Complete Solutions Guide for

CALCULUS

EIGHTH EDITION

Larson / Hostetler / Edwards

Volume II
Chapters 7–11

Bruce H. Edwards
University of Florida

Houghton Mifflin Company Boston New York

Printed in the U. S. A.

ISBN: 0-618-52794-X

6789- EB -09 08

CONTENTS

CHAPTER 7
Applications of Integration

CHAPTER 7
Applications of Integration

Section 7.1 Area of a Region Between Two Curves

1. $A = \int_0^6 [0 - (x^2 - 6x)]\,dx = -\int_0^6 (x^2 - 6x)\,dx$

2. $A = \int_{-2}^{2} [(2x + 5) - (x^2 + 2x + 1)]\,dx$
$= \int_{-2}^{2} (-x^2 + 4)\,dx$

3. $A = \int_0^3 [(-x^2 + 2x + 3) - (x^2 - 4x + 3)]\,dx$
$= \int_0^3 (-2x^2 + 6x)\,dx$

4. $A = \int_0^1 (x^2 - x^3)\,dx$

5. $A = 2\int_{-1}^{0} 3(x^3 - x)\,dx = 6\int_{-1}^{0} (x^3 - x)\,dx$
or $-6\int_0^1 (x^3 - x)\,dx$

6. $A = 2\int_0^1 [(x - 1)^3 - (x - 1)]\,dx$

7. $\int_0^4 \left[(x + 1) - \frac{x}{2}\right] dx$

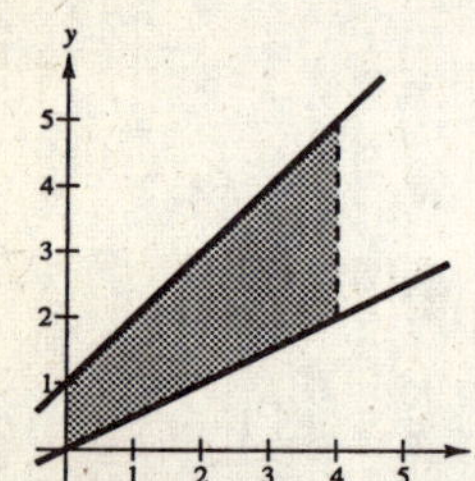

8. $\int_{-1}^{1} [(1 - x^2) - (x^2 - 1)]\,dx$

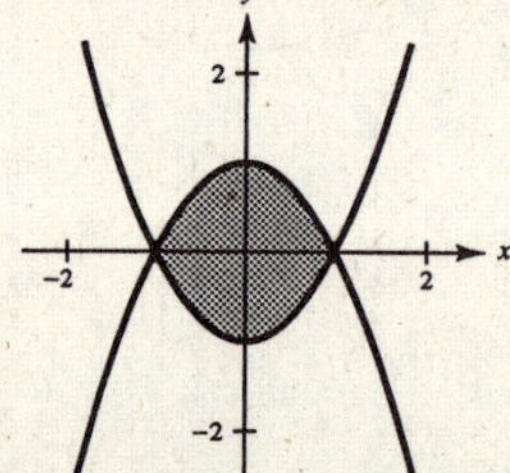

9. $\int_0^6 \left[4(2^{-x/3}) - \frac{x}{6}\right] dx$

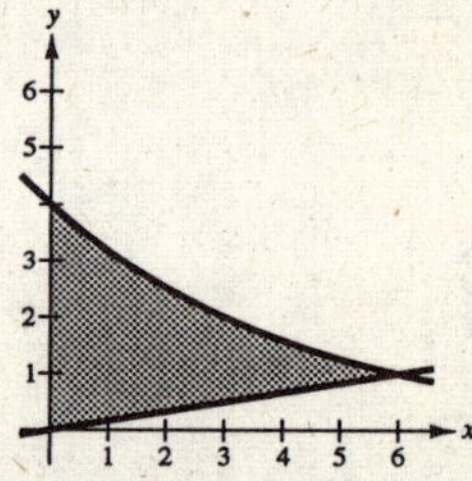

10. $\int_2^3 \left[\left(\frac{x^3}{3} - x\right) - \frac{x}{3}\right] dx$

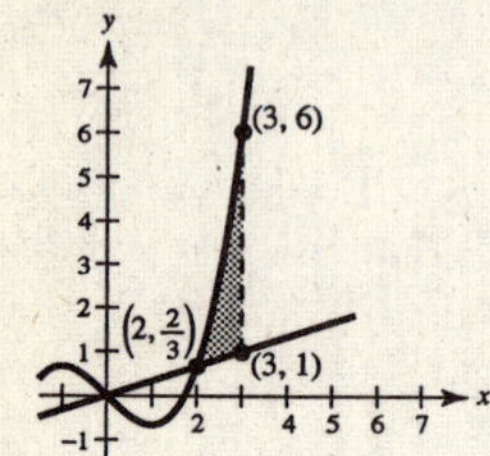

11. $\int_{-\pi/3}^{\pi/3} (2 - \sec x)\,dx$

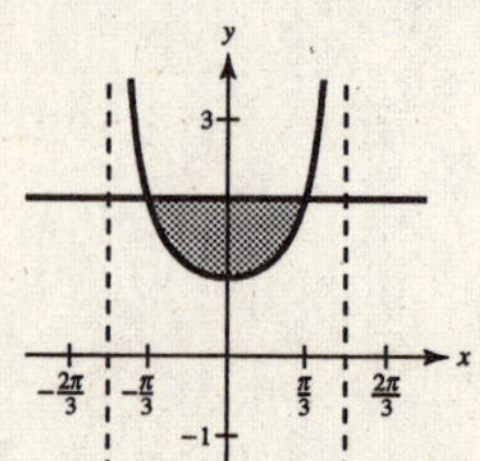

12. $\int_{-\pi/4}^{\pi/4} (\sec^2 x - \cos x)\,dx$

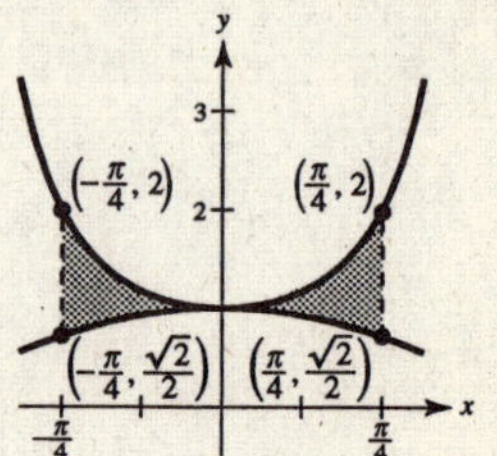

13. (a)

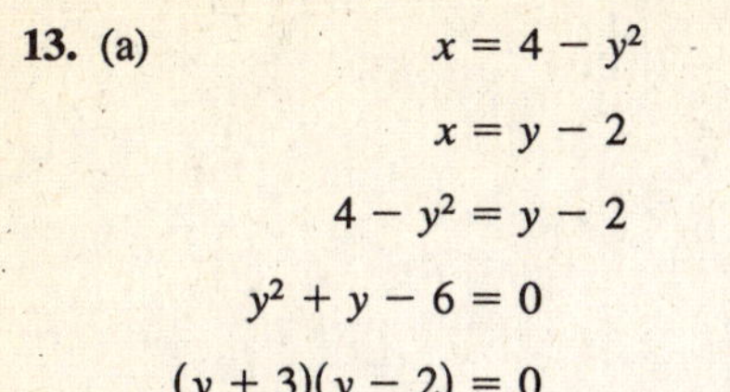

$$x = 4 - y^2$$
$$x = y - 2$$
$$4 - y^2 = y - 2$$
$$y^2 + y - 6 = 0$$
$$(y + 3)(y - 2) = 0$$

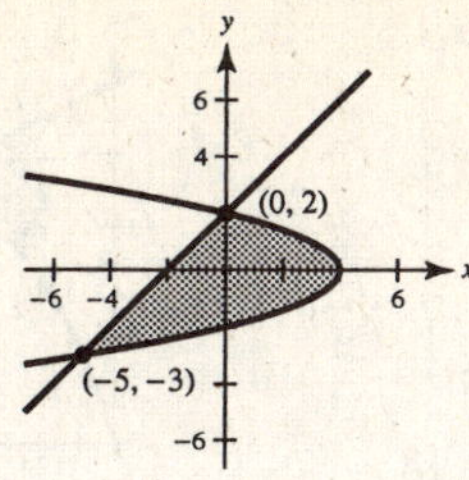

Intersection points: $(0, 2)$ and $(-5, -3)$

$$A = \int_{-5}^{0} [(x + 2) + \sqrt{4 - x}\, dx + \int_{0}^{4} 2\sqrt{4 - x}\, dx = \frac{61}{6} + \frac{32}{3} = \frac{125}{6}$$

(b) $A = \displaystyle\int_{-3}^{2} [(4 - y^2) - (y - 2)]\, dy = \frac{125}{6}$

14. (a) $y = x^2$ and $y = 6 - x$

$x^2 = 6 - x \Rightarrow x^2 + x - 6 = 0 \Rightarrow (x + 3)(x - 2) = 0$

Intersection points: $(2, 4)$ and $(-3, 9)$

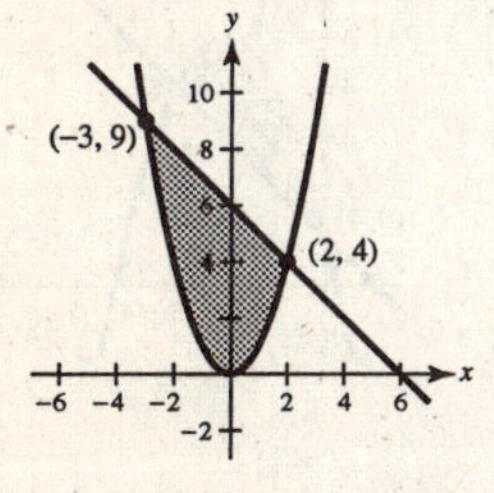

$$A = \int_{-3}^{2} [(6 - x) - x^2]\, dx = \frac{125}{6}$$

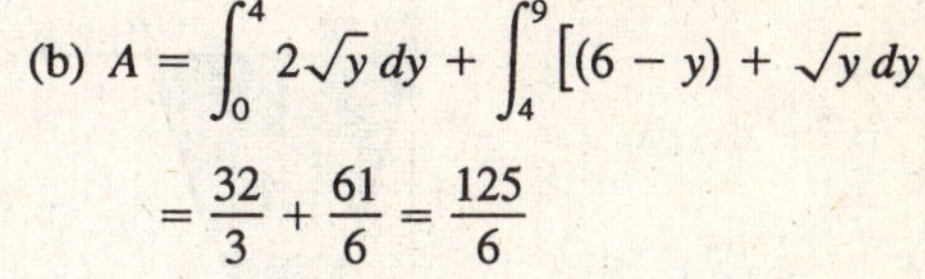

(b) $A = \displaystyle\int_{0}^{4} 2\sqrt{y}\, dy + \int_{4}^{9} [(6 - y) + \sqrt{y}\, dy$

$$= \frac{32}{3} + \frac{61}{6} = \frac{125}{6}$$

15. $f(x) = x + 1$

$g(x) = (x - 1)^2$

$A \approx 4$

Matches (d)

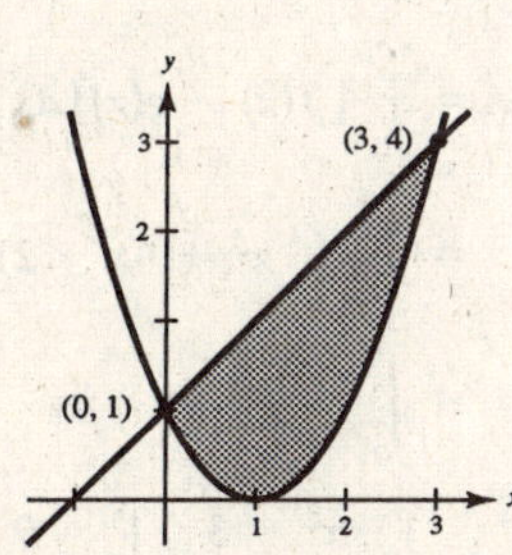

16. $f(x) = 2 - \frac{1}{2}x$

$g(x) = 2 - \sqrt{x}$

$A \approx 1$

Matches (a)

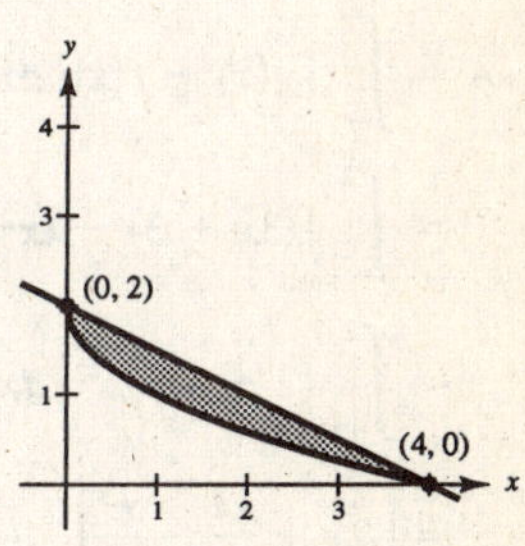

17. $A = \displaystyle\int_{0}^{2} \left[\left(\frac{1}{2}x^3 + 2\right) - (x + 1)\right] dx$

$$= \int_{0}^{2} \left(\frac{1}{2}x^3 - x + 1\right) dx$$

$$= \left[\frac{x^4}{8} - \frac{x^2}{2} + x\right]_{0}^{2}$$

$$= \left(\frac{16}{8} - \frac{4}{2} + 2\right) - 0 = 2$$

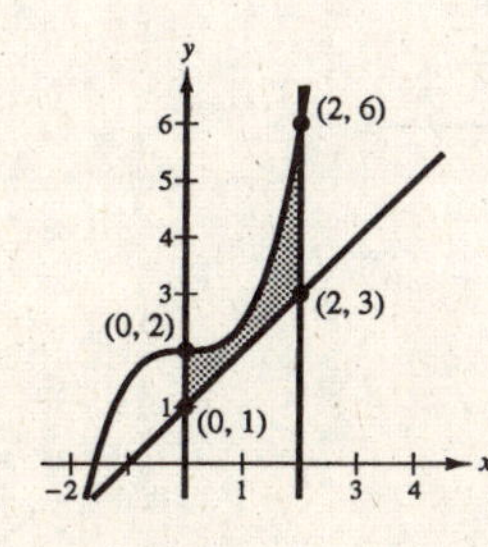

18. $A = \int_2^8 \left[\left(10 - \frac{1}{2}x\right) - \left(-\frac{3}{8}x(x-8)\right)\right] dx$

$= \int_2^8 \left(\frac{3}{8}x^2 - \frac{7}{2}x + 10\right) dx$

$= \left[\frac{x^3}{8} - \frac{7x^2}{4} + 10x\right]_2^8$

$= (64 - 112 + 80) - (1 - 7 + 20) = 18$

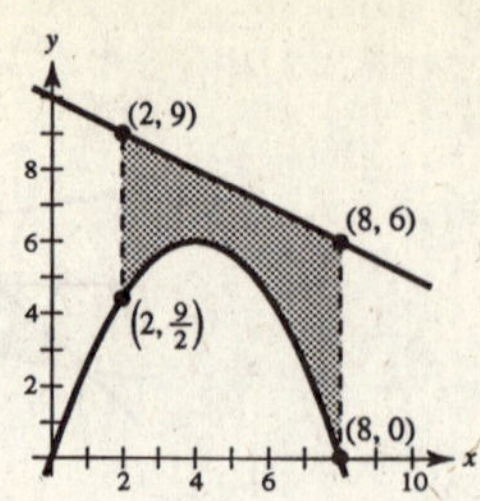

19. The points of intersection are given by:

$x^2 - 4x = 0$

$x(x-4) = 0$ when $x = 0, 4$

$A = \int_0^4 [g(x) - f(x)]\, dx$

$= -\int_0^4 (x^2 - 4x)\, dx$

$= -\left[\frac{x^3}{3} - 2x^2\right]_0^4$

$= \frac{32}{3}$

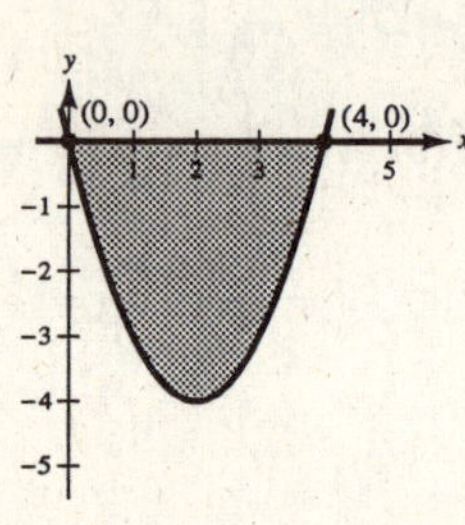

20. The points of intersection are given by:

$-x^2 + 4x + 1 = x + 1$

$-x^2 + 3x = 0$

$x^2 = 3x$ when $x = 0, 3$

$A = \int_0^3 [(-x^2 + 4x + 1) - (x + 1)]\, dx$

$= \int_0^3 (-x^2 + 3x)\, dx$

$= \left[-\frac{x^3}{3} + \frac{3x^2}{2}\right]_0^3$

$= -9 + \frac{27}{2} = \frac{9}{2}$

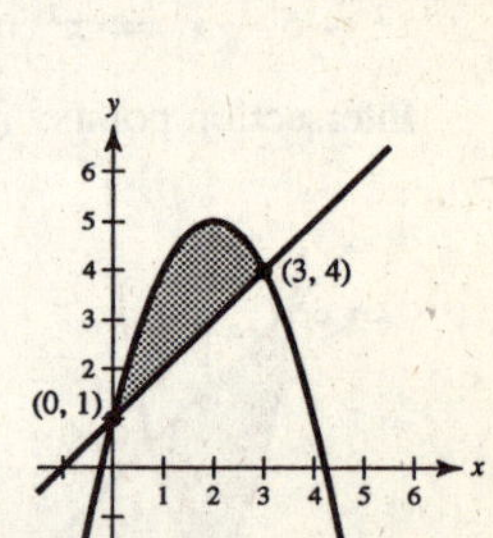

21. The points of intersection are given by:

$x^2 + 2x + 1 = 3x + 3$

$(x-2)(x+1) = 0$ when $x = -1, 2$

$A = \int_{-1}^2 [g(x) - f(x)]\, dx$

$= \int_{-1}^2 [(3x + 3) - (x^2 + 2x + 1)]\, dx$

$= \int_{-1}^2 (2 + x - x^2)\, dx$

$= \left[2x + \frac{x^2}{2} - \frac{x^3}{3}\right]_{-1}^2 = \frac{9}{2}$

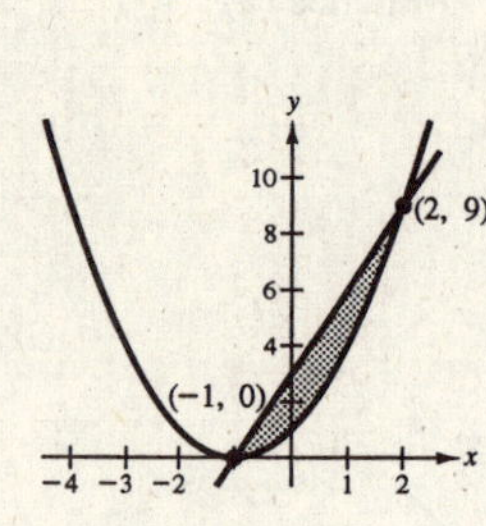

22. The points of intersection are given by:

$-x^2 + 4x + 2 = x + 2$

$x(3 - x) = 0$ when $x = 0, 3$

$A = \int_0^3 [f(x) - g(x)]\, dx$

$= \int_0^3 [(-x^2 + 4x + 2) - (x + 2)]\, dx$

$= \int_0^3 (-x^2 + 3x)\, dx$

$= \left[\frac{-x^3}{3} + \frac{3}{2}x^2\right]_0^3 = \frac{9}{2}$

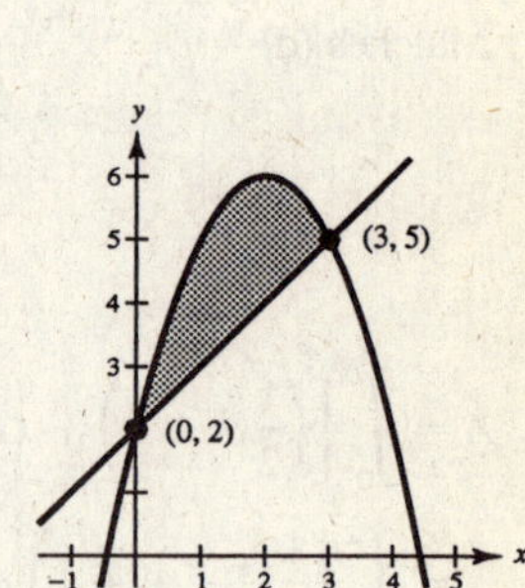

23. The points of intersection are given by:

$x = 2 - x$ and $x = 0$ and $2 - x = 0$

$x = 1$ $\qquad x = 0$ $\qquad x = 2$

$A = \int_0^1 [(2 - y) - (y)]\, dy = \left[2y - y^2\right]_0^1 = 1$

Note that if we integrate with respect to x, we need two integrals. Also, note that the region is a triangle.

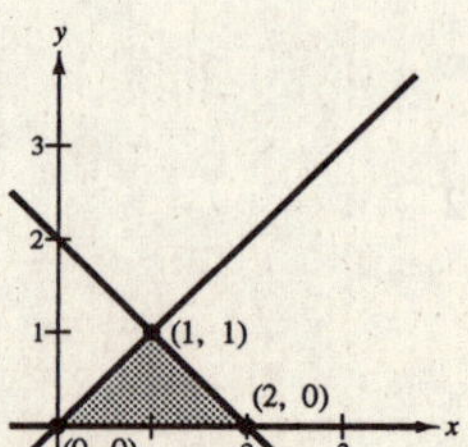

24. $A = \int_1^5 \left(\frac{1}{x^2} - 0\right) dx = \left[-\frac{1}{x}\right]_1^5 = \frac{4}{5}$

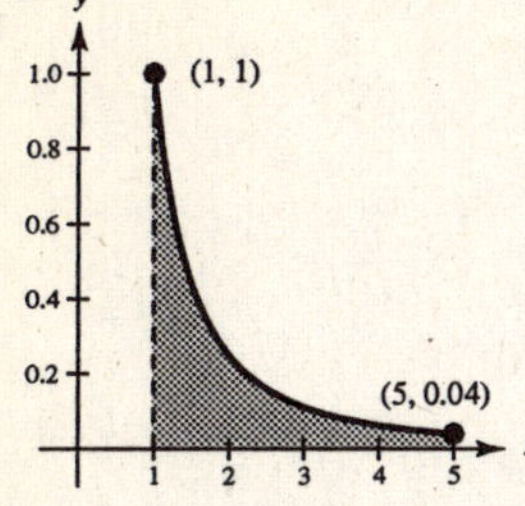

25. The points of intersection are given by:

$$\sqrt{3x} + 1 = x + 1$$

$$\sqrt{3x} = x \quad \text{when } x = 0, 3$$

$$A = \int_0^3 [f(x) - g(x)]\,dx$$

$$= \int_0^3 \left[\left(\sqrt{3x} + 1\right) - (x + 1)\right] dx$$

$$= \int_0^3 [(3x)^{1/2} - x]\,dx$$

$$= \left[\frac{2}{9}(3x)^{3/2} - \frac{x^2}{2}\right]_0^3 = \frac{3}{2}$$

26. The points of intersection are given by:

$$\sqrt[3]{x - 1} = x - 1$$

$$x - 1 = (x - 1)^3 = x^3 - 3x^2 + 3x - 1$$

$$x^3 - 3x^2 + 2x = 0$$

$$x(x^2 - 3x + 2) = 0$$

$$x(x - 2)(x - 1) = 0 \implies x = 0, 1, 2$$

$$A = 2\int_0^1 \left[(x - 1) - \sqrt[3]{x - 1}\right] dx$$

$$= 2\left[\frac{x^2}{2} - x - \frac{3}{4}(x - 1)^{4/3}\right]_0^1$$

$$= 2\left[\left(\frac{1}{2} - 1 - 0\right) - \left(-\frac{3}{4}\right)\right] = \frac{1}{2}$$

27. The points of intersection are given by:

$$y^2 = y + 2$$

$$(y - 2)(y + 1) = 0 \quad \text{when } y = -1, 2$$

$$A = \int_{-1}^2 [g(y) - f(y)]\,dy$$

$$= \int_{-1}^2 [(y + 2) - y^2]\,dy$$

$$= \left[2y + \frac{y^2}{2} - \frac{y^3}{3}\right]_{-1}^2 = \frac{9}{2}$$

28. The points of intersection are given by:

$$2y - y^2 = -y$$

$$y(y - 3) = 0 \quad \text{when } y = 0, 3$$

$$A = \int_0^3 [f(y) - g(y)]\,dy$$

$$= \int_0^3 [(2y - y^2) - (-y)]\,dy$$

$$= \int_0^3 (3y - y^2)\,dy$$

$$= \left[\frac{3}{2}y^2 - \frac{1}{3}y^3\right]_0^3 = \frac{9}{2}$$

29. $A = \int_{-1}^{2} [f(y) - g(y)]\,dy$

$= \int_{-1}^{2} [(y^2 + 1) - 0]\,dy$

$= \left[\frac{y^3}{3} + y\right]_{-1}^{2} = 6$

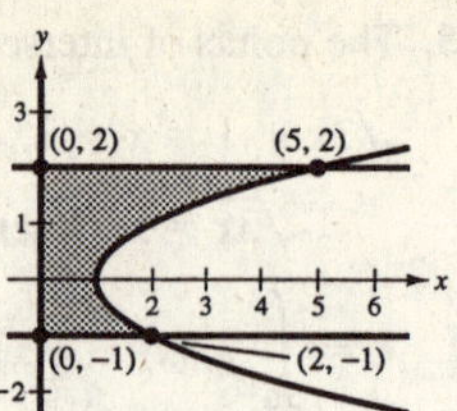

30. $A = \int_{0}^{3} [f(y) - g(y)]\,dy$

$= \int_{0}^{3} \left[\frac{y}{\sqrt{16 - y^2}} - 0\right] dy$

$= -\frac{1}{2}\int_{0}^{3} (16 - y^2)^{-1/2}(-2y)\,dy$

$= \left[-\sqrt{16 - y^2}\right]_{0}^{3} = 4 - \sqrt{7} \approx 1.354$

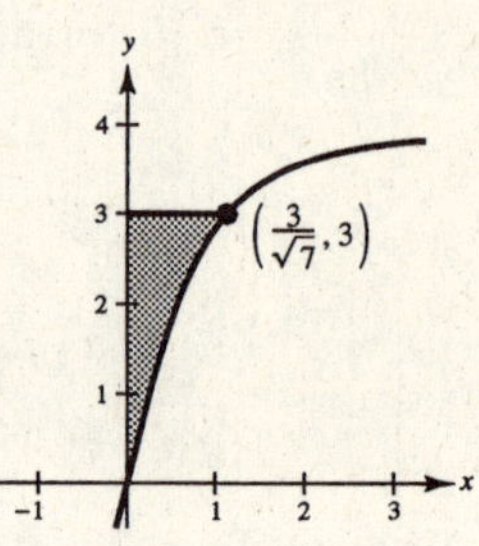

31. $y = \frac{10}{x} \Rightarrow x = \frac{10}{y}$

$A = \int_{2}^{10} \frac{10}{y}\,dy$

$= \left[10 \ln y\right]_{2}^{10}$

$= 10(\ln 10 - \ln 2)$

$= 10 \ln 5 \approx 16.0944$

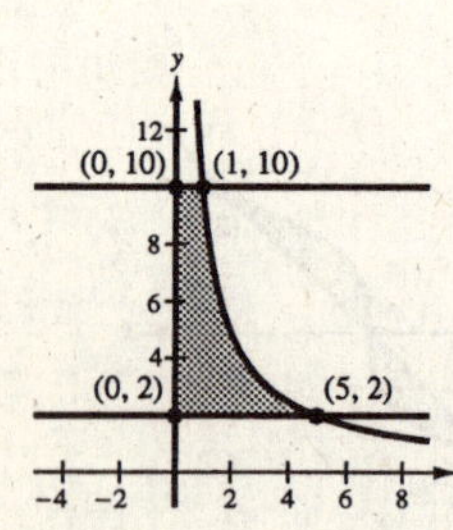

32. $A = \int_{0}^{1} \left(4 - \frac{4}{2 - x}\right) dx$

$= \left[4x + 4 \ln|2 - x|\right]_{0}^{1}$

$= 4 - 4 \ln 2$

≈ 1.227

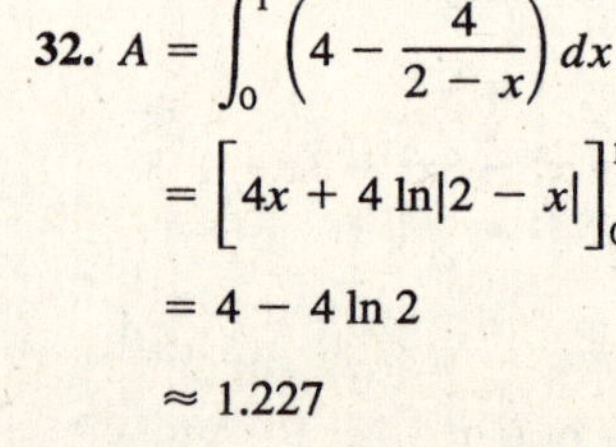

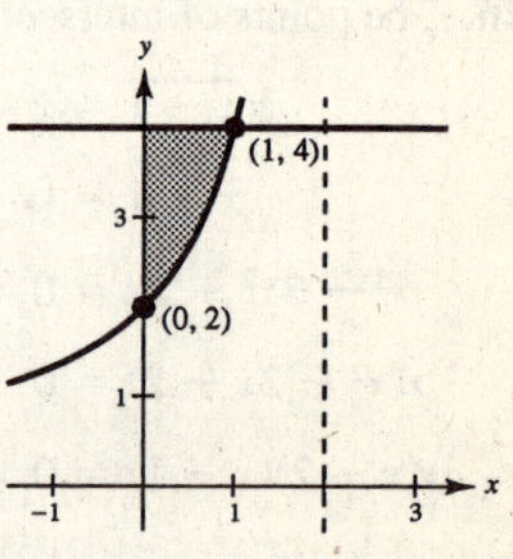

33. (a)

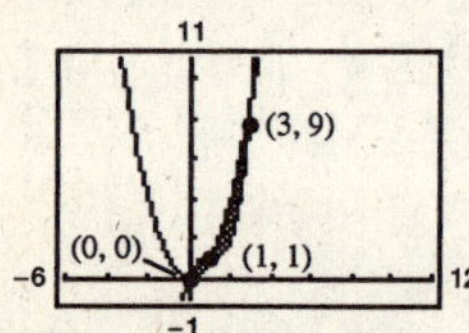

(b) The points of intersection are given by:

$x^3 - 3x^2 + 3x = x^2$

$x(x - 1)(x - 3) = 0$ when $x = 0, 1, 3$

$A = \int_{0}^{1} [f(x) - g(x)]\,dx + \int_{1}^{3} [g(x) - f(x)]\,dx$

$= \int_{0}^{1} [(x^3 - 3x^2 + 3x) - x^2]\,dx + \int_{1}^{3} [x^2 - (x^3 - 3x^2 + 3x)]\,dx$

$= \int_{0}^{1} (x^3 - 4x^2 + 3x)\,dx + \int_{1}^{3} (-x^3 + 4x^2 - 3x)\,dx$

$= \left[\frac{x^4}{4} - \frac{4}{3}x^3 + \frac{3}{2}x^2\right]_{0}^{1} + \left[\frac{-x^4}{4} + \frac{4}{3}x^3 - \frac{3}{2}x^2\right]_{1}^{3} = \frac{5}{12} + \frac{8}{3} = \frac{37}{12}$

(c) Numerical approximation: $0.417 + 2.667 \approx 3.083$

34. (a)

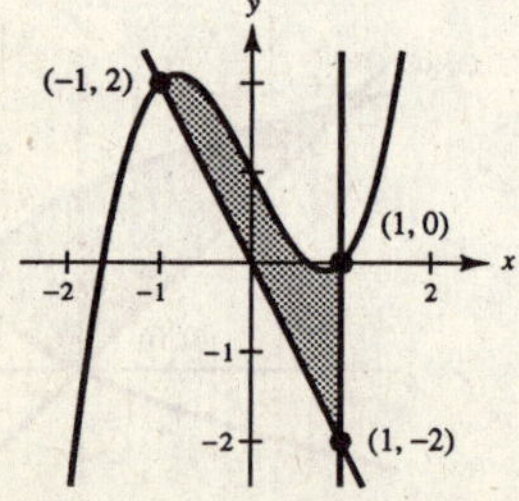

(b) The point of intersection is given by:

$x^3 - 2x + 1 = -2x$

$x^3 + 1 = 0$ when $x = -1$

$A = \int_{-1}^{1} [f(x) - g(x)]\,dx$

$= \int_{-1}^{1} [(x^3 - 2x + 1) - (-2x)]\,dx$

$= \int_{-1}^{1} (x^3 + 1)\,dx = \left[\frac{x^4}{4} + x\right]_{-1}^{1} = 2$

(c) Numerical approximation: 2.0

35. (a)

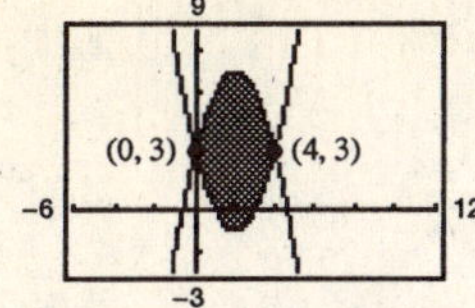

(b) The points of intersection are given by:

$$x^2 - 4x + 3 = 3 + 4x - x^2$$

$$2x(x - 4) = 0 \quad \text{when} \quad x = 0, 4$$

$$A = \int_0^4 [(3 + 4x - x^2) - (x^2 - 4x + 3)]\,dx$$

$$= \int_0^4 (-2x^2 + 8x)\,dx$$

$$= \left[-\frac{2x^3}{3} + 4x^2\right]_0^4 = \frac{64}{3}$$

(c) Numerical approximation: 21.333

36. (a)

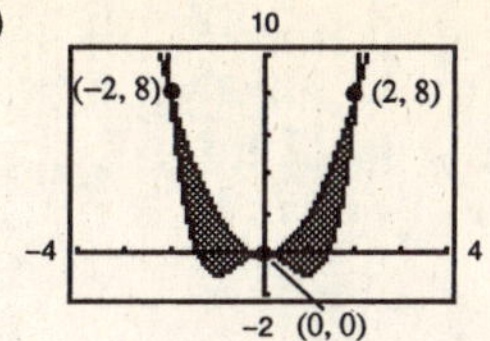

(b) The points of intersection are given by:

$$x^4 - 2x^2 = 2x^2$$

$$x^2(x^2 - 4) = 0 \quad \text{when} \quad x = 0, \pm 2$$

$$A = 2\int_0^2 [2x^2 - (x^4 - 2x^2)]\,dx$$

$$= 2\int_0^2 (4x^2 - x^4)\,dx$$

$$= 2\left[\frac{4x^3}{3} - \frac{x^5}{5}\right]_0^2 = \frac{128}{15}$$

(c) Numerical approximation: 8.533

37. (a) $f(x) = x^4 - 4x^2, \quad g(x) = x^2 - 4$

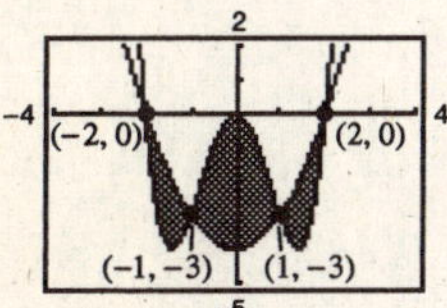

(c) Numerical approximation:
$5.067 + 2.933 = 8.0$

(b) The points of intersection are given by:

$$x^4 - 4x^2 = x^2 - 4$$

$$x^4 - 5x^2 + 4 = 0$$

$$(x^2 - 4)(x^2 - 1) = 0 \quad \text{when} \quad x = \pm 2, \pm 1$$

By symmetry:

$$A = 2\int_0^1 [(x^4 - 4x^2) - (x^2 - 4)]\,dx + 2\int_1^2 [(x^2 - 4) - (x^4 - 4x^2)]\,dx$$

$$= 2\int_0^1 (x^4 - 5x^2 + 4)\,dx + 2\int_1^2 (-x^4 + 5x^2 - 4)\,dx$$

$$= 2\left[\frac{x^5}{5} - \frac{5x^3}{3} + 4x\right]_0^1 + 2\left[-\frac{x^5}{5} + \frac{5x^3}{3} - 4x\right]_1^2$$

$$= 2\left[\frac{1}{5} - \frac{5}{3} + 4\right] + 2\left[\left(-\frac{32}{5} + \frac{40}{3} - 8\right) - \left(-\frac{1}{5} + \frac{5}{3} - 4\right)\right] = 8$$

38. (a) $f(x) = x^4 - 4x^2, \quad g(x) = x^3 - 4x$

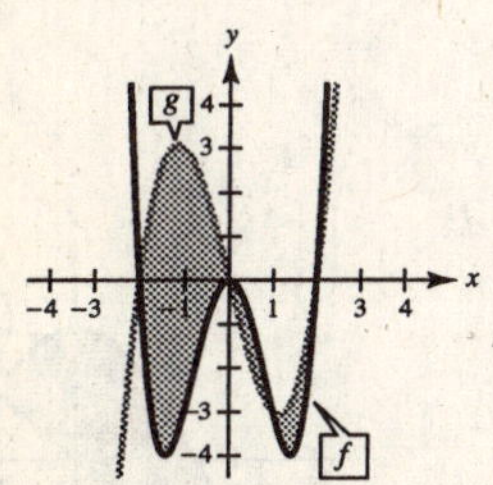

(c) Numerical approximation:
$8.267 + 0.617 + 0.883 \approx 9.767$

(b) The points of intersection are given by:

$$x^4 - 4x^2 = x^3 - 4x$$

$$x^4 - x^3 - 4x^2 + 4x = 0$$

$$x(x - 1)(x + 2)(x - 2) = 0 \quad \text{when} \quad x = -2, 0, 1, 2$$

$$A = \int_{-2}^0 [(x^3 - 4x) - (x^4 - 4x^2)]\,dx + \int_0^1 [(x^4 - 4x^2) - (x^3 - 4x)]\,dx$$

$$+ \int_1^2 [(x^3 - 4x) - (x^4 - 4x^2)]\,dx$$

$$= \frac{248}{30} + \frac{37}{60} + \frac{53}{60} = \frac{293}{30}$$

39. (a)

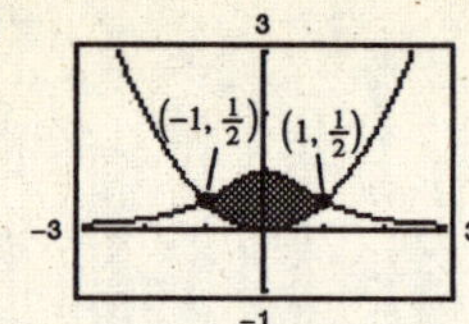

(b) The points of intersection are given by:

$$\frac{1}{1+x^2} = \frac{x^2}{2}$$

$$x^4 + x^2 - 2 = 0$$

$$(x^2+2)(x^2-1) = 0$$

$$x = \pm 1$$

$$A = 2\int_0^1 [f(x) - g(x)]\,dx$$

$$= 2\int_0^1 \left[\frac{1}{1+x^2} - \frac{x^2}{2}\right]dx$$

$$= 2\left[\arctan x - \frac{x^3}{6}\right]_0^1$$

$$= 2\left(\frac{\pi}{4} - \frac{1}{6}\right) = \frac{\pi}{2} - \frac{1}{3} \approx 1.237$$

(c) Numerical approximation: 1.237

40. (a)

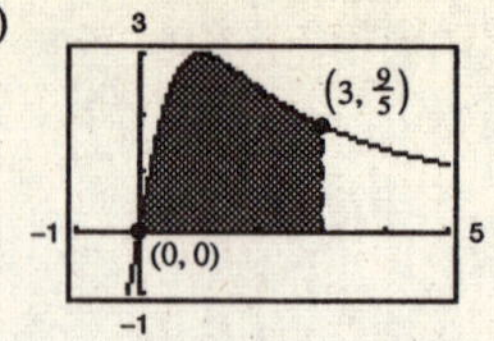

(b) $A = \int_0^3 \left[\frac{6x}{x^2+1} - 0\right] dx$

$$= \left[3\ln(x^2+1)\right]_0^3$$

$$= 3\ln 10$$

$$\approx 6.908$$

(c) Numerical approximation: 6.908

41. (a)

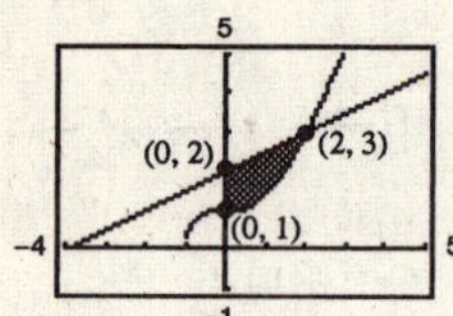

(b) and (c) $\sqrt{1+x^3} \le \frac{1}{2}x + 2$ on $[0, 2]$

You must use numerical integration because $y = \sqrt{1+x^3}$ does not have an elementary antiderivative.

$$A = \int_0^2 \left[\frac{1}{2}x + 2 - \sqrt{1+x^3}\right] dx \approx 1.759$$

42. (a)

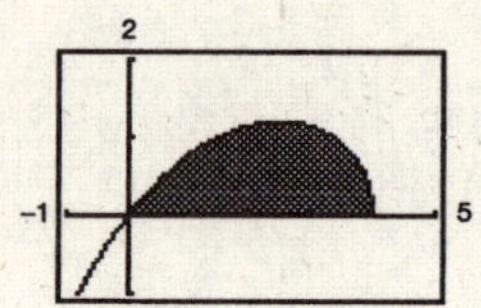

(b) and (c) You must use numerical integration:

$$A = \int_0^4 x\sqrt{\frac{4-x}{4+x}}\,dx \approx 3.434$$

43. $A = 2\int_0^{\pi/3} [f(x) - g(x)]\,dx$

$$= 2\int_0^{\pi/3} (2\sin x - \tan x)\,dx$$

$$= 2\Big[-2\cos x + \ln|\cos x|\Big]_0^{\pi/3}$$

$$= 2(1 - \ln 2) \approx 0.614$$

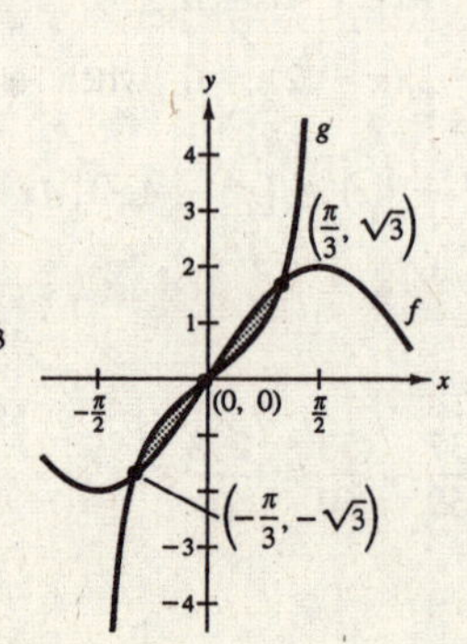

44. $A = \int_{-\pi/2}^{\pi/6} (\cos 2x - \sin x)\,dx$

$$= \left[\frac{1}{2}\sin 2x + \cos x\right]_{-\pi/2}^{\pi/6}$$

$$= \left(\frac{\sqrt{3}}{4} + \frac{\sqrt{3}}{2}\right) - (0)$$

$$= \frac{3\sqrt{3}}{4} \approx 1.299$$

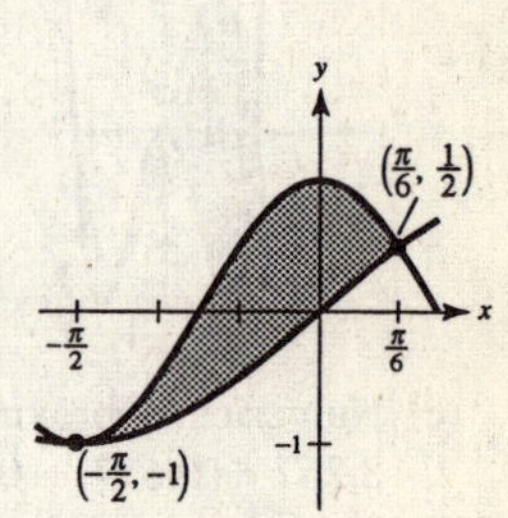

45. $A = \int_0^{2\pi} [(2 - \cos x) - \cos x]\,dx$

$= 2\int_0^{2\pi} (1 - \cos x)\,dx$

$= 2\Big[x - \sin x\Big]_0^{2\pi} = 4\pi \approx 12.566$

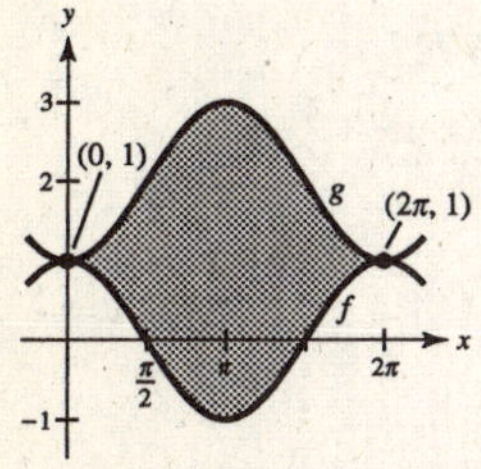

46. $A = \int_0^1 \left[\left(\sqrt{2} - 4\right)x + 4 - \sec\frac{\pi x}{4}\tan\frac{\pi x}{4}\right] dx$

$= \left[\frac{\sqrt{2} - 4}{2}x^2 + 4x - \frac{4}{\pi}\sec\frac{\pi x}{4}\right]_0^1$

$= \left(\frac{\sqrt{2} - 4}{2} + 4 - \frac{4}{\pi}\sqrt{2}\right) - \left(-\frac{4}{\pi}\right)$

$= \frac{\sqrt{2}}{2} + 2 + \frac{4}{\pi}\left(1 - \sqrt{2}\right) \approx 2.1797$

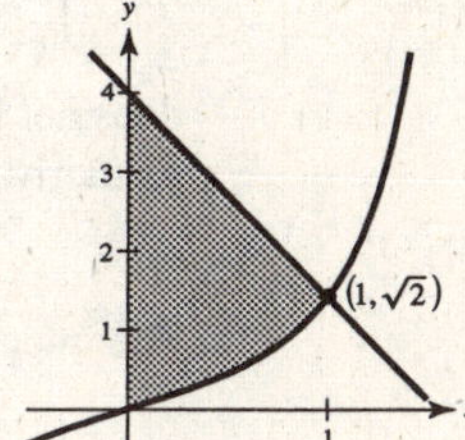

47. $A = \int_0^1 \left[xe^{-x^2} - 0\right] dx$

$= \left[-\frac{1}{2}e^{-x^2}\right]_0^1 = \frac{1}{2}\left(1 - \frac{1}{e}\right) \approx 0.316$

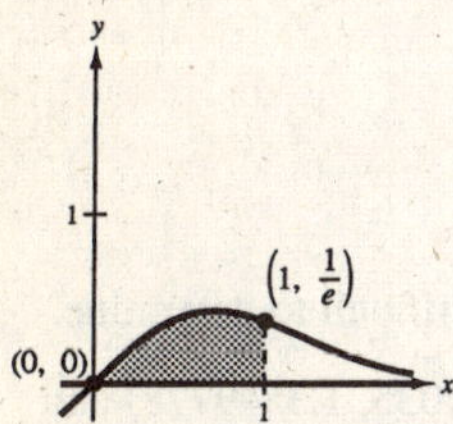

48. From the graph we see that f and g intersect twice at $x = 0$ and $x = 1$.

$A = \int_0^1 [g(x) - f(x)]\,dx$

$= \int_0^1 [(2x + 1) - 3^x]\,dx$

$= \left[x^2 + x - \frac{1}{\ln 3}(3^x)\right]_0^1$

$= 2\left(1 - \frac{1}{\ln 3}\right) \approx 0.180$

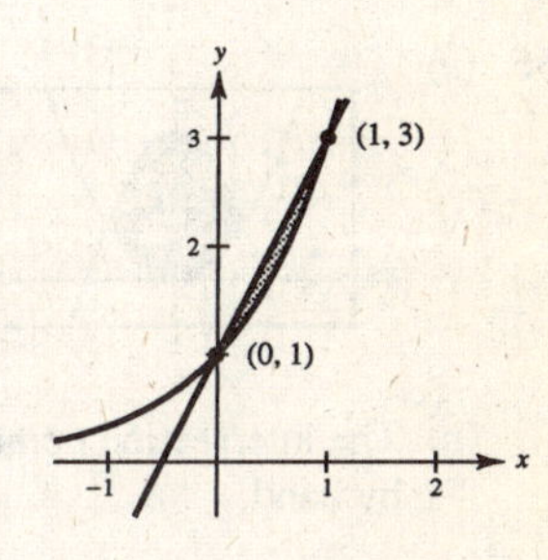

49. (a)

(b) $A = \int_0^{\pi} (2\sin x + \sin 2x)\,dx$

$= \left[-2\cos x - \frac{1}{2}\cos 2x\right]_0^{\pi}$

$= \left(2 - \frac{1}{2}\right) - \left(-2 - \frac{1}{2}\right) = 4$

(c) Numerical approximation: 4.0

50. (a)

(b) $A = \int_0^{\pi} (2\sin x + \cos 2x)\,dx$

$= \left[-2\cos x + \frac{1}{2}\sin 2x\right]_0^{\pi} = 4$

(c) Numerical approximation: 4

51. (a)

(b) $A = \int_1^3 \frac{1}{x^2}e^{1/x}\,dx$

$= \left[-e^{-1/x}\right]_1^3$

$= e - e^{1/3}$

(c) Numerical approximation: 1.323

52. (a)

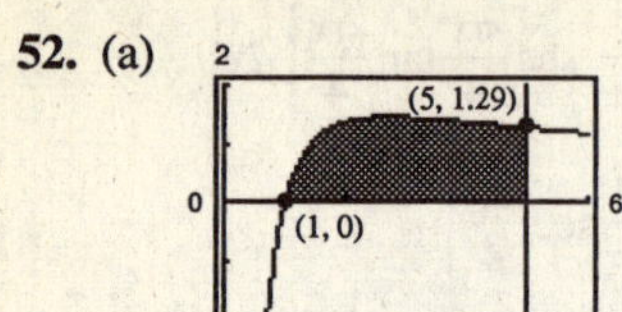

(b) $A = \int_1^5 \frac{4 \ln x}{x}\, dx$

$= \left[2(\ln x)^2\right]_1^5$

$= 2(\ln 5)^2$

(c) Numerical approximation: 5.181

53. (a)

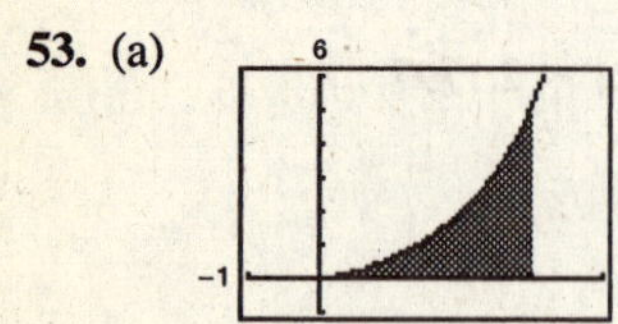

(b) The integral

$$A = \int_0^3 \sqrt{\frac{x^3}{4-x}}\, dx$$

does not have an elementary antiderivative.

(c) $A \approx 4.7721$

54. (a)

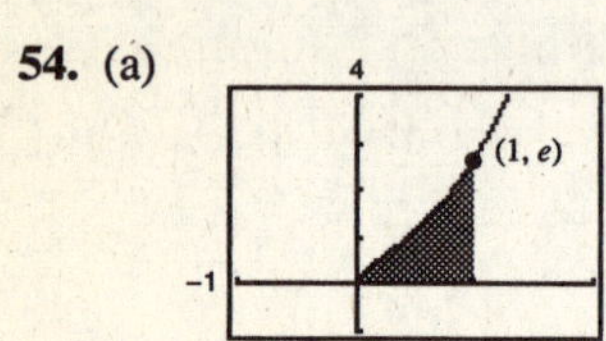

(b) The integral

$$A = \int_0^1 \sqrt{x}e^x\, dx$$

does not have an elementary antiderivative.

(c) 1.2556

55. (a)

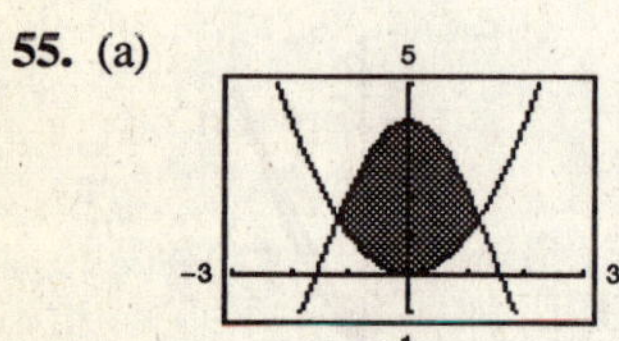

(b) The intersection points are difficult to determine by hand.

(c) Area $= \int_{-c}^{c} [4 \cos x - x^2]\, dx \approx 6.3043$ where $c \approx 1.201538$.

56. (a)

(b) The intersection points are difficult to determine.

(c) Intersection points: $(-1.164035, 1.3549778)$ and $(1.4526269, 2.1101248)$

$$A = \int_{-1.164035}^{1.4526269} \left[\sqrt{3+x} - x^2\right] dx \approx 3.0578$$

57. $F(x) = \int_0^x \left(\frac{1}{2}t + 1\right) dt = \left[\frac{t^2}{4} + t\right]_0^x = \frac{x^2}{4} + x$

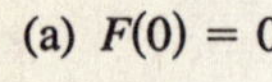

(a) $F(0) = 0$

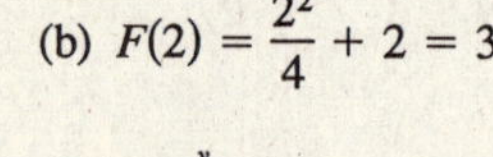

(b) $F(2) = \frac{2^2}{4} + 2 = 3$

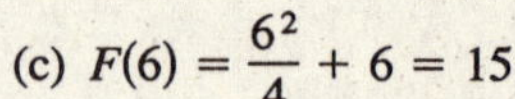

(c) $F(6) = \frac{6^2}{4} + 6 = 15$

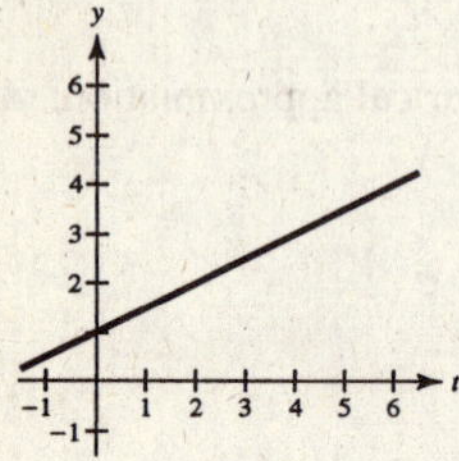

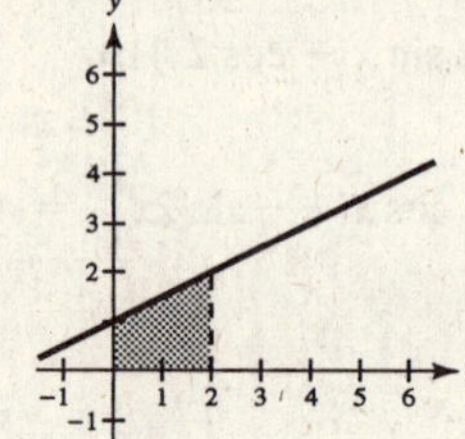

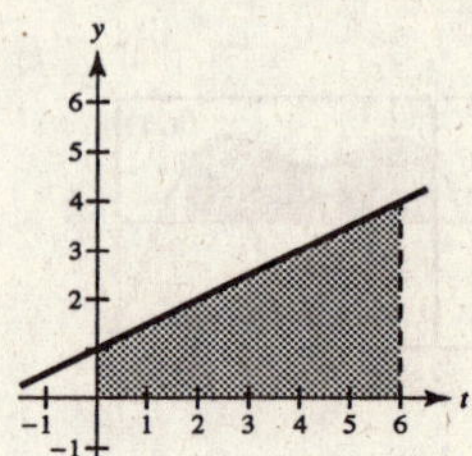

58. $F(x) = \int_0^x \left(\frac{1}{2}t^2 + 2\right) dt = \left[\frac{1}{6}t^3 + 2t\right]_0^x = \frac{x^3}{6} + 2x$

(a) $F(0) = 0$

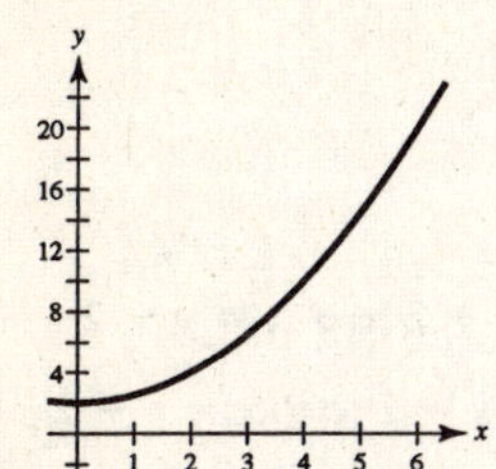

(b) $F(4) = \frac{4^3}{6} + 2(4) = \frac{56}{3}$

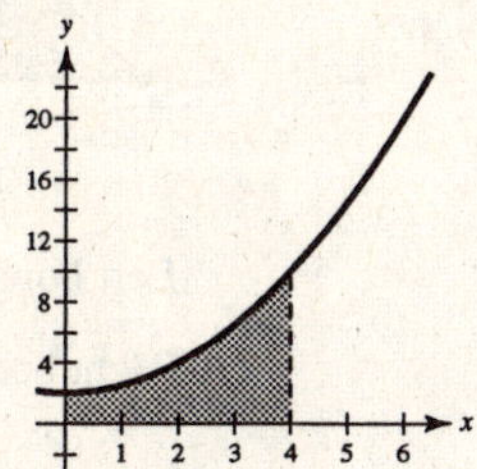

(c) $F(6) = 36 + 12 = 48$

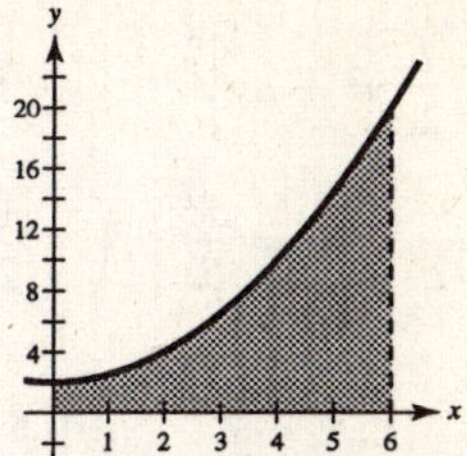

59. $F(\alpha) = \int_{-1}^{\alpha} \cos\frac{\pi\theta}{2}\, d\theta = \left[\frac{2}{\pi}\sin\frac{\pi\theta}{2}\right]_{-1}^{\alpha} = \frac{2}{\pi}\sin\frac{\pi\alpha}{2} + \frac{2}{\pi}$

(a) $F(-1) = 0$

(b) $F(0) = \frac{2}{\pi} \approx 0.6366$

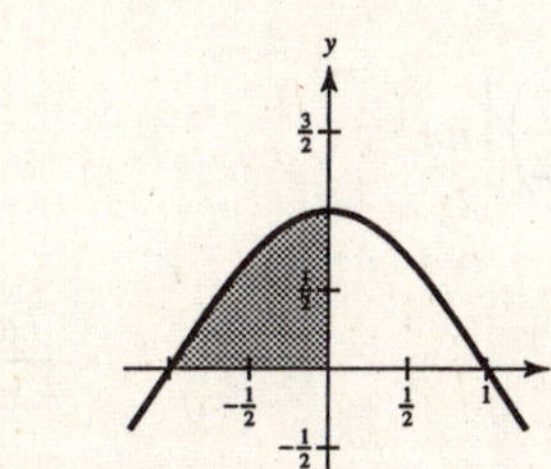

(c) $F\left(\frac{1}{2}\right) = \frac{2 + \sqrt{2}}{\pi} \approx 1.0868$

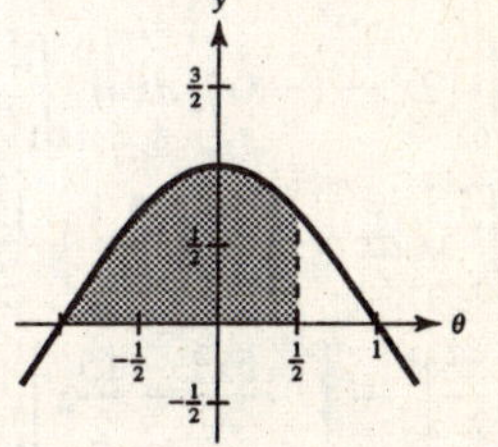

60. $F(y) = \int_{-1}^{y} 4e^{x/2}\, dx = \left[8e^{x/2}\right]_{-1}^{y} = 8e^{y/2} - 8e^{-1/2}$

(a) $F(-1) = 0$

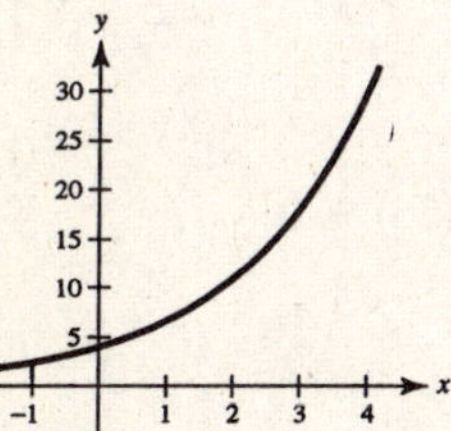

(b) $F(0) = 8 - 8e^{-1/2} \approx 3.1478$

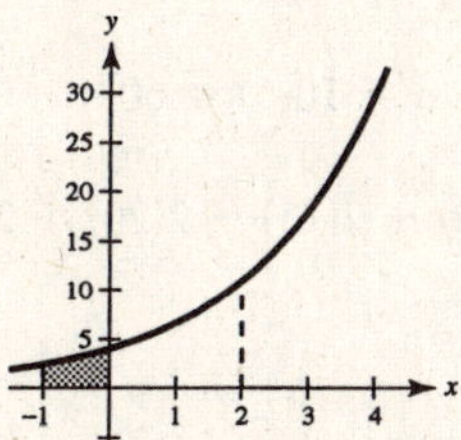

(c) $F(4) = 8e^2 - 8e^{-1/2} \approx 54.2602$

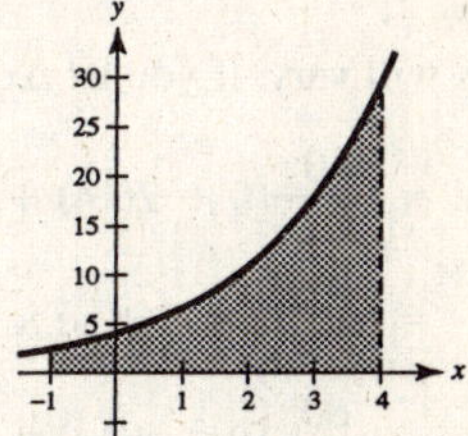

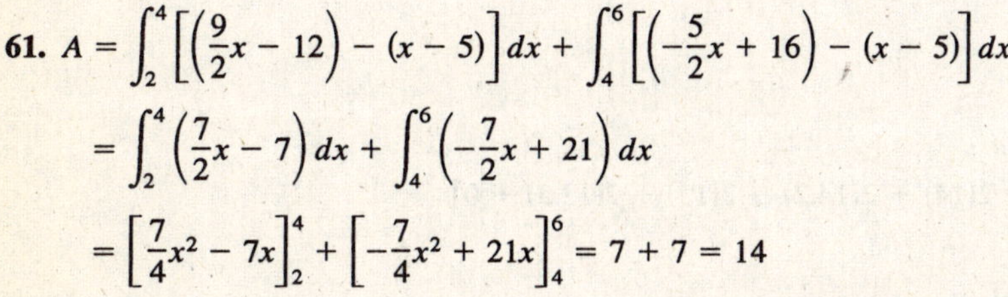

61. $A = \int_2^4 \left[\left(\frac{9}{2}x - 12\right) - (x - 5)\right] dx + \int_4^6 \left[\left(-\frac{5}{2}x + 16\right) - (x - 5)\right] dx$

$= \int_2^4 \left(\frac{7}{2}x - 7\right) dx + \int_4^6 \left(-\frac{7}{2}x + 21\right) dx$

$= \left[\frac{7}{4}x^2 - 7x\right]_2^4 + \left[-\frac{7}{4}x^2 + 21x\right]_4^6 = 7 + 7 = 14$

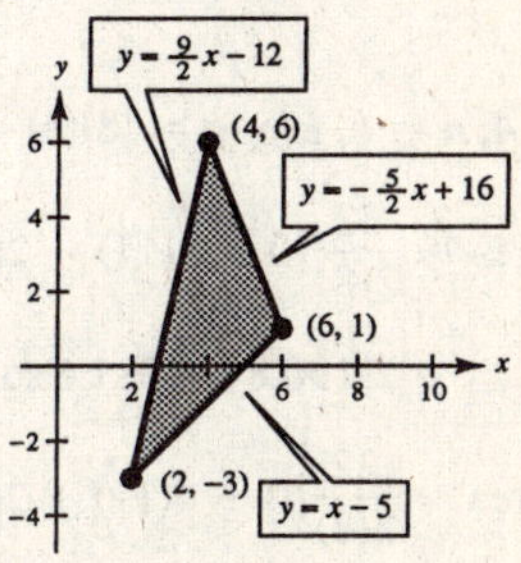

62. $A = \int_0^c \left[\left(\frac{b-a}{c}y + a\right) - \frac{b}{c}y\right] dy$

$= \int_0^c \left(-\frac{a}{c}y + a\right) dy$

$= \left[-\frac{a}{2c}y^2 + ay\right]_0^c$

$= -\frac{ac}{2} + ac = \frac{ac}{2} \quad \left(= \frac{1}{2}(\text{base})(\text{height})\right)$

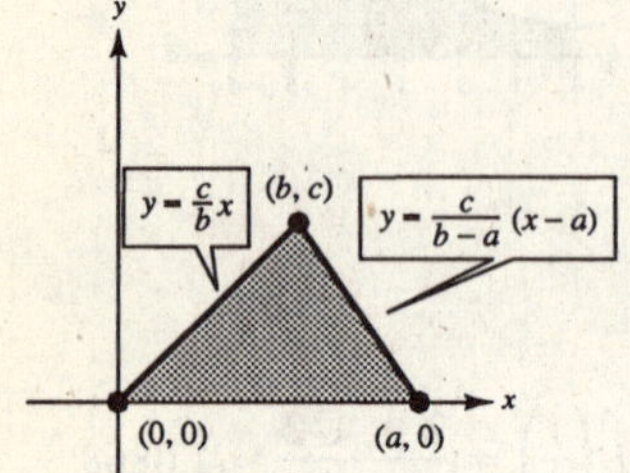

63.

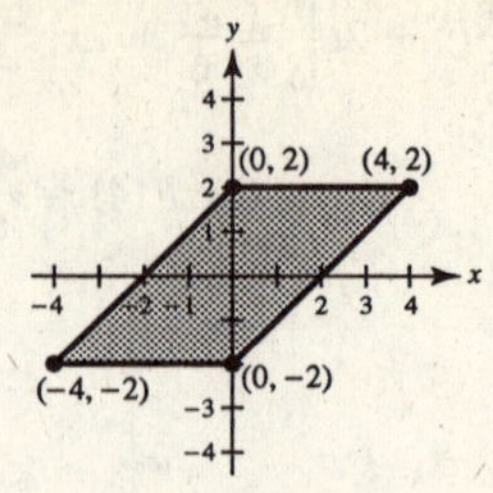

Left boundary line: $y = x + 2 \Leftrightarrow x = y - 2$

Right boundary line: $y = x - 2 \Leftrightarrow x = y + 2$

$A = \int_{-2}^{2} [(y + 2) - (y - 2)]\, dy$

$= \int_{-2}^{2} 4\, dy = 4y\Big]_{-2}^{2} = 8 - (-8) = 16$

64. $A = \int_0^1 [2x - (-3x)]\, dx + \int_1^3 \left[(-2x + 4) - \left(\frac{1}{2}x - \frac{7}{2}\right)\right] dx$

$= \int_0^1 5x\, dx + \int_1^3 \left(-\frac{5}{2}x + \frac{15}{2}\right) dx$

$= \frac{5x^2}{2}\Big]_0^1 + \left[-\frac{5x^2}{4} + \frac{15}{2}x\right]_1^3$

$= \frac{5}{2} + \left[-\frac{45}{4} + \frac{45}{2} + \frac{5}{4} - \frac{15}{2}\right]$

$= \frac{15}{2}$

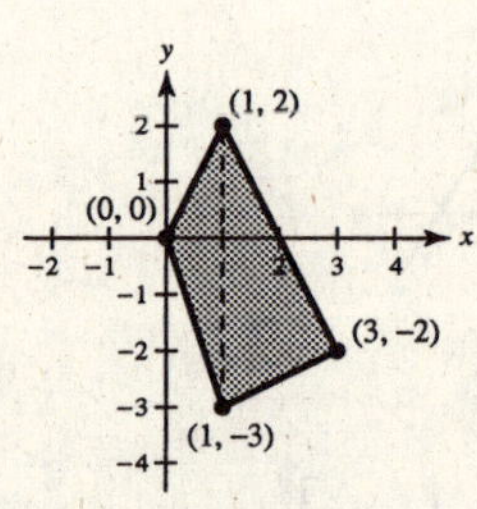

65. Answers will vary. If you let $\Delta x = 6$ and $n = 10$, $b - a = 10(6) = 60$.

(a) Area $\approx \frac{60}{2(10)}[0 + 2(14) + 2(14) + 2(12) + 2(12) + 2(15) + 2(20) + 2(23) + 2(25) + 2(26) + 0]$

$= 3[322] = 966$ sq ft

(b) Area $\approx \frac{60}{3(10)}[0 + 4(14) + 2(14) + 4(12) + 2(12) + 4(15) + 2(20) + 4(23) + 2(25) + 4(26) + 0]$

$= 2[502] = 1004$ sq ft

66. $\Delta x = 4$, $n = 8$, $b - a = (8)(4) = 32$

(a) Area $\approx \frac{32}{2(8)}[0 + 2(11) + 2(13.5) + 2(14.2) + 2(14) + 2(14.2) + 2(15) + 2(13.5) + 0]$

$= 2[190.8] = 381.6$ sq mi

(b) Area $\approx \frac{32}{3(8)}[0 + 4(11) + 2(13.5) + 4(14.2) + 2(14) + 4(14.2) + 2(15) + 4(13.5) + 0]$

$= \frac{4}{3}[296.6] \approx 395.5$ sq mi

67. $f(x) = x^3$

$f'(x) = 3x^2$

At $(1, 1)$, $f'(1) = 3$.

Tangent line: $y - 1 = 3(x - 1)$ or $y = 3x - 2$

The tangent line intersects $f(x) = x^3$ at $x = -2$.

$$A = \int_{-2}^{1} [x^3 - (3x - 2)]\,dx = \left[\frac{x^4}{4} - \frac{3x^2}{2} + 2x\right]_{-2}^{1} = \frac{27}{4}$$

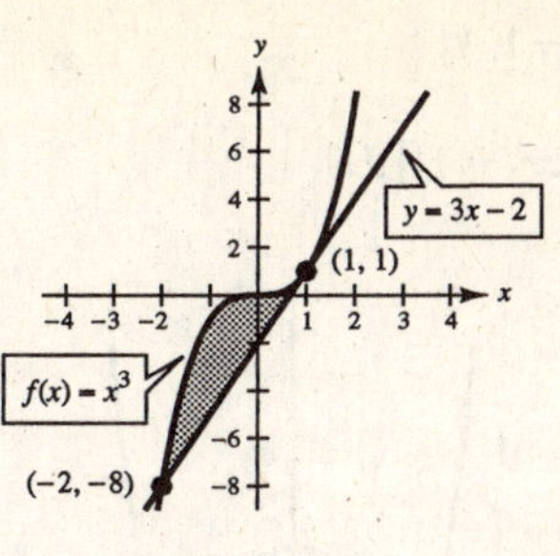

68. $y = x^3 - 2x, \quad (-1, 1)$

$y' = 3x^2 - 2$

$y'(-1) = 3 - 2 = 1$

Tangent line: $y - 1 = 1(x + 1) \implies y = x + 2$

Intersection points: $(-1, 1)$ and $(2, 4)$

$$A = \int_{-1}^{2} [(x + 2) - (x^3 - 2x)]\,dx = \int_{-1}^{2} (-x^3 + 3x + 2)\,dx$$

$$= \left[-\frac{x^4}{4} + \frac{3x^2}{2} + 2x\right]_{-1}^{2} = \left[(-4 + 6 + 4) - \left(-\frac{1}{4} + \frac{3}{2} - 2\right)\right] = \frac{27}{4}$$

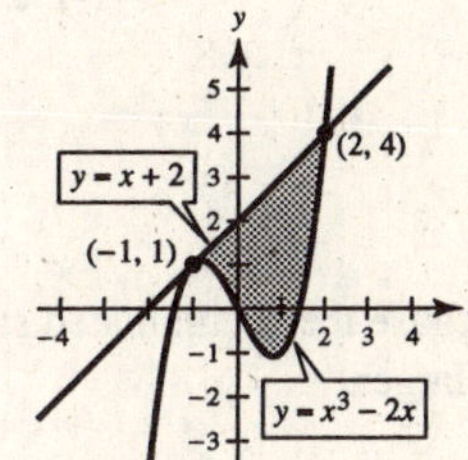

69. $f(x) = \dfrac{1}{x^2 + 1}$

$f'(x) = -\dfrac{2x}{(x^2 + 1)^2}$

At $\left(1, \dfrac{1}{2}\right)$, $f'(1) = -\dfrac{1}{2}$.

Tangent line: $y - \dfrac{1}{2} = -\dfrac{1}{2}(x - 1)$ or $y = -\dfrac{1}{2}x + 1$

The tangent line intersects $f(x) = \dfrac{1}{x^2 + 1}$ at $x = 0$.

$$A = \int_{0}^{1} \left[\frac{1}{x^2 + 1} - \left(-\frac{1}{2}x + 1\right)\right] dx = \left[\arctan x + \frac{x^2}{4} - x\right]_{0}^{1} = \frac{\pi - 3}{4} \approx 0.0354$$

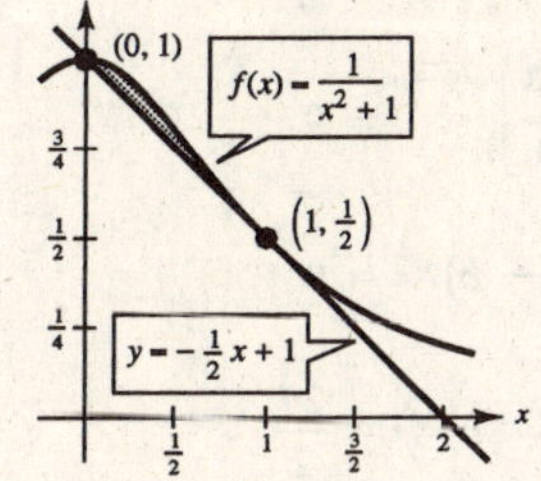

70. $y = \dfrac{2}{1 + 4x^2}, \quad \left(\dfrac{1}{2}, 1\right)$

$y' = \dfrac{-16x}{(1 + 4x^2)^2}$

$y'\left(\dfrac{1}{2}\right) = \dfrac{-8}{2^2} = -2$

Tangent line: $y - 1 = -2\left(x - \dfrac{1}{2}\right)$

$y = -2x + 2$

Intersection points: $\left(\dfrac{1}{2}, 1\right)$, $(0, 2)$

$$A = \int_{0}^{1/2} \left[\frac{2}{1 + 4x^2} - (-2x + 2)\right] dx = \left[\arctan(2x) + x^2 - 2x\right]_{0}^{1/2} = \arctan(1) + \frac{1}{4} - 1 = \frac{\pi}{4} - \frac{3}{4} \approx 0.0354$$

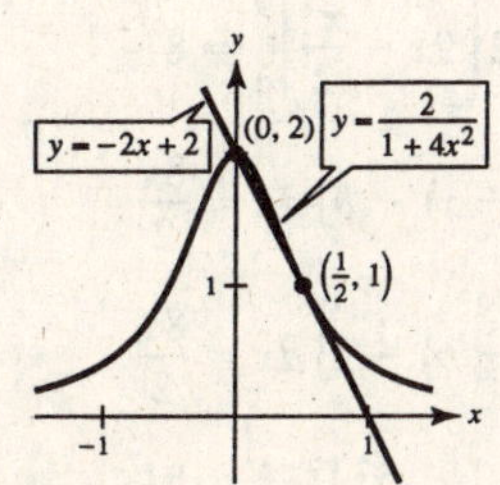

71. $x^4 - 2x^2 + 1 \le 1 - x^2$ on $[-1, 1]$

$$A = \int_{-1}^{1} [(1 - x^2) - (x^4 - 2x^2 + 1)]\,dx$$

$$= \int_{-1}^{1} (x^2 - x^4)\,dx$$

$$= \left[\frac{x^3}{3} - \frac{x^5}{5}\right]_{-1}^{1} = \frac{4}{15}$$

You can use a single integral because $x^4 - 2x^2 + 1 \le 1 - x^2$ on $[-1, 1]$.

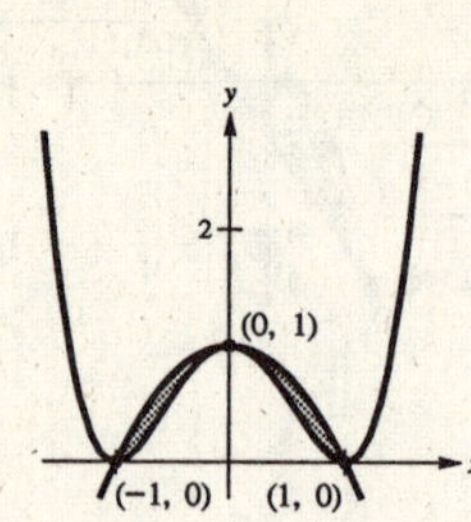

72. $x^3 \ge x$ on $[-1, 0]$, $x^3 \le x$ on $[0, 1]$

Both functions symmetric to origin.

$$\int_{-1}^{0} (x^3 - x)\,dx = -\int_{0}^{1} (x^3 - x)\,dx$$

Thus, $\displaystyle\int_{-1}^{1} (x^3 - x)\,dx = 0.$

$$A = 2\int_{0}^{1} (x - x^3)\,dx$$

$$= 2\left[\frac{x^2}{2} - \frac{x^4}{4}\right]_{0}^{1} = \frac{1}{2}$$

73. Offer 2 is better because the accumulated salary (area under the curve) is larger.

74. Proposal 2 is better since the cumulative deficit (the area under the curve) is less.

75.

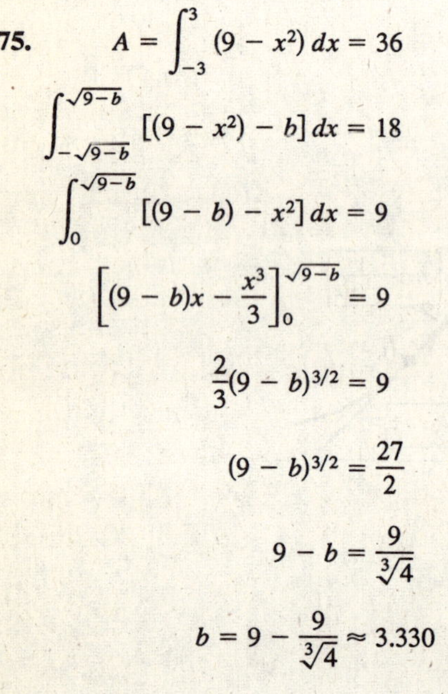

$$A = \int_{-3}^{3} (9 - x^2)\,dx = 36$$

$$\int_{-\sqrt{9-b}}^{\sqrt{9-b}} [(9 - x^2) - b]\,dx = 18$$

$$\int_{0}^{\sqrt{9-b}} [(9 - b) - x^2]\,dx = 9$$

$$\left[(9 - b)x - \frac{x^3}{3}\right]_{0}^{\sqrt{9-b}} = 9$$

$$\frac{2}{3}(9 - b)^{3/2} = 9$$

$$(9 - b)^{3/2} = \frac{27}{2}$$

$$9 - b = \frac{9}{\sqrt[3]{4}}$$

$$b = 9 - \frac{9}{\sqrt[3]{4}} \approx 3.330$$

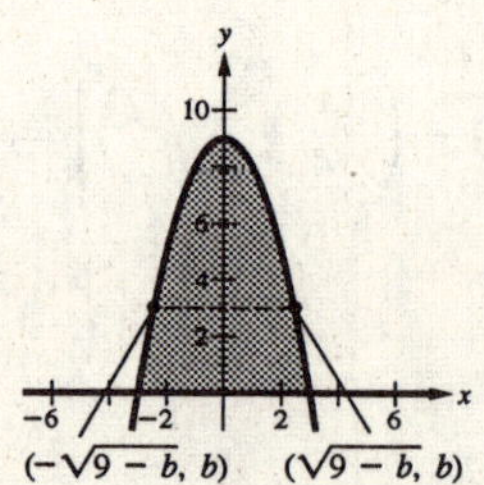

76. $$A = 2\int_{0}^{9} (9 - x)\,dx = 2\left[9x - \frac{x^2}{2}\right]_{0}^{9} = 81$$

$$2\int_{0}^{9-b} [(9 - x) - b]\,dx = \frac{81}{2}$$

$$2\int_{0}^{9-b} [(9 - b) - x]\,dx = \frac{81}{2}$$

$$2\left[(9 - b)x - \frac{x^2}{2}\right]_{0}^{9-b} = \frac{81}{2}$$

$$(9 - b)(9 - b) = \frac{81}{2}$$

$$9 - b = \frac{9}{\sqrt{2}}$$

$$b = 9 - \frac{9}{\sqrt{2}} \approx 2.636$$

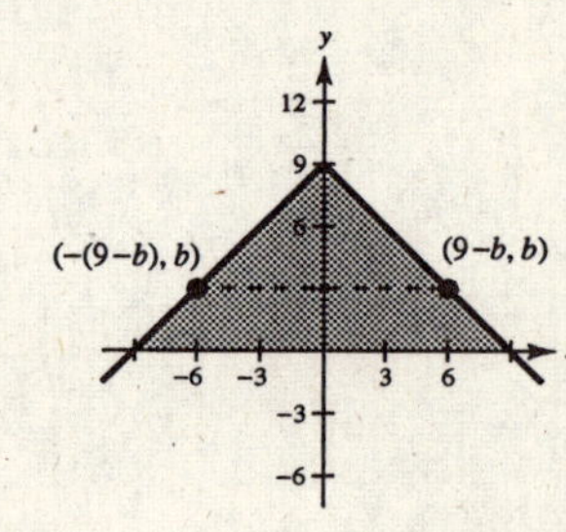

77. Area of triangle OAB is $\frac{1}{2}(4)(4) = 8$.

$$4 = \int_0^a (4 - x)\,dx = \left[4x - \frac{x^2}{2}\right]_0^a = 4a - \frac{a^2}{2}$$

$$a^2 - 8a + 8 = 0$$

$$a = 4 \pm 2\sqrt{2}$$

Since $0 < a < 4$, select $a = 4 - 2\sqrt{2} \approx 1.172$.

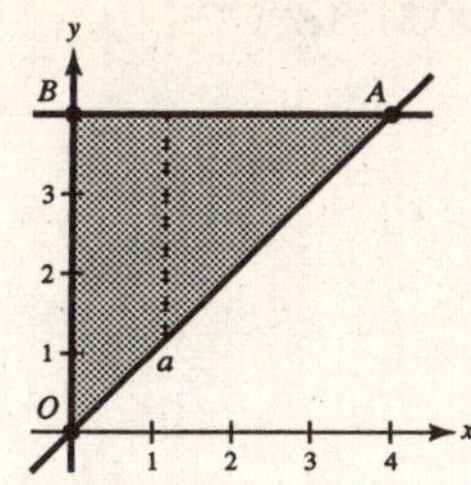

78. Total area $= \displaystyle\int_{-2}^{2} (4 - y^2)\,dy = 2\int_0^2 (4 - y^2)\,dy$

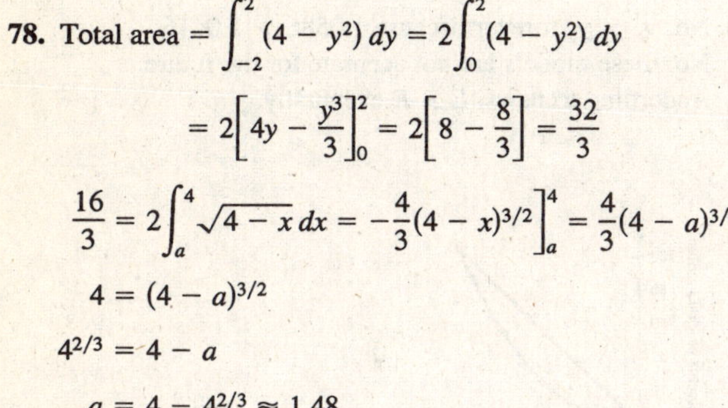

$$= 2\left[4y - \frac{y^3}{3}\right]_0^2 = 2\left[8 - \frac{8}{3}\right] = \frac{32}{3}$$

$$\frac{16}{3} = 2\int_a^4 \sqrt{4 - x}\,dx = -\frac{4}{3}(4 - x)^{3/2}\Big]_a^4 = \frac{4}{3}(4 - a)^{3/2}$$

$$4 = (4 - a)^{3/2}$$

$$4^{2/3} = 4 - a$$

$$a = 4 - 4^{2/3} \approx 1.48$$

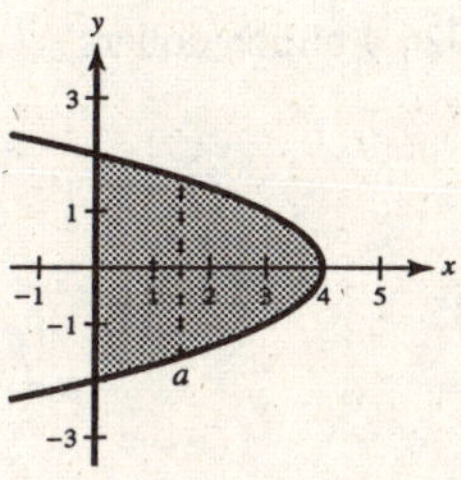

79. $\displaystyle\lim_{\|\Delta\| \to 0} \sum_{i=1}^{n} (x_i - x_i^2)\,\Delta x$

where $x_i = \dfrac{i}{n}$ and $\Delta x = \dfrac{1}{n}$ is the same as

$$\int_0^1 (x - x^2)\,dx = \left[\frac{x^2}{2} - \frac{x^3}{3}\right]_0^1 = \frac{1}{6}.$$

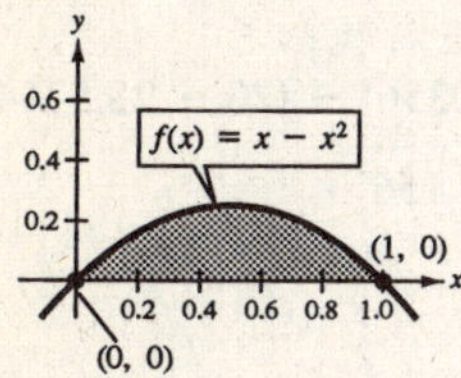

80. $\displaystyle\lim_{\|\Delta\| \to 0} \sum_{i=1}^{n} (4 - x_i^2)\,\Delta x$

where $x_i = -2 + \dfrac{4i}{n}$ and $\Delta x = \dfrac{4}{n}$ is the same as

$$\int_{-2}^{2} (4 - x^2)\,dx = \left[4x - \frac{x^3}{3}\right]_{-2}^{2} = \frac{32}{3}.$$

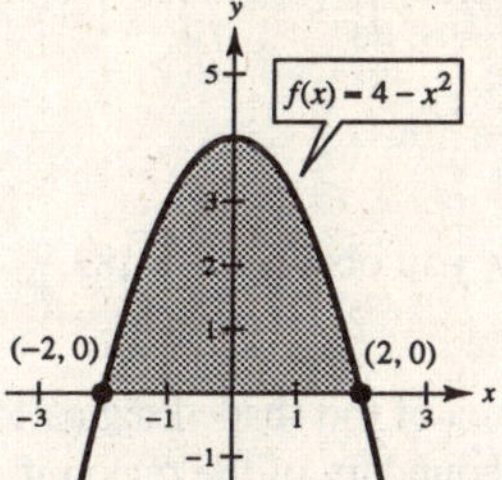

81. $\displaystyle\int_0^5 [(7.21 + 0.58t) - (7.21 + 0.45t)]\,dt = \int_0^5 0.13t\,dt = \left[\frac{0.13t^2}{2}\right]_0^5 = \$1.625$ billion

82. $\displaystyle\int_0^5 [(7.21 + 0.26t + 0.02t^2) - (7.21 + 0.1t + 0.01t^2)]\,dt = \int_0^5 (0.01t^2 + 0.16t)\,dt$

$$= \left[\frac{0.01t^3}{3} + \frac{0.16t^2}{2}\right]_0^5$$

$$= \frac{29}{12} \text{ billion} \approx \$2.417 \text{ billion}$$

83. (a) $y_1 = (270.3151)(1.0586)^t = 270.3151e^{0.05695t}$

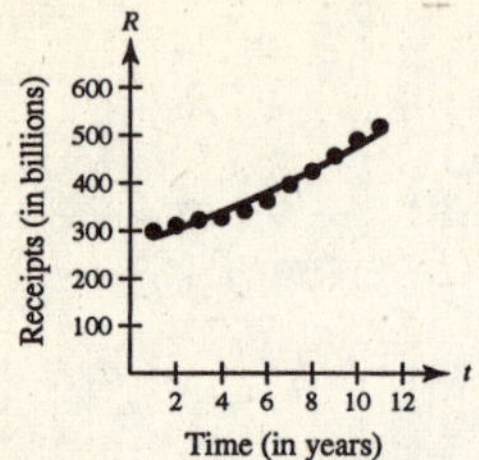

(b) $y_2 = (239.9704)(1.0416)^t = 239.9704e^{0.04074t}$

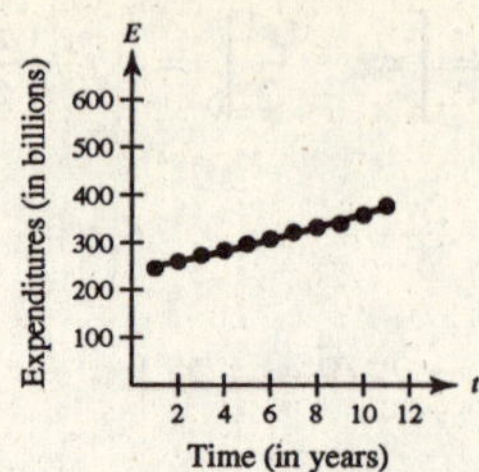

(c) Surplus $= \int_{12}^{17} (y_1 - y_2)\, dt \approx 926.4$ billion dollars

(Answers will vary.)

(d) No, $y_1 > y_2$ forever because $1.0586 > 1.0416$.
No, these models are not accurate for the future.
According to news, $E > R$ eventually.

84. (a) $y_1 = 0.0124x^2 - 0.385x + 7.85$

(b)

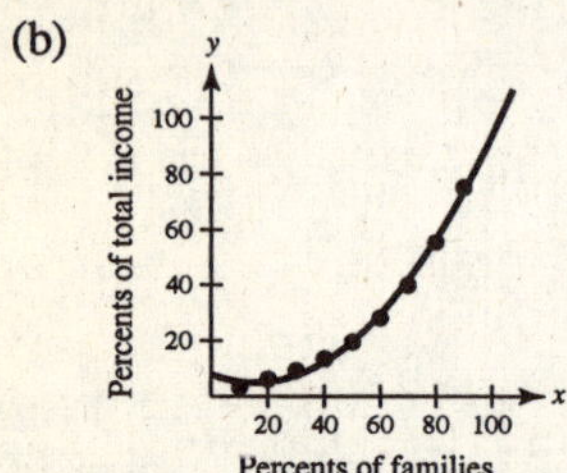

(c)

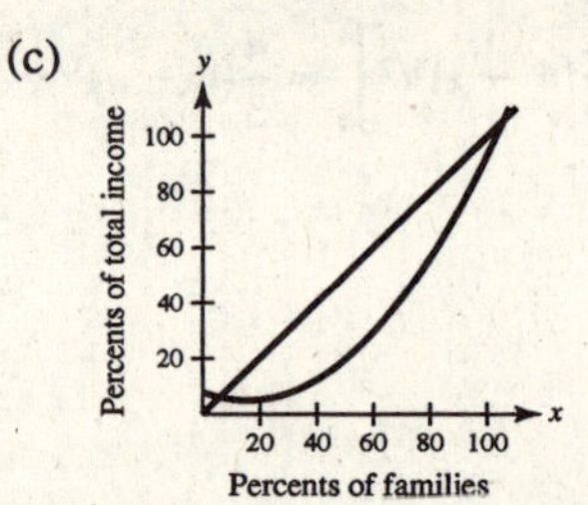

(d) Income inequality $= \int_0^{100} [x - y_1]\, dx \approx 2006.7$

85. 5%: $P_1 = 893{,}000e^{(0.05)t}$

$3\frac{1}{2}$%: $P_2 = 893{,}000e^{(0.035)t}$

Difference in profits over 5 years:
$$\int_0^5 [893{,}000e^{0.05t} - 893{,}000e^{0.035t}]\, dt = 893{,}000\left[\frac{e^{0.05t}}{0.05} - \frac{e^{0.035t}}{0.035}\right]_0^5$$
$$\approx 893{,}000[(25.6805 - 34.0356) - (20 - 28.5714)]$$
$$\approx 893{,}000(0.2163) \approx \$193{,}156$$

Note: Using a graphing utility, you obtain \$193,183.

86. The total area is 8 times the area of the shaded region to the right.
A point (x, y) is on the upper boundary of the region if

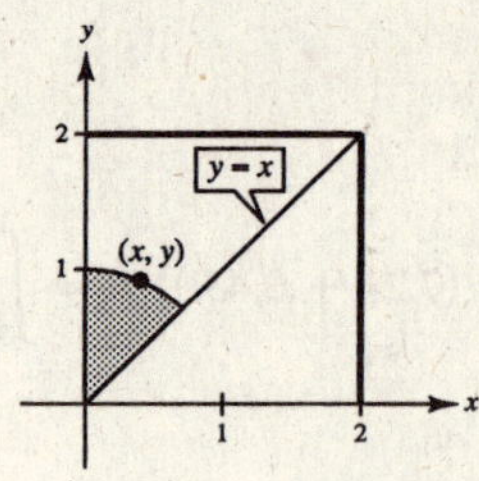

$$\sqrt{x^2 + y^2} = 2 - y$$
$$x^2 + y^2 = 4 - 4y + y^2$$
$$x^2 = 4 - 4y$$
$$4y = 4 - x^2$$
$$y = 1 - \frac{x^2}{4}.$$

We now determine where this curve intersects the line $y = x$.

$$x = 1 - \frac{x^2}{4}$$
$$x^2 + 4x - 4 = 0$$
$$x = \frac{-4 \pm \sqrt{16 + 16}}{2} = -2 \pm 2\sqrt{2} \Rightarrow x = -2 + 2\sqrt{2}$$

$$\text{Total area} = 8\int_0^{-2+2\sqrt{2}} \left(1 - \frac{x^2}{4} - x\right) dx = 8\left[x - \frac{x^3}{12} - \frac{x^2}{2}\right]_0^{-2+2\sqrt{2}} = \frac{16}{3}(4\sqrt{2} - 5) \approx 8(0.4379) = 3.503$$

87. The curves intersect at the point where the slope of y_2 equals that of y_1, 1.

$$y_2 = 0.08x^2 + k \Rightarrow y'_2 = 0.16x = 1 \Rightarrow x = \frac{1}{0.16} = 6.25$$

(a) The value of k is given by

$$y_1 = y_2$$
$$6.25 = (0.08)(6.25)^2 + k$$
$$k = 3.125.$$

(b) $\text{Area} = 2\int_0^{6.25} (y_2 - y_1)\,dx$

$$= 2\int_0^{6.25} (0.08x^2 + 3.125 - x)\,dx$$
$$= 2\left[\frac{0.08x^3}{3} + 3.125x - \frac{x^2}{2}\right]_0^{6.25}$$
$$= 2(6.510417) \approx 13.02083$$

88. (a) $A = 2\left[\int_0^5 \left(1 - \frac{1}{3}\sqrt{5-x}\right)dx + \int_5^{5.5} (1 - 0)\,dx\right]$

$$= 2\left(\left[x + \frac{2}{9}(5-x)^{3/2}\right]_0^5 + \left[x\right]_5^{5.5}\right)$$
$$= 2\left(5 - \frac{10\sqrt{5}}{9} + 5.5 - 5\right) \approx 6.031 \text{ m}^2$$

(b) $V = 2A \approx 2(6.031) \approx 12.062 \text{ m}^3$

(c) $5000\,V \approx 5000(12.062) = 60{,}310$ pounds

89. (a) $A \approx 6.031 - 2\left[\pi\left(\frac{1}{16}\right)^2\right] - 2\left[\pi\left(\frac{1}{8}\right)^2\right] \approx 5.908$

(b) $V = 2A \approx 2(5.908) \approx 11.816 \text{ m}^3$

(c) $5000V \approx 5000(11.816) = 59{,}082$ pounds

90. True

91. True

92. False. Let $f(x) = x$ and $g(x) = 2x - x^2$. f and g intersect at $(1, 1)$, the midpoint of $[0, 2]$. But

$$\int_a^b [f(x) - g(x)]\,dx = \int_0^2 [x - (2x - x^2)]\,dx = \frac{2}{3} \neq 0.$$

93. Line: $y = \dfrac{-3}{7\pi}x$

$$A = \int_0^{7\pi/6} \left[\sin x + \frac{3x}{7\pi}\right]dx$$
$$= \left[-\cos x + \frac{3x^2}{14\pi}\right]_0^{7\pi/6}$$
$$= \frac{\sqrt{3}}{2} + \frac{7\pi}{24} + 1$$
$$\approx 2.7823$$

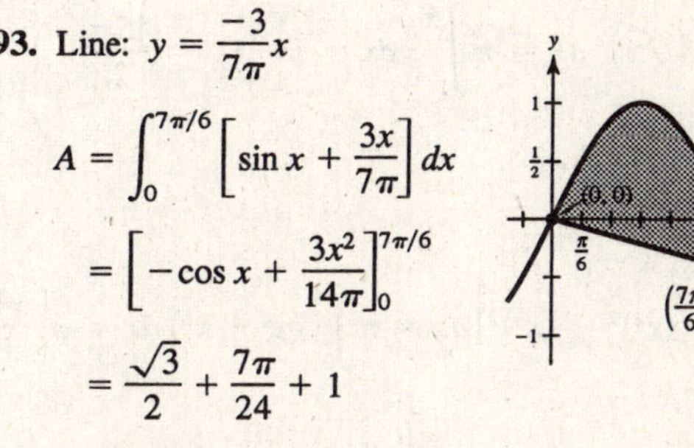

94. $A = 4\int_0^a b\sqrt{1 - \frac{x^2}{a^2}}\,dx = \frac{4b}{a}\int_0^a \sqrt{a^2 - x^2}\,dx$

$\int_0^a \sqrt{a^2 - x^2}\,dx$ is the area of $\frac{1}{4}$ of a circle $= \frac{\pi a^2}{4}$.

Hence, $A = \frac{4b}{a}\left(\frac{\pi a^2}{4}\right) = \pi ab.$

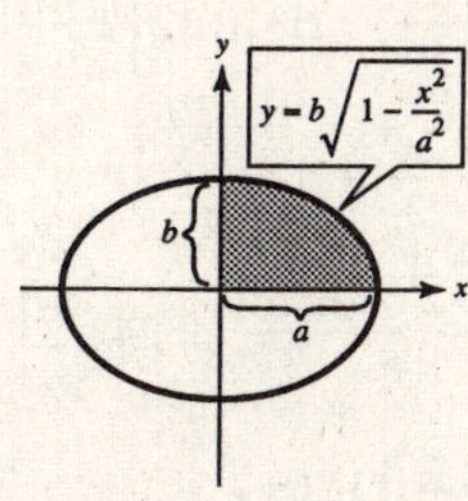

95. We want to find c such that:

$$\int_0^b [(2x - 3x^3) - c]\,dx = 0$$

$$\left[x^2 - \frac{3}{4}x^4 - cx\right]_0^b = 0$$

$$b^2 - \frac{3}{4}b^4 - cb = 0$$

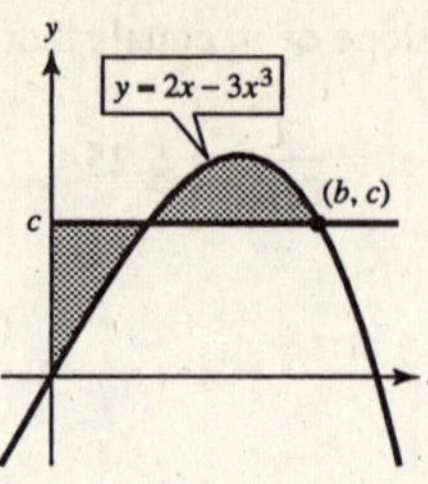

But, $c = 2b - 3b^3$ because (b, c) is on the graph.

$$b^2 - \frac{3}{4}b^4 - (2b - 3b^3)b = 0$$

$$4 - 3b^2 - 8 + 12b^2 = 0$$

$$9b^2 = 4$$

$$b = \frac{2}{3}$$

$$c = \frac{4}{9}$$

Section 7.2 Volume: The Disk Method

1. $V = \pi \int_0^1 (-x + 1)^2\,dx = \pi \int_0^1 (x^2 - 2x + 1)\,dx = \pi\left[\frac{x^3}{3} - x^2 + x\right]_0^1 = \frac{\pi}{3}$

2. $V = \pi \int_0^2 (4 - x^2)^2\,dx = \pi \int_0^2 (x^4 - 8x^2 + 16)\,dx = \pi\left[\frac{x^5}{5} - \frac{8x^3}{3} + 16x\right]_0^2 = \frac{256\pi}{15}$

3. $V = \pi \int_1^4 (\sqrt{x})^2\,dx = \pi \int_1^4 x\,dx = \pi\left[\frac{x^2}{2}\right]_1^4 = \frac{15\pi}{2}$

4. $V = \pi \int_0^3 (\sqrt{9 - x^2})^2\,dx = \pi \int_0^3 (9 - x^2)\,dx$

$= \pi\left[9x - \frac{x^3}{3}\right]_0^3 = 18\pi$

5. $V = \pi \int_0^1 [(x^2)^2 - (x^3)^2]\,dx = \pi \int_0^1 (x^4 - x^6)\,dx = \pi\left[\frac{x^5}{5} - \frac{x^7}{7}\right]_0^1 = \frac{2\pi}{35}$

6. $2 = 4 - \frac{x^2}{4}$

$8 = 16 - x^2$

$x^2 = 8$

$x = \pm 2\sqrt{2}$

$V = \pi \int_{-2\sqrt{2}}^{2\sqrt{2}} \left[\left(4 - \frac{x^2}{4}\right)^2 - (2)^2\right] dx$

$= 2\pi \int_0^{2\sqrt{2}} \left[\frac{x^4}{16} - 2x^2 + 12\right] dx$

$= 2\pi\left[\frac{x^5}{80} - \frac{2x^3}{3} + 12x\right]_0^{2\sqrt{2}}$

$= 2\pi\left[\frac{128\sqrt{2}}{80} - \frac{32\sqrt{2}}{3} + 24\sqrt{2}\right]$

$= \frac{448\sqrt{2}}{15}\pi \approx 132.69$

7. $y = x^2 \Rightarrow x = \sqrt{y}$

$V = \pi \int_0^4 (\sqrt{y})^2\,dy = \pi \int_0^4 y\,dy$

$= \pi\left[\frac{y^2}{2}\right]_0^4 = 8\pi$

8. $y = \sqrt{16 - x^2} \Rightarrow x = \sqrt{16 - y^2}$

$V = \pi \int_0^4 (\sqrt{16 - y^2})^2\,dy = \pi \int_0^4 (16 - y^2)\,dy$

$= \pi\left[16y - \frac{y^3}{3}\right]_0^4 = \frac{128\pi}{3}$

9. $y = x^{2/3} \Rightarrow x = y^{3/2}$

$V = \pi \int_0^1 (y^{3/2})^2\,dy = \pi \int_0^1 y^3\,dy = \pi\left[\frac{y^4}{4}\right]_0^1 = \frac{\pi}{4}$

10. $V = \pi\int_1^4 (-y^2 + 4y)^2\,dy = \pi\int_1^4 (y^4 - 8y^3 + 16y^2)\,dy$

$= \pi\left[\frac{y^5}{5} - 2y^4 + \frac{16y^3}{3}\right]_1^4 = \frac{459\pi}{15} = \frac{153\pi}{5}$

11. $y = \sqrt{x},\ y = 0,\ x = 4$

(a) $R(x) = \sqrt{x},\ r(x) = 0$

$V = \pi\int_0^4 (\sqrt{x})^2\,dx$

$= \pi\int_0^4 x\,dx = \left[\frac{\pi}{2}x^2\right]_0^4 = 8\pi$

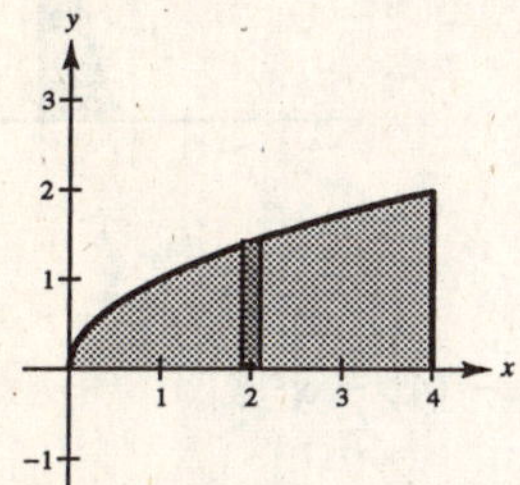

(b) $R(y) = 4,\ r(y) = y^2$

$V = \pi\int_0^2 (16 - y^4)\,dy$

$= \pi\left[16y - \frac{1}{5}y^5\right]_0^2 = \frac{128\pi}{5}$

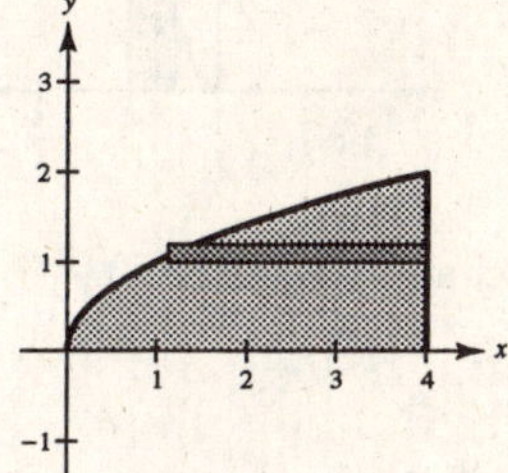

(c) $R(y) = 4 - y^2,\ r(y) = 0$

$V = \pi\int_0^2 (4 - y^2)^2\,dy$

$= \pi\int_0^2 (16 - 8y^2 + y^4)\,dy$

$= \pi\left[16y - \frac{8}{3}y^3 + \frac{1}{5}y^5\right]_0^2 = \frac{256\pi}{15}$

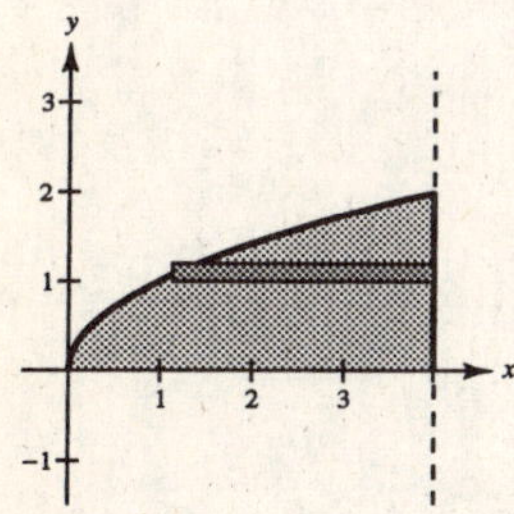

(d) $R(y) = 6 - y^2,\ r(y) = 2$

$V = \pi\int_0^2 [(6 - y^2)^2 - 4]\,dy$

$= \pi\int_0^2 (32 - 12y^2 + y^4)\,dy$

$= \pi\left[32y - 4y^3 + \frac{1}{5}y^5\right]_0^2 = \frac{192\pi}{5}$

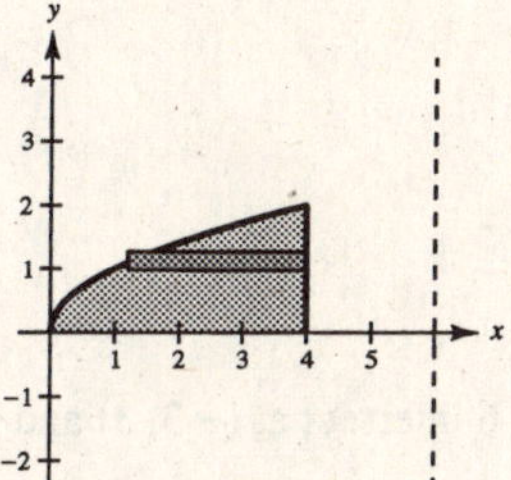

12. $y = 2x^2,\ y = 0,\ x = 2$

(a) $R(y) = 2,\ r(y) = \sqrt{y/2}$

$V = \pi\int_0^8 \left(4 - \frac{y}{2}\right)dy = \pi\left[4y - \frac{y^2}{4}\right]_0^8 = 16\pi$

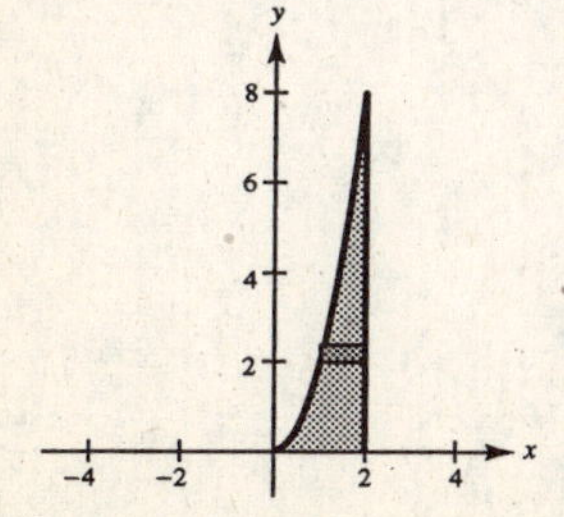

(b) $R(x) = 2x^2,\ r(x) = 0$

$V = \pi\int_0^2 4x^4\,dx = \pi\left[\frac{4x^5}{5}\right]_0^2 = \frac{128\pi}{5}$

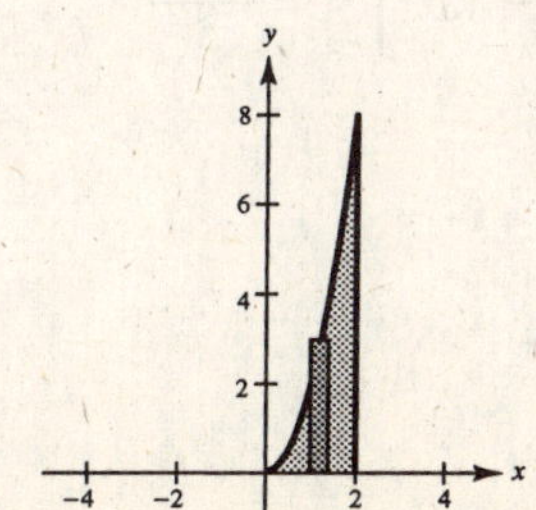

—CONTINUED—

12. —CONTINUED—

(c) $R(x) = 8,\ r(x) = 8 - 2x^2$

$$V = \pi\int_0^2 [64 - (64 - 32x^2 + 4x^4)]\,dx$$

$$= \pi\int_0^2 (32x^2 - 4x^4)\,dx = 4\pi\int_0^2 (8x^2 - x^4)\,dx$$

$$= 4\pi\left[\frac{8}{3}x^3 - \frac{1}{5}x^5\right]_0^2$$

$$= \frac{896\pi}{15}$$

(d) $R(y) = 2 - \sqrt{y/2},\ r(y) = 0$

$$V = \pi\int_0^8 \left(2 - \sqrt{\frac{y}{2}}\right)^2 dy$$

$$= \pi\int_0^8 \left(4 - 4\sqrt{\frac{y}{2}} + \frac{y}{2}\right) dy$$

$$= \pi\left[4y - \frac{4\sqrt{2}}{3}y^{3/2} + \frac{y^2}{4}\right]_0^8$$

$$= \frac{16\pi}{3}$$

13. $y = x^2,\ y = 4x - x^2$ intersect at $(0, 0)$ and $(2, 4)$.

(a) $R(x) = 4x - x^2,\ r(x) = x^2$

$$V = \pi\int_0^2 [(4x - x^2)^2 - x^4]\,dx$$

$$= \pi\int_0^2 (16x^2 - 8x^3)\,dx$$

$$= \pi\left[\frac{16}{3}x^3 - 2x^4\right]_0^2 = \frac{32\pi}{3}$$

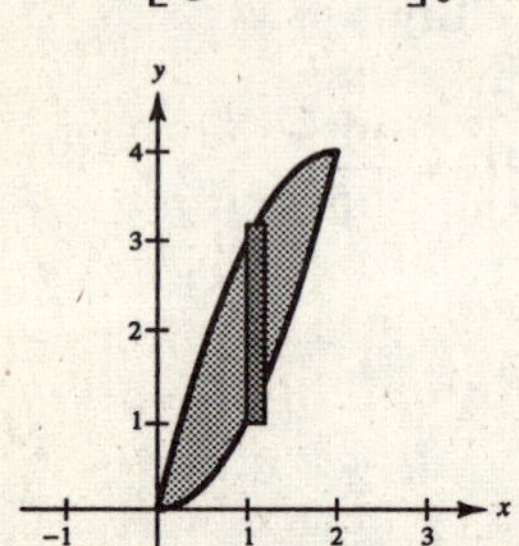

(b) $R(x) = 6 - x^2,\ r(x) = 6 - (4x - x^2)$

$$V = \pi\int_0^2 [(6 - x^2)^2 - (6 - 4x + x^2)^2]\,dx$$

$$= 8\pi\int_0^2 (x^3 - 5x^2 + 6x)\,dx$$

$$= 8\pi\left[\frac{x^4}{4} - \frac{5}{3}x^3 + 3x^2\right]_0^2 = \frac{64\pi}{3}$$

14. $y = 6 - 2x - x^2,\ y = x + 6$ intersect at $(-3, 3)$ and $(0, 6)$.

(a) $R(x) = 6 - 2x - x^2,\ r(x) = x + 6$

$$V = \pi\int_{-3}^0 [(6 - 2x - x^2)^2 - (x + 6)^2]\,dx$$

$$= \pi\int_{-3}^0 (x^4 + 4x^3 - 9x^2 - 36x)\,dx$$

$$= \pi\left[\frac{1}{5}x^5 + x^4 - 3x^3 - 18x^2\right]_{-3}^0 = \frac{243\pi}{5}$$

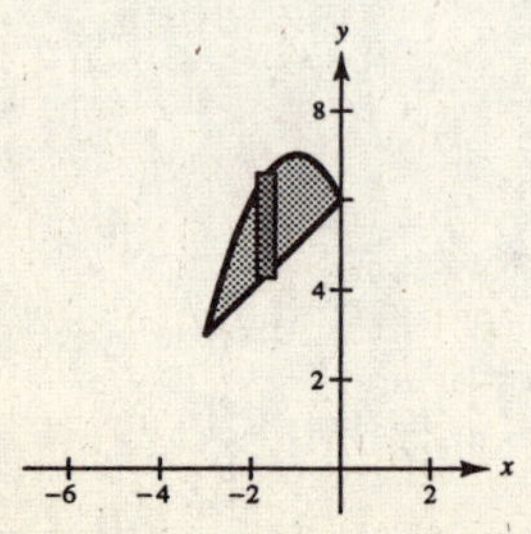

(b) $R(x) = (6 - 2x - x^2) - 3,\ r(x) = (x + 6) - 3$

$$V = \pi\int_{-3}^0 [(3 - 2x - x^2)^2 - (x + 3)^2]\,dx$$

$$= \pi\int_{-3}^0 (x^4 + 4x^3 - 3x^2 - 18x)\,dx$$

$$= \pi\left[\frac{1}{5}x^5 + x^4 - x^3 - 9x^2\right]_{-3}^0 = \frac{108\pi}{5}$$

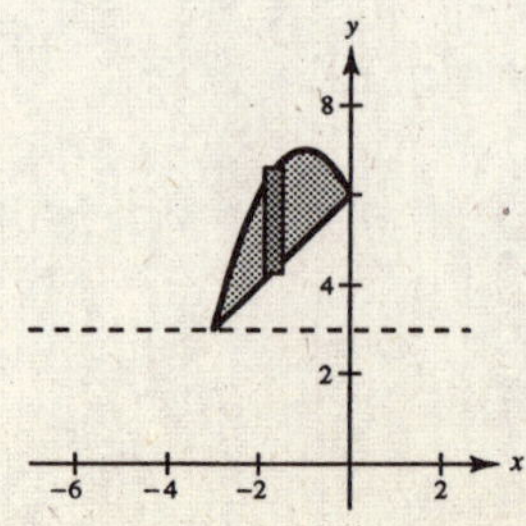

29. $x^2 + 1 = -x^2 + 2x + 5$

$2x^2 - 2x - 4 = 0$

$x^2 - x - 2 = 0$

$(x - 2)(x + 1) = 0$

The curves intersect at $(-1, 2)$ and $(2, 5)$.

$$V = \pi\int_0^2 [(5 + 2x - x^2)^2 - (x^2 + 1)^2)]\,dx + \pi\int_2^3 [(x^2 + 1)^2 - (5 + 2x - x^2)^2]\,dx$$

$$= \pi\int_0^2 (-4x^3 - 8x^2 + 20x + 24)\,dx + \pi\int_2^3 (4x^3 + 8x^2 - 20x - 24)\,dx$$

$$= \pi\left[-x^4 - \frac{8}{3}x^3 + 10x^2 + 24x\right]_0^2 + \pi\left[x^4 + \frac{8}{3}x^3 - 10x^2 - 24x\right]_2^3$$

$$= \pi\frac{152}{3} + \pi\frac{125}{3} = \frac{277\pi}{3}$$

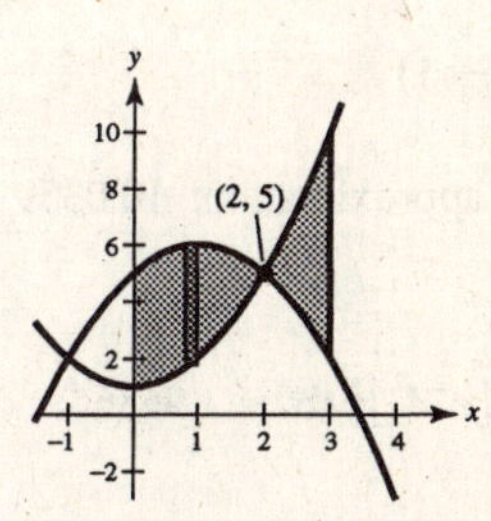

30. $$V = \pi\int_0^4 \left[\left(4 - \frac{1}{2}x\right)^2 - \left(\sqrt{x}\right)^2\right] dx + \pi\int_4^8 \left[\left(\sqrt{x}\right)^2 - \left(4 - \frac{1}{2}x\right)^2\right] dx$$

$$= \pi\int_0^4 \left(\frac{x^2}{4} - 5x + 16\right) dx + \pi\int_4^8 \left(-\frac{x^2}{4} + 5x - 16\right) dx$$

$$= \pi\left[\frac{x^3}{12} - \frac{5x^2}{2} + 16x\right]_0^4 + \pi\left[-\frac{x^3}{12} + \frac{5x^2}{2} - 16x\right]_4^8$$

$$= \frac{88}{3}\pi + \frac{56}{3}\pi = 48\pi$$

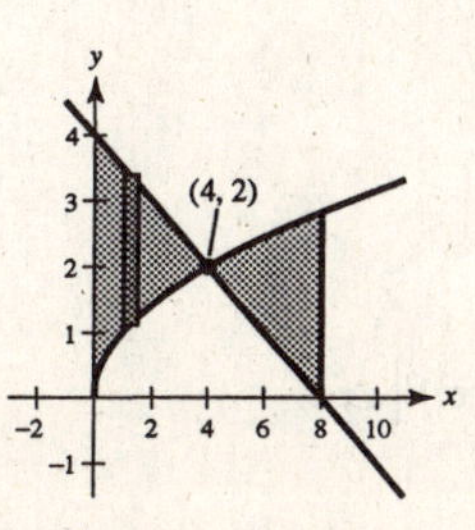

31. $y = 6 - 3x \Rightarrow x = \frac{1}{3}(6 - y)$

$$V = \pi\int_0^6 \left[\frac{1}{3}(6 - y)\right]^2 dy$$

$$= \frac{\pi}{9}\int_0^6 [36 - 12y + y^2]\,dy$$

$$= \frac{\pi}{9}\left[36y - 6y^2 + \frac{y^3}{3}\right]_0^6$$

$$= \frac{\pi}{9}\left[216 - 216 + \frac{216}{3}\right]$$

$$= 8\pi = \frac{1}{3}\pi r^2 h, \quad \text{Volume of cone}$$

32. $y = 9 - x^2,\ y = 0,\ x = 2,\ x = 3$

$x = \sqrt{9 - y}$

$$V = \pi\int_0^5 \left(\sqrt{9 - y}\right)^2 - 2]^2\,dy$$

$$= \pi\int_0^5 (5 - y)\,dy$$

$$= \pi\left[5y - \frac{y^2}{2}\right]_0^5$$

$$= \pi\left(25 - \frac{25}{2}\right) = \frac{25\pi}{2}$$

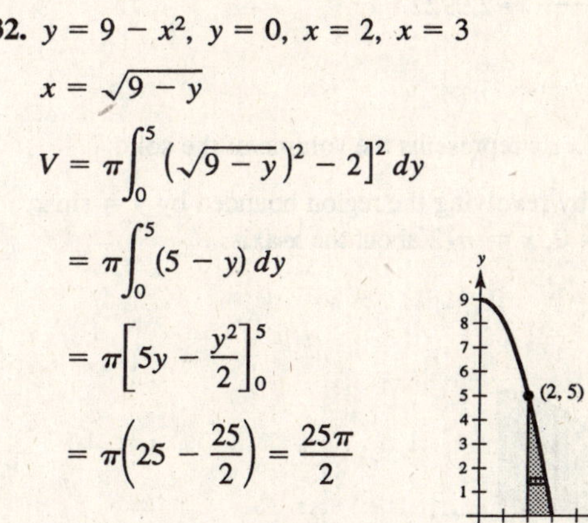

33. $$V = \pi\int_0^\pi (\sin x)^2\,dx$$

$$= \pi\int_0^\pi \frac{1 - \cos 2x}{2}\,dx$$

$$= \frac{\pi}{2}\left[x - \frac{1}{2}\sin 2x\right]_0^\pi$$

$$= \frac{\pi}{2}[\pi] = \frac{\pi^2}{2}$$

Numerical approximation: 4.9348

34. $$V = \pi\int_0^{\pi/2} [\cos x]^2\,dx$$

$$= \pi\int_0^{\pi/2} \frac{1 + \cos 2x}{2}\,dx$$

$$= \frac{\pi}{2}\left[x + \frac{1}{2}\sin 2x\right]_0^{\pi/2}$$

$$= \frac{\pi}{2}\left[\frac{\pi}{2}\right] = \frac{\pi^2}{4}$$

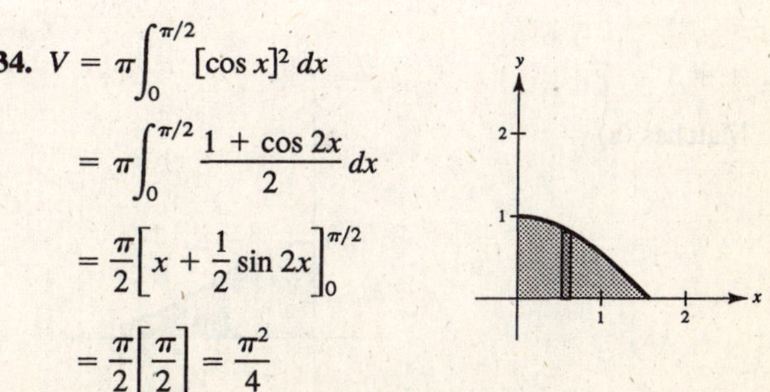

Numerical approximation: 2.4674

35. $V = \pi\int_1^2 (e^{x-1})^2\,dx$

$= \pi\int_1^2 e^{2x-2}\,dx$

$= \frac{\pi}{2}e^{2x-2}\Big]_1^2$

$= \frac{\pi}{2}(e^2 - 1)$

Numerical approximation: 10.0359

36. $V = \pi\int_{-1}^2 [e^{x/2} + e^{-x/2}]^2\,dx$

$= \pi\int_{-1}^2 [e^x + e^{-x} + 2]\,dx$

$= \pi\Big[e^x - e^{-x} + 2x\Big]_{-1}^2$

$= \pi[(e^2 - e^{-2} + 4) - (e^{-1} - e - 2)]$

$= \pi[e^2 + e + 6 - e^{-2} - e^{-1}]$

Numerical approximation: 49.0218

37. $V = \pi\int_0^2 [e^{-x^2}]^2\,dx \approx 1.9686$

38. $V = \pi\int_1^3 [\ln x]^2\,dx \approx 3.2332$

39. $V = \pi\int_0^5 [2\arctan(0.2x)]^2\,dx$

≈ 15.4115

40. $x^2 = \sqrt{2x}$

$x^4 = 2x$

$x^3 = 2$

$x = 2^{1/3} \approx 1.2599$

$V = \pi\int_0^{2^{1/3}} \left[(\sqrt{2x})^2 - (x^2)^2\right] dx$

$= \pi\int_0^{2^{1/3}} (2x - x^4)\,dx$

$= \frac{3 \cdot 2^{2/3}\pi}{5} \approx 2.9922$

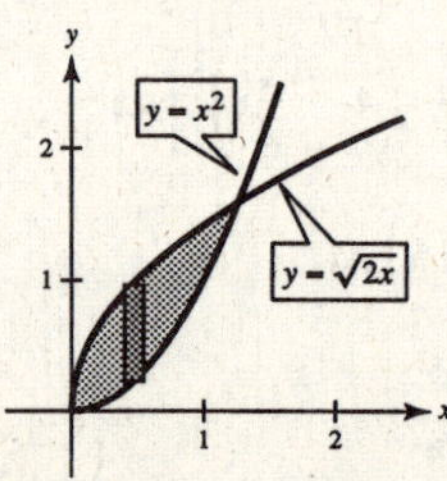

41. $\pi\int_0^{\pi/2} \sin^2 x\,dx$ represents the volume of the solid generated by revolving the region bounded by $y = \sin x$, $y = 0$, $x = 0$, $x = \pi/2$ about the x-axis.

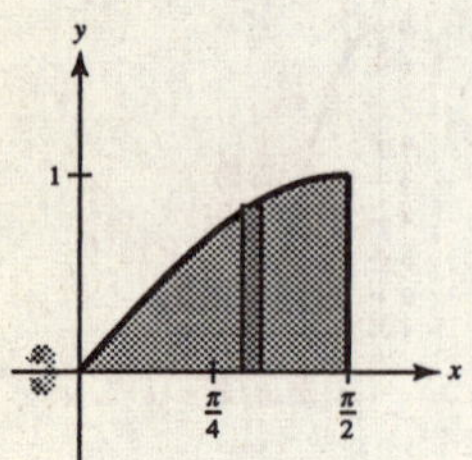

42. $\pi\int_2^4 y^4\,dy$ represents the volume of the solid generated by revolving the region bounded by $x = y^2$, $x = 0$, $y = 2$, $y = 4$ about the y-axis.

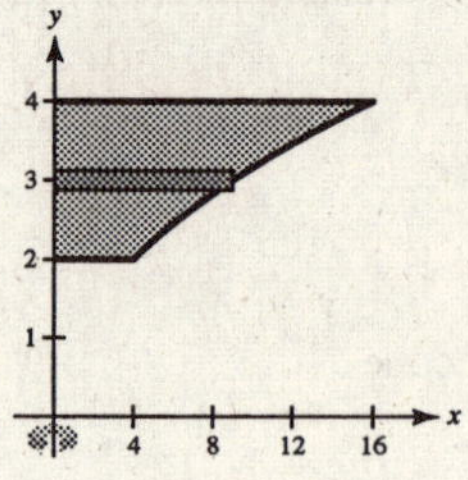

43. $A \approx 3$

Matches (a)

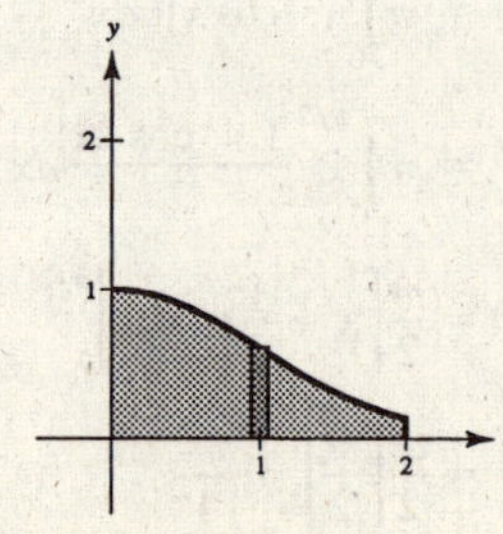

44. $A \approx \frac{3}{4}$

Matches (b)

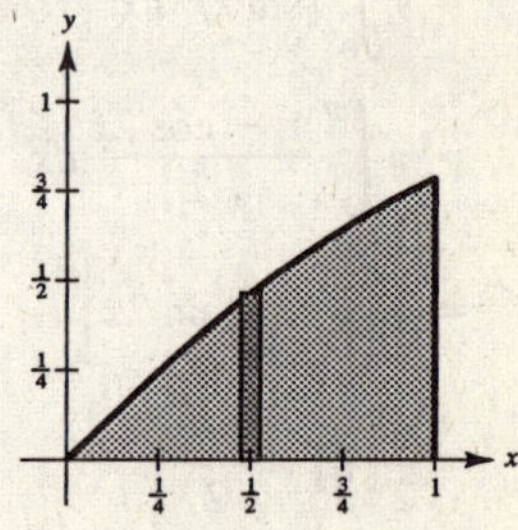

45.

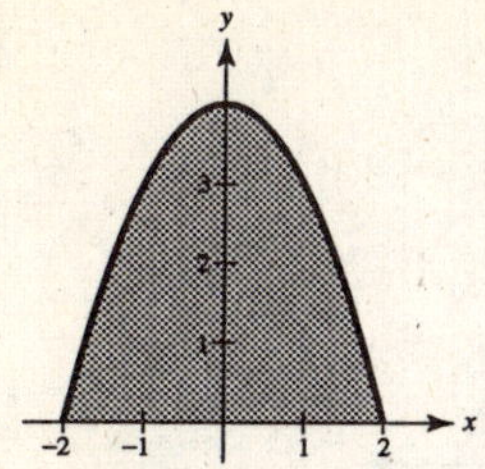
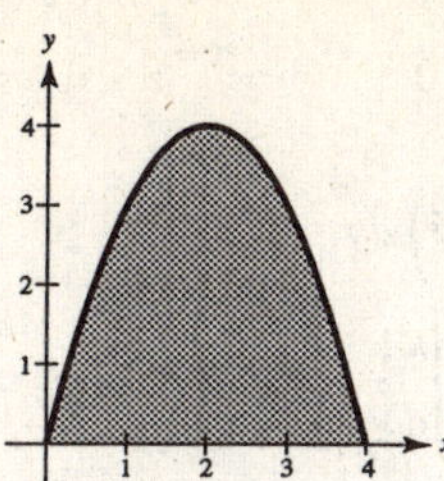

The volumes are the same because the solid has been translated horizontally. $(4x - x^2 = 4 - (x - 2)^2)$

46. (a)

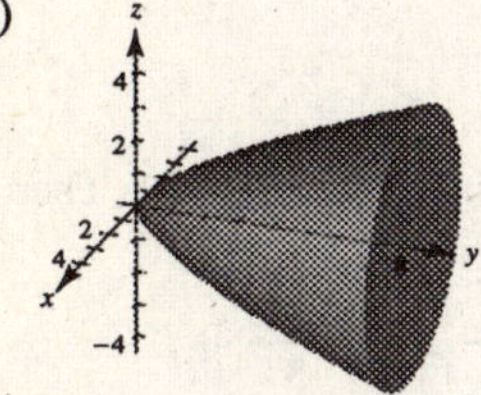

(b)

(c)

$a < c < b$

47. $R(x) = \frac{1}{2}x,\ r(x) = 0$

$$V = \pi\int_0^6 \frac{1}{4}x^2\,dx$$

$$= \left[\frac{\pi}{12}x^3\right]_0^6 = 18\pi$$

Note: $V = \frac{1}{3}\pi r^2 h$

$$= \frac{1}{3}\pi(3^2)6$$

$$= 18\pi$$

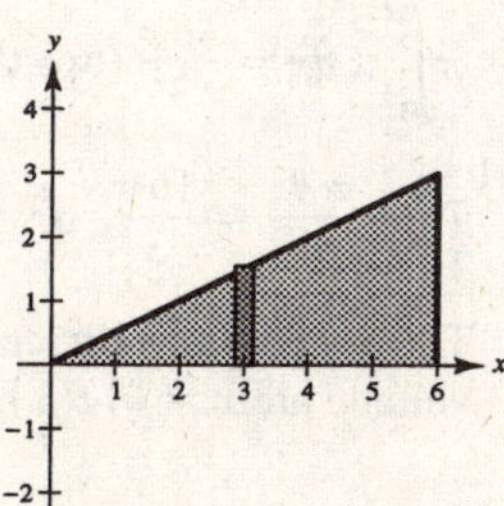

48. $R(x) = \frac{r}{h}x,\ r(x) = 0$

$$V = \pi\int_0^h \frac{r^2}{h^2}x^2\,dx$$

$$= \left[\frac{r^2\pi}{3h^2}x^3\right]_0^h$$

$$= \frac{r^2\pi}{3h^2}h^3 = \frac{1}{3}\pi r^2 h$$

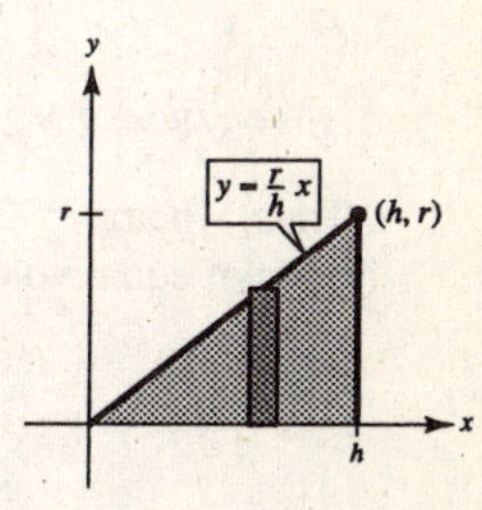

49. $R(x) = \sqrt{r^2 - x^2},\ r(x) = 0$

$$V = \pi\int_{-r}^{r} (r^2 - x^2)\,dx$$

$$= 2\pi\int_0^r (r^2 - x^2)\,dx$$

$$= 2\pi\left[r^2x - \frac{1}{3}x^3\right]_0^r$$

$$= 2\pi\left(r^3 - \frac{1}{3}r^3\right) = \frac{4}{3}\pi r^3$$

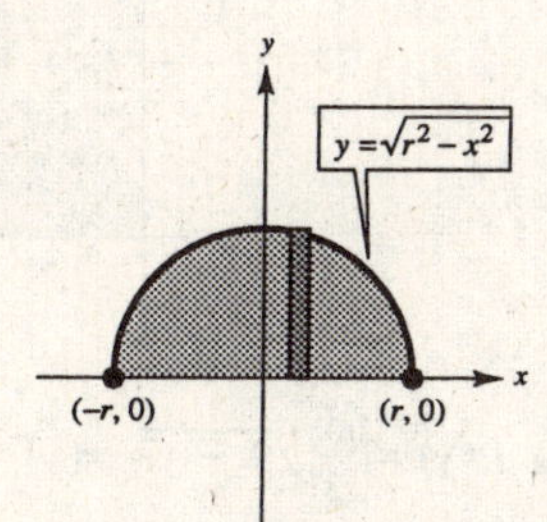

50. $x = \sqrt{r^2 - y^2},\ R(y) = \sqrt{r^2 - y^2},\ r(y) = 0$

$$V = \pi\int_h^r \left(\sqrt{r^2 - y^2}\right)^2 dy$$

$$= \pi\int_h^r (r^2 - y^2)\,dy$$

$$= \pi\left[r^2y - \frac{y^3}{3}\right]_h^r$$

$$= \pi\left[\left(r^3 - \frac{r^3}{3}\right) - \left(r^2h - \frac{h^3}{3}\right)\right]$$

$$= \pi\left(\frac{2r^3}{3} - r^2h + \frac{h^3}{3}\right)$$

$$= \frac{\pi}{3}(2r^3 - 3r^2h + h^3)$$

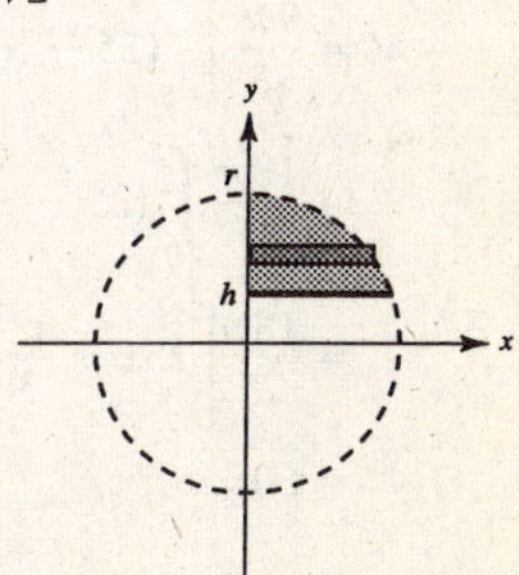

51. $x = r - \frac{r}{H}y = r\left(1 - \frac{y}{H}\right)$, $R(y) = r\left(1 - \frac{y}{H}\right)$, $r(y) = 0$

$$V = \pi\int_0^h \left[r\left(1 - \frac{y}{H}\right)\right]^2 dy = \pi r^2 \int_0^h \left(1 - \frac{2}{H}y + \frac{1}{H^2}y^2\right) dy$$

$$= \pi r^2\left[y - \frac{1}{H}y^2 + \frac{1}{3H^2}y^3\right]_0^h$$

$$= \pi r^2\left(h - \frac{h^2}{H} + \frac{h^3}{3H^2}\right)$$

$$= \pi r^2 h\left(1 - \frac{h}{H} + \frac{h^2}{3H^2}\right)$$

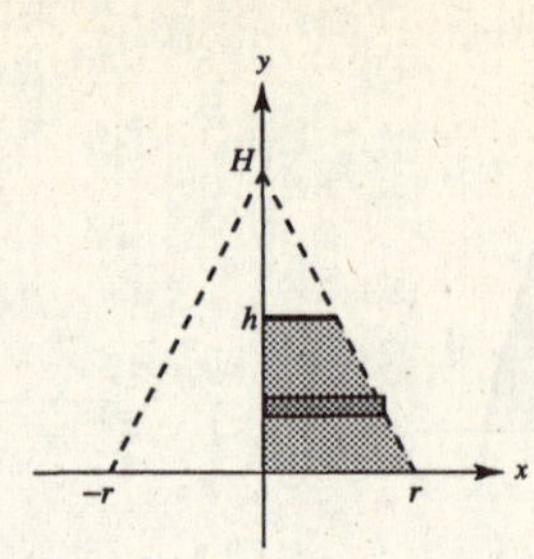

52. (a) $V = \pi\int_0^4 \left(\sqrt{x}\right)^2 dx = \pi\int_0^4 x\,dx = \left[\frac{\pi x^2}{2}\right]_0^4 = 8\pi$

Let $0 < c < 4$ and set

$$\pi\int_0^c x\,dx = \left[\frac{\pi x^2}{2}\right]_0^c = \frac{\pi c^2}{2} = 4\pi.$$

$c^2 = 8$

$c = \sqrt{8} = 2\sqrt{2}$

Thus, when $x = 2\sqrt{2}$, the solid is divided into two parts of equal volume.

(b) Set $\pi\int_0^c x\,dx = \frac{8\pi}{3}$ (one third of the volume). Then

$$\frac{\pi c^2}{2} = \frac{8\pi}{3},\ c^2 = \frac{16}{3},\ c = \frac{4}{\sqrt{3}} = \frac{4\sqrt{3}}{3}.$$

To find the other value, set

$$\pi\int_0^d x\,dx = \frac{16\pi}{3} \text{ (two thirds of the volume).}$$

Then $\frac{\pi d^2}{2} = \frac{16\pi}{3}$, $d^2 = \frac{32}{3}$, $d = \frac{\sqrt{32}}{\sqrt{3}} = \frac{4\sqrt{6}}{3}$.

The x-values that divide the solid into three parts of equal volume are $x = \left(4\sqrt{3}\right)/3$ and $x = \left(4\sqrt{6}\right)/3$.

53. $V = \pi\int_0^2 \left(\frac{1}{8}x^2\sqrt{2 - x}\right)^2 dx = \frac{\pi}{64}\int_0^2 x^4(2 - x)\,dx = \frac{\pi}{64}\left[\frac{2x^5}{5} - \frac{x^6}{6}\right]_0^2 = \frac{\pi}{30}$

54. $y = \begin{cases} \sqrt{0.1x^3 - 2.2x^2 + 10.9x + 22.2}, & 0 \le x \le 11.5 \\ 2.95, & 11.5 < x \le 15 \end{cases}$

$$V = \pi\int_0^{11.5} \left(\sqrt{0.1x^3 - 2.2x^2 + 10.9x + 22.2}\right)^2 dx + \pi\int_{11.5}^{15} 2.95^2\,dx$$

$$= \pi\left[\frac{0.1x^4}{4} - \frac{2.2x^3}{3} + \frac{10.9x^2}{2} + 22.2x\right]_0^{11.5} + \pi\left[2.95^2 x\right]_{11.5}^{15}$$

≈ 1031.9016 cubic centimeters

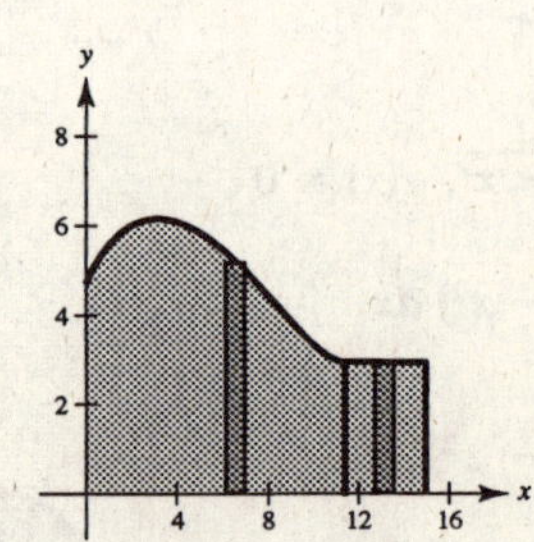

55. (a) $R(x) = \frac{3}{5}\sqrt{25 - x^2}$, $r(x) = 0$

$$V = \frac{9\pi}{25}\int_{-5}^5 (25 - x^2)\,dx$$

$$= \frac{18\pi}{25}\int_0^5 (25 - x^2)\,dx$$

$$= \frac{18\pi}{25}\left[25x - \frac{x^3}{3}\right]_0^5$$

$$= 60\pi$$

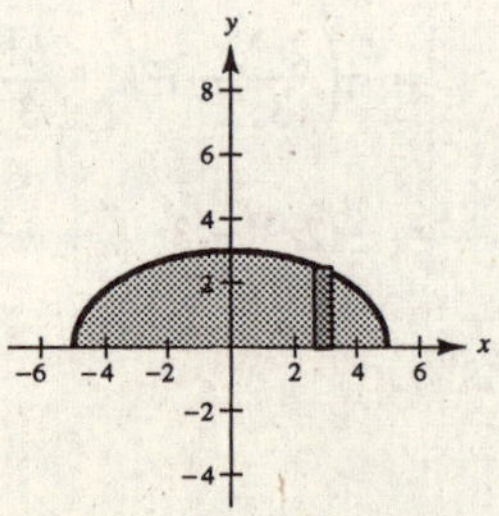

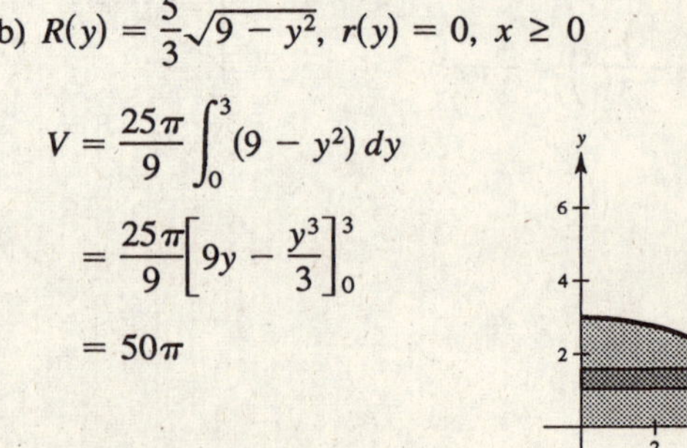

(b) $R(y) = \frac{5}{3}\sqrt{9 - y^2}$, $r(y) = 0$, $x \ge 0$

$$V = \frac{25\pi}{9}\int_0^3 (9 - y^2)\,dy$$

$$= \frac{25\pi}{9}\left[9y - \frac{y^3}{3}\right]_0^3$$

$$= 50\pi$$

56. (a) First find where $y = b$ intersects the parabola:

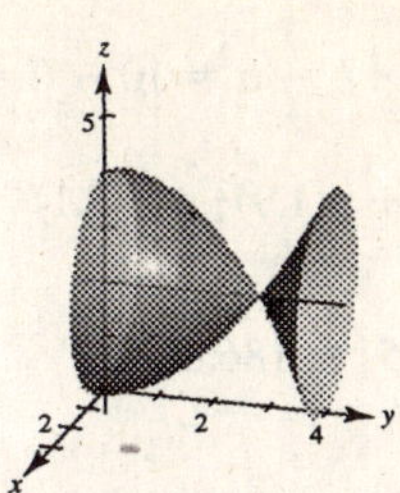

$$b = 4 - \frac{x^2}{4}$$

$$x^2 = 16 - 4b = 4(4 - b)$$

$$x = 2\sqrt{4 - b}$$

$$V = \int_0^{2\sqrt{4-b}} \pi\left[4 - \frac{x^2}{4} - b\right]^2 dx + \int_{2\sqrt{4-b}}^4 \pi\left[b - 4 + \frac{x^2}{4}\right]^2 dx$$

$$= \int_0^4 \pi\left[4 - \frac{x^2}{4} - b\right]^2 dx$$

$$= \pi\int_0^4 \left[\frac{x^4}{16} - 2x^2 + \frac{bx^2}{2} + b^2 - 8b + 16\right] dx$$

$$= \pi\left[\frac{x^5}{80} - \frac{2x^3}{3} + \frac{bx^3}{6} + b^2x - 8bx + 16x\right]_0^4$$

$$= \pi\left[\frac{64}{5} - \frac{128}{3} + \frac{32}{3}b + 4b^2 - 32b + 64\right] = \pi\left[4b^2 - \frac{64}{3}b + \frac{512}{15}\right]$$

(b) Graph of $V(b) = \pi\left[4b^2 - \dfrac{64}{3}b + \dfrac{512}{15}\right]$

(c) $V'(b) = \pi\left[8b - \dfrac{64}{3}\right] = 0 \Rightarrow b = \dfrac{64/3}{8} = \dfrac{8}{3} = 2\dfrac{2}{3}$

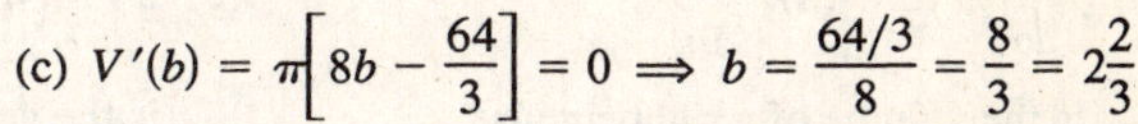

$V''(b) = 8\pi > 0 \Rightarrow b = \dfrac{8}{3}$ is a relative minimum.

Minimum volume is 17.87 for $b = 2.67$.

57. Total volume: $V = \dfrac{4\pi(50)^3}{3} = \dfrac{500{,}000\pi}{3}\ \text{ft}^3$

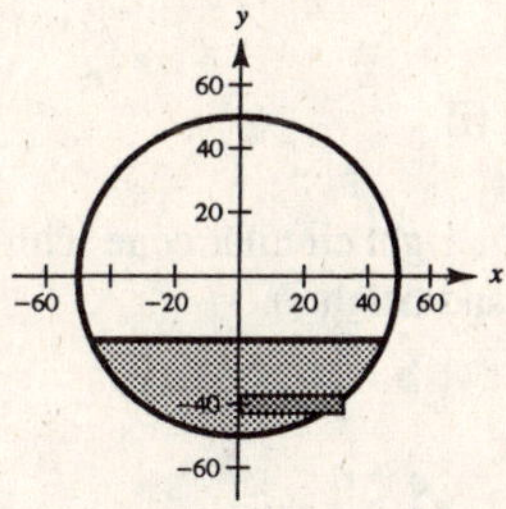

Volume of water in the tank:

$$\pi\int_{-50}^{y_0} \left(\sqrt{2500 - y^2}\right)^2 dy = \pi\int_{-50}^{y_0} (2500 - y^2)\, dy$$

$$= \pi\left[2500y - \frac{y^3}{3}\right]_{-50}^{y_0}$$

$$= \pi\left(2500y_0 - \frac{y_0^3}{3} + \frac{250{,}000}{3}\right)$$

When the tank is one-fourth of its capacity:

$$\frac{1}{4}\left(\frac{500{,}000\pi}{3}\right) = \pi\left(2500y_0 - \frac{y_0^3}{3} + \frac{250{,}000}{3}\right)$$

$$125{,}000 = 7500y_0 - y_0^3 + 250{,}000$$

$$y_0^3 - 7500y_0 - 125{,}000 = 0$$

$$y_0 \approx -17.36$$

Depth: $-17.36 - (-50) = 32.64$ feet

When the tank is three-fourths of its capacity the depth is $100 - 32.64 = 67.36$ feet.

58. (a) $V = \int_0^{10} \pi[f(x)]^2\,dx$

Simpson's Rule: $b - a = 10 - 0 = 10,\ n = 10$

$$V \approx \frac{\pi}{3}[(2.1)^2 + 4(1.9)^2 + 2(2.1)^2 + 4(2.35)^2 + 2(2.6)^2 + 4(2.85)^2 + 2(2.9)^2 + 4(2.7)^2 + 2(2.45)^2 + 4(2.2)^2 + (2.3)^2]$$

$$\approx \frac{\pi}{3}[178.405] \approx 186.83 \text{ cm}^3$$

(b) $f(x) = 0.00249x^4 - 0.0529x^3 + 0.3314x^2 - 0.4999x + 2.112$

(c) 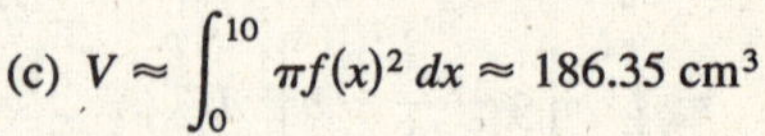
$V \approx \int_0^{10} \pi f(x)^2\,dx \approx 186.35 \text{ cm}^3$

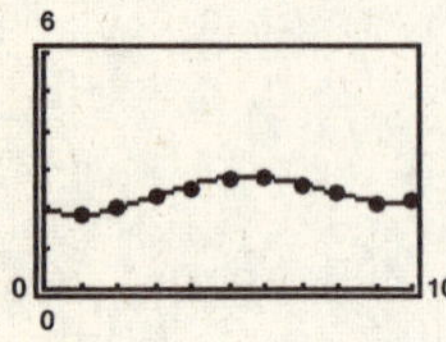

59. (a) $\pi\int_0^h r^2\,dx$ (ii)

is the volume of a right circular cylinder with radius r and height h.

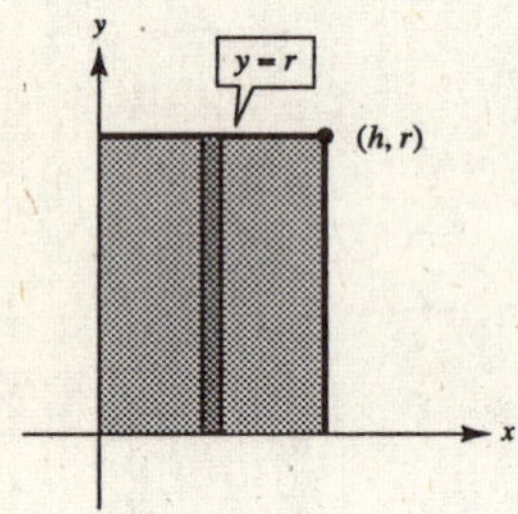

(b) 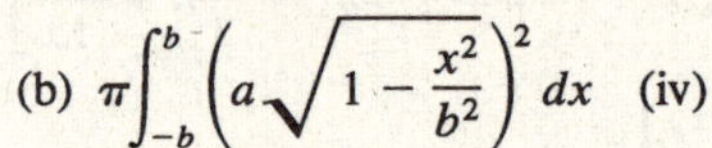
$\pi\int_{-b}^{b}\left(a\sqrt{1 - \frac{x^2}{b^2}}\right)^2 dx$ (iv)

is the volume of an ellipsoid with axes $2a$ and $2b$.

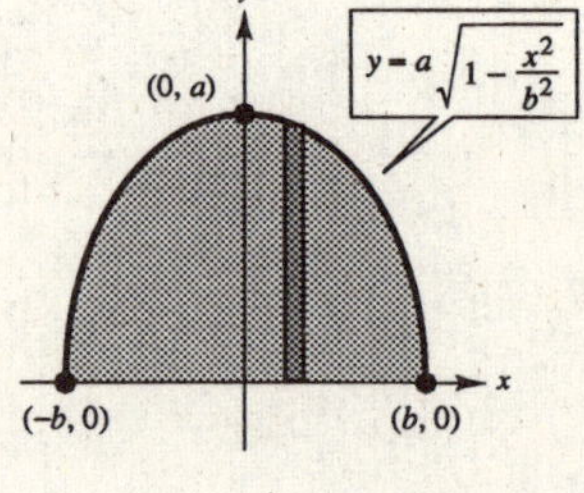

(c) 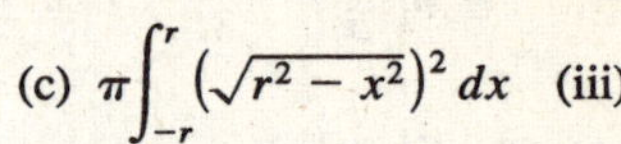
$\pi\int_{-r}^{r}\left(\sqrt{r^2 - x^2}\right)^2 dx$ (iii)

is the volume of a sphere with radius r.

(d) $\pi\int_0^h \left(\frac{rx}{h}\right)^2 dx$ (i)

is the volume of a right circular cone with the radius of the base as r and height h.

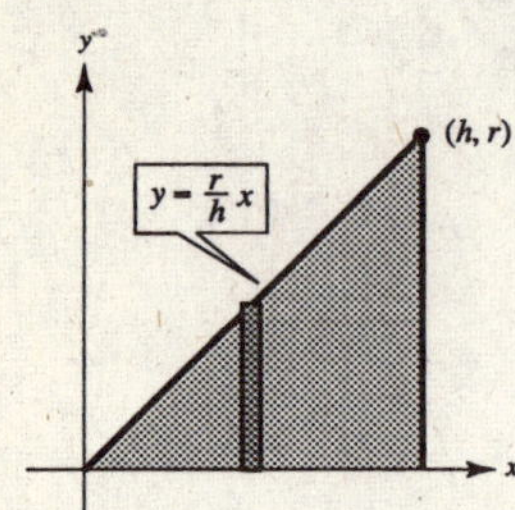

(e) $\pi\int_{-r}^{r}\left[\left(R + \sqrt{r^2 - x^2}\right)^2 - \left(R - \sqrt{r^2 - x^2}\right)^2\right] dx$ (v)

is the volume of a torus with the radius of its circular cross section as r and the distance from the axis of the torus to the center of its cross section as R.

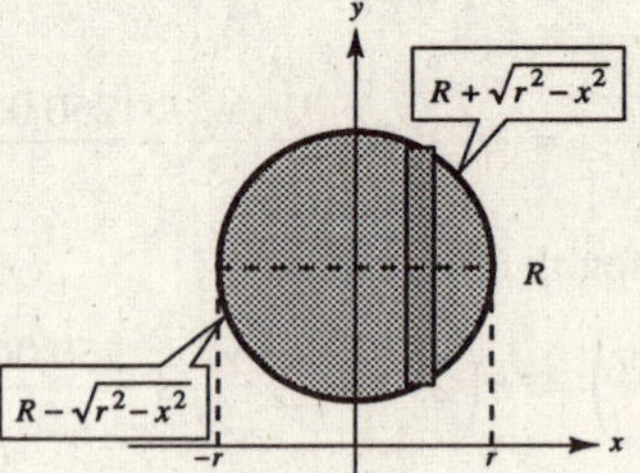

60. Let $A_1(x)$ and $A_2(x)$ equal the areas of the cross sections of the two solids for $a \le x \le b$. Since $A_1(x) = A_2(x)$, we have

$$V_1 = \int_a^b A_1(x)\,dx = \int_a^b A_2(x)\,dx = V_2.$$

Thus, the volumes are the same.

61.

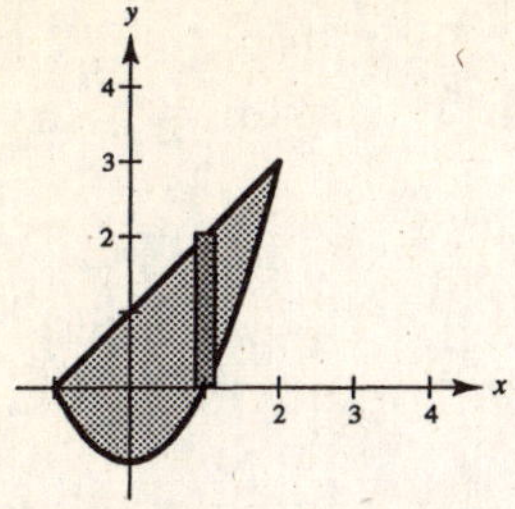

Base of cross section $= (x + 1) - (x^2 - 1) = 2 + x - x^2$

(a) $A(x) = b^2 = (2 + x - x^2)^2$

$= 4 + 4x - 3x^2 - 2x^3 + x^4$

$$V = \int_{-1}^{2} (4 + 4x - 3x^2 - 2x^3 + x^4)\,dx$$

$$= \left[4x + 2x^2 - x^3 - \frac{1}{2}x^4 + \frac{1}{5}x^5\right]_{-1}^{2} = \frac{81}{10}$$

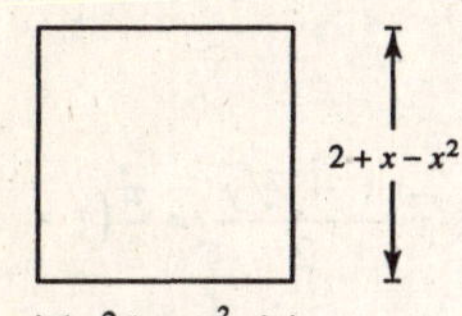

(b) $A(x) = bh = (2 + x - x^2)1$

$$V = \int_{-1}^{2} (2 + x - x^2)\,dx = \left[2x + \frac{x^2}{2} - \frac{x^3}{3}\right]_{-1}^{2} = \frac{9}{2}$$

62.

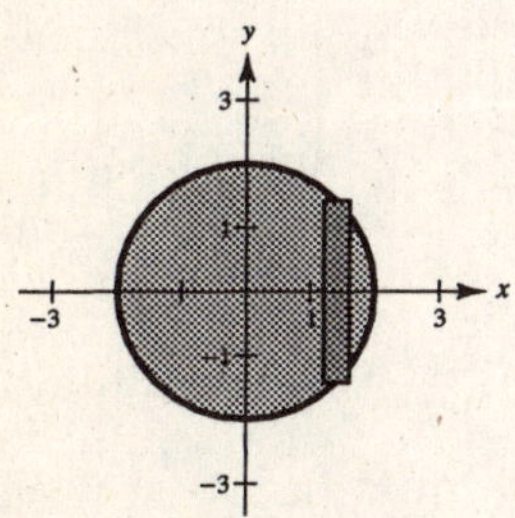

Base of cross section $= 2\sqrt{4 - x^2}$

(a) $A(x) = b^2 = \left(2\sqrt{4 - x^2}\right)^2$

$$V = \int_{-2}^{2} 4(4 - x^2)\,dx$$

$$= 4\left[4x - \frac{x^3}{3}\right]_{-2}^{2} = \frac{128}{3}$$

(b) $A(x) = \frac{1}{2}bh = \frac{1}{2}\left(2\sqrt{4 - x^2}\right)\left(\sqrt{3}\sqrt{4 - x^2}\right)$

$= \sqrt{3}(4 - x^2)$

$$V = \sqrt{3}\int_{-2}^{2} (4 - x^2)\,dx$$

$$= \sqrt{3}\left[4x - \frac{x^3}{3}\right]_{-2}^{2}$$

$$= \frac{32\sqrt{3}}{3}$$

(c) $A(x) = \frac{1}{2}\pi r^2 = \frac{\pi}{2}\left(\sqrt{4 - x^2}\right)^2 = \frac{\pi}{2}(4 - x^2)$

$$V = \frac{\pi}{2}\int_{-2}^{2} (4 - x^2)\,dx = \frac{\pi}{2}\left[4x - \frac{x^3}{3}\right]_{-2}^{2} = \frac{16\pi}{3}$$

(d) $A(x) = \frac{1}{2}bh = \frac{1}{2}\left(2\sqrt{4 - x^2}\right)\left(\sqrt{4 - x^2}\right) = 4 - x^2$

$$V = \int_{-2}^{2} (4 - x^2)\,dx = \left[4x - \frac{x^3}{3}\right]_{-2}^{2} = \frac{32}{3}$$

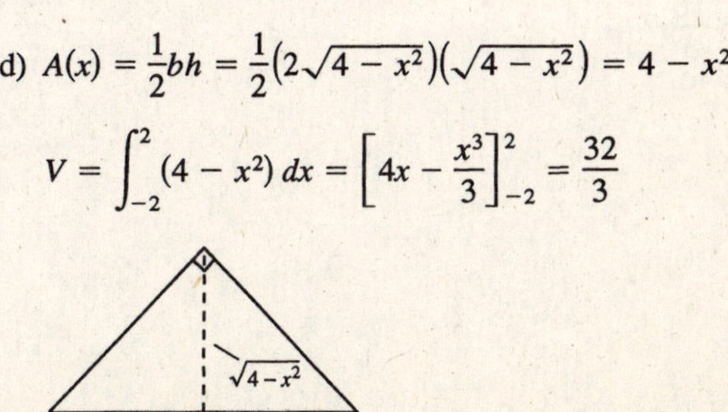

63.

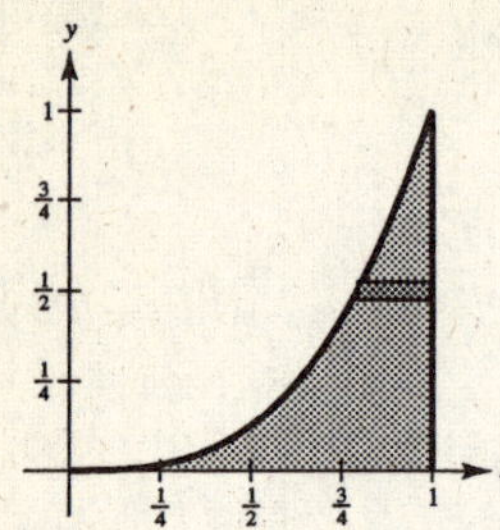

Base of cross section $= 1 - \sqrt[3]{y}$

(a) $A(y) = b^2 = \left(1 - \sqrt[3]{y}\right)^2$

$$V = \int_0^1 \left(1 - \sqrt[3]{y}\right)^2\,dy$$

$$= \int_0^1 (1 - 2y^{1/3} + y^{2/3})\,dy$$

$$= \left[y - \frac{3}{2}y^{4/3} + \frac{3}{5}y^{5/3}\right]_0^1 = \frac{1}{10}$$

(b) $A(y) = \dfrac{1}{2}\pi r^2 = \dfrac{1}{2}\pi\left(\dfrac{1 - \sqrt[3]{y}}{2}\right)^2 = \dfrac{1}{8}\pi\left(1 - \sqrt[3]{y}\right)^2$

$$V = \frac{1}{8}\pi\int_0^1 \left(1 - \sqrt[3]{y}\right)^2\,dy = \frac{\pi}{8}\left(\frac{1}{10}\right) = \frac{\pi}{80}$$

(c) $A(y) = \dfrac{1}{2}bh = \dfrac{1}{2}\left(1 - \sqrt[3]{y}\right)\left(\dfrac{\sqrt{3}}{2}\right)\left(1 - \sqrt[3]{y}\right)$

$$= \frac{\sqrt{3}}{4}\left(1 - \sqrt[3]{y}\right)^2$$

$$V = \frac{\sqrt{3}}{4}\int_0^1 \left(1 - \sqrt[3]{y}\right)^2\,dy = \frac{\sqrt{3}}{4}\left(\frac{1}{10}\right) = \frac{\sqrt{3}}{40}$$

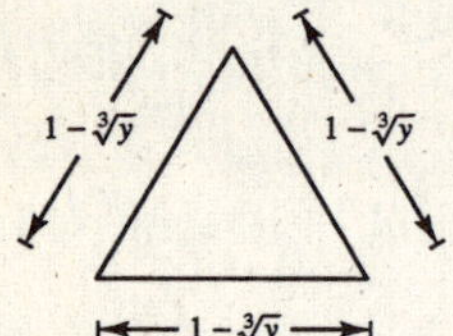

(d) $A(y) = \dfrac{1}{2}\pi ab = \dfrac{\pi}{2}(2)\left(1 - \sqrt[3]{y}\right)\dfrac{1 - \sqrt[3]{y}}{2} = \dfrac{\pi}{2}\left(1 - \sqrt[3]{y}\right)^2$

$$V = \frac{\pi}{2}\int_0^1 \left(1 - \sqrt[3]{y}\right)^2\,dy = \frac{\pi}{2}\left(\frac{1}{10}\right) = \frac{\pi}{20}$$

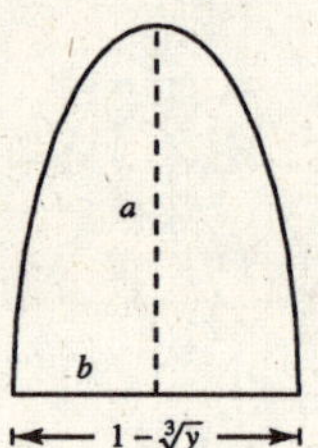

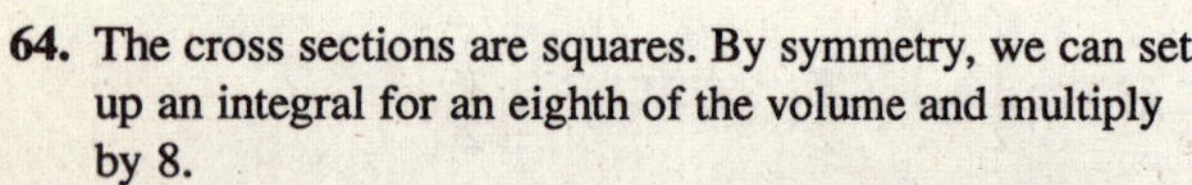

64. The cross sections are squares. By symmetry, we can set up an integral for an eighth of the volume and multiply by 8.

$A(y) = b^2 = \left(\sqrt{r^2 - y^2}\right)^2$

$$V = 8\int_0^r (r^2 - y^2)\,dy$$

$$= 8\left[r^2y - \frac{1}{3}y^3\right]_0^r$$

$$= \frac{16}{3}r^3$$

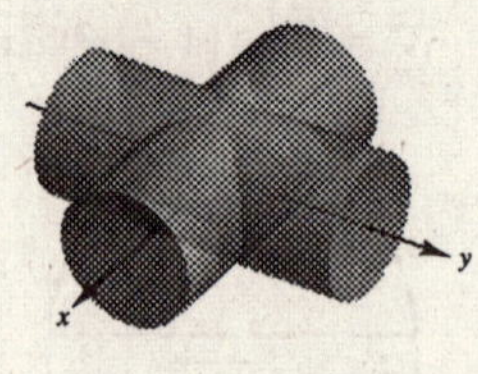

65. $V = \pi\displaystyle\int_{-\sqrt{R^2 - r^2}}^{\sqrt{R^2 - r^2}} \left[\left(\sqrt{R^2 - x^2}\right)^2 - r^2\right]dx$

$$= 2\pi\int_0^{\sqrt{R^2 - r^2}} (R^2 - r^2 - x^2)\,dx$$

$$= 2\pi\left[(R^2 - r^2)x - \frac{x^3}{3}\right]_0^{\sqrt{R^2 - r^2}}$$

$$= 2\pi\left[(R^2 - r^2)^{3/2} - \frac{(R^2 - r^2)^{3/2}}{3}\right]$$

$$= \frac{4}{3}\pi(R^2 - r^2)^{3/2}$$

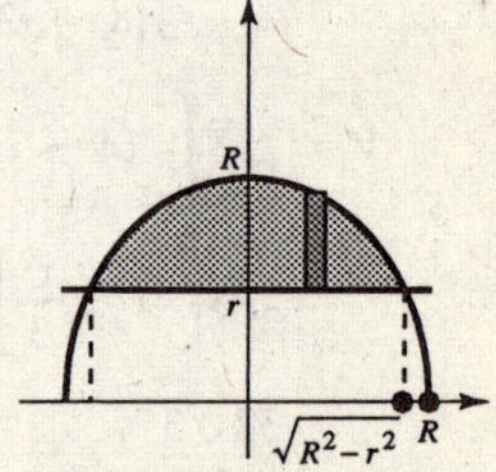

66. $\frac{4}{3}\pi(25 - r^2)^{3/2} = \frac{1}{2}\left(\frac{4}{3}\right)\pi(125)$

$$(25 - r^2)^{3/2} = \frac{125}{2}$$

$$25 - r^2 = \left(\frac{125}{2}\right)^{2/3}$$

$$25 - \frac{25}{(2^{2/3})} = r^2$$

$$25(1 - 2^{-2/3}) = r^2$$

$$r = 5\sqrt{1 - 2^{-2/3}} \approx 3.0415$$

67. $V = \pi\int_0^1 y^2\,dy = \pi\frac{y^3}{3}\Big]_0^1 = \frac{\pi}{3}$

68. $V = \pi\int_0^1 [1^2 - (1 - y)^2]\,dy$

$$= \pi\int_0^1 [2y - y^2]\,dy$$

$$= \pi\left[y^2 - \frac{y^3}{3}\right]_0^1$$

$$= \pi\left[1 - \frac{1}{3}\right] = \frac{2}{3}\pi$$

69. $V = \pi\int_0^1 (x^2 - x^4)\,dx$

$$= \pi\left[\frac{x^3}{3} - \frac{x^5}{5}\right]_0^1$$

$$= \pi\left[\frac{1}{3} - \frac{1}{5}\right]$$

$$= \frac{2\pi}{15}$$

70. $V = \pi\int_0^1 [(1 - x^2)^2 - (1 - x)^2]\,dx$

$$= \pi\int_0^1 [1 - 2x^2 + x^4 - 1 + 2x - x^2]\,dx$$

$$= \pi\int_0^1 [2x - 3x^2 + x^4]\,dx$$

$$= \pi\left[x^2 - x^3 + \frac{x^5}{5}\right]_0^1$$

$$= \pi\left[\frac{1}{5}\right] = \frac{\pi}{5}$$

71. $V = \pi\int_0^1 (1 - y)\,dy$

$$= \pi\left[y - \frac{y^2}{2}\right]_0^1$$

$$= \pi\left[1 - \frac{1}{2}\right]$$

$$= \frac{\pi}{2}$$

72. $V = \pi\int_0^1 \left(1 - \sqrt{y}\right)^2 dy$

$$= \pi\int_0^1 \left(1 - 2\sqrt{y} + y\right) dy$$

$$= \pi\left[y - \frac{4}{3}y^{3/2} + \frac{y^2}{2}\right]_0^1$$

$$= \pi\left[1 - \frac{4}{3} + \frac{1}{2}\right]$$

$$= \frac{\pi}{6}$$

73. $V = \pi\int_0^1 (y - y^2)\,dy$

$$= \pi\left[\frac{y^2}{2} - \frac{y^3}{3}\right]_0^1$$

$$= \pi\left[\frac{1}{2} - \frac{1}{3}\right]$$

$$= \frac{\pi}{6}$$

74. $V = \pi\int_0^1 \left[(1-y)^2 - \left(1-\sqrt{y}\right)^2\right] dy$

$= \pi\int_0^1 \left[1 - 2y + y^2 - 1 + 2\sqrt{y} - y\right] dy$

$= \pi\int_0^1 \left[2\sqrt{y} - 3y + y^2\right] dy$

$= \pi\left[\frac{4}{3}y^{3/2} - \frac{3y^2}{2} + \frac{y^3}{3}\right]_0^1$

$= \pi\left[\frac{4}{3} - \frac{3}{2} + \frac{1}{3}\right]$

$= \frac{\pi}{6}$

75. (a) When $a = 1$: $|x| + |y| = 1$ represents a square.

When $a = 2$: $|x|^2 + |y|^2 = 1$ represents a circle.

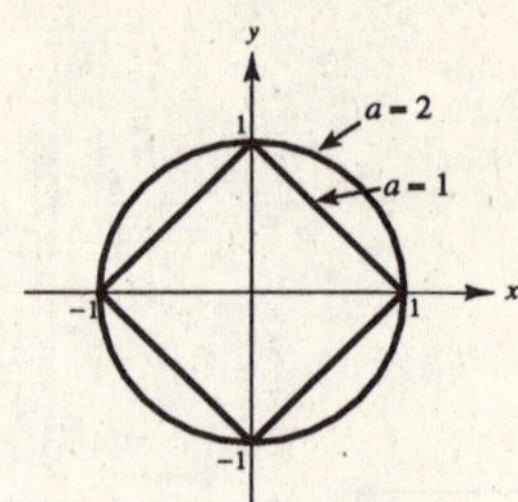

(b) $|y| = (1 - |x|^a)^{1/a}$

$A = 2\int_{-1}^{1} (1 - |x|^a)^{1/a}\, dx = 4\int_0^1 (1 - x^a)^{1/a}\, dx$

To approximate the volume of the solid, form n slices, each of whose area is approximated by the integral above. Then sum the volumes of these n slices.

76. (a) Since the cross sections are isosceles right triangles:

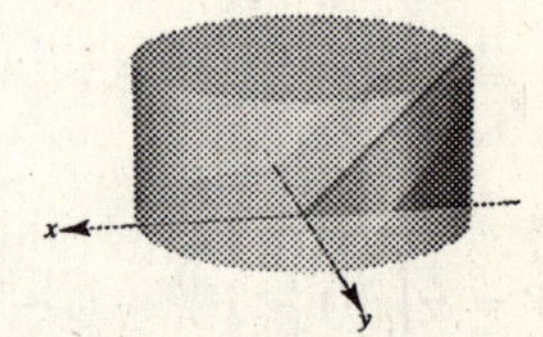

$A(x) = \frac{1}{2}bh = \frac{1}{2}\left(\sqrt{r^2 - y^2}\right)\left(\sqrt{r^2 - y^2}\right) = \frac{1}{2}(r^2 - y^2)$

$V = \frac{1}{2}\int_{-r}^{r} (r^2 - y^2)\, dy = \int_0^r (r^2 - y^2)\, dy = \left[r^2y - \frac{y^3}{3}\right]_0^r = \frac{2}{3}r^3$

(b) $A(x) = \frac{1}{2}bh = \frac{1}{2}\sqrt{r^2 - y^2}\left(\sqrt{r^2 - y^2}\tan\theta\right) = \frac{\tan\theta}{2}(r^2 - y^2)$

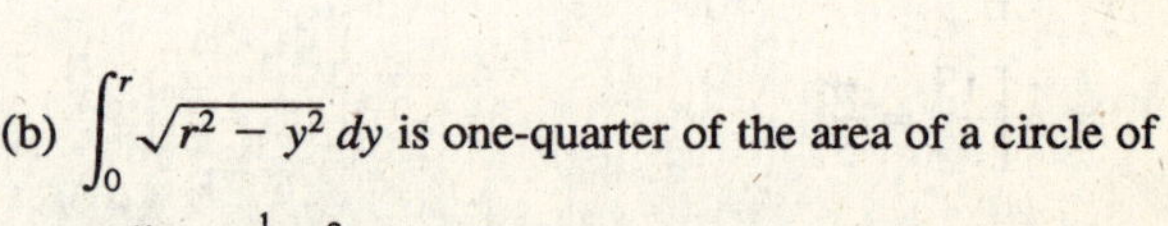

$V = \frac{\tan\theta}{2}\int_{-r}^{r} (r^2 - y^2)\, dy = \tan\theta\int_0^r (r^2 - y^2)\, dy = \tan\theta\left[r^2y - \frac{y^3}{3}\right]_0^r = \frac{2}{3}r^3\tan\theta$

As $\theta \to 90°$, $V \to \infty$.

77. (a) $(x - R)^2 + y^2 = r^2$

$x = R \pm \sqrt{r^2 - y^2}$

$V = 2\pi\int_0^r \left(\left[R + \sqrt{r^2 - y^2}\right]^2 - \left[R - \sqrt{r^2 - y^2}\right]^2\right) dy$

$= 2\pi\int_0^r 4R\sqrt{r^2 - y^2}\, dy$

$= 8\pi R\int_0^r \sqrt{r^2 - y^2}\, dy$

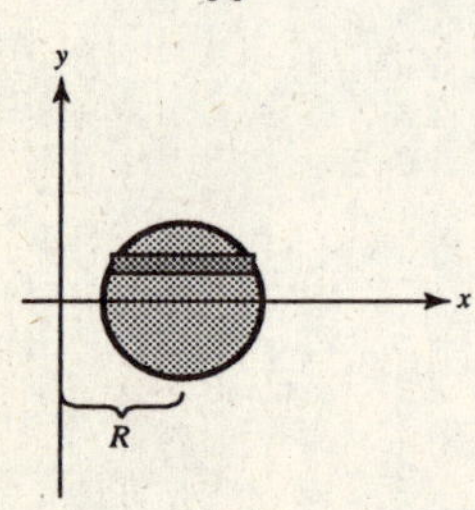

(b) $\int_0^r \sqrt{r^2 - y^2}\, dy$ is one-quarter of the area of a circle of radius r, $\frac{1}{4}\pi r^2$.

$V = 8\pi R\left(\frac{1}{4}\pi r^2\right) = 2\pi^2 r^2 R$

Section 7.3 Volume: The Shell Method

1. $p(x) = x, \ h(x) = x$

$$V = 2\pi\int_0^2 x(x)\,dx$$

$$= \left[\frac{2\pi x^3}{3}\right]_0^2 = \frac{16\pi}{3}$$

2. $p(x) = x, \ h(x) = 1 - x$

$$V = 2\pi\int_0^1 x(1 - x)\,dx$$

$$= 2\pi\int_0^1 (x - x^2)\,dx$$

$$= 2\pi\left[\frac{x^2}{2} - \frac{x^3}{3}\right]_0^1 = \frac{\pi}{3}$$

3. $p(x) = x, \ h(x) = \sqrt{x}$

$$V = 2\pi\int_0^4 x\sqrt{x}\,dx$$

$$= 2\pi\int_0^4 x^{3/2}\,dx$$

$$= \left[\frac{4\pi}{5}x^{5/2}\right]_0^4 = \frac{128\pi}{5}$$

4. $p(x) = x, \ h(x) = 8 - (x^2 + 4) = 4 - x^2$

$$V = 2\pi\int_0^2 x(4 - x^2)\,dx$$

$$= 2\pi\int_0^2 (4x - x^3)\,dx$$

$$= 2\pi\left[2x^2 - \frac{x^4}{4}\right]_0^2 = 8\pi$$

5. $p(x) = x, \ h(x) = x^2$

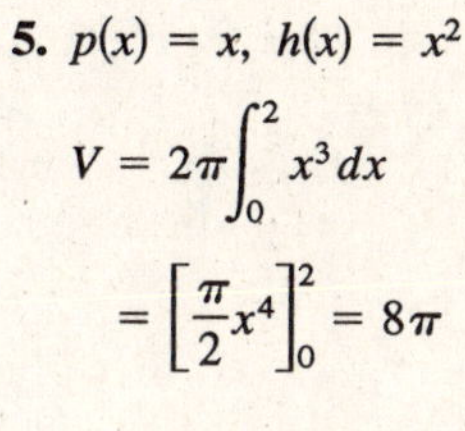

$$V = 2\pi\int_0^2 x^3\,dx$$

$$= \left[\frac{\pi}{2}x^4\right]_0^2 = 8\pi$$

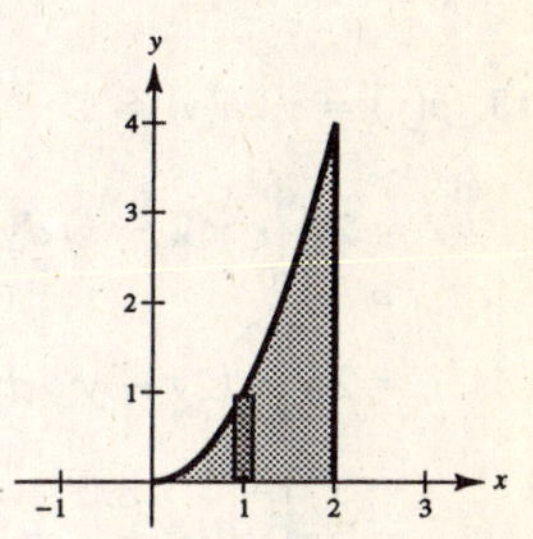

6. $p(x) = x, \ h(x) = \frac{1}{2}x^2$

$$V = 2\pi\int_0^6 \frac{1}{2}x^3\,dx$$

$$= \left[\pi\frac{x^4}{4}\right]_0^6 = 324\pi$$

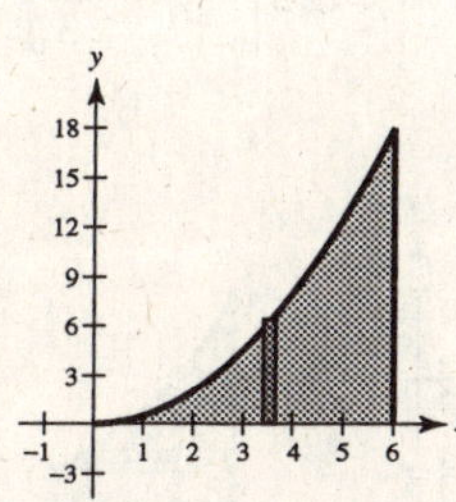

7. $p(x) = x, \ h(x) = (4x - x^2) - x^2 = 4x - 2x^2$

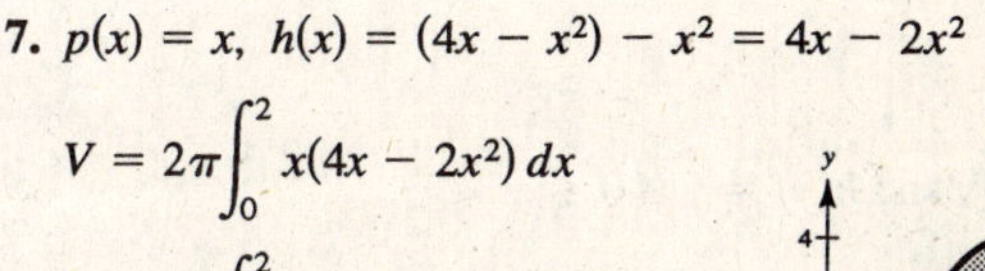

$$V = 2\pi\int_0^2 x(4x - 2x^2)\,dx$$

$$= 4\pi\int_0^2 (2x^2 - x^3)\,dx$$

$$= 4\pi\left[\frac{2}{3}x^3 - \frac{1}{4}x^4\right]_0^2 = \frac{16\pi}{3}$$

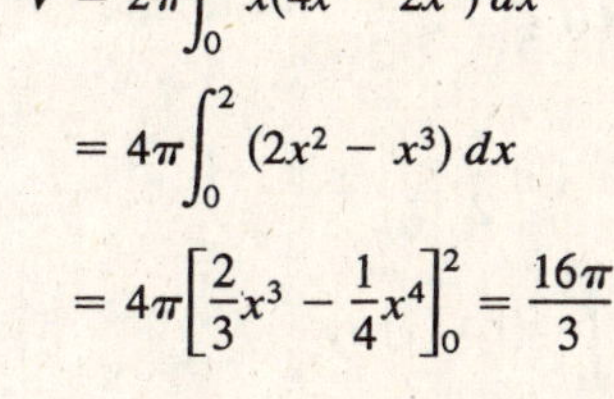

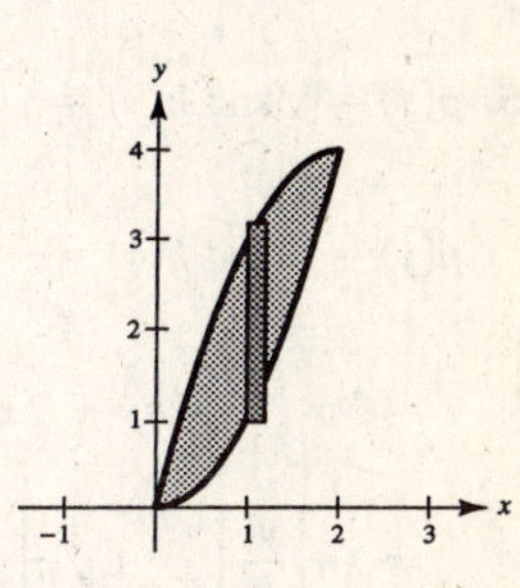

8. $p(x) = x, \ h(x) = 4 - x^2$

$$V = 2\pi\int_0^2 (4x - x^3)\,dx$$

$$= 2\pi\left[2x^2 - \frac{1}{4}x^4\right]_0^2 = 8\pi$$

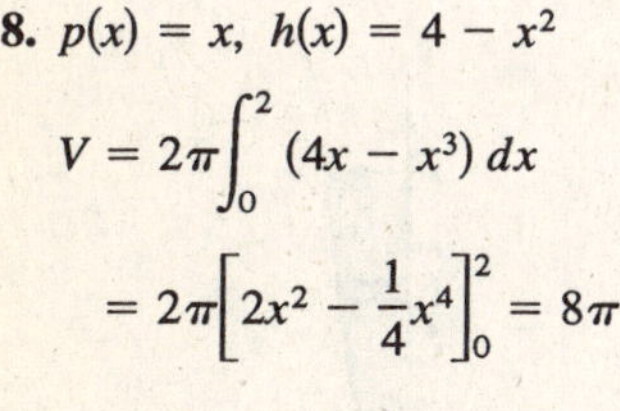

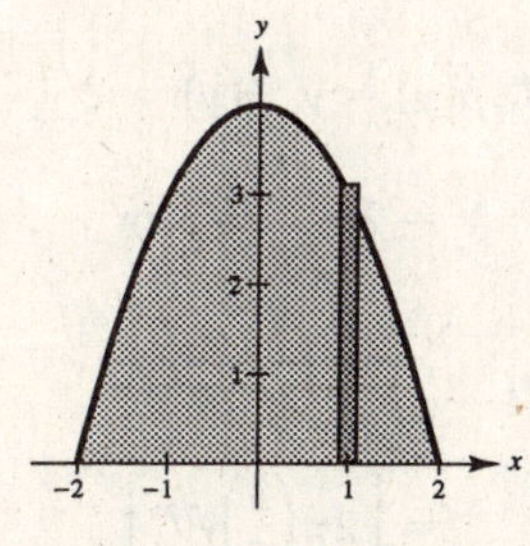

9. $p(x) = x$

$$h(x) = 4 - (4x - x^2)$$

$$= x^2 - 4x + 4$$

$$V = 2\pi\int_0^2 (x^3 - 4x^2 + 4x)\,dx$$

$$= 2\pi\left[\frac{x^4}{4} - \frac{4}{3}x^3 + 2x^2\right]_0^2$$

$$= \frac{8\pi}{3}$$

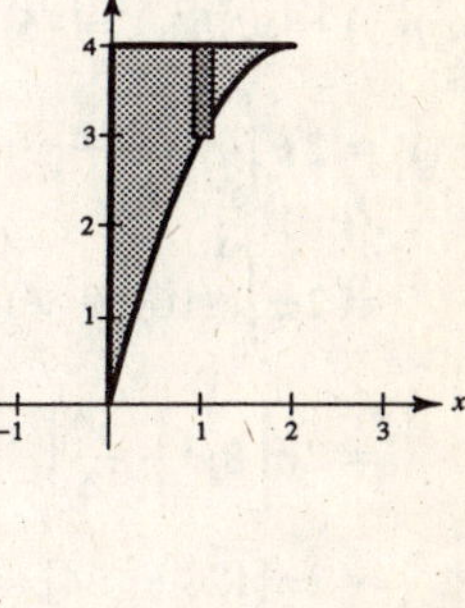

10. $p(x) = x, \ h(x) = 4 - 2x$

$$V = 2\pi\int_0^2 x(4 - 2x)\,dx$$

$$= 2\pi\int_0^2 (4x - 2x^2)\,dx$$

$$= 2\pi\left[2x^2 - \frac{2}{3}x^3\right]_0^2 = \frac{16\pi}{3}$$

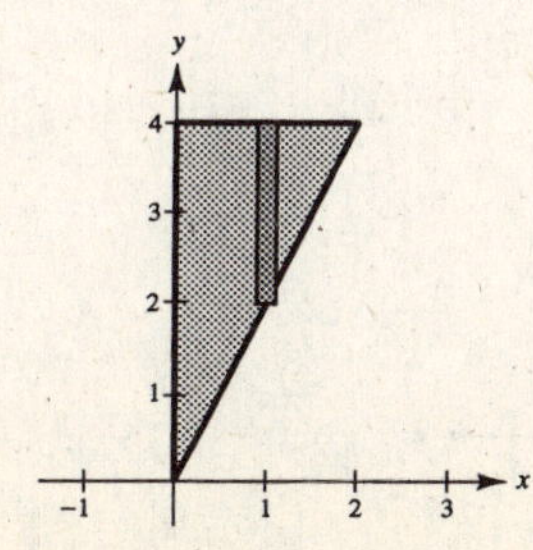

11. $p(x) = x,\ h(x) = \dfrac{1}{\sqrt{2\pi}}e^{-x^2/2}$

$$V = 2\pi\int_0^1 x\left(\frac{1}{\sqrt{2\pi}}e^{-x^2/2}\right)dx$$

$$= \sqrt{2\pi}\int_0^1 e^{-x^2/2}x\,dx$$

$$= \Big[-\sqrt{2\pi}e^{-x^2/2}\Big]_0^1$$

$$= \sqrt{2\pi}\left(1 - \frac{1}{\sqrt{e}}\right)$$

$$\approx 0.986$$

12. $p(x) = x,\ h(x) = \dfrac{\sin x}{x}$

$$V = 2\pi\int_0^{\pi} x\left[\frac{\sin x}{x}\right]dx$$

$$= 2\pi\int_0^{\pi}\sin x\,dx$$

$$= \Big[-2\pi\cos x\Big]_0^{\pi} = 4\pi$$

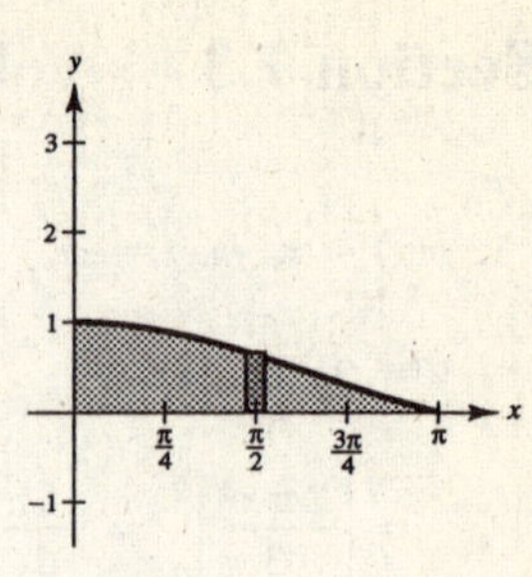

13. $p(y) = y,\ h(y) = 2 - y$

$$V = 2\pi\int_0^2 y(2 - y)\,dy$$

$$= 2\pi\int_0^2 (2y - y^2)\,dy$$

$$= 2\pi\left[y^2 - \frac{y^3}{3}\right]_0^2 = \frac{8\pi}{3}$$

14. $p(y) = -y,\quad (p(y) \ge 0 \text{ on } [-2, 0])$

$h(y) = 4 - (2 - y) = 2 + y$

$$V = 2\pi\int_{-2}^{0}(-y)(2 + y)\,dy$$

$$= 2\pi\int_{-2}^{0}(-2y - y^2)\,dy$$

$$= 2\pi\left[-y^2 - \frac{y^3}{3}\right]_{-2}^{0} = \frac{8\pi}{3}$$

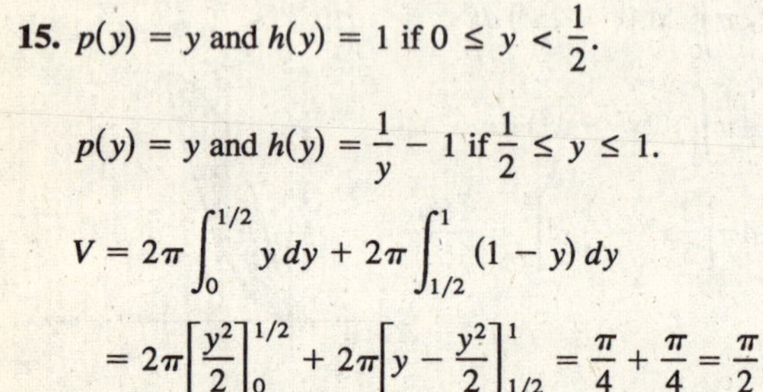

15. $p(y) = y$ and $h(y) = 1$ if $0 \le y < \dfrac{1}{2}$.

$p(y) = y$ and $h(y) = \dfrac{1}{y} - 1$ if $\dfrac{1}{2} \le y \le 1$.

$$V = 2\pi\int_0^{1/2} y\,dy + 2\pi\int_{1/2}^{1}(1 - y)\,dy$$

$$= 2\pi\left[\frac{y^2}{2}\right]_0^{1/2} + 2\pi\left[y - \frac{y^2}{2}\right]_{1/2}^{1} = \frac{\pi}{4} + \frac{\pi}{4} = \frac{\pi}{2}$$

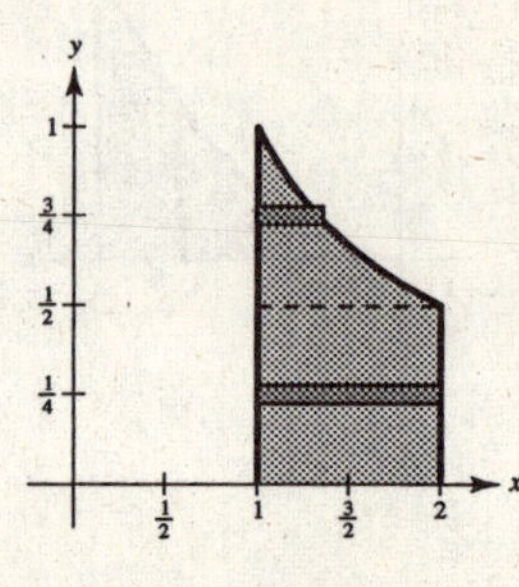

16. $p(y) = y,\ h(y) = 16 - y^2$

$$V = 2\pi\int_0^4 y(16 - y^2)\,dy$$

$$= 2\pi\int_0^4 (16y - y^3)\,dy$$

$$= 2\pi\left[8y^2 - \frac{y^4}{4}\right]_0^4$$

$$= 2\pi[128 - 64] = 128\pi$$

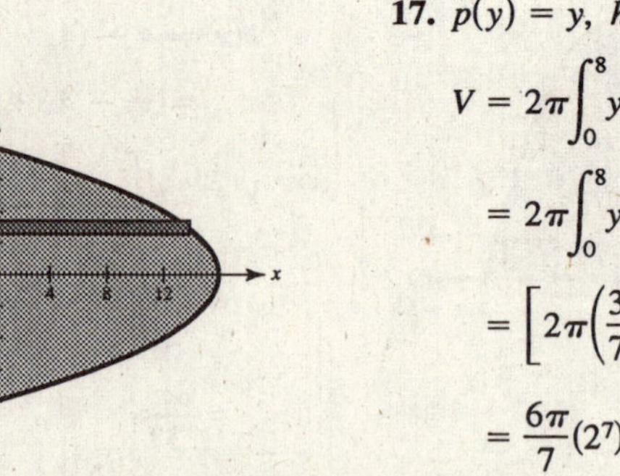

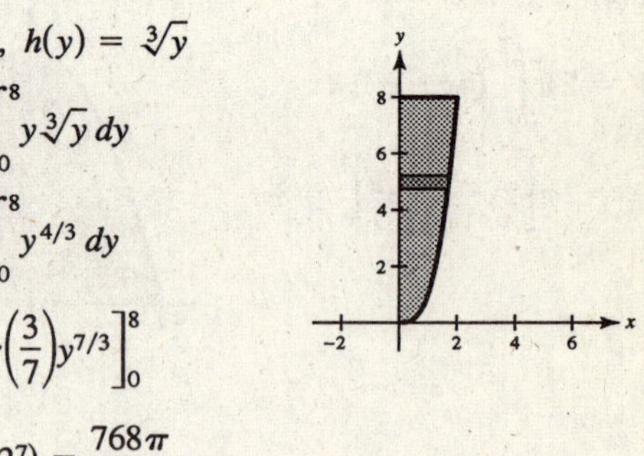

17. $p(y) = y,\ h(y) = \sqrt[3]{y}$

$$V = 2\pi\int_0^8 y\sqrt[3]{y}\,dy$$

$$= 2\pi\int_0^8 y^{4/3}\,dy$$

$$= \left[2\pi\left(\frac{3}{7}\right)y^{7/3}\right]_0^8$$

$$= \frac{6\pi}{7}(2^7) = \frac{768\pi}{7}$$

18. $p(y) = y,\ h(y) = \sqrt{y}$

$$V = 2\pi\int_0^9 y\sqrt{y}\,dy = 2\pi\int_0^9 y^{3/2}\,dy$$

$$= 2\pi\left(\frac{2}{5}\right)y^{5/2}\Big]_0^9$$

$$= \frac{4\pi}{5}(9)^{5/2} = \frac{4\pi}{5}(243) = \frac{972}{5}\pi$$

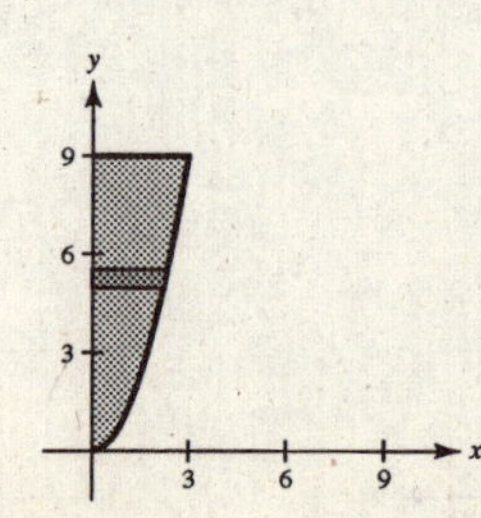

19. $p(y) = y,\ h(y) = (4 - y) - (y) = 4 - 2y$

$$V = 2\pi \int_0^2 y(4 - 2y)\, dy$$
$$= 2\pi \int_0^2 (4y - 2y^2)\, dy$$
$$= 2\pi \left[2y^2 - \frac{2}{3}y^3 \right]_0^2$$
$$= 2\pi \left[8 - \frac{16}{3} \right] = \frac{16\pi}{3}$$

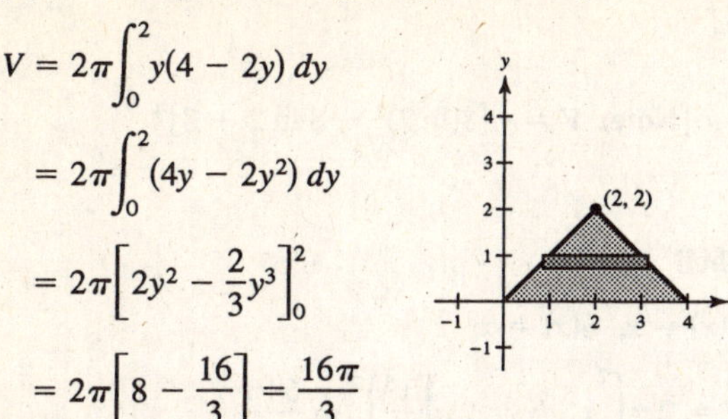

20. $p(y) = y,\ h(y) = y - (y^2 - 2) = 2 + y - y^2$

$$V = 2\pi \int_0^2 y(2 + y - y^2)\, dy$$
$$= 2\pi \int_0^2 (2y + y^2 - y^3)\, dy$$
$$= 2\pi \left[y^2 + \frac{y^3}{3} - \frac{y^4}{4} \right]_0^2$$
$$= 2\pi \left[4 + \frac{8}{3} - 4 \right] = \frac{16\pi}{3}$$

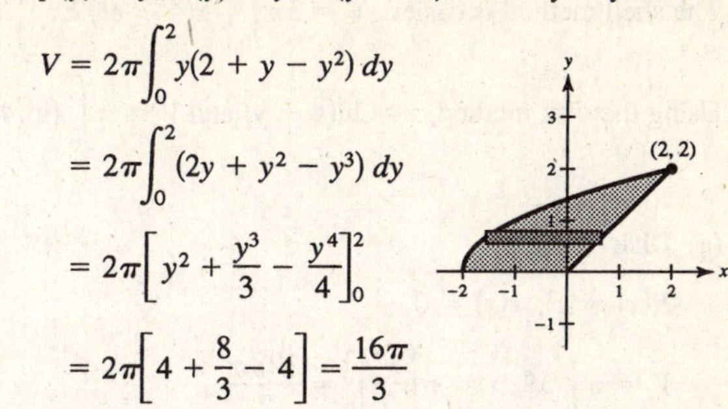

21. $p(x) = 4 - x,\ h(x) = 4x - x^2 - x^2 = 4x - 2x^2$

$$V = 2\pi \int_0^2 (4 - x)(4x - 2x^2)\, dx$$
$$= 2\pi(2) \int_0^2 (x^3 - 6x^2 + 8x)\, dx$$
$$= 4\pi \left[\frac{x^4}{4} - 2x^3 + 4x^2 \right]_0^2 = 16\pi$$

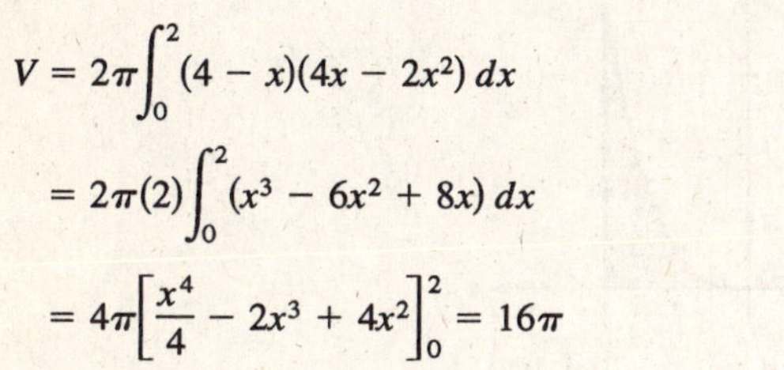

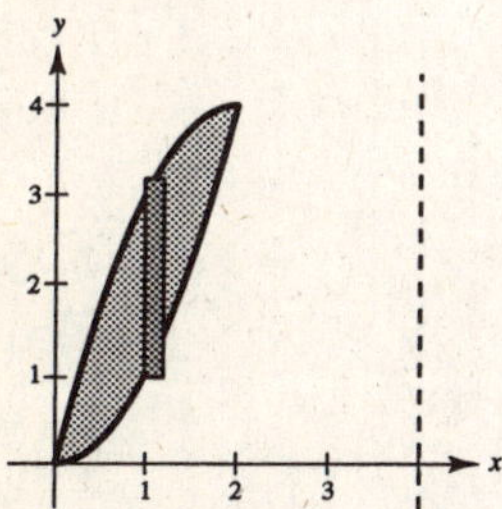

22. $p(x) = 2 - x,\ h(x) = 4x - x^2 - x^2 = 4x - 2x^2$

$$V = 2\pi \int_0^2 (2 - x)(4x - 2x^2)\, dx$$
$$= 2\pi \int_0^2 (8x - 8x^2 + 2x^3)\, dx$$
$$= 2\pi \left[4x^2 - \frac{8}{3}x^3 + \frac{1}{2}x^4 \right]_0^2 = \frac{16\pi}{3}$$

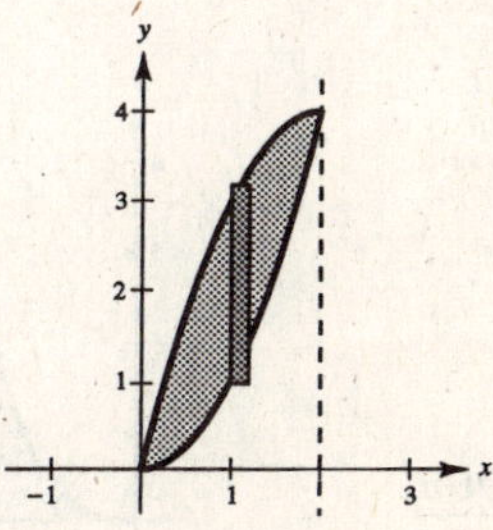

23. $p(x) = 5 - x,\ h(x) = 4x - x^2$

$$V = 2\pi \int_0^4 (5 - x)(4x - x^2)\, dx$$
$$= 2\pi \int_0^4 (x^3 - 9x^2 + 20x)\, dx$$
$$= 2\pi \left[\frac{x^4}{4} - 3x^3 + 10x^2 \right]_0^4 = 64\pi$$

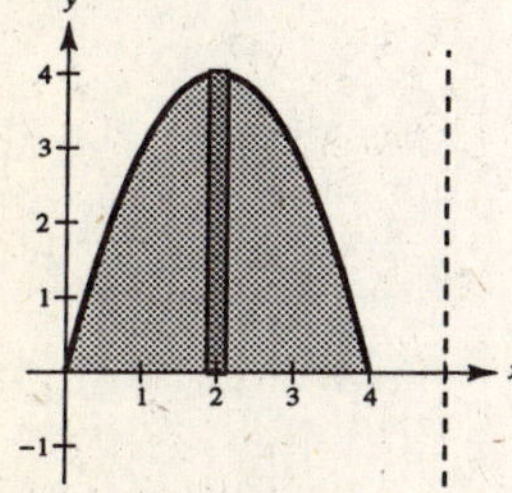

24. $p(x) = 6 - x,\ h(x) = \sqrt{x}$

$$V = 2\pi \int_0^4 (6 - x)\sqrt{x}\, dx$$
$$= 2\pi \int_0^4 (6x^{1/2} - x^{3/2})\, dx$$
$$= 2\pi \left[4x^{3/2} - \frac{2}{5}x^{5/2} \right]_0^4 = \frac{192\pi}{5}$$

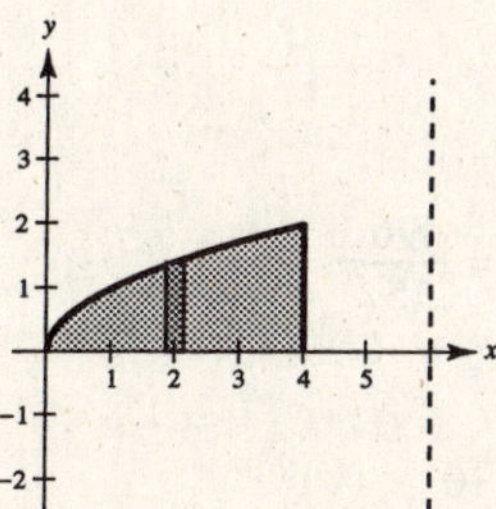

25. The shell method would be easier: $V = 2\pi \int_0^4 [4 - (y - 2)^2] y\, dy$ shells

Using the disk method: $V = \pi \int_0^4 \left[\left(2 + \sqrt{4 - x}\right)^2 - \left(2 - \sqrt{4 - x}\right)^2 \right] dx$ $\left[\textbf{Note: } V = \frac{128\pi}{3} \right]$

26. The shell method is easier: $V = 2\pi\int_0^{\ln 4} x(4 - e^x)\,dx$

Using the disk method, $x = \ln(4 - y)$ and $V = \pi\int_0^3 (\ln(4 - y))^2\,dy$. [**Note:** $V = \pi[8(\ln 2)^2 - 8\ln 2 + 3]$]

27. (a) **Disk**

$R(x) = x^3,\ r(x) = 0$

$$V = \pi\int_0^2 x^6\,dx = \pi\left[\frac{x^7}{7}\right]_0^2 = \frac{128\pi}{7}$$

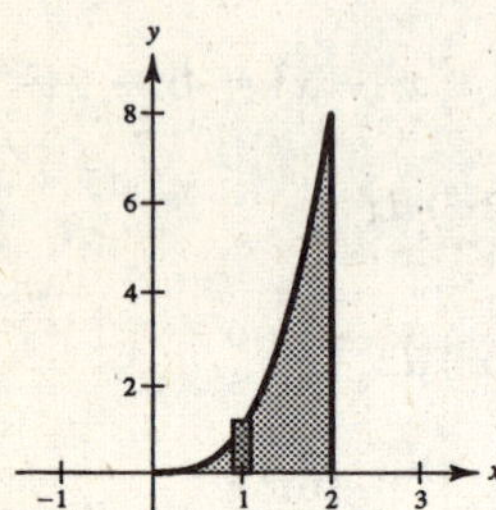

(b) **Shell**

$p(x) = x,\ h(x) = x^3$

$$V = 2\pi\int_0^2 x^4\,dx = 2\pi\left[\frac{x^5}{5}\right]_0^2 = \frac{64\pi}{5}$$

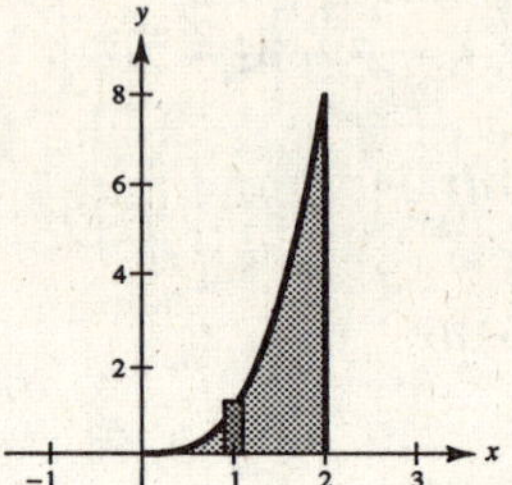

(c) **Shell**

$p(x) = 4 - x,\ h(x) = x^3$

$$V = 2\pi\int_0^2 (4 - x)x^3\,dx$$

$$= 2\pi\int_0^2 (4x^3 - x^4)\,dx$$

$$= 2\pi\left[x^4 - \frac{1}{5}x^5\right]_0^2 = \frac{96\pi}{5}$$

28. (a) **Disk**

$R(x) = \dfrac{10}{x^2},\ r(x) = 0$

$$V = \pi\int_1^5 \left(\frac{10}{x^2}\right)^2 dx$$

$$= 100\pi\int_1^5 x^{-4}\,dx$$

$$= 100\pi\left[\frac{x^{-3}}{-3}\right]_1^5$$

$$= -\frac{100\pi}{3}\left[\frac{1}{125} - 1\right] = \frac{496}{15}\pi$$

(b) **Shell**

$R(x) = x,\ r(x) = 0$

$$V = 2\pi\int_1^5 x\left(\frac{10}{x^2}\right) dx$$

$$= 20\pi\int_1^5 \frac{1}{x}\,dx$$

$$= 20\pi\Big[\ln|x|\Big]_1^5 = 20\pi\ln 5$$

(c) **Disk**

$R(x) = 10,\ r(x) = 10 - \dfrac{10}{x^2}$

$$V = \pi\int_1^5 \left[10^2 - \left(10 - \frac{10}{x^2}\right)^2\right] dx$$

$$= \pi\left[\frac{100}{3x^3} - \frac{200}{x}\right]_1^5 = \frac{1904}{15}\pi$$

29. (a) **Shell**

$p(y) = y,\ h(y) = (a^{1/2} - y^{1/2})^2$

$$V = 2\pi\int_0^a y(a - 2a^{1/2}y^{1/2} + y)\,dy$$

$$= 2\pi\int_0^a (ay - 2a^{1/2}y^{3/2} + y^2)\,dy$$

$$= 2\pi\left[\frac{a}{2}y^2 - \frac{4a^{1/2}}{5}y^{5/2} + \frac{y^3}{3}\right]_0^a$$

$$= 2\pi\left[\frac{a^3}{2} - \frac{4a^3}{5} + \frac{a^3}{3}\right] = \frac{\pi a^3}{15}$$

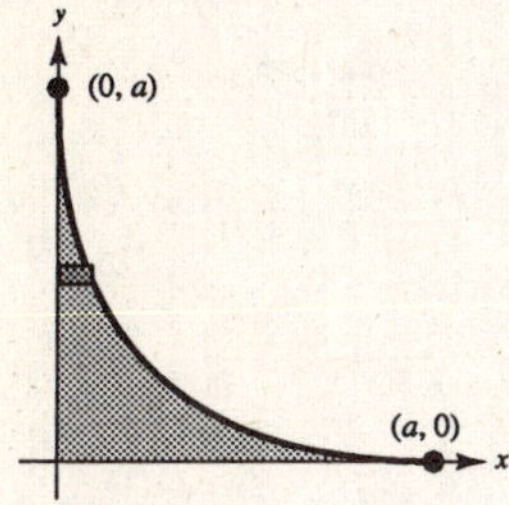

(b) Same as part (a) by symmetry

(c) **Shell**

$p(x) = a - x,\ h(x) = (a^{1/2} - x^{1/2})^2$

$$V = 2\pi\int_0^a (a - x)(a^{1/2} - x^{1/2})^2\,dx$$

$$= 2\pi\int_0^a (a^2 - 2a^{3/2}x^{1/2} + 2a^{1/2}x^{3/2} - x^2)\,dx$$

$$= 2\pi\left[a^2x - \frac{4}{3}a^{3/2}x^{3/2} + \frac{4}{5}a^{1/2}x^{5/2} - \frac{1}{3}x^3\right]_0^a = \frac{4\pi a^3}{15}$$

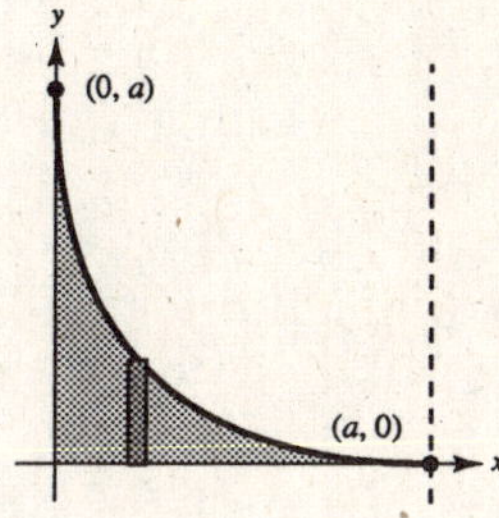

30. (a) **Disk**

$R(x) = (a^{2/3} - x^{2/3})^{3/2},\ r(x) = 0$

$$V = \pi\int_{-a}^a (a^{2/3} - x^{2/3})^3\,dx$$

$$= 2\pi\int_0^a (a^2 - 3a^{4/3}x^{2/3} + 3a^{2/3}x^{4/3} - x^2)\,dx$$

$$= 2\pi\left[a^2x - \frac{9}{5}a^{4/3}x^{5/3} + \frac{9}{7}a^{2/3}x^{7/3} - \frac{1}{3}x^3\right]_0^a$$

$$= 2\pi\left(a^3 - \frac{9}{5}a^3 + \frac{9}{7}a^3 - \frac{1}{3}a^3\right) = \frac{32\pi a^3}{105}$$

(b) Same as part (a) by symmetry

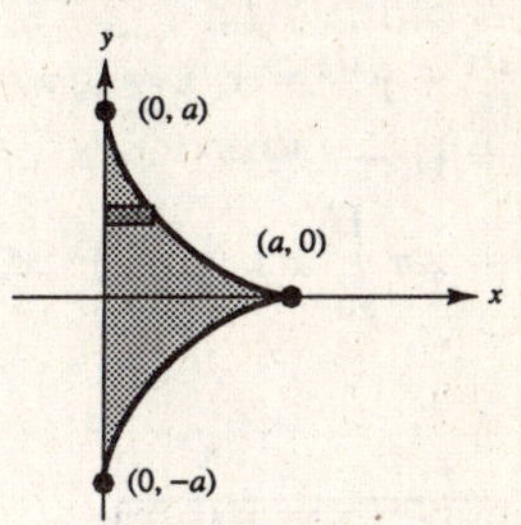

31. Answers will vary.

(a) The rectangles would be vertical.

(b) The rectangles would be horizontal.

32. (a)

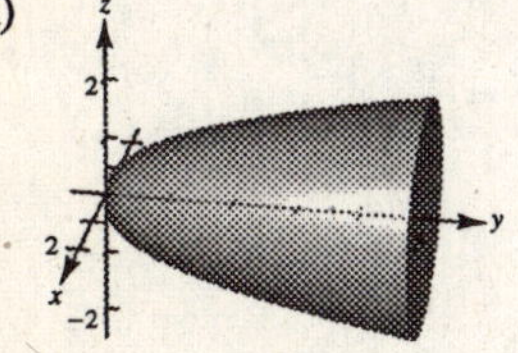

(b)

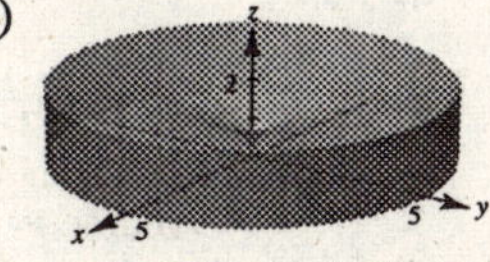

(c)

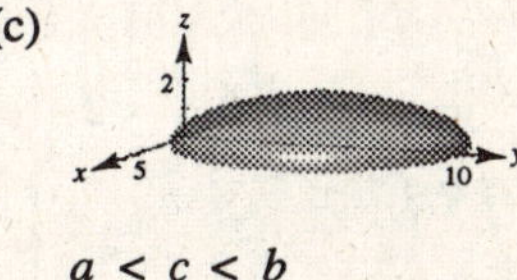

$a < c < b$

33. $\pi\int_1^5 (x-1)\,dx = \pi\int_1^5 (\sqrt{x-1})^2\,dx$

This integral represents the volume of the solid generated by revolving the region bounded by $y = \sqrt{x-1}$, $y = 0$, and $x = 5$ about the x-axis by using the disk method.

$$2\pi\int_0^2 y[5-(y^2+1)]\,dy$$

represents this same volume by using the shell method.

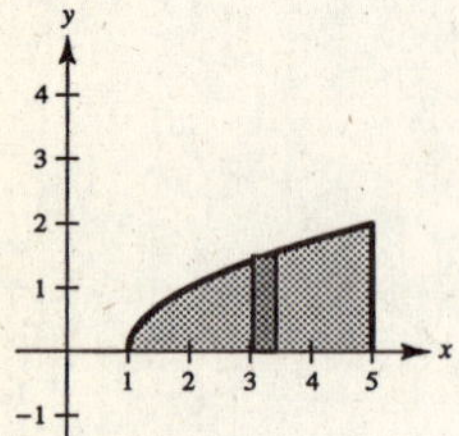

Disk method

34. $2\pi\int_0^4 x\left(\frac{x}{2}\right)dx$

represents the volume of the solid generated by revolving the region bounded by $y = x/2$, $y = 0$, and $x = 4$ about the y-axis by using the shell method.

$$\pi\int_0^2 [16-(2y)^2]\,dy = \pi\int_0^2 [(4)^2-(2y)^2]\,dy$$

represents this same volume by using the disk method.

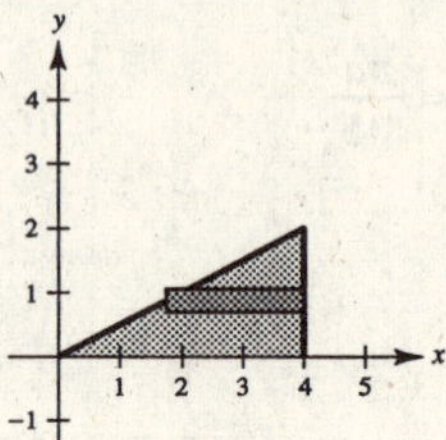

Disk method

35. (a)

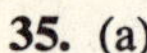

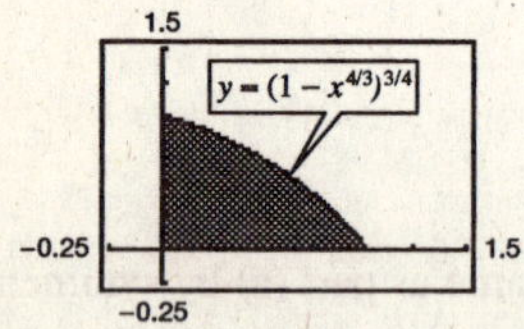

(b) $x^{4/3} + y^{4/3} = 1$, $x = 0$, $y = 0$

$y = (1 - x^{4/3})^{3/4}$

$$V = 2\pi\int_0^1 x(1-x^{4/3})^{3/4}\,dx \approx 1.5056$$

36. (a)

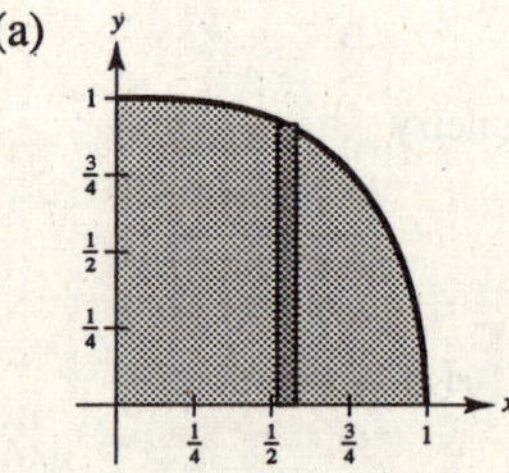

(b) $V = 2\pi\int_0^1 x\sqrt{1-x^3}\,dx \approx 2.3222$

37. (a)

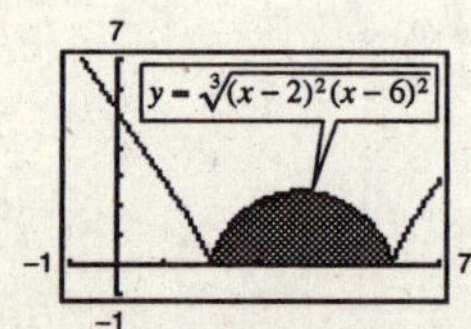

(b) $V = 2\pi\int_2^6 x\sqrt[3]{(x-2)^2(x-6)^2}\,dx \approx 187.249$

38. (a)

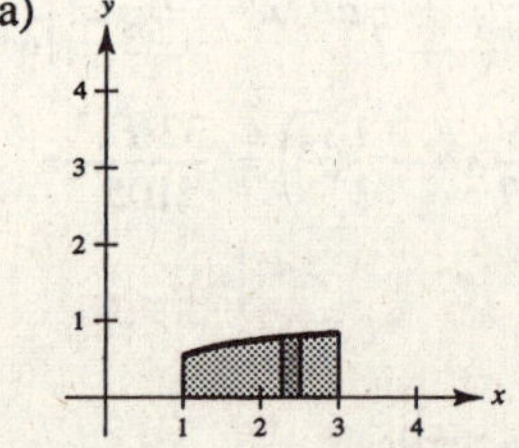

(b) $V = 2\pi\int_1^3 \frac{2x}{1+e^{1/x}}\,dx \approx 19.0162$

39. $y = 2e^{-x}$, $y = 0$, $x = 0$, $x = 2$

Volume ≈ 7.5

Matches (d)

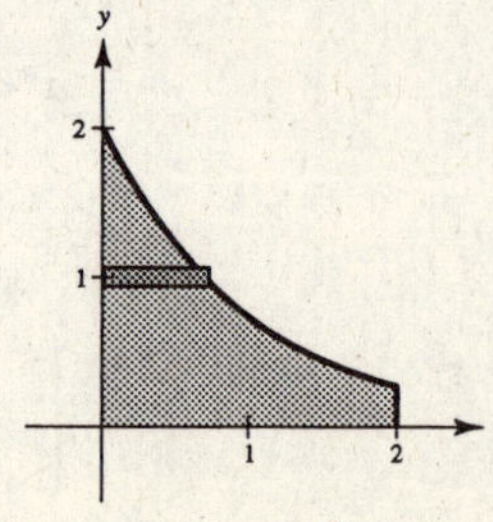

40. $y = \tan x$, $y = 0$, $x = 0$, $x = \frac{\pi}{4}$

Volume ≈ 1

Matches (e)

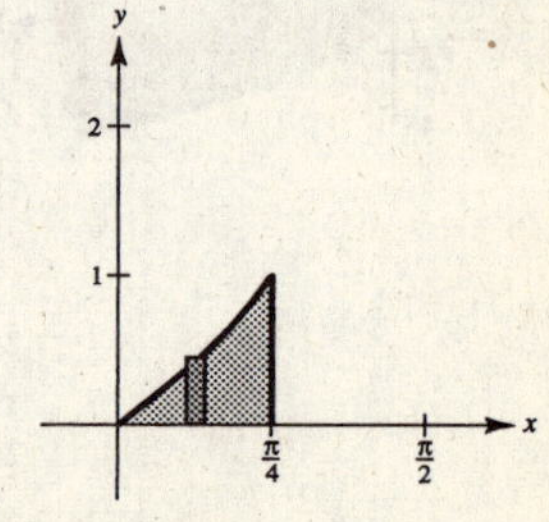

41. $p(x) = x,\ h(x) = 2 - \frac{1}{2}x^2$

$$V = 2\pi\int_0^2 x\left(2 - \frac{1}{2}x^2\right)dx = 2\pi\int_0^2\left(2x - \frac{1}{2}x^3\right)dx = 2\pi\left[x^2 - \frac{1}{8}x^4\right]_0^2 = 4\pi \quad \text{(total volume)}$$

Now find x_0 such that:

$$\pi = 2\pi\int_0^{x_0}\left(2x - \frac{1}{2}x^3\right)dx$$

$$1 = 2\left[x^2 - \frac{1}{8}x^4\right]_0^{x_0}$$

$$1 = 2x_0^2 - \frac{1}{4}x_0^4$$

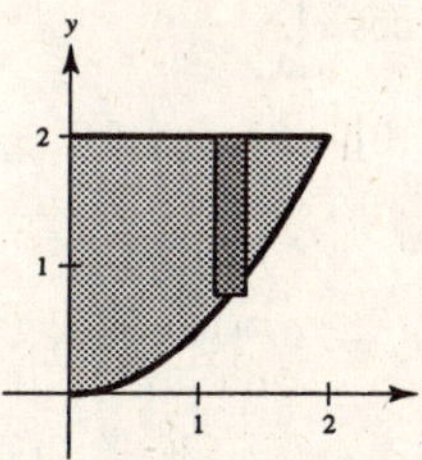

$$x_0^4 - 8x_0^2 + 4 = 0$$

$$x_0^2 = 4 \pm 2\sqrt{3} \qquad \text{(Quadratic Formula)}$$

Take $x_0 = \sqrt{4 - 2\sqrt{3}} \approx 0.73205$, since the other root is too large.

Diameter: $2\sqrt{4 - 2\sqrt{3}} \approx 1.464$

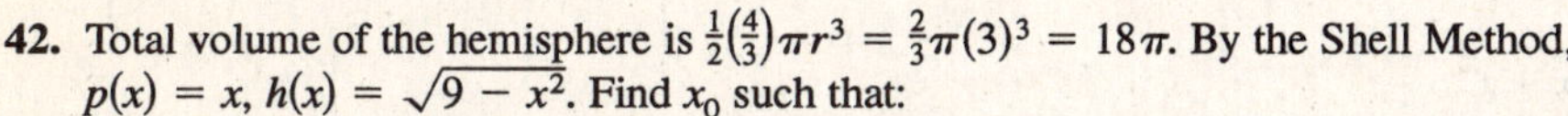

42. Total volume of the hemisphere is $\frac{1}{2}\left(\frac{4}{3}\right)\pi r^3 = \frac{2}{3}\pi(3)^3 = 18\pi$. By the Shell Method, $p(x) = x,\ h(x) = \sqrt{9 - x^2}$. Find x_0 such that:

$$6\pi = 2\pi\int_0^{x_0} x\sqrt{9 - x^2}\,dx$$

$$6 = -\int_0^{x_0}(9 - x^2)^{1/2}(-2x)\,dx$$

$$= \left[-\frac{2}{3}(9 - x^2)^{3/2}\right]_0^{x_0} = 18 - \frac{2}{3}(9 - x_0^2)^{3/2}$$

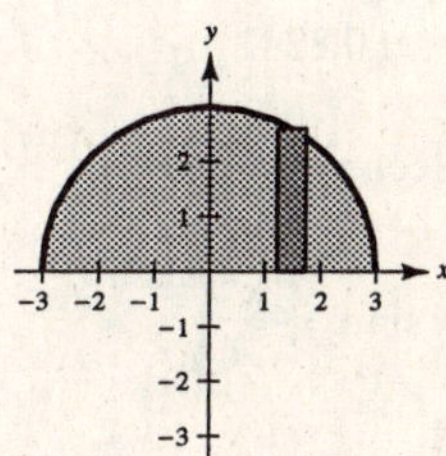

$$(9 - x_0^2)^{3/2} = 18$$

$$x_0 = \sqrt{9 - 18^{2/3}} \approx 1.460$$

Diameter: $2\sqrt{9 - 18^{2/3}} \approx 2.920$

43. $V = 4\pi\int_{-1}^{1}(2 - x)\sqrt{1 - x^2}\,dx$

$$= 8\pi\int_{-1}^{1}\sqrt{1 - x^2}\,dx - 4\pi\int_{-1}^{1}x\sqrt{1 - x^2}\,dx$$

$$= 8\pi\left(\frac{\pi}{2}\right) + 2\pi\int_{-1}^{1}x(1 - x^2)^{1/2}(-2)\,dx$$

$$= 4\pi^2 + \left[2\pi\left(\frac{2}{3}\right)(1 - x^2)^{3/2}\right]_{-1}^{1} = 4\pi^2$$

44. $V = 4\pi\int_{-r}^{r}(R - x)\sqrt{r^2 - x^2}\,dx$

$$= 4\pi R\int_{-r}^{r}\sqrt{r^2 - x^2}\,dx - 4\pi\int_{-r}^{r}x\sqrt{r^2 - x^2}\,dx$$

$$= 4\pi R\left(\frac{\pi r^2}{2}\right) + \left[2\pi\left(\frac{2}{3}\right)(r^2 - x^2)^{3/2}\right]_{-r}^{r}$$

$$= 2\pi^2 r^2 R$$

45. (a) $\frac{d}{dx}[\sin x - x\cos x + C] = \cos x + x\sin x - \cos x = x\sin x$

Hence, $\int x\sin x\,dx = \sin x - x\cos x + C$.

—CONTINUED—

45. —CONTINUED—

(b) (i) $p(x) = x,\ h(x) = \sin x$

$$V = 2\pi\int_0^{\pi/2} x \sin x\,dx$$

$$= 2\pi\Big[\sin x - x\cos x\Big]_0^{\pi/2}$$

$$= 2\pi[(1-0)-0] = 2\pi$$

(ii) $p(x) = x,\ h(x) = 2\sin x - (-\sin x) = 3\sin x$

$$V = 2\pi\int_0^{\pi} x(3\sin x)\,dx$$

$$= 6\pi\int_0^{\pi} x\sin x\,dx$$

$$= 6\pi\Big[\sin x - x\cos x\Big]_0^{\pi}$$

$$= 6\pi[\pi] = 6\pi^2$$

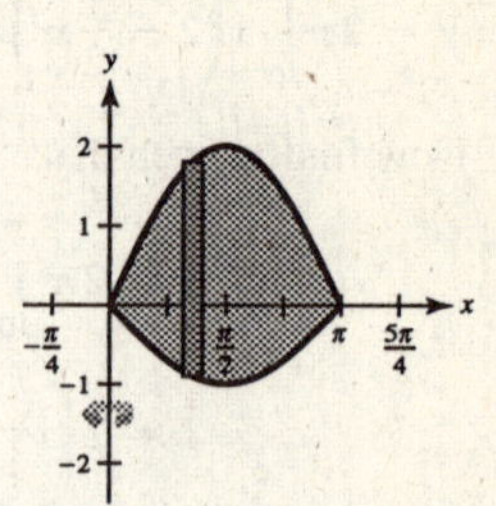

46. (a) $\dfrac{d}{dx}[\cos x + x\sin x + C] = -\sin x + \sin x + x\cos x = x\cos x$

Hence, $\displaystyle\int x\cos x\,dx = \cos x + x\sin x + C.$

(b) (i) $x^2 = \cos x \implies x \approx \pm 0.8241$

$$V \approx 2(2\pi)\int_0^{0.8241} x[\cos x - x^2]\,dx$$

$$= 4\pi\left[\cos x + x\sin x - \frac{x^4}{4}\right]_0^{0.8241}$$

$$\approx 2.1205$$

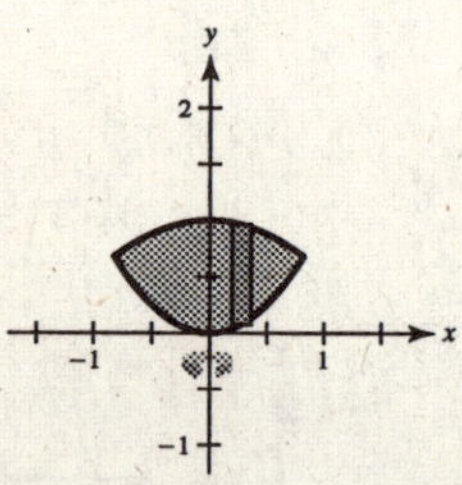

(ii) $4\cos x = (x-2)^2 \implies x = 0,\ 1.5110$

$$V \approx 2\pi\int_0^{1.511} x[4\cos x - (x-2)^2]\,dx$$

$$= 2\pi\int_0^{1.511}\left[4\cos x + 4x\sin x - \frac{(x-2)^3}{3}\right]_0^{1.511}$$

$$= 6.2993$$

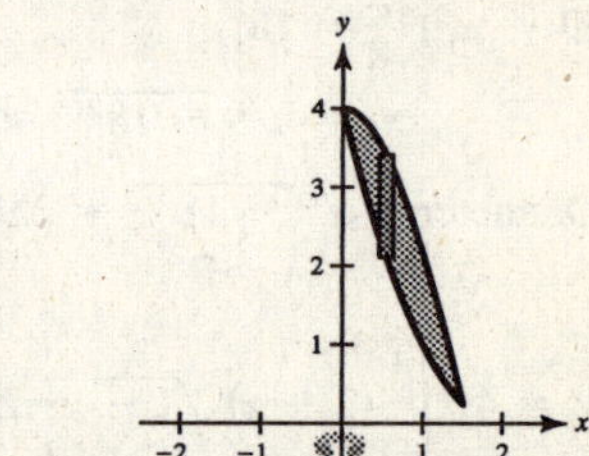

47. $\displaystyle 2\pi\int_0^2 x^3\,dx = 2\pi\int_0^2 x(x^2)\,dx$

(a) Plane region bounded by $y = x^2,\ y = 0,\ x = 0,\ x = 2$

(b) Revolved about the y-axis

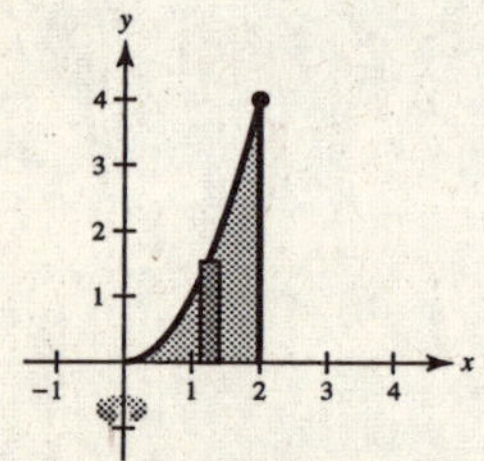

Other answers possible

48. $\displaystyle 2\pi\int_0^1 (y - y^{3/2})\,dy = 2\pi\int_0^1 y(1-\sqrt{y})\,dy$

(a) Plane region bounded by $x = \sqrt{y},\ x = 1,\ y = 0$

(b) Revolved about the x-axis

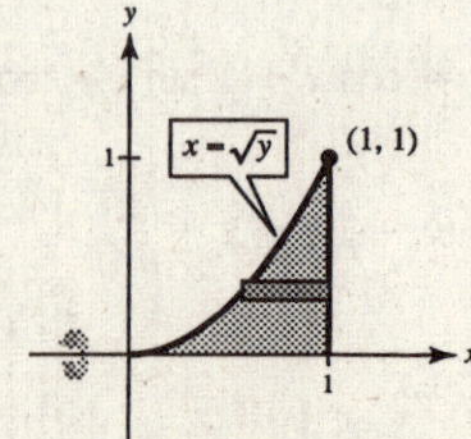

Other answers possible

49. $2\pi \int_0^6 (y + 2)\sqrt{6 - y}\, dy$

(a) Plane region bounded by $x = \sqrt{6 - y}, x = 0, y = 0$

(b) Revolved around line $y = -2$

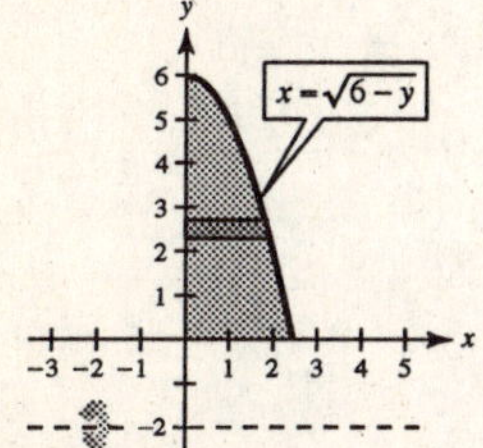

Other answers possible

50. $2\pi \int_0^1 (4 - x)e^x\, dx$

(a) Plane region bounded by $y = e^x, y = 0, x = 0, x = 1$

(b) Revolved about the line $x = 4$

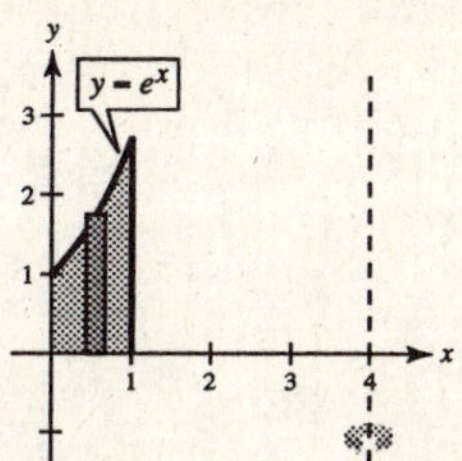

51. Disk Method

$R(y) = \sqrt{r^2 - y^2}$

$r(y) = 0$

$$V = \pi \int_{r-h}^{r} (r^2 - y^2)\, dy$$

$$= \pi\left[r^2y - \frac{y^3}{3}\right]_{r-h}^{r} = \frac{1}{3}\pi h^2(3r - h)$$

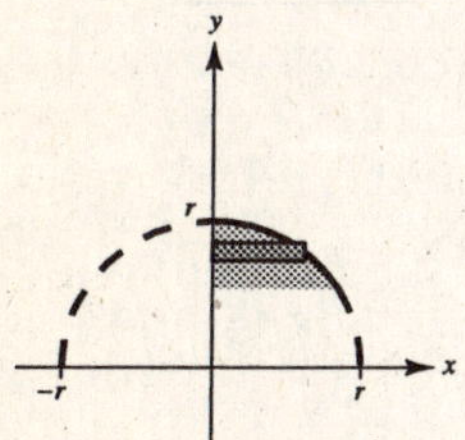

52. $\frac{x^2}{a^2} + \frac{y^2}{b^2} = 1$

$$\frac{y^2}{b^2} = 1 - \frac{x^2}{a^2}$$

$$y = \pm b\sqrt{1 - \frac{x^2}{a^2}}$$

$$p(x) = x,\ h(x) = b\sqrt{1 - \frac{x^2}{a^2}}$$

$$V = 2(2\pi)\int_0^a xb\sqrt{1 - \frac{x^2}{a^2}}\, dx$$

$$= \frac{4\pi b}{a}\int_0^a \sqrt{a^2 - x^2}\,x\, dx$$

$$= \frac{4\pi b}{a}\left(\frac{-(a^2 - x^2)^{3/2}}{3}\right)\Bigg]_0^a$$

$$= \frac{4\pi b}{3a}a^3 = \frac{4}{3}\pi a^2 b$$

Note: If $a = b$, then volume is that of a sphere.

53. (a) Area region $= \int_0^b [ab^n - ax^n]\, dx$

$$= \left[ab^n x - a\frac{x^{n+1}}{n+1}\right]_0^b$$

$$= ab^{n+1} - a\frac{b^{n+1}}{n+1}$$

$$= ab^{n+1}\left(1 - \frac{1}{n+1}\right) = ab^{n+1}\left(\frac{n}{n+1}\right)$$

$$R_1(n) = \frac{ab^{n+1}[n/(n+1)]}{(ab^n)b} = \frac{n}{n+1}$$

(b) $\lim_{n\to\infty} R_1(n) = \lim_{n\to\infty} \frac{n}{n+1} = 1$

$\lim_{n\to\infty} (ab^n)b = \infty$

(c) **Disk Method:**

$$V = 2\pi \int_0^b x(ab^n - ax^n)\, dx$$

$$= 2\pi a \int_0^b (xb^n - x^{n+1})\, dx$$

$$= 2\pi a\left[\frac{b^n}{2}x^2 - \frac{x^{n+2}}{n+2}\right]_0^b$$

$$= 2\pi a\left[\frac{b^{n+2}}{2} - \frac{b^{n+2}}{n+2}\right] = \pi ab^{n+2}\left(\frac{n}{n+2}\right)$$

$$R_2(n) = \frac{\pi ab^{n+2}[n/(n+2)]}{(\pi b^2)(ab^n)} = \left(\frac{n}{n+2}\right)$$

(d) $\lim_{n\to\infty} R_2(n) = \lim_{n\to\infty}\left(\frac{n}{n+2}\right) = 1$

$\lim_{n\to\infty} (\pi b^2)(ab^n) = \infty$

(e) As $n \to \infty$, the graph approaches the line $x = 1$.

54. (a) $2\pi \int_0^r hx\left(1 - \frac{x}{r}\right) dx$ (ii)

is the volume of a right circular cone with the radius of the base as r and height h.

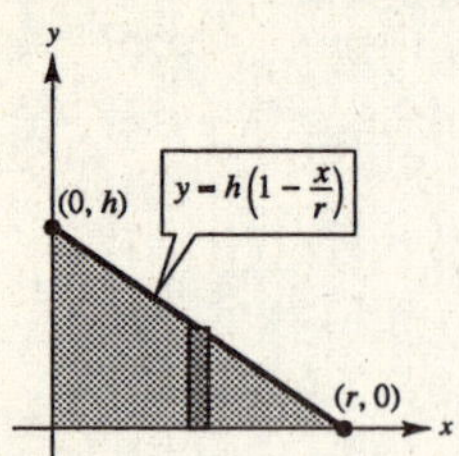

(b) $2\pi \int_{-r}^r (R - x)\left(2\sqrt{r^2 - x^2}\right) dx$ (v)

is the volume of a torus with the radius of its circular cross section as r and the distance from the axis of the torus to the center of its cross section as R.

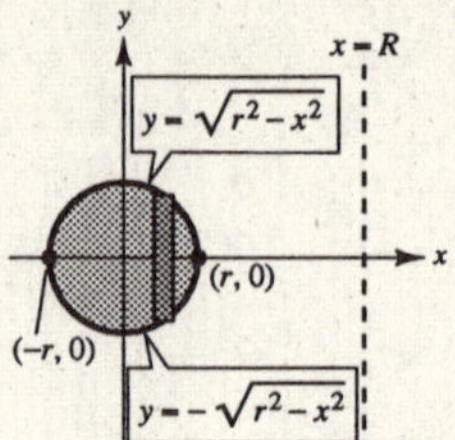

(c) $2\pi \int_0^r 2x\sqrt{r^2 - x^2}\, dx$ (iii)

is the volume of a sphere with radius r.

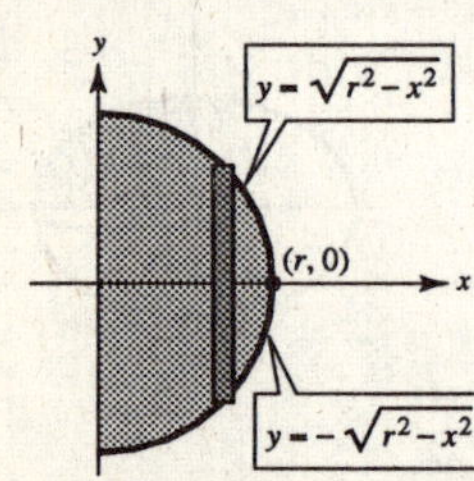

(d) $2\pi \int_0^r hx\, dx$ (i)

is the volume of a right circular cylinder with a radius of r and a height of h.

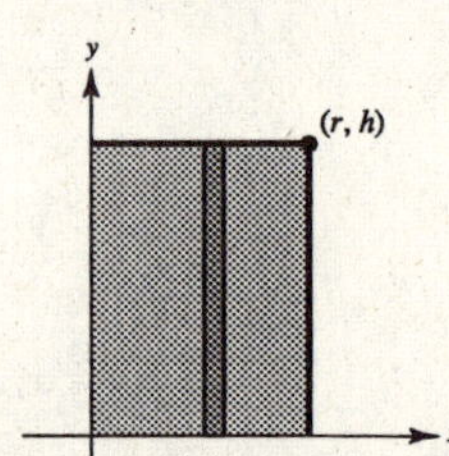

(e) $2\pi \int_0^b 2ax\sqrt{1 - (x^2/b^2)}\, dx$ (iv)

is the volume of an ellipsoid with axes $2a$ and $2b$.

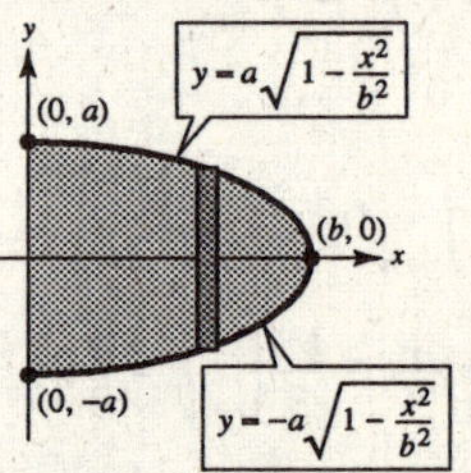

55. (a) $V = 2\pi \int_0^4 xf(x)\, dx$

$$= \frac{2\pi(40)}{3(4)}[0 + 4(10)(45) + 2(20)(40) + 4(30)(20) + 0]$$

$$= \frac{20\pi}{3}[5800] \approx 121{,}475 \text{ cubic feet}$$

(b) Top line: $y - 50 = \dfrac{40 - 50}{20 - 0}(x - 0) = -\dfrac{1}{2}x \Rightarrow y = -\dfrac{1}{2}x + 50$

Bottom line: $y - 40 = \dfrac{0 - 40}{40 - 20}(x - 20) = -2(x - 20) \Rightarrow y = -2x + 80$

$$V = 2\pi \int_0^{20} x\left(-\frac{1}{2}x + 50\right) dx + 2\pi \int_{20}^{40} x(-2x + 80)\, dx$$

$$= 2\pi \int_0^{20} \left(-\frac{1}{2}x^2 + 50x\right) dx + 2\pi \int_{20}^{40} (-2x^2 + 80x)\, dx$$

$$= 2\pi\left[-\frac{x^3}{6} + 25x^2\right]_0^{20} + 2\pi\left[-\frac{2x^3}{3} + 40x^2\right]_{20}^{40}$$

$$= 2\pi\left[\frac{26{,}000}{3}\right] + 2\pi\left[\frac{32{,}000}{3}\right]$$

$$\approx 121{,}475 \text{ cubic feet}$$

(Note that Simpson's Rule is exact for this problem.)

56. (a) $V = 2\pi\int_0^{200} xf(x)\,dx$

$\approx \dfrac{2\pi(200)}{3(8)}[0 + 4(25)(19) + 2(50)(19) + 4(75)(17) + 2(100)15 + 4(125)(14) + 2(150)(10) + 4(175)(6) + 0]$

$\approx 1{,}366{,}593$ cubic feet

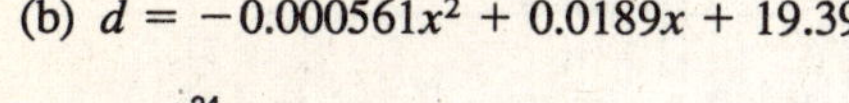

(b) $d = -0.000561x^2 + 0.0189x + 19.39$

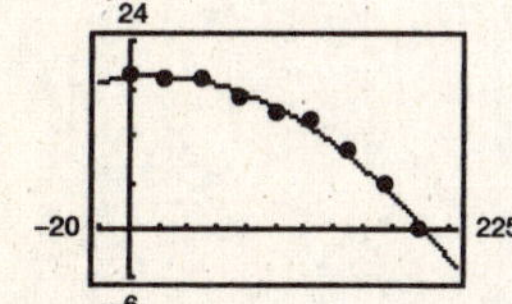

(c) $V \approx 2\pi\int_0^{200} xd(x)\,dx \approx 2\pi(213{,}800)$

$= 1{,}343{,}345$ cubic feet

(d) Number gallons $\approx V(7.48) = 10{,}048{,}221$ gallons

57. $y^2 = x(4 - x)^2, \quad 0 \le x \le 4$

$y_1 = \sqrt{x(4 - x)^2} = (4 - x)\sqrt{x}$

$y_2 = -\sqrt{x(4 - x)^2} = -(4 - x)\sqrt{x}$

(a) $V = \pi\int_0^4 x(4 - x)^2\,dx$

$= \pi\int_0^4 (x^3 - 8x^2 + 16x)\,dx$

$= \pi\left[\dfrac{x^4}{4} - \dfrac{8x^3}{3} + 8x^2\right]_0^4 = \dfrac{64\pi}{3}$

(b) $V = 4\pi\int_0^4 x(4 - x)\sqrt{x}\,dx$

$= 4\pi\int_0^4 (4x^{3/2} - x^{5/2})\,dx$

$= 4\pi\left[\dfrac{8}{5}x^{5/2} - \dfrac{2}{7}x^{7/2}\right]_0^4 = \dfrac{2048\pi}{35}$

(c) $V = 4\pi\int_0^4 (4 - x)(4 - x)\sqrt{x}\,dx$

$= 4\pi\int_0^4 \left(16\sqrt{x} - 8x^{3/2} + x^{5/2}\right)dx$

$= 4\pi\left[\dfrac{32}{3}x^{3/2} - \dfrac{16}{5}x^{5/2} + \dfrac{2}{7}x^{7/2}\right]_0^4 = \dfrac{8192\pi}{105}$

58. $y^2 = x^2(x + 5), \quad -5 \le x \le 0$

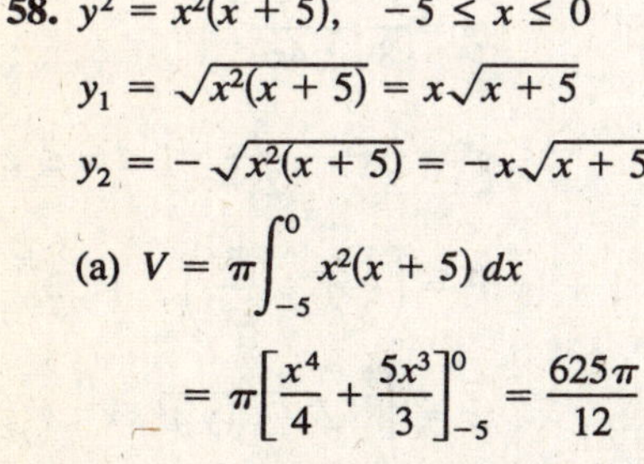

$y_1 = \sqrt{x^2(x + 5)} = x\sqrt{x + 5}$

$y_2 = -\sqrt{x^2(x + 5)} = -x\sqrt{x + 5}$

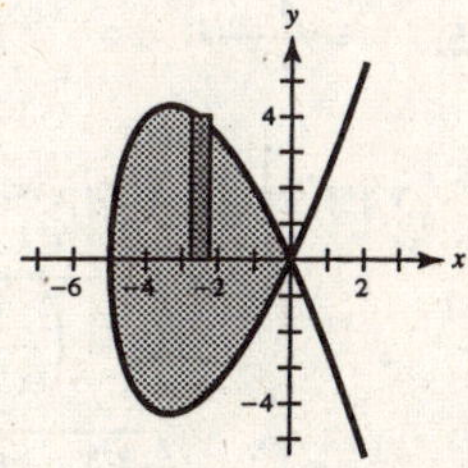

(a) $V = \pi\int_{-5}^0 x^2(x + 5)\,dx$

$= \pi\left[\dfrac{x^4}{4} + \dfrac{5x^3}{3}\right]_{-5}^0 = \dfrac{625\pi}{12}$

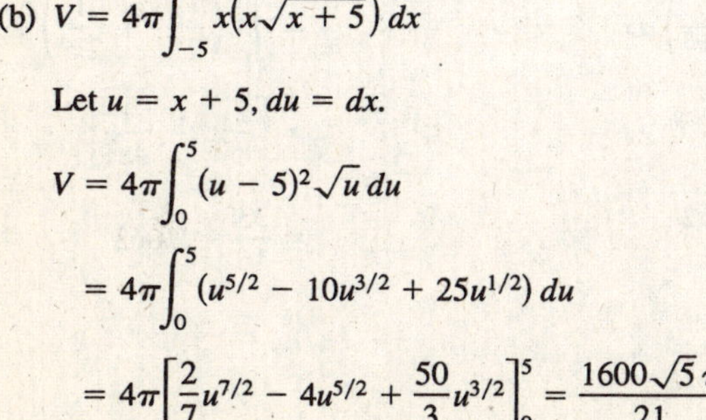

(b) $V = 4\pi\int_{-5}^0 x\left(x\sqrt{x + 5}\right)dx$

Let $u = x + 5$, $du = dx$.

$V = 4\pi\int_0^5 (u - 5)^2\sqrt{u}\,du$

$= 4\pi\int_0^5 (u^{5/2} - 10u^{3/2} + 25u^{1/2})\,du$

$= 4\pi\left[\dfrac{2}{7}u^{7/2} - 4u^{5/2} + \dfrac{50}{3}u^{3/2}\right]_0^5 = \dfrac{1600\sqrt{5}\,\pi}{21}$

(c) $V = 4\pi\int_{-5}^0 (-5 - x)x\sqrt{x + 5}\,dx$

Let $u = x + 5$, $du = dx$.

$V = 4\pi\int_0^5 (-u)(u - 5)\sqrt{u}\,du$

$= 4\pi\int_0^5 (-u^{5/2} + 5u^{3/2})\,du$

$= 4\pi\left[-\dfrac{2}{7}u^{7/2} + 2u^{5/2}\right]_0^5 = \dfrac{400\sqrt{5}\,\pi}{7}$

59. $V_1 = \pi\int_{1/4}^{c} \frac{1}{x^2}\,dx = \pi\left[-\frac{1}{x}\right]_{1/4}^{c} = \pi\left[-\frac{1}{c} + 4\right] = \frac{4c-1}{c}\pi$

$V_2 = 2\pi\int_{1/4}^{c} x\left(\frac{1}{x}\right)dx = 2\pi x\Big]_{1/4}^{c} = 2\pi\left(c - \frac{1}{4}\right)$

$V_1 = V_2 \Rightarrow \frac{4c-1}{c}\pi = 2\pi\left(c - \frac{1}{4}\right)$

$4c - 1 = 2c\left(c - \frac{1}{4}\right)$

$4c^2 - 9c + 2 = 0$

$(4c - 1)(c - 2) = 0$

$c = 2 \quad \left(c = \frac{1}{4} \text{ yields no volume.}\right)$

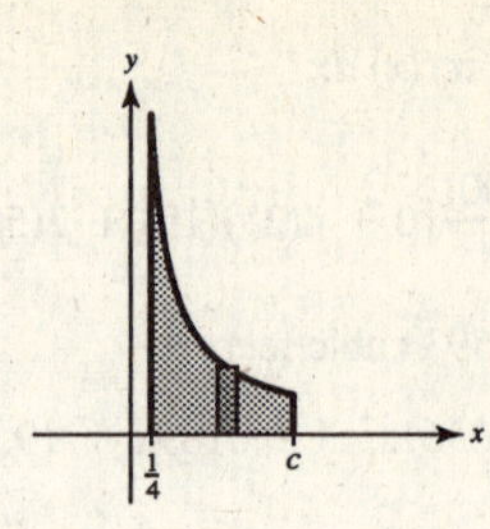

Section 7.4 Arc Length and Surfaces of Revolution

1. (0, 0), (5, 12)

(a) $d = \sqrt{(5-0)^2 + (12-0)^2}$

$= 13$

(b) $y = \frac{12}{5}x$

$y' = \frac{12}{5}$

$s = \int_0^5 \sqrt{1 + \left(\frac{12}{5}\right)^2}\,dx$

$= \left[\frac{13}{5}x\right]_0^5 = 13$

2. (1, 2), (7, 10)

(a) $d = \sqrt{(7-1)^2 + (10-2)^2}$

$= 10$

(b) $y = \frac{4}{3}x + \frac{2}{3}$

$y' = \frac{4}{3}$

$s = \int_1^7 \sqrt{1 + \left(\frac{4}{3}\right)^2}\,dx$

$= \left[\frac{5}{3}x\right]_1^7 = 10$

3. $y = \frac{2}{3}x^{3/2} + 1$

$y' = x^{1/2}, \quad 0 \le x \le 1$

$s = \int_0^1 \sqrt{1+x}\,dx$

$= \left[\frac{2}{3}(1+x)^{3/2}\right]_0^1$

$= \frac{2}{3}(\sqrt{8} - 1) \approx 1.219$

4. $y = 2x^{3/2} + 3$

$y' = 3x^{1/2}, \quad 0 \le x \le 9$

$s = \int_0^9 \sqrt{1+9x}\,dx$

$= \left[\frac{2}{27}(1+9x)^{3/2}\right]_0^9$

$= \frac{2}{27}(82^{3/2} - 1) \approx 54.929$

5. $y = \frac{3}{2}x^{2/3}$

$y' = \frac{1}{x^{1/3}}, \quad 1 \le x \le 8$

$s = \int_1^8 \sqrt{1 + \left(\frac{1}{x^{1/3}}\right)^2}\,dx$

$= \int_1^8 \sqrt{\frac{x^{2/3}+1}{x^{2/3}}}\,dx$

$= \frac{3}{2}\int_1^8 \sqrt{x^{2/3}+1}\left(\frac{2}{3x^{1/3}}\right)dx$

$= \frac{3}{2}\left[\frac{2}{3}(x^{2/3}+1)^{3/2}\right]_1^8$

$= 5\sqrt{5} - 2\sqrt{2} \approx 8.352$

6. $y = \frac{x^4}{8} + \frac{1}{4x^2}$

$y' = \frac{1}{2}x^3 - \frac{1}{2x^3}, \quad 1 \le x \le 2$

$1 + (y')^2 = \left(\frac{1}{2}x^3 + \frac{1}{2x^3}\right)^2, \quad [1, 2]$

$s = \int_a^b \sqrt{1 + (y')^2}\,dx$

$= \int_1^2 \left(\frac{1}{2}x^3 + \frac{1}{2x^3}\right)dx$

$= \left[\frac{1}{8}x^4 - \frac{1}{4x^2}\right]_1^2$

$= \frac{33}{16} \approx 2.063$

7. $y = \dfrac{x^5}{10} + \dfrac{1}{6x^3}$

$$y' = \frac{1}{2}x^4 - \frac{1}{2x^4}$$

$$1 + (y')^2 = \left(\frac{1}{2}x^4 + \frac{1}{2x^4}\right)^2, \quad 1 \le x \le 2$$

$$s = \int_a^b \sqrt{1 + (y')^2}\,dx$$

$$= \int_1^2 \sqrt{\left(\frac{1}{2}x^4 + \frac{1}{2x^4}\right)^2}\,dx$$

$$= \int_1^2 \left(\frac{1}{2}x^4 + \frac{1}{2x^4}\right)dx$$

$$= \left[\frac{1}{10}x^5 - \frac{1}{6x^3}\right]_1^2 = \frac{779}{240} \approx 3.2458$$

8. $y = \dfrac{3}{2}x^{2/3} + 4$

$$y' = x^{-1/3}, \quad 1 \le x \le 27$$

$$s = \int_1^{27} \sqrt{1 + \left(\frac{1}{x^{1/3}}\right)^2}\,dx$$

$$= \int_1^{27} \sqrt{\frac{x^{2/3} + 1}{x^{2/3}}}\,dx$$

$$= \frac{3}{2}\int_1^{27} \sqrt{x^{2/3} + 1}\left(\frac{2}{3x^{1/3}}\right)dx$$

$$= \left[\frac{3}{2} \cdot \frac{2}{3}(x^{2/3} + 1)^{3/2}\right]_1^{27}$$

$$= 10^{3/2} - 2^{3/2} \approx 28.794$$

9. $y = \ln(\sin x), \quad \left[\dfrac{\pi}{4}, \dfrac{3\pi}{4}\right]$

$$y' = \frac{1}{\sin x}\cos x = \cot x$$

$$1 + (y')^2 = 1 + \cot^2 x = \csc^2 x$$

$$s = \int_{\pi/4}^{3\pi/4} \csc x\,dx$$

$$= \Big[\ln|\csc x - \cot x|\Big]_{\pi/4}^{3\pi/4}$$

$$= \ln(\sqrt{2} + 1) - \ln(\sqrt{2} - 1) \approx 1.763$$

10. $y = \ln(\cos x), \quad 0 \le x \le \dfrac{\pi}{3}$

$$y' = \frac{-\sin x}{\cos x} = -\tan x$$

$$1 + (y')^2 = 1 + \tan^2 x = \sec^2 x$$

$$s = \int_0^{\pi/3} \sqrt{\sec^2 x}\,dx$$

$$= \int_0^{\pi/3} \sec x\,dx$$

$$= \ln|\sec x + \tan x|\Big]_0^{\pi/3}$$

$$= \ln(2 + \sqrt{3}) \approx 1.3170$$

11. $y = \dfrac{1}{2}(e^x + e^{-x})$

$$y' = \frac{1}{2}(e^x - e^{-x}), \quad [0, 2]$$

$$1 + (y')^2 = \left[\frac{1}{2}(e^x + e^{-x})\right]^2, \quad [0, 2]$$

$$s = \int_0^2 \sqrt{\left[\frac{1}{2}(e^x + e^{-x})\right]^2}\,dx$$

$$= \frac{1}{2}\int_0^2 (e^x + e^{-x})\,dx$$

$$= \frac{1}{2}\Big[e^x - e^{-x}\Big]_0^2 = \frac{1}{2}\left(e^2 - \frac{1}{e^2}\right) \approx 3.627$$

12. $y = \ln\left(\dfrac{e^x + 1}{e^x - 1}\right) = \ln(e^x + 1) - \ln(e^x - 1)$

$$\frac{dy}{dx} = \frac{e^x}{e^x + 1} - \frac{e^x}{e^x - 1} = \frac{-2e^x}{e^{2x} - 1} = \frac{2e^x}{1 - e^{2x}}$$

$$1 + \left(\frac{dy}{dx}\right)^2 = 1 + \frac{4e^{2x}}{1 - 2e^{2x} + e^{4x}}$$

$$= \frac{1 + 2e^{2x} + e^{4x}}{(1 - e^{2x})^2} = \left(\frac{1 + e^{2x}}{1 - e^{2x}}\right)^2$$

$$s = \int_a^b \sqrt{1 + \left(\frac{dy}{dx}\right)^2}\,dx = \int_{\ln 2}^{\ln 3} \frac{1 + e^{2x}}{e^{2x} - 1}\,dx$$

$$= \int_{\ln 2}^{\ln 3} \frac{e^x + e^{-x}}{e^x - e^{-x}}\,dx = \int_{\ln 2}^{\ln 3} \coth x\,dx$$

$$= \ln(\sinh(x))\Big]_{\ln 2}^{\ln 3} = \ln\left(\frac{4}{3}\right) - \ln\left(\frac{3}{4}\right)$$

$$= \ln\left(\frac{4/3}{3/4}\right) = \ln\frac{16}{9} - 2\ln\left(\frac{4}{3}\right) \approx 0.57536$$

13. $x = \frac{1}{3}(y^2 + 2)^{3/2}, \quad 0 \le y \le 4$

$$\frac{dx}{dy} = y(y^2 + 2)^{1/2}$$

$$s = \int_0^4 \sqrt{1 + y^2(y^2 + 2)}\, dy$$

$$= \int_0^4 \sqrt{y^4 + 2y^2 + 1}\, dy$$

$$= \int_0^4 (y^2 + 1)\, dy$$

$$= \left[\frac{y^3}{3} + y\right]_0^4 = \frac{64}{3} + 4 = \frac{76}{3}$$

14. $x = \frac{1}{3}\sqrt{y}(y - 3), \quad 1 \le y \le 4$

$$x = \frac{1}{3}(y^{3/2} - 3y^{1/2})$$

$$\frac{dx}{dy} = \frac{1}{2}y^{1/2} - \frac{1}{2}y^{-1/2}$$

$$1 + \left(\frac{dx}{dy}\right)^2 = 1 + \frac{1}{4}y + \frac{1}{4}y^{-1} - \frac{1}{2}$$

$$= \frac{1}{4}(y + 2 + y^{-1}) = \frac{1}{4}\left(\sqrt{y} + \frac{1}{\sqrt{y}}\right)^2$$

$$s = \int_1^4 \frac{1}{2}\left(\sqrt{y} + \frac{1}{\sqrt{y}}\right) dy$$

$$= \left[\frac{1}{2}\left(\frac{2}{3}y^{3/2} + 2y^{1/2}\right)\right]_1^4$$

$$= \frac{1}{2}\left(\frac{16}{3} + 4\right) - \frac{1}{2}\left(\frac{2}{3} + 2\right) = \frac{10}{3}$$

15. (a) $y = 4 - x^2, \quad 0 \le x \le 2$

(b) $y' = -2x$

$$1 + (y')^2 = 1 + 4x^2$$

$$L = \int_0^2 \sqrt{1 + 4x^2}\, dx$$

(c) $L \approx 4.647$

16. (a) $y = x^2 + x - 2, \quad -2 \le x \le 1$

(b) $y' = 2x + 1$

$$1 + (y')^2 = 1 + 4x^2 + 4x + 1$$

$$L = \int_{-2}^1 \sqrt{2 + 4x + 4x^2}\, dx$$

(c) $L \approx 5.653$

17. (a) $y = \frac{1}{x}, \quad 1 \le x \le 3$

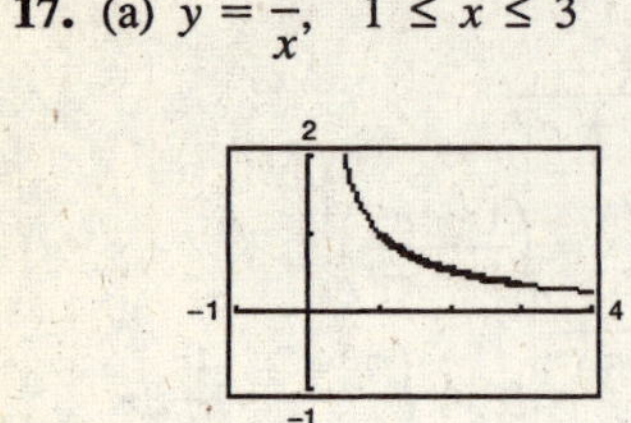

(b) $y' = -\frac{1}{x^2}$

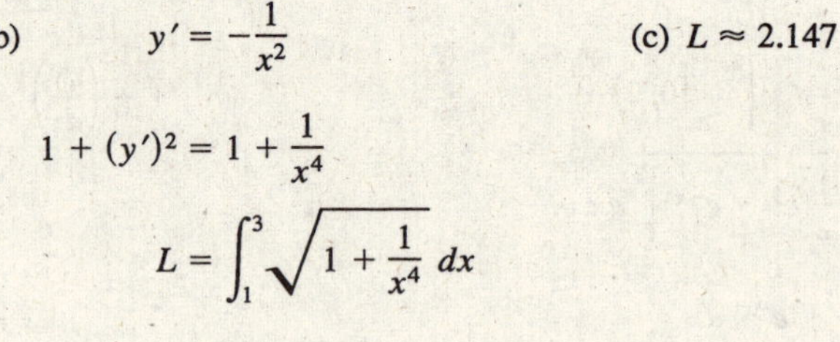

$$1 + (y')^2 = 1 + \frac{1}{x^4}$$

$$L = \int_1^3 \sqrt{1 + \frac{1}{x^4}}\, dx$$

(c) $L \approx 2.147$

18. (a) $y = \frac{1}{1 + x}, \quad 0 \le x \le 1$

(b) $y' = -\frac{1}{(1 + x)^2}$

$$1 + (y')^2 = 1 + \frac{1}{(1 + x)^4}$$

$$L = \int_0^1 \sqrt{1 + \frac{1}{(1 + x)^4}}\, dx$$

(c) $L \approx 1.132$

19. (a) $y = \sin x, \quad 0 \le x \le \pi$

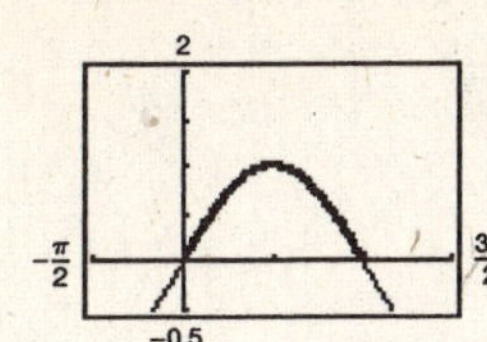

(b) $$y' = \cos x$$
$$1 + (y')^2 = 1 + \cos^2 x$$
$$L = \int_0^{\pi} \sqrt{1 + \cos^2 x}\, dx$$

(c) $L \approx 3.820$

20. (a) $y = \cos x, \quad -\frac{\pi}{2} \le x \le \frac{\pi}{2}$

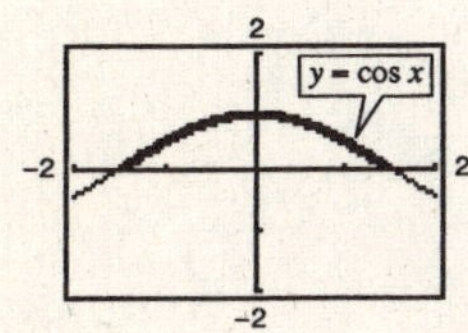

(b) $$y' = -\sin x$$
$$1 + (y')^2 = 1 + \sin^2 x$$
$$L = \int_{-\pi/2}^{\pi/2} \sqrt{1 + \sin^2 x}\, dx$$

(c) 3.820

21. (a) $x = e^{-y}, \quad 0 \le y \le 2$

$y = -\ln x$

$1 \ge x \ge e^{-2} \approx 0.135$

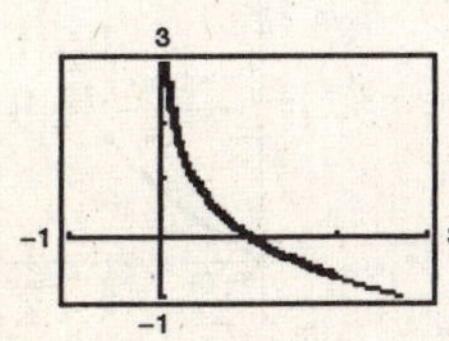

(b) $$y' = -\frac{1}{x}$$
$$1 + (y')^2 = 1 + \frac{1}{x^2}$$
$$L = \int_{e^{-2}}^{1} \sqrt{1 + \frac{1}{x^2}}\, dx$$

(c) $L \approx 2.221$

Alternatively, you can do all the computations with respect to y.

(a) $x = e^{-y}, \quad 0 \le y \le 2$

(b) $$\frac{dx}{dy} = -e^{-y}$$
$$1 + \left(\frac{dx}{dy}\right)^2 = 1 + e^{-2y}$$
$$L = \int_0^2 \sqrt{1 + e^{-2y}}\, dy$$

(c) $L \approx 2.221$

. (a) $y = \ln x, \quad 1 \le x \le 5$

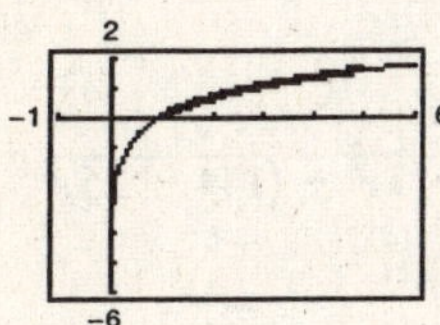

(b) $$y' = \frac{1}{x}$$
$$1 + (y')^2 = 1 + \frac{1}{x^2}$$
$$L = \int_1^5 \sqrt{1 + \frac{1}{x^2}}\, dx$$

(c) $L \approx 4.367$

23. (a) $y = 2 \arctan x, \quad 0 \le x \le 1$

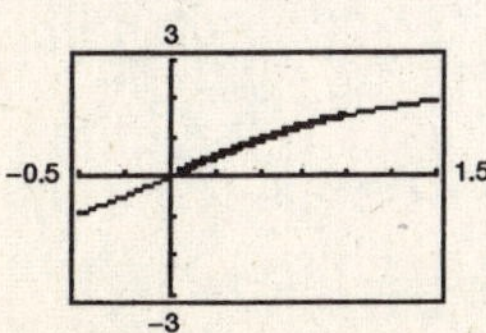

(b) $y' = \frac{2}{1 + x^2}$

$$L = \int_0^1 \sqrt{1 + \frac{4}{(1 + x^2)^2}}\, dx$$

(c) $L \approx 1.871$

24. (a) $x = \sqrt{36 - y^2}, \quad 0 \le y \le 3$

$y = \sqrt{36 - x^2}, \quad 3\sqrt{3} \le x \le 6$

(b) $\dfrac{dx}{dy} = \dfrac{1}{2}(36 - y^2)^{-1/2}(-2y)$

$= \dfrac{-y}{\sqrt{36 - y^2}}$

$L = \displaystyle\int_0^3 \sqrt{1 + \frac{y^2}{36 - y^2}}\, dy$

$= \displaystyle\int_0^3 \frac{6}{\sqrt{36 - y^2}}\, dy$

(c) $L \approx 3.142 \quad (\pi!)$

Alternatively, you can convert to a function of x.

$y = \sqrt{36 - x^2}$

$y' = \dfrac{dy}{dx} = -\dfrac{x}{\sqrt{36 - x^2}}$

$L = \displaystyle\int_{3\sqrt{3}}^6 \sqrt{1 + \frac{x^2}{36 - x^2}}\, dx = \int_{3\sqrt{3}}^6 \frac{6}{\sqrt{36 - x^2}}\, dx$

Although this integral is undefined at $x = 0$, a graphing utility still gives $L \approx 3.142$.

25. $\displaystyle\int_0^2 \sqrt{1 + \left[\frac{d}{dx}\left(\frac{5}{x^2 + 1}\right)\right]^2}\, dx$

$s \approx 5$

Matches (b)

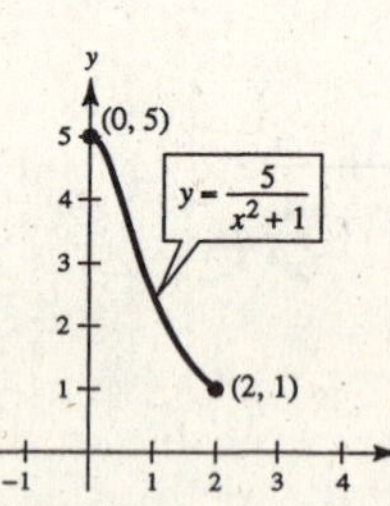

26. $\displaystyle\int_0^{\pi/4} \sqrt{1 + \left[\frac{d}{dx}(\tan x)\right]^2}\, dx$

$s \approx 1$

Matches (e)

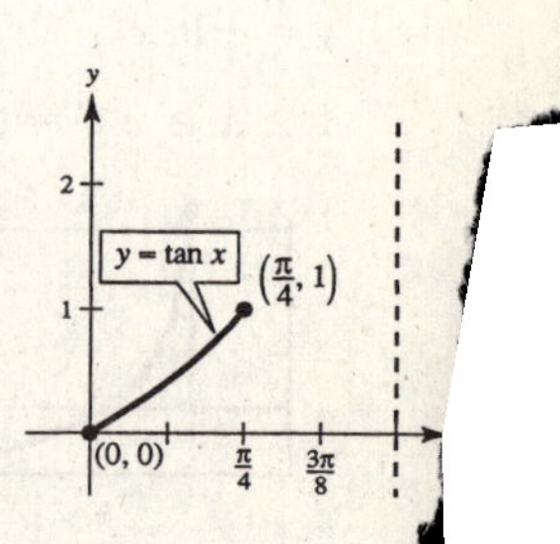

27. $y = x^3, \quad [0, 4]$

(a) $d = \sqrt{(4 - 0)^2 + (64 - 0)^2} \approx 64.125$

(b) $d = \sqrt{(1 - 0)^2 + (1 - 0)^2} + \sqrt{(2 - 1)^2 + (8 - 1)^2} + \sqrt{(3 - 2)^2 + (27 - 8)^2} + \sqrt{(4 - 3)^2 + (64 - 27)^2}$

≈ 64.525

(c) $s = \displaystyle\int_0^4 \sqrt{1 + (3x^2)^2}\, dx = \int_0^4 \sqrt{1 + 9x^4}\, dx \approx 64.666$ (Simpson's Rule, $n = 10$)

(d) 64.672

28. $f(x) = (x^2 - 4)^2, \quad [0, 4]$

(a) $d = \sqrt{(4 - 0)^2 + (144 - 16)^2} \approx 128.062$

(b) $d = \sqrt{(1 - 0)^2 + (9 - 16)^2} + \sqrt{(2 - 1)^2 + (0 - 9)^2} + \sqrt{(3 - 2)^2 + (25 - 0)^2} + \sqrt{(4 - 3)^2 + (144 - 25)^2}$

≈ 160.151

(c) $s = \displaystyle\int_0^4 \sqrt{1 + [4x(x^2 - 4)]^2}\, dx \approx 159.087$

(d) 160.287

29. (a) $f(x) = x^{2/3}$

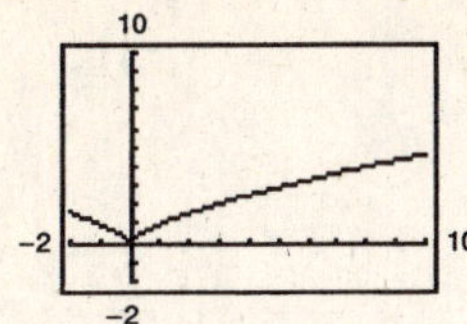

(b) No, $f'(0)$ is not defined.

(c) $$f'(x) = \frac{2}{3}x^{-1/3}$$

$$1 + f'(x)^2 = 1 + \frac{4}{9x^{2/3}} = \frac{9x^{2/3} + 4}{9x^{2/3}}$$

Divide $[-1, 8]$ into two intervals.

$$[-1, 0]:\ s_1 = \int_{-1}^{0} \sqrt{\frac{9x^{2/3} + 4}{9x^{2/3}}}\,dx$$

$$= \frac{-1}{3}\int_{-1}^{0} \sqrt{9x^{2/3} + 4}\frac{1}{x^{1/3}}\,dx, \quad (x0)$$

$$= -\frac{1}{18}\int_{-1}^{0} (9x^{2/3} + 4)^{1/2}\left(\frac{6}{x^{1/3}}\right) dx$$

$$= -\frac{1}{27}(9x^{2/3} + 4)^{3/2}\Big]_{-1}^{0}$$

$$= -\frac{1}{27}(4^{3/2} - 13^{3/2})$$

$$= -\frac{1}{27}(8 - 13^{3/2}) \approx 1.4397$$

$$[0, 8]:\ s_2 = \int_{0}^{8} \sqrt{\frac{9x^{2/3} + 4}{9x^{2/3}}}\,dx$$

$$= \frac{1}{3}\int_{0}^{8} \sqrt{9x^{2/3} + 4}\frac{1}{x^{1/3}}\,dx, \quad (x \geq 0)$$

$$= \frac{1}{27}(9x^{2/3} + 4)^{3/2}\Big]_{0}^{8}$$

$$= \frac{1}{27}(40^{3/2} - 4^{3/2})$$

$$= \frac{1}{27}(40^{3/2} - 8) \approx 9.0734$$

$$s_1 + s_2 = \frac{1}{27}[40^{3/2} - 8 - 8 + 13^{3/2}]$$

$$= \frac{1}{27}[40^{3/2} + 13^{3/2} - 16] \approx 10.5131$$

30. $x^{2/3} + y^{2/3} = 4$

$$y^{2/3} = 4 - x^{2/3}$$

$$y = (4 - x^{2/3})^{3/2}, \quad 0 \leq x \leq 8$$

$$y' = \frac{3}{2}(4 - x^{2/3})^{1/2}\left(-\frac{2}{3}x^{-1/3}\right) = \frac{-(4 - x^{2/3})^{1/2}}{x^{1/3}}$$

$$1 + (y')^2 = 1 + \frac{4 - x^{2/3}}{x^{2/3}} = \frac{4}{x^{2/3}}$$

$$\frac{1}{4}s = \int_0^8 \sqrt{\frac{4}{x^{2/3}}}\,dx$$

$$= 2\int_0^8 x^{-2/3}\,dx = 6x^{1/3}\Big]_0^8 = 12$$

Total length: $s = 4(12) = 48$

31. (a)

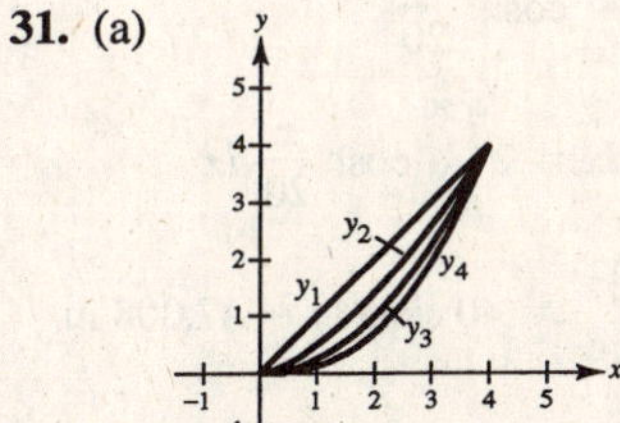

(b) y_1, y_2, y_3, y_4

(c) $y_1' = 1,\ L_1 = \int_0^4 \sqrt{2}\,dx \approx 5.657$

$$y_2' = \frac{3}{4}x^{1/2},\ L_2 = \int_0^4 \sqrt{1 + \frac{9x}{16}}\,dx \approx 5.759$$

$$y_3' = \frac{1}{2}x,\ L_3 = \int_0^4 \sqrt{1 + \frac{x^2}{4}}\,dx \approx 5.916$$

$$y_4' = \frac{5}{16}x^{3/2},\ L_4 = \int_0^4 \sqrt{1 + \frac{25}{256}x^3}\,dx \approx 6.063$$

32. Let $y = \ln x$, $1 \le x \le e$, $y' = \dfrac{1}{x}$ and $L_1 = \displaystyle\int_1^e \sqrt{1 + \frac{1}{x^2}}\, dx.$

Equivalently, $x = e^y$, $0 \le y \le 1$, $\dfrac{dx}{dy} = e^y$, and $L_2 = \displaystyle\int_0^1 \sqrt{1 + e^{2y}}\, dy = \int_0^1 \sqrt{1 + e^{2x}}\, dx.$

Numerically, both integrals yield $L = 2.0035$.

33. $y = \dfrac{1}{3}[x^{3/2} - 3x^{1/2} + 2]$

When $x = 0$, $y = \frac{2}{3}$. Thus, the fleeing object has traveled $\frac{2}{3}$ units when it is caught.

$$y' = \frac{1}{3}\left[\frac{3}{2}x^{1/2} - \frac{3}{2}x^{-1/2}\right] = \left(\frac{1}{2}\right)\frac{x-1}{x^{1/2}}$$

$$1 + (y')^2 = 1 + \frac{(x-1)^2}{4x} = \frac{(x+1)^2}{4x}$$

$$s = \int_0^1 \frac{x+1}{2x^{1/2}}\, dx = \frac{1}{2}\int_0^1 (x^{1/2} + x^{-1/2})\, dx$$

$$= \frac{1}{2}\left[\frac{2}{3}x^{3/2} + 2x^{1/2}\right]_0^1 = \frac{4}{3} = 2\left(\frac{2}{3}\right)$$

The pursuer has traveled twice the distance that the fleeing object has traveled when it is caught.

34. $y = 31 - 10(e^{x/20} + e^{-x/20})$

$$y' = -\frac{1}{2}(e^{x/20} - e^{-x/20})$$

$$1 + (y')^2 = 1 + \frac{1}{4}(e^{x/10} - 2 + e^{-x/10})$$

$$= \left[\frac{1}{2}(e^{x/20} + e^{-x/20})\right]^2$$

$$s = \int_{-20}^{20} \sqrt{\left[\frac{1}{2}(e^{x/20} + e^{-x/20})\right]^2}\, dx$$

$$= \frac{1}{2}\int_{-20}^{20} (e^{x/20} + e^{-x/20})\, dx$$

$$= \Big[10(e^{x/20} - e^{-x/20})\Big]_{-20}^{20}$$

$$= 20\left(e - \frac{1}{e}\right) \approx 47 \text{ ft}$$

Thus, there are $100(47) = 4700$ square feet of roofing on the barn.

35. $y = 20\cosh\dfrac{x}{20}, \quad -20 \le x \le 20$

$$y' = \sinh\frac{x}{20}$$

$$1 + (y')^2 = 1 + \sinh^2\frac{x}{20} = \cosh^2\frac{x}{20}$$

$$L = \int_{-20}^{20} \cosh\frac{x}{20}\, dx = 2\int_0^{20} \cosh\frac{x}{20}\, dx$$

$$= 2(20)\sinh\frac{x}{20}\Big]_0^{20} = 40\sinh(1) \approx 47.008 \text{ m}$$

36. $y = 693.8597 - 68.7672\cosh 0.0100333x$

$$y' = -0.6899619478\sinh 0.0100333x$$

$$s = \int_{-299.2239}^{299.2239} \sqrt{1 + (-0.6899619478\sinh 0.0100333x)^2}\, dx$$

$$\approx 1480$$

(Use Simpson's Rule with $n = 100$ or a graphing utility.)

37. $y = \sqrt{9 - x^2}$

$$y' = \frac{-x}{\sqrt{9 - x^2}}$$

$$1 + (y')^2 = \frac{9}{9 - x^2}$$

$$s = \int_0^2 \sqrt{\frac{9}{9 - x^2}}\, dx = \int_0^2 \frac{3}{\sqrt{9 - x^2}}\, dx$$

$$= \left[3\arcsin\frac{x}{3}\right]_0^2 = 3\left(\arcsin\frac{2}{3} - \arcsin 0\right)$$

$$= 3\arcsin\frac{2}{3} \approx 2.1892$$

38. $y = \sqrt{25 - x^2}$

$$y' = \frac{-x}{\sqrt{25 - x^2}}$$

$$1 + (y')^2 = \frac{25}{25 - x^2}$$

$$s = \int_{-3}^4 \sqrt{\frac{25}{25 - x^2}}\, dx = \int_{-3}^4 \frac{5}{\sqrt{25 - x^2}}\, dx$$

$$= \left[5\arcsin\frac{x}{5}\right]_{-3}^4 = 5\left[\arcsin\frac{4}{5} - \arcsin\left(-\frac{3}{5}\right)\right]$$

$$\approx 7.8540$$

$$\frac{1}{4}[2\pi(5)] \approx 7.8540 = s$$

39. $y = \dfrac{x^3}{3}$

$y' = x^2, \quad [0, 3]$

$$S = 2\pi \int_0^3 \frac{x^3}{3}\sqrt{1 + x^4}\,dx$$

$$= \frac{\pi}{6}\int_0^3 (1 + x^4)^{1/2}(4x^3)\,dx$$

$$= \left[\frac{\pi}{9}(1 + x^4)^{3/2}\right]_0^3$$

$$= \frac{\pi}{9}\left(82\sqrt{82} - 1\right) \approx 258.85$$

40. $y = 2\sqrt{x}$

$y' = \dfrac{1}{\sqrt{x}}, \quad [4, 9]$

$$S = 2\pi\int_4^9 2\sqrt{x}\sqrt{1 + \frac{1}{x}}\,dx$$

$$= 4\pi\int_4^9 \sqrt{x + 1}\,dx$$

$$= \left[\frac{8}{3}\pi(x + 1)^{3/2}\right]_4^9$$

$$= \frac{8\pi}{3}(10^{3/2} - 5^{3/2}) \approx 171.258$$

41. $y = \dfrac{x^3}{6} + \dfrac{1}{2x}$

$y' = \dfrac{x^2}{2} - \dfrac{1}{2x^2}$

$1 + (y')^2 = \left(\dfrac{x^2}{2} + \dfrac{1}{2x^2}\right)^2, \quad [1, 2]$

$$S = 2\pi\int_1^2 \left(\frac{x^3}{6} + \frac{1}{2x}\right)\left(\frac{x^2}{2} + \frac{1}{2x^2}\right)dx$$

$$= 2\pi\int_1^2 \left(\frac{x^5}{12} + \frac{x}{3} + \frac{1}{4x^3}\right)dx$$

$$= 2\pi\left[\frac{x^6}{72} + \frac{x^2}{6} - \frac{1}{8x^2}\right]_1^2 = \frac{47\pi}{16}$$

42. $y = \dfrac{x}{2}$

$y' = \dfrac{1}{2}$

$1 + (y')^2 = \dfrac{5}{4}, \quad [0, 6]$

$$S = 2\pi\int_0^6 \frac{x}{2}\sqrt{\frac{5}{4}}\,dx$$

$$= \left[\frac{2\pi\sqrt{5}}{8}x^2\right]_0^6 = 9\sqrt{5}\pi$$

43. $y = \sqrt[3]{x} + 2$

$y' = \dfrac{1}{3x^{2/3}}, \quad [1, 8]$

$$S = 2\pi\int_1^8 x\sqrt{1 + \frac{1}{9x^{4/3}}}\,dx$$

$$= \frac{2\pi}{3}\int_1^8 x^{1/3}\sqrt{9x^{4/3} + 1}\,dx$$

$$= \frac{\pi}{18}\int_1^8 (9x^{4/3} + 1)^{1/2}(12x^{1/3})\,dx$$

$$= \left[\frac{\pi}{27}(9x^{4/3} + 1)^{3/2}\right]_1^8$$

$$= \frac{\pi}{27}\left(145\sqrt{145} - 10\sqrt{10}\right) \approx 199.48$$

44. $y = 9 - x^2, \quad [0, 3]$

$y' = -2x$

$$S = 2\pi\int_0^3 x\sqrt{1 + 4x^2}\,dx$$

$$= \frac{\pi}{4}\int_0^3 (1 + 4x^2)^{1/2}(8x)\,dx$$

$$= \left[\frac{\pi}{6}(1 + 4x^2)^{3/2}\right]_0^3$$

$$= \frac{\pi}{6}(37^{3/2} - 1) \approx 117.319$$

45. $y = \sin x$

$y' = \cos x, \quad [0, \pi]$

$$S = 2\pi\int_0^{\pi} \sin x\sqrt{1 + \cos^2 x}\,dx$$

$$\approx 14.4236$$

46. $y = \ln x$

$y' = \dfrac{1}{x}$

$1 + (y')^2 = \dfrac{x^2 + 1}{x^2}, \quad [1, e]$

$$S = 2\pi\int_1^e x\sqrt{\frac{x^2 + 1}{x^2}}\,dx = 2\pi\int_1^e \sqrt{x^2 + 1}\,dx$$

$$\approx 22.943$$

47. A rectifiable curve is one that has a finite arc length.

48. The precalculus formula is the distance formula between two points. The representative element is

$$\sqrt{(\Delta x_i)^2 + (\Delta y_i)^2} = \sqrt{1 + \left(\frac{\Delta y_i}{\Delta x_i}\right)^2}\,\Delta xi.$$

49. The precalculus formula is the surface area formula for the lateral surface of the frustum of a right circular cone. The representative element is

$$2\pi f(d_i)\sqrt{\Delta x_i^2 + \Delta y_i^2} = 2\pi f(d_i)\sqrt{1 + \left(\frac{\Delta y_i}{\Delta x_i}\right)^2}\,\Delta x_i.$$

50. The surface of revolution given by f_1 will be larger. $r(x)$ is larger for f_1.

51.
$$y = \frac{hx}{r}$$
$$y' = \frac{h}{r}$$
$$1 + (y')^2 = \frac{r^2 + h^2}{r^2}$$
$$S = 2\pi\int_0^r x\sqrt{\frac{r^2 + h^2}{r^2}}\,dx$$
$$= \left[\frac{2\pi\sqrt{r^2 + h^2}}{r}\left(\frac{x^2}{2}\right)\right]_0^r = \pi r\sqrt{r^2 + h^2}$$

52.
$$y = \sqrt{r^2 - x^2}$$
$$y' = \frac{-x}{\sqrt{r^2 - x^2}}$$
$$1 + (y')^2 = \frac{r^2}{r^2 - x^2}$$
$$S = 2\pi\int_{-r}^r \sqrt{r^2 - x^2}\sqrt{\frac{r^2}{r^2 - x^2}}\,dx$$
$$= 2\pi\int_{-r}^r r\,dx = \Big[2\pi rx\Big]_{-r}^r = 4\pi r^2$$

53.
$$y = \sqrt{9 - x^2}$$
$$y' = \frac{-x}{\sqrt{9 - x^2}}$$
$$\sqrt{1 + (y')^2} = \frac{3}{\sqrt{9 - x^2}}$$
$$S = 2\pi\int_0^2 \frac{3x}{\sqrt{9 - x^2}}\,dx$$
$$= -3\pi\int_0^2 \frac{-2x}{\sqrt{9 - x^2}}\,dx$$
$$= \left[-6\pi\sqrt{9 - x^2}\right]_0^2$$
$$= 6\pi\left(3 - \sqrt{5}\right) \approx 14.40$$

See figure in Exercise 54.

54. From Exercise 53 we have:

$$S = 2\pi\int_0^a \frac{rx}{\sqrt{r^2 - x^2}}\,dx$$
$$= -r\pi\int_0^a \frac{-2x\,dx}{\sqrt{r^2 - x^2}}$$
$$= \left[-2r\pi\sqrt{r^2 - x^2}\right]_0^a$$
$$= 2r^2\pi - 2r\pi\sqrt{r^2 - a^2}$$
$$= 2r\pi\left(r - \sqrt{r^2 - a^2}\right)$$
$= 2\pi rh$ (where h is the height of the zone)

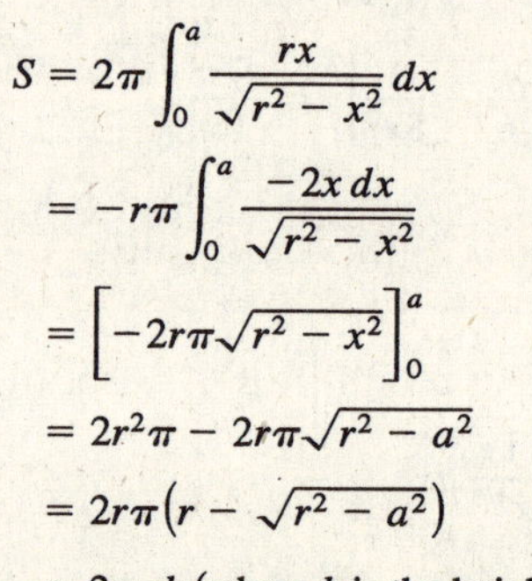

55.
$$y = \frac{1}{3}x^{1/2} - x^{3/2}$$
$$y' = \frac{1}{6}x^{-1/2} - \frac{3}{2}x^{1/2} = \frac{1}{6}(x^{-1/2} - 9x^{1/2})$$
$$1 + (y')^2 = 1 + \frac{1}{36}(x^{-1} - 18 + 81x) = \frac{1}{36}(x^{-1/2} + 9x^{1/2})^2$$
$$S = 2\pi\int_0^{1/3}\left(\frac{1}{3}x^{1/2} - x^{3/2}\right)\sqrt{\frac{1}{36}(x^{-1/2} + 9x^{1/2})^2}\,dx = \frac{2\pi}{6}\int_0^{1/3}\left(\frac{1}{3}x^{1/2} - x^{3/2}\right)(x^{-1/2} + 9x^{1/2})\,dx$$
$$= \frac{\pi}{3}\int_0^{1/3}\left(\frac{1}{3} + 2x - 9x^2\right)dx = \frac{\pi}{3}\left[\frac{1}{3}x + x^2 - 3x^3\right]_0^{1/3} = \frac{\pi}{27}\text{ ft}^2 \approx 0.1164\text{ ft}^2 \approx 16.8\text{ in.}^2$$

Amount of glass needed: $V = \dfrac{\pi}{27}\left(\dfrac{0.015}{12}\right) \approx 0.00015\text{ ft}^3 \approx 0.25\text{ in.}^3$

56. (a) $\frac{x^2}{9} + \frac{y^2}{4} = 1$

Ellipse: $y_1 = 2\sqrt{1 - \frac{x^2}{9}}$

$y_2 = -2\sqrt{1 - \frac{x^2}{9}}$

(b) $y = 2\sqrt{1 - \frac{x^2}{9}}, \quad 0 \le x \le 3$

$$y' = 2\left(\frac{1}{2}\right)\left(1 - \frac{x^2}{9}\right)^{-1/2}\left(\frac{-2x}{9}\right)$$

$$= \frac{-2x}{9\sqrt{1 - (x^2/9)}} = \frac{-2x}{3\sqrt{9 - x^2}}$$

$$L = \int_0^3 \sqrt{1 + \frac{4x^2}{81 - 9x^2}}\, dx$$

(c) You cannot evaluate this definite integral, since the integrand is not defined at $x = 3$. Simpson's Rule will not work for the same reason. Also, the integrand does not have an elementary antiderivative.

57. (a) We approximate the volume by summing six disks of thickness 3 and circumference C_i equal to the average of the given circumferences:

$$V \approx \sum_{i=1}^{6} \pi r_i^2(3) = \sum_{i=1}^{6} \pi\left(\frac{C_i}{2\pi}\right)^2(3) = \frac{3}{4\pi}\sum_{i=1}^{6} C_i^2$$

$$= \frac{3}{4\pi}\left[\left(\frac{50 + 65.5}{2}\right)^2 + \left(\frac{65.5 + 70}{2}\right)^2 + \left(\frac{70 + 66}{2}\right)^2 + \left(\frac{66 + 58}{2}\right)^2 + \left(\frac{58 + 51}{2}\right)^2 + \left(\frac{51 + 48}{2}\right)^2\right]$$

$$= \frac{3}{4\pi}[57.75^2 + 67.75^2 + 68^2 + 62^2 + 54.5^2 + 49.5^2]$$

$$= \frac{3}{4\pi}[21813.625] = 5207.62 \text{ cubic inches}$$

(b) The lateral surface area of a frustum of a right circular cone is $\pi s(R + r)$. For the first frustum:

$$S_1 \approx \pi\left[3^2 + \left(\frac{65.5 - 50}{2\pi}\right)^2\right]^{1/2}\left[\frac{50}{2\pi} + \frac{65.5}{2\pi}\right]$$

$$= \left(\frac{50 + 65.5}{2}\right)\left[9 + \left(\frac{65.5 - 50}{2\pi}\right)^2\right]^{1/2}.$$

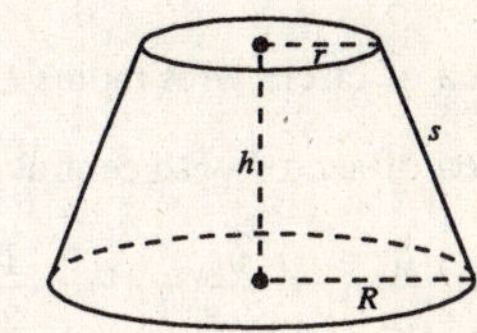

Adding the six frustums together:

$$S \approx \left(\frac{50 + 65.5}{2}\right)\left[9 + \left(\frac{15.5}{2\pi}\right)^2\right]^{1/2} + \left(\frac{65.5 + 70}{2}\right)\left[9 + \left(\frac{4.5}{2\pi}\right)^2\right]^{1/2} +$$

$$\left(\frac{70 + 66}{2}\right)\left[9 + \left(\frac{4}{2\pi}\right)^2\right]^{1/2} + \left(\frac{66 + 58}{2}\right)\left[9 + \left(\frac{8}{2\pi}\right)^2\right]^{1/2} +$$

$$\left(\frac{58 + 51}{2}\right)\left[9 + \left(\frac{7}{2\pi}\right)^2\right]^{1/2} + \left(\frac{51 + 48}{2}\right)\left[9 + \left(\frac{3}{2\pi}\right)^2\right]^{1/2}$$

$$\approx 224.30 + 208.96 + 208.54 + 202.06 + 174.41 + 150.37$$

$$= 1168.64$$

(c) $r = 0.00401y^3 - 0.1416y^2 + 1.232y + 7.943$

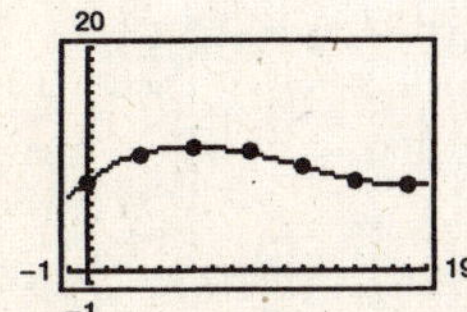

(d) $V = \int_0^{18} \pi r^2\, dy \approx 5275.9$ cubic inches

$$S = \int_0^{18} 2\pi r(y)\sqrt{1 + r'(y)^2}\, dy$$

≈ 1179.5 square inches

58. (a) $y = f(x) = 0.0000001953x^4 - 0.0001804x^3 + 0.0496x^2 - 4.8323x + 536.9270$

(b) Area $= \int_0^{400} f(x)\,dx \approx 131{,}734.5$ square feet ≈ 3.0 acres (1 acre = 43,560 square feet)

(Answers will vary.)

(c) $L = \int_0^{400} \sqrt{1 + f'(x)^2}\,dx \approx 794.9$ feet

(Answers will vary.)

59. (a) $V = \pi \int_1^b \frac{1}{x^2}\,dx = \left[-\frac{\pi}{x}\right]_1^b = \pi\left(1 - \frac{1}{b}\right)$

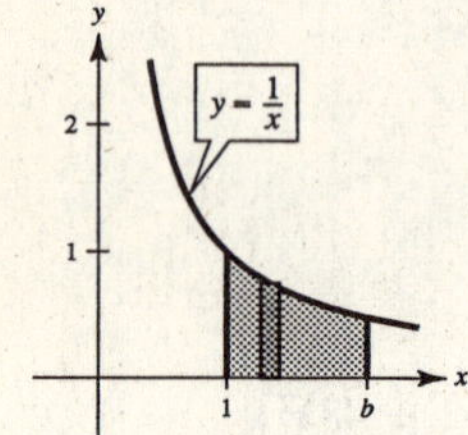

(b) $S = 2\pi \int_1^b \frac{1}{x}\sqrt{1 + \left(-\frac{1}{x^2}\right)^2}\,dx$

$= 2\pi \int_1^b \frac{1}{x}\sqrt{1 + \frac{1}{x^4}}\,dx$

$= 2\pi \int_1^b \frac{\sqrt{x^4+1}}{x^3}\,dx$

(c) $\lim\limits_{b\to\infty} V = \lim\limits_{b\to\infty} \pi\left(1 - \frac{1}{b}\right) = \pi$

(d) Since

$$\frac{\sqrt{x^4+1}}{x^3} > \frac{\sqrt{x^4}}{x^3} = \frac{1}{x} > 0 \text{ on } [1, b]$$

we have

$$\int_1^b \frac{\sqrt{x^4+1}}{x^3}\,dx > \int_1^b \frac{1}{x}\,dx = \left[\ln x\right]_1^b = \ln b$$

and $\lim\limits_{b\to\infty} \ln b \to \infty$. Thus,

$$\lim_{b\to\infty} 2\pi \int_1^b \frac{\sqrt{x^4+1}}{x^3}\,dx = \infty.$$

60. (a) Area of circle with radius L: $A = \pi L^2$

Area of sector with central angle θ (in radians):

$$S = \frac{\theta}{2\pi}A = \frac{\theta}{2\pi}(\pi L^2) = \frac{1}{2}L^2\theta$$

(b) Let s be the arc length of the sector, which is the circumference of the base of the cone. Here, $s = L\theta = 2\pi r$, and you have

$$S = \frac{1}{2}L^2\theta = \frac{1}{2}L^2\left(\frac{s}{L}\right) = \frac{1}{2}Ls = \frac{1}{2}L(2\pi r) = \pi r L.$$

(c) The lateral surface area of the frustum is the difference of the large cone and the small one.

$S = \pi r_2(L + L_1) - \pi r_1 L_1$

$= \pi r_2 L + \pi L_1(r_2 - r_1)$

By similar triangles, $\frac{L + L_1}{r_2} = \frac{L_1}{r_1} \Rightarrow L r_1 = L_1(r_2 - r_1)$. Hence,

$S = \pi r_2 L + \pi L_1(r_2 - r_1) = \pi r_2 L + \pi L r_1$

$= \pi L(r_1 + r_2)$.

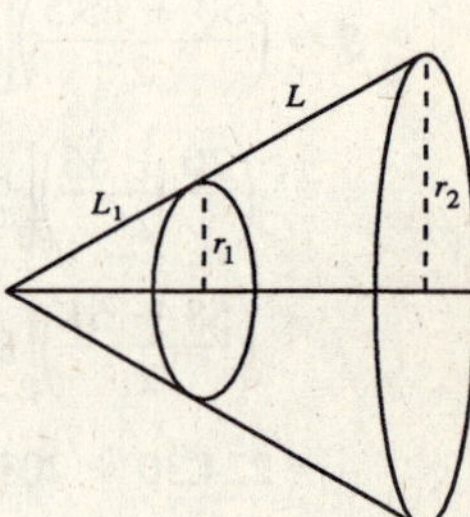

61. Individual project

62. Essay

63. $x^{2/3} + y^{2/3} = 4$

$$y^{2/3} = 4 - x^{2/3}$$

$$y = (4 - x^{2/3})^{3/2}, \quad 0 \le x \le 8$$

$$y' = \frac{3}{2}(4 - x^{2/3})^{1/2}\left(-\frac{2}{3}x^{-1/3}\right) = \frac{-(4 - x^{2/3})^{1/2}}{x^{1/3}}$$

$$1 + (y')^2 = 1 + \frac{4 - x^{2/3}}{x^{2/3}} = \frac{4}{x^{2/3}}$$

$$S = 2\pi\int_0^8 (4 - x^{2/3})^{3/2}\sqrt{\frac{4}{x^{2/3}}}\,dx$$

$$= 4\pi\int_0^8 \frac{(4 - x^{2/3})^{3/2}}{x^{1/3}}\,dx$$

$$= \left[-\frac{12\pi}{5}(4 - x^{2/3})^{5/2}\right]_0^8 = \frac{192\pi}{5}$$

[Surface area of portion above the x-axis]

64. $y^2 = \frac{1}{12}x(4 - x)^2, \quad 0 \le x \le 4$

$$y = \frac{(4 - x)\sqrt{x}}{\sqrt{12}}$$

$$y' = \frac{(4 - 3x)\sqrt{3}}{12\sqrt{x}}$$

$$1 + (y')^2 = 1 + \frac{(4 - 3x)^2}{48x}$$

$$= \frac{48x + 16 - 24x + 9x^2}{48x} = \frac{(4 + 3x)^2}{48x}, \quad x \ne 0$$

$$S = 2\pi\int_0^4 \frac{(4 - x)\sqrt{x}}{\sqrt{12}} \cdot \frac{(4 + 3x)}{\sqrt{48x}}\,dx$$

$$= 2\pi\int_0^4 \frac{(4 - x)(4 + 3x)}{24}\,dx$$

$$= \frac{\pi}{12}\int_0^4 (16 + 8x - 3x^2)\,dx$$

$$= \frac{\pi}{12}\left[16x + 4x^2 - x^3\right]_0^4$$

$$= \frac{\pi}{12}[64 + 64 - 64] = \frac{16\pi}{3}$$

65. $y = kx^2, y' = 2kx$

$$1 + (y')^2 = 1 + 4k^2x^2$$

$$h = kw^2 \Rightarrow k = \frac{h}{w^2} \Rightarrow 1 + (y') = 1 + \frac{4h^2}{w^4}x^2$$

By symmetry, $C = 2\int_0^w \sqrt{1 + \frac{4h^2}{w^4}x^2}\,dx.$

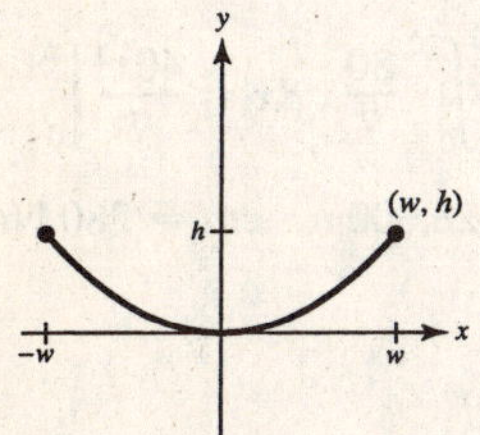

66. $C = 2\int_0^w \sqrt{1 + \frac{4h^2}{w^4}x^2}\,dx$

$$= 2\int_0^{700} \sqrt{1 + \frac{4(155)^2x^2}{700^4}}\,dx$$

≈ 1444.5 meters

67. Let (x_0, y_0) be the point on the graph of $y^2 = x^3$ where the tangent line makes an angle of 45° with the x-axis.

$$y = x^{3/2}$$

$$y' = \tfrac{3}{2}x^{1/2} = 1$$

$$x_0 = \tfrac{4}{9}$$

$$L = \int_0^{4/9} \sqrt{1 + \tfrac{9}{4}x}\,dx = \tfrac{8}{27}(2\sqrt{2} - 1)$$

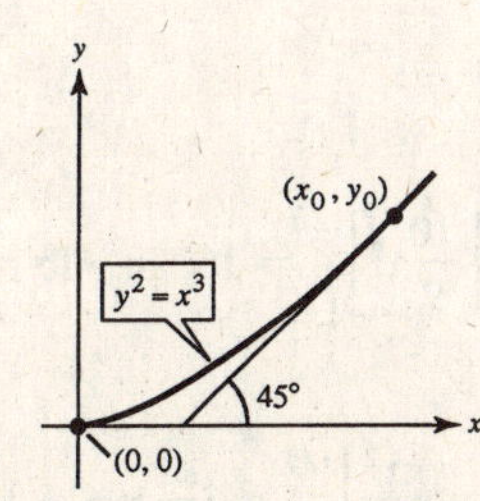

Section 7.5 Work

1. $W = Fd = (100)(10)$

$= 1000 \text{ ft} \cdot \text{lb}$

2. $W = Fd = (2800)(4)$

$= 11{,}200 \text{ ft} \cdot \text{lb}$

3. $W = Fd = (112)(4)$

$= 448$ joules (newton-meters)

4. $W = Fd = [9(2000)][\frac{1}{2}(5280)] = 47{,}520{,}000 \text{ ft} \cdot \text{lb}$

5. Work equals force times distance, $W = FD$.

6. $W = \int_a^b F(x)\,dx$ is the work done by a force F moving an object along a straight line from $x = a$ to $x = b$.

7. Since the work equals the area under the force function, you have $(c) < (d) < (a) < (b)$.

8. (a) $W = \int_0^9 6\,dx = 54 \text{ ft} \cdot \text{lbs}$

(b) $W = \int_0^7 20\,dx + \int_7^9 (-10x + 90)\,dx = 140 + 20$

$= 160 \text{ ft} \cdot \text{lbs}$

(c) $W = \int_0^9 \frac{1}{27}x^2\,dx = \left.\frac{x^3}{81}\right]_0^9 = 9 \text{ ft} \cdot \text{lbs}$

(d) $W = \int_0^9 \sqrt{x}\,dx = \left.\frac{2}{3}x^{3/2}\right]_0^9 = \frac{2}{3}(27) = 18 \text{ ft} \cdot \text{lbs}$

9. $F(x) = kx$

$5 = k(4)$

$k = \frac{5}{4}$

$W = \int_0^7 \frac{5}{4}x\,dx = \left[\frac{5}{8}x^2\right]_0^7$

$= \frac{245}{8} \text{ in.} \cdot \text{lb}$

$= 30.625 \text{ in.} \cdot \text{lb} \approx 2.55 \text{ ft} \cdot \text{lb}$

10. $W = \int_5^9 \frac{5}{4}x\,dx = \left[\frac{5}{8}x^2\right]_5^9$

$= 35 \text{ in.} \cdot \text{lb}$

11. $F(x) = kx$

$250 = k(30) \Rightarrow k = \frac{25}{3}$

$W = \int_{20}^{50} F(x)\,dx$

$= \int_{20}^{50} \frac{25}{3}x\,dx = \left.\frac{25x^2}{6}\right]_{20}^{50}$

$= 8750 \text{ n} \cdot \text{cm}$

$= 87.5$ joules or Nm

12. $F(x) = kx$

$800 = k(70) \Rightarrow k = \frac{80}{7}$

$W = \int_0^{70} F(x)\,dx$

$= \int_0^{70} \frac{80}{7}x\,dx = \left.\frac{40x^2}{7}\right]_0^{70}$

$= 28{,}000 \text{ n} \cdot \text{cm} = 280 \text{ Nm}$

13. $F(x) = kx$

$20 = k(9)$

$k = \frac{20}{9}$

$W = \int_0^{12} \frac{20}{9}x\,dx = \left[\frac{10}{9}x^2\right]_0^{12} = 160 \text{ in.} \cdot \text{lb} = \frac{40}{3} \text{ ft} \cdot \text{lb}$

14. $F(x) = kx$

$15 = k(1) = k$

$W = 2\int_0^4 15x\,dx = \left[15x^2\right]_0^4 = 240 \text{ ft} \cdot \text{lb}$

15. $W = 18 = \int_0^{1/3} kx\,dx = \left.\frac{kx^2}{2}\right]_0^{1/3} = \frac{k}{18} \Rightarrow k = 324$

$W = \int_{1/3}^{7/12} 324x\,dx = \left.162x^2\right]_{1/3}^{7/12} = 37.125 \text{ ft} \cdot \text{lbs}$

[**Note:** 4 inches $= \frac{1}{3}$ foot]

16. $W = 7.5 = \int_0^{1/6} kx\,dx = \left.\frac{kx^2}{2}\right]_0^{1/6} = \frac{k}{72} \Rightarrow k = 540$

$W = \int_{1/6}^{5/24} 540x\,dx = \left.270x^2\right]_{1/6}^{5/24} = 4.21875 \text{ ft} \cdot \text{lbs}$

17. Assume that Earth has a radius of 4000 miles.

$$F(x) = \frac{k}{x^2}$$

$$5 = \frac{k}{(4000)^2}$$

$$k = 80{,}000{,}000$$

$$F(x) = \frac{80{,}000{,}000}{x^2}$$

(a) $W = \int_{4000}^{4100} \frac{80{,}000{,}000}{x^2}\,dx = \left[\frac{-80{,}000{,}000}{x}\right]_{4000}^{4100}$

$\approx 487.8 \text{ mi} \cdot \text{tons} \approx 5.15 \times 10^9 \text{ ft} \cdot \text{lb}$

(b) $W = \int_{4000}^{4300} \frac{80{,}000{,}000}{x^2}\,dx$

$\approx 1395.3 \text{ mi} \cdot \text{ton} \approx 1.47 \times 10^{10} \text{ ft} \cdot \text{ton}$

18. $W = \int_{4000}^{h} \frac{80{,}000{,}000}{x^2}\,dx = \left[-\frac{80{,}000{,}000}{x}\right]_{4000}^{h} = \frac{-80{,}000{,}000}{h} + 20{,}000$

$\lim_{h\to\infty} W = 20{,}000 \text{ mi/ton} \approx 2.1 \times 10^{11} \text{ ft} \cdot \text{lb}$

19. Assume that Earth has a radius of 4000 miles.

$$F(x) = \frac{k}{x^2}$$

$$10 = \frac{k}{(4000)^2}$$

$$k = 160{,}000{,}000$$

$$F(x) = \frac{160{,}000{,}000}{x^2}$$

(a) $W = \int_{4000}^{15{,}000} \frac{160{,}000{,}000}{x^2}\,dx = \left[-\frac{160{,}000{,}000}{x}\right]_{4000}^{15{,}000} \approx -10{,}666.667 + 40{,}000$

$= 29{,}333.333 \text{ mi} \cdot \text{ton}$

$\approx 2.93 \times 10^4 \text{ mi} \cdot \text{ton}$

$\approx 3.10 \times 10^{11} \text{ ft} \cdot \text{lb}$

(b) $W = \int_{4000}^{26{,}000} \frac{160{,}000{,}000}{x^2}\,dx = \left[-\frac{160{,}000{,}000}{x}\right]_{4000}^{26{,}000} \approx -6{,}153.846 + 40{,}000$

$= 33{,}846.154 \text{ mi} \cdot \text{ton}$

$\approx 3.38 \times 10^4 \text{ mi} \cdot \text{ton}$

$\approx 3.57 \times 10^{11} \text{ ft} \cdot \text{lb}$

20. Weight on surface of moon: $\frac{1}{6}(12) = 2$ tons

Weight varies inversely as the square of distance from the center of the moon. Therefore:

$$F(x) = \frac{k}{x^2}$$

$$2 = \frac{k}{(1100)^2}$$

$$k = 2.42 \times 10^6$$

$$W = \int_{1100}^{1150} \frac{2.42 \times 10^6}{x^2}\,dx = \left[\frac{-2.42 \times 10^6}{x}\right]_{1100}^{1150} = 2.42 \times 10^6\left(\frac{1}{1100} - \frac{1}{1150}\right)$$

$\approx 95.652 \text{ mi} \cdot \text{ton} \approx 1.01 \times 10^9 \text{ ft} \cdot \text{lb}$

21. Weight of each layer: $62.4(20)\,\Delta y$

Distance: $4 - y$

(a) $W = \int_2^4 62.4(20)(4 - y)\,dy = \left[4992y - 624y^2\right]_2^4 = 2496 \text{ ft} \cdot \text{lb}$

(b) $W = \int_0^4 62.4(20)(4 - y)\,dy = \left[4992y - 624y^2\right]_0^4 = 9984 \text{ ft} \cdot \text{lb}$

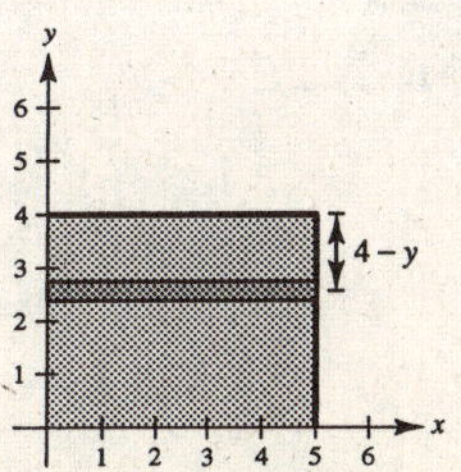

22. The bottom half had to be pumped a greater distance than the top half.

23. Volume of disk: $\pi(2)^2\,\Delta y = 4\pi\,\Delta y$

Weight of disk of water: $9800(4\pi)\,\Delta y$

Distance the disk of water is moved: $5 - y$

$$W = \int_0^4 (5-y)(9800)4\pi\,dy = 39{,}200\pi\int_0^4 (5-y)\,dy$$

$$= 39{,}200\pi\left[5y - \frac{y^2}{2}\right]_0^4$$

$$= 39{,}200\pi(12) = 470{,}400\pi \text{ newton–meters}$$

24. Volume of disk: $4\pi\,\Delta y$

Weight of disk: $9800(4\pi)\,\Delta y$

Distance the disk of water is moved: y

$$W = \int_{10}^{12} y(9800)(4\pi)\,dy = 39{,}200\pi\left[\frac{y^2}{2}\right]_{10}^{12}$$

$$= 39{,}200\pi(22)$$

$$= 862{,}400\pi \text{ newton–meters}$$

25. Volume of disk: $\pi\left(\frac{2}{3}y\right)^2\Delta y$

Weight of disk: $62.4\pi\left(\frac{2}{3}y\right)^2\Delta y$

Distance: $6 - y$

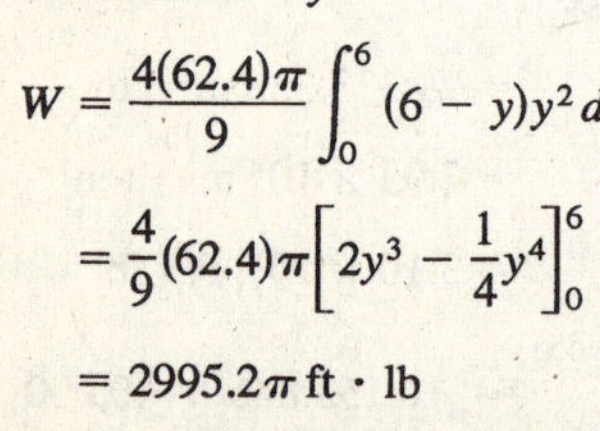

$$W = \frac{4(62.4)\pi}{9}\int_0^6 (6-y)y^2\,dy$$

$$= \frac{4}{9}(62.4)\pi\left[2y^3 - \frac{1}{4}y^4\right]_0^6$$

$$= 2995.2\pi \text{ ft} \cdot \text{lb}$$

26. Volume of disk: $\pi\left(\frac{2}{3}y\right)^2\Delta y$

Weight of disk: $62.4\pi\left(\frac{2}{3}y\right)^2\Delta y$

Distance: y

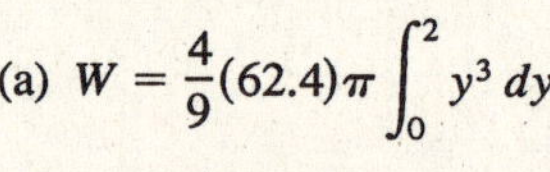

(a) $W = \frac{4}{9}(62.4)\pi\int_0^2 y^3\,dy$

$$= \left[\frac{4}{9}(62.4)\pi\left(\frac{1}{4}y^4\right)\right]_0^2 \approx 110.9\pi \text{ ft} \cdot \text{lb}$$

(b) $W = \frac{4}{9}(62.4)\pi\int_4^6 y^3\,dy$

$$= \left[\frac{4}{9}(62.4)\pi\left(\frac{1}{4}y^4\right)\right]_4^6 \approx 7210.7\pi \text{ ft} \cdot \text{lb}$$

27. Volume of disk: $\pi\left(\sqrt{36-y^2}\right)^2\Delta y$

Weight of disk: $62.4\pi(36 - y^2)\,\Delta y$

Distance: y

$$W = 62.4\pi\int_0^6 y(36-y^2)\,dy$$

$$= 62.4\pi\int_0^6 (36y - y^3)\,dy = 62.4\pi\left[18y^2 - \frac{1}{4}y^4\right]_0^6$$

$$= 20{,}217.6\pi \text{ ft} \cdot \text{lb}$$

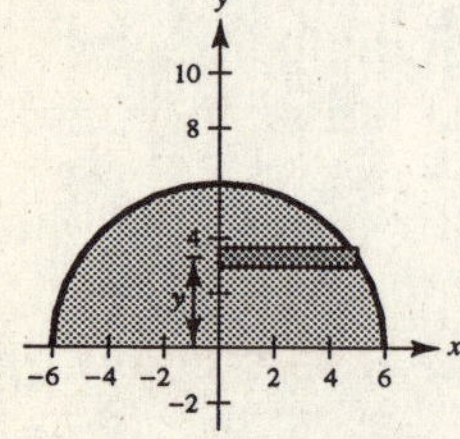

28. Volume of each layer: $\frac{y+3}{3}(3)\,\Delta y = (y+3)\,\Delta y$

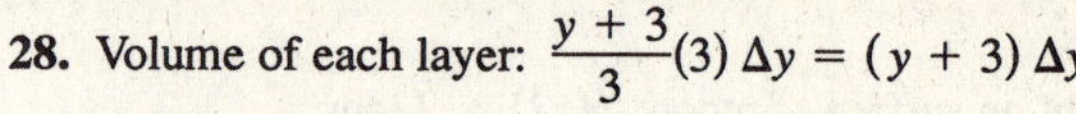

Weight of each layer: $53.1(y+3)\,\Delta y$

Distance: $6 - y$

$$W = \int_0^3 53.1(6-y)(y+3)\,dy$$

$$= 53.1\int_0^3 (18 + 3y - y^2)\,dy$$

$$= 53.1\left[18y + \frac{3y^2}{2} - \frac{y^3}{3}\right]_0^3$$

$$= 53.1\left[\frac{117}{2}\right]$$

$$= 3106.35 \text{ ft} \cdot \text{lb}$$

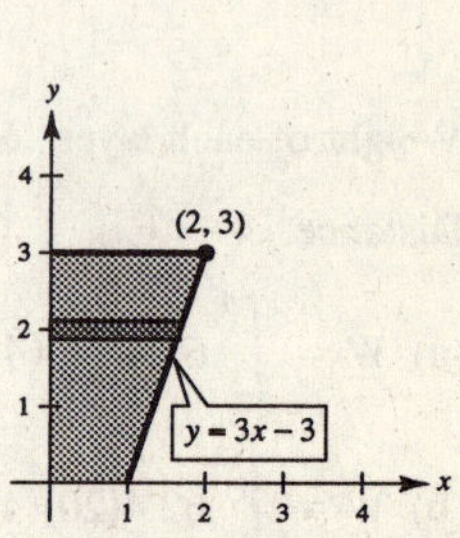

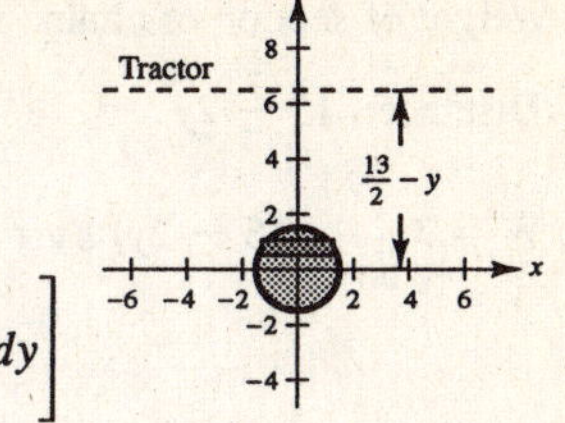

29. Volume of layer: $V = lwh = 4(2)\sqrt{(9/4) - y^2}\,\Delta y$

Weight of layer: $W = 42(8)\sqrt{(9/4) - y^2}\,\Delta y$

Distance: $\frac{13}{2} - y$

$$W = \int_{-1.5}^{1.5} 42(8)\sqrt{\frac{9}{4} - y^2}\left(\frac{13}{2} - y\right) dy = 336\left[\frac{13}{2}\int_{-1.5}^{1.5}\sqrt{\frac{9}{4} - y^2}\,dy - \int_{-1.5}^{1.5}\sqrt{\frac{9}{4} - y^2}\,y\,dy\right]$$

The second integral is zero since the integrand is odd and the limits of integration are symmetric to the origin. The first integral represents the area of a semicircle of radius $\frac{3}{2}$. Thus, the work is

$$W = 336\left(\frac{13}{2}\right)\pi\left(\frac{3}{2}\right)^2\left(\frac{1}{2}\right) = 2457\pi \text{ ft} \cdot \text{lb}.$$

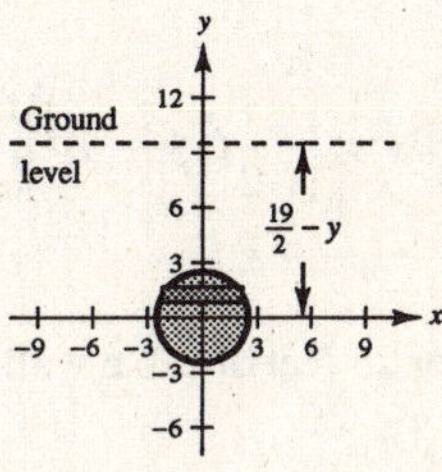

30. Volume of layer: $V = 12(2)\sqrt{(25/4) - y^2}\,\Delta y$

Weight of layer: $W = 42(24)\sqrt{(25/4) - y^2}\,\Delta y$

Distance: $\frac{19}{2} - y$

$$W = \int_{-2.5}^{2.5} 42(24)\sqrt{\frac{25}{4} - y^2}\left(\frac{19}{2} - y\right) dy$$

$$= 1008\left[\frac{19}{2}\int_{-2.5}^{2.5}\sqrt{\frac{25}{4} - y^2}\,dy + \int_{-2.5}^{2.5}\sqrt{\frac{25}{4} - y^2}(-y)\,dy\right]$$

The second integral is zero since the integrand is odd and the limits of integration are symmetric to the origin. The first integral represents the area of a semicircle of radius $\frac{5}{2}$. Thus, the work is

$$W = 1008\left(\frac{19}{2}\right)\pi\left(\frac{5}{2}\right)^2\left(\frac{1}{2}\right) = 29{,}925\pi \text{ ft} \cdot \text{lb} \approx 94{,}012.16 \text{ ft} \cdot \text{lb}.$$

31. Weight of section of chain: $3\,\Delta y$

Distance: $15 - y$

$$W = 3\int_0^{15}(15 - y)\,dy$$

$$= \left[-\frac{3}{2}(15 - y)^2\right]_0^{15}$$

$$= 337.5 \text{ ft} \cdot \text{lb}$$

32. The lower 10 feet of chain are raised 5 feet with a constant force.

$W_1 = 3(10)5 = 150 \text{ ft} \cdot \text{lb}$

The top 5 feet will be raised with variable force.

Weight of section: $3\,\Delta y$

Distance: $5 - y$

$$W_2 = 3\int_0^5 (5 - y)\,dy = \left[-\frac{3}{2}(5 - y)^2\right]_0^5 = \frac{75}{2} \text{ ft} \cdot \text{lb}$$

$$W = W_1 + W_2 = 150 + \frac{75}{2} = \frac{375}{2} \text{ ft} \cdot \text{lb}$$

33. The lower 5 feet of chain are raised 10 feet with a constant force.

$W_1 = 3(5)(10) = 150 \text{ ft} \cdot \text{lb}$

The top 10 feet of chain are raised with a variable force.

Weight per section: $3\,\Delta y$

Distance: $10 - y$

$$W_2 = 3\int_0^{10}(10 - y)\,dy = \left[-\frac{3}{2}(10 - y)^2\right]_0^{10} = 150 \text{ ft} \cdot \text{lb}$$

$$W = W_1 + W_2 = 300 \text{ ft} \cdot \text{lb}$$

34. The work required to lift the chain is 337.5 ft · lb (from Exercise 31). The work required to lift the 500-pound load is $W = (500)(15) = 7500$. The work required to lift the chain with a 100-pound load attached is

$W = 337.5 + 7500 = 7837.5$ ft · lbs.

35. Weight of section of chain: $3\,\Delta y$

Distance: $15 - 2y$

$$W = 3\int_0^{7.5} (15 - 2y)\,dy = \left[-\frac{3}{4}(15 - 2y)^2\right]_0^{7.5}$$

$$= \frac{3}{4}(15)^2 = 168.75 \text{ ft} \cdot \text{lb}$$

36. $W = 3\int_0^6 (12 - 2y)\,dy = \left[-\frac{3}{4}(12 - 2y)^2\right]_0^6$

$$= \frac{3}{4}(12)^2 = 108 \text{ ft} \cdot \text{lb}$$

37. Work to pull up the ball: $W_1 = 500(15) = 7500$ ft · lb

Work to wind up the top 15 feet of cable: force is variable

Weight per section: $1\,\Delta y$

Distance: $15 - x$

$$W_2 = \int_0^{15} (15 - x)\,dx = \left[-\frac{1}{2}(15 - x)^2\right]_0^{15}$$

$$= 112.5 \text{ ft} \cdot \text{lb}$$

Work to lift the lower 25 feet of cable with a constant force:

$$W_3 = (1)(25)(15) = 375 \text{ ft} \cdot \text{lb}$$

$$W = W_1 + W_2 + W_3 = 7500 + 112.5 + 375$$

$$= 7987.5 \text{ ft} \cdot \text{lb}$$

38. Work to pull up the ball: $W_1 = 500(40) = 20{,}000$ ft · lb

Work to pull up the cable: force is variable

Weight per section: $1\,\Delta y$

Distance: $40 - x$

$$W_2 = \int_0^{40} (40 - x)\,dx = \left[-\frac{1}{2}(40 - x)^2\right]_0^{40}$$

$$= 800 \text{ ft} \cdot \text{lb}$$

$$W = W_1 + W_2 = 20{,}000 + 800 = 20{,}800 \text{ ft} \cdot \text{lb}$$

39. $p = \frac{k}{V}$

$$1000 = \frac{k}{2}$$

$$k = 2000$$

$$W = \int_2^3 \frac{2000}{V}\,dV$$

$$= \Big[2000 \ln|V|\Big]_2^3$$

$$= 2000 \ln\left(\frac{3}{2}\right) \approx 810.93 \text{ ft} \cdot \text{lb}$$

40. $p = \frac{k}{V}$

$$2500 = \frac{k}{1} \Rightarrow k = 2500$$

$$W = \int_1^3 \frac{2500}{V}\,dV$$

$$= \Big[2500 \ln V\Big]_1^3$$

$$= 2500 \ln 3$$

$$\approx 2746.53 \text{ ft} \cdot \text{lb}$$

41. $F(x) = \frac{k}{(2 - x)^2}$

$$W = \int_{-2}^{1} \frac{k}{(2 - x)^2}\,dx$$

$$= \left[\frac{k}{2 - x}\right]_{-2}^{1} = k\left(1 - \frac{1}{4}\right)$$

$$= \frac{3k}{4} \text{ (units of work)}$$

42. (a) $W = FD = (8000\pi)(2) = 16{,}000\pi$ ft · lbs

(b) $W \approx \frac{2 - 0}{3(6)}[0 + 4(20{,}000) + 2(22{,}000) + 4(15{,}000) + 2(10{,}000) + 4(5000) + 0] \approx 24{,}88.889$ ft · lb

(c) $F(x) = -16{,}261.36x^4 + 85{,}295.45x^3 - 157{,}738.64x^2 + 104{,}386.36x - 32.4675$

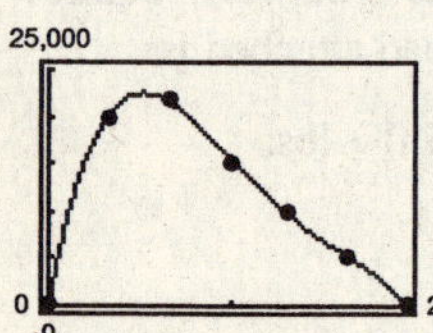

(d) $F(x) = 0$ when $x \approx 0.524$ feet. $F(x)$ is a maximum when $x \approx 0.524$ feet.

(e) $W = \int_0^2 F(x)\,dx \approx 25{,}180.5$ ft · lbs

43. $W = \int_0^5 1000[1.8 - \ln(x + 1)]\, dx \approx 3249.44 \text{ ft} \cdot \text{lb}$

44. $W = \int_0^4 \left(\frac{e^{x^2} - 1}{100}\right) dx \approx 11{,}494 \text{ ft} \cdot \text{lb}$

45. $W = \int_0^5 100x\sqrt{125 - x^3}\, dx \approx 10{,}330.3 \text{ ft} \cdot \text{lb}$

46. $W = \int_0^2 1000 \sinh x\, dx \approx 2762.2 \text{ ft} \cdot \text{lb}$

Section 7.6 Moments, Centers of Mass, and Centroids

1. $\bar{x} = \frac{6(-5) + 3(1) + 5(3)}{6 + 3 + 5} = -\frac{6}{7}$

2. $\bar{x} = \frac{7(-3) + 4(-2) + 3(5) + 8(6)}{7 + 4 + 3 + 8} = \frac{17}{11}$

3. $\bar{x} = \frac{1(7) + 1(8) + 1(12) + 1(15) + 1(18)}{1 + 1 + 1 + 1 + 1} = 12$

4. $\bar{x} = \frac{12(-6) + 1(-4) + 6(-2) + 3(0) + 11(8)}{12 + 1 + 6 + 3 + 11} = 0$

5. (a) $\bar{x} = \frac{(7 + 5) + (8 + 5) + (12 + 5) + (15 + 5) + (18 + 5)}{5} = 17 = 12 + 5$

(b) $\bar{x} = \frac{12(-6 - 3) + 1(-4 - 3) + 6(-2 - 3) + 3(0 - 3) + 11(8 - 3)}{12 + 1 + 6 + 3 + 11} = \frac{-99}{33} = -3$

6. The center of mass is translated k units as well.

7. $50x = 75(L - x) = 75(10 - x)$

$50x = 750 - 75x$

$125x = 750$

$x = 6$ feet

8. $200x = 550(5 - x)$ (Person on left)

$200x = 2750 - 550x$

$750x = 2750$

$x = 3\frac{2}{3}$ feet

9. $\bar{x} = \frac{5(2) + 1(-3) + 3(1)}{5 + 1 + 3} = \frac{10}{9}$

$\bar{y} = \frac{5(2) + 1(1) + 3(-4)}{5 + 1 + 3} = -\frac{1}{9}$

$(\bar{x}, \bar{y}) = \left(\frac{10}{9}, -\frac{1}{9}\right)$

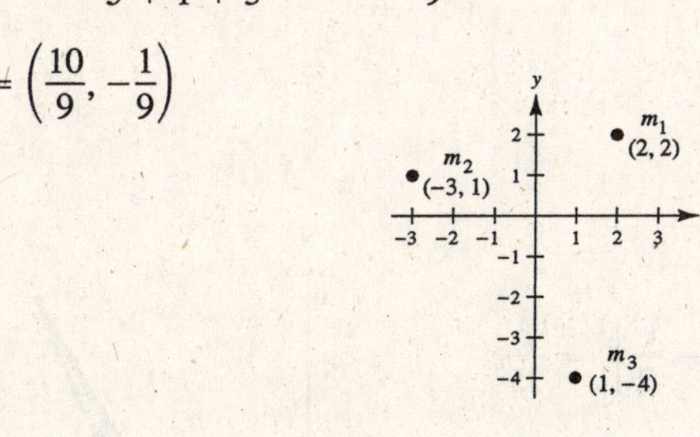

10.

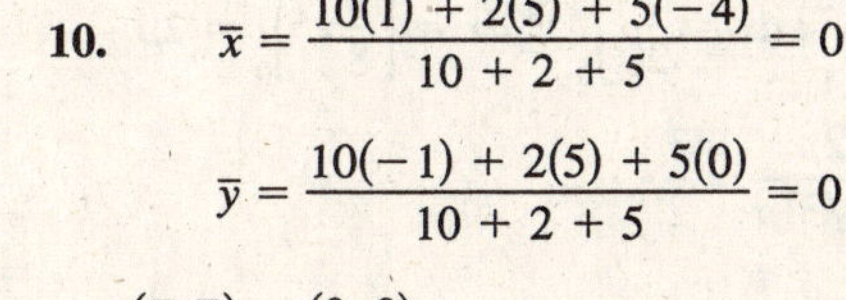

$\bar{x} = \frac{10(1) + 2(5) + 5(-4)}{10 + 2 + 5} = 0$

$\bar{y} = \frac{10(-1) + 2(5) + 5(0)}{10 + 2 + 5} = 0$

$(\bar{x}, \bar{y}) = (0, 0)$

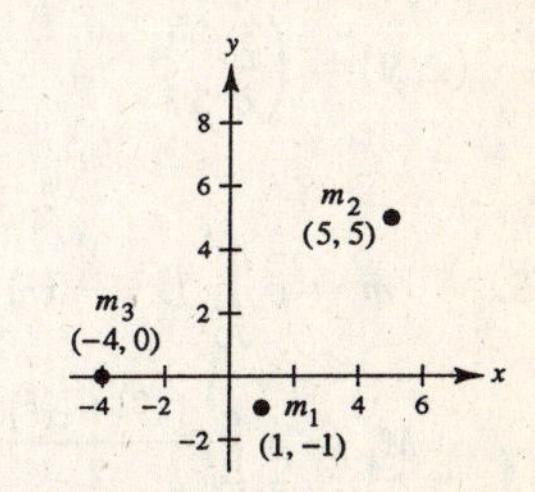

11. $\bar{x} = \frac{3(-2) + 4(5) + 2(7) + 1(0) + 6(-3)}{3 + 4 + 2 + 1 + 6} = \frac{5}{8}$

$\bar{y} = \frac{3(-3) + 4(5) + 2(1) + 1(0) + 6(0)}{3 + 4 + 2 + 1 + 6} = \frac{13}{16}$

$(\bar{x}, \bar{y}) = \left(\frac{5}{8}, \frac{13}{16}\right)$

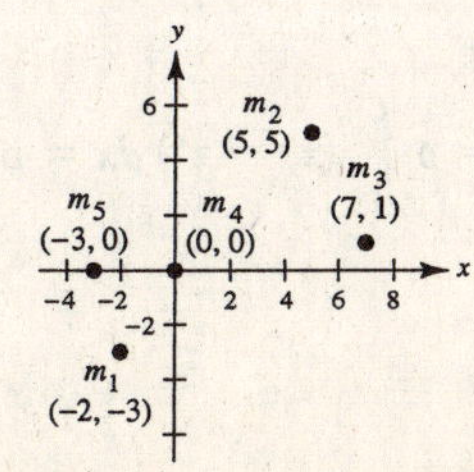

12. $\bar{x} = \dfrac{12(2) + 6(-1) + (15/2)(6) + 15(2)}{12 + 6 + (15/2) + 15} = \dfrac{93}{40.5} = \dfrac{62}{27}$

$\bar{y} = \dfrac{12(3) + 6(5) + (15/2)(8) + 15(-2)}{12 + 6 + (15/2) + 15} = \dfrac{96}{40.5} = \dfrac{64}{27}$

$(\bar{x}, \bar{y}) = \left(\dfrac{62}{27}, \dfrac{64}{27}\right)$

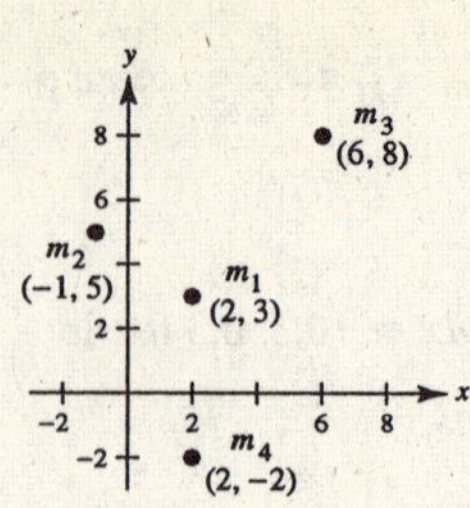

13.

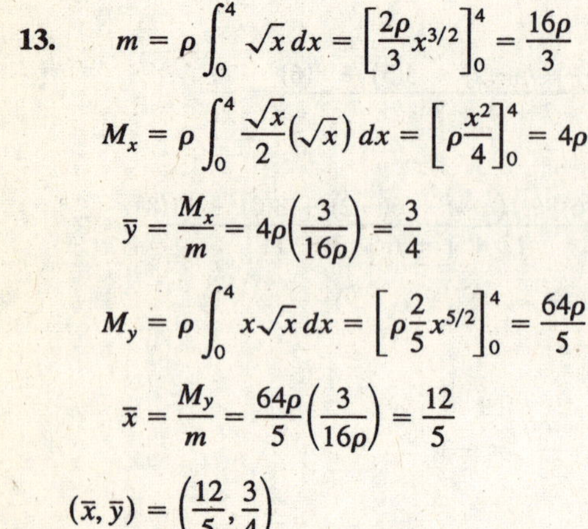

$$m = \rho\int_0^4 \sqrt{x}\,dx = \left[\frac{2\rho}{3}x^{3/2}\right]_0^4 = \frac{16\rho}{3}$$

$$M_x = \rho\int_0^4 \frac{\sqrt{x}}{2}(\sqrt{x})\,dx = \left[\rho\frac{x^2}{4}\right]_0^4 = 4\rho$$

$$\bar{y} = \frac{M_x}{m} = 4\rho\left(\frac{3}{16\rho}\right) = \frac{3}{4}$$

$$M_y = \rho\int_0^4 x\sqrt{x}\,dx = \left[\rho\frac{2}{5}x^{5/2}\right]_0^4 = \frac{64\rho}{5}$$

$$\bar{x} = \frac{M_y}{m} = \frac{64\rho}{5}\left(\frac{3}{16\rho}\right) = \frac{12}{5}$$

$$(\bar{x}, \bar{y}) = \left(\frac{12}{5}, \frac{3}{4}\right)$$

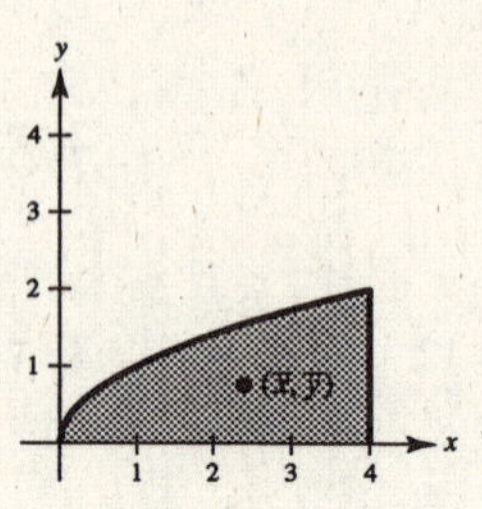

14.

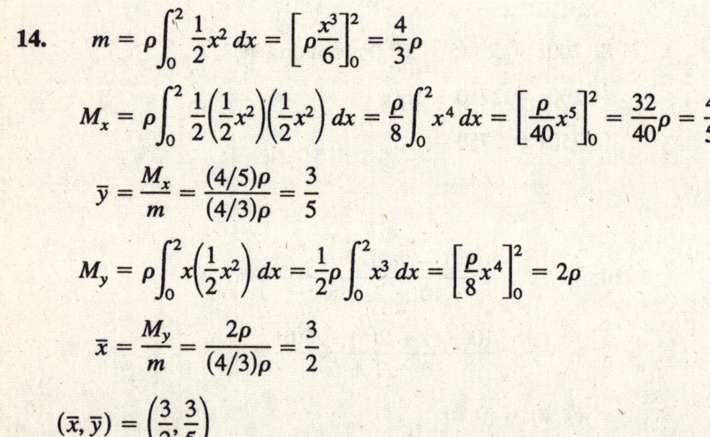

$$m = \rho\int_0^2 \frac{1}{2}x^2\,dx = \left[\rho\frac{x^3}{6}\right]_0^2 = \frac{4}{3}\rho$$

$$M_x = \rho\int_0^2 \frac{1}{2}\left(\frac{1}{2}x^2\right)\left(\frac{1}{2}x^2\right)dx = \frac{\rho}{8}\int_0^2 x^4\,dx = \left[\frac{\rho}{40}x^5\right]_0^2 = \frac{32}{40}\rho = \frac{4}{5}\rho$$

$$\bar{y} = \frac{M_x}{m} = \frac{(4/5)\rho}{(4/3)\rho} = \frac{3}{5}$$

$$M_y = \rho\int_0^2 x\left(\frac{1}{2}x^2\right)dx = \frac{1}{2}\rho\int_0^2 x^3\,dx = \left[\frac{\rho}{8}x^4\right]_0^2 = 2\rho$$

$$\bar{x} = \frac{M_y}{m} = \frac{2\rho}{(4/3)\rho} = \frac{3}{2}$$

$$(\bar{x}, \bar{y}) = \left(\frac{3}{2}, \frac{3}{5}\right)$$

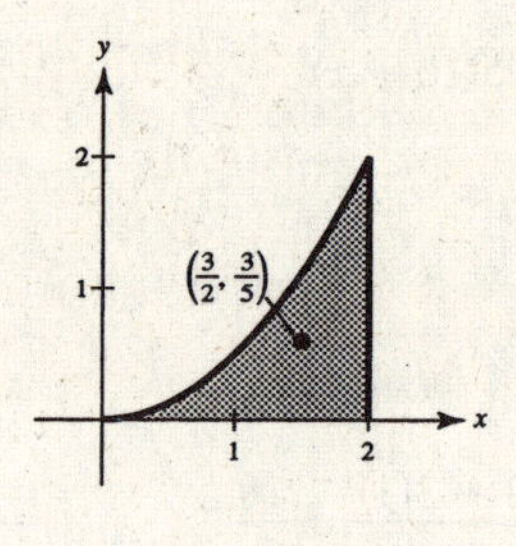

15.

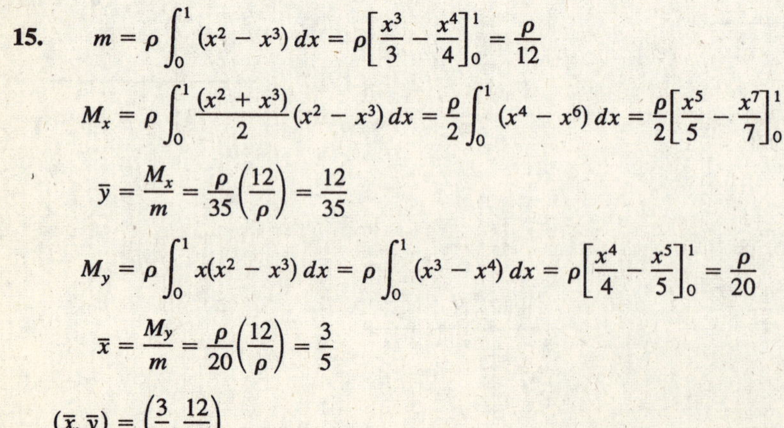

$$m = \rho\int_0^1 (x^2 - x^3)\,dx = \rho\left[\frac{x^3}{3} - \frac{x^4}{4}\right]_0^1 = \frac{\rho}{12}$$

$$M_x = \rho\int_0^1 \frac{(x^2 + x^3)}{2}(x^2 - x^3)\,dx = \frac{\rho}{2}\int_0^1 (x^4 - x^6)\,dx = \frac{\rho}{2}\left[\frac{x^5}{5} - \frac{x^7}{7}\right]_0^1 = \frac{\rho}{35}$$

$$\bar{y} = \frac{M_x}{m} = \frac{\rho}{35}\left(\frac{12}{\rho}\right) = \frac{12}{35}$$

$$M_y = \rho\int_0^1 x(x^2 - x^3)\,dx = \rho\int_0^1 (x^3 - x^4)\,dx = \rho\left[\frac{x^4}{4} - \frac{x^5}{5}\right]_0^1 = \frac{\rho}{20}$$

$$\bar{x} = \frac{M_y}{m} = \frac{\rho}{20}\left(\frac{12}{\rho}\right) = \frac{3}{5}$$

$$(\bar{x}, \bar{y}) = \left(\frac{3}{5}, \frac{12}{35}\right)$$

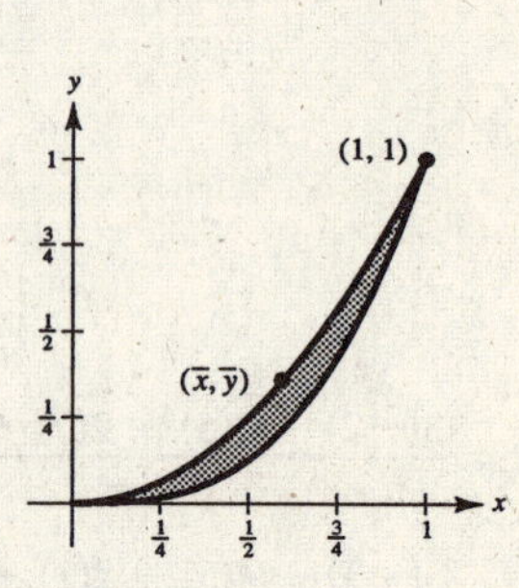

16. $m = \rho\int_0^1 \left(\sqrt{x} - x\right)dx = \rho\left[\frac{2}{3}x^{3/2} - \frac{x^2}{2}\right]_0^1 = \frac{\rho}{6}$

$M_x = \rho\int_0^1 \frac{\left(\sqrt{x} + x\right)}{2}\left(\sqrt{x} - x\right)dx = \frac{\rho}{2}\int_0^1 (x - x^2)\,dx = \frac{\rho}{2}\left[\frac{x^2}{2} - \frac{x^3}{3}\right]_0^1 = \frac{\rho}{12}$

$\bar{y} = \frac{M_x}{m} = \frac{\rho}{12}\left(\frac{6}{\rho}\right) = \frac{1}{2}$

$M_y = \rho\int_0^1 x\left(\sqrt{x} - x\right)dx = \rho\int_0^1 (x^{3/2} - x^2)\,dx = \rho\left[\frac{2}{5}x^{5/2} - \frac{x^3}{3}\right]_0^1 = \frac{\rho}{15}$

$\bar{x} = \frac{M_y}{m} = \frac{\rho}{15}\left(\frac{6}{\rho}\right) = \frac{2}{5}$

$(\bar{x}, \bar{y}) = \left(\frac{2}{5}, \frac{1}{2}\right)$

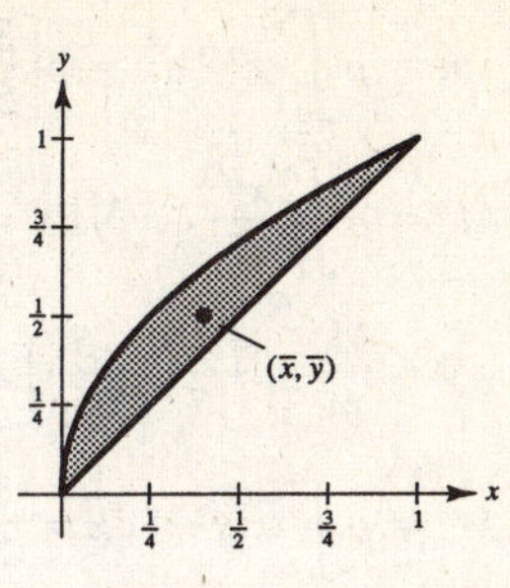

17. $m = \rho\int_0^3 [(-x^2 + 4x + 2) - (x + 2)]\,dx = -\rho\left[\frac{x^3}{3} + \frac{3x^2}{2}\right]_0^3 = \frac{9\rho}{2}$

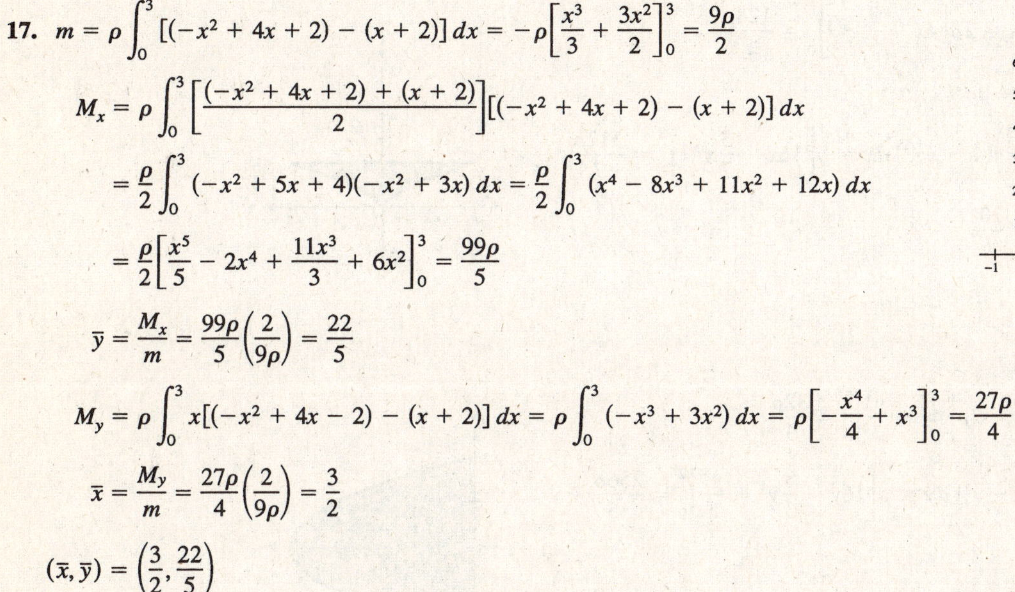

$M_x = \rho\int_0^3 \left[\frac{(-x^2 + 4x + 2) + (x + 2)}{2}\right][(-x^2 + 4x + 2) - (x + 2)]\,dx$

$= \frac{\rho}{2}\int_0^3 (-x^2 + 5x + 4)(-x^2 + 3x)\,dx = \frac{\rho}{2}\int_0^3 (x^4 - 8x^3 + 11x^2 + 12x)\,dx$

$= \frac{\rho}{2}\left[\frac{x^5}{5} - 2x^4 + \frac{11x^3}{3} + 6x^2\right]_0^3 = \frac{99\rho}{5}$

$\bar{y} = \frac{M_x}{m} = \frac{99\rho}{5}\left(\frac{2}{9\rho}\right) = \frac{22}{5}$

$M_y = \rho\int_0^3 x[(-x^2 + 4x - 2) - (x + 2)]\,dx = \rho\int_0^3 (-x^3 + 3x^2)\,dx = \rho\left[-\frac{x^4}{4} + x^3\right]_0^3 = \frac{27\rho}{4}$

$\bar{x} = \frac{M_y}{m} = \frac{27\rho}{4}\left(\frac{2}{9\rho}\right) = \frac{3}{2}$

$(\bar{x}, \bar{y}) = \left(\frac{3}{2}, \frac{22}{5}\right)$

18. $m = \rho\int_0^9 \left[\left(\sqrt{x} + 1\right) - \left(\frac{1}{3}x + 1\right)\right]dx = \rho\int_0^9 \left(\sqrt{x} - \frac{1}{3}x\right)dx = \rho\left[\frac{2}{3}x^{3/2} - \frac{x^2}{6}\right]_0^9 = \rho\left(18 - \frac{27}{2}\right) = \frac{9}{2}\rho$

$M_x = \rho\int_0^9 \frac{\sqrt{x} + 1 + (1/3)x + 1}{2}\left(\sqrt{x} + 1 - \frac{1}{3}x - 1\right)dx = \frac{\rho}{2}\int_0^9 \left(\sqrt{x} + \frac{1}{3}x + 2\right)\left(\sqrt{x} - \frac{1}{3}x\right)dx$

$= \frac{\rho}{2}\int_0^9 \left(x - \frac{1}{3}x^{3/2} + \frac{1}{3}x^{3/2} - \frac{1}{9}x^2 + 2\sqrt{x} - \frac{2}{3}x\right)dx = \frac{\rho}{2}\int_0^9 \left(\frac{1}{3}x - \frac{1}{9}x^2 + 2\sqrt{x}\right)dx$

$= \frac{\rho}{2}\left[\frac{x^2}{6} - \frac{x^3}{27} + \frac{4}{3}x^{3/2}\right]_0^9 = \frac{\rho}{2}\left[\frac{27}{2} - 27 + 36\right] = \frac{45}{4}\rho$

$M_y = \rho\int_0^9 x\left[\sqrt{x} + 1 - \frac{1}{3}x - 1\right]dx = \rho\int_0^9 \left(x^{3/2} - \frac{1}{3}x^2\right)dx = \rho\left[\frac{2}{5}x^{5/2} - \frac{1}{9}x^3\right]_0^9 = \rho\left[\frac{486}{5} - 81\right] = \frac{81}{5}\rho$

$\bar{x} = \frac{M_y}{m} = \frac{(81/5)\rho}{(9/2)\rho} = \frac{18}{5};\ \bar{y} = \frac{M_x}{m} = \frac{(45/4)\rho}{(9/2)\rho} = \frac{5}{2}$

$(\bar{x}, \bar{y}) = \left(\frac{18}{5}, \frac{5}{2}\right)$

19. $m = \rho\int_0^8 x^{2/3}\,dx = \rho\left[\frac{3}{5}x^{5/3}\right]_0^8 = \frac{96\rho}{5}$

$M_x = \rho\int_0^8 \frac{x^{2/3}}{2}(x^{2/3})\,dx = \frac{\rho}{2}\left[\frac{3}{7}x^{7/3}\right]_0^8 = \frac{192\rho}{7}$

$\bar{y} = \frac{M_x}{m} = \frac{192\rho}{7}\left(\frac{5}{96\rho}\right) = \frac{10}{7}$

$M_y = \rho\int_0^8 x(x^{2/3})\,dx = \rho\left[\frac{3}{8}x^{8/3}\right]_0^8 = 96\rho$

$\bar{x} = \frac{M_y}{m} = 96\rho\left(\frac{5}{96\rho}\right) = 5$

$(\bar{x}, \bar{y}) = \left(5, \frac{10}{7}\right)$

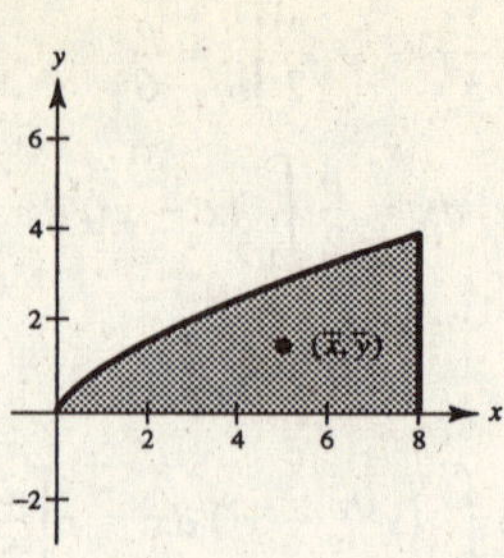

20. $m = 2\rho\int_0^8 (4 - x^{2/3})\,dx = 2\rho\left[4x - \frac{3}{5}x^{5/3}\right]_0^8 = \frac{128\rho}{5}$

By symmetry, M_y and $\bar{x} = 0$.

$M_x = 2\rho\int_0^8 \left(\frac{4 + x^{2/3}}{2}\right)(4 - x^{2/3})\,dx = \rho\left[16x - \frac{3}{7}x^{7/3}\right]_0^8 = \frac{512\rho}{7}$

$\bar{y} = \frac{512\rho}{7}\left(\frac{5}{128\rho}\right) = \frac{20}{7}$

$(\bar{x}, \bar{y}) = \left(0, \frac{20}{7}\right)$

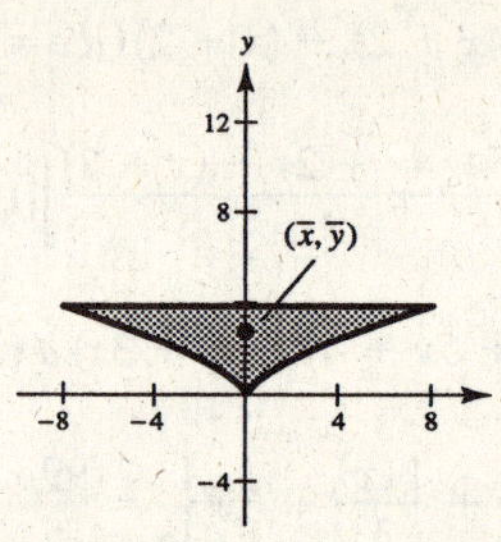

21. $m = 2\rho\int_0^2 (4 - y^2)\,dy = 2\rho\left[4y - \frac{y^3}{3}\right]_0^2 = \frac{32\rho}{3}$

$M_y = 2\rho\int_0^2 \left(\frac{4 - y^2}{2}\right)(4 - y^2)\,dy = \rho\left[16y - \frac{8}{3}y^3 + \frac{y^5}{5}\right]_0^2 = \frac{256\rho}{15}$

$\bar{x} = \frac{M_y}{m} = \frac{256\rho}{15}\left(\frac{3}{32\rho}\right) = \frac{8}{5}$

By symmetry, M_x and $\bar{y} = 0$.

$(\bar{x}, \bar{y}) = \left(\frac{8}{5}, 0\right)$

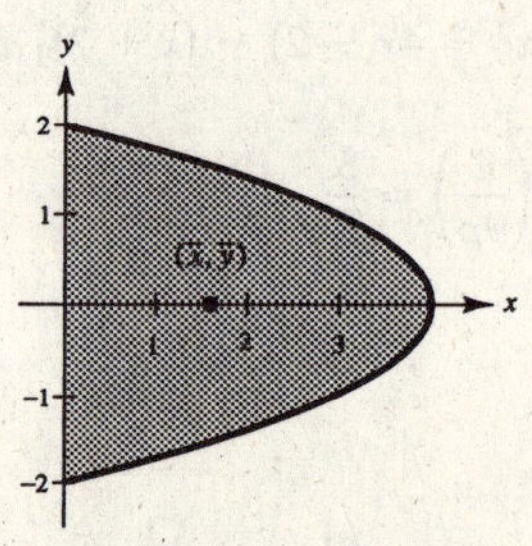

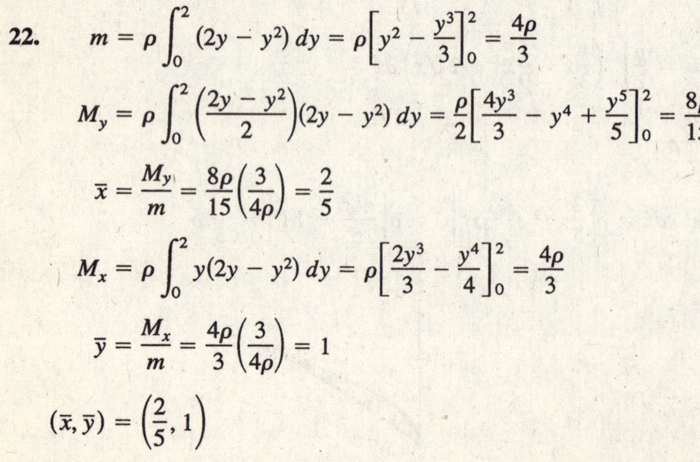

22. $m = \rho\int_0^2 (2y - y^2)\,dy = \rho\left[y^2 - \frac{y^3}{3}\right]_0^2 = \frac{4\rho}{3}$

$M_y = \rho\int_0^2 \left(\frac{2y - y^2}{2}\right)(2y - y^2)\,dy = \frac{\rho}{2}\left[\frac{4y^3}{3} - y^4 + \frac{y^5}{5}\right]_0^2 = \frac{8\rho}{15}$

$\bar{x} = \frac{M_y}{m} = \frac{8\rho}{15}\left(\frac{3}{4\rho}\right) = \frac{2}{5}$

$M_x = \rho\int_0^2 y(2y - y^2)\,dy = \rho\left[\frac{2y^3}{3} - \frac{y^4}{4}\right]_0^2 = \frac{4\rho}{3}$

$\bar{y} = \frac{M_x}{m} = \frac{4\rho}{3}\left(\frac{3}{4\rho}\right) = 1$

$(\bar{x}, \bar{y}) = \left(\frac{2}{5}, 1\right)$

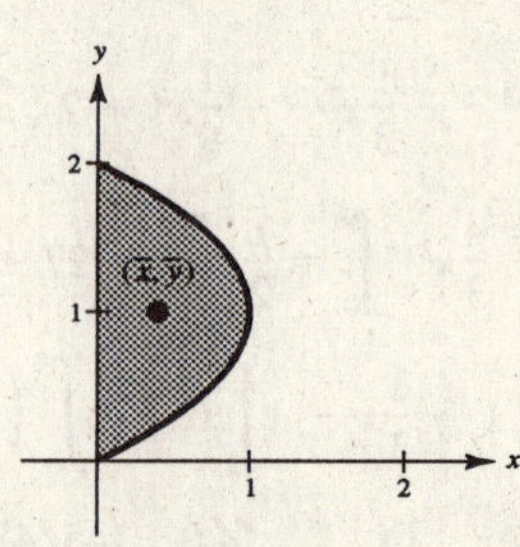

23. $m = \rho \int_0^3 [(2y - y^2) - (-y)]\,dy = \rho\left[\frac{3y^2}{2} - \frac{y^3}{3}\right]_0^3 = \frac{9\rho}{2}$

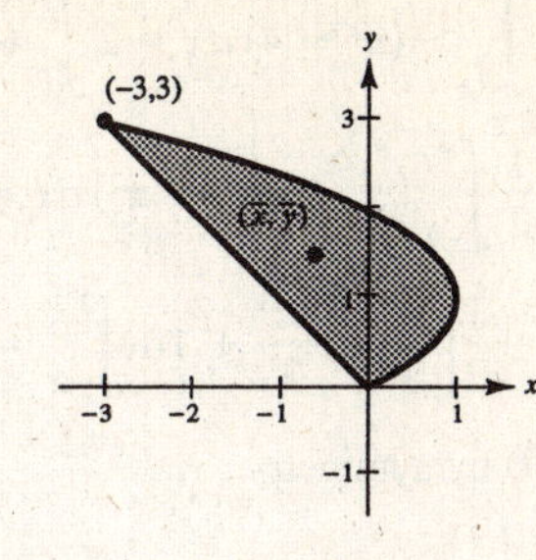

$$M_y = \rho \int_0^3 \frac{[(2y - y^2) + (-y)]}{2}[(2y - y^2) - (-y)]\,dy = \frac{\rho}{2}\int_0^3 (y - y^2)(3y - y^2)\,dy$$

$$= \frac{\rho}{2}\int_0^3 (y^4 - 4y^3 + 3y^2)\,dy = \frac{\rho}{2}\left[\frac{y^5}{5} - y^4 + y^3\right]_0^3 = -\frac{27\rho}{10}$$

$$\bar{x} = \frac{M_y}{m} = -\frac{27\rho}{10}\left(\frac{2}{9\rho}\right) = -\frac{3}{5}$$

$$M_x = \rho \int_0^3 y[(2y - y^2) - (-y)]\,dy = \rho \int_0^3 (3y^2 - y^3)\,dy = \rho\left[y^3 - \frac{y^4}{4}\right]_0^3 = \frac{27\rho}{4}$$

$$\bar{y} = \frac{M_x}{m} = \frac{27\rho}{4}\left(\frac{2}{9\rho}\right) = \frac{3}{2}$$

$$(\bar{x}, \bar{y}) = \left(-\frac{3}{5}, \frac{3}{2}\right)$$

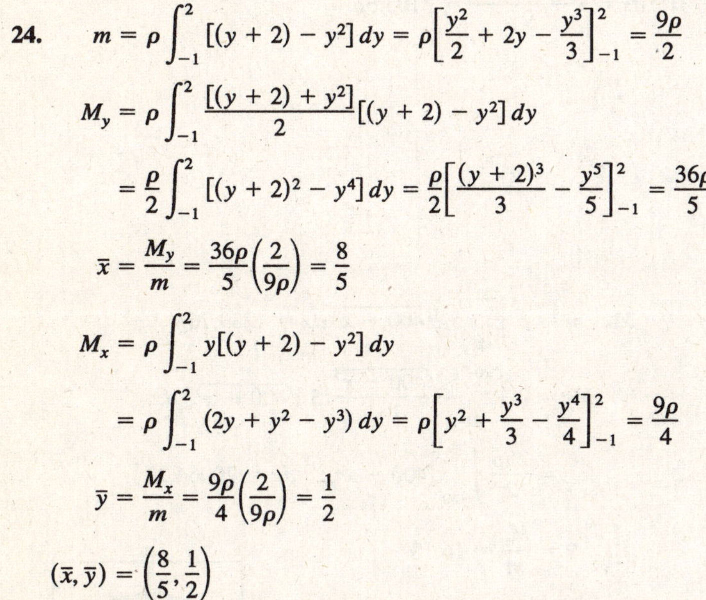

24. $m = \rho \int_{-1}^2 [(y + 2) - y^2]\,dy = \rho\left[\frac{y^2}{2} + 2y - \frac{y^3}{3}\right]_{-1}^2 = \frac{9\rho}{2}$

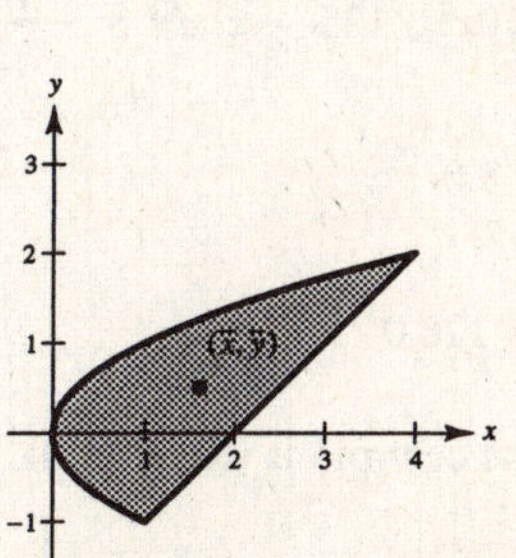

$$M_y = \rho \int_{-1}^2 \frac{[(y + 2) + y^2]}{2}[(y + 2) - y^2]\,dy$$

$$= \frac{\rho}{2}\int_{-1}^2 [(y + 2)^2 - y^4]\,dy = \frac{\rho}{2}\left[\frac{(y + 2)^3}{3} - \frac{y^5}{5}\right]_{-1}^2 = \frac{36\rho}{5}$$

$$\bar{x} = \frac{M_y}{m} = \frac{36\rho}{5}\left(\frac{2}{9\rho}\right) = \frac{8}{5}$$

$$M_x = \rho \int_{-1}^2 y[(y + 2) - y^2]\,dy$$

$$= \rho \int_{-1}^2 (2y + y^2 - y^3)\,dy = \rho\left[y^2 + \frac{y^3}{3} - \frac{y^4}{4}\right]_{-1}^2 = \frac{9\rho}{4}$$

$$\bar{y} = \frac{M_x}{m} = \frac{9\rho}{4}\left(\frac{2}{9\rho}\right) = \frac{1}{2}$$

$$(\bar{x}, \bar{y}) = \left(\frac{8}{5}, \frac{1}{2}\right)$$

25. $A = \int_0^1 (x - x^2)\,dx = \left[\frac{1}{2}x^2 - \frac{x^3}{3}\right]_0^1 = \frac{1}{6}$

$$M_x = \frac{1}{2}\int_0^1 (x^2 - x^4)\,dx = \frac{1}{2}\left[\frac{x^3}{3} - \frac{x^5}{5}\right]_0^1 = \frac{1}{2}\left(\frac{1}{3} - \frac{1}{5}\right) = \frac{1}{15}$$

$$M_y = \int_0^1 (x^2 - x^3)\,dx = \left[\frac{x^3}{3} - \frac{x^4}{4}\right]_0^1 = \left(\frac{1}{3} - \frac{1}{4}\right) = \frac{1}{12}$$

26. $A = \int_1^4 \frac{1}{x}\,dx = \Big[\ln|x|\Big]_1^4 = \ln 4$

$$M_x = \frac{1}{2}\int_1^4 \frac{1}{x^2}\,dx = \left[\frac{1}{2}\left(-\frac{1}{x}\right)\right]_1^4 = \left(-\frac{1}{8} + \frac{1}{2}\right) = \frac{3}{8}$$

$$M_y = \int_1^4 x\left(\frac{1}{x}\right)dx = \Big[x\Big]_1^4 = 3$$

27. $A = \int_0^3 (2x + 4)\,dx = \Big[x^2 + 4x\Big]_0^3 = 9 + 12 = 21$

$$M_x = \frac{1}{2}\int_0^3 (2x + 4)^2\,dx = \int_0^3 (2x^2 + 8x + 8)\,dx$$

$$= \left[\frac{2x^3}{3} + 4x^2 + 8x\right]_0^3 = 18 + 36 + 24 = 78$$

$$M_y = \int_0^3 (2x^2 + 4x)\,dx = \left[\frac{2x^3}{3} + 2x^2\right]_0^3 = 18 + 18 = 36$$

28. $A = \int_{-2}^{2} -(x^2 - 4)\,dx = 2\int_{0}^{2} (4 - x^2)\,dx = \left[8x - \frac{2x^3}{3}\right]_0^2 = 16 - \frac{16}{3} = \frac{32}{3}$

$M_x = \frac{1}{2}\int_{-2}^{2} (x^2 - 4)(4 - x^2)\,dx = -\frac{1}{2}\int_{-2}^{2} (x^4 - 8x^2 + 16)\,dx$

$= -\frac{1}{2}\left[\frac{x^5}{5} - \frac{8x^3}{3} + 16x\right]_{-2}^{2} = -\left[\frac{32}{5} - \frac{64}{3} + 32\right] = -\frac{256}{15}$

$M_y = 0$ by symmetry.

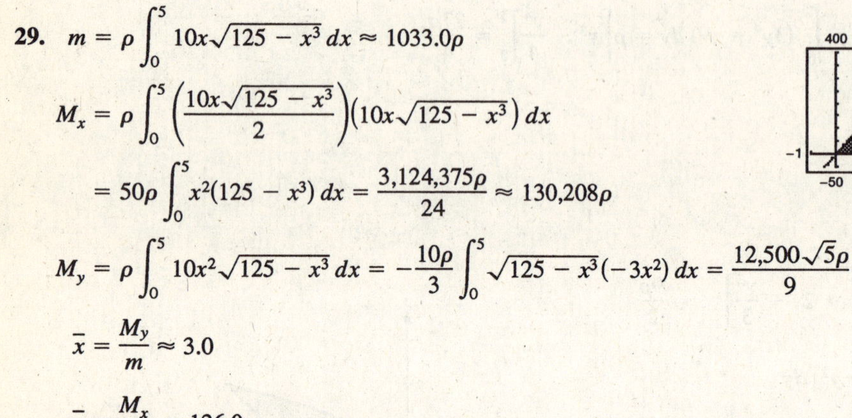

29. $m = \rho\int_0^5 10x\sqrt{125 - x^3}\,dx \approx 1033.0\rho$

$M_x = \rho\int_0^5 \left(\frac{10x\sqrt{125 - x^3}}{2}\right)\left(10x\sqrt{125 - x^3}\right)dx$

$= 50\rho\int_0^5 x^2(125 - x^3)\,dx = \frac{3{,}124{,}375\rho}{24} \approx 130{,}208\rho$

$M_y = \rho\int_0^5 10x^2\sqrt{125 - x^3}\,dx = -\frac{10\rho}{3}\int_0^5 \sqrt{125 - x^3}(-3x^2)\,dx = \frac{12{,}500\sqrt{5}\rho}{9} \approx 3105.6\rho$

$\bar{x} = \frac{M_y}{m} \approx 3.0$

$\bar{y} = \frac{M_x}{m} \approx 126.0$

Therefore, the centroid is (3.0, 126.0).

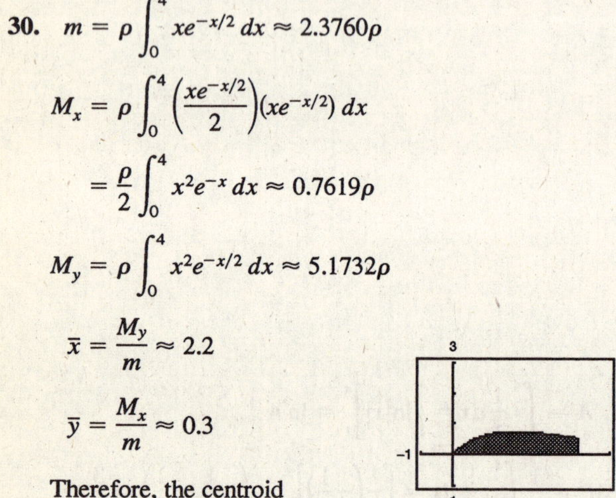

30. $m = \rho\int_0^4 xe^{-x/2}\,dx \approx 2.3760\rho$

$M_x = \rho\int_0^4 \left(\frac{xe^{-x/2}}{2}\right)(xe^{-x/2})\,dx$

$= \frac{\rho}{2}\int_0^4 x^2e^{-x}\,dx \approx 0.7619\rho$

$M_y = \rho\int_0^4 x^2e^{-x/2}\,dx \approx 5.1732\rho$

$\bar{x} = \frac{M_y}{m} \approx 2.2$

$\bar{y} = \frac{M_x}{m} \approx 0.3$

Therefore, the centroid is (2.2, 0.3).

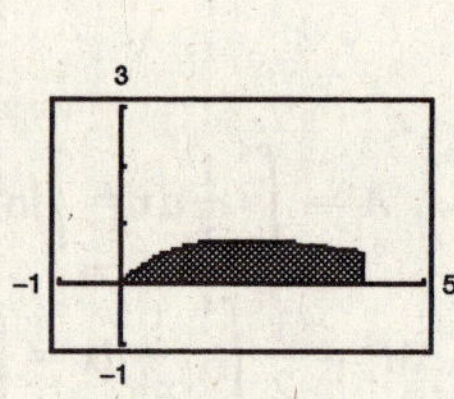

31. $m = \rho\int_{-20}^{20} 5\sqrt[3]{400 - x^2}\,dx \approx 1239.76\rho$

$M_x = \rho\int_{-20}^{20} \frac{5\sqrt[3]{400 - x^2}}{2}\left(5\sqrt[3]{400 - x^2}\right)dx$

$= \frac{25\rho}{2}\int_{-20}^{20} (400 - x^2)^{2/3}\,dx \approx 20064.27$

$\bar{y} = \frac{M_x}{m} \approx 16.18$

$\bar{x} = 0$ by symmetry. Therefore, the centroid is (0, 16.2).

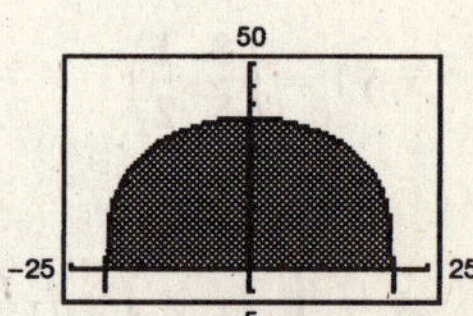

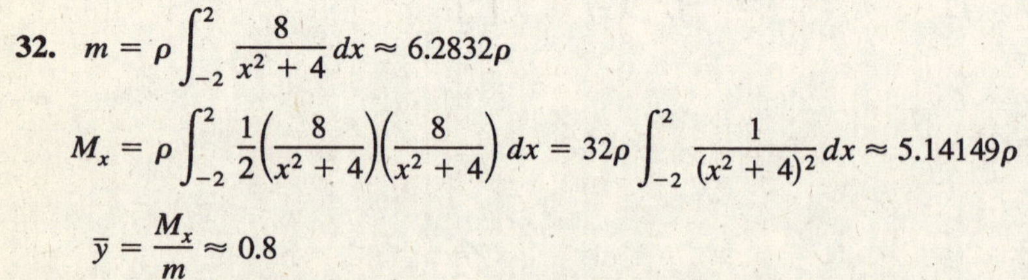

32. $m = \rho\int_{-2}^{2} \frac{8}{x^2 + 4}\,dx \approx 6.2832\rho$

$M_x = \rho\int_{-2}^{2} \frac{1}{2}\left(\frac{8}{x^2 + 4}\right)\left(\frac{8}{x^2 + 4}\right)dx = 32\rho\int_{-2}^{2} \frac{1}{(x^2 + 4)^2}\,dx \approx 5.14149\rho$

$\bar{y} = \frac{M_x}{m} \approx 0.8$

$\bar{x} = 0$ by symmetry. Therefore, the centroid is (0, 0.8).

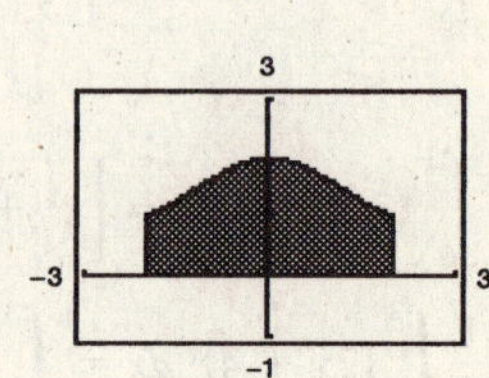

33. $A = \frac{1}{2}(2a)c = ac$

$\frac{1}{A} = \frac{1}{ac}$

$\bar{x} = \left(\frac{1}{ac}\right)\frac{1}{2}\int_0^c \left[\left(\frac{b-a}{c}y + a\right)^2 - \left(\frac{b+a}{c}y - a\right)^2\right] dy$

$= \frac{1}{2ac}\int_0^c \left[\frac{4ab}{c}y - \frac{4ab}{c^2}y^2\right] dy$

$= \frac{1}{2ac}\left[\frac{2ab}{c}y^2 - \frac{4ab}{3c^2}y^3\right]_0^c = \frac{1}{2ac}\left(\frac{2}{3}abc\right) = \frac{b}{3}$

$\bar{y} = \frac{1}{ac}\int_0^c y\left[\left(\frac{b-a}{c}y + a\right) - \left(\frac{b+a}{c}y - a\right)\right] dy$

$= \frac{1}{ac}\int_0^c y\left(-\frac{2a}{c}y + 2a\right) dy = \frac{2}{c}\int_0^c \left(y - \frac{y^2}{c}\right) dy$

$= \frac{2}{c}\left[\frac{y^2}{2} - \frac{y^3}{3c}\right]_0^c = \frac{c}{3}$

$(\bar{x}, \bar{y}) = \left(\frac{b}{3}, \frac{c}{3}\right)$

From elementary geometry, $(b/3, c/3)$ is the point of intersection of the medians.

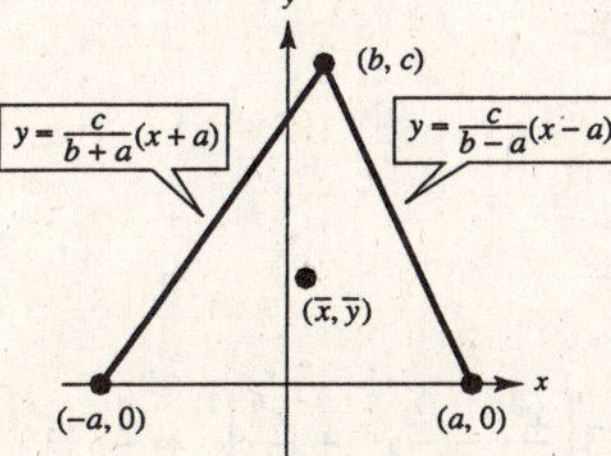

34. $A = bh = ac$

$\frac{1}{A} = \frac{1}{ac}$

$\bar{x} = \frac{1}{ac}\frac{1}{2}\int_0^c \left[\left(\frac{b}{c}y + a\right)^2 - \left(\frac{b}{c}y\right)^2\right] dy$

$= \frac{1}{2ac}\int_0^c \left(\frac{2ab}{c}y + a^2\right) dy$

$= \frac{1}{2ac}\left[\frac{ab}{c}y^2 + a^2y\right]_0^c$

$= \frac{1}{2ac}[abc + a^2c] = \frac{1}{2}(b + a)$

$\bar{y} = \frac{1}{ac}\int_0^c y\left[\left(\frac{b}{c}y + a\right) - \left(\frac{b}{c}y\right)\right] dy = \left[\frac{1}{c}\frac{y^2}{2}\right]_0^c = \frac{c}{2}$

$(\bar{x}, \bar{y}) = \left(\frac{b+a}{2}, \frac{c}{2}\right)$

This is the point of intersection of the diagonals.

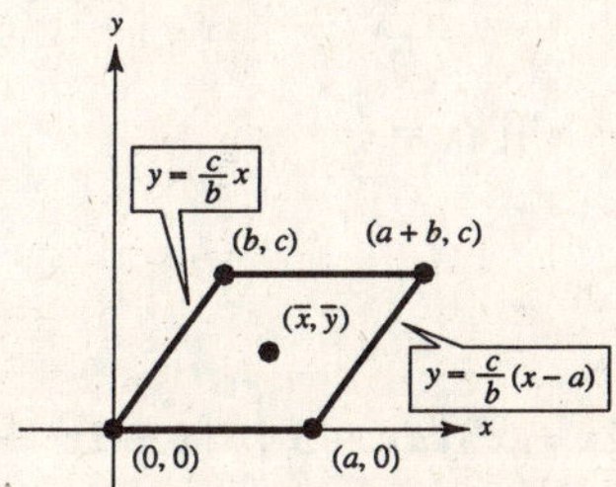

35. $A = \frac{c}{2}(a + b)$

$\frac{1}{A} = \frac{2}{c(a + b)}$

$\bar{x} = \frac{2}{c(a+b)}\int_0^c x\left(\frac{b-a}{c}x + a\right) dx = \frac{2}{c(a+b)}\int_0^c \left(\frac{b-a}{c}x^2 + ax\right) dx = \frac{2}{c(a+b)}\left[\frac{b-a}{c}\frac{x^3}{3} + \frac{ax^2}{2}\right]_0^c$

$= \frac{2}{c(a+b)}\left[\frac{(b-a)c^2}{3} + \frac{ac^2}{2}\right] = \frac{2}{c(a+b)}\left[\frac{2bc^2 - 2ac^2 + 3ac^2}{6}\right] = \frac{c(2b+a)}{3(a+b)} = \frac{(a+2b)c}{3(a+b)}$

$\bar{y} = \frac{2}{c(a+b)}\frac{1}{2}\int_0^c \left(\frac{b-a}{c}x + a\right)^2 dx = \frac{1}{c(a+b)}\int_0^c \left[\left(\frac{b-a}{c}\right)^2 x^2 + \frac{2a(b-a)}{c}x + a^2\right] dx$

$= \frac{1}{c(a+b)}\left[\left(\frac{b-a}{c}\right)^2\frac{x^3}{3} + \frac{2a(b-a)}{c}\frac{x^2}{2} + a^2x\right]_0^c = \frac{1}{c(a+b)}\left[\frac{(b-a)^2c}{3} + ac(b-a) + a^2c\right]$

$= \frac{1}{3c(a+b)}[(b^2 - 2ab + a^2)c + 3ac(b-a) + 3a^2c]$

$= \frac{1}{3(a+b)}[b^2 - 2ab + a^2 + 3ab - 3a^2 + 3a^2] = \frac{a^2 + ab + b^2}{3(a+b)}$

Thus, $(\bar{x}, \bar{y}) = \left(\frac{(a+2b)c}{3(a+b)}, \frac{a^2 + ab + b^2}{3(a+b)}\right)$.

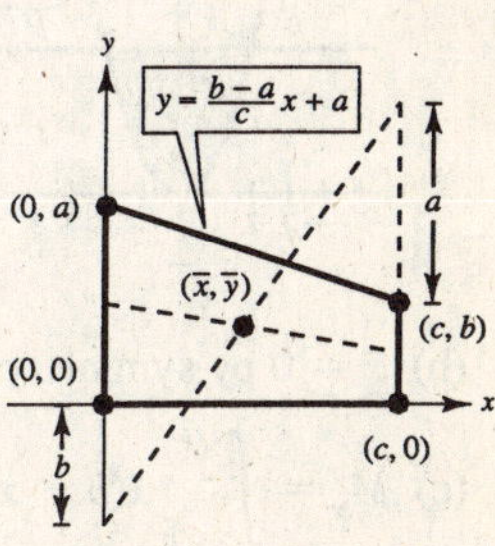

The one line passes through $(0, a/2)$ and $(c, b/2)$. It's equation is $y = \frac{b-a}{2c}x + \frac{a}{2}$. The other line passes through $(0, -b)$ and $(c, a + b)$. It's equation is $y = \frac{a+2b}{c}x - b$. $(\bar{x}, \bar{y})$ is the point of intersection of these two lines.

36. $\bar{x} = 0$ by symmetry.

$$A = \frac{1}{2}\pi r^2$$

$$\frac{1}{A} = \frac{2}{\pi r^2}$$

$$\bar{y} = \frac{2}{\pi r^2}\frac{1}{2}\int_{-r}^{r}\left(\sqrt{r^2 - x^2}\right)^2 dx$$

$$= \frac{1}{\pi r^2}\left[r^2x - \frac{x^3}{3}\right]_{-r}^{r} = \frac{1}{\pi r^2}\left[\frac{4r^3}{3}\right] = \frac{4r}{3\pi}$$

$$(\bar{x}, \bar{y}) = \left(0, \frac{4r}{3\pi}\right)$$

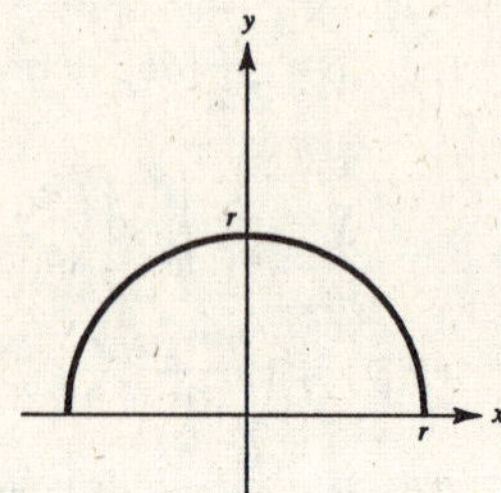

37. $\bar{x} = 0$ by symmetry.

$$A = \frac{1}{2}\pi ab$$

$$\frac{1}{A} = \frac{2}{\pi ab}$$

$$\bar{y} = \frac{2}{\pi ab}\frac{1}{2}\int_{-a}^{a}\left(\frac{b}{a}\sqrt{a^2 - x^2}\right)^2 dx$$

$$= \frac{1}{\pi ab}\left(\frac{b^2}{a^2}\right)\left[a^2x - \frac{x^3}{3}\right]_{-a}^{a} = \frac{b}{\pi a^3}\left[\frac{4a^3}{3}\right] = \frac{4b}{3\pi}$$

$$(\bar{x}, \bar{y}) = \left(0, \frac{4b}{3\pi}\right)$$

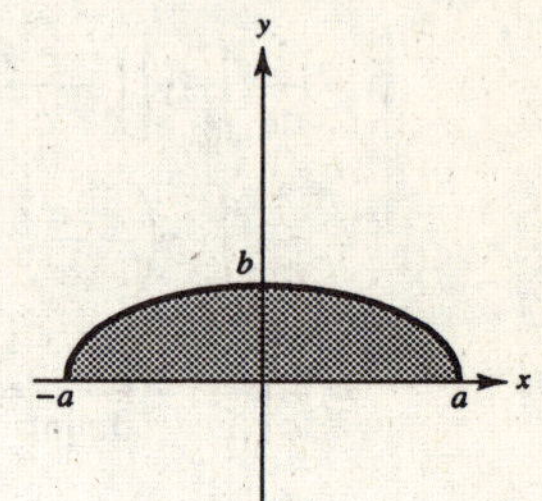

38. $$A = \int_0^1 [1 - (2x - x^2)]\,dx = \frac{1}{3}$$

$$\frac{1}{A} = 3$$

$$\bar{x} = 3\int_0^1 x[1 - (2x - x^2)]\,dx = 3\int_0^1 [x - 2x^2 + x^3]\,dx = 3\left[\frac{x^2}{2} - \frac{2}{3}x^3 + \frac{x^4}{4}\right]_0^1 = \frac{1}{4}$$

$$\bar{y} = 3\int_0^1 \frac{[1 + (2x - x^2)]}{2}[1 - (2x - x^2)]\,dx \qquad = \frac{3}{2}\int_0^1 [1 - (2x - x^2)^2]\,dx$$

$$= \frac{3}{2}\int_0^1 [1 - 4x^2 + 4x^3 - x^4]\,dx = \frac{3}{2}\left[x - \frac{4}{3}x^3 + x^4 - \frac{x^5}{5}\right]_0^1 = \frac{7}{10}$$

$$(\bar{x}, \bar{y}) = \left(\frac{1}{4}, \frac{7}{10}\right)$$

39. (a)

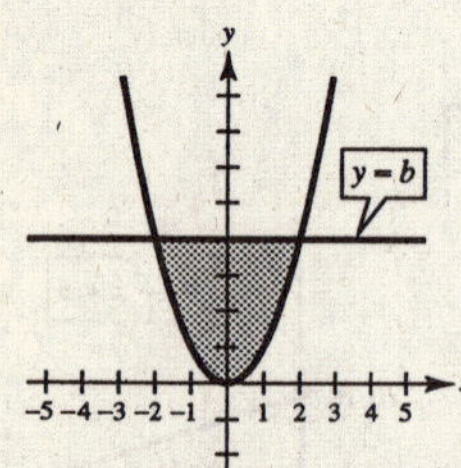

(b) $\bar{x} = 0$ by symmetry.

(c) $M_y = \int_{-\sqrt{b}}^{\sqrt{b}} x(b - x^2)\,dx = 0$ because $bx - x^3$ is odd.

(d) $\bar{y} > \frac{b}{2}$ since there is more area above $y = \frac{b}{2}$ than below.

(e) $$M_x = \int_{-\sqrt{b}}^{\sqrt{b}} \frac{(b + x^2)(b - x^2)}{2}\,dx$$

$$= \int_{-\sqrt{b}}^{\sqrt{b}} \frac{b^2 - x^4}{2}\,dx = \frac{1}{2}\left[b^2x - \frac{x^5}{5}\right]_{-\sqrt{b}}^{\sqrt{b}}$$

$$= b^2\sqrt{b} - \frac{b^2\sqrt{b}}{5} = \frac{4b^2\sqrt{b}}{5}$$

$$A = \int_{-\sqrt{b}}^{\sqrt{b}} (b - x^2)\,dx = \left[bx - \frac{x^3}{3}\right]_{-\sqrt{b}}^{\sqrt{b}}$$

$$= \left(b\sqrt{b} - \frac{b\sqrt{b}}{3}\right)2 = 4\frac{b\sqrt{b}}{3}$$

$$\bar{y} = \frac{M_x}{A} = \frac{4b^2\sqrt{b}/5}{4b\sqrt{b}/3} = \frac{3}{5}b$$

40. (a) $M_y = 0$ by symmetry.

$$M_y = \int_{-\sqrt[2n]{b}}^{\sqrt[2n]{b}} x(b - x^{2n})\,dx = 0$$

because $bx - x^{2n+1}$ is an odd function.

(c) $$M_x = \int_{-\sqrt[2n]{b}}^{\sqrt[2n]{b}} \frac{(b + x^{2n})(b - x^{2n})}{2}\,dx = \int_{-\sqrt[2n]{b}}^{\sqrt[2n]{b}} \frac{1}{2}(b^2 - x^{4n})\,dx$$

$$= \frac{1}{2}\left(b^2x - \frac{x^{4n+1}}{4n+1}\right)\Bigg]_{-\sqrt[2n]{b}}^{\sqrt[2n]{b}}$$

$$= b^2b^{1/2n} - \frac{b^{(4n+1)/2n}}{4n+1} = \frac{4n}{4n+1}b^{(4n+1)/2n}$$

$$A = \int_{-\sqrt[2n]{b}}^{\sqrt[2n]{b}} (b - x^{2n})\,dx = 2\left[bx - \frac{x^{2n+1}}{2n+1}\right]_0^{\sqrt[2n]{b}}$$

$$= 2\left[b \cdot b^{1/2n} - \frac{b^{(2n+1)/2n}}{2n+1}\right] = \frac{4n}{2n+1}b^{(2n+1)/2n}$$

$$\bar{y} = \frac{M_x}{A} = \frac{4n\,b^{(4n+1)/2n}/(4n+1)}{4n\,b^{(24n+1)/2n}/(2n+1)} = \frac{2n+1}{4n+1}b$$

(b) $\bar{y} > \dfrac{b}{2}$ because there is more area above $y = \dfrac{b}{2}$ than below.

(d)

n	1	2	3	4
$\bar{y}$	$\frac{3}{5}b$	$\frac{5}{9}b$	$\frac{7}{13}b$	$\frac{9}{17}b$

(e) $$\lim_{n\to\infty} \bar{y} = \lim_{n\to\infty} \frac{2n+1}{4n+1}b = \frac{1}{2}b$$

(f) As $n \to \infty$, the figure gets narrower.

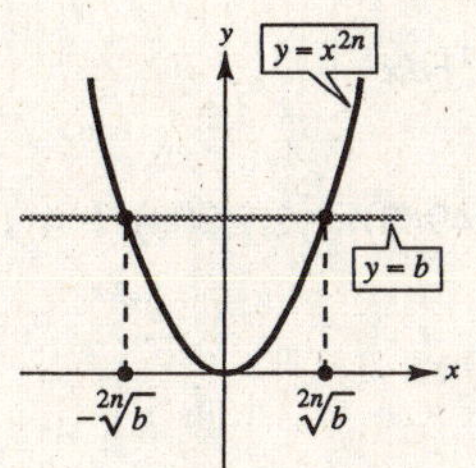

41. (a) $\bar{x} = 0$ by symmetry.

$$A = 2\int_0^{40} f(x)\,dx = \frac{2(40)}{3(4)}[30 + 4(29) + 2(26) + 4(20) + 0] = \frac{20}{3}(278) = \frac{5560}{3}$$

$$M_x = \int_{-40}^{40} \frac{f(x)^2}{2}\,dx = \frac{40}{3(4)}[30^2 + 4(29)^2 + 2(26)^2 + 4(20)^2 + 0] = \frac{10}{3}(7216) = \frac{72{,}160}{3}$$

$$\bar{y} = \frac{M_x}{A} = \frac{72{,}160/3}{5560/3} = \frac{72{,}160}{5560} \approx 12.98$$

$(\bar{x}, \bar{y}) = (0, 12.98)$

(b) $y = (-1.02 \times 10^{-5})x^4 - 0.0019x^2 + 29.28$ (Use nine data points.)

(c) $$\bar{y} = \frac{M_x}{A} \approx \frac{23{,}697.68}{1843.54} \approx 12.85$$

$(\bar{x}, \bar{y}) = (0, 12.85)$

42. Let $f(x)$ be the top curve, given by $l + d$. The bottom curve is $d(x)$.

x	0	0.5	1.0	1.5	2.0
f	2.0	1.93	1.73	1.32	0
d	0.50	0.48	0.43	0.33	0

—CONTINUED—

42. —CONTINUED—

(a) Area $= 2\int_0^2 [f(x) - d(x)]\,dx$

$$\approx 2\frac{2}{3(4)}[1.50 + 4(1.45) + 2(1.30) + 4(.99) + 0]$$

$$= \frac{1}{3}[13.86] = 4.62$$

$$M_x = \int_{-2}^{2} \frac{f(x) + d(x)}{2}(f(x) - d(x))\,dx$$

$$= \int_0^2 [f(x)^2 - d(x)^2]\,dx$$

$$= \frac{2}{3(4)}[3.75 + 4(3.4945) + 2(2.808) + 4(1.6335) + 0]$$

$$= \frac{1}{6}[29.878] = 4.9797$$

$$\bar{y} = \frac{M_x}{A} = \frac{4.9797}{4.62} = 1.078$$

$(\bar{x}, \bar{y}) = (0, 1.078)$

(b) $f(x) = -0.1061x^4 - 0.06126x^2 + 1.9527$

$d(x) = -0.02648x^4 - 0.01497x^2 + .4862$

(c) $\bar{y} = \frac{M_x}{A} \approx \frac{4.9133}{4.59998} = 1.068$

$(\bar{x}, \bar{y}) = (0, 1.068)$

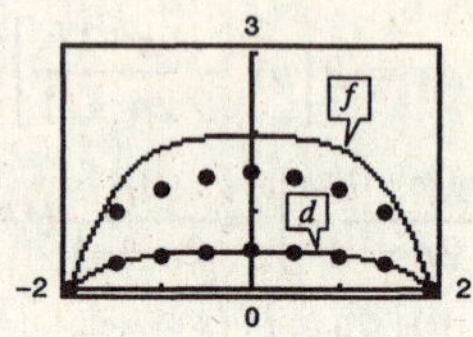

43. Centroids of the given regions: (1, 0) and (3, 0)

Area: $A = 4 + \pi$

$$\bar{x} = \frac{4(1) + \pi(3)}{4 + \pi} = \frac{4 + 3\pi}{4 + \pi}$$

$$\bar{y} = \frac{4(0) + \pi(0)}{4 + \pi} = 0$$

$$(\bar{x}, \bar{y}) = \left(\frac{4 + 3\pi}{4 + \pi}, 0\right) \approx (1.88, 0)$$

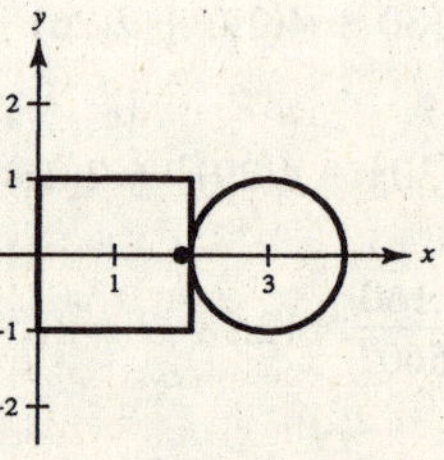

44. Centroids of the given regions: $\left(\frac{1}{2}, \frac{3}{2}\right)$, $\left(2, \frac{1}{2}\right)$, and $\left(\frac{7}{2}, 1\right)$

Area: $A = 3 + 2 + 2 = 7$

$$\bar{x} = \frac{3(1/2) + 2(2) + 2(7/2)}{7} = \frac{25/2}{7} = \frac{25}{14}$$

$$\bar{y} = \frac{3(3/2) + 2(1/2) + 2(1)}{7} = \frac{15/2}{7} = \frac{15}{14}$$

$$(\bar{x}, \bar{y}) = \left(\frac{25}{14}, \frac{15}{14}\right)$$

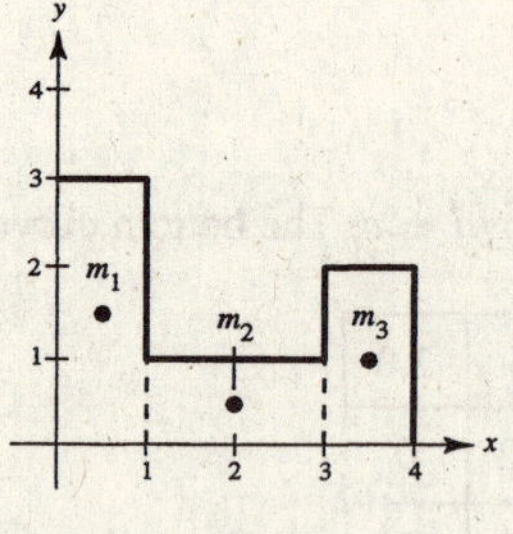

45. Centroids of the given regions: $\left(0, \frac{3}{2}\right)$, (0, 5), and $\left(0, \frac{15}{2}\right)$

Area: $A = 15 + 12 + 7 = 34$

$$\bar{x} = \frac{15(0) + 12(0) + 7(0)}{34} = 0$$

$$\bar{y} = \frac{15(3/2) + 12(5) + 7(15/2)}{34} = \frac{135}{34}$$

$$(\bar{x}, \bar{y}) = \left(0, \frac{135}{34}\right) \approx (0, 3.97)$$

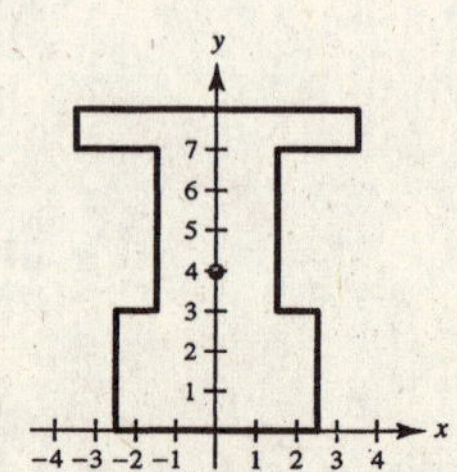

46. $m_1 = \frac{7}{8}(2) = \frac{7}{4}, P_1 = \left(0, \frac{7}{16}\right)$

$m_2 = \frac{7}{8}\left(6 - \frac{7}{8}\right) = \frac{287}{64}, P_2 = \left(0, \frac{55}{16}\right)$

By symmetry, $\bar{x} = 0$.

$$\bar{y} = \frac{(7/4)(7/16) + (287/64)(55/16)}{(7/4) + (287/64)} = \frac{16{,}569}{6384} = \frac{789}{304}$$

$$(\bar{x}, \bar{y}) = \left(0, \frac{789}{304}\right) \approx (0, 2.595)$$

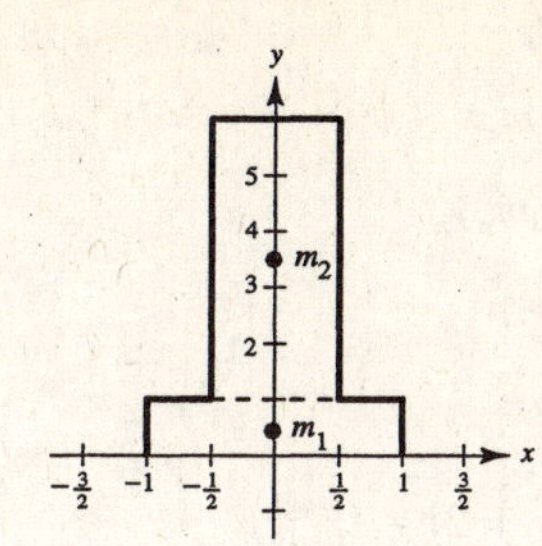

47. Centroids of the given regions: (1, 0) and (3, 0)

Mass: $4 + 2\pi$

$$\bar{x} = \frac{4(1) + 2\pi(3)}{4 + 2\pi} = \frac{2 + 3\pi}{2 + \pi}$$

$$\bar{y} = 0$$

$$(\bar{x}, \bar{y}) = \left(\frac{2 + 3\pi}{2 + \pi}, 0\right) \approx (2.22, 0)$$

48. Centroids of the given regions: (3, 0) and (1, 0)

Mass: $8 + \pi$

$$\bar{y} = 0$$

$$\bar{x} = \frac{8(1) + \pi(3)}{8 + \pi} = \frac{8 + 3\pi}{8 + \pi}$$

$$(\bar{x}, \bar{y}) = \left(\frac{8 + 3\pi}{8 + \pi}, 0\right) \approx (1.56, 0)$$

49. $r = 5$ is distance between center of circle and y-axis. $A \approx \pi(4)^2 = 16\pi$ is area of circle. Hence,

$$V = 2\pi rA = 2\pi(5)(16\pi) = 160\pi^2 \approx 1579.14.$$

50. $V = 2\pi rA = 2\pi(3)(4\pi) = 24\pi^2$

51. $A = \frac{1}{2}(4)(4) = 8$

$$\bar{y} = \left(\frac{1}{8}\right)\frac{1}{2}\int_0^4 (4 + x)(4 - x)\, dx = \frac{1}{16}\left[16x - \frac{x^3}{3}\right]_0^4 = \frac{8}{3}$$

$$r = \bar{y} = \frac{8}{3}$$

$$V = 2\pi rA = 2\pi\left(\frac{8}{3}\right)(8) = \frac{128\pi}{3} \approx 134.04$$

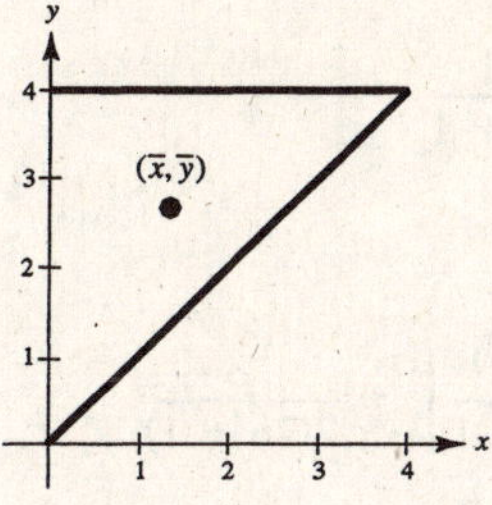

52. $A = \int_2^6 2\sqrt{x - 2}\, dx = \frac{4}{3}(x - 2)^{3/2}\Big]_2^6 = \frac{32}{3}$

$$M_y = \int_2^6 (x)2\sqrt{x - 2}\, dx = 2\int_2^6 x\sqrt{x - 2}\, dx$$

Let $u = x - 2, x = u + 2, du = dx$:

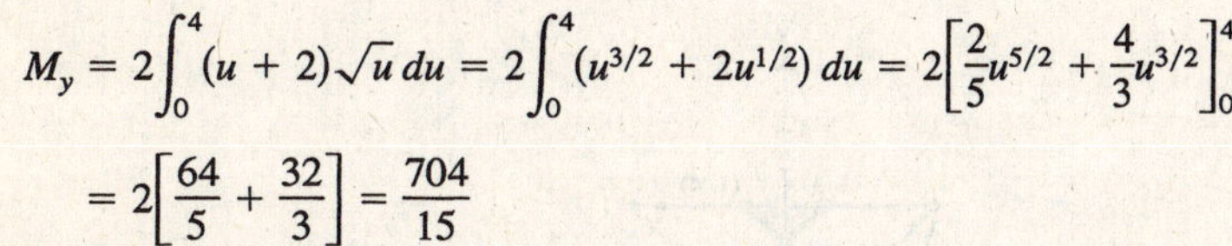

$$M_y = 2\int_0^4 (u + 2)\sqrt{u}\, du = 2\int_0^4 (u^{3/2} + 2u^{1/2})\, du = 2\left[\frac{2}{5}u^{5/2} + \frac{4}{3}u^{3/2}\right]_0^4$$

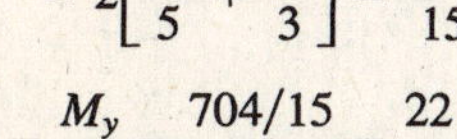

$$= 2\left[\frac{64}{5} + \frac{32}{3}\right] = \frac{704}{15}$$

$$\bar{x} = \frac{M_y}{A} = \frac{704/15}{32/3} = \frac{22}{5}$$

$$r = \bar{x} = \frac{22}{5}$$

$$V = 2\pi rA = 2\pi\left(\frac{22}{5}\right)\left(\frac{32}{3}\right) = \frac{1408\pi}{15} \approx 294.89$$

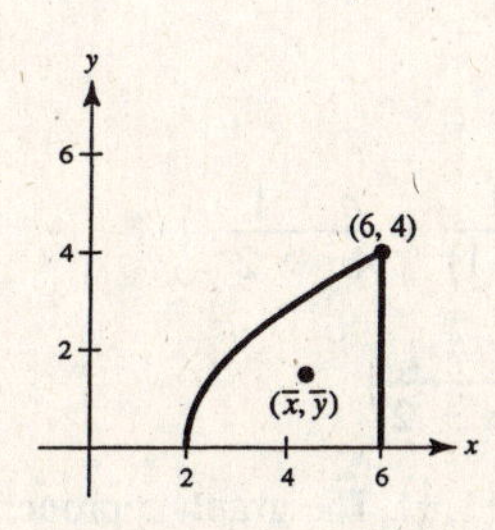

53. $m = m_1 + \cdots + m_n$

$M_y = m_1x_1 + \cdots + m_nx_n$

$M_x = m_1y_1 + \cdots + m_ny_n$

$\bar{x} = \dfrac{M_y}{m}, \bar{y} = \dfrac{M_x}{m}$

54. A planar lamina is a thin flat plate of constant density. The center of mass $(\bar{x}, \bar{y})$ is the balancing point on the lamina.

55. (a) Yes. $(\bar{x}, \bar{y}) = \left(\frac{5}{6}, \frac{5}{18} + 2\right) = \left(\frac{5}{6}, \frac{41}{18}\right)$

(b) Yes. $(\bar{x}, \bar{y}) = \left(\frac{5}{6} + 2, \frac{5}{18}\right) = \left(\frac{17}{6}, \frac{5}{18}\right)$

(c) Yes. $(\bar{x}, \bar{y}) = \left(\frac{5}{6}, -\frac{5}{18}\right)$

(d) No

56. Let R be a region in a plane and let L be a line such that L does not intersect the interior of R. If r is the distance between the centroid of R and L, then the volume V of the solid of revolution formed by revolving R about L is $V = 2\pi rA$ where A is the area of R.

57. The surface area of the sphere is $S = 4\pi r^2$. The arc length of C is $s = \pi r$. The distance traveled by the centroid is

$$d = \frac{S}{s} = \frac{4\pi r^2}{\pi r} = 4r.$$

This distance is also the circumference of the circle of radius y.

$$d = 2\pi y$$

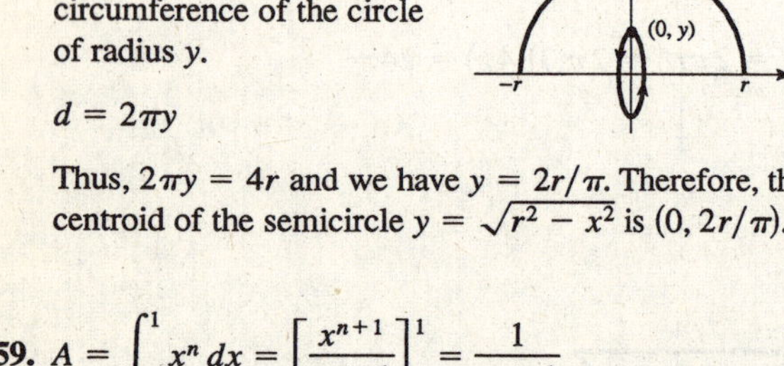

Thus, $2\pi y = 4r$ and we have $y = 2r/\pi$. Therefore, the centroid of the semicircle $y = \sqrt{r^2 - x^2}$ is $(0, 2r/\pi)$.

58. The centroid of the circle is $(1, 0)$. The distance traveled by the centroid is 2π. The arc length of the circle is also 2π. Therefore, $S = (2\pi)(2\pi) = 4\pi^2$.

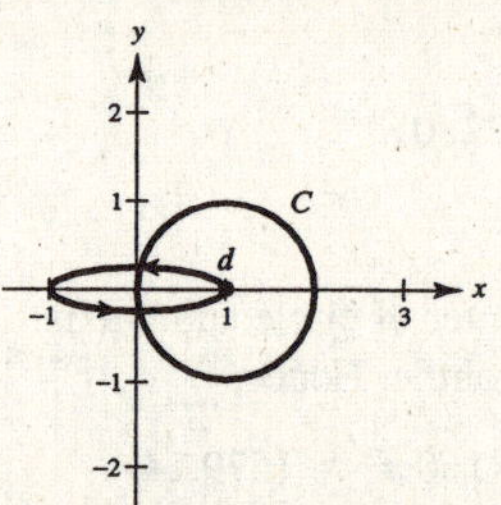

59. $A = \displaystyle\int_0^1 x^n\,dx = \left[\frac{x^{n+1}}{n+1}\right]_0^1 = \frac{1}{n+1}$

$m = \rho A = \dfrac{\rho}{n+1}$

$M_x = \dfrac{\rho}{2}\displaystyle\int_0^1 (x^n)^2\,dx = \left[\frac{\rho}{2} \cdot \frac{x^{2n+1}}{2n+1}\right]_0^1 = \frac{\rho}{2(2n+1)}$

$M_y = \rho\displaystyle\int_0^1 x(x^n)\,dx = \left[\rho \cdot \frac{x^{n+2}}{n+2}\right]_0^1 = \frac{\rho}{n+2}$

$\bar{x} = \dfrac{M_y}{m} = \dfrac{n+1}{n+2}$

$\bar{y} = \dfrac{M_x}{m} = \dfrac{n+1}{2(2n+1)} = \dfrac{n+1}{4n+2}$

Centroid: $\left(\dfrac{n+1}{n+2}, \dfrac{n+1}{4n+2}\right)$

As $n \to \infty$, $(\bar{x}, \bar{y}) \to \left(1, \frac{1}{4}\right)$. The graph approaches the x-axis and the line $x = 1$ as $n \to \infty$.

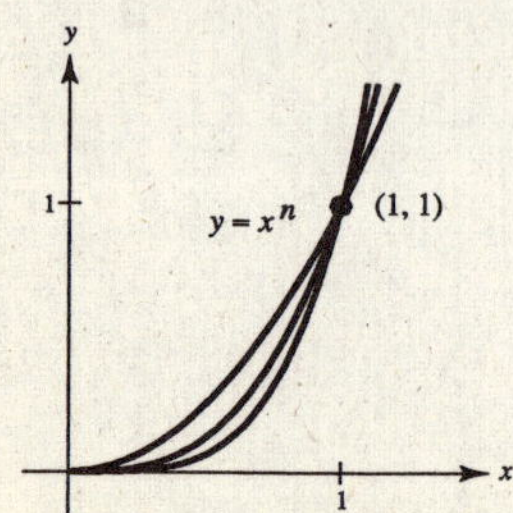

60. Let T be the shaded triangle with vertices $(-1, 4)$, $(1, 4)$, and $(0, 3)$. Let U be the large triangle with vertices $(-4, 4)$, $(4, 4)$, and $(0, 0)$. V consists of the region U minus the region T.

Centroid of T: $\left(0, \frac{11}{3}\right)$; Area $= 1$

Centroid of U: $\left(0, \frac{8}{3}\right)$; Area $= 16$

Area: $V = 16 - 1 = 15$

$\bar{x} = 0$ by symmetry.

$15\bar{y} + 1\left(\frac{11}{3}\right) = 16\left(\frac{8}{3}\right)$

$15\bar{y} = \frac{117}{3}$

$\bar{y} = \frac{13}{5}$

$(\bar{x}, \bar{y}) = \left(0, \frac{13}{5}\right)$

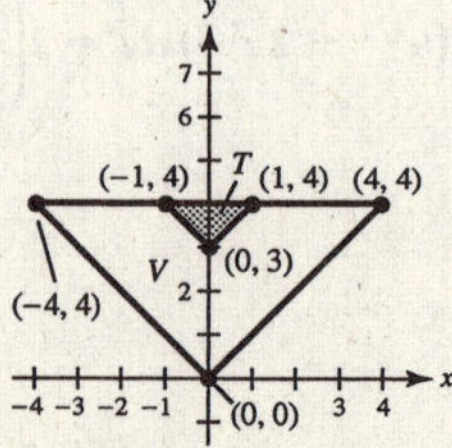

Section 7.7 Fluid Pressure and Fluid Force

1. $F = PA = [62.4(5)](3) = 936 \text{ lb}$

2. $F = PA = [62.4(5)](16) = 4992 \text{ lb}$

3. $F = 62.4(h + 2)(6) - (62.4)(h)(6)$

$= 62.4(2)(6) = 748.8 \text{ lb}$

4. $F = 62.4(h + 4)(48) - (62.4)(h)(48)$

$= 62.4(4)(48) = 11{,}980.8 \text{ lb}$

5. $h(y) = 3 - y$

$L(y) = 4$

$$F = 62.4 \int_0^3 (3 - y)(4)\,dy$$

$$= 249.6 \int_0^3 (3 - y)\,dy$$

$$= 249.6\left[3y - \frac{y^2}{2}\right]_0^3$$

$= 1123.2 \text{ lb}$

6. $h(y) = 3 - y$

$L(y) = \frac{4}{3}y$

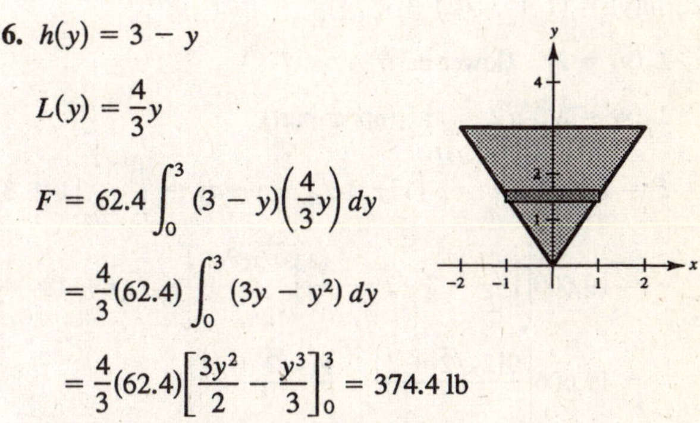

$$F = 62.4 \int_0^3 (3 - y)\left(\frac{4}{3}y\right) dy$$

$$= \frac{4}{3}(62.4) \int_0^3 (3y - y^2)\,dy$$

$$= \frac{4}{3}(62.4)\left[\frac{3y^2}{2} - \frac{y^3}{3}\right]_0^3 = 374.4 \text{ lb}$$

Force is one-third that of Exercise 5.

7. $h(y) = 3 - y$

$L(y) = 2\left(\frac{y}{3} + 1\right)$

$$F = 2(62.4) \int_0^3 (3 - y)\left(\frac{y}{3} + 1\right) dy$$

$$= 124.8 \int_0^3 \left(3 - \frac{y^2}{3}\right) dy$$

$$= 124.8\left[3y - \frac{y^3}{9}\right]_0^3$$

$= 748.8 \text{ lb}$

8. $h(y) = -y$

$L(y) = 2\sqrt{4 - y^2}$

$$F = 62.4 \int_{-2}^0 (-y)(2)\sqrt{4 - y^2}\,dy$$

$$= \left[62.4\left(\frac{2}{3}\right)(4 - y^2)^{3/2}\right]_{-2}^0 = 332.8 \text{ lb}$$

9. $h(y) = 4 - y$

$L(y) = 2\sqrt{y}$

$$F = 2(62.4) \int_0^4 (4 - y)\sqrt{y}\,dy$$

$$= 124.8 \int_0^4 (4y^{1/2} - y^{3/2})\,dy$$

$$= 124.8\left[\frac{8y^{3/2}}{3} - \frac{2y^{5/2}}{5}\right]_0^4$$

$= 1064.96 \text{ lb}$

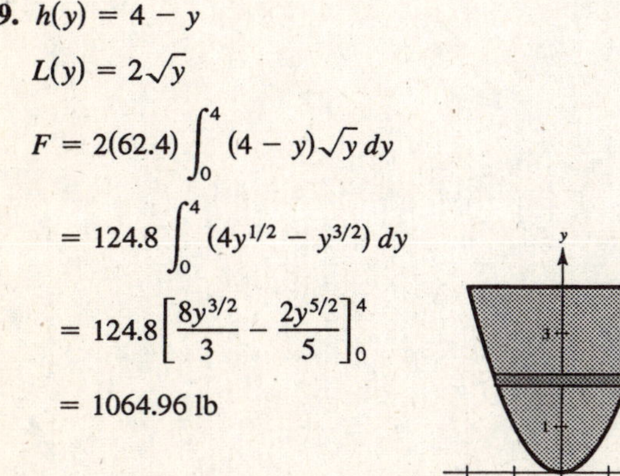

10. $h(y) = -y$

$L(y) = \frac{4}{3}\sqrt{9 - y^2}$

$$F = 62.4 \int_{-3}^0 (-y)\frac{4}{3}\sqrt{9 - y^2}\,dy$$

$$= 62.4\left(\frac{2}{3}\right) \int_{-3}^0 (9 - y^2)^{1/2}(-2y)\,dy$$

$$= \left[62.4\left(\frac{4}{9}\right)(9 - y^2)^{3/2}\right]_{-3}^0$$

$= 748.8 \text{ lb}$

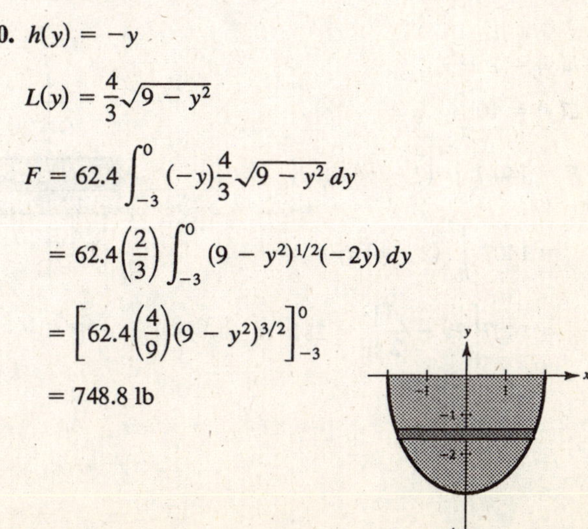

11. $h(y) = 4 - y$

$L(y) = 2$

$$F = 9800 \int_0^2 2(4 - y)\, dy$$

$$= 9800\Big[8y - y^2\Big]_0^2 = 117{,}600 \text{ newtons}$$

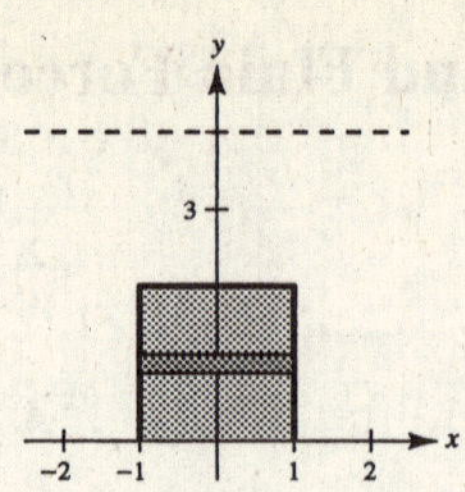

12. $h(y) = \left(1 + 3\sqrt{2}\right) - y$

$L_1(y) = 2y$ (lower part)

$L_2(y) = 2\left(3\sqrt{2} - y\right)$ (upper part)

$$F = 2(9800)\left[\int_0^{3\sqrt{2}/2} \left(1 + 3\sqrt{2} - y\right)y\, dy + \int_{3\sqrt{2}/2}^{3\sqrt{2}} \left(1 + 3\sqrt{2} - y\right)\left(3\sqrt{2} - y\right) dy\right]$$

$$= 19{,}600\left[\left[\frac{y^2}{2} - 3\sqrt{2}y - \frac{y^3}{3}\right]_0^{3\sqrt{2}/2} + \left[3\sqrt{2}y + 18y + \frac{y^3}{3} - \frac{6\sqrt{2} + 1}{2}y\right]_{3\sqrt{2}/2}^{3\sqrt{2}}\right]$$

$$= 19{,}600\left[\frac{9\left(2\sqrt{2} + 1\right)}{4} + \frac{9\left(\sqrt{2} + 1\right)}{4}\right]$$

$$= 44{,}100\left(3\sqrt{2} + 2\right) \text{ newtons}$$

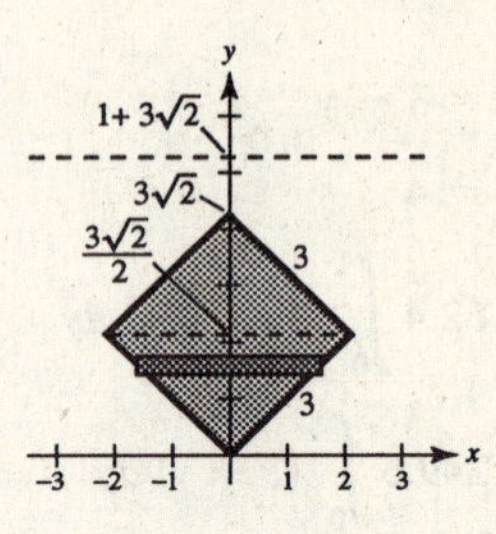

13. $h(y) = 12 - y$

$$L(y) = 6 - \frac{2y}{3}$$

$$F = 9800 \int_0^9 (12 - y)\left(6 - \frac{2y}{3}\right) dy$$

$$= 9800\left[72y - 7y^2 + \frac{2y^3}{9}\right]_0^9$$

$$= 2{,}381{,}400 \text{ newtons}$$

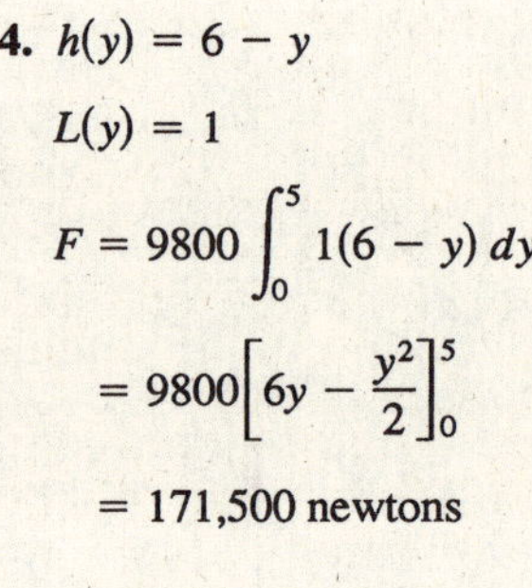

14. $h(y) = 6 - y$

$L(y) = 1$

$$F = 9800 \int_0^5 1(6 - y)\, dy$$

$$= 9800\left[6y - \frac{y^2}{2}\right]_0^5$$

$$= 171{,}500 \text{ newtons}$$

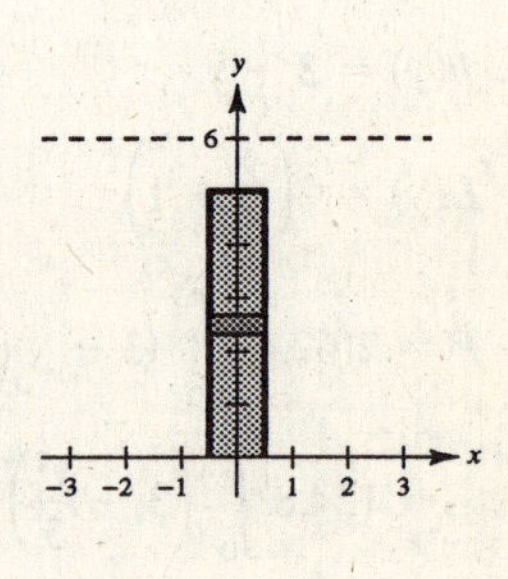

15. $h(y) = 2 - y$

$L(y) = 10$

$$F = 140.7 \int_0^2 (2 - y)(10)\, dy$$

$$= 1407 \int_0^2 (2 - y)\, dy$$

$$= 1407\left[2y - \frac{y^2}{2}\right]_0^2 = 2814 \text{ lb}$$

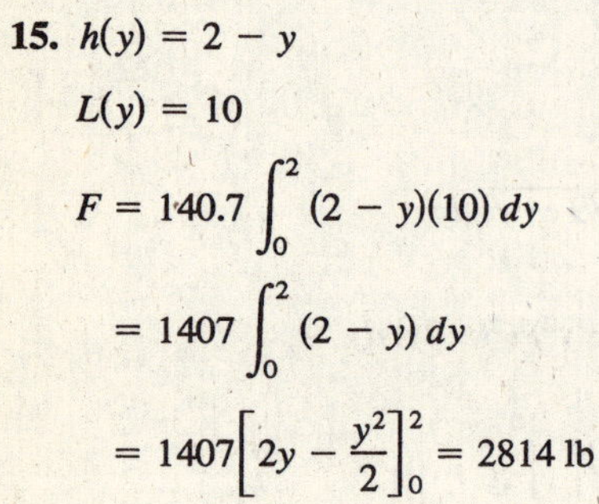

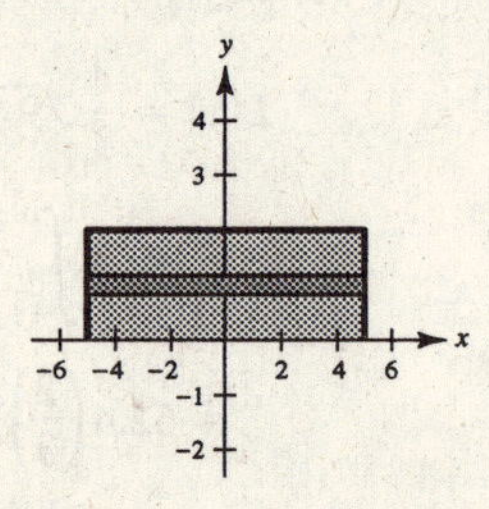

16. $h(y) = -y$

$L(y) = 2\left(\frac{4}{3}\sqrt{9 - y^2}\right)$

$$F = 140.7 \int_{-3}^{0} (-y)(2)\left(\frac{4}{3}\sqrt{9 - y^2}\right) dy$$

$$= \frac{(140.7)(4)}{3} \int_{-3}^{0} \sqrt{9 - y^2}\,(-2y)\, dy$$

$$= \left[\frac{(140.7)(4)}{3}\left(\frac{2}{3}\right)(9 - y^2)^{3/2}\right]_{-3}^{0}$$

$$= 3376.8 \text{ lb}$$

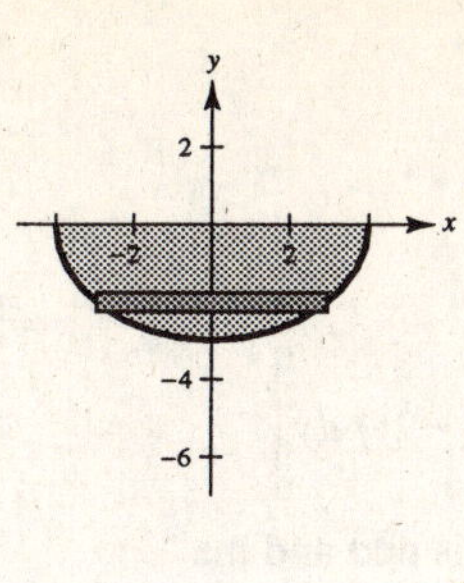

17. $h(y) = 4 - y$

$L(y) = 6$

$$F = 140.7 \int_{0}^{4} (4 - y)(6)\, dy$$

$$= 844.2 \int_{0}^{4} (4 - y)\, dy$$

$$= 844.2\left[4y - \frac{y^2}{2}\right]_{0}^{4} = 6753.6 \text{ lb}$$

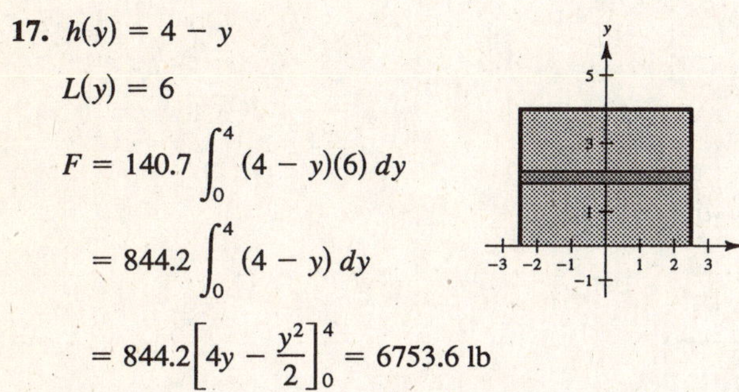

18. $h(y) = -y$

$L(y) = 5 + \frac{5}{3}y$

$$F = 140.7 \int_{-3}^{0} (-y)\left(5 + \frac{5}{3}y\right) dy$$

$$= 140.7 \int_{-3}^{0} \left(-5y - \frac{5}{3}y^2\right) dy$$

$$= 140.7\left[-\frac{5}{2}y^2 - \frac{5}{9}y^3\right]_{-3}^{0}$$

$$= 140.7\left[\frac{45}{2} - 15\right]$$

$$= 1055.25 \text{ lb}$$

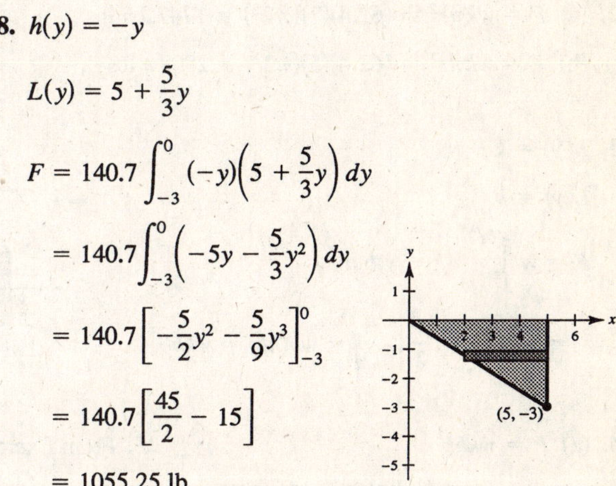

19. $h(y) = -y$

$L(y) = 2\left(\frac{1}{2}\right)\sqrt{9 - 4y^2}$

$$F = 42 \int_{-3/2}^{0} (-y)\sqrt{9 - 4y^2}\, dy$$

$$= \frac{42}{8} \int_{-3/2}^{0} (9 - 4y^2)^{1/2}(-8y)\, dy$$

$$= \left[\left(\frac{21}{4}\right)\left(\frac{2}{3}\right)(9 - 4y^2)^{3/2}\right]_{-3/2}^{0} = 94.5 \text{ lb}$$

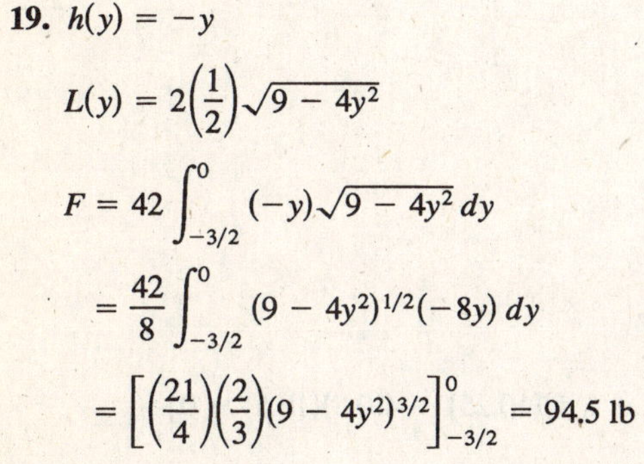

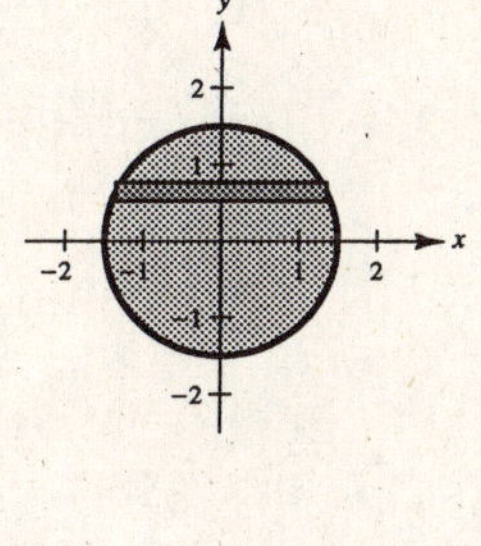

20. $h(y) = \frac{3}{2} - y$

$L(y) = 2\left(\frac{1}{2}\right)\sqrt{9 - 4y^2}$

$$F = 42 \int_{-3/2}^{3/2} \left(\frac{3}{2} - y\right)\sqrt{9 - 4y^2}\, dy = 63 \int_{-3/2}^{3/2} \sqrt{9 - 4y^2}\, dy + \frac{21}{4} \int_{-3/2}^{3/2} \sqrt{9 - 4y^2}\,(-8y)\, dy$$

The second integral is zero since it is an odd function and the limits of integration are symmetric to the origin. The first integral is twice the area of a semicircle of radius $\frac{3}{2}$.

$\left(\sqrt{9 - 4y^2} = 2\sqrt{(9/4) - y^2}\right)$

Thus, the force is $63\left(\frac{9}{4}\pi\right) = 141.75\pi \approx 445.32$ lb.

21. $h(y) = k - y$

$L(y) = 2\sqrt{r^2 - y^2}$

$$F = w\int_{-r}^{r} (k - y)\sqrt{r^2 - y^2}\,(2)\,dy$$

$$= w\left[2k\int_{-r}^{r} \sqrt{r^2 - y^2}\,dy + \int_{-r}^{r} \sqrt{r^2 - y^2}\,(-2y)\,dy\right]$$

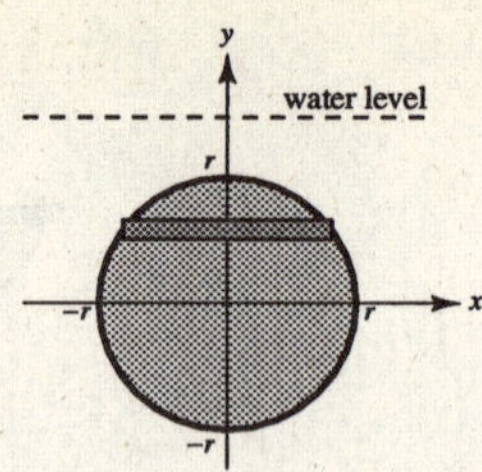

The second integral is zero since its integrand is odd and the limits of integration are symmetric to the origin. The first integral is the area of a semicircle with radius r.

$$F = w\left[(2k)\frac{\pi r^2}{2} + 0\right] = wk\pi r^2$$

22. (a) $F = wk\pi r^2 = (62.4)(7)(\pi 2^2) = 1747.2\pi$ lbs

(b) $F = wk\pi r^2 = (62.4)(5)(\pi 3^2) = 2808\pi$ lbs

23. $h(y) = k - y$

$L(y) = b$

$$F = w\int_{-h/2}^{h/2} (k - y)b\,dy$$

$$= wb\left[ky - \frac{y^2}{2}\right]_{-h/2}^{h/2} = wb(hk) = wkhb$$

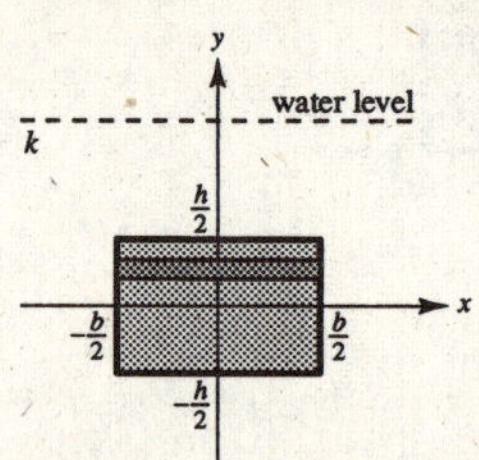

24. (a) $F = wkhb$

$= (62.4)\left(\frac{11}{2}\right)(3)(5) = 5148$ lbs

(b) $F = wkhb$

$= (62.4)\left(\frac{17}{2}\right)(5)(10) = 26{,}520$ lbs

25. From Exercise 23:

$F = 64(15)(1)(1) = 960$ lb

26. From Exercise 21:

$F = 64(15)\pi\left(\frac{1}{2}\right)^2 \approx 753.98$ lb

27. $h(y) = 4 - y$

$$F = 62.4\int_0^4 (4 - y)L(y)\,dy$$

Using Simpson's Rule with $n = 8$ we have:

$$F \approx 62.4\left(\frac{4 - 0}{3(8)}\right)[0 + 4(3.5)(3) + 2(3)(5) + 4(2.5)(8) + 2(2)(9) + 4(1.5)(10) + 2(1)(10.25) + 4(0.5)(10.5) + 0]$$

$= 3010.8$ lb

28. $h(y) = 3 - y$

Solving $y = 5x^2/(x^2 + 4)$ for x, you obtain

$x = \sqrt{4y/(5 - y)}$.

$$L(y) = 2\sqrt{\frac{4y}{5 - y}}$$

$$F = 62.4(2)\int_0^3 (3 - y)\sqrt{\frac{4y}{5 - y}}\,dy$$

$$= 2(124.8)\int_0^3 (3 - y)\sqrt{\frac{y}{5 - y}}\,dy \approx 546.265 \text{ lb}$$

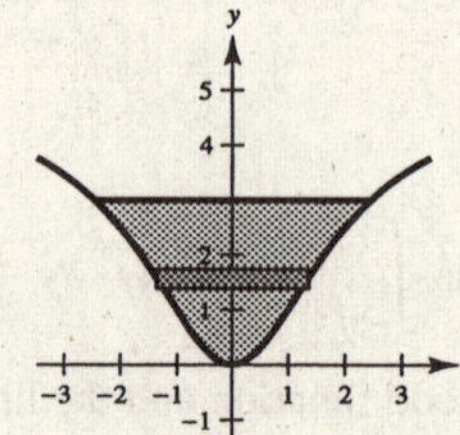

29. $h(y) = 12 - y$

$L(y) = 2(4^{2/3} - y^{2/3})^{3/2}$

$F = 62.4\int_0^4 2(12 - y)(4^{2/3} - y^{2/3})^{3/2}\,dy$

≈ 6448.73 lb

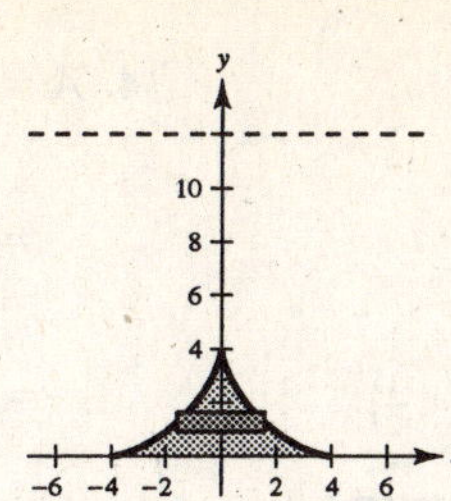

30. $h(y) = 12 - y$

$L(y) = 2\dfrac{\sqrt{7(16 - y^2)}}{2} = \sqrt{7(16 - y^2)}$

$F = 62.4\int_0^4 (12 - y)\sqrt{7(16 - y^2)}\,dy$

$= 62.4\sqrt{7}\int_0^4 (12 - y)\sqrt{16 - y^2}\,dy \approx 21373.7$ lb

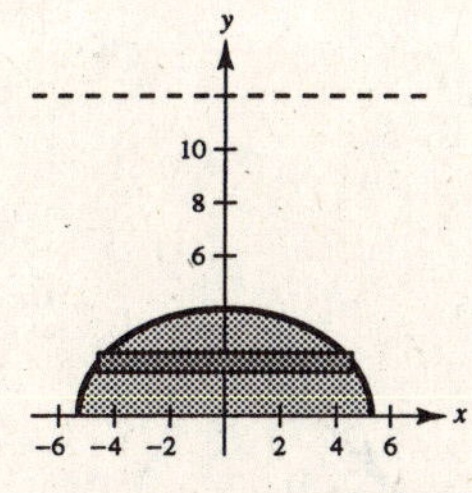

31. (a) If the fluid force is one-half of 1123.2 lb, and the height of the water is b, then

$h(y) = b - y$

$L(y) = 4$

$F = 62.4\int_0^b (b - y)(4)\,dy = \frac{1}{2}(1123.2)$

$\int_0^b (b - y)\,dy = 2.25$

$\left[by - \frac{y^2}{2}\right]_0^b = 2.25$

$b^2 - \frac{b^2}{2} = 2.25$

$b^2 = 4.5 \implies b \approx 2.12$ ft.

(b) The pressure increases with increasing depth.

32. Fluid pressure is the force per unit of area exerted by a fluid over the surface of a body.

33. $F = Fw = w\int_c^d h(y)L(y)\,dy$, see page 508.

34. The left window experiences the greater fluid force because its centroid is lower.

Review Exercises for Chapter 7

1. $A = \int_1^5 \frac{1}{x^2}\,dx = \left[-\frac{1}{x}\right]_1^5 = \frac{4}{5}$

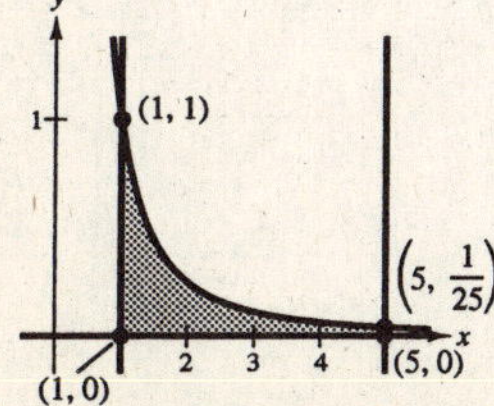

2. $A = \int_{1/2}^5 \left(4 - \frac{1}{x^2}\right)dx$

$= \left[4x + \frac{1}{x}\right]_{1/2}^5 = \frac{81}{5}$

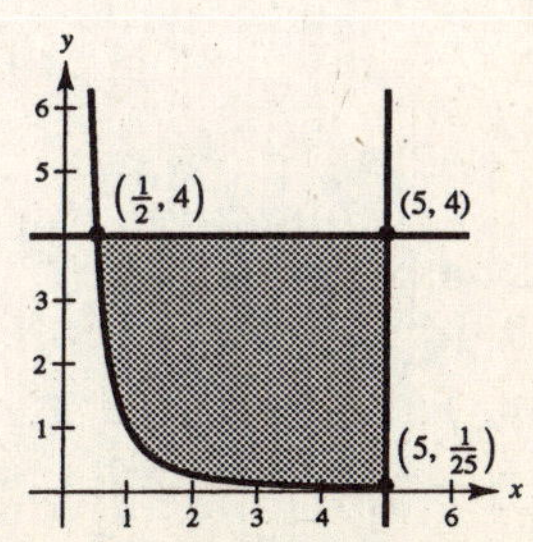

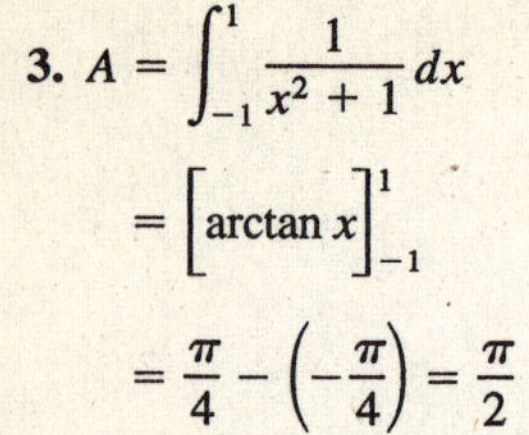

3. $A = \int_{-1}^{1} \frac{1}{x^2+1}\,dx$

$= \Big[\arctan x\Big]_{-1}^{1}$

$= \frac{\pi}{4} - \left(-\frac{\pi}{4}\right) = \frac{\pi}{2}$

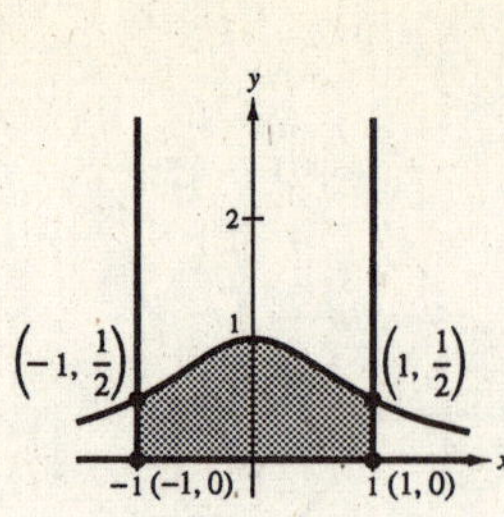

4. $A = \int_{0}^{1} [(y^2 - 2y) - (-1)]\,dy$

$= \int_{0}^{1} (y^2 - 2y + 1)\,dy$

$= \int_{0}^{1} (y-1)^2\,dy$

$= \left[\frac{(y-1)^3}{3}\right]_0^1 = \frac{1}{3}$

5. $A = 2\int_{0}^{1} (x - x^3)\,dx$

$= 2\left[\frac{1}{2}x^2 - \frac{1}{4}x^4\right]_0^1$

$= \frac{1}{2}$

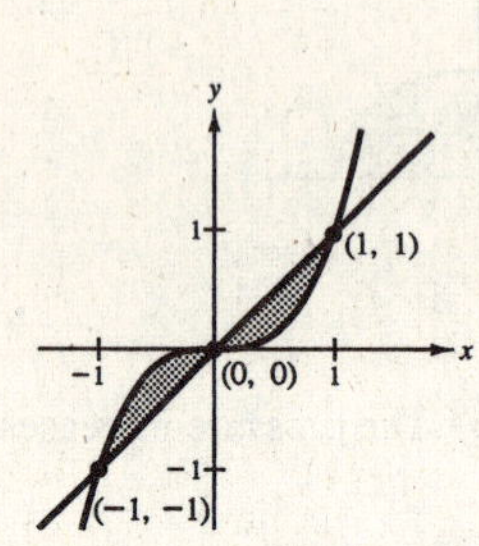

6. $A = \int_{-1}^{2} [(y+3) - (y^2+1)]\,dy$

$= \int_{-1}^{2} (2 + y - y^2)\,dy$

$= \left[2y + \frac{1}{2}y^2 - \frac{1}{3}y^3\right]_{-1}^{2}$

$= \frac{9}{2}$

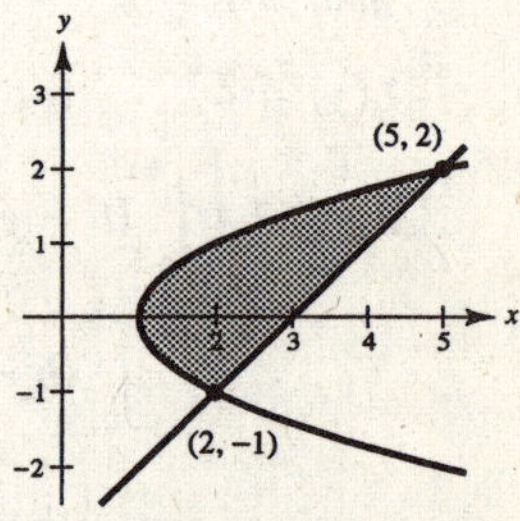

7. $A = \int_{0}^{2} (e^2 - e^x)\,dx$

$= \Big[xe^2 - e^x\Big]_0^2$

$= e^2 + 1$

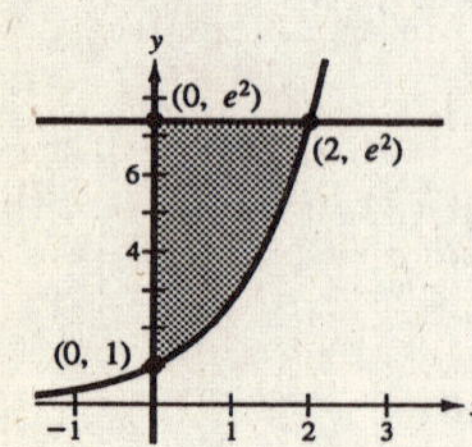

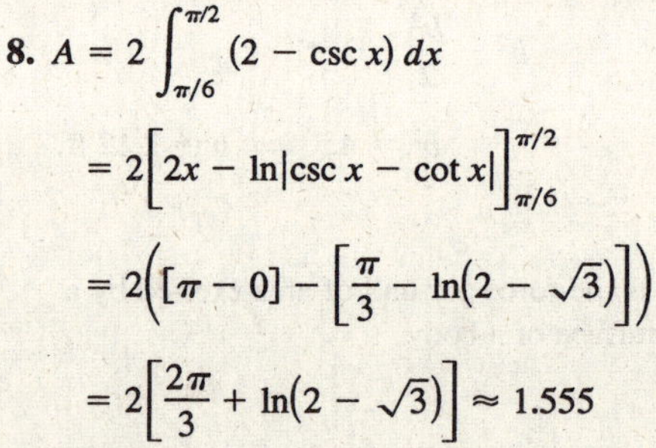

8. $A = 2\int_{\pi/6}^{\pi/2} (2 - \csc x)\,dx$

$= 2\Big[2x - \ln|\csc x - \cot x|\Big]_{\pi/6}^{\pi/2}$

$= 2\left([\pi - 0] - \left[\frac{\pi}{3} - \ln(2 - \sqrt{3})\right]\right)$

$= 2\left[\frac{2\pi}{3} + \ln(2 - \sqrt{3})\right] \approx 1.555$

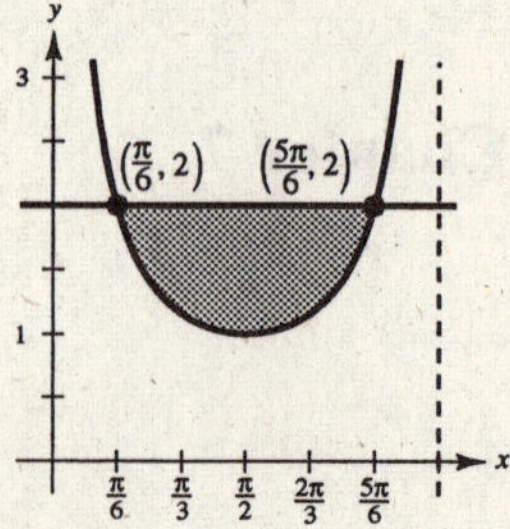

9. $A = \int_{\pi/4}^{5\pi/4} (\sin x - \cos x)\, dx$

$= \left[-\cos x - \sin x\right]_{\pi/4}^{5\pi/4}$

$= \left(\frac{1}{\sqrt{2}} + \frac{1}{\sqrt{2}}\right) - \left(-\frac{1}{\sqrt{2}} - \frac{1}{\sqrt{2}}\right)$

$= \frac{4}{\sqrt{2}} = 2\sqrt{2}$

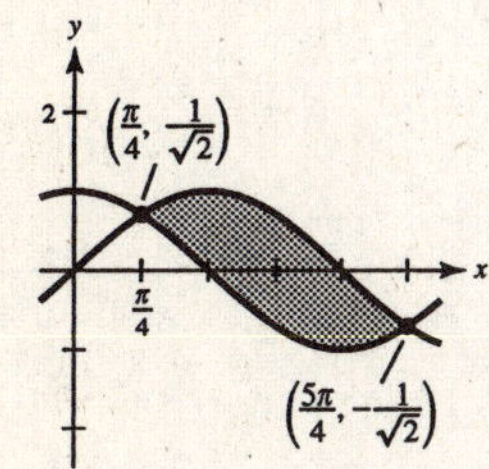

10. $A = \int_{\pi/3}^{5\pi/3} \left(\frac{1}{2} - \cos y\right) dy + \int_{5\pi/3}^{7\pi/3} \left(\cos y - \frac{1}{2}\right) dy$

$= \left[\frac{y}{2} - \sin y\right]_{\pi/3}^{5\pi/3} + \left[\sin y - \frac{y}{2}\right]_{5\pi/3}^{7\pi/3}$

$= \frac{\pi}{3} + 2\sqrt{3}$

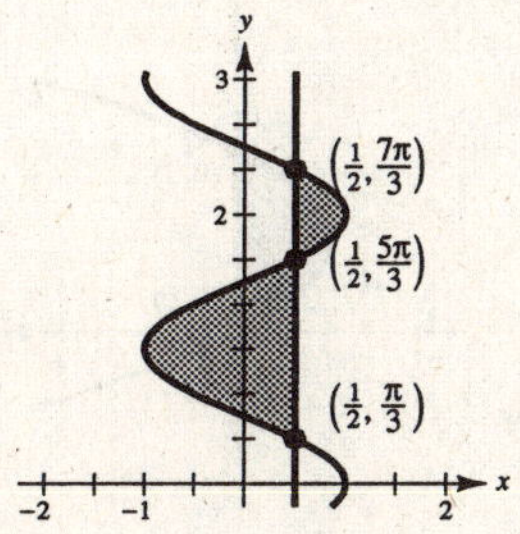

11. $A = \int_0^8 [(3 + 8x - x^2) - (x^2 - 8x + 3)]\, dx$

$= \int_0^8 (16x - 2x^2)\, dx$

$= \left[8x^2 - \frac{2}{3}x^3\right]_0^8 = \frac{512}{3} \approx 170.667$

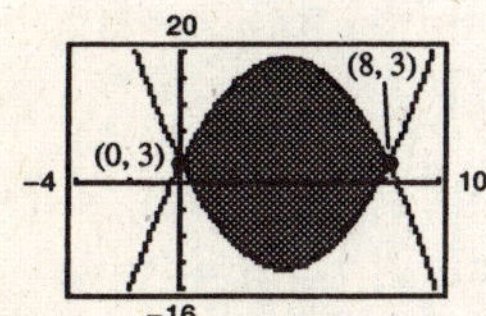

12. Point of intersection is given by:

$x^3 - x^2 + 4x - 3 = 0 \implies x \approx 0.783.$

$A \approx \int_0^{0.783} (3 - 4x + x^2 - x^3)\, dx$

$= \left[3x - 2x^2 + \frac{1}{3}x^3 - \frac{1}{4}x^4\right]_0^{0.783}$

≈ 1.189

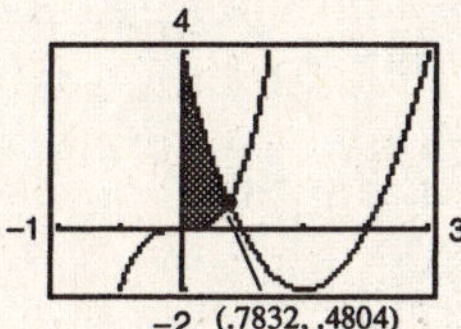

13. $y = (1 - \sqrt{x})^2$

$A = \int_0^1 (1 - \sqrt{x})^2\, dx$

$= \int_0^1 (1 - 2x^{1/2} + x)\, dx$

$= \left[x - \frac{4}{3}x^{3/2} + \frac{1}{2}x^2\right]_0^1 = \frac{1}{6} \approx 0.1667$

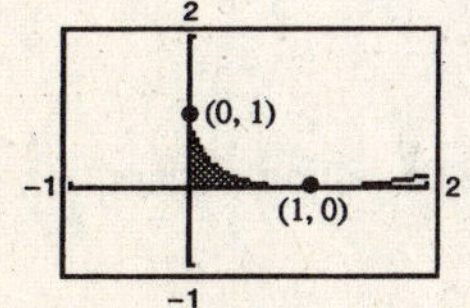

14. $A = 2\int_0^2 [2x^2 - (x^4 - 2x^2)]\, dx$

$= 2\int_0^2 (4x^2 - x^4)\, dx$

$= 2\left[\frac{4}{3}x^3 - \frac{1}{5}x^5\right]_0^2 = \frac{128}{15} \approx 8.5333$

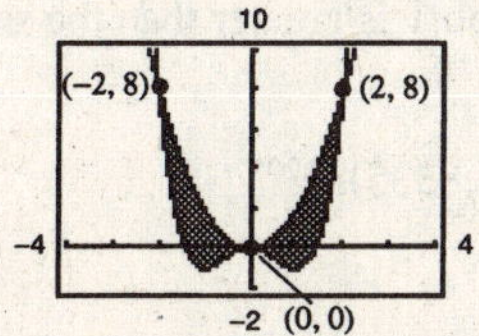

15. $x = y^2 - 2y \Rightarrow x + 1 = (y-1)^2 \Rightarrow y = 1 \pm \sqrt{x+1}$

$$A = \int_{-1}^{0} \left[\left(1 + \sqrt{x+1}\right) - \left(1 - \sqrt{x+1}\right)\right] dx$$

$$= \int_{-1}^{0} 2\sqrt{x+1}\, dx$$

$$A = \int_{0}^{2} [0 - (y^2 - 2y)]\, dy$$

$$= \int_{0}^{2} (2y - y^2)\, dy$$

$$= \left[y^2 - \frac{1}{3}y^3\right]_0^2$$

$$= \frac{4}{3}$$

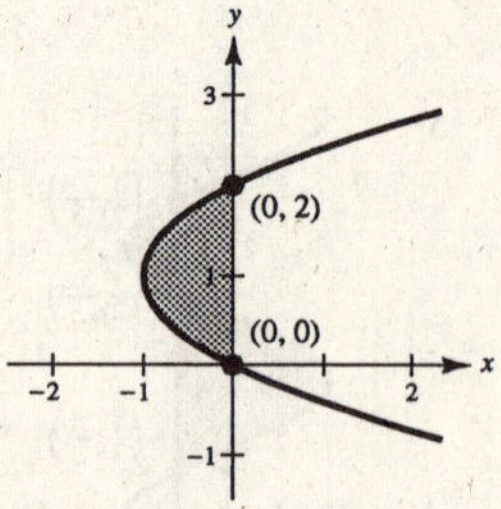

16. $y = \sqrt{x-1} \Rightarrow x = y^2 + 1$

$y = \dfrac{x-1}{2} \Rightarrow x = 2y + 1$

$$A = \int_{0}^{2} [(2y + 1) - (y^2 + 1)]\, dy$$

$$= \int_{1}^{5} \left[\sqrt{x-1} - \frac{x-1}{2}\right] dx$$

$$= \left[\frac{2}{3}(x-1)^{3/2} - \frac{1}{4}(x-1)^2\right]_1^5 = \frac{4}{3}$$

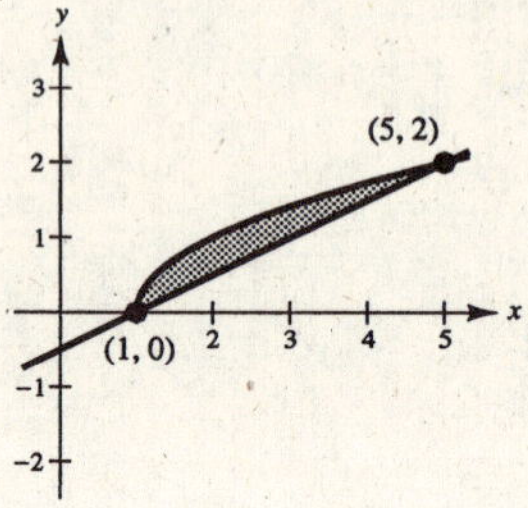

17. $$A = \int_{0}^{2} \left[1 - \left(1 - \frac{x}{2}\right)\right] dx + \int_{2}^{3} [1 - (x - 2)]\, dx$$

$$= \int_{0}^{2} \frac{x}{2}\, dx + \int_{2}^{3} (3 - x)\, dx$$

$y = 1 - \dfrac{x}{2} \Rightarrow x = 2 - 2y$

$y = x - 2 \Rightarrow x = y + 2,\ y = 1$

$$A = \int_{0}^{1} [(y + 2) - (2 - 2y)]\, dy$$

$$= \int_{0}^{1} 3y\, dy = \left[\frac{3}{2}y^2\right]_0^1 = \frac{3}{2}$$

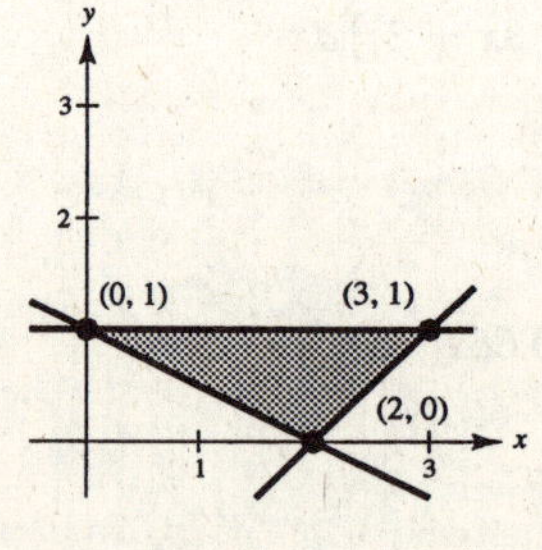

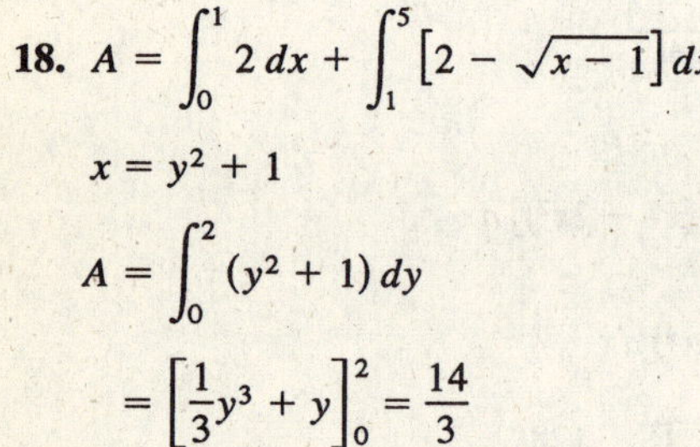

18. $$A = \int_{0}^{1} 2\, dx + \int_{1}^{5} \left[2 - \sqrt{x-1}\right] dx$$

$x = y^2 + 1$

$$A = \int_{0}^{2} (y^2 + 1)\, dy$$

$$= \left[\frac{1}{3}y^3 + y\right]_0^2 = \frac{14}{3}$$

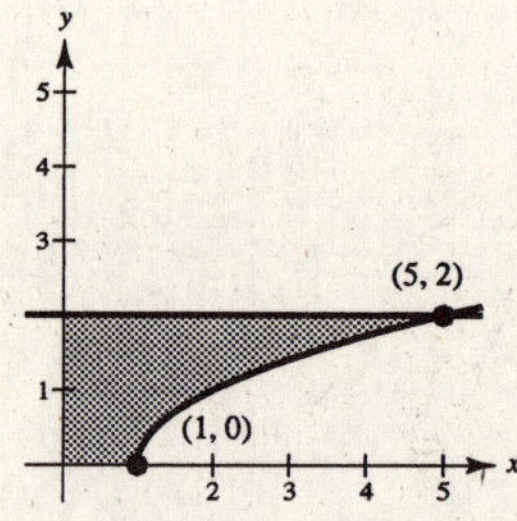

19. Job 1 is better. The salary for Job 1 is greater than the salary for Job 2 for all the years except the first and 10th years.

20. (a) $y = (6.8335)(1.2235)^t = (6.8335)e^{0.2017t}$

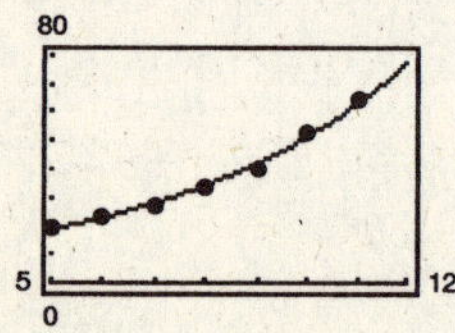

(b) $R_2 = 5 + 6.83e^{0.2t}$

Difference: $\displaystyle\int_{15}^{20} (R_2 - y)\, dt \approx 12.06$ billion dollars

21. (a) **Disk**

$$V = \pi \int_0^4 x^2\,dx = \left[\frac{\pi x^3}{3}\right]_0^4 = \frac{64\pi}{3}$$

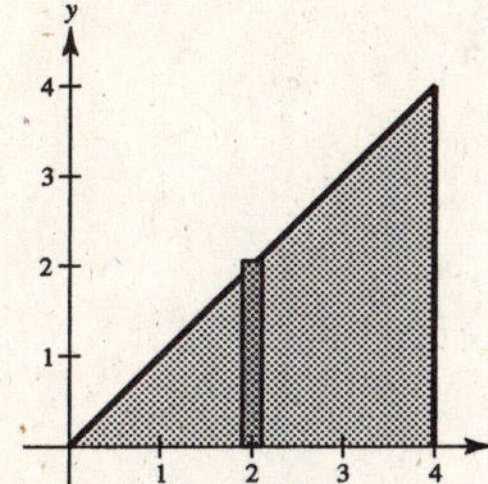

(b) **Shell**

$$V = 2\pi \int_0^4 x^2\,dx = \left[\frac{2\pi}{3}x^3\right]_0^4 = \frac{128\pi}{3}$$

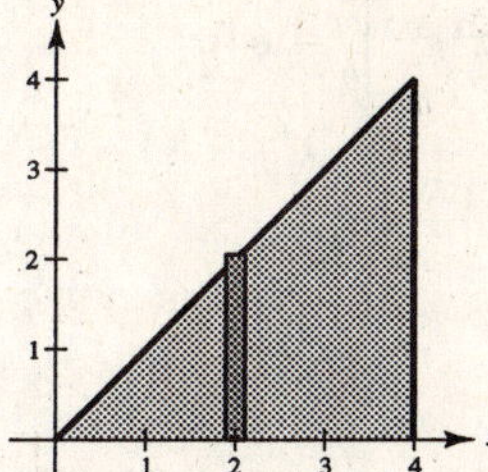

(c) **Shell**

$$V = 2\pi \int_0^4 (4 - x)x\,dx$$

$$= 2\pi \int_0^4 (4x - x^2)\,dx$$

$$= 2\pi\left[2x^2 - \frac{x^3}{3}\right]_0^4 = \frac{64\pi}{3}$$

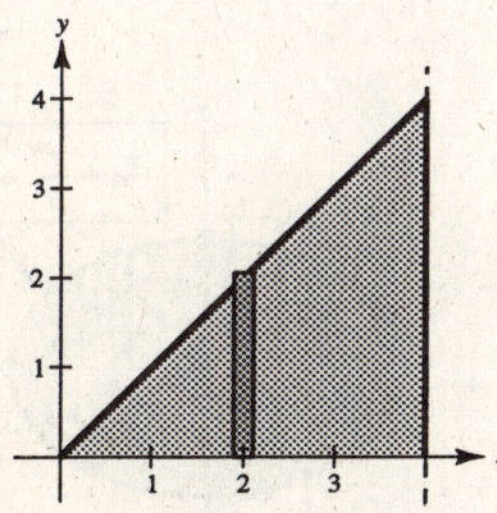

(d) **Shell**

$$V = 2\pi \int_0^4 (6 - x)x\,dx$$

$$= 2\pi \int_0^4 (6x - x^2)\,dx$$

$$= 2\pi\left[3x^2 - \frac{1}{3}x^3\right]_0^4 = \frac{160\pi}{3}$$

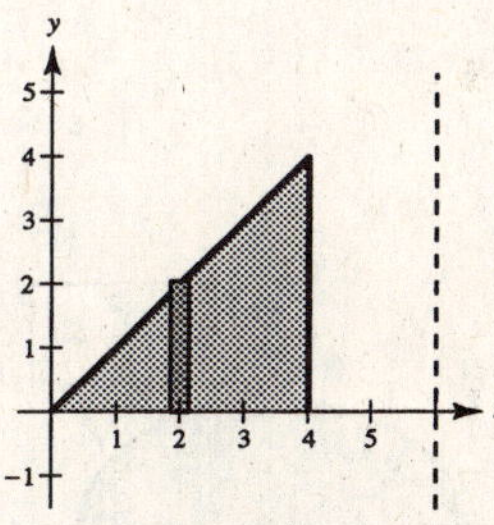

22. (a) **Shell**

$$V = 2\pi \int_0^2 y^3\,dy = \left[\frac{\pi}{2}y^4\right]_0^2 = 8\pi$$

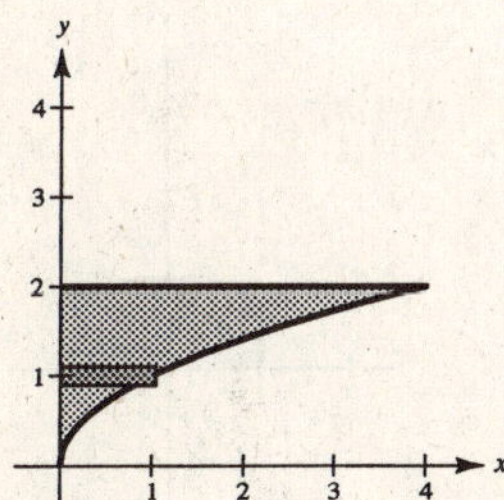

(b) **Shell**

$$V = 2\pi \int_0^2 (2 - y)y^2\,dy$$

$$= 2\pi \int_0^2 (2y^2 - y^3)\,dy$$

$$= 2\pi\left[\frac{2}{3}y^3 - \frac{1}{4}y^4\right]_0^2$$

$$= \frac{8\pi}{3}$$

(c) **Disk**

$$V = \pi \int_0^2 y^4\,dy = \left[\frac{\pi}{5}y^5\right]_0^2 = \frac{32\pi}{5}$$

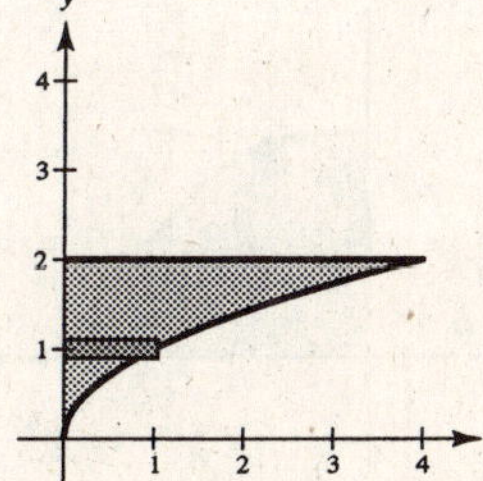

(d) **Disk**

$$V = \pi \int_0^2 [(y^2 + 1)^2 - 1^2]\,dy$$

$$= \pi \int_0^2 (y^4 + 2y^2)\,dy$$

$$= \pi\left[\frac{1}{5}y^5 + \frac{2}{3}y^3\right]_0^2$$

$$= \frac{176\pi}{15}$$

23. (a) **Shell**

$$V = 4\pi\int_0^4 x\left(\frac{3}{4}\right)\sqrt{16-x^2}\,dx$$

$$= \left[3\pi\left(-\frac{1}{2}\right)\left(\frac{2}{3}\right)(16-x^2)^{3/2}\right]_0^4 = 64\pi$$

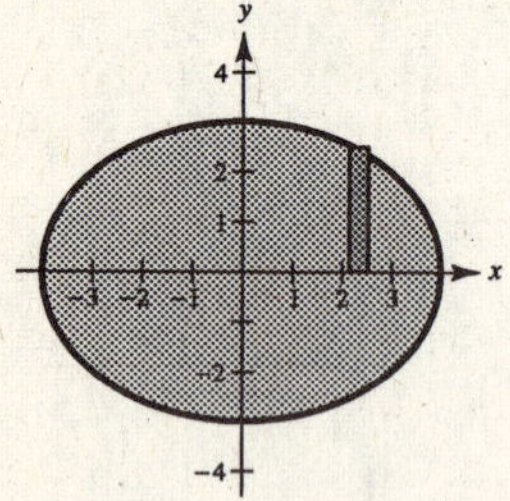

(b) **Disk**

$$V = 2\pi\int_0^4 \left[\frac{3}{4}\sqrt{16-x^2}\right]^2 dx$$

$$= \frac{9\pi}{8}\left[16x - \frac{x^3}{3}\right]_0^4 = 48\pi$$

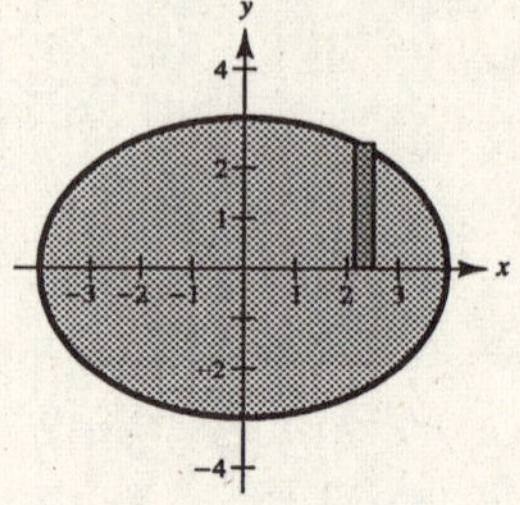

24. (a) **Shell**

$$V = 4\pi\int_0^a (x)\frac{b}{a}\sqrt{a^2-x^2}\,dx$$

$$= \frac{-2\pi b}{a}\int_0^a (a^2-x^2)^{1/2}(-2x)\,dx$$

$$= \left[\frac{-4\pi b}{3a}(a^2-x^2)^{3/2}\right]_0^a$$

$$= \frac{4}{3}\pi a^2 b$$

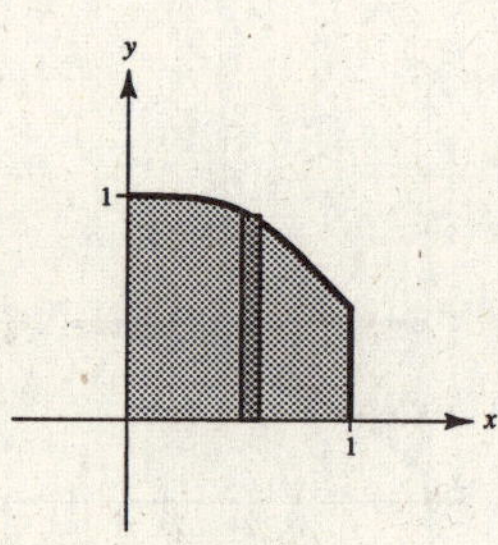

(b) **Disk**

$$V = 2\pi\int_0^a \frac{b^2}{a^2}(a^2-x^2)\,dx$$

$$= \frac{2\pi b^2}{a^2}\left[a^2x - \frac{1}{3}x^3\right]_0^a$$

$$= \frac{4}{3}\pi ab^2$$

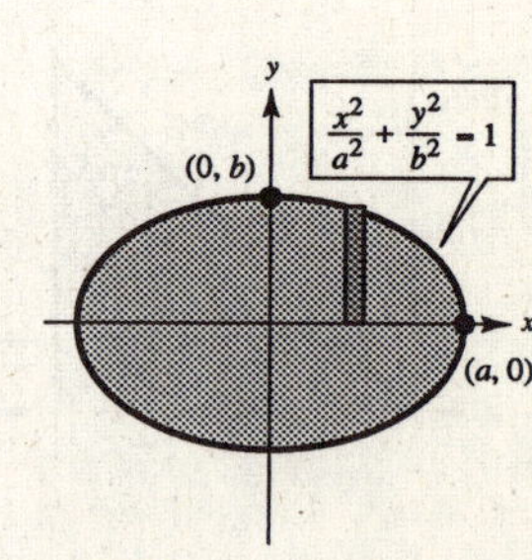

25. Shell

$$V = 2\pi\int_0^1 \frac{x}{x^4+1}\,dx$$

$$= \pi\int_0^1 \frac{(2x)}{(x^2)^2+1}\,dx$$

$$= \left[\pi\arctan(x^2)\right]_0^1$$

$$= \pi\left[\frac{\pi}{4} - 0\right] = \frac{\pi^2}{4}$$

26. Disk

$$V = 2\pi\int_0^1 \left[\frac{1}{\sqrt{1+x^2}}\right]^2 dx$$

$$= \left[2\pi\arctan x\right]_0^1$$

$$= 2\pi\left(\frac{\pi}{4} - 0\right)$$

$$= \frac{\pi^2}{2}$$

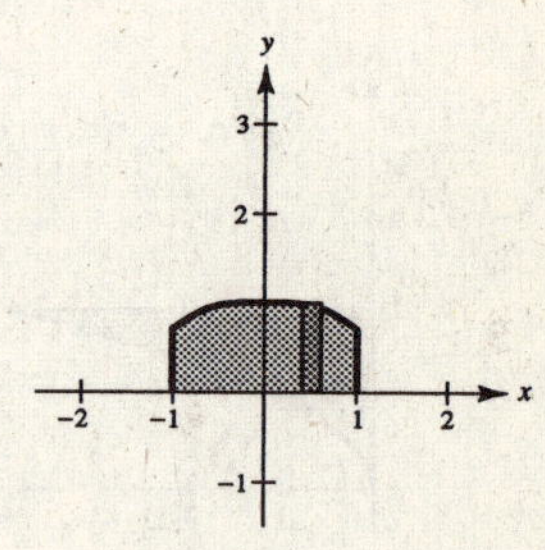

27. Shell: $V = 2\pi\int_2^6 \frac{x}{1+\sqrt{x-2}}\,dx$

$u = \sqrt{x-2}$

$x = u^2 + 2$

$dx = 2u\,du$

$$V = 2\pi\int_2^6 \frac{x}{1+\sqrt{x-2}}\,dx = 4\pi\int_0^2 \frac{(u^2+2)u}{1+u}\,du$$

$$= 4\pi\int_0^2 \frac{u^3+2u}{1+u}\,du = 4\pi\int_0^2\left(u^2 - u + 3 - \frac{3}{1+u}\right)du$$

$$= 4\pi\left[\frac{1}{3}u^3 - \frac{1}{2}u^2 + 3u - 3\ln(1+u)\right]_0^2 = \frac{4\pi}{3}(20 - 9\ln 3) \approx 42.359$$

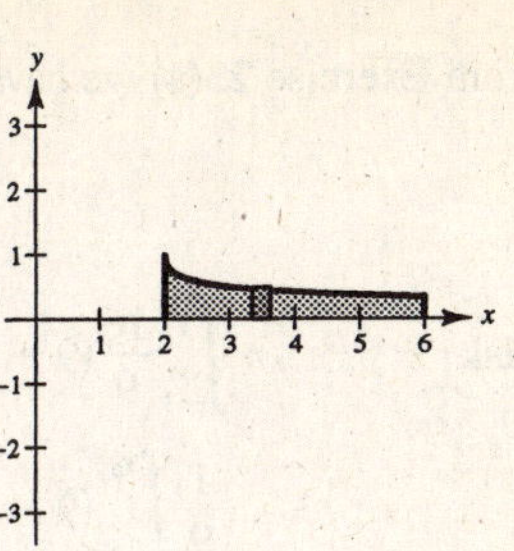

28. Disk

$$V = \pi\int_0^1 (e^{-x})^2\,dx$$

$$= \pi\int_0^1 e^{-2x}\,dx = \left[-\frac{\pi}{2}e^{-2x}\right]_0^1$$

$$= \left(\frac{-\pi}{2e^2} + \frac{\pi}{2}\right) = \frac{\pi}{2}\left(1 - \frac{1}{e^2}\right)$$

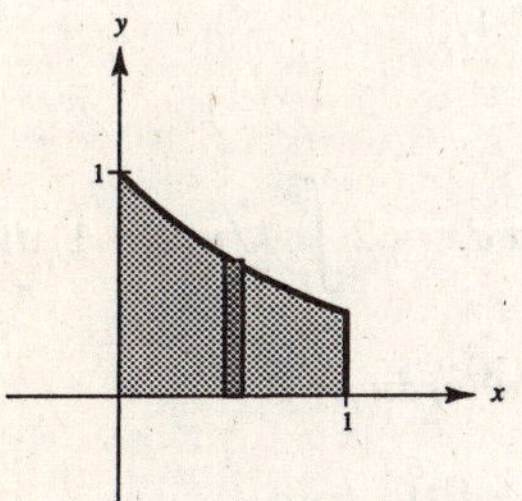

29. Since $y \le 0$, $A = -\int_{-1}^0 x\sqrt{x+1}\,dx$.

$u = x + 1$

$x = u - 1$

$dx = du$

$$A = -\int_0^1 (u-1)\sqrt{u}\,du = -\int_0^1 (u^{3/2} - u^{1/2})\,du$$

$$= -\left[\frac{2}{5}u^{5/2} - \frac{2}{3}u^{3/2}\right]_0^1 = \frac{4}{15}$$

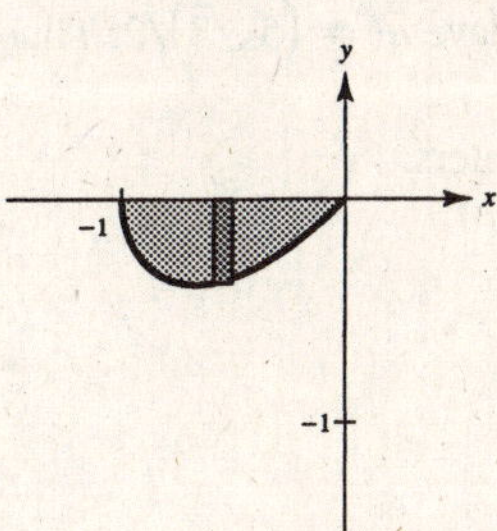

30. (a) **Disk**

$$V = \pi\int_{-1}^0 x^2(x+1)\,dx$$

$$= \pi\int_{-1}^0 (x^3 + x^2)\,dx$$

$$= \pi\left[\frac{x^4}{4} + \frac{x^3}{3}\right]_{-1}^0 = \frac{\pi}{12}$$

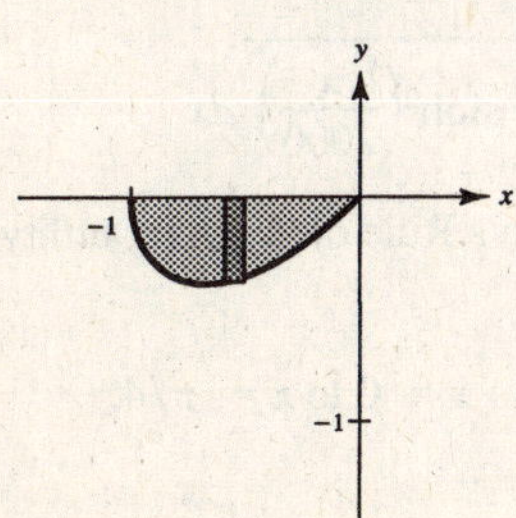

(b) **Shell**

$u = \sqrt{x+1}$

$x = u^2 - 1$

$dx = 2u\,du$

$$V = 2\pi\int_{-1}^0 x^2\sqrt{x+1}\,dx$$

$$= 4\pi\int_0^1 (u^2-1)^2u^2\,du$$

$$= 4\pi\int_0^1 (u^6 - 2u^4 + u^2)\,du$$

$$= 4\pi\left[\frac{1}{7}u^7 - \frac{2}{5}u^5 + \frac{1}{3}u^3\right]_0^1 = \frac{32\pi}{105}$$

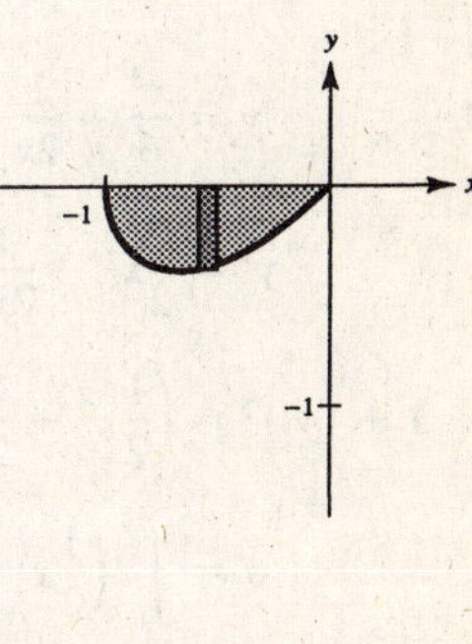

31. From Exercise 23(a) we have: $V = 64\pi \text{ ft}^3$

$$\frac{1}{4}V = 16\pi$$

Disk: $$\pi\int_{-3}^{y_0}\frac{16}{9}(9-y^2)\,dy = 16\pi$$

$$\frac{1}{9}\int_{-3}^{y_0}(9-y^2)\,dy = 1$$

$$\left[9y-\frac{1}{3}y^3\right]_{-3}^{y_0} = 9$$

$$\left(9y_0-\frac{1}{3}y_0^3\right)-(-27+9) = 9$$

$$y_0^3-27y_0-27 = 0$$

By Newton's Method, $y_0 \approx -1.042$ and the depth of the gasoline is $3 - 1.042 = 1.958$ ft.

32. $$A(x) = \frac{1}{2}bh = \frac{1}{2}\left(2\sqrt{a^2-x^2}\right)\left(\sqrt{3}\sqrt{a^2-x^2}\right)$$

$$= \sqrt{3}(a^2-x^2)$$

$$V = \sqrt{3}\int_{-a}^{a}(a^2-x^2)\,dx = \sqrt{3}\left[a^2x-\frac{x^3}{3}\right]_{-a}^{a}$$

$$= \sqrt{3}\left(\frac{4a^3}{3}\right)$$

Since $\left(4\sqrt{3}\,a^3\right)/3 = 10$, we have $a^3 = \left(5\sqrt{3}\right)/2$. Thus,

$$a = \sqrt[3]{\frac{5\sqrt{3}}{2}} \approx 1.630 \text{ meters.}$$

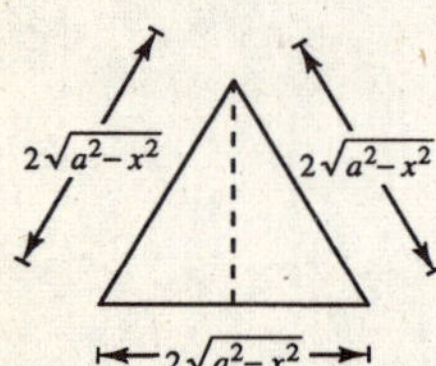

33. $$f(x) = \frac{4}{5}x^{5/4}$$

$$f'(x) = x^{1/4}$$

$$1+[f'(x)]^2 = 1+\sqrt{x}$$

$$u = 1+\sqrt{x}$$

$$x = (u-1)^2$$

$$dx = 2(u-1)\,du$$

$$s = \int_0^4\sqrt{1+\sqrt{x}}\,dx = 2\int_1^3\sqrt{u}(u-1)\,du$$

$$= 2\int_1^3\left(u^{3/2}-u^{1/2}\right)du$$

$$= 2\left[\frac{2}{5}u^{5/2}-\frac{2}{3}u^{3/2}\right]_1^3 = \frac{4}{15}\left[u^{3/2}(3u-5)\right]_1^3$$

$$= \frac{8}{15}\left(1+6\sqrt{3}\right) \approx 6.076$$

34. $$y = \frac{x^3}{6}+\frac{1}{2x}$$

$$y' = \frac{1}{2}x^2-\frac{1}{2x^2}$$

$$1+(y')^2 = \left(\frac{1}{2}x^2+\frac{1}{2x^2}\right)^2$$

$$s = \int_1^3\left(\frac{1}{2}x^2+\frac{1}{2x^2}\right)dx = \left[\frac{1}{6}x^3-\frac{1}{2x}\right]_1^3 = \frac{14}{3}$$

35. $$y = 300\cosh\left(\frac{x}{2000}\right)-280, \; -2000 \le x \le 2000$$

$$y' = \frac{3}{20}\sinh\left(\frac{x}{2000}\right)$$

$$s = \int_{-2000}^{2000}\sqrt{1+\left[\frac{3}{20}\sinh\left(\frac{x}{2000}\right)\right]^2}\,dx$$

$$= \frac{1}{20}\int_{-2000}^{2000}\sqrt{400+9\sinh^2\left(\frac{x}{2000}\right)}\,dx$$

≈ 4018.2 ft (by Simpson's Rule or graphing utility)

36. Since $f(x) = \tan x$ has $f'(x) = \sec^2 x$, this integral represents the length of the graph of $\tan x$ from $x = 0$ to $x = \pi/4$. This length is a little over 1 unit. Answers (b).

37. $y = \frac{3}{4}x$

$$y' = \frac{3}{4}$$

$$1 + (y')^2 = \frac{25}{16}$$

$$S = 2\pi \int_0^4 \left(\frac{3}{4}x\right)\sqrt{\frac{25}{16}}\,dx = \left[\left(\frac{15\pi}{8}\right)\frac{x^2}{2}\right]_0^4 = 15\pi$$

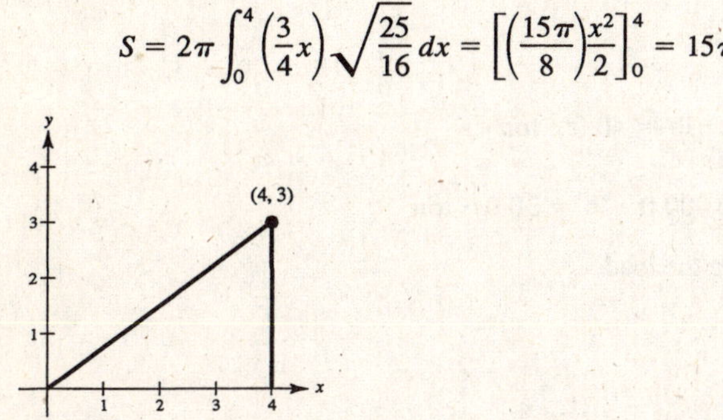

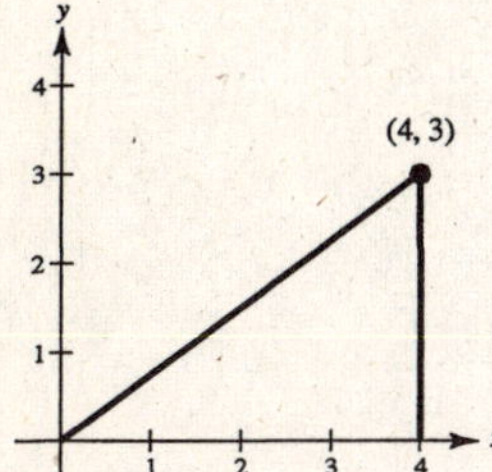

38. $y = 2\sqrt{x}$

$$y' = \frac{1}{\sqrt{x}}$$

$$1 + (y')^2 = 1 + \frac{1}{x} = \frac{x+1}{x}$$

$$S = 2\pi \int_0^3 2\sqrt{x}\sqrt{\frac{x+1}{x}}\,dx = 4\pi \int_0^3 \sqrt{x+1}\,dx$$

$$= 4\pi\left[\left(\frac{2}{3}\right)(x+1)^{3/2}\right]_0^3 = \frac{56\pi}{3}$$

39. $F = kx$

$$4 = k(1)$$

$$F = 4x$$

$$W = \int_0^5 4x\,dx = \left[2x^2\right]_0^5$$

$$= 50 \text{ in.} \cdot \text{lb} \approx 4.167 \text{ ft} \cdot \text{lb}$$

40. $F = kx$

$$50 = k(9) \Rightarrow k = \frac{50}{9}$$

$$F = \frac{50}{9}x$$

$$W = \int_0^9 \frac{50}{9}x\,dx = \left[\frac{25}{9}x^2\right]_0^9$$

$$= 225 \text{ in.} \cdot \text{lb} = 18.75 \text{ ft} \cdot \text{lb}$$

41. Volume of disk: $\pi\left(\frac{1}{3}\right)^2 \Delta y$

Weight of disk: $62.4\pi\left(\frac{1}{3}\right)^2 \Delta y$

Distance: $175 - y$

$$W = \frac{62.4\pi}{9}\int_0^{150} (175 - y)\,dy = \frac{62.4\pi}{9}\left[175y - \frac{y^2}{2}\right]_0^{150}$$

$$= 104{,}000\pi \text{ ft} \cdot \text{lb} \approx 163.4 \text{ ft} \cdot \text{ton}$$

42. We know that

$$\frac{dV}{dt} = \frac{4 \text{ gal/min} - 12 \text{ gal/min}}{7.481 \text{ gal/ft}^3} = -\frac{8}{7.481} \text{ ft}^3\text{/min}$$

$$V = \pi r^2 h = \pi\left(\frac{1}{9}\right)h$$

$$\frac{dV}{dt} = \frac{\pi}{9}\left(\frac{dh}{dt}\right)$$

$$\frac{dh}{dt} = \frac{9}{\pi}\left(\frac{dV}{dt}\right) = \frac{9}{\pi}\left(-\frac{8}{7.481}\right) \approx -3.064 \text{ ft/min.}$$

Depth of water: $-3.064t + 150$

Time to drain well: $t = \frac{150}{3.064} \approx 49$ minutes

$(49)(12) = 588$ gallons pumped

Volume of water pumped in Exercise 41: 391.7 gallons

$$\frac{391.7}{52\pi} = \frac{588}{x\pi}$$

$$x = \frac{588(52)}{391.7} \approx 78$$

Work $\approx 78\pi$ ft · ton

43. Weight of section of chain: $5\,\Delta x$

Distance moved: $10 - x$

$$W = 5\int_0^{10} (10 - x)\,dx = \left[-\frac{5}{2}(10 - x)^2\right]_0^{10} = 250 \text{ ft} \cdot \text{lb}$$

44. (a) Weight of section of cable: $4\,\Delta x$

Distance: $200 - x$

$$W = 4\int_0^{200} (200 - x)\,dx = \left[-2(200 - x)^2\right]_0^{200} = 80{,}000 \text{ ft} \cdot \text{lb} = 40 \text{ ft} \cdot \text{ton}$$

(b) Work to move 300 pounds 200 feet vertically: $200(300) = 60{,}000 \text{ ft} \cdot \text{lb} = 30 \text{ ft} \cdot \text{ton}$

Total work = work for drawing up the cable + work of lifting the load

$= 40 \text{ ft} \cdot \text{ton} + 30 \text{ ft} \cdot \text{ton} = 70 \text{ ft} \cdot \text{ton}$

45. $W = \displaystyle\int_a^b F(x)\,dx$

$$80 = \int_0^4 ax^2\,dx = \frac{ax^3}{3}\Big]_0^4 = \frac{64}{3}a$$

$$a = \frac{3(80)}{64} = \frac{15}{4} = 3.75$$

46. $W = \displaystyle\int_a^b F(x)\,dx$

$$F(x) = \begin{cases} -(2/9)x + 6, & 0 \le x \le 9 \\ -(4/3)x + 16, & 9 \le x \le 12 \end{cases}$$

$$W = \int_0^9 \left(-\frac{2}{9}x + 6\right)dx + \int_9^{12}\left(-\frac{4}{3}x + 16\right)dx$$

$$= \left[-\frac{1}{9}x^2 + 6x\right]_0^9 + \left[-\frac{2}{3}x^2 + 16x\right]_9^{12}$$

$$= (-9 + 54) + (-96 + 192 + 54 - 144)$$

$$= 51 \text{ ft} \cdot \text{lbs}$$

47. $A = \displaystyle\int_0^a (\sqrt{a} - \sqrt{x})^2\,dx = \int_0^a \left(a - 2\sqrt{a}x^{1/2} + x\right)dx = \left[ax - \frac{4}{3}\sqrt{a}x^{3/2} + \frac{1}{2}x^2\right]_0^a = \frac{a^2}{6}$

$$\frac{1}{A} = \frac{6}{a^2}$$

$$\bar{x} = \frac{6}{a^2}\int_0^a x(\sqrt{a} - \sqrt{x})^2\,dx = \frac{6}{a^2}\int_0^a \left(ax - 2\sqrt{a}x^{3/2} + x^2\right)dx = \frac{a}{5}$$

$$\bar{y} = \left(\frac{6}{a^2}\right)\frac{1}{2}\int_0^a (\sqrt{a} - \sqrt{x})^4\,dx$$

$$= \frac{3}{a^2}\int_0^a \left(a^2 - 4a^{3/2}x^{1/2} + 6ax - 4a^{1/2}x^{3/2} + x^2\right)dx$$

$$= \frac{3}{a^2}\left[a^2x - \frac{8}{3}a^{3/2}x^{3/2} + 3ax^2 - \frac{8}{5}a^{1/2}x^{5/2} + \frac{1}{3}x^3\right]_0^a = \frac{a}{5}$$

$$(\bar{x}, \bar{y}) = \left(\frac{a}{5}, \frac{a}{5}\right)$$

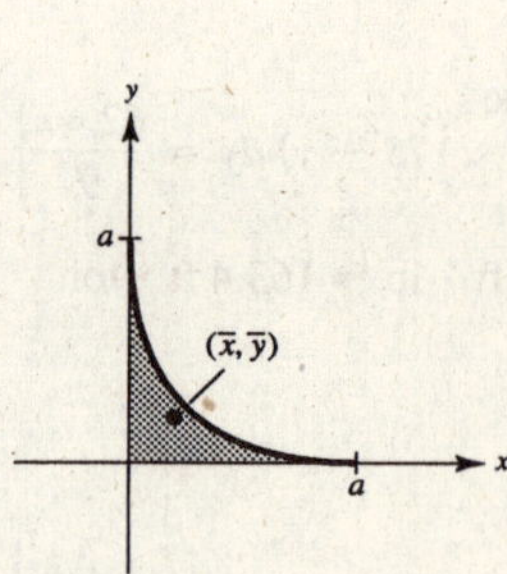

48. $A = \int_{-1}^{3} [(2x + 3) - x^2]\, dx = \left[x^2 + 3x - \frac{1}{3}x^3\right]_{-1}^{3} = \frac{32}{3}$

$$\frac{1}{A} = \frac{3}{32}$$

$$\bar{x} = \frac{3}{32}\int_{-1}^{3} x(2x + 3 - x^2)\, dx = \frac{3}{32}\int_{-1}^{3} (3x + 2x^2 - x^3)\, dx = \frac{3}{32}\left[\frac{3}{2}x^2 + \frac{2}{3}x^3 - \frac{1}{4}x^4\right]_{-1}^{3} = 1$$

$$\bar{y} = \left(\frac{3}{32}\right)\frac{1}{2}\int_{-1}^{3} [(2x + 3)^2 - x^4]\, dx = \frac{3}{64}\int_{-1}^{3} (9 + 12x + 4x^2 - x^4)\, dx$$

$$= \frac{3}{64}\left[9x + 6x^2 + \frac{4}{3}x^3 - \frac{1}{5}x^5\right]_{-1}^{3} = \frac{17}{5}$$

$$(\bar{x}, \bar{y}) = \left(1, \frac{17}{5}\right)$$

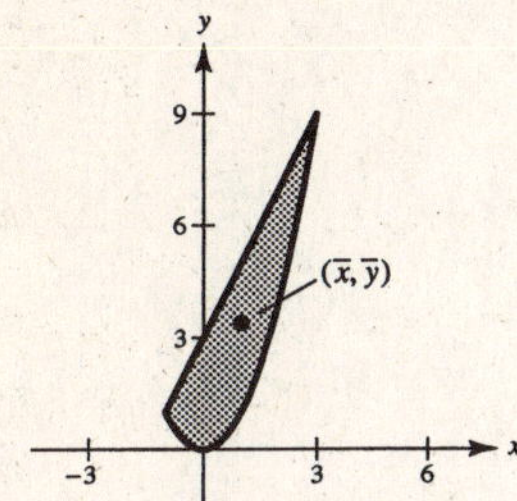

49. By symmetry, $x = 0$.

$$A = 2\int_0^1 (a^2 - x^2)\, dx = 2\left[a^2x - \frac{x^3}{3}\right]_0^a = \frac{4a^3}{3}$$

$$\frac{1}{A} = \frac{3}{4a^3}$$

$$\bar{y} = \left(\frac{3}{4a^3}\right)\frac{1}{2}\int_{-a}^{a} (a^2 - x^2)^2\, dx$$

$$= \frac{6}{8a^3}\int_0^a (a^4 - 2a^2x^2 + x^4)\, dx$$

$$= \frac{6}{8a^3}\left[a^4x - \frac{2a^2}{3}x^3 + \frac{1}{5}x^5\right]_0^a$$

$$= \frac{6}{8a^3}\left(a^5 - \frac{2}{3}a^5 + \frac{1}{5}a^5\right) = \frac{2a^2}{5}$$

$$(\bar{x}, \bar{y}) = \left(0, \frac{2a^2}{5}\right)$$

50. $A = \int_0^8 \left(x^{2/3} - \frac{1}{2}x\right) dx = \left[\frac{3}{5}x^{5/3} - \frac{1}{4}x^2\right]_0^8 = \frac{16}{5}$

$$\frac{1}{A} = \frac{5}{16}$$

$$\bar{x} = \frac{5}{16}\int_0^8 x\left(x^{2/3} - \frac{1}{2}x\right) dx$$

$$= \frac{5}{16}\left[\frac{3}{8}x^{8/3} - \frac{1}{6}x^3\right]_0^8 = \frac{10}{3}$$

$$\bar{y} = \left(\frac{5}{16}\right)\frac{1}{2}\int_0^8 \left(x^{4/3} - \frac{1}{4}x^2\right) dx$$

$$= \frac{1}{2}\left(\frac{5}{16}\right)\left[\frac{3}{7}x^{7/3} - \frac{1}{12}x^3\right]_0^8 = \frac{40}{21}$$

$$(\bar{x}, \bar{y}) = \left(\frac{10}{3}, \frac{40}{21}\right)$$

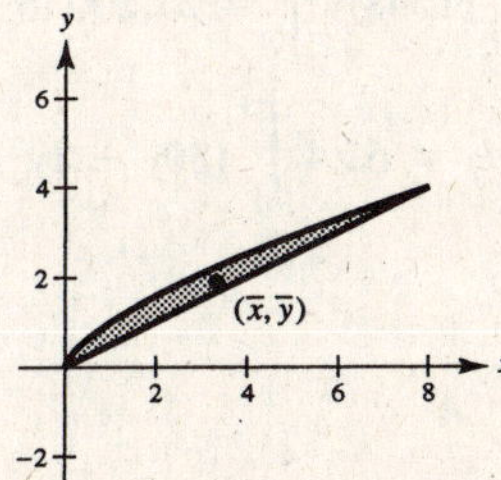

51. $\bar{y} = 0$ by symmetry.

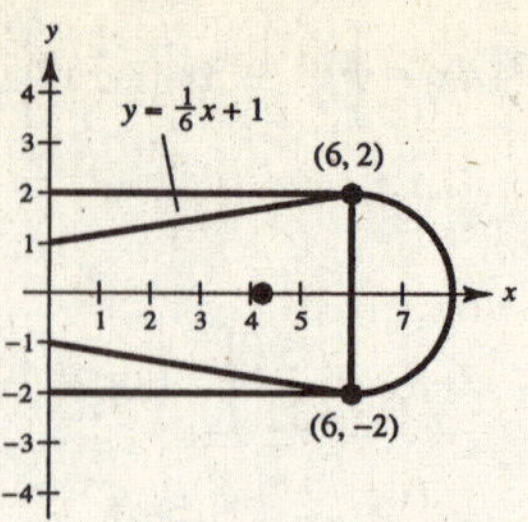

For the trapezoid:

$$m = [(4)(6) - (1)(6)]\rho = 18\rho$$

$$M_y = \rho\int_0^6 x\left[\left(\frac{1}{6}x + 1\right) - \left(-\frac{1}{6}x - 1\right)\right]dx$$

$$= \rho\int_0^6 \left(\frac{1}{3}x^2 + 2x\right)dx = \rho\left[\frac{x^3}{9} + x^2\right]_0^6 = 60\rho$$

For the semicircle:

$$m = \left(\frac{1}{2}\right)(\pi)(2)^2\rho = 2\pi\rho$$

$$M_y = \rho\int_6^8 x\left[\sqrt{4-(x-6)^2} - \left(-\sqrt{4-(x-6)^2}\right)\right]dx = 2\rho\int_6^8 x\sqrt{4-(x-6)^2}\,dx$$

Let $u = x - 6$, then $x = u + 6$ and $dx = du$. When $x = 6$, $u = 0$. When $x = 8$, $u = 2$.

$$M_y = 2\rho\int_0^2 (u+6)\sqrt{4-u^2}\,du = 2\rho\int_0^2 u\sqrt{4-u^2}\,du + 12\rho\int_0^2 \sqrt{4-u^2}\,du$$

$$= 2\rho\left[\left(-\frac{1}{2}\right)\left(\frac{2}{3}\right)(4-u^2)^{3/2}\right]_0^2 + 12\rho\left[\frac{\pi(2)^2}{4}\right] = \frac{16\rho}{3} + 12\pi\rho = \frac{4\rho(4+9\pi)}{3}$$

Thus, we have:

$$\bar{x}(18\rho + 2\pi\rho) = 60\rho + \frac{4\rho(4+9\pi)}{3}$$

$$\bar{x} = \frac{180\rho + 4\rho(4+9\pi)}{3}\cdot\frac{1}{2\rho(9+\pi)} = \frac{2(9\pi+49)}{3(\pi+9)}$$

The centroid of the blade is $\left(\frac{2(9\pi+49)}{3(\pi+9)}, 0\right)$.

52. Wall at shallow end:

$$F = 62.4\int_0^5 y(20)\,dy = \left[(1248)\frac{y^2}{2}\right]_0^5 = 15{,}600 \text{ lb}$$

Wall at deep end:

$$F = 62.4\int_0^{10} y(20)\,dy = \left[(624)y^2\right]_0^{10} = 62{,}400 \text{ lb}$$

Side wall:

$$F_1 = 62.4\int_0^5 y(40)\,dy = \left[(1248)y^2\right]_0^5 = 31{,}200 \text{ lb}$$

$$F_2 = 62.4\int_0^5 (10-y)8y\,dy = 62.4\int_0^5 (80y - 8y^2)\,dy$$

$$F = F_1 + F_2 = 72{,}800 \text{ lb}$$

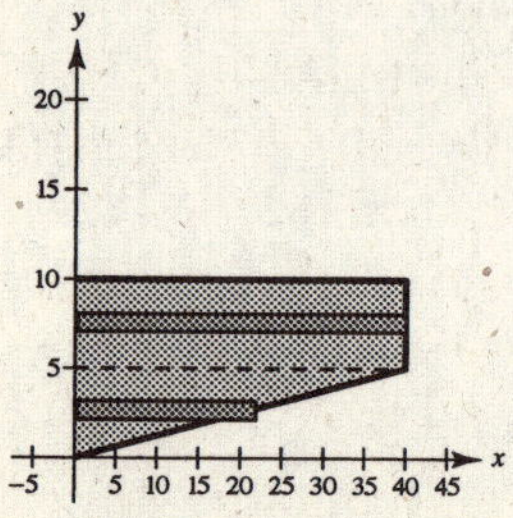

53. Let D = surface of liquid; ρ = weight per cubic volume.

$$F = \rho\int_c^d (D-y)[f(y) - g(y)]\,dy$$

$$= \rho\left[\int_c^d D[f(y)-g(y)]\,dy - \int_c^d y[f(y)-g(y)]\,dy\right]$$

$$= \rho\left[\int_c^d [f(y)-g(y)]\,dy\right]\left[D - \frac{\int_c^d y[f(y)-g(y)]\,dy}{\int_c^d [f(y)-g(y)]\,dy}\right]$$

$$= \rho(\text{Area})(D - \bar{y})$$

$$= \rho(\text{Area})(\text{depth of centroid})$$

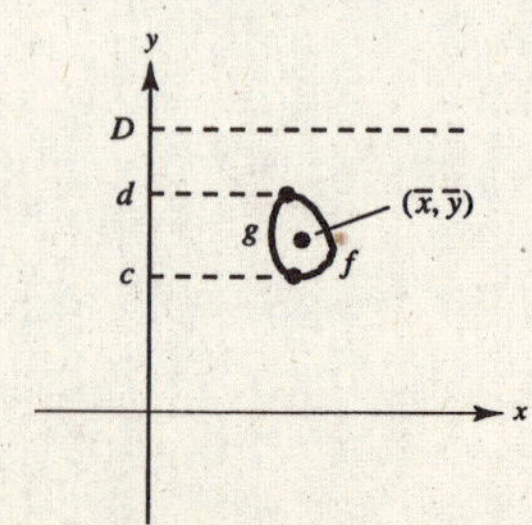

54. $F = 62.4(16\pi)5 = 4992\pi$ lb

Problem Solving for Chapter 7

1. $T = \frac{1}{2}c(c^2) = \frac{1}{2}c^3$

$$R = \int_0^c (cx - x^2)\,dx = \left[\frac{cx^2}{2} - \frac{x^3}{3}\right]_0^c = \frac{c^3}{2} - \frac{c^3}{3} = \frac{c^3}{6}$$

$$\lim_{c\to 0^+} \frac{T}{R} = \lim_{c\to 0^+} \frac{\frac{1}{2}c^3}{\frac{1}{6}c^3} = 3$$

2. $R = \int_0^1 x(1 - x)\,dx = \left[\frac{x^2}{2} - \frac{x^3}{3}\right]_0^1 = \frac{1}{2} - \frac{1}{3} = \frac{1}{6}$

Let (c, mc) be the intersection of the line and the parabola.

Then, $mc = c(1 - c) \Rightarrow m = 1 - c$ or $c = 1 - m$.

$$\frac{1}{2}\left(\frac{1}{6}\right) = \int_0^{1-m} (x - x^2 - mx)\,dx$$

$$\frac{1}{12} = \left[\frac{x^2}{2} - \frac{x^3}{3} - m\frac{x^2}{2}\right]_0^{1-m}$$

$$= \frac{(1 - m)^2}{2} - \frac{(1 - m)^3}{3} - m\frac{(1 - m)^2}{2}$$

$$1 = 6(1 - m)^2 - 4(1 - m)^3 - 6m(1 - m)^2$$

$$= (1 - m)^2(6 - 4(1 - m) - 6m)$$

$$= (1 - m)^2(2 - 2m)$$

$$\frac{1}{2} = (1 - m)^3$$

$$\left(\frac{1}{2}\right)^{1/3} = 1 - m$$

$$m = 1 - \left(\frac{1}{2}\right)^{1/3} \approx 0.2063$$

3. (a) $\frac{1}{2}V = \int_0^1 \left[\pi\left(2 + \sqrt{1 - y^2}\right)^2 - \pi\left(2 - \sqrt{1 - y^2}\right)^2\right] dy$

$$= \pi\int_0^1 \left[\left(4 + 4\sqrt{1 - y^2} + (1 - y^2)\right) - \left(4 - 4\sqrt{1 - y^2} + (1 - y^2)\right)\right] dy$$

$= 8\pi\int_0^1 \sqrt{1 - y^2}\,dy$ (Integral represents 1/4 (area of circle))

$$= 8\pi\left(\frac{\pi}{4}\right) = 2\pi^2 \Rightarrow V = 4\pi^2$$

(b) $(x - R)^2 + y^2 = r^2 \Rightarrow x = R \pm \sqrt{r^2 - y^2}$

$$\frac{1}{2}V = \int_0^r \left[\pi\left(R + \sqrt{r^2 - y^2}\right)^2 - \pi\left(R - \sqrt{r^2 - y^2}\right)^2\right] dy$$

$$= \pi\int_0^r 4R\sqrt{r^2 - y^2}\,dy$$

$$= \pi(4R)\frac{1}{4}\pi r^2 = \pi^2 r^2 R$$

$$V = 2\pi^2 r^2 R$$

4. $8y^2 = x^2(1 - x^2)$

$$y = \pm\frac{|x|\sqrt{1-x^2}}{2\sqrt{2}}$$

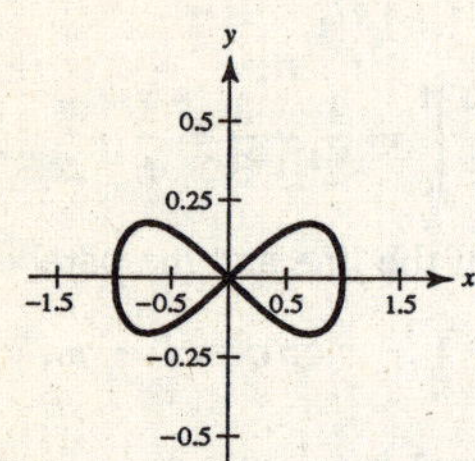

For $x > 0$, $y' = \dfrac{1 - 2x^2}{2\sqrt{2}\sqrt{1-x^2}}$

$$S = 2(2\pi)\int_0^1 x\sqrt{1 + \left(\frac{1-2x^2}{2\sqrt{2}\sqrt{1-x^2}}\right)^2}\,dx$$

$$= \frac{5\sqrt{2}\pi}{3}$$

5. $V = 2(2\pi)\displaystyle\int_{\sqrt{r^2-(h^2/4)}}^{r} x\sqrt{r^2 - x^2}\,dx$

$$= -2\pi\left[\frac{2}{3}(r^2 - x^2)^{3/2}\right]_{\sqrt{r^2-(h^2/4)}}^{r}$$

$$= \frac{-4\pi}{3}\left[-\frac{h^3}{8}\right] = \frac{\pi h^3}{6} \text{ which does not depend on } r!$$

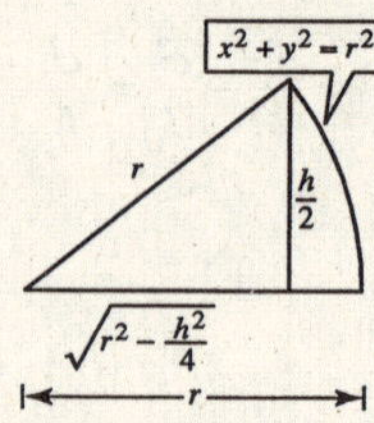

6. By the Theorem of Pappus,

$$V = 2\pi r A$$

$$= 2\pi\left[d + \frac{1}{2}\sqrt{w^2 + l^2}\right]lw$$

7. (a) Tangent at A: $y = x^3$, $y' = 3x^2$

$$y - 1 = 3(x - 1)$$

$$y = 3x - 2$$

To find point B: $x^3 = 3x - 2$

$$x^3 - 3x + 2 = 0$$

$$(x - 1)^2(x + 2) = 0 \Longrightarrow B = (-2, -8)$$

Tangent at B: $y = x^3$, $y' = 3x^2$

$$y + 8 = 12(x + 2)$$

$$y = 12x + 16$$

To find point C: $x^3 = 12x + 16$

$$x^3 - 12x - 16 = 0$$

$$(x + 2)^2(x - 4) = 0 \Longrightarrow C = (4, 64)$$

Area of $R = \displaystyle\int_{-2}^{1} (x^3 - 3x + 2)\,dx = \frac{27}{4}$

Area of $S = \displaystyle\int_{-2}^{4} (12x + 16 - x^3)\,dx = 108$

Area of $S = 16$(area of R) $\left[\dfrac{\text{area } S}{\text{area } R} = 16\right]$

(b) Tangent at $A(a, a^3)$: $y - a^3 = 3a^2(x - a)$

$$y = 3a^2x - 2a^3$$

To find point B: $x^3 - 3a^2x + 2a^3 = 0$

$$(x - a)^2(x + 2a) = 0$$

$$\Longrightarrow B = (-2a, -8a^3)$$

Tangent at B: $y + 8a^3 = 12a^2(x + 2a)$

$$y = 12a^2x + 16a^3$$

To find point C: $x^3 - 12a^2x - 16a^3 = 0$

$$(x + 2a)^2(x - 4a) = 0$$

$$\Longrightarrow C = (4a, 64a^3)$$

Area of $R = \displaystyle\int_{-2a}^{a} [x^3 - 3a^2x + 2a^3]\,dx = \frac{27}{4}a^4$

Area of $S = \displaystyle\int_{-2a}^{4a} [12a^2x + 16a^3 - x^3]\,dx = 108a^4$

Area of $S = 16$(area of R)

8. $f'(x)^2 = e^x$

$f'(x) = e^{x/2}$

$f(x) = 2e^{x/2} + C$

$f(0) = 0 \Rightarrow C = -2$

$f(x) = 2e^{x/2} - 2$

9. $s(x) = \int_{\alpha}^{x} \sqrt{1 + f'(t)^2}\,dt$

(a) $s'(x) = \dfrac{ds}{dx} = \sqrt{1 + f'(x)^2}$

(b) $ds = \sqrt{1 + f'(x)^2}\,dx$

$$(ds)^2 = [1 + f'(x)^2](dx)^2 = \left[1 + \left(\frac{dy}{dx}\right)^2\right](dx)^2 = (dx)^2 + (dy)^2$$

(c) $s(x) = \displaystyle\int_1^x \sqrt{1 + \left(\frac{3}{2}t^{1/2}\right)^2}\,dt = \int_1^x \sqrt{1 + \frac{9}{4}t}\,dt$

(d) $s(2) = \displaystyle\int_1^2 \sqrt{1 + \frac{9}{4}t}\,dt = \left[\frac{8}{27}\left(1 + \frac{9}{4}t\right)^{3/2}\right]_1^2 = \frac{22}{27}\sqrt{22} - \frac{13}{27}\sqrt{13} \approx 2.0858$

This is the length of the curve $y = x^{3/2}$ from $x = 1$ to $x = 2$.

10. Let ρ_f be the density of the fluid and ρ_0 the density of the iceberg. The buoyant force is

$$F = \rho_f g \int_{-h}^{0} A(y)\,dy$$

where $A(y)$ is a typical cross section and g is the acceleration due to gravity. The weight of the object is

$$W = \rho_0 g \int_{-h}^{L-h} A(y)\,dy.$$

$F = W$

$$\rho_f g \int_{-h}^{0} A(y)\,dy = \rho_0 g \int_{-h}^{L-h} A(y)\,dy$$

$$\frac{\rho_0}{\rho_f} = \frac{\text{submerged volume}}{\text{total volume}} = \frac{0.92 \times 10^3}{1.03 \times 10^3} = 0.893 \text{ or } 89.3\%$$

11. (a) $\bar{y} = 0$ by symmetry

$$M_y = \int_1^6 x\left(\frac{1}{x^3} - \left(-\frac{1}{x^3}\right)\right)dx = \int_1^6 \frac{2}{x^2}\,dx = \left[-2\frac{1}{x}\right]_1^6 = \frac{5}{3}$$

$$m = 2\int_1^6 \frac{1}{x^3}\,dx = \left[-\frac{1}{x^2}\right]_1^6 = \frac{35}{36}$$

$$\bar{x} = \frac{5/3}{35/36} = \frac{12}{7} \qquad (\bar{x}, \bar{y}) = \left(\frac{12}{7}, 0\right)$$

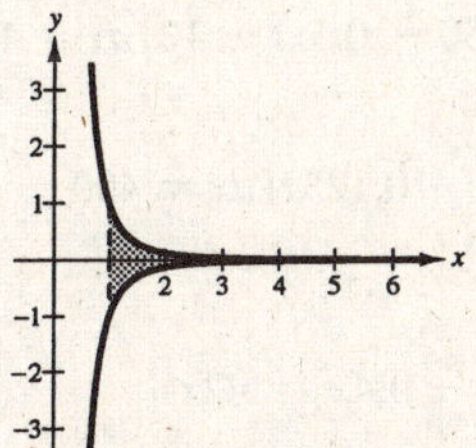

(b) $m = 2\displaystyle\int_1^b \frac{1}{x^3}\,dx = \frac{b^2 - 1}{b^2}$

$$M_y = 2\int_1^6 \frac{1}{x^2}\,dx = \frac{2(b-1)}{b}$$

$$\bar{x} = \frac{2(b-1)/b}{(b^2-1)/b^2} = \frac{2b}{b+1} \qquad (\bar{x}, \bar{y}) = \left(\frac{2b}{b+1}, 0\right)$$

(c) $\displaystyle\lim_{b\to\infty} \bar{x} = \lim_{b\to\infty} \frac{2b}{b+1} = 2 \qquad (\bar{x}, \bar{y}) = (2, 0)$

12. (a) $\bar{y} = 0$ by symmetry

$$M_y = 2\int_1^6 x\frac{1}{x^4}\,dx = 2\int_1^6 \frac{1}{x^3}\,dx = \frac{35}{36}$$

$$m = 2\int_1^6 \frac{1}{x^4}\,dx = \frac{215}{324}$$

$$\bar{x} = \frac{35/36}{215/324} = \frac{63}{43} \qquad (\bar{x}, \bar{y}) = \left(\frac{63}{43}, 0\right)$$

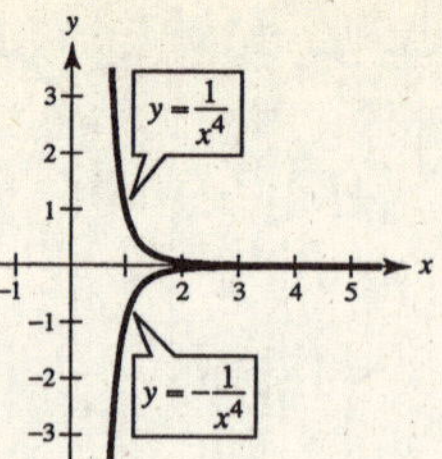

(b) $$M_y = 2\int_1^b \frac{1}{x^3}\,dx = \frac{b^2 - 1}{b^2}$$

$$m = 2\int_1^b \frac{1}{x^4}\,dx = \frac{2(b^3 - 1)}{3b^3}$$

$$\bar{x} = \frac{(b^2 - 1)/b^2}{2(b^3 - 1)/3b^3} = \frac{3b(b + 1)}{2(b^2 + b + 1)} \qquad (\bar{x}, \bar{y}) = \left(\frac{3b(b + 1)}{2(b^2 + b + 1)}, 0\right)$$

$$\lim_{b\to\infty} \bar{x} = \frac{3}{2} \qquad (\bar{x}, \bar{y}) = \left(\frac{3}{2}, 0\right)$$

13. (a) $W = \text{area} = 2 + 4 + 6 = 12$

(b) $W = \text{area} = 3 + (1 + 1) + 2 + \frac{1}{2} = 7\frac{1}{2}$

14. (a) Trapezoidal: Area $\approx \frac{160}{2(8)}[0 + 2(50) + 2(54) + 2(82) + 2(82) + 2(73) + 2(75) + 2(80) + 0] = 9920$ sq ft

(b) Simpson's: Area $\approx \frac{160}{3(8)}[0 + 4(50) + 2(54) + 4(82) + 2(82) + 4(73) + 2(75) + 4(80) + 0] = 10{,}413\frac{1}{3}$ sq ft

15. Point of equilibrium: $50 - 0.5x = 0.125x$

$x = 80, p = 10$

$(P_0, x_0) = (10, 80)$

Consumer surplus $= \displaystyle\int_0^{80} [(50 - 0.5x) - 10]\,dx = 1600$

Producer surplus $= \displaystyle\int_0^{80} [10 - 0.125x]\,dx = 400$

16. Point of equilibrium: $1000 - 0.4x^2 = 42x$

$x = 20, p = 840$

$(P_0, x_0) = (840, 20)$

Consumer surplus $= \displaystyle\int_0^{20} [(1000 - 0.4x^2) - 840]\,dx = 2133.33$

Producer surplus $= \displaystyle\int_0^{20} [840 - 42x]\,dx = 8400$

17. We use Exercise 23, Section 7.7, which gives $F = wkhb$ for a rectangle plate.

Wall at shallow end

From Exercise 23: $F = 62.4(2)(4)(20) = 9984$ lb

Wall at deep end

From Exercise 23: $F = 62.4(4)(8)(20) = 39{,}936$ lb

Side wall

From Exercise 23: $F_1 = 62.4(2)(4)(40) = 19{,}968$ lb

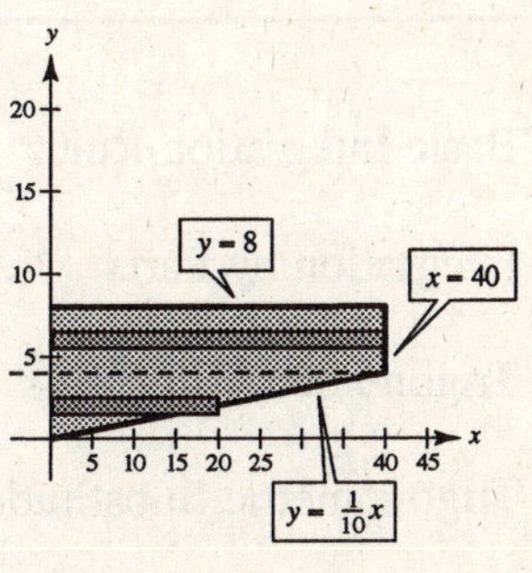

$$F_2 = 62.4\int_0^4 (8 - y)(10y)\,dy$$

$$= 624\int_0^4 (8y - y^2)\,dy = 624\left[4y^2 - \frac{y^3}{3}\right]_0^4$$

$$= 26{,}624 \text{ lb}$$

Total force: $F_1 + F_2 = 46{,}592$ lb

18. (a) Answers will vary.

$f_1(x) = 6(x - x^2)$

$f_2(x) = \frac{\pi}{2}\sin(\pi x)$

(b) f_1 arc length ≈ 3.2490

f_2 arc length ≈ 3.3655

(c) See the article by Professor Larson Riddle at http://ecademy.agnesscott.edu/lriddle/arc/contest.htm One such function is

$f_3(x) = \frac{8}{\pi}\sqrt{x - x^2}$ (arc length ≈ 2.9195)

CHAPTER 8
Integration Techniques, L'Hôpital's Rule, and Improper Integrals

CHAPTER 8
Integration Techniques, L'Hôpital's Rule, and Improper Integrals

Section 8.1 Basic Integration Rules

1. (a) $\frac{d}{dx}\left[2\sqrt{x^2+1}+C\right] = 2\left(\frac{1}{2}\right)(x^2+1)^{-1/2}(2x)$

$= \frac{2x}{\sqrt{x^2+1}}$

(b) $\frac{d}{dx}\left[\sqrt{x^2+1}+C\right] = \frac{1}{2}(x^2+1)^{-1/2}(2x) = \frac{x}{\sqrt{x^2+1}}$

(c) $\frac{d}{dx}\left[\frac{1}{2}\sqrt{x^2+1}+C\right] = \frac{1}{2}\left(\frac{1}{2}\right)(x^2+1)^{-1/2}(2x)$

$= \frac{x}{2\sqrt{x^2+1}}$

(d) $\frac{d}{dx}\left[\ln(x^2+1)+C\right] = \frac{2x}{x^2+1}$

$\int \frac{x}{\sqrt{x^2+1}}\,dx$ matches (b).

2. (a) $\frac{d}{dx}\left[\ln\sqrt{x^2+1}+C\right] = \frac{1}{2}\left(\frac{2x}{x^2+1}\right) = \frac{x}{x^2+1}$

(b) $\frac{d}{dx}\left[\frac{2x}{(x^2+1)^2}+C\right] = \frac{(x^2+1)^2(2)-(2x)(2)(x^2+1)(2x)}{(x^2+1)^4}$

$= \frac{2(1-3x^2)}{(x^2+1)^3}$

(c) $\frac{d}{dx}\left[\arctan x + C\right] = \frac{1}{1+x^2}$

(d) $\frac{d}{dx}\left[\ln(x^2+1)+C\right] = \frac{2x}{x^2+1}$

$\int \frac{x}{x^2+1}\,dx$ matches (a).

3. (a) $\frac{d}{dx}\left[\ln\sqrt{x^2+1}+C\right] = \frac{1}{2}\left(\frac{2x}{x^2+1}\right) = \frac{x}{x^2+1}$

(b) $\frac{d}{dx}\left[\frac{2x}{(x^2+1)^2}+C\right] = \frac{(x^2+1)^2(2)-(2x)(2)(x^2+1)(2x)}{(x^2+1)^4} = \frac{2(1-3x^2)}{(x^2+1)^3}$

(c) $\frac{d}{dx}\left[\arctan x + C\right] = \frac{1}{1+x^2}$

(d) $\frac{d}{dx}\left[\ln(x^2+1)+C\right] = \frac{2x}{x^2+1}$

$\int \frac{1}{x^2+1}\,dx$ matches (c).

4. (a) $\frac{d}{dx}\left[2x\sin(x^2+1)+C)\right] = 2x[\cos(x^2+1)(2x)] + 2\sin(x^2+1) = 2[2x^2\cos(x^2+1)+\sin(x^2+1)]$

(b) $\frac{d}{dx}\left[-\frac{1}{2}\sin(x^2+1)+C\right] = -\frac{1}{2}\cos(x^2+1)(2x) = -x\cos(x^2+1)$

(c) $\frac{d}{dx}\left[\frac{1}{2}\sin(x^2+1)+C\right] = \frac{1}{2}\cos(x^2+1)(2x) = x\cos(x^2+1)$

(d) $\frac{d}{dx}\left[-2x\sin(x^2+1)+C\right] = -2x[\cos(x^2+1)(2x)] - 2\sin(x^2+1) = -2[2x^2\cos(x^2+1)+\sin(x^2+1)]$

$\int x\cos(x^2+1)\,dx$ matches (c).

5. $\int (3x - 2)^4\, dx$

$u = 3x - 2,\ du = 3\, dx,\ n = 4$

Use $\int u^n\, du$.

6. $\int \frac{2t - 1}{t^2 - t + 2}\, dt$

$u = t^2 - t + 2,\ du = (2t - 1)\, dt$

Use $\int \frac{du}{u}$.

7. $\int \frac{1}{\sqrt{x}(1 - 2\sqrt{x})}\, dx$

$u = 1 - 2\sqrt{x},\ du = -\frac{1}{\sqrt{x}}\, dx$

Use $\int \frac{du}{u}$.

8. $\int \frac{2}{(2t - 1)^2 + 4}\, dt$

$u = 2t - 1,\ du = 2\, dt,\ a = 2$

Use $\int \frac{du}{u^2 + a^2}$.

9. $\int \frac{3}{\sqrt{1 - t^2}}\, dt$

$u = t,\ du = dt,\ a = 1$

Use $\int \frac{du}{\sqrt{a^2 - u^2}}$.

10. $\int \frac{-2x}{\sqrt{x^2 - 4}}\, dx$

$u = x^2 - 4,\ du = 2x\, dx,\ n = -\frac{1}{2}$

Use $\int u^n\, du$.

11. $\int t \sin t^2\, dt$

$u = t^2,\ du = 2t\, dt$

Use $\int \sin u\, du$.

12. $\int \sec 3x \tan 3x\, dx$

$u = 3x,\ du = 3\, dx$

Use $\int \sec u \tan u\, du$.

13. $\int (\cos x) e^{\sin x}\, dx$

$u = \sin x,\ du = \cos x\, dx$

Use $\int e^u\, du$.

14. $\int \frac{1}{x\sqrt{x^2 - 4}}\, dx$

$u = x,\ du = dx,\ a = 2$

Use $\int \frac{du}{u\sqrt{u^2 - a^2}}$.

15. Let $u = x - 4,\ du = dx$.

$$\int 6(x - 4)^5\, dx = 6\int (x - 4)^5\, dx = 6\frac{(x - 4)^6}{6} + C = (x - 4)^6 + C$$

16. Let $u = t - 9,\ du = dt$.

$$\int \frac{2}{(t - 9)^2}\, dt = 2\int (t - 9)^{-2}\, dt = \frac{-2}{t - 9} + C$$

17. Let $u = z - 4,\ du = dz$.

$$\int \frac{5}{(z - 4)^5}\, dz = 5\int (z - 4)^{-5}\, dz = 5\frac{(z - 4)^{-4}}{-4} + C = \frac{-5}{4(z - 4)^4} + C$$

18. Let $u = t^3 - 1,\ du = 3t^2\, dt$.

$$\int t^2 \sqrt[3]{t^3 - 1}\, dt = \frac{1}{3}\int (t^3 - 1)^{1/3}(3t^2)\, dt = \frac{1}{3}\frac{(t^3 - 1)^{4/3}}{4/3} + C = \frac{(t^3 - 1)^{4/3}}{4} + C$$

19. $$\int \left[v + \frac{1}{(3v - 1)^3}\right] dv = \int v\, dv + \frac{1}{3}\int (3v - 1)^{-3}(3)\, dv = \frac{1}{2}v^2 - \frac{1}{6(3v - 1)^2} + C$$

20. $$\int \left[x - \frac{3}{(2x + 3)^2}\right] dx = \int x\, dx - \frac{3}{2}\int (2x + 3)^{-2}(2)\, dx = \frac{x^2}{2} - \frac{3}{2}\frac{(2x + 3)^{-1}}{-1} + C = \frac{x^2}{2} + \frac{3}{2(2x + 3)} + C$$

21. Let $u = -t^3 + 9t + 1,\ du = (-3t^2 + 9)\, dt = -3(t^2 - 3)\, dt$.

$$\int \frac{t^2 - 3}{-t^3 + 9t + 1}\, dt = -\frac{1}{3}\int \frac{-3(t^2 - 3)}{-t^3 + 9t + 1}\, dt = -\frac{1}{3}\ln\left|-t^3 + 9t + 1\right| + C$$

22. Let $u = x^2 + 2x - 4$, $du = 2(x + 1)\,dx$.

$$\int \frac{x+1}{\sqrt{x^2+2x-4}}\,dx = \frac{1}{2}\int (x^2 + 2x - 4)^{-1/2}(2)(x+1)\,dx$$
$$= \sqrt{x^2 + 2x - 4} + C$$

23. $$\int \frac{x^2}{x-1}\,dx = \int (x+1)\,dx + \int \frac{1}{x-1}\,dx$$
$$= \frac{1}{2}x^2 + x + \ln|x-1| + C$$

24. $$\int \frac{2x}{x-4}\,dx = \int 2\,dx + \int \frac{8}{x-4}\,dx$$
$$= 2x + 8\ln|x-4| + C$$

25. Let $u = 1 + e^x$, $du = e^x\,dx$.

$$\int \frac{e^x}{1+e^x}\,dx = \ln(1 + e^x) + C$$

26. $$\int \left(\frac{1}{3x-1} - \frac{1}{3x+1}\right)dx = \frac{1}{3}\int \frac{1}{3x-1}(3)\,dx - \frac{1}{3}\int \frac{1}{3x+1}(3)\,dx$$
$$= \frac{1}{3}\ln|3x-1| - \frac{1}{3}\ln|3x+1| + C = \frac{1}{3}\ln\left|\frac{3x-1}{3x+1}\right| + C$$

27. $$\int (1 + 2x^2)^2\,dx = \int (4x^4 + 4x^2 + 1)\,dx = \frac{4}{5}x^5 + \frac{4}{3}x^3 + x + C = \frac{x}{15}(12x^4 + 20x^2 + 15) + C$$

28. $$\int x\left(1 + \frac{1}{x}\right)^3 = \int x\left(1 + \frac{3}{x} + \frac{3}{x^2} + \frac{1}{x^3}\right)dx = \int \left(x + 3 + \frac{3}{x} + \frac{1}{x^2}\right)dx = \frac{1}{2}x^2 + 3x + 3\ln|x| - \frac{1}{x} + C$$

29. Let $u = 2\pi x^2$, $du = 4\pi x\,dx$.

$$\int x(\cos 2\pi x^2)\,dx = \frac{1}{4\pi}\int (\cos 2\pi x^2)(4\pi x)\,dx$$
$$= \frac{1}{4\pi}\sin 2\pi x^2 + C$$

30. $$\int \sec 4x\,dx = \frac{1}{4}\int \sec(4x)(4)\,dx$$
$$= \frac{1}{4}\ln|\sec 4x + \tan 4x| + C$$

31. Let $u = \pi x$, $du = \pi\,dx$.

$$\int \csc(\pi x)\cot(\pi x)\,dx = \frac{1}{\pi}\int \csc(\pi x)\cot(\pi x)\pi\,dx$$
$$= -\frac{1}{\pi}\csc(\pi x) + C$$

32. Let $u = \cos x$, $du = -\sin x\,dx$.

$$\int \frac{\sin x}{\sqrt{\cos x}}\,dx = -\int (\cos x)^{-1/2}(-\sin x)\,dx$$
$$= -2\sqrt{\cos x} + C$$

33. Let $u = 5x$, $du = 5\,dx$.

$$\int e^{5x}\,dx = \frac{1}{5}\int e^{5x}(5)\,dx = \frac{1}{5}e^{5x} + C$$

34. Let $u = \cot x$, $du = -\csc^2 x\,dx$.

$$\int \csc^2 x e^{\cot x}\,dx = -\int e^{\cot x}(-\csc^2 x)\,dx = -e^{\cot x} + C$$

35. Let $u = 1 + e^x$, $du = e^x\,dx$.

$$\int \frac{2}{e^{-x}+1}\,dx = 2\int \left(\frac{1}{e^{-x}+1}\right)\left(\frac{e^x}{e^x}\right)dx$$
$$= 2\int \frac{e^x}{1+e^x}\,dx$$
$$= 2\ln(1 + e^x) + C$$

36. $$\int \frac{5}{3e^x - 2}\,dx = 5\int \left(\frac{1}{3e^x-2}\right)\left(\frac{e^{-x}}{e^{-x}}\right)dx$$
$$= 5\int \frac{e^{-x}}{3 - 2e^{-x}}\,dx$$
$$= \frac{5}{2}\int \frac{1}{3 - 2e^{-x}}(2e^{-x})\,dx$$
$$= \frac{5}{2}\ln|3 - 2e^{-x}| + C$$

37. $\displaystyle\int \frac{\ln x^2}{x}\,dx = 2\int (\ln x)\frac{1}{x}\,dx = 2\frac{(\ln x)^2}{2} + C = (\ln x)^2 + C$

38. Let $u = \ln(\cos x)$, $du = \dfrac{-\sin x}{\cos x}\,dx = -\tan x\,dx$.

$$\int (\tan x)(\ln \cos x)\,dx = -\int (\ln \cos x)(-\tan x)\,dx = \frac{-[\ln(\cos x)]^2}{2} + C$$

39.
$$\begin{aligned}\int \frac{1 + \sin x}{\cos x}\,dx &= \int \frac{1 + \sin x}{\cos x}\cdot\frac{1 - \sin x}{1 - \sin x}\,dx\\ &= \int \frac{1 - \sin^2 x}{\cos x(1 - \sin x)}\,dx\\ &= \int \frac{\cos^2 x}{\cos x(1 - \sin x)}\,dx\\ &= -\int \frac{-\cos x}{1 - \sin x}\,dx\\ &= -\ln|1 - \sin x| + C, \quad (u = 1 - \sin x)\end{aligned}$$

Alternate Solution:

$$\begin{aligned}\int \frac{1 + \sin x}{\cos x}\,dx &= \int (\sec x + \tan x)\,dx\\ &= \ln|\sec x + \tan x| + \ln|\sec x| + C\\ &= \ln|\sec x(\sec x + \tan x)| + C\end{aligned}$$

40.
$$\begin{aligned}\int \frac{1 + \cos\alpha}{\sin\alpha}\,d\alpha &= \int \csc\alpha\,d\alpha + \int \cot\alpha\,d\alpha\\ &= -\ln|\csc\alpha + \cot\alpha| + \ln|\sin\alpha| + C\end{aligned}$$

41.
$$\begin{aligned}\frac{1}{\cos\theta - 1} &= \frac{1}{\cos\theta - 1}\cdot\frac{\cos\theta + 1}{\cos\theta + 1} = \frac{\cos\theta + 1}{\cos^2\theta - 1}\\ &= \frac{\cos\theta + 1}{-\sin^2\theta} = -\csc\theta\cdot\cot\theta - \csc^2\theta\end{aligned}$$

$$\begin{aligned}\int \frac{1}{\cos\theta - 1}\,d\theta &= \int (-\csc\theta\cot\theta - \csc^2\theta)\,d\theta\\ &= \csc\theta + \cot\theta + C\\ &= \frac{1}{\sin\theta} + \frac{\cos\theta}{\sin\theta} + C\\ &= \frac{1 + \cos\theta}{\sin\theta} + C\end{aligned}$$

42.
$$\begin{aligned}\int \frac{2}{3(\sec x - 1)}\,dx &= \frac{2}{3}\int \frac{1}{\sec x - 1}\cdot\left(\frac{\sec x + 1}{\sec x + 1}\right)dx\\ &= \frac{2}{3}\int \frac{\sec x + 1}{\tan^2 x}\,dx\\ &= \frac{2}{3}\int \frac{\sec x}{\tan^2 x}\,dx + \frac{2}{3}\int \cot^2 x\,dx\\ &= \frac{2}{3}\int \frac{\cos x}{\sin^2 x}\,dx + \frac{2}{3}\int (\csc^2 x - 1)\,dx\\ &= \frac{2}{3}\left(-\frac{1}{\sin x}\right) - \frac{2}{3}\cot x - \frac{2}{3}x + C\\ &= -\frac{2}{3}[\csc x + \cot x + x] + C\end{aligned}$$

43. Let $u = 2t - 1$, $du = 2\,dt$.

$$\begin{aligned}\int \frac{-1}{\sqrt{1 - (2t - 1)^2}}\,dt &= -\frac{1}{2}\int \frac{2}{\sqrt{1 - (2t - 1)^2}}\,dt\\ &= -\frac{1}{2}\arcsin(2t - 1) + C\end{aligned}$$

44. Let $u = \sqrt{3}x$, $du = \sqrt{3}\,dx$.

$$\begin{aligned}\int \frac{1}{4 + 3x^2}\,dx &= \frac{1}{\sqrt{3}}\int \frac{\sqrt{3}}{4 + (\sqrt{3}x)^2}\,dx\\ &= \frac{1}{2\sqrt{3}}\arctan\left(\frac{\sqrt{3}x}{2}\right) + C\end{aligned}$$

45. Let $u = \cos\left(\dfrac{2}{t}\right)$, $du = \dfrac{2\sin(2/t)}{t^2}\,dt$.

$$\begin{aligned}\int \frac{\tan(2/t)}{t^2}\,dt &= \frac{1}{2}\int \frac{1}{\cos(2/t)}\left[\frac{2\sin(2/t)}{t^2}\right]dt\\ &= \frac{1}{2}\ln\left|\cos\left(\frac{2}{t}\right)\right| + C\end{aligned}$$

46. Let $u = \dfrac{1}{t}$, $du = \dfrac{-1}{t^2}\,dt$.

$$\int \frac{e^{1/t}}{t^2}\,dt = -\int e^{1/t}\left(\frac{-1}{t^2}\right)dt = -e^{1/t} + C$$

47. $\displaystyle\int \frac{3}{\sqrt{6x - x^2}}\,dx = 3\int \frac{1}{\sqrt{9 - (x-3)^2}}\,dx = 3\arcsin\left(\frac{x-3}{3}\right) + C$

48. $\displaystyle\int \frac{1}{(x-1)\sqrt{4x^2 - 8x + 3}}\,dx = \int \frac{2}{[2(x-1)]\sqrt{[2(x-1)]^2 - 1}}\,dx = \operatorname{arcsec}|2(x-1)| + C$

49. $\displaystyle\int \frac{4}{4x^2 + 4x + 65}\,dx = \int \frac{1}{[x + (1/2)]^2 + 16}\,dx = \frac{1}{4}\arctan\left[\frac{x + (1/2)}{4}\right] + C = \frac{1}{4}\arctan\left(\frac{2x+1}{8}\right) + C$

50. $\displaystyle\int \frac{1}{\sqrt{1 - 4x - x^2}}\,dx = \int \frac{1}{\sqrt{5 - (x^2 + 4x + 4)}}\,dx = \int \frac{1}{\sqrt{5 - (x+2)^2}}\,dx = \arcsin\left(\frac{x+2}{\sqrt{5}}\right) + C, \quad (a = \sqrt{5})$

51. $\displaystyle\frac{ds}{dt} = \frac{t}{\sqrt{1 - t^4}}, \quad \left(0, -\frac{1}{2}\right)$

(a)

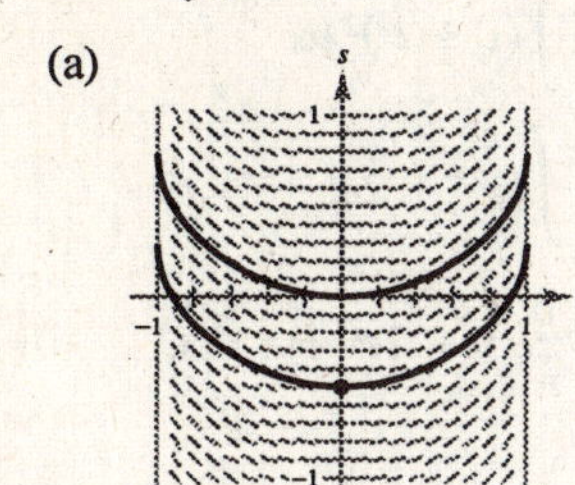

(b) $u = t^2,\ du = 2t\,dt$

$$\int \frac{t}{\sqrt{1 - t^4}}\,dt = \frac{1}{2}\int \frac{2t}{\sqrt{1 - (t^2)^2}}\,dt$$

$$= \frac{1}{2}\arcsin t^2 + C$$

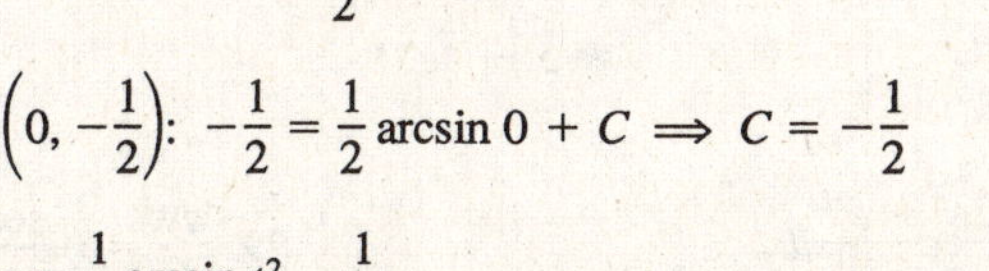

$$\left(0, -\frac{1}{2}\right):\ -\frac{1}{2} = \frac{1}{2}\arcsin 0 + C \Rightarrow C = -\frac{1}{2}$$

$$s = \frac{1}{2}\arcsin t^2 - \frac{1}{2}$$

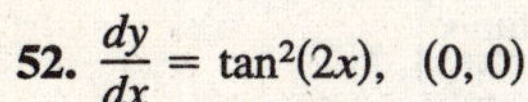

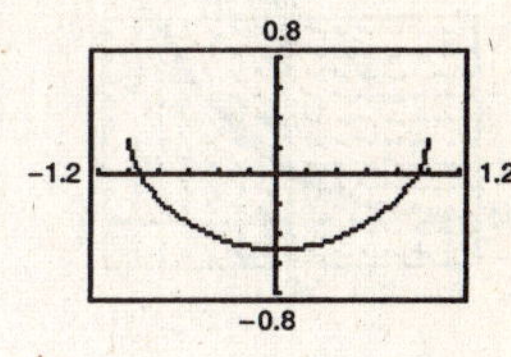

52. $\displaystyle\frac{dy}{dx} = \tan^2(2x), \quad (0, 0)$

(a)

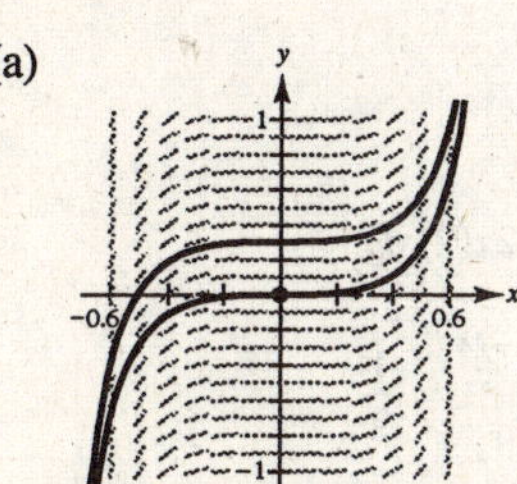

(b) $\displaystyle\int \tan^2(2x)\,dx = \int (\sec^2(2x) - 1)\,dx$

$$= \frac{1}{2}\tan(2x) - x + C$$

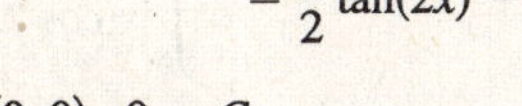

$(0, 0)$: $0 = C$

$$y = \frac{1}{2}\tan(2x) - x$$

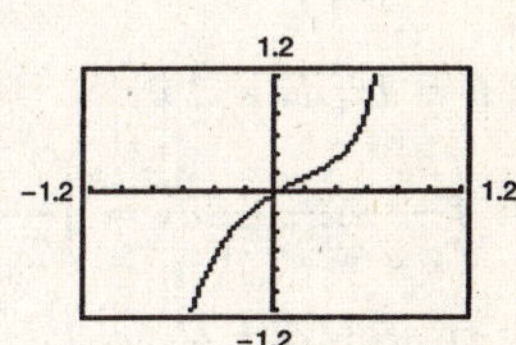

53. (a)

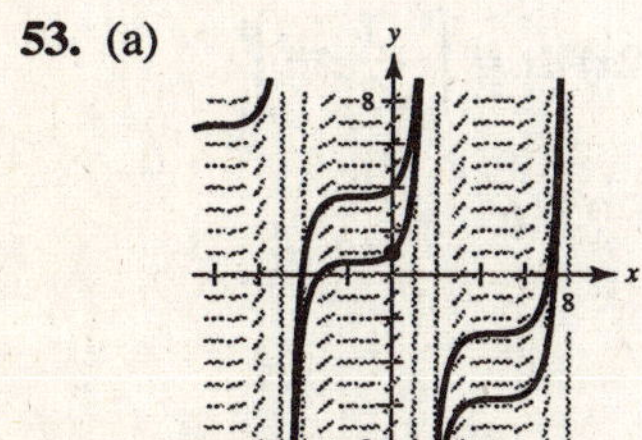

(b)

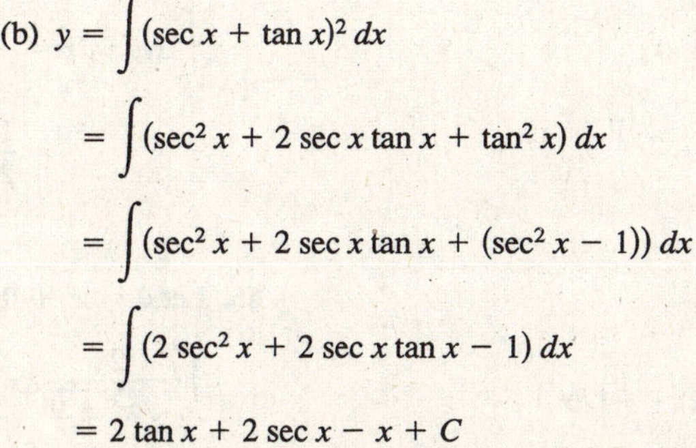

$$y = \int (\sec x + \tan x)^2\,dx$$

$$= \int (\sec^2 x + 2\sec x \tan x + \tan^2 x)\,dx$$

$$= \int (\sec^2 x + 2\sec x \tan x + (\sec^2 x - 1))\,dx$$

$$= \int (2\sec^2 x + 2\sec x \tan x - 1)\,dx$$

$$= 2\tan x + 2\sec x - x + C$$

At $(0, 1)$: $1 = 0 + 2 - 0 + C \Rightarrow C = -1$

$$y = 2\tan x + 2\sec x - x - 1$$

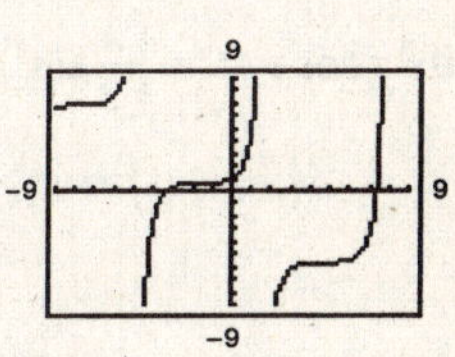

54. (a)

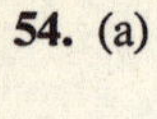

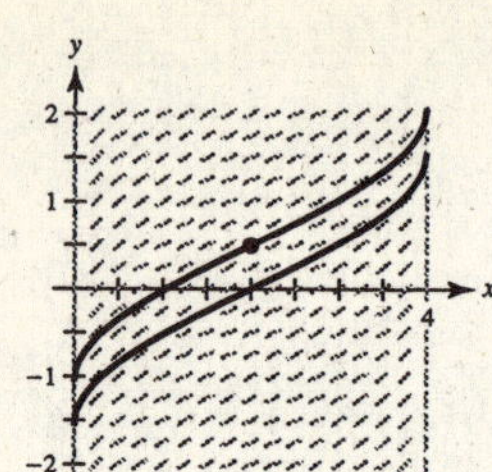

(b) $y = \int \frac{1}{\sqrt{4x - x^2}}\,dx$

$= \int \frac{1}{\sqrt{4 - (x^2 - 4x + 4)}}\,dx$

$= \int \frac{1}{\sqrt{4 - (x-2)^2}}\,dx$

$= \arcsin\left(\frac{x-2}{2}\right) + C$

At $\left(2, \frac{1}{2}\right)$: $\frac{1}{2} = \arcsin(0) + C \Rightarrow C = \frac{1}{2}$

$y = \arcsin\left(\frac{x-2}{2}\right) + \frac{1}{2}$

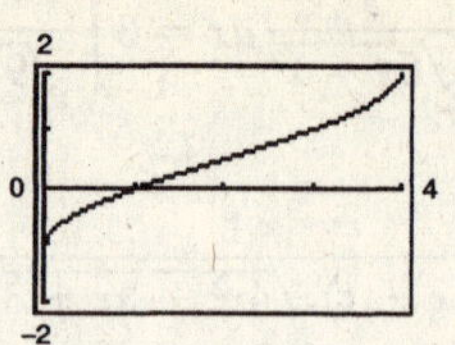

55.

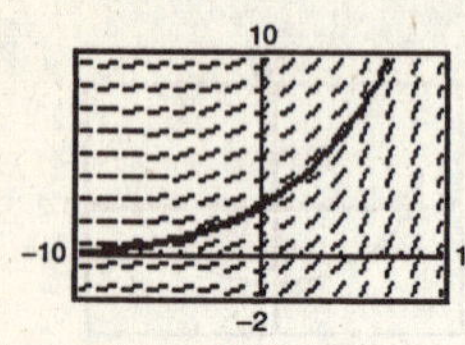

$y = 3e^{0.2x}$

56.

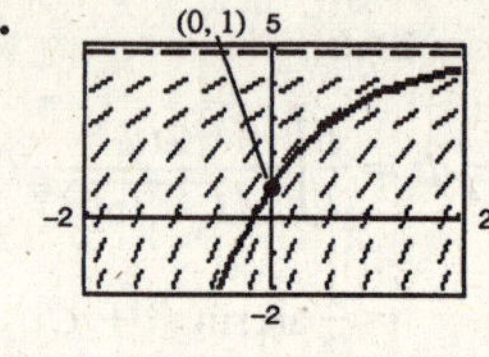

$y = 5 - 4e^{-x}$

57. $y = \int (1 + e^x)^2\,dx$

$= \int (e^{2x} + 2e^x + 1)\,dx$

$= \frac{1}{2}e^{2x} + 2e^x + x + C$

58. $r = \int \frac{(1 + e^t)^2}{e^t}\,dt = \int \frac{1 + 2e^t + e^{2t}}{e^t}\,dt$

$= \int (e^{-t} + 2 + e^t)\,dt = -e^{-t} + 2t + e^t + C$

59. $\frac{dy}{dx} = \frac{\sec^2 x}{4 + \tan^2 x}$

Let $u = \tan x$, $du = \sec^2 x\,dx$.

$y = \int \frac{\sec^2 x}{4 + \tan^2 x}\,dx = \frac{1}{2}\arctan\left(\frac{\tan x}{2}\right) + C$

60. Let $u = 2x$, $du = 2\,dx$.

$y = \int \frac{1}{x\sqrt{4x^2 - 1}}\,dx = \int \frac{2}{2x\sqrt{(2x)^2 - 1}}\,dx$

$= \operatorname{arcsec}|2x| + C$

61. Let $u = 2x$, $du = 2\,dx$.

$\int_0^{\pi/4} \cos 2x\,dx = \frac{1}{2}\int_0^{\pi/4} \cos 2x(2)\,dx$

$= \left[\frac{1}{2}\sin 2x\right]_0^{\pi/4} = \frac{1}{2}$

62. Let $u = \sin t$, $du = \cos t\,dt$.

$\int_0^{\pi} \sin^2 t \cos t\,dt = \left[\frac{1}{3}\sin^3 t\right]_0^{\pi} = 0$

63. Let $u = -x^2$, $du = -2x\,dx$.

$\int_0^1 xe^{-x^2}\,dx = -\frac{1}{2}\int_0^1 e^{-x^2}(-2x)\,dx = \left[-\frac{1}{2}e^{-x^2}\right]_0^1$

$= \frac{1}{2}(1 - e^{-1}) \approx 0.316$

64. Let $u = 1 - \ln x$, $du = \frac{-1}{x}\,dx$.

$\int_1^e \frac{1 - \ln x}{x}\,dx = -\int_1^e (1 - \ln x)\left(\frac{-1}{x}\right)dx$

$= \left[-\frac{1}{2}(1 - \ln x)^2\right]_1^e = \frac{1}{2}$

65. Let $u = x^2 + 9$, $du = 2x\,dx$.

$\int_0^4 \frac{2x}{\sqrt{x^2 + 9}}\,dx = \int_0^4 (x^2 + 9)^{-1/2}(2x)\,dx$

$= \left[2\sqrt{x^2 + 9}\right]_0^4 = 4$

66. $\displaystyle\int_1^2 \frac{x-2}{x}\,dx = \int_1^2 \left(1 - \frac{2}{x}\right) dx$

$\displaystyle = \Big[x - 2\ln x\Big]_1^2 = 1 - \ln 4 \approx -0.386$

67. Let $u = 3x$, $du = 3\,dx$.

$\displaystyle\int_0^{2/\sqrt{3}} \frac{1}{4+9x^2}\,dx = \frac{1}{3}\int_0^{2/\sqrt{3}} \frac{3}{4+(3x)^2}\,dx$

$\displaystyle = \left[\frac{1}{6}\arctan\left(\frac{3x}{2}\right)\right]_0^{2/\sqrt{3}}$

$\displaystyle = \frac{\pi}{18} \approx 0.175$

68. $\displaystyle\int_0^4 \frac{1}{\sqrt{25-x^2}}\,dx = \left[\arcsin\frac{x}{5}\right]_0^4 = \arcsin\frac{4}{5} \approx 0.927$

69. $\displaystyle A = \int_0^{5/2} (-2x+5)^{3/2}\,dx$

$\displaystyle = -\frac{1}{2}\int_0^{5/2} (5-2x)^{3/2}(-2)\,dx$

$\displaystyle = -\frac{1}{5}(5-2x)^{5/2}\Big]_0^{5/2}$

$\displaystyle = 0 + \frac{1}{5}(5)^{5/2} = 5^{3/2}$

$= 5\sqrt{5} \approx 11.1803$

70. $\displaystyle A = \int_0^2 x\sqrt{8-2x^2}\,dx$

$\displaystyle = -\frac{1}{4}\int_0^2 (8-2x^2)^{1/2}(-4x)\,dx$

$\displaystyle = -\frac{1}{6}(8-2x^2)^{3/2}\Big]_0^2$

$\displaystyle = 0 + \frac{1}{6}(8)^{3/2}$

$\displaystyle = \frac{8\sqrt{2}}{3} \approx 3.7712$

71. $\displaystyle A = \int_0^5 \frac{3x+2}{x^2+9}\,dx$

$\displaystyle = \int_0^5 \frac{3x}{x^2+9}\,dx + \int_0^5 \frac{2}{x^2+9}\,dx$

$\displaystyle = \left[\frac{3}{2}\ln|x^2+9| + \frac{2}{3}\arctan\left(\frac{x}{3}\right)\right]_0^5$

$\displaystyle = \frac{3}{2}\ln(34) + \frac{2}{3}\arctan\left(\frac{5}{3}\right) - \frac{3}{2}\ln 9$

$\displaystyle = \frac{3}{2}\ln\left(\frac{34}{9}\right) + \frac{2}{3}\arctan\left(\frac{5}{3}\right)$

≈ 2.6806

72. $\displaystyle A = \int_{-3}^3 \frac{3}{x^2+1}\,dx$

$\displaystyle = 2\int_0^3 \frac{3}{x^2+1}\,dx$

$\displaystyle = 6\arctan(x)\Big]_0^3$

$= 6\arctan(3)$

≈ 7.4943

73. $y^2 = x^2(1-x^2)$

$y = \pm\sqrt{x^2(1-x^2)}$

$\displaystyle A = 4\int_0^1 x\sqrt{1-x^2}\,dx$

$\displaystyle = -2\int_0^1 (1-x^2)^{1/2}(-2x)\,dx$

$\displaystyle = -\frac{4}{3}(1-x)^{3/2}\Big]_0^1$

$\displaystyle = -\frac{4}{3}(0-1) = \frac{4}{3}$

74. $\displaystyle A = \int_0^{\pi/2} \sin 2x\,dx$

$\displaystyle = -\frac{1}{2}\cos 2x\Big]_0^{\pi/2}$

$\displaystyle = -\frac{1}{2}(-1-1) = 1$

75. $\displaystyle\int \frac{1}{x^2 + 4x + 13}\,dx = \frac{1}{3}\arctan\left(\frac{x+2}{3}\right) + C$

The antiderivatives are vertical translations of each other.

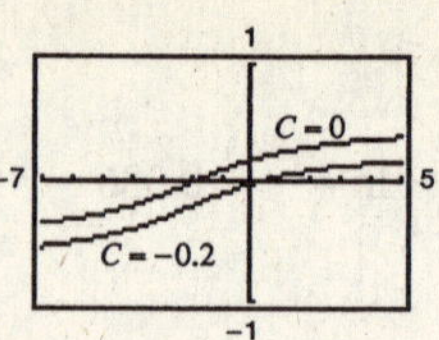

76. $\displaystyle\int \frac{x-2}{x^2 + 4x + 13}\,dx = \frac{1}{2}\ln(x^2 + 4x + 13) - \frac{4}{3}\arctan\left(\frac{x+2}{3}\right) + C$

The antiderivatives are vertical translations of each other.

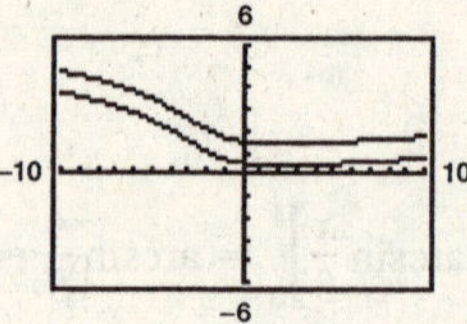

77. $\displaystyle\int \frac{1}{1+\sin\theta}\,d\theta = \tan\theta - \sec\theta + C \quad \left(\text{or } \frac{-2}{1+\tan(\theta/2)}\right)$

The antiderivatives are vertical translations of each other.

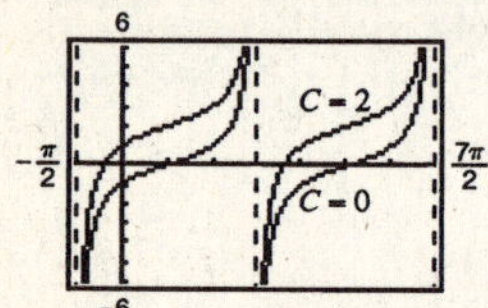

78. 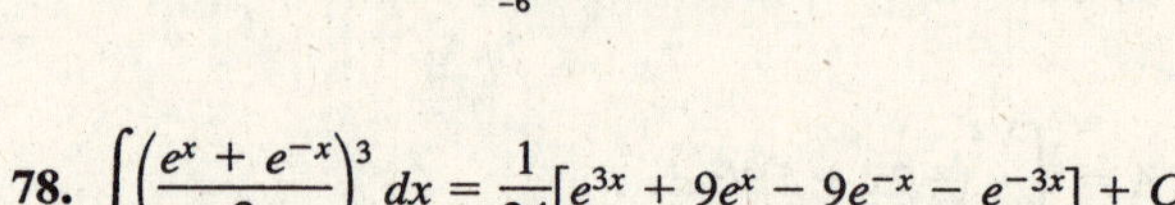 $\displaystyle\int \left(\frac{e^x + e^{-x}}{2}\right)^3 dx = \frac{1}{24}[e^{3x} + 9e^x - 9e^{-x} - e^{-3x}] + C$

The antiderivatives are vertical translations of each other.

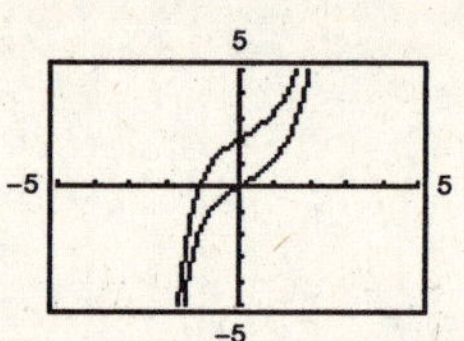

79. Power Rule: $\displaystyle\int u^n\,du = \frac{u^{n+1}}{n+1} + C,\ n \neq -1$

$u = x^2 + 1, n = 3$

80. $\displaystyle\int \sec u \tan u\,du = \sec u + C$

81. Log Rule: $\displaystyle\int \frac{du}{u} = \ln|u| + C,\ \ u = x^2 + 1$

82. Arctan Rule: $\displaystyle\int \frac{du}{a^2 + u^2} = \frac{1}{a}\arctan\left(\frac{u}{a}\right) + C$

83. They are equivalent because

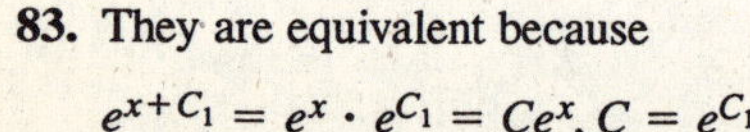

$e^{x+C_1} = e^x \cdot e^{C_1} = Ce^x, C = e^{C_1}.$

84. They differ by a constant.

$\sec^2 x + C_1 = (\tan^2 x + 1) + C_1 = \tan^2 x + C$

85. $\sin x + \cos x = a\sin(x + b)$

$\sin x + \cos x = a\sin x\cos b + a\cos x\sin b$

$\sin x + \cos x = (a\cos b)\sin x + (a\sin b)\cos x$

Equate coefficients of like terms to obtain the following.

$1 = a\cos b \quad \text{and} \quad 1 = a\sin b$

Thus, $a = 1/\cos b$. Now, substitute for a in $1 = a\sin b$.

$$1 = \left(\frac{1}{\cos b}\right)\sin b$$

$$1 = \tan b \Rightarrow b = \frac{\pi}{4}$$

Since $b = \dfrac{\pi}{4}$, $a = \dfrac{1}{\cos(\pi/4)} = \sqrt{2}$. Thus, $\sin x + \cos x = \sqrt{2}\sin\left(x + \dfrac{\pi}{4}\right)$.

$$\int \frac{dx}{\sin x + \cos x} = \int \frac{dx}{\sqrt{2}\sin(x + (\pi/4))} = \frac{1}{\sqrt{2}}\int \csc\left(x + \frac{\pi}{4}\right)dx = -\frac{1}{\sqrt{2}}\ln\left|\csc\left(x + \frac{\pi}{4}\right) + \cot\left(x + \frac{\pi}{4}\right)\right| + C$$

86. $\int_0^{1/a} (x - ax^2)\,dx = \left[\frac{1}{2}x^2 - \frac{a}{3}x^3\right]_0^{1/a} = \frac{1}{6a^2}$

Let $\frac{1}{6a^2} = \frac{2}{3}$, $12a^2 = 3$, $a = \frac{1}{2}$.

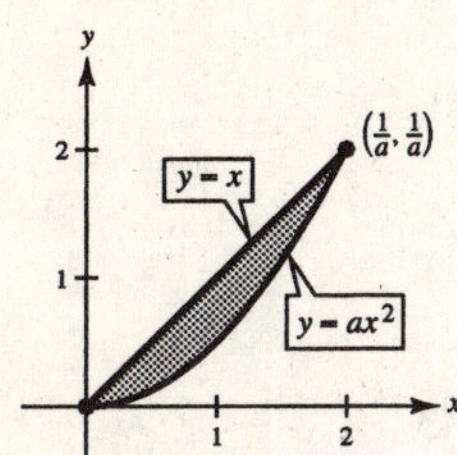

87. $f(x) = \frac{1}{5}(x^3 - 7x^2 + 10x)$

$\int_0^5 f(x)\,dx < 0$ because more area is below the x-axis than above.

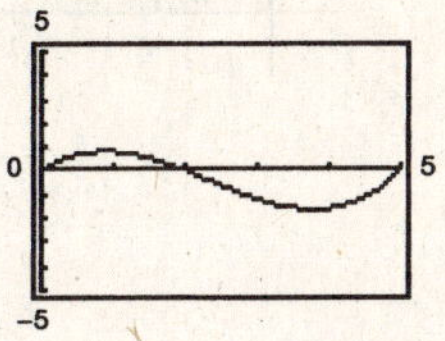

88. No. When $u = x^2$, it does not follow that $x = \sqrt{u}$ since x is negative on $[-1, 0)$.

89. $\int_0^2 \frac{4x}{x^2 + 1}\,dx \approx 3$

Matches (a).

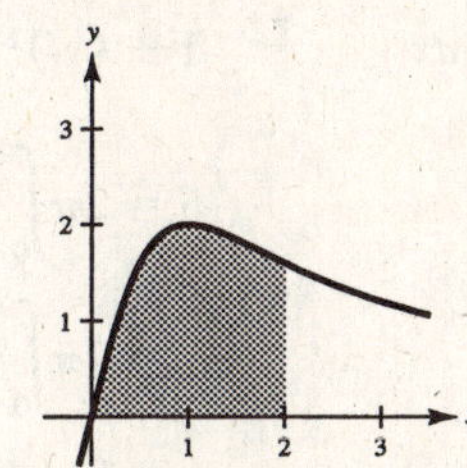

90. $\int_0^2 \frac{4}{x^2 + 1}\,dx \approx 4$

Matches (d).

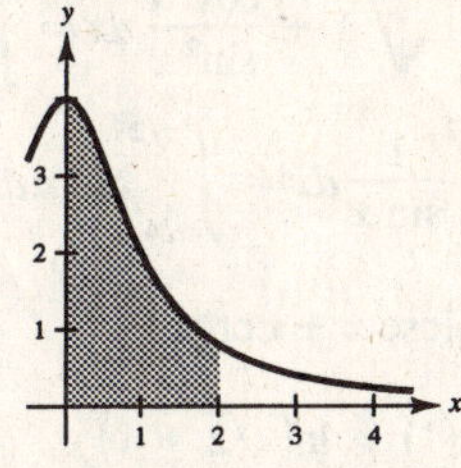

91. (a) $y = 2\pi x^2, \quad 0 \le x \le 2$

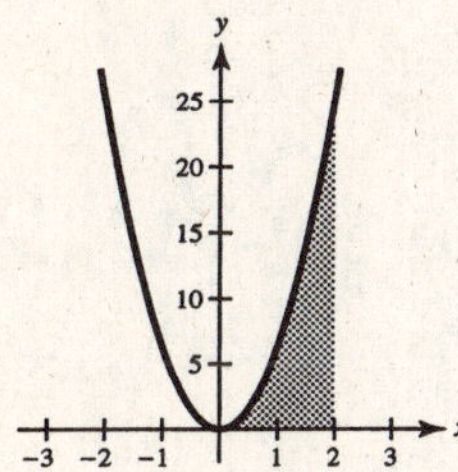

(b) $y = \sqrt{2}x, \quad 0 \le x \le 2$

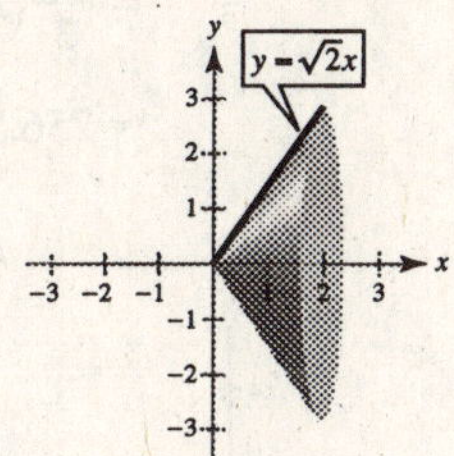

(c) $y = x, \quad 0 \le x \le 2$

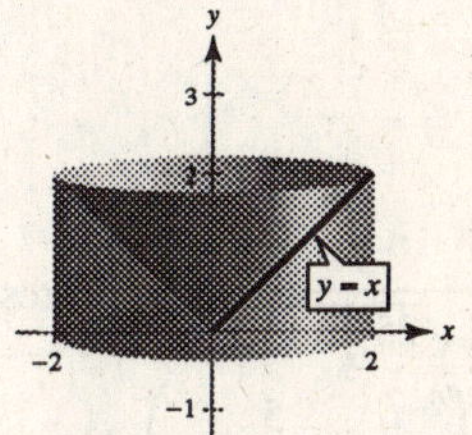

92. (a) $x = \pi y, \quad 0 \le y \le 4$

$y = \frac{1}{\pi}x, \quad 0 \le x \le 4\pi$

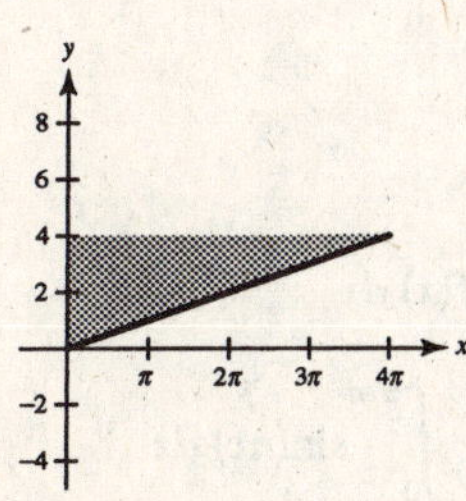

(b) $x = \sqrt{y}, \quad 0 \le y \le 4$

$y = x^2, \quad 0 \le x \le 2$

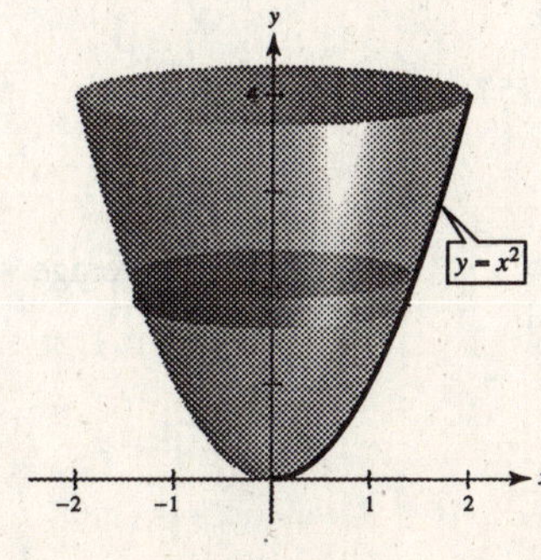

(c) $x = \frac{1}{2}, \quad 0 \le y \le 4$

$2\pi\int_0^4 y\left(\frac{1}{2}\right) dy$

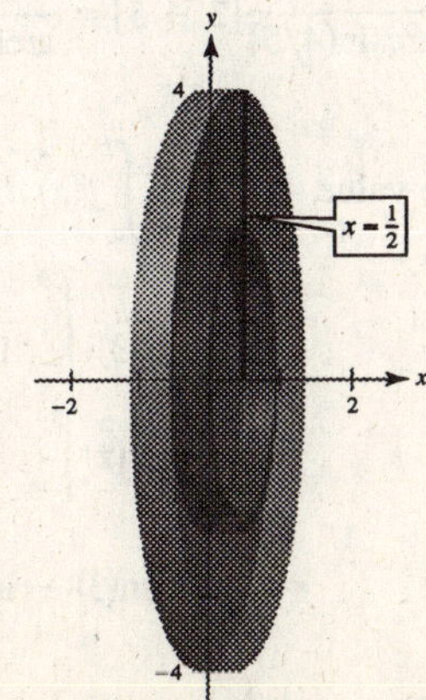

93. (a) **Shell Method:**

Let $u = -x^2$, $du = -2x\,dx$.

$$V = 2\pi\int_0^1 xe^{-x^2}\,dx$$
$$= -\pi\int_0^1 e^{-x^2}(-2x)\,dx$$
$$= \Big[-\pi e^{-x^2}\Big]_0^1$$
$$= \pi(1 - e^{-1}) \approx 1.986$$

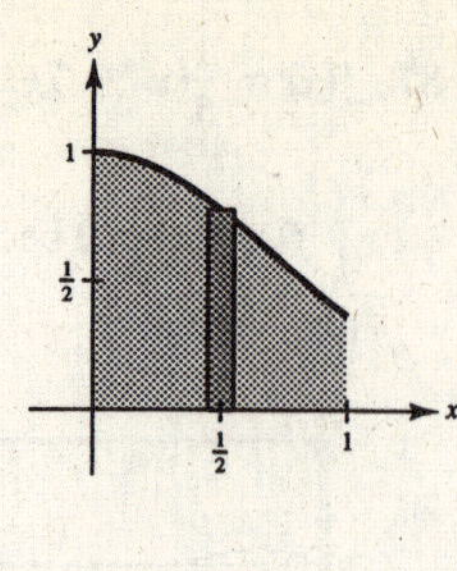

(b) **Shell Method:**

$$V = 2\pi\int_0^b xe^{-x^2}\,dx$$
$$= \Big[-\pi e^{-x^2}\Big]_0^b$$
$$= \pi(1 - e^{-b^2}) = \frac{4}{3}$$
$$e^{-b^2} = \frac{3\pi - 4}{3\pi}$$
$$b = \sqrt{\ln\left(\frac{3\pi}{3\pi - 4}\right)} \approx 0.743$$

94. $y = f(x) = \ln(\sin x)$

$$f'(x) = \frac{\cos x}{\sin x}$$
$$s = \int_{\pi/4}^{\pi/2}\sqrt{1 + \frac{\cos^2 x}{\sin^2 x}}\,dx = \int_{\pi/4}^{\pi/2}\sqrt{\frac{\sin^2 x + \cos^2 x}{\sin^2 x}}\,dx$$
$$= \int_{\pi/4}^{\pi/2}\frac{1}{\sin x}\,dx = \int_{\pi/4}^{\pi/2}\csc x\,dx$$
$$= -\ln|\csc x + \cot x|\Big]_{\pi/4}^{\pi/2}$$
$$= -\ln(1) + \ln(\sqrt{2} + 1)$$
$$= \ln(\sqrt{2} + 1) \approx 0.8814$$

95. $y = 2\sqrt{x}$

$$y' = \frac{1}{\sqrt{x}}$$
$$1 + (y')^2 = 1 + \frac{1}{x} = \frac{x + 1}{x}$$
$$S = 2\pi\int_0^9 2\sqrt{x}\sqrt{\frac{x + 1}{x}}\,dx$$
$$= 2\pi\int_0^9 2\sqrt{x + 1}\,dx$$
$$= \left[4\pi\left(\frac{2}{3}\right)(x + 1)^{3/2}\right]_0^9$$
$$= \frac{8\pi}{3}(10\sqrt{10} - 1)$$
$$\approx 256.545$$

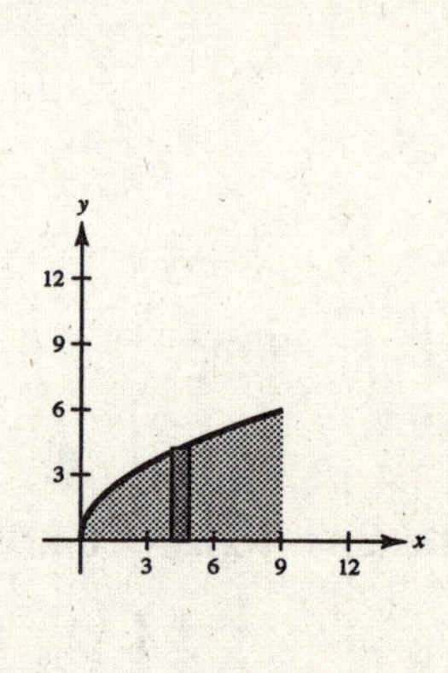

96. $A = \displaystyle\int_0^4 \frac{5}{\sqrt{25 - x^2}}\,dx = \left[5\arcsin\frac{x}{5}\right]_0^4 = 5\arcsin\frac{4}{5}$

$$\bar{x} = \frac{1}{A}\int_0^4 x\left(\frac{5}{\sqrt{25 - x^2}}\right)dx$$
$$= \frac{1}{5\arcsin(4/5)}\left(-\frac{5}{2}\right)\int_0^4 (25 - x^2)^{-1/2}(-2x)\,dx$$
$$= \frac{1}{5\arcsin(4/5)}(-5)\Big[(25 - x^2)^{1/2}\Big]_0^4$$
$$= -\frac{1}{\arcsin(4/5)}[3 - 5] = \frac{2}{\arcsin(4/5)} \approx 2.157$$

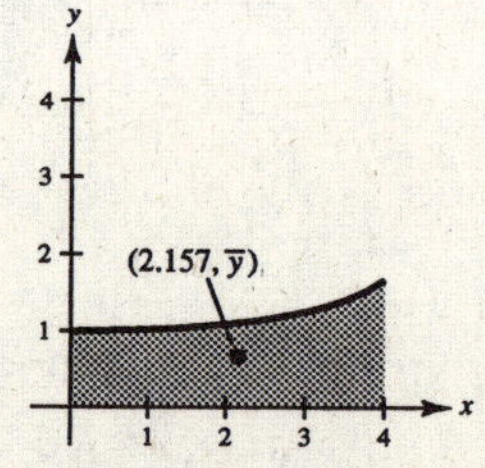

97. Average value $= \displaystyle\frac{1}{b - a}\int_a^b f(x)\,dx$

$$= \frac{1}{3 - (-3)}\int_{-3}^3 \frac{1}{1 + x^2}\,dx$$
$$= \frac{1}{6}\arctan(x)\Big]_{-3}^3$$
$$= \frac{1}{6}[\arctan(3) - \arctan(-3)]$$
$$= \frac{1}{3}\arctan(3) \approx 0.4163$$

98. Average value $= \displaystyle\frac{1}{b - a}\int_a^b f(x)\,dx$

$$= \frac{1}{(\pi/n) - 0}\int_0^{\pi/n}\sin(nx)\,dx$$
$$= \frac{n}{\pi}\left[\frac{-1}{n}\cos(nx)\right]_0^{\pi/n}$$
$$= -\frac{1}{\pi}[\cos(\pi) - \cos(0)]$$
$$= \frac{2}{\pi}$$

99. $y = \tan(\pi x)$

$y' = \pi \sec^2(\pi x)$

$1 + (y')^2 = 1 + \pi^2 \sec^4(\pi x)$

$s = \int_0^{1/4} \sqrt{1 + \pi^2 \sec^4(\pi x)}\, dx$

≈ 1.0320

100. $y = x^{2/3}$

$y' = \dfrac{2}{3x^{1/3}}$

$1 + (y')^2 = 1 + \dfrac{4}{9x^{2/3}}$

$s = \int_1^8 \sqrt{1 + \dfrac{4}{9x^{2/3}}}\, dx \approx 7.6337$

101. (a) $\int \cos^3 x\, dx = \int (1 - \sin^2 x) \cos x\, dx$

$= \sin x - \dfrac{\sin^3 x}{3} + C$

(b) $\int \cos^5 x\, dx = \int (1 - \sin^2 x)^2 \cos x\, dx$

$= \int (1 - 2\sin^2 x + \sin^4 x) \cos x\, dx$

$= \sin x - \dfrac{2}{3}\sin^3 x + \dfrac{\sin^5 x}{5} + C$

(c) $\int \cos^7 x\, dx = \int (1 - \sin^2 x)^3 \cos x\, dx$

$= \int (1 - 3\sin^2 x + 3\sin^4 x - \sin^6 x) \cos x\, dx$

$= \sin x - \sin^3 x + \dfrac{3}{5}\sin^5 x - \dfrac{1}{7}\sin^7 x + C$

(d) $\int \cos^{15} x\, dx = \int (1 - \cos^2 x)^7 \cos x\, dx$

You would expand $(1 - \cos^2 x)^7$.

102. (a) $\int \tan^3 x\, dx = \int (\sec^2 x - 1) \tan x\, dx$

$= \int \sec^2 x \tan x\, dx - \int \tan x\, dx$

$= \dfrac{\tan^2 x}{2} - \int \tan x\, dx$

$\int \tan^3 x\, dx = \dfrac{\tan^2 x}{2} + \ln|\cos x| + C$

(b) $\int \tan^5 x\, dx = \int (\sec^2 x - 1) \tan^3 x\, dx$

$= \dfrac{\tan^4 x}{4} - \int \tan^3 x\, dx$

(c) $\int \tan^{2k+1} x\, dx = \int (\sec^2 x - 1) \tan^{2k-1} x\, dx$

$= \dfrac{\tan^{2k} x}{2k} - \int \tan^{2k-1} x\, dx$

(d) You would use these formulas recursively.

103. Let $f(x) = \frac{1}{2}\left(x\sqrt{x^2 + 1} + \ln\left|x + \sqrt{x^2 + 1}\right|\right) + C.$

$$f'(x) = \frac{1}{2}\left(x\frac{1}{2}(x^2 + 1)^{-1/2}(2x) + \sqrt{x^2 + 1} + \frac{1}{x + \sqrt{x^2 + 1}}\left(1 + \frac{1}{2}(x^2 + 1)^{-1/2}(2x)\right)\right)$$

$$= \frac{1}{2}\left(\frac{x^2}{\sqrt{x^2 + 1}} + \sqrt{x^2 + 1} + \frac{1}{x + \sqrt{x^2 + 1}}\left(1 + \frac{x}{\sqrt{x^2 + 1}}\right)\right)$$

$$= \frac{1}{2}\left(\frac{x^2 + (x^2 + 1)}{\sqrt{x^2 + 1}} + \frac{1}{x + \sqrt{x^2 + 1}}\left(\frac{\sqrt{x^2 + 1} + x}{\sqrt{x^2 + 1}}\right)\right)$$

$$= \frac{1}{2}\left(\frac{2x^2 + 1}{\sqrt{x^2 + 1}} + \frac{1}{\sqrt{x^2 + 1}}\right) = \frac{1}{2}\left(\frac{2(x^2 + 1)}{\sqrt{x^2 + 1}}\right) = \sqrt{x^2 + 1}$$

Thus, $\int \sqrt{x^2 + 1}\, dx = \frac{1}{2}\left(x\sqrt{x^2 + 1} + \ln\left|x + \sqrt{x^2 + 1}\right|\right) + C.$

—CONTINUED—

103. —CONTINUED—

Let $g(x) = \frac{1}{2}\left(x\sqrt{x^2+1} + \operatorname{arcsinh}(x)\right)$.

$$g'(x) = \frac{1}{2}\left(x\frac{1}{2}(x^2+1)^{-1/2}(2x) + \sqrt{x^2+1} + \frac{1}{\sqrt{x^2+1}}\right)$$

$$= \frac{1}{2}\left(\frac{x^2}{\sqrt{x^2+1}} + \sqrt{x^2+1} + \frac{1}{\sqrt{x^2+1}}\right)$$

$$= \frac{1}{2}\left(\frac{x^2 + (x^2+1) + 1}{\sqrt{x^2+1}}\right)$$

$$= \frac{1}{2}\left(\frac{2(x^2+1)}{\sqrt{x^2+1}}\right) = \sqrt{x^2+1}$$

Thus, $\int \sqrt{x^2+1}\,dx = \frac{1}{2}\left(x\sqrt{x^2+1} + \operatorname{arcsinh}(x)\right) + C.$

104. Let $I = \int_2^4 \frac{\sqrt{\ln(9-x)}}{\sqrt{\ln(9-x)} + \sqrt{\ln(x+3)}}\,dx.$

I is defined and continuous on $[2, 4]$. Note the symmetry: as x goes from 2 to 4, $9 - x$ goes from 7 to 5 and $x + 3$ goes from 5 to 7. So, let $y = 6 - x$, $dy = -dx$.

$$I = \int_4^2 \frac{\sqrt{\ln(3+y)}}{\sqrt{\ln(3+y)} + \sqrt{\ln(9-y)}}(-dy) = \int_2^4 \frac{\sqrt{\ln(3+y)}}{\sqrt{\ln(3+y)} + \sqrt{\ln(9-y)}}\,dy$$

Adding:

$$2I = \int_2^4 \frac{\sqrt{\ln(9-x)}}{\sqrt{\ln(9-x)} + \sqrt{\ln(x+3)}}\,dx + \int_2^4 \frac{\sqrt{\ln(3+x)}}{\sqrt{\ln(3+x)} + \sqrt{\ln(9-x)}}\,dx = \int_2^4 dx = 2 \Rightarrow I = 1$$

You can easily check this result numerically.

Section 8.2 Integration by Parts

1. $\frac{d}{dx}[\sin x - x\cos x] = \cos x - (-x\sin x + \cos x) = x\sin x$

Matches (b)

2. $\frac{d}{dx}[x^2\sin x + 2x\cos x - 2\sin x] = x^2\cos x + 2x\sin x - 2x\sin x + 2\cos x - 2\cos x = x^2\cos x$

Matches (d)

3. $\frac{d}{dx}[x^2e^x - 2xe^x + 2e^x] = x^2e^x + 2xe^x - 2xe^x - 2e^x + 2e^x$

$= x^2e^x$

Matches (c)

4. $\frac{d}{dx}[-x + x\ln x] = -1 + x\left(\frac{1}{x}\right) + \ln x = \ln x$

Matches (a)

5. $\int xe^{2x}\,dx$

$u = x,\ dv = e^{2x}\,dx$

6. $\int x^2e^{2x}\,dx$

$u = x^2,\ dv = e^{2x}\,dx$

7. $\int (\ln x)^2\,dx$

$u = (\ln x)^2,\ dv = dx$

8. $\int \ln 3x\,dx$

$u = \ln 3x,\ dv = dx$

9. $\int x \sec^2 x\,dx$

$u = x,\ dv = \sec^2 x\,dx$

10. $\int x^2 \cos x\,dx$

$u = x^2,\ dv = \cos x\,dx$

11. $dv = e^{-2x}\,dx \Rightarrow v = \int e^{-2x}\,dx = -\frac{1}{2}e^{-2x}$

$u = x \Rightarrow du = dx$

$$\int xe^{-2x}\,dx = -\frac{1}{2}xe^{-2x} - \int -\frac{1}{2}e^{-2x}\,dx$$
$$= -\frac{1}{2}xe^{-2x} - \frac{1}{4}e^{-2x} + C$$
$$= \frac{-1}{4e^{2x}}(2x+1) + C$$

12. $dv = e^{-x}\,dx \Rightarrow v = \int e^{-x}\,dx = -e^{-x}$

$u = x \Rightarrow du = dx$

$$2\int \frac{x}{e^x}\,dx = 2\int xe^{-x}\,dx$$
$$= 2\left[-xe^{-x} - \int -e^{-x}\,dx\right]$$
$$= 2[-xe^{-x} - e^{-x}] + C$$
$$= -2xe^{-x} - 2e^{-x} + C$$

13. Use integration by parts three times.

(1) $dv = e^x\,dx \Rightarrow v = \int e^x\,dx = e^x$

$u = x^3 \Rightarrow du = 3x^2\,dx$

(2) $dv = e^x\,dx \Rightarrow v = \int e^x\,dx = e^x$

$u = x^2 \Rightarrow du = 2x\,dx$

(3) $dv = e^x\,dx \Rightarrow v = \int e^x\,dx = e^x$

$u = x \Rightarrow du = dx$

$$\int x^3e^x\,dx = x^3e^x - 3\int x^2e^x\,dx = x^3e^x - 3x^2e^x + 6\int xe^x\,dx$$
$$= x^3e^x - 3x^2e^x + 6xe^x - 6e^x + C = e^x(x^3 - 3x^2 + 6x - 6) + C$$

14. $\int \frac{e^{1/t}}{t^2}\,dt = -\int e^{1/t}\left(\frac{-1}{t^2}\right)dt = -e^{1/t} + C$

15. $\int x^2e^{x^3}\,dx = \frac{1}{3}\int e^{x^3}(3x^2)\,dx = \frac{1}{3}e^{x^3} + C$

16. $dv = x^4\,dx \Rightarrow v = \frac{x^5}{5}$

$u = \ln x \Rightarrow du = \frac{1}{x}\,dx$

$$\int x^4 \ln x\,dx = \frac{x^5}{5}\ln x - \int \frac{x^5}{5}\left(\frac{1}{x}\right)dx$$
$$= \frac{x^5}{5}\ln x - \frac{1}{5}\int x^4\,dx$$
$$= \frac{x^5}{5}\ln x - \frac{1}{25}x^5 + C$$
$$= \frac{x^5}{25}(5\ln x - 1) + C$$

17. $dv = t\,dt \Rightarrow v = \int t\,dt = \frac{t^2}{2}$

$u = \ln(t+1) \Rightarrow du = \frac{1}{t+1}\,dt$

$$\int t\ln(t+1)\,dt = \frac{t^2}{2}\ln(t+1) - \frac{1}{2}\int \frac{t^2}{t+1}\,dt$$
$$= \frac{t^2}{2}\ln(t+1) - \frac{1}{2}\int \left(t - 1 + \frac{1}{t+1}\right)dt$$
$$= \frac{t^2}{2}\ln(t+1) - \frac{1}{2}\left[\frac{t^2}{2} - t + \ln(t+1)\right] + C$$
$$= \frac{1}{4}[2(t^2-1)\ln|t+1| - t^2 + 2t] + C$$

18. Let $u = \ln x$, $du = \frac{1}{x}\,dx$.

$$\int \frac{1}{x(\ln x)^3}\,dx = \int (\ln x)^{-3}\left(\frac{1}{x}\right)dx = \frac{-1}{2(\ln x)^2} + C$$

19. Let $u = \ln x$, $du = \frac{1}{x}\,dx$.

$$\int \frac{(\ln x)^2}{x}\,dx = \int (\ln x)^2\left(\frac{1}{x}\right)dx = \frac{(\ln x)^3}{3} + C$$

20. $dv = \frac{1}{x^2}\,dx \implies v = \int \frac{1}{x^2}\,dx = -\frac{1}{x}$

$u = \ln x \implies du = \frac{1}{x}\,dx$

$$\int \frac{\ln x}{x^2}\,dx = -\frac{\ln x}{x} + \int \frac{1}{x^2}\,dx = -\frac{\ln x}{x} - \frac{1}{x} + C$$

21. $dv = \frac{1}{(2x+1)^2}\,dx \implies v = \int (2x+1)^{-2}\,dx = -\frac{1}{2(2x+1)}$

$u = xe^{2x} \implies du = (2xe^{2x} + e^{2x})\,dx = e^{2x}(2x+1)\,dx$

$$\int \frac{xe^{2x}}{(2x+1)^2}\,dx = -\frac{xe^{2x}}{2(2x+1)} + \int \frac{e^{2x}}{2}\,dx = \frac{-xe^{2x}}{2(2x+1)} + \frac{e^{2x}}{4} + C = \frac{e^{2x}}{4(2x+1)} + C$$

22. $dv = \frac{x}{(x^2+1)^2}\,dx \implies v = \int (x^2+1)^{-2}\,x\,dx = -\frac{1}{2(x^2+1)}$

$u = x^2e^{x^2} \implies du = (2x^3e^{x^2} + 2xe^{x^2})\,dx = 2xe^{x^2}(x^2+1)\,dx$

$$\int \frac{x^3e^{x^2}}{(x^2+1)^2}\,dx = -\frac{x^2e^{x^2}}{2(x^2+1)} + \int xe^{x^2}\,dx = -\frac{x^2e^{x^2}}{2(x^2+1)} + \frac{e^{x^2}}{2} + C = \frac{e^{x^2}}{2(x^2+1)} + C$$

23. Use integration by parts twice.

(1) $dv = e^x\,dx \implies v = \int e^x\,dx = e^x$

$u = x^2 \implies du = 2x\,dx$

(2) $dv = e^x\,dx \implies v = \int e^x\,dx = e^x$

$u = x \implies du = dx$

$$\int (x^2-1)e^x\,dx = \int x^2e^x\,dx - \int e^x\,dx = x^2e^x - 2\int xe^x\,dx - e^x$$

$$= x^2e^x - 2\left[xe^x - \int e^x\,dx\right] - e^x = x^2e^x - 2xe^x + e^x + C = (x-1)^2e^x + C$$

24. $dv = \frac{1}{x^2}\,dx \implies v = \int \frac{1}{x^2}\,dx = -\frac{1}{x}$

$u = \ln 2x \implies du = \frac{1}{x}\,dx$

$$\int \frac{\ln(2x)}{x^2}\,dx = -\frac{\ln(2x)}{x} + \int \frac{1}{x^2}\,dx = -\frac{\ln(2x)}{x} - \frac{1}{x} + C = -\frac{\ln(2x)+1}{x} + C$$

25. $dv = \sqrt{x-1}\,dx \implies v = \int (x-1)^{1/2}\,dx = \frac{2}{3}(x-1)^{3/2}$

$u = x \implies du = dx$

$$\int x\sqrt{x-1}\,dx = \frac{2}{3}x(x-1)^{3/2} - \frac{2}{3}\int (x-1)^{3/2}\,dx$$

$$= \frac{2}{3}x(x-1)^{3/2} - \frac{4}{15}(x-1)^{5/2} + C$$

$$= \frac{2(x-1)^{3/2}}{15}(3x+2) + C$$

26. $dv = \frac{1}{\sqrt{2+3x}}\,dx \implies v = \int (2+3x)^{-1/2}\,dx = \frac{2}{3}\sqrt{2+3x}$

$u = x \implies du = dx$

$$\int \frac{x}{\sqrt{2+3x}}\,dx = \frac{2x\sqrt{2+3x}}{3} - \frac{2}{3}\int \sqrt{2+3x}\,dx$$

$$= \frac{2x\sqrt{2+3x}}{3} - \frac{4}{27}(2+3x)^{3/2} + C = \frac{2\sqrt{2+3x}}{27}[9x - 2(2+3x)] + C = \frac{2\sqrt{2+3x}}{27}(3x-4) + C$$

27. $dv = \cos x\,dx \Rightarrow v = \int \cos x\,dx = \sin x$

$u = x \Rightarrow du = dx$

$$\int x \cos x\,dx = x \sin x - \int \sin x\,dx = x \sin x + \cos x + C$$

28. $dv = \sin x\,dx \Rightarrow v = -\cos x$

$u = x \Rightarrow du = dx$

$$\int x \sin dx = -x \cos x - \int -\cos x\,dx$$
$$= -x \cos x + \sin x + C$$

29. Use integration by parts three times.

(1) $u = x^3, du = 3x^2\,dx, dv = \sin x\,dx, v = -\cos x$

$$\int x^3 \sin dx = -x^3 \cos x + 3\int x^2 \cos x\,dx$$

(2) $u = x^2, du = 2x\,dx, dv = \cos x\,dx, v = \sin x$

$$\int x^3 \sin x\,dx = -x^3 \cos x + 3\left[x^2 \sin x - 2\int x \sin x\,dx\right]$$
$$= -x^3 \cos x + 3x^2 \sin x - 6\int x \sin x\,dx$$

(3) $u = x, du = dx, dv = \sin x\,dx, v = -\cos x$

$$\int x^3 \sin x\,dx = -x^3 \cos x + 3x^2 \sin x - 6\left[-x \cos x + \int \cos x\,dx\right]$$
$$= -x^3 \cos x + 3x^2 \sin x + 6x \cos x - 6 \sin x + C$$

30. Use integration by parts twice.

(1) $u = x^2, du = 2x\,dx, dv = \cos x\,dx, v = \sin x$

$$\int x^2 \cos x\,dx = x^2 \sin x - 2\int x \sin x\,dx$$

(2) $u = x, du = dx, dv = \sin x\,dx, v = -\cos x$

$$\int x^2 \cos x\,dx = x^2 \sin x - 2\left[-x \cos x + \int \cos x\,dx\right]$$
$$= x^2 \sin x + 2x \cos x - 2 \sin x + C$$

31. $u = t, du = dt, dv = \csc t \cot dt, v = -\csc t$

$$\int t \csc t \cot t\,dt = -t \csc t + \int \csc t\,dt$$
$$= -t \csc t - \ln|\csc t + \cot t| + C$$

32. $dv = \sec \theta \tan \theta\,d\theta \Rightarrow v = \int \sec \theta \tan \theta\,d\theta = \sec \theta$

$u = \theta \Rightarrow du = d\theta$

$$\int \theta \sec \theta \tan \theta\,d\theta = \theta \sec \theta - \int \sec \theta\,d\theta$$
$$= \theta \sec \theta - \ln|\sec \theta + \tan \theta| + C$$

33. $dv = dx \Rightarrow v = \int dx = x$

$u = \arctan x \Rightarrow du = \frac{1}{1 + x^2}\,dx$

$$\int \arctan x\,dx = x \arctan x - \int \frac{x}{1 + x^2}\,dx$$
$$= x \arctan x - \frac{1}{2}\ln(1 + x^2) + C$$

34. $dv = dx \Rightarrow v = \int dx = x$

$u = \arccos x \Rightarrow du = -\frac{1}{\sqrt{1 - x^2}}\,dx$

$$4\int \arccos x\,dx = 4\left[x \arccos x + \int \frac{x}{\sqrt{1 - x^2}}\,dx\right]$$
$$= 4\left[x \arccos x - \sqrt{1 - x^2}\right] + C$$

35. Use integration by parts twice.

(1) $dv = e^{2x}\,dx \Rightarrow v = \int e^{2x}\,dx = \frac{1}{2}e^{2x}$

$u = \sin x \Rightarrow du = \cos x\,dx$

(2) $dv = e^{2x}\,dx \Rightarrow v = \int e^{2x}\,dx = \frac{1}{2}e^{2x}$

$u = \cos x \Rightarrow du = -\sin x\,dx$

—CONTINUED—

35. —CONTINUED—

$$\int e^{2x}\sin x\,dx = \frac{1}{2}e^{2x}\sin x - \frac{1}{2}\int e^{2x}\cos x\,dx = \frac{1}{2}e^{2x}\sin x - \frac{1}{2}\left(\frac{1}{2}e^{2x}\cos x + \frac{1}{2}\int e^{2x}\sin x\,dx\right)$$

$$\frac{5}{4}\int e^{2x}\sin x\,dx = \frac{1}{2}e^{2x}\sin x - \frac{1}{4}e^{2x}\cos x$$

$$\int e^{2x}\sin x\,dx = \frac{1}{5}e^{2x}(2\sin x - \cos x) + C$$

36. Use integration by parts twice.

(1) $dv = e^x\,dx \implies v = \int e^x\,dx = e^x$

$u = \cos 2x \implies du = -2\sin 2x\,dx$

(2) $dv = e^x\,dx \implies v = \int e^x\,dx = e^x$

$u = \sin 2x \implies du = 2\cos 2x\,dx$

$$\int e^x\cos 2x\,dx = e^x\cos 2x + 2\int e^x\sin 2x\,dx = e^x\cos 2x + 2\left(e^x\sin 2x - 2\int e^x\cos 2x\,dx\right)$$

$$5\int e^x\cos 2x\,dx = e^x\cos 2x + 2e^x\sin 2x$$

$$\int e^x\cos 2x\,dx = \frac{e^x}{5}(\cos 2x + 2\sin 2x) + C$$

37. $y' = xe^{x^2}$

$$y = \int xe^{x^2}\,dx = \frac{1}{2}e^{x^2} + C$$

38. $dv = dx \implies v = x$

$u = \ln x \implies du = \frac{1}{x}\,dx$

$y' = \ln x$

$$y = \int \ln x\,dx = x\ln x - \int x\left(\frac{1}{x}\right)dx$$

$$= x\ln x - x + C = x(-1 + \ln x) + C$$

39. Use integration by parts twice.

(1) $dv = \frac{1}{\sqrt{2+3t}}\,dt \implies v = \int (2+3t)^{-1/2}\,dt = \frac{2}{3}\sqrt{2+3t}$

$u = t^2 \implies du = 2t\,dt$

(2) $dv = \sqrt{2+3t}\,dt \implies v = \int (2+3t)^{1/2}\,dt = \frac{2}{9}(2+3t)^{3/2}$

$u = t \implies du = dt$

$$y = \int \frac{t^2}{\sqrt{2+3t}}\,dt = \frac{2t^2\sqrt{2+3t}}{3} - \frac{4}{3}\int t\sqrt{2+3t}\,dt$$

$$= \frac{2t^2\sqrt{2+3t}}{3} - \frac{4}{3}\left[\frac{2t}{9}(2+3t)^{3/2} - \frac{2}{9}\int (2+3t)^{3/2}\,dt\right]$$

$$= \frac{2t^2\sqrt{2+3t}}{3} - \frac{8t}{27}(2+3t)^{3/2} + \frac{16}{405}(2+3t)^{5/2} + C$$

$$= \frac{2\sqrt{2+3t}}{405}(27t^2 - 24t + 32) + C$$

40. Use integration by parts twice.

(1) $dv = \sqrt{x-1}\,dx \Rightarrow v = \int (x-1)^{1/2}\,dx = \frac{2}{3}(x-1)^{3/2}$

$u = x^2 \Rightarrow du = 2x\,dx$

(2) $dv = (x-1)^{3/2}\,dx \Rightarrow v = \int (x-1)^{3/2}\,dx = \frac{2}{5}(x-1)^{5/2}$

$u = x \Rightarrow du = dx$

$$y = \int x^2\sqrt{x-1}\,dx$$

$$= \frac{2}{3}x^2(x-1)^{3/2} - \frac{4}{3}\int x(x-1)^{3/2}\,dx = \frac{2}{3}x^2(x-1)^{3/2} - \frac{4}{3}\left[\frac{2}{5}x(x-1)^{5/2} - \frac{2}{5}\int (x-1)^{5/2}\,dx\right]$$

$$= \frac{2}{3}x^2(x-1)^{3/2} - \frac{8}{15}x(x-1)^{5/2} + \frac{16}{105}(x-1)^{7/2} + C = \frac{2(x-1)^{3/2}}{105}(15x^2 + 12x + 8) + C$$

41. $(\cos y)y' = 2x$

$$\int \cos y\,dy = \int 2x\,dx$$

$$\sin y = x^2 + C$$

42. $dv = dx \Rightarrow v = \int dx = x$

$u = \arctan\frac{x}{2} \Rightarrow du = \frac{1}{1 + (x/2)^2}\left(\frac{1}{2}\right)dx = \frac{2}{4 + x^2}\,dx$

$$y = \int \arctan\frac{x}{2}\,dx = x\arctan\frac{x}{2} - \int \frac{2x}{4 + x^2}\,dx = x\arctan\frac{x}{2} - \ln(4 + x^2) + C$$

43. (a)

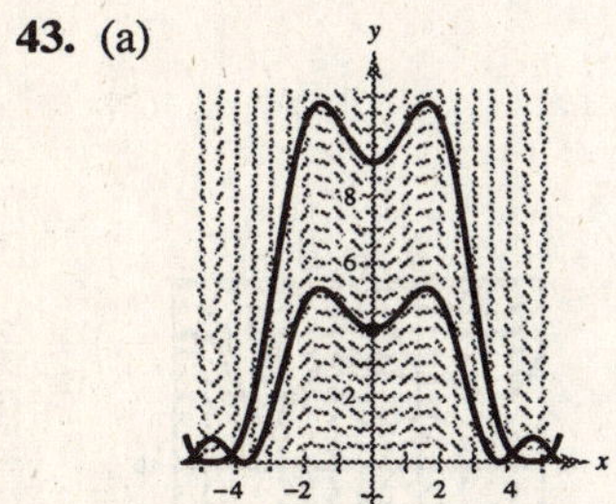

(b)

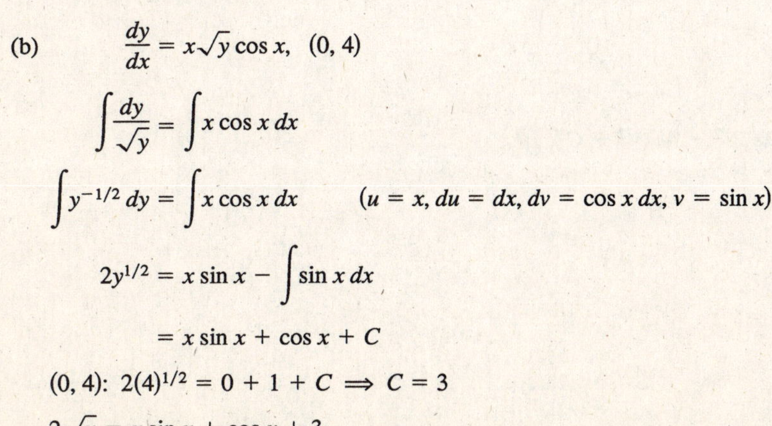

$$\frac{dy}{dx} = x\sqrt{y}\cos x, \quad (0, 4)$$

$$\int \frac{dy}{\sqrt{y}} = \int x\cos x\,dx$$

$$\int y^{-1/2}\,dy = \int x\cos x\,dx \qquad (u = x,\ du = dx,\ dv = \cos x\,dx,\ v = \sin x)$$

$$2y^{1/2} = x\sin x - \int \sin x\,dx$$

$$= x\sin x + \cos x + C$$

$(0, 4)$: $2(4)^{1/2} = 0 + 1 + C \Rightarrow C = 3$

$2\sqrt{y} = x\sin x + \cos x + 3$

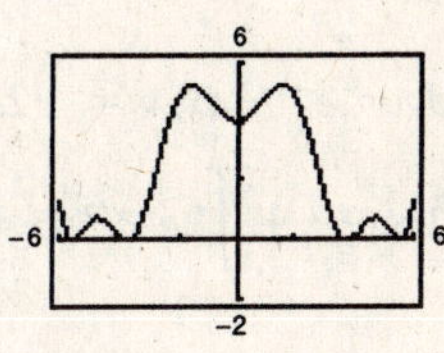

44. (a)

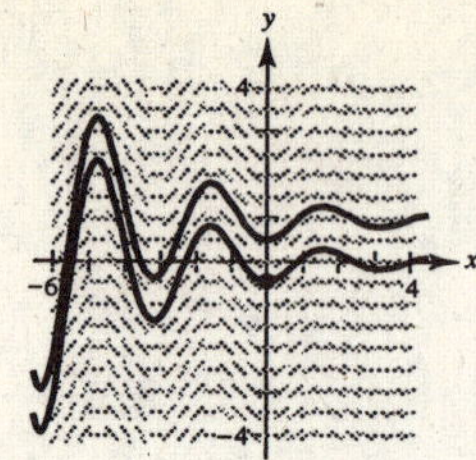

(b) $\dfrac{dy}{dx} = e^{-x/3} \sin 2x, \quad \left(0, -\dfrac{18}{37}\right)$

$$y = \int e^{-x/3} \sin 2x \, dx$$

Use integration by parts twice.

(1) $u = \sin 2x, \ du = 2 \cos 2x$

$dv = e^{-x/3} \, dx, v = -3e^{-x/3}$

$$\int e^{-x/3} \sin 2x \, dx = -3e^{-x/3} \sin 2x + \int 6e^{-x/3} \cos 2x \, dx$$

(2) $u = \cos 2x, \ du = -2 \sin 2x$

$dv = e^{-x/3} \, dx, v = -3e^{-x/3}$

$$\int e^{-x/3} \sin 2x \, dx = -3e^{-x/3} \sin 2x + 6\left[-3e^{-x/3} \cos 2x - \int 6e^{-x/3} \sin 2x \, dx\right] + C$$

$$37\int e^{-x/3} \sin 2x \, dx = -3e^{-x/3} \sin 2x - 18e^{-x/3} \cos 2x + C$$

$$y = \int e^{-x/3} \sin 2x \, dx = \frac{1}{37}\left[-3e^{-x/3} \sin 2x - 18e^{-x/3} \cos 2x\right] + C$$

$$\left(0, \frac{-18}{37}\right): \ \frac{-18}{37} = \frac{1}{37}[0 - 18] + C \Rightarrow C = 0$$

$$y = \frac{-1}{37}[3e^{-x/3} \sin 2x + 18e^{-x/3} \cos 2x]$$

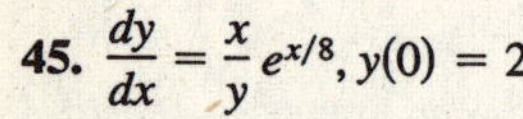

45. $\dfrac{dy}{dx} = \dfrac{x}{y} e^{x/8}, y(0) = 2$

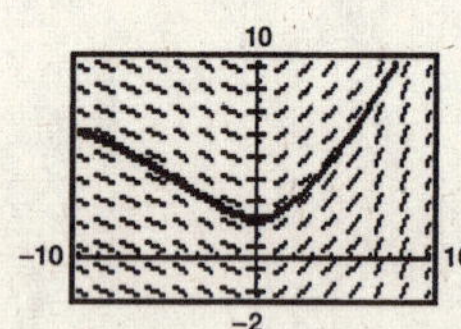

46. $\dfrac{dy}{dx} = \dfrac{x}{y} \sin x, y(0) = 4$

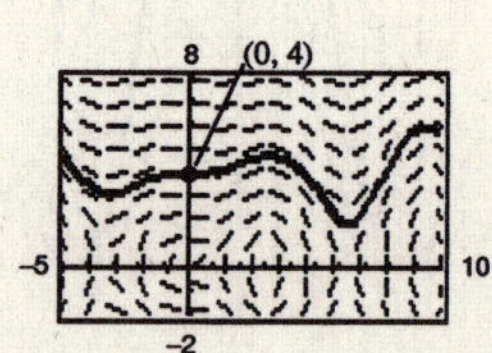

47. $u = x, du = dx, dv = e^{-x/2} \, dx, v = -2e^{-x/2}$

$$\int xe^{-x/2} \, dx = -2xe^{-x/2} + \int 2e^{-x/2} \, dx = -2xe^{-x/2} - 4e^{-x/2} + C$$

Thus, $\displaystyle\int_0^4 xe^{-x/2} \, dx = \left[-2xe^{-x/2} - 4e^{-x/2}\right]_0^4$

$$= -8e^{-2} - 4e^{-2} + 4$$

$$= -12e^{-2} + 4 \approx 2.376.$$

48. See Exercise 3.

$$\int_0^1 x^2e^x\,dx = \Big[x^2e^x - 2xe^x + 2e^x\Big]_0^1 = e - 2 \approx 0.718$$

49. See Exercise 27.

$$\int_0^{\pi/2} x\cos x\,dx = \Big[x\sin x + \cos x\Big]_0^{\pi/2} = \frac{\pi}{2} - 1$$

50. $dv = \sin 2x\,dx \Rightarrow \quad v = \int \sin 2x\,dx = -\frac{1}{2}\cos 2x$

$u = x \qquad \Rightarrow du = dx$

$$\int x\sin 2x\,dx = \frac{-1}{2}x\cos 2x + \frac{1}{2}\int \cos 2x\,dx$$

$$= \frac{-1}{2}x\cos 2x + \frac{1}{4}\sin 2x + C$$

$$= \frac{1}{4}(\sin 2x - 2x\cos 2x) + C$$

Thus, $\displaystyle\int_0^{\pi} x\sin 2x\,dx = \left[\frac{1}{4}(\sin 2x - 2x\cos 2x)\right]_0^{\pi} = -\frac{\pi}{2}.$

51. $u = \arccos x,\ du = -\dfrac{1}{\sqrt{1-x^2}}dx,\ dv = dx,\ v = x$

$$\int \arccos x\,dx = x\arccos x + \int \frac{x}{\sqrt{1-x^2}}dx$$

$$= x\arccos x - \sqrt{1-x^2} + C$$

Thus, $\displaystyle\int_0^{1/2} \arccos x = \Big[x\arccos x - \sqrt{1-x^2}\Big]_0^{1/2}$

$$= \frac{1}{2}\arccos\left(\frac{1}{2}\right) - \sqrt{\frac{3}{4}} + 1$$

$$= \frac{\pi}{6} - \frac{\sqrt{3}}{2} + 1 \approx 0.658.$$

52. $dv = x\,dx \qquad \Rightarrow \quad v = \int x\,dx = \dfrac{x^2}{2}$

$u = \arcsin x^2 \Rightarrow du = \dfrac{2x}{\sqrt{1-x^4}}dx$

$$\int x\arcsin x^2\,dx = \frac{x^2}{2}\arcsin x^2 - \int \frac{x^3}{\sqrt{1-x^4}}dx$$

$$= \frac{x^2}{2}\arcsin x^2 + \frac{1}{4}(2)(1-x^4)^{1/2} + C$$

$$= \frac{1}{2}\Big[x^2\arcsin x^2 + \sqrt{1-x^4}\Big] + C$$

Thus, $\displaystyle\int_0^1 x\arcsin x^2\,dx = \frac{1}{2}\Big[x^2\arcsin x^2 + \sqrt{1-x^4}\Big]_0^1 = \frac{1}{4}(\pi - 2).$

53. Use integration by parts twice.

(1) $dv = e^x\,dx \Rightarrow \quad v = \int e^x\,dx = e^x$

$u = \sin x \Rightarrow du = \cos x\,dx$

(2) $dv = e^x\,dx \Rightarrow \quad v = \int e^x\,dx = e^x$

$u = \cos x \Rightarrow du = -\sin x\,dx$

$$\int e^x\sin x\,dx = e^x\sin x - \int e^x\cos x\,dx = e^x\sin x - e^x\cos x - \int e^x\sin x\,dx$$

$$2\int e^x\sin x\,dx = e^x(\sin x - \cos x)$$

$$\int e^x\sin x\,dx = \frac{e^x}{2}(\sin x - \cos x) + C$$

Thus, $\displaystyle\int_0^1 e^x\sin x\,dx = \left[\frac{e^x}{2}(\sin x - \cos x)\right]_0^1 = \frac{e}{2}(\sin 1 - \cos 1) + \frac{1}{2} = \frac{e(\sin 1 - \cos 1) + 1}{2} \approx 0.909.$

54. Use integration by parts twice.

(1) $dv = e^{-x}$, $v = -e^{-x}$, $u = \cos x$, $du = -\sin x\,dx$

$$\int e^{-x}\cos x\,dx = -e^{-x}\cos x - \int e^{-x}\sin x\,dx$$

(2) $dv = e^{-x}\,dx$, $v = -e^{-x}$, $u = \sin x$, $du = \cos x\,dx$

$$\int e^{-x}\cos x\,dx = -e^{-x}\cos x - \left[-e^{-x}\sin x + \int e^{-x}\cos x\,dx\right] \Rightarrow 2\int e^{-x}\cos x\,dx = e^{-x}\sin x - e^{-x}\cos x$$

Thus, $\displaystyle\int_0^2 e^{-x}\cos x\,dx = \left[\frac{e^{-x}\sin x - e^{-x}\cos x}{2}\right]_0^2 = \frac{-e^{-2}}{2}[\sin 2 - \cos 2] + \frac{1}{2}.$

55. $dv = x^2\,dx$, $v = \dfrac{x^3}{3}$, $u = \ln x$, $du = \dfrac{1}{x}\,dx$

$$\int x^2 \ln x\,dx = \frac{x^3}{3}\ln x - \int \frac{x^3}{3}\left(\frac{1}{x}\right)dx$$

$$= \frac{x^3}{3}\ln x - \frac{1}{3}\int x^2\,dx$$

Hence, $\displaystyle\int_1^2 x^2\ln x\,dx = \left[\frac{x^3}{3}\ln x - \frac{1}{9}x^3\right]_1^2$

$$= \frac{8}{3}\ln 2 - \frac{8}{9} + \frac{1}{9}$$

$$= \frac{8}{3}\ln 2 - \frac{7}{9} \approx 1.071.$$

56. $dv = dx \quad \Rightarrow \quad v = \displaystyle\int dx = x$

$$u = \ln(1 + x^2) \Rightarrow du = \frac{2x}{1 + x^2}\,dx$$

$$\int \ln(1 + x^2)\,dx = x\ln(1 + x^2) - \int \frac{2x^2}{1 + x^2}\,dx$$

$$= x\ln(1 + x^2) - 2\int\left[1 - \frac{1}{1 + x^2}\right]dx$$

$$= x\ln(1 + x^2) - 2x + 2\arctan x + C$$

Thus,

$$\int_0^1 \ln(1 + x^2)\,dx = \left[x\ln(1 + x^2) - 2x + 2\arctan x\right]_0^1$$

$$= \ln 2 - 2 + \frac{\pi}{2}.$$

57. $dv = x\,dx$, $v = \dfrac{x^2}{2}$, $u = \operatorname{arcsec} x$, $du = \dfrac{1}{x\sqrt{x^2 - 1}}\,dx$

$$\int x \operatorname{arcsec} x\,dx = \frac{x^2}{2}\operatorname{arcsec} x - \int \frac{x^2/2}{x\sqrt{x^2 - 1}}\,dx$$

$$= \frac{x^2}{2}\operatorname{arcsec} x - \frac{1}{4}\int \frac{2x}{\sqrt{x^2 - 1}}\,dx$$

$$= \frac{x^2}{2}\operatorname{arcsec} x - \frac{1}{2}\sqrt{x^2 - 1} + C$$

Hence,

$$\int_2^4 x\operatorname{arcsec} x\,dx = \left[\frac{x^2}{2}\operatorname{arcsec} x - \frac{1}{2}\sqrt{x^2 - 1}\right]_2^4$$

$$= \left(8\operatorname{arcsec} 4 - \frac{\sqrt{15}}{2}\right) - \left(\frac{2\pi}{3} - \frac{\sqrt{3}}{2}\right)$$

$$= 8\operatorname{arcsec} 4 - \frac{\sqrt{15}}{2} + \frac{\sqrt{3}}{2} - \frac{2\pi}{3}$$

$$\approx 7.380.$$

58. $u = x$, $du = dx$, $dv = \sec^2 x\,dx$, $v = \tan x$

$$\int x\sec^2 x\,dx = x\tan x - \int \tan x\,dx$$

Hence,

$$\int_0^{\pi/4} x\sec^2 x\,dx = \left[x\tan x + \ln|\cos x|\right]_0^{\pi/4}$$

$$= \left(\frac{\pi}{4} + \ln\frac{\sqrt{2}}{2}\right) - 0$$

$$= \frac{\pi}{4} - \frac{1}{2}\ln 2.$$

59. $\int x^2e^{2x}\,dx = x^2\left(\frac{1}{2}e^{2x}\right) - (2x)\left(\frac{1}{4}e^{2x}\right) + 2\left(\frac{1}{8}e^{2x}\right) + C$

$= \frac{1}{2}x^2e^{2x} - \frac{1}{2}xe^{2x} + \frac{1}{4}e^{2x} + C$

$= \frac{1}{4}e^{2x}(2x^2 - 2x + 1) + C$

Alternate signs	u and its derivatives	v' and its antiderivatives
+	x^2	e^{2x}
−	$2x$	$\frac{1}{2}e^{2x}$
+	2	$\frac{1}{4}e^{2x}$
−	0	$\frac{1}{8}e^{2x}$

60. $\int x^3e^{-2x}\,dx = x^3\left(-\frac{1}{2}e^{-2x}\right) - 3x^2\left(\frac{1}{4}e^{-2x}\right) + 6x\left(-\frac{1}{8}e^{-2x}\right) - 6\left(\frac{1}{16}e^{-2x}\right) + C$

$= -\frac{1}{8}e^{-2x}(4x^3 + 6x^2 + 6x + 3) + C$

Alternate signs	u and its derivatives	v' and its antiderivatives
+	x^3	e^{-2x}
−	$3x^2$	$-\frac{1}{2}e^{-2x}$
+	$6x$	$\frac{1}{4}e^{-2x}$
−	6	$-\frac{1}{8}e^{-2x}$
+	0	$\frac{1}{16}e^{-2x}$

61. $\int x^3 \sin x\,dx = x^3(-\cos x) - 3x^2(-\sin x) + 6x\cos x - 6\sin x + C$

$= -x^3\cos x + 3x^2\sin x + 6x\cos x - 6\sin x + C$

$= (3x^2 - 6)\sin x - (x^3 - 6x)\cos x + C$

Alternate signs	u and its derivatives	v' and its antiderivatives
+	x^3	$\sin x$
−	$3x^2$	$-\cos x$
+	$6x$	$-\sin x$
−	6	$\cos x$
+	0	$\sin x$

62. $\int x^3\cos 2x\,dx = x^3\left(\frac{1}{2}\sin 2x\right) - 3x^2\left(-\frac{1}{4}\cos 2x\right) + 6x\left(-\frac{1}{8}\sin 2x\right) - 6\left(\frac{1}{16}\cos 2x\right) + C$

$= \frac{1}{2}x^3\sin 2x + \frac{3}{4}x^2\cos 2x - \frac{3}{4}x\sin 2x - \frac{3}{8}\cos 2x + C$

$= \frac{1}{8}[4x^3\sin 2x + 6x^2\cos 2x - 6x\sin 2x - 3\cos 2x] + C$

Alternate signs	u and its derivatives	v' and its antiderivatives
+	x^3	$\cos 2x$
−	$3x^2$	$\frac{1}{2}\sin 2x$
+	$6x$	$-\frac{1}{4}\cos 2x$
−	6	$-\frac{1}{8}\sin 2x$
+	0	$\frac{1}{16}\cos 2x$

63. $\int x\sec^2 x\,dx = x\tan x + \ln|\cos x| + C$

Alternate signs	u and its derivatives	v' and its antiderivatives
+	x	$\sec^2 x$
−	1	$\tan x$
+	0	$-\ln|\cos x|$

64. $\int x^2(x-2)^{3/2}\,dx = \frac{2}{5}x^2(x-2)^{5/2} - \frac{8}{35}x(x-2)^{7/2} + \frac{16}{315}(x-2)^{9/2} + C$

$= \frac{2}{315}(x-2)^{5/2}(35x^2 + 40x + 32) + C$

Alternate signs	u and its derivatives	v' and its antiderivatives
+	x^2	$(x-2)^{3/2}$
−	$2x$	$\frac{2}{5}(x-2)^{5/2}$
+	2	$\frac{4}{35}(x-2)^{7/2}$
−	0	$\frac{8}{315}(x-2)^{9/2}$

65. $u = \sqrt{x} \Rightarrow u^2 = x \Rightarrow 2u\,du = dx$

$\int \sin\sqrt{x}\,dx = \int \sin u(2u\,du) = 2\int u\sin u\,du$

Integration by parts: $w = u, dw = du, dv = \sin u\,du,$ $v = -\cos u$

$$2\int u\sin u\,du = 2\left(-u\cos u + \int \cos u\,du\right)$$
$$= 2(-u\cos u + \sin u) + C$$
$$= 2(-\sqrt{x}\cos\sqrt{x} + \sin\sqrt{x}) + C$$

66. $u = x^2, du = 2x\,dx$

$\int 2x^3\cos(x^2)\,dx = \int x^2\cos(x^2)(2x)\,dx = \int u\cos u\,du$

Integration by parts: $w = u, dw = du, dv = \cos u\,du,$ $v = \sin u$

$$\int u\cos u\,du = u\sin u - \int \sin u\,du$$
$$= u\sin u + \cos u + C$$
$$= x^2\sin(x^2) + \cos(x^2) + C$$

67. Let $u = 4 - x, du = -dx, x = 4 - u.$

$$\int_0^4 x\sqrt{4-x}\,dx = \int_4^0 (4-u)u^{1/2}(-du)$$
$$= \int_0^4 (4u^{1/2} - u^{3/2})\,du$$
$$= \left[\frac{8}{3}u^{3/2} - \frac{2}{5}u^{5/2}\right]_0^4$$
$$= \frac{8}{3}(8) - \frac{2}{5}(32) = \frac{128}{15}$$

68. Let $u = \sqrt{2x}, u^2 = 2x, 2u\,du = 2\,dx.$

$$\int_0^2 e^{\sqrt{2x}}\,dx = \int_0^2 e^u(u\,du)$$
$$= \left[ue^u - e^u\right]_0^2 \quad \text{(Integration by parts)}$$
$$= (2e^2 - e^2) - (0 - 1)$$
$$= e^2 + 1$$

69. Let $w = \ln x, dw = \frac{1}{x}dx, x = e^w, dx = e^w\,dw.$

$\int \cos(\ln x)\,dx = \int \cos w(e^w\,dw)$

Now use integration by parts twice.

$$\int \cos w e^w\,dw = \cos w e^w + \int \sin w e^w\,dw \qquad [u = \cos w, dv = e^w\,dw]$$
$$= \cos w e^w + \left[\sin w e^w - \int \cos w e^w\,dw\right] \qquad [u = \sin w, dv = e^w\,dw]$$
$$2\int \cos w e^w\,dw = \cos w e^w + \sin w e^w$$
$$\int \cos w e^w\,dw = \frac{1}{2}e^w[\cos w + \sin w] + C$$
$$\int \cos(\ln x)\,dx = \frac{1}{2}x[\cos(\ln x) + \sin(\ln x)] + C$$

70. Let $w = 1 + x^2$, $dw = 2x\,dx$, $x^2 = w - 1$, $x = \sqrt{w-1}$.

$$\int \ln(x^2+1)\,dx = \int \ln(w)\frac{dw}{2\sqrt{w-1}}$$

Integration by parts: $u = \ln w$, $du = \frac{1}{w}\,dw$, $dv = \frac{1}{2\sqrt{w-1}}\,dw$, $v = \sqrt{w-1}$

$$\int \ln(x^2+1)\,dx = \ln(w)\sqrt{w-1} - \int \frac{\sqrt{w-1}}{2}\,dw$$

Substitution: $z = \sqrt{w-1}$, $z^2 = w - 1$, $2z\,dz = dw$

$$\int \ln(x^2+1)\,dx = \ln(w)\sqrt{w-1} - \int \frac{z}{z^2+1}(2z\,dz)$$

$$= \ln(w)\sqrt{w-1} - 2\int\left(1 - \frac{1}{z^2+1}\right)dz$$

$$= \ln(w)\sqrt{w-1} - 2z + 2\arctan(z) + C$$

$$= \ln(1+x^2)x - 2x + 2\arctan(x) + C$$

71. Integration by parts is based on the Product Rule.

72. Answers will vary.

73. No
Substitution

74. Yes
$u = \ln x$, $dv = x\,dx$

75. Yes
$u = x^2$, $dv = e^{2x}\,dx$

76. No
Substitution

77. Yes. Let $u = x$ and
$du = \frac{1}{\sqrt{x+1}}\,dx.$
(Substitution also works. Let $u = \sqrt{x+1}$.)

78. No
Substitution

79. (a) $\int t^3 e^{-4t}\,dt = \frac{-e^{-4t}}{128}(32t^3 + 24t^2 + 12t + 3) + C$

(b)

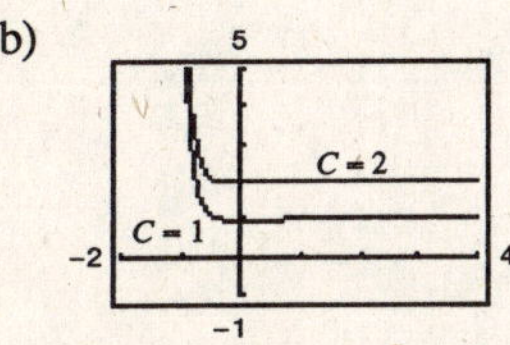

(c) The graphs are vertical translations of each other.

80. (a) $\int \alpha^4 \sin(\pi\alpha)\,d\alpha = \frac{1}{\pi^5}[-(\alpha\pi)^4\cos\pi\alpha + 4(\alpha\pi)^3\sin\pi\alpha + 12(\alpha\pi)^2\cos\pi\alpha - 24(\alpha\pi)\sin\pi\alpha - 24\cos\pi\alpha] + C$

(b)

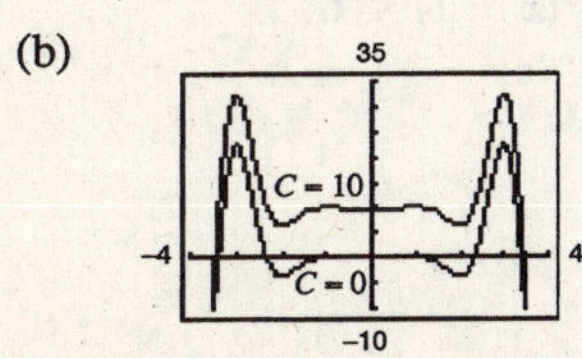

(c) The graphs are vertical translations of each other.

81. (a) $\displaystyle\int e^{-2x}\sin 3x\,dx = \frac{e^{-2x}}{13}[-2\sin 3x - 3\cos 3x] + C$

$$\int_0^{\pi/2} e^{-2x}\sin 3x\,dx = \frac{1}{13}[2e^{-\pi} + 3] \approx 0.2374$$

(b)

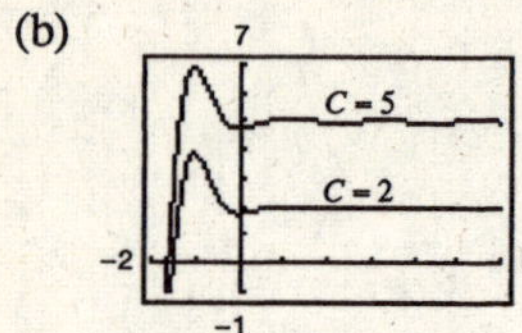

(c) The graphs are vertical translations of each other.

82. (a) $\displaystyle\int x^4(25 - x^2)^{3/2}\,dx = \frac{1{,}171{,}875\arcsin|x/5|}{128} - \frac{x(2x^2 + 25)(25 - x^2)^{5/2}}{16} + \frac{625x(25 - x^2)^{3/2}}{64} + \frac{46{,}875x\sqrt{25 - x^2}}{128} + C$

$$\int_0^5 x^4(25 - x^2)^{2/3}\,dx = \frac{1{,}171{,}875}{256}\pi \approx 14{,}381.0699$$

(b)

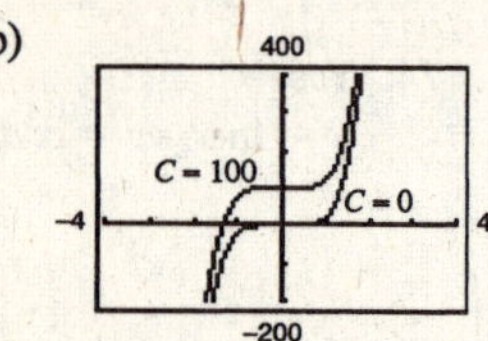

(c) The graphs are vertical translations of each other.

83. (a) $dv = \sqrt{2x - 3}\,dx \Rightarrow \quad v = \displaystyle\int (2x - 3)^{1/2}\,dx = \frac{1}{3}(2x - 3)^{3/2}$

$u = 2x \qquad\qquad \Rightarrow du = 2\,dx$

$$\int 2x\sqrt{2x - 3}\,dx = \frac{2}{3}x(2x - 3)^{3/2} - \frac{2}{3}\int (2x - 3)^{3/2}\,dx$$

$$= \frac{2}{3}x(2x - 3)^{3/2} - \frac{2}{15}(2x - 3)^{5/2} + C$$

$$= \frac{2}{15}(2x - 3)^{3/2}(3x + 3) + C = \frac{2}{5}(2x - 3)^{3/2}(x + 1) + C$$

(b) $u = 2x - 3 \Rightarrow x = \dfrac{u + 3}{2}$ and $dx = \dfrac{1}{2}\,du$

$$\int 2x\sqrt{2x - 3}\,dx = \int 2\left(\frac{u + 3}{2}\right)u^{1/2}\left(\frac{1}{2}\right)du = \frac{1}{2}\int (u^{3/2} + 3u^{1/2})\,du = \frac{1}{2}\left[\frac{2}{5}u^{5/2} + 2u^{3/2}\right] + C$$

$$= \frac{1}{5}u^{3/2}(u + 5) + C = \frac{1}{5}(2x - 3)^{3/2}[(2x - 3) + 5] + C = \frac{2}{5}(2x - 3)^{3/2}(x + 1) + C$$

84. (a) $dv = \sqrt{4+x}\,dx \Rightarrow \quad v = \int (4+x)^{1/2}\,dx = \frac{2}{3}(4+x)^{3/2}$

$u = x \qquad \Rightarrow du = dx$

$$\int x\sqrt{4+x}\,dx = \frac{2}{3}x(4+x)^{3/2} - \frac{2}{3}\int (4+x)^{3/2}\,dx$$

$$= \frac{2}{3}x(4+x)^{3/2} - \frac{4}{15}(4+x)^{5/2} + C = \frac{2}{15}(4+x)^{3/2}(3x-8) + C$$

(b) $u = 4 + x \Rightarrow x = u - 4$ and $dx = du$

$$\int x\sqrt{4+x}\,dx = \int (u-4)u^{1/2}\,du = \int (u^{3/2} - 4u^{1/2})\,du$$

$$= \frac{2}{5}u^{5/2} - \frac{8}{3}u^{3/2} + C = \frac{2}{15}u^{3/2}(3u-20) + C$$

$$= \frac{2}{15}(4+x)^{3/2}[3(4+x) - 20] + C = \frac{2}{15}(4+x)^{3/2}(3x-8) + C$$

85. (a) $dv = \frac{x}{\sqrt{4+x^2}}\,dx \Rightarrow \quad v = \int (4+x^2)^{-1/2}x\,dx = \sqrt{4+x^2}$

$u = x^2 \qquad \Rightarrow du = 2x\,dx$

$$\int \frac{x^3}{\sqrt{4+x^2}}\,dx = x^2\sqrt{4+x^2} - 2\int x\sqrt{4+x^2}\,dx$$

$$= x^2\sqrt{4+x^2} - \frac{2}{3}(4+x^2)^{3/2} + C = \frac{1}{3}\sqrt{4+x^2}\,(x^2-8) + C$$

(b) $u = 4 + x^2 \Rightarrow x^2 = u - 4$ and $2x\,dx = du \Rightarrow x\,dx = \frac{1}{2}\,du$

$$\int \frac{x^3}{\sqrt{4+x^2}}\,dx = \int \frac{x^2}{\sqrt{4+x^2}}\,x\,dx = \int \frac{u-4}{\sqrt{u}}\frac{1}{2}\,du$$

$$= \frac{1}{2}\int (u^{1/2} - 4u^{-1/2})\,du = \frac{1}{2}\left(\frac{2}{3}u^{3/2} - 8u^{1/2}\right) + C$$

$$= \frac{1}{3}u^{1/2}(u-12) + C = \frac{1}{3}\sqrt{4+x^2}\,[(4+x^2) - 12] + C = \frac{1}{3}\sqrt{4+x^2}\,(x^2-8) + C$$

86. (a) $dv = \sqrt{4-x}\,dx \Rightarrow \quad v = \int (4-x)^{1/2}\,dx$

$$= -\frac{2}{3}(4-x)^{3/2}$$

$u = x \qquad \Rightarrow du = dx$

$$\int x\sqrt{4-x}\,dx = -\frac{2}{3}x(4-x)^{3/2} + \frac{2}{3}\int (4-x)^{3/2}\,dx$$

$$= -\frac{2}{3}x(4-x)^{3/2} - \frac{4}{15}(4-x)^{5/2} + C$$

$$= -\frac{2}{15}(4-x)^{3/2}[5x + 2(4-x)] + C$$

$$= -\frac{2}{15}(4-x)^{3/2}(3x+8) + C$$

(b) $u = 4 - x \Rightarrow x = 4 - u$ and $dx = -du$

$$\int x\sqrt{4-x}\,dx = -\int (4-u)\sqrt{u}\,du$$

$$= -\int (4u^{1/2} - u^{3/2})\,du$$

$$= -\frac{8}{3}u^{3/2} + \frac{2}{5}u^{5/2} + C$$

$$= -\frac{2}{15}u^{3/2}(20-3u) + C$$

$$= -\frac{2}{15}(4-x)^{3/2}[20 - 3(4-x)] + C$$

$$= -\frac{2}{15}(4-x)^{3/2}(3x+8) + C$$

87. $n = 0$: $\int \ln x\,dx = x(\ln x - 1) + C$

$n = 1$: $\int x \ln x\,dx = \frac{x^2}{4}(2 \ln x - 1) + C$

$n = 2$: $\int x^2 \ln x\,dx = \frac{x^3}{9}(3 \ln x - 1) + C$

$n = 3$: $\int x^3 \ln x\,dx = \frac{x^4}{16}(4 \ln x - 1) + C$

$n = 4$: $\int x^4 \ln x\,dx = \frac{x^5}{25}(5 \ln x - 1) + C$

In general, $\int x^n \ln x\,dx = \frac{x^{n+1}}{(n+1)^2}[(n+1)\ln x - 1] + C.$

88. $n = 0$: $\int e^x\,dx = e^x + C$

$n = 1$: $\int xe^x\,dx = xe^x - e^x + C = xe^x - \int e^x\,dx$

$n = 2$: $\int x^2e^x\,dx = x^2e^x - 2xe^x + 2e^x + C = x^2e^x - 2\int xe^x\,dx$

$n = 3$: $\int x^3e^x\,dx = x^3e^x - 3x^2e^x + 6xe^x - 6e^x + C = x^3e^x - 3\int x^2e^x\,dx$

$n = 4$: $\int x^4e^x\,dx = x^4e^x - 4x^3e^x + 12x^2e^x - 24xe^x + 24e^x + C = x^4e^x - 4\int x^3e^x\,dx$

In general, $\int x^ne^x\,dx = x^ne^x - n\int x^{n-1}e^x\,dx.$

89. $dv = \sin x\,dx \implies v = -\cos x$

$u = x^n \implies du = nx^{n-1}\,dx$

$\int x^n \sin x\,dx = -x^n \cos x + n\int x^{n-1} \cos x\,dx$

90. $dv = \cos x\,dx \implies v = \sin x$

$u = x^n \implies du = nx^{n-1}\,dx$

$\int x^n \cos x\,dx = x^n \sin x - n\int x^{n-1} \sin x\,dx$

91. $dv = x^n\,dx \implies v = \frac{x^{n+1}}{n+1}$

$u = \ln x \implies du = \frac{1}{x}\,dx$

$$\int x^n \ln x\,dx = \frac{x^{n+1}}{n+1}\ln x - \int \frac{x^n}{n+1}\,dx$$

$$= \frac{x^{n+1}}{n+1}\ln x - \frac{x^{n+1}}{(n+1)^2} + C$$

$$= \frac{x^{n+1}}{(n+1)^2}[(n+1)\ln x - 1] + C$$

92. $dv = e^{ax}\,dx \implies v = \frac{1}{a}e^{ax}$

$u = x^n \implies du = nx^{n-1}\,dx$

$\int x^ne^{ax}\,dx = \frac{x^ne^{ax}}{a} - \frac{n}{a}\int x^{n-1}e^{ax}\,dx$

93. Use integration by parts twice.

(1) $dv = e^{ax}\,dx \implies v = \frac{1}{a}e^{ax}$ (2) $dv = e^{ax}\,dx \implies v = \frac{1}{a}e^{ax}$

$u = \sin bx \implies du = b\cos bx\,dx$ $u = \cos bx \implies du = -b\sin bx\,dx$

$$\int e^{ax}\sin bx\,dx = \frac{e^{ax}\sin bx}{a} - \frac{b}{a}\int e^{ax}\cos bx\,dx$$

$$= \frac{e^{ax}\sin bx}{a} - \frac{b}{a}\left[\frac{e^{ax}\cos bx}{a} + \frac{b}{a}\int e^{ax}\sin bx\,dx\right] = \frac{e^{ax}\sin bx}{a} - \frac{b}{a^2}e^{ax}\cos bx - \frac{b^2}{a^2}\int e^{ax}\sin bx\,dx$$

Therefore, $\left(1 + \frac{b^2}{a^2}\right)\int e^{ax}\sin bx\,dx = \frac{e^{ax}(a\sin bx - b\cos bx)}{a^2}$

$$\int e^{ax}\sin bx\,dx = \frac{e^{ax}(a\sin bx - b\cos bx)}{a^2 + b^2} + C.$$

94. Use integration by parts twice.

(1) $dv = e^{ax}\,dx \implies v = \frac{1}{a}e^{ax}$ (2) $dv = e^{ax}\,dx \implies v = \frac{1}{a}e^{ax}$

$u = \cos bx \implies du = -b\sin bx$ $u = \sin bx \implies du = b\cos bx$

$$\int e^{ax}\cos bx\,dx = \frac{e^{ax}\cos bx}{a} + \frac{b}{a}\int e^{ax}\sin bx\,dx = \frac{e^{ax}\cos bx}{a} + \frac{b}{a}\left[\frac{e^{ax}\sin bx}{a} - \frac{b}{a}\int e^{ax}\cos bx\,dx\right]$$

$$= \frac{e^{ax}\cos bx}{a} + \frac{be^{ax}\sin bx}{a^2} - \frac{b^2}{a^2}\int e^{ax}\cos bx\,dx$$

Therefore, $\left(1 + \frac{b^2}{a^2}\right)\int e^{ax}\cos bx\,dx = \frac{e^{ax}(a\cos bx + b\sin bx)}{a^2}$

$$\int e^{ax}\cos bx\,dx = \frac{e^{ax}(a\cos bx + b\sin bx)}{a^2 + b^2} + C.$$

95. $n = 3$, (Use formula in Exercise 91.)

$$\int x^3\ln x\,dx = \frac{x^4}{16}[4\ln x - 1] + C$$

96. $n = 2$, (Use formula in Exercise 90.)

$$\int x^2\cos x\,dx = x^2\sin x - 2\int x\sin x\,dx,$$ (Use formula in Exercise 83.) $(n = 1)$

$$= x^2\sin x - 2\left[-x\cos x + \int\cos x\,dx\right] = x^2\sin x + 2x\cos x - 2\sin x + C$$

97. $a = 2, b = 3$, (Use formula in Exercise 94.)

$$\int e^{2x}\cos 3x\,dx = \frac{e^{2x}(2\cos 3x + 3\sin 3x)}{13} + C$$

98. $n = 3, a = 2,$ (Use formula in Exercise 92 three times.)

$$\int x^3e^{2x}\,dx = \frac{x^3e^{2x}}{2} - \frac{3}{2}\int x^2e^{2x}\,dx, \quad (n = 3, a = 2)$$

$$= \frac{x^3e^{2x}}{2} - \frac{3}{2}\left[\frac{x^2e^{2x}}{2} - \int xe^{2x}\,dx\right], \quad (n = 2, a = 2)$$

$$= \frac{x^3e^{2x}}{2} - \frac{3x^2e^{2x}}{4} + \frac{3}{2}\left[\frac{xe^{2x}}{2} - \frac{1}{2}\int e^{2x}\,dx\right]$$

$$= \frac{x^3e^{2x}}{2} - \frac{3x^2e^{2x}}{4} + \frac{3xe^{2x}}{4} - \frac{3e^{2x}}{8} + C, \quad (n = 1, a = 2)$$

$$= \frac{e^{2x}}{8}(4x^3 - 6x^2 + 6x - 3) + C$$

99. $dv = e^{-x}\,dx \implies v = -e^{-x}$

$u = x \implies du = dx$

$$A = \int_0^4 xe^{-x}\,dx = \Big[-xe^{-x}\Big]_0^4 + \int_0^4 e^{-x}\,dx = \frac{-4}{e^4} - \Big[e^{-x}\Big]_0^4$$

$$= 1 - \frac{5}{e^4} \approx 0.908$$

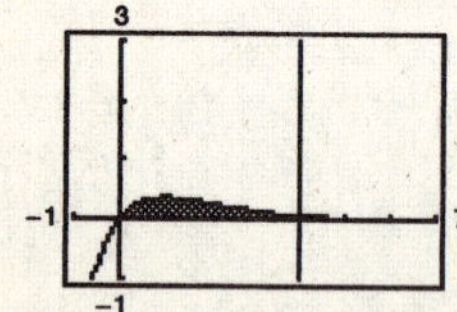

100. $dv = e^{-x/3}\,dx \implies v = -3e^{-x/3}$

$u = x \implies du = dx$

$$A = \frac{1}{9}\int_0^3 xe^{-x/3}\,dx$$

$$= \frac{1}{9}\left(\Big[-3xe^{-x/3}\Big]_0^3 + 3\int_0^3 e^{-x/3}\,dx\right)$$

$$= \frac{1}{9}\left(\frac{-9}{e} - \Big[9e^{-x/3}\Big]_0^3\right)$$

$$= -\frac{1}{e} - \frac{1}{e} + 1$$

$$= 1 - \frac{2}{e} \approx 0.264$$

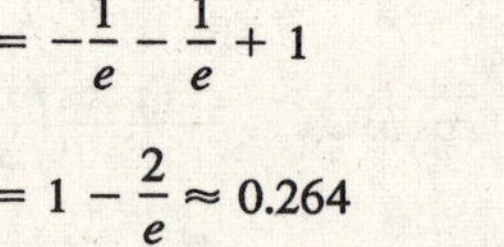

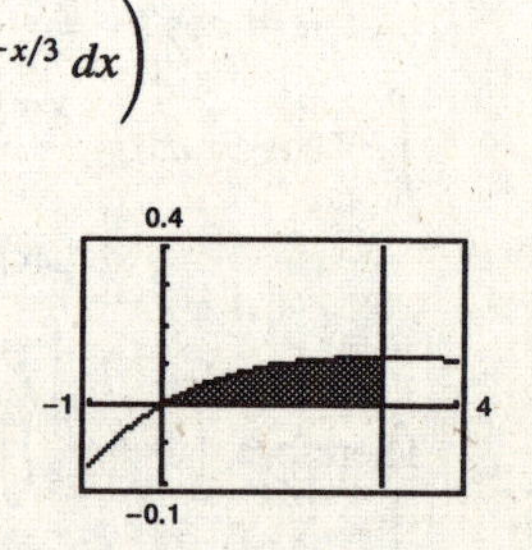

101. $$A = \int_0^1 e^{-x}\sin(\pi x)\,dx$$

$$= \left[\frac{e^{-x}(-\sin \pi x - \pi \cos \pi x)}{1 + \pi^2}\right]_0^1$$

$$= \frac{1}{1 + \pi^2}\left(\frac{\pi}{e} + \pi\right)$$

$$= \frac{\pi}{1 + \pi^2}\left(\frac{1}{e} + 1\right)$$

≈ 0.395 (See Exercise 93.)

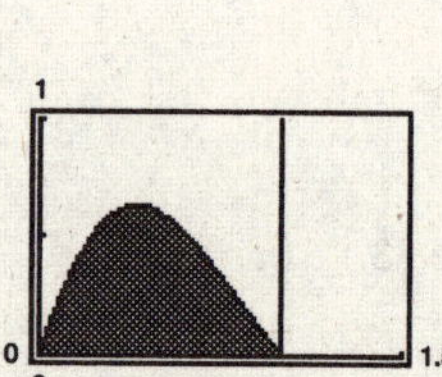

102. $$A = \int_0^{\pi} x \sin x\,dx = \Big[-x\cos x + \sin x\Big]_0^{\pi}$$

$= \pi$ (See Exercise 89.)

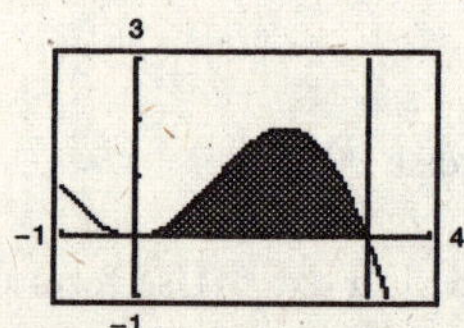

103. (a) $$A = \int_1^e \ln x\,dx = \Big[-x + x\ln x\Big]_1^e = 1$$ (See Exercise 4.)

(b) $R(x) = \ln x, r(x) = 0$

$$V = \pi\int_1^e (\ln x)^2\,dx$$

$$= \pi\Big[x(\ln x)^2 - 2x\ln x + 2x\Big]_1^e$$ (Use integration by parts twice, see Exercise 7.)

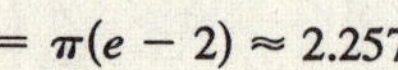

$$= \pi(e - 2) \approx 2.257$$

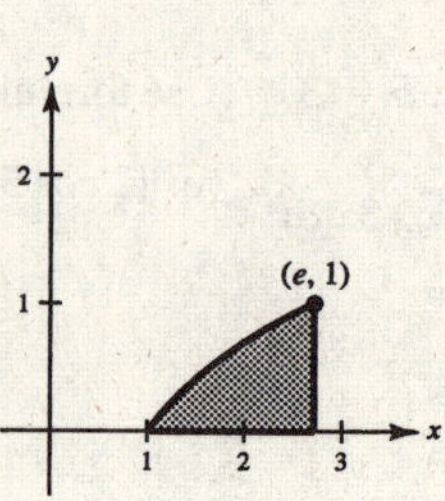

—CONTINUED—

103. —CONTINUED—

(c) $p(x) = x,\ h(x) = \ln x$

$$V = 2\pi\int_1^e x\ln x\,dx = 2\pi\left[\frac{x^2}{4}(-1 + 2\ln x)\right]_1^e$$

$$= \frac{(e^2 + 1)\pi}{2} \approx 13.177 \quad \text{(See Exercise 91.)}$$

(d)

$$\bar{x} = \frac{\int_1^e x\ln x\,dx}{1} = \frac{e^2 + 1}{4} \approx 2.097$$

$$\bar{y} = \frac{\frac{1}{2}\int_1^e (\ln x)^2\,dx}{1} = \frac{e - 2}{2} \approx 0.359$$

$$(\bar{x}, \bar{y}) = \left(\frac{e^2 + 1}{4}, \frac{e - 2}{2}\right) \approx (2.097, 0.359)$$

104. $y = x\sin x,\quad 0 \le x \le \pi$

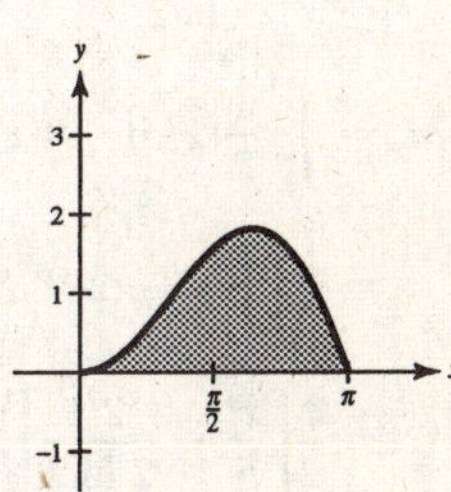

(a) $V = \displaystyle\int_0^\pi \pi[x\sin x]^2\,dx = \pi\int_0^\pi x^2\sin^2 x\,dx$

Let $u = x^2,\ du = 2x\,dx,\ dv = \sin^2 x\,dx = \dfrac{1 - \cos 2x}{2}\,dx,\ v = \dfrac{1}{2}x - \dfrac{\sin 2x}{4}$.

$$\int x^2\sin^2 x\,dx = x^2\left[\frac{1}{2}x - \frac{\sin 2x}{4}\right] - \int\left(\frac{1}{2}x - \frac{\sin 2x}{4}\right)(2x\,dx)$$

$$= \frac{1}{2}x^3 - \frac{x^2\sin 2x}{4} - \int\left(x^2 - \frac{x\sin 2x}{2}\right)dx$$

$$= \frac{1}{2}x^3 - \frac{x^2\sin 2x}{4} - \frac{x^3}{3} + \int\frac{x\sin 2x}{2}\,dx$$

$$= \frac{1}{6}x^3 - \frac{1}{4}x^2\sin 2x + \frac{1}{8}(\sin 2x - 2x\cos 2x) + C \quad \text{(Integration by Parts)}$$

$$V = \pi\int_0^\pi x^2\sin^2 x\,dx = \pi\left[\frac{1}{6}x^3 - \frac{1}{4}x^2\sin 2x + \frac{1}{8}(\sin 2x - 2x\cos 2x)\right]_0^\pi = \frac{1}{6}\pi^4 - \frac{1}{4}\pi^2$$

(b) $V = \displaystyle\int_0^\pi 2\pi x(x\sin x)\,dx = 2\pi\Big[2\cos x + 2x\sin x - x^2\cos x\Big]_0^\pi = 2\pi[\pi^2 - 4] = 2\pi^3 - 8\pi$

(c) $m = \displaystyle\int_0^\pi x\sin(x)\,dx = \Big[\sin x - x\cos x\Big]_0^\pi = \pi$

$$M_x = \int_0^\pi \frac{1}{2}(x\sin x)^2\,dx$$

$$= \frac{1}{2}\left[\frac{1}{6}\pi^3 - \frac{1}{4}\pi\right] \quad \text{(See part (a).)}$$

$$= \frac{1}{12}\pi^3 - \frac{1}{8}\pi$$

$$M_y = \int_0^\pi x(x\sin x)\,dx = \pi^2 - 4 \quad \text{(See part (b).)}$$

$$\bar{x} = \frac{M_y}{m} = \frac{\pi^2 - 4}{\pi} \approx 1.8684,\quad \bar{y} = \frac{M_x}{m} = \frac{(1/12)\pi^3 - (1/8)\pi}{\pi} = \frac{1}{2}\pi^2 - \frac{1}{8} \approx 0.6975$$

105. In Example 6, we showed that the centroid of an equivalent region was $(1, \pi/8)$. By symmetry, the centroid of this region is $(\pi/8, 1)$. You can also solve this problem directly.

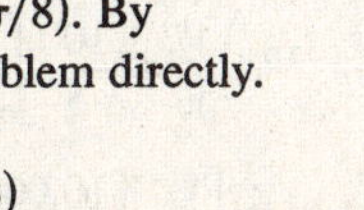

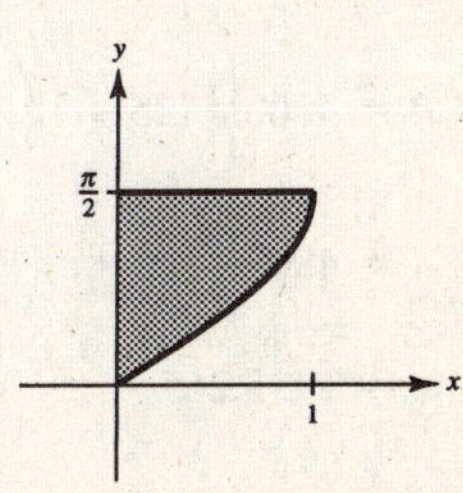

$$A = \int_0^1\left(\frac{\pi}{2} - \arcsin x\right)dx = \left[\frac{\pi}{2}x - x\arcsin x - \sqrt{1 - x^2}\right]_0^1 \quad \text{(Example 3)}$$

$$= \left(\frac{\pi}{2} - \frac{\pi}{2} - 0\right) - (-1) = 1$$

$$\bar{x} = \frac{M_y}{A} = \int_0^1 x\left[\frac{\pi}{2} - \arcsin x\right]dx = \frac{\pi}{8},\quad \bar{y} = \frac{M_x}{A} = \int_0^1 \frac{(\pi/2) + \arcsin x}{2}\left[\frac{\pi}{2} - \arcsin x\right]dx = 1$$

106. $f(x) = x^2, g(x) = 2^x$

$f(2) = g(2) = 4,\ f(4) = g(4) = 16$

$$m = \int_2^4 (x^2 - 2^x)\,dx = \left[\frac{x^3}{3} - \frac{1}{\ln 2}2^x\right]_2^4$$

$$= \left(\frac{64}{3} - \frac{16}{\ln 2}\right) - \left(\frac{8}{3} - \frac{4}{\ln 2}\right)$$

$$= \frac{56}{3} - \frac{12}{\ln 2} \approx 1.3543$$

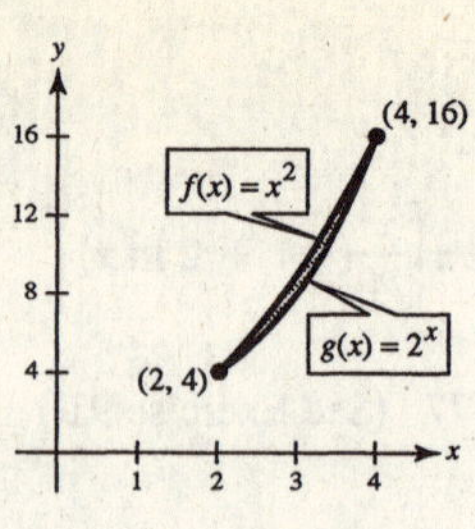

$$M_x = \int_2^4 \frac{1}{2}(x^2 + 2^x)(x^2 - 2^x)\,dx$$

$$= \frac{1}{2}\int_2^4 (x^4 - 2^{2x})\,dx$$

$$= \frac{1}{2}\left[\frac{x^5}{5} - \frac{2^{2x}}{2\ln 2}\right]_2^4$$

$$= \frac{1}{2}\left[\left(\frac{1024}{5} - \frac{128}{\ln 2}\right) - \left(\frac{32}{5} - \frac{8}{\ln 2}\right)\right]$$

$$= \frac{496}{5} - \frac{60}{\ln 2} \approx 12.6383$$

$$M_y = \int_2^4 x[x^2 - 2^x]\,dx$$

$$= -\frac{56}{\ln 2} + \frac{12}{(\ln 2)^2} \approx 4.1855$$

$$(\bar{x}, \bar{y}) = \left(\frac{M_y}{m}, \frac{M_x}{m}\right) \approx (3.0905, 9.3318)$$

107. Average value $= \dfrac{1}{\pi}\displaystyle\int_0^{\pi} e^{-4t}(\cos 2t + 5\sin 2t)\,dt$

$$= \frac{1}{\pi}\left[e^{-4t}\left(\frac{-4\cos 2t + 2\sin 2t}{20}\right) + 5e^{-4t}\left(\frac{-4\sin 2t - 2\cos 2t}{20}\right)\right]_0^{\pi} \quad \text{(From Exercises 93 and 94)}$$

$$= \frac{7}{10\pi}(1 - e^{-4\pi}) \approx 0.223$$

108. (a) Average $= \displaystyle\int_1^2 (1.6t\ln t + 1)\,dt = \Big[0.8t^2\ln t - 0.4t^2 + t\Big]_1^2 = 3.2(\ln 2) - 0.2 \approx 2.018$

(b) Average $= \displaystyle\int_3^4 (1.6t\ln t + 1)\,dt = \Big[0.8t^2\ln t - 0.4t^2 + t\Big]_3^4 = 12.8(\ln 4) - 7.2(\ln 3) - 1.8 \approx 8.035$

109. $c(t) = 100{,}000 + 4000t, r = 5\%, t_1 = 10$

$$P = \int_0^{10} (100{,}000 + 4000t)e^{-0.05t}\,dt$$

$$= 4000\int_0^{10} (25 + t)e^{-0.05t}\,dt$$

Let $u = 25 + t, dv = e^{-0.05t}dt, du = dt, v = -\dfrac{100}{5}e^{-0.05t}$.

$$P = 4000\left\{\left[(25 + t)\left(-\frac{100}{5}e^{-0.05t}\right)\right]_0^{10} + \frac{100}{5}\int_0^{10} e^{-0.05t}\,dt\right\}$$

$$= 4000\left\{\left[(25 + t)\left(-\frac{100}{5}e^{-0.05t}\right)\right]_0^{10} - \left[\frac{10{,}000}{25}e^{-0.05t}\right]_0^{10}\right\}$$

$\approx \$931{,}265$

110. $c(t) = 30{,}000 + 500t, r = 7\%, t_1 = 5$

$$P\int_0^5 (30{,}000 + 500t)e^{-0.07t}\,dt = 500\int_0^5 (60 + t)e^{-0.07t}\,dt$$

Let $u = 60 + t, dv = e^{-0.07t}\,dt, du = dt, v = -\dfrac{100}{7}e^{-0.07t}$.

$$P = 500\left\{\left[(60 + t)\left(-\frac{100}{7}e^{-0.07t}\right)\right]_0^5 + \frac{100}{7}\int_0^5 e^{-0.07t}\,dt\right\}$$

$$= 500\left\{\left[(60 + t)\left(-\frac{100}{7}e^{-0.07t}\right)\right]_0^5 - \left[\frac{10{,}000}{49}e^{-0.07t}\right]_0^5\right\}$$

$\approx \$131{,}528.68$

111. $\displaystyle\int_{-\pi}^{\pi} x\sin nx\,dx = \left[-\frac{x}{n}\cos nx + \frac{1}{n^2}\sin nx\right]_{-\pi}^{\pi}$

$$= -\frac{\pi}{n}\cos \pi n - \frac{\pi}{n}\cos(-\pi n)$$

$$= -\frac{2\pi}{n}\cos \pi n$$

$$= \begin{cases} -(2\pi/n), & \text{if } n \text{ is even} \\ (2\pi/n), & \text{if } n \text{ is odd} \end{cases}$$

112. $\displaystyle\int_{-\pi}^{\pi} x^2\cos nx\,dx = \left[\frac{x^2}{n}\sin nx + \frac{2x}{n^2}\cos nx - \frac{2}{n^3}\sin nx\right]_{-\pi}^{\pi}$

$$= \frac{2\pi}{n^2}\cos n\pi + \frac{2\pi}{n^2}\cos(-n\pi)$$

$$= \frac{4\pi}{n^2}\cos n\pi$$

$$= \begin{cases} (4\pi/n^2), & \text{if } n \text{ is even} \\ -(4\pi/n^2), & \text{if } n \text{ is odd} \end{cases}$$

$$= \frac{(-1)^n 4\pi}{n^2}$$

113. Let $u = x$, $dv = \sin\left(\frac{n\pi}{2}x\right)dx$, $du = dx$, $v = -\frac{2}{n\pi}\cos\left(\frac{n\pi}{2}x\right)$.

$$I_1 = \int_0^1 x\sin\left(\frac{n\pi}{2}x\right)dx = \left[\frac{-2x}{n\pi}\cos\left(\frac{n\pi}{2}x\right)\right]_0^1 + \frac{2}{n\pi}\int_0^1 \cos\left(\frac{n\pi}{2}x\right)dx$$

$$= -\frac{2}{n\pi}\cos\left(\frac{n\pi}{2}\right) + \left[\left(\frac{2}{n\pi}\right)^2\sin\left(\frac{n\pi}{2}x\right)\right]_0^1$$

$$= -\frac{2}{n\pi}\cos\left(\frac{n\pi}{2}\right) + \left(\frac{2}{n\pi}\right)^2\sin\left(\frac{n\pi}{2}\right)$$

Let $u = (-x + 2)$, $dv = \sin\left(\frac{n\pi}{2}x\right)dx$, $du = -dx$, $v = -\frac{2}{n\pi}\cos\left(\frac{n\pi}{2}x\right)$.

$$I_2 = \int_1^2 (-x+2)\sin\left(\frac{n\pi}{2}x\right)dx = \left[\frac{-2(-x+2)}{n\pi}\cos\left(\frac{n\pi}{2}x\right)\right]_1^2 - \frac{2}{n\pi}\int_1^2 \cos\left(\frac{n\pi}{2}x\right)dx$$

$$= \frac{2}{n\pi}\cos\left(\frac{n\pi}{2}\right) - \left[\left(\frac{2}{n\pi}\right)^2\sin\left(\frac{n\pi}{2}x\right)\right]_1^2$$

$$= \frac{2}{n\pi}\cos\left(\frac{n\pi}{2}\right) + \left(\frac{2}{n\pi}\right)^2\sin\left(\frac{n\pi}{2}\right)$$

$$h(I_1 + I_2) = b_n = h\left[\left(\frac{2}{n\pi}\right)^2\sin\left(\frac{n\pi}{2}\right) + \left(\frac{2}{n\pi}\right)^2\sin\left(\frac{n\pi}{2}\right)\right] = \frac{8h}{(n\pi)^2}\sin\left(\frac{n\pi}{2}\right)$$

114. For any integrable function, $\int f(x)\,dx = C + \int f(x)\,dx$, but this cannot be used to imply that $C = 0$.

115. Shell Method:

$$V = 2\pi\int_a^b x f(x)\,dx$$

$$dv = x\,dx \implies v = \frac{x^2}{2}$$

$$u = f(x) \implies du = f'(x)\,dx$$

$$V = 2\pi\left[\frac{x^2}{2}f(x) - \int\frac{x^2}{2}f'(x)\,dx\right]_a^b$$

$$= \pi\left[(b^2f(b) - a^2f(a)) - \int_a^b x^2f'(x)\,dx\right]$$

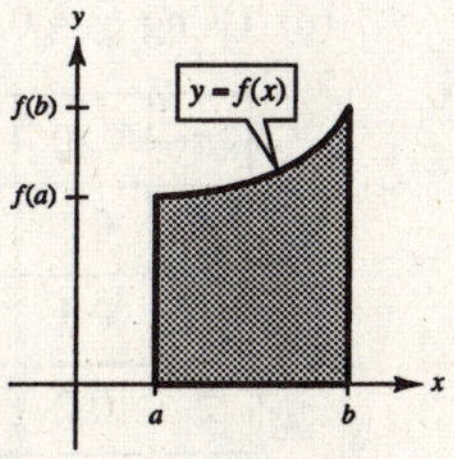

Disk Method:

$$V = \pi\int_0^{f(a)}(b^2 - a^2)\,dy + \pi\int_{f(a)}^{f(b)}\left[b^2 - [f^{-1}(y)]^2\right]dy$$

$$= \pi(b^2 - a^2)f(a) + \pi b^2(f(b) - f(a)) - \pi\int_{f(a)}^{f(b)}[f^{-1}(y)]^2\,dy$$

$$= \pi\left[(b^2f(b) - a^2f(a)) - \int_{f(a)}^{f(b)}[f^{-1}(y)]^2\,dy\right]$$

Since $x = f^{-1}(y)$, we have $f(x) = y$ and $f'(x)\,dx = dy$. When $y = f(a)$, $x = a$. When $y = f(b)$, $x = b$. Thus,

$$\int_{f(a)}^{f(b)}[f^{-1}(y)]^2\,dy = \int_a^b x^2f'(x)\,dx$$

and the volumes are the same.

116. $f'(x) = xe^{-x}$

(a) $f(x) = \int xe^{-x}\,dx = -xe^{-x} - e^{-x} + C$

(Parts: $u = x$, $dv = e^{-x}\,dx$)

$f(0) = 0 = -1 + C \Rightarrow C = 1$

$f(x) = -xe^{-x} - e^{-x} + 1$

(b)

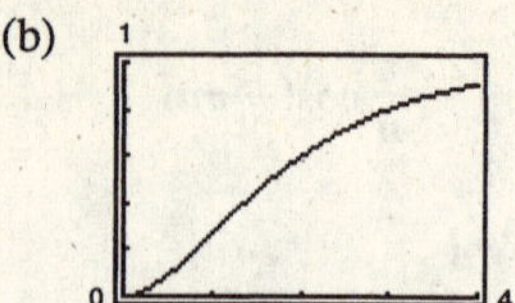

(c) You obtain the points:

n	x_n	y_n
0	0	0
1	0.05	0
2	0.10	2.378×10^{-3}
3	0.15	0.0069
4	0.20	0.0134
⋮	⋮	⋮
80	4.0	0.9064

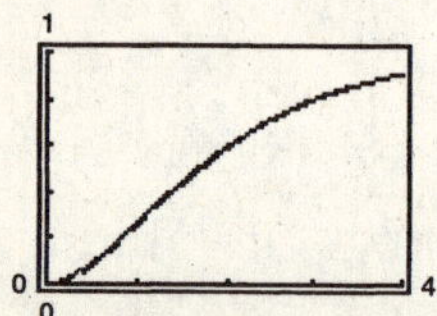

(d) You obtain the points:

n	x_n	y_n
0	0	0
1	0.1	0
2	0.2	0.0090484
3	0.3	0.025423
4	0.4	0.047648
⋮	⋮	⋮
40	4.0	0.9039

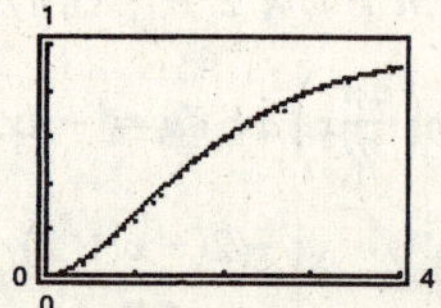

(e) The result in part (c) is better because h is smaller.

117. $f'(x) = 3x\sin(2x)$, $f(0) = 0$

(a) $f(x) = \int 3x\sin 2x\,dx = -\frac{3}{4}(2x\cos 2x - \sin 2x) + C$

(Parts: $u = 3x$, $dv = \sin 2x\,dx$)

$f(0) = 0 = -\frac{3}{4}(0) + C \Rightarrow C = 0$

$f(x) = -\frac{3}{4}(2x\cos 2x - \sin 2x)$

(b)

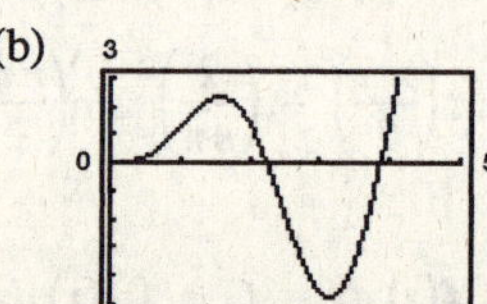

(c) Using $h = 0.05$, you obtain the points:

n	x_n	y_n
0	0	0
1	0.05	0.05
2	0.10	7.4875×10^{-4}
3	0.15	0.0037
4	0.20	0.0104
⋮	⋮	⋮
80	4.0	1.3181

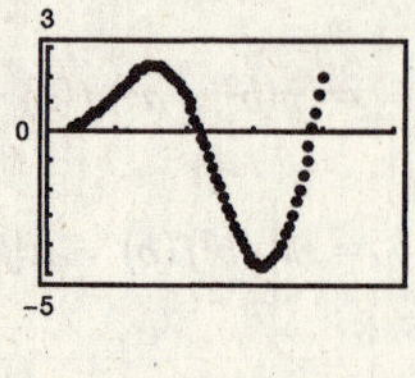

(d) Using $h = 0.1$, you obtain the points:

n	x_n	y_n
0	0	0
1	0.1	0
2	0.2	0.0060
3	0.3	0.0293
4	0.4	0.0801
⋮	⋮	⋮
40	4.0	1.0210

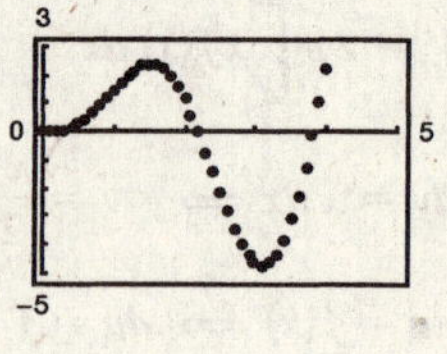

118. $f'(x) = \cos\sqrt{x},\ f(0) = 1$

(a) Let $w = \sqrt{x}$, $w^2 = x$, $2w\,dw = dx$.

$$\int \cos\sqrt{x}\,dx = \int \cos w(2w\,dw)$$

Now use parts: $u = 2w$, $dv = \cos w\,dw$.

$$\int \cos\sqrt{x}\,dx = 2w\sin w + 2\cos w + C$$

$$= 2\sqrt{x}\sin\sqrt{x} + 2\cos\sqrt{x} + C$$

$f(0) = 1 = 2 + C \Rightarrow C = -1$

$f(x) = 2\sqrt{x}\sin\sqrt{x} + 2\cos\sqrt{x} - 1$

(b)

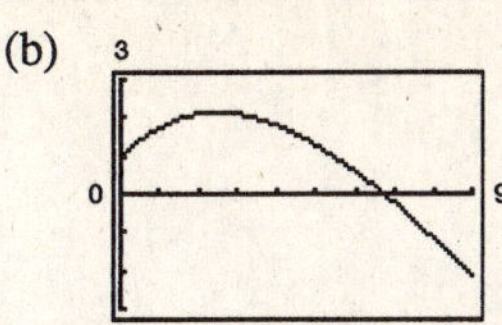

(c) Using $h = 0.05$, you obtain the points:

n	x_n	y_n
0	0	1
1	0.05	1.05
2	0.1	1.0988
3	0.15	1.1463
4	0.2	1.1926
⋮	⋮	⋮
80	4.0	1.8404

(d) Using $h = 0.1$, you obtain the points:

n	x_n	y_n
0	0	1
1	0.1	1.1
2	0.2	1.1950
3	0.3	1.2852
4	0.4	1.3706
⋮	⋮	⋮
80	4.0	1.8759

119. On $\left[0, \frac{\pi}{2}\right]$, $\sin x \le 1 \Rightarrow x\sin x \le x \Rightarrow \int_0^{\pi/2} x\sin x\,dx \le \int_0^{\pi/2} x\,dx.$

120. (a) $A = \int_0^{\pi} x\sin x\,dx = \left[\sin x - x\cos x\right]_0^{\pi} = \pi$

(b) $\int_{\pi}^{2\pi} x\sin x\,dx = \left[\sin x - x\cos x\right]_{\pi}^{2\pi} = -2\pi - \pi = -3\pi$

$A = 3\pi$

(c) $\int_{2\pi}^{3\pi} x\sin x\,dx = \left[\sin x - x\cos x\right]_{2\pi}^{3\pi} = 3\pi + 2\pi = 5\pi$

$A = 5\pi$

The area between $y = x\sin x$ and $y = 0$ on $[n\pi, (n+1)\pi]$ is $(2n+1)\pi$:

$$\int_{n\pi}^{(n+1)\pi} x\sin x\,dx = \left[\sin x - x\cos x\right]_{n\pi}^{(n+1)\pi} = \pm(n+1)\pi \pm n\pi = \pm(2n+1)\pi$$

$A = |\pm(2n+1)\pi| = (2n+1)\pi$

Section 8.3 Trigonometric Integrals

1. $y = \sec x$

$y' = \sec x \tan x = \sin x \sec^2 x$

$\int \sin x \sec^2 x\, dx = \sec x + C$

Matches (c)

2. $y = \cos x + \sec x$

$y' = -\sin x + \sec x \tan x$

$= -\sin x + \sin x \sec^2 x$

$= -\sin x(1 - \sec^2 x)$

$= \sin x \tan^2 x$

$\int \sin x \tan^2 x\, dx = \cos x + \sec x + C$

Matches (a)

3. $y = x - \tan x + \frac{1}{3}\tan^3 x$

$y' = 1 - \sec^2 x + \tan^2 x(\sec^2 x)$

$= -\tan^2 x + \tan^2 x(1 + \tan^2 x)$

$= \tan^4 x$

$\int \tan^4 x\, dx = x - \tan x + \frac{1}{3}\tan^3 x + C$

Matches (d)

4.

$$y = 3x + 2\sin x \cos^3 x + 3\sin x \cos x$$

$$y' = 3 + 2\cos^4 x - 6\sin^2 x \cos^2 x + 3\cos^2 x - 3\sin^2 x$$

$$= 3 + 2\cos^4 x - 6\cos^2 x(1 - \cos^2 x) + 3\cos^2 x - 3(1 - \cos^2 x) = 8\cos^4 x$$

$\int 8\cos^4 x\, dx = 3x + 2\sin x \cos^3 x + 3\sin x \cos x + C$

Matches (b)

5. Let $u = \cos x$, $du = -\sin x\, dx$.

$$\int \cos^3 x \sin x\, dx = -\int \cos^3 x(-\sin x)\, dx$$

$$= -\frac{1}{4}\cos^4 x + C$$

6. $$\int \cos^3 x \sin^4 x\, dx = \int \cos x(1 - \sin^2 x)\sin^4 x\, dx$$

$$= \int (\sin^4 x - \sin^6 x)\cos x\, dx$$

$$= \frac{\sin^5 x}{5} - \frac{\sin^7 x}{7} + C$$

7. Let $u = \sin 2x$, $du = 2\cos 2x\, dx$.

$$\int \sin^5 2x \cos 2x\, dx = \frac{1}{2}\int \sin^5 2x(2\cos 2x)\, dx$$

$$= \frac{1}{12}\sin^6 2x + C$$

8. Let $u = \cos x$, $du = -\sin x\, dx$.

$$\int \sin^3 x\, dx = \int \sin x(1 - \cos^2 x)\, dx$$

$$= \int \cos^2 x(-\sin x)\, dx + \int \sin x\, dx$$

$$= \frac{1}{3}\cos^3 x - \cos x + C$$

9. Let $u = \cos x$, $du = -\sin x\, dx$.

$$\int \sin^5 x \cos^2 x\, dx = \int \sin x(1 - \cos^2 x)^2 \cos^2 x\, dx$$

$$= -\int (\cos^2 x - 2\cos^4 x + \cos^6 x)(-\sin x)\, dx = \frac{-1}{3}\cos^3 x + \frac{2}{5}\cos^5 x - \frac{1}{7}\cos^7 x + C$$

10. Let $u = \sin\frac{x}{3}$, $du = \frac{1}{3}\cos\frac{x}{3}\,dx$.

$$\int \cos^3\frac{x}{3}\,dx = \int\left(\cos\frac{x}{3}\right)\left(1 - \sin^2\frac{x}{3}\right)dx$$
$$= 3\int\left(1 - \sin^2\frac{x}{3}\right)\left(\frac{1}{3}\cos\frac{x}{3}\right)dx$$
$$= 3\left(\sin\frac{x}{3} - \frac{1}{3}\sin^3\frac{x}{3}\right) + C$$
$$= 3\sin\frac{x}{3} - \sin^3\frac{x}{3} + C$$

11. $$\int \cos^3\theta\sqrt{\sin\theta}\,d\theta = \int \cos\theta(1 - \sin^2\theta)(\sin\theta)^{1/2}\,d\theta$$
$$= \int\left[(\sin\theta)^{1/2} - (\sin\theta)^{5/2}\right]\cos\theta\,d\theta$$
$$= \frac{2}{3}(\sin\theta)^{3/2} - \frac{2}{7}(\sin\theta)^{7/2} + C$$

12. $$\int \frac{\sin^5 t}{\sqrt{\cos t}}\,dt = \int \sin t(1 - \cos^2 t)^2(\cos t)^{-1/2}\,dt$$
$$= \int \sin t(1 - 2\cos^2 t + \cos^4 t)(\cos t)^{-1/2}\,dt$$
$$= \int\left[(\cos t)^{-1/2} - 2(\cos t)^{3/2} + (\cos t)^{7/2}\right]\sin t\,dt = -2(\cos t)^{1/2} + \frac{4}{5}(\cos t)^{5/2} - \frac{2}{9}(\cos t)^{9/2} + C$$

13. $$\int \cos^2 3x\,dx = \int \frac{1 + \cos 6x}{2}\,dx$$
$$= \frac{1}{2}\left(x + \frac{1}{6}\sin 6x\right) + C$$
$$= \frac{1}{12}(6x + \sin 6x) + C$$

14. $$\int \sin^2 2x\,dx = \int \frac{1 - \cos 4x}{2}\,dx = \frac{1}{2}\left(x - \frac{1}{4}\sin 4x\right) + C$$
$$= \frac{1}{8}(4x - \sin 4x) + C$$

15. $$\int \sin^2\alpha\cdot\cos^2\alpha\,d\alpha = \int \frac{1 - \cos 2\alpha}{2}\cdot\frac{1 + \cos 2\alpha}{2}\,d\alpha$$
$$= \frac{1}{4}\int(1 - \cos^2 2\alpha)\,d\alpha$$
$$= \frac{1}{4}\int\left(1 - \frac{1 + \cos 4\alpha}{2}\right)d\alpha$$
$$= \frac{1}{8}\int(1 - \cos 4\alpha)\,d\alpha$$
$$= \frac{1}{8}\left[\alpha - \frac{1}{4}\sin 4\alpha\right] + C$$
$$= \frac{1}{32}[4\alpha - \sin 4\alpha] + C$$

16. $$\int \sin^4 2\theta\,d\theta = \int \frac{1 - \cos 4\theta}{2}\cdot\frac{1 - \cos 4\theta}{2}\,d\theta$$
$$= \frac{1}{4}\int(1 - 2\cos 4\theta + \cos^2 4\theta)\,d\theta$$
$$= \frac{1}{4}\int\left(1 - 2\cos 4\theta + \frac{1 + \cos 8\theta}{2}\right)d\theta$$
$$= \frac{1}{4}\int\left(\frac{3}{2} - 2\cos 4\theta + \frac{1}{2}\cos 8\theta\right)d\theta$$
$$= \frac{1}{4}\left[\frac{3}{2}\theta - \frac{1}{2}\sin 4\theta + \frac{1}{16}\sin 8\theta\right] + C$$
$$= \frac{3}{8}\theta - \frac{1}{8}\sin 4\theta + \frac{1}{64}\sin 8\theta + C$$

17. Integration by parts:

$$dv = \sin^2 x\,dx = \frac{1 - \cos 2x}{2} \implies v = \frac{x}{2} - \frac{\sin 2x}{4} = \frac{1}{4}(2x - \sin 2x)$$

$$u = x \implies du = dx$$

$$\int x\sin^2 x\,dx = \frac{1}{4}x(2x - \sin 2x) - \frac{1}{4}\int(2x - \sin 2x)\,dx$$
$$= \frac{1}{4}x(2x - \sin 2x) - \frac{1}{4}\left(x^2 + \frac{1}{2}\cos 2x\right) + C = \frac{1}{8}(2x^2 - 2x\sin 2x - \cos 2x) + C$$

18. Use integration by parts twice.

$$dv = \sin^2 x\,dx = \frac{1-\cos 2x}{2} \Rightarrow v = \frac{x}{2} - \frac{\sin 2x}{4} = \frac{1}{4}(2x - \sin 2x)$$

$$u = x^2 \Rightarrow du = 2x\,dx$$

$$dv = \sin 2x\,dx \Rightarrow \quad v = -\frac{1}{2}\cos 2x$$

$$u = x \qquad \Rightarrow du = dx$$

$$\begin{aligned}\int x^2 \sin^2 x\,dx &= \frac{1}{4}x^2(2x - \sin 2x) - \frac{1}{2}\int (2x^2 - x\sin 2x)\,dx\\ &= \frac{1}{2}x^3 - \frac{1}{4}x^2\sin 2x - \frac{1}{3}x^3 + \frac{1}{2}\int x\sin 2x\,dx\\ &= \frac{1}{6}x^3 - \frac{1}{4}x^2\sin 2x + \frac{1}{2}\left[-\frac{1}{2}x\cos 2x + \frac{1}{2}\int \cos 2x\,dx\right]\\ &= \frac{1}{6}x^3 - \frac{1}{4}x^2\sin 2x - \frac{1}{4}x\cos 2x + \frac{1}{8}\sin 2x + C\\ &= \frac{1}{24}(4x^3 - 6x^2\sin 2x - 6x\cos 2x + 3\sin 2x) + C\end{aligned}$$

19. $\displaystyle\int_0^{\pi/2} \cos^3 x\,dx = \frac{2}{3}, \quad (n = 3)$

20. $\displaystyle\int_0^{\pi/2} \cos^5 x\,dx = \left(\frac{2}{3}\right)\left(\frac{4}{5}\right) = \frac{8}{15}, \quad (n = 5)$

21. $\displaystyle\int_0^{\pi/2} \cos^7 x\,dx = \left(\frac{2}{3}\right)\left(\frac{4}{5}\right)\left(\frac{6}{7}\right) = \frac{16}{35}, \quad (n = 7)$

22. $\displaystyle\int_0^{\pi/2} \sin^2 x\,dx = \left(\frac{1}{2}\right)\frac{\pi}{2} = \frac{\pi}{4}, \quad (n = 2)$

23. $\displaystyle\int_0^{\pi/2} \sin^6 x\,dx = \left(\frac{1}{2}\right)\left(\frac{3}{4}\right)\left(\frac{5}{6}\right)\frac{\pi}{2} = \frac{5\pi}{32}, \quad (n = 6)$

24. $\displaystyle\int_0^{\pi/2} \sin^7 x\,dx = \left(\frac{2}{3}\right)\left(\frac{4}{5}\right)\left(\frac{6}{7}\right) = \frac{16}{35}, \quad (n = 7)$

25. $\displaystyle\int \sec(3x)\,dx = \frac{1}{3}\ln|\sec 3x + \tan 3x| + C$

26. $\displaystyle\int \sec^2(2x - 1)\,dx = \frac{1}{2}\tan(2x - 1) + C$

27. $$\begin{aligned}\int \sec^4 5x\,dx &= \int (1 + \tan^2 5x)\sec^2 5x\,dx\\ &= \frac{1}{5}\left(\tan 5x + \frac{\tan^3 5x}{3}\right) + C\\ &= \frac{\tan 5x}{15}(3 + \tan^2 5x) + C\end{aligned}$$

28. $$\begin{aligned}\int \sec^6 3x\,dx &= \int (1 + \tan^2 3x)^2 \sec^2 3x\,dx\\ &= \int (1 + 2\tan^2 3x + \tan^4 3x)\sec^2 3x\,dx\\ &= \frac{1}{3}\tan 3x + \frac{2}{9}\tan^3 3x + \frac{1}{15}\tan^5 3x + C\end{aligned}$$

29. $dv = \sec^2 \pi x\,dx \Rightarrow \quad v = \dfrac{1}{\pi}\tan \pi x$

$$u = \sec \pi x \qquad \Rightarrow du = \pi \sec \pi x \tan \pi x\,dx$$

$$\int \sec^3 \pi x\,dx = \frac{1}{\pi}\sec \pi x \tan \pi x - \int \sec \pi x \tan^2 \pi x\,dx = \frac{1}{\pi}\sec \pi x \tan \pi x - \int \sec \pi x(\sec^2 \pi x - 1)\,dx$$

$$2\int \sec^3 \pi x\,dx = \frac{1}{\pi}(\sec \pi x \tan \pi x + \ln|\sec \pi x + \tan \pi x|) + C_1$$

$$\int \sec^3 \pi x\,dx = \frac{1}{2\pi}(\sec \pi x \tan \pi x + \ln|\sec \pi x + \tan \pi x|) + C$$

30. $\int \tan^2 x\,dx = \int (\sec^2 x - 1)\,dx = \tan x - x + C$

31. $\int \tan^5 \frac{x}{4}\,dx = \int \left(\sec^2 \frac{x}{4} - 1\right) \tan^3 \frac{x}{4}\,dx$

$= \int \tan^3 \frac{x}{4} \sec^2 \frac{x}{4}\,dx - \int \tan^3 \frac{x}{4}\,dx$

$= \tan^4 \frac{x}{4} - \int \left(\sec^2 \frac{x}{4} - 1\right) \tan \frac{x}{4}\,dx$

$= \tan^4 \frac{x}{4} - 2\tan^2 \frac{x}{4} - 4 \ln\left|\cos \frac{x}{4}\right| + C$

32. $\int \tan^3 \frac{\pi x}{2} \sec^2 \frac{\pi x}{2}\,dx = \frac{1}{2\pi} \tan^4 \frac{\pi x}{2} + C$

33. $u = \tan x$, $du = \sec^2 x\,dx$

$\int \sec^2 x \tan x\,dx = \frac{1}{2}\tan^2 x + C$

[or, $u = \sec x$, $du = \sec x \tan x\,dx$,

$\int \sec^2 x \tan x\,dx = \frac{1}{2}\sec^2 x + C$.]

34. Let $u = \sec 2t$, $du = 2 \sec 2t \tan 2t$.

$$\int \tan^3 2t \cdot \sec^3 2t\,dt = \int (\sec^2 2t - 1)\sec^3 2t \cdot \tan 2t\,dt = \int (\sec^4 2t - \sec^2 2t)(\sec 2t \tan 2t)\,dt = \frac{\sec^5 2t}{10} - \frac{\sec^3 2t}{6} + C$$

35. $\int \tan^2 x \sec^2 x\,dx = \frac{\tan^3 x}{3} + C$

36. $\int \tan^5 2x \sec^2 2x\,dx = \frac{1}{12}\tan^6 2x + C$

37. $\int \sec^6 4x \tan 4x\,dx = \frac{1}{4}\int \sec^5 4x(4 \sec 4x \tan 4x)\,dx$

$= \frac{\sec^6 4x}{24} + C$

38. $\int \sec^2 \frac{x}{2} \tan \frac{x}{2}\,dx = 2\int \sec \frac{x}{2}\left(\frac{1}{2}\sec \frac{x}{2} \tan \frac{x}{2}\right)dx$

$= \sec^2 \frac{x}{2} + C$ or

$\int \sec^2 \frac{x}{2} \tan \frac{x}{2}\,dx = 2\int \tan \frac{x}{2}\left(\frac{1}{2}\sec^2 \frac{x}{2}\right)dx = \tan^2 \frac{x}{2} + C$

39. Let $u = \sec x$, $du = \sec x \tan x\,dx$.

$\int \sec^3 x \tan x\,dx = \int \sec^2 x(\sec x \tan x)\,dx$

$= \frac{1}{3}\sec^3 x + C$

40. $\int \tan^3 3x\,dx = \int (\sec^2 3x - 1)\tan 3x\,dx$

$= \frac{1}{3}\int \tan 3x(3\sec^2 3x)\,dx + \frac{1}{3}\int \frac{-3\sin 3x}{\cos 3x}\,dx$

$= \frac{1}{6}\tan^2 3x + \frac{1}{3}\ln|\cos 3x| + C$

41. $\int \frac{\tan^2 x}{\sec x}\,dx = \int \frac{(\sec^2 x - 1)}{\sec x}\,dx$

$= \int (\sec x - \cos x)\,dx$

$= \ln|\sec x + \tan x| - \sin x + C$

42. $\int \frac{\tan^2 x}{\sec^5 x} = \int \frac{\sin^2 x}{\cos^2 x} \cdot \cos^5 x\,dx$

$= \int \sin^2 x \cdot \cos^3 x\,dx$

$= \int \sin^2 x(1 - \sin^2 x)\cos x\,dx$

$= \int (\sin^2 x - \sin^4 x)\cos x\,dx$

$= \frac{\sin^3 x}{3} - \frac{\sin^5 x}{5} + C$

43. $r = \int \sin^4(\pi\theta)\,d\theta = \frac{1}{4}\int [1 - \cos(2\pi\theta)]^2\,d\theta$

$= \frac{1}{4}\int [1 - 2\cos(2\pi\theta) + \cos^2(2\pi\theta)]\,d\theta$

$= \frac{1}{4}\int \left[1 - 2\cos(2\pi\theta) + \frac{1 + \cos(4\pi\theta)}{2}\right] d\theta$

$= \frac{1}{4}\left[\theta - \frac{1}{\pi}\sin(2\pi\theta) + \frac{\theta}{2} + \frac{1}{8\pi}\sin(4\pi\theta)\right] + C$

$= \frac{1}{32\pi}[12\pi\theta - 8\sin(2\pi\theta) + \sin(4\pi\theta)] + C$

44. $s = \int \sin^2\frac{\alpha}{2}\cos^2\frac{\alpha}{2}\,d\alpha$

$= \int \left(\frac{1 - \cos\alpha}{2}\right)\left(\frac{1 + \cos\alpha}{2}\right) d\alpha = \int \frac{1 - \cos^2\alpha}{4}\,d\alpha$

$= \frac{1}{4}\int \sin^2\alpha\,d\alpha = \frac{1}{8}\int (1 - \cos 2\alpha)\,d\alpha$

$= \frac{1}{8}\left[\theta - \frac{\sin 2\alpha}{2}\right] + C$

$= \frac{1}{16}(2\alpha - \sin 2\alpha) + C$

45. $y = \int \tan^3 3x \sec 3x\,dx$

$= \int (\sec^2 3x - 1)\sec 3x \tan 3x\,dx$

$= \frac{1}{3}\int \sec^2 3x(3\sec 3x \tan 3x)\,dx - \frac{1}{3}\int 3\sec 3x \tan 3x\,dx$

$= \frac{1}{9}\sec^3 3x - \frac{1}{3}\sec 3x + C$

46. $y = \int \sqrt{\tan x}\sec^4 x\,dx$

$= \int \tan^{1/2} x(\tan^2 x + 1)\sec^2 x\,dx$

$= \int (\tan^{5/2} x + \tan^{1/2} x)\sec^2 x\,dx$

$= \frac{2}{7}\tan^{7/2} x + \frac{2}{3}\tan^{3/2} x + C$

47. (a)

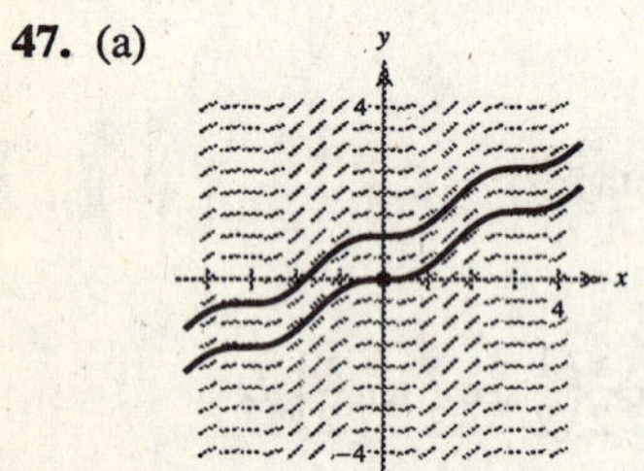

(b) $\frac{dy}{dx} = \sin^2 x,\ (0, 0)$

$y = \int \sin^2 x\,dx$

$= \int \frac{1 - \cos 2x}{2}\,dx$

$= \frac{1}{2}x - \frac{\sin 2x}{4} + C$

$(0, 0)$: $0 = C,\ y = \frac{1}{2}x - \frac{\sin 2x}{4}$

48. (a)

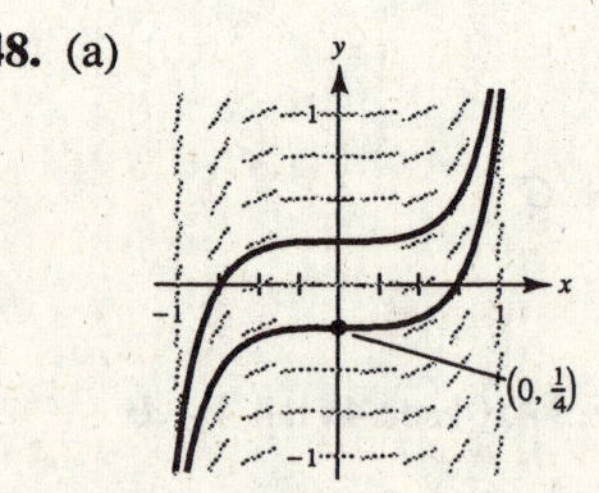

(b) $\frac{dy}{dx} = \sec^2 x \tan^2 x,\ \left(0, -\frac{1}{4}\right)$

$y = \int \sec^2 x \tan^2 x\,dx \quad u = \tan x,\ du = \sec^2 x\,dx$

$y = \frac{\tan^3 x}{3} + C$

$\left(0, -\frac{1}{4}\right)$: $-\frac{1}{4} = C \Rightarrow y = \frac{1}{3}\tan^3 x - \frac{1}{4}$

49. $\frac{dy}{dx} = \frac{3\sin x}{y},\ y(0) = 2$

50. $\frac{dy}{dx} = 3\sqrt{y}\tan^2 x,\ y(0) = 3$

51. $\displaystyle\int \sin 3x \cos 2x\,dx = \frac{1}{2}\int (\sin 5x + \sin x)\,dx$

$\displaystyle = \frac{-1}{2}\left(\frac{1}{5}\cos 5x + \cos x\right) + C$

$\displaystyle = \frac{-1}{10}(\cos 5x + 5\cos x) + C$

52. $\displaystyle\int \cos 4\theta \cos(-3\theta)\,d\theta = \int \cos 4\theta \cos 3\theta\,d\theta$

$\displaystyle = \frac{1}{2}\int (\cos 7\theta + \cos\theta)\,d\theta$

$\displaystyle = \frac{\sin 7\theta}{14} + \frac{\sin\theta}{2} + C$

53. $\displaystyle\int \sin\theta \sin 3\theta\,d\theta = \frac{1}{2}\int (\cos 2\theta - \cos 4\theta)\,d\theta$

$\displaystyle = \frac{1}{2}\left(\frac{1}{2}\sin 2\theta - \frac{1}{4}\sin 4\theta\right) + C$

$\displaystyle = \frac{1}{8}(2\sin 2\theta - \sin 4\theta) + C$

54. $\displaystyle\int \sin(-4x)\cos 3x\,dx = -\int \sin 4x \cos 3x\,dx$

$\displaystyle = -\frac{1}{2}\int (\sin x + \sin 7x)\,dx$

$\displaystyle = -\frac{1}{2}\left[-\cos x - \frac{1}{7}\cos 7x\right] + C$

$\displaystyle = \frac{1}{14}[7\cos x + \cos 7x] + C$

55. $\displaystyle\int \cot^3 2x\,dx = \int (\csc^2 2x - 1)\cot 2x\,dx$

$\displaystyle = -\frac{1}{2}\int \cot 2x(-2\csc^2 2x)\,dx - \frac{1}{2}\int \frac{2\cos 2x}{\sin 2x}\,dx$

$\displaystyle = -\frac{1}{4}\cot^2 2x - \frac{1}{2}\ln|\sin 2x| + C$

$\displaystyle = \frac{1}{4}(\ln|\csc^2 2x| - \cot^2 2x) + C$

56. Let $u = \tan\dfrac{x}{2}$, $du = \dfrac{1}{2}\sec^2\dfrac{x}{2}\,dx$.

$\displaystyle\int \tan^4\frac{x}{2}\sec^4\frac{x}{2}\,dx = \int \tan^4\frac{x}{2}\left(\tan^2\frac{x}{2} + 1\right)\sec^2\frac{x}{2}\,dx$

$\displaystyle = 2\int\left(\tan^6\frac{x}{2} + \tan^4\frac{x}{2}\right)\left(\frac{1}{2}\sec^2\frac{x}{2}\right)dx$

$\displaystyle = \frac{2}{7}\tan^7\frac{x}{2} + \frac{2}{5}\tan^5\frac{x}{2} + C$

57. Let $u = \cot\theta$, $du = -\csc^2\theta\,d\theta$.

$\displaystyle\int \csc^4\theta\,d\theta = \int \csc^2\theta(1 + \cot^2\theta)\,d\theta$

$\displaystyle = \int \csc^2\theta\,d\theta + \int \csc^2\theta\cot^2\theta\,d\theta$

$\displaystyle = -\cot\theta - \frac{1}{3}\cot^3\theta + C$

58. $u = \cot 3x$, $du = -3\csc^2 3x\,dx$

$\displaystyle\int \csc^2 3x \cot 3x\,dx = -\frac{1}{3}\int \cot 3x(-3\csc^2 3x)\,dx$

$\displaystyle = -\frac{1}{6}\cot^2 3x + C$

59. $\displaystyle\int \frac{\cot^2 t}{\csc t}\,dt = \int \frac{\csc^2 t - 1}{\csc t}\,dt$

$\displaystyle = \int (\csc t - \sin t)\,dt$

$\displaystyle = \ln|\csc t - \cot t| + \cos t + C$

60. $\displaystyle\int \frac{\cot^3 t}{\csc t}\,dt = \int \frac{\cos^3 t}{\sin^2 t}\,dt = \int \frac{(1 - \sin^2 t)\cos t}{\sin^2 t}\,dt$

$\displaystyle = \int \frac{\cos t}{\sin^2 t}\,dt - \int \cos t\,dt$

$\displaystyle = \frac{-1}{\sin t} - \sin t + C = -\csc t - \sin t + C$

61. $\displaystyle\int \frac{1}{\sec x \tan x}\,dx = \int \frac{\cos^2 x}{\sin x}\,dx = \int \frac{1 - \sin^2 x}{\sin x}\,dx$

$\displaystyle = \int (\csc x - \sin x)\,dx$

$\displaystyle = \ln|\csc x - \cot x| + \cos x + C$

62. $\displaystyle\int \frac{\sin^2 x - \cos^2 x}{\cos x}\,dx = \int \frac{1 - 2\cos^2 x}{\cos x}\,dx$

$\displaystyle = \int (\sec x - 2\cos x)\,dx$

$\displaystyle = \ln|\sec x + \tan x| - 2\sin x + C$

63. $\displaystyle\int(\tan^4 t - \sec^4 t)\,dt = \int(\tan^2 t + \sec^2 t)(\tan^2 t - \sec^2 t)\,dt, \qquad (\tan^2 t - \sec^2 t = -1)$

$$= -\int(\tan^2 t + \sec^2 t)\,dt = -\int(2\sec^2 t - 1)\,dt = -2\tan t + t + C$$

64. $\displaystyle\int\frac{1 - \sec t}{\cos t - 1}\,dt = \int\frac{\cos t - 1}{(\cos t - 1)\cos t}\,dt$

$$= \int \sec t\,dt = \ln|\sec t + \tan t| + C$$

65. $\displaystyle\int_{-\pi}^{\pi}\sin^2 x\,dx = 2\int_0^{\pi}\frac{1 - \cos 2x}{2}\,dx$

$$= \left[x - \frac{1}{2}\sin 2x\right]_0^{\pi} = \pi$$

66. $\displaystyle\int_0^{\pi/3}\tan^2 x\,dx = \int_0^{\pi/3}(\sec^2 x - 1)\,dx$

$$= \left[\tan x - x\right]_0^{\pi/3} = \sqrt{3} - \frac{\pi}{3}$$

67. $\displaystyle\int_0^{\pi/4}\tan^3 x\,dx = \int_0^{\pi/4}(\sec^2 x - 1)\tan x\,dx$

$$= \int_0^{\pi/4}\sec^2 x\tan x\,dx - \int_0^{\pi/4}\frac{\sin x}{\cos x}\,dx$$

$$= \left[\frac{1}{2}\tan^2 x + \ln|\cos x|\right]_0^{\pi/4}$$

$$= \frac{1}{2}(1 - \ln 2)$$

68. Let $u = \tan t$, $du = \sec^2 t\,dt$.

$$\int_0^{\pi/4}\sec^2 t\sqrt{\tan t}\,dt = \left[\frac{2}{3}\tan^{3/2} t\right]_0^{\pi/4} = \frac{2}{3}$$

69. Let $u = 1 + \sin t$, $du = \cos t\,dt$.

$$\int_0^{\pi/2}\frac{\cos t}{1 + \sin t}\,dt = \left[\ln|1 + \sin t|\right]_0^{\pi/2} = \ln 2$$

70. $\displaystyle\int_{-\pi}^{\pi}\sin 3\theta\cos\theta\,d\theta = \frac{1}{2}\int_{-\pi}^{\pi}(\sin 4\theta + \sin 2\theta)\,d\theta$

$$= -\frac{1}{2}\left[\frac{1}{4}\cos 4\theta + \frac{1}{2}\cos 2\theta\right]_{-\pi}^{\pi} = 0$$

71. Let $u = \sin x$, $du = \cos x\,dx$.

$$\int_{-\pi/2}^{\pi/2}\cos^3 x\,dx = 2\int_0^{\pi/2}(1 - \sin^2 x)\cos x\,dx$$

$$= 2\left[\sin x - \frac{1}{3}\sin^3 x\right]_0^{\pi/2} = \frac{4}{3}$$

72. $\displaystyle\int_{-\pi/2}^{\pi/2}(\sin^2 x + 1)\,dx = \int_{-\pi/2}^{\pi/2}\left(\frac{1 - \cos 2x}{2} + 1\right)dx$

$$= \int_{-\pi/2}^{\pi/2}\left(\frac{3}{2} - \frac{1}{2}\cos 2x\right)dx$$

$$= \left[\frac{3}{2}x - \frac{1}{4}\sin 2x\right]_{-\pi/2}^{\pi/2} = \frac{3\pi}{2}$$

73. $\displaystyle\int\cos^4\frac{x}{2}\,dx = \frac{1}{16}[6x + 8\sin x + \sin 2x] + C$

$$= \frac{1}{8}\left[4\sin\frac{x}{2}\cos^3\frac{x}{2} + 6\sin\frac{x}{2}\cos\frac{x}{2} + 3x\right] + C$$

74. $\displaystyle\int\sin^2 x\cos^2 x\,dx = \frac{1}{32}[4x - \sin 4x] + C$

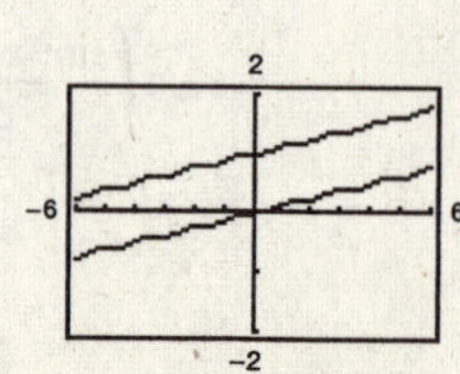

75. $\displaystyle\int \sec^5 \pi x\,dx = \frac{1}{4\pi}\left\{\sec^3 \pi x \tan \pi x + \frac{3}{2}[\sec \pi x \tan \pi x + \ln|\sec \pi x + \tan \pi x|]\right\} + C$

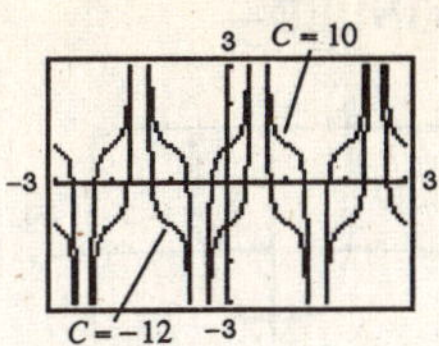

76. $\displaystyle\int \tan^3(1 - x)\,dx = -\frac{\tan^2(1 - x)}{2} - \ln|\cos(1 - x)| + C$

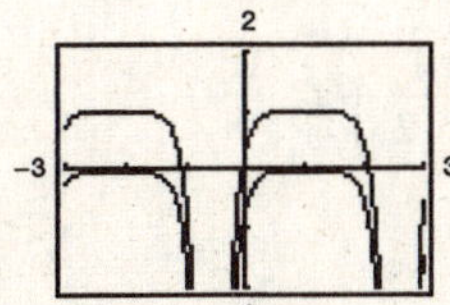

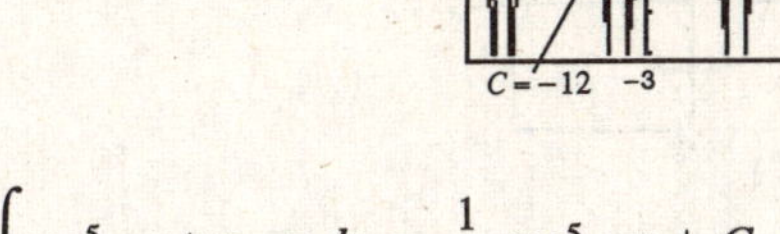

77. $\displaystyle\int \sec^5 \pi x \tan \pi x\,dx = \frac{1}{5\pi}\sec^5 \pi x + C$

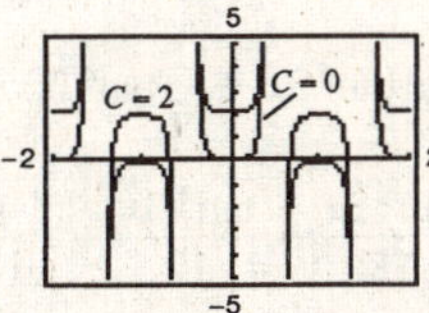

78. $\displaystyle\int \sec^4(1 - x)\tan(1 - x)\,dx = -\frac{\sec^4(1 - x)}{4} + C$

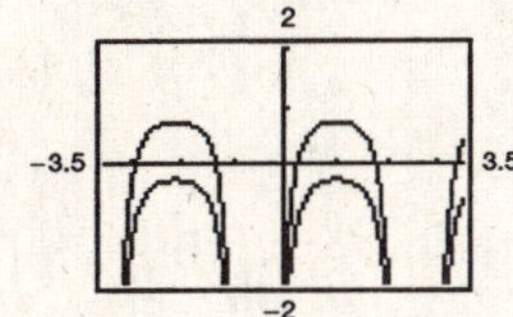

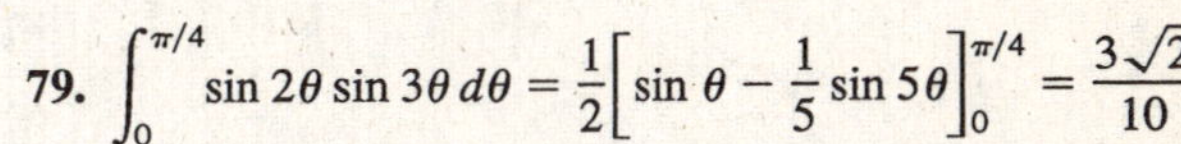

79. $\displaystyle\int_0^{\pi/4} \sin 2\theta \sin 3\theta\,d\theta = \frac{1}{2}\left[\sin \theta - \frac{1}{5}\sin 5\theta\right]_0^{\pi/4} = \frac{3\sqrt{2}}{10}$

80. $\displaystyle\int_0^{\pi/2} (1 - \cos \theta)^2\,d\theta = \left[\frac{3}{2}\theta - 2\sin \theta + \frac{1}{4}\sin 2\theta\right]_0^{\pi/2}$

$$= \frac{3\pi}{4} - 2$$

81. $\displaystyle\int_0^{\pi/2} \sin^4 x\,dx = \frac{1}{4}\left[\frac{3x}{2} - \sin 2x + \frac{1}{8}\sin 4x\right]_0^{\pi/2}$

$$= \frac{3\pi}{16}$$

82. $\displaystyle\int_0^{\pi/2} \sin^6 x\,dx = \frac{1}{8}\left[\frac{5x}{2} - 2\sin 2x + \frac{3}{8}\sin 4x + \frac{1}{6}\sin^3 2x\right]_0^{\pi/2} = \frac{5\pi}{32}$

83. (a) Save one sine factor and convert the remaining sine factors to cosine. Then expand and integrate.

(b) Save one cosine factor and convert the remaining cosine factors to sine. Then expand and integrate.

(c) Make repeated use of the power reducing formula to convert the integrand to odd powers of the cosine.

84. See guidelines on page 537.

85. (a) Let $u = \tan 3x$, $du = 3\sec^2 3x\,dx$.

$$\int \sec^4 3x \tan^3 3x\,dx = \int \sec^2 3x \tan^3 3x \sec^2 3x\,dx = \frac{1}{3}\int (\tan^2 3x + 1)\tan^3 3x(3\sec^2 3x)\,dx$$

$$= \frac{1}{3}\int (\tan^5 3x + \tan^3 3x)(3\sec^2 3x)\,dx = \frac{\tan^6 3x}{18} + \frac{\tan^4 3x}{12} + C_1$$

Or let $u = \sec 3x$, $du = 3\sec 3x \tan 3x\,dx$.

$$\int \sec^4 3x \tan^3 3x\,dx = \int \sec^3 3x \tan^2 3x \sec 3x \tan 3x\,dx$$

$$= \frac{1}{3}\int \sec^3 3x(\sec^2 3x - 1)(3\sec 3x \tan 3x)\,dx = \frac{\sec^6 3x}{18} - \frac{\sec^4 3x}{12} + C$$

—CONTINUED—

85. —CONTINUED—

(b)

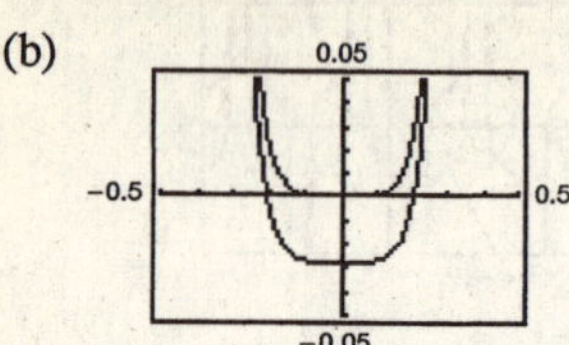

(c) $$\frac{\sec^6 3x}{18} - \frac{\sec^4 3x}{12} + C = \frac{(1+\tan^2 3x)^3}{18} - \frac{(1+\tan^2 3x)^2}{12} + C$$

$$= \frac{1}{18}\tan^6 3x + \frac{1}{6}\tan^4 3x + \frac{1}{6}\tan^2 3x + \frac{1}{18} - \frac{1}{12}\tan^4 3x - \frac{1}{6}\tan^2 3x - \frac{1}{12} + C$$

$$= \frac{\tan^6 3x}{18} + \frac{\tan^4 3x}{12} + \left(\frac{1}{18} - \frac{1}{12}\right) + C$$

$$= \frac{\tan^6 3x}{18} + \frac{\tan^4 3x}{12} + C_2$$

86. (a) Let $u = \tan x$, $du = \sec^2 x\,dx$.

$$\int \sec^2 x \tan x\,dx = \frac{1}{2}\tan^2 x + C_1$$

Or let $u = \sec x$, $du = \sec x \tan x\,dx$.

$$\int \sec x(\sec x \tan x)\,dx = \frac{1}{2}\sec^2 x + C$$

(b)

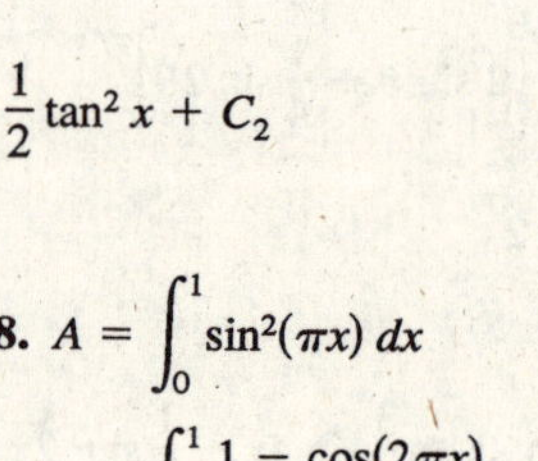

(c) $$\frac{1}{2}\sec^2 x + C = \frac{1}{2}(\tan^2 x + 1) + C = \frac{1}{2}\tan^2 x + \left(\frac{1}{2} + C\right) = \frac{1}{2}\tan^2 x + C_2$$

87. $$A = \int_0^{\pi/2} (\sin x - \sin^3 x)\,dx$$

$$= \int_0^{\pi/2} \sin x\,dx - \int_0^{\pi/2} \sin^3 x\,dx$$

$$= \Big[-\cos x\Big]_0^{\pi/2} - \frac{2}{3} \quad \text{(Wallis's Formula)}$$

$$= 1 - \frac{2}{3} = \frac{1}{3}$$

88. $$A = \int_0^1 \sin^2(\pi x)\,dx$$

$$= \int_0^1 \frac{1 - \cos(2\pi x)}{2}\,dx$$

$$= \left[\frac{1}{2}x - \frac{\sin 2\pi x}{4\pi}\right]_0^1$$

$$= \frac{1}{2}$$

89. $$A = \int_{-\pi/4}^{\pi/4} [\cos^2 x - \sin^2 x]\,dx$$

$$= \int_{-\pi/4}^{\pi/4} \cos 2x\,dx$$

$$= \left[\frac{\sin 2x}{2}\right]_{-\pi/4}^{\pi/4}$$

$$= \frac{1}{2} + \frac{1}{2} = 1$$

90. $$A = \int_{-\pi/2}^{\pi/4} [\cos^2 x - \sin x \cos x]\,dx$$

$$= \int_{-\pi/2}^{\pi/4} \left[\frac{1 + \cos 2x}{2} - \sin x \cos x\right] dx$$

$$= \left[\frac{1}{2}x + \frac{\sin 2x}{4} - \frac{\sin^2 x}{2}\right]_{-\pi/2}^{\pi/4}$$

$$= \left(\frac{\pi}{8} + \frac{1}{4} - \frac{1}{4}\right) - \left(-\frac{\pi}{4} - \frac{1}{2}\right)$$

$$= \frac{3\pi}{8} + \frac{1}{2}$$

91. Disks

$R(x) = \tan x, r(x) = 0$

$$V = 2\pi\int_0^{\pi/4} \tan^2 x\,dx$$

$$= 2\pi\int_0^{\pi/4} (\sec^2 x - 1)\,dx$$

$$= 2\pi\Big[\tan x - x\Big]_0^{\pi/4}$$

$$= 2\pi\left(1 - \frac{\pi}{4}\right) \approx 1.348$$

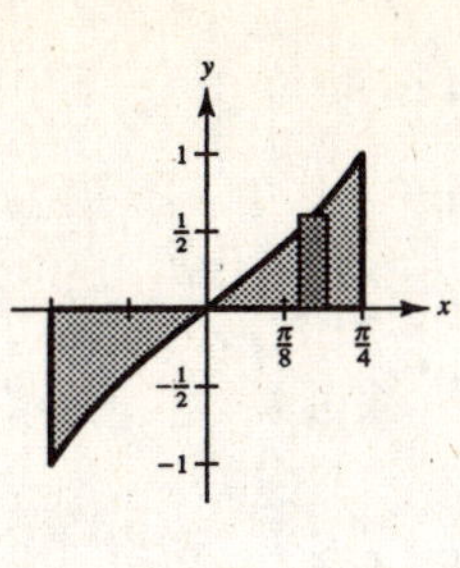

92. $V = \pi\int_0^{\pi/2} \left[\cos^2\left(\frac{x}{2}\right) - \sin^2\left(\frac{x}{2}\right)\right] dx$

$$= \pi\int_0^{\pi/2} \cos x\,dx$$

$$= \pi\Big[\sin x\Big]_0^{\pi/2} = \pi$$

93. (a) $V = \pi\int_0^{\pi} \sin^2 x\,dx = \frac{\pi}{2}\int_0^{\pi}(1 - \cos 2x)\,dx = \frac{\pi}{2}\left[x - \frac{1}{2}\sin 2x\right]_0^{\pi} = \frac{\pi^2}{2}$

(b) $A = \int_0^{\pi} \sin x\,dx = \Big[-\cos x\Big]_0^{\pi} = 1 + 1 = 2$

Let $u = x$, $dv = \sin x\,dx$, $du = dx$, $v = -\cos x$.

$$\bar{x} = \frac{1}{A}\int_0^{\pi} x\sin x\,dx = \frac{1}{2}\left[\Big[-x\cos x\Big]_0^{\pi} + \int_0^{\pi}\cos x\,dx\right] = \frac{1}{2}\Big[-x\cos x + \sin x\Big]_0^{\pi} = \frac{\pi}{2}$$

$$\bar{y} = \frac{1}{2A}\int_0^{\pi} \sin^2 x\,dx = \frac{1}{8}\int_0^{\pi}(1 - \cos 2x)\,dx = \frac{1}{8}\left[x - \frac{1}{2}\sin 2x\right]_0^{\pi} = \frac{\pi}{8}$$

$$(\bar{x}, \bar{y}) = \left(\frac{\pi}{2}, \frac{\pi}{8}\right)$$

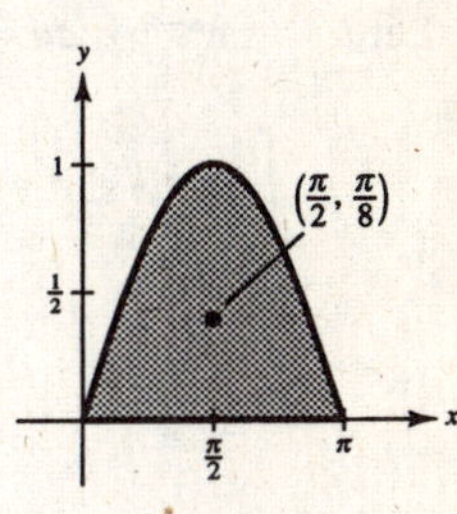

94. (a) $V = \pi\int_0^{\pi/2} \cos^2 x\,dx = \frac{\pi}{2}\int_0^{\pi/2}(1 + \cos 2x)\,dx = \frac{\pi}{2}\left[x + \frac{1}{2}\sin 2x\right]_0^{\pi/2} = \frac{\pi^2}{4}$

(b) $A = \int_0^{\pi/2} \cos x\,dx = \Big[\sin x\Big]_0^{\pi/2} = 1$

Let $u = x$, $dv = \cos x\,dx$, $du = dx$, $v = \sin x$.

$$\bar{x} = \int_0^{\pi/2} x\cos x\,dx = \Big[x\sin x\Big]_0^{\pi/2} - \int_0^{\pi/2}\sin x\,dx = \Big[x\sin x + \cos x\Big]_0^{\pi/2} = \frac{\pi}{2} - 1 = \frac{\pi - 2}{2}$$

$$\bar{y} = \frac{1}{2}\int_0^{\pi/2} \cos^2 x\,dx = \frac{1}{4}\int_0^{\pi/2}(1 + \cos 2x)\,dx = \frac{1}{4}\left[x + \frac{1}{2}\sin 2x\right]_0^{\pi/2} = \frac{\pi}{8}$$

$$(\bar{x}, \bar{y}) = \left(\frac{\pi - 2}{2}, \frac{\pi}{8}\right)$$

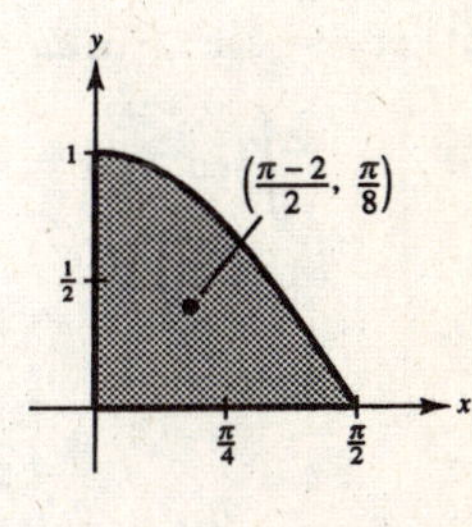

95. $dv = \sin x\,dx \implies v = -\cos x$

$u = \sin^{n-1} x \implies du = (n - 1)\sin^{n-2} x\cos x\,dx$

$$\int \sin^n x\,dx = -\sin^{n-1} x\cos x + (n - 1)\int \sin^{n-2} x\cos^2 x\,dx = -\sin^{n-1} x\cos x + (n - 1)\int \sin^{n-2} x(1 - \sin^2 x)\,dx$$

$$= -\sin^{n-1} x\cos x + (n - 1)\int \sin^{n-2} x\,dx - (n - 1)\int \sin^n x\,dx$$

Therefore, $n\int \sin^n x\,dx = -\sin^{n-1} x\cos x + (n - 1)\int \sin^{n-2} x\,dx$

$$\int \sin^n x\,dx = \frac{-\sin^{n-1} x\cos x}{n} + \frac{n - 1}{n}\int \sin^{n-2} x\,dx.$$

96. $dv = \cos x\,dx \implies v = \sin x$

$u = \cos^{n-1} x \implies du = -(n-1)\cos^{n-2} x \sin x\,dx$

$$\int \cos^n x\,dx = \cos^{n-1} x \sin x + (n-1)\int \cos^{n-2} x \sin^2 x\,dx$$

$$= \cos^{n-1} x \sin x + (n-1)\int \cos^{n-2} x(1 - \cos^2 x)\,dx$$

$$= \cos^{n-1} x \sin x + (n-1)\int \cos^{n-2} x\,dx - (n-1)\int \cos^n x\,dx$$

Therefore, $n\int \cos^n x\,dx = \cos^{n-1} x \sin x + (n-1)\int \cos^{n-2} x\,dx$

$$\int \cos^n x\,dx = \frac{\cos^{n-1} x \sin x}{n} + \frac{n-1}{n}\int \cos^{n-2} x\,dx.$$

97. Let $u = \sin^{n-1} x$, $du = (n-1)\sin^{n-2} x \cos x\,dx$, $dv = \cos^m x \sin x\,dx$, $v = \dfrac{-\cos^{m+1} x}{m+1}$.

$$\int \cos^m x \sin^n x\,dx = \frac{-\sin^{n-1} x \cos^{m+1} x}{m+1} + \frac{n-1}{m+1}\int \sin^{n-2} x \cos^{m+2} x\,dx$$

$$= \frac{-\sin^{n-1} x \cos^{m+1} x}{m+1} + \frac{n-1}{m+1}\int \sin^{n-2} x \cos^m x(1 - \sin^2 x)\,dx$$

$$= \frac{-\sin^{n-1} x \cos^{m+1} x}{m+1} + \frac{n-1}{m+1}\int \sin^{n-2} x \cos^m x\,dx - \frac{n-1}{m+1}\int \sin^n x \cos^m x\,dx$$

$$\frac{m+n}{m+1}\int \cos^m x \sin^n x\,dx = \frac{-\sin^{n-1} x \cos^{m+1} x}{m+1} + \frac{n-1}{m+1}\int \sin^{n-2} x \cos^m x\,dx$$

$$\int \cos^m x \sin^n x\,dx = \frac{-\cos^{m+1} x \sin^{n-1} x}{m+n} + \frac{n-1}{m+n}\int \cos^m x \sin^{n-2} x\,dx$$

98. Let $u = \sec^{n-2} x$, $du = (n-2)\sec^{n-2} x \tan x\,dx$, $dv = \sec^2 x\,dx$, $v = \tan x$.

$$\int \sec^n x\,dx = \sec^{n-2} x \tan x - \int (n-2)\sec^{n-2} x \tan^2 x\,dx$$

$$= \sec^{n-2} x \tan x - (n-2)\int \sec^{n-2} x(\sec^2 x - 1)\,dx$$

$$= \sec^{n-2} x \tan x - (n-2)\left[\int \sec^n x\,dx - \int \sec^{n-2} x\,dx\right]$$

$$(n-1)\int \sec^n x\,dx = \sec^{n-2} x \tan x + (n-2)\int \sec^{n-2} x\,dx$$

$$\int \sec^n x\,dx = \frac{1}{n-1}\sec^{n-2} x \tan x + \frac{n-2}{n-1}\int \sec^{n-2} x\,dx$$

99. $$\int \sin^5 x\,dx = -\frac{\sin^4 x \cos x}{5} + \frac{4}{5}\int \sin^3 x\,dx$$

$$= -\frac{\sin^4 x \cos x}{5} + \frac{4}{5}\left[-\frac{\sin^2 x \cos x}{3} + \frac{2}{3}\int \sin x\,dx\right]$$

$$= -\frac{1}{5}\sin^4 x \cos x - \frac{4}{15}\sin^2 x \cos x - \frac{8}{15}\cos x + C$$

$$= -\frac{\cos x}{15}[3\sin^4 x + 4\sin^2 x + 8] + C$$

100. $\displaystyle\int \cos^4 x\,dx = \frac{\cos^3 x \sin x}{4} + \frac{3}{4}\int \cos^2 x\,dx$

$\displaystyle = \frac{\cos^3 x \sin x}{4} + \frac{3}{4}\left[\frac{\cos x \sin x}{2} + \frac{1}{2}\int dx\right]$

$\displaystyle = \frac{1}{4}\cos^3 x \sin x + \frac{3}{8}\cos x \sin x + \frac{3}{8}x + C$

$\displaystyle = \frac{1}{8}[2\cos^3 x \sin x + 3\cos x \sin x + 3x] + C$

101. $\displaystyle\int \sec^4 \frac{2\pi x}{5}\,dx = \frac{5}{2\pi}\int \sec^4\left(\frac{2\pi x}{5}\right)\frac{2\pi}{5}\,dx$

$\displaystyle = \frac{5}{2\pi}\left[\frac{1}{3}\sec^2\left(\frac{2\pi x}{5}\right)\tan\left(\frac{2\pi x}{5}\right) + \frac{2}{3}\int \sec^2\left(\frac{2\pi x}{5}\right)\frac{2\pi}{5}\,dx\right]$

$\displaystyle = \frac{5}{6\pi}\left[\sec^2\left(\frac{2\pi x}{5}\right)\tan\left(\frac{2\pi x}{5}\right) + 2\tan\left(\frac{2\pi x}{5}\right)\right] + C$

$\displaystyle = \frac{5}{6\pi}\tan\left(\frac{2\pi x}{5}\right)\left[\sec^2\left(\frac{2\pi x}{5}\right) + 2\right] + C$

102. $\displaystyle\int \sin^4 x \cos^2 x\,dx = -\frac{\cos^3 x \sin^3 x}{6} + \frac{1}{2}\int \cos^2 x \sin^2 x\,dx$

$\displaystyle = -\frac{\cos^3 x \sin^3 x}{6} + \frac{1}{2}\left[-\frac{\cos^3 x \sin x}{4} + \frac{1}{4}\int \cos^2 x\,dx\right]$

$\displaystyle = -\frac{1}{6}\cos^3 x \sin^3 x - \frac{1}{8}\cos^3 x \sin x + \frac{1}{8}\left[\frac{\cos x \sin x}{2} + \frac{x}{2}\right] + C$

$\displaystyle = -\frac{1}{48}[8\cos^3 x \sin^3 x + 6\cos^3 x \sin x - 3\cos x \sin x - 3x] + C$

103. $\displaystyle f(t) = a_0 + a_1\cos\frac{\pi t}{6} + b_1\sin\frac{\pi t}{6}$

$\displaystyle a_0 = \frac{1}{12}\int_0^{12} f(t)\,dt,\quad a_1 = \frac{1}{6}\int_0^{12} f(t)\cos\frac{\pi t}{6}\,dt,\quad b_1 = \frac{1}{6}\int_0^{12} f(t)\sin\frac{\pi t}{6}\,dt$

(a) $\displaystyle a_0 \approx \frac{1}{12}\cdot\frac{(12-0)}{3(12)}[33.5 + 4(35.4) + 2(44.7) + 4(55.6) + 2(67.4) + 4(76.2) + 2(80.4) + 4(79.0) + 2(72.0)$

$\displaystyle + 4(61.0) + 2(49.3) + 4(38.6) + 33.5]$

≈ 57.72

$a_1 \approx -23.36$

$b_1 \approx -2.75$ (Answers will vary.)

$\displaystyle H(t) \approx 57.72 - 23.36\cos\left(\frac{\pi t}{6}\right) - 2.75\sin\left(\frac{\pi t}{6}\right)$

(b) $\displaystyle L(t) \approx 42.04 - 20.91\cos\left(\frac{\pi t}{6}\right) - 4.33\sin\left(\frac{\pi t}{6}\right)$

(c)

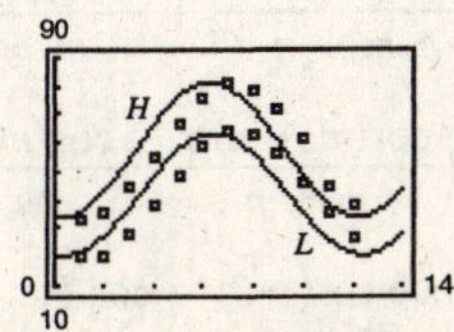

Temperature difference is greatest in the summer ($t \approx 4.9$ or end of May).

104. (a) n is odd and $n \geq 3$.

$$\int_0^{\pi/2} \cos^n x\,dx = \left[\frac{\cos^{n-1} x \sin x}{n}\right]_0^{\pi/2} + \frac{n-1}{n}\int_0^{\pi/2} \cos^{n-2} x\,dx$$

$$= \frac{n-1}{n}\left[\left[\frac{\cos^{n-3} x \sin x}{n-2}\right]_0^{\pi/2} + \frac{n-3}{n-2}\int_0^{\pi/2} \cos^{n-4} x\,dx\right]$$

$$= \frac{n-1}{n}\cdot\frac{n-3}{n-2}\left[\left[\frac{\cos^{n-5} x \sin x}{n-4}\right]_0^{\pi/2} + \frac{n-5}{n-4}\int_0^{\pi/2} \cos^{n-6} x\,dx\right]$$

$$= \frac{n-1}{n}\cdot\frac{n-3}{n-2}\cdot\frac{n-5}{n-4}\int_0^{\pi/2} \cos^{n-6} x\,dx$$

$$= \frac{n-1}{n}\cdot\frac{n-3}{n-2}\cdot\frac{n-5}{n-4}\cdots\int_0^{\pi/2} \cos x\,dx$$

$$= \left[\frac{n-1}{n}\cdot\frac{n-3}{n-2}\cdot\frac{n-5}{n-4}\cdots(\sin x)\right]_0^{\pi/2}$$

$$= \frac{n-1}{n}\cdot\frac{n-3}{n-2}\cdot\frac{n-5}{n-4}\cdots 1 \quad \text{(Reverse the order.)}$$

$$= (1)\left(\frac{2}{3}\right)\left(\frac{4}{5}\right)\left(\frac{6}{7}\right)\cdots\left(\frac{n-1}{n}\right)$$

$$= \left(\frac{2}{3}\right)\left(\frac{4}{5}\right)\left(\frac{6}{7}\right)\cdots\left(\frac{n-1}{n}\right)$$

(b) n is even and $n \geq 2$.

$$\int_0^{\pi/2} \cos^n x\,dx = \frac{n-1}{n}\cdot\frac{n-3}{n-2}\cdot\frac{n-5}{n-4}\cdots\int_0^{\pi/2} \cos^2 x\,dx \quad \text{(From part (a))}$$

$$= \left[\frac{n-1}{n}\cdot\frac{n-3}{n-2}\cdot\frac{n-5}{n-4}\cdots\left(\frac{x}{2}+\frac{1}{4}\sin 2x\right)\right]_0^{\pi/2}$$

$$= \frac{n-1}{n}\cdot\frac{n-3}{n-2}\cdot\frac{n-5}{n-4}\cdots\frac{\pi}{4} \quad \text{(Reverse the order.)}$$

$$= \left(\frac{\pi}{2}\cdot\frac{1}{2}\right)\left(\frac{3}{4}\right)\left(\frac{5}{6}\right)\cdots\left(\frac{n-1}{n}\right)$$

$$= \left(\frac{1}{2}\right)\left(\frac{3}{4}\right)\left(\frac{5}{6}\right)\cdots\left(\frac{n-1}{n}\right)\left(\frac{\pi}{2}\right)$$

105. $$\int_{-\pi}^{\pi} \cos(mx)\cos(nx)\,dx = \frac{1}{2}\left[\frac{\sin(m+n)x}{m+n} + \frac{\sin(m-n)x}{m-n}\right]_{-\pi}^{\pi} = 0, \quad (m \neq n)$$

$$\int_{-\pi}^{\pi} \sin(mx)\sin(nx)\,dx = \frac{1}{2}\int_{-\pi}^{\pi} [\cos(m-n)x - \cos(m+n)x]\,dx$$

$$= \frac{1}{2}\left[\frac{\sin(m-n)x}{m-n} - \frac{\sin(m+n)x}{m+n}\right]_{-\pi}^{\pi} = 0, \quad (m \neq n)$$

$$\int_{-\pi}^{\pi} \sin(mx)\cos(nx)\,dx = \frac{1}{2}\int_{-\pi}^{\pi} [\sin(m+n)x + \sin(m-n)x]\,dx$$

$$= -\frac{1}{2}\left[\frac{\cos(m+n)x}{m+n} + \frac{\cos(m-n)x}{m-n}\right]_{-\pi}^{\pi}, \quad (m \neq n)$$

$$= -\frac{1}{2}\left[\left(\frac{\cos(m+n)\pi}{m+n} + \frac{\cos(m-n)\pi}{m-n}\right) - \left(\frac{\cos(m+n)(-\pi)}{m+n} + \frac{\cos(m-n)(-\pi)}{m-n}\right)\right]$$

$$= 0, \quad \text{since } \cos(-\theta) = \cos\theta.$$

$$\int_{-\pi}^{\pi} \sin(mx)\cos(mx)\,dx = \frac{1}{m}\left[\frac{\sin^2(mx)}{2}\right]_{-\pi}^{\pi} = 0$$

106. $f(x) = \sum_{i=1}^{N} a_i \sin(ix)$

(a)
$$f(x)\sin(nx) = \left[\sum_{i=1}^{N} a_i \sin(ix)\right]\sin(nx)$$

$$\int_{-\pi}^{\pi} f(x)\sin(nx)\,dx = \int_{-\pi}^{\pi}\left[\sum_{i=1}^{N} a_i \sin(ix)\right]\sin(nx)\,dx$$
$$= \int_{-\pi}^{\pi} a_n \sin^2(nx)\,dx \quad \text{(by Exercise 106)}$$
$$= \int_{-\pi}^{\pi} a_n \frac{1-\cos(2nx)}{2}\,dx$$
$$= \left[\frac{a_n}{2}\left(x - \frac{\sin(2nx)}{2n}\right)\right]_{-\pi}^{\pi}$$
$$= \frac{a_n}{2}(\pi + \pi) = a_n\pi$$

Hence, $a_n = \dfrac{1}{\pi}\displaystyle\int_{-\pi}^{\pi} f(x)\sin(nx)\,dx.$

(b) $f(x) = x$

$$a_1 = \frac{1}{\pi}\int_{-\pi}^{\pi} x\sin x\,dx = 2$$
$$a_2 = \frac{1}{\pi}\int_{-\pi}^{\pi} x\sin 2x\,dx = -1$$
$$a_3 = \frac{1}{\pi}\int_{-\pi}^{\pi} x\sin 3x\,dx = \frac{2}{3}$$

Section 8.4 Trigonometric Substitution

1. $$\frac{d}{dx}\left[4\ln\left|\frac{\sqrt{x^2+16}-4}{x}\right| + \sqrt{x^2+16} + C\right] = \frac{d}{dx}\left[4\ln\left|\sqrt{x^2+16}-4\right| - 4\ln|x| + \sqrt{x^2+16} + C\right]$$
$$= 4\left[\frac{x/\sqrt{x^2+16}}{\sqrt{x^2+16}-4}\right] - \frac{4}{x} + \frac{x}{\sqrt{x^2+16}}$$
$$= \frac{4x}{\sqrt{x^2+16}\left(\sqrt{x^2+16}-4\right)} - \frac{4}{x} + \frac{x}{\sqrt{x^2+16}}$$
$$= \frac{4x^2 - 4\sqrt{x^2+16}\left(\sqrt{x^2+16}-4\right) + x^2\left(\sqrt{x^2+16}-4\right)}{x\sqrt{x^2+16}\left(\sqrt{x^2+16}-4\right)}$$
$$= \frac{4x^2 - 4(x^2+16) + 16\sqrt{x^2+16} + x^2\sqrt{x^2+16} - 4x^2}{x\sqrt{x^2+16}\left(\sqrt{x^2+16}-4\right)}$$
$$= \frac{\sqrt{x^2+16}(x^2+16) - 4(x^2+16)}{x\sqrt{x^2+16}\left(\sqrt{x^2+16}-4\right)}$$
$$= \frac{(x^2+16)\left(\sqrt{x^2+16}-4\right)}{x\sqrt{x^2+16}(}$$

Indefinite integral: $\displaystyle\int \frac{\sqrt{x^2+16}}{x}\,dx$, matches (b).

2. $$\frac{d}{dx}\left[8\ln\left|\sqrt{x^2-16}+x\right| + \frac{1}{2}x\sqrt{x^2-16} + C\right] = 8\left[\frac{\left(x/\sqrt{x^2-16}\right)+1}{\sqrt{x^2-16}+x}\right] + \frac{1}{2}x\left(\frac{x}{\sqrt{x^2-16}}\right) + \frac{1}{2}\sqrt{x^2-16}$$
$$= \frac{8\left(x+\sqrt{x^2-16}\right)}{\sqrt{x^2-16}\left(\sqrt{x^2-16}+x\right)} + \frac{x^2}{2\sqrt{x^2+16}} + \frac{\sqrt{x^2-16}}{2}$$
$$= \frac{16 + x^2 + x^2 - 16}{2\sqrt{x^2-16}}$$
$$= \frac{x^2}{\sqrt{x^2-16}}$$

Indefinite integral: $\displaystyle\int \frac{x^2}{\sqrt{x^2-16}}$, matches (d).

3. $$\frac{d}{dx}\left[8\arcsin\frac{x}{4} - \frac{x\sqrt{16-x^2}}{2} + C\right] = 8\frac{1/4}{\sqrt{1-(x/4)^2}} - \frac{x(1/2)(16-x^2)^{-1/2}(-2x) + \sqrt{16-x^2}}{2}$$

$$= \frac{8}{\sqrt{16-x^2}} + \frac{x^2}{2\sqrt{16-x^2}} - \frac{\sqrt{16-x^2}}{2}$$

$$= \frac{16}{2\sqrt{16-x^2}} + \frac{x^2}{2\sqrt{16-x^2}} - \frac{(16-x^2)}{2\sqrt{16-x^2}} = \frac{x^2}{\sqrt{16-x^2}}$$

Matches (a)

4. $$\frac{d}{dx}\left[8\arcsin\frac{x-3}{4} + \frac{(x-3)\sqrt{7+6x-x^2}}{2} + C\right] = 8\left[\frac{1}{\sqrt{1-[(x-3)/4]^2}}\cdot\frac{1}{4}\right] + \frac{1}{2}(x-3)\frac{3-x}{\sqrt{7+6x-x^2}} + \frac{1}{2}\sqrt{7+6x-x^2}$$

$$= \frac{8}{\sqrt{16-(x-3)^2}} - \frac{(x-3)^2}{2\sqrt{16-(x-3)^2}} + \frac{\sqrt{16-(x-3)^2}}{2}$$

$$= \frac{16-(x^2-6x+9)+16-(x^2-6x+9)}{2\sqrt{16-(x-3)^2}}$$

$$= \frac{2[16-(x-3)^2]}{2\sqrt{16-(x-3)^2}}$$

$$= \sqrt{16-(x-3)^2}$$

$$= \sqrt{7+6x-x^2}$$

Indefinite integral: $\int \sqrt{7+6x-x^2}\,dx$, matches (c).

5. Let $x = 5\sin\theta$, $dx = 5\cos\theta\,d\theta$, $\sqrt{25-x^2} = 5\cos\theta$.

$$\int \frac{1}{(25-x^2)^{3/2}}\,dx = \int \frac{5\cos\theta}{(5\cos\theta)^3}\,d\theta$$

$$= \frac{1}{25}\int \sec^2\theta\,d\theta$$

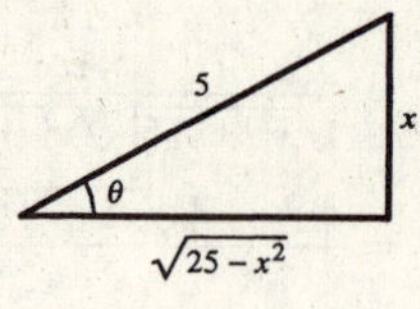

$$= \frac{1}{25}\tan\theta + C$$

$$= \frac{x}{25\sqrt{25-x^2}} + C$$

6. Same substitution as in Exercise 5

$$\int \frac{10}{x^2\sqrt{25-x^2}}\,dx = 10\int \frac{5\cos\theta\,d\theta}{(25\sin^2\theta)(5\cos\theta)} = \frac{2}{5}\int \csc^2\theta\,d\theta = -\frac{2}{5}\cot\theta + C = \frac{-2\sqrt{25-x^2}}{5x} + C$$

7. Same substitution as in Exercise 5

$$\int \frac{\sqrt{25-x^2}}{x}\,dx = \int \frac{25\cos^2\theta\,d\theta}{5\sin\theta} = 5\int \frac{1-\sin^2\theta}{\sin\theta}\,d\theta = 5\int (\csc\theta - \sin\theta)\,d\theta$$

$$= 5[\ln|\csc\theta - \cot\theta| + \cos\theta] + C = 5\ln\left|\frac{5-\sqrt{25-x^2}}{x}\right| + \sqrt{25-x^2} + C$$

8. Same substitution as in Exercise 5

$$\int \frac{x^2}{\sqrt{25-x^2}}\,dx = \int \frac{25\sin^2\theta}{5\cos\theta}(5\cos\theta)\,d\theta = \frac{25}{2}\int (1-\cos 2\theta)\,d\theta$$

$$= \frac{25}{2}\left(\theta - \frac{1}{2}\sin 2\theta\right) + C = \frac{25}{2}(\theta - \sin\theta\cos\theta) + C$$

$$= \frac{25}{2}\left[\arcsin\left(\frac{x}{5}\right) - \left(\frac{x}{5}\right)\left(\frac{\sqrt{25-x^2}}{5}\right)\right] + C = \frac{1}{2}\left[25\arcsin\left(\frac{x}{5}\right) - x\sqrt{25-x^2}\right] + C$$

9. Let $x = 2\sec\theta$, $dx = 2\sec\theta\tan\theta\,d\theta$, $\sqrt{x^2-4} = 2\tan\theta$.

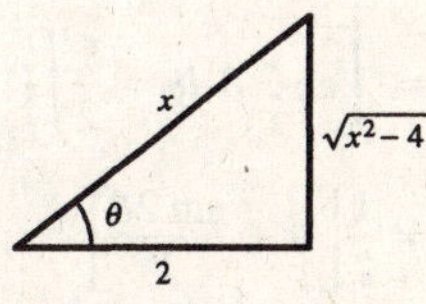

$$\int \frac{1}{\sqrt{x^2-4}}\,dx = \int \frac{2\sec\theta\tan\theta\,d\theta}{2\tan\theta} = \int \sec\theta\,d\theta = \ln|\sec\theta + \tan\theta| + C_1$$

$$= \ln\left|\frac{x}{2} + \frac{\sqrt{x^2-4}}{2}\right| + C_1$$

$$= \ln\left|x + \sqrt{x^2-4}\right| - \ln 2 + C_1 = \ln\left|x + \sqrt{x^2-4}\right| + C$$

10. Same substitution as in Exercise 9

$$\int \frac{\sqrt{x^2-4}}{x}\,dx = \int \frac{2\tan\theta}{2\sec\theta}(2\sec\theta\tan\theta)\,d\theta = 2\int \tan^2\theta\,d\theta = 2\int (\sec^2\theta - 1)\,d\theta$$

$$= 2(\tan\theta - \theta) + C = 2\left[\frac{\sqrt{x^2-4}}{2} - \operatorname{arcsec}\left(\frac{x}{2}\right)\right] + C = \sqrt{x^2-4} - 2\operatorname{arcsec}\left(\frac{x}{2}\right) + C$$

11. Same substitution as in Exercise 9

$$\int x^3\sqrt{x^2-4}\,dx = \int (8\sec^3\theta)(2\tan\theta)(2\sec\theta\tan\theta)\,d\theta = 32\int \tan^2\theta\sec^4\theta\,d\theta$$

$$= 32\int \tan^2\theta(1+\tan^2\theta)\sec^2\theta\,d\theta = 32\left(\frac{\tan^3\theta}{3} + \frac{\tan^5\theta}{5}\right) + C$$

$$= \frac{32}{15}\tan^3\theta[5 + 3\tan^2\theta] + C = \frac{32}{15}\frac{(x^2-4)^{3/2}}{8}\left[5 + 3\frac{(x^2-4)}{4}\right] + C$$

$$= \frac{1}{15}(x^2-4)^{3/2}[20 + 3(x^2-4)] + C = \frac{1}{15}(x^2-4)^{3/2}(3x^2+8) + C$$

12. Same substitution as in Exercise 9

$$\int \frac{x^3}{\sqrt{x^2-4}}\,dx = \int \frac{8\sec^3\theta}{2\tan\theta}(2\sec\theta\tan\theta)\,d\theta = 8\int \sec^4\theta\,d\theta$$

$$= 8\int (1+\tan^2\theta)\sec^2\theta\,d\theta = 8\left(\tan\theta + \frac{\tan^3\theta}{3}\right) + C = \frac{8}{3}\tan\theta(3+\tan^2\theta) + C$$

$$= \frac{8}{3}\left(\frac{\sqrt{x^2-4}}{2}\right)\left(3 + \frac{x^2-4}{4}\right) + C = \frac{1}{3}\sqrt{x^2-4}\,(12 + x^2 - 4) + C = \frac{1}{3}\sqrt{x^2-4}\,(x^2+8) + C$$

13. Let $x = \tan\theta$, $dx = \sec^2\theta\,d\theta$, $\sqrt{1+x^2} = \sec\theta$.

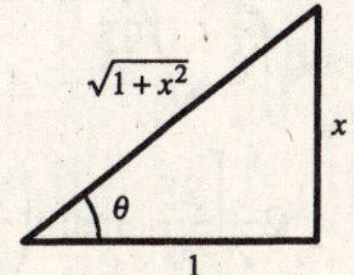

$$\int x\sqrt{1+x^2}\,dx = \int \tan\theta(\sec\theta)\sec^2\theta\,d\theta = \frac{\sec^3\theta}{3} + C = \frac{1}{3}(1+x^2)^{3/2} + C$$

Note: This integral could have been evaluated with the Power Rule.

14. Same substitution as in Exercise 13

$$\int \frac{9x^3}{\sqrt{1+x^2}}\,dx = 9\int \frac{\tan^3\theta}{\sec\theta}\sec^2\theta\,d\theta = 9\int(\sec^2\theta - 1)\sec\theta\tan\theta\,d\theta = 9\left[\frac{\sec^3\theta}{3} - \sec\theta\right] + C$$

$$= 3\sec\theta(\sec^2\theta - 3) + C = 3\sqrt{1+x^2}[(1+x^2) - 3] + C = 3\sqrt{1+x^2}(x^2-2) + C$$

15. Same substitution as in Exercise 13

$$\int \frac{1}{(1+x^2)^2}\,dx = \int \frac{1}{\left(\sqrt{1+x^2}\right)^4}\,dx = \int \frac{\sec^2\theta\,d\theta}{\sec^4\theta}$$

$$= \int \cos^2\theta\,d\theta = \frac{1}{2}\int(1+\cos 2\theta)\,d\theta$$

$$= \frac{1}{2}\left[\theta + \frac{\sin 2\theta}{2}\right]$$

$$= \frac{1}{2}[\theta + \sin\theta\cos\theta] + C$$

$$= \frac{1}{2}\left[\arctan x + \left(\frac{x}{\sqrt{1+x^2}}\right)\left(\frac{1}{\sqrt{1+x^2}}\right)\right] + C$$

$$= \frac{1}{2}\left[\arctan x + \frac{x}{1+x^2}\right] + C$$

16. Same substitution as in Exercise 13

$$\int \frac{x^2}{(1+x^2)^2}\,dx = \int \frac{x^2}{\left(\sqrt{1+x^2}\right)^4}\,dx = \int \frac{\tan^2\theta\sec^2\theta\,d\theta}{\sec^4\theta} = \int \sin^2\theta\,d\theta$$

$$= \frac{1}{2}\int(1-\cos 2\theta)\,d\theta = \frac{1}{2}\left[\theta - \frac{\sin 2\theta}{2}\right] = \frac{1}{2}[\theta - \sin\theta\cos\theta] + C$$

$$= \frac{1}{2}\left[\arctan x - \left(\frac{x}{\sqrt{1+x^2}}\right)\left(\frac{1}{\sqrt{1+x^2}}\right)\right] + C = \frac{1}{2}\left[\arctan x - \frac{x}{1+x^2}\right] + C$$

17. Let $u = 3x$, $a = 2$, and $du = 3\,dx$.

$$\int \sqrt{4+9x^2}\,dx = \frac{1}{3}\int \sqrt{(2)^2 + (3x)^2}\,3\,dx$$

$$= \frac{1}{3}\left(\frac{1}{2}\right)\left(3x\sqrt{4+9x^2} + 4\ln\left|3x + \sqrt{4+9x^2}\right|\right) + C$$

$$= \frac{1}{2}x\sqrt{4+9x^2} + \frac{2}{3}\ln\left|3x + \sqrt{4+9x^2}\right| + C$$

18. Let $u = x$, $a = 1$, and $du = dx$.

$$\int \sqrt{1+x^2}\,dx = \frac{1}{2}\left(x\sqrt{1+x^2} + \ln\left|x + \sqrt{1+x^2}\right|\right) + C$$

19. $\displaystyle\int \sqrt{25-4x^2}\,dx = \int 2\sqrt{\frac{25}{4} - x^2}\,dx, \quad a = \frac{5}{2}$

$$= 2\frac{1}{2}\left[\frac{25}{4}\arcsin\left(\frac{2x}{5}\right) + x\sqrt{\frac{25}{4} - x^2}\right] + C$$

$$= \frac{25}{4}\arcsin\left(\frac{2x}{5}\right) + \frac{x}{2}\sqrt{25-4x^2} + C$$

20. $\int \sqrt{2x^2 - 1}\, dx = \int \sqrt{(\sqrt{2}x)^2 - 1}\, dx, \quad u = \sqrt{2}x, du = \sqrt{2}\, dx$

$$= \frac{1}{\sqrt{2}}\left(\frac{1}{2}\right)\left[\sqrt{2}x\sqrt{2x^2 - 1} - \ln\left|\sqrt{2}x + \sqrt{2x^2 - 1}\right|\right] + C$$

$$= \frac{x}{2}\sqrt{2x^2 - 1} - \frac{\sqrt{2}}{4}\ln\left|\sqrt{2}x + \sqrt{2x^2 - 1}\right| + C$$

21. $\int \frac{x}{\sqrt{x^2 + 9}}\, dx = \frac{1}{2}\int (x^2 + 9)^{-1/2}(2x)\, dx$

$$= \sqrt{x^2 + 9} + C$$

(Power Rule)

22. $\int \frac{x}{\sqrt{9 - x^2}}\, dx = -\frac{1}{2}\int (9 - x^2)^{-1/2}(-2x)\, dx$

$$= -(9 - x^2)^{1/2} + C$$

(Power Rule)

23. $\int \frac{1}{\sqrt{16 - x^2}}\, dx = \arcsin\left(\frac{x}{4}\right) + C$

24. $\int \frac{1}{\sqrt{25 - x^2}}\, dx = \arcsin\frac{x}{5} + C$

25. Let $x = 2\sin\theta$, $dx = 2\cos\theta\, d\theta$, $\sqrt{4 - x^2} = 2\cos\theta$.

$$\int \sqrt{16 - 4x^2}\, dx = 2\int \sqrt{4 - x^2}\, dx$$

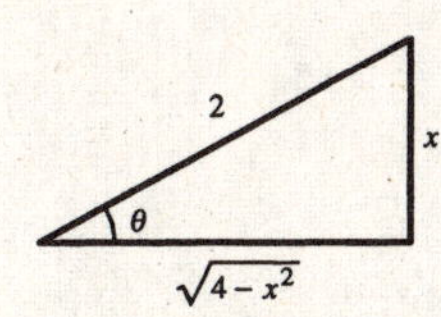

$$= 2\int 2\cos\theta(2\cos\theta\, d\theta)$$

$$= 8\int \cos^2\theta\, d\theta$$

$$= 4\int (1 + \cos 2\theta)\, d\theta$$

$$= 4\left[\theta + \frac{1}{2}\sin 2\theta\right] + C$$

$$= 4\theta + 4\sin\theta\cos\theta + C$$

$$= 4\arcsin\left(\frac{x}{2}\right) + x\sqrt{4 - x^2} + C$$

26. Let $u = 16 - 4x^2$, $du = -8x\, dx$.

$$\int x\sqrt{16 - 4x^2}\, dx = -\frac{1}{8}\int (16 - 4x^2)^{1/2}(-8x)\, dx = \left[-\frac{1}{12}(16 - 4x^2)^{3/2}\right] + C = -\frac{2}{3}(4 - x^2)^{3/2} + C$$

27. Let $x = 3\sec\theta$, $dx = 3\sec\theta\tan\theta\, d\theta$, $\sqrt{x^2 - 9} = 3\tan\theta$.

$$\int \frac{1}{\sqrt{x^2 - 9}}\, dx = \int \frac{3\sec\theta\tan\theta\, d\theta}{3\tan\theta}$$

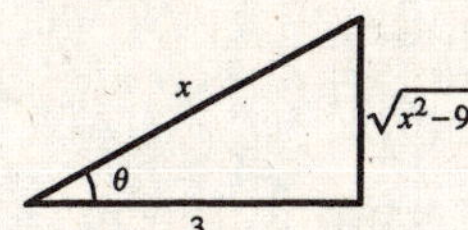

$$= \int \sec\theta\, d\theta$$

$$= \ln|\sec\theta + \tan\theta| + C_1$$

$$= \ln\left|\frac{x}{3} + \frac{\sqrt{x^2 - 9}}{3}\right| + C_1$$

$$= \ln\left|x + \sqrt{x^2 - 9}\right| + C$$

28. Let $u = 1 - t^2$, $du = -2t\,dt$.

$$\int \frac{t}{(1-t^2)^{3/2}}\,dt = -\frac{1}{2}\int (1-t^2)^{-3/2}(-2t)\,dt = \frac{1}{\sqrt{1-t^2}} + C$$

29. Let $x = \sin\theta$, $dx = \cos\theta\,d\theta$, $\sqrt{1-x^2} = \cos\theta$.

$$\int \frac{\sqrt{1-x^2}}{x^4}\,dx = \int \frac{\cos\theta(\cos\theta\,d\theta)}{\sin^4\theta}$$

$$= \int \cot^2\theta\csc^2\theta\,d\theta$$

$$= -\frac{1}{3}\cot^3\theta + C$$

$$= \frac{-(1-x^2)^{3/2}}{3x^3} + C$$

30. Let $2x = 3\tan\theta$, $dx = \frac{3}{2}\sec^2\theta\,d\theta$, $\sqrt{4x^2+9} = 3\sec\theta$.

$$\int \frac{\sqrt{4x^2+9}}{x^4}\,dx = \int \frac{3\sec\theta[(3/2)\sec^2\theta\,d\theta]}{(3/2)^4\tan^4\theta}$$

$$= \frac{8}{9}\int \frac{\cos\theta}{\sin^4\theta}\,d\theta$$

$$= \frac{-8}{27\sin^3\theta} + C$$

$$= -\frac{8}{27}\csc^3\theta + C$$

$$= \frac{-(4x^2+9)^{3/2}}{27x^3} + C$$

31. Same substitution as in Exercise 30

$$x = \frac{3}{2}\tan\theta,\ dx = \frac{3}{2}\sec^2\theta\,d\theta$$

$$\int \frac{1}{x\sqrt{4x^2+9}}\,dx = \int \frac{(3/2)\sec^2\theta\,d\theta}{(3/2)\tan\theta\, 3\sec\theta}$$

$$= \frac{1}{3}\int \csc\theta\,d\theta$$

$$= -\frac{1}{3}\ln|\csc\theta + \cot\theta| + C$$

$$= -\frac{1}{3}\ln\left|\frac{\sqrt{4x^2+9}+3}{2x}\right| + C$$

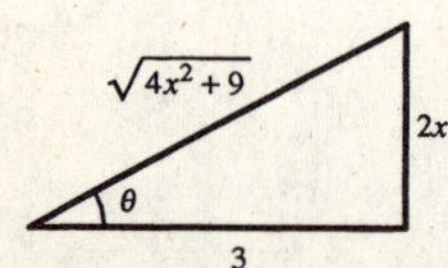

32. Let $2x = 4\tan\theta$, $dx = 2\sec^2\theta\,d\theta$, $\sqrt{4x^2+16} = 4\sec\theta$.

$$\int \frac{1}{x\sqrt{4x^2+16}}\,dx = \int \frac{2\sec^2\theta\,d\theta}{2\tan\theta(4\sec\theta)}$$

$$= \frac{1}{4}\int \frac{\sec\theta}{\tan\theta}\,d\theta = \frac{1}{4}\int \csc\theta\,d\theta$$

$$= -\frac{1}{4}\ln|\csc\theta + \cot\theta| + C$$

$$= -\frac{1}{4}\ln\left|\frac{\sqrt{x^2+4}+2}{x}\right| + C$$

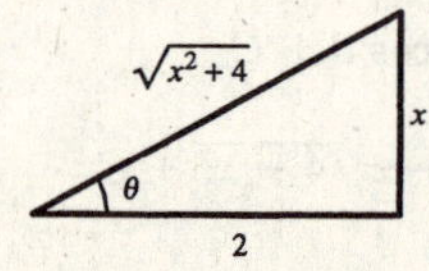

33. Let $x = \sqrt{5}\tan\theta$, $dx = \sqrt{5}\sec^2\theta\,d\theta$, $x^2 + 5 = 5\sec^2\theta$.

$$\int \frac{-5x}{(x^2+5)^{3/2}}\,dx = \int \frac{-5\sqrt{5}\tan\theta}{(5\sec^2\theta)^{3/2}}\sqrt{5}\sec^2\theta\,d\theta$$

$$= -\sqrt{5}\int \frac{\tan\theta}{\sec\theta}\,d\theta$$

$$= -\sqrt{5}\int \sin\theta\,d\theta$$

$$= \sqrt{5}\cos\theta + C$$

$$= \sqrt{5}\,\frac{\sqrt{5}}{\sqrt{x^2+5}} + C$$

$$= \frac{5}{\sqrt{x^2+5}} + C$$

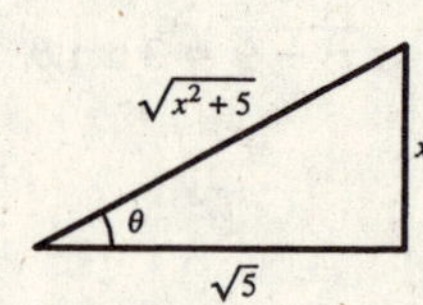

34. Let $x = \sqrt{3}\tan\theta$, $dx = \sqrt{3}\sec^2\theta\, d\theta$, $x^2 + 3 = 3\sec^2\theta$.

$$\int \frac{1}{(x^2+3)^{3/2}}\,dx = \int \frac{\sqrt{3}\sec^2\theta\, d\theta}{3\sqrt{3}\sec^3\theta}$$
$$= \frac{1}{3}\int \cos\theta\, d\theta = \frac{1}{3}\sin\theta + C = \frac{x}{3\sqrt{x^2+3}} + C$$

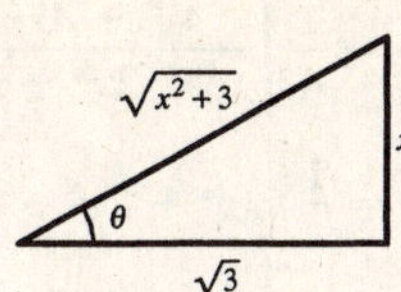

35. Let $u = 1 + e^{2x}$, $du = 2e^{2x}\,dx$.

$$\int e^{2x}\sqrt{1+e^{2x}}\,dx = \frac{1}{2}\int (1+e^{2x})^{1/2}(2e^{2x})\,dx = \frac{1}{3}(1+e^{2x})^{3/2} + C$$

36. Let $u = x^2 + 2x + 2$, $du = (2x+2)\,dx$.

$$\int (x+1)\sqrt{x^2+2x+2}\,dx = \frac{1}{2}\int (x^2+2x+2)^{1/2}(2x+2)\,dx = \frac{1}{3}(x^2+2x+2)^{3/2} + C$$

37. Let $e^x = \sin\theta$, $e^x\,dx = \cos\theta\, d\theta$, $\sqrt{1-e^{2x}} = \cos\theta$.

$$\int e^x\sqrt{1-e^{2x}}\,dx = \int \cos^2\theta\, d\theta$$
$$= \frac{1}{2}\int (1+\cos 2\theta)\,d\theta$$
$$= \frac{1}{2}\left[\theta + \frac{\sin 2\theta}{2}\right]$$
$$= \frac{1}{2}(\theta + \sin\theta\cos\theta) + C$$
$$= \frac{1}{2}\left(\arcsin e^x + e^x\sqrt{1-e^{2x}}\right) + C$$

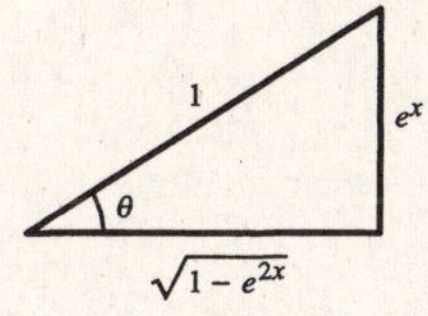

38. Let $\sqrt{x} = \sin\theta$, $x = \sin^2\theta$, $dx = 2\sin\theta\cos\theta\, d\theta$, $\sqrt{1-x} = \cos\theta$.

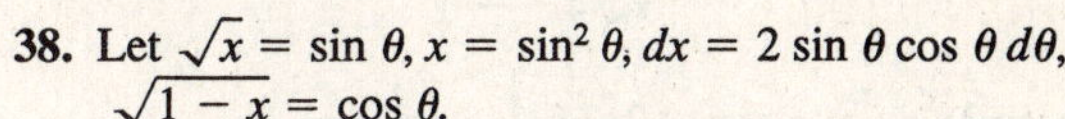

$$\int \frac{\sqrt{1-x}}{\sqrt{x}}\,dx = \int \frac{\cos\theta(2\sin\theta\cos\theta\, d\theta)}{\sin\theta}$$
$$= 2\int \cos^2\theta\, d\theta$$
$$= \int (1+\cos 2\theta)\,d\theta$$
$$= (\theta + \sin\theta\cos\theta) + C$$
$$= \arcsin\sqrt{x} + \sqrt{x}\sqrt{1-x} + C$$

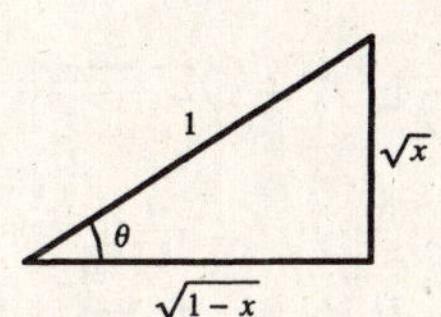

39. Let $x = \sqrt{2}\tan\theta$, $dx = \sqrt{2}\sec^2\theta\, d\theta$, $x^2 + 2 = 2\sec^2\theta$.

$$\int \frac{1}{4+4x^2+x^4}\,dx = \int \frac{1}{(x^2+2)^2}\,dx = \int \frac{\sqrt{2}\sec^2\theta\, d\theta}{4\sec^4\theta}$$
$$= \frac{\sqrt{2}}{4}\int \cos^2\theta\, d\theta$$
$$= \frac{\sqrt{2}}{4}\left(\frac{1}{2}\right)\int (1+\cos 2\theta)\,d\theta$$
$$= \frac{\sqrt{2}}{8}\left(\theta + \frac{1}{2}\sin 2\theta\right) + C$$
$$= \frac{\sqrt{2}}{8}(\theta + \sin\theta\cos\theta) + C$$
$$= \frac{\sqrt{2}}{8}\left(\arctan\frac{x}{\sqrt{2}} + \frac{x}{\sqrt{x^2+2}}\cdot\frac{\sqrt{2}}{\sqrt{x^2+2}}\right)$$
$$= \frac{1}{4}\left[\frac{x}{x^2+2} + \frac{1}{\sqrt{2}}\arctan\frac{x}{\sqrt{2}}\right] + C$$

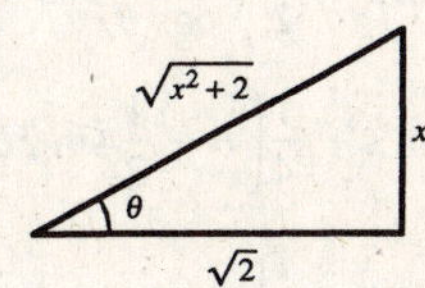

40. Let $x = \tan\theta$, $dx = \sec^2\theta\, d\theta$, $x^2 + 1 = \sec^2\theta$.

$$\int \frac{x^3 + x + 1}{x^4 + 2x^2 + 1}\,dx = \frac{1}{4}\int \frac{4x^3 + 4x}{x^4 + 2x^2 + 1}\,dx + \int \frac{1}{(x^2+1)^2}\,dx$$

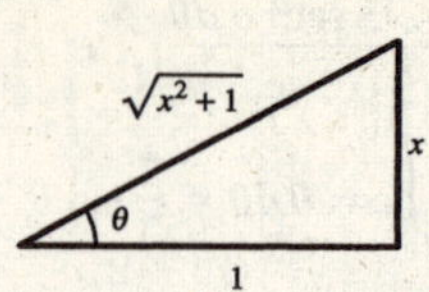

$$= \frac{1}{4}\ln(x^4 + 2x^2 + 1) + \int \frac{\sec^2\theta\, d\theta}{\sec^4\theta}$$

$$= \frac{1}{2}\ln(x^2 + 1) + \frac{1}{2}\int (1 + \cos 2\theta)\, d\theta$$

$$= \frac{1}{2}\ln(x^2 + 1) + \frac{1}{2}(\theta + \sin\theta\cos\theta) + C$$

$$= \frac{1}{2}\left[\ln(x^2 + 1) + \arctan x + \frac{x}{x^2 + 1}\right] + C$$

41. Use integration by parts. Since $x > \frac{1}{2}$,

$$u = \operatorname{arcsec} 2x \implies du = \frac{1}{x\sqrt{4x^2 - 1}}\,dx,\ dv = dx \implies v = x$$

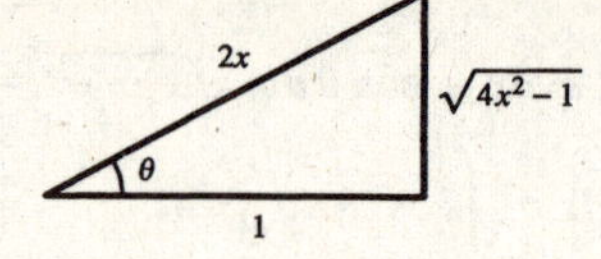

$$\int \operatorname{arcsec} 2x\, dx = x \operatorname{arcsec} 2x - \int \frac{1}{\sqrt{4x^2 - 1}}\,dx$$

$$2x = \sec\theta,\ dx = \frac{1}{2}\sec\theta\tan\theta\, d\theta,\ \sqrt{4x^2 - 1} = \tan\theta$$

$$\int \operatorname{arcsec} 2x\, dx = x \operatorname{arcsec} 2x - \int \frac{(1/2)\sec\theta\tan\theta\, d\theta}{\tan\theta}$$

$$= x \operatorname{arcsec} 2x - \frac{1}{2}\int \sec\theta\, d\theta$$

$$= x \operatorname{arcsec} 2x - \frac{1}{2}\ln|\sec\theta + \tan\theta| + C$$

$$= x \operatorname{arcsec} 2x - \frac{1}{2}\ln\left|2x + \sqrt{4x^2 - 1}\right| + C.$$

42. $u = \arcsin x \implies du = \dfrac{1}{\sqrt{1 - x^2}}\,dx,\ dv = x\,dx \implies v = \dfrac{x^2}{2}$

$$\int x \arcsin x\, dx = \frac{x^2}{2}\arcsin x - \frac{1}{2}\int \frac{x^2}{\sqrt{1 - x^2}}\,dx$$

$$x = \sin\theta,\ dx = \cos\theta\, d\theta,\ \sqrt{1 - x^2} = \cos\theta$$

$$\int x \arcsin x\, dx = \frac{x^2}{2}\arcsin x = \frac{1}{2}\int \frac{\sin^2\theta}{\cos\theta}\cos\theta\, d\theta = \frac{x^2}{2}\arcsin x - \frac{1}{4}\int (1 - \cos 2\theta)\, d\theta$$

$$= \frac{x^2}{2}\arcsin x - \frac{1}{4}\left[\theta - \frac{1}{2}\sin 2\theta\right] + C = \frac{x^2}{2}\arcsin x - \frac{1}{4}[\theta - \sin\theta\cos\theta] + C$$

$$= \frac{x^2}{2}\arcsin x - \frac{1}{4}\left[\arcsin x - x\sqrt{1 - x^2}\right] + C = \frac{1}{4}\left[(2x^2 - 1)\arcsin x + x\sqrt{1 - x^2}\right] + C$$

43. $\displaystyle\int \frac{1}{\sqrt{4x - x^2}}\,dx = \int \frac{1}{\sqrt{4 - (x - 2)^2}}\,dx = \arcsin\left(\frac{x - 2}{2}\right) + C$

44. Let $x - 1 = \sin\theta$, $dx = \cos\theta\,d\theta$, $\sqrt{1-(x-1)^2} = \sqrt{2x - x^2} = \cos\theta$.

$$\begin{aligned}
\int \frac{x^2}{\sqrt{2x-x^2}}\,dx &= \int \frac{x^2}{\sqrt{1-(x-1)^2}}\,dx \\
&= \int \frac{(1+\sin\theta)^2(\cos\theta\,d\theta)}{\cos\theta} \\
&= \int (1 + 2\sin\theta + \sin^2\theta)\,d\theta \\
&= \int\left(\frac{3}{2} + 2\sin\theta - \frac{1}{2}\cos 2\theta\right)d\theta \\
&= \frac{3}{2}\theta - 2\cos\theta - \frac{1}{4}\sin 2\theta + C \\
&= \frac{3}{2}\theta - 2\cos\theta - \frac{1}{2}\sin\theta\cos\theta + C \\
&= \frac{3}{2}\arcsin(x-1) - 2\sqrt{2x-x^2} - \frac{1}{2}(x-1)\sqrt{2x-x^2} + C \\
&= \frac{3}{2}\arcsin(x-1) - \frac{1}{2}\sqrt{2x-x^2}(x+3) + C
\end{aligned}$$

45. Let $x + 2 = 2\tan\theta$, $dx = 2\sec^2\theta\,d\theta$, $\sqrt{(x+2)^2+4} = 2\sec\theta$.

$$\begin{aligned}
\int \frac{x}{\sqrt{x^2+4x+8}}\,dx = \int \frac{x}{\sqrt{(x+2)^2+4}}\,dx &= \int \frac{(2\tan\theta - 2)(2\sec^2\theta)\,d\theta}{2\sec\theta} \\
&= 2\int(\tan\theta - 1)(\sec\theta)\,d\theta \\
&= 2[\sec\theta - \ln|\sec\theta + \tan\theta|] + C_1 \\
&= 2\left[\frac{\sqrt{(x+2)^2+4}}{2} - \ln\left|\frac{\sqrt{(x+2)^2+4}}{2} + \frac{x+2}{2}\right|\right] + C_1 \\
&= \sqrt{x^2+4x+8} - 2\left[\ln\left|\sqrt{x^2+4x+8} + (x+2)\right| - \ln 2\right] + C_1 \\
&= \sqrt{x^2+4x+8} - 2\ln\left|\sqrt{x^2+4x+8} + (x+2)\right| + C
\end{aligned}$$

46. Let $x - 3 = 2\sec\theta$, $dx = 2\sec\theta\tan\theta\,d\theta$, $\sqrt{(x-3)^2-4} = 2\tan\theta$.

$$\begin{aligned}
\int \frac{x}{\sqrt{x^2-6x+5}}\,dx &= \int \frac{x}{\sqrt{(x-3)^2-4}}\,dx \\
&= \int \frac{(2\sec\theta + 3)}{2\tan\theta}(2\sec\theta\tan\theta)\,d\theta \\
&= \int (2\sec^2\theta + 3\sec\theta)\,d\theta \\
&= 2\tan\theta + 3\ln|\sec\theta + \tan\theta| + C_1 \\
&= 2\left(\frac{\sqrt{(x-3)^3-4}}{2}\right) + 3\ln\left|\frac{x-3}{2} + \frac{\sqrt{(x-3)^2-4}}{2}\right| + C_1 \\
&= \sqrt{x^2-6x+5} + 3\ln\left|(x-3) + \sqrt{x^2-6x+5}\right| + C
\end{aligned}$$

47. Let $t = \sin\theta$, $dt = \cos\theta\,d\theta$, $1 - t^2 = \cos^2\theta$.

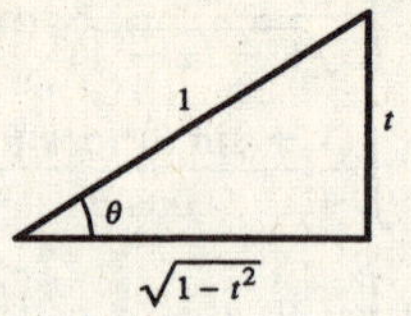

(a) $$\int \frac{t^2}{(1-t^2)^{3/2}}\,dt = \int \frac{\sin^2\theta\cos\theta\,d\theta}{\cos^3\theta}$$
$$= \int \tan^2\theta\,d\theta = \int (\sec^2\theta - 1)\,d\theta$$
$$= \tan\theta - \theta + C$$
$$= \frac{t}{\sqrt{1-t^2}} - \arcsin t + C$$

Thus, $$\int_0^{\sqrt{3}/2} \frac{t^2}{(1-t^2)^{3/2}}\,dt = \left[\frac{t}{\sqrt{1-t^2}} - \arcsin t\right]_0^{\sqrt{3}/2} = \frac{\sqrt{3}/2}{\sqrt{1/4}} - \arcsin\frac{\sqrt{3}}{2} = \sqrt{3} - \frac{\pi}{3} \approx 0.685.$$

(b) When $t = 0$, $\theta = 0$. When $t = \sqrt{3}/2$, $\theta = \pi/3$. Thus,
$$\int_0^{\sqrt{3}/2} \frac{t^2}{(1-t^2)^{3/2}}\,dt = \Big[\tan\theta - \theta\Big]_0^{\pi/3} = \sqrt{3} - \frac{\pi}{3} \approx 0.685.$$

48. Same substitution as in Exercise 47

(a) $$\int \frac{1}{(1-t^2)^{5/2}}\,dt = \int \frac{\cos\theta\,d\theta}{\cos^5\theta} = \int \sec^4\theta\,d\theta = \int (\tan^2\theta + 1)\sec^2\theta\,d\theta$$
$$= \frac{1}{3}\tan^3\theta + \tan\theta + C = \frac{1}{3}\left(\frac{t}{\sqrt{1-t^2}}\right)^3 + \frac{t}{\sqrt{1-t^2}} + C$$

Thus, $$\int_0^{\sqrt{3}/2} \frac{1}{(1-t^2)^{5/2}}\,dt = \left[\frac{t^3}{3(1-t^2)^{3/2}} + \frac{t}{\sqrt{1-t^2}}\right]_0^{\sqrt{3}/2}$$
$$= \frac{3\sqrt{3}/8}{3(1/4)^{3/2}} + \frac{\sqrt{3}/2}{\sqrt{1/4}} = \sqrt{3} + \sqrt{3} = 2\sqrt{3} \approx 3.464.$$

(b) When $t = 0$, $\theta = 0$. When $t = \sqrt{3}/2$, $\theta = \pi/3$. Thus,
$$\int_0^{\sqrt{3}/2} \frac{1}{(1-t^2)^{5/2}}\,dt = \left[\frac{1}{3}\tan^3\theta + \tan\theta\right]_0^{\pi/3} = \frac{1}{3}\left(\sqrt{3}\right)^3 + \sqrt{3} = 2\sqrt{3} \approx 3.464.$$

49. (a) Let $x = 3\tan\theta$, $dx = 3\sec^2\theta\,d\theta$, $\sqrt{x^2+9} = 3\sec\theta$.
$$\int \frac{x^3}{\sqrt{x^2+9}}\,dx = \int \frac{(27\tan^3\theta)(3\sec^2\theta\,d\theta)}{3\sec\theta}$$
$$= 27\int (\sec^2\theta - 1)\sec\theta\tan\theta\,d\theta$$
$$= 27\left[\frac{1}{3}\sec^3\theta - \sec\theta\right] + C = 9[\sec^3\theta - 3\sec\theta] + C$$
$$= 9\left[\left(\frac{\sqrt{x^2+9}}{3}\right)^3 - 3\left(\frac{\sqrt{x^2+9}}{3}\right)\right] + C = \frac{1}{3}(x^2+9)^{3/2} - 9\sqrt{x^2+9} + C$$

Thus, $$\int_0^3 \frac{x^3}{\sqrt{x^2+9}}\,dx = \left[\frac{1}{3}(x^2+9)^{3/2} - 9\sqrt{x^2+9}\right]_0^3$$
$$= \left(\frac{1}{3}(54\sqrt{2}) - 27\sqrt{2}\right) - (9 - 27)$$
$$= 18 - 9\sqrt{2} = 9\left(2 - \sqrt{2}\right) \approx 5.272.$$

(b) When $x = 0$, $\theta = 0$. When $x = 3$, $\theta = \pi/4$. Thus,
$$\int_0^3 \frac{x^3}{\sqrt{x^2+9}}\,dx = 9\Big[\sec^3\theta - 3\sec\theta\Big]_0^{\pi/4} = 9\left(2\sqrt{2} - 3\sqrt{2}\right) - 9(1 - 3) = 9\left(2 - \sqrt{2}\right) \approx 5.272.$$

50. (a) Let $5x = 3\sin\theta$, $dx = \frac{3}{5}\cos\theta\,d\theta$, $\sqrt{9-25x^2} = 3\cos\theta$.

$$\begin{aligned}\int\sqrt{9-25x^2}\,dx &= \int(3\cos\theta)\frac{3}{5}\cos\theta\,d\theta\\ &= \frac{9}{5}\int\frac{1+\cos 2\theta}{2}\,d\theta\\ &= \frac{9}{10}\left[\theta+\frac{1}{2}\sin 2\theta\right]+C\\ &= \frac{9}{10}[\theta+\sin\theta\cos\theta]+C\\ &= \frac{9}{10}\left[\arcsin\frac{5x}{3}+\frac{5x}{3}\cdot\frac{\sqrt{9-25x^2}}{3}\right]+C\end{aligned}$$

Thus, $\displaystyle\int_0^{3/5}\sqrt{9-25x^2}\,dx = \left[\frac{9}{10}\arcsin\frac{5x}{3}+\frac{5x\sqrt{9-25x^2}}{9}\right]_0^{3/5} = \frac{9}{10}\left[\frac{\pi}{2}\right] = \frac{9\pi}{20}.$

(b) When $x = 0$, $\theta = 0$. When $x = \frac{3}{5}$, $\theta = \frac{\pi}{2}$.

Thus, $\displaystyle\int_0^{3/5}\sqrt{9-25x^2}\,dx = \left[\frac{9}{10}(\theta+\sin\theta\cos\theta)\right]_0^{\pi/2} = \frac{9}{10}\left(\frac{\pi}{2}\right) = \frac{9\pi}{20}.$

51. (a) Let $x = 3\sec\theta$, $dx = 3\sec\theta\tan\theta\,d\theta$, $\sqrt{x^2-9} = 3\tan\theta$.

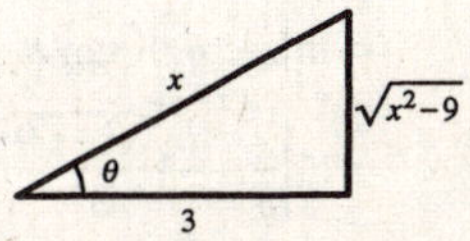

$$\begin{aligned}\int\frac{x^2}{\sqrt{x^2-9}}\,dx &= \int\frac{9\sec^2\theta}{3\tan\theta}3\sec\theta\tan\theta\,d\theta\\ &= 9\int\sec^3\theta\,d\theta\\ &= 9\left[\frac{1}{2}\sec\theta\tan\theta+\frac{1}{2}\int\sec\theta\,d\theta\right] \quad \text{(8.3 Exercise 98 or Example 5, Section 8.2)}\\ &= \frac{9}{2}[\sec\theta\tan\theta+\ln|\sec\theta+\tan\theta|]\\ &= \frac{9}{2}\left[\frac{x}{3}\cdot\frac{\sqrt{x^2-9}}{3}+\ln\left|\frac{x}{3}+\frac{\sqrt{x^2-9}}{3}\right|\right]\end{aligned}$$

Hence,

$$\begin{aligned}\int_4^6\frac{x^2}{\sqrt{x^2-9}}\,dx &= \frac{9}{2}\left[\frac{x\sqrt{x^2-9}}{9}+\ln\left|\frac{x}{3}+\frac{\sqrt{x^2-9}}{3}\right|\right]_4^6\\ &= \frac{9}{2}\left[\left(\frac{6\sqrt{27}}{9}+\ln\left|2+\frac{\sqrt{27}}{3}\right|\right)-\left(\frac{4\sqrt{7}}{9}+\ln\left|\frac{4}{3}+\frac{\sqrt{7}}{3}\right|\right)\right]\\ &= 9\sqrt{3}-2\sqrt{7}+\frac{9}{2}\left(\ln\left(\frac{6+\sqrt{27}}{3}\right)-\ln\left(\frac{4+\sqrt{7}}{3}\right)\right)\\ &= 9\sqrt{3}-2\sqrt{7}+\frac{9}{2}\ln\left(\frac{6+3\sqrt{3}}{4+\sqrt{7}}\right)\\ &= 9\sqrt{3}-2\sqrt{7}+\frac{9}{2}\ln\left(\frac{(4-\sqrt{7})(2+\sqrt{3})}{3}\right)\approx 12.644.\end{aligned}$$

—CONTINUED—

51. —CONTINUED—

(b) When $x = 4$, $\theta = \text{arcsec}\left(\frac{4}{3}\right)$. When $x = 6$, $\theta = \text{arcsec}(2) = \frac{\pi}{3}$.

$$\int_4^6 \frac{x^2}{\sqrt{x^2-9}}\,dx = \frac{9}{2}\Big[\sec\theta\tan\theta + \ln|\sec\theta + \tan\theta|\Big]_{\text{arcsec}(4/3)}^{\pi/3}$$

$$= \frac{9}{2}\big[2\cdot\sqrt{3} + \ln|2+\sqrt{3}|\big] - \frac{9}{2}\left[\frac{4}{3}\frac{\sqrt{7}}{3} + \ln\left|\frac{4}{3} + \frac{\sqrt{7}}{3}\right|\right]$$

$$= 9\sqrt{3} - 2\sqrt{7} + \frac{9}{2}\ln\left(\frac{6+3\sqrt{3}}{4+\sqrt{7}}\right) \approx 12.644$$

52. (a) Let $x = 3\sec\theta$, $dx = 3\sec\theta\tan\theta\,d\theta$, $\sqrt{x^2-9} = 3\tan\theta$.

$$\int \frac{\sqrt{x^2-9}}{x^2}\,dx = \int \frac{3\tan\theta}{9\sec^2\theta}3\sec\theta\tan\theta\,d\theta$$

$$= \int \frac{\tan^2\theta}{\sec\theta}\,d\theta = \int \frac{\sin^2\theta}{\cos\theta}\,d\theta$$

$$= \int \frac{1-\cos^2\theta}{\cos\theta}\,d\theta$$

$$= \int (\sec\theta - \cos\theta)\,d\theta$$

$$= \ln|\sec\theta + \tan\theta| - \sin\theta + C$$

$$= \ln\left|\frac{x}{3} + \frac{\sqrt{x^2-9}}{3}\right| - \frac{\sqrt{x^2-9}}{x} + C$$

Hence,

$$\int_3^6 \frac{\sqrt{x^2-9}}{x^2}\,dx = \left[\ln\left|\frac{x}{3} + \frac{\sqrt{x^2-9}}{3}\right| - \frac{\sqrt{x^2-9}}{x}\right]_3^6$$

$$= \ln\left|2+\sqrt{3}\right| - \frac{\sqrt{3}}{2}.$$

(b) When $x = 3$, $\theta = 0$; when $x = 6$, $\theta = \frac{\pi}{3}$. Hence,

$$\int_3^6 \frac{\sqrt{x^2-9}}{x^2}\,dx = \Big[\ln|\sec\theta + \tan\theta| - \sin\theta\Big]_0^{\pi/3}$$

$$= \ln\left|2+\sqrt{3}\right| - \frac{\sqrt{3}}{2}.$$

53. $x\frac{dy}{dx} = \sqrt{x^2-9}$, $x \ge 3$, $y(3) = 1$

$$y = \int \frac{\sqrt{x^2-9}}{x}\,dx$$

Let $x = 3\sec\theta$, $dx = 3\sec\theta\tan\theta\,d\theta$, $\sqrt{x^2-9} = 3\tan\theta$.

$$y = \int \frac{3\tan\theta}{3\sec\theta}3\sec\theta\tan\theta\,d\theta = 3\int \tan^2\theta\,d\theta$$

$$= 3\int (\sec^2\theta - 1)\,d\theta = 3[\tan\theta - \theta] + C$$

$$= 3\left[\frac{\sqrt{x^2-9}}{3} - \arctan\left(\frac{\sqrt{x^2-9}}{3}\right)\right] + C$$

$$= \sqrt{x^2-9} - 3\arctan\left(\frac{\sqrt{x^2-9}}{3}\right) + C$$

$y(3) = 1$: $1 = 0 - 3(0) + C \Rightarrow C = 1$

$$y = \sqrt{x^2-9} - 3\arctan\left(\frac{\sqrt{x^2-9}}{3}\right) + 1$$

54. $\sqrt{x^2+4}\frac{dy}{dx} = 1$, $x \ge -2$, $y(0) = 4$

$$\frac{dy}{dx} = \frac{1}{\sqrt{x^2+4}}$$

$$y = \int \frac{1}{\sqrt{x^2+4}}\,dx$$

Let $x = 2\tan\theta$, $x^2 + 4 = 4\sec^2\theta$, $dx = 2\sec^2\theta\,d\theta$.

$$y = \int \frac{1}{2\sec\theta}2\sec^2\theta\,d\theta = \int \sec\theta\,d\theta$$

$$= \ln|\sec\theta + \tan\theta| + C$$

$$= \ln\left|\frac{\sqrt{x^2+4}}{2} + \frac{x}{2}\right| + C$$

$$= \ln\left|\sqrt{x^2+4} + x\right| + C_1$$

$y(0) = 4 \Rightarrow 4 = \ln|2| + C_1 \Rightarrow C_1 = 4 - \ln 2$

$$y = \ln\left|\sqrt{x^2+4} + x\right| + 4 - \ln 2$$

55. $\displaystyle\int \frac{x^2}{\sqrt{x^2+10x+9}}\,dx = \frac{1}{2}\sqrt{x^2+10x+9}\,(x-15) + 33\ln\left|(x+5)+\sqrt{x^2+10x+9}\right| + C$

56. $\displaystyle\int (x^2+2x+11)^{3/2}\,dx = \frac{1}{4}(x+1)(x^2+2x+26)\sqrt{x^2+2x+11} + \frac{75}{2}\ln\left|\sqrt{x^2+2x+11}+(x+1)\right| + C$

57. $\displaystyle\int \frac{x^2}{\sqrt{x^2-1}}\,dx = \frac{1}{2}\left(x\sqrt{x^2-1} + \ln\left|x+\sqrt{x^2-1}\right|\right) + C$

58. $\displaystyle\int x^2\sqrt{x^2-4}\,dx = \frac{1}{4}x^3\sqrt{x^2-4} - \frac{1}{2}x\sqrt{x^2-4} - 2\ln\left|x+\sqrt{x^2-4}\right| + C$

59. (a) $u = a\sin\theta$

(b) $u = a\tan\theta$

(c) $u = a\sec\theta$

60. (a) Substitution: $u = x^2+1$, $du = 2x\,dx$

(b) Trigonometric substitution: $x = \sec\theta$

61. (a) $u = x^2+9$, $du = 2x\,dx$

$$\int \frac{x}{x^2+9}\,dx = \frac{1}{2}\int\frac{du}{u} = \frac{1}{2}\ln|u| + C = \frac{1}{2}\ln(x^2+9) + C$$

(b) Let $x = 3\tan\theta$, $x^2+9 = 9\sec^2\theta$, $dx = 3\sec^2\theta\,d\theta$.

$$\begin{aligned}\int \frac{x}{x^2+9}\,dx &= \int\frac{3\tan\theta}{9\sec^2\theta}3\sec^2\theta\,d\theta = \int\tan\theta\,d\theta\\ &= -\ln|\cos\theta| + C_1\\ &= -\ln\left|\frac{3}{\sqrt{x^2+9}}\right| + C_1\\ &= -\ln 3 + \ln\sqrt{x^2+9} + C_1\\ &= \frac{1}{2}\ln(x^2+9) + C_2\end{aligned}$$

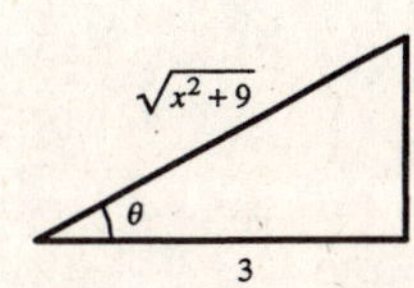

The answers are equivalent.

62. (a) $\displaystyle\int\frac{x^2}{x^2+9}\,dx = \int\frac{x^2+9-9}{x^2+9}\,dx = \int\left(1-\frac{9}{x^2+9}\right)dx = x - 3\arctan\left(\frac{x}{3}\right) + C$

(b) Let $x = 3\tan\theta$, $x^2+9 = 9\sec^2\theta$, $dx = 3\sec^2\theta\,d\theta$.

$$\begin{aligned}\int\frac{x^2}{x^2+9}\,dx &= \int\frac{9\tan^2\theta}{9\sec^2\theta}3\sec^2\theta\,d\theta\\ &= 3\int\tan^2\theta\,d\theta = 3\int(\sec^2\theta - 1)\,d\theta\\ &= 3\tan\theta - 3\theta + C_1\\ &= x - 3\arctan\left(\frac{x}{3}\right) + C_1\end{aligned}$$

The answers are equivalent.

63. True

$$\int\frac{dx}{\sqrt{1-x^2}} = \int\frac{\cos\theta\,d\theta}{\cos\theta} = \int d\theta$$

64. False

$$\int\frac{\sqrt{x^2-1}}{x}\,dx = \int\frac{\tan\theta}{\sec\theta}(\sec\theta\tan\theta\,d\theta) = \int\tan^2\theta\,d\theta$$

65. False

$$\int_0^{\sqrt{3}} \frac{dx}{(\sqrt{1+x^2})^3} = \int_0^{\pi/3} \frac{\sec^2\theta\, d\theta}{\sec^3\theta} = \int_0^{\pi/3} \cos\theta\, d\theta$$

66. True

$$\int_{-1}^{1} x^2\sqrt{1-x^2}\,dx = 2\int_0^1 x^2\sqrt{1-x^2}\,dx = 2\int_0^{\pi/2} (\sin^2\theta)(\cos\theta)(\cos\theta\,d\theta) = 2\int_0^{\pi/2} \sin^2\theta\cos^2\theta\,d\theta$$

67. $A = 4\int_0^a \frac{b}{a}\sqrt{a^2-x^2}\,dx$

$= \frac{4b}{a}\int_0^a \sqrt{a^2-x^2}\,dx$

$= \left[\frac{4b}{a}\left(\frac{1}{2}\right)\left(a^2\arcsin\frac{x}{a} + x\sqrt{a^2-x^2}\right)\right]_0^a$

$= \frac{2b}{a}\left(a^2\left(\frac{\pi}{2}\right)\right) = \pi ab$

Note: See Theorem 8.2 for $\int\sqrt{a^2-x^2}\,dx$.

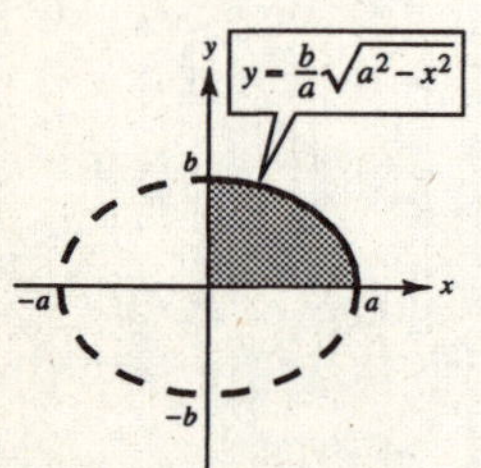

68. $x^2 + y^2 = a^2$

$x = \pm\sqrt{a^2-y^2}$

$A = 2\int_h^a \sqrt{a^2-y^2}\,dy$

$= \left[a^2\arcsin\left(\frac{y}{a}\right) + y\sqrt{a^2-y^2}\right]_h^a$ (Theorem 8.2)

$= \left(a^2\frac{\pi}{2}\right) - \left(a^2\arcsin\left(\frac{h}{a}\right) + h\sqrt{a^2-h^2}\right)$

$= \frac{a^2\pi}{2} - a^2\arcsin\left(\frac{h}{a}\right) - h\sqrt{a^2-h^2}$

69. (a) $x^2 + (y-k)^2 = 25$

Radius of circle $= 5$

$k^2 = 5^2 + 5^2 = 50$

$k = 5\sqrt{2}$

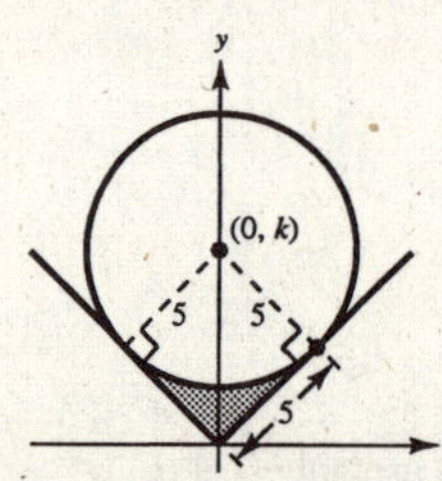

(b) Area $=$ square $- \frac{1}{4}$(circle)

$= 25 - \frac{1}{4}\pi(5)^2 = 25\left(1 - \frac{\pi}{4}\right)$

(c) Area $= r^2 - \frac{1}{4}\pi r^2 = r^2\left(1 - \frac{\pi}{4}\right)$

70. (a) Place the center of the circle at $(0, 1)$; $x^2 + (y-1)^2 = 1$. The depth d satisfies $0 \le d \le 2$. The volume is

$V = 3\cdot 2\int_0^d \sqrt{1-(y-1)^2}\,dy = 6\cdot\frac{1}{2}\left[\arcsin(y-1) + (y-1)\sqrt{1-(y-1)^2}\right]_0^d$ (Theorem 8.2 (1))

$= 3\left[\arcsin(d-1) + (d-1)\sqrt{1-(d-1)^2} - \arcsin(-1)\right]$

$= \frac{3\pi}{2} + 3\arcsin(d-1) + 3(d-1)\sqrt{2d-d^2}.$

(b)

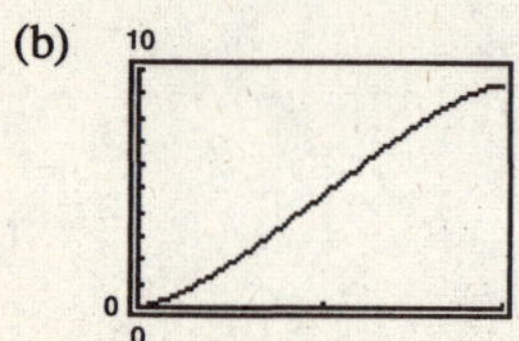

(c) The full tank holds $3\pi \approx 9.4248$ cubic meters. The horizontal lines

$$y = \frac{3\pi}{4},\ y = \frac{3\pi}{2},\ y = \frac{9\pi}{4}$$

intersect the curve at $d = 0.596, 1.0, 1.404$. The dipstick would have these markings on it.

—CONTINUED—

70. —CONTINUED—

(d) $V = 6\int_0^d \sqrt{1-(y-1)^2}\,dy$

$$\frac{dV}{dt} = \frac{dV}{dd}\cdot\frac{dd}{dt} = 6\sqrt{1-(d-1)^2}\cdot d'(t)$$

$$= \frac{1}{4} \Rightarrow d'(t) = \frac{1}{24\sqrt{1-(d-1)^2}}$$

(e) 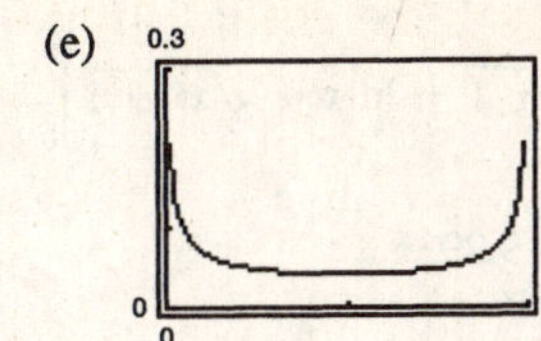

The minimum occurs at $d = 1$, which is the widest part of the tank.

71. Let $x - 3 = \sin\theta$, $dx = \cos\theta\,d\theta$, $\sqrt{1-(x-3)^2} = \cos\theta$.

Shell Method:

$$V = 4\pi\int_2^4 x\sqrt{1-(x-3)^2}\,dx$$

$$= 4\pi\int_{-\pi/2}^{\pi/2} (3+\sin\theta)\cos^2\theta\,d\theta$$

$$= 4\pi\left[\frac{3}{2}\int_{-\pi/2}^{\pi/2}(1+\cos 2\theta)\,d\theta + \int_{-\pi/2}^{\pi/2}\cos^2\theta\sin\theta\,d\theta\right]$$

$$= 4\pi\left[\frac{3}{2}\left(\theta+\frac{1}{2}\sin 2\theta\right) - \frac{1}{3}\cos^3\theta\right]_{-\pi/2}^{\pi/2} = 6\pi^2$$

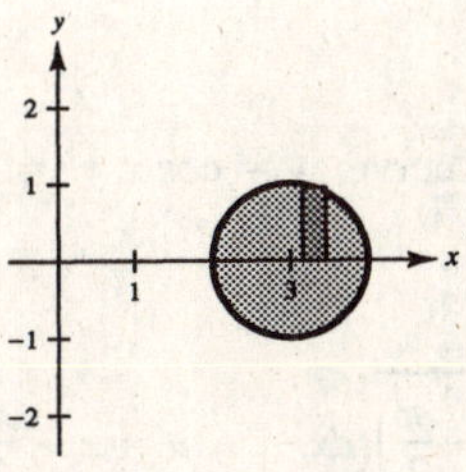

72. Let $x - h = r\sin\theta$, $dx = r\cos\theta\,d\theta$, $\sqrt{r^2-(x-h)^2} = r\cos\theta$.

Shell Method:

$$V = 4\pi\int_{h-r}^{h+r} x\sqrt{r^2-(x-h)^2}\,dx$$

$$= 4\pi\int_{-\pi/2}^{\pi/2}(h+r\sin\theta)r\cos\theta(r\cos\theta)\,d\theta = 4\pi r^2\int_{-\pi/2}^{\pi/2}(h+r\sin\theta)\cos^2\theta\,d\theta$$

$$= 4\pi r^2\left[\frac{h}{2}\int_{-\pi/2}^{\pi/2}(1+\cos 2\theta)\,d\theta + r\int_{-\pi/2}^{\pi/2}\sin\theta\cos^2\theta\,d\theta\right]$$

$$= 2\pi r^2 h\left[\theta+\frac{1}{2}\sin 2\theta\right]_{-\pi/2}^{\pi/2} - \left[4\pi r^3\left(\frac{\cos^3\theta}{3}\right)\right]_{-\pi/2}^{\pi/2} = 2\pi^2 r^2 h$$

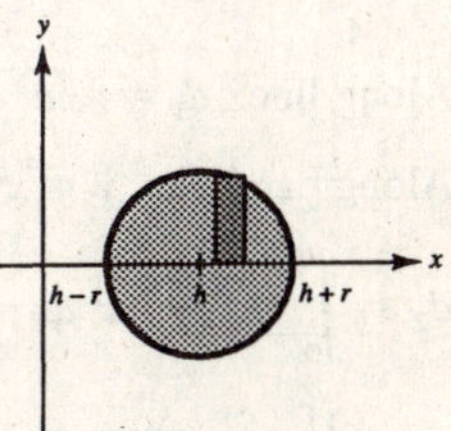

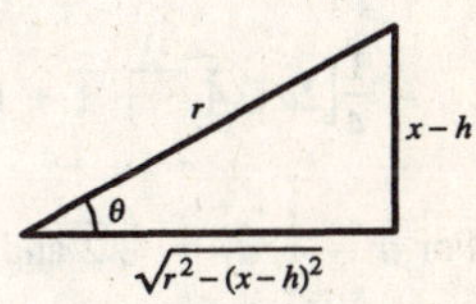

73. $y = \ln x$, $y' = \frac{1}{x}$, $1 + (y')^2 = 1 + \frac{1}{x^2} = \frac{x^2+1}{x^2}$

Let $x = \tan\theta$, $dx = \sec^2\theta\,d\theta$, $\sqrt{x^2+1} = \sec\theta$.

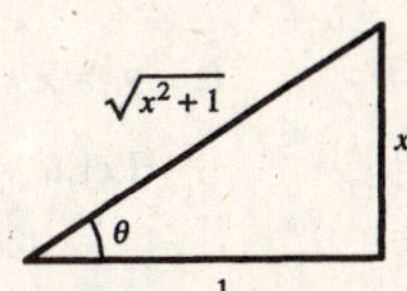

$$s = \int_1^5 \sqrt{\frac{x^2+1}{x^2}}\,dx = \int_1^5 \frac{\sqrt{x^2+1}}{x}\,dx$$

$$= \int_a^b \frac{\sec\theta}{\tan\theta}\sec^2\theta\,d\theta = \int_a^b \frac{\sec\theta}{\tan\theta}(1+\tan^2\theta)\,d\theta$$

$$= \int_a^b (\csc\theta + \sec\theta\tan\theta)\,d\theta = \Big[-\ln|\csc\theta+\cot\theta| + \sec\theta\Big]_a^b$$

$$= \left[-\ln\left|\frac{\sqrt{x^2+1}}{x}+\frac{1}{x}\right| + \sqrt{x^2+1}\right]_1^5$$

$$= \left[-\ln\left(\frac{\sqrt{26}+1}{5}\right) + \sqrt{26}\right] - \left[-\ln(\sqrt{2}+1) + \sqrt{2}\right]$$

$$= \ln\left[\frac{5(\sqrt{2}+1)}{\sqrt{26}+1}\right] + \sqrt{26} - \sqrt{2} \approx 4.367 \text{ or } \ln\left[\frac{\sqrt{26}-1}{5(\sqrt{2}-1)}\right] + \sqrt{26} - \sqrt{2}$$

74. $y = \frac{1}{2}x^2, y' = x, 1 + (y')^2 = 1 + x^2$

$$s = \int_0^4 \sqrt{1 + x^2}\,dx = \left[\frac{1}{2}\left(x\sqrt{x^2 + 1} + \ln\left|x + \sqrt{x^2 + 1}\right|\right)\right]_0^4 \quad \text{(Theorem 8.2)}$$

$$= \frac{1}{2}\left[4\sqrt{17} + \ln\left(4 + \sqrt{17}\right)\right] \approx 9.2936$$

75. Length of one arch of sine curve: $y = \sin x, y' = \cos x$

$$L_1 = \int_0^{\pi} \sqrt{1 + \cos^2 x}\,dx$$

Length of one arch of cosine curve: $y = \cos x, y' = -\sin x$

$$L_2 = \int_{-\pi/2}^{\pi/2} \sqrt{1 + \sin^2 x}\,dx$$

$$= \int_{-\pi/2}^{\pi/2} \sqrt{1 + \cos^2\left(x - \frac{\pi}{2}\right)}\,dx, \qquad u = x - \frac{\pi}{2}, du = dx$$

$$= \int_{-\pi}^{0} \sqrt{1 + \cos^2 u}\,du$$

$$= \int_0^{\pi} \sqrt{1 + \cos^2 u}\,du = L_1$$

76. (a) Along line: $d_1 = \sqrt{a^2 + a^4} = a\sqrt{1 + a^2}$

Along parabola: $y = x^2, y' = 2x$

$$d_2 = \int_0^a \sqrt{1 + 4x^2}\,dx$$

$$= \frac{1}{4}\left[2x\sqrt{4x^2 + 1} + \ln\left|2x + \sqrt{4x^2 + 1}\right|\right]_0^a \quad \text{(Theorem 8.2)}$$

$$= \frac{1}{4}\left[2a\sqrt{4a^2 + 1} + \ln\left(2a + \sqrt{4a^2 + 1}\right)\right]$$

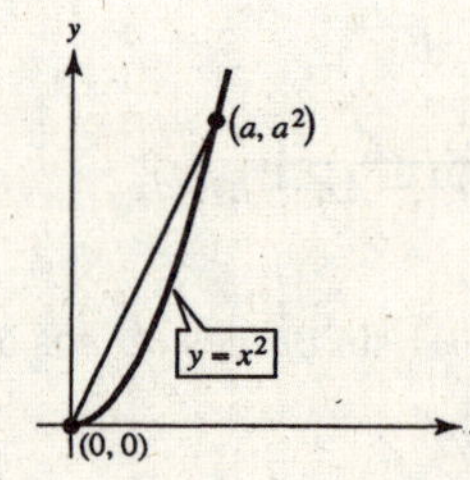

(b) For $a = 1$, $d_1 = \sqrt{2}$ and $d_2 = \frac{\sqrt{5}}{2} + \frac{1}{4}\ln\left(2 + \sqrt{5}\right) \approx 1.4789$.

For $a = 10$, $d_1 = 10\sqrt{101} \approx 100.4988$, $d_2 \approx 101.0473$.

(c) As a increases, $d_2 - d_1 \to 0$.

77. (a)

(b) $y = 0$ for $x = 200$ (range)

(c) $y = x - 0.005x^2, y' = 1 - 0.01x, 1 + (y')^2 = 1 + (1 - 0.01x)^2$

Let $u = 1 - 0.01x$, $du = -0.01\,dx$, $a = 1$. (See Theorem 8.2.)

$$s = \int_0^{200} \sqrt{1 + (1 - 0.01x)^2}\,dx = -100\int_0^{200} \sqrt{(1 - 0.01x)^2 + 1}\,(-0.01)\,dx$$

$$= -50\left[(1 - 0.01x)\sqrt{(1 - 0.01x)^2 + 1} + \ln\left|(1 - 0.01x) + \sqrt{(1 - 0.01x)^2 + 1}\right|\right]_0^{200}$$

$$= -50\left[\left(-\sqrt{2} + \ln\left|-1 + \sqrt{2}\right|\right) - \left(\sqrt{2} + \ln\left|1 + \sqrt{2}\right|\right)\right]$$

$$= 100\sqrt{2} + 50\ln\left(\frac{\sqrt{2} + 1}{\sqrt{2} - 1}\right) \approx 229.559$$

78. (a)

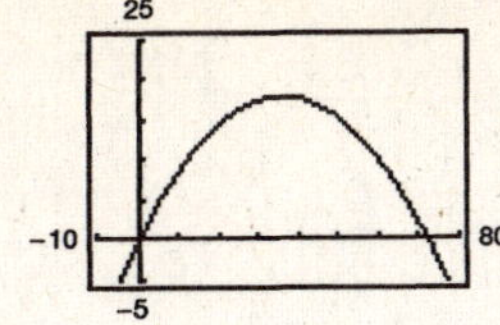

(b) $y = 0$ for $x = 72$

(c) $y = x - \dfrac{x^2}{72}, y' = 1 - \dfrac{x}{36}, 1 + (y')^2 = 1 + \left(1 - \dfrac{x}{36}\right)^2$

$$s = \int_0^{72} \sqrt{1 + \left(1 - \frac{x}{36}\right)^2}\, dx = -36\int_0^{72} \sqrt{1 + \left(1 - \frac{x}{36}\right)^2}\left(-\frac{1}{36}\right)dx$$

$$= -\frac{36}{2}\left[\left(1 - \frac{x}{36}\right)\sqrt{1 + \left(1 - \frac{x}{36}\right)^2} + \ln\left|\left(1 - \frac{x}{36}\right) + \sqrt{1 + \left(1 - \frac{x}{36}\right)^2}\right|\right]_0^{72}$$

$$= -18\left[\left(-\sqrt{2} + \ln\left|-1 + \sqrt{2}\right|\right) - \left(\sqrt{2} + \ln\left|1 + \sqrt{2}\right|\right)\right] = 36\sqrt{2} + 18\ln\left(\frac{\sqrt{2}+1}{\sqrt{2}-1}\right) \approx 82.641$$

79. Let $x = 3\tan\theta, dx = 3\sec^2\theta\, d\theta, \sqrt{x^2 + 9} = 3\sec\theta$.

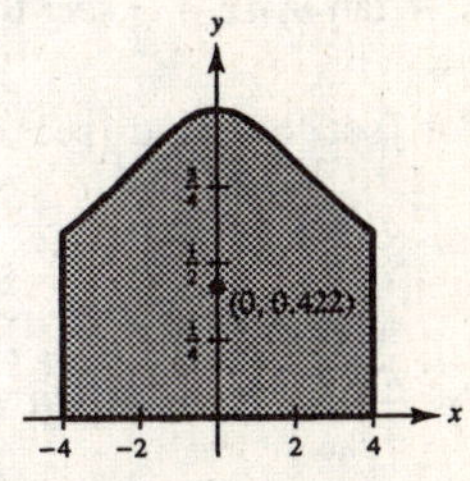

$$A = 2\int_0^4 \frac{3}{\sqrt{x^2+9}}\, dx = 6\int_0^4 \frac{dx}{\sqrt{x^2+9}} = 6\int_a^b \frac{3\sec^2\theta\, d\theta}{3\sec\theta}$$

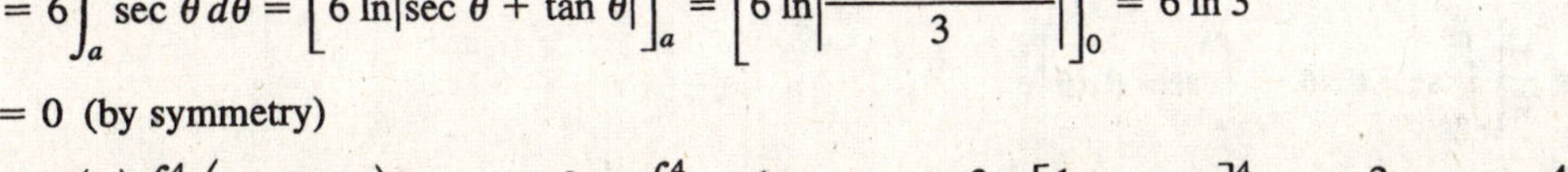

$$= 6\int_a^b \sec\theta\, d\theta = \Big[6\ln|\sec\theta + \tan\theta|\Big]_a^b = \left[6\ln\left|\frac{\sqrt{x^2+9} + x}{3}\right|\right]_0^4 = 6\ln 3$$

$\bar{x} = 0$ (by symmetry)

$$\bar{y} = \frac{1}{2}\left(\frac{1}{A}\right)\int_{-4}^{4}\left(\frac{3}{\sqrt{x^2+9}}\right)^2 dx = \frac{9}{12\ln 3}\int_{-4}^{4}\frac{1}{x^2+9}\, dx = \frac{3}{4\ln 3}\left[\frac{1}{3}\arctan\frac{x}{3}\right]_{-4}^{4} = \frac{2}{4\ln 3}\arctan\frac{4}{3} \approx 0.422$$

$$(\bar{x}, \bar{y}) = \left(0, \frac{1}{2\ln 3}\arctan\frac{4}{3}\right) \approx (0, 0.422)$$

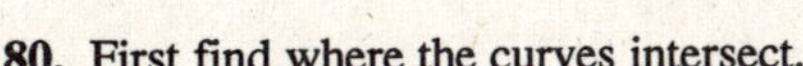

80. First find where the curves intersect.

$$y^2 = 16 - (x-4)^2 = \frac{1}{16}x^4$$

$$16^2 - 16(x-4)^2 = x^4$$

$$16^2 - 16x^2 + 128x - 16^2 = x^4$$

$$x^4 + 16x^2 - 128x = 0$$

$$x(x-4)(x^2 + 4x + 32) \Longrightarrow x = 0, 4$$

$$A = \int_0^4 \frac{1}{4}x^2\, dx + \frac{1}{4}\pi(4)^2 = \left[\frac{1}{12}x^3\right]_0^4 + 4\pi = \frac{16}{3} + 4\pi$$

$$M_y = \int_0^4 x\left[\frac{1}{4}x^2\right]dx + \int_4^8 x\sqrt{16 - (x-4)^2}\, dx$$

$$= \left[\frac{x^4}{16}\right]_0^4 + \int_4^8 (x-4)\sqrt{16-(x-4)^2}\, dx + \int_4^8 4\sqrt{16-(x-4)^2}\, dx$$

$$= 16 + \left[\frac{-1}{3}(16 - (x-4)^2)^{3/2}\right]_4^8 + 2\left[16\arcsin\frac{x-4}{4} + (x-4)\sqrt{16-(x-4)^2}\right]_4^8$$

$$= 16 + \frac{1}{3}16^{3/2} + 2\left[16\left(\frac{\pi}{2}\right)\right] = 16 + \frac{64}{3} + 16\pi = \frac{112}{3} + 16\pi$$

—CONTINUED—

80. —CONTINUED—

$$M_x = \int_0^4 \frac{1}{2}\left(\frac{1}{4}x^2\right)^2 dx + \int_4^8 \frac{1}{2}(16 - (x-4)^2)\,dx$$

$$= \left[\frac{1}{32} \cdot \frac{x^5}{5}\right]_0^4 + \left[8x - \frac{(x-4)^3}{6}\right]_4^8$$

$$= \frac{32}{5} + \left(64 - \frac{64}{6}\right) - 32 = \frac{416}{15}$$

$$\bar{x} = \frac{M_y}{A} = \frac{112/3 + 16\pi}{16/3 + 4\pi} = \frac{112 + 48\pi}{16 + 12\pi} = \frac{28 + 12\pi}{4 + 3\pi} \approx 4.89$$

$$\bar{y} = \frac{M_x}{A} = \frac{416/15}{(16/3) + 4\pi} = \frac{104}{5(4 + 3\pi)} \approx 1.55$$

$(\bar{x}, \bar{y}) \approx (4.89, 1.55)$

81. $y = x^2,\ y' = 2x,\ 1 + (y') = 1 + 4x^2$

$2x = \tan\theta,\ dx = \frac{1}{2}\sec^2\theta\,d\theta,\ \sqrt{1 + 4x^2} = \sec\theta$

(For $\int \sec^5\theta\,d\theta$ and $\int \sec^3\theta\,d\theta$, see Exercise 98 in Section 8.3.)

$$S = 2\pi\int_0^{\sqrt{2}} x^2\sqrt{1 + 4x^2}\,dx = 2\pi\int_a^b \left(\frac{\tan\theta}{2}\right)^2(\sec\theta)\left(\frac{1}{2}\sec^2\theta\right)d\theta$$

$$= \frac{\pi}{4}\int_a^b \sec^3\theta\tan^2\theta\,d\theta = \frac{\pi}{4}\left[\int_a^b \sec^5\theta\,d\theta - \int_a^b \sec^3\theta\,d\theta\right]$$

$$= \frac{\pi}{4}\left\{\frac{1}{4}\left[\sec^3\theta\tan\theta + \frac{3}{2}(\sec\theta\tan\theta + \ln|\sec\theta + \tan\theta|)\right] - \frac{1}{2}(\sec\theta\tan\theta + \ln|\sec\theta + \tan\theta|)\right\}_a^b$$

$$= \frac{\pi}{4}\left[\frac{1}{4}[(1 + 4x^2)^{3/2}(2x)] - \frac{1}{8}\left[(1 + 4x^2)^{1/2}(2x) + \ln\left|\sqrt{1 + 4x^2} + 2x\right|\right]\right]_0^{\sqrt{2}}$$

$$= \frac{\pi}{4}\left[\frac{54\sqrt{2}}{4} - \frac{6\sqrt{2}}{8} - \frac{1}{8}\ln\left(3 + 2\sqrt{2}\right)\right]$$

$$= \frac{\pi}{4}\left(\frac{51\sqrt{2}}{4} - \frac{\ln\left(3 + 2\sqrt{2}\right)}{8}\right) = \frac{\pi}{32}\left[102\sqrt{2} - \ln\left(3 + 2\sqrt{2}\right)\right] \approx 13.989$$

82. Let $r = L\tan\theta,\ dr = L\sec^2\theta\,d\theta,\ r^2 + L^2 = L^2\sec^2\theta.$

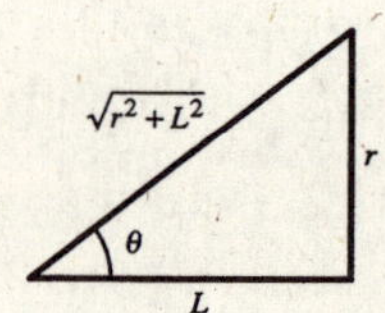

$$\frac{1}{R}\int_0^R \frac{2mL}{(r^2 + L^2)^{3/2}}\,dr = \frac{2mL}{R}\int_a^b \frac{L\sec^2\theta\,d\theta}{L^3\sec^3\theta}$$

$$= \frac{2m}{RL}\int_a^b \cos\theta\,d\theta$$

$$= \left[\frac{2m}{RL}\sin\theta\right]_a^b$$

$$= \left[\frac{2m}{RL}\frac{r}{\sqrt{r^2 + L^2}}\right]_0^R$$

$$= \frac{2m}{L\sqrt{R^2 + L^2}}$$

83. (a) Area of representative rectangle: $2\sqrt{1-y^2}\,\Delta y$

Force: $2(62.4)(3-y)\sqrt{1-y^2}\,\Delta y$

$$F = 124.8\int_{-1}^{1}(3-y)\sqrt{1-y^2}\,dy$$

$$= 124.8\left[3\int_{-1}^{1}\sqrt{1-y^2}\,dy - \int_{-1}^{1}y\sqrt{1-y^2}\,dy\right]$$

$$= 124.8\left[\frac{3}{2}\left(\arcsin y + y\sqrt{1-y^2}\right) + \frac{1}{2}\left(\frac{2}{3}\right)(1-y^2)^{3/2}\right]_{-1}^{1}$$

$$= (62.4)3[\arcsin 1 - \arcsin(-1)] = 187.2\pi \text{ lb}$$

(b) $$F = 124.8\int_{-1}^{1}(d-y)\sqrt{1-y^2}\,dy = 124.8d\int_{-1}^{1}\sqrt{1-y^2}\,dy - 124.8\int_{-1}^{1}y\sqrt{1-y^2}\,dy$$

$$= 124.8\left(\frac{d}{2}\right)\left[\arcsin y + y\sqrt{1-y^2}\right]_{-1}^{1} - 124.8(0) = 62.4\pi d \text{ lb}$$

84. (a) $$F_{\text{inside}} = 48\int_{-1}^{0.8}(0.8-y)(2)\sqrt{1-y^2}\,dy$$

$$= 96\left[0.8\int_{-1}^{0.8}\sqrt{1-y^2}\,dy - \int_{-1}^{0.8}y\sqrt{1-y^2}\,dy\right]$$

$$= 96\left[\frac{0.8}{2}\left(\arcsin y + y\sqrt{1-y^2}\right) + \frac{1}{3}(1-y^2)^{3/2}\right]_{-1}^{0.8} \approx 96(1.263) \approx 121.3 \text{ lbs}$$

(b) $$F_{\text{outside}} = 64\int_{-1}^{0.4}(0.4-y)(2)\sqrt{1-y^2}\,dy$$

$$= 128\left[0.4\int_{-1}^{0.4}\sqrt{1-y^2}\,dy - \int_{-1}^{0.4}y\sqrt{1-y^2}\,dy\right] = 128\left[\frac{0.4}{2}\left(\arcsin y + y\sqrt{1-y^2}\right) + \frac{1}{3}(1-y^2)^{3/2}\right]_{-1}^{0.4} \approx 92.98$$

85. Let $u = a\sin\theta$, $du = a\cos\theta\,d\theta$, $\sqrt{a^2-u^2} = a\cos\theta$.

$$\int\sqrt{a^2-u^2}\,du = \int a^2\cos^2\theta\,d\theta = a^2\int\frac{1+\cos 2\theta}{2}\,d\theta$$

$$= \frac{a^2}{2}\left(\theta + \frac{1}{2}\sin 2\theta\right) + C = \frac{a^2}{2}(\theta + \sin\theta\cos\theta) + C$$

$$= \frac{a^2}{2}\left[\arcsin\frac{u}{a} + \left(\frac{u}{a}\right)\left(\frac{\sqrt{a^2+u^2}}{a}\right)\right] + C = \frac{1}{2}\left[a^2\arcsin\frac{u}{a} + u\sqrt{a^2-u^2}\right] + C$$

Let $u = a\sec\theta$, $du = a\sec\theta\tan\theta\,d\theta$, $\sqrt{u^2-a^2} = a\tan\theta$.

$$\int\sqrt{u^2-a^2}\,du = \int a\tan\theta(a\sec\theta\tan\theta)\,d\theta = a^2\int\tan^2\theta\sec\theta\,d\theta$$

$$= a^2\int(\sec^2\theta - 1)\sec\theta\,d\theta = a^2\int(\sec^3\theta - \sec\theta)\,d\theta$$

$$= a^2\left[\frac{1}{2}\sec\theta\tan\theta + \frac{1}{2}\int\sec\theta\,d\theta\right] - a^2\int\sec\theta\,d\theta = a^2\left[\frac{1}{2}\sec\theta\tan\theta - \frac{1}{2}\ln|\sec\theta+\tan\theta|\right]$$

$$= \frac{a^2}{2}\left[\frac{u}{a}\cdot\frac{\sqrt{u^2-a^2}}{a} - \ln\left|\frac{u}{a} + \frac{\sqrt{u^2-a^2}}{a}\right|\right] + C_1$$

$$= \frac{1}{2}\left[u\sqrt{u^2-a^2} - a^2\ln\left|u + \sqrt{u^2-a^2}\right|\right] + C$$

—CONTINUED—

85. —CONTINUED—

Let $u = a\tan\theta$, $du = a\sec^2\theta\,d\theta$, $\sqrt{u^2+a^2} = a\sec\theta$.

$$\int \sqrt{u^2+a^2}\,du = \int (a\sec\theta)(a\sec^2\theta)\,d\theta$$

$$= a^2\int \sec^3\theta\,d\theta = a^2\left[\frac{1}{2}\sec\theta\tan\theta + \frac{1}{2}\ln|\sec\theta + \tan\theta|\right] + C_1$$

$$= \frac{a^2}{2}\left[\frac{\sqrt{u^2+a^2}}{a}\cdot\frac{u}{a} + \ln\left|\frac{\sqrt{u^2+a^2}}{a} + \frac{u}{a}\right|\right] + C_1 = \frac{1}{2}\left[u\sqrt{u^2+a^2} + a^2\ln\left|u + \sqrt{u^2+a^2}\right|\right] + C$$

86. $y = \sin x$ on $[0, 2]$

$y' = \cos x$

$s_1 = 2\int_0^{\pi} \sqrt{1+\cos^2 x}\,dx \quad (\approx 3.820197789)$

Ellipse: $x^2 + 2y^2 = 2$

Upper half: $y = \sqrt{1 - \frac{1}{2}x^2}, \quad -\sqrt{2} \le x \le \sqrt{2}$

$$y' = \frac{-x}{2\sqrt{1-(1/2)x^2}}$$

$$s_2 = 2\int_{-\sqrt{2}}^{\sqrt{2}} \sqrt{1 + \frac{x^2}{4(1-(1/2)x^2)}}\,dx = 2\int_{-\sqrt{2}}^{\sqrt{2}} \sqrt{1 + \frac{x^2}{4-2x^2}}\,dx$$

Let $x = \sqrt{2}\sin\theta$, $dx = \sqrt{2}\cos\theta\,d\theta$, $x^2 = 2\sin^2\theta$, $4 - 2x^2 = 4 - 4\sin^2\theta = 4\cos^2\theta$.

$$s_2 = 2\int_{-\pi/2}^{\pi/2} \sqrt{1 + \frac{2\sin^2\theta}{4\cos^2\theta}}\sqrt{2}\cos\theta\,d\theta$$

$$= 2\int_{-\pi/2}^{\pi/2} \frac{\sqrt{4\cos^2\theta + 2\sin^2\theta}}{2\cos\theta}\sqrt{2}\cos\theta\,d\theta$$

$$= 2\int_{-\pi/2}^{\pi/2} \frac{\sqrt{2+2\cos^2\theta}}{\sqrt{2}}\,d\theta$$

$$= 2\int_{-\pi/2}^{\pi/2} \sqrt{1+\cos^2\theta}\,d\theta$$

$$= 2\int_0^{\pi} \sqrt{1+\cos^2\theta}\,d\theta = s_1$$

87. Large circle: $x^2 + y^2 = 25$

$y = \sqrt{25 - x^2}$, upper half

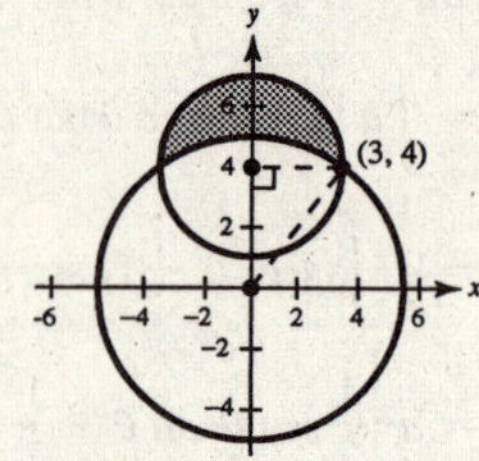

From the right triangle, the center of the small circle is $(0, 4)$.

$x^2 + (y-4)^2 = 9$

$y = 4 + \sqrt{9 - x^2}$, upper half

$$A = 2\int_0^3 \left[\left(4 + \sqrt{9-x^2}\right) - \sqrt{25-x^2}\right]dx$$

$$= 2\left[4x + \frac{1}{2}\left[9\arcsin\left(\frac{x}{3}\right) + x\sqrt{9-x^2}\right] - \frac{1}{2}\left[25\arcsin\left(\frac{x}{5}\right) + x\sqrt{25-x^2}\right]\right]_0^3$$

$$= 2\left[12 + \frac{9}{2}\arcsin(1) - \frac{25}{2}\arcsin\frac{3}{5} - 6\right]$$

$$= 12 + \frac{9\pi}{2} - 25\arcsin\frac{3}{5} \approx 10.050$$

Section 8.5 Partial Fractions

1. $\dfrac{5}{x^2 - 10x} = \dfrac{5}{x(x - 10)} = \dfrac{A}{x} + \dfrac{B}{x - 10}$

2. $\dfrac{4x^2 + 3}{(x - 5)^3} = \dfrac{A}{x - 5} + \dfrac{B}{(x - 5)^2} + \dfrac{C}{(x - 5)^3}$

3. $\dfrac{2x - 3}{x^3 + 10x} = \dfrac{2x - 3}{x(x^2 + 10)} = \dfrac{A}{x} + \dfrac{Bx + C}{x^2 + 10}$

4. $\dfrac{x - 2}{x^2 + 4x + 3} = \dfrac{x - 2}{(x + 1)(x + 3)} = \dfrac{A}{x + 1} + \dfrac{B}{x + 3}$

5. $\dfrac{16}{x(x - 10)} = \dfrac{A}{x} + \dfrac{B}{x - 10}$

6. $\dfrac{2x - 1}{x(x^2 + 1)^2} = \dfrac{A}{x} + \dfrac{Bx + C}{x^2 + 1} + \dfrac{Dx + E}{(x^2 + 1)^2}$

7. $\dfrac{1}{x^2 - 1} = \dfrac{1}{(x + 1)(x - 1)} = \dfrac{A}{x + 1} + \dfrac{B}{x - 1}$

$1 = A(x - 1) + B(x + 1)$

When $x = -1$, $1 = -2A$, $A = -\frac{1}{2}$.

When $x = 1$, $1 = 2B$, $B = \frac{1}{2}$.

$$\int \frac{1}{x^2 - 1}\,dx = -\frac{1}{2}\int \frac{1}{x + 1}\,dx + \frac{1}{2}\int \frac{1}{x - 1}\,dx$$

$$= -\frac{1}{2}\ln|x + 1| + \frac{1}{2}\ln|x - 1| + C$$

$$= \frac{1}{2}\ln\left|\frac{x - 1}{x + 1}\right| + C$$

8. $\dfrac{1}{4x^2 - 9} = \dfrac{1}{(2x - 3)(2x + 3)} = \dfrac{A}{2x - 3} + \dfrac{B}{2x + 3}$

$1 = A(2x + 3) + B(2x - 3)$

When $x = \frac{3}{2}$, $1 = 6A$, $A = \frac{1}{6}$.

When $x = -\frac{3}{2}$, $1 = -6B$, $B = -\frac{1}{6}$.

$$\int \frac{1}{4x^2 - 9}\,dx = \frac{1}{6}\left[\int \frac{1}{2x - 3}\,dx - \int \frac{1}{2x + 3}\,dx\right]$$

$$= \frac{1}{12}[\ln|2x - 3| - \ln|2x + 3|] + C$$

$$= \frac{1}{12}\ln\left|\frac{2x - 3}{2x + 3}\right| + C$$

9. $\dfrac{3}{x^2 + x - 2} = \dfrac{3}{(x - 1)(x + 2)} = \dfrac{A}{x - 1} + \dfrac{B}{x + 2}$

$3 = A(x + 2) + B(x - 1)$

When $x = 1$, $3 = 3A$, $A = 1$.

When $x = -2$, $3 = -3B$, $B = -1$.

$$\int \frac{3}{x^2 + x - 2}\,dx = \int \frac{1}{x - 1}\,dx - \int \frac{1}{x + 2}\,dx$$

$$= \ln|x - 1| - \ln|x + 2| + C$$

$$= \ln\left|\frac{x - 1}{x + 2}\right| + C$$

10. $$\int \frac{x + 1}{x^2 + 4x + 3}\,dx = \int \frac{(x + 1)}{(x + 1)(x + 3)}\,dx$$

$$= \int \frac{1}{x + 3}\,dx = \ln|x + 3| + C$$

11. $\dfrac{5 - x}{2x^2 + x - 1} = \dfrac{5 - x}{(2x - 1)(x + 1)} = \dfrac{A}{2x - 1} + \dfrac{B}{x + 1}$

$5 - x = A(x + 1) + B(2x - 1)$

When $x = \frac{1}{2}$, $\frac{9}{2} = \frac{3}{2}A$, $A = 3$.

When $x = -1$, $6 = -3B$, $B = -2$.

$$\int \frac{5 - x}{2x^2 + x - 1}\,dx = 3\int \frac{1}{2x - 1}\,dx - 2\int \frac{1}{x + 1}\,dx$$

$$= \frac{3}{2}\ln|2x - 1| - 2\ln|x + 1| + C$$

12. $\dfrac{5x^2 - 12x - 12}{x(x - 2)(x + 2)} = \dfrac{A}{x} + \dfrac{B}{x - 2} + \dfrac{C}{x + 2}$

$5x^2 - 12x - 12 = A(x^2 - 4) + Bx(x + 2) + Cx(x - 2)$

When $x = 0$, $-12 = -4A \Rightarrow A = 3$. When $x = 2$, $-16 = 8B \Rightarrow B = -2$. When $x = -2$, $32 = 8C \Rightarrow C = 4$.

$$\int \frac{5x^2 - 12x - 12}{x^3 - 4x}\,dx$$

$$= \int \frac{3}{x}\,dx + \int \frac{-2}{x - 2}\,dx + \int \frac{4}{x + 2}\,dx$$

$$= 3\ln|x| - 2\ln|x - 2| + 4\ln|x + 2| + C$$

13. $\dfrac{x^2 + 12x + 12}{x(x+2)(x-2)} = \dfrac{A}{x} + \dfrac{B}{x+2} + \dfrac{C}{x-2}$

$x^2 + 12x + 12 = A(x+2)(x-2) + Bx(x-2) + Cx(x+2)$

When $x = 0$, $12 = -4A$, $A = -3$. When $x = -2$, $-8 = 8B$, $B = -1$. When $x = 2$, $40 = 8C$, $C = 5$.

$$\int \frac{x^2 + 12x + 12}{x^3 - 4x}\,dx = 5\int \frac{1}{x-2}\,dx - \int \frac{1}{x+2}\,dx - 3\int \frac{1}{x}\,dx$$

$$= 5\ln|x-2| - \ln|x+2| - 3\ln|x| + C$$

14. $\dfrac{x^3 - x + 3}{x^2 + x - 2} = x - 1 + \dfrac{2x+1}{(x+2)(x-1)} = x - 1 + \dfrac{A}{x+2} + \dfrac{B}{x-1}$

$2x + 1 = A(x-1) + B(x+2)$

When $x = -2$, $-3 = -3A$, $A = 1$. When $x = 1$, $3 = 3B$, $B = 1$.

$$\int \frac{x^3 - x + 3}{x^2 + x - 2}\,dx = \int \left[x - 1 + \frac{1}{x+2} + \frac{1}{x-1}\right] dx$$

$$= \frac{x^2}{2} - x + \ln|x+2| + \ln|x-1| + C = \frac{x^2}{2} - x + \ln|x^2 + x - 2| + C$$

15. $\dfrac{2x^3 - 4x^2 - 15x + 5}{x^2 - 2x - 8} = 2x + \dfrac{x+5}{(x-4)(x+2)} = 2x + \dfrac{A}{x-4} + \dfrac{B}{x+2}$

$x + 5 = A(x+2) + B(x-4)$

When $x = 4$, $9 = 6A$, $A = \frac{3}{2}$. When $x = -2$, $3 = -6B$, $B = -\frac{1}{2}$.

$$\int \frac{2x^3 - 4x^2 - 15x + 5}{x^2 - 2x - 8}\,dx = \int \left[2x + \frac{3/2}{x-4} - \frac{1/2}{x+2}\right] dx$$

$$= x^2 + \frac{3}{2}\ln|x-4| - \frac{1}{2}\ln|x+2| + C$$

16. $\dfrac{x+2}{x(x-4)} = \dfrac{A}{x-4} + \dfrac{B}{x}$

$x + 2 = Ax + B(x-4)$

When $x = 4$, $6 = 4A$, $A = \frac{3}{2}$.

When $x = 0$, $2 = -4B$, $B = -\frac{1}{2}$.

$$\int \frac{x+2}{x^2 - 4x}\,dx = \int \left[\frac{3/2}{x-4} - \frac{1/2}{x}\right] dx$$

$$= \frac{3}{2}\ln|x-4| - \frac{1}{2}\ln|x| + C$$

17. $\dfrac{4x^2 + 2x - 1}{x^2(x+1)} = \dfrac{A}{x} + \dfrac{B}{x^2} + \dfrac{C}{x+1}$

$4x^2 + 2x - 1 = Ax(x+1) + B(x+1) + Cx^2$

When $x = 0$, $B = -1$. When $x = -1$, $C = 1$. When $x = 1$, $A = 3$.

$$\int \frac{4x^2 + 2x - 1}{x^3 + x^2}\,dx = \int \left[\frac{3}{x} - \frac{1}{x^2} + \frac{1}{x+1}\right] dx$$

$$= 3\ln|x| + \frac{1}{x} + \ln|x+1| + C$$

$$= \frac{1}{x} + \ln|x^4 + x^3| + C$$

18. $\dfrac{2x-3}{(x-1)^2} = \dfrac{A}{x-1} + \dfrac{B}{(x-1)^2}$

$2x - 3 = A(x-1) + B$

When $x = 1$, $B = -1$. When $x = 0$, $A = 2$.

$$\int \frac{2x-3}{(x-1)^2}\,dx = \int \left[\frac{2}{x-1} - \frac{1}{(x-1)^2}\right] dx = 2\ln|x-1| + \frac{1}{x-1} + C$$

19. $\dfrac{x^2 + 3x - 4}{x^3 - 4x^2 + 4x} = \dfrac{x^2 + 3x - 4}{x(x - 2)^2} = \dfrac{A}{x} + \dfrac{B}{(x - 2)} + \dfrac{C}{(x - 2)^2}$

$x^2 + 3x - 4 = A(x - 2)^2 + Bx(x - 2) + Cx$

When $x = 0, -4 = 4A \implies A = -1$. When $x = 2, 6 = 2C \implies C = 3$. When $x = 1, 0 = -1 - B + 3 \implies B = 2$.

$$\int \frac{x^2 + 3x - 4}{x^3 - 4x^2 + 4x}\,dx = \int \frac{-1}{x}\,dx + \int \frac{2}{(x - 2)}\,dx + \int \frac{3}{(x - 2)^2}\,dx$$

$$= -\ln|x| + 2\ln|x - 2| - \frac{3}{(x - 2)} + C$$

20. $\dfrac{4x^2}{x^3 + x^2 - x - 1} = \dfrac{4x^2}{x^2(x + 1) - (x + 1)} = \dfrac{4x^2}{(x^2 - 1)(x + 1)} = \dfrac{A}{x - 1} + \dfrac{B}{x + 1} + \dfrac{C}{(x + 1)^2}$

$4x^2 = A(x + 1)^2 + B(x - 1)(x + 1) + C(x - 1)$

When $x = -1, 4 = -2C \implies C = -2$. When $x = 1, 4 = 4A \implies A = 1$. When $x = 0, 0 = 1 - B + 2 \implies B = 3$.

$$\int \frac{4x^2}{x^3 + x^2 - x - 1}\,dx = \int \frac{1}{x - 1}\,dx + \int \frac{3}{x + 1}\,dx - \int \frac{2}{(x + 1)^2}\,dx$$

$$= \ln|x - 1| + 3\ln|x + 1| + \frac{2}{(x + 1)} + C$$

21. $\dfrac{x^2 - 1}{x(x^2 + 1)} = \dfrac{A}{x} + \dfrac{Bx + C}{x^2 + 1}$

$x^2 - 1 = A(x^2 + 1) + (Bx + C)x$

When $x = 0, A = -1$. When $x = 1, 0 = -2 + B + C$. When $x = -1, 0 = -2 + B - C$.
Solving these equations we have $A = -1, B = 2, C = 0$.

$$\int \frac{x^2 - 1}{x^3 + x}\,dx = -\int \frac{1}{x}\,dx + \int \frac{2x}{x^2 + 1}\,dx$$

$$= -\ln|x| + \ln|x^2 + 1| + C$$

$$= \ln\left|\frac{x^2 + 1}{x}\right| + C$$

22. $\dfrac{6x}{x^3 - 8} = \dfrac{6x}{(x - 2)(x^2 + 2x + 4)} = \dfrac{A}{x - 2} + \dfrac{Bx + C}{x^2 + 2x + 4}$

$6x = A(x^2 + 2x + 4) + (Bx + C)(x - 2)$

When $x = 2, 12 = 12A \implies A = 1$. When $x = 0, 0 = 4 - 2C \implies C = 2$.
When $x = 1, 6 = 7 + (B + 2)(-1) \implies B = -1$.

$$\int \frac{6x}{x^3 - 8}\,dx = \int \frac{1}{x - 2}\,dx + \int \frac{-x + 2}{x^2 + 2x + 4}\,dx$$

$$= \int \frac{1}{x - 2}\,dx + \int \frac{-x - 1}{x^2 + 2x + 4}\,dx + \int \frac{3}{(x^2 + 2x + 1) + 3}\,dx$$

$$= \ln|x - 2| - \frac{1}{2}\ln|x^2 + 2x + 4| + \frac{3}{\sqrt{3}}\arctan\left(\frac{x + 1}{\sqrt{3}}\right) + C$$

$$= \ln|x - 2| - \frac{1}{2}\ln|x^2 + 2x + 4| + \sqrt{3}\arctan\left(\frac{\sqrt{3}(x + 1)}{3}\right) + C$$

23. $\dfrac{x^2}{x^4 - 2x^2 - 8} = \dfrac{A}{x - 2} + \dfrac{B}{x + 2} + \dfrac{Cx + D}{x^2 + 2}$

$$x^2 = A(x + 2)(x^2 + 2) + B(x - 2)(x^2 + 2) + (Cx + D)(x + 2)(x - 2)$$

When $x = 2$, $4 = 24A$. When $x = -2$, $4 = -24B$. When $x = 0$, $0 = 4A - 4B - 4D$, and when $x = 1$, $1 = 9A - 3B - 3C - 3D$. Solving these equations we have $A = \frac{1}{6}$, $B = -\frac{1}{6}$, $C = 0$, $D = \frac{1}{3}$.

$$\int \frac{x^2}{x^4 - 2x^2 - 8}\,dx = \frac{1}{6}\left[\int \frac{1}{x - 2}\,dx - \int \frac{1}{x + 2}\,dx + 2\int \frac{1}{x^2 + 2}\,dx\right]$$
$$= \frac{1}{6}\left[\ln\left|\frac{x - 2}{x + 2}\right| + \sqrt{2}\arctan\frac{x}{\sqrt{2}}\right] + C$$

24. $\dfrac{x^2 - x + 9}{(x^2 + 9)^2} = \dfrac{Ax + B}{x^2 + 9} + \dfrac{Cx + D}{(x^2 + 9)^2}$

$$x^2 - x + 9 = (Ax + B)(x^2 + 9) + Cx + D$$
$$= Ax^3 + Bx^2 + (9A + C)x + (9B + D)$$

By equating coefficients of like terms, we have $A = 0$, $B = 1$, $D = 0$, and $C = -1$.

$$\int \frac{x^2 - x - 9}{(x^2 + 9)^2} = \int \frac{1}{x^2 + 9}\,dx - \int \frac{x}{(x^2 + 9)^2}\,dx = \frac{1}{3}\arctan\left(\frac{x}{3}\right) + \frac{1}{2(x^2 + 9)} + C$$

25. $\dfrac{x}{(2x - 1)(2x + 1)(4x^2 + 1)} = \dfrac{A}{2x - 1} + \dfrac{B}{2x + 1} + \dfrac{Cx + D}{4x^2 + 1}$

$$x = A(2x + 1)(4x^2 + 1) + B(2x - 1)(4x^2 + 1) + (Cx + D)(2x - 1)(2x + 1)$$

When $x = \frac{1}{2}$, $\frac{1}{2} = 4A$. When $x = -\frac{1}{2}$, $-\frac{1}{2} = -4B$. When $x = 0$, $0 = A - B - D$, and when $x = 1$, $1 = 15A + 5B + 3C + 3D$. Solving these equations we have $A = \frac{1}{8}$, $B = \frac{1}{8}$, $C = -\frac{1}{2}$, $D = 0$.

$$\int \frac{x}{16x^4 - 1}\,dx = \frac{1}{8}\left[\int \frac{1}{2x - 1}\,dx + \int \frac{1}{2x + 1}\,dx - 4\int \frac{x}{4x^2 + 1}\,dx\right] = \frac{1}{16}\ln\left|\frac{4x^2 - 1}{4x^2 + 1}\right| + C$$

26. $\dfrac{x^2 - 4x + 7}{(x + 1)(x^2 - 2x + 3)} = \dfrac{A}{x + 1} + \dfrac{Bx + C}{x^2 - 2x + 3}$

$$x^2 - 4x + 7 = A(x^2 - 2x + 3) + (Bx + C)(x + 1)$$

When $x = -1$, $12 = 6A$. When $x = 0$, $7 = 3A + C$. When $x = 1$, $4 = 2A + 2B + 2C$. Solving these equations we have $A = 2$, $B = -1$, $C = 1$.

$$\int \frac{x^2 - 4x + 7}{x^3 - x^2 + x + 3}\,dx = 2\int \frac{1}{x + 1}\,dx + \int \frac{-x + 1}{x^2 - 2x + 3}\,dx$$
$$= 2\ln|x + 1| - \frac{1}{2}\ln|x^2 - 2x + 3| + C$$

27. $\dfrac{x^2 + 5}{(x + 1)(x^2 - 2x + 3)} = \dfrac{A}{x + 1} + \dfrac{Bx + C}{x^2 - 2x + 3}$

$$x^2 + 5 = A(x^2 - 2x + 3) + (Bx + C)(x + 1)$$
$$= (A + B)x^2 + (-2A + B + C)x + (3A + C)$$

When $x = -1$, $A = 1$. By equating coefficients of like terms, we have $A + B = 1$, $-2A + B + C = 0$, $3A + C = 5$. Solving these equations we have $A = 1$, $B = 0$, $C = 2$.

$$\int \frac{x^2 + 5}{x^3 - x^2 + x + 3}\,dx = \int \frac{1}{x + 1}\,dx + 2\int \frac{1}{(x - 1)^2 + 2}\,dx$$
$$= \ln|x + 1| + \sqrt{2}\arctan\left(\frac{x - 1}{\sqrt{2}}\right) + C$$

28. $\dfrac{x^2+x+3}{(x^2+3)^2} = \dfrac{Ax+B}{x^2+3} + \dfrac{Cx+D}{(x^2+3)^2}$

$$x^2+x+3 = (Ax+B)(x^2+3) + Cx + D$$
$$= Ax^3 + Bx^2 + (3A+C)x + (3B+D)$$

By equating coefficients of like terms, we have $A = 0$, $B = 1$, $3A + C = 1$, $3B + D = 3$. Solving these equations we have $A = 0$, $B = 1$, $C = 1$, $D = 0$.

$$\int \frac{x^2+x+3}{x^4+6x^2+9}\,dx = \int \left[\frac{1}{x^2+3} + \frac{x}{(x^2+3)^2}\right] dx$$
$$= \frac{1}{\sqrt{3}} \arctan\frac{x}{\sqrt{3}} - \frac{1}{2(x^2+3)} + C$$

29. $\dfrac{3}{(2x+1)(x+2)} = \dfrac{A}{2x+1} + \dfrac{B}{x+2}$

$$3 = A(x+2) + B(2x+1)$$

When $x = -\frac{1}{2}$, $A = 2$. When $x = -2$, $B = -1$.

$$\int_0^1 \frac{3}{2x^2+5x+2}\,dx = \int_0^1 \frac{2}{2x+1}\,dx - \int_0^1 \frac{1}{x+2}\,dx$$
$$= \Big[\ln|2x+1| - \ln|x+2|\Big]_0^1$$
$$= \ln 2$$

30. $\dfrac{x-1}{x^2(x+1)} = \dfrac{A}{x} + \dfrac{B}{x^2} + \dfrac{C}{x+1}$

$$x - 1 = Ax(x+1) + B(x+1) + Cx^2$$

When $x = 0$, $B = -1$. When $x = -1$, $C = -2$. When $x = 1$, $0 = 2A + 2B + C$. Solving these equations we have $A = 2$, $B = -1$, $C = -2$.

$$\int_1^5 \frac{x-1}{x^2(x+1)}\,dx = 2\int_1^5 \frac{1}{x}\,dx - \int_1^5 \frac{1}{x^2}\,dx - 2\int_1^5 \frac{1}{x+1}\,dx$$
$$= \left[2\ln|x| + \frac{1}{x} - 2\ln|x+1|\right]_1^5$$
$$= \left[2\ln\left|\frac{x}{x+1}\right| + \frac{1}{x}\right]_1^5$$
$$= 2\ln\frac{5}{3} - \frac{4}{5}$$

31. $\dfrac{x+1}{x(x^2+1)} = \dfrac{A}{x} + \dfrac{Bx+C}{x^2+1}$

$$x + 1 = A(x^2+1) + (Bx+C)x$$

When $x = 0$, $A = 1$. When $x = 1$, $2 = 2A + B + C$. When $x = -1$, $0 = 2A + B - C$. Solving these equations we have $A = 1$, $B = -1$, $C = 1$.

$$\int_1^2 \frac{x+1}{x(x^2+1)}\,dx = \int_1^2 \frac{1}{x}\,dx - \int_1^2 \frac{x}{x^2+1}\,dx + \int_1^2 \frac{1}{x^2+1}\,dx$$
$$= \left[\ln|x| - \frac{1}{2}\ln(x^2+1) + \arctan x\right]_1^2$$
$$= \frac{1}{2}\ln\frac{8}{5} - \frac{\pi}{4} + \arctan 2$$
$$\approx 0.557$$

32. $$\int_0^1 \frac{x^2-x}{x^2+x+1}\,dx = \int_0^1 dx - \int_0^1 \frac{2x+1}{x^2+x+1}\,dx$$
$$= \Big[x - \ln|x^2+x+1|\Big]_0^1$$
$$= 1 - \ln 3$$

33. $\displaystyle\int \frac{3x\,dx}{x^2-6x+9} = 3\ln|x-3| - \frac{9}{x-3} + C$

$(4, 0)$: $\;3\ln|4-3| - \dfrac{9}{4-3} + C = 0 \Rightarrow C = 9$

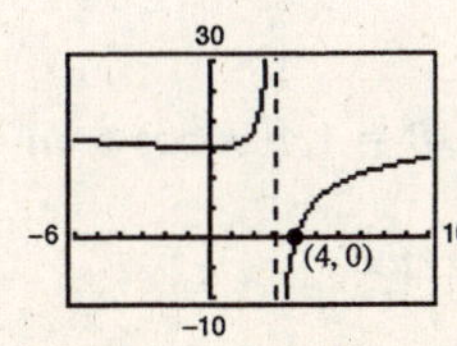

34. $\displaystyle\int \frac{6x^2+1}{x^2(x-1)^3}\,dx = 3\ln\left|\frac{x-1}{x}\right| + \frac{1}{x} + \frac{2}{x-1} - \frac{7}{2(x-1)^2} + C$

(2, 1): $3\ln\left|\frac{1}{2}\right| + \frac{1}{2} + \frac{2}{1} - \frac{7}{2} + C = 1 \Rightarrow C = 2 - 3\ln\frac{1}{2}$

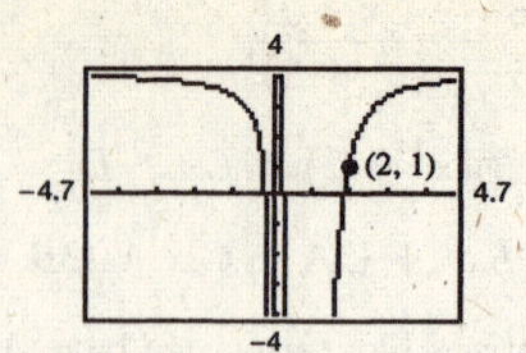

35. $\displaystyle\int \frac{x^2+x+2}{(x^2+2)^2}\,dx = \frac{\sqrt{2}}{2}\arctan\frac{x}{\sqrt{2}} - \frac{1}{2(x^2+2)} + C$

(0, 1): $0 - \frac{1}{4} + C = 1 \Rightarrow C = \frac{5}{4}$

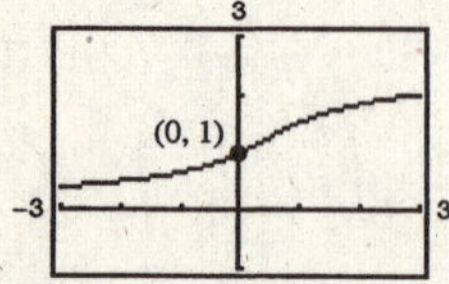

36. $\displaystyle\int \frac{x^3}{(x^2-4)^2}\,dx = \frac{1}{2}\ln|x^2-4| - \frac{2}{x^2-4} + C$

(3, 4): $\frac{1}{2}\ln 5 - \frac{2}{5} + C = 4 \Rightarrow C = \frac{22}{5} - \frac{1}{2}\ln 5$

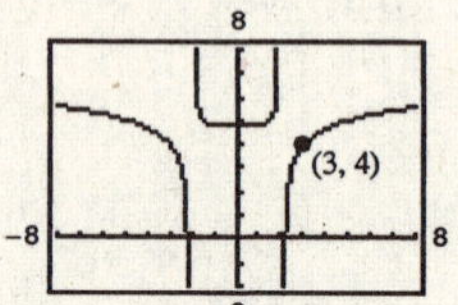

37. $\displaystyle\int \frac{2x^2-2x+3}{x^3-x^2-x-2}\,dx = \ln|x-2| + \frac{1}{2}\ln|x^2+x+1| - \sqrt{3}\arctan\left(\frac{2x+1}{\sqrt{3}}\right) + C$

(3, 10): $0 + \frac{1}{2}\ln 13 - \sqrt{3}\arctan\frac{7}{\sqrt{3}} + C = 10 \Rightarrow C = 10 - \frac{1}{2}\ln 13 + \sqrt{3}\arctan\frac{7}{\sqrt{3}}$

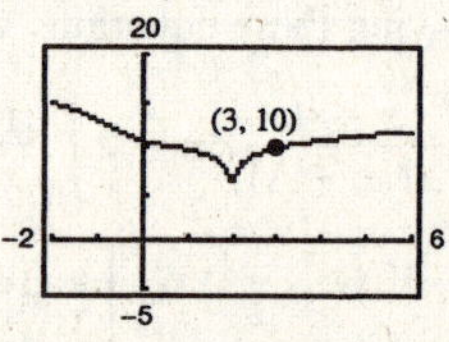

38. $\displaystyle\int \frac{x(2x-9)}{x^3-6x^2+12x-8}\,dx = 2\ln|x-2| + \frac{1}{x-2} + \frac{5}{(x-2)^2} + C$

(3, 2): $0 + 1 + 5 + C = 2 \Rightarrow C = -4$

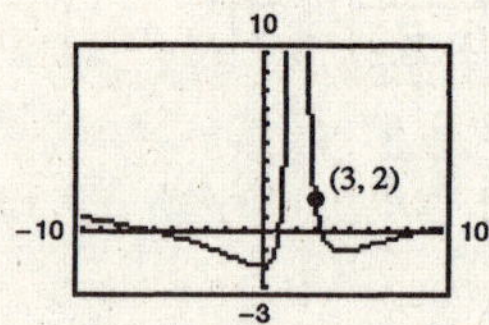

39. $\displaystyle\int \frac{1}{x^2-4}\,dx = \frac{1}{4}\ln\left|\frac{x-2}{x+2}\right| + C$

(6, 4): $\frac{1}{4}\ln\left|\frac{4}{8}\right| + C = 4 \Rightarrow C = 4 - \frac{1}{4}\ln\frac{1}{2} = 4 + \frac{1}{4}\ln 2$

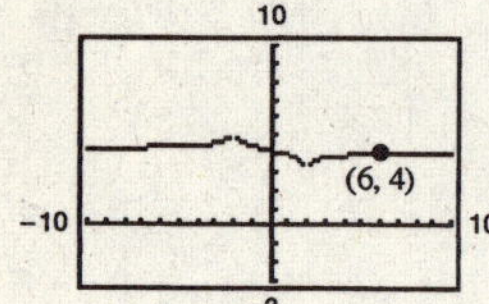

40. $\displaystyle\int \frac{x^2-x+2}{x^3-x^2+x-1}\,dx = -\arctan x + \ln|x-1| + C$

(2, 6): $-\arctan 2 + 0 + C = 6 \Rightarrow C = 6 + \arctan 2$

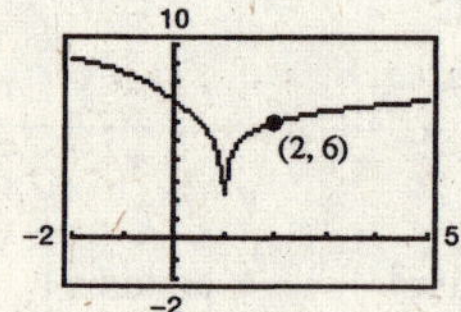

41. Let $u = \cos x$ $du = -\sin x\,dx$.

$$\frac{1}{u(u-1)} = \frac{A}{u} + \frac{B}{u-1}$$

$$1 = A(u-1) + Bu$$

When $u = 0, A = -1$. When $u = 1, B = 1, u = \cos x, du = -\sin x\,dx$.

$$\int \frac{\sin x}{\cos x(\cos x - 1)}\,dx = -\int \frac{1}{u(u-1)}\,du$$

$$= \int \frac{1}{u}\,du - \int \frac{1}{u-1}\,du = \ln|u| - \ln|u-1| + C = \ln\left|\frac{u}{u-1}\right| + C = \ln\left|\frac{\cos x}{\cos x - 1}\right| + C$$

42. Let $u = \cos x,\ du = \sin x\,dx.$

$$\frac{1}{u(u+1)} = \frac{A}{u} + \frac{B}{u+1}$$

$$1 = A(u+1) + Bu$$

When $u = 0, A = 1$. When $u = -1, B = -1, u = \cos x,$ $du = -\sin dx.$

$$\int \frac{\sin x}{\cos x + \cos^2 x}\,dx = -\int \frac{1}{u(u+1)}\,du$$

$$= \int \frac{1}{u+1}\,du - \int \frac{1}{u}\,du$$

$$= \ln|u+1| - \ln|u| + C$$

$$= \ln\left|\frac{u+1}{u}\right| + C$$

$$= \ln\left|\frac{\cos x + 1}{\cos x}\right| + C$$

$$= \ln|1 + \sec x| + C$$

43. $$\int \frac{3\cos x}{\sin^2 x + \sin x - 2}\,dx = 3\int \frac{1}{u^2 + u - 2}\,du$$

$$= \ln\left|\frac{u-1}{u+2}\right| + C$$

$$= \ln\left|\frac{-1 + \sin x}{2 + \sin x}\right| + C$$

(From Exercise 9 with $u = \sin x,\ du = \cos x\,dx$)

44. $\dfrac{1}{u(u+1)} = \dfrac{A}{u} + \dfrac{B}{u+1},\quad u = \tan x,\ du = \sec^2 x\,dx$

$$1 = A(u+1) + Bu$$

When $u = 0, A = 1.$
When $u = -1, 1 = -B \Rightarrow B = -1.$

$$\int \frac{\sec^2 x\,dx}{\tan x(\tan x + 1)} = \int \frac{1}{u(u+1)}\,du$$

$$= \int \left(\frac{1}{u} - \frac{1}{u+1}\right) du$$

$$= \ln|u| - \ln|u+1| + C$$

$$= \ln\left|\frac{u}{u+1}\right| + C$$

$$= \ln\left|\frac{\tan x}{\tan x + 1}\right| + C$$

45. Let $u = e^x,\ du = e^x\,dx.$

$$\frac{1}{(u-1)(u+4)} = \frac{A}{u-1} + \frac{B}{u+4}$$

$$1 = A(u+4) + B(u-1)$$

When $u = 1, A = \frac{1}{5}$. When $u = -4, B = -\frac{1}{5}, u = e^x,$ $du = e^x\,dx.$

$$\int \frac{e^x}{(e^x - 1)(e^x + 4)}\,dx = \int \frac{1}{(u-1)(u+4)}\,du$$

$$= \frac{1}{5}\left(\int \frac{1}{u-1}\,du - \int \frac{1}{u+4}\,du\right)$$

$$= \frac{1}{5}\ln\left|\frac{u-1}{u+4}\right| + C$$

$$= \frac{1}{5}\ln\left|\frac{e^x - 1}{e^x + 4}\right| + C$$

46. Let $u = e^x,\ du = e^x\,dx.$

$$\frac{1}{(u^2+1)(u-1)} = \frac{A}{u-1} + \frac{Bu + C}{u^2 + 1}$$

$$1 = A(u^2 + 1) + (Bu + C)(u - 1)$$

When $u = 1, A = \frac{1}{2}$. When $u = 0, 1 = A - C$. When $u = -1, 1 = 2A + 2B - 2C$. Solving these equations we have $A = \frac{1}{2}, B = -\frac{1}{2}, C = -\frac{1}{2}, u = e^x, du = e^x\,dx.$

$$\int \frac{e^x}{(e^{2x} + 1)(e^x - 1)}\,dx = \int \frac{1}{(u^2+1)(u-1)}\,du$$

$$= \frac{1}{2}\left(\int \frac{1}{u-1}\,du - \int \frac{u+1}{u^2+1}\,du\right)$$

$$= \frac{1}{2}\left(\ln|u-1| - \frac{1}{2}\ln|u^2 + 1| - \arctan u\right) + C$$

$$= \frac{1}{4}(2\ln|e^x - 1| - \ln|e^{2x} + 1| - 2\arctan e^x) + C$$

47. $\dfrac{1}{x(a+bx)} = \dfrac{A}{x} + \dfrac{B}{a+bx}$

$$1 = A(a+bx) + Bx$$

When $x = 0$, $1 = aA \Rightarrow A = 1/a$.
When $x = -a/b$, $1 = -(a/b)B \Rightarrow B = -b/a$.

$$\int \frac{1}{x(a+bx)}\,dx = \frac{1}{a}\int\left(\frac{1}{x} - \frac{b}{a+bx}\right)dx$$

$$= \frac{1}{a}(\ln|x| - \ln|a+bx|) + C$$

$$= \frac{1}{a}\ln\left|\frac{x}{a+bx}\right| + C$$

48. $\dfrac{1}{a^2-x^2} = \dfrac{A}{a-x} + \dfrac{B}{a+x}$

$$1 = A(a+x) + B(a-x)$$

When $x = a$, $A = 1/2a$.
When $x = -a$, $B = 1/2a$.

$$\int \frac{1}{a^2-x^2}\,dx = \frac{1}{2a}\int\left(\frac{1}{a-x} + \frac{1}{a+x}\right)dx$$

$$= \frac{1}{2a}(-\ln|a-x| + \ln|a+x|) + C$$

$$= \frac{1}{2a}\ln\left|\frac{a+x}{a-x}\right| + C$$

49. $\dfrac{x}{(a+bx)^2} = \dfrac{A}{a+bx} + \dfrac{B}{(a+bx)^2}$

$$x = A(a+bx) + B$$

When $x = -a/b$, $B = -a/b$.
When $x = 0$, $0 = aA + B \Rightarrow A = 1/b$.

$$\int \frac{x}{(a+bx)^2}\,dx = \int\left(\frac{1/b}{a+bx} + \frac{-a/b}{(a+bx)^2}\right)dx$$

$$= \frac{1}{b}\int\frac{1}{a+bx}\,dx - \frac{a}{b}\int\frac{1}{(a+bx)^2}\,dx$$

$$= \frac{1}{b^2}\ln|a+bx| + \frac{a}{b^2}\left(\frac{1}{a+bx}\right) + C$$

$$= \frac{1}{b^2}\left(\frac{a}{a+bx} + \ln|a+bx|\right) + C$$

50. $\dfrac{1}{x^2(a+bx)} = \dfrac{A}{x} + \dfrac{B}{x^2} + \dfrac{C}{a+bx}$

$$1 = Ax(a+bx) + B(a+bx) + Cx^2$$

When $x = 0$, $1 = Ba \Rightarrow B = 1/a$. When $x = -a/b$, $1 = C(a^2/b^2) \Rightarrow C = b^2/a^2$. When $x = 1$, $1 = (a+b)A + (a+b)B + C \Rightarrow A = -b/a^2$.

$$\int \frac{1}{x^2(a+bx)}\,dx = \int\left(\frac{-b/a^2}{x} + \frac{1/a}{x^2} + \frac{b^2/a^2}{a+bx}\right)dx$$

$$= -\frac{b}{a^2}\ln|x| - \frac{1}{ax} + \frac{b}{a^2}\ln|a+bx| + C$$

$$= -\frac{1}{ax} + \frac{b}{a^2}\ln\left|\frac{a+bx}{x}\right| + C$$

$$= -\frac{1}{ax} - \frac{b}{a^2}\ln\left|\frac{x}{a+bx}\right| + C$$

51. 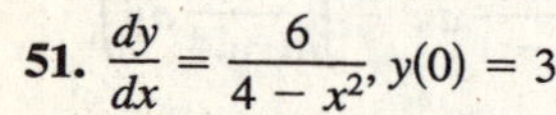

$\dfrac{dy}{dx} = \dfrac{6}{4-x^2}$, $y(0) = 3$

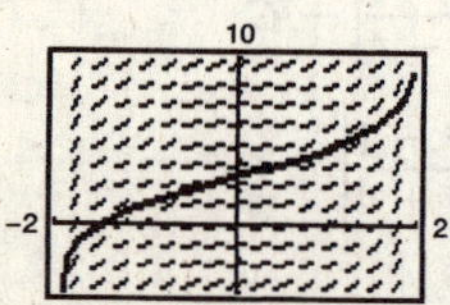

52. $\dfrac{dy}{dx} = \dfrac{4}{(x^2-2x-3)}$, $y(0) = 5$

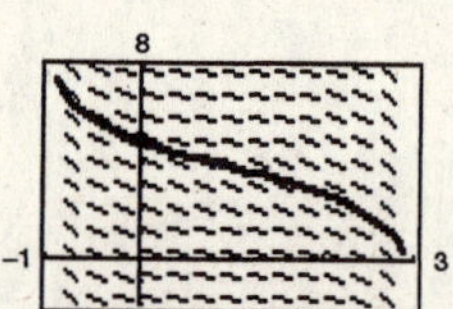

53. Dividing x^3 by $x - 5$

54. (a) $\dfrac{N(x)}{D(x)} = \dfrac{A_1}{px+q} + \dfrac{A_2}{(px+q)^2} + \cdots + \dfrac{A_m}{(px+q)^m}$

(b) $\dfrac{N(x)}{D(x)} = \dfrac{A_1 + B_1x}{(ax^2+bx+c)} + \cdots + \dfrac{A_n + B_nx}{(ax^2+bx+c)^n}$

55. (a) Substitution: $u = x^2 + 2x - 8$

(b) Partial fractions

(c) Trigonometric substitution (tan) or inverse tangent rule

56. $A = \int_1^3 \frac{10}{x(x^2+1)}\,dx \approx 3$, matches (c).

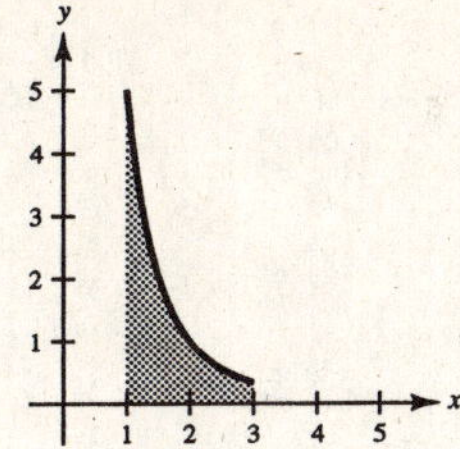

57. $$A = \int_0^1 \frac{12}{x^2+5x+6}\,dx$$

$$\frac{12}{x^2+5x+6} = \frac{12}{(x+2)(x+3)} = \frac{A}{x+2} + \frac{B}{x+3}$$

$$12 = A(x+3) + B(x+2)$$

Let $x = -3$: $12 = B(-1) \Rightarrow B = -12$

Let $x = -2$: $12 = A(1) \Rightarrow A = 12$

$$A = \int_0^1 \left(\frac{12}{x+2} - \frac{12}{x+3}\right) dx$$

$$= \Big[12\ln|x+2| - 12\ln|x+3|\Big]_0^1$$

$$= 12[\ln 3 - \ln 4 - \ln 2 + \ln 3]$$

$$= 12\ln\left(\frac{9}{8}\right) \approx 1.4134$$

58. $A = 2\int_0^3 \left(1 - \frac{7}{16-x^2}\right) dx = 2\int_0^3 dx - 14\int_0^3 \frac{1}{16-x^2}\,dx$

$= \left[2x - \frac{14}{8}\ln\left|\frac{4+x}{4-x}\right|\right]_0^3$ (From Exercise 48)

$= 6 - \frac{7}{4}\ln 7 \approx 2.595$

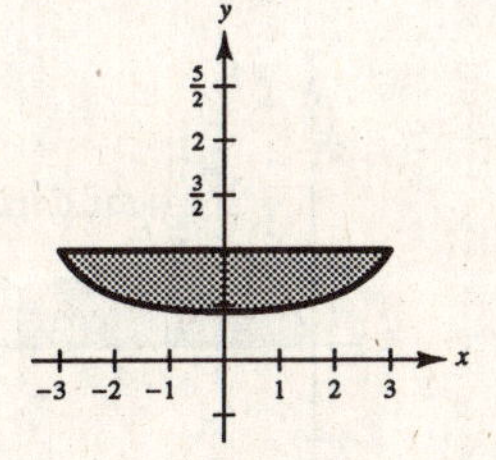

59. Average cost $= \dfrac{1}{80-75}\displaystyle\int_{75}^{80} \frac{124p}{(10+p)(100-p)}\,dp$

$$= \frac{1}{5}\int_{75}^{80}\left(\frac{-124}{(10+p)11} + \frac{1240}{(100-p)11}\right) dp$$

$$= \frac{1}{5}\left[\frac{-124}{11}\ln(10+p) - \frac{1240}{11}\ln(100-p)\right]_{75}^{80}$$

$$\approx \frac{1}{5}(24.51) = 4.9$$

Approximately $490,000

60. (a)

(b) The slope is negative because the function is decreasing.

(c) For $y > 0$, $\lim_{t\to\infty} y(t) = 3$.

(d) $$\frac{dy}{y(L-y)} = \frac{A}{y} + \frac{B}{L-y}$$

$$1 = A(L-y) + By \Rightarrow A = \frac{1}{L}, B = \frac{1}{L}$$

$$\int \frac{dy}{y(L-y)} = \int k\,dt$$

$$\frac{1}{L}\left[\int \frac{1}{y}\,dy + \int \frac{1}{L-y}\,dy\right] = \int k\,dt$$

$$\frac{1}{L}\big[\ln|y| - \ln|L-y|\big] = kt + C_1$$

$$\ln\left|\frac{y}{L-y}\right| = kLt + LC_1$$

$$C_2 e^{kLt} = \frac{y}{L-y}$$

When $t = 0$, $\dfrac{y_0}{L-y_0} = C_2 \Rightarrow \dfrac{y}{L-y} = \dfrac{y_0}{L-y_0}e^{kLt}$.

Solving for y, you obtain $y = \dfrac{y_0 L}{y_0 + (L-y_0)e^{-kLt}}$.

(e) $k = 1, L = 3$

(*i*) $y(0) = 5$: $y = \dfrac{15}{5 - 2e^{-3t}}$

(*ii*) $y(0) = \dfrac{1}{2}$:

$$y = \frac{3/2}{(1/2) + (5/2)e^{-3t}}$$

$$= \frac{3}{1 + 5e^{-3t}}$$

—CONTINUED—

60. —CONTINUED—

(f) $\dfrac{dy}{dt} = ky(L - y)$

$$\frac{d^2y}{dt^2} = k\left[y\left(\frac{-dy}{dt}\right) + (L - y)\frac{dy}{dt}\right] = 0$$

$$\Rightarrow y\frac{dy}{dt} = (L - y)\frac{dy}{dt}$$

$$\Rightarrow \quad y = \frac{L}{2}$$

From the first derivative test, this is a maximum.

61. $V = \pi\displaystyle\int_0^3 \left(\frac{2x}{x^2 + 1}\right)^2 dx = 4\pi\int_0^3 \frac{x^2}{(x^2 + 1)^2}\,dx$

$$= 4\pi\int_0^3 \left(\frac{1}{x^2 + 1} - \frac{1}{(x^2 + 1)^2}\right) dx \qquad \text{(partial fractions)}$$

$$= 4\pi\left[\arctan x - \frac{1}{2}\left(\arctan x + \frac{x}{x^2 + 1}\right)\right]_0^3 \qquad \text{(trigonometric substitution)}$$

$$= 2\pi\left[\arctan x - \frac{x}{x^2 + 1}\right]_0^3 = 2\pi\left[\arctan 3 - \frac{3}{10}\right] \approx 5.963$$

$$A = \int_0^3 \frac{2x}{x^2 + 1}\,dx = \Big[\ln(x^2 + 1)\Big]_0^3 = \ln 10$$

$$\bar{x} = \frac{1}{A}\int_0^3 \frac{2x^2}{x^2 + 1}\,dx = \frac{1}{\ln 10}\int_0^3 \left(2 - \frac{2}{x^2 + 1}\right) dx$$

$$= \frac{1}{\ln 10}\Big[2x - 2\arctan x\Big]_0^3 = \frac{2}{\ln 10}[3 - \arctan 3] \approx 1.521$$

$$\bar{y} = \frac{1}{A}\left(\frac{1}{2}\right)\int_0^3 \left(\frac{2x}{x^2 + 1}\right)^2 dx = \frac{2}{\ln 10}\int_0^3 \frac{x^2}{(x^2 + 1)^2}\,dx$$

$$= \frac{2}{\ln 10}\int_0^3 \left(\frac{1}{x^2 + 1} - \frac{1}{(x^2 + 1)^2}\right) dx \qquad \text{(partial fractions)}$$

$$= \frac{2}{\ln 10}\left[\arctan x - \frac{1}{2}\left(\arctan x + \frac{x}{x^2 + 1}\right)\right]_0^3 \qquad \text{(trigonometric substitution)}$$

$$= \frac{2}{\ln 10}\left[\frac{1}{2}\arctan x - \frac{x}{2(x^2 + 1)}\right]_0^3 = \frac{1}{\ln 10}\left[\arctan x - \frac{x}{x^2 + 1}\right]_0^3 = \frac{1}{\ln 10}\left[\arctan 3 - \frac{3}{10}\right] \approx 0.412$$

$(\bar{x}, \bar{y}) \approx (1.521, 0.412)$

62. $y^2 = \dfrac{(2 - x)^2}{(1 + x)^2}, \quad [0, 1]$

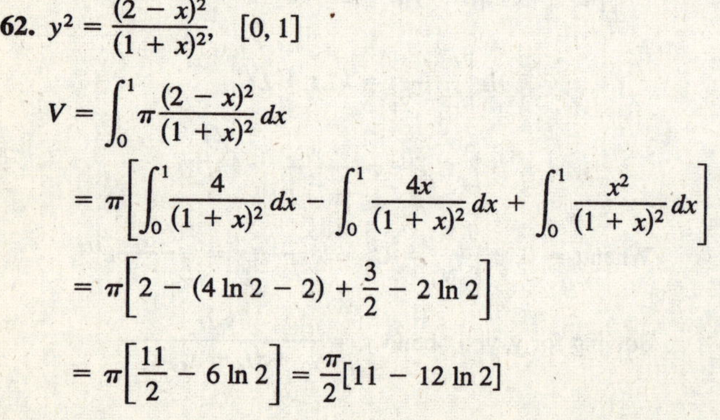

$$V = \int_0^1 \pi\frac{(2 - x)^2}{(1 + x)^2}\,dx$$

$$= \pi\left[\int_0^1 \frac{4}{(1 + x)^2}\,dx - \int_0^1 \frac{4x}{(1 + x)^2}\,dx + \int_0^1 \frac{x^2}{(1 + x)^2}\,dx\right]$$

$$= \pi\left[2 - (4\ln 2 - 2) + \frac{3}{2} - 2\ln 2\right]$$

$$= \pi\left[\frac{11}{2} - 6\ln 2\right] = \frac{\pi}{2}[11 - 12\ln 2]$$

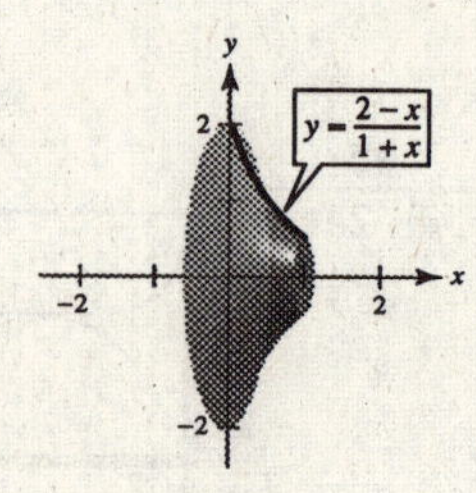

63. $$\frac{1}{(x+1)(n-x)} = \frac{A}{x+1} + \frac{B}{n-x}, A = B = \frac{1}{n+1}$$

$$\frac{1}{n+1}\int\left(\frac{1}{x+1} + \frac{1}{n-x}\right)dx = kt + C$$

$$\frac{1}{n+1}\ln\left|\frac{x+1}{n-x}\right| = kt + C$$

When $t = 0, x = 0, C = \frac{1}{n+1}\ln\frac{1}{n}$.

$$\frac{1}{n+1}\ln\left|\frac{x+1}{n-x}\right| = kt + \frac{1}{n+1}\ln\frac{1}{n}$$

$$\frac{1}{n+1}\left[\ln\left|\frac{x+1}{n-x}\right| - \ln\frac{1}{n}\right] = kt$$

$$\ln\frac{nx+n}{n-x} = (n+1)kt$$

$$\frac{nx+n}{n-x} = e^{(n+1)kt}$$

$$x = \frac{n[e^{(n+1)kt} - 1]}{n + e^{(n+1)kt}}$$ **Note:** $\lim_{t\to\infty} x = n$

64. (a) $$\frac{1}{(y_0 - x)(z_0 - x)} = \frac{A}{y_0 - x} + \frac{B}{z_0 - x},$$

$A = \frac{1}{z_0 - y_0}, B = -\frac{1}{z_0 - y_0}$, (Assume $y_0 \neq z_0$.)

$$\frac{1}{z_0 - y_0}\int\left(\frac{1}{y_0 - x} - \frac{1}{z_0 - x}\right)dx = kt + C$$

$$\frac{1}{z_0 - y_0}\ln\left|\frac{z_0 - x}{y_0 - x}\right| = kt + C, \text{ when } t = 0, x = 0$$

$$C = \frac{1}{z_0 - y_0}\ln\frac{z_0}{y_0}$$

$$\frac{1}{z_0 - y_0}\left[\ln\left|\frac{z_0 - x}{y_0 - x}\right| - \ln\left(\frac{z_0}{y_0}\right)\right] = kt$$

$$\ln\left[\frac{y_0(z_0 - x)}{z_0(y_0 - x)}\right] = (z_0 - y_0)kt$$

$$\frac{y_0(z_0 - x)}{z_0(y_0 - x)} = e^{(z_0 - y_0)kt}$$

$$x = \frac{y_0 z_0[e^{(z_0 - y_0)kt} - 1]}{z_0 e^{(z_0 - y_0)kt} - y_0}$$

(b) (1) If $y_0 < z_0$, $\lim_{t\to\infty} x = y_0$.

(2) If $y_0 > z_0$, $\lim_{t\to\infty} x = z_0$.

(3) If $y_0 = z_0$, then the original equation is:

$$\int\frac{1}{(y_0 - x)^2}\,dx = \int k\,dt$$

$$(y_0 - x)^{-1} = kt + C_1$$

$$x = 0 \text{ when } t = 0 \Rightarrow \frac{1}{y_0} = C_1$$

$$\frac{1}{y_0 - x} = kt + \frac{1}{y_0} = \frac{kty_0 + 1}{y_0}$$

$$y_0 - x = \frac{y_0}{kty_0 + 1}$$

$$x = y_0 - \frac{y_0}{kty_0 + 1}$$

As $t \to \infty$, $x \to y_0 = x_0$.

65. $\dfrac{x}{1+x^4} = \dfrac{Ax+B}{x^2+\sqrt{2}x+1} + \dfrac{Cx+D}{x^2-\sqrt{2}x+1}$

$$x = (Ax+B)(x^2-\sqrt{2}x+1) + (Cx+D)(x^2+\sqrt{2}x+1)$$

$$= (A+C)x^3 + (B+D-\sqrt{2}A+\sqrt{2}C)x^2 + (A+C-\sqrt{2}B+\sqrt{2}D)x + (B+D)$$

$0 = A + C \Rightarrow C = -A$

$0 = B + D - \sqrt{2}A + \sqrt{2}C \qquad -2\sqrt{2}A = 0 \Rightarrow A = 0$ and $C = 0$

$1 = A + C - \sqrt{2}B + \sqrt{2}D \qquad -2\sqrt{2}B = 1 \Rightarrow B = -\dfrac{\sqrt{2}}{4}$ and $D = \dfrac{\sqrt{2}}{4}$

$0 = B + D \Rightarrow D = -B$

Thus,

$$\int_0^1 \frac{x}{1+x^4}\,dx = \int_0^1 \left[\frac{-\sqrt{2}/4}{x^2+\sqrt{2}x+1} + \frac{\sqrt{2}/4}{x^2-\sqrt{2}x+1}\right] dx$$

$$= \frac{\sqrt{2}}{4}\int_0^1 \left[\frac{-1}{\left[x+(\sqrt{2}/2)\right]^2+(1/2)} + \frac{1}{\left[x-(\sqrt{2}/2)\right]^2+(1/2)}\right] dx$$

$$= \frac{\sqrt{2}}{4}\cdot\frac{1}{1/\sqrt{2}}\left[-\arctan\left(\frac{x+(\sqrt{2}/2)}{1/\sqrt{2}}\right) + \arctan\left(\frac{x-(\sqrt{2}/2)}{1/\sqrt{2}}\right)\right]_0^1$$

$$= \frac{1}{2}\left[-\arctan(\sqrt{2}x+1) + \arctan(\sqrt{2}x-1)\right]_0^1$$

$$= \frac{1}{2}\left[(-\arctan(\sqrt{2}+1) + \arctan(\sqrt{2}-1)) - (-\arctan 1 + \arctan(-1)\right]$$

$$= \frac{1}{2}\left[\arctan(\sqrt{2}-1) - \arctan(\sqrt{2}+1) + \frac{\pi}{4} + \frac{\pi}{4}\right].$$

Since $\arctan x - \arctan y = \arctan[(x-y)/(1+xy)]$, we have:

$$\int_0^1 \frac{x}{1+x^4}\,dx = \frac{1}{2}\left[\arctan\left(\frac{(\sqrt{2}-1)-(\sqrt{2}+1)}{1+(\sqrt{2}-1)(\sqrt{2}+1)}\right) + \frac{\pi}{2}\right] = \frac{1}{2}\left[\arctan\left(\frac{-2}{2}\right) + \frac{\pi}{2}\right] = \frac{1}{2}\left[-\frac{\pi}{4} + \frac{\pi}{2}\right] = \frac{\pi}{8}$$

66. The partial fraction decomposition is:

$$\frac{x^4(1-x)^4}{1+x^2} = x^6 - 4x^5 + 5x^4 - 4x^2 + 4 - \frac{4}{1+x^2}$$

$$\int_0^1 \frac{x^4(1-x)^4}{1+x^2}\,dx = \left[\frac{x^7}{7} - \frac{2x^6}{3} + x^5 - \frac{4}{3}x^3 + 4x - 4\arctan x\right]_0^1$$

$$= \frac{1}{7} - \frac{2}{3} + 1 - \frac{4}{3} + 4 - 4\left(\frac{\pi}{4}\right)$$

$$= \frac{22}{7} - \pi$$

Note: You can easily verify this calculation with a graphing utility.

Section 8.6 Integration by Tables and Other Integration Techniques

1. By Formula 6: $\displaystyle\int \frac{x^2}{1+x}\,dx = -\frac{x}{2}(2-x) + \ln|1+x| + C$

2. By Formula 13: $(b = 2, a = -5)$

$$\frac{2}{3}\int \frac{1}{x^2(2x-5)^2}\,dx = \frac{2}{3}\left(\frac{-1}{25}\right)\left[\frac{-5+4x}{x(-5+2x)} + \frac{4}{-5}\ln\left|\frac{x}{2x-5}\right|\right] + C$$

$$= \frac{8}{375}\ln\left|\frac{x}{2x-5}\right| - \frac{2}{75}\frac{(4x-5)}{x(2x-5)} + C$$

3. By Formula 26: $\displaystyle\int e^x\sqrt{1+e^{2x}}\,dx = \frac{1}{2}\left[e^x\sqrt{e^{2x}+1} + \ln\left|e^x + \sqrt{e^{2x}+1}\right|\right] + C$

$u = e^x,\ du = e^x\,dx$

4. By Formula 29: $(a = 3)$

$$\frac{1}{3}\int \frac{\sqrt{x^2-9}}{x}\,dx = \frac{1}{3}\sqrt{x^2-9} - \operatorname{arcsec}\frac{|x|}{3} + C$$

5. By Formula 44: $\displaystyle\int \frac{1}{x^2\sqrt{1-x^2}}\,dx = -\frac{\sqrt{1-x^2}}{x} + C$

6. By Formula 41: $\displaystyle\int \frac{x}{\sqrt{9-x^4}}\,dx = \frac{1}{2}\int \frac{2x}{\sqrt{3^2-(x^2)^2}}\,dx$

$$= \frac{1}{2}\arcsin\frac{x^2}{3} + C$$

7. By Formulas 50 and 48: $\displaystyle\int \sin^4(2x)\,dx = \frac{1}{2}\int \sin^4(2x)(2)\,dx$

$$= \frac{1}{2}\left[\frac{-\sin^3(2x)\cos(2x)}{4} + \frac{3}{4}\int \sin^2(2x)(2)\,dx\right]$$

$$= \frac{1}{2}\left[\frac{-\sin^3(2x)\cos(2x)}{4} + \frac{3}{8}(2x - \sin 2x\cos 2x)\right] + C$$

$$= \frac{1}{16}(6x - 3\sin 2x\cos 2x - 2\sin^3 2x\cos 2x) + C$$

8. By Formulas 51 and 47: $\displaystyle\int \frac{\cos^3\sqrt{x}}{\sqrt{x}}\,dx = 2\int \cos^3\sqrt{x}\left(\frac{1}{2\sqrt{x}}\right)dx$

$$= 2\left[\frac{\cos^2\sqrt{x}\sin\sqrt{x}}{3} + \frac{2}{3}\int \cos\sqrt{x}\left(\frac{1}{2\sqrt{x}}\right)dx\right] = \frac{2}{3}\sin\sqrt{x}\left(\cos^2\sqrt{x} + 2\right) + C$$

$u = \sqrt{x},\ du = \dfrac{1}{2\sqrt{x}}\,dx$

9. By Formula 57: $\displaystyle\int \frac{1}{\sqrt{x}\left(1-\cos\sqrt{x}\right)}\,dx = 2\int \frac{1}{1-\cos\sqrt{x}}\left(\frac{1}{2\sqrt{x}}\right)dx$

$$= -2\left(\cot\sqrt{x} + \csc\sqrt{x}\right) + C$$

$u = \sqrt{x},\ du = \dfrac{1}{2\sqrt{x}}\,dx$

10. By Formula 71:

$$\int \frac{1}{1-\tan 5x}\,dx = \frac{1}{5}\int \frac{1}{1-\tan 5x}(5)\,dx$$

$$= \frac{1}{5}\left(\frac{1}{2}\right)\left(u - \ln|\cos u - \sin u|\right) + C$$

$$= \frac{1}{10}\left(5x - \ln|\cos 5x - \sin 5x|\right) + C$$

$u = 5x,\ du = 5\,dx$

11. By Formula 84:

$$\int \frac{1}{1+e^{2x}}\,dx = x - \frac{1}{2}\ln(1+e^{2x}) + C$$

12. By Formula 85: $\left(a = -\frac{1}{2}, b = 2\right)$

$$\int e^{-x/2}\sin 2x\,dx = \frac{e^{-x/2}}{(1/4)+4}\left(-\frac{1}{2}\sin 2x - 2\cos 2x\right) + C$$

$$= \frac{4}{17}e^{-x/2}\left(-\frac{1}{2}\sin 2x - 2\cos 2x\right) + C$$

13. By Formula 89:

$$\int x^3 \ln x\,dx = \frac{x^4}{16}(4\ln|x| - 1) + C$$

14. By Formulas 90 and 91: $\int (\ln x)^3\,dx = x(\ln x)^3 - 3\int (\ln x)^2\,dx$

$$= x(\ln x)^3 - 3x[2 - 2\ln x + (\ln x)^2] + C$$

$$= x[(\ln x)^3 - 3(\ln x)^2 + 6\ln x - 6] + C$$

15. (a) By Formulas 83 and 82: $\int x^2e^x\,dx = x^2e^x - 2\int xe^x\,dx$

$$= x^2e^x - 2[(x-1)e^x + C_1]$$

$$= x^2e^x - 2xe^x + 2e^x + C$$

(b) Integration by parts: $u = x^2,\ du = 2x\,dx,\ dv = e^x\,dx,\ v = e^x$

$$\int x^2e^x\,dx = x^2e^x - \int 2xe^x\,dx$$

Parts again: $u = 2x,\ du = 2\,dx,\ dv = e^x\,dx,\ v = e^x$

$$\int x^2e^x\,dx = x^2e^x - \left[2xe^x - \int 2e^x\,dx\right] = x^2e^x - 2xe^x + 2e^x + C$$

16. (a) By Formula 89: $\int x^4\ln x\,dx = \frac{x^5}{5^2}[-1 + (4+1)\ln x] + C = \frac{-x^5}{25} + \frac{1}{5}x^5\ln x + C$

(b) Integration by parts: $u = \ln x,\ du = \frac{1}{x}\,dx,\ dv = x^4\,dx,\ v = \frac{x^5}{5}$

$$\int x^4\ln x\,dx = \frac{x^5}{5}\ln x - \int \frac{x^5}{5}\frac{1}{x}\,dx = \frac{x^5}{5}\ln x - \frac{x^5}{25} + C$$

17. (a) By Formula: 12, $a = b = 1$, $u = x$, and

$$\int \frac{1}{x^2(x+1)}\,dx = \frac{-1}{1}\left(\frac{1}{x} + \frac{1}{1}\ln\left|\frac{x}{1+x}\right|\right) + C$$

$$= \frac{-1}{x} - \ln\left|\frac{x}{1+x}\right| + C$$

$$= \frac{-1}{x} + \ln\left|\frac{x+1}{x}\right| + C$$

(b) Partial fractions:

$$\frac{1}{x^2(x+1)} = \frac{A}{x} + \frac{B}{x^2} + \frac{C}{x+1}$$

$$1 = Ax(x+1) + B(x+1) + Cx^2$$

$x = 0$: $1 = B$

$x = -1$: $1 = C$

$x = 1$: $1 = 2A + 2 + 1 \Rightarrow A = -1$

$$\int \frac{1}{x^2(x+1)}\,dx = \int\left[\frac{-1}{x} + \frac{1}{x^2} + \frac{1}{x+1}\right]dx$$

$$= -\ln|x| - \frac{1}{x} + \ln|x+1| + C$$

$$= -\frac{1}{x} - \ln\left|\frac{x}{x+1}\right| + C$$

18. (a) By Formula 24: $a = \sqrt{75}$, $x = u$, and

$$\int \frac{1}{x^2 - 75}\,dx = \frac{1}{2\sqrt{75}}\ln\left|\frac{x-\sqrt{75}}{x+\sqrt{75}}\right| + C$$

$$= \frac{\sqrt{3}}{30}\ln\left|\frac{x-\sqrt{75}}{x+\sqrt{75}}\right| + C$$

(b) Partial fractions:

$$\frac{1}{x^2 - 75} = \frac{A}{x - \sqrt{75}} + \frac{B}{x + \sqrt{75}}$$

$$1 = A\left(x + \sqrt{75}\right) + B\left(x - \sqrt{75}\right)$$

$x = \sqrt{75}$: $1 = 2A\sqrt{75} \Rightarrow A = \dfrac{1}{2\sqrt{75}} = \dfrac{1}{10\sqrt{3}} = \dfrac{\sqrt{3}}{30}$

$x = -\sqrt{75}$: $1 = -2B\sqrt{75} \Rightarrow B = -\dfrac{\sqrt{3}}{30}$

$$\int \frac{1}{x^2 - 75}\,dx = \int\left[\frac{\sqrt{3}/30}{x - \sqrt{75}} - \frac{\sqrt{3}/30}{x + \sqrt{75}}\right]dx$$

$$= \frac{\sqrt{3}}{30}\ln\left|\frac{x-\sqrt{75}}{x+\sqrt{75}}\right| + C$$

19. By Formula 79: $\displaystyle\int x \operatorname{arcsec}(x^2+1)\,dx = \frac{1}{2}\int \operatorname{arcsec}(x^2+1)(2x)\,dx$

$$= \frac{1}{2}\left[(x^2+1)\operatorname{arcsec}(x^2+1) - \ln\left((x^2+1) + \sqrt{x^4+2x^2}\right)\right] + C$$

$u = x^2 + 1$, $du = 2x\,dx$

20. By Formula 79: $\displaystyle\int \operatorname{arcsec} 2x\,dx = \frac{1}{2}\left[2x \operatorname{arcsec} 2x - \ln\left|2x + \sqrt{4x^2-1}\right|\right] + C$

$u = 2x$, $du = 2\,dx$

21. By Formula 35: $\displaystyle\int \frac{1}{x^2\sqrt{x^2-4}}\,dx = \frac{\sqrt{x^2-4}}{4x} + C$

22. By Formula 14: $\displaystyle\int \frac{1}{x^2+2x+2}\,dx = \frac{2}{\sqrt{4}}\arctan\left(\frac{2x+2}{2}\right) + C = \arctan(x+1) + C$

23. By Formula 4: $\displaystyle\int \frac{2x}{(1-3x)^2}\,dx = 2\int \frac{x}{(1-3x)^2}\,dx = \frac{2}{9}\left(\ln|1-3x| + \frac{1}{1-3x}\right) + C$

24. By Formula 56:

$$\int \frac{\theta^2}{1 - \sin \theta^3}\, d\theta = \frac{1}{3}\int \frac{1}{1 - \sin \theta^3} 3\theta^2\, d\theta$$

$$= \frac{1}{3}(\tan \theta^3 + \sec \theta^3) + C$$

25. By Formula 76:

$$\int e^x \arccos e^x\, dx = e^x \arccos e^x - \sqrt{1 - e^{2x}} + C$$

$u = e^x, du = e^x\, dx$

26. By Formula 71:

$$\int \frac{e^x}{1 - \tan e^x}\, dx = \frac{1}{2}\left(e^x - \ln|\cos e^x - \sin e^x|\right) + C$$

$u = e^x, du = e^x\, dx$

27. By Formula 73:

$$\int \frac{x}{1 - \sec x^2}\, dx = \frac{1}{2}\int \frac{2x}{1 - \sec x^2}\, dx$$

$$= \frac{1}{2}(x^2 + \cot x^2 + \csc x^2) + C$$

28. By Formula 23: $\displaystyle\int \frac{1}{t[1 + (\ln t)^2]}\, dt = \int \frac{1}{1 + (\ln t)^2}\left(\frac{1}{t}\right) dt = \arctan(\ln t) + C$

$u = \ln t, du = \dfrac{1}{t}\, dt$

29. By Formula 14: $\displaystyle\int \frac{\cos \theta}{3 + 2\sin \theta + \sin^2 \theta}\, d\theta = \frac{\sqrt{2}}{2}\arctan\left(\frac{1 + \sin \theta}{\sqrt{2}}\right) + C \quad (b^2 = 4 < 12 = 4ac)$

$u = \sin \theta, du = \cos \theta\, d\theta$

30. By Formula 27: $\displaystyle\int x^2\sqrt{2 + (3x)^2}\, dx = \frac{1}{27}\int (3x)^2\sqrt{\left(\sqrt{2}\right)^2 + (3x)^2}\, 3\, dx$

$$= \frac{1}{8(27)}\left[3x(18x^2 + 2)\sqrt{2 + 9x^2} - 4\ln\left|3x + \sqrt{2 + 9x^2}\right|\right] + C$$

31. By Formula 35: $\displaystyle\int \frac{1}{x^2\sqrt{2 + 9x^2}}\, dx = 3\int \frac{3}{(3x)^2\sqrt{\left(\sqrt{2}\right)^2 + (3x)^2}}\, dx$

$$= -\frac{3\sqrt{2 + 9x^2}}{6x} + C$$

$$= -\frac{\sqrt{2 + 9x^2}}{2x} + C$$

32. By Formula 77: $\displaystyle\int \sqrt{x}\arctan(x^{3/2})\, dx = \frac{2}{3}\int \arctan(x^{3/2})\left(\frac{3}{2}\sqrt{x}\right) dx$

$$= \frac{2}{3}\left[x^{3/2}\arctan(x^{3/2}) - \ln\sqrt{1 + x^3}\right] + C$$

33. By Formula 3: $\displaystyle\int \frac{\ln x}{x(3 + 2\ln x)}\, dx = \frac{1}{4}\left(2\ln|x| - 3\ln|3 + 2\ln|x||\right) + C$

$u = \ln x, du = \dfrac{1}{x}\, dx$

34. By Formula 45: $\displaystyle\int \frac{e^x}{(1-e^{2x})^{3/2}}\,dx = \frac{e^x}{\sqrt{1-e^{2x}}} + C$

$u = e^x,\ du = e^x\,dx$

35. By Formulas 1, 25, and 33: $\displaystyle\int \frac{x}{(x^2-6x+10)^2}\,dx = \frac{1}{2}\int \frac{2x-6+6}{(x^2-6x+10)^2}\,dx$

$$= \frac{1}{2}\int (x^2-6x+10)^{-2}(2x-6)\,dx + 3\int \frac{1}{[(x-3)^2+1]^2}\,dx$$

$$= -\frac{1}{2(x^2-6x+10)} + \frac{3}{2}\left[\frac{x-3}{x^2-6x+10} + \arctan(x-3)\right] + C$$

$$= \frac{3x-10}{2(x^2-6x+10)} + \frac{3}{2}\arctan(x-3) + C$$

36. By Formula 27:

$$\int (2x-3)^2\sqrt{(2x-3)^2+4}\,dx = \frac{1}{2}\int (2x-3)^2\sqrt{(2x-3)^2+4}(2)\,dx$$

$$= \frac{1}{8}(2x-3)[(2x-3)^2+2]\sqrt{(2x-3)^2+4} - \ln\left|2x-3+\sqrt{(2x-3)^2+4}\right| + C$$

$u = 2x-3,\ du = 2\,dx$

37. By Formula 31: $\displaystyle\int \frac{x}{\sqrt{x^4-6x^2+5}}\,dx = \frac{1}{2}\int \frac{2x}{\sqrt{(x^2-3)^2-4}}\,dx$

$$= \frac{1}{2}\ln\left|x^2-3+\sqrt{x^4-6x^2+5}\right| + C$$

$u = x^2-3,\ du = 2x\,dx$

38. By Formula 31: $\displaystyle\int \frac{\cos x}{\sqrt{\sin^2 x+1}}\,dx = \ln\left|\sin x + \sqrt{\sin^2 x+1}\right| + C$

$u = \sin x,\ du = \cos x\,dx$

39. $\displaystyle\int \frac{x^3}{\sqrt{4-x^2}}\,dx = \int \frac{8\sin^3\theta(2\cos\theta\,d\theta)}{2\cos\theta}$

$$= 8\int (1-\cos^2\theta)\sin\theta\,d\theta$$

$$= 8\int [\sin\theta - \cos^2\theta(\sin\theta)]\,d\theta$$

$$= -8\cos\theta + \frac{8\cos^3\theta}{3} + C$$

$$= -8\frac{\sqrt{4-x^2}}{2} + \frac{8}{3}\left(\frac{\sqrt{4-x^2}}{2}\right)^3 + C$$

$$= \sqrt{4-x^2}\left[-4 + \frac{1}{3}(4-x^2)\right] + C$$

$$= \frac{-\sqrt{4-x^2}}{3}(x^2+8) + C$$

2

x

θ

$\sqrt{4-x^2}$

$x = 2\sin\theta,\ dx = 2\cos\theta\,d\theta,\ \sqrt{4-x^2} = 2\cos\theta$

40. $\displaystyle\int\sqrt{\frac{3-x}{3+x}}\,dx = \int\frac{3-x}{\sqrt{9-x^2}}\,dx$

$$= 3\int\frac{1}{\sqrt{9-x^2}}\,dx + \int\frac{-x}{\sqrt{9-x^2}}\,dx$$

$$= 3\arcsin\frac{x}{3} + \sqrt{9-x^2} + C$$

41. By Formula 8:

$$\int\frac{e^{3x}}{(1+e^x)^3}\,dx = \int\frac{(e^x)^2}{(1+e^x)^3}(e^x)\,dx$$

$$= \frac{2}{1+e^x} - \frac{1}{2(1+e^x)^2} + \ln|1+e^x| + C$$

$u = e^x,\ du = e^x\,dx$

42. By Formula 67:

$$\int\tan^3\theta\,d\theta = \frac{\tan^2\theta}{2} - \int\tan\theta\,d\theta$$

$$= \frac{\tan^2\theta}{2} + \ln|\cos x| + C$$

43. $\displaystyle\int_0^1 xe^{x^2}\,dx$

By Formula 81:

$$\int_0^1 xe^{x^2}\,dx = \frac{1}{2}e^{x^2}\Big]_0^1 = \frac{1}{2}(e-1)$$

44. By Formula 21:

$$\int_0^3\frac{x}{\sqrt{1+x}}\,dx = \left[\frac{-2}{3}(2-x)\sqrt{1+x}\right]_0^3$$

$$= \frac{-2}{3}(-1)(2) + \frac{2}{3}(2) = \frac{8}{3}$$

45. By Formula 89:

$$\int_1^3 x^2\ln x\,dx = \left[\frac{x^3}{9}(-1+3\ln|x|)\right]_1^3$$

$$= 3(-1+3\ln 3) + \frac{1}{9} = 9\ln 3 - \frac{26}{9}$$

46. By Formula 52:

$$\int_0^\pi x\sin x\,dx = \Big[\sin x - x\cos x\Big]_0^\pi$$

$$= \pi$$

47. By Formula 23, and letting $u = \sin x$:

$$\int_{-\pi/2}^{\pi/2}\frac{\cos x}{1+\sin^2 x}\,dx = \Big[\arctan(\sin x)\Big]_{-\pi/2}^{\pi/2}$$

$$= \arctan(1) - \arctan(-1) = \frac{\pi}{2}$$

48. By Formula 7:

$$\int_2^4\frac{x^2}{(3x-5)^2}\,dx = \left[\frac{1}{27}\left(3x - \frac{25}{3x-5} + 10\ln|3x-5|\right)\right]_2^4$$

$$= \frac{1}{27}\left[\left(12 - \frac{25}{7} + 10\ln 7\right) - (6-25)\right] = \frac{64}{63} + \frac{10}{27}\ln 7$$

49. By Formulas 54 and 55:

$$\int t^3\cos t\,dt = t^3\sin t - 3\int t^2\sin t\,dt$$

$$= t^3\sin t - 3\left[-t^2\cos t + 2\int t\cos t\,dt\right]$$

$$= t^3\sin t + 3t^2\cos t - 6\left[t\sin t - \int\sin t\,dt\right]$$

$$= t^3\sin t + 3t^2\cos t - 6t\sin t - 6\cos t + C$$

Thus,

$$\int_0^{\pi/2} t^3\cos t\,dt = \Big[t^3\sin t + 3t^2\cos t - 6t\sin t - 6\cos t\Big]_0^{\pi/2}$$

$$= \left(\frac{\pi^3}{8} - 3\pi\right) + 6 = \frac{\pi^3}{8} + 6 - 3\pi$$

50. By Formula 26:

$$\int_0^1 \sqrt{3+x^2}\,dx = \left[\frac{1}{2}\left(x\sqrt{x^2+3} + 3\ln\left|x+\sqrt{x^2+3}\right|\right)\right]_0^1 = \frac{1}{2}[(2) + 3\ln 3 - 3\ln\sqrt{3}] = 1 + \frac{3}{4}\ln 3$$

51. $$\frac{u^2}{(a+bu)^2} = \frac{1}{b^2} - \frac{(2a/b)u + (a^2/b^2)}{(a+bu)^2} = \frac{1}{b^2} + \frac{A}{a+bu} + \frac{B}{(a+bu)^2}$$

$$-\frac{2a}{b}u - \frac{a^2}{b^2} = A(a+bu) + B = (aA + B) + bAu$$

Equating the coefficients of like terms we have $aA + B = -a^2/b^2$ and $bA = -2a/b$. Solving these equations we have $A = -2a/b^2$ and $B = a^2/b^2$.

$$\int \frac{u^2}{(a+bu)^2}\,du = \frac{1}{b^2}\int du - \frac{2a}{b^2}\left(\frac{1}{b}\right)\int \frac{1}{a+bu}b\,du + \frac{a^2}{b^2}\left(\frac{1}{b}\right)\int \frac{1}{(a+bu)^2}b\,du = \frac{1}{b^2}u - \frac{2a}{b^3}\ln|a+bu| - \frac{a^2}{b^3}\left(\frac{1}{a+bu}\right) + C$$

$$= \frac{1}{b^3}\left(bu - \frac{a^2}{a+bu} - 2a\ln|a+bu|\right) + C$$

52. Integration by parts: $w = u^n$, $dw = nu^{n-1}\,du$, $dv = \dfrac{du}{\sqrt{a+bu}}$, $v = \dfrac{2}{b}\sqrt{a+bu}$

$$\int \frac{u^n}{\sqrt{a+bu}}\,du = \frac{2u^n}{b}\sqrt{a+bu} - \frac{2n}{b}\int u^{n-1}\sqrt{a+bu}\,du$$

$$= \frac{2u^n}{b}\sqrt{a+bu} - \frac{2n}{b}\int u^{n-1}\sqrt{a+bu}\cdot\frac{\sqrt{a+bu}}{\sqrt{a+bu}}\,du$$

$$= \frac{2u^n}{b}\sqrt{a+bu} - \frac{2n}{b}\int \frac{au^{n-1} + bu^n}{\sqrt{a+bu}}\,du$$

$$= \frac{2u^n}{b}\sqrt{a+bu} - \frac{2na}{b}\int \frac{u^{n-1}}{\sqrt{a+bu}}\,du - 2n\int \frac{u^n}{\sqrt{a+bu}}\,du$$

Therefore, $(2n+1)\displaystyle\int \frac{u^n}{\sqrt{a+bu}}\,du = \frac{2}{b}\left[u^n\sqrt{a+bu} - na\int \frac{u^{n-1}}{\sqrt{a+bu}}\,du\right]$ and

$$\int \frac{u^n}{\sqrt{a+bu}} = \frac{2}{(2n+1)b}\left[u^n\sqrt{a+bu} - na\int \frac{u^{n-1}}{\sqrt{a+bu}}\,du\right].$$

53. When we have $u^2 + a^2$:

$$u = a\tan\theta$$

$$du = a\sec^2\theta\,d\theta$$

$$u^2 + a^2 = a^2\sec^2\theta$$

$$\int \frac{1}{(u^2+a^2)^{3/2}}\,du = \int \frac{a\sec^2\theta\,d\theta}{a^3\sec^3\theta}$$

$$= \frac{1}{a^2}\int \cos\theta\,d\theta$$

$$= \frac{1}{a^2}\sin\theta + C$$

$$= \frac{u}{a^2\sqrt{u^2+a^2}} + C$$

When we have $u^2 - a^2$:

$$u = a\sec\theta$$

$$du = a\sec\theta\tan\theta\,d\theta$$

$$u^2 - a^2 = a^2\tan^2\theta$$

$$\int \frac{1}{(u^2-a^2)^{3/2}}\,du = \int \frac{a\sec\theta\tan\theta\,d\theta}{a^3\tan^3\theta}$$

$$= \frac{1}{a^2}\int \frac{\cos\theta}{\sin^2\theta}\,d\theta = \frac{1}{a^2}\int \csc\theta\cot\theta\,d\theta$$

$$= -\frac{1}{a^2}\csc\theta + C$$

$$= \frac{-u}{a^2\sqrt{u^2-a^2}} + C$$

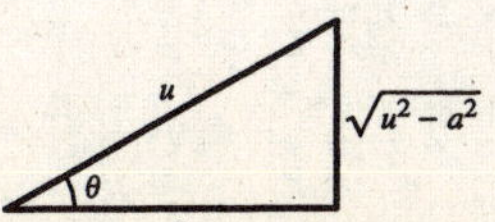

54. $\displaystyle\int u^n(\cos u)\,du = u^n \sin u - n\int u^{n-1}(\sin u)\,du$

$w = u^n,\ dv = \cos u\,du,\ dw = nu^{n-1}\,du,\ v = \sin u$

55. $\displaystyle\int (\arctan u)\,du = u\arctan u - \frac{1}{2}\int \frac{2u}{1+u^2}\,du$

$\displaystyle = u\arctan u - \frac{1}{2}\ln(1+u^2) + C$

$= u\arctan u - \ln\sqrt{1+u^2} + C$

$\displaystyle w = \arctan u,\ dv = du,\ dw = \frac{du}{1+u^2},\ v = u$

56. $\displaystyle\int (\ln u)^n\,du = u(\ln u)^n - \int n(\ln u)^{n-1}\left(\frac{1}{u}\right)u\,du = u(\ln u)^n - n\int (\ln u)^{n-1}\,du$

$\displaystyle w = (\ln u)^n,\ dv = du,\ dw = n(\ln u)^{n-1}\left(\frac{1}{u}\right)du,\ v = u$

57. $\displaystyle\int \frac{1}{x^{3/2}\sqrt{1-x}}\,dx = \frac{-2\sqrt{1-x}}{\sqrt{x}} + C$

$\displaystyle\left(\frac{1}{2}, 5\right)\text{:}\ \frac{-2\sqrt{1/2}}{\sqrt{1/2}} + C = 5 \Rightarrow C = 7$

$\displaystyle y = \frac{-2\sqrt{1-x}}{\sqrt{x}} + 7$

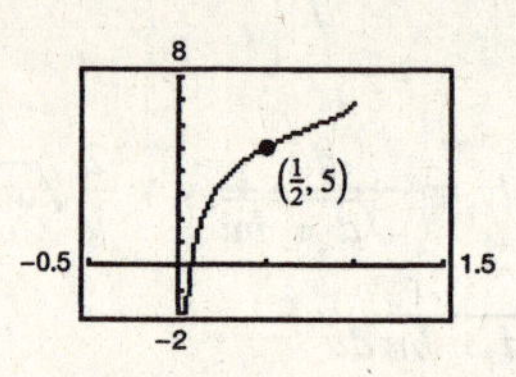

58. $\displaystyle\int x\sqrt{x^2+2x}\,dx = \frac{1}{6}\left[2(x^2+2x)^{3/2} - 3(x+1)\sqrt{x^2+2x} + 3\ln\left|x+1+\sqrt{x^2+2x}\right|\right] + C$

$\displaystyle (0, 0)\text{:}\ \frac{1}{6}[3\ln|1|] + C = 0 \Rightarrow C = 0$

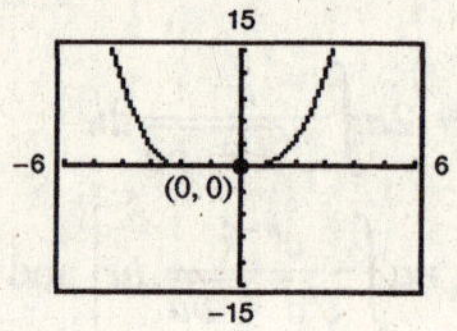

59. $\displaystyle\int \frac{1}{(x^2-6x+10)^2}\,dx = \frac{1}{2}\left[\tan^{-1}(x-3) + \frac{x-3}{x^2-6x+10}\right] + C$

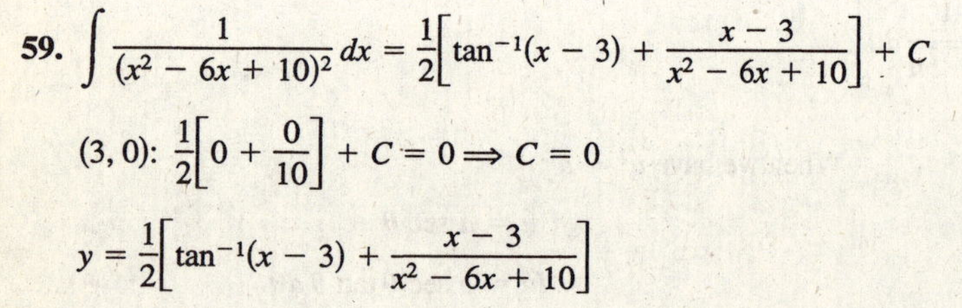

$\displaystyle (3, 0)\text{:}\ \frac{1}{2}\left[0 + \frac{0}{10}\right] + C = 0 \Rightarrow C = 0$

$\displaystyle y = \frac{1}{2}\left[\tan^{-1}(x-3) + \frac{x-3}{x^2-6x+10}\right]$

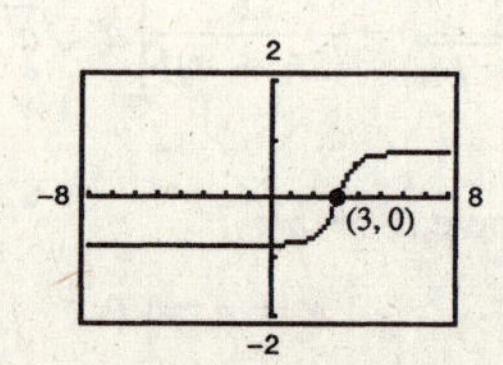

60. $\displaystyle\int \frac{\sqrt{2-2x-x^2}}{x+1}\,dx = \sqrt{2-2x-x^2} - \sqrt{3}\ln\left|\frac{\sqrt{3}+\sqrt{2-2x-x^2}}{x+1}\right| + C$

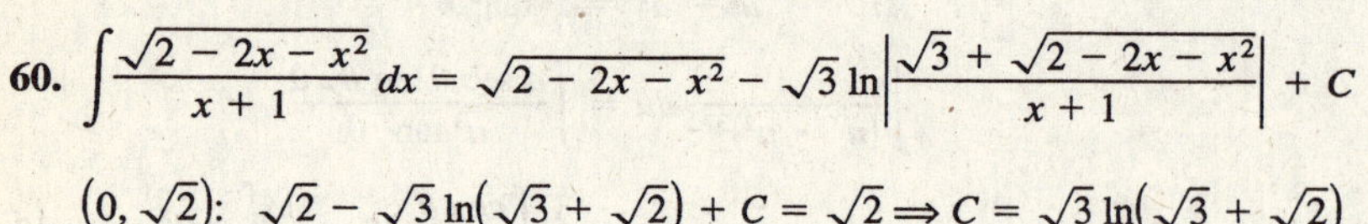

$(0, \sqrt{2})\text{:}\ \sqrt{2} - \sqrt{3}\ln(\sqrt{3}+\sqrt{2}) + C = \sqrt{2} \Rightarrow C = \sqrt{3}\ln(\sqrt{3}+\sqrt{2})$

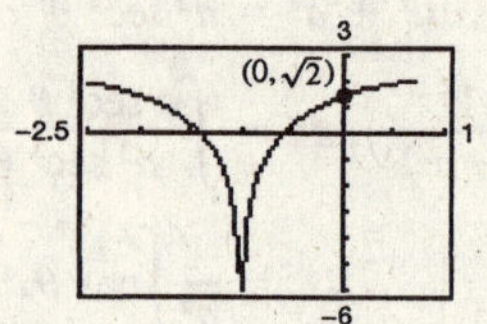

61. $\displaystyle\int \frac{1}{\sin\theta\tan\theta}\,d\theta = -\csc\theta + C$

$\displaystyle\left(\frac{\pi}{4}, 2\right)\text{:}\ -\frac{2}{\sqrt{2}} + C = 2 \Rightarrow C = 2 + \sqrt{2}$

$y = -\csc\theta + 2 + \sqrt{2}$

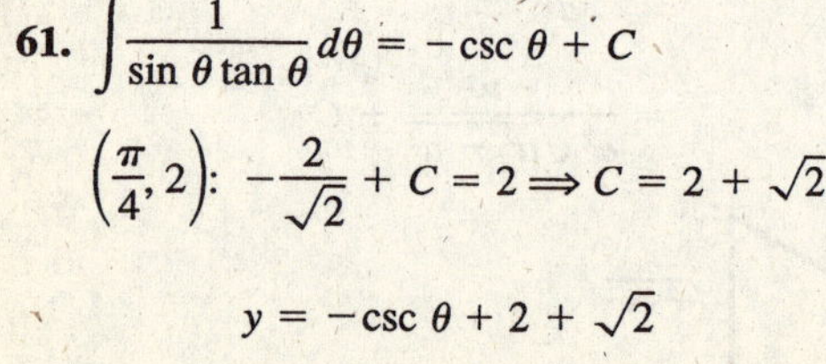

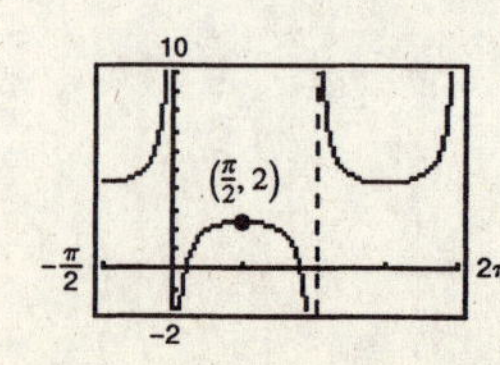

62. $\displaystyle\int \frac{\sin\theta}{(\cos\theta)(1+\sin\theta)}\,d\theta = \frac{1}{2}\left[\frac{-\sin\theta}{1+\sin\theta} + \ln\left|\frac{1+\sin\theta}{\cos\theta}\right|\right] + C$

$(0, 1)$: $C = 1 \Rightarrow y = \dfrac{1}{2}\left[\dfrac{-\sin\theta}{1+\sin\theta} + \ln\left|\dfrac{1+\sin\theta}{\cos\theta}\right|\right] + 1$

10
−8
8
(0, 1)
−2

63. $\displaystyle\int \frac{1}{2-3\sin\theta}\,d\theta = \int\left[\frac{\dfrac{2\,du}{1+u^2}}{2-3\left(\dfrac{2u}{1+u^2}\right)}\right],\ u = \tan\frac{\theta}{2}$

$$= \int \frac{2}{2(1+u^2)-6u}\,du$$

$$= \int \frac{1}{u^2-3u+1}\,du$$

$$= \int \frac{1}{\left(u-\frac{3}{2}\right)^2-\frac{5}{4}}\,du$$

$$= \frac{1}{\sqrt{5}}\ln\left|\frac{\left(u-\frac{3}{2}\right)-\frac{\sqrt{5}}{2}}{\left(u-\frac{3}{2}\right)+\frac{\sqrt{5}}{2}}\right| + C$$

$$= \frac{1}{\sqrt{5}}\ln\left|\frac{2u-3-\sqrt{5}}{2u-3+\sqrt{5}}\right| + C$$

$$= \frac{1}{\sqrt{5}}\ln\left|\frac{2\tan\left(\frac{\theta}{2}\right)-3-\sqrt{5}}{2\tan\left(\frac{\theta}{2}\right)-3+\sqrt{5}}\right| + C$$

64. $\displaystyle\int \frac{\sin\theta}{1+\cos^2\theta}\,d\theta = -\int \frac{-\sin\theta}{1+(\cos\theta)^2}\,d\theta$

$$= -\arctan(\cos\theta) + C$$

65. $\displaystyle\int_0^{\pi/2} \frac{1}{1+\sin\theta+\cos\theta}\,d\theta = \int_0^1\left[\frac{\dfrac{2\,du}{1+u^2}}{1+\dfrac{2u}{1+u^2}+\dfrac{1-u^2}{1+u^2}}\right]$

$$= \int_0^1 \frac{1}{1+u}\,du$$

$$= \Big[\ln|1+u|\Big]_0^1$$

$$= \ln 2$$

$u = \tan\dfrac{\theta}{2}$

66. $\displaystyle\int_0^{\pi/2} \frac{1}{3-2\cos\theta}\,d\theta = \int_0^1\left[\frac{\dfrac{2u}{1+u^2}}{3-\dfrac{2(1-u^2)}{1+u^2}}\right]$

$$= 2\int_0^1 \frac{1}{5u^2+1}\,du$$

$$= \left[\frac{2}{\sqrt{5}}\arctan\left(\sqrt{5}\,u\right)\right]_0^1$$

$$= \frac{2}{\sqrt{5}}\arctan\sqrt{5}$$

67. $\displaystyle\int \frac{\sin\theta}{3-2\cos\theta}\,d\theta = \frac{1}{2}\int \frac{2\sin\theta}{3-2\cos\theta}\,d\theta$

$$= \frac{1}{2}\ln|u| + C$$

$$= \frac{1}{2}\ln(3-2\cos\theta) + C$$

$u = 3 - 2\cos\theta,\ du = 2\sin\theta\,d\theta$

68. $\displaystyle\int \frac{\cos\theta}{1+\cos\theta}\,d\theta = \int \frac{\cos\theta(1-\cos\theta)}{(1+\cos\theta)(1-\cos\theta)}\,d\theta$

$$= \int \frac{\cos\theta-\cos^2\theta}{\sin^2\theta}\,d\theta$$

$$= \int(\csc\theta\cot\theta - \cot^2\theta)\,d\theta$$

$$= \int(\csc\theta\cot\theta - (\csc^2\theta - 1))\,d\theta$$

$$= -\csc\theta + \cot\theta + \theta + C$$

69. $\displaystyle\int \frac{\cos\sqrt{\theta}}{\sqrt{\theta}}\,d\theta = 2\int \cos\sqrt{\theta}\left(\frac{1}{2\sqrt{\theta}}\right)d\theta$

$\displaystyle = 2\sin\sqrt{\theta} + C$

$\displaystyle u = \sqrt{\theta},\ du = \frac{1}{2\sqrt{\theta}}\,d\theta$

70. $\displaystyle\int \frac{1}{\sec\theta - \tan\theta}\,d\theta = \int \frac{1}{(1/\cos\theta) - (\sin\theta/\cos\theta)}\,d\theta$

$\displaystyle = -\int \frac{-\cos\theta}{1 - \sin\theta}\,d\theta$

$\displaystyle = -\ln|1 - \sin\theta| + C$

$u = 1 - \sin\theta,\ du = -\cos\theta\,d\theta$

71. $\displaystyle A = \int_0^8 \frac{x}{\sqrt{x+1}}\,dx$

$\displaystyle = \left[\frac{-2(2-x)}{3}\sqrt{x+1}\right]_0^8$

$\displaystyle = 12 - \left(-\frac{4}{3}\right)$

$\displaystyle = \frac{40}{3} \approx 13.333$ square units

72. $\displaystyle A = \int_0^2 \frac{x}{1+e^{x^2}}\,dx$

$\displaystyle = \frac{1}{2}\int_0^2 \frac{2x\,dx}{1+e^{x^2}}$

$\displaystyle = \frac{1}{2}\left[x^2 - \ln(1+e^{x^2})\right]_0^2$

$\displaystyle = \frac{1}{2}\left[4 - \ln(1+e^4)\right] + \frac{1}{2}\ln 2$

≈ 0.337 square units

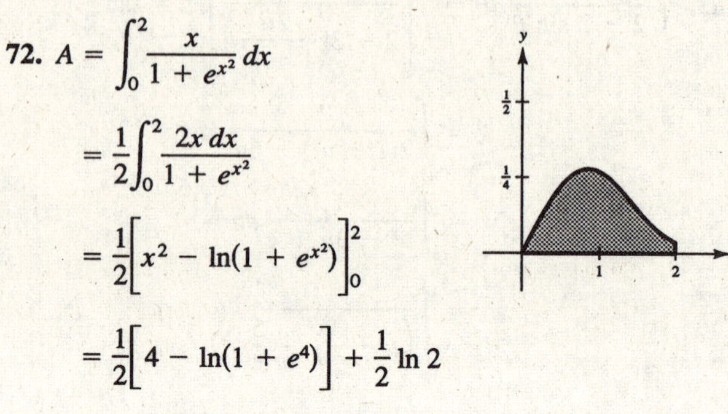

73. Arctangent Formula, Formula 23,

$\displaystyle\int \frac{1}{u^2+1}\,du,\ u = e^x$

74. Log Rule: $\displaystyle\int \frac{1}{u}\,du,\ u = e^x + 1$

75. Substitution: $u = x^2,\ du = 2x\,dx$

Then Formula 81.

76. Integration by parts

77. Cannot be integrated.

78. Formula 16 with $u = e^{2x}$

79. (a) $n = 1$: $\displaystyle u = \ln x,\ du = \frac{1}{x}\,dx,\ dv = x\,dx,\ v = \frac{x^2}{2}$

$$\int x\ln x\,dx = \frac{x^2}{2}\ln x - \int \frac{x^2}{2}\frac{1}{x}\,dx = \frac{x^2}{2}\ln x - \frac{x^2}{4} + C$$

$n = 2$: $\displaystyle u = \ln x,\ du = \frac{1}{x}\,dx,\ dv = x^2\,dx,\ v = \frac{x^3}{3}$

$$\int x^2\ln x\,dx = \frac{x^3}{3}\ln x - \int \frac{x^3}{3}\frac{1}{x}\,dx = \frac{x^3}{3}\ln x - \frac{x^3}{9} + C$$

$n = 3$: $\displaystyle u = \ln x,\ du = \frac{1}{x}\,dx,\ dv = x^3\,dx,\ v = \frac{x^4}{4}$

$$\int x^3\ln x\,dx = \frac{x^4}{4}\ln x - \int \frac{x^4}{4}\frac{1}{x}\,dx = \frac{x^4}{4}\ln x - \frac{x^4}{16} + C$$

(b) $\displaystyle\int x^n\ln x\,dx = \frac{x^{n+1}}{n+1}\ln x - \frac{x^{n+1}}{(n+1)^2} + C$

80. A reduction formula reduces an integral to the sum of a function and a simpler integral. For example, see Formula 50, 54.

81. False. You might need to convert your integral using substitution or algebra.

82. True

83. $W = \int_0^5 2000xe^{-x}\,dx$

$= -2000\int_0^5 -xe^{-x}\,dx$

$= 2000\int_0^5 (-x)e^{-x}(-1)\,dx$

$= 2000\Big[(-x)e^{-x} - e^{-x}\Big]_0^5$

$= 2000\left(-\dfrac{6}{e^5} + 1\right)$

≈ 1919.145 ft · lbs

84. $W = \int_0^5 \dfrac{500x}{\sqrt{26 - x^2}}\,dx$

$= -250\int_0^5 (26 - x^2)^{-1/2}(-2x)\,dx$

$= \Big[-500\sqrt{26 - x^2}\Big]_0^5$

$= 500\left(\sqrt{26} - 1\right)$

≈ 2049.51 ft · lbs

85. $V = 20(2)\int_0^3 \dfrac{2}{\sqrt{1 + y^2}}\,dy$

$= \Big[80\ln\left|y + \sqrt{1 + y^2}\right|\Big]_0^3$

$= 80\ln\left(3 + \sqrt{10}\right)$

≈ 145.5 cubic feet

$W = 148\left(80\ln\left(3 + \sqrt{10}\right)\right)$

$= 11{,}840\ln\left(3 + \sqrt{10}\right)$

$\approx 21{,}530.4$ lb

By symmetry, $\bar{x} = 0$.

$$M = \rho(2)\int_0^3 \frac{2}{\sqrt{1 + y^2}}\,dy = \Big[4\rho\ln\left|y + \sqrt{1 + y^2}\right|\Big]_0^3 = 4\rho\ln\left(3 + \sqrt{10}\right)$$

$$M_x = 2\rho\int_0^3 \frac{2y}{\sqrt{1 + y^2}}\,dy = \Big[4\rho\sqrt{1 + y^2}\Big]_0^3 = 4\rho\left(\sqrt{10} - 1\right)$$

$$\bar{y} = \frac{M_x}{M} = \frac{4\rho\left(\sqrt{10} - 1\right)}{4\rho\ln\left(3 + \sqrt{10}\right)} \approx 1.19$$

Centroid: $(\bar{x}, \bar{y}) \approx (0, 1.19)$

86. $\dfrac{1}{2 - 0}\int_0^2 \dfrac{5000}{1 + e^{4.8 - 1.9t}}\,dt = \dfrac{2500}{-1.9}\int_0^2 \dfrac{-1.9\,dt}{1 + e^{4.8 - 1.9t}}$

$= -\dfrac{2500}{1.9}\Big[(4.8 - 1.9t) - \ln(1 + e^{4.8 - 1.9t})\Big]_0^2$

$= -\dfrac{2500}{1.9}[(1 - \ln(1 + e)) - (4.8 - \ln(1 + e^{4.8}))]$

$= \dfrac{2500}{1.9}\left[3.8 + \ln\left(\dfrac{1 + e}{1 + e^{4.8}}\right)\right] \approx 401.4$

87. (a) $\int_0^4 \dfrac{k}{2 + 3x}\,dx = 10$

$k = \dfrac{10}{\int_0^4 \dfrac{1}{2 + 3x}\,dx} = \dfrac{10}{\frac{1}{3}\ln 7} \approx \dfrac{10}{0.6486}$

$= 15.417 \quad \left(= \dfrac{30}{\ln 7}\right)$

(b) $\int_0^4 \dfrac{15.417}{2 + 3x}\,dx$

88. (a) $\displaystyle\int_0^k 6x^2e^{-x/2}\,dx = 50$

By trial and error, $k = 5.51897$.

(b) $\displaystyle\int_0^{5.51897} 6x^2e^{-x/2}\,dx$

89. Let $I = \displaystyle\int_0^{\pi/2} \frac{dx}{1 + (\tan x)^{\sqrt{2}}}$.

For $x = \dfrac{\pi}{2} - u$, $dx = -du$, and

$$I = \int_{\pi/2}^{0} \frac{-du}{1 + (\tan(\pi/2 - u))^{\sqrt{2}}} = \int_0^{\pi/2} \frac{du}{1 + (\cot u)^{\sqrt{2}}} = \int_0^{\pi/2} \frac{(\tan u)^{\sqrt{2}}}{(\tan u)^{\sqrt{2}} + 1}\,du.$$

$$2I = \int_0^{\pi/2} \frac{dx}{1 + (\tan x)^{\sqrt{2}}} + \int_0^{\pi/2} \frac{(\tan x)^{\sqrt{2}}}{(\tan x)^{\sqrt{2}} + 1}\,dx = \int_0^{\pi/2} dx = \frac{\pi}{2}$$

Thus, $I = \dfrac{\pi}{4}$.

Section 8.7 Indeterminate Forms and L'Hôpital's Rule

1. $\displaystyle\lim_{x\to 0} \frac{\sin 5x}{\sin 2x} \approx 2.5 \left(\text{exact: } \frac{5}{2}\right)$

x	-0.1	-0.01	-0.001	0.001	0.01	0.1
$f(x)$	2.4132	2.4991	2.500	2.500	2.4991	2.4132

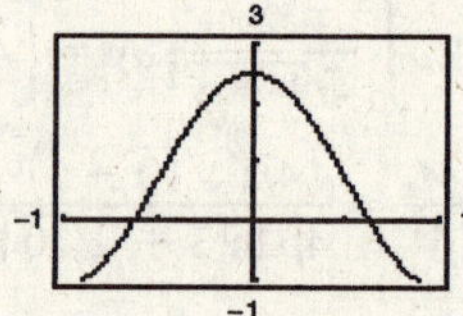

2. $\displaystyle\lim_{x\to 0} \frac{1 - e^x}{x} \approx -1$

x	-0.1	-0.01	-0.001	0.001	0.01	0.1
$f(x)$	-0.9516	-0.9950	-0.9995	-1.00005	-1.005	-1.0517

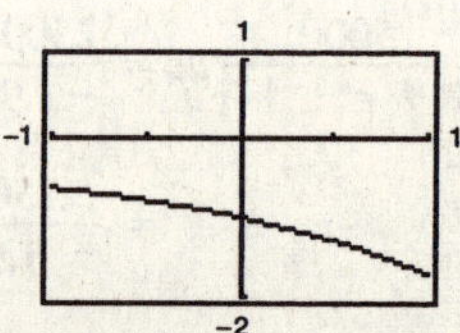

3. $\displaystyle\lim_{x\to\infty} x^5e^{-x/100} \approx 0$

x	1	10	10^2	10^3	10^4	10^5
$f(x)$	0.9900	90,484	3.7×10^9	4.5×10^{10}	0	0

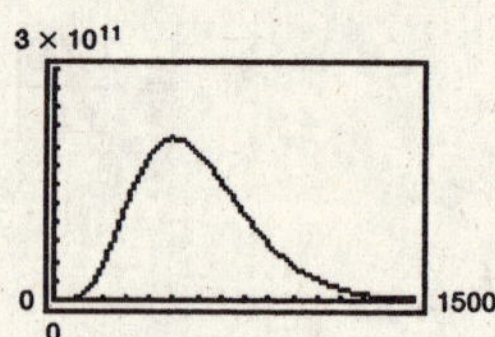

4. $\displaystyle\lim_{x\to\infty} \frac{6x}{\sqrt{3x^2 - 2x}} \approx 3.4641 \left(\text{exact: } \frac{6}{\sqrt{3}}\right)$

x	1	10	10^2	10^3	10^4	10^5
$f(x)$	6	3.5857	3.4757	3.4653	3.4642	3.4641

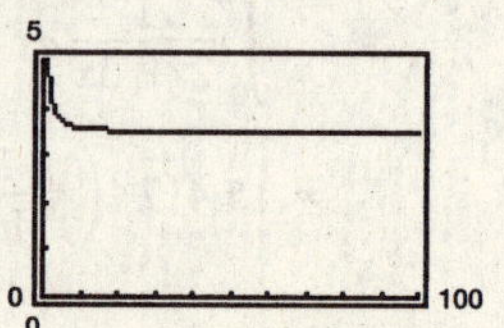

5. (a) $\lim_{x\to3} \frac{2(x-3)}{x^2-9} = \lim_{x\to3} \frac{2(x-3)}{(x+3)(x-3)} = \lim_{x\to3} \frac{2}{x+3} = \frac{1}{3}$

(b) $\lim_{x\to3} \frac{2(x-3)}{x^2-9} = \lim_{x\to3} \frac{(d/dx)[2(x-3)]}{(d/dx)[x^2-9]} = \lim_{x\to3} \frac{2}{2x} = \frac{2}{6} = \frac{1}{3}$

6. (a) $\lim_{x\to-1} \frac{2x^2-x-3}{x+1} = \lim_{x\to-1} \frac{(2x-3)(x+1)}{x+1} = \lim_{x\to-1} (2x-3) = -5$

(b) $\lim_{x\to-1} \frac{2x^2-x-3}{x+1} = \lim_{x\to-1} \frac{(d/dx)[2x^2-x-3]}{(d/dx)[x+1]} = \lim_{x\to-1} \frac{4x-1}{1} = -5$

7. (a) $\lim_{x\to3} \frac{\sqrt{x+1}-2}{x-3} = \lim_{x\to3} \frac{\sqrt{x+1}-2}{x-3} \cdot \frac{\sqrt{x+1}+2}{\sqrt{x+1}+2} = \lim_{x\to3} \frac{(x+1)-4}{(x-3)\left[\sqrt{x+1}+2\right]} = \lim_{x\to3} \frac{1}{\sqrt{x+1}+2} = \frac{1}{4}$

(b) $\lim_{x\to3} \frac{\sqrt{x+1}-2}{x-3} = \lim_{x\to3} \frac{(d/dx)\left[\sqrt{x+1}-2\right]}{(d/dx)[x-3]} = \lim_{x\to3} \frac{1/\left(2\sqrt{x+1}\right)}{1} = \frac{1}{4}$

8. (a) $\lim_{x\to0} \frac{\sin 4x}{2x} = \lim_{x\to0} 2\left(\frac{\sin 4x}{4x}\right) = 2(1) = 2$

(b) $\lim_{x\to0} \frac{\sin 4x}{2x} = \lim_{x\to0} \frac{(d/dx)[\sin 4x]}{(d/dx)[2x]} = \lim_{x\to0} \frac{4\cos 4x}{2} = 2$

9. (a) $\lim_{x\to\infty} \frac{5x^2-3x+1}{3x^2-5} = \lim_{x\to\infty} \frac{5-(3/x)+(1/x^2)}{3-(5/x^2)} = \frac{5}{3}$

(b) $\lim_{x\to\infty} \frac{5x^2-3x+1}{3x^2-5} = \lim_{x\to\infty} \frac{(d/dx)[5x^2-3x+1]}{(d/dx)[3x^2-5]} = \lim_{x\to\infty} \frac{10x-3}{6x} = \lim_{x\to\infty} \frac{(d/dx)[10x-3]}{(d/dx)[6x]} = \lim_{x\to\infty} \frac{10}{6} = \frac{5}{3}$

10. (a) $\lim_{x\to\infty} \frac{2x+1}{4x^2+x} = \lim_{x\to\infty} \frac{(2/x)+(1/x^2)}{4+(1/x)} = \frac{0}{4} = 0$

(b) $\lim_{x\to\infty} \frac{2x+1}{4x^2+x} = \lim_{x\to\infty} \frac{(d/dx)[2x+1]}{(d/dx)[4x^2+x]} = \lim_{x\to\infty} \frac{2}{8x+1} = 0$

11. $\lim_{x\to2} \frac{x^2-x-2}{x-2} = \lim_{x\to2} \frac{2x-1}{1} = 3$

12. $\lim_{x\to-1} \frac{x^2-x-2}{x+1} = \lim_{x\to-1} \frac{2x-1}{1} = -3$

13. $\lim_{x\to0} \frac{\sqrt{4-x^2}-2}{x} = \lim_{x\to0} \frac{-x/\sqrt{4-x^2}}{1} = 0$

14. $\lim_{x\to2^-} \frac{\sqrt{4-x^2}}{x-2} = \lim_{x\to2^-} \frac{-x/\sqrt{4-x^2}}{1}$

$= \lim_{x\to2^-} \frac{-x}{\sqrt{4-x^2}} = -\infty$

15. $\lim_{x\to0} \frac{e^x-(1-x)}{x} = \lim_{x\to0} \frac{e^x+1}{1} = 2$

16. $\lim_{x\to1} \frac{\ln x^2}{x^2-1} = \lim_{x\to1} \frac{2\ln x}{x^2-1}$

$= \lim_{x\to1} \frac{2/x}{2x}$

$= \lim_{x\to1} \frac{1}{x^2} = 1$

17. $\lim_{x\to0^+} \frac{e^x-(1+x)}{x^3} = \lim_{x\to0^+} \frac{e^x-1}{3x^2}$

$= \lim_{x\to0^+} \frac{e^x}{6x} = \infty$

18. Case 1: $n = 1$

$$\lim_{x\to 0^+} \frac{e^x - (1 + x)}{x} = \lim_{x\to 0^+} \frac{e^x - 1}{1} = 0$$

Case 2: $n = 2$

$$\lim_{x\to 0^+} \frac{e^x - (1 + x)}{x^2} = \lim_{x\to 0^+} \frac{e^x - 1}{2x} = \lim_{x\to 0^+} \frac{e^x}{2} = \frac{1}{2}$$

Case 3: $n \geq 3$

$$\lim_{x\to 0^+} \frac{e^x - (1 + x)}{x^n} = \lim_{x\to 0^+} \frac{e^x - 1}{nx^{n-1}} = \lim_{x\to 0^+} \frac{e^x}{n(n-1)x^{n-2}} = \infty$$

19. $\lim_{x\to 0} \frac{\sin 2x}{\sin 3x} = \lim_{x\to 0} \frac{2\cos 2x}{3\cos 3x} = \frac{2}{3}$

20. $\lim_{x\to 0} \frac{\sin ax}{\sin bx} = \lim_{x\to 0} \frac{a\cos ax}{b\cos bx} = \frac{a}{b}$

21. $\lim_{x\to 0} \frac{\arcsin x}{x} = \lim_{x\to 0} \frac{1/\sqrt{1-x^2}}{1} = 1$

22. $\lim_{x\to 1} \frac{\arctan x - (\pi/4)}{x - 1} = \lim_{x\to 1} \frac{1/(1 + x^2)}{1} = \frac{1}{2}$

23. $\lim_{x\to\infty} \frac{3x^2 - 2x + 1}{2x^2 + 3} = \lim_{x\to\infty} \frac{6x - 2}{4x}$

$$= \lim_{x\to\infty} \frac{6}{4} = \frac{3}{2}$$

24. $\lim_{x\to\infty} \frac{x - 1}{x^2 + 2x + 3} = \lim_{x\to\infty} \frac{1}{2x + 2} = 0$

25. $\lim_{x\to\infty} \frac{x^2 + 2x + 3}{x - 1} = \lim_{x\to\infty} \frac{2x + 2}{1} = \infty$

26. $\lim_{x\to\infty} \frac{x^3}{x + 1} = \lim_{x\to\infty} \frac{3x^2}{1} = \infty$

27. $\lim_{x\to\infty} \frac{x^3}{e^{x/2}} = \lim_{x\to\infty} \frac{3x^2}{(1/2)e^{x/2}}$

$$= \lim_{x\to\infty} \frac{6x}{(1/4)e^{x/2}} = \lim_{x\to\infty} \frac{6}{(1/8)e^{x/2}} = 0$$

28. $\lim_{x\to\infty} \frac{x^2}{e^x} = \lim_{x\to\infty} \frac{2x}{e^x} = \lim_{x\to\infty} \frac{2}{e^x} = 0$

29. $\lim_{x\to\infty} \frac{x}{\sqrt{x^2 + 1}} = \lim_{x\to\infty} \frac{1}{\sqrt{1 + (1/x^2)}} = 1$

Note: L'Hôpital's Rule does not work on this limit. See Exercise 79.

30. $\lim_{x\to\infty} \frac{x^2}{\sqrt{x^2 + 1}} = \lim_{x\to\infty} \frac{x}{\sqrt{1 + (1/x)^2}} = \infty$

31. $\lim_{x\to\infty} \frac{\cos x}{x} = 0$ by Squeeze Theorem

$$\left(\frac{\cos x}{x} \leq \frac{1}{x}, \text{ for } x > 0\right)$$

32. $\lim_{x\to\infty} \frac{\sin x}{x - \pi} = 0$

Note: Use the Squeeze Theorem for $x > \pi$.

$$-\frac{1}{x - \pi} \leq \frac{\sin x}{x - \pi} \leq \frac{1}{x - \pi}$$

33. $\lim_{x\to\infty} \frac{\ln x}{x^2} = \lim_{x\to\infty} \frac{1/x}{2x} = \lim_{x\to\infty} \frac{1}{2x^2} = 0$

34. $\lim_{x\to\infty} \frac{\ln x^4}{x^3} = \lim_{x\to\infty} \frac{4 \ln x}{x^3} = \lim_{x\to\infty} \frac{4/x}{3x^2}$

$$= \lim_{x\to\infty} \frac{4}{3x^3} = 0$$

35. $\lim_{x\to\infty} \frac{e^x}{x^2} = \lim_{x\to\infty} \frac{e^x}{2x} = \lim_{x\to\infty} \frac{e^x}{2} = \infty$

36. $\lim_{x\to\infty} \frac{e^{x/2}}{x} = \lim_{x\to\infty} \frac{(1/2)e^{x/2}}{1} = \infty$

37. (a) $\lim_{x\to\infty} x\ln x$, not indeterminate

(b) $\lim_{x\to\infty} x\ln x = (\infty)(\infty) = \infty$

(c)

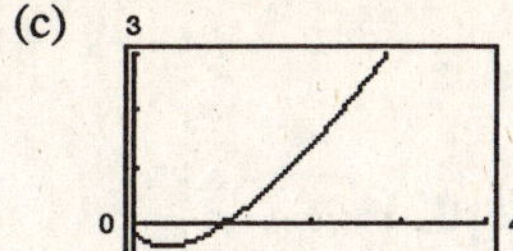

38. (a) $\lim_{x\to 0^+} x^3 \cot x = (0)(\infty)$

(b) $\lim_{x\to 0^+} x^3 \cot x = \lim_{x\to 0^+} \dfrac{x^3}{\tan x} = \lim_{x\to 0^+} \dfrac{3x^2}{\sec^2 x} = 0$

(c)

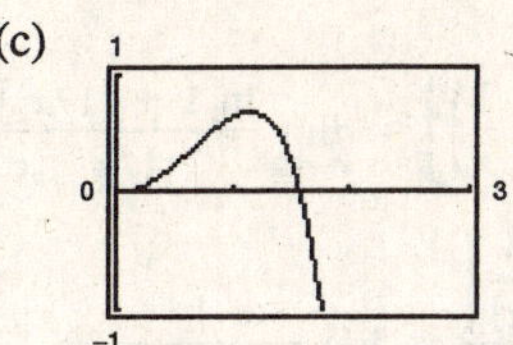

39. (a) $\lim_{x\to\infty} \left(x \sin \dfrac{1}{x}\right) = (\infty)(0)$

(b) $\lim_{x\to\infty} x \sin \dfrac{1}{x} = \lim_{x\to\infty} \dfrac{\sin(1/x)}{1/x}$

$= \lim_{x\to\infty} \dfrac{(-1/x^2)\cos(1/x)}{-1/x^2}$

$= \lim_{x\to\infty} \cos\left(\dfrac{1}{x}\right) = 1$

(c)

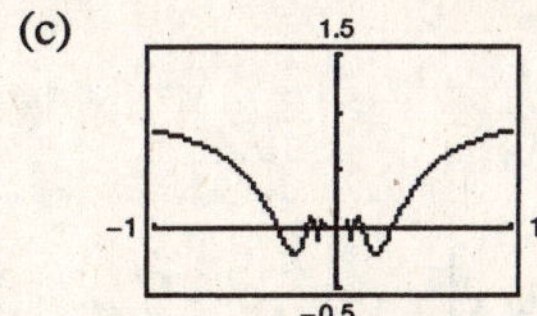

40. (a) $\lim_{x\to\infty} \left(x \tan \dfrac{1}{x}\right) = (\infty)(0)$

(b) $\lim_{x\to\infty} x \tan \dfrac{1}{x} = \lim_{x\to\infty} \dfrac{\tan(1/x)}{1/x}$

$= \lim_{x\to\infty} \dfrac{-(1/x^2)\sec^2(1/x)}{-(1/x^2)}$

$= \lim_{x\to\infty} \sec^2\left(\dfrac{1}{x}\right) = 1$

(c)

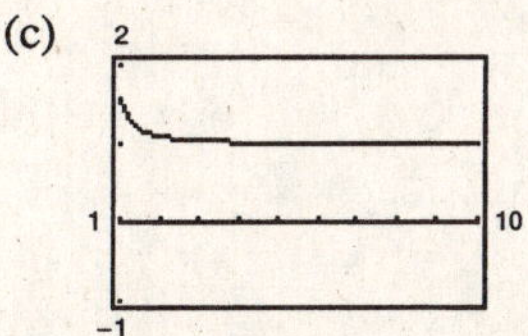

41. (a) $\lim_{x\to 0^+} x^{1/x} = 0^{\infty} = 0$, not indeterminate (See Exercise 106).

(b) Let $y = x^{1/x}$

$\ln y = \ln x^{1/x} = \dfrac{1}{x}\ln x.$

Since $x \to 0^+$, $\dfrac{1}{x}\ln x \to (\infty)(-\infty) = -\infty$. Hence,

$\ln y \to -\infty \implies y \to 0^+.$

Therefore, $\lim_{x\to 0^+} x^{1/x} = 0.$

(c)

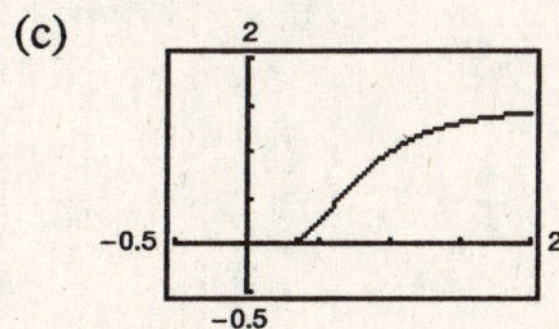

42. (a) $\lim_{x\to 0^+} (e^x + x)^{2/x} = 1^{\infty}$

(b) Let $y = \lim_{x\to 0^+} (e^x + x)^{2/x}$.

$\ln y = \lim_{x\to 0^+} \dfrac{2\ln(e^x + x)}{x}$

$= \lim_{x\to 0^+} \dfrac{2(e^x + 1)/(e^x + x)}{1} = 4$

Thus, $\ln y = 4 \implies y = e^4 \approx 54.598$.

(c)

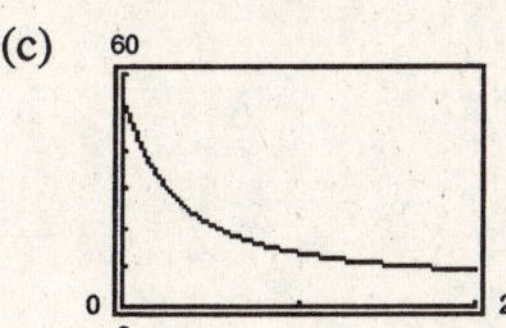

43. (a) $\lim_{x\to\infty} x^{1/x} = \infty^0$

(b) Let $y = \lim_{x\to\infty} x^{1/x}$.

$\ln y = \lim_{x\to\infty} \dfrac{\ln x}{x} = \lim_{x\to\infty} \left(\dfrac{1/x}{1}\right) = 0$

Thus, $\ln y = 0 \implies y = e^0 = 1$. Therefore,

$\lim_{x\to\infty} x^{1/x} = 1.$

(c)

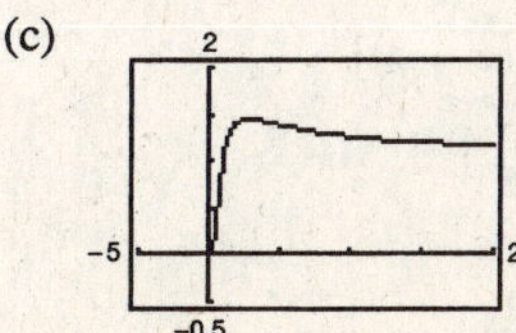

44. (a) $\lim_{x\to\infty}\left(1+\frac{1}{x}\right)^x = 1^\infty$

(b) Let $y = \lim_{x\to\infty}\left(1+\frac{1}{x}\right)^x$.

$$\ln y = \lim_{x\to\infty}\left[x\ln\left(1+\frac{1}{x}\right)\right] = \lim_{x\to\infty}\frac{\ln[1+(1/x)]}{1/x}$$

$$= \lim_{x\to\infty}\frac{\left[\frac{(-1/x^2)}{1+(1/x)}\right]}{(-1/x^2)} = \lim_{x\to\infty}\frac{1}{1+(1/x)} = 1$$

Thus, $\ln y = 1 \Rightarrow y = e^1 = e$. Therefore,

$$\lim_{x\to\infty}\left(1+\frac{1}{x}\right)^x = e.$$

(c)

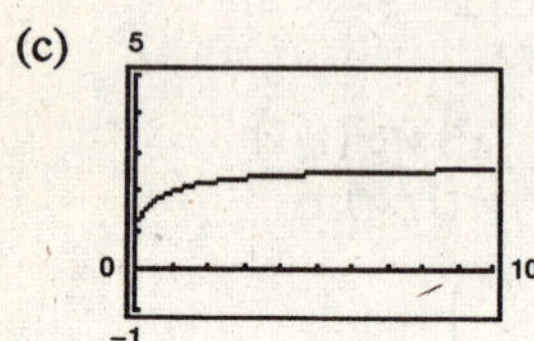

45. (a) $\lim_{x\to0^+}(1+x)^{1/x} = 1^\infty$

(b) Let $y = \lim_{x\to0^+}(1+x)^{1/x}$.

$$\ln y = \lim_{x\to0^+}\frac{\ln(1+x)}{x}$$

$$= \lim_{x\to0^+}\left(\frac{1/(1+x)}{1}\right) = 1$$

Thus, $\ln y = 1 \Rightarrow y = e^1 = e$.

Therefore, $\lim_{x\to0^+}(1+x)^{1/x} = e$.

(c)

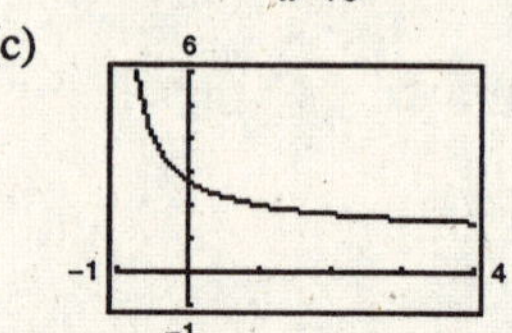

46. (a) $\lim_{x\to\infty}(1+x)^{1/x} = \infty^0$

(b) Let $y = \lim_{x\to\infty}(1+x)^{1/x}$.

$$\ln y = \lim_{x\to\infty}\frac{\ln(1+x)}{x}$$

$$= \lim_{x\to\infty}\left(\frac{1/(1+x)}{1}\right) = 0$$

Thus, $\ln y = 0 \Rightarrow y = e^0 = 1$.
Therefore, $\lim_{x\to\infty}(1+x)^{1/x} = 1$.

(c)

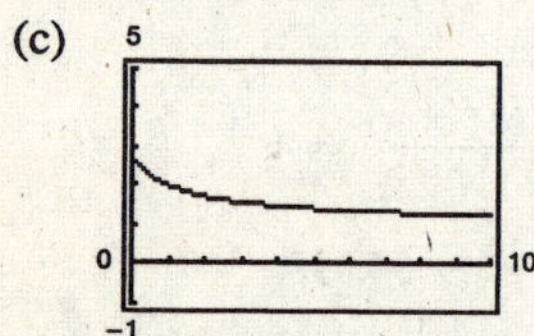

47. (a) $\lim_{x\to0^+}[3(x)^{x/2}] = 0^0$

(b) Let $y = \lim_{x\to0^+}3(x)^{x/2}$.

$$\ln y = \lim_{x\to0^+}\left[\ln 3 + \frac{x}{2}\ln x\right]$$

$$= \lim_{x\to0^+}\left[\ln 3 + \frac{\ln x}{2/x}\right]$$

$$= \lim_{x\to0^+}\ln 3 + \lim_{x\to0^+}\frac{1/x}{-2/x^2}$$

$$= \lim_{x\to0^+}\ln 3 - \lim_{x\to0^+}\frac{x}{2}$$

$$= \ln 3$$

Hence, $\lim_{x\to0^+}3(x)^{x/2} = 3$.

(c)

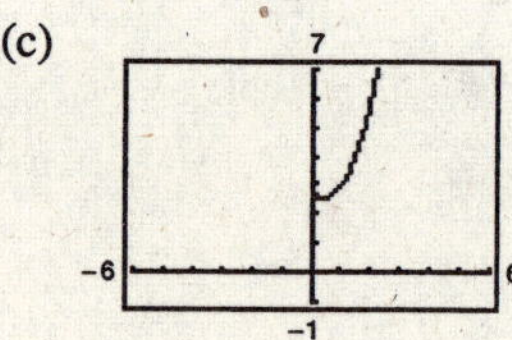

48. (a) $\lim_{x\to4^+}[3(x-4)]^{x-4} = 0^0$

(b) Let $y = \lim_{x\to4^+}[3(x-4)]^{x-4}$.

$$\ln y = \lim_{x\to4^+}(x-4)\ln[3(x-4)]$$

$$= \lim_{x\to4^+}\frac{\ln[3(x-4)]}{1/(x-4)}$$

$$= \lim_{x\to4^+}\frac{1/(x-4)}{-1/(x-4)^2}$$

$$= \lim_{x\to4^+}[-(x-4)] = 0$$

Hence, $\lim_{x\to4^+}[3(x-4)]^{x-4} = 1$.

(c)

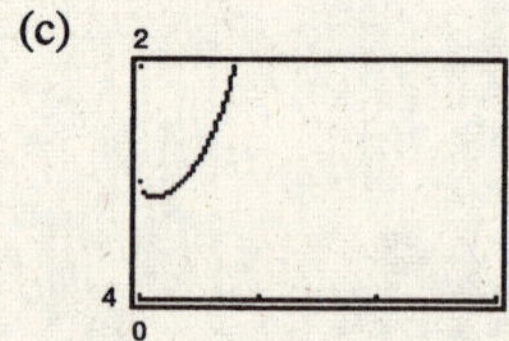

49. (a) $\lim_{x\to1^+} (\ln x)^{x-1} = 0^0$

(b) Let $y = \lim_{x\to1^+} (\ln x)^{x-1}$

$= \lim_{x\to1^+} (x-1)\ln x = 0.$

Hence, $\lim_{x\to1^+} (\ln x)^{x-1} = 1.$

(c)

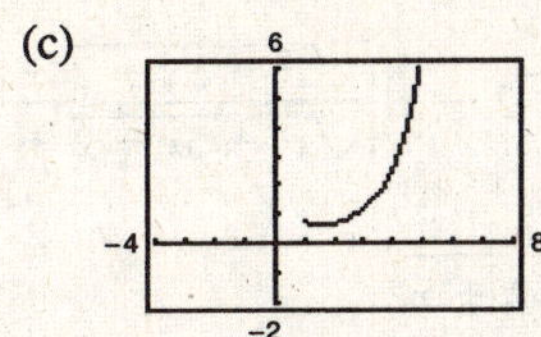

50. (a) $\lim_{x\to0^+} \left[\cos\left(\frac{\pi}{2} - x\right)\right]^x = 0^0$

(b) Let $y = \lim_{x\to0^+} \left[\cos\left(\frac{\pi}{2} - x\right)\right]^x.$

$\ln y = \lim_{x\to0^+} x \ln\left[\cos\left(\frac{\pi}{2} - x\right)\right]$

$= 0 \cdot 0 = 0$

Hence, $\lim_{x\to0^+} \left[\cos\left(\frac{\pi}{2} - x\right)\right]^x = 1.$

(c)

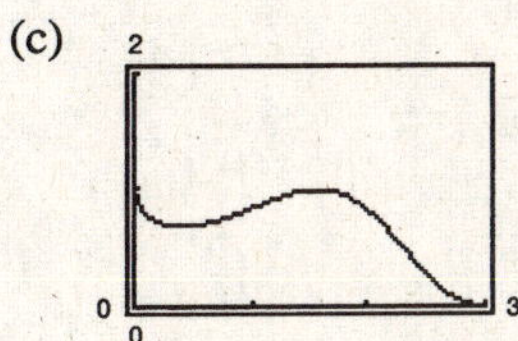

51. (a) $\lim_{x\to2^+} \left(\frac{8}{x^2-4} - \frac{x}{x-2}\right) = \infty - \infty$

(b) $\lim_{x\to2^+} \left(\frac{8}{x^2-4} - \frac{x}{x-2}\right) = \lim_{x\to2^+} \frac{8 - x(x+2)}{x^2-4}$

$= \lim_{x\to2^+} \frac{(2-x)(4+x)}{(x+2)(x-2)}$

$= \lim_{x\to2^+} \frac{-(x+4)}{x+2} = \frac{-3}{2}$

(c)

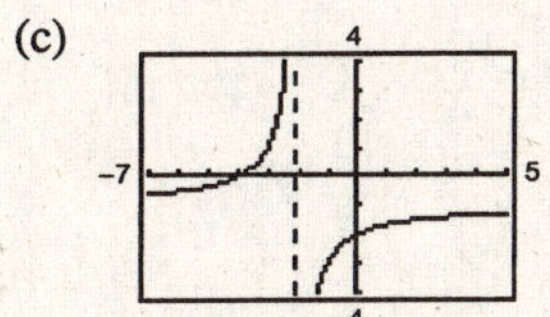

52. (a) $\lim_{x\to2^+} \left(\frac{1}{x^2-4} - \frac{\sqrt{x-1}}{x^2-4}\right) = \infty - \infty$

(b) $\lim_{x\to2^+} \left(\frac{1}{x^2-4} - \frac{\sqrt{x-1}}{x^2-4}\right) = \lim_{x\to2^+} \frac{1 - \sqrt{x-1}}{x^2-4}$

$= \lim_{x\to2^+} \frac{-1/(2\sqrt{x-1})}{2x}$

$= \lim_{x\to2^+} \frac{-1}{4x\sqrt{x-1}} = \frac{-1}{8}$

(c)

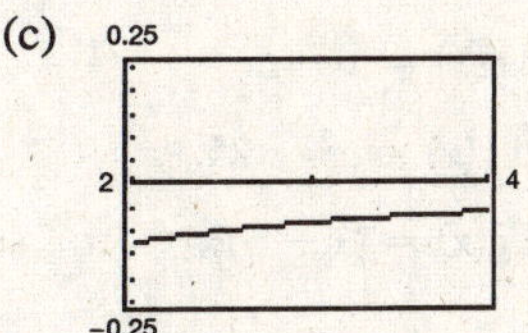

53. (a) $\lim_{x\to1^+} \left(\frac{3}{\ln x} - \frac{2}{x-1}\right) = \infty - \infty$

(b) $\lim_{x\to1^+} \left(\frac{3}{\ln x} - \frac{2}{x-1}\right) = \lim_{x\to1^+} \frac{3x - 3 - 2\ln x}{(x-1)\ln x}$

$= \lim_{x\to1^+} \frac{3 - (2/x)}{[(x-1)/x] + \ln x} = \infty$

(c)

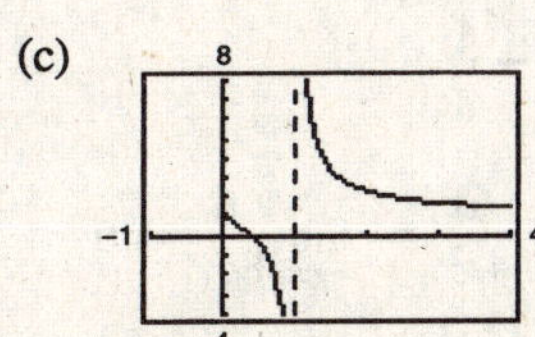

54. (a) $\lim_{x\to0^+} \left(\frac{10}{x} - \frac{3}{x^2}\right) = \infty - \infty$

(b) $\lim_{x\to0^+} \left(\frac{10}{x} - \frac{3}{x^2}\right) = \lim_{x\to0^+} \left(\frac{10x - 3}{x^2}\right) = -\infty$

(c)

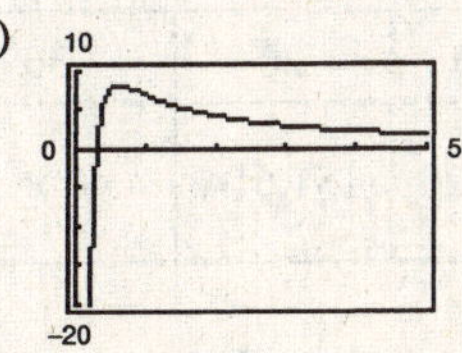

55. (a)

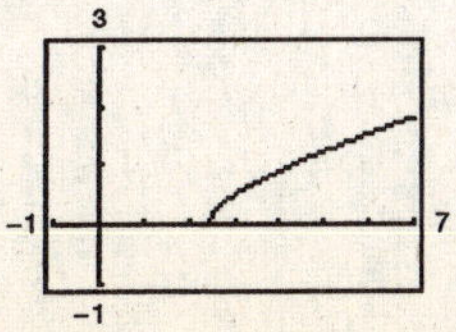

(b) $\lim_{x\to3} \frac{x-3}{\ln(2x-5)} = \lim_{x\to3} \frac{1}{2/(2x-5)}$

$= \lim_{x\to3} \frac{2x-5}{2} = \frac{1}{2}$

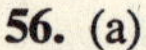

56. (a)

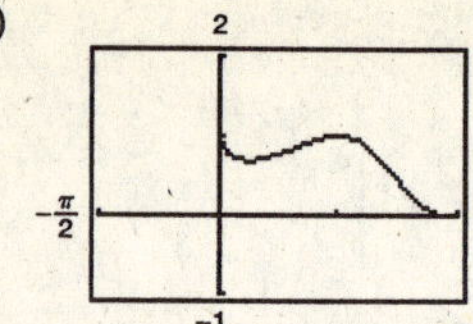

(b) Let $y = (\sin x)^x$, then $\ln y = x \ln(\sin x)$.

$$\lim_{x\to 0^+} \frac{\ln(\sin x)}{1/x} = \lim_{x\to 0^+} \frac{\cos x/\sin x}{-1/x^2} = \lim_{x\to 0^+} \frac{-x^2}{\tan x} = \lim_{x\to 0^+} \frac{-2x}{\sec^2 x} = 0$$

Therefore, since $\ln y = 0$, $y = 1$ and $\lim_{x\to 0^+} (\sin x)^x = 1$.

57. (a)

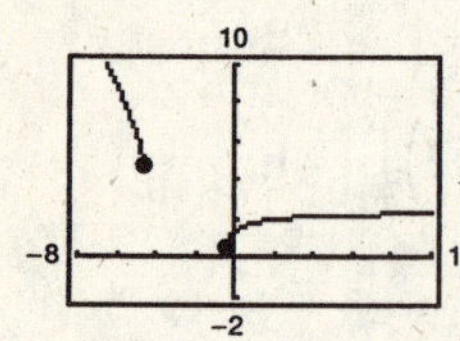

(b)
$$\lim_{x\to\infty}\left(\sqrt{x^2+5x+2} - x\right) = \lim_{x\to\infty}\left(\sqrt{x^2+5x+2} - x\right)\frac{\left(\sqrt{x^2+5x+2}+x\right)}{\left(\sqrt{x^2+5x+2}+x\right)}$$

$$= \lim_{x\to\infty} \frac{(x^2+5x+2) - x^2}{\sqrt{x^2+5x+2}+x}$$

$$= \lim_{x\to\infty} \frac{5x+2}{\sqrt{x^2+5x+2}+x}$$

$$= \lim_{x\to\infty} \frac{5+(2/x)}{\sqrt{1+(5/x)+(2/x^2)}+1} = \frac{5}{2}$$

58. (a)

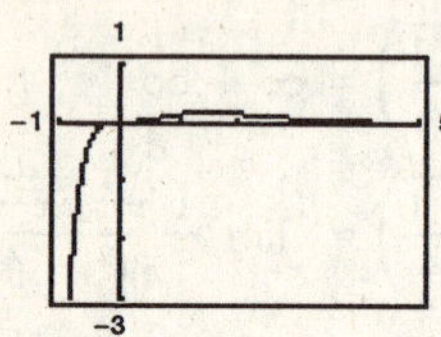

(b) $\displaystyle \lim_{x\to\infty} \frac{x^3}{e^{2x}} = \lim_{x\to\infty} \frac{3x^2}{2e^{2x}} = \lim_{x\to\infty} \frac{6x}{4e^{2x}} = \lim_{x\to\infty} \frac{6}{8e^{2x}} = 0$

59. $\frac{0}{0}, \frac{\infty}{\infty}, 0 \cdot \infty, 1^\infty, 0^0, \infty - \infty, \infty^0$

60. See Theorem 8.4.

61. (a) Let $f(x) = x^2 - 25$ and $g(x) = x - 5$.

(b) Let $f(x) = (x-5)^2$ and $g(x) = x^2 - 25$.

(c) Let $f(x) = x^2 - 25$ and $g(x) = (x-5)^3$.

62. Let $f(x) = x + 25$ and $g(x) = x$.

63.

x	10	10^2	10^4	10^6	10^8	10^{10}
$\frac{(\ln x)^4}{x}$	2.811	4.498	0.720	0.036	0.001	0.000

64.

x	1	5	10	20	30	40	50	100
$\frac{e^x}{x^5}$	2.718	0.047	0.220	151.614	4.40×10^5	2.30×10^9	1.66×10^{13}	2.69×10^{33}

65. $\displaystyle \lim_{x\to\infty} \frac{x^2}{e^{5x}} = \lim_{x\to\infty} \frac{2x}{5e^{5x}} = \lim_{x\to\infty} \frac{2}{25e^{5x}} = 0$

66. $\displaystyle \lim_{x\to\infty} \frac{x^3}{e^{2x}} = \lim_{x\to\infty} \frac{3x^2}{2e^{2x}} = \lim_{x\to\infty} \frac{6x}{4e^{2x}} = \lim_{x\to\infty} \frac{6}{8e^{2x}} = 0$

67. $\displaystyle\lim_{x\to\infty}\frac{(\ln x)^3}{x} = \lim_{x\to\infty}\frac{3(\ln x)^2(1/x)}{1}$

$\displaystyle = \lim_{x\to\infty}\frac{3(\ln x)^2}{x}$

$\displaystyle = \lim_{x\to\infty}\frac{6(\ln x)(1/x)}{1}$

$\displaystyle = \lim_{x\to\infty}\frac{6(\ln x)}{x} = \lim_{x\to\infty}\frac{6}{x} = 0$

68. $\displaystyle\lim_{x\to\infty}\frac{(\ln x)^2}{x^3} = \lim_{x\to\infty}\frac{(2\ln x)/x}{3x^2}$

$\displaystyle = \lim_{x\to\infty}\frac{2\ln x}{3x^3}$

$\displaystyle = \lim_{x\to\infty}\frac{2/x}{9x^2} = \lim_{x\to\infty}\frac{2}{9x^3} = 0$

69. $\displaystyle\lim_{x\to\infty}\frac{(\ln x)^n}{x^m} = \lim_{x\to\infty}\frac{n(\ln x)^{n-1}/x}{mx^{m-1}}$

$\displaystyle = \lim_{x\to\infty}\frac{n(\ln x)^{n-1}}{mx^m}$

$\displaystyle = \lim_{x\to\infty}\frac{n(n-1)(\ln x)^{n-2}}{m^2x^m}$

$\displaystyle = \cdots = \lim_{x\to\infty}\frac{n!}{m^nx^m} = 0$

70. $\displaystyle\lim_{x\to\infty}\frac{x^m}{e^{nx}} = \lim_{x\to\infty}\frac{mx^{m-1}}{ne^{nx}}$

$\displaystyle = \lim_{x\to\infty}\frac{m(m-1)x^{m-2}}{n^2e^{nx}}$

$\displaystyle = \cdots = \lim_{x\to\infty}\frac{m!}{n^me^{nx}} = 0$

71. $y = x^{1/x}, x > 0$

Horizontal asymptote: $y = 1$ (See Exercise 43.)

$\displaystyle \ln y = \frac{1}{x}\ln x$

$\displaystyle \frac{1}{y}\frac{dy}{dx} = \frac{1}{x}\left(\frac{1}{x}\right) + (\ln x)\left(-\frac{1}{x^2}\right)$

$\displaystyle \frac{dy}{dx} = x^{1/x}\left(\frac{1}{x^2}\right)(1-\ln x) = x^{(1/x)-2}(1-\ln x) = 0$

Critical number: $x = e$

Intervals:	$(0, e)$	(e, ∞)
Sign of dy/dx:	$+$	$-$
$y = f(x)$:	Increasing	Decreasing

Relative maximum: $(e, e^{1/e})$

72. $y = x^x, x > 0$

$\displaystyle\lim_{x\to\infty} x^x = \infty$ and $\displaystyle\lim_{x\to 0^+} x^x = 1$

No horizontal asymptotes

$\ln y = x\ln x$

$\displaystyle \frac{1}{y}\frac{dy}{dx} = x\left(\frac{1}{x}\right) + \ln x$

$\displaystyle \frac{dy}{dx} = x^x(1+\ln x) = 0$

Critical number: $x = e^{-1}$

Intervals:	$(0, e^{-1})$	$(e^{-1}, 0)$
Sign of dy/dx:	$-$	$+$
$y = f(x)$:	Decreasing	Increasing

Relative minimum: $\displaystyle (e^{-1}, (e^{-1})^{e^{-1}}) = \left(\frac{1}{e}, \left(\frac{1}{e}\right)^{1/e}\right)$

73. $y = 2xe^{-x}$

$\displaystyle\lim_{x\to\infty}\frac{2x}{e^x} = \lim_{x\to\infty}\frac{2}{e^x} = 0$

Horizontal asymptote: $y = 0$

$\displaystyle \frac{dy}{dx} = 2x(-e^{-x}) + 2e^{-x}$

$= 2e^{-x}(1-x) = 0$

Critical number: $x = 1$

Intervals:	$(-\infty, 1)$	$(1, \infty)$
Sign of dy/dx:	$+$	$-$
$y = f(x)$:	Increasing	Decreasing

Relative maximum: $\displaystyle\left(1, \frac{2}{e}\right)$

74. $\displaystyle y = \frac{\ln x}{x}$

Horizontal asymptote: $y = 0$ (See Exercise 29.)

$\displaystyle \frac{dy}{dx} = \frac{x(1/x) - (\ln x)(1)}{x^2} = \frac{1-\ln x}{x^2} = 0$

Critical number: $x = e$

Intervals:	$(0, e)$	(e, ∞)
Sign of dy/dx:	$+$	$-$
$y = f(x)$:	Increasing	Decreasing

Relative maximum: $\displaystyle\left(e, \frac{1}{e}\right)$

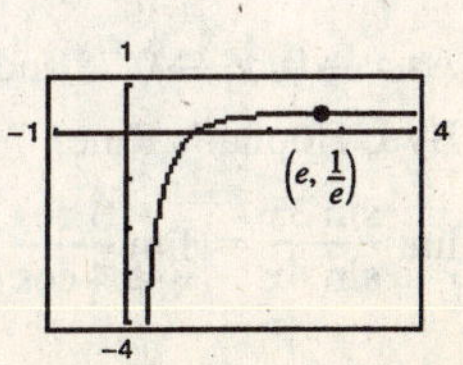

75. $\lim_{x\to 0} \frac{e^{2x}-1}{e^x} = \frac{0}{1} = 0$

Limit is not of the form $0/0$ or ∞/∞.
L'Hôpital's Rule does not apply.

76. $\lim_{x\to\infty} \frac{\sin \pi x - 1}{x} = 0$ (Numerator is bounded)

Limit is not of the form $0/0$ or ∞/∞.
L'Hôpital's Rule does not apply.

77. $\lim_{x\to\infty} x \cos \frac{1}{x} = \infty(1) = \infty$

Limit is not of the form $0/0$ or ∞/∞.
L'Hôpital's Rule does not apply.

78. $\lim_{x\to\infty} \frac{e^{-x}}{1+e^{-x}} = \frac{0}{1+0} = 0$

Limit is not of the form $0/0$ or ∞/∞.
L'Hôpital's Rule does not apply.

79. (a) Applying L'Hôpital's Rule twice results in the original limit, so L'Hôpital's Rule fails:

$$\lim_{x\to\infty} \frac{x}{\sqrt{x^2+1}} = \lim_{x\to\infty} \frac{1}{x/\sqrt{x^2+1}}$$
$$= \lim_{x\to\infty} \frac{\sqrt{x^2+1}}{x}$$
$$= \lim_{x\to\infty} \frac{x/\sqrt{x^2+1}}{1}$$
$$= \lim_{x\to\infty} \frac{x}{\sqrt{x^2+1}}$$

(b) $$\lim_{x\to\infty} \frac{x}{\sqrt{x^2+1}} = \lim_{x\to\infty} \frac{x/x}{\sqrt{x^2+1}/x}$$
$$= \lim_{x\to\infty} \frac{1}{\sqrt{1+1/x^2}}$$
$$= \frac{1}{\sqrt{1+0}} = 1$$

(c)

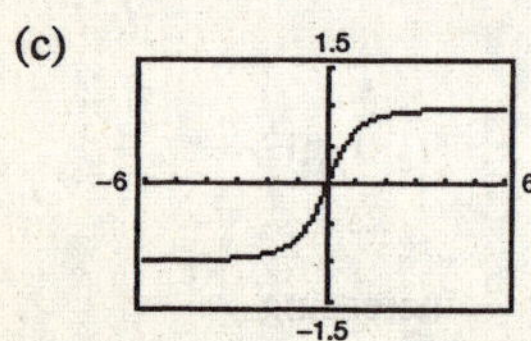

80. (a) $\lim_{x\to\pi/2^-} \frac{\tan x}{\sec x}$ is indeterminate: $\frac{\infty}{\infty}$

$$\lim_{x\to\pi/2^-} \frac{\tan x}{\sec x} = \lim_{x\to\pi/2^-} \frac{\sec^2 x}{\sec x \tan x}$$
$$= \lim_{x\to\pi/2^-} \frac{\sec x}{\tan x} \quad \left(\frac{\infty}{\infty}\right)$$
$$= \lim_{x\to\pi/2^-} \frac{\sec x \tan x}{\sec^2 x}$$
$$= \lim_{x\to\pi/2^-} \frac{\tan x}{\sec x}, \text{ the original problem!}$$

(b) $$\lim_{x\to\pi/2^-} \frac{\tan x}{\sec x} = \lim_{x\to\pi/2^-} \frac{\sin x}{\cos x}(\cos x)$$
$$= \lim_{x\to\pi/2^-} \sin x = 1$$

(c)

2
0
π
−1

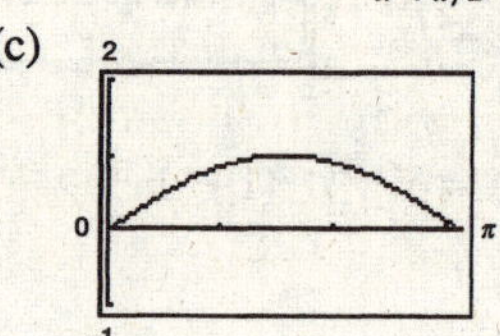

81. $f(x) = \sin(3x)$, $g(x) = \sin(4x)$

$f'(x) = 3\cos(3x)$, $g'(x) = 4\cos(4x)$

$y_1 = \frac{f(x)}{g(x)} = \frac{\sin 3x}{\sin 4x}$, $y_2 = \frac{f'(x)}{g'(x)} = \frac{3\cos 3x}{4\cos 4x}$

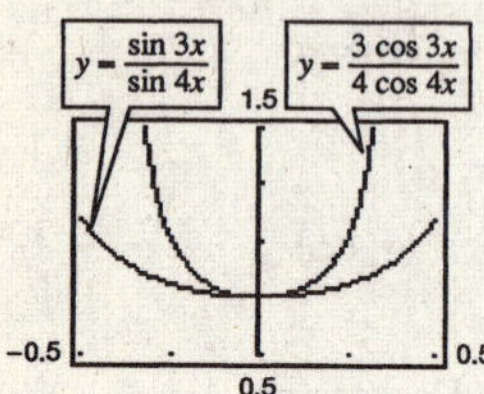

As $x \to 0$, $y_1 \to 0.75$ and $y_2 \to 0.75$
By L'Hôpital's Rule,

$$\lim_{x\to 0} \frac{\sin 3x}{\sin 4x} = \lim_{x\to 0} \frac{3\cos 3x}{4\cos 4x} = \frac{3}{4}$$

82. $f(x) = e^{3x} - 1$, $g(x) = x$

$f'(x) = 3e^{3x}$, $g'(x) = 1$

$y_1 = \frac{f(x)}{g(x)} = \frac{e^{3x}-1}{x}$, $y_2 = \frac{f'(x)}{g'(x)} = 3e^{3x}$

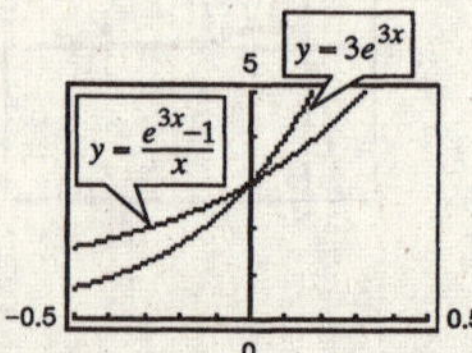

As $x \to 0$, $y_1 \to 3$ and $y_2 \to 3$
By L'Hôpital's Rule,

$$\lim_{x\to 0} \frac{e^{3x}-1}{x} = \lim_{x\to 0} \frac{3e^{3x}}{1} = 3$$

83. $\lim_{k\to 0} \dfrac{32\left(1 - e^{-kt} + \dfrac{v_0 k e^{-kt}}{32}\right)}{k} = \lim_{k\to 0} \dfrac{32(1 - e^{-kt})}{k} + \lim_{k\to 0} (v_0 e^{-kt})$

$= \lim_{k\to 0} \dfrac{32(0 + te^{-kt})}{1} + \lim_{k\to 0} \left(\dfrac{v_0}{e^{kt}}\right) = 32t + v_0$

84. $A = P\left(1 + \dfrac{r}{n}\right)^{nt}$

$\ln A = \ln P + nt \ln\left(1 + \dfrac{r}{n}\right) = \ln P + \dfrac{\ln\left(1 + \dfrac{r}{n}\right)}{\dfrac{1}{nt}}$

$$\lim_{n\to\infty} \left[\frac{\ln\left(1 + \frac{r}{n}\right)}{\frac{1}{nt}}\right] = \lim_{n\to\infty} \left[\frac{-\frac{r}{n^2}\left(\frac{1}{1 + (r/n)}\right)}{-\left(\frac{1}{n^2 t}\right)}\right] = \lim_{n\to\infty} \left[rt\left(\frac{1}{1 + \frac{r}{n}}\right)\right] = rt$$

Since $\lim_{n\to\infty} \ln A = \ln P + rt$, we have $\lim_{n\to\infty} A = e^{(\ln P + rt)} = e^{\ln P} e^{rt} = Pe^{rt}$. Alternatively,

$$\lim_{n\to\infty} A = \lim_{n\to\infty} P\left(1 + \frac{r}{n}\right)^{nt} = \lim_{n\to\infty} P\left[\left(1 + \frac{r}{n}\right)^{n/r}\right]^{rt} = Pe^{rt}.$$

85. Let N be a fixed value for n. Then

$$\lim_{x\to\infty} \frac{x^{N-1}}{e^x} = \lim_{x\to\infty} \frac{(N-1)x^{N-2}}{e^x} = \lim_{x\to\infty} \frac{(N-1)(N-2)x^{N-3}}{e^x} = \ldots = \lim_{x\to\infty} \left[\frac{(N-1)!}{e^x}\right] = 0.$$ (See Exercise 70.)

86. (a) $m = \dfrac{dy}{dx} = \dfrac{y - \left(y + \sqrt{144 - x^2}\right)}{x - 0}$

$= -\dfrac{\sqrt{144 - x^2}}{x}$

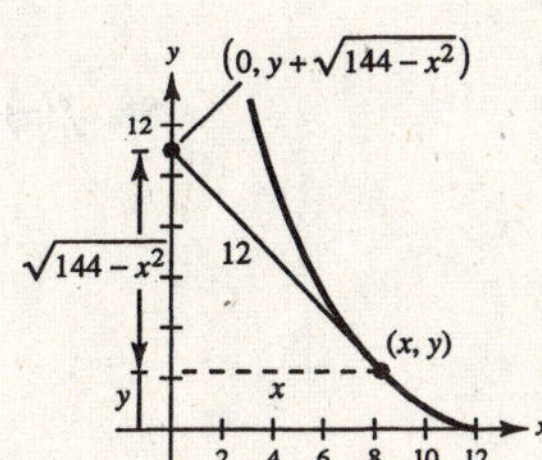

(b) $y = -\displaystyle\int \frac{\sqrt{144 - x^2}}{x}\, dx$

Let $x = 12 \sin \theta$, $dx = 12 \cos \theta \, d\theta$, $\sqrt{144 - x^2} = 12 \cos \theta$.

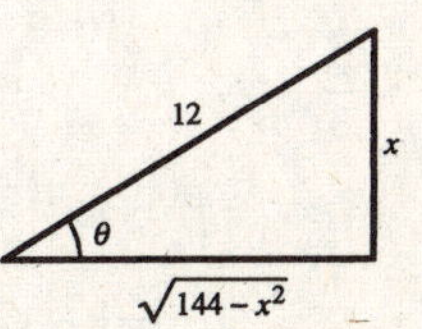

$y = -\displaystyle\int \frac{12 \cos \theta}{12 \sin \theta} 12 \cos \theta \, d\theta = -12 \int \frac{1 - \sin^2 \theta}{\sin \theta}\, d\theta$

$= -12 \displaystyle\int (\csc \theta - \sin \theta)\, d\theta = -12 \ln|\csc \theta - \cot \theta| - 12 \cos \theta + C$

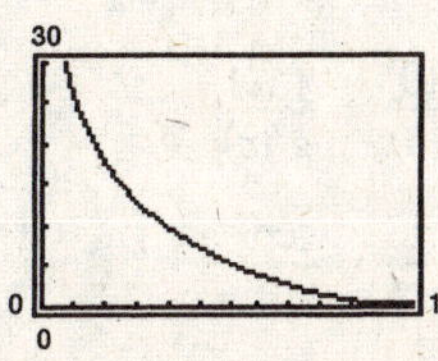

$= -12 \ln\left|\dfrac{12}{x} - \dfrac{\sqrt{144 - x^2}}{x}\right| - 12\left(\dfrac{\sqrt{144 - x^2}}{12}\right) + C$

$= -12 \ln\left|\dfrac{12 - \sqrt{144 - x^2}}{x}\right| - \sqrt{144 - x^2} + C$

When $x = 12$, $y = 0 \Rightarrow C = 0$. Thus, $y = -12 \ln\left(\dfrac{12 - \sqrt{144 - x^2}}{x}\right) - \sqrt{144 - x^2}$.

Note: $\dfrac{12 - \sqrt{144 - x^2}}{x} > 0$ for $0 < x \le 12$

—CONTINUED—

86. —CONTINUED—

(c) Vertical asymptote: $x = 0$

(d) $y + \sqrt{144 - x^2} = 12 \Rightarrow y = 12 - \sqrt{144 - x^2}$

Thus,

$$12 - \sqrt{144 - x^2} = -12 \ln\left(\frac{12 - \sqrt{144 - x^2}}{x}\right) - \sqrt{144 - x^2}$$

$$-1 = \ln\left(\frac{12 - \sqrt{144 - x^2}}{x}\right)$$

$$xe^{-1} = 12 - \sqrt{144 - x^2}$$

$$(xe^{-1} - 12)^2 = \left(-\sqrt{144 - x^2}\right)^2$$

$$x^2e^{-2} - 24xe^{-1} + 144 = 144 - x^2$$

$$x^2(e^{-2} + 1) - 24xe^{-1} = 0$$

$$x[x(e^{-2} + 1) - 24e^{-1}] = 0$$

$$x = 0 \text{ or } x = \frac{24e^{-1}}{e^{-2} + 1} \approx 7.77665.$$

Therefore,

$$s = \int_{7.77665}^{12} \sqrt{1 + \left(-\frac{\sqrt{144 - x^2}}{x}\right)^2}\, dx = \int_{7.77665}^{12} \sqrt{\frac{x^2 + (144 - x^2)}{x^2}}\, dx$$

$$= \int_{7.77665}^{12} \frac{12}{x}\, dx = \Big[12 \ln|x|\Big]_{7.77665}^{12} = 12(\ln 12 - \ln 7.77665) \approx 5.2 \text{ meters.}$$

87. $f(x) = x^3, g(x) = x^2 + 1, [0, 1]$

$$\frac{f(b) - f(a)}{g(b) - g(a)} = \frac{f'(c)}{g'(c)}$$

$$\frac{f(1) - f(0)}{g(1) - g(0)} = \frac{3c^2}{2c}$$

$$\frac{1}{1} = \frac{3c}{2}$$

$$c = \frac{2}{3}$$

88. $f(x) = \frac{1}{x}; g(x) = x^2 - 4, [1, 2]$

$$\frac{f(2) - f(1)}{g(2) - g(1)} = \frac{f'(c)}{g'(c)}$$

$$\frac{-1/2}{3} = \frac{-1/c^2}{2c}$$

$$-\frac{1}{6} = -\frac{1}{2c^3}$$

$$2c^3 = 6$$

$$c = \sqrt[3]{3}$$

89. $f(x) = \sin x, g(x) = \cos x, \left[0, \frac{\pi}{2}\right]$

$$\frac{f(\pi/2) - f(0)}{g(\pi/2) - g(0)} = \frac{f'(c)}{g'(c)}$$

$$\frac{1}{-1} = \frac{\cos c}{-\sin c}$$

$$-1 = -\cot c$$

$$c = \frac{\pi}{4}$$

90. $f(x) = \ln x, g(x) = x^3, [1, 4]$

$$\frac{f(4) - f(1)}{g(4) - g(1)} = \frac{f'(c)}{g'(c)}$$

$$\frac{\ln 4}{63} = \frac{1/c}{3c^2} = \frac{1}{3c^3}$$

$$3c^3 \ln 4 = 63$$

$$c^3 = \frac{21}{\ln 4}$$

$$c = \sqrt[3]{\frac{21}{\ln 4}} \approx 2.474$$

91. False. L'Hôpital's Rule does not apply since

$$\lim_{x\to 0}(x^2 + x + 1) \neq 0.$$

$$\lim_{x\to 0^+}\frac{x^2 + x + 1}{x} = \lim_{x\to 0^+}\left(x + 1 + \frac{1}{x}\right) = 1 + \infty = \infty$$

92. False. If $y = e^x/x^2$, then

$$y' = \frac{x^2e^x - 2xe^x}{x^4} = \frac{xe^x(x-2)}{x^4} = \frac{e^x(x-2)}{x^3}.$$

93. True

94. False. Let $f(x) = x$ and $g(x) = x + 1$. Then

$$\lim_{x\to\infty}\frac{x}{x+1} = 1, \text{ but } \lim_{x\to\infty}[x - (x+1)] = -1.$$

95. Area of triangle: $\frac{1}{2}(2x)(1 - \cos x) = x - x\cos x$

Shaded area: Area of rectangle − Area under curve

$$2x(1 - \cos x) - 2\int_0^x (1 - \cos t)\,dt = 2x(1 - \cos x) - 2\Big[t - \sin t\Big]_0^x$$

$$= 2x(1 - \cos x) - 2(x - \sin x) = 2\sin x - 2x\cos x$$

Ratio: $$\lim_{x\to 0}\frac{x - x\cos x}{2\sin x - 2x\cos x} = \lim_{x\to 0}\frac{1 + x\sin x - \cos x}{2\cos x + 2x\sin x - 2\cos x}$$

$$= \lim_{x\to 0}\frac{1 + x\sin x - \cos x}{2x\sin x}$$

$$= \lim_{x\to 0}\frac{x\cos x + \sin x + \sin x}{2x\cos x + 2\sin x}$$

$$= \lim_{x\to 0}\frac{x\cos x + 2\sin x}{2x\cos x + 2\sin x}\cdot\frac{1/\cos x}{1/\cos x}$$

$$= \lim_{x\to 0}\frac{x + 2\tan x}{2x + 2\tan x}$$

$$= \lim_{x\to 0}\frac{1 + 2\sec^2 x}{2 + 2\sec^2 x} = \frac{3}{4}$$

96. (a) $\sin\theta = BD$

$\cos\theta = DO \implies AD = 1 - \cos\theta$

Area $\triangle ABD = \frac{1}{2}bh = \frac{1}{2}(1 - \cos\theta)\sin\theta = \frac{1}{2}\sin\theta - \frac{1}{2}\sin\theta\cos\theta$

(b) Area of sector: $\frac{1}{2}\theta$

Shaded area: $\frac{1}{2}\theta - \text{Area }\triangle OBD = \frac{1}{2}\theta - \frac{1}{2}(\cos\theta)(\sin\theta) = \frac{1}{2}\theta - \frac{1}{2}\sin\theta\cos\theta$

(c) $$R = \frac{(1/2)\sin\theta - (1/2)\sin\theta\cos\theta}{(1/2)\theta - (1/2)\sin\theta\cos\theta} = \frac{\sin\theta - \sin\theta\cos\theta}{\theta - \sin\theta\cos\theta}$$

(d) $$\lim_{\theta\to 0}R = \lim_{\theta\to 0}\frac{\sin\theta - (1/2)\sin 2\theta}{\theta - (1/2)\sin 2\theta}$$

$$= \lim_{\theta\to 0}\frac{\cos\theta - \cos 2\theta}{1 - \cos 2\theta} = \lim_{\theta\to 0}\frac{-\sin\theta + 2\sin 2\theta}{2\sin 2\theta} = \lim_{\theta\to 0}\frac{-\cos\theta + 4\cos 2\theta}{4\cos 2\theta} = \frac{3}{4}$$

97. $\displaystyle\lim_{x\to0}\frac{4x-2\sin 2x}{2x^3}=\lim_{x\to0}\frac{4-4\cos 2x}{6x^2}$

$\displaystyle=\lim_{x\to0}\frac{8\sin 2x}{12x}$

$\displaystyle=\lim_{x\to0}\frac{16\cos 2x}{12}=\frac{16}{12}=\frac{4}{3}$

Let $c=\dfrac{4}{3}$.

98. Let $y=(e^x+x)^{1/x}$.

$\displaystyle\ln y=\frac{1}{x}\ln(e^x+x)=\frac{\ln(e^x+x)}{x}$

$\displaystyle\lim_{x\to0}\frac{\ln(e^x+x)}{x}=\lim_{x\to0}\frac{e^x+1}{e^x+x}=\frac{2}{1}=2$

Hence, $\lim_{x\to0}(e^x+x)^{1/x}=e^2$.

Let $c=e^2\approx 7.389$.

99. $\displaystyle\lim_{x\to0}\frac{a-\cos bx}{x^2}=2$

Near $x=0$, $\cos bx\approx 1$ and $x^2\approx 0\implies a=1$.

Using L'Hôpital's Rule,

$\displaystyle\lim_{x\to0}\frac{1-\cos bx}{x^2}=\lim_{x\to0}\frac{b\sin bx}{2x}$

$\displaystyle=\lim_{x\to0}\frac{b^2\cos bx}{2}=2.$

Hence, $b^2=4$ and $b=\pm2$.

Answer: $a=1, b=\pm2$

100. We use mathematical induction.

For $n=1$, $\displaystyle\lim_{x\to\infty}\frac{x^1}{e^x}=\lim_{x\to\infty}\frac{1}{e^x}=0.$

Assume that $\displaystyle\lim_{x\to\infty}\frac{x^k}{e^x}=0.$

Then, $\displaystyle\lim_{x\to\infty}\frac{x^{k+1}}{e^x}=\lim_{x\to\infty}\frac{(k+1)x^k}{e^x}$

$\displaystyle=(k+1)\lim_{x\to0}\frac{x^k}{e^x}$

$=(k+1)(0)=0.$

101. (a) $\displaystyle\lim_{h\to0}\frac{f(x+h)-f(x-h)}{2h}=\lim_{h\to0}\frac{f'(x+h)(1)-f'(x-h)(-1)}{2}$

$\displaystyle=\lim_{h\to0}\left[\frac{f'(x+h)+f'(x-h)}{2}\right]$

$\displaystyle=\frac{f'(x)+f'(x)}{2}=f'(x)$

(b)

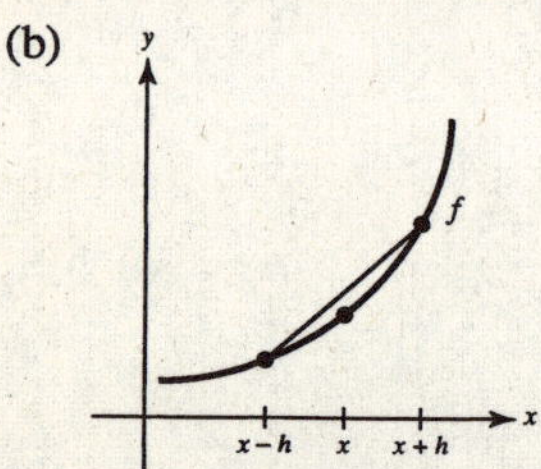

Graphically, the slope of the line joining $(x-h, f(x-h))$ and $(x+h, f(x+h))$ is approximately $f'(x)$. And, as $h\to0$,

$\displaystyle\lim_{h\to0}\frac{f(x+h)-f(x-h)}{2h}=f'(x).$

102. $\displaystyle\lim_{h\to0}\frac{f(x+h)-2f(x)+f(x-h)}{h^2}=\lim_{h\to0}\frac{f'(x+h)(1)+f'(x-h)(-1)}{2h}$

$\displaystyle=\lim_{h\to0}\frac{f'(x+h)-f'(x-h)}{2h}$

$\displaystyle=\lim_{h\to0}\frac{f''(x+h)(1)-f''(x-h)(-1)}{2}$

$\displaystyle=\lim_{h\to0}\frac{f''(x+h)+f''(x-h)}{2}$

$\displaystyle=\frac{f''(x)+f''(x)}{2}=f''(x)$

103. $g(x) = \begin{cases} e^{-1/x^2}, & x \neq 0 \\ 0, & x = 0 \end{cases}$

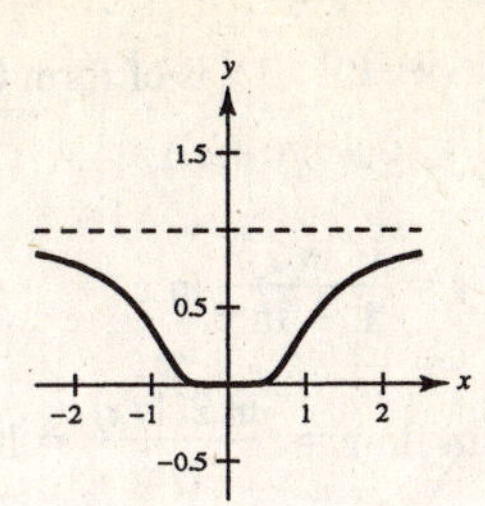

$$g'(0) = \lim_{x \to 0} \frac{g(x) - g(0)}{x - 0} = \lim_{x \to 0} \frac{e^{-1/x^2}}{x}$$

Let $y = \dfrac{e^{-1/x^2}}{x}$, then $\ln y = \ln\left(\dfrac{e^{-1/x^2}}{x}\right) = -\dfrac{1}{x^2} - \ln x = \dfrac{-1 - x^2 \ln x}{x^2}$. Since

$$\lim_{x \to 0} x^2 \ln x = \lim_{x \to 0} \frac{\ln x}{1/x^2} = \lim_{x \to 0} \frac{1/x}{-2/x^3} = \lim_{x \to 0}\left(-\frac{x^2}{2}\right) = 0$$

we have $\displaystyle\lim_{x \to 0}\left(\frac{-1 - x^2 \ln x}{x^2}\right) = -\infty$. Thus, $\displaystyle\lim_{x \to 0} y = e^{-\infty} = 0 \Rightarrow g'(0) = 0$.

Note: The graph appears to support this conclusion—the tangent line is horizontal at (0, 0).

104. $f(x) = \dfrac{x^k - 1}{k}$

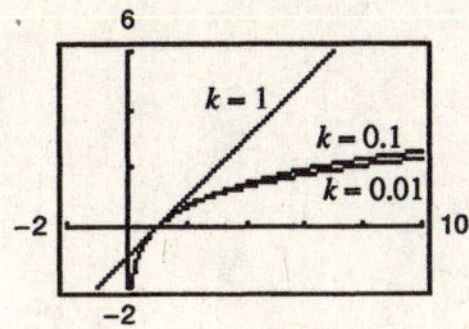

$k = 1, \quad f(x) = x - 1$

$k = 0.1, \quad f(x) = \dfrac{x^{0.1} - 1}{0.1} = 10(x^{0.1} - 1)$

$k = 0.01, \quad f(x) = \dfrac{x^{0.01} - 1}{0.01} = 100(x^{0.01} - 1)$

$$\lim_{k \to 0^+} \frac{x^k - 1}{k} = \lim_{k \to 0^+} \frac{x^k(\ln x)}{1} = \ln x$$

105. (a) $\displaystyle\lim_{x \to 0^+} (-x \ln x)$ is the form $0 \cdot \infty$.

(b) $\displaystyle\lim_{x \to 0^+} \frac{-\ln x}{1/x} = \lim_{x \to 0^+} \frac{-1/x}{-1/x^2} = \lim_{x \to 0^+} (x) = 0$

(c)

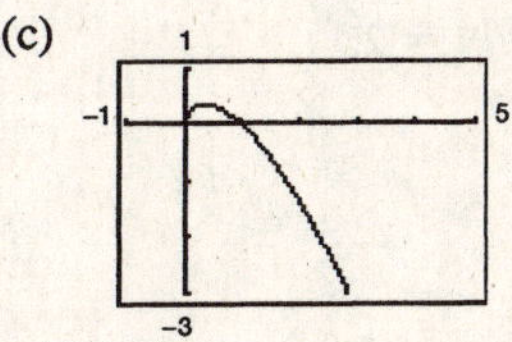

106. $\displaystyle\lim_{x \to a} f(x)^{g(x)}$

$y = f(x)^{g(x)}$

$\ln y = g(x) \ln f(x)$

$\displaystyle\lim_{x \to a} g(x) \ln f(x) = (\infty)(-\infty) = -\infty$

As $x \to a$, $\ln y \Rightarrow -\infty$, and hence $y = 0$. Thus,

$\displaystyle\lim_{x \to a} f(x)^{g(x)} = 0.$

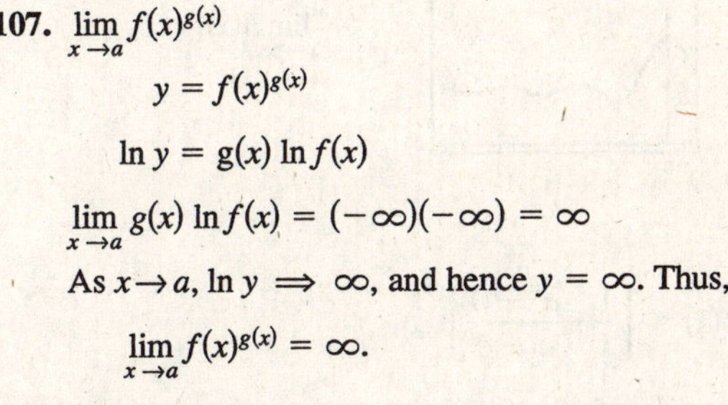

107. $\displaystyle\lim_{x \to a} f(x)^{g(x)}$

$y = f(x)^{g(x)}$

$\ln y = g(x) \ln f(x)$

$\displaystyle\lim_{x \to a} g(x) \ln f(x) = (-\infty)(-\infty) = \infty$

As $x \to a$, $\ln y \Rightarrow \infty$, and hence $y = \infty$. Thus,

$\displaystyle\lim_{x \to a} f(x)^{g(x)} = \infty.$

108. $\displaystyle f'(a)(b - a) - \int_a^b f''(t)(t - b)\,dt = f'(a)(b - a) - \left\{\Big[f'(t)(t - b)\Big]_a^b - \int_a^b f'(t)\,dt\right\}$

$$= f'(a)(b - a) + f'(a)(a - b) + \Big[f(t)\Big]_a^b = f(b) - f(a)$$

$dv = f''(t)\,dt \Rightarrow v = f'(t)$

$u = t - b \Rightarrow du = dt$

109. (a) $\lim_{x\to 0^+} x^{(\ln 2)/(1+\ln x)}$ is of form 0^0.

Let $y = x^{(\ln 2)/(1+\ln x)}$

$$\ln y = \frac{\ln 2}{1+\ln x}\ln x$$

$$\lim_{x\to 0^+} \ln y = \frac{\ln 2(1/x)}{1/x} = \ln 2.$$

Thus, $\lim_{x\to 0^+} x^{(\ln 2)/(1+\ln x)} = 2.$

(b) $\lim_{x\to\infty} x^{(\ln 2)/(1+\ln x)}$ is of form ∞^0.

Let $y = x^{(\ln 2)/(1+\ln x)}$

$$\ln y = \frac{\ln 2}{1+\ln x}\ln x$$

$$\lim_{x\to\infty} \ln y = \frac{\ln 2(1/x)}{1/x} = \ln 2.$$

Thus, $\lim_{x\to\infty} x^{(\ln 2)/(1+\ln x)} = 2.$

(c) $\lim_{x\to 0} (x+1)^{(\ln 2)/(x)}$ is of form 1^∞.

Let $y = (x+1)^{(\ln 2)/(x)}$

$$\ln y = \frac{\ln 2}{x}\ln(x+1)$$

$$\lim_{x\to 0} \ln y = \lim_{x\to 0}\frac{(\ln 2)1/(x+1)}{1} = \ln 2.$$

Thus, $\lim_{x\to 0} (x+1)^{(\ln 2)/(x)} = 2.$

110. $$\lim_{x\to a}\frac{\sqrt{2a^3x - x^4} - a\sqrt[3]{a^2x}}{a - \sqrt[4]{ax^3}}$$

$$= \lim_{x\to a}\frac{\frac{1}{2}(2a^3x - x^4)^{-1/2}(2a^3 - 4x^3) - \frac{a}{3}(a^2x)^{-2/3}a^2}{-\frac{1}{4}(ax^3)^{-3/4}}$$

$$= \frac{\frac{1}{2}(a^4)^{-1/2}(-2a^3) - \frac{a^3}{3}(a^3)^{-2/3}}{-\frac{1}{4}(ax^3)^{-3/4}(3ax^2)}$$

$$= \frac{a + \frac{a}{3}}{\frac{1}{4}(a^{-3})(3a^3)}$$

$$= \frac{\frac{4}{3}a}{\frac{3}{4}} = \frac{16}{9}a$$

111. (a) $h(x) = \dfrac{x + \sin x}{x}$

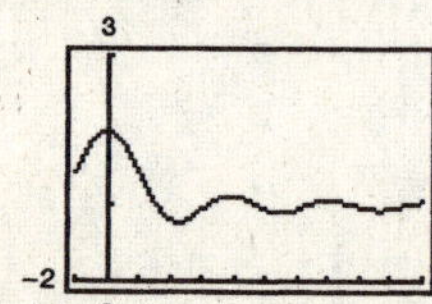

$\lim_{x\to\infty} h(x) = 1$

(b) $h(x) = \dfrac{x + \sin x}{x} = \dfrac{x}{x} + \dfrac{\sin x}{x} = 1 + \dfrac{\sin x}{x}, x > 0$

Hence, $\lim_{x\to\infty} h(x) = \lim_{x\to\infty}\left[1 + \dfrac{\sin x}{x}\right] = 1 + 0 = 1.$

(c) No. $h(x)$ is not an indeterminate form.

112. Let $f(x) = \left[\dfrac{1}{x}\cdot\dfrac{a^x - 1}{a - 1}\right]^{1/x}$.

For $a > 1$ and $x > 0$,

$$\ln f(x) = \frac{1}{x}\left[\ln\frac{1}{x} + \ln(a^x - 1) - \ln(a - 1)\right] = -\frac{\ln x}{x} + \frac{\ln(a^x - 1)}{x} - \frac{\ln(a - 1)}{x}.$$

As $x\to\infty$, $\dfrac{\ln x}{x}\to 0$, $\dfrac{\ln(a-1)}{x}\to 0$, and $\dfrac{\ln(a^x - 1)}{x} = \dfrac{\ln[(1 - a^{-x})a^x]}{x} = \dfrac{\ln(1 - a^{-x})}{x} + \ln a \to \ln a.$

Hence, $\ln f(x) \to \ln a$.

For $0 < a < 1$ and $x > 0$,

$$\ln f(x) = \frac{-\ln x}{x} + \frac{\ln(1 - a^x)}{x} - \frac{\ln(1 - a)}{x} \to 0 \text{ as } x\to\infty.$$

Combining these results, $\lim_{x\to\infty} f(x) = \begin{cases} a & \text{if } a > 1 \\ 1 & \text{if } 0 < a < 1 \end{cases}$.

Section 8.8 Improper Integrals

1. $\displaystyle\int_0^1 \frac{dx}{3x-2}$ is improper because $3x - 2 = 0$ when $x = \frac{2}{3}$.

2. $\displaystyle\int_1^2 \frac{dx}{x^2}$ is not improper because $\frac{1}{x^2}$ is continuous on $[1, 2]$.

3. $\displaystyle\int_0^1 \frac{2x-5}{x^2-5x+6}\,dx = \int_0^1 \frac{2x-5}{(x-2)(x-3)}\,dx$

is not improper because

$\displaystyle\frac{2x-5}{(x-2)(x-3)}$ is continuous on $[0, 1]$.

4. $\displaystyle\int_1^\infty \ln(x^2)dx$

is improper because the upper limit of integration is ∞.

5. Infinite discontinuity at $x = 0$.

$$\int_0^4 \frac{1}{\sqrt{x}}\,dx = \lim_{b\to 0^+} \int_b^4 \frac{1}{\sqrt{x}}\,dx$$
$$= \lim_{b\to 0^+} \Big[2\sqrt{x}\Big]_b^4$$
$$= \lim_{b\to 0^+} \left(4 - 2\sqrt{b}\right) = 4$$

Converges

6. Infinite discontinuity at $x = 3$.

$$\int_3^4 \frac{1}{(x-3)^{3/2}}\,dx = \lim_{b\to 3^+} \int_b^4 (x-3)^{-3/2}\,dx$$
$$= \lim_{b\to 3^+} \Big[-2(x-3)^{-1/2}\Big]_b^4$$
$$= \lim_{b\to 3^+} \left[-2 + \frac{2}{\sqrt{b-3}}\right] = \infty$$

Diverges

7. Infinite discontinuity at $x = 1$.

$$\int_0^2 \frac{1}{(x-1)^2}\,dx = \int_0^1 \frac{1}{(x-1)^2}\,dx + \int_1^2 \frac{1}{(x-1)^2}\,dx$$
$$= \lim_{b\to 1^-} \int_0^b \frac{1}{(x-1)^2}\,dx + \lim_{c\to 1^+} \int_c^2 \frac{1}{(x-1)^2}\,dx$$
$$= \lim_{b\to 1^-} \left[-\frac{1}{x-1}\right]_0^b + \lim_{c\to 1^+} \left[-\frac{1}{x-1}\right]_c^2 = (\infty - 1) + (-1 + \infty)$$

Diverges

8. Infinite discontinuity at $x = 1$.

$$\int_0^2 \frac{1}{(x-1)^{2/3}}\,dx = \int_0^1 \frac{1}{(x-1)^{2/3}}\,dx + \int_1^2 \frac{1}{(x-1)^{2/3}}\,dx$$
$$= \lim_{b\to 1^-} \int_0^b \frac{1}{(x-1)^{2/3}}\,dx + \lim_{c\to 1^+} \int_c^2 \frac{1}{(x-1)^{2/3}}\,dx$$
$$= \lim_{b\to 1^-} \Big[3\sqrt[3]{x-1}\Big]_0^b + \lim_{c\to 1^+} \Big[3\sqrt[3]{x-1}\Big]_c^2 = (0 + 3) + (3 - 0) = 6$$

Converges

9. Infinite limit of integration.

$$\int_0^\infty e^{-x}\,dx = \lim_{b\to\infty} \int_0^b e^{-x}\,dx$$
$$= \lim_{b\to\infty} \Big[-e^{-x}\Big]_0^b = 0 + 1 = 1$$

Converges

10. Infinite limit of integration.

$$\int_{-\infty}^0 e^{2x}\,dx = \lim_{b\to -\infty} \int_b^0 e^{2x}\,dx$$
$$= \lim_{b\to -\infty} \left[\frac{1}{2}e^{2x}\right]_b^0 = \frac{1}{2} - 0 = \frac{1}{2}$$

Converges

11. $\displaystyle\int_{-1}^{1} \frac{1}{x^2}\,dx \neq -2$

because the integrand is not defined at $x = 0$.
Diverges

12. $\displaystyle\int_{-2}^{2} \frac{-2}{(x-1)^3}\,dx \neq \frac{8}{9}$ because the integral is not defined at $x = 1$. The integral diverges.

13. $\displaystyle\int_0^{\infty} e^{-x}\,dx \neq 0$. You need to evaluate the limit.

$$\lim_{b\to\infty} \int_0^b e^{-x}\,dx = \lim_{b\to\infty}\Big[-e^{-x}\Big]_0^b$$
$$= \lim_{b\to\infty}\Big[-e^{-b} + 1\Big] = 1$$

14. $\displaystyle\int_0^{\pi} \sec x\,dx \neq 0$ because $\sec x$ is not defined at $x = \pi/2$.
The integral diverges.

15.
$$\int_1^{\infty} \frac{1}{x^2}\,dx = \lim_{b\to\infty}\int_1^b \frac{1}{x^2}\,dx$$
$$= \lim_{b\to\infty}\left[-\frac{1}{x}\right]_1^b = 1$$

16.
$$\int_1^{\infty} \frac{5}{x^3}\,dx = \lim_{b\to\infty}\int_1^b \frac{5}{x^3}\,dx$$
$$= \lim_{b\to\infty}\left[-\frac{5}{2}x^{-2}\right]_1^b = \frac{5}{2}$$

17.
$$\int_1^{\infty} \frac{3}{\sqrt[3]{x}}\,dx = \lim_{b\to\infty}\int_1^b 3x^{-1/3}\,dx$$
$$= \lim_{b\to\infty}\left[\frac{9}{2}x^{2/3}\right]_1^b = \infty$$

Diverges

18.
$$\int_1^{\infty} \frac{4}{\sqrt[4]{x}}\,dx = \lim_{b\to\infty}\int_1^b 4x^{-1/4}\,dx$$
$$= \lim_{b\to\infty}\left[\frac{16}{3}x^{3/4}\right]_1^b = \infty \quad \text{Diverges}$$

19. $\displaystyle\int_{-\infty}^{0} xe^{-2x}\,dx = \lim_{b\to-\infty}\int_b^0 xe^{-2x}\,dx = \lim_{b\to-\infty}\frac{1}{4}\Big[(-2x-1)e^{-2x}\Big]_b^0 = \lim_{b\to-\infty}\frac{1}{4}\big[-1 + (2b+1)e^{-2b}\big] = -\infty$ **(Integration by parts)**

Diverges

20. $\displaystyle\int_0^{\infty} xe^{-x/2}\,dx = \lim_{b\to\infty}\int_0^b xe^{-x/2}\,dx = \lim_{b\to\infty}\Big[e^{-x/2}(-2x-4)\Big]_0^b = \lim_{b\to\infty} e^{-b/2}(-2b-4) + 4 = 4$

21. $\displaystyle\int_0^{\infty} x^2e^{-x}\,dx = \lim_{b\to\infty}\int_0^b x^2e^{-x}\,dx = \lim_{b\to\infty}\Big[-e^{-x}(x^2+2x+2)\Big]_0^b = \lim_{b\to\infty}\left(-\frac{b^2+2b+2}{e^b} + 2\right) = 2$

Since $\displaystyle\lim_{b\to\infty}\left(-\frac{b^2+2b+2}{e^b}\right) = 0$ by L'Hôpital's Rule.

22. $\displaystyle\int_0^{\infty} (x-1)e^{-x}\,dx = \lim_{b\to\infty}\int_0^b (x-1)e^{-x}\,dx = \lim_{b\to\infty}\Big[-xe^{-x}\Big]_0^b = \lim_{b\to\infty}\left(\frac{-b}{e^b} + 0\right) = 0$ by L'Hôpital's Rule.

23.
$$\int_0^{\infty} e^{-x}\cos x\,dx = \lim_{b\to\infty}\frac{1}{2}\Big[e^{-x}(-\cos x + \sin x)\Big]_0^b$$
$$= \frac{1}{2}[0 - (-1)] = \frac{1}{2}$$

24.
$$\int_0^{\infty} e^{-ax}\sin bx\,dx = \lim_{c\to\infty}\left[\frac{e^{-ax}(-a\sin bx - b\cos bx)}{a^2+b^2}\right]_0^c$$
$$= 0 - \frac{-b}{a^2+b^2} = \frac{b}{a^2+b^2}$$

25. $$\int_4^{\infty} \frac{1}{x(\ln x)^3}\,dx = \lim_{b\to\infty}\int_4^b (\ln x)^{-3}\frac{1}{x}\,dx$$
$$= \lim_{b\to\infty}\left[-\frac{1}{2}(\ln x)^{-2}\right]_4^b$$
$$= -\frac{1}{2}(\ln b)^{-2} + \frac{1}{2}(\ln 4)^{-2}$$
$$= \frac{1}{2}\frac{1}{(2\ln 2)^2} = \frac{1}{8(\ln 2)^2}$$

26. $$\int_1^{\infty} \frac{\ln x}{x}\,dx = \lim_{b\to\infty}\int_1^b \frac{\ln x}{x}\,dx$$
$$= \lim_{b\to\infty}\left[\frac{(\ln x)^2}{2}\right]_1^b = \infty$$

Diverges

27. $$\int_{-\infty}^{\infty} \frac{2}{4+x^2}\,dx = \int_{-\infty}^{0} \frac{2}{4+x^2}\,dx + \int_0^{\infty} \frac{2}{4+x^2}\,dx$$
$$= \lim_{b\to-\infty}\int_b^0 \frac{2}{4+x^2}\,dx + \lim_{c\to\infty}\int_0^c \frac{2}{4+x^2}\,dx$$
$$= \lim_{b\to-\infty}\left[\arctan\left(\frac{x}{2}\right)\right]_b^0 + \lim_{c\to\infty}\left[\arctan\left(\frac{x}{2}\right)\right]_0^c$$
$$= \left(0 - \left(-\frac{\pi}{2}\right)\right) + \left(\frac{\pi}{2} - 0\right) = \pi$$

28. $$\int_0^{\infty} \frac{x^3}{(x^2+1)^2}\,dx = \lim_{b\to\infty}\int_0^b \frac{x}{x^2+1}\,dx - \lim_{b\to\infty}\int_0^b \frac{x}{(x^2+1)^2}\,dx$$
$$= \lim_{b\to\infty}\left[\frac{1}{2}\ln(x^2+1) + \frac{1}{2(x^2+1)}\right]_0^b$$
$$= \infty - \frac{1}{2}$$

Diverges

29. $$\int_0^{\infty} \frac{1}{e^x + e^{-x}}\,dx = \lim_{b\to\infty}\int_0^b \frac{e^x}{1+e^{2x}}\,dx$$
$$= \lim_{b\to\infty}\left[\arctan(e^x)\right]_0^b$$
$$= \frac{\pi}{2} - \frac{\pi}{4} = \frac{\pi}{4}$$

30. $$\int_0^{\infty} \frac{e^x}{1+e^x}\,dx = \lim_{b\to\infty}\left[\ln(1+e^x)\right]_0^b = \infty - \ln 2$$

Diverges

31. $$\int_0^{\infty} \cos \pi x\,dx = \lim_{b\to\infty}\left[\frac{1}{\pi}\sin \pi x\right]_0^b$$

Diverges since $\sin \pi b$ does not approach a limit as $b \to \infty$.

32. $$\int_0^{\infty} \sin\frac{x}{2}\,dx = \lim_{b\to\infty}\left[-2\cos\frac{x}{2}\right]_0^b$$

Diverges since $\cos\frac{x}{2}$ does not approach a limit as $x \to \infty$.

33. $$\int_0^1 \frac{1}{x^2}\,dx = \lim_{b\to 0^+}\left[\frac{-1}{x}\right]_b^1 = \lim_{b\to 0^+}\left[-1 + \frac{1}{b}\right] = -1 + \infty$$

Diverges

34. $$\int_0^4 \frac{8}{x}\,dx = \lim_{b\to 0^+}\int_b^4 \frac{8}{x}\,dx = \lim_{b\to 0^+}\left[8\ln x\right]_b^4 = \infty$$

Diverges

35. $$\int_0^8 \frac{1}{\sqrt[3]{8-x}}\,dx = \lim_{b\to 8^-}\int_0^b \frac{1}{\sqrt[3]{8-x}}\,dx$$
$$= \lim_{b\to 8^-}\left[\frac{-3}{2}(8-x)^{2/3}\right]_0^b = 6$$

36. $$\int_0^6 \frac{4}{\sqrt{6-x}}\,dx = \lim_{b\to 6^-}\int_0^b 4(6-x)^{-1/2}\,dx$$
$$= \lim_{b\to 6^-}\left[-8(6-x)^{1/2}\right]_0^b$$
$$= -8(0) + 8\sqrt{6}$$
$$= 8\sqrt{6}$$

37. $$\int_0^1 x\ln x\,dx = \lim_{b\to 0^+}\left[\frac{x^2}{2}\ln|x| - \frac{x^2}{4}\right]_b^1$$
$$= \lim_{b\to 0^+}\left[\frac{-1}{4} - \frac{b^2\ln b}{2} + \frac{b^2}{4}\right] = \frac{-1}{4}$$

since $\lim_{b\to 0^+}(b^2\ln b) = 0$ by L'Hôpital's Rule.

38. $$\int_0^e \ln x^2\,dx = \lim_{b\to 0^+}\int_0^e 2\ln x\,dx$$
$$= \lim_{b\to 0^+}\left[2x\ln x - 2x\right]_b^e$$
$$= \lim_{b\to 0^+}\left[(2e - 2e) - (2b\ln b - 2b)\right]$$
$$= 0$$

39. $\displaystyle\int_0^{\pi/2} \tan\theta\, d\theta = \lim_{b\to(\pi/2)^-}\Big[\ln|\sec\theta|\Big]_0^b = \infty$

Diverges

40. $\displaystyle\int_0^{\pi/2} \sec\theta\, d\theta = \lim_{b\to(\pi/2)}\Big[\ln|\sec\theta + \tan\theta|\Big]_0^b = \infty$

Diverges

41. $$\int_2^4 \frac{2}{x\sqrt{x^2-4}}\,dx = \lim_{b\to2^+}\int_b^4 \frac{2}{x\sqrt{x^2-4}}\,dx$$
$$= \lim_{b\to2^+}\left[\operatorname{arcsec}\left|\frac{x}{2}\right|\right]_b^4$$
$$= \lim_{b\to2^+}\left(\operatorname{arcsec} 2 - \operatorname{arcsec}\left(\frac{b}{2}\right)\right)$$
$$= \frac{\pi}{3} - 0 = \frac{\pi}{3}$$

42. $\displaystyle\int_0^2 \frac{1}{\sqrt{4-x^2}}\,dx = \lim_{b\to2^-}\left[\arcsin\left(\frac{x}{2}\right)\right]_0^b = \frac{\pi}{2}$

43. $\displaystyle\int_2^4 \frac{1}{\sqrt{x^2-4}} = \lim_{b\to2^+}\left[\ln\left|x + \sqrt{x^2-4}\right|\right]_b^4 = \ln(4 + 2\sqrt{3}) - \ln 2 = \ln(2+\sqrt{3}) \approx 1.317$

44. $\displaystyle\int_0^2 \frac{1}{4-x^2}\,dx = \lim_{b\to2^-}\int_0^b \frac{1}{4}\left(\frac{1}{2+x} + \frac{1}{2-x}\right)dx = \lim_{b\to2^-}\left[\frac{1}{4}\ln\left|\frac{2+x}{2-x}\right|\right]_0^b = \infty - 0$

Diverges

45. $$\int_0^2 \frac{1}{\sqrt[3]{x-1}}\,dx = \int_0^1 \frac{1}{\sqrt[3]{x-1}}\,dx + \int_1^2 \frac{1}{\sqrt[3]{x-1}}\,dx$$
$$= \lim_{b\to1^-}\left[\frac{3}{2}(x-1)^{2/3}\right]_0^b + \lim_{c\to1^+}\left[\frac{3}{2}(x-1)^{2/3}\right]_c^2 = \frac{-3}{2} + \frac{3}{2} = 0$$

46. $$\int_1^3 \frac{2}{(x-2)^{8/3}}\,dx = \int_1^2 2(x-2)^{-8/3}\,dx + \int_2^3 2(x-2)^{-8/3}\,dx$$
$$= \lim_{b\to2^-}\int_1^b 2(x-2)^{-8/3}\,dx + \lim_{c\to2^+}\int_c^3 2(x-2)^{-8/3}\,dx$$
$$= \lim_{b\to2^-}\left[-\frac{6}{5}(x-2)^{-5/3}\right]_1^b + \lim_{c\to2^+}\left[-\frac{6}{5}(x-2)^{-5/3}\right]_c^3 = \infty$$

Diverges

47. $$\int_0^\infty \frac{4}{\sqrt{x}(x+6)}\,dx = \int_0^1 \frac{4}{\sqrt{x}(x+6)}\,dx + \int_1^\infty \frac{4}{\sqrt{x}(x+6)}\,dx$$

Let $u = \sqrt{x}$, $u^2 = x$, $2u\,du = dx$.

$$\int \frac{4}{\sqrt{x}(x+6)}\,dx = \int \frac{4(2u\,du)}{u(u^2+6)} = 8\int \frac{du}{u^2+6} = \frac{8}{\sqrt{6}}\arctan\left(\frac{u}{\sqrt{6}}\right) + C = \frac{8}{\sqrt{6}}\arctan\left(\frac{\sqrt{x}}{\sqrt{6}}\right) + C$$

Thus, $$\int_0^\infty \frac{4}{\sqrt{x}(x+6)}\,dx = \lim_{b\to0^+}\left[\frac{8}{\sqrt{6}}\arctan\left(\frac{\sqrt{x}}{\sqrt{6}}\right)\right]_b^1 + \lim_{c\to\infty}\left[\frac{8}{\sqrt{6}}\arctan\left(\frac{\sqrt{x}}{\sqrt{6}}\right)\right]_1^c$$
$$= \left(\frac{8}{\sqrt{6}}\arctan\left(\frac{1}{\sqrt{6}}\right) - \frac{8}{\sqrt{6}}0\right) + \left(\frac{8}{\sqrt{6}}\frac{\pi}{2} - \frac{8}{\sqrt{6}}\arctan\left(\frac{1}{\sqrt{6}}\right)\right)$$
$$= \frac{8\pi}{2\sqrt{6}} = \frac{2\pi\sqrt{6}}{3}.$$

48. $\displaystyle\int \frac{1}{x \ln x}\,dx = \ln|\ln|x|| + C$

Thus,

$$\int_1^\infty \frac{1}{x\ln x}\,dx = \int_1^e \frac{1}{x\ln x}\,dx + \int_e^\infty \frac{1}{x\ln x}\,dx$$
$$= \lim_{b\to 1^+}\Big[\ln(\ln x)\Big]_1^e + \lim_{c\to\infty}\Big[\ln(\ln x)\Big]_e^\infty.$$

Diverges

49. If $p = 1$, $\displaystyle\int_1^\infty \frac{1}{x}\,dx = \lim_{b\to\infty}\int_1^b \frac{1}{x}\,dx = \lim_{b\to\infty} \ln x\Big]_1^b$

$$= \lim_{b\to\infty}[\ln b] = \infty.$$

Diverges. For $p \neq 1$,

$$\int_1^\infty \frac{1}{x^p}\,dx = \lim_{b\to\infty}\left[\frac{x^{1-p}}{1-p}\right]_1^b = \lim_{b\to\infty}\left[\frac{b^{1-p}}{1-p} - \frac{1}{1-p}\right].$$

This converges to $\dfrac{1}{p-1}$ if $1 - p < 0$ or $p > 1$.

50. If $p = 1$, $\displaystyle\int_0^1 \frac{1}{x}\,dx = \lim_{a\to 0^+} \ln x\Big]_a^1 = \lim_{a\to 0^+} -\ln a = \infty.$

Diverges. If $p \neq 1$,

$$\int_0^1 \frac{1}{x^p}\,dx = \lim_{a\to 0^+}\left[\frac{x^{1-p}}{1-p}\right]_a^1 = \lim_{a\to 0^+}\left[\frac{1}{1-p} - \frac{a^{1-p}}{1-p}\right].$$

This converges to $\dfrac{1}{1-p}$ if $1 - p > 0$ or $p < 1$.

51. For $n = 1$ we have

$$\int_0^\infty xe^{-x}\,dx = \lim_{b\to\infty}\int_0^b xe^{-x}\,dx$$
$$= \lim_{b\to\infty}\Big[-e^{-x}x - e^{-x}\Big]_0^b \qquad \text{(Parts: } u = x,\ dv = e^{-x}\,dx)$$
$$= \lim_{b\to\infty}\left[-e^{-b}b - e^{-b} + 1\right]$$
$$= \lim_{b\to\infty}\left[\frac{-b}{e^b} - \frac{1}{e^b} + 1\right] = 1 \qquad \text{(L'Hôpital's Rule).}$$

Assume that $\displaystyle\int_0^\infty x^n e^{-x}\,dx$ converges. Then for $n + 1$ we have

$$\int x^{n+1}e^{-x}\,dx = -x^{n+1}e^{-x} + (n+1)\int x^n e^{-x}\,dx$$

by parts ($u = x^{n+1}$, $du = (n+1)x^n\,dx$, $dv = e^{-x}\,dx$, $v = -e^{-x}$).

Thus,

$$\int_0^\infty x^{n+1}e^{-x}\,dx = \lim_{b\to\infty}\Big[-x^{n+1}e^{-x}\Big]_0^b + (n+1)\int_0^\infty x^n e^{-x}\,dx = 0 + (n+1)\int_0^\infty x^n e^{-x}\,dx, \text{ which converges.}$$

52. (a) Assume $\displaystyle\int_a^\infty g(x)\,dx = L$ (converges).

Since $0 \le f(x) \le g(x)$ on $[a, \infty)$, $\displaystyle 0 \le \int_a^\infty f(x)\,dx \le \int_a^\infty g(x)\,dx = L$ and $\displaystyle\int_a^\infty f(x)\,dx$ converges.

(b) $\displaystyle\int_a^\infty g(x)\,dx$ diverges, because otherwise, by part (a), if $\displaystyle\int_a^\infty g(x)\,dx$ converges, then so does $\displaystyle\int_a^\infty f(x)\,dx$.

53. $\displaystyle\int_0^1 \frac{1}{x^3}\,dx$ diverges.

(See Exercise 50, $p = 3 \not< 1$.)

54. $\displaystyle\int_0^1 \frac{1}{\sqrt[3]{x}}\,dx = \frac{1}{1 - (1/3)} = \frac{3}{2}$ converges.

(See Exercise 50, $p = \frac{1}{3}$.)

55. $\displaystyle\int_1^{\infty} \frac{1}{x^3}\,dx = \frac{1}{3-1} = \frac{1}{2}$ converges.

(See Exercise 49, $p = 3$.)

56. $\displaystyle\int_0^{\infty} x^4 e^{-x}\,dx$ converges.

(See Exercise 51.)

57. Since $\displaystyle\frac{1}{x^2+5} \le \frac{1}{x^2}$ on $[1, \infty)$ and $\displaystyle\int_1^{\infty} \frac{1}{x^2}\,dx$ converges by Exercise 49, $\displaystyle\int_1^{\infty} \frac{1}{x^2+5}\,dx$ converges.

58. Since $\displaystyle\frac{1}{\sqrt{x-1}} \ge \frac{1}{x}$ on $[2, \infty)$ and $\displaystyle\int_2^{\infty} \frac{1}{x}\,dx$ diverges by Exercise 49, $\displaystyle\int_2^{\infty} \frac{1}{\sqrt{x-1}}\,dx$ diverges.

59. Since $\displaystyle\frac{1}{\sqrt[3]{x(x-1)}} \ge \frac{1}{\sqrt[3]{x^2}}$ on $[2, \infty)$ and $\displaystyle\int_2^{\infty} \frac{1}{\sqrt[3]{x^2}}\,dx$ diverges by Exercise 49, $\displaystyle\int_2^{\infty} \frac{1}{\sqrt[3]{x(x-1)}}\,dx$ diverges.

60. Since $\displaystyle\frac{1}{\sqrt{x}(1+x)} \le \frac{1}{x^{3/2}}$ on $[1, \infty)$ and $\displaystyle\int_1^{\infty} \frac{1}{x^{3/2}}\,dx$ converges by Exercise 49, $\displaystyle\int_1^{\infty} \frac{1}{\sqrt{x}(1+x)}\,dx$ converges.

61. Since $e^{-x^2} \le e^{-x}$ on $[1, \infty)$ and $\displaystyle\int_0^{\infty} e^{-x}\,dx$ converges (see Exercise 9), $\displaystyle\int_0^{\infty} e^{-x^2}\,dx$ converges.

62. $\displaystyle\frac{1}{\sqrt{x}\ln x} \ge \frac{1}{x}$ since $\sqrt{x}\ln x < x$ on $[2, \infty)$. Since $\displaystyle\int_2^{\infty} \frac{1}{x}\,dx$ diverges by Exercise 49, $\displaystyle\int_2^{\infty} \frac{1}{\sqrt{x}\ln x}\,dx$ diverges.

63. Answers will vary.

64. See the definitions, pages 578, 581.

65. $\displaystyle\int_{-1}^{1} \frac{1}{x^3}\,dx = \int_{-1}^{0} \frac{1}{x^3}\,dx + \int_0^{1} \frac{1}{x^3}\,dx$

These two integrals diverge by Exercise 50.

66. $\displaystyle\frac{10}{x^2 - 2x} = \frac{10}{x(x-2)} \Rightarrow x = 0, 2.$

You must analyze three improper integrals, and each must converge in order for the original integral to converge.

$$\int_0^3 f(x)\,dx = \int_0^1 f(x)\,dx + \int_1^2 f(x)\,dx + \int_2^3 f(x)\,dx$$

67.
$$\begin{aligned} A &= \int_{-\infty}^{1} e^x\,dx \\ &= \lim_{b\to-\infty} \int_b^1 e^x\,dx \\ &= \lim_{b\to-\infty} \Big[e^x\Big]_b^1 \\ &= \lim_{b\to-\infty} \left[e - e^b\right] = e \end{aligned}$$

68.
$$\begin{aligned} A &= \int_0^1 -\ln x\,dx \\ &= -\lim_{b\to0^+} \int_b^1 \ln x\,dx \\ &= -\lim_{b\to0^+} \Big[x\ln x - x\Big]_b^1 \\ &= -\lim_{b\to0^+} [(0-1) - b\ln b + b] \\ &= 1 \end{aligned}$$

Note: $\displaystyle\lim_{b\to0^+} b\ln b = \lim_{b\to0^+} \frac{\ln b}{1/b} = \lim_{b\to0^+} \frac{1/b}{-1/b^2} = 0$

69. $A = \int_{-\infty}^{\infty} \frac{1}{x^2+1}\,dx$

$= \lim_{b\to-\infty} \int_b^0 \frac{1}{x^2+1}\,dx + \lim_{b\to\infty} \int_0^b \frac{1}{x^2+1}\,dx$

$= \lim_{b\to-\infty} \Big[\arctan(x)\Big]_b^0 + \lim_{b\to\infty} \Big[\arctan(x)\Big]_0^b$

$= \lim_{b\to-\infty} [0 - \arctan(b)] + \lim_{b\to\infty} [\arctan(b) - 0]$

$= -\left(-\frac{\pi}{2}\right) + \frac{\pi}{2} = \pi$

70. $A = \int_{-\infty}^{\infty} \frac{8}{x^2+4}\,dx$

$= \lim_{b\to-\infty} \int_b^0 \frac{8}{x^2+4}\,dx + \lim_{b\to\infty} \int_0^b \frac{8}{x^2+4}\,dx$

$= \lim_{b\to-\infty} \left[4\arctan\left(\frac{x}{2}\right)\right]_b^0 + \lim_{b\to\infty} \left[4\arctan\left(\frac{x}{2}\right)\right]_0^b$

$= \lim_{b\to-\infty} \left[0 - 4\arctan\left(\frac{b}{2}\right)\right] + \lim_{b\to\infty} \left[4\arctan\left(\frac{b}{2}\right) - 0\right]$

$= -4\left(\frac{-\pi}{2}\right) + 4\left(\frac{\pi}{2}\right) = 4\pi$

71. (a) $A = \int_0^{\infty} e^{-x}\,dx$

$= \lim_{b\to\infty} \Big[-e^{-x}\Big]_0^b = 0 - (-1) = 1$

(b) **Disk:**

$V = \pi \int_0^{\infty} (e^{-x})^2\,dx$

$= \lim_{b\to\infty} \pi\left[-\frac{1}{2}e^{-2x}\right]_0^b = \frac{\pi}{2}$

(c) **Shell:**

$V = 2\pi \int_0^{\infty} xe^{-x}\,dx$

$= \lim_{b\to\infty} \left\{2\pi\Big[-e^{-x}(x+1)\Big]_0^b\right\} = 2\pi$

72. (a) $A = \int_1^{\infty} \frac{1}{x^2}\,dx = \left[-\frac{1}{x}\right]_1^{\infty} = 1$

(b) **Disk:**

$V = \pi \int_1^{\infty} \frac{1}{x^4}\,dx = \lim_{b\to\infty} \left[-\frac{\pi}{3x^3}\right]_1^b = \frac{\pi}{3}$

(c) **Shell:**

$V = 2\pi \int_1^{\infty} x\left(\frac{1}{x^2}\right)dx = \lim_{b\to\infty} \Big[2\pi(\ln x)\Big]_1^b = \infty$

Diverges

73. $x^{2/3} + y^{2/3} = 4$

$\frac{2}{3}x^{-1/3} + \frac{2}{3}y^{-1/3}y' = 0$

$y' = \frac{-y^{1/3}}{x^{1/3}}$

$\sqrt{1+(y')^2} = \sqrt{1 + \frac{y^{2/3}}{x^{2/3}}} = \sqrt{\frac{x^{2/3}+y^{2/3}}{x^{2/3}}} = \sqrt{\frac{4}{x^{2/3}}} = \frac{2}{x^{1/3}},\quad (x > 0)$

$s = 4\int_0^8 \frac{2}{x^{1/3}}\,dx = \lim_{b\to0^+} \left[8 \cdot \frac{3}{2}x^{2/3}\right]_b^8 = 48$

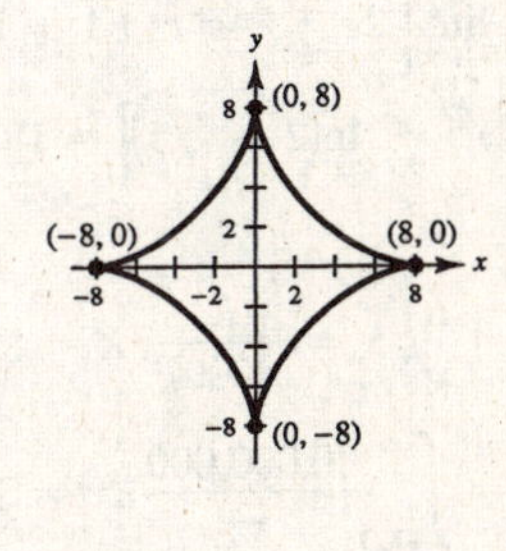

74. $y = \sqrt{16 - x^2}, 0 \le x \le 4$

$y' = \frac{-x}{\sqrt{16-x^2}}$

$s = \int_0^4 \sqrt{1 + \frac{x^2}{16-x^2}}\,dx = \int_0^4 \frac{4}{\sqrt{16-x^2}}\,dx$

$= \lim_{t\to4^-} \int_0^t \frac{4}{\sqrt{16-x^2}}\,dx$

$= \lim_{t\to4^-} \left[4\arcsin\left(\frac{x}{4}\right)\right]_0^t$

$= \lim_{t\to4^-} 4\arcsin\left(\frac{t}{4}\right) = 2\pi$

75. $(x-2)^2 + y^2 = 1$

$2(x-2) + 2yy' = 0$

$$y' = \frac{-(x-2)}{y}$$

$$\sqrt{1+(y')^2} = \sqrt{1+[(x-2)^2/y^2]} = \frac{1}{y} \text{ (Assume } y > 0.)$$

$$S = 4\pi \int_1^3 \frac{x}{y}\,dx = 4\pi \int_1^3 \frac{x}{\sqrt{1-(x-2)^2}}\,dx = 4\pi \int_1^3 \left[\frac{x-2}{\sqrt{1-(x-2)^2}} + \frac{2}{\sqrt{1-(x-2)^2}}\right] dx$$

$$= \lim_{\substack{a\to 1^+ \\ b\to 3^-}} \left\{ 4\pi\left[-\sqrt{1-(x-2)^2} + 2\arcsin(x-2)\right]_a^b \right\} = 4\pi[0 + 2\arcsin(1) - 2\arcsin(-1)] = 8\pi^2$$

76. $y = 2e^{-x}$

$y' = -2e^{-x}$

$$S = 2\pi \int_0^\infty (2e^{-x})\sqrt{1+4e^{-2x}}\,dx$$

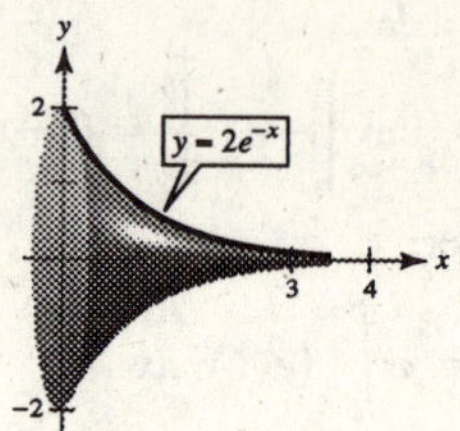

Let $u = e^{-x}$, $du = -e^{-x}\,dx$.

$$\int e^{-x}\sqrt{1+4e^{-2x}}\,dx = -\int \sqrt{1+4u^2}\,du$$

$$= -\frac{1}{4}\left[2u\sqrt{4u^2+1} + \ln\left|2u + \sqrt{4u^2+1}\right|\right] + C$$

$$= -\frac{1}{4}\left[2e^{-x}\sqrt{4e^{-2x}+1} + \ln\left|2e^{-x} + \sqrt{4e^{-2x}+1}\right|\right] + C$$

$$S = 4\pi \lim_{b\to\infty} \int_0^b (e^{-x})\sqrt{1+4e^{-2x}}\,dx$$

$$= -\pi \lim_{b\to\infty} \left[2e^{-x}\sqrt{4e^{-2x}+1} + \ln\left|2e^{-x} + \sqrt{4e^{-2x}+1}\right|\right]_0^b$$

$$= \pi\left[2\sqrt{5} + \ln(2+\sqrt{5})\right] \approx 18.5849$$

77. (a) $F(x) = \dfrac{K}{x^2}$, $5 = \dfrac{K}{(4000)^2}$, $K = 80{,}000{,}000$

$$W = \int_{4000}^\infty \frac{80{,}000{,}000}{x^2}\,dx = \lim_{b\to\infty}\left[\frac{-80{,}000{,}000}{x}\right]_{4000}^b = 20{,}000 \text{ mi-ton}$$

(b) $$\frac{W}{2} = 10{,}000 = \left[\frac{-80{,}000{,}000}{x}\right]_{4000}^b = \frac{-80{,}000{,}000}{b} + 20{,}000$$

$$\frac{80{,}000{,}000}{b} = 10{,}000$$

$$b = 8000$$

Therefore, 4000 miles *above* the earth's surface.

78. (a) $F(x) = \dfrac{k}{x^2}$, $10 = \dfrac{k}{4000^2}$, $k = 10(4000^2)$

$$W = \int_{4000}^\infty \frac{10(4000^2)}{x^2}\,dx = \lim_{b\to\infty}\left[\frac{-10(4000^2)}{x}\right]_{4000}^b$$

$$= \frac{10(4000^2)}{4000} = 40{,}000 \text{ mi-ton}$$

(b) $$\frac{W}{2} = 20{,}000 = \left[\frac{-10(4000^2)}{x}\right]_{4000}^b = \frac{-10(4000^2)}{b} + 40{,}000$$

$$\frac{10(4000^2)}{b} = 20{,}000$$

$$b = 8000$$

Therefore, 4000 miles above the earth's surface.

79. (a) $\displaystyle\int_{-\infty}^{\infty} \frac{1}{7}e^{-t/7}\,dt = \int_0^{\infty}\frac{1}{7}e^{-t/7}\,dt = \lim_{b\to\infty}\Big[-e^{-t/7}\Big]_0^b = 1$

(b) $\displaystyle\int_0^4 \frac{1}{7}e^{-t/7}\,dt = \Big[-e^{-t/7}\Big]_0^4 = -e^{-4/7} + 1$

$\approx 0.4353 = 43.53\%$

(c) $\displaystyle\int_0^{\infty} t\left[\frac{1}{7}e^{-t/7}\right]dt = \lim_{b\to\infty}\Big[-te^{-t/7} - 7e^{-t/7}\Big]_0^b$

$= 0 + 7 = 7$

80. (a) $\displaystyle\int_{-\infty}^{\infty} \frac{2}{5}e^{-2t/5}\,dt = \int_0^{\infty}\frac{2}{5}e^{-2t/5}\,dt = \lim_{b\to\infty}\Big[-e^{-2t5}\Big]_0^b = 1$

(b) $\displaystyle\int_0^4 \frac{2}{5}e^{-2t/5}\,dt = \Big[-e^{-2t/5}\Big]_0^4 = -e^{-8/5} + 1$

$\approx 0.7981 = 79.81\%$

(c) $\displaystyle\int_0^{\infty} t\left[\frac{2}{5}e^{-2t/5}\right]dt = \lim_{b\to\infty}\Big[-te^{2t/5} - \frac{5}{2}e^{-2t/5}\Big]_0^b = \frac{5}{2}$

81. (a) $\displaystyle C = 650{,}000 + \int_0^5 25{,}000\,e^{-0.06t}\,dt = 650{,}000 - \left[\frac{25{,}000}{0.06}e^{-0.06t}\right]_0^5 \approx \$757{,}992.41$

(b) $\displaystyle C = 650{,}000 + \int_0^{10} 25{,}000e^{-0.06t}\,dt \approx \$837{,}995.15$

(c) $\displaystyle C = 650{,}000 + \int_0^{\infty} 25{,}000e^{-0.06t}\,dt = 650{,}000 - \lim_{b\to\infty}\left[\frac{25{,}000}{0.06}e^{-0.06t}\right]_0^b \approx \$1{,}066{,}666.67$

82. (a) $\displaystyle C = 650{,}000 + \int_0^5 25{,}000(1 + 0.08t)e^{-0.06t}\,dt$

$\displaystyle = 650{,}000 + 25{,}000\left[-\frac{1}{0.06}e^{-0.06t} - 0.08\left(\frac{t}{0.06}e^{-0.06t} + \frac{1}{(0.06)^2}e^{-0.06t}\right)\right]_0^5 \approx \$778{,}512.58$

(b) $\displaystyle C = 650{,}000 + \int_0^{10} 25{,}000(1 + 0.08t)e^{-0.06t}\,dt$

$\displaystyle = 650{,}000 + 25{,}000\left[-\frac{1}{0.06}e^{-0.06t} - 0.08\left(\frac{t}{0.06}e^{-0.06t} + \frac{1}{(0.06)^2}e^{-0.06t}\right)\right]_0^{10} \approx \$905{,}718.14$

(c) $\displaystyle C = 650{,}000 + \int_0^{\infty} 25{,}000(1 + 0.08t)e^{-0.06t}\,dt$

$\displaystyle = 650{,}000 + 25{,}000\lim_{b\to\infty}\left[-\frac{t}{0.06}e^{-0.06t} - 0.08\left(\frac{t}{0.06}e^{-0.06t} + \frac{1}{(0.06)^2}e^{-0.06t}\right)\right]_0^b \approx \$1{,}622{,}222.22$

83. Let $K = \dfrac{2\pi NI r}{k}$. Then

$$P = K\int_c^{\infty} \frac{1}{(r^2 + x^2)^{3/2}}\,dx.$$

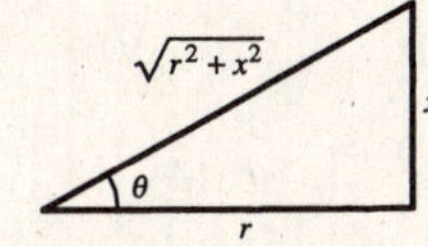

Let $x = r\tan\theta$, $dx = r\sec^2\theta\,d\theta$, $\sqrt{r^2 + x^2} = r\sec\theta$.

$$\int \frac{1}{(r^2 + x^2)^{3/2}}\,dx = \int \frac{r\sec^2\theta\,d\theta}{r^3\sec^3\theta} = \frac{1}{r^2}\int \cos\theta\,d\theta$$

$$= \frac{1}{r^2}\sin\theta + C = \frac{1}{r^2}\frac{x}{\sqrt{r^2 + x^2}} + C$$

Hence,

$$P = K\frac{1}{r^2}\lim_{b\to\infty}\left[\frac{x}{\sqrt{r^2 + x^2}}\right]_c^b$$

$$= \frac{K}{r^2}\left[1 - \frac{c}{\sqrt{r^2 + c^2}}\right]$$

$$= \frac{K\left(\sqrt{r^2 + c^2} - c\right)}{r^2\sqrt{r^2 + c^2}}$$

$$= \frac{2\pi NI\left(\sqrt{r^2 + c^2} - c\right)}{kr\sqrt{r^2 + c^2}}.$$

84. $F = \int_0^\infty \frac{GM\delta}{(a+x)^2}\,dx$

$= \lim_{b\to\infty}\left[\frac{-GM\delta}{a+x}\right]_0^b$

$= \frac{GM\delta}{a}$

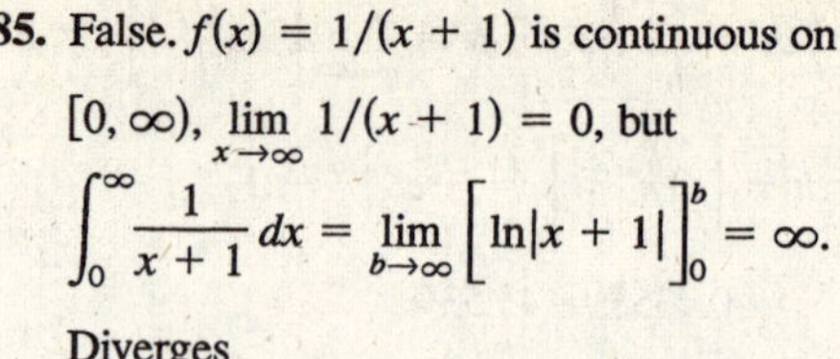

85. False. $f(x) = 1/(x+1)$ is continuous on $[0, \infty)$, $\lim_{x\to\infty} 1/(x+1) = 0$, but

$\int_0^\infty \frac{1}{x+1}\,dx = \lim_{b\to\infty}\left[\ln|x+1|\right]_0^b = \infty.$

Diverges

86. False. This is equivalent to Exercise 85.

87. True

88. True

89. (a) $\int_1^\infty \frac{1}{x}\,dx = \lim_{b\to\infty}\left[\ln|x|\right]_1^b = \infty$

$\int_1^\infty \frac{1}{x^2}\,dx = \lim_{b\to\infty}\left[-\frac{1}{x}\right]_1^b = 1$

$\int_1^\infty \frac{1}{x^n}\,dx$ will converge if $n > 1$ and will diverge if $n \le 1$.

(b) It would appear to converge.

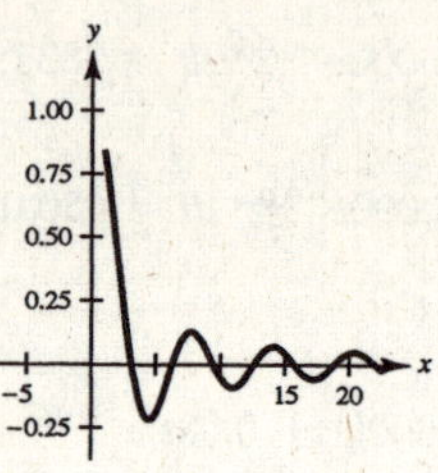

(c) Let $dv = \sin x\,dx \Rightarrow v = -\cos x$

$u = \frac{1}{x} \Rightarrow du = -\frac{1}{x^2}\,dx.$

$\int_1^\infty \frac{\sin x}{x}\,dx = \lim_{b\to 0}\left[-\frac{\cos x}{x}\right]_1^b - \int_1^\infty \frac{\cos x}{x^2}\,dx$

$= \cos 1 - \int_1^\infty \frac{\cos x}{x^2}\,dx$

Converges

90. (a) Yes, the integrand is not defined at $x = \pi/2$.

(b)

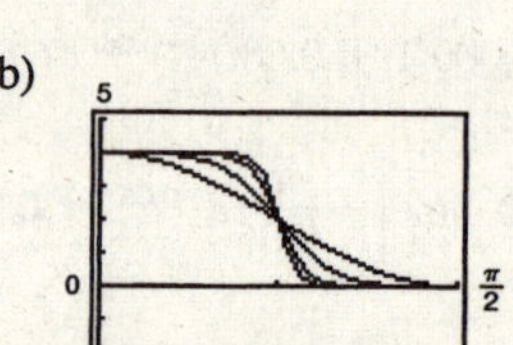

(c) As $n \to \infty$, the integral approaches $4(\pi/4) = \pi$.

(d) $I_n = \int_0^{\pi/2} \frac{4}{1 + (\tan x)^n}\,dx$

$I_2 \approx 3.14159$

$I_4 \approx 3.14159$

$I_8 \approx 3.14159$

$I_{12} \approx 3.14159$

91. $\Gamma(n) = \int_0^\infty x^{n-1}e^{-x}\,dx$

(a) $\Gamma(1) = \int_0^\infty e^{-x}\,dx = \lim_{b\to\infty}\left[-e^{-x}\right]_0^b = 1$

$\Gamma(2) = \int_0^\infty xe^{-x}\,dx = \lim_{b\to\infty}\left[-e^{-x}(x+1)\right]_0^b = 1$

$\Gamma(3) = \int_0^\infty x^2e^{-x}\,dx = \lim_{b\to\infty}\left[-x^2e^{-x} - 2xe^{-x} - 2e^{-x}\right]_0^b = 2$

(b) $\Gamma(n+1) = \int_0^\infty x^ne^{-x}\,dx = \lim_{b\to\infty}\left[-x^ne^{-x}\right]_0^b + \lim_{b\to\infty} n\int_0^b x^{n-1}e^{-x}\,dx = 0 + n\Gamma(n) \quad (u = x^n,\ dv = e^{-x}\,dx)$

(c) $\Gamma(n) = (n-1)!$

92. For $n = 1$,

$$I_1 = \int_0^{\infty} \frac{x}{(x^2+1)^4}\,dx = \lim_{b\to\infty} \frac{1}{2}\int_0^b (x^2+1)^{-4}(2x\,dx) = \lim_{b\to\infty}\left[-\frac{1}{6}\frac{1}{(x^2+1)^3}\right]_0^b = \frac{1}{6}.$$

For $n > 1$,

$$I_n = \int_0^{\infty} \frac{x^{2n-1}}{(x^2+1)^{n+3}}\,dx = \lim_{b\to\infty}\left[\frac{-x^{2n-2}}{2(n+2)(x^2+1)^{n+2}}\right]_0^b + \frac{n-1}{n+2}\int_0^{\infty}\frac{x^{2n-3}}{(x^2+1)^{n+2}}\,dx = 0 + \frac{n-1}{n+2}(I_{n-1})$$

$$\left(\text{Parts: } u = x^{2n-2},\ du = (2n-2)x^{2n-3}\,dx,\ dv = \frac{x}{(x^2+1)^{n+3}}\,dx,\ v = \frac{-1}{2(n+2)(x^2+1)^{n+2}}\right)$$

(a) $\displaystyle\int_0^{\infty}\frac{x}{(x^2+1)^4}\,dx = \lim_{b\to\infty}\left[-\frac{1}{6(x^2+1)^3}\right]_0^b = \frac{1}{6}$

(b) $\displaystyle\int_0^{\infty}\frac{x^3}{(x^2+1)^5}\,dx = \frac{1}{4}\int_0^{\infty}\frac{x}{(x^2+1)^4}\,dx = \frac{1}{4}\left(\frac{1}{6}\right) = \frac{1}{24}$

(c) $\displaystyle\int_0^{\infty}\frac{x^5}{(x^2+1)^6} = \frac{2}{5}\int_0^{\infty}\frac{x^3}{(x^2+1)^5}\,dx = \frac{2}{5}\left(\frac{1}{24}\right) = \frac{1}{60}$

93. $f(t) = 1$

$$F(s) = \int_0^{\infty} e^{-st}\,dt = \lim_{b\to\infty}\left[-\frac{1}{s}e^{-st}\right]_0^b = \frac{1}{s},\ s > 0$$

94. $f(t) = t$

$$F(s) = \int_0^{\infty} te^{-st}\,dt = \lim_{b\to\infty}\left[\frac{1}{s^2}(-st-1)e^{-st}\right]_0^b$$

$$= \frac{1}{s^2},\ s > 0$$

95. $f(t) = t^2$

$$F(s) = \int_0^{\infty} t^2e^{-st}\,dt = \lim_{b\to\infty}\left[\frac{1}{s^3}(-s^2t^2 - 2st - 2)e^{-st}\right]_0^b$$

$$= \frac{2}{s^3},\ s > 0$$

96. $f(t) = e^{at}$

$$F(s) = \int_0^{\infty} e^{at}e^{-st}\,dt = \int_0^{\infty} e^{t(a-s)}\,dt$$

$$= \lim_{b\to\infty}\left[\frac{1}{a-s}e^{t(a-s)}\right]_0^b$$

$$= 0 - \frac{1}{a-s} = \frac{1}{s-a},\ s > a$$

97. $f(t) = \cos at$

$$F(s) = \int_0^{\infty} e^{-st}\cos at\,dt$$

$$= \lim_{b\to\infty}\left[\frac{e^{-st}}{s^2+a^2}(-s\cos at + a\sin at)\right]_0^b$$

$$= 0 + \frac{s}{s^2+a^2} = \frac{s}{s^2+a^2},\ s > 0$$

98. $f(t) = \sin at$

$$F(s) = \int_0^{\infty} e^{-st}\sin at\,dt$$

$$= \lim_{b\to\infty}\left[\frac{e^{-st}}{s^2+a^2}(-s\sin at - a\cos at)\right]_0^b$$

$$= 0 + \frac{a}{s^2+a^2} = \frac{a}{s^2+a^2},\ s > 0$$

99. $f(t) = \cosh at$

$$F(s) = \int_0^{\infty} e^{-st}\cosh at\,dt = \int_0^{\infty} e^{-st}\left(\frac{e^{at}+e^{-at}}{2}\right)dt = \frac{1}{2}\int_0^{\infty}\left[e^{t(-s+a)} + e^{t(-s-a)}\right]dt$$

$$= \lim_{b\to\infty}\frac{1}{2}\left[\frac{1}{(-s+a)}e^{t(-s+a)} + \frac{1}{(-s-a)}e^{t(-s-a)}\right]_0^b = 0 - \frac{1}{2}\left[\frac{1}{(-s+a)} + \frac{1}{(-s-a)}\right]$$

$$= \frac{-1}{2}\left[\frac{1}{(-s+a)} + \frac{1}{(-s-a)}\right] = \frac{s}{s^2-a^2},\ s > |a|$$

100. $f(t) = \sinh at$

$$F(s) = \int_0^\infty e^{-st} \sinh at\, dt = \int_0^\infty e^{-st}\left(\frac{e^{at} - e^{-at}}{2}\right) dt = \frac{1}{2}\int_0^\infty \left[e^{t(-s+a)} - e^{t(-s-a)}\right] dt$$

$$= \lim_{b\to\infty} \frac{1}{2}\left[\frac{1}{(-s+a)}e^{t(-s+a)} - \frac{1}{(-s-a)}e^{t(-s-a)}\right]_0^b = 0 - \frac{1}{2}\left[\frac{1}{(-s+a)} - \frac{1}{(-s-a)}\right]$$

$$= \frac{-1}{2}\left[\frac{1}{(-s+a)} - \frac{1}{(-s-a)}\right] = \frac{a}{s^2 - a^2}, s > |a|$$

101. (a) $f(x) = \dfrac{1}{3\sqrt{2\pi}}e^{-(x-70)^2/18}$

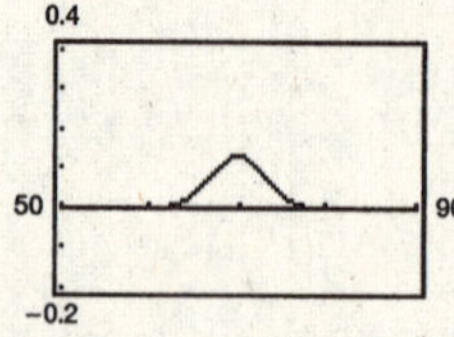

$$\int_{50}^{90} f(x)\, dx \approx 1.0$$

(b) $P(72 \le x < \infty) \approx 0.2525$

(c) $0.5 - P(70 \le x \le 72) \approx 0.5 - 0.2475 = 0.2525$

These are the same answers because by symmetry,

$P(70 \le x < \infty) = 0.5$

and

$0.5 = P(70 \le x < \infty)$

$= P(70 \le x \le 72) + P(72 \le x < \infty)$.

102. (a)

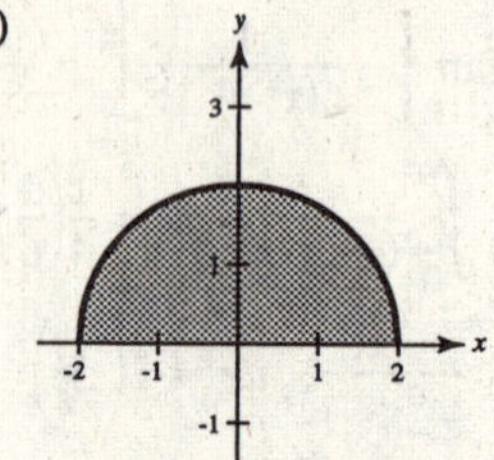

(b) Area $= \dfrac{1}{2}\pi(2)^2 = 2\pi$

Arc length is also $\dfrac{1}{2}(2\pi(2)) = 2\pi$.

Hence, the corresponding integrals are equal.

Let $y = \sqrt{4 - x^2}$, $y' = \dfrac{-x}{\sqrt{4 - x^2}}$

$$1 + (y')^2 = \frac{4}{4 - x^2} \Rightarrow \sqrt{1 + (y')^2} = \frac{2}{\sqrt{4 - x^2}}.$$

Thus, $\displaystyle\int_{-2}^{2} \sqrt{4 - x^2}\, dx = \int_{-2}^{2} \frac{2}{\sqrt{4 - x^2}}\, dx.$

(area) (arc length)

103. $$\int_0^\infty \left(\frac{1}{\sqrt{x^2+1}} - \frac{c}{x+1}\right) dx = \lim_{b\to\infty}\int_0^b \left(\frac{1}{\sqrt{x^2+1}} - \frac{c}{x+1}\right) dx$$

$$= \lim_{b\to\infty}\left[\ln\left|x + \sqrt{x^2+1}\right| - c\ln|x+1|\right]_0^b$$

$$= \lim_{b\to\infty}\left[\ln\left(b + \sqrt{b^2+1}\right) - \ln(b+1)^c\right] = \lim_{b\to\infty} \ln\left[\frac{b + \sqrt{b^2+1}}{(b+1)^c}\right]$$

This limit exists for $c = 1$, and you have

$$\lim_{b\to\infty} \ln\left[\frac{b + \sqrt{b^2+1}}{(b+1)}\right] = \ln 2.$$

104. $$\int_1^\infty \left(\frac{cx}{x^2+2} - \frac{1}{3x}\right) dx = \lim_{b\to\infty}\int_1^b \left(\frac{cx}{x^2+2} - \frac{1}{3x}\right) dx$$

$$= \lim_{b\to\infty}\left[\frac{c}{2}\ln(x^2+2) - \frac{1}{3}\ln|x|\right]_1^b = \lim_{b\to\infty} \ln\left[\frac{(x^2+2)^{c/2}}{x^{1/3}}\right]_1^b = \lim_{b\to\infty}\left[\ln\frac{(b^2+2)^{c/2}}{b^{1/3}} - \ln 3^{c/2}\right]$$

This limit exists if $c = 1/3$, and you have

$$\lim_{b\to\infty}\left[\ln\frac{(b^2+2)^{1/6}}{b^{1/3}} - \ln 3^{1/6}\right] = -\ln 3^{1/6} = \frac{-\ln 3}{6}.$$

105. $f(x) = \begin{cases} x \ln x, & 0 < x \le 2 \\ 0, & x = 0 \end{cases}$

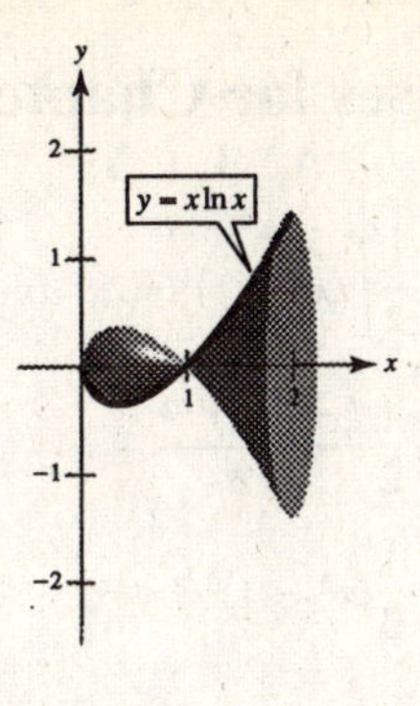

$$V = \pi \int_0^2 (x \ln x)^2 \, dx$$

Let $u = \ln x$, $e^u = x$, $e^u \, du = dx$.

$$V = \pi \int_{-\infty}^{\ln 2} e^{2u} u^2 (e^4 \, du) = \pi \int_{-\infty}^{\ln 2} e^{3u} u^2 \, du$$

$$= \lim_{b \to -\infty} \left[\pi \left[\frac{u^2}{3} - \frac{2u}{9} + \frac{2}{27} \right] e^{3u} \right]_b^{\ln 2} = \pi \left[\frac{(\ln 2)^2}{3} - \frac{2 \ln 2}{9} + \frac{2}{27} \right] 8 \approx 2.0155$$

106. $V = \pi \int_0^1 (-\ln x)^2 \, dx$

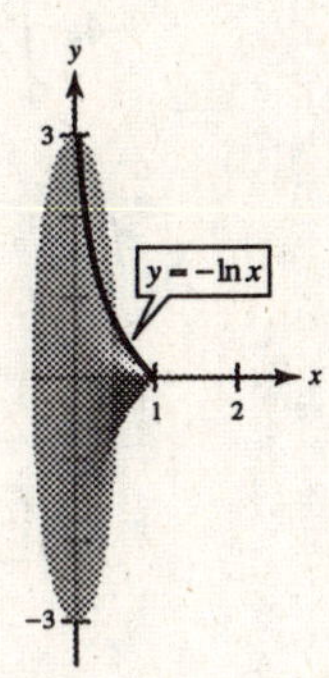

$$= \lim_{b \to 0^+} \pi \int_b^1 (\ln x)^2 \, dx$$

$$= \lim_{b \to 0^+} \pi x \left[(\ln x)^2 - 2 \ln x + 2 \right]_b^1$$

$$= \lim_{b \to 0^+} \pi [2 - b(\ln b)^2 - 2b \ln b - 2b]$$

$$= 2\pi$$

107. $u = \sqrt{x}$, $u^2 = x$, $2u \, du = dx$

$$\int_0^1 \frac{\sin x}{\sqrt{x}} \, dx = \int_0^1 \frac{\sin(u^2)}{u} (2u \, du) = \int_0^1 2 \sin(u^2) \, du$$

Trapezoidal Rule ($n = 5$): 0.6278

108. $u = \sqrt{1 - x}$, $1 - x = u^2$, $2u \, du = -dx$

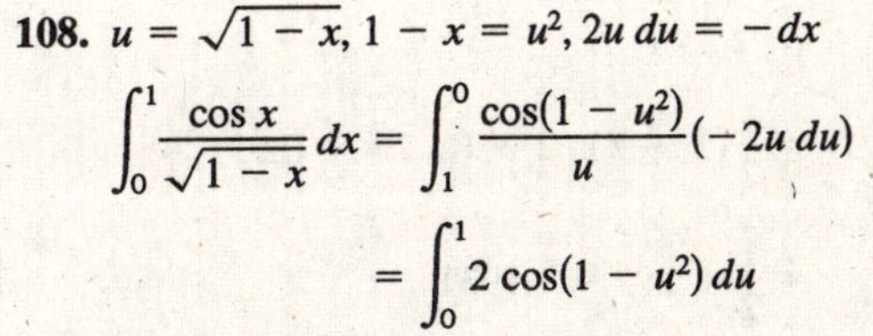

$$\int_0^1 \frac{\cos x}{\sqrt{1 - x}} \, dx = \int_1^0 \frac{\cos(1 - u^2)}{u} (-2u \, du)$$

$$= \int_0^1 2 \cos(1 - u^2) \, du$$

Trapezoidal Rule ($n = 5$): 1.4997

109. (a)

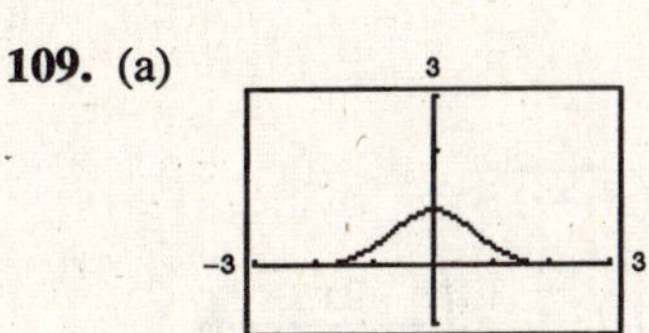

(b) Let $y = e^{-x^2}$, $0 \le x < \infty$.

$\ln y = -x^2$

$x = \sqrt{-\ln y}$ for $0 < y \le 1$

The area bounded by $y = e^{-x^2}$, $x = 0$ and $y = 0$ is

$$\int_0^\infty e^{-x^2} \, dx = \int_0^1 \sqrt{-\ln y} \, dy, \quad \left(= \frac{\sqrt{\pi}}{2} \right).$$

110. Assume $a < b$. The proof is similar if $a > b$.

$$\int_{-\infty}^a f(x) \, dx + \int_a^\infty f(x) \, dx = \lim_{c \to -\infty} \int_c^a f(x) \, dx + \lim_{d \to \infty} \int_a^d f(x) \, dx$$

$$= \lim_{c \to -\infty} \int_c^a f(x) \, dx + \lim_{d \to \infty} \left[\int_a^b f(x) \, dx + \int_b^d f(x) \, dx \right]$$

$$= \lim_{c \to -\infty} \int_c^a f(x) \, dx + \int_a^b f(x) \, dx + \lim_{d \to \infty} \int_b^d f(x) \, dx$$

$$= \lim_{c \to -\infty} \left[\int_c^a f(x) \, dx + \int_a^b f(x) \, dx \right] + \lim_{d \to \infty} \int_b^d f(x) \, dx = \lim_{c \to -\infty} \int_c^b f(x) \, dx + \lim_{d \to \infty} \int_b^d f(x) \, dx$$

$$= \int_{-\infty}^b f(x) \, dx + \int_b^\infty f(x) \, dx$$

Review Exercises for Chapter 8

1. $\displaystyle\int x\sqrt{x^2-1}\,dx = \frac{1}{2}\int (x^2-1)^{1/2}(2x)\,dx$

$\displaystyle= \frac{1}{2}\frac{(x^2-1)^{3/2}}{3/2} + C$

$\displaystyle= \frac{1}{3}(x^2-1)^{3/2} + C$

2. $\displaystyle\int xe^{x^2-1}\,dx = \frac{1}{2}\int e^{x^2-1}(2x)\,dx$

$\displaystyle= \frac{1}{2}e^{x^2-1} + C$

3. $\displaystyle\int \frac{x}{x^2-1}\,dx = \frac{1}{2}\int \frac{2x}{x^2-1}\,dx$

$\displaystyle= \frac{1}{2}\ln|x^2-1| + C$

4. $\displaystyle\int \frac{x}{\sqrt{1-x^2}}\,dx = -\frac{1}{2}\int (1-x^2)^{-1/2}(-2x)\,dx$

$\displaystyle= -\frac{1}{2}\frac{(1-x^2)^{1/2}}{1/2} + C$

$= -\sqrt{1-x^2} + C$

5. Let $u = \ln(2x)$, $du = \dfrac{1}{x}\,dx$.

$\displaystyle\int_1^e \frac{\ln(2x)}{x}\,dx = \int_{\ln 2}^{1+\ln 2} u\,du$

$\displaystyle= \left.\frac{u^2}{2}\right]_{\ln 2}^{1+\ln 2}$

$\displaystyle= \frac{1}{2}[1 + 2\ln 2 + (\ln 2)^2 - (\ln 2)^2]$

$\displaystyle= \frac{1}{2} + \ln 2 \approx 1.1931$

6. Let $u = 2x-3$, $du = 2\,dx$, $x = \dfrac{1}{2}(u+3)$.

$\displaystyle\int_{3/2}^{2} 2x\sqrt{2x-3}\,dx = \int_0^1 (u+3)u^{1/2}\frac{1}{2}\,du$

$\displaystyle= \frac{1}{2}\int_0^1 (u^{3/2} + 3u^{1/2})\,du$

$\displaystyle= \frac{1}{2}\left[\frac{2}{5}u^{5/2} + 2u^{3/2}\right]_0^1$

$\displaystyle= \frac{1}{2}\left[\frac{2}{5} + 2\right]$

$\displaystyle= \frac{6}{5}$

7. $\displaystyle\int \frac{16}{\sqrt{16-x^2}}\,dx = 16\arcsin\left(\frac{x}{4}\right) + C$

8. $\displaystyle\frac{x^4+2x^2+x+1}{x^4+2x^2+1} = 1 + \frac{x}{(x^2+1)^2}$

$\displaystyle\int \frac{x^4+2x^2+x+1}{(x^2+1)^2}\,dx = \int dx + \frac{1}{2}\int \frac{2x}{(x^2+1)^2}\,dx$

$\displaystyle= x - \frac{1}{2(x^2+1)} + C$

9. $\displaystyle\int e^{2x}\sin 3x\,dx = -\frac{1}{3}e^{2x}\cos 3x + \frac{2}{3}\int e^{2x}\cos 3x\,dx$

$\displaystyle= -\frac{1}{3}e^{2x}\cos 3x + \frac{2}{3}\left(\frac{1}{3}e^{2x}\sin 3x - \frac{2}{3}\int e^{2x}\sin 3x\,dx\right)$

$\displaystyle\frac{13}{9}\int e^{2x}\sin 3x\,dx = -\frac{1}{3}e^{2x}\cos 3x + \frac{2}{9}e^{2x}\sin 3x$

$\displaystyle\int e^{2x}\sin 3x\,dx = \frac{e^{2x}}{13}(2\sin 3x - 3\cos 3x) + C$

(1) $dv = \sin 3x\,dx \Rightarrow v = -\dfrac{1}{3}\cos 3x$

$u = e^{2x} \Rightarrow du = 2e^{2x}\,dx$

(2) $dv = \cos 3x\,dx \Rightarrow v = \dfrac{1}{3}\sin 3x$

$u = e^{2x} \Rightarrow du = 2e^{2x}\,dx$

10. $\int (x^2 - 1)e^x\,dx = (x^2 - 1)e^x - 2\int xe^x\,dx = (x^2 - 1)e^x - 2xe^x + 2\int e^x\,dx = e^x(x^2 - 2x + 1) + 1$

(1) $dv = e^x\,dx \Rightarrow v = e^x$ (2) $dv = e^x\,dx \Rightarrow v = e^x$

$u = x^2 - 1 \Rightarrow du = 2x\,dx$ $u = x \Rightarrow du = dx$

11. $u = x,\ du = dx,\ dv = (x-5)^{1/2}\,dx,\ v = \frac{2}{3}(x-5)^{3/2}$

$$\int x\sqrt{x-5}\,dx = \frac{2}{3}x(x-5)^{3/2} - \int \frac{2}{3}(x-5)^{3/2}\,dx$$
$$= \frac{2}{3}x(x-5)^{3/2} - \frac{4}{15}(x-5)^{5/2} + C$$
$$= (x-5)^{3/2}\left[\frac{2}{3}x - \frac{4}{15}(x-5)\right] + C$$
$$= (x-5)^{3/2}\left[\frac{6}{15}x + \frac{4}{3}\right] + C$$
$$= \frac{2}{15}(x-5)^{3/2}[3x + 10] + C$$

12. $u = \arctan 2x,\ du = \frac{2}{1+4x^2}\,dx,\ dv = dx, v = x$

$$\int \arctan 2x\,dx = x\arctan 2x - \int \frac{2x}{1+4x^2}\,dx$$
$$= x\arctan 2x - \frac{1}{4}\ln(1 + 4x^2) + C$$

13. $$\int x^2 \sin 2x\,dx = -\frac{1}{2}x^2\cos 2x + \int x\cos 2x\,dx$$
$$= -\frac{1}{2}x^2\cos 2x + \frac{1}{2}x\sin 2x - \frac{1}{2}\int \sin 2x\,dx$$
$$= -\frac{1}{2}x^2\cos 2x + \frac{x}{2}\sin 2x + \frac{1}{4}\cos 2x + C$$

(1) $dv = \sin 2x\,dx \Rightarrow v = -\frac{1}{2}\cos 2x$

$u = x^2 \Rightarrow du = 2x\,dx$

(2) $dv = \cos 2x\,dx \Rightarrow v = \frac{1}{2}\sin 2x$

$u = x \Rightarrow du = dx$

14. $$\int \ln\sqrt{x^2-1}\,dx = \frac{1}{2}\int \ln(x^2-1)\,dx$$
$$= \frac{1}{2}x\ln|x^2-1| - \int \frac{x^2}{x^2-1}\,dx$$
$$= \frac{1}{2}x\ln|x^2-1| - \int dx - \int \frac{1}{x^2-1}\,dx$$
$$= \frac{1}{2}x\ln|x^2-1| - x - \frac{1}{2}\ln\left|\frac{x-1}{x+1}\right| + C$$

$dv = dx \Rightarrow v = x$

$u = \ln(x^2-1) \Rightarrow du = \frac{2x}{x^2-1}\,dx$

15. $$\int x\arcsin 2x\,dx = \frac{x^2}{2}\arcsin 2x - \int \frac{x^2}{\sqrt{1-4x^2}}\,dx$$
$$= \frac{x^2}{2}\arcsin 2x - \frac{1}{8}\int \frac{2(2x)^2}{\sqrt{1-(2x)^2}}\,dx$$
$$= \frac{x^2}{2}\arcsin 2x - \frac{1}{8}\left(\frac{1}{2}\right)\left[-(2x)\sqrt{1-4x^2} + \arcsin 2x\right] + C \quad \text{(by Formula 43 of Integration Tables)}$$
$$= \frac{1}{16}\left[(8x^2-1)\arcsin 2x + 2x\sqrt{1-4x^2}\right] + C$$

$dv = x\,dx \Rightarrow v = \frac{x^2}{2}$

$u = \arcsin 2x \Rightarrow du = \frac{2}{\sqrt{1-4x^2}}\,dx$

16. $\displaystyle\int e^x \arctan(e^x)\,dx = e^x \arctan(e^x) - \int \frac{e^{2x}}{1+e^{2x}}\,dx$

$$= e^x \arctan(e^x) - \frac{1}{2}\ln(1+e^{2x}) + C$$

$dv = e^x\,dx \implies v = e^x$

$u = \arctan e^x \implies du = \dfrac{e^x}{1+e^{2x}}\,dx$

17. $\displaystyle\int \cos^3(\pi x - 1)\,dx = \int [1 - \sin^2(\pi x - 1)]\cos(\pi x - 1)\,dx$

$$= \frac{1}{\pi}\left[\sin(\pi x - 1) - \frac{1}{3}\sin^3(\pi x - 1)\right] + C$$

$$= \frac{1}{3\pi}\sin(\pi x - 1)[3 - \sin^2(\pi x - 1)] + C$$

$$= \frac{1}{3\pi}\sin(\pi x - 1)[3 - (1 - \cos^2(\pi x - 1))] + C$$

$$= \frac{1}{3\pi}\sin(\pi x - 1)[2 + \cos^2(\pi x - 1)] + C$$

18. $\displaystyle\int \sin^2\frac{\pi x}{2}\,dx = \int \frac{1}{2}(1 - \cos \pi x)\,dx = \frac{1}{2}\left[x - \frac{1}{\pi}\sin \pi x\right] + C = \frac{1}{2\pi}[\pi x - \sin \pi x] + C$

19. $\displaystyle\int \sec^4\left(\frac{x}{2}\right)dx = \int \left[\tan^2\left(\frac{x}{2}\right) + 1\right]\sec^2\left(\frac{x}{2}\right)dx$

$$= \int \tan^2\left(\frac{x}{2}\right)\sec^2\left(\frac{x}{2}\right)dx + \int \sec^2\left(\frac{x}{2}\right)dx$$

$$= \frac{2}{3}\tan^3\left(\frac{x}{2}\right) + 2\tan\left(\frac{x}{2}\right) + C = \frac{2}{3}\left[\tan^3\left(\frac{x}{2}\right) + 3\tan\left(\frac{x}{2}\right)\right] + C$$

20. $\displaystyle\int \tan\theta \sec^4\theta\,d\theta = \int (\tan^3\theta + \tan\theta)\sec^2\theta\,d\theta = \frac{1}{4}\tan^4\theta + \frac{1}{2}\tan^2\theta + C_1$

or

$$\int \tan\theta\sec^4\theta\,d\theta = \int \sec^3\theta(\sec\theta\tan\theta)\,d\theta = \frac{1}{4}\sec^4\theta + C_2$$

21. $\displaystyle\int \frac{1}{1-\sin\theta}\,d\theta = \int \frac{1}{1-\sin\theta}\cdot\frac{1+\sin\theta}{1+\sin\theta}\,d\theta = \int \frac{1+\sin\theta}{\cos^2\theta}\,d\theta = \int (\sec^2\theta + \sec\theta\tan\theta)\,d\theta = \tan\theta + \sec\theta + C$

22. $\displaystyle\int \cos 2\theta(\sin\theta + \cos\theta)^2\,d\theta = \int (\cos^2\theta - \sin^2\theta)(\sin\theta + \cos\theta)^2\,d\theta$

$$= \int (\sin\theta + \cos\theta)^3(\cos\theta - \sin\theta)\,d\theta = \frac{1}{4}(\sin\theta + \cos\theta)^4 + C$$

23. $A = \int_{\pi/4}^{3\pi/4} \sin^4 x\, dx$. Using the Table of Integrals,

$$\int \sin^4 x\, dx = -\frac{\sin^3 x \cos x}{4} + \frac{3}{4}\int \sin^2 x\, dx$$

$$= \frac{-\sin^3 x \cos x}{4} + \frac{3}{4}\left[\frac{1}{2}(x - \sin x \cos x)\right] + C$$

$$\int_{\pi/4}^{3\pi/4} \sin^4 x\, dx = \left[\frac{-\sin^3 x \cos x}{4} + \frac{3}{8}x - \frac{3}{8}\sin x \cos x\right]_{\pi/4}^{3\pi/4}$$

$$= \left(\frac{1}{16} + \frac{9\pi}{32} + \frac{3}{16}\right) - \left(\frac{-1}{16} + \frac{3\pi}{32} - \frac{3}{16}\right)$$

$$= \frac{3\pi}{16} + \frac{1}{2} \approx 1.0890$$

24. $A = \int_0^{\pi/6} \cos(3x)\cos x\, dx$

$$= \int_0^{\pi/6} \frac{1}{2}[\cos 2x + \cos 4x]\, dx$$

$$= \left[\frac{\sin 2x}{4} + \frac{\sin 4x}{8}\right]_0^{\pi/6}$$

$$= \left(\frac{\sqrt{3}}{8} + \frac{\sqrt{3}}{16}\right) - 0$$

$$= \frac{3\sqrt{3}}{16}$$

25. $\displaystyle\int \frac{-12}{x^2\sqrt{4-x^2}}\, dx = \int \frac{-24\cos\theta\, d\theta}{(4\sin^2\theta)(2\cos\theta)}$

$$= -3\int \csc^2\theta\, d\theta$$

$$= 3\cot\theta + C$$

$$= \frac{3\sqrt{4-x^2}}{x} + C$$

$x = 2\sin\theta$, $dx = 2\cos\theta\, d\theta$, $\sqrt{4-x^2} = 2\cos\theta$

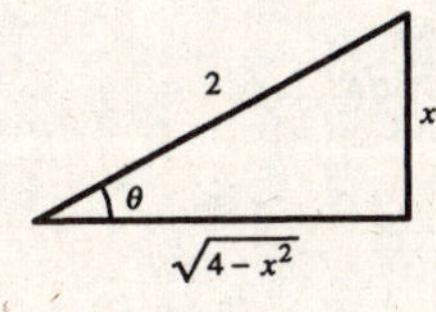

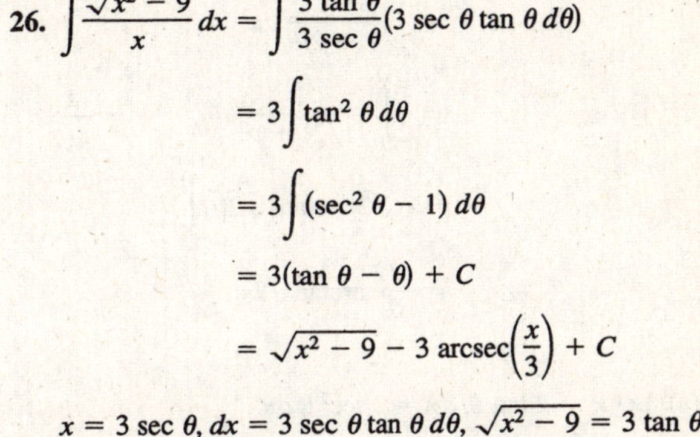

26. $\displaystyle\int \frac{\sqrt{x^2-9}}{x}\, dx = \int \frac{3\tan\theta}{3\sec\theta}(3\sec\theta\tan\theta\, d\theta)$

$$= 3\int \tan^2\theta\, d\theta$$

$$= 3\int (\sec^2\theta - 1)\, d\theta$$

$$= 3(\tan\theta - \theta) + C$$

$$= \sqrt{x^2-9} - 3\,\text{arcsec}\left(\frac{x}{3}\right) + C$$

$x = 3\sec\theta$, $dx = 3\sec\theta\tan\theta\, d\theta$, $\sqrt{x^2-9} = 3\tan\theta$

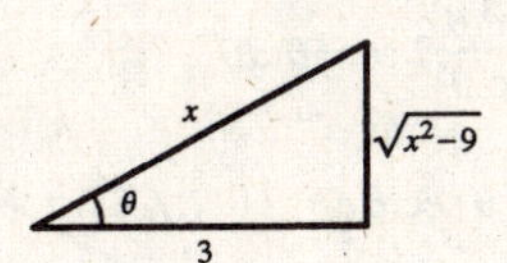

27. $x = 2\tan\theta$

$dx = 2\sec^2\theta\, d\theta$

$4 + x^2 = 4\sec^2\theta$

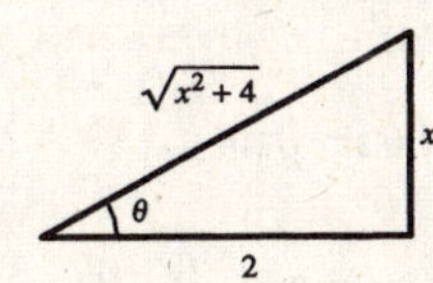

$$\int \frac{x^3}{\sqrt{4+x^2}}\, dx = \int \frac{8\tan^3\theta}{2\sec\theta} 2\sec^2\theta\, d\theta$$

$$= 8\int \tan^3\theta\sec\theta\, d\theta$$

$$= 8\int (\sec^2\theta - 1)\tan\theta\sec\theta\, d\theta$$

$$= 8\left[\frac{\sec^3\theta}{3} - \sec\theta\right] + C$$

$$= 8\left[\frac{(x^2+4)^{3/2}}{24} - \frac{\sqrt{x^2+4}}{2}\right] + C$$

$$= \sqrt{x^2+4}\left[\frac{1}{3}(x^2+4) - 4\right] + C$$

$$= \frac{1}{3}x^2\sqrt{x^2+4} - \frac{8}{3}\sqrt{x^2+4} + C$$

$$= \frac{1}{3}(x^2+4)^{1/2}(x^2-8) + C$$

28. $\displaystyle\int \sqrt{9-4x^2}\,dx = \frac{1}{2}\int \sqrt{9-(2x)^2}\,(2)\,dx$

$\displaystyle = \frac{1}{2}\cdot\frac{1}{2}\left[9\arcsin\frac{2x}{3} + 2x\sqrt{9-4x^2}\right] + C$

$\displaystyle = \frac{9}{4}\arcsin\frac{2x}{3} + \frac{x}{2}\sqrt{9-4x^2} + C$

29. $\displaystyle\int_{-2}^{0} \sqrt{4-x^2}\,dx = \frac{1}{2}\left[4\arcsin\left(\frac{x}{2}\right) + x\sqrt{4-x^2}\right]_{-2}^{0}$

$\displaystyle = \frac{1}{2}[0 - 4\arcsin(-1)]$

$\displaystyle = \frac{1}{2}\left[-4\left(\frac{-\pi}{2}\right)\right]$

$= \pi$

Note: The integral represents the area of a quarter circle of radius 2: $A = \frac{1}{4}(\pi 2^2) = \pi$.

30. Let $u = \cos\theta$, $du = -\sin\theta\,d\theta$.

$\displaystyle\int_{0}^{\pi/2} \frac{\sin\theta}{1+2\cos^2\theta}\,d\theta = \int_{1}^{0} \frac{1}{1+2u^2}(-du)$

$\displaystyle = \int_{0}^{1} \frac{1}{1+2u^2}\,du$

$\displaystyle = \frac{1}{2}\int_{0}^{1} \frac{1}{(1/2)+u^2}\,du, \quad a = \frac{1}{\sqrt{2}}$

$\displaystyle = \frac{1}{2}\sqrt{2}\arctan\left(\sqrt{2}u\right)\Big]_{0}^{1}$

$\displaystyle = \frac{\sqrt{2}}{2}\arctan\sqrt{2}$

31. (a) Let $x = 2\tan\theta$, $dx = 2\sec^2\theta\,d\theta$.

$\displaystyle\int \frac{x^3}{\sqrt{4+x^2}}\,dx = \int \frac{8\tan^3\theta}{2\sec\theta}\,2\sec^2\theta\,d\theta$

$\displaystyle = 8\int \tan^3\theta\sec\theta\,d\theta$

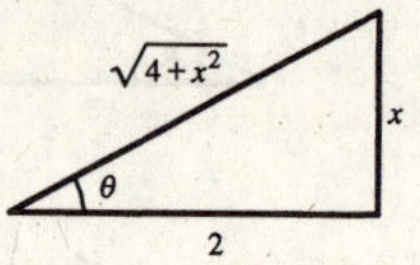

$\displaystyle = 8\int \frac{\sin^3\theta}{\cos^4\theta}\,d\theta$

$\displaystyle = 8\int (1-\cos^2\theta)\cos^{-4}\theta\sin\theta\,d\theta$

$\displaystyle = 8\int (\cos^{-4}\theta - \cos^{-2}\theta)\sin\theta\,d\theta$

$\displaystyle = 8\left[\frac{\cos^{-3}\theta}{3} - \frac{\cos^{-1}\theta}{-1}\right] + C$

$\displaystyle = \frac{8}{3}\sec\theta(\sec^2\theta - 3) + C$

$\displaystyle = \frac{8}{3}\,\frac{\sqrt{4+x^2}}{2}\left(\frac{4+x^2}{4} - 3\right) + C$

$\displaystyle = \frac{1}{3}\sqrt{4+x^2}\,(x^2-8) + C$

(b) $\displaystyle\int \frac{x^3}{\sqrt{4+x^2}}\,dx = \int \frac{x^2}{\sqrt{4+x^2}}\,x\,dx$

$\displaystyle = \int \frac{(u^2-4)u\,du}{u}$

$\displaystyle = \int (u^2-4)\,du$

$\displaystyle = \frac{1}{3}u^3 - 4u + C$

$\displaystyle = \frac{u}{3}(u^2-12) + C$

$\displaystyle = \frac{\sqrt{4+x^2}}{3}(x^2-8) + C$

$u^2 = 4+x^2$, $2u\,du = 2x\,dx$

(c) $\displaystyle\int \frac{x^3}{\sqrt{4+x^2}}\,dx = x^2\sqrt{4+x^2} - \int 2x\sqrt{4+x^2}\,dx$

$\displaystyle = x^2\sqrt{4+x^2} - \frac{2}{3}(4+x^2)^{3/2} + C = \frac{\sqrt{4+x^2}}{3}(x^2-8) + C$

$\displaystyle dv = \frac{x}{\sqrt{4+x^2}}\,dx \Rightarrow v = \sqrt{4+x^2}$

$u = x^2 \qquad \Rightarrow du = 2x\,dx$

32. (a) $\int x\sqrt{4+x}\,dx = 64\int \tan^3\theta\sec^3\theta\,d\theta$

$= 64\int (\sec^4\theta - \sec^2\theta)\sec\theta\tan\theta\,d\theta$

$= \frac{64\sec^3\theta}{15}(3\sec^3\theta - 5) + C$

$= \frac{2(4+x)^{3/2}}{15}(3x-8) + C$

$x = 4\tan^2\theta,\ dx = 8\tan\theta\sec^2\theta\,d\theta,$

$\sqrt{4+x} = 2\sec\theta$

(b) $\int x\sqrt{4+x}\,dx = 2\int (u^4 - 4u^2)\,du$

$= \frac{2u^3}{15}(3u^2 - 20) + C$

$= \frac{2(4+x)^{3/2}}{15}(3x-8) + C$

$u^2 = 4 + x,\ dx = 2u\,du$

(c) $\int x\sqrt{4+x}\,dx = \int (u^{3/2} - 4u^{1/2})\,du$

$= \frac{2u^{3/2}}{15}(3u - 20) + C$

$= \frac{2(4+x)^{3/2}}{15}(3x-8) + C$

$u = 4 + x,\ du = dx$

(d) $\int x\sqrt{4+x}\,dx = \frac{2x}{3}(4+x)^{3/2} - \frac{2}{3}\int (4+x)^{3/2}\,dx$

$= \frac{2x}{3}(4+x)^{3/2} - \frac{4}{15}(4+x)^{5/2} + C$

$= \frac{2(4+x)^{3/2}}{15}(3x-8) + C$

$dv = \sqrt{4+x}\,dx \Rightarrow\ v = \frac{2}{3}(4+x)^{3/2}$

$u = x \qquad \Rightarrow du = dx$

33. $\frac{x-28}{x^2-x-6} = \frac{A}{x-3} + \frac{B}{x+2}$

$x - 28 = A(x+2) + B(x-3)$

$x = -2 \Rightarrow -30 = B(-5) \Rightarrow B = 6$

$x = 3 \quad \Rightarrow -25 = A(5) \quad \Rightarrow A = -5$

$\int \frac{x-28}{x^2-x-6}\,dx = \int\left(\frac{-5}{x-3} + \frac{6}{x+2}\right)dx = -5\ln|x-3| + 6\ln|x+2| + C$

34. $\frac{2x^3 - 5x^2 + 4x - 4}{x^2 - x} = 2x - 3 + \frac{4}{x} - \frac{3}{x-1}$

$\int \frac{2x^3 - 5x^2 + 4x - 4}{x^2 - x}\,dx = \int\left(2x - 3 + \frac{4}{x} - \frac{3}{x-1}\right)dx = x^2 - 3x + 4\ln|x| - 3\ln|x-1| + C$

35. $\frac{x^2+2x}{(x-1)(x^2+1)} = \frac{A}{x-1} + \frac{Bx+C}{x^2+1}$

$x^2 + 2x = A(x^2+1) + (Bx+C)(x-1)$

Let $x = 1$: $3 = 2A \Rightarrow A = \frac{3}{2}$ Let $x = 0$: $0 = A - C \Rightarrow C = \frac{3}{2}$ Let $x = 2$: $8 = 5A + 2B + C \Rightarrow B = -\frac{1}{2}$

$\int \frac{x^2+2x}{x^3-x^2+x-1}\,dx = \frac{3}{2}\int \frac{1}{x-1}\,dx - \frac{1}{2}\int \frac{x-3}{x^2+1}\,dx$

$= \frac{3}{2}\int \frac{1}{x-1}\,dx - \frac{1}{4}\int \frac{2x}{x^2+1}\,dx + \frac{3}{2}\int \frac{1}{x^2+1}\,dx$

$= \frac{3}{2}\ln|x-1| - \frac{1}{4}\ln|x^2+1| + \frac{3}{2}\arctan x + C$

$= \frac{1}{4}[6\ln|x-1| - \ln(x^2+1) + 6\arctan x] + C$

36. $\dfrac{4x-2}{3(x-1)^2} = \dfrac{A}{x-1} + \dfrac{B}{(x-1)^2}$

$4x - 2 = 3A(x-1) + 3B$

Let $x = 1$: $2 = 3B \Rightarrow B = \dfrac{2}{3}$

Let $x = 2$: $6 = 3A + 3B \Rightarrow A = \dfrac{4}{3}$

$$\int \frac{4x-2}{3(x-1)^2}\,dx = \frac{4}{3}\int \frac{1}{x-1}\,dx + \frac{2}{3}\int \frac{1}{(x-1)^2}\,dx = \frac{4}{3}\ln|x-1| - \frac{2}{3(x-1)} + C = \frac{2}{3}\left(2\ln|x-1| - \frac{1}{x-1}\right) + C$$

37. $\dfrac{x^2}{x^2+2x-15} = 1 + \dfrac{15-2x}{x^2+2x-15}$

$\dfrac{15-2x}{(x-3)(x+5)} = \dfrac{A}{x-3} + \dfrac{B}{x+5}$

$15 - 2x = A(x+5) + B(x-3)$

Let $x = 3$: $9 = 8A \Rightarrow A = \dfrac{9}{8}$

Let $x = -5$: $25 = -8B \Rightarrow B = -\dfrac{25}{8}$

$$\int \frac{x^2}{x^2+2x-15}\,dx = \int dx + \frac{9}{8}\int \frac{1}{x-3}\,dx - \frac{25}{8}\int \frac{1}{x+5}\,dx = x + \frac{9}{8}\ln|x-3| - \frac{25}{8}\ln|x+5| + C$$

38. $\displaystyle\int \frac{\sec^2\theta}{\tan\theta(\tan\theta - 1)}\,d\theta = \int \frac{1}{u(u-1)}\,du = \int \frac{1}{u-1}\,du - \int \frac{1}{u}\,du$

$$= \ln|u-1| - \ln|u| + C = \ln\left|\frac{\tan\theta - 1}{\tan\theta}\right| + C = \ln|1 - \cot\theta| + C$$

$u = \tan\theta$, $du = \sec^2\theta\,d\theta$

$\dfrac{1}{u(u-1)} = \dfrac{A}{u} + \dfrac{B}{u-1}$

$1 = A(u-1) + Bu$

Let $u = 0$: $1 = -A \Rightarrow A = -1$

Let $u = 1$: $1 = B$

39. $\displaystyle\int \frac{x}{(2+3x)^2}\,dx = \frac{1}{9}\left[\frac{2}{2+3x} + \ln|2+3x|\right] + C$

(Formula 4)

40. $\displaystyle\int \frac{x}{\sqrt{2+3x}}\,dx = \frac{-2(4-3x)}{27}\sqrt{2+3x} + C$ (Formula 21)

$$= \frac{6x-8}{27}\sqrt{2+3x} + C$$

41. Let $u = x^2$, $du = 2x\,dx$.

$$\int_0^{\sqrt{\pi}/2} \frac{x}{1+\sin x^2}\,dx = \frac{1}{2}\int_0^{\pi/4} \frac{1}{1+\sin u}\,du$$

$$= \frac{1}{2}\Big[\tan u - \sec u\Big]_0^{\pi/4}$$

$$= \frac{1}{2}[(1-\sqrt{2}) - (0-1)]$$

$$= 1 - \frac{\sqrt{2}}{2}$$

42. Let $u = x^2$, $du = 2x\,dx$.

$$\int_0^1 \frac{x}{1+e^{x^2}}\,dx = \frac{1}{2}\int_0^1 \frac{1}{1+e^u}\,du$$

$$= \frac{1}{2}\Big[u - \ln(1+e^u)\Big]_0^1$$

$$= \frac{1}{2}[(1 - \ln(1+e)) + \ln 2]$$

$$= \frac{1}{2}\left[1 + \ln\left(\frac{2}{1+e}\right)\right]$$

43. $\displaystyle\int \frac{x}{x^2+4x+8}\,dx = \frac{1}{2}\left[\ln|x^2+4x+8| - 4\int \frac{1}{x^2+4x+8}\,dx\right]$ (Formula 15)

$\displaystyle= \frac{1}{2}[\ln|x^2+4x+8|] - 2\left[\frac{2}{\sqrt{32-16}}\arctan\left(\frac{2x+4}{\sqrt{32-16}}\right)\right] + C$ (Formula 14)

$\displaystyle= \frac{1}{2}\ln|x^2+4x+8| - \arctan\left(1+\frac{x}{2}\right) + C$

44. $\displaystyle\int \frac{3}{2x\sqrt{9x^2-1}}\,dx = \frac{3}{2}\int \frac{1}{3x\sqrt{(3x)^2-1}}\,3\,dx$ $(u = 3x)$

$\displaystyle= \frac{3}{2}\operatorname{arcsec}|3x| + C$ (Formula 33)

45. $\displaystyle\int \frac{1}{\sin \pi x \cos \pi x}\,dx = \frac{1}{\pi}\int \frac{1}{\sin \pi x \cos \pi x}(\pi)\,dx$ $(u = \pi x)$

$\displaystyle= \frac{1}{\pi}\ln|\tan \pi x| + C$ (Formula 58)

46. $\displaystyle\int \frac{1}{1+\tan \pi x}\,dx = \frac{1}{\pi}\int \frac{1}{1+\tan \pi x}(\pi)dx$ $(u = \pi x)$

$\displaystyle= \frac{1}{\pi}\frac{1}{2}[\pi x + \ln|\cos \pi x + \sin \pi x|] + C$ (Formula 71)

47. $dv = dx \implies v = x$

$u = (\ln x)^n \implies du = n(\ln x)^{n-1}\frac{1}{x}dx$

$\displaystyle\int (\ln x)^n\,dx = x(\ln x)^n - n\int (\ln x)^{n-1}\,dx$

48. $\displaystyle\int \tan^n x\,dx = \int \tan^{n-2} x(\sec^2 x - 1)\,dx$

$\displaystyle= \int \tan^{n-2} x \sec^2 x\,dx - \int \tan^{n-2} x\,dx$

$\displaystyle= \frac{1}{n-1}\tan^{n-1} x - \int \tan^{n-2} x\,dx$

49. $\displaystyle\int \theta \sin\theta \cos\theta\,d\theta = \frac{1}{2}\int \theta \sin 2\theta\,d\theta$

$\displaystyle= -\frac{1}{4}\theta\cos 2\theta + \frac{1}{4}\int \cos 2\theta\,d\theta = -\frac{1}{4}\theta\cos 2\theta + \frac{1}{8}\sin 2\theta + C = \frac{1}{8}(\sin 2\theta - 2\theta\cos 2\theta) + C$

$dv = \sin 2\theta\,d\theta \implies v = -\frac{1}{2}\cos 2\theta$

$u = \theta \implies du = d\theta$

50. $\displaystyle\int \frac{\csc\sqrt{2x}}{\sqrt{x}}\,dx = \sqrt{2}\int \csc\sqrt{2x}\left(\frac{1}{\sqrt{2x}}\right)dx$

$= -\sqrt{2}\ln|\csc\sqrt{2x} + \cot\sqrt{2x}| + C$

$u = \sqrt{2x},\ du = \dfrac{1}{\sqrt{2x}}\,dx$

51. $\displaystyle\int \frac{x^{1/4}}{1+x^{1/2}}\,dx = 4\int \frac{u(u^3)}{1+u^2}\,du$

$\displaystyle= 4\int\left(u^2 - 1 + \frac{1}{u^2+1}\right)du$

$\displaystyle= 4\left(\frac{1}{3}u^3 - u + \arctan u\right) + C$

$\displaystyle= \frac{4}{3}[x^{3/4} - 3x^{1/4} + 3\arctan(x^{1/4})] + C$

$u = \sqrt[4]{x},\ x = u^4,\ dx = 4u^3\,du$

52. $\displaystyle\int \sqrt{1+\sqrt{x}}\,dx = \int u(4u^3 - 4u)\,du = \int (4u^4 - 4u^2)\,du = \frac{4u^5}{5} - \frac{4u^3}{3} + C = \frac{4}{15}(1+\sqrt{x})^{3/2}(3\sqrt{x} - 2) + C$

$u = \sqrt{1+\sqrt{x}},\ x = u^4 - 2u^2 + 1,\ dx = (4u^3 - 4u)\,du$

53. $\displaystyle\int \sqrt{1+\cos x}\,dx = \int \frac{\sqrt{1+\cos x}}{1}\cdot\frac{\sqrt{1-\cos x}}{\sqrt{1-\cos x}}\,dx$

$\displaystyle = \int \frac{\sin x}{\sqrt{1-\cos x}}\,dx$

$\displaystyle = \int (1-\cos x)^{-1/2}(\sin x)\,dx$

$= 2\sqrt{1-\cos x} + C$

$u = 1 - \cos x,\ du = \sin x\,dx$

54. $\displaystyle\frac{3x^3+4x}{(x^2+1)^2} = \frac{Ax+B}{x^2+1} + \frac{Cx+D}{(x^2+1)^2}$

$3x^3 + 4x = (Ax+B)(x^2+1) + Cx + D$

$= Ax^3 + Bx^2 + (A+C)x + (B+D)$

$A = 3, B = 0, A + C = 4 \Rightarrow C = 1,$

$B + D = 0 \Rightarrow D = 0$

$\displaystyle\int \frac{3x^3+4x}{(x^2+1)^2}\,dx = 3\int\frac{x}{x^2+1}\,dx + \int\frac{x}{(x^2+1)^2}\,dx$

$\displaystyle = \frac{3}{2}\ln(x^2+1) - \frac{1}{2(x^2+1)} + C$

55. $\displaystyle\int \cos x \ln(\sin x)\,dx = \sin x\ln(\sin x) - \int \cos x\,dx = \sin x \ln(\sin x) - \sin x + C$

$dv = \cos x\,dx \Rightarrow\ v = \sin x$

$\displaystyle u = \ln(\sin x) \Rightarrow du = \frac{\cos x}{\sin x}\,dx$

56. $\displaystyle\int (\sin\theta + \cos\theta)^2\,d\theta = \int(\sin^2\theta + 2\sin\theta\cos\theta + \cos^2\theta)\,d\theta$

$\displaystyle = \int (1 + \sin 2\theta)\,d\theta = \theta - \frac{1}{2}\cos 2\theta + C = \frac{1}{2}(2\theta - \cos 2\theta) + C$

57. $\displaystyle y = \int\frac{9}{x^2-9}\,dx = \frac{3}{2}\ln\left|\frac{x-3}{x+3}\right| + C$ (by Formula 24 of Integration Tables)

58. $\displaystyle y = \int\frac{\sqrt{4-x^2}}{2x}\,dx = \int\frac{2\cos\theta(2\cos\theta)\,d\theta}{4\sin\theta}$

$\displaystyle = \int(\csc\theta - \sin\theta)\,d\theta$

$= [-\ln|\csc\theta + \cos\theta| + \cos\theta] + C$

$\displaystyle = -\ln\left|\frac{2+\sqrt{4-x^2}}{x}\right| + \frac{\sqrt{4-x^2}}{2} + C$

$x = 2\sin\theta,\ dx = 2\cos\theta\,d\theta,\ \sqrt{4-x^2} = 2\cos\theta$

59. $\displaystyle y = \int \ln(x^2+x)dx = x\ln|x^2+x| - \int\frac{2x^2+x}{x^2+x}\,dx$

$\displaystyle = x\ln|x^2+x| - \int\frac{2x+1}{x+1}\,dx$

$\displaystyle = x\ln|x^2+x| - \int 2dx + \int\frac{1}{x+1}\,dx$

$= x\ln|x^2+x| - 2x + \ln|x+1| + C$

$dv = dx \quad\Rightarrow\quad v = x$

$\displaystyle u = \ln(x^2+x) \Rightarrow du = \frac{2x+1}{x^2+x}\,dx$

60. $\displaystyle y = \int\sqrt{1-\cos\theta}\,d\theta = \int\frac{\sin\theta}{\sqrt{1+\cos\theta}}\,d\theta = -\int(1+\cos\theta)^{-1/2}(-\sin\theta)d\theta = -2\sqrt{1+\cos\theta} + C$

$u = 1 + \cos\theta,\ du = -\sin\theta\,d\theta$

61. $\displaystyle\int_2^{\sqrt{5}} x(x^2-4)^{3/2}\,dx = \left[\frac{1}{5}(x^2-4)^{5/2}\right]_2^{\sqrt{5}} = \frac{1}{5}$

62. $\displaystyle\int_0^1 \frac{x}{(x-2)(x-4)}\,dx = \Big[2\ln|x-4| - \ln|x-2|\Big]_0^1$

$= 2\ln 3 - 2\ln 4 + \ln 2$

$\displaystyle = \ln\frac{9}{8} \approx 0.118$

63. $\displaystyle\int_1^4 \frac{\ln x}{x}\,dx = \left[\frac{1}{2}(\ln x)^2\right]_1^4 = \frac{1}{2}(\ln 4)^2 = 2(\ln 2)^2 \approx 0.961$

64. $\displaystyle\int_0^2 xe^{3x}\,dx = \left[\frac{e^{3x}}{9}(3x - 1)\right]_0^2 = \frac{1}{9}(5e^6 + 1) \approx 224.238$

65. $\displaystyle\int_0^{\pi} x\sin x\,dx = \Big[-x\cos x + \sin x\Big]_0^{\pi} = \pi$

66. $\displaystyle\int_0^3 \frac{x}{\sqrt{1+x}}\,dx = \left[\frac{-2(2-x)}{3}\sqrt{1+x}\right]_0^3 = \frac{4}{3} + \frac{4}{3} = \frac{8}{3}$

67. $\displaystyle A = \int_0^4 x\sqrt{4-x}\,dx = \int_2^0 (4-u^2)u(-2u)\,du$

$\displaystyle = \int_2^0 2(u^4 - 4u^2)\,du$

$\displaystyle = \left[2\left(\frac{u^5}{5} - \frac{4u^3}{3}\right)\right]_2^0 = \frac{128}{15}$

$u = \sqrt{4-x},\ x = 4 - u^2,\ dx = -2u\,du$

68. $\displaystyle A = \int_0^4 \frac{1}{25 - x^2}\,dx$

$\displaystyle = \left[-\frac{1}{10}\ln\left|\frac{x-5}{x+5}\right|\right]_0^4 = -\frac{1}{10}\ln\frac{1}{9} = \frac{1}{10}\ln 9 \approx 0.220$

69. By symmetry, $\bar{x} = 0$, $A = \dfrac{1}{2}\pi$.

$\displaystyle \bar{y} = \frac{2}{\pi}\left(\frac{1}{2}\right)\int_{-1}^{1}\left(\sqrt{1-x^2}\right)^2 dx = \frac{1}{\pi}\left[x - \frac{1}{3}x^3\right]_{-1}^{1} = \frac{4}{3\pi}$

$\displaystyle (\bar{x}, \bar{y}) = \left(0, \frac{4}{3\pi}\right)$

70. By symmetry, $\bar{y} = 0$.

$A = \pi + 4\pi = 5\pi$

$\displaystyle \bar{x} = \frac{1(\pi) + 4(4\pi)}{\pi + 4\pi}$

$\displaystyle = \frac{17\pi}{5\pi} = 3.4$

$(\bar{x}, \bar{y}) = (3.4, 0)$

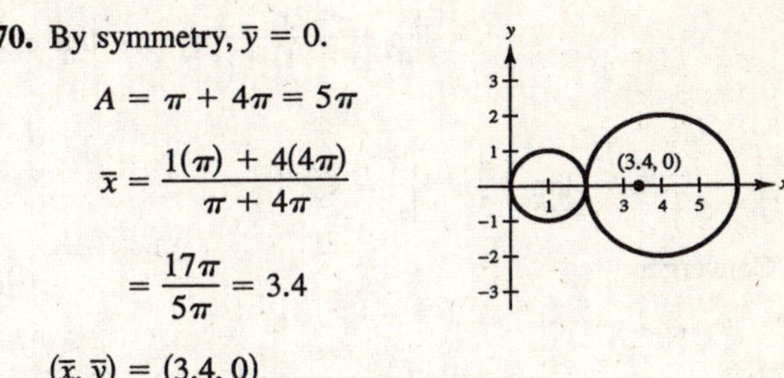

71. $\displaystyle s = \int_0^{\pi} \sqrt{1 + \cos^2 x}\,dx \approx 3.82$

72. $\displaystyle s = \int_0^{\pi} \sqrt{1 + \sin^2 2x}\,dx \approx 3.82$

73. $\displaystyle \lim_{x\to 1}\left[\frac{(\ln x)^2}{x - 1}\right] = \lim_{x\to 1}\left[\frac{2(1/x)\ln x}{1}\right] = 0$

74. $\displaystyle \lim_{x\to 0}\frac{\sin \pi x}{\sin 2\pi x} = \lim_{x\to 0}\frac{\pi\cos\pi x}{2\pi\cos 2\pi x} = \frac{\pi}{2\pi} = \frac{1}{2}$

75. $\displaystyle \lim_{x\to\infty}\frac{e^{2x}}{x^2} = \lim_{x\to\infty}\frac{2e^{2x}}{2x} = \lim_{x\to\infty}\frac{4e^{2x}}{2} = \infty$

76. $\displaystyle \lim_{x\to\infty} xe^{-x^2} = \lim_{x\to\infty}\frac{x}{e^{x^2}} = \lim_{x\to\infty}\frac{1}{2xe^{x^2}} = 0$

77. $\displaystyle y = \lim_{x\to\infty}(\ln x)^{2/x}$

$\displaystyle \ln y = \lim_{x\to\infty}\frac{2\ln(\ln x)}{x} = \lim_{x\to\infty}\left[\frac{2/(x\ln x)}{1}\right] = 0$

Since $\ln y = 0$, $y = 1$.

78. $\displaystyle y = \lim_{x\to 1^+}(x-1)^{\ln x}$

$\displaystyle \ln y = \lim_{x\to 1^+}\left[(\ln x)\ln(x-1)\right]$

$\displaystyle = \lim_{x\to 1^+}\left[\frac{\ln(x-1)}{\dfrac{1}{\ln x}}\right] = \lim_{x\to 1^+}\left[\frac{\dfrac{1}{x-1}}{\left(\dfrac{1}{x}\right)\dfrac{-1}{\ln^2 x}}\right] = \lim_{x\to 1^+}\left[\frac{-\ln^2 x}{\dfrac{x-1}{x}}\right] = \lim_{x\to 1^+}\left[\frac{-2\left(\dfrac{1}{x}\right)(\ln x)}{\dfrac{1}{x^2}}\right]$

$\displaystyle = \lim_{x\to 1^+} 2x(\ln x) = 0$

Since $\ln y = 0$, $y = 1$.

79. $\lim_{n\to\infty} 1000\left(1+\frac{0.09}{n}\right)^n = 1000 \lim_{n\to\infty}\left(1+\frac{0.09}{n}\right)^n$

Let $y = \lim_{n\to\infty}\left(1+\frac{0.09}{n}\right)^n$.

$$\ln y = \lim_{n\to\infty} n\ln\left(1+\frac{0.09}{n}\right) = \lim_{n\to\infty}\frac{\ln\left(1+\frac{0.09}{n}\right)}{\frac{1}{n}} = \lim_{n\to\infty}\left(\frac{\frac{-0.09/n^2}{1+(0.09/n)}}{-\frac{1}{n^2}}\right) = \lim_{n\to\infty}\frac{0.09}{1+\left(\frac{0.09}{n}\right)} = 0.09$$

Thus, $\ln y = 0.09 \implies y = e^{0.09}$ and $\lim_{n\to\infty} 1000\left(1+\frac{0.09}{n}\right)^n = 1000e^{0.09} \approx 1094.17$.

80. $$\lim_{x\to1^+}\left(\frac{2}{\ln x}-\frac{2}{x-1}\right) = \lim_{x\to1^+}\left[\frac{2x-2-2\ln x}{(\ln x)(x-1)}\right]$$
$$= \lim_{x\to1^+}\left[\frac{2-(2/x)}{(x-1)(1/x)+\ln x}\right]$$
$$= \lim_{x\to1^+}\frac{2x-2}{(x-1)+x\ln x} = \lim_{x\to1^+}\frac{2}{1+1+\ln x} = 1$$

81. $$\int_0^{16}\frac{1}{\sqrt[4]{x}}\,dx = \lim_{b\to0^+}\left[\frac{4}{3}x^{3/4}\right]_b^{16} = \frac{32}{3}$$

Converges

82. $$\int_0^1\frac{6}{x-1}\,dx = \lim_{b\to1^-}\left[6\ln|x-1|\right]_0^b = -\infty$$

Diverges

83. $$\int_1^\infty x^2\ln x\,dx = \lim_{b\to\infty}\left[\frac{x^3}{9}(-1+3\ln x)\right]_1^b = \infty$$

Diverges

84. $$\int_0^\infty\frac{e^{-1/x}}{x^2}\,dx = \lim_{\substack{a\to0^+\\ b\to\infty}}\left[e^{-1/x}\right]_a^b = 1-0=1$$

85. Let $u=\ln x$, $du=\frac{1}{x}dx$, $dv=x^{-2}\,dx$, $v=-x^{-1}$.

$$\int\frac{\ln x}{x^2}\,dx = \frac{-\ln x}{x}+\int\frac{1}{x^2}\,dx = \frac{-\ln x}{x}-\frac{1}{x}+C$$
$$\int_1^\infty\frac{\ln x}{x^2}\,dx = \lim_{b\to\infty}\left[\frac{-\ln x}{x}-\frac{1}{x}\right]_1^b$$
$$= \lim_{b\to\infty}\left(\frac{-\ln b}{b}-\frac{1}{b}\right)-(-1)$$
$$= 0+1=1$$

86. $$\int_1^\infty\frac{1}{\sqrt[4]{x}}\,dx = \lim_{b\to\infty}\int_1^b x^{-1/4}\,dx$$
$$= \lim_{b\to\infty}\left[\frac{4}{3}x^{3/4}\right]_1^b$$
$$= \lim_{b\to\infty}\left[\frac{4}{3}b^{3/4}-\frac{4}{3}\right]$$

Diverges

87. $$\int_0^{t_0}500{,}000e^{-0.05t}\,dt = \left[\frac{500{,}000}{-0.05}e^{-0.05t}\right]_0^{t_0}$$
$$= \frac{-500{,}000}{0.05}(e^{-0.05t_0}-1)$$
$$= 10{,}000{,}000(1-e^{-0.05t_0})$$

(a) $t_0 = 20$: \$6,321,205.59

(b) $t_0\to\infty$: \$10,000,000

88. $$V = \pi\int_0^\infty(xe^{-x})^2dx$$
$$= \pi\int_0^\infty x^2e^{-2x}\,dx$$
$$= \lim_{b\to\infty}\left[-\frac{\pi e^{-2x}}{4}(2x^2+2x+1)\right]_0^b = \frac{\pi}{4}$$

89. (a) $P(13\le x<\infty) = \frac{1}{0.95\sqrt{2\pi}}\int_{13}^\infty e^{-(x-12.9)^2/2(0.95)^2}\,dx \approx 0.4581$

(b) $P(15\le x<\infty) = \frac{1}{0.95\sqrt{2\pi}}\int_{15}^\infty e^{-(x-12.9)^2/2(0.95)^2}\,dx \approx 0.0135$

Problem Solving for Chapter 8

1. (a) $\displaystyle\int_{-1}^{1}(1-x^2)\,dx = \left[x - \frac{x^3}{3}\right]_{-1}^{1} = 2\left(1-\frac{1}{3}\right) = \frac{4}{3}$

$$\int_{-1}^{1}(1-x^2)^2\,dx = \int_{-1}^{1}(1-2x^2+x^4)\,dx = \left[x - \frac{2x^3}{3} + \frac{x^5}{5}\right]_{-1}^{1} = 2\left(1-\frac{2}{3}+\frac{1}{5}\right) = \frac{16}{15}$$

(b) Let $x = \sin u$, $dx = \cos u\,du$, $1 - x^2 = 1 - \sin^2 u = \cos^2 u$.

$$\int_{-1}^{1}(1-x^2)^n\,dx = \int_{-\pi/2}^{\pi/2}(\cos^2 u)^n \cos u\,du$$

$$= \int_{-\pi/2}^{\pi/2}\cos^{2n+1} u\,du$$

$$= 2\left[\frac{2}{3}\cdot\frac{4}{5}\cdot\frac{6}{7}\cdots\frac{(2n)}{(2n+1)}\right] \qquad \text{(Wallis's Formula)}$$

$$= 2\left[\frac{2^2\cdot 4^2\cdot 6^2\cdots(2n)^2}{2\cdot3\cdot4\cdot5\cdots(2n)(2n+1)}\right]$$

$$= \frac{2(2^{2n})(n!)^2}{(2n+1)!} = \frac{2^{2n+1}(n!)^2}{(2n+1)!}$$

2. (a) $\displaystyle\int_{0}^{1}\ln x\,dx = \lim_{b\to0^+}\left[x\ln - x\right]_{b}^{1}$

$$= (-1) - \lim_{b\to0^+}(b\ln b - b) = -1$$

Note: $\displaystyle\lim_{b\to0^+} b\ln b = \lim_{b\to0^+}\frac{\ln b}{b^{-1}} = \lim_{b\to0^+}\frac{1/b}{-1/b^2} = 0$

$$\int_{0}^{1}(\ln x)^2\,dx = \lim_{b\to0^+}\left[x(\ln x)^2 - 2x\ln x + 2x\right]_{b}^{1}$$

$$= 2 - \lim_{b\to0^+}(b(\ln b)^2 - 2b\ln b + 2b) = 2$$

(b) Note first that $\displaystyle\lim_{b\to0^+} b(\ln b)^n = 0$ (Mathematical induction).

Also, $\displaystyle\int(\ln x)^{n+1}\,dx = x(\ln x)^{n+1} - (n+1)\int(\ln x)^n\,dx.$

Assume $\displaystyle\int_{0}^{1}(\ln x)^n\,dx = (-1)^n n!.$

Then, $\displaystyle\int_{0}^{1}(\ln x)^{n+1}\,dx = \lim_{b\to0^+}\left[x(\ln x)^{n+1}\right]_{b}^{1} - (n+1)\int_{0}^{1}(\ln x)^n\,dx$

$$= 0 - (n+1)(-1)^n n! = (-1)^{n+1}(n+1)!.$$

3.
$$\lim_{x\to\infty}\left(\frac{x+c}{x-c}\right)^x = 9$$
$$\lim_{x\to\infty} x\ln\left(\frac{x+c}{x-c}\right) = \ln 9$$
$$\lim_{x\to\infty}\frac{\ln(x+c)-\ln(x-c)}{1/x} = \ln 9$$
$$\lim_{x\to\infty}\frac{\dfrac{1}{x+c}-\dfrac{1}{x-c}}{-\dfrac{1}{x^2}} = \ln 9$$
$$\lim_{x\to\infty}\frac{-2c}{(x+c)(x-c)}(-x^2) = \ln 9$$
$$\lim_{x\to\infty}\left(\frac{2cx^2}{x^2-c^2}\right) = \ln 9$$
$$2c = \ln 9$$
$$2c = 2\ln 3$$
$$c = \ln 3$$

4.
$$\lim_{x\to\infty}\left(\frac{x-c}{x+c}\right)^x = \frac{1}{4}$$
$$\lim_{x\to\infty} x\ln\left(\frac{x-c}{x+c}\right) = \ln\frac{1}{4}$$
$$\lim_{x\to\infty}\frac{\ln(x-c)-\ln(x+c)}{1/x} = -\ln 4$$
$$\lim_{x\to\infty}\frac{\dfrac{1}{x-c}-\dfrac{1}{x+c}}{-\dfrac{1}{x^2}} = -\ln 4$$
$$\lim_{x\to\infty}\frac{2c}{(x-c)(x+c)}(-x^2) = -\ln 4$$
$$\lim_{x\to\infty}\frac{2cx^2}{x^2-c^2} = \ln 4$$
$$2c = \ln 4$$
$$2x = 2\ln 2$$
$$c = \ln 2$$

5. $\sin\theta = \dfrac{PB}{OP} = PB, \cos\theta = OB$

$AQ = \widehat{AP} = \theta$

$BR = OR + OB = OR + \cos\theta$

The triangles $\triangle AQR$ and $\triangle BPR$ are similar:

$$\frac{AR}{AQ} = \frac{BR}{BP} \Rightarrow \frac{OR+1}{\theta} = \frac{OR+\cos\theta}{\sin\theta}$$
$$\sin\theta(OR) + \sin\theta = (OR)\theta + \theta\cos\theta$$
$$OR = \frac{\theta\cos\theta - \sin\theta}{\sin\theta - \theta}$$

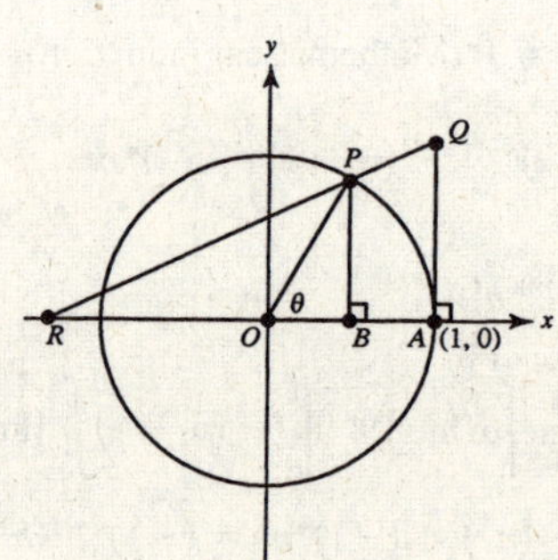

$$\lim_{\theta\to0^+} OR = \lim_{\theta\to0^+}\frac{\theta\cos\theta-\sin\theta}{\sin\theta-\theta}$$
$$= \lim_{\theta\to0^+}\frac{-\theta\sin\theta+\cos\theta-\cos\theta}{\cos\theta-1}$$
$$= \lim_{\theta\to0^+}\frac{-\theta\sin\theta}{\cos\theta-1}$$
$$= \lim_{\theta\to0^+}\frac{-\sin\theta-\theta\cos\theta}{-\sin\theta}$$
$$= \lim_{\theta\to0^+}\frac{\cos\theta+\cos\theta-\theta\sin\theta}{\cos\theta}$$
$$= 2$$

6. $\sin\theta = BD, \cos\theta = OD$

$$\text{Area } \triangle DAB = \frac{1}{2}(DA)(BD) = \frac{1}{2}(1-\cos\theta)\sin\theta$$

$$\text{Shaded area} = \frac{\theta}{2} - \frac{1}{2}(1)(BD) = \frac{\theta}{2} - \frac{1}{2}\sin\theta$$

$$R = \frac{\triangle DAB}{\text{Shaded area}} = \frac{1/2(1-\cos\theta)\sin\theta}{1/2(\theta-\sin\theta)}$$

$$\lim_{\theta\to 0^+} R = \lim_{\theta\to 0^+}\frac{(1-\cos\theta)\sin\theta}{\theta-\sin\theta} = \lim_{\theta\to 0^+}\frac{(1-\cos\theta)\cos\theta + \sin^2\theta}{1-\cos\theta}$$

$$= \lim_{\theta\to 0^+}\frac{(1-\cos\theta)(-\sin\theta) + \cos\theta\sin\theta + 2\sin\theta\cos\theta}{\sin\theta}$$

$$= \lim_{\theta\to 0^+}\frac{-\sin\theta - 4\cos\theta\sin\theta}{\sin\theta} = \lim_{\theta\to 0}\frac{4\cos\theta - 1}{1} = 3$$

7. (a)

Area ≈ 0.2986

(b) Let $x = 3\tan\theta$, $dx = 3\sec^2\theta\,d\theta$, $x^2 + 9 = 9\sec^2\theta$.

$$\int\frac{x^2}{(x^2+9)^{3/2}}\,dx = \int\frac{9\tan^2\theta}{(9\sec^2\theta)^{3/2}}(3\sec^2\theta\,d\theta)$$

$$= \int\frac{\tan^2\theta}{\sec\theta}\,d\theta$$

$$= \int\frac{\sin^2\theta}{\cos\theta}\,d\theta$$

$$= \int\frac{1-\cos^2\theta}{\cos\theta}\,d\theta$$

$$= \ln|\sec\theta + \tan\theta| - \sin\theta + C$$

$$\text{Area} = \int_0^4\frac{x^2}{(x^2+9)^{3/2}}\,dx = \Big[\ln|\sec\theta+\tan\theta| - \sin\theta\Big]_0^{\tan^{-1}(4/3)}$$

$$= \left[\ln\left(\frac{\sqrt{x^2+9}}{3} + \frac{x}{3}\right) - \frac{x}{\sqrt{x^2+9}}\right]_0^4$$

$$= \ln\left(\frac{5}{3} + \frac{4}{3}\right) - \frac{4}{5} = \ln 3 - \frac{4}{5}$$

(c) $x = 3\sinh u$, $dx = 3\cosh u\,du$, $x^2 + 9 = 9\sinh^2 u + 9 = 9\cosh^2 u$

$$A = \int_0^4\frac{x^2}{(x^2+9)^{3/2}}\,dx = \int_0^{\sinh^{-1}(4/3)}\frac{9\sinh^2 u}{(9\cosh^2 u)^{3/2}}(3\cosh u\,du) = \int_0^{\sinh^{-1}(4/3)}\tanh^2 u\,du$$

$$= \int_0^{\sinh^{-1}(4/3)}(1-\operatorname{sech}^2 u)\,du = \Big[u - \tanh u\Big]_0^{\sinh^{-1}(4/3)}$$

$$= \sinh^{-1}\left(\frac{4}{3}\right) - \tanh\left(\sinh^{-1}\left(\frac{4}{3}\right)\right) = \ln\left(\frac{4}{3} + \sqrt{\frac{16}{9}+1}\right) - \tanh\left[\ln\left(\frac{4}{3} + \sqrt{\frac{16}{9}+1}\right)\right]$$

$$= \ln\left(\frac{4}{3}+\frac{5}{3}\right) - \tanh\left(\ln\left(\frac{4}{3}+\frac{5}{3}\right)\right) = \ln 3 - \tanh(\ln 3)$$

$$= \ln 3 - \frac{3-(1/3)}{3+(1/3)} = \ln 3 - \frac{4}{5}$$

8. $u = \tan\frac{x}{2}, \cos x = \frac{1-u^2}{1+u^2}, 2 + \cos x = 2 + \frac{1-u^2}{1+u^2} = \frac{3+u^2}{1+u^2}$

$$dx = \frac{2\,du}{1+u^2}$$

$$\int_0^{\pi/2} \frac{1}{2+\cos x}\,dx = \int_0^1 \left(\frac{1+u^2}{3+u^2}\right)\left(\frac{2}{1+u^2}\right) du$$

$$= \int_0^1 \frac{2}{3+u^2}\,du$$

$$= \left[2\frac{1}{\sqrt{3}}\arctan\left(\frac{u}{\sqrt{3}}\right)\right]_0^1$$

$$= \frac{2}{\sqrt{3}}\arctan\left(\frac{1}{\sqrt{3}}\right)$$

$$= \frac{2}{\sqrt{3}}\frac{\pi}{6} = \frac{\pi\sqrt{3}}{9} \approx 0.6046$$

9. $y = \ln(1-x^2), y' = \frac{-2x}{1-x^2}$

$$1 + (y')^2 = 1 + \frac{4x^2}{(1-x^2)^2} = \frac{1 - 2x^2 + x^4 + 4x^2}{(1-x^2)^2} = \left(\frac{1+x^2}{1-x^2}\right)^2$$

$$\text{Arc length} = \int_0^{1/2} \sqrt{1+(y')^2}\,dx$$

$$= \int_0^{1/2} \left(\frac{1+x^2}{1-x^2}\right) dx$$

$$= \int_0^{1/2} \left(-1 + \frac{2}{1-x^2}\right) dx$$

$$= \int_0^{1/2} \left(-1 + \frac{1}{x+1} + \frac{1}{1-x}\right) dx$$

$$= \Big[-x + \ln(1+x) - \ln(1-x)\Big]_0^{1/2}$$

$$= \left(-\frac{1}{2} + \ln\frac{3}{2} - \ln\frac{1}{2}\right)$$

$$= -\frac{1}{2} + \ln 3 - \ln 2 + \ln 2$$

$$= \ln 3 - \frac{1}{2} \approx 0.5986$$

10. Let $u = cx$, $du = c\,dx$.

$$\int_0^b e^{-c^2x^2}\,dx = \int_0^{cb} e^{-u^2}\frac{du}{c} = \frac{1}{c}\int_0^{cb} e^{-u^2}\,du$$

As $b \to \infty$, $cb \to \infty$. Hence, $\displaystyle\int_0^\infty e^{-c^2x^2}\,dx = \frac{1}{c}\int_0^\infty e^{-x^2}\,dx.$

$\bar{x} = 0$ by symmetry.

$$\bar{y} = \frac{M_x}{m} = \frac{2\displaystyle\int_0^\infty \frac{(e^{-c^2x^2})^2}{2}\,dx}{2\displaystyle\int_0^\infty e^{-c^2x^2}\,dx}$$

$$= \frac{1}{2}\,\frac{\displaystyle\int_0^\infty e^{-2c^2x^2}\,dx}{\displaystyle\int_0^\infty e^{-c^2x^2}\,dx}$$

$$= \frac{1}{2}\,\frac{\dfrac{1}{\sqrt{2}c}\displaystyle\int_0^\infty e^{-x^2}\,dx}{\dfrac{1}{c}\displaystyle\int_0^\infty e^{-x^2}\,dx}$$

$$= \frac{1}{2\sqrt{2}} = \frac{\sqrt{2}}{4}$$

Thus, $(\bar{x}, \bar{y}) = \left(0, \dfrac{\sqrt{2}}{4}\right)$.

11. Consider $\displaystyle\int \frac{1}{\ln x}\,dx$.

Let $u = \ln x$, $du = \dfrac{1}{x}\,dx$, $x = e^u$. Then $\displaystyle\int \frac{1}{\ln x}\,dx = \int \frac{1}{u}e^u\,du = \int \frac{e^u}{u}\,du.$

If $\displaystyle\int \frac{1}{\ln x}\,dx$ were elementary, then $\displaystyle\int \frac{e^u}{u}\,du$ would be too, which is false.

Hence, $\displaystyle\int \frac{1}{\ln x}\,dx$ is not elementary.

12. (a) Let $y = f^{-1}(x)$, $f(y) = x$, $dx = f'(y)\,dy$.

$$\int f^{-1}(x)\,dx = \int y f'(y)\,dy$$

$$= yf(y) - \int f(y)\,dy \qquad \left[\begin{array}{l} u = y,\ du = dy \\ dv = f'(y)\,dy,\ v = f(y) \end{array}\right]$$

$$= xf^{-1}(x) - \int f(y)\,dy$$

(b) $f^{-1}(x) = \arcsin x = y$, $f(x) = \sin x$

$$\int \arcsin x\,dx = x\arcsin x - \int \sin y\,dy$$

$$= x\arcsin x + \cos y + C$$

$$= x\arcsin x + \sqrt{1 - x^2} + C$$

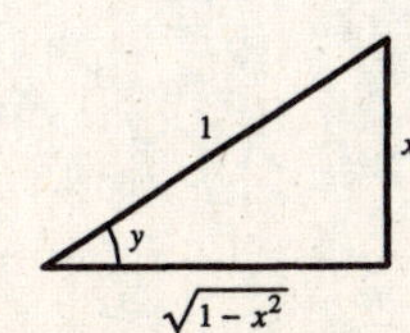

—CONTINUED—

12. —CONTINUED—

(c) $f(x) = e^x, f^{-1}(x) = \ln x = y \qquad x = 1 \Leftrightarrow y = 0; x = e \Leftrightarrow y = 1$

$$\int_1^e \ln x\,dx = \Big[x \ln x\Big]_1^e - \int_0^1 e^y\,dy$$

$$= e - \Big[e^y\Big]_0^1$$

$$= e - (e - 1) = 1$$

13. $x^4 + 1 = (x^2 + ax + b)(x^2 + cx + d)$

$$= x^4 + (a + c)x^3 + (ac + b + d)x^2 + (ad + bc)x + bd$$

$a = -c, b = d = 1, a = \sqrt{2}$

$$x^4 + 1 = \left(x^2 + \sqrt{2}x + 1\right)\left(x^2 - \sqrt{2}x + 1\right)$$

$$\int_0^1 \frac{1}{x^4 + 1}\,dx = \int_0^1 \frac{Ax + B}{x^2 + \sqrt{2}x + 1}\,dx + \int_0^1 \frac{Cx + D}{x^2 - \sqrt{2}x + 1}\,dx$$

$$= \int_0^1 \frac{\frac{1}{2} + \frac{\sqrt{2}}{4}x}{x^2 + \sqrt{2}x + 1}\,dx - \int_0^1 \frac{-\frac{1}{2} + \frac{\sqrt{2}}{4}x}{x^2 + \sqrt{2}x + 1}\,dx$$

$$= \frac{\sqrt{2}}{4}\Big[\arctan\left(\sqrt{2}x + 1\right) + \arctan\left(\sqrt{2}x - 1\right)\Big]_0^1 + \frac{\sqrt{2}}{8}\Big[\ln\left(x^2 + \sqrt{2}x + 1\right) - \ln\left(x^2 - \sqrt{2}x + 1\right)\Big]_0^1$$

$$= \frac{\sqrt{2}}{4}\left[\arctan\left(\sqrt{2} + 1\right) + \arctan\left(\sqrt{2} - 1\right)\right] + \frac{\sqrt{2}}{8}\left[\ln\left(2 + \sqrt{2}\right) - \ln\left(2 - \sqrt{2}\right)\right] - \frac{\sqrt{2}}{4}\left[\frac{\pi}{4} - \frac{\pi}{4}\right] - \frac{\sqrt{2}}{8}[0]$$

$$\approx 0.5554 + 0.3116$$

$$\approx 0.8670$$

14. (a) Let $x = \frac{\pi}{2} - u, dx = du$.

$$I = \int_0^{\pi/2} \frac{\sin x}{\cos x + \sin x}\,dx = \int_{\pi/2}^0 \frac{\sin\left(\frac{\pi}{2} - u\right)}{\cos\left(\frac{\pi}{2} - u\right) + \sin\left(\frac{\pi}{2} - u\right)}(-du)$$

$$= \int_0^{\pi/2} \frac{\cos u}{\sin u + \cos u}\,du$$

Hence,

$$2I = \int_0^{\pi/2} \frac{\sin x}{\cos x + \sin x}\,dx + \int_0^{\pi/2} \frac{\cos x}{\sin x + \cos x}\,dx$$

$$= \int_0^{\pi/2} 1\,dx = \frac{\pi}{2} \Rightarrow I = \frac{\pi}{4}.$$

(b) $$I = \int_{\pi/2}^0 \frac{\sin^n\left(\frac{\pi}{2} - u\right)}{\cos^n\left(\frac{\pi}{2} - u\right) + \sin^n\left(\frac{\pi}{2} - u\right)}(-du)$$

$$= \int_0^{\pi/2} \frac{\cos^n u}{\sin^n u + \cos^n u}\,du$$

Thus, $2I = \displaystyle\int_0^{\pi/2} 1\,dx = \frac{\pi}{2} \Rightarrow I = \frac{\pi}{4}.$

15. Using a graphing utility,

(a) $\lim_{x\to 0^+}\left(\cot x + \frac{1}{x}\right) = \infty$

(b) $\lim_{x\to 0^+}\left(\cot x - \frac{1}{x}\right) = 0$

(c) $\lim_{x\to 0^+}\left(\cot x + \frac{1}{x}\right)\left(\cot x - \frac{1}{x}\right) \approx -\frac{2}{3}.$

Analytically,

(a) $\lim_{x\to 0^+}\left(\cot x + \frac{1}{x}\right) = \infty + \infty = \infty$

(b) $$\lim_{x\to 0^+}\left(\cot x - \frac{1}{x}\right) = \lim_{x\to 0^+}\frac{x\cot x - 1}{x} = \lim_{x\to 0^+}\frac{x\cos x - \sin x}{x\sin x}$$

$$= \lim_{x\to 0^+}\frac{\cos x - x\sin x - \cos x}{\sin x + x\cos x} = \lim_{x\to 0^+}\frac{-x\sin x}{\sin x + x\cos x}$$

$$= \lim_{x\to 0^+}\frac{-\sin x - x\cos x}{\cos x + \cos x - x\sin x} = 0.$$

(c) $$\left(\cot x + \frac{1}{x}\right)\left(\cot x - \frac{1}{x}\right) = \cot^2 x - \frac{1}{x^2}$$

$$= \frac{x^2\cot^2 x - 1}{x^2}$$

$$\lim_{x\to 0^+}\frac{x^2\cot^2 x - 1}{x^2} = \lim_{x\to 0^+}\frac{2x\cot^2 x - 2x^2\cot x\csc^2 x}{2x}$$

$$= \lim_{x\to 0^+}\frac{\cot^2 x - x\cot x\csc^2 x}{1}$$

$$= \lim_{x\to 0^+}\frac{\cos^2 x\sin x - x\cos x}{\sin^3 x}$$

$$= \lim_{x\to 0^+}\frac{(1 - \sin^2 x)\sin x - x\cos x}{\sin^3 x}$$

$$= \lim_{x\to 0^+}\frac{\sin x - x\cos x}{\sin^3 x} - 1$$

Now, $$\lim_{x\to 0^+}\frac{\sin x - x\cos x}{\sin^3 x} = \lim_{x\to 0^+}\frac{\cos x - \cos x + x\sin x}{3\sin^2 x\cos x}$$

$$= \lim_{x\to 0^+}\frac{x}{3\sin x\cdot\cos x}$$

$$= \lim_{x\to 0^+}\left(\frac{x}{\sin x}\right)\frac{1}{3\cos x} = \frac{1}{3}.$$

Thus, $\lim_{x\to 0^+}\left(\cot x + \frac{1}{x}\right)\left(\cot x - \frac{1}{x}\right) = \frac{1}{3} - 1 = -\frac{2}{3}.$

The form $0\cdot\infty$ is indeterminant.

16. $\dfrac{N(x)}{D(x)} = \dfrac{P_1}{x - c_1} + \dfrac{P_2}{x - c_2} + \cdots + \dfrac{P_n}{x - c_n}$

$N(x) = P_1(x - c_2)(x - c_3)\ldots(x - c_n) + P_2(x - c_1)(x - c_3)\ldots(x - c_n) + \cdots + P_n(x - c_1)(x - c_2)\ldots(x - c_{n-1})$

Let $x = c_1$: $N(c_1) = P_1(c_1 - c_2)(c_1 - c_3)\ldots(c_1 - c_n)$

$$P_1 = \frac{N(c_1)}{(c_1 - c_2)(c_1 - c_3)\ldots(c_1 - c_n)}$$

Let $x = c_2$: $N(c_2) = P_2(c_2 - c_1)(c_2 - c_3)\ldots(c_2 - c_n)$

$$P_2 = \frac{N(c_2)}{(c_2 - c_1)(c_2 - c_3)\ldots(c_2 - c_n)}$$

$\vdots \qquad\qquad \vdots$

Let $x = c_n$: $N(c_n) = P_n(c_n - c_1)(c_n - c_2)\ldots(c_n - c_{n-1})$

$$P_n = \frac{N(c_n)}{(c_n - c_1)(c_n - c_2)\ldots(c_n - c_{n-1})}$$

If $D(x) = (x - c_1)(x - c_2)(x - c_3)\ldots(x - c_n)$, then by the Product Rule

$$D'(x) = (x - c_2)(x - c_3)\ldots(x - c_n) + (x - c_1)(x - c_3)\ldots(x - c_n) + \cdots + (x - c_1)(x - c_2)(x - c_3)\ldots(x - c_{n-1})$$

and

$$D'(c_1) = (c_1 - c_2)(c_1 - c_3)\ldots(c_1 - c_n)$$
$$D'(c_2) = (c_2 - c_1)(c_2 - c_3)\ldots(c_2 - c_n)$$
$$\vdots$$
$$D'(c_n) = (c_n - c_1)(c_n - c_2)\ldots(c_n - c_{n-1}).$$

Thus, $P_k = N(c_k)/D'(c_k)$ for $k = 1, 2, \ldots, n$.

17. $\dfrac{x^3 - 3x^2 + 1}{x^4 - 13x^2 + 12x} = \dfrac{P_1}{x} + \dfrac{P_2}{x - 1} + \dfrac{P_3}{x + 4} + \dfrac{P_4}{x - 3} \Rightarrow c_1 = 0,\ c_2 = 1,\ c_3 = -4,\ c_4 = 3$

$N(x) = x^3 - 3x^2 + 1$

$D'(x) = 4x^3 - 26x + 12$

$$P_1 = \frac{N(0)}{D'(0)} = \frac{1}{12}$$

$$P_2 = \frac{N(1)}{D'(1)} = \frac{-1}{-10} = \frac{1}{10}$$

$$P_3 = \frac{N(-4)}{D'(-4)} = \frac{-111}{-140} = \frac{111}{140}$$

$$P_4 = \frac{N(3)}{D'(3)} = \frac{1}{42}$$

Thus, $\dfrac{x^3 - 3x^2 + 1}{x^4 - 13x^2 + 12x} = \dfrac{1/12}{x} + \dfrac{1/10}{x - 1} + \dfrac{111/140}{x + 4} + \dfrac{1/42}{x - 3}$.

18. $$s(t) = \int\left[-32t + 12{,}000 \ln\frac{50{,}000}{50{,}000 - 400t}\right] dt$$

$$= -16t^2 + 12{,}000\int[\ln 50{,}000 - \ln(50{,}000 - 400t)]\, dt$$

$$= 16t^2 + 12{,}000t \ln 50{,}000 - 12{,}000\left[t\ln(50{,}000 - 400t) - \int\frac{-400t}{50{,}000 - 400t}\,dt\right]$$

$$= -16t^2 + 12{,}000t \ln\frac{50{,}000}{50{,}000 - 400t} + 12{,}000t\int\left[1 - \frac{50{,}000}{50{,}000 - 400t}\right] dt$$

$$= -16t^2 + 12{,}000t \ln\frac{50{,}000}{50{,}000 - 400t} + 12{,}000t + 1{,}500{,}000 \ln(50{,}000 - 400t) + C$$

$$s(0) = 1{,}500{,}000 \ln 50{,}000 + C = 0$$

$$C = -1{,}500{,}000 \ln 50{,}000$$

$$s(t) = -16t^2 + 12{,}000t\left[1 + \ln\frac{50{,}000}{50{,}000 - 400t}\right] + 1{,}500{,}000 \ln\frac{50{,}000 - 400t}{50{,}000}$$

When $t = 100$, $s(100) \approx 557{,}168.626$ feet.

19. By parts,

$$\int_a^b f(x)g''(x)\,dx = \Big[f(x)g'(x)\Big]_a^b - \int_a^b f'(x)g'(x)\,dx\,[u = f(x),\ dv = g''(x)\,dx]$$

$$= -\int_a^b f'(x)g'(x)\,dx$$

$$= \Big[-f'(x)g(x)\Big]_a^b + \int_a^b g(x)f''(x)\,dx\,[u = f'(x),\ dv = g'(x)\,dx]$$

$$= \int_a^b f''(x)g(x)\,dx.$$

20. Let $u = (x - a)(x - b)$, $du = [(x - a) + (x - b)]\,dx$, $dv = f''(x)\,dx$, $v = f'(x)$.

$$\int_a^b (x - a)(x - b)\,dx = \Big[(x - a)(x - b)f'(x)\Big]_a^b - \int_a^b [(x - a) + (x - b)]f'(x)\,dx$$

$$= -\int_a^b (2x - a - b)f'(x)\,dx \qquad \begin{pmatrix} u = 2x - a - b \\ dv = f'(x)\,dx \end{pmatrix}$$

$$= \Big[-(2x - a - b)f(x)\Big]_a^b + \int_a^b 2f(x)\,dx$$

$$= 2\int_a^b f(x)\,dx$$

21. $$\int_2^\infty\left[\frac{1}{x^5} + \frac{1}{x^{10}} + \frac{1}{x^{15}}\right] dx < \int_2^\infty \frac{1}{x^5 - 1}\,dx < \int_2^\infty\left[\frac{1}{x^5} + \frac{1}{x^{10}} + \frac{2}{x^{15}}\right] dx$$

$$\lim_{b\to\infty}\left[-\frac{1}{4x^4} - \frac{1}{9x^9} - \frac{1}{14x^{14}}\right]_2^b < \int_2^\infty \frac{1}{x^5 - 1}\,dx < \lim_{b\to\infty}\left[-\frac{1}{4x^4} - \frac{1}{9x^9} - \frac{1}{7x^{14}}\right]_2^b$$

$$0.015846 < \int_2^\infty \frac{1}{x^5 - 1}\,dx < 0.015851$$

CHAPTER 9
Infinite Series

CHAPTER 9
Infinite Series

Section 9.1 Sequences

1. $a_n = 2^n$

$a_1 = 2^1 = 2$

$a_2 = 2^2 = 4$

$a_3 = 2^3 = 8$

$a_4 = 2^4 = 16$

$a_5 = 2^5 = 32$

2. $a_n = \dfrac{3^n}{n!}$

$a_1 = \dfrac{3}{1!} = 3$

$a_2 = \dfrac{3^2}{2!} = \dfrac{9}{2}$

$a_3 = \dfrac{3^3}{3!} = \dfrac{27}{6} = \dfrac{9}{2}$

$a_4 = \dfrac{3^4}{4!} = \dfrac{81}{24} = \dfrac{27}{8}$

$a_5 = \dfrac{3^5}{5!} = \dfrac{243}{120} = \dfrac{81}{40}$

3. $a_n = \left(-\dfrac{1}{2}\right)^n$

$a_1 = \left(-\dfrac{1}{2}\right)^1 = -\dfrac{1}{2}$

$a_2 = \left(-\dfrac{1}{2}\right)^2 = \dfrac{1}{4}$

$a_3 = \left(-\dfrac{1}{2}\right)^3 = -\dfrac{1}{8}$

$a_4 = \left(-\dfrac{1}{2}\right)^4 = \dfrac{1}{16}$

$a_5 = \left(-\dfrac{1}{2}\right)^5 = -\dfrac{1}{32}$

4. $a_n = \left(-\dfrac{2}{3}\right)^n$

$a_1 = -\dfrac{2}{3}$

$a_2 = \dfrac{4}{9}$

$a_3 = -\dfrac{8}{27}$

$a_4 = \dfrac{16}{81}$

$a_5 = -\dfrac{32}{243}$

5. $a_n = \sin\dfrac{n\pi}{2}$

$a_1 = \sin\dfrac{\pi}{2} = 1$

$a_2 = \sin \pi = 0$

$a_3 = \sin\dfrac{3\pi}{2} = -1$

$a_4 = \sin 2\pi = 0$

$a_5 = \sin\dfrac{5\pi}{2} = 1$

6. $a_n = \dfrac{2n}{n+3}$

$a_1 = \dfrac{2}{4} = \dfrac{1}{2}$

$a_2 = \dfrac{4}{5}$

$a_3 = \dfrac{6}{6} = 1$

$a_4 = \dfrac{8}{7}$

$a_5 = \dfrac{10}{8} = \dfrac{5}{4}$

7. $a_n = \dfrac{(-1)^{n(n+1)/2}}{n^2}$

$a_1 = \dfrac{(-1)^1}{1^2} = -1$

$a_2 = \dfrac{(-1)^3}{2^2} = -\dfrac{1}{4}$

$a_3 = \dfrac{(-1)^6}{3^2} = \dfrac{1}{9}$

$a_4 = \dfrac{(-1)^{10}}{4^2} = \dfrac{1}{16}$

$a_5 = \dfrac{(-1)^{15}}{5^2} = -\dfrac{1}{25}$

8. $a_n = (-1)^{n+1}\left(\dfrac{2}{n}\right)$

$a_1 = \dfrac{2}{1} = 2$

$a_2 = -\dfrac{2}{2} = -1$

$a_3 = \dfrac{2}{3}$

$a_4 = -\dfrac{2}{4} = -\dfrac{1}{2}$

$a_5 = \dfrac{2}{5}$

9. $a_n = 5 - \dfrac{1}{n} + \dfrac{1}{n^2}$

$a_1 = 5 - 1 + 1 = 5$

$a_2 = 5 - \dfrac{1}{2} + \dfrac{1}{4} = \dfrac{19}{4}$

$a_3 = 5 - \dfrac{1}{3} + \dfrac{1}{9} = \dfrac{43}{9}$

$a_4 = 5 - \dfrac{1}{4} + \dfrac{1}{16} = \dfrac{77}{16}$

$a_5 = 5 - \dfrac{1}{5} + \dfrac{1}{25} = \dfrac{121}{25}$

10. $a_n = 10 + \frac{2}{n} + \frac{6}{n^2}$

$a_1 = 10 + 2 + 6 = 18$

$a_2 = 10 + 1 + \frac{3}{2} = \frac{25}{2}$

$a_3 = 10 + \frac{2}{3} + \frac{2}{3} = \frac{34}{3}$

$a_4 = 10 + \frac{1}{2} + \frac{3}{8} = \frac{87}{8}$

$a_5 = 10 + \frac{2}{5} + \frac{6}{25} = \frac{266}{25}$

11. $a_1 = 3, a_{k+1} = 2(a_k - 1)$

$a_2 = 2(a_1 - 1)$
$= 2(3 - 1) = 4$

$a_3 = 2(a_2 - 1)$
$= 2(4 - 1) = 6$

$a_4 = 2(a_3 - 1)$
$= 2(6 - 1) = 10$

$a_5 = 2(a_4 - 1)$
$= 2(10 - 1) = 18$

12. $a_1 = 4, a_{k+1} = \left(\frac{k+1}{2}\right)a_k$

$a_2 = \left(\frac{1+1}{2}\right)a_1 = 4$

$a_3 = \left(\frac{2+1}{2}\right)a_2 = 6$

$a_4 = \left(\frac{3+1}{2}\right)a_3 = 12$

$a_5 = \left(\frac{4+1}{2}\right)a_4 = 30$

13. $a_1 = 32, a_{k+1} = \frac{1}{2}a_k$

$a_2 = \frac{1}{2}a_1 = \frac{1}{2}(32) = 16$

$a_3 = \frac{1}{2}a_2 = \frac{1}{2}(16) = 8$

$a_4 = \frac{1}{2}a_3 = \frac{1}{2}(8) = 4$

$a_5 = \frac{1}{2}a_4 = \frac{1}{2}(4) = 2$

14. $a_1 = 6, a_{k+1} = \frac{1}{3}a_k^2$

$a_2 = \frac{1}{3}a_1^2 = \frac{1}{3}(6^2) = 12$

$a_3 = \frac{1}{3}a_2^2 = \frac{1}{3}(12^2) = 48$

$a_4 = \frac{1}{3}a_3^2 = \frac{1}{3}(48^2) = 768$

$a_5 = \frac{1}{3}a_4^2 = \frac{1}{3}(768)^2 = 196{,}608$

15. $a_n = \frac{8}{n+1}, a_1 = 4, a_2 = \frac{8}{3},$
decreases to 0; matches (f).

16. $a_n = \frac{8n}{n+1}, a_1 = 4, a_2 = \frac{16}{3},$
increases towards 8; matches (a).

17. $a_n = 4(0.5)^{n-1}, a_1 = 4, a_2 = 2,$
decreases to 0; matches (e).

18. $a_n = \frac{4^n}{n!}, a_1 = 4, a_2 = 8,$
eventually approaches 0; matches (b).

19. $a_n(-1)^n, a_1 = -1, a_2 = 1, a_3 = -1,$ etc.; matches (d).

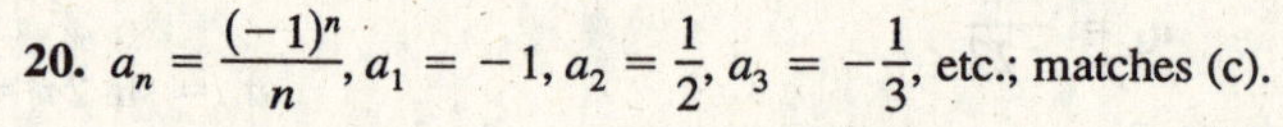

20. $a_n = \frac{(-1)^n}{n}, a_1 = -1, a_2 = \frac{1}{2}, a_3 = -\frac{1}{3},$ etc.; matches (c).

21.

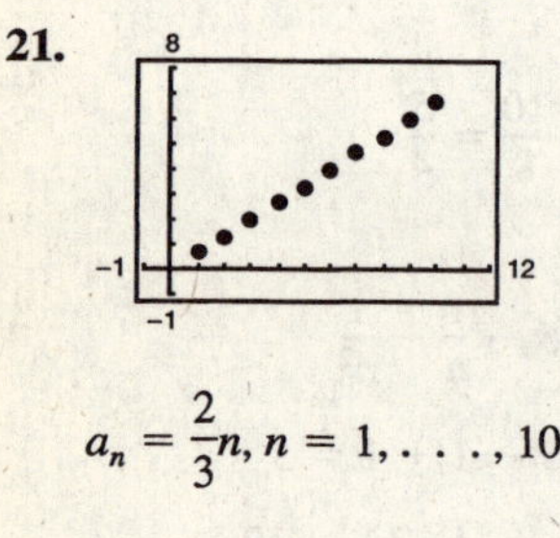

$a_n = \frac{2}{3}n, n = 1, \ldots, 10$

22.

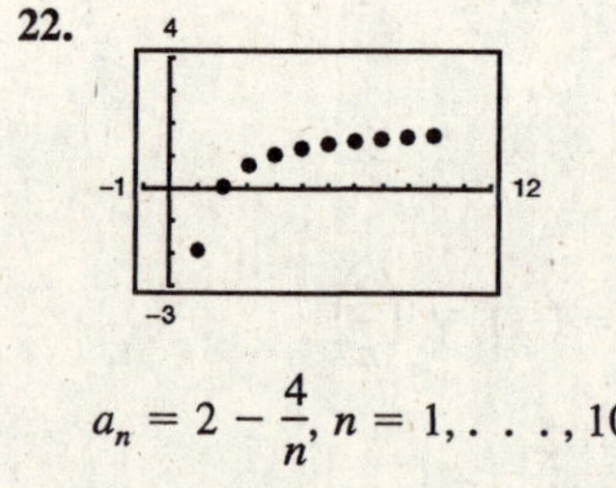

$a_n = 2 - \frac{4}{n}, n = 1, \ldots, 10$

23.

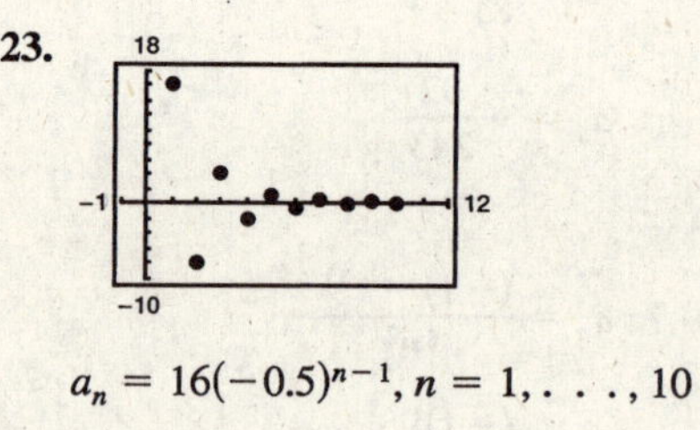

$a_n = 16(-0.5)^{n-1}, n = 1, \ldots, 10$

24.

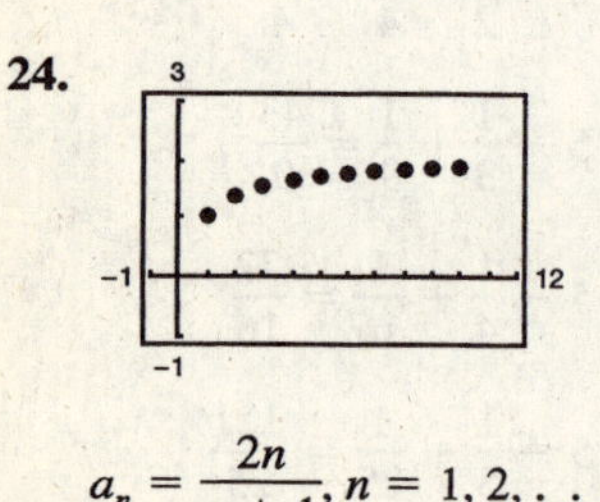

$a_n = \frac{2n}{n+1}, n = 1, 2, \ldots, 10$

25. $a_n = 3n - 1$

$a_5 = 3(5) - 1 = 14$

$a_6 = 3(6) - 1 = 17$

Add 3 to preceding term.

26. $a_n = \frac{n+6}{2}$

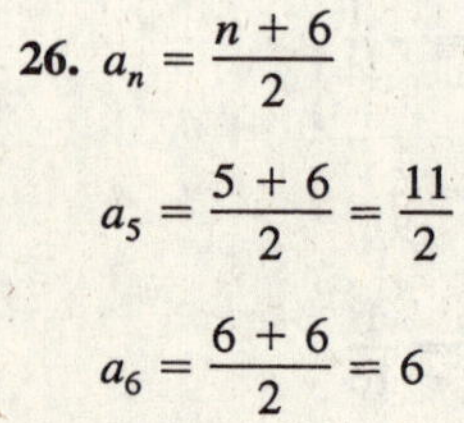

$a_5 = \frac{5+6}{2} = \frac{11}{2}$

$a_6 = \frac{6+6}{2} = 6$

27. $a_{n+1} = 2a_n,\ a_1 = 5$

$a_5 = 2(40) = 80$

$a_6 = 2(80) = 160$

28. $a_n = -\frac{1}{2}a_{n-1},\ a_1 = 1$

$a_5 = \frac{1}{16},\ a_6 = -\frac{1}{32}$

29. $a_n = \dfrac{3}{(-2)^{n-1}}$

$a_5 = \dfrac{3}{(-2)^4} = \dfrac{3}{16}$

$a_6 = \dfrac{3}{(-2)^5} = -\dfrac{3}{32}$

Multiply the preceding term by $-\frac{1}{2}$.

30. $a_n = -\frac{3}{2}a_{n-1},\ a_1 = 1$

$a_5 = \frac{81}{16},\ a_6 = -\frac{243}{32}$

31. $\dfrac{10!}{8!} = \dfrac{8!(9)(10)}{8!}$

$= (9)(10) = 90$

32. $\dfrac{25!}{23!} = \dfrac{23!(24)(25)}{23!}$

$= (24)(25) = 600$

33. $\dfrac{(n+1)!}{n!} = \dfrac{n!(n+1)}{n!} = n + 1$

34. $\dfrac{(n+2)!}{n!} = \dfrac{n!(n+1)(n+2)}{n!}$

$= (n+1)(n+2)$

35. $\dfrac{(2n-1)!}{(2n+1)!} = \dfrac{(2n-1)!}{(2n-1)!(2n)(2n+1)}$

$= \dfrac{1}{2n(2n+1)}$

36. $\dfrac{(2n+2)!}{(2n)!} = \dfrac{(2n)!(2n+1)(2n+2)}{(2n)!}$

$= (2n+1)(2n+2)$

37. $\lim\limits_{n\to\infty} \dfrac{5n^2}{n^2+2} = 5$

38. $\lim\limits_{n\to\infty} \left(5 - \dfrac{1}{n^2}\right) = 5 - 0 = 5$

39. $\lim\limits_{n\to\infty} \dfrac{2n}{\sqrt{n^2+1}} = \lim\limits_{n\to\infty} \dfrac{2}{\sqrt{1+(1/n^2)}} = \dfrac{2}{1} = 2$

40. $\lim\limits_{n\to\infty} \dfrac{5n}{\sqrt{n^2+4}} = \lim\limits_{n\to\infty} \dfrac{5}{\sqrt{1+(4/n^2)}} = \dfrac{5}{1} = 5$

41. $\lim\limits_{n\to\infty} \sin\left(\dfrac{1}{n}\right) = 0$

42. $\lim\limits_{n\to\infty} \cos\left(\dfrac{2}{n}\right) = 1$

43.

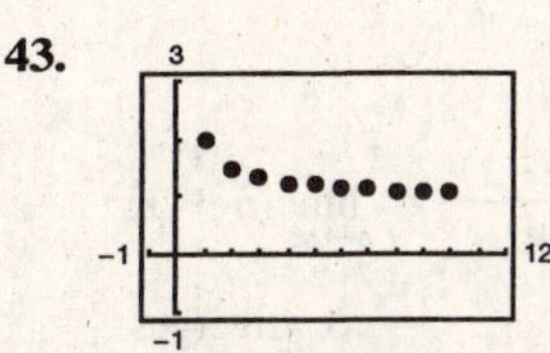

The graph seems to indicate that the sequence converges to 1. Analytically,

$$\lim_{n\to\infty} a_n = \lim_{n\to\infty} \frac{n+1}{n} = \lim_{x\to\infty} \frac{x+1}{x} = \lim_{x\to\infty} 1 = 1.$$

44.

The graph seems to indicate that the sequence converges to 0. Analytically,

$$\lim_{n\to\infty} a_n = \lim_{n\to\infty} \frac{1}{n^{3/2}} = \lim_{x\to\infty} \frac{1}{x^{3/2}} = 0.$$

45.

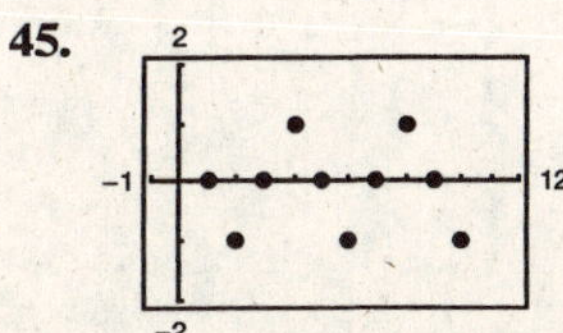

The graph seems to indicate that the sequence diverges. Analytically, the sequence is

$\{a_n\} = \{0, -1, 0, 1, 0, -1, \ldots\}$.

Hence, $\lim\limits_{n\to\infty} a_n$ does not exist.

46.

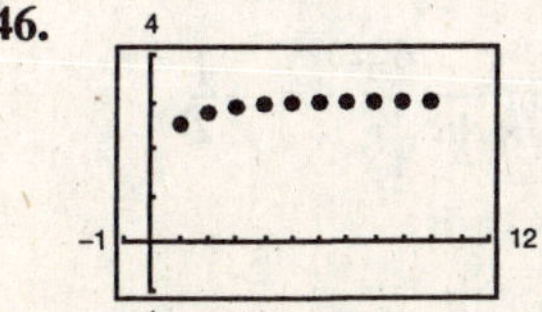

The graph seems to indicate that the sequence converges to 3. Analytically,

$$\lim_{n\to\infty} a_n = \lim_{n\to\infty} \left(3 - \frac{1}{2^n}\right) = 3 - 0 = 3.$$

47. $\lim_{n\to\infty}(-1)^n\left(\frac{n}{n+1}\right)$

does not exist (oscillates between -1 and 1), diverges.

48. $\lim_{n\to\infty}[1+(-1)^n]$

does not exist, (alternates between 0 and 2), diverges.

49. $\lim_{n\to\infty}\frac{3n^2-n+4}{2n^2+1}=\frac{3}{2}$, converges

50. $\lim_{n\to\infty}\frac{\sqrt[3]{n}}{\sqrt[3]{n}+1}=1$, converges

51. $a_n=\frac{1\cdot 3\cdot 5\cdots(2n-1)}{(2n)^n}$

$$=\frac{1}{2n}\cdot\frac{3}{2n}\cdot\frac{5}{2n}\cdots\frac{2n-1}{2n}<\frac{1}{2n}$$

Thus, $\lim_{n\to\infty}a_n=0$, converges.

52. The sequence diverges. To prove this analytically, we use mathematical induction to show that $a_n\geq\left(\frac{3}{2}\right)^{n-1}$. Clearly, $a_1=1\geq\left(\frac{3}{2}\right)^0=1$. Assume that $a_k\geq\left(\frac{3}{2}\right)^{k-1}$. Then,

$$a_{k+1}=\frac{1\cdot 3\cdot 5\cdots(2k-1)(2k+1)}{(k+1)!}=a_k\frac{2k+1}{k+1}\geq\left(\frac{3}{2}\right)^{k-1}\left(\frac{2k+1}{k+1}\right).$$

Since

$$\frac{2k+1}{k+1}\geq\frac{3}{2},$$

we have $a_{k+1}\geq\left(\frac{3}{2}\right)^k$, which shows that $a_n\geq\left(\frac{3}{2}\right)^{n-1}$ for all n.

53. $\lim_{n\to\infty}\frac{1+(-1)^n}{n}=0$, converges

54. $\lim_{n\to\infty}\frac{1+(-1)^n}{n^2}=0$, converges

55. $\lim_{n\to\infty}\frac{\ln(n^3)}{2n}=\lim_{n\to\infty}\frac{3}{2}\frac{\ln(n)}{n}$

$$=\lim_{n\to\infty}\frac{3}{2}\left(\frac{1}{n}\right)=0,\text{ converges}$$

(L'Hôpital's Rule)

56. $\lim_{n\to\infty}\frac{\ln\sqrt{n}}{n}=\lim_{n\to\infty}\frac{1/2\ln n}{n}$

$$=\lim_{n\to\infty}\frac{1}{2n}=0,\text{ converges}$$

(L'Hôpital's Rule)

57. $\lim_{n\to\infty}\left(\frac{3}{4}\right)^n=0$, converges

58. $\lim_{n\to\infty}(0.5)^n=0$, converges

59. $\lim_{n\to\infty}\frac{(n+1)!}{n!}=\lim_{n\to\infty}(n+1)$

$=\infty$, diverges

60. $\lim_{n\to\infty}\frac{(n-2)!}{n!}=\lim_{n\to\infty}\frac{1}{n(n-1)}=0$, converges

61. $\lim_{n\to\infty}\left(\frac{n-1}{n}-\frac{n}{n-1}\right)=\lim_{n\to\infty}\frac{(n-1)^2-n^2}{n(n-1)}$

$$=\lim_{n\to\infty}\frac{1-2n}{n^2-n}=0,\text{ converges}$$

62. $\lim_{n\to\infty}\left(\frac{n^2}{2n+1}-\frac{n^2}{2n-1}\right)=\lim_{n\to\infty}\frac{-2n^2}{4n^2-1}=-\frac{1}{2}$,

converges

63. $\lim_{n\to\infty}\frac{n^p}{e^n}=0$, converges

$(p>0, n\geq 2)$

64. $a_n = n \sin \frac{1}{n}$

Let $f(x) = x \sin \frac{1}{x}$.

$$\lim_{x\to\infty} x \sin \frac{1}{x} = \lim_{x\to\infty} \frac{\sin(1/x)}{1/x} = \lim_{x\to\infty} \frac{(-1/x^2)\cos(1/x)}{-1/x^2} = \lim_{x\to\infty} \cos\frac{1}{x} = \cos 0 = 1 \quad \text{(L'Hôpital's Rule)}$$

or,

$$\lim_{x\to\infty} \frac{\sin(1/x)}{1/x} = \lim_{y\to 0^+} \frac{\sin(y)}{y} = 1. \text{ Therefore } \lim_{n\to\infty} n \sin\frac{1}{n} = 1.$$

65. $a_n = \left(1 + \frac{k}{n}\right)^n$

$$\lim_{n\to\infty} \left(1 + \frac{k}{n}\right)^n = \lim_{u\to 0} [(1 + u)^{1/u}]^k = e^k$$

where $u = k/n$, converges

66. $\lim_{n\to\infty} 2^{1/n} = 2^0 = 1$, converges

67. $\lim_{n\to\infty} \frac{\sin n}{n} = \lim_{n\to\infty} (\sin n)\frac{1}{n} = 0$,

converges (because $(\sin n)$ is bounded)

68. $\lim_{n\to\infty} \frac{\cos \pi n}{n^2} = 0$, converges

69. $a_n = 3n - 2$

70. $a_n = 4n - 1$

71. $a_n = n^2 - 2$

72. $a_n = \frac{(-1)^{n-1}}{n^2}$

73. $a_n = \frac{n+1}{n+2}$

74. $a_n = \frac{(-1)^{n-1}}{2^{n-2}}$

75. $a_n = 1 + \frac{1}{n} = \frac{n+1}{n}$

76. $a_n = 1 + \frac{2^n - 1}{2^n}$

$= \frac{2^{n+1} - 1}{2^n}$

77. $a_n = \frac{n}{(n+1)(n+2)}$

78. $a_n = \frac{1}{n!}$

79. $a_n = \frac{(-1)^{n-1}}{1 \cdot 3 \cdot 5 \cdot \cdot \cdot (2n-1)}$

$= \frac{(-1)^{n-1} 2^n n!}{(2n)!}$

80. $a_n = \frac{x^{n-1}}{(n-1)!}$

81. $a_n = (2n)!, n = 1, 2, 3, \ldots$

82. $a_n = (2n - 1)!, n = 1, 2, 3, \ldots$

83. $a_n = 4 - \frac{1}{n} < 4 - \frac{1}{n+1} = a_{n+1}$,

monotonic; $|a_n| < 4$, bounded

84. Let $f(x) = \frac{3x}{x+2}$. Then $f'(x) = \frac{6}{(x+2)^2}$.

Thus, f is increasing which implies $\{a_n\}$ is increasing.

$|a_n| < 3$, bounded

85. $\frac{n}{2^{n+2}} \overset{?}{\geq} \frac{n+1}{2^{(n+1)+2}}$

$2^{n+3}n \overset{?}{\geq} 2^{n+2}(n+1)$

$2n \overset{?}{\geq} n+1$

$n \geq 1$

Hence, $n \geq 1$

$2n \geq n+1$

$2^{n+3}n \geq 2^{n+2}(n+1)$

$\frac{n}{2^{n+2}} \geq \frac{n+1}{2^{(n+1)+2}}$

$a_n \geq a_{n+1}.$

True; monotonic; $|a_n| \leq \frac{1}{8}$, bounded

86. $a_n = ne^{-n/2}$

$a_1 = 0.6065$

$a_2 = 0.7358$

$a_3 = 0.6694$

Not monotonic; $|a_n| \leq 0.7358$, bounded

87. $a_n = (-1)^n\left(\frac{1}{n}\right)$

$a_1 = -1$

$a_2 = \frac{1}{2}$

$a_3 = -\frac{1}{3}$

Not monotonic; $|a_n| \leq 1$, bounded

88. $a_n = \left(-\frac{2}{3}\right)^n$

$a_1 = -\frac{2}{3}$

$a_2 = \frac{4}{9}$

$a_3 = -\frac{8}{27}$

Not monotonic; $|a_n| \leq \frac{2}{3}$, bounded

89. $a_n = \left(\frac{2}{3}\right)^n > \left(\frac{2}{3}\right)^{n+1} = a_{n+1}$

Monotonic; $|a_n| \leq \frac{2}{3}$, bounded

90. $a_n = \left(\frac{3}{2}\right)^n < \left(\frac{3}{2}\right)^{n+1} = a_{n+1}$

Monotonic; $\lim_{n\to\infty} a_n = \infty$, not bounded

91. $a_n = \sin\left(\frac{n\pi}{6}\right)$

$a_1 = 0.500$

$a_2 = 0.8660$

$a_3 = 1.000$

$a_4 = 0.8660$

Not monotonic; $|a_n| \leq 1$, bounded

92. $a_n = \cos\left(\frac{n\pi}{2}\right)$

$a_1 = \cos\frac{\pi}{2} = 0$

$a_2 = \cos \pi = -1$

$a_3 = \cos\left(\frac{3\pi}{2}\right) = 0$

Not monotonic; $|a_n| \leq 1$, bounded

93. $a_n = \frac{\cos n}{n}$

$a_1 = 0.5403$

$a_2 = -0.2081$

$a_3 = -0.3230$

$a_4 = -0.1634$

Not monotonic; $|a_n| \leq 1$, bounded

94. $a_n = \frac{\sin\sqrt{n}}{n}$

$a_1 = \frac{\sin(1)}{1} \approx 0.8415$ $\qquad a_4 = \frac{\sin(4)}{4} \approx -0.1892$

$a_2 = \frac{\sin(2)}{2} \approx 0.4546$ $\qquad a_5 = \frac{\sin(5)}{5} \approx -0.1918$

$a_3 = \frac{\sin(3)}{3} \approx 0.0470$ $\qquad a_6 = \frac{\sin(6)}{6} \approx -0.0466$

Not monotonic, $|a_n| \leq 1$, bounded

95. (a) $a_n = 5 + \frac{1}{n}$

$\left|5 + \frac{1}{n}\right| \le 6 \Rightarrow \{a_n\}$, bounded

$a_n = 5 + \frac{1}{n} > 5 + \frac{1}{n+1}$

$= a_{n+1} \Rightarrow \{a_n\}$, monotonic

Therefore, $\{a_n\}$ converges.

(b)

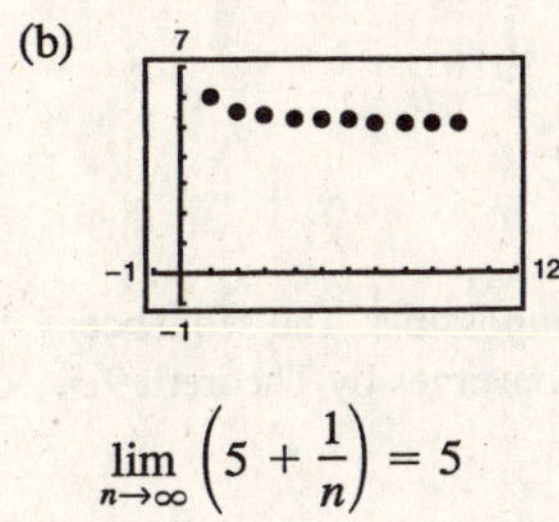

$\lim_{n\to\infty}\left(5 + \frac{1}{n}\right) = 5$

96. (a) $a_n = 4 - \frac{3}{n}$

$\left|4 - \frac{3}{n}\right| < 4 \Rightarrow$ bounded

$a_n = 4 - \frac{3}{n} < 3 - \frac{4}{n+1} = a_{n+1} \Rightarrow$ monotonic

Therefore, $\{a_n\}$ converges.

(b)

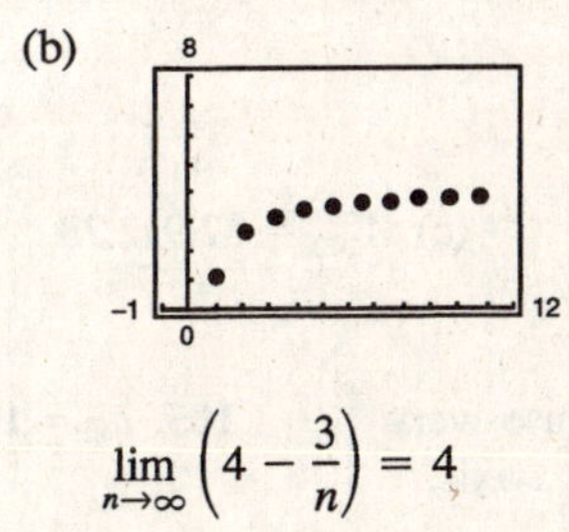

$\lim_{n\to\infty}\left(4 - \frac{3}{n}\right) = 4$

97. (a) $a_n = \frac{1}{3}\left(1 - \frac{1}{3^n}\right)$

$\left|\frac{1}{3}\left(1 - \frac{1}{3^n}\right)\right| < \frac{1}{3} \Rightarrow \{a_n\}$,bounded

$a_n = \frac{1}{3}\left(1 - \frac{1}{3^n}\right) < \frac{1}{3}\left(1 - \frac{1}{3^{n+1}}\right)$

$= a_{n+1} \Rightarrow \{a_n\}$, monotonic

Therefore, $\{a_n\}$ converges.

(b)

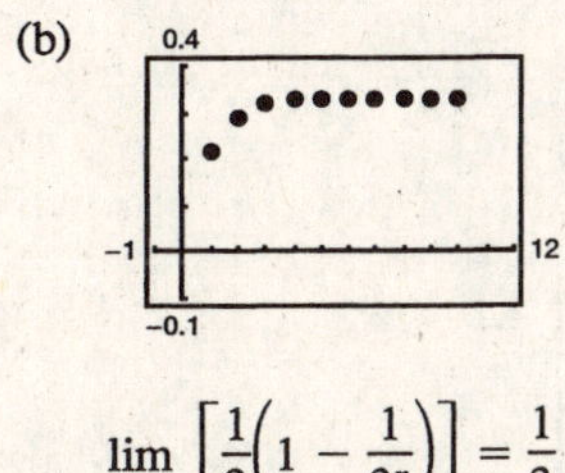

$\lim_{n\to\infty}\left[\frac{1}{3}\left(1 - \frac{1}{3^n}\right)\right] = \frac{1}{3}$

98. (a) $a_n = 4 + \frac{1}{2^n}$

$\left|4 + \frac{1}{2^n}\right| \le 4.5 \Rightarrow \{a_n\}$, bounded

$a_n = 4 + \frac{1}{2^n} > 4 + \frac{1}{2^{n+1}}$

$= a_{n+1} \Rightarrow \{a_n\}$, monotonic

Therefore, $\{a_n\}$ converges.

(b)

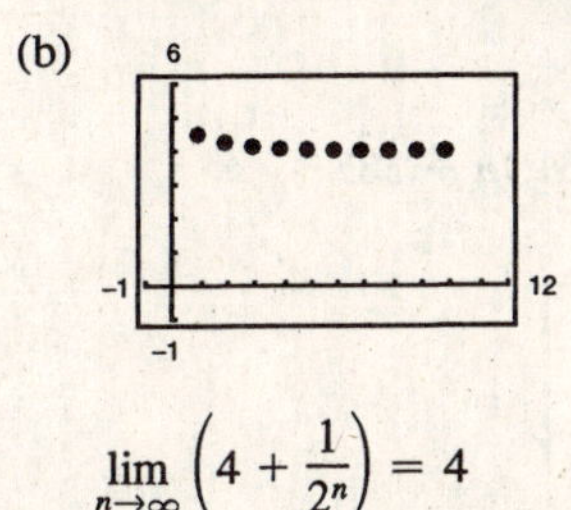

$\lim_{n\to\infty}\left(4 + \frac{1}{2^n}\right) = 4$

99. $\{a_n\}$ has a limit because it is a bounded, monotonic sequence. The limit is less than or equal to 4, and greater than or equal to 2.

$2 \le \lim_{n\to\infty} a_n \le 4$

100. The sequence $\{a_n\}$ could converge or diverge. If $\{a_n\}$ is increasing, then it converges to a limit less than or equal to 1. If $\{a_n\}$ is decreasing, then it could converge (example: $a_n = 1/n$) or diverge (example: $a_n = -n$).

101. $A_n = P\left(1 + \frac{r}{12}\right)^n$

(a) No, the sequence $\{A_n\}$ diverges, $\lim_{n\to\infty} A_n = \infty$. The amount will grow arbitrarily large over time.

(b) $P = 9000, r = 0.055, A_n = 9000\left(1 + \frac{0.055}{12}\right)^n$

$A_1 = 9041.25$ $\quad A_6 = 9250.35$

$A_2 = 9082.69$ $\quad A_7 = 9292.75$

$A_3 = 9124.32$ $\quad A_8 = 9335.34$

$A_4 = 9166.14$ $\quad A_9 = 9378.13$

$A_5 = 9208.15$ $\quad A_{10} = 9421.11$

102. (a) $A_n = 100(401)(1.0025^n - 1)$

$A_0 = 0$

$A_1 = 100.25$

$A_2 = 200.75$

$A_3 = 301.50$

$A_4 = 402.51$

$A_5 = 503.76$

$A_6 = 605.27$

(b) $A_{60} = 6480.83$ (c) $A_{240} = 32{,}912.28$

103. (a) A sequence is a function whose domain is the set of positive integers.

(b) A sequence converges if it has a limit. See the definition.

(c) A sequence is monotonic if its terms are nondecreasing, or nonincreasing.

(d) A sequence is bounded if it is bounded below ($a_n \geq N$ for some N) and bounded above ($a_n \leq M$ for some M).

104. The first sequence because every other point is below the x-axis.

105. $a_n = 10 - \frac{1}{n}$

106. Impossible. The sequence converges by Theorem 9.5.

107. $a_n = \frac{3n}{4n + 1}$

108. Impossible. An unbounded sequence diverges.

109. (a) $A_n = (0.8)^n(2.5)$ billion

(b) $A_1 = \$2$ billion

$A_2 = \$1.6$ billion

$A_3 = \$1.28$ billion

$A_4 = \$1.024$ billion

(c) $\lim_{n\to\infty} (0.8)^n(2.5) = 0$

110. $P_n = 16{,}000(1.045)^n$

$P_1 = \$16{,}720.00$

$P_2 = \$17{,}472.40$

$P_3 \approx \$18{,}258.66$

$P_4 \approx \$19{,}080.30$

$P_5 \approx \$19{,}938.91$

111. (a) $a_n = -6.60n^2 + 151.7n + 387$

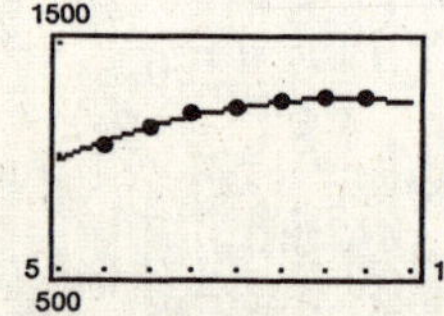

(b) In 2008, $n = 18$ and $a_{18} \approx 979$ endangered species.

112. (a) $a_n = 244.2n + 3250$

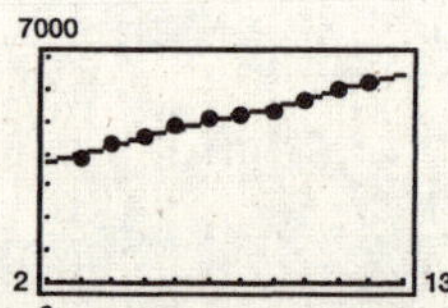

(b) For 2008, $n = 18$ and $a_{18} \approx 7646$ million.

113. $a_n = \frac{10^n}{n!}$

(a) $a_9 = a_{10} = \frac{10^9}{9!}$

$= \frac{1{,}000{,}000{,}000}{362{,}880}$

$= \frac{1{,}562{,}500}{567}$

(b) Decreasing

(c) Factorials increase more rapidly than exponentials.

114. $a_n = \left(1 + \frac{1}{n}\right)^n$

$a_1 = 2.0000$

$a_2 = 2.2500$

$a_3 \approx 2.3704$

$a_4 \approx 2.4414$

$a_5 \approx 2.4883$

$a_6 \approx 2.5216$

$\lim_{n\to\infty} \left(1 + \frac{1}{n}\right)^n = e$

115. $\{a_n\} = \{\sqrt[n]{n}\} = \{n^{1/n}\}$

$a_1 = 1^{1/1} = 1$

$a_2 = \sqrt{2} \approx 1.4142$

$a_3 = \sqrt[3]{3} \approx 1.4422$

$a_4 = \sqrt[4]{4} \approx 1.4142$

$a_5 = \sqrt[5]{5} \approx 1.3797$

$a_6 = \sqrt[6]{6} \approx 1.3480$

Let $y = \lim_{n\to\infty} n^{1/n}$.

$$\ln y = \lim_{n\to\infty}\left(\frac{1}{n}\ln n\right) = \lim_{n\to\infty}\frac{\ln n}{n} = \lim_{n\to\infty}\frac{1/n}{n} = 0$$

Since $\ln y = 0$, we have $y = e^0 = 1$. Therefore, $\lim_{n\to\infty}\sqrt[n]{n} = 1$.

116. Since

$$\lim_{n\to\infty} s_n = L > 0,$$

there exists for each $\varepsilon > 0$,

an integer N such that $|s_n - L| < \varepsilon$ for every $n > N$.

Let $\varepsilon = L > 0$ and we have,

$$|s_n - L| < L, -L < s_n - L < L, \text{ or } 0 < s_n < 2L$$

for each $n > N$.

117. True

118. True

119. True

120. True

121. $a_{n+2} = a_n + a_{n+1}$

(a) $a_1 = 1$

$a_2 = 1$

$a_3 = 1 + 1 = 2$

$a_4 = 2 + 1 = 3$

$a_5 = 3 + 2 = 5$

$a_6 = 5 + 3 = 8$

$a_7 = 8 + 5 = 13$

$a_8 = 13 + 8 = 21$

$a_9 = 21 + 13 = 34$

$a_{10} = 34 + 21 = 55$

$a_{11} = 55 + 34 = 89$

$a_{12} = 89 + 55 = 144$

(b) $b_n = \dfrac{a_{n+1}}{a_n}, n \ge 1$

$b_1 = \dfrac{1}{1} = 1$

$b_2 = \dfrac{2}{1} = 2$

$b_3 = \dfrac{3}{2}$

$b_4 = \dfrac{5}{3}$

$b_5 = \dfrac{8}{5}$

$b_6 = \dfrac{13}{8}$

$b_7 = \dfrac{21}{13}$

$b_8 = \dfrac{34}{21}$

$b_9 = \dfrac{55}{34}$

$b_{10} = \dfrac{89}{55}$

(c) $$1 + \frac{1}{b_{n-1}} = 1 + \frac{1}{a_n/a_{n-1}} = 1 + \frac{a_{n-1}}{a_n} = \frac{a_n + a_{n-1}}{a_n} = \frac{a_{n+1}}{a_n} = b_n$$

(d) If $\lim_{n\to\infty} b_n = \rho$, then $\lim_{n\to\infty}\left(1 + \dfrac{1}{b_{n-1}}\right) = \rho$.

Since $\lim_{n\to\infty} b_n = \lim_{n\to\infty} b_{n-1}$, we have

$1 + (1/\rho) = \rho$.

$\rho + 1 = \rho^2$

$0 = \rho^2 - \rho - 1$

$$\rho = \frac{1 \pm \sqrt{1+4}}{2} = \frac{1 \pm \sqrt{5}}{2}$$

Since a_n, and thus b_n, is positive, $\rho = \dfrac{1+\sqrt{5}}{2} \approx 1.6180$.

122. $x_0 = 1, x_n = \dfrac{1}{2}x_{n-1} + \dfrac{1}{x_{n-1}}, n = 1, 2, \ldots$

$x_1 = 1.5$

$x_2 = 1.41667$

$x_3 = 1.414216$

$x_4 = 1.414214$

$x_5 = 1.414214$

$x_6 = 1.414214$

$x_7 = 1.414214$

$x_8 = 1.414114$

$x_9 = 1.414214$

$x_{10} = 1.414214$

The limit of the sequence appears to be $\sqrt{2}$. In fact, this sequence is Newton's Method applied to $f(x) = x^2 - 2$.

123. (a) $a_1 = \sqrt{2} \approx 1.4142$

$a_2 = \sqrt{2 + \sqrt{2}} \approx 1.8478$

$a_3 = \sqrt{2 + \sqrt{2 + \sqrt{2}}} \approx 1.9616$

$a_4 = \sqrt{2 + \sqrt{2 + \sqrt{2 + \sqrt{2}}}} \approx 1.9904$

$a_5 = \sqrt{2 + \sqrt{2 + \sqrt{2 + \sqrt{2 + \sqrt{2}}}}} \approx 1.9976$

(b) $a_n = \sqrt{2 + a_{n-1}}, \quad n \geq 2, a_1 = \sqrt{2}$

(c) We first use mathematical induction to show that $a_n \leq 2$; clearly $a_1 \leq 2$. So assume $a_k \leq 2$. Then

$$a_k + 2 \leq 4$$
$$\sqrt{a_k + 2} \leq 2$$
$$a_{k+1} \leq 2.$$

Now we show that $\{a_n\}$ is an increasing sequence. Since $a_n \geq 0$ and $a_n \leq 2$,

$$(a_n - 2)(a_n + 1) \leq 0$$
$$a_n^2 - a_n - 2 \leq 0$$
$$a_n^2 \leq a_n + 2$$
$$a_n \leq \sqrt{a_n + 2}$$
$$a_n \leq a_{n+1}.$$

Since $\{a_n\}$ is a bounding increasing sequence, it converges to some number L, by Theorem 9.5.

$$\lim_{n\to\infty} a_n = L \Rightarrow \sqrt{2 + L} = L \Rightarrow 2 + L = L^2 \Rightarrow L^2 - L - 2 = 0$$
$$\Rightarrow (L - 2)(L + 1) = 0 \Rightarrow L = 2 \quad (L \neq -1)$$

124. (a) $a_1 = \sqrt{6} \approx 2.4495$

$a_2 = \sqrt{6 + \sqrt{6}} \approx 2.9068$

$a_3 = \sqrt{6 + \sqrt{6 + \sqrt{6}}} \approx 2.9844$

$a_4 = \sqrt{6 + \sqrt{6 + \sqrt{6 + \sqrt{6}}}} \approx 2.9974$

$a_5 = \sqrt{6 + \sqrt{6 + \sqrt{6 + \sqrt{6 + \sqrt{6}}}}} \approx 2.9996$

(b) $a_n = \sqrt{6 + a_{n-1}}, \quad n \geq 2, a_1 = \sqrt{6}$

(c) We first use mathematical induction to show that $a_n \leq 3$; clearly $a_1 \leq 3$. So assume $a_k \leq 3$. Then

$$6 + a_k \leq 9$$
$$\sqrt{6 + a_k} \leq 3$$
$$a_{k+1} \leq 3.$$

Now we show that $\{a_n\}$ is an increasing sequence. Since $a_n \geq 0$ and $a_n \leq 3$,

$$(a_n - 3)(a_n + 2) \leq 0$$
$$a_n^2 - a_n - 6 \leq 0$$
$$a_n^2 \leq a_n + 6$$
$$a_n \leq \sqrt{a_n + 6}$$
$$a_n \leq a_{n+1}.$$

Since $\{a_n\}$ is a bounded increasing sequence, it converges to some number L: $\lim_{n\to\infty} a_n = L$. Thus,

$$\sqrt{6 + L} = L \Rightarrow 6 + L = L^2 \Rightarrow L^2 - L - 6 = 0$$
$$\Rightarrow (L - 3)(L + 2) = 0 \Rightarrow L = 3 \quad (L \neq -2)$$

125. (a) We use mathematical induction to show that

$$a_n \le \frac{1 + \sqrt{1 + 4k}}{2}.$$

[Note that if $k = 2$, then $a_n \le 3$, and if $k = 6$, then $a_n \le 3$.] Clearly,

$$a_1 = \sqrt{k} \le \frac{\sqrt{1 + 4k}}{2} \le \frac{1 + \sqrt{1 + 4k}}{2}.$$

Before proceeding to the induction step, note that

$$2 + 2\sqrt{1 + 4k} + 4k = 2 + 2\sqrt{1 + 4k} + 4k$$

$$\frac{1 + \sqrt{1 + 4k}}{2} + k = \frac{1 + 2\sqrt{1 + 4k} + 1 + 4k}{4}$$

$$\frac{1 + \sqrt{1 + 4k}}{2} + k = \left[\frac{1 + \sqrt{1 + 4k}}{2}\right]^2$$

$$\sqrt{\frac{1 + \sqrt{1 + 4k}}{2} + k} = \frac{1 + \sqrt{1 + 4k}}{2}.$$

So assume $a_n \le \frac{1 + \sqrt{1 + 4k}}{2}$. Then

$$a_n + k \le \frac{1 + \sqrt{1 + 4k}}{2} + k$$

$$\sqrt{a_n + k} \le \sqrt{\frac{1 + \sqrt{1 + 4k}}{2} + k}$$

$$a_{n+1} \le \frac{1 + \sqrt{1 + 4k}}{2}.$$

$\{a_n\}$ is increasing because

$$\left(a_n - \frac{1 + \sqrt{1 + 4k}}{2}\right)\left(a_n - \frac{1 - \sqrt{1 + 4k}}{2}\right) \le 0$$

$$a_n^2 - a_n - k \le 0$$

$$a_n^2 \le a_n + k$$

$$a_n \le \sqrt{a_n + k}$$

$$a_n \le a_{n+1}.$$

(b) Since $\{a_n\}$ is bounded and increasing, it has a limit L.

(c) $\lim_{n\to\infty} a_n = L$ implies that

$$L = \sqrt{k + L} \Rightarrow L^2 = k + L$$

$$\Rightarrow L^2 - L - k = 0$$

$$\Rightarrow L = \frac{1 \pm \sqrt{1 + 4k}}{2}.$$

Since $L > 0$, $L = \frac{1 + \sqrt{1 + 4k}}{2}$.

126. (a) $a_0 = 10$, $b_0 = 3$

$a_1 = \frac{a_0 + b_0}{2} = \frac{10 + 3}{2} = 6.5$ $\qquad b_1 = \sqrt{a_0 b_0} = \sqrt{10(3)} \approx 5.4772$

$a_2 = \frac{a_1 + b_1}{2} \approx 5.9886$ $\qquad b_2 = \sqrt{a_1 b_1} \approx 5.9667$

$a_3 = \frac{a_2 + b_2}{2} \approx 5.9777$ $\qquad b_3 = \sqrt{a_2 b_2} \approx 5.9777$

$a_4 = \frac{a_3 + b_3}{2} \approx 5.9777$ $\qquad b_4 = \sqrt{a_3 b_3} \approx 5.9777$

$a_5 = \frac{a_4 + b_4}{2} \approx 5.9777$ $\qquad b_5 = \sqrt{a_4 b_4} \approx 5.9777$

The terms of $\{a_n\}$ are decreasing, and those of $\{b_n\}$ are increasing. They both seem to approach the same limit.

—CONTINUED—

126. —CONTINUED—

(b) For $n = 0$, we need to show $a_0 > a_1 > b_1 > b_0$. Since $a_0 > b_0$, $2a_0 > a_0 + b_0 \Rightarrow a_0 > \dfrac{a_0 + b_0}{2} = a_1$.

Since $(a_0 - b_0)^2 > 0$,

$$a_0^2 - 2a_0b_0 + b_0^2 > 0 \Rightarrow a_0^2 + 2a_0b_0 + b_0^2 > 4a_0b_0$$
$$\Rightarrow (a_0 + b_0)^2 > 4a_0b_0 \Rightarrow a_0 + b_0 > 2\sqrt{a_0b_0} \Rightarrow a_1 > b_1.$$

Since $a_0 > b_0$, $a_0b_0 > b_0^2 \Rightarrow \sqrt{a_0b_0} > b_0 \Rightarrow b_1 > b_0$. Thus, we have shown that $a_0 > a_1 > b_1 > b_0$.

Now assume $a_k > a_{k+1} > b_{k+1} > b_k$. Since $a_{k+1} > b_{k+1}$,

$$2a_{k+1} > a_{k+1} + b_{k+1} \Rightarrow a_{k+1} > \frac{a_{k+1} + b_{k+1}}{2} = a_{k+2}.$$

Since $(a_{k+1} - b_{k+1})^2 > 0$,

$$a_{k+1}^2 - 2a_{k+1}b_{k+1} + b_{k+1}^2 > 0 \Rightarrow a_{k+1}^2 + 2a_{k+1}b_{k+1} + b_{k+1}^2 > 4a_{k+1}b_{k+1}$$
$$\Rightarrow (a_{k+1} + b_{k+1})^2 > 4a_{k+1}b_{k+1}$$
$$\Rightarrow a_{k+1} + b_{k+1} > 2\sqrt{a_{k+1}b_{k+1}}$$
$$\Rightarrow a_{k+2} > b_{k+2}.$$

Since $a_{k+1} > b_{k+1}$, $a_{k+1}b_{k+1} > b_{k+1}^2 \Rightarrow \sqrt{a_{k+1}b_{k+1}} > b_{k+1} \Rightarrow b_{k+2} > b_{k+1}$.

Thus, we have shown that $a_{k+1} > a_{k+2} > b_{k+2} > b_{k+1}$.

(c) $\{a_n\}$ converges because it is decreasing and bounded below by 0. $\{b_n\}$ converges because it is increasing and bounded above by a_0.

(d) Let $\lim_{n\to\infty} a_n = A$ and $\lim_{n\to\infty} b_n = B$. Then $A = \dfrac{A + B}{2} \Rightarrow A = B$.

127. (a) $f(x) = \sin x$, $a_n = n \sin \dfrac{1}{n}$

$f'(x) = \cos x$, $f'(0) = 1$

$$\lim_{n\to\infty} a_n = \lim_{n\to\infty} n \sin\frac{1}{n} = \lim_{n\to\infty} \frac{\sin(1/n)}{(1/n)} = 1 = f'(0)$$

(b) $$f'(0) = \lim_{h\to 0^+} \frac{f(0 + h) - f(0)}{h}$$
$$= \lim_{h\to 0^+} \frac{f(h)}{h}$$
$$= \lim_{n\to\infty} \frac{f(1/n)}{(1/n)}$$
$$= \lim_{n\to\infty} nf\left(\frac{1}{n}\right)$$
$$= \lim_{n\to\infty} a_n$$

128. $a_n = nr^n$

(a) $r = \dfrac{1}{2}$: $a_n = n\left(\dfrac{1}{2}\right)^n = \dfrac{n}{2^n} \to 0$

(b) $r = 1$: $a_n = n$, diverges

(c) $r = \dfrac{3}{2}$: $a_n = n\left(\dfrac{3}{2}\right)^2$, diverges

(d) If $|r| \ge 1$, diverges. If $|r| < 1$,

$$nr^n = \frac{n}{r^{-n}} \qquad \text{“}\frac{\infty}{\infty}\text{”}$$
$$\to \frac{1}{-r^{-n}\ln(r)} = \frac{-r^n}{\ln(r)} \to 0.$$

The sequence converges for $|r| < 1$.

129. (a)

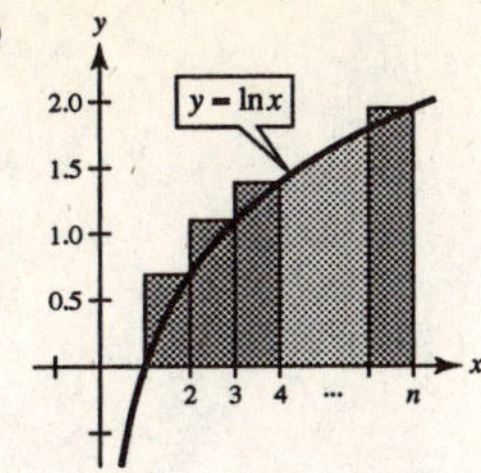

$$\int_1^n \ln x\,dx < \ln 2 + \ln 3 + \cdots + \ln n$$

$$= \ln[1 \cdot 2 \cdot 3 \cdots n] = \ln(n!)$$

(b)

$$\int_1^{n+1} \ln x\,dx > \ln 2 + \ln 3 + \cdots + \ln n$$

$$= \ln(n!)$$

(c) $\displaystyle\int \ln x\,dx = x \ln x - x + C$

$$\int_1^n \ln x\,dx = n \ln n - n + 1 = \ln n^n - n + 1$$

From part (a): $\ln n^n - n + 1 < \ln(n!)$

$$e^{\ln n^n - n + 1} < n!$$

$$\frac{n^n}{e^{n-1}} < n!$$

$$\int_1^{n+1} \ln x\,dx = (n+1)\ln(n+1) - (n+1) + 1 = \ln(n+1)^{n+1} - n$$

From part (b): $\ln(n+1)^{n+1} - n > \ln(n!)$

$$e^{\ln(n+1)^{n+1} - n} > n!$$

$$\frac{(n+1)^{n+1}}{e^n} > n!$$

(d) $$\frac{n^n}{e^{n-1}} < n! < \frac{(n+1)^{n+1}}{e^n}$$

$$\frac{n}{e^{1-(1/n)}} < \sqrt[n]{n!} < \frac{(n+1)^{(n+1)/n}}{e}$$

$$\frac{1}{e^{1-(1/n)}} < \frac{\sqrt[n]{n!}}{n} < \frac{(n+1)^{1+(1/n)}}{ne}$$

$$\lim_{n\to\infty} \frac{1}{e^{1-(1/n)}} = \frac{1}{e}$$

$$\lim_{n\to\infty} \frac{(n+1)^{1+(1/n)}}{ne} = \lim_{n\to\infty} \frac{(n+1)}{n}\,\frac{(n+1)^{1/n}}{e}$$

$$= (1)\frac{1}{e} = \frac{1}{e}$$

By the Squeeze Theorem, $\displaystyle\lim_{n\to\infty} \frac{\sqrt[n]{n!}}{n} = \frac{1}{e}$.

(e) $n = 20$: $\displaystyle\frac{\sqrt[20]{20!}}{20} \approx 0.4152$

$n = 50$: $\displaystyle\frac{\sqrt[50]{50!}}{50} \approx 0.3897$

$n = 100$: $\displaystyle\frac{\sqrt[100]{100!}}{100} \approx 0.3799$

$\displaystyle\frac{1}{e} \approx 0.3679$

130. $a_n = \frac{1}{n}\sum_{k=1}^{n}\frac{1}{1+(k/n)}$

(a) $a_1 = \frac{1}{1+1} = \frac{1}{2} = 0.5$

$$a_2 = \frac{1}{2}\left[\frac{1}{1+(1/2)} + \frac{1}{1+1}\right] = \frac{1}{2}\left[\frac{2}{3} + \frac{1}{2}\right] = \frac{7}{12} \approx 0.5833$$

$$a_3 = \frac{1}{3}\left[\frac{1}{1+(1/3)} + \frac{1}{1+(2/3)} + \frac{1}{1+1}\right] = \frac{37}{60} \approx 0.6167$$

$$a_4 = \frac{1}{4}\left[\frac{1}{1+(1/4)} + \frac{1}{1+(2/4)} + \frac{1}{1+(3/4)} + \frac{1}{1+1}\right] = \frac{533}{840} \approx 0.6345$$

$$a_5 = \frac{1}{5}\left[\frac{1}{1+(1/5)} + \frac{1}{1+(2/5)} + \frac{1}{1+(3/5)} + \frac{1}{1+(4/5)} + \frac{1}{1+1}\right] = \frac{1627}{2520} \approx 0.6456$$

(b) $\lim_{n\to\infty} a_n = \lim_{n\to\infty}\frac{1}{n}\sum_{k=1}^{n}\frac{1}{1+(k/n)} = \int_0^1 \frac{1}{1+x}\,dx = \ln|1+x|\Big]_0^1 = \ln 2$

131. For a given $\epsilon > 0$, we must find $M > 0$ such that

$$|a_n - L| = \left|\frac{1}{n^3}\right| < \varepsilon$$

whenever $n > M$. That is,

$$n^3 > \frac{1}{\varepsilon} \text{ or } n > \left(\frac{1}{\varepsilon}\right)^{1/3}.$$

So, let $\varepsilon > 0$ be given. Let M be an integer satisfying $M > (1/\varepsilon)^{1/3}$. For $n > M$, we have

$$n > \left(\frac{1}{\varepsilon}\right)^{1/3}$$

$$n^3 > \frac{1}{\varepsilon}$$

$$\varepsilon > \frac{1}{n^3} \Rightarrow \left|\frac{1}{n^3} - 0\right| < \varepsilon.$$

Thus, $\lim_{n\to\infty}\frac{1}{n^3} = 0.$

132. For a given $\varepsilon > 0$, we must find $M > 0$ such that $|a_n - L| = |r^n|\varepsilon$ whenever $n > M$. That is, $n\ln|r| < \ln(\varepsilon)$ or

$$n > \frac{\ln(\varepsilon)}{\ln|r|} \text{ (since } \ln|r| < 0 \text{ for } |r| < 1).$$

So, let $\varepsilon > 0$ be given. Let M be an integer satisfying

$$M > \frac{\ln(\varepsilon)}{\ln|r|}.$$

For $n > M$, we have

$$n > \frac{\ln(\varepsilon)}{\ln|r|}$$

$$n\ln|r| < \ln(\varepsilon)$$

$$\ln|r|^n < \ln(\varepsilon)$$

$$|r|^n < \varepsilon$$

$$|r^n - 0| < \varepsilon.$$

Thus, $\lim_{n\to\infty} r^n = 0$ for $-1 < r < 1$.

133. If $\{a_n\}$ is bounded, monotonic and nonincreasing, then $a_1 \ge a_2 \ge a_3 \ge \cdots \ge a_n \ge \cdots$. Then $-a_1 \le -a_2 \le -a_3 \le \cdots \le -a_n \le \cdots$ is a bounded, monotonic, nondecreasing sequence which converges by the first half of the theorem. Since $\{-a_n\}$ converges, then so does $\{a_n\}$.

134. Define $a_n = \frac{x_{n+1} + x_{n-1}}{x_n}$, $n \ge 1$.

$$x_{n+1}^2 - x_n x_{n+2} = 1 = x_n^2 - x_{n-1}x_{n+1} \Rightarrow$$

$$x_{n+1}(x_{n+1} + x_{n-1}) = x_n(x_n + x_{n+2})$$

$$\frac{x_{n+1} + x_{n-1}}{x_n} = \frac{x_{n+2} + x_n}{x_{n+1}}$$

$$a_n = a_{n+1}$$

Hence, $a_1 = a_2 = \ldots = a$. Thus,

$$x_{n+1} = a_n x_n - x_{n-1} = a x_n - x_{n-1}.$$

135. $T_n = n! + 2^n$

We use mathematical induction to verify the formula.

$T_0 = 1 + 1 = 2$

$T_1 = 1 + 2 = 3$

$T_2 = 2 + 4 = 6$

Assume $T_k = k! + 2^k$. Then

$$\begin{aligned} T_{k+1} &= (k + 1 + 4)T_k - 4(k + 1)T_{k-1} + (4(k + 1) - 8)T_{k-2} \\ &= (k + 5)[k! + 2^k] - 4(k + 1)((k - 1)! + 2^{k-1}) + (4k - 4)((k - 2)! + 2^{k-2}) \\ &= [(k + 5)(k)(k - 1) - 4(k + 1)(k - 1) + 4(k - 1)](k - 2)! + [(k + 5)4 - 8(k + 1) + 4(k - 1)]2^{k-2} \\ &= [k^2 + 5k - 4k - 4 + 4](k - 1)! + 8 \cdot 2^{k-2} \\ &= (k + 1)! + 2^{k+1}. \end{aligned}$$

By mathematical induction, the formula is valid for all n.

Section 9.2 Series and Convergence

1. $S_1 = 1$

$S_2 = 1 + \frac{1}{4} = 1.2500$

$S_3 = 1 + \frac{1}{4} + \frac{1}{9} \approx 1.3611$

$S_4 = 1 + \frac{1}{4} + \frac{1}{9} + \frac{1}{16} \approx 1.4236$

$S_5 = 1 + \frac{1}{4} + \frac{1}{9} + \frac{1}{16} + \frac{1}{25} \approx 1.4636$

2. $S_1 = \frac{1}{6} \approx 0.1667$

$S_2 = \frac{1}{6} + \frac{1}{6} \approx 0.3333$

$S_3 = \frac{1}{6} + \frac{1}{6} + \frac{3}{20} \approx 0.4833$

$S_4 = \frac{1}{6} + \frac{1}{6} + \frac{3}{20} + \frac{2}{15} \approx 0.6167$

$S_5 = \frac{1}{6} + \frac{1}{6} + \frac{3}{20} + \frac{2}{15} + \frac{5}{42} \approx 0.7357$

3. $S_1 = 3$

$S_2 = 3 - \frac{9}{2} = -1.5$

$S_3 = 3 - \frac{9}{2} + \frac{27}{4} = 5.25$

$S_4 = 3 - \frac{9}{2} + \frac{27}{4} - \frac{81}{8} = -4.875$

$S_5 = 3 - \frac{9}{2} + \frac{27}{4} - \frac{81}{8} + \frac{243}{16} = 10.3125$

4. $S_1 = 1$

$S_2 = 1 + \frac{1}{3} \approx 1.3333$

$S_3 = 1 + \frac{1}{3} + \frac{1}{5} \approx 1.5333$

$S_4 = 1 + \frac{1}{3} + \frac{1}{5} + \frac{1}{7} \approx 1.6762$

$S_5 = 1 + \frac{1}{3} + \frac{1}{5} + \frac{1}{7} + \frac{1}{9} \approx 1.7873$

5. $S_1 = 3$

$S_2 = 3 + \frac{3}{2} = 4.5$

$S_3 = 3 + \frac{3}{2} + \frac{3}{4} = 5.250$

$S_4 = 3 + \frac{3}{2} + \frac{3}{4} + \frac{3}{8} = 5.625$

$S_5 = 3 + \frac{3}{2} + \frac{3}{4} + \frac{3}{8} + \frac{3}{16} = 5.8125$

6. $S_1 = 1$

$S_2 = 1 - \frac{1}{2} = 0.5$

$S_3 = 1 - \frac{1}{2} + \frac{1}{6} \approx 0.6667$

$S_4 = 1 - \frac{1}{2} + \frac{1}{6} - \frac{1}{24} \approx 0.6250$

$S_5 = 1 - \frac{1}{2} + \frac{1}{6} - \frac{1}{24} + \frac{1}{120} \approx 0.6333$

7. $\sum_{n=0}^{\infty} 3\left(\frac{3}{2}\right)^n$

Geometric series

$r = \frac{3}{2} > 1$

Diverges by Theorem 9.6

8. $\sum_{n=0}^{\infty} \left(\frac{4}{3}\right)^n$

Geometric series

$r = \frac{4}{3} > 1$

Diverges by Theorem 9.6

9. $\sum_{n=0}^{\infty} 1000(1.055)^n$

Geometric series

$r = 1.055 > 1$

Diverges by Theorem 9.6

10. $\sum_{n=0}^{\infty} 2(-1.03)^n$

Geometric series

$|r| = 1.03 > 1$

Diverges by Theorem 9.6

11. $\sum_{n=1}^{\infty} \frac{n}{n+1}$

$\lim_{n\to\infty} \frac{n}{n+1} = 1 \neq 0$

Diverges by Theorem 9.9

12. $\sum_{n=1}^{\infty} \frac{n}{2n+3}$

$\lim_{n\to\infty} \frac{n}{2n+3} = \frac{1}{2} \neq 0$

Diverges by Theorem 9.9

13. $\sum_{n=1}^{\infty} \frac{n^2}{n^2+1}$

$\lim_{n\to\infty} \frac{n^2}{n^2+1} = 1 \neq 0$

Diverges by Theorem 9.9

14. $\sum_{n=1}^{\infty} \frac{n}{\sqrt{n^2+1}}$

$\lim_{n\to\infty} \frac{n}{\sqrt{n^2+1}} = \lim_{n\to\infty} \frac{1}{\sqrt{1+(1/n^2)}} = 1 \neq 0$

Diverges by Theorem 9.9

15. $\sum_{n=1}^{\infty} \frac{2^n+1}{2^{n+1}}$

$\lim_{n\to\infty} \frac{2^n+1}{2^{n+1}} = \lim_{n\to\infty} \frac{1+2^{-n}}{2} = \frac{1}{2} \neq 0$

Diverges by Theorem 9.9

16. $\sum_{n=1}^{\infty} \frac{n!}{2^n}$

$\lim_{n\to\infty} \frac{n!}{2^n} = \infty$

Diverges by Theorem 9.9

17. $\sum_{n=0}^{\infty} \frac{9}{4}\left(\frac{1}{4}\right)^n = \frac{9}{4}\left[1 + \frac{1}{4} + \frac{1}{16} + \cdots\right]$

$S_0 = \frac{9}{4}, S_1 = \frac{9}{4}\cdot\frac{5}{4} = \frac{45}{16}, S_2 = \frac{9}{4}\cdot\frac{21}{16} \approx 2.95, \ldots$

Matches graph (c). Analytically, the series is geometric:

$$\sum_{n=0}^{\infty} \left(\frac{9}{4}\right)\left(\frac{1}{4}\right)^n = \frac{9/4}{1-1/4} = \frac{9/4}{3/4} = 3$$

18. $\sum_{n=0}^{\infty} \left(\frac{2}{3}\right)^n = 1 + \frac{2}{3} + \frac{4}{9} + \cdots$

$S_0 = 1, S_1 = \frac{5}{3}, S_2 \approx 2.11, \ldots$

Matches graph (b). Analytically, the series is geometric:

$$\sum_{n=0}^{\infty} \left(\frac{2}{3}\right)^n = \frac{1}{1-2/3} = \frac{1}{1/3} = 3$$

19. $\sum_{n=0}^{\infty} \frac{15}{4}\left(-\frac{1}{4}\right)^n = \frac{15}{4}\left[1 - \frac{1}{4} + \frac{1}{16} - \cdots\right]$

$S_0 = \frac{15}{4}, S_1 = \frac{45}{16}, S_2 \approx 3.05, \ldots$

Matches graph (a). Analytically, the series is geometric:

$$\sum_{n=0}^{\infty} \frac{15}{4}\left(-\frac{1}{4}\right)^n = \frac{15/4}{1-(-1/4)} = \frac{15/4}{5/4} = 3$$

20. $\sum_{n=0}^{\infty} \frac{17}{3}\left(-\frac{8}{9}\right)^n = \frac{17}{3}\left[1 - \frac{8}{9} + \frac{64}{81} - \cdots\right]$

$S_0 = \frac{17}{3}, S_1 \approx 0.63, S_3 \approx 5.1, \ldots$

Matches (d). Analytically, the series is geometric:

$$\sum_{n=0}^{\infty} \frac{17}{3}\left(-\frac{8}{9}\right)^n = \frac{17/3}{1-(-8/9)} = \frac{17/3}{17/9} = 3$$

21. $\sum_{n=0}^{\infty} \frac{17}{3}\left(-\frac{1}{2}\right)^n = \frac{17}{3}\left[1 - \frac{1}{2} + \frac{1}{4} - \cdots\right]$

$S_0 = \frac{17}{3}, S_1 = \frac{17}{6}, \ldots$

Matches (f). The series is geometric:

$$\sum_{n=0}^{\infty} \frac{17}{3}\left(-\frac{1}{2}\right)^n = \frac{17}{3}\,\frac{1}{1-(-1/2)} = \frac{17}{3}\,\frac{2}{3} = \frac{34}{9}$$

22. $\sum_{n=0}^{\infty} \left(\frac{2}{5}\right)^n = 1 + \frac{2}{5} + \frac{4}{25} + \cdots$

$S_0 = 1, S_1 = \frac{7}{5}, \ldots$

Matches (e). The series is geometric:

$$\sum_{n=0}^{\infty} \left(\frac{2}{5}\right)^n = \frac{1}{1-(2/5)} = \frac{5}{3}$$

23. $\sum_{n=1}^{\infty} \frac{1}{n(n+1)} = \sum_{n=1}^{\infty} \left(\frac{1}{n} - \frac{1}{n+1}\right) = \left(1 - \frac{1}{2}\right) + \left(\frac{1}{2} - \frac{1}{3}\right) + \left(\frac{1}{3} - \frac{1}{4}\right) + \left(\frac{1}{4} - \frac{1}{5}\right) + \cdots, \quad S_n = 1 - \frac{1}{n+1}$

$$\sum_{n=1}^{\infty} \frac{1}{n(n+1)} = \lim_{n\to\infty} S_n = \lim_{n\to\infty} \left(1 - \frac{1}{n+1}\right) = 1$$

24. $\displaystyle\sum_{n=1}^{\infty} \frac{1}{n(n+2)} = \sum_{n=1}^{\infty}\left(\frac{1}{2n} - \frac{1}{2(n+2)}\right) = \left(\frac{1}{2} - \frac{1}{6}\right) + \left(\frac{1}{4} - \frac{1}{8}\right) + \left(\frac{1}{6} - \frac{1}{10}\right) + \left(\frac{1}{8} - \frac{1}{12}\right) + \left(\frac{1}{10} - \frac{1}{14}\right) + \cdots$

$$\sum_{n=1}^{\infty} \frac{1}{n(n+2)} = \lim_{n\to\infty} S_n = \lim_{n\to\infty}\left[\frac{1}{2} + \frac{1}{4} - \frac{1}{2(n+1)} - \frac{1}{2(n+2)}\right] = \frac{1}{2} + \frac{1}{4} = \frac{3}{4}$$

25. $\displaystyle\sum_{n=0}^{\infty} 2\left(\frac{3}{4}\right)^n$

Geometric series with $r = \frac{3}{4} < 1$

Converges by Theorem 9.6

26. $\displaystyle\sum_{n=0}^{\infty} 2\left(-\frac{1}{2}\right)^n$

Geometric series with $|r| = \left|-\frac{1}{2}\right| < 1$

Converges by Theorem 9.6

27. $\displaystyle\sum_{n=0}^{\infty} (0.9)^n$

Geometric series with $r = 0.9 < 1$

Converges by Theorem 9.6

28. $\displaystyle\sum_{n=0}^{\infty} (-0.6)^n$

Geometric series with $|r| = |-0.6| < 1$

Converges by Theorem 9.6

29. (a) $\displaystyle\sum_{n=1}^{\infty} \frac{6}{n(n+3)} = 2\sum_{n=1}^{\infty}\left(\frac{1}{n} - \frac{1}{n+3}\right)$

$$= 2\left[\left(1 - \frac{1}{4}\right) + \left(\frac{1}{2} - \frac{1}{5}\right) + \left(\frac{1}{3} - \frac{1}{6}\right) + \left(\frac{1}{4} - \frac{1}{7}\right) + \cdots\right]$$

$$\left(S_n = 2\left[1 + \frac{1}{2} + \frac{1}{3} - \left(\frac{1}{n+1} + \frac{1}{n+2} + \frac{1}{n+3}\right)\right]\right)$$

$$= 2\left[1 + \frac{1}{2} + \frac{1}{3}\right] = \frac{11}{3} \approx 3.667$$

(b)

n	5	10	20	50	100
S_n	2.7976	3.1643	3.3936	3.5513	3.6078

(c)

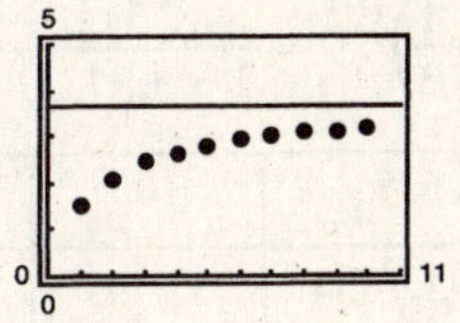

(d) The terms of the series decrease in magnitude slowly. Thus, the sequence of partial sums approaches the sum slowly.

30. (a) $\displaystyle\sum_{n=1}^{\infty} \frac{4}{n(n+4)} = \sum_{n=1}^{\infty}\left(\frac{1}{n} - \frac{1}{n+4}\right)$

$$= \left(1 - \frac{1}{5}\right) + \left(\frac{1}{2} - \frac{1}{6}\right) + \left(\frac{1}{3} - \frac{1}{7}\right) + \left(\frac{1}{4} - \frac{1}{8}\right) + \left(\frac{1}{5} - \frac{1}{9}\right) + \left(\frac{1}{6} - \frac{1}{10}\right) + \cdots$$

$$= 1 + \frac{1}{2} + \frac{1}{3} + \frac{1}{4} = \frac{25}{12} \approx 2.0833$$

(b)

n	5	10	20	50	100
S_n	1.5377	1.7607	1.9051	2.0071	2.0443

(c)

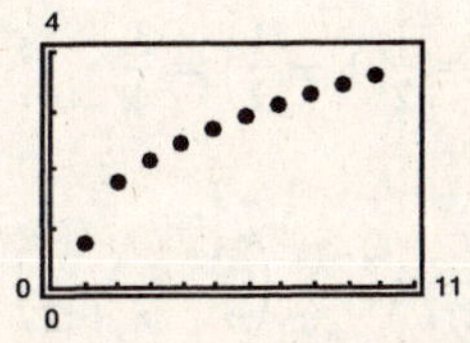

(d) The terms of the series decrease in magnitude slowly. Thus, the sequence of partial sums approaches the sum slowly.

31. (a) $\displaystyle\sum_{n=1}^{\infty} 2(0.9)^{n-1} = \sum_{n=0}^{\infty} 2(0.9)^n = \frac{2}{1-0.9} = 20$

(b)

n	5	10	20	50	100
S_n	8.1902	13.0264	17.5685	19.8969	19.9995

(c)

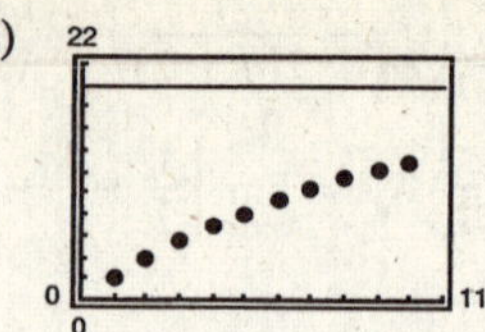

(d) The terms of the series decrease in magnitude slowly. Thus, the sequence of partial sums approaches the sum slowly.

32. (a) $\displaystyle\sum_{n=1}^{\infty} 3(0.85)^{n-1} = \frac{3}{1-0.85} = 20$ (Geometric series)

(b)

n	5	10	20	50	100
S_n	11.1259	16.0625	19.2248	19.9941	19.999998

(c)

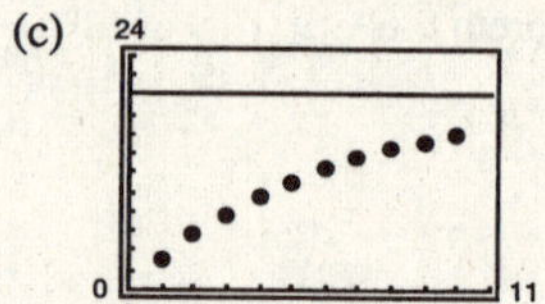

(d) The terms of the series decrease in magnitude slowly. Thus, the sequence of partial sums approaches the sum slowly.

33. (a) $\displaystyle\sum_{n=1}^{\infty} 10(0.25)^{n-1} = \frac{10}{1-0.25} = \frac{40}{3} \approx 13.3333$

(b)

n	5	10	20	50	100
S_n	13.3203	13.3333	13.3333	13.3333	13.3333

(c)

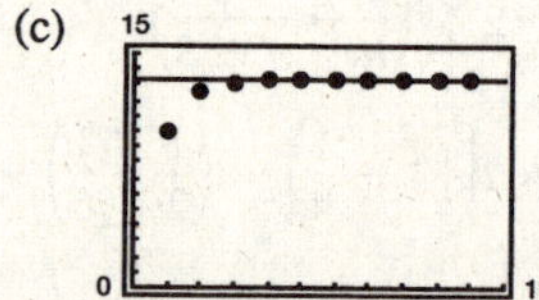

(d) The terms of the series decrease in magnitude rapidly. Thus, the sequence of partial sums approaches the sum rapidly.

34. (a) $\displaystyle\sum_{n=1}^{\infty} 5\left(-\frac{1}{3}\right)^{n-1} = \frac{5}{1-(-1/3)} = \frac{15}{4} = 3.75$

(b)

n	5	10	20	50	100
S_n	3.7654	3.7499	3.7500	3.7500	3.7500

(c)

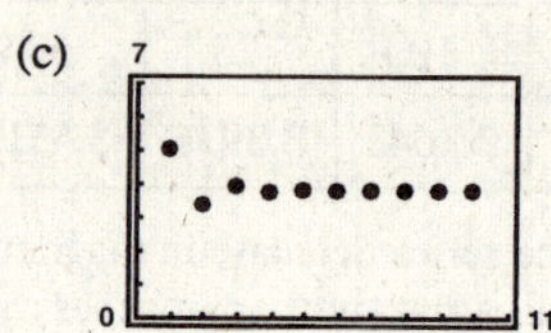

(d) The terms of the series decrease in magnitude rapidly. Thus, the sequence of partial sums approaches the sum rapidly.

35. $$\sum_{n=2}^{\infty} \frac{1}{n^2-1} = \sum_{n=2}^{\infty}\left(\frac{1/2}{n-1} - \frac{1/2}{n+1}\right) = \frac{1}{2}\sum_{n=2}^{\infty}\left(\frac{1}{n-1} - \frac{1}{n+1}\right)$$
$$= \frac{1}{2}\left[\left(1-\frac{1}{3}\right) + \left(\frac{1}{2}-\frac{1}{4}\right) + \left(\frac{1}{3}-\frac{1}{5}\right) + \left(\frac{1}{4}-\frac{1}{6}\right) + \cdots\right]$$
$$= \frac{1}{2}\left(1+\frac{1}{2}\right) = \frac{3}{4}$$

36. $$\sum_{n=1}^{\infty} \frac{4}{n(n+2)} = 2\sum_{n=1}^{\infty}\left(\frac{1}{n} - \frac{1}{n+2}\right) = 2\left[\left(1-\frac{1}{3}\right) + \left(\frac{1}{2}-\frac{1}{4}\right) + \left(\frac{1}{3}-\frac{1}{5}\right) + \cdots\right] = 2\left(1+\frac{1}{2}\right) = 3$$

37. $$\sum_{n=1}^{\infty} \frac{8}{(n+1)(n+2)} = 8\sum_{n=1}^{\infty}\left(\frac{1}{n+1} - \frac{1}{n+2}\right) = 8\left[\left(\frac{1}{2}-\frac{1}{3}\right) + \left(\frac{1}{3}-\frac{1}{4}\right) + \left(\frac{1}{4}-\frac{1}{5}\right) + \cdots\right] = 8\left(\frac{1}{2}\right) = 4$$

38. $$\sum_{n=1}^{\infty} \frac{1}{(2n+1)(2n+3)} = \frac{1}{2}\sum_{n=1}^{\infty}\left(\frac{1}{2n+1} - \frac{1}{2n+3}\right) = \frac{1}{2}\left[\left(\frac{1}{3}-\frac{1}{5}\right) + \left(\frac{1}{5}-\frac{1}{7}\right) + \left(\frac{1}{7}-\frac{1}{9}\right) + \cdots\right] = \frac{1}{2}\left(\frac{1}{3}\right) = \frac{1}{6}$$

39. $\sum_{n=0}^{\infty}\left(\frac{1}{2}\right)^n = \frac{1}{1-(1/2)} = 2$

40. $\sum_{n=0}^{\infty}6\left(\frac{4}{5}\right)^n = \frac{6}{1-(4/5)} = 30$

(Geometric)

41. $\sum_{n=0}^{\infty}\left(-\frac{1}{2}\right)^n = \frac{1}{1-(-1/2)} = \frac{2}{3}$

42. $\sum_{n=0}^{\infty}2\left(-\frac{2}{3}\right)^n = \frac{2}{1-(-2/3)} = \frac{6}{5}$

43. $\sum_{n=0}^{\infty}\left(\frac{1}{10}\right)^n = \frac{1}{1-(1/10)} = \frac{10}{9}$

44. $\sum_{n=0}^{\infty}8\left(\frac{3}{4}\right)^n = \frac{8}{1-(3/4)} = 32$

45. $\sum_{n=0}^{\infty}3\left(-\frac{1}{3}\right)^n = \frac{3}{1-(-1/3)} = \frac{9}{4}$

46. $\sum_{n=0}^{\infty}4\left(-\frac{1}{2}\right)^n = \frac{4}{1-(-1/2)} = \frac{8}{3}$

47. $\sum_{n=0}^{\infty}\left(\frac{1}{2^n}-\frac{1}{3^n}\right) = \sum_{n=0}^{\infty}\left(\frac{1}{2}\right)^n - \sum_{n=0}^{\infty}\left(\frac{1}{3}\right)^n$

$= \frac{1}{1-(1/2)} - \frac{1}{1-(1/3)}$

$= 2 - \frac{3}{2} = \frac{1}{2}$

48. $\sum_{n=1}^{\infty}[(0.7)^n + (0.9)^n] = \sum_{n=0}^{\infty}\left(\frac{7}{10}\right)^n + \sum_{n=0}^{\infty}\left(\frac{9}{10}\right)^n - 2$

$= \frac{1}{1-(7/10)} + \frac{1}{1-(9/10)} - 2$

$= \frac{10}{3} + 10 - 2 = \frac{34}{3}$

49. Note that $\sin(1) \approx 0.8415 < 1$. The series $\sum_{n=1}^{\infty}[\sin(1)]^n$ is geometric with $r = \sin(1) < 1$. Thus,

$\sum_{n=1}^{\infty}[\sin(1)]^n = \sin(1)\sum_{n=0}^{\infty}[\sin(1)]^n = \frac{\sin(1)}{1-\sin(1)} \approx 5.3080.$

50. $S_n = \sum_{k=1}^{n}\frac{1}{9k^2+3k-2} = \sum_{k=1}^{n}\frac{1}{(3k-1)(3k+2)}$

$= \sum_{k=1}^{n}\left[\frac{1}{9k-3} - \frac{1}{9k+6}\right] = \frac{1}{3}\sum_{k=1}^{n}\left[\frac{1}{3k-1} - \frac{1}{3k+2}\right]$

$= \frac{1}{3}\left[\left(\frac{1}{2}-\frac{1}{5}\right) + \left(\frac{1}{5}-\frac{1}{8}\right) + \left(\frac{1}{8}-\frac{1}{11}\right) + \cdots + \left(\frac{1}{3n-1} - \frac{1}{3n+2}\right)\right]$

$= \frac{1}{3}\left[\frac{1}{2} - \frac{1}{3n+2}\right]$

$\lim_{n\to\infty} S_n = \lim_{n\to\infty}\frac{1}{3}\left[\frac{1}{2} - \frac{1}{3n+2}\right] = \frac{1}{6}$

51. (a) $0.\overline{4} = \sum_{n=0}^{\infty}\frac{4}{10}\left(\frac{1}{10}\right)^n$

(b) Geometric series with $a = \frac{4}{10}$ and $r = \frac{1}{10}$

$S = \frac{a}{1-r} = \frac{4/10}{1-(1/10)} = \frac{4}{9}$

52. (a) $0.\overline{9} = \frac{9}{10} + \frac{9}{100} + \frac{9}{1000} + \cdots$

$= \sum_{n=1}^{\infty}9\left(\frac{1}{10}\right)^n = \frac{9}{10}\sum_{n=0}^{\infty}\left(\frac{1}{10}\right)^n$

(b) $0.\overline{9} = \frac{9}{10}\,\frac{1}{1-(1/10)} = 1$

53. (a) $0.\overline{81} = \sum_{n=0}^{\infty}\frac{81}{100}\left(\frac{1}{100}\right)^n$

(b) Geometric series with $a = \frac{81}{100}$ and $r = \frac{1}{100}$

$S = \frac{a}{1-r} = \frac{81/100}{1-(1/100)} = \frac{81}{99} = \frac{9}{11}$

54. (a) $0.\overline{01} = \sum_{n=1}^{\infty}\left(\frac{1}{100}\right)^n = \frac{1}{100}\sum_{n=0}^{\infty}\left(\frac{1}{100}\right)^n$

(b) $0.\overline{01} = \frac{1}{100}\cdot\frac{1}{1-(1/100)} = \frac{1}{100}\cdot\frac{100}{99} = \frac{1}{99}$

55. (a) $0.0\overline{75} = \sum_{n=0}^{\infty} \frac{3}{40}\left(\frac{1}{100}\right)^n$

(b) Geometric series with $a = \frac{3}{40}$ and $r = \frac{1}{100}$

$$S = \frac{a}{1-r} = \frac{3/40}{99/100} = \frac{5}{66}$$

56. (a) $0.2\overline{15} = \frac{1}{5} + \sum_{n=0}^{\infty} \frac{3}{200}\left(\frac{1}{100}\right)^n$

(b) Geometric series with $a = \frac{3}{200}$ and $r = \frac{1}{100}$

$$S = \frac{1}{5} + \frac{a}{1-r} = \frac{1}{5} + \frac{3/200}{99/100} = \frac{71}{330}$$

57. $\sum_{n=1}^{\infty} \frac{n+10}{10n+1}$

$$\lim_{n\to\infty} \frac{n+10}{10n+1} = \frac{1}{10} \neq 0$$

Diverges by Theorem 9.9

58. $\sum_{n=1}^{\infty} \frac{n+1}{2n-1}$

$$\lim_{n\to\infty} \frac{n+1}{2n-1} = \frac{1}{2} \neq 0$$

Diverges by Theorem 9.9

59. $\sum_{n=1}^{\infty}\left(\frac{1}{n} - \frac{1}{n+2}\right) = \left(1 - \frac{1}{3}\right) + \left(\frac{1}{2} - \frac{1}{4}\right) + \left(\frac{1}{3} - \frac{1}{5}\right) + \left(\frac{1}{4} - \frac{1}{6}\right) + \cdots = 1 + \frac{1}{2} = \frac{3}{2}$, converges

60.
$$\sum_{n=1}^{\infty} \frac{1}{n(n+3)} = \frac{1}{3}\sum_{n=1}^{\infty}\left(\frac{1}{n} - \frac{1}{n+3}\right)$$
$$= \frac{1}{3}\left[\left(1 - \frac{1}{4}\right) + \left(\frac{1}{2} - \frac{1}{5}\right) + \left(\frac{1}{3} - \frac{1}{6}\right) + \left(\frac{1}{4} - \frac{1}{7}\right) + \left(\frac{1}{5} - \frac{1}{8}\right) + \left(\frac{1}{6} - \frac{1}{9}\right) + \cdots\right]$$
$$= \frac{1}{3}\left(1 + \frac{1}{2} + \frac{1}{3}\right) = \frac{11}{18}, \text{ converges}$$

61. $\sum_{n=1}^{\infty} \frac{3n-1}{2n+1}$

$$\lim_{n\to\infty} \frac{3n-1}{2n+1} = \frac{3}{2} \neq 0$$

Diverges by Theorem 9.9

62. $\sum_{n=1}^{\infty} \frac{3^n}{n^3}$

$$\lim_{n\to\infty} \frac{3^n}{n^3} = \lim_{n\to\infty} \frac{(\ln 2)3^n}{3n^2}$$
$$= \lim_{n\to\infty} \frac{(\ln 2)^2 3^n}{6n} = \lim_{n\to\infty} \frac{(\ln n)^3 3^n}{6} = \infty$$

(by L'Hôpital's Rule); diverges by Theorem 9.9

63. $\sum_{n=0}^{\infty} \frac{4}{2^n} = 4\sum_{n=0}^{\infty}\left(\frac{1}{2}\right)^n$

Geometric series with $r = \frac{1}{2}$

Converges by Theorem 9.6

64. $\sum_{n=0}^{\infty} \frac{1}{4^n}$

Geometric series with $r = \frac{1}{4}$

Converges by Theorem 9.6

65. $\sum_{n=0}^{\infty} (1.075)^n$

Geometric series with $r = 1.075$

Diverges by Theorem 9.6

66. $\sum_{n=1}^{\infty} \frac{2^n}{100}$

Geometric series with $r = 2$

Diverges by Theorem 9.6

67. Since $n > \ln(n)$, the terms $a_n = \frac{n}{\ln(n)}$ do not approach 0 as $n \to \infty$. Hence, the series $\sum_{n=2}^{\infty} \frac{n}{\ln(n)}$ diverges.

68.
$$S_n = \sum_{k=1}^{n} \ln\left(\frac{1}{k}\right) = \sum_{k=1}^{n} -\ln(k)$$
$$= 0 - \ln 2 - \ln 3 - \cdots - \ln(n)$$

Since $\lim_{n\to\infty} S_n$ diverges, $\sum_{n=1}^{\infty} \ln\left(\frac{1}{n}\right)$ diverges.

69. For $k \neq 0$,

$$\lim_{n\to\infty}\left(1+\frac{k}{n}\right)^n = \lim_{n\to\infty}\left[\left(1+\frac{k}{n}\right)^{n/k}\right]^k$$

$$= e^k \neq 0.$$

For $k = 0$, $\lim_{n\to\infty} (1 + 0)^n = 1 \neq 0$.

Hence, $\sum_{n=1}^{\infty}\left[1+\frac{k}{n}\right]^n$ diverges.

70. $\sum_{n=1}^{\infty} e^{-n} = \sum_{n=1}^{\infty}\left(\frac{1}{e}\right)^n$ converges since it is geometric with

$|r| = \frac{1}{e} < 1.$

71. $\lim_{n\to\infty} \arctan n = \frac{\pi}{2} \neq 0$

Hence, $\sum_{n=1}^{\infty} \arctan n$ diverges.

72. $S_n = \sum_{k=1}^{n} \ln\left(\frac{k+1}{k}\right)$

$= \ln\left(\frac{2}{1}\right) + \ln\left(\frac{3}{2}\right) + \cdots + \ln\left(\frac{n+1}{n}\right)$

$= (\ln 2 - \ln 1) + (\ln 3 - \ln 2) + \cdots + (\ln(n + 1) - \ln n)$

$= \ln(n + 1) - \ln(1) = \ln(n + 1)$

Diverges

73. See definition on page 606.

74. $\lim_{n\to\infty} a_n = 5$ means that the limit of the sequence $\{a_n\}$ is 5.

$\sum_{n=1}^{\infty} a_n = a_1 + a_2 + \cdots = 5$ means that the limit of the partial sums is 5.

75. The series given by

$$\sum_{n=0}^{\infty} ar^n = a + ar + ar^2 + \cdots + ar^n + \cdots, \quad a \neq 0$$

is a geometric series with ratio r. When $0 < |r| < 1$, the series converges to $a/(1 - r)$. The series diverges if $|r| \geq 1$.

76. If $\lim_{n\to\infty} a_n \neq 0$, then $\sum_{n=1}^{\infty} a_n$ diverges.

77. $a_n = \dfrac{n+1}{n}$

$\lim_{n\to\infty} a_n = \lim_{n\to\infty} \dfrac{n+1}{n} = 1 \quad (\{a_n\}$ converges)

$\sum_{n=1}^{\infty} a_n$ diverges because $\lim_{n\to\infty} a_n \neq 0$.

78. (a) $\sum_{n=1}^{\infty} a_n = a_1 + a_2 + a_3 + \cdots$

(b) $\sum_{k=1}^{\infty} a_k = a_1 + a_2 + a_3 + \cdots$

These are the same. The third series is different, unless $a_1 = a_2 = \cdots = a$ is constant.

(c) $\sum_{n=1}^{\infty} a_k = a_k + a_k + \cdots$

79. $\sum_{n=1}^{\infty} \frac{x^n}{2^n} = \sum_{n=1}^{\infty}\left(\frac{x}{2}\right)^n = \frac{x}{2}\sum_{n=0}^{\infty}\left(\frac{x}{2}\right)^n$

Geometric series: converges for $\left|\frac{x}{2}\right| < 1$ or $|x| < 2$

$f(x) = \frac{x}{2}\sum_{n=0}^{\infty}\left(\frac{x}{2}\right)^n$

$= \frac{x}{2}\frac{1}{1-(x/2)} = \frac{x}{2}\frac{2}{2-x} = \frac{x}{2-x}, \quad |x| < 2$

80. $\sum_{n=1}^{\infty} (3x)^n = (3x)\sum_{n=0}^{\infty} (3x)^n$

Geometric series: converges for $|3x| < 1 \Rightarrow |x| < \frac{1}{3}$

$f(x) = (3x)\sum_{n=0}^{\infty} (3x)^n = (3x)\frac{1}{1-3x} = \frac{3x}{1-3x}, \quad |x| < \frac{1}{3}$

81. $\sum_{n=1}^{\infty}(x-1)^n = (x-1)\sum_{n=0}^{\infty}(x-1)^n$

Geometric series: converges for $|x-1| < 1 \Rightarrow 0 < x < 2$

$$f(x) = (x-1)\sum_{n=0}^{\infty}(x-1)^n$$

$$= (x-1)\frac{1}{1-(x-1)} = \frac{x-1}{2-x}, \quad 0 < x < 2$$

82. $\sum_{n=0}^{\infty} 4\left(\frac{x-3}{4}\right)^n$

Geometric series: converges for

$$\left|\frac{x-3}{4}\right| < 1 \Rightarrow |x-3| < 4 \Rightarrow -1 < x < 7$$

$$f(x) = \sum_{n=0}^{\infty} 4\left(\frac{x-3}{4}\right)^n = \frac{4}{1-[(x-3)/4]}$$

$$= \frac{4}{(4-x+3)/4} = \frac{16}{7-x}, \quad -1 < x < 7$$

83. $\sum_{n=0}^{\infty}(-1)^n x^n = \sum_{n=0}^{\infty}(-x)^n$

Geometric series: converges for

$|-x| < 1 \Rightarrow |x| < 1 \Rightarrow -1 < x < 1$

$$f(x) = \sum_{n=0}^{\infty}(-x)^n = \frac{1}{1+x}, \quad -1 < x < 1$$

84. $\sum_{n=0}^{\infty}(-1)^n x^{2n} = \sum_{n=0}^{\infty}(-x^2)^n$

Geometric series: converges for

$|-x^2| < 1 \Rightarrow -1 < x < 1$

$$f(x) = \sum_{n=0}^{\infty}(-x^2)^n = \frac{1}{1-(-x^2)} = \frac{1}{1+x^2}, \quad -1 < x < 1$$

85. $\sum_{n=0}^{\infty}\left(\frac{1}{x}\right)^n$

Geometric series: converges if $\left|\frac{1}{x}\right| < 1$

$\Rightarrow |x| > 1 \Rightarrow x < -1$ or $x > 1$

$$f(x) = \sum_{n=0}^{\infty}\left(\frac{1}{x}\right)^n = \frac{1}{1-(1/x)} = \frac{x}{x-1}, \quad x > 1 \text{ or } x < -1$$

86. $\sum_{n=1}^{\infty}\left(\frac{x^2}{x^2+4}\right)^n = \frac{x^2}{x^2+4}\sum_{n=0}^{\infty}\left(\frac{x^2}{x^2+4}\right)^n$

Geometric series: converges for $\left|\frac{x^2}{x^2+4}\right| < 1$

$\Rightarrow x^2 < x^2 + 4 \Rightarrow$ converges for all x

$$f(x) = \frac{x^2}{x^2+4}\cdot\frac{1}{1-\dfrac{x^2}{x^2+4}} = \frac{x^2}{x^2+4}\cdot\frac{x^2+4}{4} = \frac{x^2}{4}$$

87. (a) Yes, the new series will still diverge.

(b) Yes, the new series will converge.

88. Neither statement is true. The formula

$$\frac{1}{x-1} = 1 + x + x^2 + \cdots$$

holds for $-1 < x < 1$.

89. (a) x is the common ratio.

(b) $1 + x + x^2 + \cdots = \sum_{n=0}^{\infty} x^n = \frac{1}{1-x}, \quad |x| < 1$

(c) $y_1 = \frac{1}{1-x}$

$y_2 = S_3 = 1 + x + x^2$

$y_3 = S_5 = 1 + x + x^2 + x^3 + x^4$

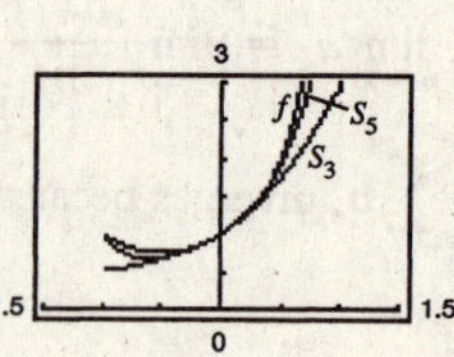

90. (a) $\left(-\frac{x}{2}\right)$ is the common ratio.

(b) $1 - \frac{x}{2} + \frac{x^2}{4} - \frac{x^3}{8} + \cdots = \sum_{n=0}^{\infty}\left(-\frac{x}{2}\right)^n$

$$= \frac{1}{1-(-x/2)}$$

$$= \frac{2}{2+x}, \quad |x| < 2$$

(c) $y_1 = \frac{2}{2+x}$

$y_2 = S_3 = 1 - \frac{x}{2} + \frac{x^2}{4}$

$y_3 = S_5 = 1 - \frac{x}{2} + \frac{x^2}{4} - \frac{x^3}{8} + \frac{x^4}{16}$

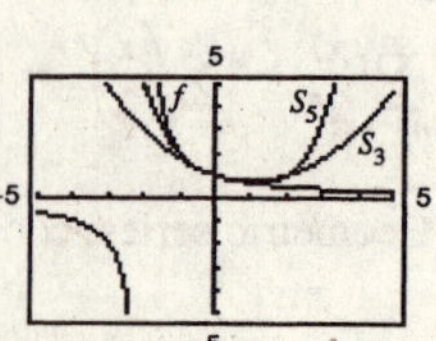

91. $f(x) = 3\left[\dfrac{1 - 0.5^x}{1 - 0.5}\right]$

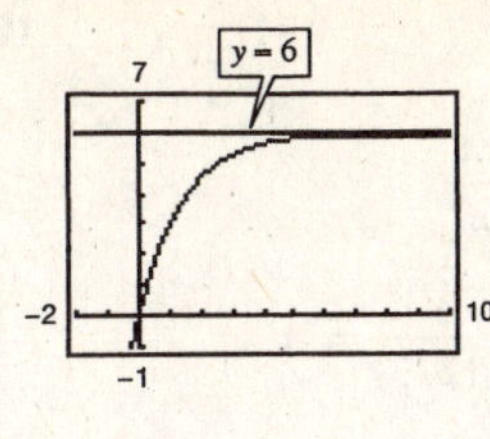

Horizontal asymptote: $y = 6$

$\sum_{n=0}^{\infty} 3\left(\dfrac{1}{2}\right)^n$

$S = \dfrac{3}{1 - (1/2)} = 6$

The horizontal asymptote is the sum of the series. $f(n)$ is the nth partial sum.

92. $f(x) = 2\left[\dfrac{1 - 0.8^x}{1 - 0.8}\right]$

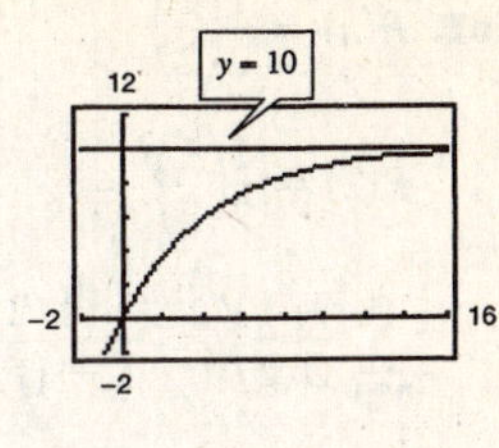

Horizontal asymptote: $y = 10$

$\sum_{n=0}^{\infty} 2\left(\dfrac{4}{5}\right)^n$

$S = \dfrac{2}{1 - (4/5)} = 10$

The horizontal asymptote is the sum of the series. $f(n)$ is the nth partial sum.

93. $\dfrac{1}{n(n + 1)} < 0.0001$

$10{,}000 < n^2 + n$

$0 < n^2 + n - 10{,}000$

$n = \dfrac{-1 \pm \sqrt{1^2 - 4(1)(-10{,}000)}}{2}$

Choosing the positive value for n we have $n \approx 99.5012$. The first *term* that is less than 0.0001 is $n = 100$.

$\left(\dfrac{1}{8}\right)^n < 0.0001$

$10{,}000 < 8^n$

This inequality is true when $n = 5$. This series converges at a faster rate.

94. $\dfrac{1}{2^n} < 0.0001$

$10{,}000 < 2^n$

This inequality is true when $n = 14$.

$(0.01)^n < 0.0001$

$10{,}000 < 10^n$

This inequality is true when $n = 5$. This series converges at a faster rate.

95. $\sum_{i=0}^{n-1} 8000(0.9)^i = \dfrac{8000[1 - (0.9)^{(n-1)+1}]}{1 - 0.9}$

$= 80{,}000(1 - 0.9^n),\ n > 0$

96. $V(t) = 225{,}000(1 - 0.3)^n = (0.7)^n(225{,}000)$

$V(5) = (0.7)^5(225{,}000) = \$37{,}815.75$

97. $\sum_{i=0}^{n-1} 100(0.75)^i = \dfrac{100[1 - 0.75^{(n-1)+1}]}{1 - 0.75}$

$= 400(1 - 0.75^n)$ million dollars

Sum $= 400$ million dollars

98. $\sum_{i=0}^{n-1} 100(0.60)^i = \dfrac{100[1 - 0.6^n]}{1 - 0.6}$

$= 250(1 - 0.6^n)$ million dollars

Sum $= 250$ million dollars

99. $D_1 = 16$

$D_2 = \underbrace{0.81(16)}_{\text{up}} + \underbrace{0.81(16)}_{\text{down}} = 32(0.81)$

$D_3 = 16(0.81)^2 + 16(0.81)^2 = 32(0.81)^2$

$\vdots$

$D = 16 + 32(0.81) + 32(0.81)^2 + \cdots$

$= -16 + \sum_{n=0}^{\infty} 32(0.81)^n = -16 + \dfrac{32}{1 - 0.81}$

≈ 152.42 feet

100. The ball in Exercise 99 takes the following times for each fall.

$s_1 = -16t^2 + 16$	$s_1 = 0$ if $t = 1$
$s_2 = -16t^2 + 16(0.81)$	$s_2 = 0$ if $t = 0.9$
$s_3 = -16t^2 + 16(0.81)^2$	$s_3 = 0$ if $t = (0.9)^2$
$\vdots$	$\vdots$
$s_n = -16t^2 + 16(0.81)^{n-1}$	$s_n = 0$ if $t = (0.9)^{n-1}$

Beginning with s_2, the ball takes the same amount of time to bounce up as it takes to fall. The total elapsed time before the ball comes to rest is

$t = 1 + 2\sum_{n=1}^{\infty} (0.9)^n = -1 + 2\sum_{n=0}^{\infty} (0.9)^n$

$= -1 + \dfrac{2}{1 - 0.9} = 19$ seconds.

101. $P(n) = \frac{1}{2}\left(\frac{1}{2}\right)^n$

$P(2) = \frac{1}{2}\left(\frac{1}{2}\right)^2 = \frac{1}{8}$

$\sum_{n=0}^{\infty} \frac{1}{2}\left(\frac{1}{2}\right)^n = \frac{1/2}{1-(1/2)} = 1$

102. $P(n) = \frac{1}{3}\left(\frac{2}{3}\right)^n$

$P(2) = \frac{1}{3}\left(\frac{2}{3}\right)^2 = \frac{4}{27}$

$\sum_{n=0}^{\infty} \frac{1}{3}\left(\frac{2}{3}\right)^n = \frac{1/3}{1-(2/3)} = 1$

103. (a) $\sum_{n=1}^{\infty}\left(\frac{1}{2}\right)^n = \sum_{n=0}^{\infty}\frac{1}{2}\left(\frac{1}{2}\right)^n$

$= \frac{1}{2}\frac{1}{(1-(1/2))} = 1$

(b) No, the series is not geometric.

(c) $\sum_{n=1}^{\infty} n\left(\frac{1}{2}\right)^n = 2$

104. Person 1: $\frac{1}{2} + \frac{1}{2^4} + \frac{1}{2^7} + \cdots = \frac{1}{2}\sum_{n=0}^{\infty}\left(\frac{1}{8}\right)^n = \frac{1}{2}\frac{1}{1-(1/8)} = \frac{4}{7}$

Person 2: $\frac{1}{2^2} + \frac{1}{2^5} + \frac{1}{2^8} + \cdots = \frac{1}{4}\sum_{n=0}^{\infty}\left(\frac{1}{8}\right)^n = \frac{1}{4}\frac{1}{1-(1/8)} = \frac{2}{7}$

Person 3: $\frac{1}{2^3} + \frac{1}{2^6} + \frac{1}{2^9} + \cdots = \frac{1}{8}\sum_{n=0}^{\infty}\left(\frac{1}{8}\right)^n = \frac{1}{8}\frac{1}{1-(1/8)} = \frac{1}{7}$

Sum: $\frac{4}{7} + \frac{2}{7} + \frac{1}{7} = 1$

105. (a) $64 + 32 + 16 + 8 + 4 + 2 = 126 \text{ in.}^2$

(b) $\sum_{n=0}^{\infty} 64\left(\frac{1}{2}\right)^n = \frac{64}{1-(1/2)} = 128 \text{ in.}^2$

Note: This is one-half of the area of the original square!

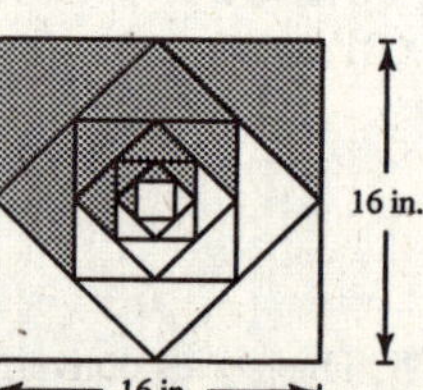

106. (a) $\sin\theta = \frac{|Yy_1|}{z} \Rightarrow |Yy_1| = z\sin\theta$

$\sin\theta = \frac{|x_1y_1|}{|Yy_1|} \Rightarrow |x_1y_1| = |Yy_1|\sin\theta = z\sin^2\theta$

$\sin\theta = \frac{|x_1y_2|}{|x_1y_1|} \Rightarrow |x_1y_2| = |x_1y_1|\sin\theta = z\sin^3\theta$

Total: $z\sin\theta + z\sin^2\theta + z\sin^3\theta + \cdots = z\frac{\sin\theta}{1-\sin\theta}$

(b) If $z = 1$ and $\theta = \frac{\pi}{6}$, then total $= \frac{1/2}{1-(1/2)} = 1$.

107. $\sum_{n=1}^{20} 50{,}000\left(\frac{1}{1.06}\right)^n = \frac{50{,}000}{1.06}\sum_{i=0}^{19}\left(\frac{1}{1.06}\right)^i$

$= \frac{50{,}000}{1.06}\left[\frac{1-1.06^{-20}}{1-1.06^{-1}}\right]$, $[n = 20, r = 1.06^{-1}]$

$\approx 573{,}496.06$

108. Surface area $= 4\pi(1)^2 + 9\left(4\pi\left(\frac{1}{3}\right)^2\right) + 9^2 \cdot 4\pi\left(\frac{1}{9}\right)^2 + \cdots = 4[\pi + \pi + \cdots] = \infty$

109. $w = \sum_{i=0}^{n-1} 0.01(2)^i = \frac{0.01(1-2^n)}{1-2} = 0.01(2^n - 1)$

(a) When $n = 29$: $w = \$5{,}368{,}709.11$

(b) When $n = 30$: $w = \$10{,}737{,}418.23$

(c) When $n = 31$: $w = \$21{,}474{,}836.47$

110. $\displaystyle\sum_{n=0}^{12t-1} P\left(1+\frac{r}{12}\right)^n = \frac{P\left[1-\left(1+\frac{r}{12}\right)^{12t}\right]}{1-\left(1+\frac{r}{12}\right)} = P\left(-\frac{12}{r}\right)\left[\left(1-\left(1+\frac{r}{12}\right)^{12t}\right] = P\left(\frac{12}{r}\right)\left[\left(1+\frac{r}{12}\right)^{12t}-1\right]$

$\displaystyle\sum_{n=0}^{12t-1} P(e^{r/12})^n = \frac{P(1-(e^{r/12})^{12t})}{1-e^{r/12}} = \frac{P(e^{rt}-1)}{e^{r/12}-1}$

111. $P = 50, r = 0.03, t = 20$

(a) $A = 50\left(\frac{12}{0.03}\right)\left[\left(1+\frac{0.03}{12}\right)^{12(20)} - 1\right] \approx \$16{,}415.10$

(b) $A = \dfrac{50(e^{0.03(20)}-1)}{e^{0.03/12}-1} \approx \$16{,}421.83$

112. $P = 75, r = 0.05, t = 25$

(a) $A = 75\left(\frac{12}{0.05}\right)\left[\left(1+\frac{0.05}{12}\right)^{12(25)} - 1\right] \approx \$44{,}663.23$

(b) $A = \dfrac{75(e^{0.05(25)}-1)}{e^{0.05/12}-1} \approx \$44{,}732.85$

113. $P = 100, r = 0.04, t = 40$

(a) $A = 100\left(\frac{12}{0.04}\right)\left[\left(1+\frac{0.04}{12}\right)^{12(40)} - 1\right] \approx \$118{,}196.13$

(b) $A = \dfrac{100(e^{0.04(40)}-1)}{e^{0.04/12}-1} \approx \$118{,}393.43$

114. $P = 20, r = 0.06, t = 50$

(a) $A = 20\left(\frac{12}{0.06}\right)\left[\left(1+\frac{0.06}{12}\right)^{12(50)} - 1\right] \approx \$75{,}743.82$

(b) $A = \dfrac{20(e^{0.06(50)}-1)}{e^{0.06/12}-1} \approx \$76{,}151.45$

115. (a) $a_n = 3484.1363(1.0502)^n = 3484.1363e^{0.04897n}$

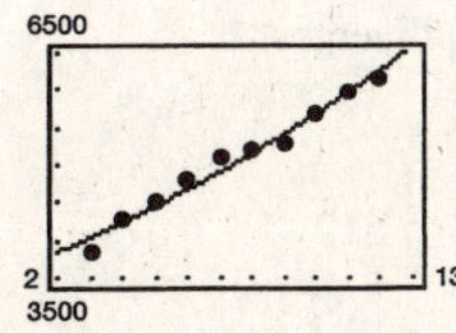

(b) Adding the ten values $a_3, a_4, \ldots, a_{12}$ you obtain

$\displaystyle\sum_{n=3}^{12} a_n = 50{,}809$, or \$50,809,000,000.

(c) $\displaystyle\sum_{n=3}^{12} 3484.1363e^{0.04897n} \approx 50{,}803$, or \$50,803,000,000

(Answers will vary.)

116. $T = 40{,}000 + 40{,}000(1.04) + \cdots + 40{,}000(1.04)^{39}$

$\displaystyle = \sum_{n=0}^{39} 40{,}000(1.04)^n = 40{,}000\left(\frac{1-1.04^{40}}{1-1.04}\right) \approx \$3{,}801{,}020$

117. False. $\displaystyle\lim_{n\to\infty}\frac{1}{n} = 0$, but $\displaystyle\sum_{n=1}^{\infty}\frac{1}{n}$ diverges.

118. True

119. False; $\displaystyle\sum_{n=1}^{\infty} ar^n = \left(\frac{a}{1-r}\right) - a$

The formula requires that the geometric series begins with $n = 0$.

120. True

$\displaystyle\lim_{n\to\infty}\frac{n}{1000(n+1)} = \frac{1}{1000} \neq 0$

121. True

$0.74999\ldots = 0.74 + \dfrac{9}{10^3} + \dfrac{9}{10^4} + \cdots$

$\displaystyle = 0.74 + \frac{9}{10^3}\sum_{n=0}^{\infty}\left(\frac{1}{10}\right)^n$

$= 0.74 + \dfrac{9}{10^3}\cdot\dfrac{1}{1-(1/10)}$

$= 0.74 + \dfrac{9}{10^3}\cdot\dfrac{10}{9}$

$= 0.74 + \dfrac{1}{100} = 0.75$

122. True

123. By letting $S_0 = 0$, we have

$$a_n = \sum_{k=1}^{n} a_k - \sum_{k=1}^{n-1} a_k = S_n - S_{n-1}. \text{ Thus,}$$

$$\begin{aligned}\sum_{n=1}^{\infty} a_n &= \sum_{n=1}^{\infty} (S_n - S_{n-1}) \\ &= \sum_{n=1}^{\infty} (S_n - S_{n-1} + c - c) \\ &= \sum_{n=1}^{\infty} [(c - S_{n-1}) - (c - S_n)].\end{aligned}$$

124. Let $\{S_n\}$ be the sequence of partial sums for the convergent series

$\sum_{n=1}^{\infty} a_n = L$. Then $\lim_{n\to\infty} S_n = L$ and since

$$R_n = \sum_{k=n+1}^{\infty} a_k = L - S_n,$$

we have

$$\begin{aligned}\lim_{n\to\infty} R_n &= \lim_{n\to\infty} (L - S_n) \\ &= \lim_{n\to\infty} L - \lim_{n\to\infty} S_n = L - L = 0.\end{aligned}$$

125. Let $\sum a_n = \sum_{n=0}^{\infty} 1$ and $\sum b_n = \sum_{n=0}^{\infty} (-1)$.

Both are divergent series.

$$\sum(a_n + b_n) = \sum_{n=0}^{\infty} [1 + (-1)] = \sum_{n=0}^{\infty} [1 - 1] = 0$$

126. If $\Sigma\,(a_n + b_n)$ converged, then $\Sigma\,(a_n + b_n) - \Sigma\, a_n = \Sigma\, b_n$ would converge, which is a contradiction. Thus, $\Sigma\,(a_n + b_n)$ diverges.

127. Suppose, on the contrary, that $\Sigma\, ca_n$ converges. Because $c \neq 0$,

$$\sum\left(\frac{1}{c}\right) ca_n = \sum a_n$$

converges. This is a contradiction since $\Sigma\, ca_n$ diverged. Hence, $\Sigma\, ca_n$ diverges.

128. If $\Sigma\, a_n$ converges, then $\lim_{n\to\infty} a_n = 0$. Hence,

$$\lim_{n\to\infty} \frac{1}{a_n} \neq 0,$$

which implies that $\Sigma\, 1/a_n$ diverges.

129. (a) $$\begin{aligned}\frac{1}{a_{n+1}a_{n+2}} - \frac{1}{a_{n+2}a_{n+3}} &= \frac{a_{n+3} - a_{n+1}}{a_{n+1}a_{n+2}a_{n+3}} \\ &= \frac{a_{n+2}}{a_{n+1}a_{n+2}a_{n+3}} \\ &= \frac{1}{a_{n+1}a_{n+3}}\end{aligned}$$

(b) $$\begin{aligned}S_n &= \sum_{k=0}^{n} \frac{1}{a_{k+1}a_{k+3}} \\ &= \sum_{k=0}^{n} \left[\frac{1}{a_{k+1}a_{k+2}} - \frac{1}{a_{k+2}a_{k+3}}\right] \\ &= \left[\frac{1}{a_1a_2} - \frac{1}{a_2a_3}\right] + \left[\frac{1}{a_2a_3} - \frac{1}{a_3a_4}\right] + \cdots + \left[\frac{1}{a_{n+1}a_{n+2}} - \frac{1}{a_{n+2}a_{n+3}}\right] \\ &= \frac{1}{a_1a_2} - \frac{1}{a_{n+2}a_{n+3}} \\ &= 1 - \frac{1}{a_{n+2}a_{n+3}}\end{aligned}$$

$$\sum_{n=0}^{\infty} \frac{1}{a_{n+1}a_{n+3}} = \lim_{n\to\infty} S_n = \lim_{n\to\infty} \left[1 - \frac{1}{a_{n+2}a_{n+3}}\right] = 1$$

130. $S_{2n} = 1 + 2x + x^2 + 2x^3 + \cdots + x^{2n} + 2x^{2n+1}$

$$= (1 + x^2 + x^4 + \cdots + x^{2n}) + (2x + 2x^3 + \cdots + 2x^{2n+1})$$

$$= \sum_{k=0}^{n} (x^2)^k + 2x \sum_{k=0}^{n} (x^2)^k$$

As $n \to \infty$, the two geometric series converge for $|x| < 1$:

$$\lim_{n\to\infty} S_{2n} = \frac{1}{1-x^2} + 2x\left(\frac{1}{1-x^2}\right) = \frac{1+2x}{1-x^2}, \quad |x| < 1$$

131. $\frac{1}{r} + \frac{1}{r^2} + \frac{1}{r^3} + \cdots = \sum_{n=0}^{\infty} \frac{1}{r}\left(\frac{1}{r}\right)^n$

$$= \frac{1/r}{1 - (1/r)}$$

$$= \frac{1}{r-1} \quad \left(\text{since } \left|\frac{1}{r}\right| < 1\right)$$

This is a geometric series which converges if

$$\left|\frac{1}{r}\right| < 1 \Leftrightarrow |r| > 1.$$

132. The entire rectangle has area 2 because the height is 1 and the base is $1 + \frac{1}{2} + \frac{1}{4} + \cdots = 2$. The squares all lie inside the rectangle, and the sum of their areas is

$$1 + \frac{1}{2^2} + \frac{1}{3^2} + \frac{1}{4^2} + \cdots.$$

Thus, $\sum_{n=1}^{\infty} \frac{1}{n^2} < 2$.

133. Let H represent the half-life of the drug. If a patient receives n equal doses of P units each of this drug, administered at equal time interval of length t, the total amount of the drug in the patient's system at the time the last dose is administered is given by $T_n = P + Pe^{kt} + Pe^{2kt} + \cdots + Pe^{(n-1)kt}$ where $k = -(\ln 2)/H$. One time interval *after* the last dose is administered is given by $T_{n+1} = Pe^{kt} + Pe^{2kt} + Pe^{3kt} + \cdots + Pe^{nkt}$. Two time intervals *after* the last dose is administered is given by $T_{n+2} = Pe^{2kt} + Pe^{3kt} + Pe^{4kt} + \cdots + Pe^{(n+1)kt}$ and so on. Since $k < 0$, $T_{n+s} \to 0$ as $s \to \infty$, where s is an integer.

134. The series is telescoping:

$$S_n = \sum_{k=1}^{n} \frac{6^k}{(3^{k+1} - 2^{k+1})(3^k - 2^k)}$$

$$= \sum_{k=1}^{n} \left[\frac{3^k}{3^k - 2^k} - \frac{3^{k+1}}{3^{k+1} - 2^{k+1}}\right]$$

$$= 3 - \frac{3^{n+1}}{3^{n+1} - 2^{n+1}}$$

$$\lim_{n\to\infty} S_n = 3 - 1 = 2$$

135. $f(1) = 0, f(2) = 1, f(3) = 2, f(4) = 4, \ldots$

In general: $f(n) = \begin{cases} n^2/4, & n \text{ even} \\ (n^2 - 1)/4, & n \text{ odd.} \end{cases}$

(See below for a proof of this.)

$x + y$ and $x - y$ are either both odd or both even. If both even, then

$$f(x + y) - f(x - y) = \frac{(x+y)^2}{4} - \frac{(x-y)^2}{4} = xy.$$

If both odd,

$$f(x + y) - f(x - y) = \frac{(x+y)^2 - 1}{4} - \frac{(x-y)^2 - 1}{4}$$

$$= xy.$$

Proof by induction that the formula for $f(n)$ is correct. It is true for $n = 1$. Assume that the formula is valid for k. If k is even, then $f(k) = k^2/4$ and

$$f(k + 1) = f(k) + \frac{k}{2} = \frac{k^2}{4} + \frac{k}{2}$$

$$= \frac{k^2 + 2k}{4} = \frac{(k+1)^2 - 1}{4}.$$

The argument is similar if k is odd.

Section 9.3 The Integral Test and p-Series

1. $\displaystyle\sum_{n=1}^{\infty} \frac{1}{n+1}$

Let $f(x) = \dfrac{1}{x+1}$.

f is positive, continuous, and decreasing for $x \geq 1$.

$$\int_1^{\infty} \frac{1}{x+1}\,dx = \Big[\ln(x+1)\Big]_1^{\infty} = \infty$$

Diverges by Theorem 9.10

2. $\displaystyle\sum_{n=1}^{\infty} \frac{2}{3n+5}$

Let $f(x) = \dfrac{2}{3x+5}$.

f is positive, continuous, and decreasing for $x \geq 1$.

$$\int_1^{\infty} \frac{2}{3x+5}\,dx = \left[\frac{2}{3}\ln(3x+5)\right]_1^{\infty} = \infty$$

Diverges by Theorem 9.10

3. $\displaystyle\sum_{n=1}^{\infty} e^{-n}$

Let $f(x) = e^{-x}$.

f is positive, continuous, and decreasing for $x \geq 1$.

$$\int_1^{\infty} e^{-x}\,dx = \Big[-e^{-x}\Big]_1^{\infty} = \frac{1}{e}$$

Converges by Theorem 9.10

4. $\displaystyle\sum_{n=1}^{\infty} ne^{-n/2}$

Let $f(x) = xe^{-x/2}$.

f is positive, continuous, and decreasing for $x \geq 3$ since

$f'(x) = \dfrac{2-x}{2e^{x/2}} < 0$ for $x \geq 3$.

$$\int_3^{\infty} xe^{-x/2}\,dx = \Big[-2(x+2)e^{-x/2}\Big]_3^{\infty} = 10e^{-3/2}$$

Converges by Theorem 9.10

5. $\displaystyle\sum_{n=1}^{\infty} \frac{1}{n^2+1}$

Let $f(x) = \dfrac{1}{x^2+1}$.

f is positive, continuous, and decreasing for $x \geq 1$.

$$\int_1^{\infty} \frac{1}{x^2+1}\,dx = \Big[\arctan x\Big]_1^{\infty} = \frac{\pi}{4}$$

Converges by Theorem 9.10

6. $\displaystyle\sum_{n=1}^{\infty} \frac{1}{2n+1}$

Let $f(x) = \dfrac{1}{2x+1}$.

f is positive, continuous, and decreasing for $x \geq 1$.

$$\int_1^{\infty} \frac{1}{2x+1}\,dx = \Big[\ln\sqrt{2x+1}\Big]_1^{\infty} = \infty$$

Diverges by Theorem 9.10

7. $\displaystyle\sum_{n=1}^{\infty} \frac{\ln(n+1)}{n+1}$

Let $f(x) = \dfrac{\ln(x+1)}{x+1}$.

f is positive, continuous, and decreasing for $x \geq 2$ since

$f'(x) = \dfrac{1-\ln(x+1)}{(x+1)^2} < 0$ for $x \geq 2$.

$$\int_1^{\infty} \frac{\ln(x+1)}{x+1}\,dx = \left[\frac{[\ln(x+1)]^2}{2}\right]_1^{\infty} = \infty$$

Diverges by Theorem 9.10

8. $\displaystyle\sum_{n=2}^{\infty} \frac{\ln n}{\sqrt{n}}$

Let $f(x) = \dfrac{\ln x}{\sqrt{x}}$, $f'(x) = \dfrac{2-\ln x}{2x^{3/2}}$.

f is positive, continuous, and decreasing for $x > e^2 \approx 7.4$.

$$\int_2^{\infty} \frac{\ln x}{\sqrt{x}}\,dx = \Big[2\sqrt{x}(\ln x - 2)\Big]_2^{\infty} = \infty, \text{ diverges}$$

Hence, the series diverges by Theorem 9.10.

9. $\sum_{n=1}^{\infty} \frac{1}{\sqrt{n}(\sqrt{n}+1)}$

Let $f(x) = \frac{1}{\sqrt{x}(\sqrt{x}+1)}$, $f'(x) = -\frac{1+2\sqrt{x}}{2x^{3/2}(\sqrt{x}+1)^2} < 0.$

f is positive, continuous, and decreasing for $x \geq 1$.

$$\int_1^{\infty} \frac{1}{\sqrt{x}(\sqrt{x}+1)}\,dx = \Big[2\ln(\sqrt{x}+1)\Big]_1^{\infty} = \infty, \text{ diverges}$$

Hence, the series diverges by Theorem 9.10.

10. $\sum_{n=1}^{\infty} \frac{n}{n^2+3}$

Let $f(x) = \frac{x}{x^2+3}$.

$f(x)$ is positive, continuous, and decreasing for $x \geq 2$ since

$$f'(x) = \frac{3-x^2}{(x^2+3)} < 0 \text{ for } x \geq 2.$$

$$\int_1^{\infty} \frac{x}{x^2+3}\,dx = \Big[\ln\sqrt{x^2+3}\Big]_1^{\infty} = \infty$$

Diverges by Theorem 9.10

11. $\sum_{n=1}^{\infty} \frac{1}{\sqrt{n+1}}$

Let $f(x) = \frac{1}{\sqrt{x+1}}$, $f'(x) = \frac{-1}{2(x+1)^{3/2}} < 0.$

f is positive, continuous, and decreasing for $x \geq 1$.

$$\int_1^{\infty} \frac{1}{\sqrt{x+1}}\,dx = \Big[2\sqrt{x+1}\Big]_1^{\infty} = \infty, \text{ diverges}$$

Hence, the series diverges by Theorem 9.10.

12. $\sum_{n=2}^{\infty} \frac{\ln n}{n^3}$

Let $f(x) = \frac{\ln x}{x^3}$, $f'(x) = \frac{1-3\ln x}{x^4}$.

f is positive, continuous, and decreasing for $x > 2$.

$$\int_2^{\infty} \frac{\ln x}{x^2}\,dx = \left[-\frac{(2\ln x+1)}{4x^4}\right]_2^{\infty}$$

$$= \frac{2\ln 2+1}{16}, \text{ converges}$$

Hence, the series converges by Theorem 9.10.

13. $\sum_{n=1}^{\infty} \frac{\ln n}{n^2}$

Let $f(x) = \frac{\ln x}{x^2}$, $f'(x) = \frac{1-2\ln x}{x^3}$.

f is positive, continuous, and decreasing for $x > e^{1/2} \approx 1.6$.

$$\int_1^{\infty} \frac{\ln x}{x^2}\,dx = \left[\frac{-(\ln x+1)}{x}\right]_1^{\infty} = 1, \text{ converges}$$

Hence, the series converges by Theorem 9.10.

14. $\sum_{n=2}^{\infty} \frac{1}{n\sqrt{\ln n}}$

Let $f(x) = \frac{1}{x\sqrt{\ln x}}$, $f'(x) = -\frac{2\ln x+1}{2x^2(\ln x)^{3/2}}$.

f is positive, continuous, and decreasing for $x \geq 2$.

$$\int_2^{\infty} \frac{1}{x\sqrt{\ln x}}\,dx = \Big[2\sqrt{\ln x}\Big]_2^{\infty} = \infty, \text{ diverges}$$

Hence, the series diverges by Theorem 9.10.

15. $\sum_{n=1}^{\infty} \frac{\arctan n}{n^2+1}$

Let $f(x) = \frac{\arctan x}{x^2+1}$, $f'(x) = \frac{1-2x\arctan x}{(x^2+1)^2} < 0 \text{ for } x \geq 1.$

f is positive, continuous, and decreasing for $x \geq 1$.

$$\int_1^{\infty} \frac{\arctan x}{x^2+1}\,dx = \left[\frac{(\arctan x)^2}{2}\right]_1^{\infty} = \frac{3\pi^2}{32}, \text{ converges}$$

Hence, the series converges by Theorem 9.10.

16. $\sum_{n=3}^{\infty} \frac{1}{n\ln n\ln(\ln n)}$

Let $f(x) = \frac{1}{x\ln x\ln(\ln x)}$,

$$f'(x) = \frac{-[(\ln x+1)\ln(\ln x)+1]}{x^2(\ln x)^2(\ln(\ln x))^2}.$$

f is positive, continuous, and decreasing for $x \geq 3$.

$$\int_3^{\infty} \frac{1}{x\ln x\ln(\ln x)}\,dx = \Big[\ln(\ln(\ln x))\Big]_3^{\infty} = \infty, \text{ diverges}$$

Hence, the series diverges by Theorem 9.10.

17. $\sum_{n=1}^{\infty} \frac{2n}{n^2+1}$

Let $f(x) = \frac{2x}{x^2+1}$, $f'(x) = 2\frac{1-x^2}{(x^2+1)^2} < 0$ for $x > 1$.

f is positive, continuous, and decreasing for $x > 1$.

$$\int_1^{\infty} \frac{2x}{x^2+1}\,dx = \Big[\ln(x^2+1)\Big]_1^{\infty} = \infty, \text{ diverges}$$

Hence, the series diverges by Theorem 9.10.

18. $\sum_{n=1}^{\infty} \frac{n}{n^4+1}$

Let $f(x) = \frac{x}{x^4+1}$, $f'(x) = \frac{1-3x^4}{(x^4+1)^2} < 0$ for $x > 1$.

f is positive, continuous, and decreasing for $x > 1$.

$$\int_1^{\infty} \frac{x}{x^4+1}\,dx = \left[\frac{1}{2}\arctan(x^2)\right]_1^{\infty} = \frac{\pi}{8}, \text{ converges}$$

Hence, the series converges by Theorem 9.10.

19. $\sum_{n=1}^{\infty} \frac{n^{k-1}}{n^k+c}$

Let $f(x) = \frac{x^{k-1}}{x^k+c}$.

f is positive, continuous, and decreasing for $x > \sqrt[k]{c(k-1)}$ since

$$f'(x) = \frac{x^{k-2}[c(k-1)-x^k]}{(x^k+c)^2} < 0$$

for $x > \sqrt[k]{c(k-1)}$.

$$\int_1^{\infty} \frac{x^{k-1}}{x^k+c}\,dx = \left[\frac{1}{k}\ln(x^k+c)\right]_1^{\infty} = \infty$$

Diverges by Theorem 9.10

20. $\sum_{n=1}^{\infty} n^k e^{-n}$

Let $f(x) = \frac{x^k}{e^x}$.

f is positive, continuous, and decreasing for $x > k$ since

$$f'(x) = \frac{x^{k-1}(k-x)}{e^x} < 0 \text{ for } x > k.$$

We use integration by parts.

$$\int_1^{\infty} x^k e^{-x}\,dx = \Big[-x^k e^{-x}\Big]_1^{\infty} + k\int_1^{\infty} x^{k-1}e^{-x}\,dx$$

$$= \frac{1}{e} + \frac{k}{e} + \frac{k(k-1)}{e} + \cdots + \frac{k!}{e}$$

Converges by Theorem 9.10

21. Let $f(x) = \frac{(-1)^x}{x}$, $f(n) = a_n$.

The function f is not positive for $x \geq 1$.

22. Let $f(x) = e^{-x}\cos x$, $f(n) = a_n$.

The function f is not positive for $x \geq 1$.

23. Let $f(x) = \frac{2+\sin x}{x}$, $f(n) = a_n$.

The function f is not decreasing for $x \geq 1$.

24. Let $f(x) = \left(\frac{\sin x}{x}\right)^2$, $f(n) = a_n$.

The function f is not decreasing for $x \geq 1$.

25. $\sum_{n=1}^{\infty} \frac{1}{n^3}$

Let $f(x) = \frac{1}{x^3}$.

f is positive, continuous, and decreasing for $x \geq 1$.

$$\int_1^{\infty} \frac{1}{x^3}\,dx = \left[-\frac{1}{2x^2}\right]_1^{\infty} = \frac{1}{2}$$

Converges by Theorem 9.10

26. $\sum_{n=1}^{\infty} \frac{1}{n^{1/3}}$

Let $f(x) = \frac{1}{x^{1/3}}$.

f is positive, continuous, and decreasing for $x \geq 1$.

$$\int_1^{\infty} \frac{1}{x^{1/3}}\,dx = \left[\frac{3}{2}x^{2/3}\right]_1^{\infty} = \infty$$

Diverges by Theorem 9.10

27. $\sum_{n=1}^{\infty} \frac{1}{\sqrt{n}}$

Let $f(x) = \frac{1}{\sqrt{x}}$, $f'(x) = \frac{-1}{2x^{3/2}} < 0$ for $x \geq 1$.

f is positive, continuous, and decreasing for $x \geq 1$.

$$\int_1^{\infty} \frac{1}{\sqrt{x}}\,dx = \left[2\sqrt{x}\right]_1^{\infty} = \infty, \text{ diverges}$$

Hence, the series diverges by Theorem 9.10.

28. $\sum_{n=1}^{\infty} \frac{1}{n^2}$

Let $f(x) = \frac{1}{x^2}$, $f'(x) = \frac{-2}{x^3} < 0$ for $x \geq 1$.

f is positive, continuous, and decreasing for $x \geq 1$.

$$\int_1^{\infty} \frac{1}{x^2}\,dx = \left[-\frac{1}{x}\right]_1^{\infty} = 1, \text{ converges}$$

Hence, the series converges by Theorem 9.10.

29. $\sum_{n=1}^{\infty} \frac{1}{\sqrt[5]{n}} = \sum_{n=1}^{\infty} \frac{1}{n^{1/5}}$

Divergent p-series with $p = \frac{1}{5} < 1$

30. $\sum_{n=1}^{\infty} \frac{3}{n^{5/3}}$

Convergent p-series with $p = \frac{5}{3} > 1$

31. $\sum_{n=1}^{\infty} \frac{1}{n^{1/2}}$

Divergent p-series with $p = \frac{1}{2} < 1$

32. $\sum_{n=1}^{\infty} \frac{1}{n^2}$

Convergent p-series with $p = 2 > 1$

33. $\sum_{n=1}^{\infty} \frac{1}{n^{3/2}}$

Convergent p-series with $p = \frac{3}{2} > 1$

34. $\sum_{n=1}^{\infty} \frac{1}{n^{2/3}}$

Divergent p-series with $p = \frac{2}{3} < 1$

35. $\sum_{n=1}^{\infty} \frac{1}{n^{1.04}}$

Convergent p-series with $p = 1.04 > 1$

36. $\sum_{n=1}^{\infty} \frac{1}{n^{\pi}}$

Convergent p-series with $p = \pi > 1$

37. $\sum_{n=1}^{\infty} \frac{2}{n^{3/4}}$

$S_1 = 2$

$S_2 \approx 3.1892$

$S_3 \approx 4.0666$

$S_4 \approx 4.7740$

Matches (c), diverges

38. $\sum_{n=1}^{\infty} \frac{2}{n}$

$S_1 = 2$

$S_2 = 3$

$S_3 \approx 3.6667$

$S_4 \approx 4.1667$

Matches (f), diverges

39. $\sum_{n=1}^{\infty} \frac{2}{\sqrt{n^{\pi}}} = \sum_{n=1}^{\infty} \frac{2}{n^{\pi/2}}$

$S_1 = 2$

$S_2 \approx 2.6732$

$S_3 \approx 3.0293$

$S_4 \approx 3.2560$

Matches (b), converges

Note: The partial sums for 39 and 41 are very similar because $\pi/2 \approx 3/2$.

40. $\sum_{n=1}^{\infty} \frac{2}{n^{2/5}}$

$S_1 = 2$

$S_2 \approx 3.5157$

$S_3 \approx 4.8045$

Matches (a), diverges

41. $\sum_{n=1}^{\infty} \frac{2}{n\sqrt{n}} = \sum_{n=1}^{\infty} \frac{2}{n^{3/2}}$

$S_1 = 2$

$S_2 \approx 2.7071$

$S_3 \approx 3.0920$

$S_4 \approx 3.3420$

Matches (d), converges

Note: The partial sums for 39 and 41 are very similar because $\pi/2 \approx 3/2$.

42. $\sum_{n=1}^{\infty} \frac{2}{n^2}$

$S_1 = 2$

$S_2 = 2.5$

$S_3 \approx 2.7222$

Matches (e), converges

43. (a)

n	5	10	20	50	100
S_n	3.7488	3.75	3.75	3.75	3.75

The partial sums approach the sum 3.75 very rapidly.

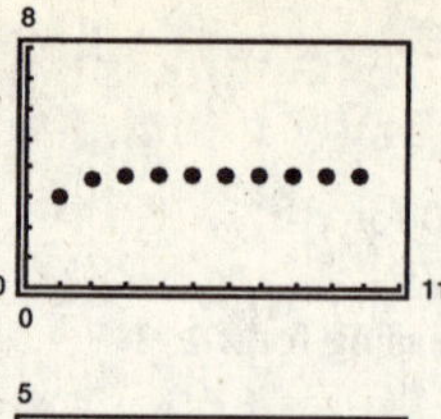

(b)

n	5	10	20	50	100
S_n	1.4636	1.5498	1.5962	1.6251	1.635

The partial sums approach the sum $\pi^2/6 \approx 1.6449$ slower than the series in part (a).

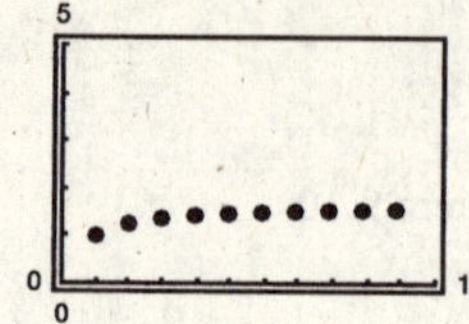

44. 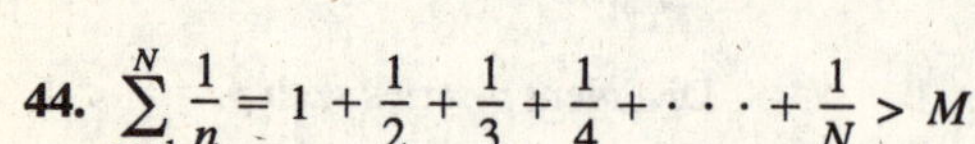

$$\sum_{n=1}^{N} \frac{1}{n} = 1 + \frac{1}{2} + \frac{1}{3} + \frac{1}{4} + \cdots + \frac{1}{N} > M$$

(a)

M	2	4	6	8
N	4	31	227	1674

(b) No. Since the terms are decreasing (approaching zero), more and more terms are required to increase the partial sum by 2.

45. Let f be positive, continuous, and decreasing for $x \geq 1$ and $a_n = f(n)$. Then,

$$\sum_{n=1}^{\infty} a_n \text{ and } \int_1^{\infty} f(x)\,dx$$

either both converge or both diverge (Theorem 9.10). See Example 1, page 618.

46. A series of the form $\displaystyle\sum_{n=1}^{\infty} \frac{1}{n^p}$ is a p-series, $p > 0$.

The p-series converges if $p > 1$ and diverges if $0 < p \leq 1$.

47. Your friend is not correct. The series

$$\sum_{n=10,000}^{\infty} \frac{1}{n} = \frac{1}{10,000} + \frac{1}{10,001} + \cdots$$

is the harmonic series, starting with the 10,000th term, and hence diverges.

48. No. Theorem 9.9 says that if the series converges, then the terms a_n tend to zero. Some of the series in Exercises 37–42 converge because the terms tend to 0 very rapidly.

49. The area under the rectangles is greater than the area under the graph of $y = 1/\sqrt{x}, x \geq 1$.

$$\sum_{n=1}^{\infty} \frac{1}{\sqrt{n}} > \int_1^{\infty} \frac{1}{\sqrt{x}}\,dx$$

Since $\displaystyle\int_1^{\infty} \frac{1}{\sqrt{x}}\,dx = \Big[2\sqrt{x}\Big]_1^{\infty} = \infty$ diverges, the series $\displaystyle\sum_{n=1}^{\infty} \frac{1}{\sqrt{n}}$ diverges.

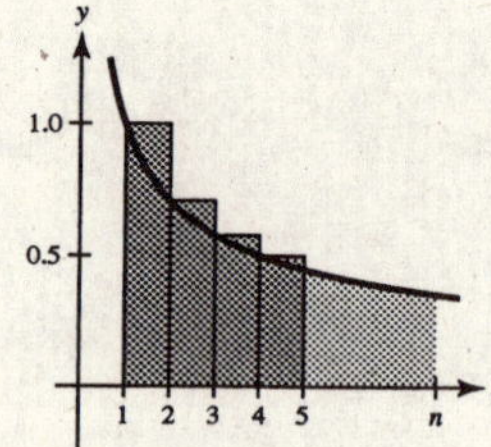

50. $\displaystyle\sum_{n=1}^{6} a_n \geq \int_1^7 f(x)\,dx \geq \sum_{n=2}^{7} a_n$

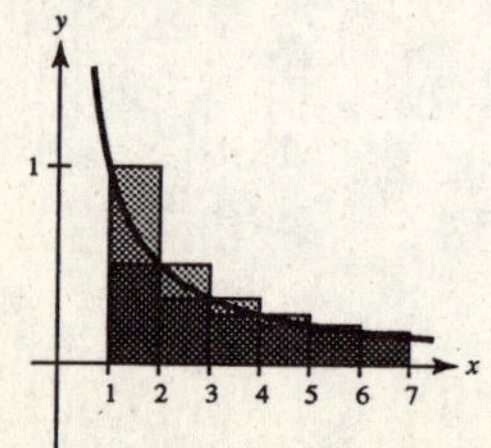

51. $\displaystyle\sum_{n=2}^{\infty} \frac{1}{n(\ln n)^p}$

If $p = 1$, then the series diverges by the Integral Test. If $p \neq 1$,

$$\int_2^{\infty} \frac{1}{x(\ln x)^p}\,dx = \int_2^{\infty} (\ln x)^{-p} \frac{1}{x}\,dx = \left[\frac{(\ln x)^{-p+1}}{-p+1}\right]_2^{\infty}.$$

Converges for $-p + 1 < 0$ or $p > 1$

52. $\displaystyle\sum_{n=2}^{\infty} \frac{\ln n}{n^p}$

If $p = 1$, then the series diverges by the Integral Test. If $p \neq 1$,

$$\int_2^{\infty} \frac{\ln x}{x^p}\,dx = \int_2^{\infty} x^{-p} \ln x\,dx = \left[\frac{x^{-p+1}}{(-p+1)^2}[-1 + (-p+1)\ln x]\right]_2^{\infty}. \text{ (Use integration by parts.)}$$

Converges for $-p + 1 < 0$ or $p > 1$

53. $\displaystyle\sum_{n=1}^{\infty} \frac{n}{(1+n^2)^p}$

If $p = 1$, $\displaystyle\sum_{n=1}^{\infty} \frac{n}{1+n^2}$ diverges (see Example 1). Let

$$f(x) = \frac{x}{(1+x^2)^p},\ p \neq 1$$

$$f'(x) = \frac{1-(2p-1)x^2}{(1+x^2)^{p+1}}.$$

For a fixed $p > 0$, $p \neq 1$, $f'(x)$ is eventually negative. f is positive, continuous, and eventually decreasing.

$$\int_1^{\infty} \frac{x}{(1+x^2)^p}\,dx = \left[\frac{1}{(x^2+1)^{p-1}(2-2p)}\right]_1^{\infty}$$

For $p > 1$, this integral converges. For $0 < p < 1$, it diverges.

54. $\displaystyle\sum_{n=1}^{\infty} n(1+n^2)^p$

Since $p > 0$, the series diverges for all values of p.

55. $\displaystyle\sum_{n=2}^{\infty} \frac{1}{n(\ln n)^p}$ converges for $p > 1$.

Hence, $\displaystyle\sum_{n=2}^{\infty} \frac{1}{n \ln n}$, diverges.

56. $\displaystyle\sum_{n=2}^{\infty} \frac{1}{n(\ln n)^p}$ converges for $p > 1$.

Hence, $\displaystyle\sum_{n=2}^{\infty} \frac{1}{n\sqrt[3]{(\ln n)^2}} = \sum_{n=2}^{\infty} \frac{1}{n(\ln n)^{2/3}}$, diverges.

57. $\displaystyle\sum_{n=2}^{\infty} \frac{1}{n(\ln n)^p}$ converges for $p > 1$.

Hence, $\displaystyle\sum_{n=2}^{\infty} \frac{1}{n(\ln n)^2}$, converges.

58. $\displaystyle\sum_{n=2}^{\infty} \frac{1}{n(\ln n)^p}$ converges for $p > 1$.

Hence, $\displaystyle\sum_{n=2}^{\infty} \frac{1}{n\ln(n^2)} = \sum_{n=2}^{\infty} \frac{1}{2n \ln n}$

$$= \frac{1}{2}\sum_{n=2}^{\infty} \frac{1}{n \ln(n)}, \text{ diverges}$$

59. Since f is positive, continuous, and decreasing for $x \geq 1$ and $a_n = f(n)$, we have,

$$R_N = S - S_N = \sum_{n=1}^{\infty} a_n - \sum_{n=1}^{N} a_n = \sum_{n=N+1}^{\infty} a_n > 0.$$

Also,

$$R_N = S - S_N = \sum_{n=N+1}^{\infty} a_n \leq a_{N+1} + \int_{N+1}^{\infty} f(x)\,dx$$

$$\leq \int_N^{\infty} f(x)\,dx.$$

Thus, $0 \leq R_N \leq \displaystyle\int_N^{\infty} f(x)\,dx$.

60. From Exercise 59, we have:

$$0 \le S - S_N \le \int_N^\infty f(x)\,dx$$

$$S_N \le S \le S_N + \int_N^\infty f(x)\,dx$$

$$\sum_{n=1}^{N} a_n \le S \le \sum_{n=1}^{N} a_n + \int_N^\infty f(x)\,dx$$

61. $S_6 = 1 + \frac{1}{2^4} + \frac{1}{3^4} + \frac{1}{4^4} + \frac{1}{5^4} + \frac{1}{6^4} \approx 1.0811$

$$R_6 \le \int_6^\infty \frac{1}{x^4}\,dx = \left[-\frac{1}{3x^3}\right]_6^\infty \approx 0.0015$$

$$1.0811 \le \sum_{n=1}^\infty \frac{1}{n^4} \le 1.0811 + 0.0015 = 1.0826$$

62. $S_4 = 1 + \frac{1}{2^5} + \frac{1}{3^5} + \frac{1}{4^5} \approx 1.0363$

$$R_4 \le \int_4^\infty \frac{1}{x^5}\,dx = \left[-\frac{1}{4x^4}\right]_4^\infty \approx 0.0010$$

$$1.0363 \le \sum_{n=1}^\infty \frac{1}{n^5} \le 1.0363 + 0.0010 = 1.0373$$

63. $S_{10} = \frac{1}{2} + \frac{1}{5} + \frac{1}{10} + \frac{1}{17} + \frac{1}{26} + \frac{1}{37} + \frac{1}{50} + \frac{1}{65} + \frac{1}{82} + \frac{1}{101} \approx 0.9818$

$$R_{10} \le \int_{10}^\infty \frac{1}{x^2+1}\,dx = \Big[\arctan x\Big]_{10}^\infty = \frac{\pi}{2} - \arctan 10 \approx 0.0997$$

$$0.9818 \le \sum_{n=1}^\infty \frac{1}{n^2+1} \le 0.9818 + 0.0997 = 1.0815$$

64. $S_{10} = \frac{1}{2(\ln 2)^3} + \frac{1}{3(\ln 3)^3} + \frac{1}{4(\ln 4)^3} + \cdots + \frac{1}{11(\ln 11)^3} \approx 1.9821$

$$R_{10} \le \int_{10}^\infty \frac{1}{(x+1)[\ln(x+1)]^3}\,dx = \left[-\frac{1}{2[\ln(x+1)]^2}\right]_{10}^\infty = \frac{1}{2(\ln 11)^3} \approx 0.0870$$

$$1.9821 \le \sum_{n=1}^\infty \frac{1}{(n+1)[\ln(n+1)]^3} \le 1.9821 + 0.0870 = 2.0691$$

65. $S_4 = \frac{1}{e} + \frac{2}{e^4} + \frac{3}{e^9} + \frac{4}{e^{16}} \approx 0.4049$

$$R_4 \le \int_4^\infty xe^{-x^2}\,dx = \left[-\frac{1}{2}e^{-x^2}\right]_4^\infty = \frac{e^{-16}}{2} \approx 5.6 \times 10^{-8}$$

$$0.4049 \le \sum_{n=1}^\infty ne^{-n^2} \le 0.4049 + 5.6 \times 10^{-8}$$

66. $S_4 = \frac{1}{e} + \frac{1}{e^2} + \frac{1}{e^3} + \frac{1}{e^4} \approx 0.5713$

$$R_4 \le \int_4^\infty e^{-x}\,dx = \Big[-e^{-x}\Big]_4^\infty \approx 0.0183$$

$$0.5713 \le \sum_{n=0}^\infty e^{-n} \le 0.5713 + 0.0183 = 0.5896$$

67. $0 \le R_N \le \int_N^\infty \frac{1}{x^4}\,dx = \left[-\frac{1}{3x^3}\right]_N^\infty = \frac{1}{3N^3} < 0.001$

$$\frac{1}{N^3} < 0.003$$

$$N^3 > 333.33$$

$$N > 6.93$$

$$N \ge 7$$

68. $0 \le R_N \le \int_N^\infty \frac{1}{x^{3/2}}\,dx = \left[-\frac{2}{x^{1/2}}\right]_N^\infty = \frac{2}{\sqrt{N}} < 0.001$

$$N^{-1/2} < 0.0005$$

$$\sqrt{N} > 2000$$

$$N \ge 4{,}000{,}000$$

69. $R_N \le \int_N^\infty e^{-5x}\,dx = \left[-\frac{1}{5}e^{-5x}\right]_N^\infty = \frac{e^{-5N}}{5} < 0.001$

$\frac{1}{e^{5N}} < 0.005$

$e^{5N} > 200$

$5N > \ln 200$

$N > \frac{\ln 200}{5}$

$N > 1.0597$

$N \ge 2$

70. $R_N \le \int_N^\infty e^{-x/2}\,dx = \left[-2e^{-x/2}\right]_N^\infty = \frac{2}{e^{N/2}} < 0.001$

$\frac{2}{e^{N/2}} < 0.001$

$e^{N/2} > 2000$

$\frac{N}{2} > \ln 2000$

$N > 2\ln 2000 \approx 15.2$

$N \ge 16$

71. $R_N \le \int_N^\infty \frac{1}{x^2+1}\,dx = \left[\arctan x\right]_N^\infty$

$= \frac{\pi}{2} - \arctan N < 0.001$

$-\arctan N < -1.5698$

$\arctan N > 1.5698$

$N > \tan 1.5698$

$N \ge 1004$

72. $R_n \le \int_N^\infty \frac{2}{x^2+5}\,dx = 2\left[\frac{1}{\sqrt{5}}\arctan\left(\frac{x}{\sqrt{5}}\right)\right]_N^\infty$

$= \frac{2}{\sqrt{5}}\left(\frac{\pi}{2} - \arctan\left(\frac{N}{\sqrt{5}}\right)\right) < 0.001$

$\frac{\pi}{2} - \arctan\left(\frac{N}{\sqrt{5}}\right) < 0.001118$

$1.56968 < \arctan\left(\frac{N}{\sqrt{5}}\right)$

$\frac{N}{\sqrt{5}} > \tan 1.56968$

$N \ge 2004$

73. (a) $\sum_{n=2}^{\infty} \frac{1}{n^{1.1}}$. This is a convergent p-series with $p = 1.1 > 1$. $\sum_{n=2}^{\infty} \frac{1}{n \ln n}$ is a divergent series. Use the Integral Test.

$f(x) = \frac{1}{x \ln x}$ is positive, continuous, and decreasing for $x \ge 2$.

$\int_2^\infty \frac{1}{x\ln x}\,dx = \left[\ln|\ln x|\right]_2^\infty = \infty$

(b) $\sum_{n=2}^{6} \frac{1}{n^{1.1}} = \frac{1}{2^{1.1}} + \frac{1}{3^{1.1}} + \frac{1}{4^{1.1}} + \frac{1}{5^{1.1}} + \frac{1}{6^{1.1}} \approx 0.4665 + 0.2987 + 0.2176 + 0.1703 + 0.1393$

$\sum_{n=2}^{6} \frac{1}{n\ln n} = \frac{1}{2\ln 2} + \frac{1}{3\ln 3} + \frac{1}{4\ln 4} + \frac{1}{5\ln 5} + \frac{1}{6\ln 6} \approx 0.7213 + 0.3034 + 0.1803 + 0.1243 + 0.0930$

For $n \ge 4$, the terms of the convergent series **seem** to be larger than those of the divergent series!

(c) $\frac{1}{n^{1.1}} < \frac{1}{n\ln n}$

$n \ln n < n^{1.1}$

$\ln n < n^{0.1}$

This inequality holds when $n \ge 3.5 \times 10^{15}$. Or, $n > e^{40}$. Then $\ln e^{40} = 40 < (e^{40})^{0.1} = e^4 \approx 55$.

74. (a) $\int_{10}^\infty \frac{1}{x^p}\,dx = \left[\frac{x^{-p+1}}{-p+1}\right]_{10}^\infty = \frac{1}{(p-1)10^{p-1}},\ p > 1$

(b) $f(x) = \frac{1}{x^p}$

$R_{10}(p) = \sum_{n=11}^{\infty} \frac{1}{n^p}$

$\le$ Area under the graph of f over the interval $[10, \infty)$

(c) The horizontal asymptote is $y = 0$. As n increases, the error decreases.

75. (a) Let $f(x) = 1/x$. f is positive, continuous, and decreasing on $[1, \infty)$.

$$S_n - 1 \le \int_1^n \frac{1}{x}\,dx$$

$$S_n - 1 \le \ln n$$

Hence, $S_n \le 1 + \ln n$. Similarly,

$$S_n \ge \int_1^{n+1} \frac{1}{x}\,dx = \ln(n+1).$$

Thus, $\ln(n+1) \le S_n \le 1 + \ln n$.

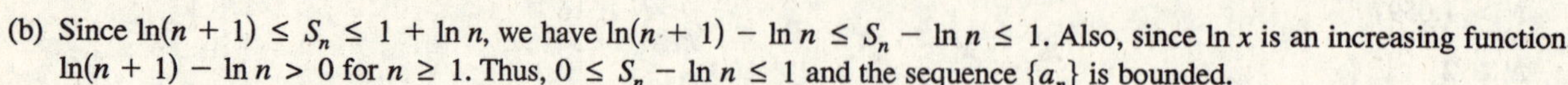

(b) Since $\ln(n+1) \le S_n \le 1 + \ln n$, we have $\ln(n+1) - \ln n \le S_n - \ln n \le 1$. Also, since $\ln x$ is an increasing function, $\ln(n+1) - \ln n > 0$ for $n \ge 1$. Thus, $0 \le S_n - \ln n \le 1$ and the sequence $\{a_n\}$ is bounded.

(c) $a_n - a_{n+1} = [S_n - \ln n] - [S_{n+1} - \ln(n+1)] = \displaystyle\int_n^{n+1} \frac{1}{x}\,dx - \frac{1}{n+1} \ge 0$

Thus, $a_n \ge a_{n+1}$ and the sequence is decreasing.

(d) Since the sequence is bounded and monotonic, it converges to a limit, γ.

(e) $a_{100} = S_{100} - \ln 100 \approx 0.5822$ (Actually $\gamma \approx 0.577216$.)

76. $$\sum_{n=2}^{\infty} \ln\left(1 - \frac{1}{n^2}\right) = \sum_{n=2}^{\infty} \ln\left(\frac{n^2-1}{n^2}\right) = \sum_{n=2}^{\infty} \ln\frac{(n+1)(n-1)}{n^2} = \sum_{n=2}^{\infty} [\ln(n+1) + \ln(n-1) - 2\ln n]$$

$$= \ln 3 + \ln 1 - 2\ln 2) + (\ln 4 + \ln 2 - 2\ln 3) + (\ln 5 + \ln 3 - 2\ln 4) + (\ln 6 + \ln 4 - 2\ln 5)$$

$$+ (\ln 7 + \ln 5 - 2\ln 6) + (\ln 8 + \ln 6 - 2\ln 7) + (\ln 9 + \ln 7 - 2\ln 8) + \cdots = -\ln 2$$

77. $\displaystyle\sum_{n=2}^{\infty} x^{\ln n}$

(a) $x = 1$: $\displaystyle\sum_{n=2}^{\infty} 1^{\ln n} = \sum_{n=2}^{\infty} 1$, diverges

(b) $x = \dfrac{1}{e}$: $\displaystyle\sum_{n=2}^{\infty}\left(\frac{1}{e}\right)^{\ln n} = \sum_{n=2}^{\infty} e^{-\ln n} = \sum_{n=2}^{\infty}\frac{1}{n}$, diverges

(c) Let x be given, $x > 0$. Put $x = e^{-p} \Leftrightarrow \ln x = -p$.

$$\sum_{n=2}^{\infty} x^{\ln n} = \sum_{n=2}^{\infty} e^{-p\ln n} = \sum_{n=2}^{\infty} n^{-p} = \sum_{n=2}^{\infty}\frac{1}{n^p}$$

This series converges for $p > 1 \Rightarrow x < \dfrac{1}{e}$.

78. $\xi(x) = \displaystyle\sum_{n=1}^{\infty} n^{-x} = \sum_{n=1}^{\infty}\frac{1}{n^x}$

Converges for $x > 1$ by Theorem 9.11

79. $\displaystyle\sum_{n=1}^{\infty}\frac{1}{2n-1}$

Let $f(x) = \dfrac{1}{2x-1}$.

f is positive, continuous, and decreasing for $x \ge 1$.

$$\int_1^{\infty}\frac{1}{2x-1}\,dx = \left[\ln\sqrt{2x-1}\right]_1^{\infty} = \infty$$

Diverges by Theorem 9.10

80. $\displaystyle\sum_{n=2}^{\infty}\frac{1}{n\sqrt{n^2-1}}$

Let $f(x) = \dfrac{1}{x\sqrt{x^2-1}}$.

f is positive, continuous, and decreasing for $x \ge 2$.

$$\int_2^{\infty}\frac{1}{x\sqrt{x^2-1}}\,dx = \left[\operatorname{arcsec} x\right]_2^{\infty} = \frac{\pi}{2} - \frac{\pi}{3}$$

Converges by Theorem 9.10

81. $\displaystyle\sum_{n=1}^{\infty} \frac{1}{n\sqrt[4]{n}} = \sum_{n=1}^{\infty} \frac{1}{n^{5/4}}$

p-series with $p = \frac{5}{4}$

Converges by Theorem 9.11

82. $\displaystyle 3\sum_{n=1}^{\infty} \frac{1}{n^{0.95}}$

p-series with $p = 0.95$

Diverges by Theorem 9.11

83. $\displaystyle\sum_{n=0}^{\infty} \left(\frac{2}{3}\right)^n$

Geometric series with $r = \frac{2}{3}$

Converges by Theorem 9.6

84. $\displaystyle\sum_{n=0}^{\infty} (1.075)^n$

Geometric series with $r = 1.075$

Diverges by Theorem 9.6

85. $\displaystyle\sum_{n=1}^{\infty} \frac{n}{\sqrt{n^2+1}}$

$$\lim_{n\to\infty} \frac{n}{\sqrt{n^2+1}} = \lim_{n\to\infty} \frac{1}{\sqrt{1+(1/n^2)}} = 1 \neq 0$$

Diverges by Theorem 9.9

86. $\displaystyle\sum_{n=1}^{\infty} \left(\frac{1}{n^2} - \frac{1}{n^3}\right) = \sum_{n=1}^{\infty} \frac{1}{n^2} - \sum_{n=1}^{\infty} \frac{1}{n^3}$

Since these are both convergent p-series, the difference is convergent.

87. $\displaystyle\sum_{n=1}^{\infty} \left(1 + \frac{1}{n}\right)^n$

$$\lim_{n\to\infty} \left(1 + \frac{1}{n}\right)^n = e \neq 0$$

Fails nth-Term Test

Diverges by Theorem 9.9

88. $\displaystyle\sum_{n=2}^{\infty} \ln(n)$

$$\lim_{n\to\infty} \ln(n) = \infty$$

Diverges by Theorem 9.9

89. $\displaystyle\sum_{n=2}^{\infty} \frac{1}{n(\ln n)^3}$

Let $f(x) = \dfrac{1}{x(\ln x)^3}$.

f is positive, continuous and decreasing for $x \geq 2$.

$$\int_2^{\infty} \frac{1}{x(\ln x)^3}\,dx = \int_2^{\infty} (\ln x)^{-3}\frac{1}{x}\,dx = \left[\frac{(\ln x)^{-2}}{-2}\right]_2^{\infty} = \left[-\frac{1}{2(\ln x)^2}\right]_2^{\infty} = \frac{1}{2(\ln 2)^2}$$

Converges by Theorem 9.10. See Exercise 51.

90. $\displaystyle\sum_{n=2}^{\infty} \frac{\ln n}{n^3}$

Let $f(x) = \dfrac{\ln x}{x^3}$.

f is positive, continuous, and decreasing for $x \geq 2$ since $f'(x) = \dfrac{1 - 3\ln x}{x^4} < 0$ for $x \geq 2$.

$$\int_2^{\infty} \frac{\ln x}{x^3}\,dx = \left[-\frac{\ln x}{2x^2}\right]_2^{\infty} + \frac{1}{2}\int_2^{\infty} \frac{1}{x^3}\,dx = \frac{\ln 2}{8} + \left[-\frac{1}{4x^2}\right]_2^{\infty} = \frac{\ln 2}{8} + \frac{1}{16} \quad \text{(Use integration by parts.)}$$

Converges by Theorem 9.10. See Exercise 30.

Section 9.4 Comparisons of Series

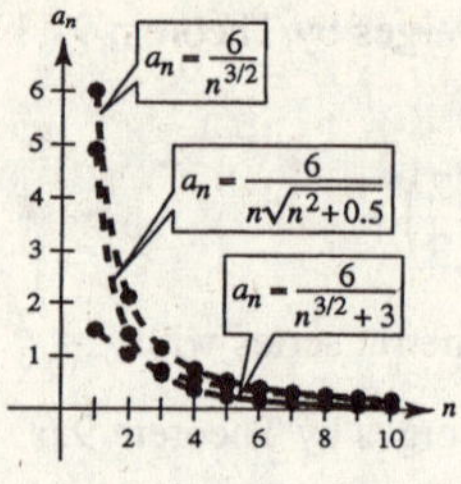

1. (a) $\sum_{n=1}^{\infty} \frac{6}{n^{3/2}} = \frac{6}{1} + \frac{6}{2^{3/2}} + \cdots; \; S_1 = 6$

$\sum_{n=1}^{\infty} \frac{6}{n^{3/2} + 3} = \frac{6}{4} + \frac{6}{2^{3/2} + 3} + \cdots; \; S_1 = \frac{3}{2}$

$\sum_{n=1}^{\infty} \frac{6}{n\sqrt{n^2 + 0.5}} = \frac{6}{1\sqrt{1.5}} + \frac{6}{2\sqrt{4.5}} + \cdots; \; S_1 = \frac{6}{\sqrt{1.5}} \approx 4.9$

(b) The first series is a p-series. It converges $\left(p = \frac{3}{2} > 1\right)$.

(c) The magnitude of the terms of the other two series are less than the corresponding terms at the convergent p-series. Hence, the other two series converge.

(d) The smaller the magnitude of the terms, the smaller the magnitude of the terms of the sequence of partial sums.

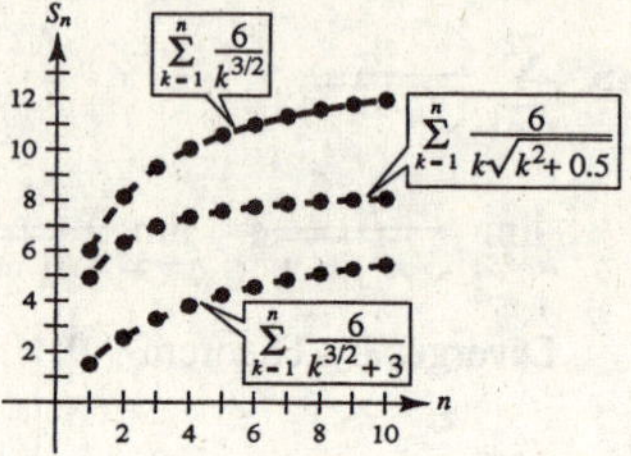

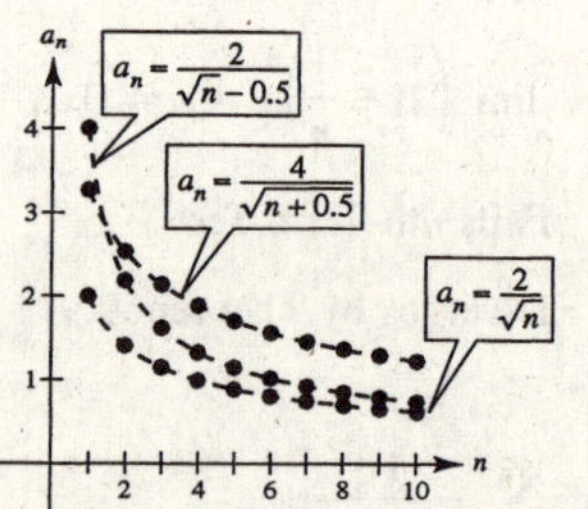

2. (a) $\sum_{n=1}^{\infty} \frac{2}{\sqrt{n}} = 2 + \frac{2}{\sqrt{2}} + \cdots \; S_1 = 2$

$\sum_{n=1}^{\infty} \frac{2}{\sqrt{n} - 0.5} = \frac{2}{0.5} + \frac{2}{\sqrt{2} - 0.5} + \cdots \; S_1 = 4$

$\sum_{n=1}^{\infty} \frac{4}{\sqrt{n} + 0.5} = \frac{4}{\sqrt{1.5}} + \frac{4}{\sqrt{2.5}} + \cdots \; S_1 \approx 3.3$

(b) The first series is a p-series. It diverges $\left(p = \frac{1}{2} < 1\right)$.

(c) The magnitude of the terms of the other two series are greater than the corresponding terms of the divergent p-series. Hence, the other two series diverge.

(d) The larger the magnitude of the terms, the larger the magnitude of the terms of the sequence of partial sums.

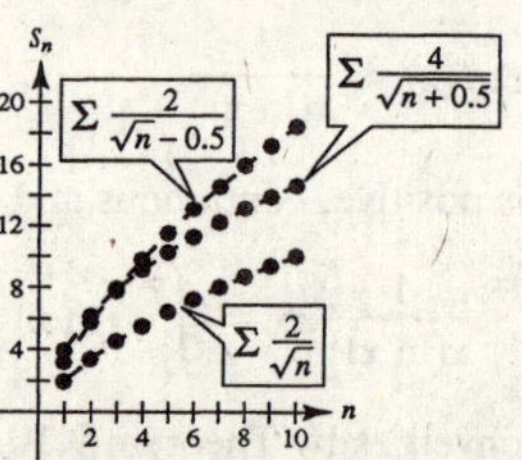

3. $0 < \frac{1}{n^2 + 1} < \frac{1}{n^2}$

Therefore,

$$\sum_{n=1}^{\infty} \frac{1}{n^2 + 1}$$

converges by comparison with the convergent p-series

$$\sum_{n=1}^{\infty} \frac{1}{n^2}.$$

4. $\frac{1}{3n^2 + 2} < \frac{1}{3n^2}$

Therefore,

$$\sum_{n=1}^{\infty} \frac{1}{3n^2 + 2}$$

converges by comparison with the convergent p-series

$$\frac{1}{3}\sum_{n=1}^{\infty} \frac{1}{n^2}.$$

5. $\frac{1}{n - 1} > \frac{1}{n} > 0$ for $n \geq 2$

Therefore,

$$\sum_{n=2}^{\infty} \frac{1}{n - 1}$$

diverges by comparison with the divergent p-series

$$\sum_{n=2}^{\infty} \frac{1}{n}.$$

6. $\dfrac{1}{\sqrt{n}-1} > \dfrac{1}{\sqrt{n}}$ for $n \geq 2$

Therefore,

$$\sum_{n=2}^{\infty} \frac{1}{\sqrt{n}-1}$$

diverges by comparison with the divergent p-series

$$\sum_{n=2}^{\infty} \frac{1}{\sqrt{n}}.$$

7. $0 < \dfrac{1}{3^n+1} < \dfrac{1}{3^n}$

Therefore,

$$\sum_{n=0}^{\infty} \frac{1}{3^n+1}$$

converges by comparison with the convergent geometric series

$$\sum_{n=0}^{\infty} \left(\frac{1}{3}\right)^n.$$

8. $\dfrac{3^n}{4^n+5} < \left(\dfrac{3}{4}\right)^n$

Therefore,

$$\sum_{n=0}^{\infty} \frac{3^n}{4^n+5}$$

converges by comparison with the convergent geometric series

$$\sum_{n=0}^{\infty} \left(\frac{3}{4}\right)^n.$$

9. For $n \geq 3$, $\dfrac{\ln n}{n+1} > \dfrac{1}{n+1} > 0.$

Therefore,

$$\sum_{n=1}^{\infty} \frac{\ln n}{n+1}$$

diverges by comparison with the divergent series

$$\sum_{n=1}^{\infty} \frac{1}{n+1}.$$

Note: $\displaystyle\sum_{n=1}^{\infty} \frac{1}{n+1}$ diverges by the Integral Test.

10. $\dfrac{1}{\sqrt{n^3+1}} < \dfrac{1}{n^{3/2}}$

Therefore,

$$\sum_{n=1}^{\infty} \frac{1}{\sqrt{n^3+1}}$$

converges by comparison with the convergent p-series

$$\sum_{n=1}^{\infty} \frac{1}{n^{3/2}}.$$

11. For $n > 3$, $\dfrac{1}{n^2} > \dfrac{1}{n!} > 0.$

Therefore,

$$\sum_{n=0}^{\infty} \frac{1}{n!}$$

converges by comparison with the convergent p-series

$$\sum_{n=1}^{\infty} \frac{1}{n^2}.$$

12. $\dfrac{1}{4\sqrt[3]{n}-1} > \dfrac{1}{4\sqrt[4]{n}}$

Therefore,

$$\sum_{n=1}^{\infty} \frac{1}{4\sqrt[3]{n}-1}$$

diverges by comparison with the divergent p-series

$$\frac{1}{4}\sum_{n=1}^{\infty} \frac{1}{\sqrt[4]{n}}.$$

13. $0 < \dfrac{1}{e^{n^2}} \leq \dfrac{1}{e^n}$

Therefore,

$$\sum_{n=0}^{\infty} \frac{1}{e^{n^2}}$$

converges by comparison with the convergent geometric series

$$\sum_{n=0}^{\infty} \left(\frac{1}{e}\right)^n.$$

14. $\dfrac{4^n}{3^n-1} > \dfrac{4^n}{3^n}$

Therefore,

$$\sum_{n=1}^{\infty} \frac{4^n}{3^n-1}$$

diverges by comparison with the divergent geometric series

$$\sum_{n=1}^{\infty} \left(\frac{4}{3}\right)^n.$$

15. $\displaystyle\lim_{n\to\infty} \frac{n/(n^2+1)}{1/n} = \lim_{n\to\infty} \frac{n^2}{n^2+1} = 1$

Therefore,

$$\sum_{n=1}^{\infty} \frac{n}{n^2+1}$$

diverges by a limit comparison with the divergent p-series

$$\sum_{n=1}^{\infty} \frac{1}{n}.$$

16. $\displaystyle\lim_{n\to\infty} \frac{2/(3^n-5)}{1/3^n} = \lim_{n\to\infty} \frac{2 \cdot 3^n}{3^n-5} = 2$

Therefore,

$$\sum_{n=1}^{\infty} \frac{2}{3^n-5}$$

converges by a limit comparison with the convergent geometric series

$$\sum_{n=1}^{\infty} \left(\frac{1}{3}\right)^n.$$

17. $\lim_{n\to\infty} \frac{1/\sqrt{n^2+1}}{1/n} = \lim_{n\to\infty} \frac{n}{\sqrt{n^2+1}} = 1$

Therefore,

$$\sum_{n=0}^{\infty} \frac{1}{\sqrt{n^2+1}}$$

diverges by a limit comparison with the divergent p-series

$$\sum_{n=1}^{\infty} \frac{1}{n}.$$

18. $\lim_{n\to\infty} \frac{3/\sqrt{n^2-4}}{1/n} = \lim_{n\to\infty} \frac{3n}{\sqrt{n^2-4}} = 3$

Therefore,

$$\sum_{n=3}^{\infty} \frac{3}{\sqrt{n^2-4}}$$

diverges by a limit comparison with the divergent harmonic series

$$\sum_{n=3}^{\infty} \frac{1}{n}.$$

19. $\lim_{n\to\infty} \frac{\frac{2n^2-1}{3n^5+2n+1}}{1/n^3} = \lim_{n\to\infty} \frac{2n^5-n^3}{3n^5+2n+1} = \frac{2}{3}$

Therefore,

$$\sum_{n=1}^{\infty} \frac{2n^2-1}{3n^5+2n+1}$$

converges by a limit comparison with the convergent p-series

$$\sum_{n=1}^{\infty} \frac{1}{n^3}.$$

20. $\lim_{n\to\infty} \frac{\frac{5n-3}{n^2-2n+5}}{1/n} = \lim_{n\to\infty} \frac{5n^2-3n}{n^2-2n+5} = 5$

Therefore,

$$\sum_{n=1}^{\infty} \frac{5n-3}{n^2-2n+5}$$

diverges by a limit comparison with the divergent p-series

$$\sum_{n=1}^{\infty} \frac{1}{n}.$$

21. $\lim_{n\to\infty} \frac{\frac{n+3}{n(n+2)}}{1/n} = \lim_{n\to\infty} \frac{n^2+3n}{n^2+2n} = 1$

Therefore,

$$\sum_{n=1}^{\infty} \frac{n+3}{n(n+2)}$$

diverges by a limit comparison with the divergent p-series

$$\sum_{n=1}^{\infty} \frac{1}{n}.$$

22. $\lim_{n\to\infty} \frac{\frac{1}{n(n^2+1)}}{1/n^3} = \lim_{n\to\infty} \frac{n^3}{n^3+n} = 1$

Therefore,

$$\sum_{n=1}^{\infty} \frac{1}{n(n^2+1)}$$

converges by a limit comparison with the convergent p-series

$$\sum_{n=1}^{\infty} \frac{1}{n^3}.$$

23. $\lim_{n\to\infty} \frac{1/\left(n\sqrt{n^2+1}\right)}{1/n^2} = \lim_{n\to\infty} \frac{n^2}{n\sqrt{n^2+1}} = 1$

Therefore,

$$\sum_{n=1}^{\infty} \frac{1}{n\sqrt{n^2+1}}$$

converges by a limit comparison with the convergent p-series

$$\sum_{n=1}^{\infty} \frac{1}{n^2}.$$

24. $\lim_{n\to\infty} \frac{n/[(n+1)2^{n-1}]}{1/(2^{n-1})} = \lim_{n\to\infty} \frac{n}{n+1} = 1$

Therefore,

$$\sum_{n=1}^{\infty} \frac{n}{(n+1)2^{n-1}}$$

converges by a limit comparison with the convergent geometric series

$$\sum_{n=1}^{\infty} \left(\frac{1}{2}\right)^{n-1}.$$

25. $\lim_{n\to\infty} \frac{(n^{k-1})/(n^k+1)}{1/n} = \lim_{n\to\infty} \frac{n^k}{n^k+1} = 1$

Therefore,

$$\sum_{n=1}^{\infty} \frac{n^{k-1}}{n^k+1}$$

diverges by a limit comparison with the divergent p-series

$$\sum_{n=1}^{\infty} \frac{1}{n}.$$

26. $\lim_{n\to\infty} \frac{5/(n+\sqrt{n^2+4})}{1/n} = \lim_{n\to\infty} \frac{5n}{n+\sqrt{n^2+4}} = \frac{5}{2}$

Therefore,

$$\sum_{n=1}^{\infty} \frac{5}{n+\sqrt{n^2+4}}$$

diverges by a limit comparison with the divergent harmonic series

$$\sum_{n=1}^{\infty} \frac{1}{n}.$$

27. $\lim_{n\to\infty} \frac{\sin(1/n)}{1/n} = \lim_{n\to\infty} \frac{(-1/n^2)\cos(1/n)}{-1/n^2}$

$= \lim_{n\to\infty} \cos\left(\frac{1}{n}\right) = 1$

Therefore,

$$\sum_{n=1}^{\infty} \sin\left(\frac{1}{n}\right)$$

diverges by a limit comparison with the divergent p-series

$$\sum_{n=1}^{\infty} \frac{1}{n}.$$

28. $\lim_{n\to\infty} \frac{\tan(1/n)}{1/n} = \lim_{n\to\infty} \frac{(-1/n^2)\sec^2(1/n)}{-1/n^2}$

$= \lim_{n\to\infty} \sec^2\left(\frac{1}{n}\right) = 1$

Therefore,

$$\sum_{n=1}^{\infty} \tan\left(\frac{1}{n}\right)$$

diverges by a limit comparison with the divergent p-series

$$\sum_{n=1}^{\infty} \frac{1}{n}.$$

29. $\sum_{n=1}^{\infty} \frac{\sqrt{n}}{n} = \sum_{n=1}^{\infty} \frac{1}{\sqrt{n}}$

Diverges

p-series with $p = \frac{1}{2}$

30. $\sum_{n=0}^{\infty} 5\left(-\frac{1}{5}\right)^n$

Converges

Geometric series with $r = -\frac{1}{5}$

31. $\sum_{n=1}^{\infty} \frac{1}{3^n+2}$

Converges

Direct comparison with $\sum_{n=1}^{\infty} \left(\frac{1}{3}\right)^n$

32. $\sum_{n=4}^{\infty} \frac{1}{3n^2-2n-15}$

Converges

Limit comparison with $\sum_{n=1}^{\infty} \frac{1}{n^2}$

33. $\sum_{n=1}^{\infty} \frac{n}{2n+3}$

Diverges; nth-Term Test

$\lim_{n\to\infty} \frac{n}{2n+3} = \frac{1}{2} \neq 0$

34. $\sum_{n=1}^{\infty} \left(\frac{1}{n+1} - \frac{1}{n+2}\right) = \left(\frac{1}{2} - \frac{1}{3}\right) + \left(\frac{1}{3} - \frac{1}{4}\right) + \left(\frac{1}{4} - \frac{1}{5}\right) + \cdots = \frac{1}{2}$

Converges; telescoping series

35. $\sum_{n=1}^{\infty} \frac{n}{(n^2+1)^2}$

Converges; Integral Test

36. $\sum_{n=1}^{\infty} \frac{3}{n(n+3)}$

Converges; telescoping series

$$\sum_{n=1}^{\infty} \left(\frac{1}{n} - \frac{1}{n+3}\right)$$

37. $\lim_{n\to\infty} \frac{a_n}{1/n} = \lim_{n\to\infty} na_n$. By given conditions $\lim_{n\to\infty} na_n$ is finite and nonzero. Therefore,

$$\sum_{n=1}^{\infty} a_n$$

diverges by a limit comparison with the p-series

$$\sum_{n=1}^{\infty} \frac{1}{n}.$$

38. If $j < k - 1$, then $k - j > 1$. The p-series with $p = k - j$ converges and since

$$\lim_{n\to\infty} \frac{P(n)/Q(n)}{1/n^{k-j}} = L > 0, \text{ the series } \sum_{n=1}^{\infty} \frac{P(n)}{Q(n)}$$

converges by the Limit Comparison Test. Similarly, if $j \geq k - 1$, then $k - j \leq 1$ which implies that

$$\sum_{n=1}^{\infty} \frac{P(n)}{Q(n)}$$

diverges by the Limit Comparison Test.

39. $\frac{1}{2} + \frac{2}{5} + \frac{3}{10} + \frac{4}{17} + \frac{5}{26} + \cdots = \sum_{n=1}^{\infty} \frac{n}{n^2+1}$,

which diverges since the degree of the numerator is only one less than the degree of the denominator.

40. $\frac{1}{3} + \frac{1}{8} + \frac{1}{15} + \frac{1}{24} + \frac{1}{35} + \cdots = \sum_{n=2}^{\infty} \frac{1}{n^2-1}$,

which converges since the degree of the numerator is two less than the degree of the denominator.

41. $\sum_{n=1}^{\infty} \frac{1}{n^3+1}$

converges since the degree of the numerator is three less than the degree of the denominator.

42. $\sum_{n=1}^{\infty} \frac{n^2}{n^3+1}$

diverges since the degree of the numerator is only one less than the degree of the denominator.

43. $\lim_{n\to\infty} n\left(\frac{n^3}{5n^4+3}\right) = \lim_{n\to\infty} \frac{n^4}{5n^4+3} = \frac{1}{5} \neq 0$

Therefore, $\sum_{n=1}^{\infty} \frac{n^3}{5n^4+3}$ diverges.

44. $\lim_{n\to\infty} \frac{n}{\ln n} = \lim_{n\to\infty} \frac{1}{1/n} = \lim_{n\to\infty} n = \infty \neq 0$

Therefore, $\sum_{n=2}^{\infty} \frac{1}{\ln n}$ diverges.

45. $\frac{1}{200} + \frac{1}{400} + \frac{1}{600} + \cdots = \sum_{n=1}^{\infty} \frac{1}{200n}$

diverges, (harmonic)

46. $\frac{1}{200} + \frac{1}{210} + \frac{1}{220} + \cdots = \sum_{n=0}^{\infty} \frac{1}{200+10n}$

diverges

47. $\frac{1}{201} + \frac{1}{204} + \frac{1}{209} + \frac{1}{216} = \sum_{n=1}^{\infty} \frac{1}{200+n^2}$

converges

48. $\frac{1}{201} + \frac{1}{208} + \frac{1}{227} + \frac{1}{264} + \cdots = \sum_{n=1}^{\infty} \frac{1}{200+n^3}$

converges

49. Some series diverge or converge very slowly. You cannot decide convergence or divergence of a series by comparing the first few terms.

50. See Theorem 9.12, page 624. One example is

$\sum_{n=1}^{\infty} \frac{1}{n^2+1}$ converges because $\frac{1}{n^2+1} < \frac{1}{n^2}$ and

$\sum_{n=1}^{\infty} \frac{1}{n^2}$ converges (p-series).

51. See Theorem 9.13, page 626. One example is

$\sum_{n=2}^{\infty} \frac{1}{\sqrt{n}-1}$ diverges because $\lim_{n\to\infty} \frac{1/\sqrt{n}-1}{1/\sqrt{n}} = 1$ and

$\sum_{n=2}^{\infty} \frac{1}{\sqrt{n}}$ diverges (p-series).

52. This is not correct. The beginning terms do not affect the convergence or divergence of a series. In fact,

$\frac{1}{1000} + \frac{1}{1001} + \cdots = \sum_{n=1000}^{\infty} \frac{1}{n}$ diverges (harmonic)

and $1 + \frac{1}{4} + \frac{1}{9} + \cdots = \sum_{n=1}^{\infty} \frac{1}{n^2}$ converges (p-series).

53.

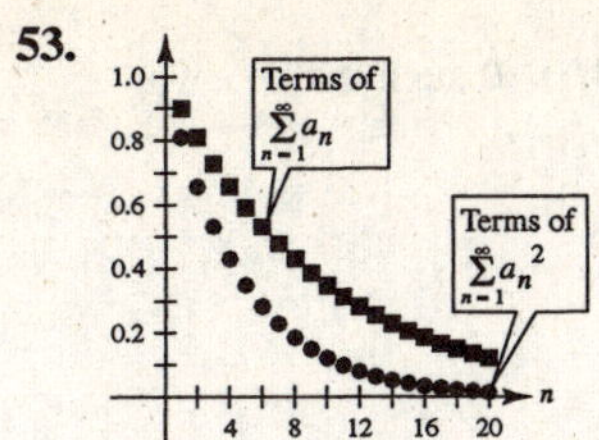

For $0 < a_n < 1, 0 < a_n^2 < a_n < 1$. Hence, the lower terms are those of $\Sigma\, a_n^2$.

54. (a) $\displaystyle\sum_{n=1}^{\infty} \frac{1}{(2n-1)^2} = \sum_{n=1}^{\infty} \frac{1}{4n^2 - 4n + 1}$

converges since the degree of the numerator is two less than the degree of the denominator. (See Exercise 38.)

(b)

n	5	10	20	50	100
S_n	1.1839	1.02087	1.2212	1.2287	1.2312

(c) $\displaystyle\sum_{n=3}^{\infty} \frac{1}{(2n-1)^2} = \frac{\pi^2}{8} - S_2 \approx 0.1226$

(d) $\displaystyle\sum_{n=10}^{\infty} \frac{1}{(2n-1)^2} = \frac{\pi^2}{8} - S_9 \approx 0.0277$

55. False. Let $a_n = \dfrac{1}{n^3}$ and $b_n = \dfrac{1}{n^2}$. $0 < a_n \le b_n$ and both $\displaystyle\sum_{n=1}^{\infty} \frac{1}{n^3}$ and $\displaystyle\sum_{n=1}^{\infty} \frac{1}{n^2}$ converge.

56. True

57. True

58. False. Let $a_n = 1/n, b_n = 1/n, c_n = 1/n^2$. Then,

$a_n \le b_n + c_n$, but $\displaystyle\sum_{n=1}^{\infty} c_n$ converges.

59. True

60. False. $\displaystyle\sum_{n=1}^{\infty} a_n$ could converge or diverge.

For example, let $\displaystyle\sum_{n=1}^{\infty} b_n = \sum_{n=1}^{\infty} \frac{1}{\sqrt{n}}$, which diverges.

$0 < \dfrac{1}{n} < \dfrac{1}{\sqrt{n}}$ and $\displaystyle\sum_{n=1}^{\infty} \frac{1}{n}$ diverges, but

$0 < \dfrac{1}{n^2} < \dfrac{1}{\sqrt{n}}$ and $\displaystyle\sum_{n=1}^{\infty} \frac{1}{n^2}$ converges.

61. Since $\displaystyle\sum_{n=1}^{\infty} b_n$ converges, $\displaystyle\lim_{n\to\infty} b_n = 0$. There exists N such that $b_n < 1$ for $n > N$. Thus, $a_n b_n < a_n$ for $n > N$ and

$\displaystyle\sum_{n=1}^{\infty} a_n b_n$ converges by comparison to the convergent

series $\displaystyle\sum_{i=1}^{\infty} a_n$.

62. Since $\displaystyle\sum_{n=1}^{\infty} a_n$ converges, then $\displaystyle\sum_{n=1}^{\infty} a_n a_n = \sum_{n=1}^{\infty} a_n^2$

converges by Exercise 61.

63. $\displaystyle\sum \frac{1}{n^2}$ and $\displaystyle\sum \frac{1}{n^3}$ both converge, and hence so does

$\displaystyle\sum\left(\frac{1}{n^2}\right)\left(\frac{1}{n^3}\right) = \sum \frac{1}{n^5}$.

64. $\displaystyle\sum \frac{1}{n^2}$ converge, and hence so does $\displaystyle\sum\left(\frac{1}{n^2}\right)^2 = \sum \frac{1}{n^4}$.

65. Suppose $\displaystyle\lim_{n\to\infty} \frac{a_n}{b_n} = 0$ and $\Sigma\, b_n$ converges.

From the definition of limit of a sequence, there exists $M > 0$ such that

$$\left|\frac{a_n}{b_n} - 0\right| < 1$$

whenever $n > M$. Hence, $a_n < b_n$ for $n > M$. From the Comparison Test, $\Sigma\, a_n$ converges.

66. Suppose $\lim_{n\to\infty} \frac{a_n}{b_n} = \infty$ and $\Sigma\, b_n$ diverges. From the definition of limit of a sequence, there exists $M > 0$ such that

$$\frac{a_n}{b_n} > 1$$

for $n > M$. Thus, $a_n > b_n$ for $n > M$. By the Comparison Test, $\Sigma\, a_n$ diverges.

67. (a) Let $\sum a_n = \sum \frac{1}{(n+1)^3}$, and $\sum b_n = \sum \frac{1}{n^2}$, converges.

$$\lim_{n\to\infty} \frac{a_n}{b_n} = \lim_{n\to\infty} \frac{1/[(n+1)^3]}{1/(n^2)} = \lim_{n\to\infty} \frac{n^2}{(n+1)^3} = 0$$

By Exercise 65, $\sum_{n=1}^{\infty} \frac{1}{(n+1)^3}$ converges.

(b) Let $\sum a_n = \sum \frac{1}{\sqrt{n}\pi^n}$, and $\sum b_n = \sum \frac{1}{\pi^n}$, converges.

$$\lim_{n\to\infty} \frac{a_n}{b_n} = \lim_{n\to\infty} \frac{1/(\sqrt{n}\pi^n)}{1/(\pi^n)} = \lim_{n\to\infty} \frac{1}{\sqrt{n}} = 0$$

By Exercise 65, $\sum_{n=1}^{\infty} \frac{1}{\sqrt{n}\pi^n}$ converges.

68. (a) Let $\sum a_n = \sum \frac{\ln n}{n}$, and $\sum b_n = \sum \frac{1}{n}$, diverges.

$$\lim_{n\to\infty} \frac{a_n}{b_n} = \lim_{n\to\infty} \frac{(\ln n)/n}{1/n} = \lim_{n\to\infty} \ln n = \infty$$

By Exercise 66, $\sum_{n=1}^{\infty} \frac{\ln n}{n}$ diverges.

(b) Let $\sum a_n = \sum \frac{1}{\ln n}$, and $\sum b_n = \sum \frac{1}{n}$, diverges.

$$\lim_{n\to\infty} \frac{a_n}{b_n} = \lim_{n\to\infty} \frac{n}{\ln n} = \infty$$

By Exercise 66, $\sum \frac{1}{\ln n}$ diverges.

69. Since $\lim_{n\to\infty} a_n = 0$, the terms of $\Sigma \sin(a_n)$ are positive for sufficiently large n. Since

$$\lim_{n\to\infty} \frac{\sin(a_n)}{a_n} = 1 \text{ and } \sum a_n$$

converges, so does $\Sigma \sin(a_n)$.

70.
$$\sum_{n=1}^{\infty} \frac{1}{1+2+\cdots+n} = \sum_{n=1}^{\infty} \frac{1}{[n(n+1)]/2} = \sum_{n=1}^{\infty} \frac{2}{n(n+1)}$$

Since $\Sigma\, 1/n^2$ converges, and

$$\lim_{n\to\infty} \frac{2/[n(n+1)]}{1/(n^2)} = \lim_{n\to\infty} \frac{2n^2}{n(n+1)} = 2,$$

$\sum \frac{1}{1+2+\cdots+n}$ converges.

71. The series diverges. For $n > 1$,

$$n < 2^n$$
$$n^{1/n} < 2$$
$$\frac{1}{n^{1/n}} > \frac{1}{2}$$
$$\frac{1}{n^{(n+1)/n}} > \frac{1}{2n}$$

Since $\sum \frac{1}{2n}$ diverges, so does $\sum \frac{1}{n^{(n+1)/n}}$.

72. Consider two cases:

If $a_n \ge \frac{1}{2^{n+1}}$, then $a_n^{1/(n+1)} \ge \left(\frac{1}{2^{n+1}}\right)^{1/(n+1)} = \frac{1}{2}$, and

$$a_n^{n/(n+1)} = \frac{a_n}{a_n^{1/(n+1)}} \le 2a_n.$$

If $a_n \le \frac{1}{2^{n+1}}$, then $a_n^{n/(n+1)} \le \left(\frac{1}{2^{n+1}}\right)^{n/(n+1)} = \frac{1}{2^n}$, and

combining, $a_n^{n/(n+1)} \le 2a_n + \frac{1}{2^n}$.

Since $\sum_{n=1}^{\infty}\left(2a_n + \frac{1}{2^n}\right)$ converges, so does $\sum_{n=1}^{\infty} a_n^{n/(n+1)}$ by the Comparison Test.

Section 9.5 Alternating Series

1. $\sum_{n=1}^{\infty} \frac{6}{n^2}$

$S_1 = 6$

$S_2 = 7.5$

$S_3 \approx 8.1667$

Matches (d).

2. $\sum_{n=1}^{\infty} \frac{(-1)^{n-1}6}{n^2}$

$S_1 = 6$

$S_2 = 4.4$

$S_3 \approx 5.1667$

Matches (f).

3. $\sum_{n=1}^{\infty} \frac{3}{n!}$

$S_1 = 3$

$S_2 = 4.5$

$S_3 = 5.0$

Matches (a).

4. $\sum_{n=1}^{\infty} \frac{(-1)^{n-1}3}{n!}$

$S_1 = 3$

$S_2 = 1.5$

$S_3 = 2.0$

Matches (b).

5. $\sum_{n=1}^{\infty} \frac{10}{n2^n}$

$S_1 = 5$

$S_2 = 6.25$

Matches (e).

6. $\sum_{n=1}^{\infty} \frac{(-1)^n 10}{n2^n}$

$S_1 = 5$

$S_2 = 3.75$

Matches (c).

7. $\sum_{n=1}^{\infty} \frac{(-1)^{n-1}}{2n-1} = \frac{\pi}{4} \approx 0.7854$

(a)

n	1	2	3	4	5	6	7	8	9	10
S_n	1	0.6667	0.8667	0.7238	0.8349	0.7440	0.8209	0.7543	0.8131	0.7605

(b)

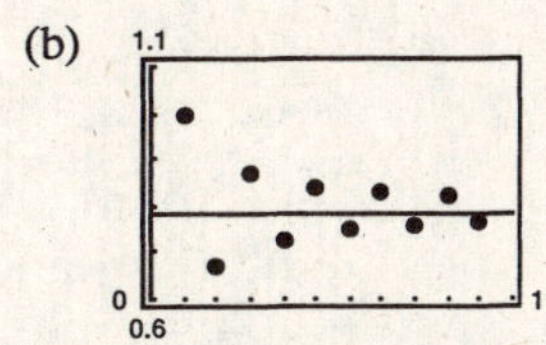

(c) The points alternate sides of the horizontal line that represents the sum of the series. The distance between successive points and the line decreases.

(d) The distance in part (c) is always less than the magnitude of the next term of the series.

8. $\sum_{n=1}^{\infty} \frac{(-1)^{n-1}}{(n-1)!} = \frac{1}{e} \approx 0.3679$

(a)

n	1	2	3	4	5	6	7	8	9	10
S_n	1	0	0.5	0.3333	0.375	0.3667	0.3681	0.3679	0.3679	0.3679

(b)

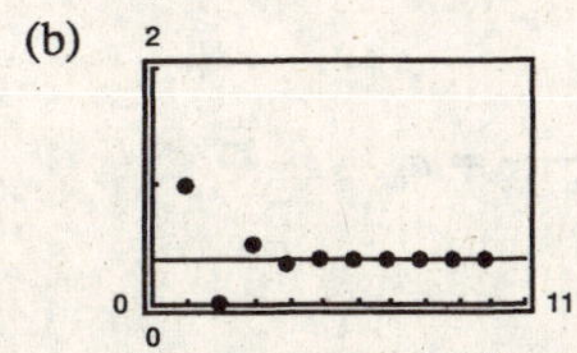

(c) The points alternate sides of the horizontal line that represents the sum of the series. The distance between successive points and the line decreases.

(d) The distance in part (c) is always less than the magnitude of the next series.

9. $\displaystyle\sum_{n=1}^{\infty} \frac{(-1)^{n-1}}{n^2} = \frac{\pi^2}{12} \approx 0.8225$

(a)

n	1	2	3	4	5	6	7	8	9	10
S_n	1	0.75	0.8611	0.7986	0.8386	0.8108	0.8312	0.8156	0.8280	0.8180

(b)

1.1

0 11

0.6

(c) The points alternate sides of the horizontal line that represents the sum of the series. The distance between successive points and the line decreases.

(d) The distance in part (c) is always less than the magnitude of the next term in the series.

10. $\displaystyle\sum_{n=1}^{\infty} \frac{(-1)^{n-1}}{(2n-1)!} = \sin(1) \approx 0.8415$

(a)

n	1	2	3	4	5	6	7	8	9	10
S_n	1	0.8333	0.8417	0.8415	0.8415	0.8415	0.8415	0.8415	0.8415	0.8415

(b)

2

0 11

0

(c) The points alternate sides of the horizontal line that represents the sum of the series. The distance between successive points and the line decreases.

(d) The distance in part (c) is always less than the magnitude of the next series.

11. $\displaystyle\sum_{n=1}^{\infty} \frac{(-1)^{n+1}}{n}$

$a_{n+1} = \dfrac{1}{n+1} < \dfrac{1}{n} = a_n$

$\displaystyle\lim_{n\to\infty} \frac{1}{n} = 0$

Converges by Theorem 9.14

12. $\displaystyle\sum_{n=1}^{\infty} \frac{(-1)^{n+1} n}{2n-1}$

$\displaystyle\lim_{n\to\infty} \frac{n}{2n-1} = \frac{1}{2}$

Diverges by the nth-Term Test

13. $\displaystyle\sum_{n=1}^{\infty} \frac{(-1)^{n+1}}{2n-1}$

$a_{n+1} = \dfrac{1}{2(n+1)-1} < \dfrac{1}{2n-1} = a_n$

$\displaystyle\lim_{n\to\infty} \frac{1}{2n-1} = 0$

Converges by Theorem 9.14

14. $\displaystyle\sum_{n=1}^{\infty} \frac{(-1)^n}{\ln(n+1)}$

$a_{n+1} = \dfrac{1}{\ln(n+2)} < \dfrac{1}{\ln(n+1)} = a_n$

$\displaystyle\lim_{n\to\infty} \frac{1}{\ln(n+1)} = 0$

Converges by Theorem 9.14

15. $\sum_{n=1}^{\infty} \frac{(-1)^n n^2}{n^2+1}$

$\lim_{n\to\infty} \frac{n^2}{n^2+1} = 1$. Thus, $\lim_{n\to\infty} a_n \neq 0$.

Diverges by the nth-Term Test

16. $\sum_{n=1}^{\infty} \frac{(-1)^{n+1} n}{n^2+1}$

$$a_{n+1} = \frac{n+1}{(n+1)^2+1} < \frac{n}{n^2+1} = a_n$$

$$\lim_{n\to\infty} \frac{n}{n^2+1} = 0$$

Converges by Theorem 9.14

17. $\sum_{n=1}^{\infty} \frac{(-1)^n}{\sqrt{n}}$

$$a_{n+1} = \frac{1}{\sqrt{n+1}} < \frac{1}{\sqrt{n}} = a_n$$

$$\lim_{n\to\infty} \frac{1}{\sqrt{n}} = 0$$

Converges by Theorem 9.14

18. $\sum_{n=1}^{\infty} \frac{(-1)^{n+1} n^2}{n^2+5}$

$$\lim_{n\to\infty} \frac{n^2}{n^2+5} = 1$$

Diverges by nth-Term Test

19. $\sum_{n=1}^{\infty} \frac{(-1)^{n+1}(n+1)}{\ln(n+1)}$

$$\lim_{n\to\infty} \frac{n+1}{\ln(n+1)} = \lim_{n\to\infty} \frac{1}{1/(n+1)} = \lim_{n\to\infty} (n+1) = \infty$$

Diverges by the nth-Term Test

20. $\sum_{n=1}^{\infty} \frac{(-1)^{n+1} \ln(n+1)}{n+1}$

$$a_{n+1} = \frac{\ln[(n+1)+1]}{(n+1)+1} < \frac{\ln(n+1)}{n+1} \text{ for } n \geq 2$$

$$\lim_{n\to\infty} \frac{\ln(n+1)}{n+1} = \lim_{n\to\infty} \frac{1/(n+1)}{1} = 0$$

Converges by Theorem 9.14

21. $\sum_{n=1}^{\infty} \sin\left[\frac{(2n-1)\pi}{2}\right] = \sum_{n=1}^{\infty} (-1)^{n+1}$

Diverges by the nth-Term Test

22. $\sum_{n=1}^{\infty} \frac{1}{n} \sin\left[\frac{(2n-1)\pi}{2}\right] = \sum_{n=1}^{\infty} \frac{(-1)^{n+1}}{n}$

Converges; (See Exercise 11.)

23. $\sum_{n=1}^{\infty} \cos n\pi = \sum_{n=1}^{\infty} (-1)^n$

Diverges by the nth-Term Test

24. $\sum_{n=1}^{\infty} \frac{1}{n} \cos n\pi = \sum_{n=1}^{\infty} \frac{(-1)^n}{n}$

Converges; (See Exercise 11.)

25. $\sum_{n=0}^{\infty} \frac{(-1)^n}{n!}$

$$a_{n+1} = \frac{1}{(n+1)!} < \frac{1}{n!} = a_n$$

$$\lim_{n\to\infty} \frac{1}{n!} = 0$$

Converges by Theorem 9.14

26. $\sum_{n=0}^{\infty} \frac{(-1)^n}{(2n+1)!}$

$$a_{n+1} = \frac{1}{(2n+3)!} < \frac{1}{(2n+1)!} = a_n$$

$$\lim_{n\to\infty} \frac{1}{(2n+1)!} = 0$$

Converges by Theorem 9.14

27. $\sum_{n=1}^{\infty} \frac{(-1)^{n+1}\sqrt{n}}{n+2}$

$$a_{n+1} = \frac{\sqrt{n+1}}{(n+1)+2} < \frac{\sqrt{n}}{n+2} \text{ for } n \geq 2$$

$$\lim_{n\to\infty} \frac{\sqrt{n}}{n+2} = 0$$

Converges by Theorem 9.14

28. $\sum_{n=1}^{\infty} \frac{(-1)^{n+1}\sqrt{n}}{\sqrt[3]{n}}$

$$\lim_{n\to\infty} \frac{n^{1/2}}{n^{1/3}} = \lim_{n\to\infty} n^{1/6} = \infty$$

Diverges by the nth-Term Test

29. $\displaystyle\sum_{n=1}^{\infty} \frac{(-1)^{n+1}n!}{1 \cdot 3 \cdot 5 \cdots (2n-1)}$

$$a_{n+1} = \frac{(n+1)!}{1 \cdot 3 \cdot 5 \cdots (2n-1)(2n+1)} = \frac{n!}{1 \cdot 3 \cdot 5 \cdots (2n-1)} \cdot \frac{n+1}{2n+1} = a_n\left(\frac{n+1}{2n+1}\right) < a_n$$

$$\lim_{n\to\infty} a_n = \lim_{n\to\infty} \frac{n!}{1 \cdot 3 \cdot 5 \cdots (2n-1)}$$

$$= \lim_{n\to\infty} \frac{1 \cdot 2 \cdot 3 \cdots n}{1 \cdot 3 \cdot 5 \cdots (2n-1)}$$

$$= \lim_{n\to\infty} 2\left[\frac{3}{3} \cdot \frac{4}{5} \cdot \frac{5}{7} \cdots \frac{n}{2n-3}\right] \cdot \frac{1}{2n-1} = 0$$

Converges by Theorem 9.14

30. $\displaystyle\sum_{n=1}^{\infty} (-1)^{n+1} \frac{1 \cdot 3 \cdot 5 \cdots (2n-1)}{1 \cdot 4 \cdot 7 \cdots (3n-2)}$

$$a_{n+1} = \frac{1 \cdot 3 \cdot 5 \cdots (2n-1)(2n+1)}{1 \cdot 4 \cdot 7 \cdots (3n-2)(3n+1)} = a_n\left(\frac{2n+1}{3n+1}\right) < a_n$$

$$\lim_{n\to\infty} a_n = \lim_{n\to\infty} 3\left[\frac{5}{4} \cdot \frac{7}{7} \cdot \frac{9}{10} \cdots \frac{2n-1}{3n-5}\right] \frac{1}{3n-2} = 0$$

Converges by Theorem 9.14

31. $\displaystyle\sum_{n=1}^{\infty} \frac{(-1)^{n+1}(2)}{e^n - e^{-n}} = \sum_{n=1}^{\infty} \frac{(-1)^{n+1}(2e^n)}{e^{2n} - 1}$

Let $f(x) = \dfrac{2e^x}{e^{2x} - 1}$. Then

$$f'(x) = \frac{-2e^x(e^{2x}+1)}{(e^{2x}-1)^2} < 0.$$

Thus, $f(x)$ is decreasing. Therefore, $a_{n+1} < a_n$, and

$$\lim_{n\to\infty} \frac{2e^n}{e^{2n}-1} = \lim_{n\to\infty} \frac{2e^n}{2e^{2n}} = \lim_{n\to\infty} \frac{1}{e^n} = 0.$$

The series converges by Theorem 9.14.

32. $\displaystyle\sum_{n=1}^{\infty} \frac{2(-1)^{n+1}}{e^n + e^{-n}} = \sum_{n=1}^{\infty} \frac{(-1)^{n+1}(2e^n)}{e^{2n} + 1}$

Let $f(x) = \dfrac{2e^x}{e^{2x} + 1}$. Then

$$f'(x) = \frac{2e^{2x}(1 - e^{2x})}{(e^{2x}+1)^2} < 0 \text{ for } x > 0.$$

Thus, $f(x)$ is decreasing for $x > 0$ which implies $a_{n+1} < a_n$.

$$\lim_{n\to\infty} \frac{2e^n}{e^{2n}+1} = \lim_{n\to\infty} \frac{2e^n}{2e^{2n}} = \lim_{n\to\infty} \frac{1}{e^n} = 0$$

The series converges by Theorem 9.14.

33. $S_6 = \displaystyle\sum_{n=1}^{6} \frac{3(-1)^{n+1}}{n^2} = 2.4325$

$2.4325 - 0.0612 \le S \le 2.4325 + 0.0612$

$|R_6| = |S - S_6| \le a_7 = \dfrac{3}{49} \approx 0.0612$

$2.3713 \le S \le 2.4937$

34. $S_6 = \displaystyle\sum_{n=1}^{6} \frac{4(-1)^{n+1}}{\ln(n+1)} \approx 2.7067$

$|R_6| = |S - S_6| \le a_7 = \dfrac{4}{\ln 8} \approx 1.9236$

$0.7831 \le S \le 4.6303$

35. $S_6 = \displaystyle\sum_{n=0}^{5} \frac{2(-1)^n}{n!} \approx 0.7333$

$0.7333 - 0.002778 \le S \le 0.7333 + 0.002778$

$|R_6| = |S - S_6| \le a_7 = \dfrac{2}{6!} = 0.002778$

$0.7305 \le S \le 0.7361$

36. $S_6 = \displaystyle\sum_{n=1}^{6} \frac{(-1)^{n+1}n}{2^n} = 0.1875$

$|R_6| = |S - S_6| \le a_7 = \dfrac{7}{2^7} \approx 0.05469$

$0.1328 \le S \le 0.2422$

37. $\sum_{n=0}^{\infty} \frac{(-1)^n}{n!}$

(a) By Theorem 9.15,

$$|R_N| \le a_{N+1} = \frac{1}{(N+1)!} < 0.001.$$

This inequality is valid when $N = 6$.

(b) We may approximate the series by

$$\sum_{n=0}^{6} \frac{(-1)^n}{n!} = 1 - 1 + \frac{1}{2} - \frac{1}{6} + \frac{1}{24} - \frac{1}{120} + \frac{1}{720}$$

$$\approx 0.368.$$

(7 terms. Note that the sum begins with $n = 0$.)

38. $\sum_{n=0}^{\infty} \frac{(-1)^n}{2^n n!}$

(a) By Theorem 9.15,

$$|R_n| \le a_{N+1} = \frac{1}{2^{N+1}(N+1)!} < 0.001.$$

This inequality is valid when $N = 4$.

(b) We may approximate the series by

$$\sum_{n=0}^{4} \frac{(-1)^n}{2^n n!} = 1 - \frac{1}{2} + \frac{1}{8} - \frac{1}{48} + \frac{1}{348} \approx 0.607.$$

(5 terms. Note that the sum begins with $n = 0$.)

39. $\sum_{n=0}^{\infty} \frac{(-1)^n}{(2n+1)!}$

(a) By Theorem 9.15,

$$|R_N| \le a_{N+1} = \frac{1}{[2(N+1)+1]!} < 0.001.$$

This inequality is valid when $N = 2$.

(b) We may approximate the series by

$$\sum_{n=0}^{2} \frac{(-1)^n}{(2n+1)!} = 1 - \frac{1}{6} + \frac{1}{120} \approx 0.842.$$

(3 terms. Note that the sum begins with $n = 0$.)

40. $\sum_{n=0}^{\infty} \frac{(-1)^n}{(2n)!}$

(a) By Theorem 9.15,

$$|R_N| \le a_{N+1} = \frac{1}{(2N+2)!} < 0.001.$$

This inequality is valid when $N = 3$.

(b) We may approximate the series by

$$\sum_{n=0}^{3} \frac{(-1)^n}{(2n)!} = 1 - \frac{1}{2} + \frac{1}{24} - \frac{1}{720} \approx 0.540.$$

(4 terms. Note that the sum begins with $n = 0$.)

41. $\sum_{n=1}^{\infty} \frac{(-1)^{n+1}}{n}$

(a) By Theorem 9.15,

$$|R_N| \le a_{N+1} = \frac{1}{N+1} < 0.001.$$

This inequality is valid when $N = 1000$.

(b) We may approximate the series by

$$\sum_{n=1}^{1000} \frac{(-1)^{n+1}}{n} = 1 - \frac{1}{2} + \frac{1}{3} - \frac{1}{4} + \cdots - \frac{1}{1000}$$

$$\approx 0.693.$$

(1000 terms)

42. $\sum_{n=1}^{\infty} \frac{(-1)^{n+1}}{4^n n}$

(a) By Theorem 9.15,

$$|R_N| \le a_{N+1} = \frac{1}{4^{N+1}(N+1)} < 0.001.$$

This inequality is valid when $N = 3$.

(b) We may approximate the series by

$$\sum_{n=1}^{3} \frac{(-1)^{n+1}}{4^n n} = \frac{1}{4} - \frac{1}{32} + \frac{1}{192} \approx 0.224.$$

(3 terms)

43. $\sum_{n=1}^{\infty} \frac{(-1)^{n+1}}{n^3}$

By Theorem 9.15,

$$|R_N| \le a_{N+1} = \frac{1}{(N+1)^3} < 0.001$$

$$\Rightarrow (N+1)^3 > 1000 \Rightarrow N + 1 > 10.$$

Use 10 terms.

44. $\sum_{n=1}^{\infty} \frac{(-1)^{n+1}}{n^2}$

By Theorem 9.15,

$$|R_N| \le a_{N+1} = \frac{1}{(N+1)^2} < 0.001$$

$$\Rightarrow (N+1)^2 > 1000 \Rightarrow N = 31.$$

Use 31 terms.

45. $\displaystyle\sum_{n=1}^{\infty} \frac{(-1)^{n+1}}{2n^3 - 1}$

By Theorem 9.15,

$$|R_N| \le a_{N+1} = \frac{1}{2(N+1)^3 - 1} < 0.001.$$

This inequality is valid when $N = 7$. Use 7 terms.

46. $\displaystyle\sum_{n=1}^{\infty} \frac{(-1)^{n+1}}{n^4}$

By Theorem 9.15, $|R_N| \le a_{N+1} = \dfrac{1}{(N+1)^4} < 0.001.$

This inequality is valid when $N = 5$.

47. $\displaystyle\sum_{n=1}^{\infty} \frac{(-1)^{n+1}}{(n+1)^2}$

$\displaystyle\sum_{n=1}^{\infty} \frac{1}{(n+1)^2}$ converges by comparison to the p-series

$\displaystyle\sum_{n=1}^{\infty} \frac{1}{n^2}.$

Therefore, the given series converges absolutely.

48. $\displaystyle\sum_{n=1}^{\infty} \frac{(-1)^{n+1}}{n+1}$

The given series converges by the Alternating Series Test, but does not converge absolutely since the series

$\displaystyle\sum_{n=1}^{\infty} \frac{1}{n+1}$

diverges by the Integral Test. Therefore, the series converges conditionally.

49. $\displaystyle\sum_{n=1}^{\infty} \frac{(-1)^{n+1}}{\sqrt{n}}$

The given series converges by the Alternating Series Test, but does not converge absolutely since

$\displaystyle\sum_{n=1}^{\infty} \frac{1}{\sqrt{n}}$

is a divergent p-series. Therefore, the series converges conditionally.

50. $\displaystyle\sum_{n=1}^{\infty} \frac{(-1)^{n+1}}{n\sqrt{n}}$

$\displaystyle\sum_{n=1}^{\infty} \frac{1}{n\sqrt{n}} = \sum_{n=1}^{\infty} \frac{1}{n^{3/2}}$ which is a convergent p-series.

Therefore, the given series converges absolutely.

51. $\displaystyle\sum_{n=1}^{\infty} \frac{(-1)^{n+1} n^2}{(n+1)^2}$

$$\lim_{n\to\infty} \frac{n^2}{(n+1)^2} = 1$$

Therefore, the series diverges by the nth-Term Test.

52. $\displaystyle\sum_{n=1}^{\infty} \frac{(-1)^{n+1}(2n+3)}{n+10}$

$$\lim_{n\to\infty} \frac{2n+3}{n+10} = 2$$

Therefore, the series diverges by the nth-Term Test.

53. $\displaystyle\sum_{n=2}^{\infty} \frac{(-1)^n}{\ln n}$

The given series converges by the Alternating Series Test but does not converge absolutely since the series

$\displaystyle\sum_{n=2}^{\infty} \frac{1}{\ln n}$ diverges by comparison to the harmonic series

$\displaystyle\sum_{n=1}^{\infty} \frac{1}{n}$. Therefore, the series converges conditionally.

54. $\displaystyle\sum_{n=0}^{\infty} \frac{(-1)^n}{e^{n^2}}$

$\displaystyle\sum_{n=0}^{\infty} \frac{1}{e^{n^2}}$ converges by a comparison to

the convergent geometric series $\displaystyle\sum_{n=0}^{\infty} \left(\frac{1}{e}\right)^n$. Therefore, the given series converges absolutely.

55. $\displaystyle\sum_{n=2}^{\infty} \frac{(-1)^n n}{n^3 - 1}$

$\displaystyle\sum_{n=2}^{\infty} \frac{n}{n^3 - 1}$ converges by a limit comparison to

the convergent p-series $\displaystyle\sum_{n=2}^{\infty} \frac{1}{n^2}$. Therefore, the given series converges absolutely.

56. $\displaystyle\sum_{n=1}^{\infty} \frac{(-1)^{n+1}}{n^{1.5}}$

$\displaystyle\sum_{n=1}^{\infty} \frac{1}{n^{1.5}}$ is a convergent p-series.

Therefore, the given series converges absolutely.

57. $\sum_{n=0}^{\infty} \frac{(-1)^n}{(2n+1)!}$

$$\sum_{n=0}^{\infty} \frac{1}{(2n+1)!}$$

is convergent by comparison to the convergent geometric series

$$\sum_{n=0}^{\infty} \left(\frac{1}{2}\right)^n$$

since

$$\frac{1}{(2n+1)!} < \frac{1}{2^n} \text{ for } n > 0.$$

Therefore, the given series converges absolutely.

58. $\sum_{n=0}^{\infty} \frac{(-1)^n}{\sqrt{n+4}}$

The given series converges by the Alternating Series Test, but

$$\sum_{n=0}^{\infty} \frac{1}{\sqrt{n+4}}$$

diverges by a limit comparison to the divergent p-series

$$\sum_{n=1}^{\infty} \frac{1}{\sqrt{n}}.$$

Therefore, the given series converges conditionally.

59. $\sum_{n=0}^{\infty} \frac{\cos n\pi}{n+1} = \sum_{n=0}^{\infty} \frac{(-1)^n}{n+1}$

The given series converges by the Alternating Series Test, but

$$\sum_{n=0}^{\infty} \frac{|\cos n\pi|}{n+1} = \sum_{n=0}^{\infty} \frac{1}{n+1}$$

diverges by a limit comparison to the divergent harmonic series,

$$\sum_{n=1}^{\infty} \frac{1}{n}.$$

$\lim_{n\to\infty} \frac{|\cos n\pi|/(n+1)}{1/n} = 1$, therefore the series converges conditionally.

60. $\sum_{n=1}^{\infty} (-1)^{n+1} \arctan n$

$$\lim_{n\to\infty} \arctan n = \frac{\pi}{2} \neq 0$$

Therefore, the series diverges by the nth-Term Test.

61. $\sum_{n=1}^{\infty} \frac{\cos n\pi}{n^2} = \sum_{n=1}^{\infty} \frac{(-1)^n}{n^2}$

$\sum_{n=1}^{\infty} \frac{1}{n^2}$ is a convergent p-series.

Therefore, the given series converges absolutely.

62. $\sum_{n=1}^{\infty} \frac{\sin[(2n-1)\pi/2]}{n} = \sum_{n=1}^{\infty} \frac{(-1)^{n+1}}{n}$

The given series converges by the Alternating Series Test, but

$$\sum_{n=1}^{\infty} \left|\frac{\sin[(2n-1)\pi/2]}{n}\right| = \sum_{n=1}^{\infty} \frac{1}{n}$$

is a divergent p-series. Therefore, the series converges conditionally.

63. An alternating series is a series whose terms alternate in sign. See Theorem 9.14.

64. $|S - S_n| = |R_n| \leq a_{n+1}$ (Theorem 9.15)

65. $\Sigma\, a_n$ is absolutely convergent if $\Sigma |a_n|$ converges. $\Sigma\, a_n$ is conditionally convergent if $\Sigma |a_n|$ diverges, but $\Sigma\, a_n$ converges.

66. (b). The partial sums alternate above and below the horizontal line representing the sum.

67. False. Let $a_n = \frac{(-1)^n}{n}$.

Then $\sum a_n$ converges and $\sum (-a_n)$ converges.

But, $\sum |a_n| = \sum \frac{1}{n}$ diverges.

68. True. If $\Sigma\, |a_n|$ converged, then so would $\Sigma\, a_n$ by Theorem 9.16.

69. True. $S_{100} = -1 + \frac{1}{2} - \frac{1}{3} + \cdots + \frac{1}{100}$

Since the next term $-\frac{1}{101}$ is negative, S_{100} is an overestimate of the sum.

70. False. Let

$$\sum a_n = \sum b_n = \sum \frac{(-1)^n}{\sqrt{n}}.$$

Then both converge by the Alternating Series Test. But,

$\sum a_n b_n = \sum \frac{1}{n}$, which diverges.

71. $\sum_{n=1}^{\infty} (-1)^n \frac{1}{n^p}$

If $p = 0$, then $\sum_{n=1}^{\infty} (-1)^n$ diverges.

If $p < 0$, then $\sum_{n=1}^{\infty} (-1)^n n^{-p}$ diverges.

If $p > 0$, then $\lim_{n\to\infty} \frac{1}{n^p} = 0$ and

$$a_{n+1} = \frac{1}{(n+1)^p} < \frac{1}{n^p} = a_n.$$

Therefore, the series converges for $p > 0$.

72. $\sum_{n=1}^{\infty} (-1)^n \frac{1}{n+p}$

Assume that $n + p \neq 0$ so that $a_n = 1/(n+p)$ are defined for all n. For all p,

$$\lim_{n\to\infty} a_n = \lim_{n\to\infty} \frac{1}{n+p} = 0$$

$$a_{n+1} = \frac{1}{n+1+p} < \frac{1}{n+p} < a_n.$$

Therefore, the series converges for all p.

73. Since

$$\sum_{n=1}^{\infty} |a_n|$$

converges we have $\lim_{n\to\infty} |a_n| = 0$. Thus, there must exist an $N > 0$ such that $|a_N| < 1$ for all $n > N$ and it follows that $a_n^2 \leq |a_n|$ for all $n > N$. Hence, by the Comparison Test,

$$\sum_{n=1}^{\infty} a_n^2$$

converges. Let $a_n = 1/n$ to see that the converse is false.

74. $\sum_{n=1}^{\infty} \frac{(-1)^{n-1}}{n}$ converges, but $\sum_{n=1}^{\infty} \frac{1}{n}$ diverges.

75. $\sum_{n=1}^{\infty} \frac{1}{n^2}$ converges, hence so does $\sum_{n=1}^{\infty} \frac{1}{n^4}$.

76. (a) $\sum_{n=1}^{\infty} \frac{x^n}{n}$

converges absolutely (by comparison) for $-1 < x < 1$, since

$$\left|\frac{x^n}{n}\right| < |x^n| \text{ and } \sum x^n$$

is a convergent geometric series for $-1 < x < 1$.

(b) When $x = -1$, we have the convergent alternating series

$$\sum_{n=1}^{\infty} \frac{(-1)^n}{n}.$$

When $x = 1$, we have the divergent harmonic series $1/n$. Therefore,

$\sum_{n=1}^{\infty} \frac{x^n}{n}$ converges conditionally for $x = -1$.

77. (a) No, the series does not satisfy $a_{n+1} \leq a_n$ for all n. For example, $\frac{1}{9} < \frac{1}{8}$.

(b) Yes, the series converges.

$$S_{2n} = \frac{1}{2} - \frac{1}{3} + \cdots + \frac{1}{2^n} - \frac{1}{3^n}$$

$$= \left(\frac{1}{2} + \cdots + \frac{1}{2^n}\right) - \left(\frac{1}{3} + \cdots + \frac{1}{3^n}\right)$$

$$= \left(1 + \frac{1}{2} + \cdots + \frac{1}{2^n}\right) - \left(1 + \frac{1}{3} + \cdots + \frac{1}{3^n}\right)$$

As $n \to \infty$,

$$S_{2n} \to \frac{1}{1-(1/2)} - \frac{1}{1-(1/3)} = 2 - \frac{3}{2} = \frac{1}{2}.$$

78. (a) No, the series does not satisfy $a_{n+1} \le a_n$:

$$\sum_{n=1}^{\infty}(-1)^{n+1}a_n = 1 - \frac{1}{8} + \frac{1}{\sqrt{3}} - \frac{1}{64} + \cdots \text{ and}$$

$$\frac{1}{8} < \frac{1}{\sqrt{3}}.$$

(b) No, the series diverges because $\sum \frac{1}{\sqrt{n}}$ diverges.

79. $\displaystyle\sum_{n=1}^{\infty} \frac{10}{n^{3/2}} = 10\sum_{n=1}^{\infty}\frac{1}{n^{3/2}},$

convergent p-series

80. $\displaystyle\sum_{n=1}^{\infty} \frac{3}{n^2+5}$

converges by limit comparison to convergent p-series

$\displaystyle\sum \frac{1}{n^2}.$

81. Diverges by nth-Term Test

$\displaystyle\lim_{n\to\infty} a_n = \infty$

82. Converges by limit comparison to convergent geometric series $\sum \frac{1}{2^n}$.

83. Convergent geometric series

$\left(r = \frac{7}{8} < 1\right)$

84. Diverges by nth-Term Test

$\displaystyle\lim_{n\to\infty} a_n = \frac{3}{2}$

85. Convergent geometric series $\left(r = 1/\sqrt{e}\right)$ or Integral Test

86. Converges (conditionally) by Alternating Series Test

87. Converges (absolutely) by Alternating Series Test

88. Diverges by comparison to Divergent Harmonic Series:

$\dfrac{\ln n}{n} > \dfrac{1}{n}$ for $n \ge 3$

89. The first term of the series is zero, not one. You cannot regroup series terms arbitrarily.

90. Rearranging the terms of an alternating series can change the sum.

91. $s = 1 - \frac{1}{2} + \frac{1}{3} - \frac{1}{4} + \cdots$

$$S = 1 + \frac{1}{3} - \frac{1}{2} + \frac{1}{5} + \frac{1}{7} - \frac{1}{4} + \frac{1}{9} + \frac{1}{11} - \frac{1}{6} + \cdots$$

(i) $s_{4n} = 1 - \frac{1}{2} + \frac{1}{3} - \frac{1}{4} + \frac{1}{5} - \frac{1}{6} + \cdots + \frac{1}{4n-1} - \frac{1}{4n}$

$$\frac{1}{2}s_{2n} = \frac{1}{2} - \frac{1}{4} + \frac{1}{6} - \frac{1}{8} + \frac{1}{10} - \cdots + \frac{1}{4n-2} - \frac{1}{4n}$$

Adding: $s_{4n} + \frac{1}{2}s_{2n} = 1 + \frac{1}{3} - \frac{1}{2} + \frac{1}{5} + \frac{1}{7} - \frac{1}{4} + \cdots + \frac{1}{4n-3} + \frac{1}{4n-1} - \frac{1}{2n} = S_{3n}$

(ii) $\displaystyle\lim_{n\to\infty} s_n = s$ (In fact, $s = \ln 2$.)

$s \ne 0$ since $s > \frac{1}{2}$.

$$S = \lim_{n\to\infty} S_{3n} = s_{4n} + \frac{1}{2}s_{2n} = s + \frac{1}{2}s = \frac{3}{2}s$$

Thus, $S \ne s$.

Section 9.6 The Ratio and Root Tests

1. $\dfrac{(n+1)!}{(n-2)!} = \dfrac{(n+1)(n)(n-1)(n-2)!}{(n-2)!}$

$= (n+1)(n)(n-1)$

2. $\dfrac{(2k-2)!}{(2k)!} = \dfrac{(2k-2)!}{(2k)(2k-1)(2k-2)!}$

$= \dfrac{1}{(2k)(2k-1)}$

3. Use the Principle of Mathematical Induction. When $k = 1$, the formula is valid since $1 = \dfrac{(2(1))!}{2^1 \cdot 1!}$. Assume that

$$1 \cdot 3 \cdot 5 \cdots (2n-1) = \frac{(2n)!}{2^n n!}$$

and show that

$$1 \cdot 3 \cdot 5 \cdots (2n-1)(2n+1) = \frac{(2n+2)!}{2^{n+1}(n+1)!}.$$

To do this, note that:

$$\begin{aligned}
1 \cdot 3 \cdot 5 \cdots (2n-1)(2n+1) &= [1 \cdot 3 \cdot 5 \cdots (2n-1)](2n+1) \\
&= \frac{(2n)!}{2^n n!} \cdot (2n+1) \quad \text{(Induction hypothesis)} \\
&= \frac{(2n)!(2n+1)}{2^n n!} \cdot \frac{(2n+2)}{2(n+1)} \\
&= \frac{(2n)!(2n+1)(2n+2)}{2^{n+1} n!(n+1)} \\
&= \frac{(2n+2)!}{2^{n+1}(n+1)!}
\end{aligned}$$

The formula is valid for all $n \geq 1$.

4. Use the Principle of Mathematical Induction. When $k = 3$, the formula is valid since $\dfrac{1}{1} = \dfrac{2^3 3!(3)(5)}{6!} = 1$. Assume that

$$\frac{1}{1 \cdot 3 \cdot 5 \cdots (2n-5)} = \frac{2^n n!(2n-3)(2n-1)}{(2n)!}$$

and show that

$$\frac{1}{1 \cdot 3 \cdot 5 \cdots (2n-5)(2n-3)} = \frac{2^{n+1}(n+1)!(2n-1)(2n+1)}{(2n+2)!}.$$

To do this, note that:

$$\begin{aligned}
\frac{1}{1 \cdot 3 \cdot 5 \cdots (2n-5)(2n-3)} &= \frac{1}{1 \cdot 3 \cdot 5 \cdots (2n-5)} \cdot \frac{1}{(2n-3)} \\
&= \frac{2^n n!\cancel{(2n-3)}(2n-1)}{(2n)!} \cdot \frac{1}{\cancel{(2n-3)}} \\
&= \frac{2^n n!(2n-1)}{(2n)!} \cdot \frac{(2n+1)(2n+2)}{(2n+1)(2n+2)} \\
&= \frac{2^n (2)(n+1)n!(2n-1)(2n+1)}{(2n)!(2n+1)(2n+2)} \\
&= \frac{2^{n+1}(n+1)!(2n-1)(2n+1)}{(2n+2)!}
\end{aligned}$$

The formula is valid for all $n \geq 3$.

5. $\sum_{n=1}^{\infty} n\left(\frac{3}{4}\right)^n = 1\left(\frac{3}{4}\right) + 2\left(\frac{9}{16}\right) + \cdots$

$S_1 = \frac{3}{4}, S_2 \approx 1.875$

Matches (d).

6. $\sum_{n=1}^{\infty} \left(\frac{3}{4}\right)^n\left(\frac{1}{n!}\right) = \frac{3}{4} + \frac{9}{16}\left(\frac{1}{2}\right) + \cdots$

$S_1 = \frac{3}{4}, S_2 \approx 1.03$

Matches (c).

7. $\sum_{n=1}^{\infty} \frac{(-3)^{n+1}}{n!} = 9 - \frac{3^3}{2} + \cdots$

$S_1 = 9$

Matches (f).

8. $\sum_{n=1}^{\infty} \frac{(-1)^{n-1} 4}{(2n)!} = \frac{4}{2} - \frac{4}{24} + \cdots$

$S_1 = 2$

Matches (b).

9. $\sum_{n=1}^{\infty} \left(\frac{4n}{5n-3}\right)^n = \frac{4}{2} + \left(\frac{8}{7}\right)^2 + \cdots$

$S_1 = 2, S_2 = 3.31$

Matches (a).

10. $\sum_{n=0}^{\infty} 4e^{-n} = 4 + \frac{4}{e} + \cdots$

$S_1 = 4$

Matches (e).

11. (a) Ratio Test: $\lim_{n\to\infty} \left|\frac{a_{n+1}}{a_n}\right| = \lim_{n\to\infty} \frac{(n+1)^2(5/8)^{n+1}}{n^2(5/8)^n} = \lim_{n\to\infty} \left(\frac{n+1}{n}\right)^2 \frac{5}{8} = \frac{5}{8} < 1.$ Converges

(b)

n	5	10	15	20	25
S_n	9.2104	16.7598	18.8016	19.1878	19.2491

(c) 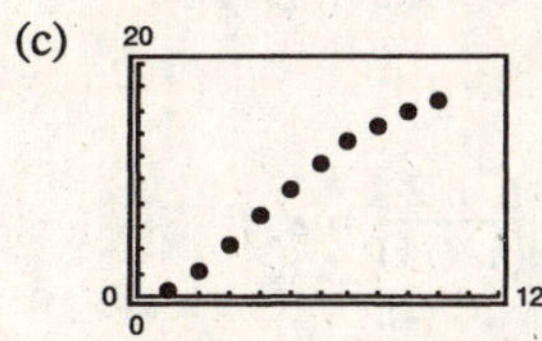

(d) The sum is approximately 19.26.

(e) The more rapidly the terms of the series approach 0, the more rapidly the sequence of the partial sums approaches the sum of the series.

12. (a) Ratio Test: $\lim_{n\to\infty} \left|\frac{a_{n+1}}{a_n}\right| = \lim_{n\to\infty} \frac{\frac{(n+1)^2+1}{(n+1)!}}{\frac{n^2+1}{n!}} = \lim_{n\to\infty} \left(\frac{n^2+2n+2}{n^2+1}\right)\left(\frac{1}{n+1}\right) = 0 < 1.$ Converges

(b)

n	5	10	15	20	25
S_n	7.0917	7.1548	7.1548	7.1548	7.1548

(c) 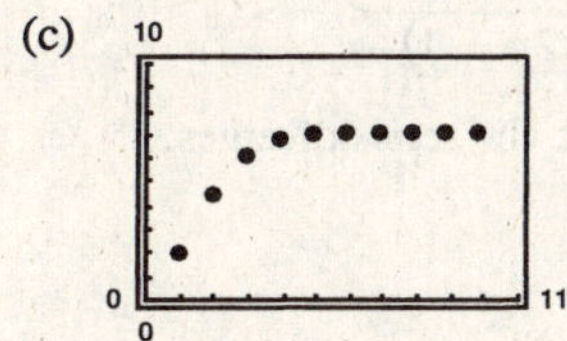

(d) The sum is approximately 7.15485

(e) The more rapidly the terms of the series approach 0, the more rapidly the sequence of the partial sums approaches the sum of the series.

13. $\sum_{n=0}^{\infty} \frac{n!}{3^n}$

$$\lim_{n\to\infty} \left|\frac{a_{n+1}}{a_n}\right| = \lim_{n\to\infty} \left|\frac{(n+1)!}{3^{n+1}} \cdot \frac{3^n}{n!}\right|$$

$$= \lim_{n\to\infty} \frac{n+1}{3} = \infty$$

Therefore, by the Ratio Test, the series diverges.

14. $\sum_{n=0}^{\infty} \frac{3^n}{n!}$

$$\lim_{n\to\infty} \left|\frac{a_{n+1}}{a_n}\right| = \lim_{n\to\infty} \left|\frac{3^{n+1}}{(n+1)!} \cdot \frac{n!}{3^n}\right|$$

$$= \lim_{n\to\infty} \frac{3}{n+1} = 0$$

Therefore, by the Ratio Test, the series converges.

15. $\sum_{n=1}^{\infty} n\left(\frac{3}{4}\right)^n$

$$\lim_{n\to\infty}\left|\frac{a_{n+1}}{a_n}\right| = \lim_{n\to\infty}\left|\frac{(n+1)(3/4)^{n+1}}{n(3/4)^n}\right| = \lim_{n\to\infty}\left|\frac{3(n+1)}{4n}\right| = \frac{3}{4}$$

Therefore, by the Ratio Test, the series converges.

16. $\sum_{n=1}^{\infty} n\left(\frac{3}{2}\right)^n$

$$\lim_{n\to\infty}\left|\frac{a_{n+1}}{a_n}\right| = \lim_{n\to\infty}\left|\frac{(n+1)3^{n+1}}{2^{n+1}}\cdot\frac{2^n}{n3^n}\right| = \lim_{n\to\infty}\frac{3(n+1)}{2n} = \frac{3}{2}$$

Therefore, by the Ratio Test, the series diverges.

17. $\sum_{n=1}^{\infty} \frac{n}{2^n}$

$$\lim_{n\to\infty}\left|\frac{a_{n+1}}{a_n}\right| = \lim_{n\to\infty}\left|\frac{n+1}{2^{n+1}}\cdot\frac{2^n}{n}\right| = \lim_{n\to\infty}\frac{n+1}{2n} = \frac{1}{2}$$

Therefore, by the Ratio Test, the series converges.

18. $\sum_{n=1}^{\infty} \frac{n^3}{2^n}$

$$\lim_{n\to\infty}\left|\frac{a_n+1}{a_n}\right| = \lim_{n\to\infty}\left|\frac{(n+1)^3/2^{n+1}}{n^3/2^n}\right| = \lim_{n\to\infty}\left|\frac{(n+1)^3}{2n^3}\right| = \frac{1}{2}$$

Therefore, by the Ratio Test, the series converges.

19. $\sum_{n=1}^{\infty} \frac{2^n}{n^2}$

$$\lim_{n\to\infty}\left|\frac{a_{n+1}}{a_n}\right| = \lim_{n\to\infty}\left|\frac{2^{n+1}}{(n+1)^2}\cdot\frac{n^2}{2^n}\right| = \lim_{n\to\infty}\frac{2n^2}{(n+1)^2} = 2$$

Therefore, by the Ratio Test, the series diverges.

20. $\sum_{n=1}^{\infty} \frac{(-1)^{n+1}(n+2)}{n(n+1)}$

$$a_{n+1} = \frac{n+3}{(n+1)(n+2)} \le \frac{n+2}{n(n+1)} = a_n$$

$$\lim_{n\to\infty}\frac{n+2}{n(n+1)} = 0$$

Therefore, by Theorem 9.14, the series converges.

Note: The Ratio Test is inconclusive since $\lim_{n\to\infty}\left|\frac{a_{n+1}}{a_n}\right| = 1.$

The series converges conditionally.

21. $\sum_{n=0}^{\infty} \frac{(-1)^n 2^n}{n!}$

$$\lim_{n\to\infty}\left|\frac{a_{n+1}}{a_n}\right| = \lim_{n\to\infty}\left|\frac{2^{n+1}}{(n+1)!}\cdot\frac{n!}{2^n}\right| = \lim_{n\to\infty}\frac{2}{n+1} = 0$$

Therefore, by the Ratio Test, the series converges.

22. $\sum_{n=1}^{\infty} \frac{(-1)^{n-1}(3/2)^n}{n^2}$

$$\lim_{n\to\infty}\left|\frac{a_{n+1}}{a_n}\right| = \lim_{n\to\infty}\left|\frac{(3/2)^{n+1}}{n^2+2n+1}\cdot\frac{n^2}{(3/2)^n}\right| = \lim_{n\to\infty}\frac{3n^2}{2(n^2+2n+1)} = \frac{3}{2} > 1$$

Therefore, by the Ratio Test, the series diverges.

23. $\sum_{n=1}^{\infty} \frac{n!}{n3^n}$

$$\lim_{n\to\infty}\left|\frac{a_{n+1}}{a_n}\right| = \lim_{n\to\infty}\left|\frac{(n+1)!}{(n+1)3^{n+1}}\cdot\frac{n3^n}{n!}\right| = \lim_{n\to\infty}\frac{n}{3} = \infty$$

Therefore, by the Ratio Test, the series diverges.

24. $\sum_{n=1}^{\infty} \frac{(2n)!}{n^5}$

$$\lim_{n\to\infty}\left|\frac{a_{n+1}}{a_n}\right| = \lim_{n\to\infty}\left|\frac{(2n+2)!}{(n+1)^5}\cdot\frac{n^5}{(2n)!}\right| = \lim_{n\to\infty}\frac{(2n+2)(2n+1)n^5}{(n+1)^5} = \infty$$

Therefore, by the Ratio Test, the series diverges.

25. $\sum_{n=0}^{\infty} \frac{4^n}{n!}$

$$\lim_{n\to\infty}\left|\frac{a_{n+1}}{a_n}\right| = \lim_{n\to\infty}\left|\frac{4^{n+1}}{(n+1)!}\cdot\frac{n!}{4^n}\right|$$

$$= \lim_{n\to\infty}\frac{4}{n+1} = 0$$

Therefore, by the Ratio Test, the series converges.

26. $\sum_{n=1}^{\infty} \frac{n^n}{n!}$

$$\lim_{n\to\infty}\left|\frac{a_{n+1}}{a_n}\right| = \lim_{n\to\infty}\left|\frac{(n+1)^{n+1}}{(n+1)!}\cdot\frac{n!}{n^n}\right|$$

$$= \lim_{n\to\infty}\frac{(n+1)(n+1)^n n!}{(n+1)n!n^n}$$

$$= \lim_{n\to\infty}\left(\frac{n+1}{n}\right)^n = e > 1$$

Therefore, by the Ratio Test, the series diverges.

27. $\sum_{n=0}^{\infty} \frac{3^n}{(n+1)^n}$

$$\lim_{n\to\infty}\left|\frac{a_{n+1}}{a_n}\right| = \lim_{n\to\infty}\left|\frac{3^{n+1}}{(n+2)^{n+1}}\cdot\frac{(n+1)^n}{3^n}\right| = \lim_{n\to\infty}\frac{3(n+1)^n}{(n+2)^{n+1}} = \lim_{n\to\infty}\frac{3}{n+2}\left(\frac{n+1}{n+2}\right)^n = (0)\left(\frac{1}{e}\right) = 0$$

To find $\lim_{n\to\infty}\left(\frac{n+1}{n+2}\right)^n$, let $y = \lim_{n\to\infty}\left(\frac{n+1}{n+2}\right)^n$. Then,

$$\ln y = \lim_{n\to\infty} n\ln\left(\frac{n+1}{n+2}\right) = \lim_{n\to\infty}\frac{\ln[(n+1)/(n+2)]}{1/n} = \frac{0}{0}$$

$$\ln y = \lim_{n\to\infty}\frac{[(1)/(n+1)] - [(1)/(n+2)]}{-(1/n^2)} = -1 \text{ by L'Hôpital's Rule}$$

$$y = e^{-1} = \frac{1}{e}.$$

Therefore, by the Ratio Test, the series converges.

28. $\sum_{n=0}^{\infty} \frac{(n!)^2}{(3n)!}$

$$\lim_{n\to\infty}\left|\frac{a_{n+1}}{a_n}\right| = \lim_{n\to\infty}\left|\frac{[(n+1)!]^2}{(3n+3)!}\cdot\frac{(3n)!}{(n!)^2}\right|$$

$$= \lim_{n\to\infty}\frac{(n+1)^2}{(3n+3)(3n+2)(3n+1)} = 0$$

Therefore, by the Ratio Test, the series converges.

29. $\sum_{n=0}^{\infty} \frac{4^n}{3^n+1}$

$$\lim_{n\to\infty}\left|\frac{a_{n+1}}{a_n}\right| = \lim_{n\to\infty}\left|\frac{4^{n+1}}{3^{n+1}+1}\cdot\frac{3^n+1}{4^n}\right|$$

$$= \lim_{n\to\infty}\frac{4(3^n+1)}{3^{n+1}+1}$$

$$= \lim_{n\to\infty}\frac{4(1+1/3^n)}{3+1/3^n} = \frac{4}{3}$$

Therefore, by the Ratio Test, the series diverges.

30. $\sum_{n=0}^{\infty} \frac{(-1)^n 2^{4n}}{(2n+1)!}$

$$\lim_{n\to\infty}\left|\frac{a_{n+1}}{a_n}\right| = \lim_{n\to\infty}\left|\frac{2^{4n+4}}{(2n+3)!}\cdot\frac{(2n+1)!}{2^{4n}}\right| = \lim_{n\to\infty}\frac{2^4}{(2n+3)(2n+2)} = 0$$

Therefore, by the Ratio Test, the series converges.

31. $\sum_{n=0}^{\infty} \frac{(-1)^{n+1}n!}{1\cdot 3\cdot 5\cdots(2n+1)}$

$$\lim_{n\to\infty}\left|\frac{a_{n+1}}{a_n}\right| = \lim_{n\to\infty}\left|\frac{(n+1)!}{1\cdot 3\cdot 5\cdots(2n+1)(2n+3)}\cdot\frac{1\cdot 3\cdot 5\cdots(2n+1)}{n!}\right| = \lim_{n\to\infty}\frac{n+1}{2n+3} = \frac{1}{2}$$

Therefore, by the Ratio Test, the series converges.

Note: The first few terms of this series are $-1 + \frac{1}{1\cdot 3} - \frac{2!}{1\cdot 3\cdot 5} + \frac{3!}{1\cdot 3\cdot 5\cdot 7} - \cdots$.

32. $\sum_{n=1}^{\infty} \frac{(-1)^n 2 \cdot 4 \cdot 6 \cdots 2n}{2 \cdot 5 \cdot 8 \cdots (3n-1)}$

$$\lim_{n\to\infty} \left|\frac{a_{n+1}}{a_n}\right| = \lim_{n\to\infty} \left|\frac{2 \cdot 4 \cdots 2n(2n+2)}{2 \cdot 5 \cdots (3n-1)(3n+2)} \cdot \frac{2 \cdot 5 \cdots (3n-1)}{2 \cdot 4 \cdots 2n}\right| = \lim_{n\to\infty} \frac{2n+2}{3n+2} = \frac{2}{3}$$

Therefore, by the Ratio Test, the series converges.

Note: The first few terms of this series are $-\frac{2}{2} + \frac{2 \cdot 4}{2 \cdot 5} - \frac{2 \cdot 4 \cdot 6}{2 \cdot 5 \cdot 8} + \cdots$.

33. $\sum_{n=1}^{\infty} \frac{1}{n^{3/2}}$

$$\lim_{n\to\infty} \left|\frac{a_{n+1}}{a_n}\right| = \lim_{n\to\infty} \left|\frac{1}{(n+1)^{3/2}} \cdot \frac{n^{3/2}}{1}\right|$$

$$= \lim_{n\to\infty} \left(\frac{n}{n+1}\right)^{3/2} = 1$$

Ratio Test is inconclusive.

34. $\sum_{n=1}^{\infty} \frac{1}{n^{1/2}}$

$$\lim_{n\to\infty} \left|\frac{a_{n+1}}{a_n}\right| = \lim_{n\to\infty} \left|\frac{1}{(n+1)^{1/2}} \cdot \frac{n^{1/2}}{1}\right|$$

$$= \lim_{n\to\infty} \left(\frac{n}{n+1}\right)^{1/2} = 1$$

Ratio Test is inconclusive.

35. $\sum_{n=1}^{\infty} \frac{1}{n^4}$

$$\lim_{n\to\infty} \left|\frac{a_{n+1}}{a_n}\right| = \lim_{n\to\infty} \left|\frac{1}{(n+1)^4} \cdot \frac{n^4}{1}\right| = \lim_{n\to\infty} \left(\frac{n}{n+1}\right)^4 = 1$$

36. $\sum_{n=1}^{\infty} \frac{1}{n^p}$

$$\lim_{n\to\infty} \left|\frac{a_{n+1}}{a_n}\right| = \lim_{n\to\infty} \left|\frac{1}{(n+1)^p} \cdot \frac{n^p}{1}\right| = \lim_{n\to\infty} \left(\frac{n}{n+1}\right)^p = 1$$

37. $\sum_{n=1}^{\infty} \left(\frac{n}{2n+1}\right)^n$

$$\lim_{n\to\infty} \sqrt[n]{|a_n|} = \lim_{n\to\infty} \sqrt[n]{\left(\frac{n}{2n+1}\right)^n}$$

$$= \lim_{n\to\infty} \frac{n}{2n+1} = \frac{1}{2}$$

Therefore, by the Root Test, the series converges.

38. $\sum_{n=1}^{\infty} \left(\frac{2n}{n+1}\right)^n$

$$\lim_{n\to\infty} \sqrt[n]{|a_n|} = \lim_{n\to\infty} \sqrt[n]{\left(\frac{2n}{n+1}\right)^n}$$

$$= \lim_{n\to\infty} \frac{2n}{n+1} = 2$$

Therefore, by the Root Test, the series diverges.

39. $\sum_{n=2}^{\infty} \left(\frac{2n+1}{n-1}\right)^n$

$$\lim_{n\to\infty} \sqrt[n]{|a_n|} = \lim_{n\to\infty} \sqrt[n]{\left(\frac{2n+1}{n-1}\right)^n}$$

$$= \lim_{n\to\infty} \left(\frac{2n+1}{n-1}\right) = 2$$

Since $2 > 1$, the series diverges.

40. $\sum_{n=1}^{\infty} \left(\frac{4n+3}{2n-1}\right)^n$

$$\lim_{n\to\infty} \sqrt[n]{|a_n|} = \lim_{n\to\infty} \sqrt[n]{\left(\frac{4n+3}{2n-1}\right)^n}$$

$$= \lim_{n\to\infty} \frac{4n+3}{2n-1} = 2$$

Since $2 > 1$, the series diverges.

41. $\sum_{n=2}^{\infty} \frac{(-1)^n}{(\ln n)^n}$

$$\lim_{n\to\infty} \sqrt[n]{|a_n|} = \lim_{n\to\infty} \sqrt[n]{\left|\frac{(-1)^n}{(\ln n)^n}\right|}$$

$$= \lim_{n\to\infty} \frac{1}{|\ln n|} = 0$$

Therefore, by the Root Test, the series converges.

42. $\sum_{n=1}^{\infty} \left(\frac{-3n}{2n+1}\right)^{3n}$

$$\lim_{n\to\infty} \sqrt[n]{|a_n|} = \lim_{n\to\infty} \sqrt[n]{\left|\left(\frac{-3n}{2n+1}\right)^{3n}\right|}$$

$$= \lim_{n\to\infty} \left(\frac{3n}{2n+1}\right)^3 = \left(\frac{3}{2}\right)^3 = \frac{27}{8}$$

Therefore, by the Root Test, the series diverges.

43. $\sum_{n=1}^{\infty} (2\sqrt[n]{n} + 1)^n$

$\lim_{n\to\infty} \sqrt[n]{|a_n|} = \lim_{n\to\infty} \sqrt[n]{(2\sqrt[n]{n} + 1)^n} = \lim_{n\to\infty} (2\sqrt[n]{n} + 1)$

To find $\lim_{n\to\infty} \sqrt[n]{n}$, let $y = \lim_{n\to\infty} \sqrt[n]{n}$. Then

$$\ln y = \lim_{n\to\infty} \left(\ln \sqrt[n]{n}\right) = \lim_{n\to\infty} \frac{1}{n} \ln n = \lim_{n\to\infty} \frac{\ln n}{n} = \lim_{n\to\infty} \frac{1/n}{1} = 0.$$

Thus, $\ln y = 0$, so $y = e^0 = 1$ and $\lim_{n\to\infty} \left(2\sqrt[n]{n} + 1\right) = 2(1) + 1 = 3$. Therefore, by the Root Test, the series diverges.

44. $\sum_{n=0}^{\infty} e^{-n}$

$\lim_{n\to\infty} \sqrt[n]{|a_n|} = \lim_{n\to\infty} \sqrt[n]{\frac{1}{e^n}} = \frac{1}{e}$

Therefore, by the Root Test, the series converges.

45. $\sum_{n=1}^{\infty} \frac{n}{4^n}$

$\lim_{n\to\infty} \sqrt[n]{|a_n|} = \lim_{n\to\infty} \sqrt[n]{\frac{n}{4^n}}$

$= \lim_{n\to\infty} \frac{n^{1/n}}{4} = \frac{1}{4}$

Since $\frac{1}{4} < 1$, the series converges.

Note: You can use L'Hôpital's Rule to show $\lim_{n\to\infty} n^{1/n} = 1$.

Let $y = n^{1/n} \Rightarrow \ln y = \frac{\ln n}{n}$

$\lim_{n\to\infty} \frac{\ln n}{n} = \lim_{n\to\infty} \frac{1/n}{1} = 0.$

Hence, $\ln y \to 0 \Rightarrow y = n^{1/n} \to 1$.

46. $\sum_{n=1}^{\infty} \left(\frac{n}{500}\right)^n$

$\lim_{n\to\infty} \sqrt[n]{|a_n|} = \lim_{n\to\infty} \sqrt[n]{\left(\frac{n}{500}\right)^n}$

$= \lim_{n\to\infty} \left(\frac{n}{500}\right) = \infty$

The series diverges.

47. $\sum_{n=1}^{\infty} \left(\frac{1}{n} - \frac{1}{n^2}\right)^n$

$\lim_{n\to\infty} \sqrt[n]{|a_n|} = \lim_{n\to\infty} \sqrt[n]{\left(\frac{1}{n} - \frac{1}{n^2}\right)^n}$

$= \lim_{n\to\infty} \left(\frac{1}{n} - \frac{1}{n^2}\right) = 0 - 0 = 0 < 1$

Hence, the series converges.

48. $\sum_{n=1}^{\infty} \left(\frac{\ln n}{n}\right)^n$

$\lim_{n\to\infty} \sqrt[n]{|a_n|} = \lim_{n\to\infty} \sqrt[n]{\left(\frac{\ln n}{n}\right)^n}$

$= \lim_{n\to\infty} \frac{\ln n}{n} = 0 < 1$

By the Root Test, the series converges.

49. $\sum_{n=2}^{\infty} \frac{n}{(\ln n)^n}$

$\lim_{n\to\infty} \sqrt[n]{|a_n|} = \lim_{n\to\infty} \sqrt[n]{\frac{n}{(\ln n)^n}}$

$= \lim_{n\to\infty} \frac{n^{1/n}}{\ln n} = 0$

Since $0 < 1$, the series converges by the Root Test.

50. $\sum_{n=1}^{\infty} \frac{(n!)^n}{(n^n)^2} = \sum_{n=1}^{\infty} \frac{(n!)^n}{(n^2)^n}$

$\lim_{n\to\infty} \sqrt[n]{|a_n|} = \lim_{n\to\infty} \sqrt[n]{\frac{(n!)}{(n^2)^n}}$

$= \lim_{n\to\infty} \frac{n!}{n^2} = \infty$

The series diverges.

51. $\sum_{n=1}^{\infty} \frac{(-1)^{n+1} 5}{n}$

$a_{n+1} = \frac{5}{n+1} < \frac{5}{n} = a_n$

$\lim_{n\to\infty} \frac{5}{n} = 0$

Therefore, by the Alternating Series Test, the series converges (conditional convergence).

52. $\sum_{n=1}^{\infty} \frac{5}{n} = 5\sum_{n=1}^{\infty} \frac{1}{n}$

This is the divergent harmonic series.

53. $\sum_{n=1}^{\infty} \frac{3}{n\sqrt{n}} = 3\sum_{n=1}^{\infty} \frac{1}{n^{3/2}}$

This is convergent p-series.

54. $\sum_{n=1}^{\infty} \left(\frac{\pi}{4}\right)^n$

Since $\pi/4 < 1$,this is convergent geometric series.

55. $\sum_{n=1}^{\infty} \frac{2n}{n+1}$

$$\lim_{n\to\infty} \frac{2n}{n+1} = 2 \neq 0$$

This diverges by the nth-Term Test for Divergence.

56. $\sum_{n=1}^{\infty} \frac{n}{2n^2+1}$

$$\lim_{n\to\infty} \frac{n/(2n^2+1)}{1/n} = \lim_{n\to\infty} \frac{n^2}{2n^2+1} = \frac{1}{2} > 0$$

This series diverges by limit comparison to the divergent harmonic series

$$\sum_{n=1}^{\infty} \frac{1}{n}.$$

57. $\sum_{n=1}^{\infty} \frac{(-1)^n 3^{n-2}}{2^n} = \sum_{n=1}^{\infty} \frac{(-1)^n 3^n 3^{-2}}{2^n} = \sum_{n=1}^{\infty} \frac{1}{9}\left(-\frac{3}{2}\right)^n$

Since $|r| = \frac{3}{2} > 1$, this is a divergent geometric series.

58. $\sum_{n=1}^{\infty} \frac{10}{3\sqrt{n^3}}$

$$\lim_{n\to\infty} \frac{10/3n^{3/2}}{1/n^{3/2}} = \frac{10}{3}$$

Therefore, the series converges by a Limit Comparison Test with the p-series

$$\sum_{n=1}^{\infty} \frac{1}{n^{3/2}}.$$

59. $\sum_{n=1}^{\infty} \frac{10n+3}{n2^n}$

$$\lim_{n\to\infty} \frac{(10n+3)/n2^n}{1/2^n} = \lim_{n\to\infty} \frac{10n+3}{n} = 10$$

Therefore, the series converges by a Limit Comparison Test with the geometric series

$$\sum_{n=0}^{\infty} \left(\frac{1}{2}\right)^n.$$

60. $\sum_{n=1}^{\infty} \frac{2^n}{4n^2-1}$

$$\lim_{n\to\infty} \frac{2^n}{4n^2-1} = \lim_{n\to\infty} \frac{(\ln 2)2^n}{8n} = \lim_{n\to\infty} \frac{(\ln 2)^2 2^n}{8} = \infty$$

Therefore, the series diverges by the nth-Term Test for Divergence.

61. $\sum_{n=1}^{\infty} \frac{\cos n}{2^n}$

$$\left|\frac{\cos n}{2^n}\right| \leq \frac{1}{2^n}$$

Therefore, the series

$$\sum_{n=1}^{\infty} \left|\frac{\cos n}{2^n}\right|$$

converges by comparison with the geometric series

$$\sum_{n=0}^{\infty} \left(\frac{1}{2}\right)^n.$$

62. $\sum_{n=2}^{\infty} \frac{(-1)^n}{n \ln n}$

$$a_{n+1} = \frac{1}{(n+1)\ln(n+1)} \leq \frac{1}{n\ln(n)} = a_n$$

$$\lim_{n\to\infty} \frac{1}{n\ln(n)} = 0$$

Therefore, by the Alternating Series Test, the series converges.

63. $\sum_{n=1}^{\infty} \frac{n7^n}{n!}$

$$\lim_{n\to\infty} \left|\frac{a_{n+1}}{a_n}\right| = \lim_{n\to\infty} \left|\frac{(n+1)7^{n+1}}{(n+1)!} \cdot \frac{n!}{n7^n}\right| = \lim_{n\to\infty} \frac{7}{n} = 0$$

Therefore, by the Ratio Test, the series converges.

64. $\sum_{n=1}^{\infty} \frac{\ln(n)}{n^2}$

$$\frac{\ln(n)}{n^2} \leq \frac{1}{n^{3/2}}$$

Therefore, the series converges by comparison with the p-series

$$\sum_{n=1}^{\infty} \frac{1}{n^{3/2}}.$$

65. $\sum_{n=1}^{\infty} \frac{(-1)^n 3^{n-1}}{n!}$

$$\lim_{n\to\infty} \left|\frac{a_{n+1}}{a_n}\right| = \lim_{n\to\infty} \left|\frac{3^n}{(n+1)!} \cdot \frac{n!}{3^{n-1}}\right| = \lim_{n\to\infty} \frac{3}{n+1} = 0$$

Therefore, by the Ratio Test, the series converges.

(Absolutely)

66. $\sum_{n=1}^{\infty} \frac{(-1)^n 3^n}{n2^n}$

$$\lim_{n\to\infty} \left|\frac{a_{n+1}}{a_n}\right| = \lim_{n\to\infty} \left|\frac{3^{n+1}}{(n+1)2^{n+1}} \cdot \frac{n2^n}{3^n}\right| = \lim_{n\to\infty} \frac{3n}{2(n+1)} = \frac{3}{2}$$

Therefore, by the Ratio Test, the series diverges.

67. $\sum_{n=1}^{\infty} \frac{(-3)^n}{3 \cdot 5 \cdot 7 \cdots (2n+1)}$

$$\lim_{n\to\infty} \left|\frac{a_{n+1}}{a_n}\right| = \lim_{n\to\infty} \left|\frac{(-3)^{n+1}}{3 \cdot 5 \cdot 7 \cdots (2n+1)(2n+3)} \cdot \frac{3 \cdot 5 \cdot 7 \cdots (2n+1)}{(-3)^n}\right| = \lim_{n\to\infty} \frac{3}{2n+3} = 0$$

Therefore, by the Ratio Test, the series converges.

68. $\sum_{n=1}^{\infty} \frac{3 \cdot 5 \cdot 7 \cdots (2n+1)}{18^n (2n-1)n!}$

$$\lim_{n\to\infty} \left|\frac{a_{n+1}}{a_n}\right| = \lim_{n\to\infty} \left|\frac{3 \cdot 5 \cdot 7 \cdots (2n+1)(2n+3)}{18^{n+1}(2n+1)(2n-1)n!} \cdot \frac{18^n (2n-1)n!}{3 \cdot 5 \cdot 7 \cdots (2n+1)}\right| = \lim_{n\to\infty} \frac{(2n+3)(2n-1)}{18(2n+1)(2n-1)} = \frac{1}{18}$$

Therefore, by the Ratio Test, the series converge.

69. (a) and (c)

$$\sum_{n=1}^{\infty} \frac{n5^n}{n!} = \sum_{n=0}^{\infty} \frac{(n+1)5^{n+1}}{(n+1)!}$$

$$= 5 + \frac{(2)(5)^2}{2!} + \frac{(3)(5)^3}{3!} + \frac{(4)(5)^4}{4!} + \cdots$$

70. (b) and (c)

$$\sum_{n=0}^{\infty} (n+1)\left(\frac{3}{4}\right)^n = \sum_{n=1}^{\infty} n\left(\frac{3}{4}\right)^{n-1}$$

$$= 1 + 2\left(\frac{3}{4}\right) + 3\left(\frac{3}{4}\right)^2 + 4\left(\frac{3}{4}\right)^3 + \cdots$$

71. (a) and (b) are the same.

72. (a) and (b) are the same.

$$\sum_{n=2}^{\infty} \frac{(-1)^n}{(n-1)2^{n-1}} = \frac{1}{2} - \frac{1}{2 \cdot 2^2} + \frac{1}{3 \cdot 2^3} - \cdots$$

$$\sum_{n=1}^{\infty} \frac{(-1)^{n+1}}{n2^n} = \frac{1}{2} - \frac{1}{2 \cdot 2^2} + \frac{1}{3 \cdot 2^3} - \cdots$$

73. Replace n with $n + 1$.

$$\sum_{n=1}^{\infty} \frac{n}{4^n} = \sum_{n=0}^{\infty} \frac{n+1}{4^{n+1}}$$

74. Replace n with $n + 2$.

$$\sum_{n=2}^{\infty} \frac{2^n}{(n-2)!} = \sum_{n=0}^{\infty} \frac{2^{n+2}}{n!}$$

75. Since

$$\frac{3^{10}}{2^{10}\,10!} \approx 1.59 \times 10^{-5},$$

use 9 terms.

$$\sum_{k=1}^{9} \frac{(-3)^k}{2^k\,k!} \approx -0.7769$$

76. $$\sum_{k=0}^{\infty} \frac{(-3)^k}{1 \cdot 3 \cdot 5 \ldots (2k+1)} = \sum_{k=0}^{\infty} \frac{(-3)^k 2^k k!}{(2k)!(2k+1)}$$

$$= \sum_{k=0}^{\infty} \frac{(-6)^k k!}{(2k+1)!}$$

$$\approx 0.40967$$

(See Exercise 3 and use 10 terms, $k = 9$.)

77. $$\lim_{n\to\infty} \left|\frac{a_{n+1}}{a_n}\right| = \lim_{n\to\infty} \left|\frac{(4n-1)/(3n+2)a_n}{a_n}\right|$$

$$= \lim_{n\to\infty} \frac{4n-1}{3n+2} = \frac{4}{3} > 1$$

The series diverges by the Ratio Test.

78. $$\lim_{n\to\infty} \left|\frac{a_{n+1}}{a_n}\right| = \lim_{n\to\infty} \left|\frac{(2n+1)/(5n-4)a_n}{a_n}\right|$$

$$= \lim_{n\to\infty} \frac{2n+1}{5n-4} = \frac{2}{5} < 1$$

The series converges by the Ratio Test.

79. $$\lim_{n\to\infty} \left|\frac{a_{n+1}}{a_n}\right| = \lim_{n\to\infty} \left|\frac{(\sin n+1)/(\sqrt{n})a_n}{a_n}\right|$$

$$= \lim_{n\to\infty} \frac{\sin n + 1}{\sqrt{n}} = 0 < 1$$

The series converges by the Ratio Test.

80. $$\lim_{n\to\infty} \left|\frac{a_{n+1}}{a_n}\right| = \lim_{n\to\infty} \left|\frac{(\cos n+1)/(n)a_n}{a_n}\right|$$

$$= \lim_{n\to\infty} \frac{\cos n + 1}{n} = 0 < 1$$

The series converges by the Ratio Test.

81. $$\lim_{n\to\infty} \left|\frac{a_{n+1}}{a_n}\right| = \lim_{n\to\infty} \left|\frac{(1+(1)/(n))a_n}{a_n}\right|$$

$$= \lim_{n\to\infty} \left(1 + \frac{1}{n}\right) = 1$$

The Ratio Test is inconclusive.

But, $\lim_{n\to\infty} a_n \neq 0$, so the series diverges.

82. The series diverges because $\lim_{n\to\infty} a_n \neq 0$.

$$a_1 = \frac{1}{4}$$

$$a_2 = \left(\frac{1}{4}\right)^{1/2} = \frac{1}{2}$$

$$a_3 = \left(\frac{1}{2}\right)^{1/3} \approx 0.7937$$

In general, $a_{n+1} > a_n > 0$.

83. $$\lim_{n\to\infty} \left|\frac{a_{n+1}}{a_n}\right| = \lim_{n\to\infty} \left|\frac{\dfrac{1 \cdot 2 \cdots n(n+1)}{1 \cdot 3 \cdots (2n-1)(2n+1)}}{\dfrac{1 \cdot 2 \cdots n}{1 \cdot 3 \cdots (2n-1)}}\right|$$

$$= \lim_{n\to\infty} \frac{n+1}{2n+1} = \frac{1}{2} < 1$$

The series converges by the Ratio Test.

84. $$\sum_{n=0}^{\infty} \frac{n+1}{3^n}$$

$$\lim_{n\to\infty} \sqrt[n]{|a_n|} = \lim_{n\to\infty} \sqrt[n]{\frac{n+1}{3^n}} = \lim_{n\to\infty} \frac{\sqrt[n]{n+1}}{3}$$

Let $y = \lim_{n\to\infty} \sqrt[n]{n+1}$

$$\ln y = \lim_{n\to\infty} \left(\ln \sqrt[n]{n+1}\right)$$

$$= \lim_{n\to\infty} \frac{1}{n} \ln(n+1)$$

$$= \lim_{n\to\infty} \frac{\ln(n+1)}{n} = \frac{1}{n+1} = 0.$$

Since $\ln y = 0$, $y = e^0 = 1$, so

$$\lim_{n\to\infty} \frac{\sqrt{n+1}}{3} = \frac{1}{3}.$$

Therefore, by the Root Test, the series converges.

85. $$\sum_{n=3}^{\infty} \frac{1}{(\ln n)^n}$$

$$\lim_{n\to\infty} \sqrt[n]{|a_n|} = \lim_{n\to\infty} \sqrt[n]{\frac{1}{(\ln n)^n}} = \lim_{n\to\infty} \frac{1}{\ln n} = 0$$

Therefore, by the Root Test, the series converges.

86. $\lim_{n\to\infty}\left|\frac{a_{n+1}}{a_n}\right| = \lim_{n\to\infty}\left|\frac{\frac{1\cdot 3\cdot 5\cdots(2n-1)(2n+1)}{1\cdot 2\cdot 3\cdots(2n-1)(2n)(2n+1)}}{\frac{1\cdot 3\cdot 5\cdots(2n-1)}{1\cdot 2\cdot 3\cdots(2n-1)}}\right|$

$= \lim_{n\to\infty}\frac{2n+1}{(2n)(2n+1)} = 0 < 1$

The series converges by the Ratio Test.

87. $\lim_{n\to\infty}\left|\frac{a_{n+1}}{a_n}\right| = \lim_{n\to\infty}\left|\frac{2(x/3)^{n+1}}{2(x/3)^n}\right|$

$= \lim_{n\to\infty}\left|\frac{x}{3}\right| = \left|\frac{x}{3}\right|$

For the series to converge: $\left|\frac{x}{3}\right| < 1 \Rightarrow -3 < x < 3$.

For $x = 3$, the series diverges.

For $x = -3$, the series diverges.

Answer: $-3 < x < 3$

88. $\lim_{n\to\infty}\sqrt[n]{|a_n|} = \lim_{n\to\infty}\sqrt[n]{\left|\frac{x+1}{4}\right|^n}$

$= \lim_{n\to\infty}\left|\frac{x+1}{4}\right| = \left|\frac{x+1}{4}\right|$

For the series to converge,

$\left|\frac{x+1}{4}\right| < 1 \Rightarrow -4 < x+1 < 4$

$\Rightarrow -5 < x < 3.$

For $x = 3$, $\sum_{n=0}^{\infty}(1)^n$, diverges.

For $x = -5$, $\sum_{n=0}^{\infty}(-1)^n$, diverges.

Answer: $-5 < x < 3$

89. $\lim_{n\to\infty}\left|\frac{a_{n+1}}{a_n}\right| = \lim_{n\to\infty}\left|\frac{(x+1)^{n+1}/(n+1)}{x^n/n}\right|$

$= \lim_{n\to\infty}\left|\frac{n}{n+1}(x+1)\right| = |x+1|$

For the series to converge,

$|x+1| < 1 \Rightarrow -1 < x+1 < 1$

$\Rightarrow -2 < x < 0.$

For $x = 0$, $\sum_{n=1}^{\infty}\frac{(-1)^n}{n}$, converges.

For $x = -2$, $\sum_{n=1}^{\infty}\frac{(-1)^n(-1)^n}{n} = \sum_{n=1}^{\infty}\frac{1}{n}$, diverges.

Answer: $-2 < x \le 0$

90. $\lim_{n\to\infty}\left|\frac{a_{n+1}}{a_n}\right| = \lim_{n\to\infty}\left|\frac{2|x-1|^{n+1}}{2|x-1|^n}\right|$

$= \lim_{n\to\infty}|x-1| = |x-1|$

For the series to converge,

$|x-1| < 1 \Rightarrow -1 < x-1 < 1$

$\Rightarrow 0 < x < 2.$

For $x = 2$, $\sum_{n=0}^{\infty}2(1)^n$, diverges.

For $x = 0$, $\sum_{n=0}^{\infty}2(-1)^n$, diverges.

Answer: $0 < x < 2$

91. $\lim_{n\to\infty}\left|\frac{a_{n+1}}{a_n}\right| = \lim_{n\to\infty}\frac{(n+1)!\left|\frac{x}{2}\right|^{n+1}}{n!\left|\frac{x}{2}\right|^n}$

$= \lim_{n\to\infty}(n+1)\left|\frac{x}{2}\right| = \infty$

The series converges only at $x = 0$.

92. $\lim_{n\to\infty}\left|\frac{a_{n+1}}{a_n}\right| = \lim_{n\to\infty}\frac{|x+1|^{n+1}}{(n+1)!}\bigg/\frac{|x+1|^n}{n!} = \lim_{n\to\infty}\frac{|x+1|}{n+1} = 0$

The series converges for all x.

93. See Theorem 9.17, page 597.

94. See Theorem 9.18.

95. No. Let $a_n = \dfrac{1}{n + 10{,}000}$.

The series $\displaystyle\sum_{n=1}^{\infty} \frac{1}{n + 10{,}000}$ diverges.

96. One example is $\displaystyle\sum_{n=1}^{\infty}\left(-100 + \frac{1}{n}\right)$.

97. The series converges absolutely. See Theorem 9.17.

98. Assume that

$\lim_{n\to\infty} |a_{n+1}/a_n| = L > 1$ or that $\lim_{n\to\infty} |a_{n+1}/a_n| = \infty$.

Then there exists $N > 0$ such that $|a_{n+1}/a_n| > 1$ for all $n > N$. Therefore,

$$|a_{n+1}| > |a_n|, \quad n > N \Longrightarrow \lim_{n\to\infty} a_n \neq 0 \Longrightarrow \sum a_n \text{ diverges.}$$

99. First, let

$$\lim_{n\to\infty} \sqrt[n]{|a_n|} = r < 1$$

and choose R such that $0 \le r < R < 1$. There must exist some $N > 0$ such that $\sqrt[n]{|a_n|} < R$ for all $n > N$. Thus, for $n > N$ $|a_n| < R^n$ and since the geometric series

$$\sum_{n=0}^{\infty} R^n$$

converges, we can apply the Comparison Test to conclude that

$$\sum_{n=1}^{\infty} |a_n|$$

converges which in turn implies that $\displaystyle\sum_{n=1}^{\infty} a_n$ converges.

Second, let

$$\lim_{n\to\infty} \sqrt[n]{|a_n|} = r > R > 1.$$

Then there must exist some $M > 0$ such that $\sqrt[n]{|a_n|} > R$ for infinitely many $n > M$. Thus, for infinitely many $n > M$, we have $|a_n| > R^n > 1$ which implies that $\lim_{n\to\infty} a_n \neq 0$ which in turn implies that

$$\sum_{n=1}^{\infty} a_n \text{ diverges.}$$

100. $\displaystyle\sum_{n=1}^{\infty} \frac{1}{n^p}$, p-series

$$\lim_{n\to\infty} \sqrt[n]{|a_n|} = \lim_{n\to\infty} \sqrt[n]{\frac{1}{n^p}} = \lim_{n\to\infty} \frac{1}{n^{p/n}} = 1$$

Hence, the Root Test is inconclusive.

Note: $\lim_{n\to\infty} n^{p/n} = 1$ because if $y = n^{p/n}$, then

$\ln y = \dfrac{p}{n} \ln n$ and $\dfrac{p}{n} \ln n \to 0$ as $n \to \infty$.

Hence $y \to 1$ as $n \to \infty$.

101. Ratio Test:

$$\lim_{n\to\infty} \left|\frac{a_{n+1}}{a_n}\right| = \lim_{n\to\infty} \frac{n(\ln n)^p}{(n+1)(\ln(n+1))^p} = 1, \text{ inconclusive.}$$

Root Test:

$$\lim_{n\to\infty} \sqrt[n]{|a_n|} = \lim_{n\to\infty} \sqrt[n]{\frac{1}{n(\ln n)^p}} = \lim_{n\to\infty} \frac{1}{n^{1/n}(\ln n)^{p/n}}$$

$\lim_{n\to\infty} n^{1/n} = 1$. Furthermore, let $y = (\ln n)^{p/n} \Longrightarrow$

$$\ln y = \frac{p}{n} \ln(\ln n). \quad \lim_{n\to\infty} \ln y = \lim_{n\to\infty} \frac{p \ln(\ln n)}{n}$$

$$= \lim_{n\to\infty} \frac{p}{\ln(n)(1/n)} = 0 \Longrightarrow \lim_{n\to\infty} (\ln n)^{p/n} = 1.$$

Thus, $\displaystyle\lim_{n\to\infty} \frac{1}{n^{1/n}(\ln n)^{p/n}} = 1$, inconclusive.

102. $\sum_{n=1}^{\infty} \frac{(n!)^2}{(xn)!}$, x positive integer

(a) $x = 1$: $\sum \frac{(n!)^2}{n!} = \sum n!$, diverges

(b) $x = 2$: $\sum \frac{(n!)^2}{(2n)!}$ converges by the Ratio Test:

$$\lim_{n\to\infty} \frac{[(n+1)!]^2}{(2n+2)!} \bigg/ \frac{(n!)^2}{(2n)!} = \lim_{n\to\infty} \frac{(n+1)^2}{(2n+2)(2n+1)} = \frac{1}{4} < 1$$

(c) $x = 3$: $\sum \frac{(n!)^2}{(3n)!}$ converges by the Ratio Test:

$$\lim_{n\to\infty} \frac{[(n+1)!]^2}{(3n+3)!} \bigg/ \frac{(n!)^2}{(3n)!} = \lim_{n\to\infty} \frac{(n+1)^2}{(3n+3)(3n+2)(3n+1)} = 0 < 1$$

(d) Use the Ratio Test:

$$\lim_{n\to\infty} \frac{[(n+1)!]^2}{[x(n+1)]!} \bigg/ \frac{(n!)^2}{(xn)!} = \lim_{n\to\infty} (n+1)^2 \frac{(xn)!}{(xn+x)!}$$

The cases $x = 1, 2, 3$ were solved above. For $x > 3$, the limit is 0. Hence, the series converges for all integers $x \geq 2$.

103. For $n = 1, 2, 3, \ldots, -|a_n| \leq a_n \leq |a_n| \Rightarrow -\sum_{n=1}^{k} |a_n| \leq \sum_{n=1}^{k} a_n \leq \sum_{n=1}^{k} |a_n|$.

Taking limits as $k \to \infty$, $-\sum_{n=1}^{\infty} |a_n| \leq \sum_{n=1}^{\infty} a_n \leq \sum_{n=1}^{\infty} |a_n| \Rightarrow \left|\sum_{n=1}^{\infty} a_n\right| \leq \sum_{n=1}^{\infty} |a_n|$.

104. The differentiation test states that if

$$\sum_{n=1}^{\infty} U_n$$

is an infinite series with real terms and $f(x)$ is a real function such that $f(1/n) = U_n$ for all positive integers n and d^2f/dx^2 exists at $x = 0$, then

$$\sum_{n=1}^{\infty} U_n$$

converges absolutely if $f(0) = f'(0) = 0$ and diverges otherwise. Below are some examples.

Convergent Series	Divergent Series
$\sum \frac{1}{n^3}, f(x) = x^3$	$\sum \frac{1}{n}, f(x) = x$
$\sum \left(1 - \cos\frac{1}{n}\right), f(x) = 1 - \cos x$	$\sum \sin\frac{1}{n}, f(x) = \sin x$

105. Using the Ratio Test,

$$\begin{aligned}
\lim_{n\to\infty}\left|\frac{a_{n+1}}{a_n}\right| &= \lim_{n\to\infty}\left[\frac{n!}{(n+1)^n}\left(\frac{19}{7}\right)^n \bigg/ \frac{(n-1)!}{n^{n-1}}\left(\frac{19}{7}\right)^{n-1}\right] \\
&= \lim_{n\to\infty}\left[\frac{n \cdot n^{n-1}}{(n+1)^n}\left(\frac{19}{7}\right)\right] \\
&= \lim_{n\to\infty}\left[\frac{1}{(1+(1/n))^n}\left(\frac{19}{7}\right)\right] \\
&= \frac{19}{7} \cdot \frac{1}{e} < 1
\end{aligned}$$

Hence, the series converges.

106. We first prove Abel's Summation Theorem:

If the partial sums of $\sum a_n$ are bounded and if $\{b_n\}$ decreases to zero, then $\sum a_n b_n$ converges.

Let $S_k = \sum_{i=1}^{k} a_i$. Let M be a bound for $\{|S_k|\}$.

$$\begin{aligned}a_1b_1 + a_2b_2 + \cdots + a_nb_n &= S_1b_1 + (S_2 - S_1)b_2 + \cdots + (S_n - S_{n-1})b_n\\ &= S_1(b_1 - b_2) + S_2(b_2 - b_3) + \cdots + S_{n-1}(b_{n-1} - b_n) + S_nb_n\\ &= \sum_{i=1}^{n-1} S_i(b_i - b_{i+1}) + S_nb_n\end{aligned}$$

The series $\sum_{i=1}^{\infty} S_i(b_i - b_{i+1})$ is absolutely convergent because $|S_i(b_i - b_{i+1})| \le M(b_i - b_{i+1})$ and $\sum_{i=1}^{\infty} (b_i - b_{i+1})$ converges to b_1.

Also, $\lim_{n\to\infty} S_nb_n = 0$ because $\{S_n\}$ bounded and $b_n \to 0$. Thus, $\sum_{n=1}^{\infty} a_nb_n = \lim_{n\to\infty} \sum_{i=1}^{n} a_ib_i$ converges.

Now let $b_n = \dfrac{1}{n}$ to finish the problem.

Section 9.7 Taylor Polynomials and Approximations

1. $y = -\frac{1}{2}x^2 + 1$

Parabola

Matches (d)

2. $y = \frac{1}{8}x^4 - \frac{1}{2}x^2 + 1$

y-axis symmetry

Three relative extrema

Matches (c)

3. $y = e^{-1/2}[(x + 1) + 1]$

Linear

Matches (a)

4. $y = e^{-1/2}[\frac{1}{3}(x - 1)^3 - (x - 1) + 1]$

Cubic

Matches (b)

5. $f(x) = \dfrac{4}{\sqrt{x}} = 4x^{-1/2}$ $\qquad f(1) = 4$

$f'(x) = -2x^{-3/2}$ $\qquad f'(1) = -2$

$P_1(x) = f(1) + f'(1)(x - 1)$

$= 4 + (-2)(x - 1)$

$P_1(x) = -2x + 6$

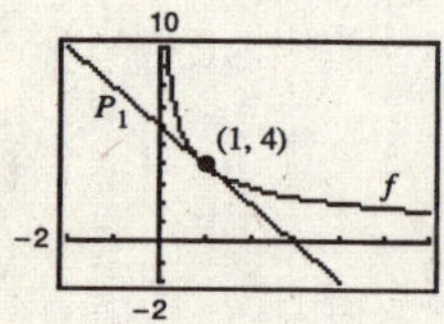

P_1 is called the first degree Taylor polynomial for f at c.

6. $f(x) = \dfrac{4}{\sqrt[3]{x}} = 4x^{-1/3}$ $\qquad f(8) = 2$

$f'(x) = -\dfrac{4}{3}x^{-4/3}$ $\qquad f'(8) = -\dfrac{1}{12}$

$P_1(x) = f(8) + f'(8)(x - 8)$

$= 2 + \left(-\dfrac{1}{12}\right)(x - 8)$

$P_1(x) = -\dfrac{1}{12}x + \dfrac{8}{3}$

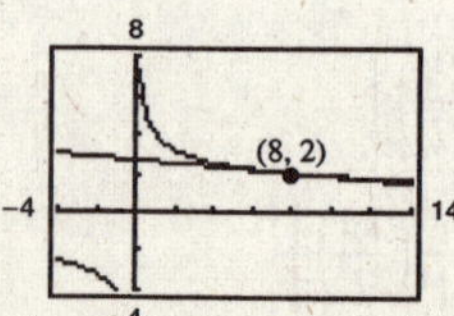

P_1 is called the first degree Taylor polynomial for f at c.

7. $f(x) = \sec x \qquad f\left(\frac{\pi}{4}\right) = \sqrt{2}$

$f'(x) = \sec x \tan x \qquad f'\left(\frac{\pi}{4}\right) = \sqrt{2}$

$$P_1(x) = f\left(\frac{\pi}{4}\right) + f'\left(\frac{\pi}{4}\right)\left(x - \frac{\pi}{4}\right)$$

$$P_1(x) = \sqrt{2} + \sqrt{2}\left(x - \frac{\pi}{4}\right)$$

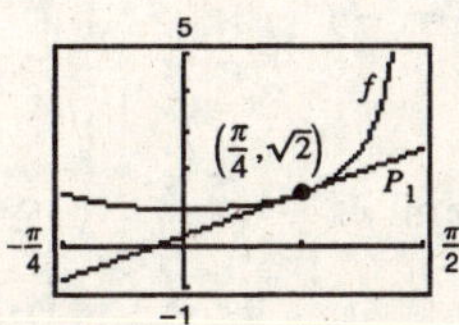

P_1 is called the first degree Taylor polynomial for f at c.

8. $f(x) = \tan x \qquad f\left(\frac{\pi}{4}\right) = 1$

$f'(x) = \sec^2 x \qquad f'\left(\frac{\pi}{4}\right) = 2$

$$P_1 = f\left(\frac{\pi}{4}\right) + f'\left(\frac{\pi}{4}\right)\left(x - \frac{\pi}{4}\right)$$

$$= 1 + 2\left(x - \frac{\pi}{4}\right)$$

$$P_1(x) = 2x + 1 - \frac{\pi}{2}$$

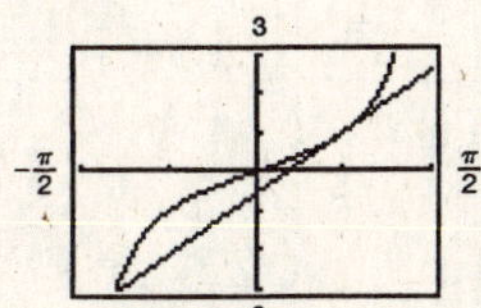

P_1 is called the first degree Taylor polynomial for f at c.

9. $f(x) = \frac{4}{\sqrt{x}} = 4x^{-1/2} \qquad f(1) = 4$

$f'(x) = -2x^{-3/2} \qquad f'(1) = -2$

$f''(x) = 3x^{-5/2} \qquad f''(1) = 3$

$$P_2 = f(1) + f'(1)(x - 1) + \frac{f''(1)}{2}(x - 1)^2$$

$$= 4 - 2(x - 1) + \frac{3}{2}(x - 1)^2$$

x	0	0.8	0.9	1.0	1.1	1.2	2
$f(x)$	Error	4.4721	4.2164	4.0	3.8139	3.6515	2.8284
$P_2(x)$	7.5	4.46	4.215	4.0	3.815	3.66	3.5

10. $f(x) = \sec x \qquad f\left(\frac{\pi}{4}\right) = \sqrt{2}$

$f'(x) = \sec x \tan x \qquad f'\left(\frac{\pi}{4}\right) = \sqrt{2}$

$f''(x) = \sec^3 x + \sec x \tan^2 x \qquad f''\left(\frac{\pi}{4}\right) = 3\sqrt{2}$

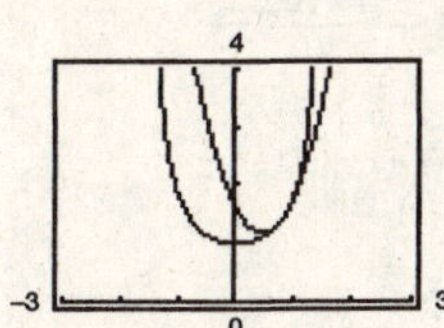

$$P_2(x) = f\left(\frac{\pi}{4}\right) + f'\left(\frac{\pi}{4}\right)\left(x - \frac{\pi}{4}\right) + \frac{f''(\pi/4)}{2}\left(x - \frac{\pi}{4}\right)^2$$

$$P_2(x) = \sqrt{2} + \sqrt{2}\left(x - \frac{\pi}{4}\right) + \frac{3}{2}\sqrt{2}\left(x - \frac{\pi}{4}\right)^2$$

x	−2.15	0.585	0.685	$\pi/4$	0.885	0.985	1.785
$f(x)$	−1.8270	1.1995	1.2913	1.4142	1.5791	1.8088	−4.7043
$P_2(x)$	15.5414	1.2160	1.2936	1.4142	1.5761	1.7810	4.9475

11. $f(x) = \cos x$

$P_2(x) = 1 - \frac{1}{2}x^2$

$P_4(x) = 1 - \frac{1}{2}x^2 + \frac{1}{24}x^4$

$P_6(x) = 1 - \frac{1}{2}x^2 + \frac{1}{24}x^4 - \frac{1}{720}x^6$

(a)

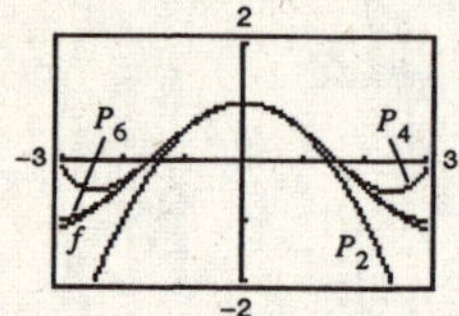

(b) $f'(x) = -\sin x \qquad P_2'(x) = -x$

$f''(x) = -\cos x \qquad P_2''(x) = -1$

$f''(0) = P_2''(0) = -1$

$f'''(x) = \sin x \qquad P_4'''(x) = x$

$f^{(4)}(x) = \cos x \qquad P_4^{(4)}(x) = 1$

$f^{(4)}(0) = 1 = P_4^{(4)}(0)$

$f^{(5)}(x) = -\sin x \qquad P_6^{(5)}(x) = -x$

$f^{(6)}(x) = -\cos x \qquad P^{(6)}(x) = -1$

$f^{(6)}(0) = -1 = P_6^{(6)}(0)$

(c) In general, $f^{(n)}(0) = P_n^{(n)}(0)$ for all n.

12. $f(x) = x^2e^x, f(0) = 0$

(a) $f'(x) = (x^2 + 2x)e^x \qquad f'(0) = 0$

$f''(x) = (x^2 + 4x + 2)e^x \qquad f''(0) = 2$

$f'''(x) = (x^2 + 6x + 6)e^x \qquad f'''(0) = 6$

$f^{(4)}(x) = (x^2 + 8x + 12)e^x \qquad f^{(4)}(0) = 12$

$$P_2(x) = \frac{2x^2}{2!} = x^2$$

$$P_3(x) = x^2 + \frac{6x^3}{3!} = x^2 + x^3$$

$$P_4(x) = x^2 + x^3 + \frac{12x^4}{4!} = x^2 + x^3 + \frac{x^4}{2}$$

(b)

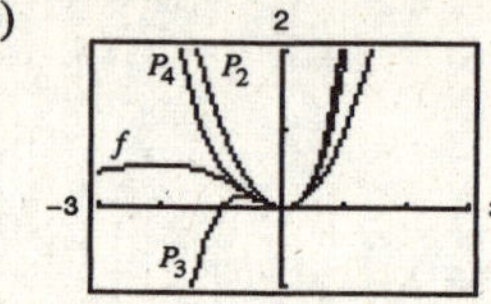

(c) $f''(0) = 2 = P_2''(0)$

$f'''(0) = 6 = P_3'''(0)$

$f^{(4)}(0) = 12 = P_4^{(4)}(0)$

(d) $f^{(n)}(0) = P_n^{(n)}(0)$

13. $f(x) = e^{-x} \qquad f(0) = 1$

$f'(x) = -e^{-x} \qquad f'(0) = -1$

$f''(x) = e^{-x} \qquad f''(0) = 1$

$f'''(x) = -e^{-x} \qquad f'''(0) = -1$

$$P_3(x) = f(0) + f'(0)x + \frac{f''(0)}{2!}x^2 + \frac{f'''(0)}{3!}x^3$$

$$= 1 - x + \frac{x^2}{2} - \frac{x^3}{6}$$

14. $f(x) = e^{-x} \qquad f(0) = 1$

$f'(x) = -e^{-x} \qquad f'(0) = -1$

$f''(x) = e^{-x} \qquad f''(0) = 1$

$f'''(x) = -e^{-x} \qquad f'''(0) = -1$

$f^{(4)}(x) = e^{-x} \qquad f^{(4)}(0) = 1$

$f^{(5)}(x) = -e^{-x} \qquad f^{(5)}(0) = -1$

$$P_5(x) = f(0) + f'(0)x + \frac{f''(0)}{2!}x^2 + \frac{f'''(0)}{3!}x^3 + \frac{f^{(4)}(0)}{4!}x^4$$

$$+ \frac{f^{(5)}(0)}{5!}x^5 = 1 - x + \frac{x^2}{2} - \frac{x^3}{6} + \frac{x^4}{24} - \frac{x^5}{120}$$

15. $f(x) = e^{2x} \qquad f(0) = 1$

$f'(x) = 2e^{2x} \qquad f'(0) = 2$

$f''(x) = 4e^{2x} \qquad f''(0) = 4$

$f'''(x) = 8e^{2x} \qquad f'''(0) = 8$

$f^{(4)}(x) = 16^{2x} \qquad f^{(4)}(0) = 16$

$$P_4(x) = 1 + 2x + \frac{4}{2!}x^2 + \frac{8}{3!}x^3 + \frac{16}{4!}x^4$$

$$= 1 + 2x + 2x^2 + \frac{4}{3}x^3 + \frac{2}{3}x^4$$

16. $f(x) = e^{3x} \qquad f(0) = 1$

$f'(x) = 3e^{3x} \qquad f'(0) = 3$

$f''(x) = 9e^{3x} \qquad f''(0) = 9$

$f'''(x) = 27e^{3x} \qquad f'''(0) = 27$

$f^{(4)}(x) = 81e^{3x} \qquad f^{(4)}(0) = 81$

$$P_4(x) = 1 + 3x + \frac{9}{2!}x^2 + \frac{27}{3!}x^3 + \frac{81}{4!}x^4$$

$$= 1 + 3x + \frac{9}{2}x^2 + \frac{9}{2}x^3 + \frac{27}{8}x^4$$

17. $f(x) = \sin x \qquad f(0) = 0$

$f'(x) = \cos x \qquad f'(0) = 1$

$f''(x) = -\sin x \qquad f''(0) = 0$

$f'''(x) = -\cos x \qquad f'''(0) = -1$

$f^{(4)}(x) = \sin x \qquad f^{(4)}(0) = 0$

$f^{(5)}(x) = \cos x \qquad f^{(5)}(0) = 1$

$$P_5(x) = 0 + (1)x + \frac{0}{2!}x^2 + \frac{-1}{3!}x^3 + \frac{0}{4!}x^4 + \frac{1}{5!}x^5$$

$$= x - \frac{1}{6}x^3 + \frac{1}{120}x^5$$

18. $f(x) = \sin \pi x \qquad f(0) = 0$

$f'(x) = \pi \cos \pi x \qquad f'(0) = \pi$

$f''(x) = -\pi^2 \sin \pi x \qquad f''(0) = 0$

$f'''(x) = -\pi^3 \cos \pi x \qquad f'''(0) = -\pi^3$

$$P_3(x) = 0 + \pi x + \frac{0}{2!}x^2 + \frac{-\pi^3}{3!}x^3 = \pi x - \frac{\pi^3}{6}x^3$$

19. $f(x) = xe^x \qquad f(0) = 0$

$f'(x) = xe^x + e^x \qquad f'(0) = 1$

$f''(x) = xe^x + 2e^x \qquad f''(0) = 2$

$f'''(x) = xe^x + 3e^x \qquad f'''(0) = 3$

$f^{(4)}(x) = xe^x + 4e^x \qquad f^{(4)}(0) = 4$

$$P_4(x) = 0 + x + \frac{2}{2!}x^2 + \frac{3}{3!}x^3 + \frac{4}{4!}x^4$$

$$= x + x^2 + \frac{1}{2}x^3 + \frac{1}{6}x^4$$

20. $f(x) = x^2e^{-x} \qquad f(0) = 0$

$f'(x) = 2xe^{-x} - x^2e^{-x} \qquad f'(0) = 0$

$f''(x) = 2e^{-x} - 4xe^{-x} + x^2e^{-x} \qquad f''(0) = 2$

$f'''(x) = -6e^{-x} + 6xe^{-x} - x^2e^{-x} \qquad f'''(0) = -6$

$f^{(4)}(x) = 12e^{-x} - 8xe^{-x} + x^2e^{-x} \qquad f^{(4)}(0) = 12$

$$P_4(x) = 0 + 0x + \frac{2}{2!}x^2 + \frac{-6}{3!}x^3 + \frac{12}{4!}x^4$$

$$= x^2 - x^3 + \frac{1}{2}x^4$$

21. $f(x) = \dfrac{1}{x+1} \qquad f(0) = 1$

$f'(x) = -\dfrac{1}{(x+1)^2} \qquad f'(0) = -1$

$f''(x) = \dfrac{2}{(x+1)^3} \qquad f''(0) = 2$

$f'''(x) = \dfrac{-6}{(x+1)^4} \qquad f'''(0) = -6$

$f^{(4)}(x) = \dfrac{24}{(x+1)^5} \qquad f^{(4)}(0) = 24$

$$P_4(x) = 1 - x + \frac{2}{2!}x^2 + \frac{-6}{3!}x^3 + \frac{24}{4!}x^4$$

$$= 1 - x + x^2 - x^3 + x^4$$

22. $f(x) = \dfrac{x}{x+1} = \dfrac{x+1-1}{x+1} \qquad f(0) = 0$

$= 1 - (x+1)^{-1}$

$f'(x) = (x+1)^{-2} \qquad f'(0) = 1$

$f''(x) = -2(x+1)^{-3} \qquad f''(0) = -2$

$f'''(x) = 6(x+1)^{-4} \qquad f'''(0) = 6$

$f^{(4)}(x) = -24(x+1)^{-5} \qquad f^{(4)}(0) = -24$

$$P_4(x) = 0 + 1(x) - \frac{2}{2}x^2 + \frac{6}{6}x^3 - \frac{24}{24}x^4$$

$$= x - x^2 + x^3 - x^4$$

23. $f(x) = \sec x \qquad f(0) = 1$

$f'(x) = \sec x \tan x \qquad f'(0) = 0$

$f''(x) = \sec^3 x + \sec x \tan^2 x \qquad f''(0) = 1$

$$P_2(x) = 1 + 0x + \frac{1}{2!}x^2 = 1 + \frac{1}{2}x^2$$

24. $f(x) = \tan x \qquad f(0) = 0$

$f'(x) = \sec^2 x \qquad f'(0) = 1$

$f''(x) = 2\sec^2 x \tan x \qquad f''(0) = 0$

$f'''(x) = 4\sec^2 x \tan^2 x + 2\sec^4 x \qquad f'''(0) = 2$

$$P_3(x) = 0 + 1(x) + 0 + \frac{2}{6}x^3 = x + \frac{1}{3}x^3$$

25.
$f(x) = \frac{1}{x}$ $\quad f(1) = 1$

$f'(x) = -\frac{1}{x^2}$ $\quad f'(1) = -1$

$f''(x) = \frac{2}{x^3}$ $\quad f''(1) = 2$

$f'''(x) = -\frac{6}{x^4}$ $\quad f'''(1) = -6$

$f^{(4)}(x) = \frac{24}{x^5}$ $\quad f^{(4)}(1) = 24$

$$P_4(x) = 1 - (x-1) + \frac{2}{2!}(x-1)^2 + \frac{-6}{3!}(x-1)^3 + \frac{24}{4!}(x-1)^4$$
$$= 1 - (x-1) + (x-1)^2 - (x-1)^3 + (x-1)^4$$

26.
$f(x) = 2x^{-2}$ $\quad f(2) = \frac{1}{2}$

$f'(x) = -4x^{-3}$ $\quad f'(2) = -\frac{1}{2}$

$f''(x) = 12x^{-4}$ $\quad f''(2) = \frac{3}{4}$

$f'''(x) = -48x^{-5}$ $\quad f'''(x) = -\frac{3}{2}$

$f^{(4)}(x) = 240x^{-6}$ $\quad f^{(4)}(x) = \frac{15}{4}$

$$P_4(x) = \frac{1}{2} - \frac{1}{2}(x-2) + \frac{3}{8}(x-2)^2 - \frac{1}{4}(x-2)^3 + \frac{5}{32}(x-2)^4$$

27.
$f(x) = \sqrt{x}$ $\quad f(1) = 1$

$f'(x) = \frac{1}{2\sqrt{x}}$ $\quad f'(1) = \frac{1}{2}$

$f''(x) = -\frac{1}{4x\sqrt{x}}$ $\quad f''(1) = -\frac{1}{4}$

$f'''(x) = \frac{3}{8x^2\sqrt{x}}$ $\quad f'''(1) = \frac{3}{8}$

$f^{(4)}(x) = -\frac{15}{16x^3\sqrt{x}}$ $\quad f^{(4)}(1) = -\frac{15}{16}$

$$P_4(x) = 1 + \frac{1}{2}(x-1) - \frac{1}{8}(x-1)^2 + \frac{1}{16}(x-1)^3 - \frac{5}{128}(x-1)^4$$

28.
$f(x) = x^{1/3}$ $\quad f(8) = 2$

$f'(x) = \frac{1}{3}x^{-2/3}$ $\quad f'(8) = \frac{1}{12}$

$f''(x) = -\frac{2}{9}x^{-5/3}$ $\quad f''(8) = -\frac{1}{144}$

$f'''(x) = \frac{10}{27}x^{-8/3}$ $\quad f'''(8) = \frac{10}{27} \cdot \frac{1}{2^8} = \frac{5}{3456}$

$$P_3(x) = 2 + \frac{1}{12}(x-8) - \frac{1}{288}(x-8)^2 + \frac{5}{20{,}736}(x-8)^3$$

29.
$f(x) = \ln x$ $\quad f(1) = 0$

$f'(x) = \frac{1}{x}$ $\quad f'(1) = 1$

$f''(x) = -\frac{1}{x^2}$ $\quad f''(1) = -1$

$f'''(x) = \frac{2}{x^3}$ $\quad f'''(1) = 2$

$f^{(4)}(x) = -\frac{6}{x^4}$ $\quad f^{(4)}(1) = -6$

$$P_4(x) = 0 + (x-1) - \frac{1}{2}(x-1)^2 + \frac{1}{3}(x-1)^3 - \frac{1}{4}(x-1)^4$$

30.
$f(x) = x^2 \cos x$ $\quad f(\pi) = -\pi^2$

$f'(x) = \cos x - x^2 \sin x$ $\quad f'(\pi) = -2\pi$

$f''(x) = 2\cos x - 4x\sin x - x^2\cos x$ $\quad f''(\pi) = -2 + \pi^2$

$$P_2(x) = -\pi^2 - 2\pi(x-\pi) + \frac{(\pi^2-2)}{2}(x-\pi)^2$$

31. $f(x) = \tan x$

$f'(x) = \sec^2 x$

$f''(x) = 2\sec^2 x \tan x$

$f'''(x) = 4\sec^2 x \tan^2 x + 2\sec^4 x$

$f^{(4)}(x) = 8\sec^2 x \tan^3 x + 16\sec^4 x \tan x$

$f^{(5)}(x) = 16\sec^2 x \tan^4 x + 88\sec^4 x \tan^2 x + 16\sec^6 x$

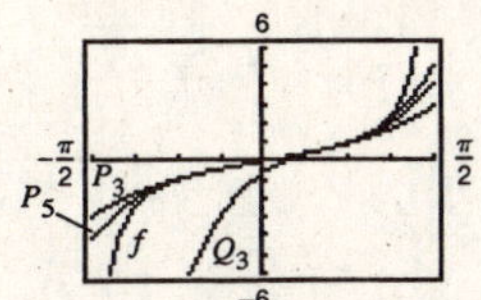

(a) $n = 3, c = 0$

$$P_3(x) = 0 + x + \frac{0}{2!}x^2 + \frac{2}{3!}x^3 = x + \frac{1}{3}x^3$$

(b) $n = 3, c = \frac{\pi}{4}$

$$Q_3(x) = 1 + 2\left(x - \frac{\pi}{4}\right) + \frac{4}{2!}\left(x - \frac{\pi}{4}\right)^2 + \frac{16}{3!}\left(x - \frac{\pi}{4}\right)^3$$

$$= 1 + 2\left(x - \frac{\pi}{4}\right) + 2\left(x - \frac{\pi}{4}\right)^2 + \frac{8}{3}\left(x - \frac{\pi}{4}\right)^3$$

32. $f(x) = \dfrac{1}{x^2 + 1}$

$f'(x) = \dfrac{-2x}{(x^2 + 1)^2}$

$f''(x) = \dfrac{2(3x^2 - 1)}{(x^2 + 1)^3}$

$f'''(x) = \dfrac{24x(1 - x^2)}{(x^2 + 1)^4}$

$f^{(4)}(x) = \dfrac{24(5x^4 - 10x^2 + 1)}{(x^2 + 1)^5}$

(a) $n = 4, c = 0$

$$P_4(x) = 1 + 0x + \frac{-2}{2!}x^2 + \frac{0}{3!}x^3 + \frac{24}{4!}x^4 = 1 - x^2 + x^4$$

(b) $n = 4, c = 1$

$$Q_4(x) = \frac{1}{2} + \left(-\frac{1}{2}\right)(x - 1) + \frac{1/2}{2!}(x - 1)^2 + \frac{0}{3!}(x - 1)^3 + \frac{-3}{4!}(x - 1)^4 = \frac{1}{2} - \frac{1}{2}(x - 1) + \frac{1}{4}(x - 1)^2 - \frac{1}{8}(x - 1)^4$$

33. $f(x) = \sin x$

$P_1(x) = x$

$P_3(x) = x - \frac{1}{6}x^3$

$P_5(x) = x - \frac{1}{6}x^3 + \frac{1}{120}x^5$

(a)

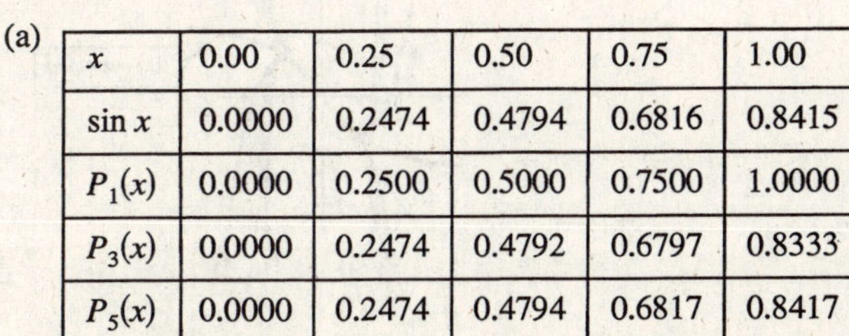

x	0.00	0.25	0.50	0.75	1.00
$\sin x$	0.0000	0.2474	0.4794	0.6816	0.8415
$P_1(x)$	0.0000	0.2500	0.5000	0.7500	1.0000
$P_3(x)$	0.0000	0.2474	0.4792	0.6797	0.8333
$P_5(x)$	0.0000	0.2474	0.4794	0.6817	0.8417

(b)

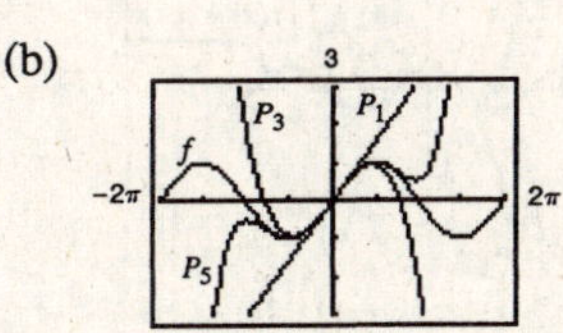

(c) As the distance increases, the accuracy decreases.

34. $f(x) = \ln x, c = 1$

$P_1(x) = x - 1$

$P_2(x) = (x - 1) - \frac{1}{2}(x - 1)^2$

$P_4(x) = (x - 1) - \frac{1}{2}(x - 1)^2 + \frac{1}{3}(x - 1)^3 - \frac{1}{4}(x - 1)^4$

(a)

x	1.00	1.25	1.50	1.75	2.00
$\ln x$	0	0.2231	0.4055	0.5596	0.6931
$P_1(x)$	0	0.2500	0.5000	0.7500	1.0000
$P_2(x)$	0	0.2188	0.375	0.4688	0.5000
$P_4(x)$	0	0.2230	0.4010	0.5303	0.5833

(b)

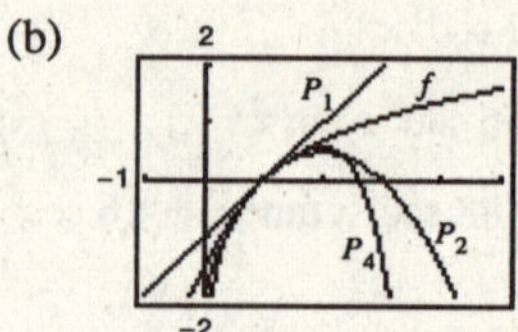

(c) As the degree increases, the accuracy increases. As the distance from x to 1 increases, the accuracy decreases.

35. $f(x) = \arcsin x$

(a) $P_3(x) = x + \dfrac{x^3}{6}$

(b)

x	−0.75	−0.50	−0.25	0	0.25	0.50	0.75
$f(x)$	−0.848	−0.524	−0.253	0	0.253	0.524	0.848
$P_3(x)$	−0.820	−0.521	−0.253	0	0.253	0.521	0.820

(c)

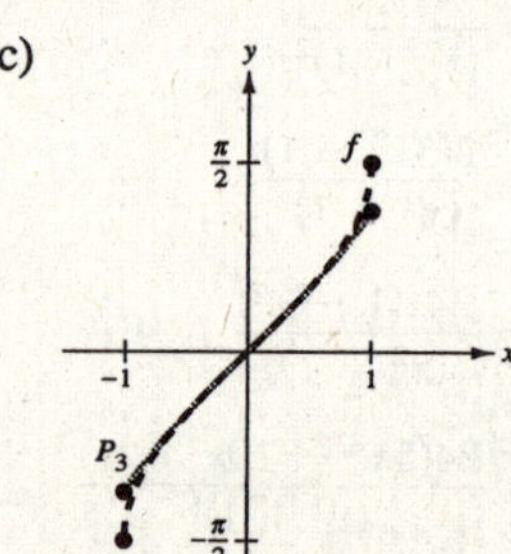

36. (a) $f(x) = \arctan x$

$P_3(x) = x - \dfrac{x^3}{3}$

(b)

x	−0.75	−0.50	−0.25	0	0.25	0.50	0.75
$f(x)$	−0.6435	−0.4636	−0.2450	0	0.2450	0.4636	0.6435
$P_3(x)$	−0.6094	−0.4583	−0.2448	0	0.2448	0.4583	0.6094

(c)

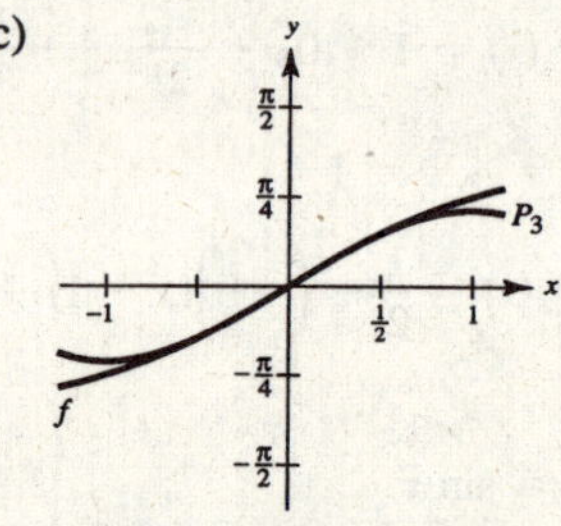

37. $f(x) = \cos x$

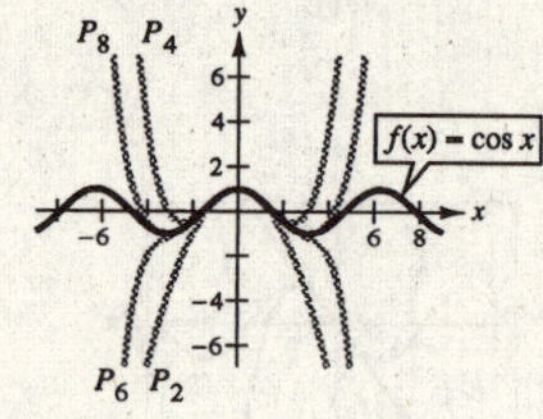

38. $f(x) = \arctan x$

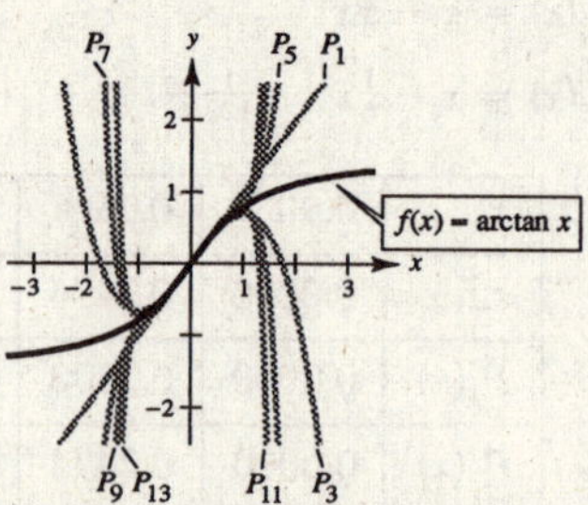

39. $f(x) = \ln(x^2 + 1)$

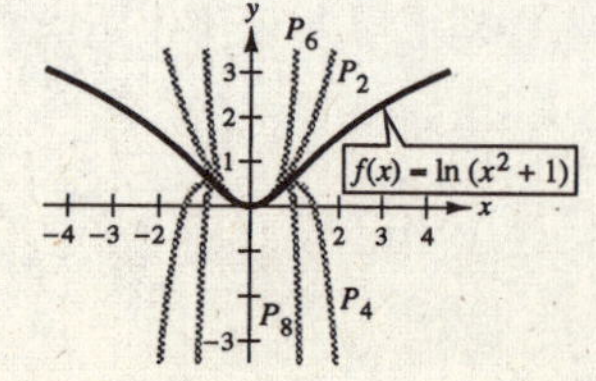

40. $f(x) = 4xe^{-x^2/4}$

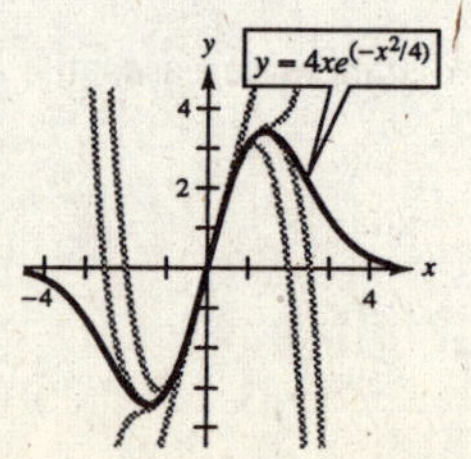

41. $f(x) = e^{-x} \approx 1 - x + \dfrac{x^2}{2} - \dfrac{x^3}{6}$

$f\left(\dfrac{1}{2}\right) \approx 0.6042$

42. $f(x) = x^2e^{-x} \approx x^2 - x^3 + \dfrac{1}{2}x^4$

$f\left(\dfrac{1}{5}\right) \approx 0.0328$

43. $f(x) = \ln x \approx (x-1) - \frac{1}{2}(x-1)^2 + \frac{1}{3}(x-1)^3 - \frac{1}{4}(x-1)^4$

$f(1.2) \approx 0.1823$

44. $f(x) = x^2 \cos x \approx -\pi^2 - 2\pi(x-\pi) + \left(\dfrac{\pi^2-2}{2}\right)(x-\pi)^2$

$f\left(\dfrac{7\pi}{8}\right) \approx -6.7954$

45. $f(x) = \cos x; f^{(5)}(x) = -\sin x \implies$ Max on $[0, 0.3]$ is 1.

$R_4(x) \le \dfrac{1}{5!}(0.3)^5 = 2.025 \times 10^{-5}$

Note: you could use $R_5(x)$: $f^{(6)}(x) = -\cos x$, max on $[0, 0.3]$ is 1.

$R_5(x) \le \dfrac{1}{6!}(0.3)^6 = 1.0125 \times 10^{-6}$

Exact error: $0.000001 = 1.0 \times 10^{-6}$

46. $f(x) = e^x; f^{(6)}(x) = e^x \implies$ Max on $[0, 1]$ is e^1.

$R_5(x) \le \dfrac{e^1}{6!}(1)^6 \approx 0.00378 = 3.78 \times 10^{-3}$

47. $f(x) = \arcsin x; f^{(4)}(x) = \dfrac{x(6x^2+9)}{(1-x^2)^{7/2}} \implies$ Max on $[0, 0.4]$ is $f^{(4)}(0.4) \approx 7.3340$.

$R_3(x) \le \dfrac{7.3340}{4!}(0.4)^4 \approx 0.00782 = 7.82 \times 10^{-3}$. The exact error is 8.5×10^{-4}. [Note: You could use R_4.]

48. $f(x) = \arctan x; f^{(4)}(x) = \dfrac{24x(x^2+1)}{(1-x^2)^4}$

$\implies$Max on $[0, 0.4]$ is $f^{(4)}(0.4) \approx 22.3672$.

$R_3(x) \le \dfrac{22.3672}{4!}(0.4)^4 \approx 0.0239$

49. $g(x) = \sin x$

$|g^{(n+1)}(x)| \le 1$ for all x.

$R_n(x) \le \dfrac{1}{(n+1)!}(0.3)^{n+1} < 0.001$

By trial and error, $n = 3$.

50. $f(x) = \cos x$

$|f^{(n+1)}(x)| \le 1$ for all x and all n.

$$|R_n(x)| = \left|\frac{f^{(n+1)}(z)x^{n+1}}{(n+1)!}\right| \le \frac{(0.1)^{n+1}}{(n+1)!} < 0.001$$

By trial and error, $n = 2$.

51. $f(x) = e^x$

$f^{(n+1)}(x) = e^x$

Max on $[0, 0.6]$ is $e^{0.6} \approx 1.8221$.

$R_n \le \dfrac{1.8221}{(n+1)!}(0.6)^{n+1} < 0.001$

By trial and error, $n = 5$.

52. $f(x) = e^x$

$f^{(n+1)}(x) = e^x$

The maximum value on $[0, 0.3]$ is $e^{0.3} \approx 1.3499$.

$$|R_n| \le \frac{1.3499}{(n+1)!}(0.3)^{n+1} < 0.001$$

By trial and error, $n = 3$.

53. $f(x) = \ln(x + 1)$

$$f^{(n+1)}(x) = \frac{(-1)^n n!}{(x+1)^{n+1}} \Rightarrow \text{Max on } [0, 0.5] \text{ is } n!.$$

$$R_n \le \frac{n!}{(n+1)!}(0.5)^{n+1} = \frac{(0.5)^{n+1}}{n+1} < 0.0001$$

By trial and error, $n = 9$. (See Example 9.) Using 9 terms, $\ln(1.5) \approx 0.4055$.

54. $f(x) = \cos(\pi x^2)$

$$g(x) = \cos x = 1 - \frac{x^2}{2!} + \frac{x^4}{4!} - \frac{x^6}{6!} + \cdots$$

$$f(x) = g(\pi x^2)$$

$$= 1 - \frac{(\pi x^2)^2}{2!} + \frac{(\pi x^2)^4}{4!} - \frac{(\pi x^2)^6}{6!} + \cdots$$

$$= 1 - \frac{\pi^2 x^4}{2!} + \frac{\pi^4 x^8}{4!} - \frac{\pi^6 x^{12}}{6!} + \cdots$$

$$f(0.6) = 1 - \frac{\pi^2}{2!}(0.6)^4 + \frac{\pi^4}{4!}(0.6)^8 - \frac{\pi^6}{6!}(0.6)^{12} + \cdots$$

Since this is an alternating series,

$$R_n \le a_{n+1} = \frac{\pi^{2n}}{(2n)!}(0.6)^{4n} < 0.0001.$$

By trial and error, $n = 4$. Using 4 terms $f(0.6) \approx 0.4257$.

55. $f(x) = e^{-\pi x}$, $f(1.3)$

$$f'(x) = (-\pi)e^{-\pi x}$$

$$f^{(n+1)}(x) = (-\pi)^{n+1}e^{-\pi x} \le |(-\pi)^{n+1}| \text{ on } [0, 1.3]$$

$$|R_n| \le \frac{(\pi)^{n+1}}{(n+1)!}(1.3)^{n+1} < 0.0001$$

By trial and error, $n = 16$.

56. $f(x) = e^{-x}$

$$f'(x) = -e^{-x}$$

$$f^{(n+1)}(x) = (-1)^{n+1}e^{-x} \le 1 \text{ on } [0, 1]$$

$$|R_n| \le \frac{1}{(n+1)!}1^{n+1} \le 0.0001$$

By trial and error, $n = 7$.

57. $f(x) = e^x \approx 1 + x + \frac{x^2}{2} + \frac{x^3}{6},\ x < 0$

$$R_3(x) = \frac{e^z}{4!}x^4 < 0.001$$

$$e^z x^4 < 0.024$$

$$|xe^{z/4}| < 0.3936$$

$$|x| < \frac{0.3936}{e^{z/4}} < 0.3936,\ z < 0$$

$$-0.3936 < x < 0$$

58. $f(x) = \sin x \approx x - \frac{x^3}{3!}$

$$|R_3(x)| = \left|\frac{\sin z}{4!}x^4\right| \le \frac{|x^4|}{4!} < 0.001$$

$$x^4 < 0.024$$

$$|x| < 0.3936$$

$$-0.3936 < x < 0.3936$$

59. $f(x) = \cos x \approx 1 - \frac{x^2}{2!} + \frac{x^4}{4!}$, fifth degree polynomial

$|f^{(n+1)}(x)| \le 1$ for all x and all n.

$$|R_5(x)| \le \frac{1}{6!}|x|^6 < 0.001$$

$$|x|^6 < 0.72$$

$$|x| < 0.9467$$

$$-0.9467 < x < 0.9467$$

Note: Use a graphing utility to graph $y = \cos x - (1 - x^2/2 + x^4/24)$ in the viewing window $[-0.9467, 0.9467] \times [-0.001, 0.001]$ to verify the answer.

60. $f(x) = e^{-2x} \approx 1 - 2x + 2x^2 - \dfrac{4}{3}x^3$

$f'(x) = -2e^{-2x}, f''(x) = 4e^{-2x}, f'''(x) = -8e^{-2x}, f^{(4)}(x) = 16e^{-2x}$

$R_3(x) = \dfrac{f^4(z)}{4!}(x-0)^4 = \dfrac{16e^{-2z}}{24}x^4 = \dfrac{2}{3}e^{-2z}x^4 < 0.001$

$e^{-2z}x^4 < 0.0015$

$x < \left[\dfrac{0.0015}{e^{-2z}}\right]^{1/4} \approx 0.1970e^{2z} < 0.1970$, for $z < 0$.

Thus $0 < x < 0.1970$.

In fact, by graphing $f(x) = e^{-2x}$ and $y = 1 - 2x + 2x^2 - \frac{4}{3}x^3$, you can verify that $|f(x) - y| < 0.001$ on $(-0.19294, 0.20068)$

61. The graph of the approximating polynomial P and the elementary function f both pass through the point $(c, f(c))$ and the slopes of P and f agree at $(c, f(c))$. Depending on the degree of P, the nth derivatives of P and f agree at $(c, f(c))$.

62. $f(c) = P_2(c), f'(c) = P_2'(c)$, and $f''(c) = P_2''(c)$

63. See definition on page 650.

64. See Theorem 9.19, page 654.

65. The accuracy increases as the degree increases (for values within the interval of convergence).

66.

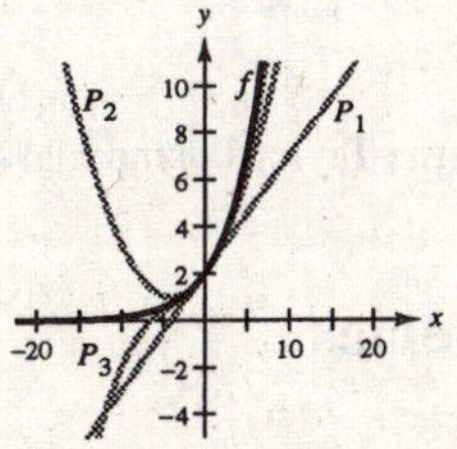

67. (a) $f(x) = e^x$

$P_4(x) = 1 + x + \dfrac{1}{2}x^2 + \dfrac{1}{6}x^3 + \dfrac{1}{24}x^4$

$g(x) = xe^x$

$Q_5(x) = x + x^2 + \dfrac{1}{2}x^3 + \dfrac{1}{6}x^4 + \dfrac{1}{24}x^5$

$Q_5(x) = x\,P_4(x)$

(b) $f(x) = \sin x$

$P_5(x) = x - \dfrac{x^3}{3!} + \dfrac{x^5}{5!}$

$g(x) = x \sin x$

$Q_6(x) = x\,P_5(x) = x^2 - \dfrac{x^4}{3!} + \dfrac{x^6}{5!}$

(c) $g(x) = \dfrac{\sin x}{x} = \dfrac{1}{x}P_5(x) = 1 - \dfrac{x^2}{3!} + \dfrac{x^4}{5!}$

68. (a) $P_5(x) = x - \dfrac{x^3}{3!} + \dfrac{x^5}{5!}$ for $f(x) = \sin x$

$P_5'(x) = 1 - \dfrac{x^2}{2!} + \dfrac{x^4}{4!}$

This is the Maclaurin polynomial of degree 4 for $g(x) = \cos x$.

(b) $Q_6(x) = 1 - \dfrac{x^2}{2} + \dfrac{x^4}{4!} - \dfrac{x^6}{6!}$ for $\cos x$

$Q_6'(x) = -x + \dfrac{x^3}{3!} - \dfrac{x^5}{5!} = -P_5(x)$

(c) $R(x) = 1 + x + \dfrac{x^2}{2!} + \dfrac{x^3}{3!} + \dfrac{x^4}{4!}$

$R'(x) = 1 + x + \dfrac{x^2}{2!} + \dfrac{x^3}{3!}$

The first four terms are the same!

69. (a) $Q_2(x) = -1 + \dfrac{\pi^2(x+2)^2}{32}$ (b) $R_2(x) = -1 + \dfrac{\pi^2(x-6)^2}{32}$ (c) No. The polynomial will be linear. Translations are possible at $x = -2 + 8n$.

70. Let f be an odd function and P_n be the nth Maclaurin polynomial for f. Since f is odd, f' is even:

$$f'(-x) = \lim_{h\to 0}\frac{f(-x+h) - f(-x)}{h} = \lim_{h\to 0}\frac{-f(x-h) + f(x)}{h} = \lim_{h\to 0}\frac{f(x+(-h)) - f(x)}{-h} = f'(x).$$

Similarly, f'' is odd, f''' is even, etc. Therefore, $f, f'', f^{(4)}$, etc. are all odd functions, which implies that $f(0) = f''(0) = \cdots = 0$. Hence, in the formula

$$P_n(x) = f(0) + f'(0)x + \frac{f''(0)x^2}{2!} + \cdots$$ all the coefficients of the even power of x are zero.

71. Let f be an even function and P_n be the nth Maclaurin polynomial for f. Since f is even, f' is odd, f'' is even, f''' is odd, etc. All of the odd derivatives of f are odd and thus, all of the odd powers of x will have coefficients of zero. P_n will only have terms with even powers of x.

72. Let $P_n(x) = a_0 + a_1(x-c) + a_2(x-c)^2 + \cdots + a_n(x-c)^n$ where $a_i = \dfrac{f^{(i)}(c)}{i!}$.

$$P_n(c) = a_0 = f(c)$$

For $1 \le k \le n$, $P_n^{(k)}(c) = a_n k! = \left(\dfrac{f^{(k)}(c)}{k!}\right)k! = f^{(k)}(c)$.

73. As you move away from $x = c$, the Taylor Polynomial becomes less and less accurate.

Section 9.8 Power Series

1. Centered at 0

2. Centered at 0

3. Centered at 2

4. Centered at π

5. $\displaystyle\sum_{n=0}^{\infty}(-1)^n\frac{x^n}{n+1}$

$$L = \lim_{n\to\infty}\left|\frac{u_{n+1}}{u_n}\right| = \lim_{n\to\infty}\left|\frac{(-1)^{n+1}x^{n+1}}{n+2}\cdot\frac{n+1}{(-1)^n x^n}\right|$$

$$= \lim_{n\to\infty}\left|\frac{n+1}{n+2}\right||x| = |x|$$

$|x| < 1 \Rightarrow R = 1$

6. $\displaystyle\sum_{n=0}^{\infty}(2x)^n$

$$L = \lim_{n\to\infty}\left|\frac{u_n + 1}{u_n}\right| = \lim_{n\to\infty}\left|\frac{(2x)^{n+1}}{(2x)^n}\right| = 2|x|$$

$2|x| < 1 \Rightarrow R = \dfrac{1}{2}$

7. $\displaystyle\sum_{n=1}^{\infty}\frac{(2x)^n}{n^2}$

$$L = \lim_{n\to\infty}\left|\frac{u_{n+1}}{u_n}\right| = \lim_{n\to\infty}\left|\frac{(2x)^{n+1}}{(n+1)^2}\cdot\frac{n^2}{(2x)^n}\right|$$

$$= \lim_{n\to\infty}\left|\frac{2n^2x}{(n+1)^2}\right| = 2|x|$$

$2|x| < 1 \Rightarrow R = \dfrac{1}{2}$

8. $\displaystyle\sum_{n=0}^{\infty}\frac{(-1)^n x^n}{2^n}$

$$L = \lim_{n\to\infty}\left|\frac{u_{n+1}}{u_n}\right| = \lim_{n\to\infty}\left|\frac{(-1)^{n+1}x^{n+1}}{2^{n+1}}\cdot\frac{2^n}{(-1)^n x^n}\right|$$

$$= \frac{1}{2}|x|$$

$\dfrac{1}{2}|x| < 1 \Rightarrow R = 2$

9. $\sum_{n=0}^{\infty} \frac{(2x)^{2n}}{(2n)!}$

$$L = \lim_{n\to\infty}\left|\frac{u_{n+1}}{u_n}\right| = \lim_{n\to\infty}\left|\frac{(2x)^{2n+2}/(2n+2)!}{(2x)^{2n}/(2n)!}\right|$$

$$= \lim_{n\to\infty}\left|\frac{(2x)^2}{(2n+2)(2n+1)}\right| = 0$$

Thus, the series converges for all x. $R = \infty$.

10. $\sum_{n=0}^{\infty} \frac{(2n)!x^{2n}}{n!}$

$$L = \lim_{n\to\infty}\left|\frac{u_n+1}{u_n}\right| = \lim_{n\to\infty}\left|\frac{(2n+2)!x^{2n+2}/(n+1)!}{(2n)!x^{2n}/n!}\right|$$

$$= \lim_{n\to\infty}\left|\frac{(2n+2)(2n+1)x^2}{(n+1)}\right| = \infty$$

The series only converges at $x = 0$. $R = 0$.

11. $\sum_{n=0}^{\infty}\left(\frac{x}{2}\right)^n$

Since the series is geometric, it converges only if $|x/2| < 1$ or $-2 < x < 2$.

12. $\sum_{n=0}^{\infty}\left(\frac{x}{5}\right)^n$

is geometric, so it converges if $|x/5| < 1 \Rightarrow |x| < 5 \Rightarrow -5 < x < 5$.

13. $\sum_{n=1}^{\infty} \frac{(-1)^n x^n}{n}$

$$\lim_{n\to\infty}\left|\frac{u_{n+1}}{u_n}\right| = \lim_{n\to\infty}\left|\frac{(-1)^{n+1}x^{n+1}}{n+1}\cdot\frac{n}{(-1)^n x^n}\right|$$

$$= \lim_{n\to\infty}\left|\frac{nx}{n+1}\right| = |x|$$

Interval: $-1 < x < 1$

When $x = 1$, the alternating series $\sum_{n=1}^{\infty} \frac{(-1)^n}{n}$ converges.

When $x = -1$, the p-series $\sum_{n=1}^{\infty} \frac{1}{n}$ diverges.

Therefore, the interval of convergence is $-1 < x \le 1$.

14. $\sum_{n=0}^{\infty} (-1)^{n+1}(n+1)x^n$

$$\lim_{n\to\infty}\left|\frac{u_n+1}{u_n}\right| = \lim_{n\to\infty}\left|\frac{(-1)^{n+2}(n+2)x^{n+1}}{(-1)^n(n+1)x^n}\right|$$

$$= \lim_{n\to\infty}\left|\frac{(n+2)x}{n+1}\right| = |x|$$

Interval: $-1 < x < 1$

When $x = 1$, the series $\sum_{n=0}^{\infty} (-1)^{n+1}(n+1)$ diverges.

When $x = -1$, the series $\sum_{n=0}^{\infty} -(n+1)$ diverges.

Therefore, the interval of convergence is $-1 < x < 1$.

15. $\sum_{n=0}^{\infty} \frac{x^n}{n!}$

$$\lim_{n\to\infty}\left|\frac{u_{n+1}}{u_n}\right| = \lim_{n\to\infty}\left|\frac{x^{n+1}}{(n+1)!}\cdot\frac{n!}{x^n}\right|$$

$$= \lim_{n\to\infty}\left|\frac{x}{n+1}\right| = 0$$

The series converges for all x. Therefore, the interval of convergence is $-\infty < x < \infty$.

16. $\sum_{n=0}^{\infty} \frac{(3x)^n}{(2n)!}$

$$\lim_{n\to\infty}\left|\frac{u_{n+1}}{u_n}\right| = \lim_{n\to\infty}\left|\frac{(3x)^{n+1}}{(2n+1)!}\cdot\frac{(2n)!}{(3x)^n}\right|$$

$$= \lim_{n\to\infty}\left|\frac{3x}{(2n+2)(2n+1)}\right| = 0$$

Therefore, the interval of convergence is $-\infty < x < \infty$.

17. $\sum_{n=0}^{\infty} (2n)!\left(\frac{x}{2}\right)^n$

$$\lim_{n\to\infty}\left|\frac{u_{n+1}}{u_n}\right| = \lim_{n\to\infty}\left|\frac{(2n+2)!x^{n+1}}{2^{n+1}}\cdot\frac{2^n}{(2n)!x^n}\right| = \lim_{n\to\infty}\left|\frac{(2n+2)(2n+1)x}{2}\right| = \infty$$

Therefore, the series converges only for $x = 0$.

18. $\displaystyle\sum_{n=0}^{\infty} \frac{(-1)^n x^n}{(n+1)(n+2)}$

$$\lim_{n\to\infty}\left|\frac{u_{n+1}}{u_n}\right| = \lim_{n\to\infty}\left|\frac{(-1)^{n+1}x^{n+1}}{(n+2)(n+3)}\cdot\frac{(n+1)(n+2)}{(-1)^n x^n}\right| = \lim_{n\to\infty}\left|\frac{(n+1)x}{n+3}\right| = |x|$$

Interval: $-1 < x < 1$

When $x = 1$, the alternating series $\displaystyle\sum_{n=0}^{\infty}\frac{(-1)^n}{(n+1)(n+2)}$ converges.

When $x = -1$, the series $\displaystyle\sum_{n=0}^{\infty}\frac{1}{(n+1)(n+2)}$ converges by limit comparison to $\displaystyle\sum_{n=1}^{\infty}\frac{1}{n^2}$.

Therefore, the interval of convergence is $-1 \le x \le 1$.

19. $\displaystyle\sum_{n=1}^{\infty}\frac{(-1)^{n+1}x^n}{4^n}$

Since the series is geometric, it converges only if $|x/4| < 1$ or $-4 < x < 4$.

20. $\displaystyle\sum_{n=0}^{\infty}\frac{(-1)^n n!(x-4)^n}{3^n}$

$$\lim_{n\to\infty}\left|\frac{u_{n+1}}{u_n}\right| = \lim_{n\to\infty}\left|\frac{(-1)^{n+1}(n+1)!(x-4)^{n+1}}{3^{n+1}}\cdot\frac{3^n}{(-1)^n n!(x-4)^n}\right| = \lim_{n\to\infty}\left|\frac{(n+1)(x-4)}{3}\right| = \infty$$

$R = 0$

Center: $x = 4$

Therefore, the series converges only for $x = 4$.

21. $\displaystyle\sum_{n=1}^{\infty}\frac{(-1)^{n+1}(x-5)^n}{n5^n}$

$$\lim_{n\to\infty}\left|\frac{u_{n+1}}{u_n}\right| = \lim_{n\to\infty}\left|\frac{(-1)^{n+2}(x-5)^{n+1}}{(n+1)5^{n+1}}\cdot\frac{n5^n}{(-1)^{n+1}(x-5)^n}\right| = \lim_{n\to\infty}\left|\frac{n(x-5)}{5(n+1)}\right| = \frac{1}{5}|x-5|$$

$R = 5$

Center: $x = 5$

Interval: $-5 < x - 5 < 5$ or $0 < x < 10$

When $x = 0$, the p-series $\displaystyle\sum_{n=1}^{\infty}\frac{-1}{n}$ diverges. When $x = 10$, the alternating series $\displaystyle\sum_{n=1}^{\infty}\frac{(-1)^{n+1}}{n}$ converges.

Therefore, the interval of convergence is $0 < x \le 10$.

22. $\displaystyle\sum_{n=0}^{\infty}\frac{(x-2)^{n+1}}{(n+1)4^{n+1}}$

$$\lim_{n\to\infty}\left|\frac{u_{n+1}}{u_n}\right| = \lim_{n\to\infty}\left|\frac{(x-2)^{n+2}}{(n+2)4^{n+2}}\cdot\frac{(n+1)4^{n+1}}{(x-2)^{n+1}}\right| = \lim_{n\to\infty}\left|\frac{(x-2)(n+1)}{4(n+2)}\right| = \frac{1}{4}|x-2|$$

$R = 4$

Center: $x = 2$

Interval: $-4 < x - 2 < 4$ or $-2 < x < 6$

When $x = -2$, the alternating series $\displaystyle\sum_{n=0}^{\infty}\frac{(-1)^{n+1}}{(n+1)}$ converges.

When $x = 6$, the series $\displaystyle\sum_{n=0}^{\infty}\frac{1}{n+1}$ diverges.

Therefore, the interval of convergence is $-2 \le x < 6$.

23. $\displaystyle\sum_{n=0}^{\infty} \frac{(-1)^{n+1}(x-1)^{n+1}}{n+1}$

$$\lim_{n\to\infty}\left|\frac{u_{n+1}}{u_n}\right| = \lim_{n\to\infty}\left|\frac{(-1)^{n+2}(x-1)^{n+2}}{n+2}\cdot\frac{n+1}{(-1)^{n+1}(x-1)^{n+1}}\right| = \lim_{n\to\infty}\left|\frac{(n+1)(x-1)}{n+2}\right| = |x-1|$$

$R = 1$

Center: $x = 1$

Interval: $-1 < x - 1 < 1$ or $0 < x < 2$

When $x = 0$, the series $\displaystyle\sum_{n=0}^{\infty}\frac{1}{n+1}$ diverges by the integral test.

When $x = 2$, the alternating series $\displaystyle\sum_{n=0}^{\infty}\frac{(-1)^{n+1}}{n+1}$ converges.

Therefore, the interval of convergence is $0 < x \le 2$.

24. $\displaystyle\sum_{n=1}^{\infty}\frac{(-1)^{n+1}(x-2)^n}{n2^n}$

$$\lim_{n\to\infty}\left|\frac{a_{n+1}}{a_n}\right| = \lim_{n\to\infty}\left|\frac{(-1)^{n+2}(x-2)^{n+1}}{(n+1)2^{n+1}}\Big/\frac{(-1)^{n+1}(x-2)^n}{n2^n}\right|$$

$$= \lim_{n\to\infty}\left|\frac{x-2}{2}\,\frac{n}{n+1}\right| = \left|\frac{x-2}{2}\right|$$

$$\left|\frac{x-2}{2}\right| < 1 \Rightarrow -2 < x-2 < 2 \Rightarrow 0 < x < 4$$

At $x = 0$,

$$\sum_{n=1}^{\infty}\frac{(-1)^{n+1}(-2)^n}{n2^n} = \sum_{n=1}^{\infty}\frac{(-1)(2^n)}{n2^n} = \sum_{n=1}^{\infty}\frac{(-1)}{n}, \text{ diverges.}$$

At $x = 4$,

$$\sum_{n=1}^{\infty}\frac{(-1)^{n+1}2^n}{n2^n} = \sum_{n=1}^{\infty}\frac{(-1)^{n+1}}{n}, \text{ converges.}$$

Interval of convergence: $0 < x \le 4$

25. $\displaystyle\sum_{n=1}^{\infty}\left(\frac{x-3}{3}\right)^{n-1}$ is geometric. It converges if

$$\left|\frac{x-3}{3}\right| < 1 \Rightarrow |x-3| < 3 \Rightarrow 0 < x < 6.$$

Interval convergence: $0 < x < 6$

26. $\displaystyle\sum_{n=0}^{\infty}\frac{(-1)^n x^{2n+1}}{2n+1}$

$$\lim_{n\to\infty}\left|\frac{u_{n+1}}{u_n}\right| = \lim_{n\to\infty}\left|\frac{(-1)^{n+1}x^{2n+3}}{(2n+3)}\cdot\frac{(2n+1)}{(-1)^n x^{2n+1}}\right| = \lim_{n\to\infty}\left|\frac{(2n+1)}{(2n+3)}x^2\right| = |x^2|$$

$R = 1$

Interval: $-1 < x < 1$

When $x = 1$, $\displaystyle\sum_{n=0}^{\infty}\frac{(-1)^n}{2n+1}$ converges.

When $x = -1$, $\displaystyle\sum_{n=0}^{\infty}\frac{(-1)^{n+1}}{2n+1}$ converges.

Therefore, the interval of convergence is $-1 \le x \le 1$.

27. $\sum_{n=1}^{\infty} \frac{n}{n+1}(-2x)^{n-1}$

$$\lim_{n\to\infty}\left|\frac{u_{n+1}}{u_n}\right| = \lim_{n\to\infty}\left|\frac{(n+1)(-2x)^n}{n+2}\cdot\frac{n+1}{n(-2x)^{n-1}}\right|$$

$$= \lim_{n\to\infty}\left|\frac{(-2x)(n+1)^2}{n(n+2)}\right| = 2|x|$$

$R = \frac{1}{2}$

Interval: $-\frac{1}{2} < x < \frac{1}{2}$

When $x = -\frac{1}{2}$, the series $\sum_{n=1}^{\infty} \frac{n}{n+1}$ diverges by the nth Term Test.

When $x = \frac{1}{2}$, the alternating series $\sum_{n=1}^{\infty} \frac{(-1)^{n-1}n}{n+1}$ diverges.

Therefore, the interval of convergence is $-\frac{1}{2} < x < \frac{1}{2}$.

28. $\sum_{n=0}^{\infty} \frac{(-1)^n x^{2n}}{n!}$

$$\lim_{n\to\infty}\left|\frac{u_{n+1}}{u_n}\right| = \lim_{n\to\infty}\left|\frac{(-1)^{n+1}x^{2n+2}}{(n+1)!}\cdot\frac{n!}{(-1)^n x^{2n}}\right|$$

$$= \lim_{n\to\infty}\left|\frac{x^2}{n+1}\right| = 0$$

Therefore, the interval of convergence is $-\infty < x < \infty$.

29. $\sum_{n=0}^{\infty} \frac{x^{2n+1}}{(2n+1)!}$

$$\lim_{n\to\infty}\left|\frac{u_{n+1}}{u_n}\right| = \lim_{n\to\infty}\left|\frac{x^{2n+3}}{(2n+3)!}\cdot\frac{(2n+1)!}{x^{2n+1}}\right|$$

$$= \lim_{n\to\infty}\left|\frac{x^2}{(2n+2)(2n+3)}\right| = 0$$

Therefore, the interval of convergence is $-\infty < x < \infty$.

30. $\sum_{n=1}^{\infty} \frac{n!x^n}{(2n)!}$

$$\lim_{n\to\infty}\left|\frac{u_{n+1}}{u_n}\right| = \lim_{n\to\infty}\left|\frac{(n+1)!x^{n+1}}{(2n+2)!}\cdot\frac{(2n)!}{n!x^n}\right|$$

$$= \lim_{n\to\infty}\left|\frac{(n+1)x}{(2n+2)(2n+1)}\right| = 0$$

Therefore, the interval of convergence is $-\infty < x < \infty$.

31. $\sum_{n=1}^{\infty} \frac{2\cdot 3\cdot 4\cdots(n+1)x^n}{n!} = \sum_{n=1}^{\infty}(n+1)x^n$

$$\lim_{n\to\infty}\left|\frac{a_{n+1}}{a_n}\right| = \lim_{n\to\infty}\left|\frac{(n+2)x^{n+1}}{(n+1)x^n}\right| = \lim_{n\to\infty}\left|\frac{n+2}{n+1}x\right| = |x|$$

Converges if $|x| < 1 \Rightarrow -1 < x < 1$.

At $x = \pm 1$, diverges.

Interval of convergence: $-1 < x < 1$

32. $\sum_{n=1}^{\infty} \frac{2\cdot 4\cdot 6\cdots(2n)}{3\cdot 5\cdot 7\cdots(2n+1)}(x^{2n+1})$

$$\lim_{n\to\infty}\left|\frac{u_{n+1}}{u_n}\right| = \lim_{n\to\infty}\left|\frac{2\cdot 4\cdots(2n)(2n+2)x^{2n+3}}{3\cdot 5\cdot 7\cdots(2n+1)(2n+3)}\cdot\frac{3\cdot 5\cdots(2n+1)}{2\cdot 4\cdots(2n)x^{2n+1}}\right| = \lim_{n\to\infty}\left|\frac{(2n+2)x^2}{(2n+3)}\right| = |x^2|$$

$R = 1$

When $x = \pm 1$, the series diverges by comparing it to

$$\sum_{n=1}^{\infty}\frac{1}{2n+1}$$

which diverges. Therefore, the interval of convergence is $-1 < x < 1$.

33. $\displaystyle\sum_{n=1}^{\infty} \frac{(-1)^{n+1} 3 \cdot 7 \cdot 11 \cdots (4n-1)(x-3)^n}{4^n}$

$$\lim_{n\to\infty}\left|\frac{u_{n+1}}{u_n}\right| = \lim_{n\to\infty}\left|\frac{(-1)^{n+2}\cdot 3\cdot 7\cdot 11\cdots(4n-1)(4n+3)(x-3)^{n+1}}{4^{n+1}}\cdot\frac{4^n}{(-1)^{n+1}\cdot 3\cdot 7\cdot 11\cdots(4n-1)(x-3)^n}\right|$$

$$= \lim_{n\to\infty}\left|\frac{(4n+3)(x-3)}{4}\right| = \infty$$

$R = 0$

Center: $x = 3$

Therefore, the series converges only for $x = 3$.

34. $\displaystyle\sum_{n=1}^{\infty} \frac{n!(x+1)^n}{1\cdot 3\cdot 5\cdots(2n-1)}$

$$\lim_{n\to\infty}\left|\frac{a_{n+1}}{a_n}\right| = \lim_{n\to\infty}\left|\frac{(n+1)!(x+1)^{n+1}}{1\cdot 3\cdot 5\cdots(2n-1)(2n+1)}\bigg/\frac{(n)!(x+1)^n}{1\cdot 3\cdot 5\cdots(2n-1)}\right|$$

$$= \lim_{n\to\infty}\left|\frac{(n+1)(x+1)}{2n+1}\right| = \frac{1}{2}|x+1|$$

Converges if $\frac{1}{2}|x+1| < 1 \Rightarrow -2 < x+1 < 2 \Rightarrow -3 < x < 1$.

At $x = 1$, $a_n = \dfrac{n!2^n}{1\cdot 3\cdot 5\cdots(2n-1)} = \dfrac{2\cdot 4\cdot 6\cdots 2n}{1\cdot 3\cdot 5\cdots(2n-1)} > 1$, diverges.

At $x = -3$, $a_n = \dfrac{n!(-2)^n}{1\cdot 3\cdots(2n-1)} = (-1)^n\dfrac{2\cdot 4\cdots 2n}{1\cdot 3\cdots(2n-1)}$, diverges.

Interval of convergence: $-3 < x < 1$

35. $\displaystyle\sum_{n=1}^{\infty} \frac{(x-c)^{n-1}}{c^{n-1}}$

$$\lim_{n\to\infty}\left|\frac{u_{n+1}}{u_n}\right| = \lim_{n\to\infty}\left|\frac{(x-c)^n}{c^n}\cdot\frac{c^{n-1}}{(x-c)^{n-1}}\right| = \frac{1}{c}|x-c|$$

$R = c$

Center: $x = c$

Interval: $-c < x - c < c$ or $0 < x < 2c$

When $x = 0$, the series $\displaystyle\sum_{n=1}^{\infty}(-1)^{n-1}$ diverges.

When $x = 2c$, the series $\displaystyle\sum_{n=1}^{\infty} 1$ diverges.

Therefore, the interval of convergence is $0 < x < 2c$.

36. $\displaystyle\sum_{n=0}^{\infty} \frac{(n!)^k x^n}{(kn)!}$, k is a positive integer.

$$\lim_{n\to\infty}\left|\frac{a_{n+1}}{a_n}\right| = \lim_{n\to\infty}\left|\frac{[(n+1)!]^k x^{n+1}}{[k(n+1)]!}\bigg/\frac{(n!)^k x^n}{(kn)!}\right|$$

$$= \lim_{n\to\infty}\left|\frac{(n+1)^k x}{(k+nk)(k-1+nk)\cdots(1+nk)}\right|$$

$$= \frac{|x|}{k^k}$$

Converges if $\dfrac{|x|}{k^k} < 1 \Rightarrow R = k^k$.

37. $\displaystyle\sum_{n=0}^{\infty}\left(\frac{x}{k}\right)^n$

Since the series is geometric, it converges only if $|x/k| < 1$ or $-k < x < k$.

38. $\displaystyle\sum_{n=1}^{\infty} \frac{(-1)^{n+1}(x-c)^n}{nc^n}$

$$\lim_{n\to\infty}\left|\frac{u_{n+1}}{u_n}\right| = \lim_{n\to\infty}\left|\frac{(-1)^{n+2}(x-c)^{n+1}}{(n+1)c^{n+1}} \cdot \frac{nc^n}{(-1)^{n+1}(x-c)^n}\right| = \lim_{n\to\infty}\left|\frac{n(x-c)}{c(n+1)}\right| = \frac{1}{c}|x-c|$$

$R = c$

Center: $x = c$

Interval: $-c < x - c < c$ or $0 < x < 2c$

When $x = 0$, the p-series $\displaystyle\sum_{n=1}^{\infty} \frac{-1}{n}$ diverges.

When $x = 2c$, the alternating series $\displaystyle\sum_{n=1}^{\infty} \frac{(-1)^{n+1}}{n}$ converges. Therefore, the interval of convergence is $0 < x \le 2c$.

39. $\displaystyle\sum_{n=1}^{\infty} \frac{k(k+1)\cdots(k+n-1)x^n}{n!}$

$$\lim_{n\to\infty}\left|\frac{u_{n+1}}{u_n}\right| = \lim_{n\to\infty}\left|\frac{k(k+1)\cdots(k+n-1)(k+n)x^{n+1}}{(n+1)!} \cdot \frac{n!}{k(k+1)\cdots(k+n-1)x^n}\right| = \lim_{n\to\infty}\left|\frac{(k+n)x}{n+1}\right| = |x|$$

$R = 1$

When $x = \pm 1$, the series diverges and the interval of convergence is $-1 < x < 1$.

$$\left[\frac{k(k+1)\cdots(k+n-1)}{1\cdot 2\cdots n} \ge 1\right]$$

40. $\displaystyle\sum_{n=1}^{\infty} \frac{n!(x-c)^n}{1\cdot 3\cdot 5\cdots(2n-1)}$

$$\lim_{n\to\infty}\left|\frac{u_{n+1}}{u_n}\right| = \lim_{n\to\infty}\left|\frac{(n+1)!(x-c)^{n+1}}{1\cdot 3\cdot 5\cdots(2n-1)(2n+1)} \cdot \frac{1\cdot 3\cdot 5\cdots(2n-1)}{n!(x-c)}\right| = \lim_{n\to\infty}\left|\frac{(n+1)(x-c)}{2n+1}\right| = \frac{1}{2}|x-c|$$

$R = 2$

Interval: $-2 < x - c < 2$ or $c - 2 < x < c + 2$

The series diverges at the endpoints. Therefore, the interval of convergence is $c - 2 < x < c + 2$.

$$\left[\frac{n!(c+2-c)^n}{1\cdot 3\cdot 5\cdots(2n-1)} = \frac{n!\,2^2}{1\cdot 3\cdot 5\cdots(2n-1)} = \frac{2\cdot 4\cdot 6\cdots(2n)}{1\cdot 3\cdot 5\cdots(2n-1)} > 1\right]$$

41. $\displaystyle\sum_{n=0}^{\infty} \frac{x^n}{n!} = 1 + \frac{x}{1} + \frac{x^2}{2} + \cdots = \sum_{n=1}^{\infty} \frac{x^{n-1}}{(n-1)!}$

42. $\displaystyle\sum_{n=0}^{\infty} (-1)^{n+1}(n+1)x^n = \sum_{n=1}^{\infty} (-1)^n(n)x^{n-1}$

43. $\displaystyle\sum_{n=0}^{\infty} \frac{x^{2n+1}}{(2n+1)!} = \sum_{n=1}^{\infty} \frac{x^{2n-1}}{(2n-1)!}$

Replace n with $n - 1$.

44. $\displaystyle\sum_{n=0}^{\infty} \frac{(-1)^n x^{2n+1}}{2n+1} = \sum_{n=1}^{\infty} \frac{(-1)^{n-1}x^{2n-1}}{2n-1}$

45. (a) $\displaystyle f(x) = \sum_{n=0}^{\infty} \left(\frac{x}{2}\right)^n, -2 < x < 2$ (Geometric)

(b) $\displaystyle f'(x) = \sum_{n=1}^{\infty} \left(\frac{n}{2}\right)\left(\frac{x}{2}\right)^{n-1}, -2 < x < 2$

(c) $\displaystyle f''(x) = \sum_{n=2}^{\infty} \left(\frac{n}{2}\right)\left(\frac{n-1}{2}\right)\left(\frac{x}{2}\right)^{n-2}, -2 < x < 2$

(d) $\displaystyle \int f(x)\,dx = \sum_{n=0}^{\infty} \frac{2}{n+1}\left(\frac{x}{2}\right)^{n+1}, -2 \le x < 2$

46. (a) $\displaystyle f(x) = \sum_{n=1}^{\infty} \frac{(-1)^{n+1}(x-5)^n}{n5^n}, 0 < x \le 10$

(b) $\displaystyle f'(x) = \sum_{n=1}^{\infty} \frac{(-1)^{n+1}(x-5)^{n-1}}{5^n}, 0 < x < 10$

(c) $\displaystyle f''(x) = \sum_{n=2}^{\infty} \frac{(-1)^{n+1}(n-1)(x-5)^{n-2}}{5^n}, 0 < x < 10$

(d) $\displaystyle \int f(x)\,dx = \sum_{n=1}^{\infty} \frac{(-1)^{n+1}(x-5)^{n+1}}{n(n+1)5^n}, 0 \le x \le 10$

47. (a) $f(x) = \sum_{n=0}^{\infty} \frac{(-1)^{n+1}(x-1)^{n+1}}{n+1}, 0 < x \le 2$

(b) $f'(x) = \sum_{n=0}^{\infty} (-1)^{n+1}(x-1)^n,\ 0 < x < 2$

(c) $f''(x) = \sum_{n=1}^{\infty} (-1)^{n+1} n(x-1)^{n-1}, 0 < x < 2$

(d) $\int f(x)\,dx = \sum_{n=1}^{\infty} \frac{(-1)^{n+1}(x-1)^{n+2}}{(n+1)(n+2)}, 0 \le x \le 2$

48. (a) $f(x) = \sum_{n=1}^{\infty} \frac{(-1)^{n+1}(x-2)^n}{n}, 1 < x \le 3$

(b) $f'(x) = \sum_{n=1}^{\infty} (-1)^{n+1}(x-2)^{n-1}, 1 < x < 3$

(c) $f''(x) = \sum_{n=2}^{\infty} (-1)^{n+1}(n-1)(x-2)^{n-2}, 1 < x < 3$

(d) $\int f(x)\,dx = \sum_{n=1}^{\infty} \frac{(-1)^{n+1}(x-2)^{n+1}}{n(n+1)}, 1 \le x \le 3$

49. $g(1) = \sum_{n=0}^{\infty} \left(\frac{1}{3}\right)^n = 1 + \frac{1}{3} + \frac{1}{9} + \cdots$

$S_1 = 1, S_2 = 1.33$. Matches (c).

50. $g(2) = \sum_{n=0}^{\infty} \left(\frac{2}{3}\right)^n = 1 + \frac{2}{3} + \frac{4}{9} + \cdots$

$S_1 = 1, S_2 = 1.67$. Matches (a).

51. $g(3.1) = \sum_{n=0}^{\infty} \left(\frac{3.1}{3}\right)^n$ diverges. Matches (b).

52. $g(-2) = \sum_{n=0}^{\infty} \left(-\frac{2}{3}\right)^n$ alternating. Matches (d).

53. $g\left(\frac{1}{8}\right) = \sum_{n=0}^{\infty} \left[2\left(\frac{1}{8}\right)\right]^n = \sum_{n=0}^{\infty} \left(\frac{1}{4}\right)^n = 1 + \frac{1}{4} + \frac{1}{16} + \cdots$, converges

$S_1 = 1, S_2 = 1.25, S_3 = 1.3125$ Matches (b).

54. $g\left(-\frac{1}{8}\right) = \sum_{n=0}^{\infty} \left[2\left(-\frac{1}{8}\right)\right]^n = \sum_{n=0}^{\infty} \left(-\frac{1}{4}\right)^n = 1 - \frac{1}{4} + \frac{1}{16} - \cdots$, converges

$S_1 = 1, S_2 = 0.75, S_3 = 0.8125$ Matches (c).

55. $g\left(\frac{9}{16}\right) = \sum_{n=0}^{\infty} \left[2\left(\frac{9}{16}\right)\right]^n = \sum_{n=0}^{\infty} \left(\frac{9}{8}\right)^n$, diverges

$S_1 = 1, S_2 = \frac{17}{8}$ Matches (d).

56. $g\left(\frac{3}{8}\right) = \sum_{n=0}^{\infty} \left[2\left(\frac{3}{8}\right)\right]^n = \sum_{n=0}^{\infty} \left(\frac{3}{4}\right)^n$, converges

$S_1 = 1, S_2 = 1.75$ Matches (a).

57. A series of the form

$$\sum_{n=0}^{\infty} a_n(x-c)^n$$

is called a power series centered at c.

58. The set of all values of x for which the power series converges is the interval of convergence. If the power series converges for all x, then the radius of convergence is $R = \infty$. If the power series converges at only c, then $R = 0$. Otherwise, according to Theorem 8.20, there exists a real number $R > 0$ (radius of convergence) such that the series converges absolutely for $|x - c| < R$ and diverges for $|x - c| > R$.

59. A single point, an interval, or the entire real line.

60. You differentiate and integrate the power series term by term. The radius of convergence remains the same. However, the interval of convergence might change.

61. Answers will vary.

$\sum_{n=1}^{\infty} \frac{x^n}{n}$ converges for $-1 \le x < 1$. At $x = -1$, the convergence is conditional because $\sum \frac{1}{n}$ diverges.

$\sum_{n=1}^{\infty} \frac{x^n}{n^2}$ converges for $-1 \le x \le 1$. At $x = \pm 1$, the convergence is absolute.

62. Many answers possible.

(a) $\sum_{n=1}^{\infty}\left(\frac{x}{2}\right)^n$ Geometric: $\left|\frac{x}{2}\right| < 1 \Rightarrow |x| < 2$

(b) $\sum_{n=1}^{\infty}\frac{(-1)^n x^n}{n}$ converges for $-1 < x \le 1$

(c) $\sum_{n=1}^{\infty}(2x+1)^n$ Geometric: $|2x+1| < 1 \Rightarrow -1 < x < 0$

(d) $\sum_{n=1}^{\infty}\frac{(x-2)^n}{n4^n}$ converges for $-2 \le x < 6$

63. (a) $f(x) = \sum_{n=0}^{\infty}\frac{(-1)^n x^{2n+1}}{(2n+1)!}, -\infty < x < \infty$

(See Exercise 29.)

$g(x) = \sum_{n=0}^{\infty}\frac{(-1)^n x^{2n}}{(2n)!}, -\infty < x < \infty$

(b) $f'(x) = \sum_{n=0}^{\infty}\frac{(-1)^n x^{2n}}{(2n)!} = g(x)$

(c) $g'(x) = \sum_{n=1}^{\infty}\frac{(-1)^n x^{2n-1}}{(2n-1)!} = \sum_{n=0}^{\infty}\frac{(-1)^{n+1}x^{2n+1}}{(2n+1)!}$

$= -\sum_{n=0}^{\infty}\frac{(-1)^n x^{2n+1}}{(2n+1)!} = -f(x)$

(d) $f(x) = \sin x$ and $g(x) = \cos x$

64. (a) $f(x) = \sum_{n=0}^{\infty}\frac{x^n}{n!}, -\infty < x < \infty$ (See Exercise 11)

(b) $f'(x) = \sum_{n=1}^{\infty}\frac{nx^{n-1}}{n!} = \sum_{n=1}^{\infty}\frac{x^{n-1}}{(n-1)!} = \sum_{n=0}^{\infty}\frac{x^n}{n!} = f(x)$

(c) $f(x) = \sum_{n=1}^{\infty}\frac{x^n}{n!} = 1 + x + \frac{x^2}{2!} + \frac{x^3}{3!} + \frac{x^4}{4!} + \cdots$

$f(0) = 1$

(d) $f(x) = e^x$

65. $y = \sum_{n=0}^{\infty}\frac{(-1)^n x^{2n+1}}{(2n+1)!} = \sum_{n=1}^{\infty}\frac{(-1)^{n-1}x^{2n-1}}{(2n-1)!}$

$y' = \sum_{n=0}^{\infty}\frac{(-1)^n(2n+1)x^{2n}}{(2n+1)!} = \sum_{n=0}^{\infty}\frac{(-1)^n x^{2n}}{(2n)!}$

$y'' = \sum_{n=1}^{\infty}\frac{(-1)^n(2n)x^{2n-1}}{(2n)!} = \sum_{n=1}^{\infty}\frac{(-1)x^{2n-1}}{(2n-1)!}$

$y'' + y = \sum_{n=1}^{\infty}\frac{(-1)^n x^{2n-1}}{(2n-1)!} + \sum_{n=1}^{\infty}\frac{(-1)^{n-1}x^{2n-1}}{(2n-1)!} = 0$

66. $y = \sum_{n=0}^{\infty}\frac{(-1)^n x^{2n}}{(2n)!} = \sum_{n=1}^{\infty}\frac{(-1)^{n-1}x^{2n-2}}{(2n-2)!}$

$y' = \sum_{n=1}^{\infty}\frac{(-1)^n(2n)x^{2n-1}}{(2n)!} = \sum_{n=1}^{\infty}\frac{(-1)^n x^{2n-1}}{(2n-1)!}$

$y'' = \sum_{n=1}^{\infty}\frac{(-1)^n(2n-1)x^{2n-2}}{(2n-1)!} = \sum_{n=1}^{\infty}\frac{(-1)^n x^{2n-2}}{(2n-2)!}$

$y'' + y = \sum_{n=1}^{\infty}\frac{(-1)^n x^{2n-2}}{(2n-2)!} + \sum_{n=1}^{\infty}\frac{(-1)^{n-1}x^{2n-2}}{(2n-2)!} = 0$

67. $y = \sum_{n=0}^{\infty}\frac{x^{2n+1}}{(2n+1)!} = \sum_{n=1}^{\infty}\frac{x^{2n-1}}{(2n-1)!}$

$y' = \sum_{n=0}^{\infty}\frac{(2n+1)x^{2n}}{(2n+1)!} = \sum_{n=0}^{\infty}\frac{x^{2n}}{(2n)!}$

$y'' = \sum_{n=1}^{\infty}\frac{(2n)x^{2n-1}}{(2n)!} = \sum_{n=1}^{\infty}\frac{x^{2n-1}}{(2n-1)!} = y$

$y'' - y = 0$

68. $y = \sum_{n=0}^{\infty}\frac{x^{2n}}{(2n)!} = \sum_{n=1}^{\infty}\frac{x^{2n-2}}{(2n-2)!}$

$y' = \sum_{n=1}^{\infty}\frac{(2n-2)x^{2n-1}}{(2n-2)!} = \sum_{n=1}^{\infty}\frac{x^{2n-1}}{(2n-1)!}$

$y'' = \sum_{n=1}^{\infty}\frac{(2n-1)x^{2n-2}}{(2n-1)!} = \sum_{n=1}^{\infty}\frac{x^{2n-2}}{(2n-2)!} = y$

$y'' - y = 0$

69. $y = \sum_{n=0}^{\infty}\frac{x^{2n}}{2^n n!}$ $\quad y' = \sum_{n=1}^{\infty}\frac{2nx^{2n-1}}{2^n n!}$ $\quad y'' = \sum_{n=1}^{\infty}\frac{2n(2n-1)x^{2n-2}}{2^n n!}$

$$y'' - xy' - y = \sum_{n=1}^{\infty}\frac{2n(2n-1)x^{2n-2}}{2^n n!} - \sum_{n=1}^{\infty}\frac{2nx^{2n}}{2^n n!} - \sum_{n=0}^{\infty}\frac{x^{2n}}{2^n n!}$$

$$= \sum_{n=1}^{\infty}\frac{2n(2n-1)x^{2n-2}}{2^n n!} - \sum_{n=0}^{\infty}\frac{(2n+1)x^{2n}}{2^n n!}$$

$$= \sum_{n=0}^{\infty}\left[\frac{(2n+2)(2n+1)x^{2n}}{2^{n+1}(n+1)!} - \frac{(2n+1)x^{2n}}{2^n n!}\cdot\frac{2(n+1)}{2(n+1)}\right]$$

$$= \sum_{n=0}^{\infty}\frac{2(n+1)x^{2n}\,[(2n+1)-(2n+1)]}{2^{n+1}(n+1)!} = 0$$

70. $$y = 1 + \sum_{n=1}^{\infty} \frac{(-1)^n x^{4n}}{2^{2n} n! \cdot 3 \cdot 7 \cdot 11 \cdot \cdot \cdot (4n - 1)}$$

$$y' = \sum_{n=1}^{\infty} \frac{(-1)^n 4nx^{4n-1}}{2^{2n} n! \cdot 3 \cdot 7 \cdot 11 \cdot \cdot \cdot (4n - 1)}$$

$$y'' = \sum_{n=1}^{\infty} \frac{(-1)^n 4n\cancel{(4n-1)}x^{4n-2}}{2^{2n} n! \cdot 3 \cdot 7 \cdot 11 \cdot \cdot \cdot \cancel{(4n-1)}} = -x^2 + \sum_{n=2}^{\infty} \frac{(-1)^n 4nx^{4n-2}}{2^{2n} n! \cdot 3 \cdot 7 \cdot 11 \cdot \cdot \cdot (4n - 5)}$$

$$y'' + x^2 y = -x^2 + \sum_{n=2}^{\infty} \frac{(-1)^n 4nx^{4n-2}}{2^{2n} n! \cdot 3 \cdot 7 \cdot 11 \cdot \cdot \cdot (4n - 5)} + \sum_{n=1}^{\infty} \frac{(-1)^n x^{4n+2}}{2^{2n} n! \cdot 3 \cdot 7 \cdot 11 \cdot \cdot \cdot (4n - 1)} + x^2$$

$$= \sum_{n=1}^{\infty} \frac{(-1)^{n+1} 4(n+1)x^{4n+2}}{2^{2n+2}(n+1)! \cdot 3 \cdot 7 \cdot 11 \cdot \cdot \cdot (4n - 1)} - \sum_{n=1}^{\infty} \frac{(-1)^{n+1} x^{4n+2}}{2^{2n} n! \cdot 3 \cdot 7 \cdot 11 \cdot \cdot \cdot (4n - 1)} \frac{2^2(n+1)}{2^2(n+1)} = 0$$

71. $J_0(x) = \displaystyle\sum_{k=0}^{\infty} \frac{(-1)^k x^{2k}}{2^{2k}(k!)^2}$

(a) $$\lim_{k \to \infty} \left| \frac{u_{k+1}}{u_k} \right| = \lim_{k \to \infty} \left| \frac{(-1)^{k+1} x^{2k+2}}{2^{2k+2}[(k+1)!]^2} \cdot \frac{2^{2k}(k!)^2}{(-1)^k x^{2k}} \right| = \lim_{k \to \infty} \left| \frac{(-1)x^2}{2^2(k+1)^2} \right| = 0$$

Therefore, the interval of convergence is $-\infty < x < \infty$.

(b) $$J_0 = \sum_{k=0}^{\infty} (-1)^k \frac{x^{2k}}{4^k (k!)^2}$$

$$J_0' = \sum_{k=1}^{\infty} (-1)^k \frac{2kx^{2k-1}}{4^k (k!)^2} = \sum_{k=0}^{\infty} (-1)^{k+1} \frac{(2k+2)x^{2k+1}}{4^{k+1}[(k+1)!]^2}$$

$$J_0'' = \sum_{k=1}^{\infty} (-1)^k \frac{2k(2k-1)x^{2k-2}}{4^k (k!)^2} = \sum_{k=0}^{\infty} (-1)^{k+1} \frac{(2k+2)(2k+1)x^{2k}}{4^{k+1}[(k+1)!]^2}$$

$$x^2 J_0'' + xJ_0' + x^2 J_0 = \sum_{k=0}^{\infty} (-1)^{k+1} \frac{2(2k+1)x^{2k+2}}{4^{k+1}(k+1)!k!} + \sum_{k=0}^{\infty} (-1)^{k+1} \frac{2x^{2k+2}}{4^{k+1}(k+1)!k!} + \sum_{k=0}^{\infty} (-1)^k \frac{x^{2k+2}}{4^k (k!)^2}$$

$$= \sum_{k=0}^{\infty} \frac{(-1)^k x^{2k+2}}{4^k (k!)^2} \left[(-1)\frac{2(2k+1)}{4(k+1)} + (-1)\frac{2}{4(k+1)} + 1 \right]$$

$$= \sum_{k=0}^{\infty} \frac{(-1)^k x^{2k+2}}{4^k (k!)^2} \left[\frac{-4k-2}{4k+4} - \frac{2}{4k+4} + \frac{4k+4}{4k+4} \right] = 0$$

(c) $P_6(x) = 1 - \dfrac{x^2}{4} + \dfrac{x^4}{64} - \dfrac{x^6}{2304}$

(d) $$\int_0^1 J_0 dx = \int_0^1 \sum_{k=0}^{\infty} \frac{(-1)^k x^{2k}}{4^k (k!)^2} dx$$

$$= \left[\sum_{k=0}^{\infty} \frac{(-1)^k x^{2k+1}}{4^k (k!)^2 (2k+1)} \right]_0^1$$

$$= \sum_{k=0}^{\infty} \frac{(-1)^k}{4^k (k!)^2 (2k+1)}$$

$$= 1 - \frac{1}{12} + \frac{1}{320} \approx 0.92$$

(integral is approximately 0.9197304101)

72. $J_1(x) = x \displaystyle\sum_{k=0}^{\infty} \frac{(-1)^k x^{2k}}{2^{2k+1}k!(k+1)!} = \sum_{k=0}^{\infty} \frac{(-1)^k x^{2k+1}}{2^{2k+1}k!(k+1)!}$

(a) $$\lim_{k \to \infty} \left| \frac{u_{k+1}}{u_k} \right| = \lim_{k \to \infty} \left| \frac{(-1)^{k+1} x^{2k+3}}{2^{2k+3}(k+1)!(k+2)!} \cdot \frac{2^{2k+1}k!(k+1)!}{(-1)^k x^{2k+1}} \right| = \lim_{k \to \infty} \left| \frac{(-1)x^2}{2^2 (k+2)(k+1)} \right| = 0$$

Therefore, the interval of convergence is $-\infty < x < \infty$.

—CONTINUED—

72. —CONTINUED

(b) $J_1(x) = \sum_{k=0}^{\infty} \frac{(-1)^k x^{2k+1}}{2^{2k+1}k!(k+1!)}$

$J_1'(x) = \sum_{k=0}^{\infty} \frac{(-1)^k (2k+1)x^{2k}}{2^{2k+1}k!(k+1)!}$

$J_1''(x) = \sum_{k=1}^{\infty} \frac{(-1)^k (2k+1)(2k)x^{2k-1}}{2^{2k+1}k!(k+1)!}$

$$x^2J_1'' + xJ_1' + (x^2-1)J_1 = \sum_{k=1}^{\infty} \frac{(-1)^k(2k+1)(2k)x^{2k+1}}{2^{2k+1}k!(k+1)!} + \sum_{k=0}^{\infty} \frac{(-1)^k(2k+1)x^{2k+1}}{2^{2k+1}k!(k+1)!}$$

$$+ \sum_{k=0}^{\infty} \frac{(-1)^k x^{2k+3}}{2^{2k+1}k!(k+1)!} - \sum_{k=0}^{\infty} \frac{(-1)^k x^{2k+1}}{2^{2k+1}k!(k+1)!}$$

$$= \left[\sum_{k=1}^{\infty} \frac{(-1)^k(2k+1)(2k)x^{2k+1}}{2^{2k+1}k!(k+1)!} + \frac{x}{2} + \sum_{k=1}^{\infty} \frac{(-1)^k(2k+1)x^{2k+1}}{2^{2k+1}k!(k+1)!}\right.$$

$$\left. - \frac{x}{2} - \sum_{k=1}^{\infty} \frac{(-1)^k x^{2k+1}}{2^{2k+1}k!(k+1)!}\right] + \sum_{k=0}^{\infty} \frac{(-1)^k x^{2k+3}}{2^{2k+1}k!(k+1)!}$$

$$= \sum_{k=1}^{\infty} \frac{(-1)^k x^{2k+1}[(2k+1)(2k) + (2k+1) - 1]}{2^{2k+1}k!(k+1)!} + \sum_{k=0}^{\infty} \frac{(-1)^k x^{2k+3}}{2^{2k+1}k!(k+1)!}$$

$$= \sum_{k=1}^{\infty} \frac{(-1)^k x^{2k+1} 4k(k+1)}{2^{2k+1}k!(k+1)!} + \sum_{k=0}^{\infty} \frac{(-1)^k x^{2k+3}}{2^{2k+1}k!(k+1)!}$$

$$= \sum_{k=1}^{\infty} \frac{(-1)^k x^{2k+1}}{2^{2k-1}(k-1)!k!} + \sum_{k=0}^{\infty} \frac{(-1)^k x^{2k+3}}{2^{2k+1}k!(k+1)!}$$

$$= \sum_{k=0}^{\infty} \frac{(-1)^{k+1} x^{2k+3}}{2^{2k+1}k!(k+1)!} + \sum_{k=0}^{\infty} \frac{(-1)^k x^{2k+3}}{2^{2k+1}k!(k+1)!}$$

$$= \sum_{k=0}^{\infty} \frac{(-1)^k x^{2k+3}[(-1)+1]}{2^{2k+1}k!(k+1)!} = 0$$

(c) $P_7(x) = \frac{x}{2} - \frac{1}{16}x^3 + \frac{1}{384}x^5 - \frac{1}{18{,}432}x^7$

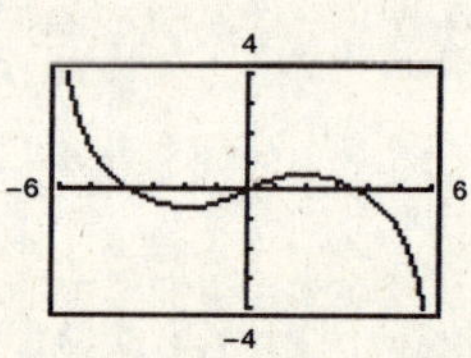

(d) $J_0'(x) = \sum_{k=0}^{\infty} \frac{(-1)^{k+1}2(k+1)x^{2k+1}}{2^{2k+2}(k+1)!(k+1)!} = \sum_{k=0}^{\infty} \frac{(-1)^{k+1}x^{2k+1}}{2^{2k+1}k!(k+1)!}$

$-J_1(x) = -\sum_{k=0}^{\infty} \frac{(-1)^k x^{2k+1}}{2^{2k+1}k!(k+1)!} = \sum_{k=0}^{\infty} \frac{(-1)^{k+1}x^{2k+1}}{2^{2k+1}k!(k+1)!}$ **Note:** $J_0'(x) = -J_1(x)$

73. $f(x) = \sum_{n=0}^{\infty} (-1)^n \frac{x^{2n}}{(2n)!} = \cos x$

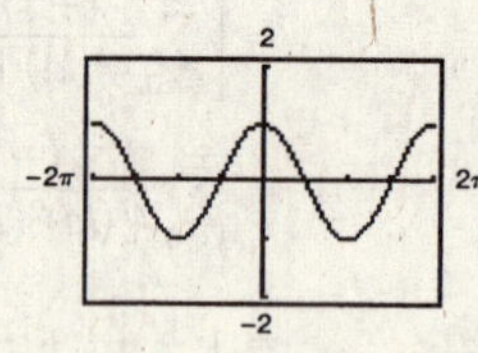

74. $f(x) = \sum_{n=0}^{\infty} (-1)^n \frac{x^{2n+1}}{(2n+1)!}$

$= \sin x$

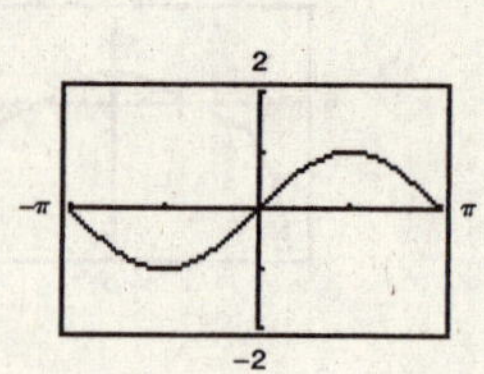

75. $f(x) = \sum_{n=0}^{\infty} (-1)^n x^n = \sum_{n=0}^{\infty} (-x)^n$ Geometric

$= \frac{1}{1-(-x)} = \frac{1}{1+x}$ for $-1 < x < 1$

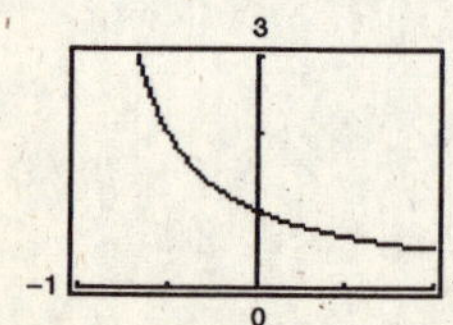

76. $f(x) = \sum_{n=0}^{\infty} (-1)^n \frac{x^{2n+1}}{2n+1} = \arctan x, \ -1 \le x \le 1$

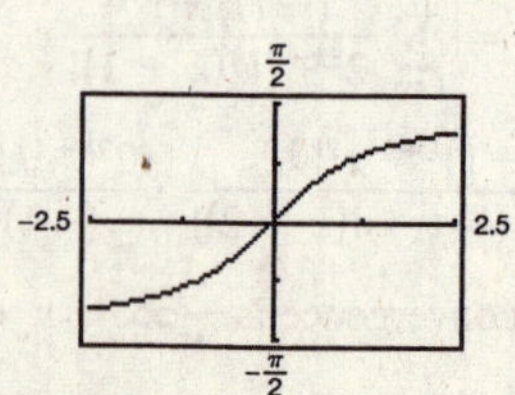

77. $\displaystyle\sum_{n=0}^{\infty}\left(\frac{x}{2}\right)^n$

(a) $\displaystyle\sum_{n=0}^{\infty}\left(\frac{3/4}{2}\right)^n = \sum_{n=0}^{\infty}\left(\frac{3}{8}\right)^n = \frac{1}{1-(3/8)} = \frac{8}{5} = 1.6$

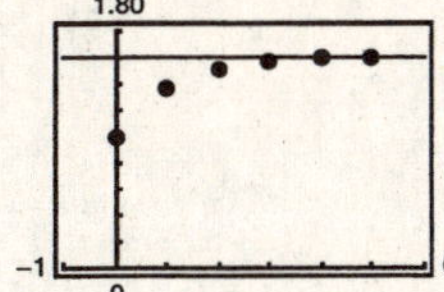

(b) 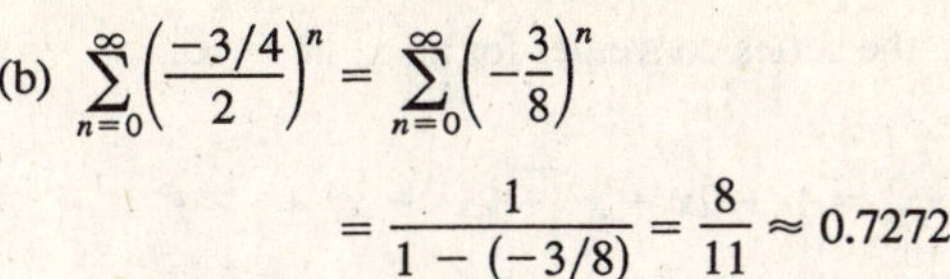

$\displaystyle\sum_{n=0}^{\infty}\left(\frac{-3/4}{2}\right)^n = \sum_{n=0}^{\infty}\left(-\frac{3}{8}\right)^n$

$\displaystyle = \frac{1}{1-(-3/8)} = \frac{8}{11} \approx 0.7272$

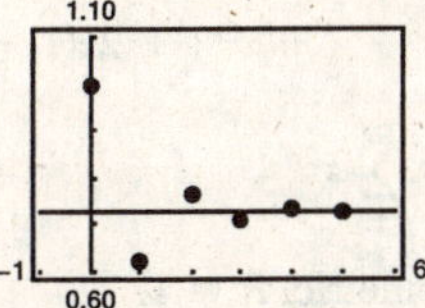

(c) The alternating series converges more rapidly. The partial sums of the series of positive terms approach the sum from below. The partial sums of the alternating series alternate sides of the horizontal line representing the sum.

(d) $\displaystyle\sum_{n=0}^{N}\left(\frac{3}{2}\right)^n > M$

M	10	100	1000	10,000
N	4	9	15	21

78. $\displaystyle\sum_{n=0}^{\infty}(3x)^n$ converges on $\left(-\frac{1}{3}, \frac{1}{3}\right)$.

(a) $x = \frac{1}{6}$: $\displaystyle\sum_{n=0}^{\infty}\left(3\left(\frac{1}{6}\right)\right)^n = \sum_{n=0}^{\infty}\left(\frac{1}{2}\right)^n = \frac{1}{1-(1/2)} = 2$

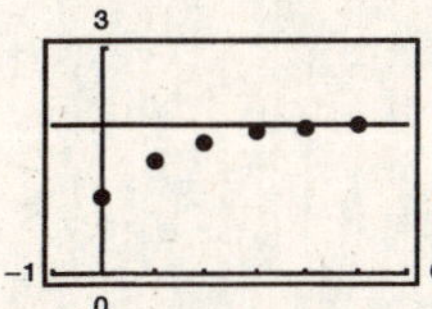

(b) $x = -\frac{1}{6}$: $\displaystyle\sum_{n=0}^{\infty}\left(3\left(-\frac{1}{6}\right)\right)^n = \sum_{n=0}^{\infty}\left(-\frac{1}{2}\right)^n$

$\displaystyle = \frac{1}{1+(1/2)} = \frac{2}{3}$

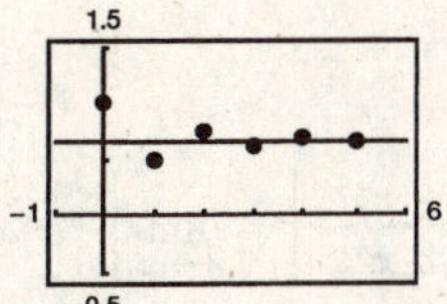

(c) The alternating series converges more rapidly. The partial sums in (a) approach the sum 2 from below. The partial sums in (b) alternate sides of the horizontal line $y = \frac{2}{3}$.

(d) $\displaystyle\sum_{n=0}^{N}\left(3\cdot\frac{2}{3}\right)^n = \sum_{n=0}^{N}2^n > M$

M	10	100	1000	10,000
N	3	6	9	13

79. False;

$$\sum_{n=0}^{\infty}\frac{(-1)^n x^n}{n2^n}$$

converges for $x = 2$ but diverges for $x = -2$.

80. True; if

$$\sum_{n=0}^{\infty}a_n x^n$$

converges for $x = 2$, then we know that it must converge on $(-2, 2]$.

81. True; the radius of convergence is $R = 1$ for both series.

82. True

$$\int_0^1 f(x)\,dx = \int_0^1\left(\sum_{n=0}^{\infty}a_n x^n\right)dx = \left[\sum_{n=0}^{\infty}\frac{a_n x^{n+1}}{n+1}\right]_0^1 = \sum_{n=0}^{\infty}\frac{a_n}{n+1}$$

83. $\lim_{n\to\infty}\left|\frac{a_{n+1}}{a_n}\right| = \lim_{n\to\infty}\left|\frac{(n+1+p)!}{(n+1)!(n+1+q)!}x^{n+1}\bigg/\frac{(n+p)!}{n!(n+q)!}x^n\right| = \lim_{n\to\infty}\left|\frac{(n+1+p)x}{(n+1)(n+1+q)}\right| = 0$

Thus, the series converges for all x: $R = \infty$.

84. (a) $g(x) = 1 + 2x + x^2 + 2x^3 + x^4 + \cdots$

$$S_{2n} = 1 + 2x + x^2 + 2x^3 + x^4 + \cdots + x^{2n} + 2x^{2n+1}$$
$$= (1 + x^2 + x^4 + \cdots + x^{2n}) + 2x(1 + x^2 + x^4 + \cdots + x^{2n})$$

$$\lim_{n\to\infty} S_{2n} = \sum_{n=0}^{\infty} x^{2n} + 2x\sum_{n=0}^{\infty} x^{2n}$$

Since each series is geometric, $R = 1$.

(b) For $|x| < 1$, $g(x) = \frac{1}{1-x^2} + 2x\frac{1}{1-x^2} = \frac{1+2x}{1-x^2}$.

85. (a) $f(x) = \sum_{n=0}^{\infty} c_n x^n$, $c_{n+3} = c_n$

$$= c_0 + c_1x + c_2x^2 + c_0x^3 + c_1x^4 + c_2x^5 + c_0x^6 + \cdots$$

$$S_{3n} = c_0(1 + x^3 + \cdots + x^{3n}) + c_1x(1 + x^3 + \cdots + x^{3n}) + c_2x^2(1 + x^3 + \cdots + x^{3n})$$

$$\lim_{n\to\infty} S_{3n} = c_0\sum_{n=0}^{\infty} x^{3n} + c_1x\sum_{n=0}^{\infty} x^{3n} + c_2x^2\sum_{n=0}^{\infty} x^{3n}$$

Each series is geometric, $R = 1$, and the interval of convergence is $(-1, 1)$.

(b) For $|x| < 1$, $f(x) = c_0\frac{1}{1-x^3} + c_1x\frac{1}{1-x^3} + c_2x^2\frac{1}{1-x^3} = \frac{c_0 + c_1x + c_2x^2}{1-x^3}$.

86. For the series $\sum c_n x^n$,

$$\lim_{n\to\infty}\left|\frac{a_{n+1}}{a_n}\right| = \lim_{n\to\infty}\left|\frac{c_{n+1}x^{n+1}}{c_nx^n}\right| = \lim_{n\to\infty}\left|\frac{c_{n+1}}{c_n}x\right| < 1 \Rightarrow |x| < \left|\frac{c_n}{c_{n+1}}\right| = R$$

For the series $\sum c_n x^{2n}$,

$$\lim_{n\to\infty}\left|\frac{b_{n+1}}{b_n}\right| = \lim_{n\to\infty}\left|\frac{c_{n+1}x^{2n+2}}{c_nx^{2n}}\right| = \lim_{n\to\infty}\left|\frac{c_{n+1}}{c_n}x^2\right| < 1 \Rightarrow |x^2| < \left|\frac{c_n}{c_{n+1}}\right| = R \Rightarrow |x| < \sqrt{R}.$$

87. At $x_0 + R$, the series is $\sum_{n=0}^{\infty} c_n(x_0 - x_0 - R)^n = \sum_{n=0}^{\infty} c_n(-R)^n$ which converges.

At $x_0 - R$, the series is $\sum_{n=0}^{\infty} c_nR^n$, which diverges.

Hence, at $x_0 + R$, we have $\sum_{n=0}^{\infty} c_n(-R)^n$ converges, $\sum_{n=0}^{\infty} |c_n(-R)^n| = \sum_{n=0}^{\infty} c_nR^n$ diverges.

$\Rightarrow$ The series converges conditionally.

Section 9.9 Representation of Functions by Power Series

1. (a) $\dfrac{1}{2-x} = \dfrac{1/2}{1-(x/2)} = \dfrac{a}{1-r}$

$$= \sum_{n=0}^{\infty} \frac{1}{2}\left(\frac{x}{2}\right)^n = \sum_{n=0}^{\infty} \frac{x^n}{2^{n+1}}$$

This series converges on $(-2, 2)$.

(b)
$$\begin{array}{r l}
 & \frac{1}{2} + \frac{x}{4} + \frac{x^2}{8} + \frac{x^3}{16} + \cdots \\
2 - x \,\big) & 1 \\
 & 1 - \frac{x}{2} \\ \hline
 & \frac{x}{2} \\
 & \frac{x}{2} - \frac{x^2}{4} \\ \hline
 & \frac{x^2}{4} \\
 & \frac{x^2}{4} - \frac{x^3}{8} \\ \hline
 & \frac{x^3}{8} \\
 & \frac{x^3}{8} - \frac{x^4}{16} \\ \hline
 & \vdots
\end{array}$$

2. (a) $f(x) = \dfrac{4}{5-x} = \dfrac{4/5}{1-x/5} = \dfrac{a}{1-r}$

$$= \sum_{n=0}^{\infty} \frac{4}{5}\left(\frac{x}{5}\right)^n = \sum_{n=0}^{\infty} \frac{4x^n}{5^{n+1}}$$

This series converges on $(-5, 5)$.

(b)
$$\begin{array}{r l}
 & \frac{4}{5} + \frac{4}{25}x + \frac{4}{125}x^2 + \frac{4x^3}{625} + \cdots \\
5 - x \,\big) & 4 \\
 & 4 - \frac{4}{5}x \\ \hline
 & \frac{4}{5}x \\
 & \frac{4}{5}x - \frac{4}{25}x^2 \\ \hline
 & \frac{4}{25}x^2 \\
 & \frac{4}{25}x^2 - \frac{4x^3}{125} \\ \hline
 & \frac{4x^3}{125} \\
 & \frac{4x^3}{125} - \frac{4x^4}{625} \\ \hline
 & \vdots
\end{array}$$

3. (a) $\dfrac{1}{2+x} = \dfrac{1/2}{1-(-x/2)} = \dfrac{a}{1-r}$

$$= \sum_{n=0}^{\infty} \frac{1}{2}\left(-\frac{x}{2}\right)^n = \sum_{n=0}^{\infty} \frac{(-1)^n x^n}{2^{n+1}}$$

This series converges on $(-2, 2)$.

(b)
$$\begin{array}{r l}
 & \frac{1}{2} - \frac{x}{4} + \frac{x^2}{8} - \frac{x^3}{16} + \cdots \\
2 + x \,\big) & 1 \\
 & 1 + \frac{x}{2} \\ \hline
 & -\frac{x}{2} \\
 & -\frac{x}{2} - \frac{x^2}{4} \\ \hline
 & \frac{x^2}{4} \\
 & \frac{x^2}{4} + \frac{x^3}{8} \\ \hline
 & -\frac{x^3}{8} \\
 & -\frac{x^3}{8} - \frac{x^4}{16} \\ \hline
 & \vdots
\end{array}$$

4. (a) $\dfrac{1}{1+x} = \dfrac{1}{1-(-x)} = \dfrac{a}{1-r}$

$$= \sum_{n=0}^{\infty} (-x)^n = \sum_{n=0}^{\infty} (-1)^n x^n$$

This series converges on $(-1, 1)$.

(b)
$$\begin{array}{r l}
 & 1 - x + x^2 - x^3 + \cdots \\
1 + x \,\big) & 1 \\
 & 1 + x \\ \hline
 & -x \\
 & -x - x^2 \\ \hline
 & x^2 \\
 & x^2 + x^3 \\ \hline
 & -x^3 \\
 & -x^3 - x^4 \\ \hline
 & \vdots
\end{array}$$

5. Writing $f(x)$ in the form $a/(1 - r)$, we have

$$\frac{1}{2-x} = \frac{1}{-3-(x-5)} = \frac{-1/3}{1+(1/3)(x-5)}$$

which implies that $a = -1/3$ and $r = (-1/3)(x - 5)$. Therefore, the power series for $f(x)$ is given by

$$\frac{1}{2-x} = \sum_{n=0}^{\infty} ar^n = \sum_{n=0}^{\infty} -\frac{1}{3}\left[-\frac{1}{3}(x-5)\right]^n$$

$$= \sum_{n=0}^{\infty} \frac{(x-5)^n}{(-3)^{n+1}}, \ |x-5| < 3 \text{ or } 2 < x < 8.$$

6. Writing $f(x)$ in the form $a/(1 - r)$, we have

$$\frac{4}{5-x} = \frac{4}{7-(x+2)} = \frac{4/7}{1-1/7(x+2)} = \frac{a}{1-r}.$$

Therefore, the power series for $f(x)$ is given by

$$\frac{4}{5-x} = \sum_{n=0}^{\infty} ar^n = \sum_{n=0}^{\infty} \frac{4}{7}\left(\frac{1}{7}(x+2)\right)^n$$

$$= \sum_{n=0}^{\infty} \frac{4(x+2)^n}{7^{n+1}}.$$

$|x + 2| < 7$ or $-9 < x < 5$

7. Writing $f(x)$ in the form $a/(1 - r)$, we have

$$\frac{3}{2x-1} = \frac{-3}{1-2x} = \frac{a}{1-r}$$

which implies that $a = -3$ and $r = 2x$. Therefore, the power series for $f(x)$ is given by

$$\frac{3}{2x-1} = \sum_{n=0}^{\infty} ar^n = \sum_{n=0}^{\infty} (-3)(2x)^n$$

$$= -3\sum_{n=0}^{\infty} (2x)^n, \ |2x| < 1 \text{ or } -\frac{1}{2} < x < \frac{1}{2}.$$

8. Writing $f(x)$ in the form $a/(1 - r)$, we have

$$\frac{3}{2x-1} = \frac{3}{3+2(x-2)} = \frac{1}{1+(2/3)(x-2)} = \frac{a}{1-r}$$

which implies that $a = 1$ and $r = (-2/3)(x - 2)$. Therefore, the power series for $f(x)$ is given by

$$\frac{3}{2x-1} = \sum_{n=0}^{\infty} ar^n = \sum_{n=0}^{\infty}\left[-\frac{2}{3}(x-2)\right]^n,$$

$$= \sum_{n=0}^{\infty} \frac{(-2)^n(x-2)^n}{3^n},$$

$$|x-2| < \frac{3}{2} \text{ or } \frac{1}{2} < x < \frac{7}{2}.$$

9. Writing $f(x)$ in the form $a/(1 - r)$, we have

$$\frac{1}{2x-5} = \frac{-1}{11-2(x+3)}$$

$$= \frac{-1/11}{1-(2/11)(x+3)} = \frac{a}{1-r}$$

which implies that $a = -1/11$ and $r = (2/11)(x + 3)$. Therefore, the power series for $f(x)$ is given by

$$\frac{1}{2x-5} = \sum_{n=0}^{\infty} ar^n = \sum_{n=0}^{\infty}\left(-\frac{1}{11}\right)\left[\frac{2}{11}(x+3)\right]^n$$

$$= -\sum_{n=0}^{\infty} \frac{2^n(x+3)^n}{11^{n+1}},$$

$$|x+3| < \frac{11}{2} \text{ or } -\frac{17}{2} < x < \frac{5}{2}.$$

10. Writing $f(x)$ in the form $a/(1 - r)$, we have

$$\frac{1}{2x-5} = \frac{1}{-5+2x} = \frac{-1/5}{1-(2/5)x} = \frac{a}{1-r}$$

which implies that $a = -1/5$ and $r = (2/5)x$. Therefore, the power series for $f(x)$ is given by

$$\frac{1}{2x-5} = \sum_{n=0}^{\infty} ar^n = \sum_{n=0}^{\infty}\left(-\frac{1}{5}\right)\left(\frac{2}{5}x\right)^n = -\sum_{n=0}^{\infty} \frac{2^n x^n}{5^{n+1}},$$

$$|x| < \frac{5}{2} \text{ or } -\frac{5}{2} < x < \frac{5}{2}.$$

11. Writing $f(x)$ in the form $a/(1 - r)$, we have

$$\frac{3}{x+2} = \frac{3}{2+x} = \frac{3/2}{1+(1/2)x} = \frac{a}{1-r}$$

which implies that $a = 3/2$ and $r = (-1/2)x$. Therefore, the power series for $f(x)$ is given by

$$\frac{3}{x+2} = \sum_{n=0}^{\infty} ar^n = \sum_{n=0}^{\infty} \frac{3}{2}\left(-\frac{1}{2}x\right)^n$$

$$= 3\sum_{n=0}^{\infty} \frac{(-1)^n x^n}{2^{n+1}} = \frac{3}{2}\sum_{n=0}^{\infty}\left(-\frac{x}{2}\right)^n,$$

$|x| < 2$ or $-2 < x < 2$.

12. Writing $f(x)$ in the form $a/(1 - r)$, we have

$$\frac{4}{3x+2} = \frac{4}{8+3(x-2)} = \frac{1/2}{1+(3/8)(x-2)} = \frac{a}{1-r}$$

which implies that $a = 1/2$ and $r = (-3/8)(x - 2)$. Therefore, the power series for $f(x)$ is given by

$$\frac{4}{3x+2} = \sum_{n=0}^{\infty} ar^n = \sum_{n=0}^{\infty} \frac{1}{2}\left[-\frac{3}{8}(x-2)\right]^n$$

$$= \frac{1}{2}\sum_{n=0}^{\infty} \frac{(-3)^n(x-2)^n}{8^n},$$

$$|x-2| < \frac{8}{3} \text{ or } -\frac{2}{3} < x < \frac{14}{3}.$$

13. $\dfrac{3x}{x^2+x-2} = \dfrac{2}{x+2} + \dfrac{1}{x-1} = \dfrac{2}{2+x} + \dfrac{1}{-1+x} = \dfrac{1}{1+(1/2)x} + \dfrac{-1}{1-x}$

Writing $f(x)$ as a sum of two geometric series, we have

$$\frac{3x}{x^2+x-2} = \sum_{n=0}^{\infty}\left(-\frac{1}{2}x\right)^n + \sum_{n=0}^{\infty}(-1)(x)^n = \sum_{n=0}^{\infty}\left[\frac{1}{(-2)^n} - 1\right]x^n.$$

The interval of convergence is $-1 < x < 1$ since

$$\lim_{n\to\infty}\left|\frac{u_{n+1}}{u_n}\right| = \lim_{n\to\infty}\left|\frac{(1-(-2)^{n+1})x^{n+1}}{(-2)^{n+1}} \cdot \frac{(-2)^n}{(1-(-2)^n)x^n}\right| = \lim_{n\to\infty}\left|\frac{(1-(-2)^{n+1})x}{-2-(-2)^{n+1}}\right| = |x|.$$

14. $\dfrac{4x-7}{2x^2+3x-2} = \dfrac{3}{x+2} - \dfrac{2}{2x-1} = \dfrac{3}{2+x} - \dfrac{2}{-1+2x} = \dfrac{3/2}{1+(1/2)x} + \dfrac{2}{1-2x}$

Writing $f(x)$ as a sum of two geometric series, we have

$$\frac{4x-7}{2x^2+3x-2} = \sum_{n=0}^{\infty}\left(\frac{3}{2}\right)\left(-\frac{1}{2}x\right)^n + \sum_{n=0}^{\infty} 2(2x)^n = \sum_{n=0}^{\infty}\left[\frac{3(-1)^n}{2^{n+1}} + 2^{n+1}\right]x^n,\ |x| < \frac{1}{2} \text{ or } -\frac{1}{2} < x < \frac{1}{2}.$$

15. $\dfrac{2}{1-x^2} = \dfrac{1}{1-x} + \dfrac{1}{1+x}$

Writing $f(x)$ as a sum of two geometric series, we have

$$\frac{2}{1-x^2} = \sum_{n=0}^{\infty} x^n + \sum_{n=0}^{\infty}(-x)^n = \sum_{n=0}^{\infty}(1+(-1)^n)x^n = \sum_{n=0}^{\infty} 2x^{2n}.$$

The interval of convergence is $|x^2| < 1$ or $-1 < x < 1$ since $\displaystyle\lim_{n\to\infty}\left|\frac{u_{n+1}}{u_n}\right| = \lim_{n\to\infty}\left|\frac{2x^{2n+2}}{2x^{2n}}\right| = |x^2|$.

16. First finding the power series for $4/(4+x)$, we have

$$\frac{1}{1+(1/4)x} = \sum_{n=0}^{\infty}\left(-\frac{1}{4}x\right)^n = \sum_{n=0}^{\infty}\frac{(-1)^n x^n}{4^n}$$

Now replace x with x^2.

$$\frac{4}{4+x^2} = \sum_{n=0}^{\infty}\frac{(-1)^n x^{2n}}{4^n}.$$

The interval of convergence is $|x^2| < 4$ or $-2 < x < 2$ since

$$\lim_{n\to\infty}\left|\frac{u_{n+1}}{u_n}\right| = \lim_{n\to\infty}\left|\frac{(-1)^{n+1}x^{2n+2}}{4^{n+1}} \cdot \frac{4^n}{(-1)^n x^{2n}}\right| = \left|-\frac{x^2}{4}\right| = \frac{|x^2|}{4}.$$

17. $\dfrac{1}{1+x} = \displaystyle\sum_{n=0}^{\infty}(-1)^n x^n$

$$\frac{1}{1-x} = \sum_{n=0}^{\infty}(-1)^n(-x)^n = \sum_{n=0}^{\infty}(-1)^{2n}x^n = \sum_{n=0}^{\infty} x^n$$

$$h(x) = \frac{-2}{x^2-1} = \frac{1}{1+x} + \frac{1}{1-x} = \sum_{n=0}^{\infty}(-1)^n x^n + \sum_{n=0}^{\infty} x^n = \sum_{n=0}^{\infty}[(-1)^n + 1]x^n$$

$$= 2 + 0x + 2x^2 + 0x^3 + 2x^4 + 0x^5 + 2x^6 + \cdots = \sum_{n=0}^{\infty} 2x^{2n},\ -1 < x < 1 \text{ (See Exercise 15.)}$$

18. $h(x) = \dfrac{x}{x^2 - 1} = \dfrac{1}{2(1 + x)} - \dfrac{1}{2(1 - x)} = \dfrac{1}{2}\sum_{n=0}^{\infty}(-1)^n x^n - \dfrac{1}{2}\sum_{n=0}^{\infty} x^n$

$$= \frac{1}{2}\sum_{n=0}^{\infty}[(-1)^n - 1]x^n = \frac{1}{2}[0 - 2x + 0x^2 - 2x^3 + 0x^4 - 2x^5 + \cdots]$$

$$= \frac{1}{2}\sum_{n=0}^{\infty}(-2)x^{2n+1} = -\sum_{n=0}^{\infty} x^{2n+1}, 1 < x < 1$$

19. By taking the first derivative, we have $\dfrac{d}{dx}\left[\dfrac{1}{x + 1}\right] = \dfrac{-1}{(x + 1)^2}$. Therefore,

$$\frac{-1}{(x + 1)^2} = \frac{d}{dx}\left[\sum_{n=0}^{\infty}(-1)^n x^n\right] = \sum_{n=1}^{\infty}(-1)^n n x^{n-1}$$

$$= \sum_{n=0}^{\infty}(-1)^{n+1}(n + 1)x^n, -1 < x < 1.$$

20. By taking the second derivative, we have $\dfrac{d^2}{dx^2}\left[\dfrac{1}{x + 1}\right] = \dfrac{2}{(x + 1)^3}$. Therefore,

$$\frac{2}{(x + 1)^3} = \frac{d^2}{dx^2}\left[\sum_{n=0}^{\infty}(-1)^n x^n\right]$$

$$= \frac{d}{dx}\left[\sum_{n=1}^{\infty}(-1)^n n x^{n-1}\right] = \sum_{n=2}^{\infty}(-1)^n n(n - 1)x^{n-2} = \sum_{n=0}^{\infty}(-1)^n(n + 2)(n + 1)x^n, -1 < x < 1.$$

21. By integrating, we have $\displaystyle\int \frac{1}{x + 1}\,dx = \ln(x + 1)$. Therefore,

$$\ln(x + 1) = \int\left[\sum_{n=0}^{\infty}(-1)^n x^n\right]dx = C + \sum_{n=0}^{\infty}\frac{(-1)^n x^{n+1}}{n + 1}, -1 < x \le 1.$$

To solve for C, let $x = 0$ and conclude that $C = 0$. Therefore,

$$\ln(x + 1) = \sum_{n=0}^{\infty}\frac{(-1)^n x^{n+1}}{n + 1}, -1 < x \le 1.$$

22. By integrating, we have

$$\int \frac{1}{1 + x}\,dx = \ln(1 + x) + C_1 \text{ and } \int \frac{1}{1 - x}\,dx = -\ln(1 - x) + C_2.$$

$f(x) = \ln(1 - x^2) = \ln(1 + x) - [-\ln(1 - x)]$. Therefore,

$$\ln(1 - x^2) = \int \frac{1}{1 + x}\,dx - \int \frac{1}{1 - x}\,dx$$

$$= \int\left[\sum_{n=0}^{\infty}(-1)^n x^n\right]dx - \int\left[\sum_{n=0}^{\infty} x^n\right]dx = \left[C_1 + \sum_{n=0}^{\infty}\frac{(-1)^n x^{n+1}}{n + 1}\right] - \left[C_2 + \sum_{n=0}^{\infty}\frac{x^{n+1}}{n + 1}\right]$$

$$= C + \sum_{n=0}^{\infty}\frac{[(-1)^n - 1]x^{n+1}}{n + 1} = C + \sum_{n=0}^{\infty}\frac{-2x^{2n+2}}{2n + 2} = C + \sum_{n=0}^{\infty}\frac{(-1)x^{2n+2}}{n + 1}$$

To solve for C, let $x = 0$ and conclude that $C = 0$. Therefore,

$$\ln(1 - x^2) = -\sum_{n=0}^{\infty}\frac{x^{2n+2}}{n + 1}, -1 < x < 1.$$

23. $\dfrac{1}{x^2 + 1} = \displaystyle\sum_{n=0}^{\infty}(-1)^n(x^2)^n = \sum_{n=0}^{\infty}(-1)^n x^{2n}, -1 < x < 1$

24. $\dfrac{2x}{x^2+1} = 2x\sum_{n=0}^{\infty}(-1)^n x^{2n}$ (See Exercise 23.)

$$= \sum_{n=0}^{\infty}(-1)^n 2x^{2n+1}$$

Since $\dfrac{d}{dx}(\ln(x^2+1)) = \dfrac{2x}{x^2+1}$, we have

$$\ln(x^2+1) = \int\left[\sum_{n=0}^{\infty}(-1)^n 2x^{2n+1}\right]dx = C + \sum_{n=0}^{\infty}\frac{(-1)^n x^{2n+2}}{n+1}, \quad -1 \le x \le 1.$$

To solve for C, let $x = 0$ and conclude that $C = 0$. Therefore,

$$\ln(x^2+1) = \sum_{n=0}^{\infty}\frac{(-1)^n x^{2n+2}}{n+1}, \quad -1 \le x \le 1.$$

25. Since, $\dfrac{1}{x+1} = \sum_{n=0}^{\infty}(-1)^n x^n$, we have $\dfrac{1}{4x^2+1} = \sum_{n=0}^{\infty}(-1)^n(4x^2)^n = \sum_{n=0}^{\infty}(-1)^n 4^n x^{2n} = \sum_{n=0}^{\infty}(-1)^n(2x)^{2n}$, $-\dfrac{1}{2} < x < \dfrac{1}{2}$.

26. Since $\displaystyle\int\frac{1}{4x^2+1}\,dx = \frac{1}{2}\arctan(2x)$, we can use the result of Exercise 25 to obtain

$$\arctan(2x) = 2\int\frac{1}{4x^2+1}\,dx = 2\int\left[\sum_{n=0}^{\infty}(-1)^n 4^n x^{2n}\right]dx = C + 2\sum_{n=0}^{\infty}\frac{(-1)^n 4^n x^{2n+1}}{2n+1}, \quad -\frac{1}{2} < x \le \frac{1}{2}.$$

To solve for C, let $x = 0$ and conclude that $C = 0$. Therefore,

$$\arctan(2x) = 2\sum_{n=0}^{\infty}\frac{(-1)^n 4^n x^{2n+1}}{2n+1}, \quad -\frac{1}{2} < x \le \frac{1}{2}.$$

27. $x - \dfrac{x^2}{2} \le \ln(x+1) \le x - \dfrac{x^2}{2} + \dfrac{x^3}{3}$

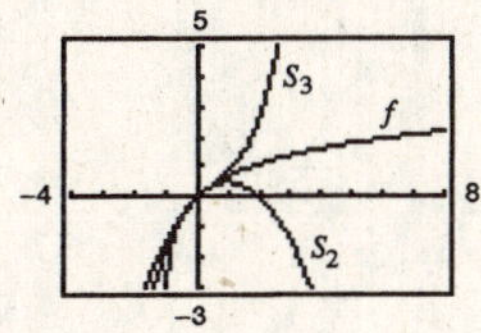

x	0.0	0.2	0.4	0.6	0.8	1.0
$x - \frac{x^2}{2}$	0.000	0.180	0.320	0.420	0.480	0.500
$\ln(x+1)$	0.000	0.182	0.336	0.470	0.588	0.693
$x - \frac{x^2}{2} + \frac{x^3}{3}$	0.000	0.183	0.341	0.492	0.651	0.833

28. $x - \dfrac{x^2}{2} + \dfrac{x^3}{3} - \dfrac{x^4}{4} \le \ln(x+1)$

$$\le x - \frac{x^2}{2} + \frac{x^3}{3} - \frac{x^4}{4} + \frac{x^5}{5}$$

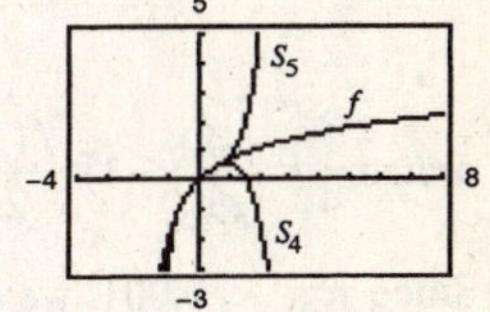

x	0.0	0.2	0.4	0.6	0.8	1.0
$x - \frac{x^2}{2} + \frac{x^3}{3} - \frac{x^4}{4}$	0.0	0.18227	0.33493	0.45960	0.54827	0.58333
$\ln(x+1)$	0.0	0.18232	0.33647	0.47000	0.58779	0.69315
$x - \frac{x^2}{2} + \frac{x^3}{3} - \frac{x^4}{4} + \frac{x^5}{5}$	0.0	0.18233	0.33698	0.47515	0.61380	0.78333

29. $\displaystyle\sum_{n=1}^{\infty} \frac{(-1)^{n+1}(x-1)^n}{n} = \frac{(x-1)}{1} - \frac{(x-1)^2}{2} + \frac{(x-1)^3}{3} - \cdots$

(a)

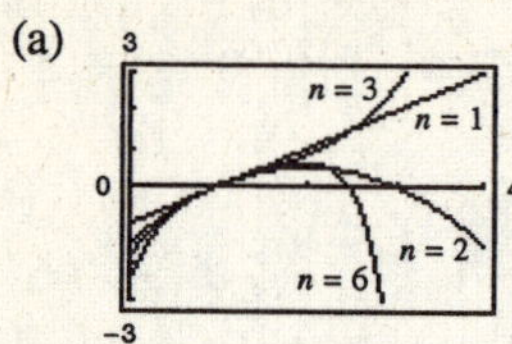

(b) From Example 4,

$$\sum_{n=1}^{\infty} \frac{(-1)^{n+1}(x-1)^n}{n} = \sum_{n=0}^{\infty} \frac{(-1)^n(x-1)^{n+1}}{n+1}$$

$$= \ln x, \quad 0 < x \le 2, \quad R = 1.$$

(c) $x = 0.5$:

$$\sum_{n=1}^{\infty} \frac{(-1)^{n+1}(-1/2)^n}{n} = \sum_{n=1}^{\infty} \frac{-(1/2)^n}{n} \approx -0.693147$$

(d) This is an approximation of $\ln\left(\frac{1}{2}\right)$. The error is approximately 0. $\left[\text{The error is less than the first omitted term, } 1/(51 \cdot 2^{51}) \approx 8.7 \times 10^{-18}\right]$

30. $\displaystyle\sum_{n=0}^{\infty} \frac{(-1)^n x^{2n+1}}{(2n+1)!} = x - \frac{x^3}{3!} + \frac{x^5}{5!} - \cdots$

(a)

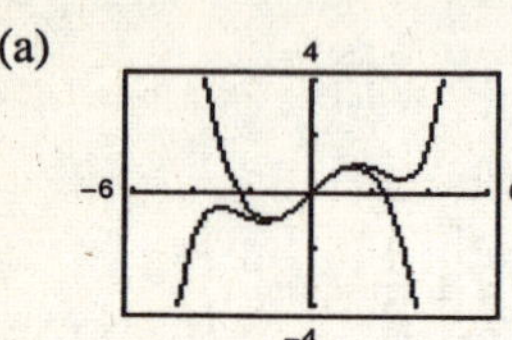

(b) $\displaystyle\sum_{n=0}^{\infty} \frac{(-1)^n x^{2n+1}}{(2n+1)!} = \sin x, R = \infty$

(c) $\displaystyle\sum_{n=0}^{\infty} \frac{(-1)^n(1/2)^{2n+1}}{(2n+1)!} \approx 0.4794255386$

(d) This is an approximation of $\sin\left(\frac{1}{2}\right)$.

The error is approximately 0.

31. $g(x) = x$ line

Matches (c)

32. $g(x) = x - \dfrac{x^3}{3}$, cubic with 3 zeros.

Matches (d)

33. $g(x) = x - \dfrac{x^3}{3} + \dfrac{x^5}{5}$

Matches (a)

34. $g(x) = x - \dfrac{x^3}{3} + \dfrac{x^5}{5} - \dfrac{x^7}{7}$,

Matches (b)

In Exercises 35-38, arctan $\displaystyle x = \sum_{n=0}^{\infty} (-1)^n \frac{x^{2n+1}}{2n+1}$.

35. $\displaystyle \arctan\frac{1}{4} = \sum_{n=0}^{\infty} (-1)^n \frac{(1/4)^{2n+1}}{2n+1} = \sum_{n=0}^{\infty} \frac{(-1)^n}{(2n+1)4^{2n+1}} = \frac{1}{4} - \frac{1}{192} + \frac{1}{5120} + \cdots$

Since $\frac{1}{5120} < 0.001$, we can approximate the series by its first two terms: $\arctan\frac{1}{4} \approx \frac{1}{4} - \frac{1}{192} \approx 0.245$.

36.

$$\arctan x^2 = \sum_{n=0}^{\infty} (-1)^n \frac{x^{4n+2}}{2n+1}$$

$$\int \arctan x^2\,dx = \sum_{n=0}^{\infty} (-1)^n \frac{x^{4n+3}}{(4n+3)(2n+1)} + C, C = 0$$

$$\int_0^{3/4} \arctan x^2\,dx = \sum_{n=0}^{\infty} (-1)^n \frac{(3/4)^{4n+3}}{(4n+3)(2n+1)} = \sum_{n=0}^{\infty} (-1)^n \frac{3^{4n+3}}{(4n+3)(2n+1)4^{4n+3}}$$

$$= \frac{27}{192} - \frac{2187}{344{,}064} + \frac{177{,}147}{230{,}686{,}720}$$

Since $177{,}147/230{,}686{,}720 < 0.001$, we can approximate the series by its first two terms: 0.134.

37. $\dfrac{\arctan x^2}{x} = \dfrac{1}{x}\displaystyle\sum_{n=0}^{\infty}(-1)^n\frac{(x^2)^{2n+1}}{2n+1} = \sum_{n=0}^{\infty}(-1)^2\frac{x^{4n+1}}{2n+1}$

$$\int \frac{\arctan x^2}{x}\,dx = \sum_{n=0}^{\infty}(-1)^n\frac{x^{4n+2}}{(4n+2)(2n+1)} + C \text{ (Note: } C = 0)$$

$$\int_0^{1/2} \frac{\arctan x^2}{x}\,dx = \sum_{n=0}^{\infty}(-1)^n\frac{1}{(4n+2)(2n+1)2^{4n+2}} = \frac{1}{8} - \frac{1}{1152} + \cdots$$

Since $\dfrac{1}{1152} < 0.001$, we can approximate the series by its first term: $\displaystyle\int_0^{1/2}\frac{\arctan x^2}{x}\,dx \approx 0.125$.

38. $x^2 \arctan x = \displaystyle\sum_{n=0}^{\infty}(-1)^n\frac{x^{2n+3}}{2n+1}$

$$\int x^2 \arctan x\,dx = \sum_{n=0}^{\infty}(-1)^n\frac{x^{2n+4}}{(2n+4)(2n+1)}$$

$$\int_0^{1/2} x^2 \arctan x\,dx = \sum_{n=0}^{\infty}(-1)^n\frac{1}{(2n+4)(2n+1)2^{2n+4}} = \frac{1}{64} - \frac{1}{1152} + \cdots$$

Since $\dfrac{1}{1152} < 0.001$, we can approximate the series by its first term: $\displaystyle\int_0^{1/2} x^2 \arctan x\,dx \approx 0.016$.

In Exercises 39–43, use $\dfrac{1}{1-x} = \displaystyle\sum_{n=0}^{\infty} x^n,\ |x| < 1$.

39. $\dfrac{1}{(1-x)^2} = \dfrac{d}{dx}\left[\dfrac{1}{1-x}\right] = \dfrac{d}{dx}\left[\displaystyle\sum_{n=0}^{\infty} x^n\right] = \displaystyle\sum_{n=1}^{\infty} nx^{n-1},\ |x| < 1$

40. $\dfrac{x}{(1-x)^2} = x\displaystyle\sum_{n=1}^{\infty} nx^{n-1} = \sum_{n=1}^{\infty} nx^n,\ |x| < 1$

41. $\dfrac{1+x}{(1-x)^2} = \dfrac{1}{(1-x)^2} + \dfrac{x}{(1-x)^2}$

$$= \sum_{n=1}^{\infty} n(x^{n-1} + x^n),\quad |x| < 1$$

$$= \sum_{n=0}^{\infty}(2n+1)x^n,\quad |x| < 1$$

42. $\dfrac{x(1+x)}{(1-x)^2} = x\displaystyle\sum_{n=0}^{\infty}(2n+1)x^n = \sum_{n=0}^{\infty}(2n+1)x^{n+1},\ |x| < 1$

(See Exercise 41.)

43. $P(n) = \left(\dfrac{1}{2}\right)^n$

$$E(n) = \sum_{n=1}^{\infty} nP(n) = \sum_{n=1}^{\infty} n\left(\frac{1}{2}\right)^n = \frac{1}{2}\sum_{n=1}^{\infty} n\left(\frac{1}{2}\right)^{n-1}$$

$$= \frac{1}{2}\,\frac{1}{[1-(1/2)]^2} = 2$$

Since the probability of obtaining a head on a single toss is $\frac{1}{2}$, it is expected that, on average, a head will be obtained in two tosses.

In Exercise 44, $\dfrac{1}{1-x} = \displaystyle\sum_{n=0}^{\infty} x^n,\ |x|1$.

44. (a) $\dfrac{1}{3}\displaystyle\sum_{n=1}^{\infty} n\left(\frac{2}{3}\right)^n = \frac{2}{9}\sum_{n=1}^{\infty} n\left(\frac{2}{3}\right)^{n-1} = \frac{2}{9}\,\frac{1}{[1-(2/3)]^2} = 2$

(b) $\dfrac{1}{10}\displaystyle\sum_{n=1}^{\infty} n\left(\frac{9}{10}\right)^n = \frac{9}{100}\sum_{n=1}^{\infty} n\left(\frac{9}{10}\right)^{n-1} = \frac{9}{100}\cdot\frac{1}{[1-(9/10)]^2} = 9$

45. Replace x with $(-x)$.

46. Replace x with x^2.

47. Replace x with $(-x)$ and multiply the series by 5.

48. Integrate the series and multiply by (-1).

49. Let $\arctan x + \arctan y = \theta$. Then,

$$\tan(\arctan x + \arctan y) = \tan\theta$$

$$\frac{\tan(\arctan x) + \tan(\arctan y)}{1 - \tan(\arctan x)\tan(\arctan y)} = \tan\theta$$

$$\frac{x+y}{1-xy} = \tan\theta$$

$$\arctan\left(\frac{x+y}{1-xy}\right) = \theta. \text{ Therefore, } \arctan x + \arctan y = \arctan\left(\frac{x+y}{1-xy}\right) \text{ for } xy \neq 1.$$

50. (a) From Exercise 49, we have

$$\arctan\frac{120}{119} - \arctan\frac{1}{239} = \arctan\frac{120}{119} + \arctan\left(-\frac{1}{239}\right)$$

$$= \arctan\left[\frac{(120/119) + (-1/239)}{1 - (120/119)(-1/239)}\right] = \arctan\left(\frac{28{,}561}{28{,}561}\right) = \arctan 1 = \frac{\pi}{4}$$

(b) $2\arctan\frac{1}{5} = \arctan\frac{1}{5} + \arctan\frac{1}{5} = \arctan\left[\frac{2(1/5)}{1-(1/5)^2}\right] = \arctan\frac{10}{24} = \arctan\frac{5}{12}$

$$4\arctan\frac{1}{5} = 2\arctan\frac{1}{5} + 2\arctan\frac{1}{5} = \arctan\frac{5}{12} + \arctan\frac{5}{12} = \arctan\left[\frac{2(5/12)}{1-(5/12)^2}\right] = \arctan\frac{120}{119}$$

$$4\arctan\frac{1}{5} - \arctan\frac{1}{239} = \arctan\frac{120}{119} - \arctan\frac{1}{239} = \frac{\pi}{4} \text{ (see part (a).)}$$

51. (a) $2\arctan\frac{1}{2} = \arctan\frac{1}{2} + \arctan\frac{1}{2} = \arctan\left[\frac{\frac{1}{2}+\frac{1}{2}}{1-(1/2)^2}\right] = \arctan\frac{4}{3}$

$$2\arctan\frac{1}{2} - \arctan\frac{1}{7} = \arctan\frac{4}{3} + \arctan\left(-\frac{1}{7}\right) = \arctan\left[\frac{(4/3)-(1/7)}{1+(4/3)(1/7)}\right] = \arctan\frac{25}{25} = \arctan 1 = \frac{\pi}{4}$$

(b) $\pi = 8\arctan\frac{1}{2} - 4\arctan\frac{1}{7} \approx 8\left[\frac{1}{2} - \frac{(0.5)^3}{3} + \frac{(0.5)^5}{5} - \frac{(0.5)^7}{7}\right] - 4\left[\frac{1}{7} - \frac{(1/7)^3}{3} + \frac{(1/7)^5}{5} - \frac{(1/7)^7}{7}\right] \approx 3.14$

52. (a) $\arctan\frac{1}{2} + \arctan\frac{1}{3} = \arctan\left[\frac{(1/2)+(1/3)}{1-(1/2)(1/3)}\right] = \arctan\left(\frac{5/6}{5/6}\right) = \frac{\pi}{4}$

(b) $\pi = 4\left[\arctan\frac{1}{2} + \arctan\frac{1}{3}\right]$

$$= 4\left[\frac{1}{2} - \frac{(1/2)^3}{3} + \frac{(1/2)^5}{5} - \frac{(1/2)^7}{7}\right] + 4\left[\frac{1}{3} - \frac{(1/3)^3}{3} + \frac{(1/3)^5}{5} - \frac{(1/3)^7}{7}\right] \approx 4(0.4635) + 4(0.3217) \approx 3.14$$

53. From Exercise 21, we have

$$\ln(x+1) = \sum_{n=0}^{\infty}\frac{(-1)^n x^{n+1}}{n+1} = \sum_{n=1}^{\infty}\frac{(-1)^{n-1}x^n}{n}$$

$$= \sum_{n=1}^{\infty}\frac{(-1)^{n+1}x^n}{n}.$$

Thus, $\displaystyle\sum_{n=1}^{\infty}(-1)^{n+1}\frac{1}{2^n n} = \sum_{n=1}^{\infty}\frac{(-1)^{n+1}(1/2)^n}{n}$

$$= \ln\left(\frac{1}{2}+1\right) = \ln\frac{3}{2} \approx 0.4055.$$

54. From Exercise 53, we have

$$\sum_{n=1}^{\infty}(-1)^{n+1}\frac{1}{3^n n} = \sum_{n=1}^{\infty}\frac{(-1)^{n+1}(1/3)^n}{n}$$

$$= \ln\left(\frac{1}{3}+1\right) = \ln\frac{4}{3} \approx 0.2877.$$

55. From Exercise 53, we have

$$\sum_{n=1}^{\infty}(-1)^{n+1}\frac{2^n}{5^n n} = \sum_{n=1}^{\infty}\frac{(-1)^{n+1}(2/5)^n}{n}$$

$$= \ln\left(\frac{2}{5}+1\right) = \ln\frac{7}{5} \approx 0.3365.$$

56. From Example 5, we have $\arctan x = \sum_{n=0}^{\infty}(-1)^n\frac{x^{2n+1}}{2n+1}$.

$$\sum_{n=0}^{\infty}(-1)^n\frac{1}{2n+1} = \sum_{n=0}^{\infty}(-1)^n\frac{(1)^{2n+1}}{2n+1}$$

$$= \arctan 1 = \frac{\pi}{4} \approx 0.7854$$

57. From Exercise 56, we have

$$\sum_{n=0}^{\infty}(-1)^n\frac{1}{2^{2n+1}(2n+1)} = \sum_{n=0}^{\infty}(-1)^n\frac{(1/2)^{2n+1}}{2n+1}$$

$$= \arctan\frac{1}{2} \approx 0.4636.$$

58. From Exercise 56, we have

$$\sum_{n=1}^{\infty}(-1)^{n+1}\frac{1}{3^{2n-1}(2n-1)} = \sum_{n=0}^{\infty}(-1)^n\frac{1}{3^{2n+1}(2n+1)}$$

$$= \sum_{n=0}^{\infty}(-1)^n\frac{(1/3)^{2n+1}}{2n+1}$$

$$= \arctan\frac{1}{3} \approx 0.3218.$$

59. $f(x) = \arctan x$ is an odd function (symmetric to the origin).

60. The approximations of degree $3, 7, 11, \ldots,$ $(4n-1, n = 1, 2, \ldots)$ have relative extrema.

61. The series in Exercise 56 converges to its sum at a slower rate because its terms approach 0 at a much slower rate.

62. Because $\frac{d}{dx}\left[\sum_{n=0}^{\infty}a_n x^n\right] = \sum_{n=1}^{\infty}na_n x^{n-1}$, the radius of convergence is the same, 3.

63. Because the first series is the derivative of the second series, the second series converges for $|x+1| < 4$ (and perhaps at the endpoints, $x = 3$ and $x = -5$.)

64. Using a graphing utility, you obtain the following partial sums for the left hand side. Note that $1/\pi \approx 0.3183098862$.

$n = 0$: $S_0 \approx 0.3183098784$

$n = 1$: $S_1 \approx 0.3183098862$

65. $\sum_{n=0}^{\infty}\frac{(-1)^n}{3^n(2n+1)}$

From Example 5 we have $\arctan x = \sum_{n=0}^{\infty}(-1)^n\frac{x^{2n+1}}{2n+1}$.

$$\sum_{n=0}^{\infty}\frac{(-1)^n}{3^n(2n+1)} = \sum_{n=0}^{\infty}\frac{(-1)^n}{(\sqrt{3})^{2n}(2n+1)}\frac{\sqrt{3}}{\sqrt{3}}$$

$$= \sqrt{3}\sum_{n=0}^{\infty}\frac{(-1)^n(1/\sqrt{3})^{2n+1}}{2n+1}$$

$$= \sqrt{3}\arctan\left(\frac{1}{\sqrt{3}}\right)$$

$$= \sqrt{3}\left(\frac{\pi}{6}\right) \approx 0.9068997$$

66. $$\sum_{n=0}^{\infty}\frac{(-1)^n\pi^{2n+1}}{3^{2n+1}(2n+1)!} = \sum_{n=0}^{\infty}(-1)^n\frac{(\pi/3)^{2n+1}}{(2n+1)!}$$

$$= \sin\left(\frac{\pi}{3}\right) = \frac{\sqrt{3}}{2} \approx 0.866025$$

Section 9.10 Taylor and Maclaurin Series

1. For $c = 0$, we have:

$$f(x) = e^{2x}$$

$$f^{(n)}(x) = 2^n e^{2x} \Rightarrow f^{(n)}(0) = 2^n$$

$$e^{2x} = 1 + 2x + \frac{4x^2}{2!} + \frac{8x^3}{3!} + \frac{16x^4}{4!} + \cdots = \sum_{n=0}^{\infty}\frac{(2x)^n}{n!}.$$

2. For $c = 0$, we have:

$$f(x) = e^{3x}$$

$$f^{(n)}(x) = 3^n e^{3x} \Rightarrow f^{(n)}(0) = 3^n$$

$$e^{3x} = 1 + 3x + \frac{9x^2}{2!} + \frac{27x^3}{3!} + \cdots = \sum_{n=0}^{\infty} \frac{(3x)^n}{n!}.$$

3. For $c = \pi/4$, we have:

$$f(x) = \cos(x) \qquad f\left(\frac{\pi}{4}\right) = \frac{\sqrt{2}}{2}$$

$$f'(x) = -\sin(x) \qquad f'\left(\frac{\pi}{4}\right) = -\frac{\sqrt{2}}{2}$$

$$f''(x) = -\cos(x) \qquad f''\left(\frac{\pi}{4}\right) = -\frac{\sqrt{2}}{2}$$

$$f'''(x) = \sin(x) \qquad f'''\left(\frac{\pi}{4}\right) = \frac{\sqrt{2}}{2}$$

$$f^{(4)}(x) = \cos(x) \qquad f^{(4)}\left(\frac{\pi}{4}\right) = \frac{\sqrt{2}}{2}$$

and so on. Therefore, we have:

$$\cos x = \sum_{n=0}^{\infty} \frac{f^{(n)}(\pi/4)[x - (\pi/4)]^n}{n!}$$

$$= \frac{\sqrt{2}}{2}\left[1 - \left(x - \frac{\pi}{4}\right) - \frac{[x - (\pi/4)]^2}{2!} + \frac{[x - (\pi/4)]^3}{3!} + \frac{[x - (\pi/4)]^4}{4!} - \cdots\right]$$

$$= \frac{\sqrt{2}}{2} \sum_{n=0}^{\infty} \frac{(-1)^{n(n+1)/2}[x - (\pi/4)]^n}{n!}.$$

[**Note:** $(-1)^{n(n+1)/2} = 1, -1, -1, 1, 1, -1, -1, 1, \ldots$]

4. For $c = \pi/4$, we have:

$$f(x) = \sin x \qquad f\left(\frac{\pi}{4}\right) = \frac{\sqrt{2}}{2}$$

$$f'(x) = \cos x \qquad f'\left(\frac{\pi}{4}\right) = \frac{\sqrt{2}}{2}$$

$$f''(x) = -\sin x \qquad f''\left(\frac{\pi}{4}\right) = -\frac{\sqrt{2}}{2}$$

$$f'''(x) = -\cos x \qquad f'''\left(\frac{\pi}{4}\right) = -\frac{\sqrt{2}}{2}$$

$$f^{(4)}(x) = \sin x \qquad f^{(4)}\left(\frac{\pi}{4}\right) = \frac{\sqrt{2}}{2}$$

and so on. Therefore we have:

$$\sin x = \sum_{n=0}^{\infty} \frac{f^{(n)}(\pi/4)[x - (\pi/4)]^n}{n!}$$

$$= \frac{\sqrt{2}}{2}\left[1 + \left(x - \frac{\pi}{4}\right) - \frac{[x - (\pi/4)]^2}{2!} - \frac{[x - (\pi/4)]^3}{3!} + \frac{[x - (\pi/4)]^4}{4!} + \cdots\right]$$

$$= \frac{\sqrt{2}}{2}\left\{\sum_{n=0}^{\infty} \frac{(-1)^{n(n-1)/2}[x - (\pi/4)]^{n+1}}{(n+1)!} + 1\right\}.$$

5. For $c = 1$, we have,

$$f(x) = \ln x \qquad f(1) = 0$$

$$f'(x) = \frac{1}{x} \qquad f'(1) = 1$$

$$f''(x) = -\frac{1}{x^2} \qquad f''(1) = -1$$

$$f'''(x) = \frac{2}{x^3} \qquad f'''(1) = 2$$

$$f^{(4)}(x) = -\frac{6}{x^4} \qquad f^{(4)}(1) = -6$$

$$f^{(5)}(x) = \frac{24}{x^5} \qquad f^{(5)}(1) = 24$$

and so on. Therefore, we have:

$$\ln x = \sum_{n=0}^{\infty} \frac{f^{(n)}(1)(x-1)^n}{n!}$$

$$= 0 + (x-1) - \frac{(x-1)^2}{2!} + \frac{2(x-1)^3}{3!} - \frac{6(x-1)^4}{4!} + \frac{24(x-1)^5}{5!} - \cdots$$

$$= (x-1) - \frac{(x-1)^2}{2} + \frac{(x-1)^3}{3} - \frac{(x-1)^4}{4} + \frac{(x-1)^5}{5} - \cdots$$

$$= \sum_{n=0}^{\infty} (-1)^n \frac{(x-1)^{n+1}}{n+1}.$$

6. For $c = 1$, we have:

$$f(x) = e^x$$

$$f^{(n)}(x) = e^x \Rightarrow f^{(n)}(1) = e$$

$$e^x = \sum_{n=0}^{\infty} \frac{f^{(n)}(1)(x-1)^n}{n!} = e\left[1 + (x-1) + \frac{(x-1)^2}{2!} + \frac{(x-1)^3}{3!} + \frac{(x-1)^4}{4!} + \cdots\right] = e\sum_{n=0}^{\infty} \frac{(x-1)^n}{n!}.$$

7. For $c = 0$, we have:

$$f(x) = \sin 2x \qquad f(0) = 0$$

$$f'(x) = 2\cos 2x \qquad f'(0) = 2$$

$$f''(x) = -4\sin 2x \qquad f''(0) = 0$$

$$f'''(x) = -8\cos 2x \qquad f'''(0) = -8$$

$$f^{(4)}(x) = 16\sin 2x \qquad f^{(4)}(0) = 0$$

$$f^{(5)}(x) = 32\cos 2x \qquad f^{(5)}(0) = 32$$

$$f^{(6)}(x) = -64\sin 2x \qquad f^{(6)}(0) = 0$$

$$f^{(7)}(x) = -128\cos 2x \qquad f^{(7)}(0) = -128$$

and so on. Therefore, we have:

$$\sin 2x = \sum_{n=0}^{\infty} \frac{f^{(n)}(0)x^n}{n!} = 0 + 2x + \frac{0x^2}{2!} - \frac{8x^3}{3!} + \frac{0x^4}{4!} + \frac{32x^5}{5!} + \frac{0x^6}{6!} - \frac{128x^7}{7!} + \cdots$$

$$= 2x - \frac{8x^3}{3!} + \frac{32x^5}{5!} - \frac{128x^7}{7!} + \cdots = \sum_{n=0}^{\infty} \frac{(-1)^n(2x)^{2n+1}}{(2n+1)!}.$$

8. For $c = 0$, we have:

$$f(x) = \ln(x^2 + 1) \qquad f(0) = 0$$

$$f'(x) = \frac{2x}{x^2 + 1} \qquad f'(0) = 0$$

$$f''(x) = \frac{2 - 2x^2}{(x^2 + 1)^2} \qquad f''(0) = 2$$

$$f'''(x) = \frac{4x(x^2 - 3)}{(x^2 + 1)^3} \qquad f'''(0) = 0$$

$$f^{(4)}(x) = \frac{12(-x^4 + 6x^2 - 1)}{(x^2 + 1)^4} \qquad f^{(4)}(0) = -12$$

$$f^{(5)}(x) = \frac{48x(x^4 - 10x^2 + 5)}{(x^2 + 1)^5} \qquad f^{(5)}(0) = 0$$

$$f^{(6)}(x) = \frac{-240(5x^6 - 15x^4 + 15x^2 - 1)}{(x^2 + 1)^6} \qquad f^{(6)}(0) = 240$$

and so on. Therefore, we have:

$$\ln(x^2 + 1) = \sum_{n=0}^{\infty} \frac{f^{(n)}(0)x^n}{n!} = 0 + 0x + \frac{2x^2}{2!} + \frac{0x^3}{3!} - \frac{12x^4}{4!} + \frac{0x^5}{5!} + \frac{240x^6}{6!} + \cdots$$

$$= x^2 - \frac{x^4}{2} + \frac{x^6}{3} - \cdots = \sum_{n=0}^{\infty} \frac{(-1)^n x^{2n+2}}{n+1}.$$

9. For $c = 0$, we have:

$$f(x) = \sec(x) \qquad f(0) = 1$$

$$f'(x) = \sec(x)\tan(x) \qquad f'(0) = 0$$

$$f''(x) = \sec^3(x) + \sec(x)\tan^2(x) \qquad f''(0) = 1$$

$$f'''(x) = 5\sec^3(x)\tan(x) + \sec(x)\tan^3(x) \qquad f'''(0) = 0$$

$$f^{(4)}(x) = 5\sec^5(x) + 18\sec^3(x)\tan^2(x) + \sec(x)\tan^4(x) \qquad f^{(4)}(0) = 5$$

$$\sec(x) = \sum_{n=0}^{\infty} \frac{f^{(n)}(0)x^n}{n!} = 1 + \frac{x^2}{2!} + \frac{5x^4}{4!} + \cdots.$$

10. For $c = 0$, we have;

$$f(x) = \tan(x) \qquad f(0) = 0$$

$$f'(x) = \sec^2(x) \qquad f'(0) = 1$$

$$f''(x) = 2\sec^2(x)\tan(x) \qquad f''(0) = 0$$

$$f'''(x) = 2[\sec^4(x) + 2\sec^2(x)\tan^2(x)] \qquad f'''(0) = 2$$

$$f^{(4)}(x) = 8[\sec^4(x)\tan(x) + \sec^2(x)\tan^3(x)] \qquad f^{(4)}(0) = 0$$

$$f^{(5)}(x) = 8[2\sec^6(x) + 11\sec^4(x)\tan^2(x) + 2\sec^2(x)\tan^4(x)] \qquad f^{(5)}(0) = 16$$

$$\tan(x) = \sum_{n=0}^{\infty} \frac{f^{(n)}(0)x^n}{n!} = x + \frac{2x^3}{3!} + \frac{16x^5}{5!} + \cdots = x + \frac{x^3}{3} + \frac{2}{15}x^5 + \cdots.$$

11. The Maclaurin series for $f(x) = \cos x$ is $\displaystyle\sum_{n=0}^{\infty} \frac{(-1)^n x^{2n}}{(2n)!}$.

Because $f^{(n+1)}(x) = \pm\sin x$ or $\pm\cos x$, we have $|f^{(n+1)}(z)| \le 1$ for all z. Hence by Taylor's Theorem,

$$0 \le |R_n(x)| = \left|\frac{f^{(n+1)}(z)}{(n+1)!}x^{n+1}\right| \le \frac{|x|^{n+1}}{(n+1)!}.$$

Since $\displaystyle\lim_{n\to\infty} \frac{|x|^{n+1}}{(n+1)!} = 0$, it follows that $R_n(x) \to 0$ as $n \to \infty$. Hence, the Maclaurin series for $\cos x$ converges to $\cos x$ for all x.

12. The Maclaurin series for $f(x) = e^{-2x}$ is $\displaystyle\sum_{n=0}^{\infty} \frac{(-2x)^n}{n!}$.

$f^{(n+1)}(x) = (-2)^{n+1}e^{-2x}$. Hence, by Taylor's Theorem,

$$0 \le |Rn(x)| = \left|\frac{f^{(n+1)}(z)}{(n+1)!}x^{n+1}\right| = \left|\frac{(-2)^{n+1}e^{-2z}}{(n+1)!}x^{n+1}\right|.$$

Since $\displaystyle\lim_{n\to\infty}\left|\frac{(-2)^{n+1}x^{n+1}}{(n+1)!}\right| = \lim_{n\to\infty}\left|\frac{(2x)^{n+1}}{(n+1)!}\right| = 0$, it follows that $R_n(x) \to 0$ as $n \to \infty$.

Hence, the Maclaurin Series for e^{-2x} converges to e^{-2x} for all x.

13. The Maclaurin series for $f(x) = \sinh x$ is $\displaystyle\sum_{n=0}^{\infty} \frac{x^{2n+1}}{(2n+1)!}$.

$f^{(n+1)}(x) = \sinh x$ (or $\cosh x$). For fixed x,

$$0 \le |R_n(x)| = \left|\frac{f^{(n+1)}(z)}{(n+1)!}x^{n+1}\right| = \left|\frac{\sinh(z)}{(n+1)!}x^{n+1}\right| \to 0 \text{ as } n \to \infty.$$

(The argument is the same if $f^{(n+1)}(x) = \cosh x$). Hence, the Maclaurin series for $\sinh x$ converges to $\sinh x$ for all x.

14. The Maclaurin series for $f(x) = \cosh x$ is $\displaystyle\sum_{n=0}^{\infty} \frac{x^{2n}}{(2n)!}$.

$f^{(n+1)}(x) = \sinh x$ (or $\cosh x$). For fixed x,

$$0 \le |R_n(x)| = \left|\frac{f^{(n+1)}(z)}{(n+1)!}x^{n+1}\right| = \left|\frac{\sinh(z)}{(n+1)!}x^{n+1}\right| \to 0 \text{ as } n \to \infty.$$

(The argument is the same if $f^{(n+1)}(x) = \cosh x$). Hence, the Maclaurin series for $\cosh x$ converges to $\cosh x$ for all x.

15. Since $(1+x)^{-k} = 1 - kx + \dfrac{k(k+1)x^2}{2!} - \dfrac{k(k+1)(k+2)x^3}{3!} + \cdots$, we have

$$(1+x)^{-2} = 1 - 2x + \frac{2(3)x^2}{2!} - \frac{2(3)(4)x^3}{3!} + \frac{2(3)(4)(5)x^4}{4!} - \cdots = 1 - 2x + 3x^2 - 4x^3 + 5x^4 - \cdots$$

$$= \sum_{n=0}^{\infty} (-1)^n (n+1)x^n.$$

16. Since $(1+x)^{-k} = 1 - kx + \dfrac{k(k+1)x^2}{2!} - \dfrac{k(k+1)(k+2)x^3}{3!} + \cdots$, we have

$$\left[1 + (-x)\right]^{-1/2} = 1 + \left(\frac{1}{2}\right)x + \frac{(1/2)(3/2)x^2}{2!} + \frac{(1/2)(3/2)(5/2)x^3}{3!} + \cdots$$

$$= 1 + \frac{x}{2} + \frac{(1)(3)x^2}{2^2 2!} + \frac{(1)(3)(5)x^3}{2^3 3!} + \cdots$$

$$= 1 + \sum_{n=1}^{\infty} \frac{1 \cdot 3 \cdot 5 \cdots (2n-1)x^n}{2^n n!}.$$

17. $\dfrac{1}{\sqrt{4+x^2}} = \left(\dfrac{1}{2}\right)\left[1+\left(\dfrac{x}{2}\right)^2\right]^{-1/2}$ and since $(1+x)^{-1/2} = 1 + \displaystyle\sum_{n=1}^{\infty} \frac{(-1)^n\, 1\cdot 3\cdot 5\cdots(2n-1)x^n}{2^n n!}$, we have

$$\frac{1}{\sqrt{4+x^2}} = \frac{1}{2}\left[1 + \sum_{n=1}^{\infty} \frac{(-1)^n\, 1\cdot 3\cdot 5\cdots(2n-1)(x/2)^{2n}}{2^n\, n!}\right] = \frac{1}{2} + \sum_{n=1}^{\infty} \frac{(-1)^n\, 1\cdot 3\cdot 5\cdots(2n-1)x^{2n}}{2^{3n+1}n!}.$$

18. $$(1+x)^{1/4} = 1 + \frac{1}{4}x + \frac{(1/4)(-3/4)}{2!}x^2 + \frac{(1/4)(-3/4)(-7/4)}{3!}x^3 + \cdots$$

$$= 1 + \frac{1}{4}x - \frac{3}{4^2 2!}x^2 + \frac{3\cdot 7}{4^3 3!}x^3 - \frac{3\cdot 7\cdot 11}{4^4 4!}x^4 + \cdots$$

$$= 1 + \frac{1}{4}x + \sum_{n=2}^{\infty} \frac{(-1)^{n+1} 3\cdot 7\cdot 11\cdots(4n-5)}{4^n n!}x^n$$

19. Since $(1+x)^{1/2} = 1 + \dfrac{x}{2} + \displaystyle\sum_{n=2}^{\infty} \frac{(-1)^{n+1}\, 1\cdot 3\cdot 5\cdots(2n-3)x^n}{2^n\, n!}$

we have $(1+x^2)^{1/2} = 1 + \dfrac{x^2}{2} + \displaystyle\sum_{n=2}^{\infty} \frac{(-1)^{n+1}\, 1\cdot 3\cdot 5\cdots(2n-3)x^{2n}}{2^n\, n!}$.

20. Since $(1+x)^{1/2} = 1 + \dfrac{x}{2} + \displaystyle\sum_{n=2}^{\infty} \frac{(-1)^{n+1}\, 1\cdot 3\cdot 5\cdots(2n-3)x^n}{2^n\, n!}$

we have $(1+x^3)^{1/2} = 1 + \dfrac{x^3}{2} + \displaystyle\sum_{n=2}^{\infty} \frac{(-1)^{n+1}\, 1\cdot 3\cdot 5\cdots(2n-3)x^{3n}}{2^n\, n!}$.

21. $$e^x = \sum_{n=0}^{\infty} \frac{x^n}{n!} = 1 + x + \frac{x^2}{2!} + \frac{x^3}{3!} + \frac{x^4}{4!} + \frac{x^5}{5!} + \cdots$$

$$e^{x^2/2} = \sum_{n=0}^{\infty} \frac{(x^2/2)^n}{n!} = \sum_{n=0}^{\infty} \frac{x^{2n}}{2^n n!} = 1 + \frac{x^2}{2} + \frac{x^4}{2^2 2!} + \frac{x^6}{2^3 3!} + \frac{x^8}{2^4 4!} + \cdots$$

22. $$e^x = \sum_{n=0}^{\infty} \frac{x^n}{n!} = 1 + x + \frac{x^2}{2!} + \frac{x^3}{3!} + \frac{x^4}{4!} + \frac{x^5}{5!} + \cdots$$

$$e^{-3x} = \sum_{n=0}^{\infty} \frac{(-3x)^n}{n!} = \sum_{n=0}^{\infty} \frac{(-1)^n 3^n x^n}{n!} = 1 - 3x + \frac{9x^2}{2!} - \frac{27x^3}{3!} + \frac{81x^4}{4!} - \frac{243x^5}{5!} + \cdots$$

23. $$\sin x = \sum_{n=0}^{\infty} \frac{(-1)^n x^{2n+1}}{(2n+1)!}$$

$$\sin 3x = \sum_{n=0}^{\infty} \frac{(-1)^n (3x)^{2n+1}}{(2n+1)!}$$

24. $$\cos x = \sum_{n=0}^{\infty} \frac{(-1)^n x^{2n}}{(2n)!} = 1 - \frac{x^2}{2!} + \frac{x^4}{4!} - \frac{x^6}{6!} + \cdots$$

$$\cos 4x = \sum_{n=0}^{\infty} \frac{(-1)^n (4x)^{2n}}{(2n)!} = \sum_{n=0}^{\infty} \frac{(-1)^n 4^{2n} x^{2n}}{(2n)!}$$

$$= 1 - \frac{16x^2}{2!} + \frac{256x^4}{4!} - \cdots$$

25. $$\cos x = \sum_{n=0}^{\infty} \frac{(-1)^n x^{2n}}{(2n)!} = 1 - \frac{x^2}{2!} + \frac{x^4}{4!} - \cdots$$

$$\cos x^{3/2} = \sum_{n=0}^{\infty} \frac{(-1)^n (x^{3/2})^{2n}}{(2n)!}$$

$$= \sum_{n=0}^{\infty} \frac{(-1)^n x^{3n}}{(2n)!}$$

$$= 1 - \frac{x^3}{2!} + \frac{x^6}{4!} - \cdots$$

26. $$\sin x = \sum_{n=0}^{\infty} \frac{(-1)^n x^{2n+1}}{(2n+1)!}$$

$$2\sin x^3 = 2\sum_{n=0}^{\infty} \frac{(-1)^n (x^3)^{2n+1}}{(2n+1)!}$$

$$= 2\left[x^3 - \frac{x^9}{3!} + \frac{x^{15}}{5!} - \cdots\right]$$

$$= 2x^3 - \frac{2x^9}{3!} + \frac{2x^{15}}{5!} - \cdots$$

27.

$$e^x = 1 + x + \frac{x^2}{2!} + \frac{x^3}{3!} + \frac{x^4}{4!} + \frac{x^5}{5!} + \cdots$$

$$e^{-x} = 1 - x + \frac{x^2}{2!} - \frac{x^3}{3!} + \frac{x^4}{4!} - \frac{x^5}{5!} + \cdots$$

$$e^x - e^{-x} = 2x + \frac{2x^3}{3!} + \frac{2x^5}{5!} + \frac{2x^7}{7!} + \cdots$$

$$\sinh(x) = \frac{1}{2}(e^x - e^{-x})$$

$$= x + \frac{x^3}{3!} + \frac{x^5}{5!} + \frac{x^7}{7!} + \cdots = \sum_{n=0}^{\infty} \frac{x^{2n+1}}{(2n+1)!}$$

28.

$$e^x = 1 + x + \frac{x^2}{2!} + \frac{x^3}{3!} + \cdots$$

$$e^{-x} = 1 - x + \frac{x^2}{2!} - \frac{x^3}{3!} + \cdots$$

$$e^x + e^{-x} = 2 + \frac{2x^2}{2!} + \frac{2x^4}{4!} + \cdots$$

$$2 \cos h(x) = e^x + e^{-x} = \sum_{n=0}^{\infty} 2\frac{x^{2n}}{(2n)!}$$

29. $\cos^2(x) = \frac{1}{2}[1 + \cos(2x)]$

$$= \frac{1}{2}\left[1 + 1 - \frac{(2x)^2}{2!} + \frac{(2x)^4}{4!} - \frac{(2x)^6}{6!} - \cdots\right] = \frac{1}{2}\left[1 + \sum_{n=0}^{\infty} \frac{(-1)^n(2x)^{2n}}{(2n)!}\right]$$

30. The formula for the binomial series gives $(1 + x)^{-1/2} = 1 + \sum_{n=1}^{\infty} \frac{(-1)^n\, 1 \cdot 3 \cdot 5 \cdots (2n-1)x^n}{2^n\, n!}$, which implies that

$$(1 + x^2)^{-1/2} = 1 + \sum_{n=1}^{\infty} \frac{(-1)^n\, 1 \cdot 3 \cdot 5 \cdots (2n-1)x^{2n}}{2^n\, n!}$$

$$\ln\left(x + \sqrt{x^2+1}\right) = \int \frac{1}{\sqrt{x^2+1}}\, dx$$

$$= x + \sum_{n=1}^{\infty} \frac{(-1)^n 1 \cdot 3 \cdot 5 \ldots (2n-1)x^{2n+1}}{2^n(2n+1)n!}$$

$$= x - \frac{x^3}{2 \cdot 3} + \frac{1 \cdot 3x^5}{2 \cdot 4 \cdot 5} - \frac{1 \cdot 3 \cdot 5x^7}{2 \cdot 4 \cdot 6 \cdot 7} + \cdots.$$

31. $x \sin x = x\left(x - \frac{x^3}{3!} + \frac{x^5}{5!} - \cdots\right)$

$$= x^2 - \frac{x^4}{3!} + \frac{x^6}{5!} - \cdots$$

$$= \sum_{n=0}^{\infty} \frac{(-1)^n x^{2n+2}}{(2n+1)!}$$

32. $x \cos x = x\left(1 - \frac{x^2}{2!} + \frac{x^4}{4!} - \cdots\right)$

$$= x - \frac{x^3}{2!} + \frac{x^5}{4!} - \cdots$$

$$= \sum_{n=0}^{\infty} \frac{(-1)^n x^{2n+1}}{(2n)!}$$

33. $\frac{\sin x}{x} = \frac{x - (x^3/3!) + (x^5/5!) - \cdots}{x}$

$$= 1 - \frac{x^2}{2!} + \frac{x^4}{4!} - \cdots = \sum_{n=0}^{\infty} \frac{(-1)^n x^{2n}}{(2n+1)!}, x \neq 0$$

34. $\frac{\arcsin x}{x} = \sum_{n=0}^{\infty} \frac{(2n)!x^{2n+1}}{(2^n n!)^2(2n+1)} \cdot \frac{1}{x}$

$$= \sum_{n=0}^{\infty} \frac{(2n)!x^{2n}}{(2^n n!)^2(2n+1)}, x \neq 0$$

35.

$$e^{ix} = 1 + ix + \frac{(ix)^2}{2!} + \frac{(ix)^3}{3!} + \frac{(ix)^4}{4!} + \cdots = 1 + ix - \frac{x^2}{2!} - \frac{ix^3}{3!} + \frac{x^4}{4!} + \frac{ix^5}{5!} - \frac{x^6}{6!} - \cdots$$

$$e^{-ix} = 1 - ix + \frac{(-ix)^2}{2!} + \frac{(-ix)^3}{3!} + \frac{(-ix)^4}{4!} + \cdots = 1 - ix - \frac{x^2}{2!} + \frac{ix^3}{3!} + \frac{x^4}{4!} - \frac{ix^5}{5!} - \frac{x^6}{6!} + \cdots$$

$$e^{ix} - e^{-ix} = 2ix - \frac{2ix^3}{3!} + \frac{2ix^5}{5!} - \frac{2ix^7}{7!} + \cdots$$

$$\frac{e^{ix} - e^{-ix}}{2i} = x - \frac{x^3}{3!} + \frac{x^5}{5!} - \frac{x^7}{7!} + \cdots = \sum_{n=0}^{\infty} \frac{(-1)^n x^{2n+1}}{(2n+1)!} = \sin(x)$$

36. $e^{ix} + e^{-ix} = 2 - \dfrac{2x^2}{2!} + \dfrac{2x^4}{4!} - \dfrac{2x^6}{6!} + \cdots$ (See Exercise 35.)

$$\frac{e^{ix} + e^{-ix}}{2} = 1 - \frac{x^2}{2!} + \frac{x^4}{4!} - \frac{x^6}{6!} + \cdots = \sum_{n=0}^{\infty} \frac{(-1)^n x^{2n}}{(2n)!} = \cos(x)$$

37. $f(x) = e^x \sin x$

$$= \left(1 + x + \frac{x^2}{2} + \frac{x^3}{6} + \frac{x^4}{24} + \cdots\right)\left(x - \frac{x^3}{6} + \frac{x^5}{120} - \cdots\right)$$

$$= x + x^2 + \left(\frac{x^3}{2} - \frac{x^3}{6}\right) + \left(\frac{x^4}{6} - \frac{x^4}{6}\right) + \left(\frac{x^5}{120} - \frac{x^5}{12} + \frac{x^5}{24}\right) + \cdots$$

$$= x + x^2 + \frac{x^3}{3} - \frac{x^5}{30} + \cdots$$

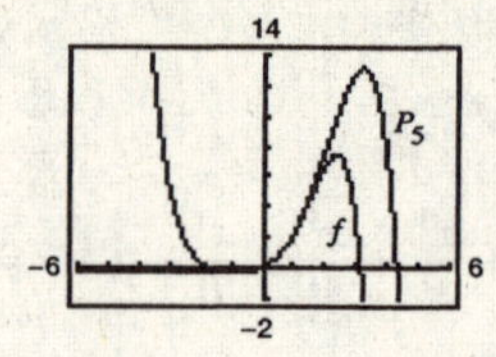

38. $g(x) = e^x \cos x$

$$= \left(1 + x + \frac{x^2}{2} + \frac{x^4}{6} + \frac{x^4}{24} + \cdots\right)\left(1 - \frac{x^2}{2} + \frac{x^4}{24} - \cdots\right)$$

$$= 1 + x + \left(\frac{x^2}{2} - \frac{x^2}{2}\right) + \left(\frac{x^3}{6} - \frac{x^3}{2}\right) + \left(\frac{x^4}{24} - \frac{x^4}{4} + \frac{x^4}{24}\right) + \cdots$$

$$= 1 + x - \frac{x^3}{3} - \frac{x^4}{6} + \cdots$$

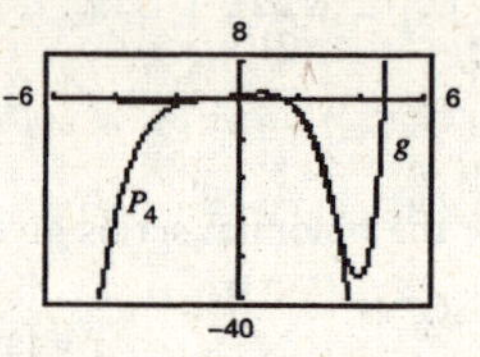

39. $h(x) = \cos x \ln(1 + x)$

$$= \left(1 - \frac{x^2}{2} + \frac{x^4}{24} - \cdots\right)\left(x - \frac{x^2}{2} + \frac{x^3}{3} - \frac{x^4}{4} + \frac{x^5}{5} - \cdots\right)$$

$$= x - \frac{x^2}{2} + \left(\frac{x^3}{3} - \frac{x^3}{2}\right) + \left(\frac{x^4}{4} - \frac{x^4}{4}\right) + \left(\frac{x^5}{5} - \frac{x^5}{6} + \frac{x^5}{24}\right) + \cdots$$

$$= x - \frac{x^2}{2} - \frac{x^3}{6} + \frac{3x^5}{40} + \cdots$$

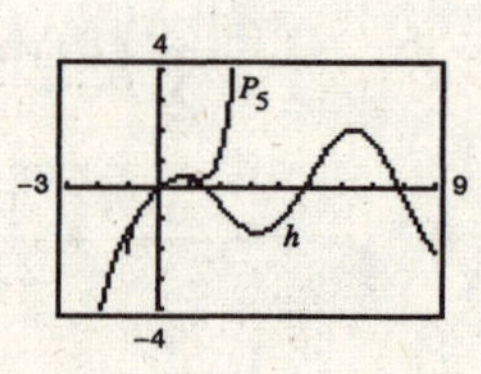

40. $f(x) = e^x \ln(1 + x)$

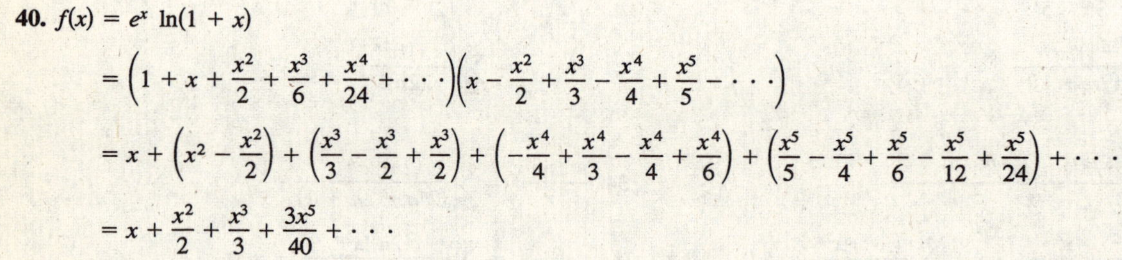

$$= \left(1 + x + \frac{x^2}{2} + \frac{x^3}{6} + \frac{x^4}{24} + \cdots\right)\left(x - \frac{x^2}{2} + \frac{x^3}{3} - \frac{x^4}{4} + \frac{x^5}{5} - \cdots\right)$$

$$= x + \left(x^2 - \frac{x^2}{2}\right) + \left(\frac{x^3}{3} - \frac{x^3}{2} + \frac{x^3}{2}\right) + \left(-\frac{x^4}{4} + \frac{x^4}{3} - \frac{x^4}{4} + \frac{x^4}{6}\right) + \left(\frac{x^5}{5} - \frac{x^5}{4} + \frac{x^5}{6} - \frac{x^5}{12} + \frac{x^5}{24}\right) + \cdots$$

$$= x + \frac{x^2}{2} + \frac{x^3}{3} + \frac{3x^5}{40} + \cdots$$

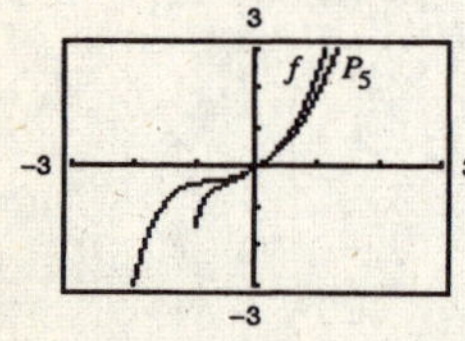

41. $g(x) = \dfrac{\sin x}{1+x}$. Divide the series for $\sin x$ by $(1 + x)$.

$$
\begin{array}{r|l}
 & x - x^2 + \frac{5x^3}{6} - \frac{5x^4}{6} + \cdots \\
\hline
1 + x & x + 0x^2 - \frac{x^3}{6} + 0x^4 + \frac{x^5}{120} + \cdots \\
 & \underline{x + x^2} \\
 & -x^2 - \frac{x^3}{6} \\
 & \underline{-x^2 - x^3} \\
 & \frac{5x^3}{6} + 0x^4 \\
 & \underline{\frac{5x^3}{6} + \frac{5x^4}{6}} \\
 & -\frac{5x^4}{6} + \frac{x^5}{120} \\
 & \underline{-\frac{5x^4}{6} - \frac{5x^5}{6}} \\
 & \vdots
\end{array}
$$

$$g(x) = x - x^2 + \frac{5x^3}{6} - \frac{5x^4}{6} + \cdots$$

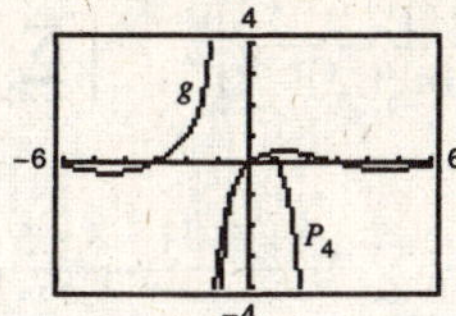

42. $f(x) = \dfrac{e^x}{1+x}$. Divide the series for e^x by $(1 + x)$.

$$
\begin{array}{r|l}
 & 1 + \frac{x^2}{2} - \frac{x^3}{3} + \frac{3x^4}{8} + \cdots \\
\hline
1 + x & 1 + x + \frac{x^2}{2} + \frac{x^3}{6} + \frac{x^4}{24} + \frac{x^5}{120} + \cdots \\
 & \underline{1 + x} \\
 & 0 + \frac{x^2}{2} + \frac{x^3}{6} \\
 & \underline{\frac{x^2}{2} + \frac{x^3}{2}} \\
 & -\frac{x^3}{3} + \frac{x^4}{24} \\
 & \underline{-\frac{x^3}{3} - \frac{x^4}{3}} \\
 & \frac{3x^4}{8} + \frac{x^5}{120} \\
 & \underline{\frac{3x^4}{8} + \frac{3x^5}{8}} \\
 & \vdots
\end{array}
$$

$$f(x) = 1 + \frac{x^2}{2} - \frac{x^3}{3} + \frac{3x^4}{8} - \cdots$$

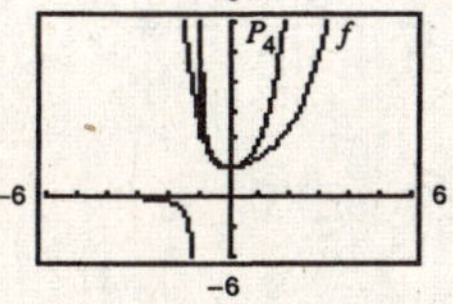

43. $y = x^2 - \dfrac{x^4}{3!} = x\left(x - \dfrac{x^3}{3!}\right)$

$f(x) = x \sin x$

Matches (c)

44. $y = x - \dfrac{x^3}{2!} + \dfrac{x^5}{4!} = x\left(1 - \dfrac{x^2}{2!} + \dfrac{x^4}{4!}\right)$

$f(x) = x \cos x$

Matches (d)

45. $y = x + x^2 + \dfrac{x^3}{2!} = x\left(1 + x + \dfrac{x^2}{2!}\right)$

$f(x) = xe^x$

Matches (a)

46. $y = x^2 - x^3 + x^4 = x^2(1 - x + x^2)$

$f(x) = x^2 \dfrac{1}{1+x}$

Matches (b)

47. $\displaystyle\int_0^x (e^{-t^2} - 1)\,dt = \int_0^x \left[\left(\sum_{n=0}^{\infty} \frac{(-1)^n t^{2n}}{n!}\right) - 1\right] dt$

$$= \int_0^x \left[\sum_{n=0}^{\infty} \frac{(-1)^{n+1} t^{2n+2}}{(n+1)!}\right] dt = \left[\sum_{n=0}^{\infty} \frac{(-1)^{n+1} t^{2n+3}}{(2n+3)(n+1)!}\right]_0^x = \sum_{n=0}^{\infty} \frac{(-1)^{n+1} x^{2n+3}}{(2n+3)(n+1)!}$$

48. $\displaystyle\int_0^x \sqrt{1+t^3}\,dt = \int_0^x \left[1 + \frac{t^3}{2} + \sum_{n=2}^{\infty} \frac{(-1)^{n-1} 1 \cdot 3 \cdot 5 \cdots (2n-3) t^{3n}}{2^n n!}\right] dt$

$$= \left[t + \frac{t^4}{8} + \sum_{n=2}^{\infty} \frac{(-1)^{n-1} 1 \cdot 3 \cdot 5 \cdots (2n-3) t^{3n+1}}{(3n+1)2^n n!}\right]_0^x$$

$$= x + \frac{x^4}{8} + \sum_{n=2}^{\infty} \frac{(-1)^{n-1} 1 \cdot 3 \cdot 5 \cdots (2n-3) x^{3n+1}}{(3n+1)2^n n!}$$

49. Since $\displaystyle\ln x = \sum_{n=0}^{\infty} \frac{(-1)^n (x-1)^{n+1}}{n+1} = (x-1) - \frac{(x-1)^2}{2} + \frac{(x-1)^3}{3} - \frac{(x-1)^4}{4} + \cdots, \quad (0 < x \le 2)$

we have $\displaystyle\ln 2 = 1 - \frac{1}{2} + \frac{1}{3} - \frac{1}{4} + \cdots = \sum_{n=1}^{\infty} (-1)^{n+1} \frac{1}{n} \approx 0.6931.$ (10,001 terms)

50. Since $\displaystyle\sin(x) = \sum_{n=0}^{\infty} \frac{(-1)^n x^{2n+1}}{(2n+1)!} = x - \frac{x^3}{3!} + \frac{x^5}{5!} - \frac{x^7}{7!} + \cdots$, we have

$\displaystyle\sin(1) = \sum_{n=0}^{\infty} \frac{(-1)^n}{(2n+1)!} = 1 - \frac{1}{3!} + \frac{1}{5!} - \frac{1}{7!} + \cdots \approx 0.8415.$ (4 terms)

51. Since $\displaystyle e^x = \sum_{n=0}^{\infty} \frac{x^n}{n!} = 1 + x + \frac{x^2}{2!} + \frac{x^3}{3!} + \cdots,$

we have $\displaystyle e^2 = 1 + 2 + \frac{2^2}{2!} + \frac{2^3}{3!} + \cdots = \sum_{n=0}^{\infty} \frac{2^n}{n!} \approx 7.3891.$ (12 terms)

52. Since $\displaystyle e^x = \sum_{n=0}^{\infty} \frac{x^n}{n!} = 1 + x + \frac{x^2}{2!} + \frac{x^3}{3!} + \frac{x^4}{4!} + \frac{x^5}{5!} + \ldots$, we have $\displaystyle e^{-1} = 1 - 1 + \frac{1}{2!} - \frac{1}{3!} + \frac{1}{4!} - \frac{1}{5!} + \cdots$

and $\displaystyle\frac{e-1}{e} = 1 - e^{-1} = 1 - \frac{1}{2!} + \frac{1}{3!} - \frac{1}{4!} + \frac{1}{5!} - \frac{1}{7!} + \cdots = \sum_{n=1}^{\infty} \frac{(-1)^{n-1}}{n!} \approx 0.6321.$ (6 terms)

53. Since

$$\cos x = \sum_{n=0}^{\infty} \frac{(-1)^n x^{2n}}{(2n)!} = 1 - \frac{x^2}{2!} + \frac{x^4}{4!} - \frac{x^6}{6!} + \frac{x^8}{8!} - \cdots$$

$$1 - \cos x = \frac{x^2}{2!} - \frac{x^4}{4!} + \frac{x^6}{6!} - \frac{x^8}{8!} + \cdots = \sum_{n=0}^{\infty} \frac{(-1)^n x^{2n+2}}{(2n+2)!}$$

$$\frac{1 - \cos}{x} = \frac{x}{2!} - \frac{x^3}{4!} + \frac{x^5}{6!} - \frac{x^7}{8!} + \cdots = \sum_{n=0}^{\infty} \frac{(-1)^n x^{2n+1}}{(2n+2)!}$$

we have $\displaystyle\lim_{x\to 0} \frac{1 - \cos x}{x} = \lim_{x\to 0} \sum_{n=0}^{\infty} \frac{(-1)x^{2n+1}}{(2n+2)!} = 0.$

54. Since

$$\sin x = \sum_{n=0}^{\infty} \frac{(-1)^n x^{2n+1}}{(2n+1)!} = x - \frac{x^3}{3!} + \frac{x^5}{5!} - \frac{x^7}{7!} + \cdots$$

$$\frac{\sin x}{x} = 1 - \frac{x^2}{3!} + \frac{x^4}{5!} - \frac{x^6}{7!} + \cdots = \sum_{n=0}^{\infty} \frac{(-1)^n x^{2n}}{(2n+1)!}$$

we have $\displaystyle\lim_{x\to 0} \frac{\sin x}{x} = \lim_{x\to 0} \sum_{n=0}^{\infty} \frac{(-1)^n x^{2n}}{(2n+1)!} = 1.$

55. $\displaystyle\int_0^1 \frac{\sin x}{x}\,dx = \int_0^1 \left[\sum_{n=0}^{\infty} \frac{(-1)^n x^{2n}}{(2n+1)!}\right] dx = \left[\sum_{n=0}^{\infty} \frac{(-1)^n x^{2n+1}}{(2n+1)(2n+1)!}\right]_0^1 = \sum_{n=0}^{\infty} \frac{(-1)^n}{(2n+1)(2n+1)!}$

Since $1/(7 \cdot 7!) < 0.0001$, we need three terms:

$$\int_0^1 \frac{\sin x}{x}\,dx = 1 - \frac{1}{3 \cdot 3!} + \frac{1}{5 \cdot 5!} - \cdots \approx 0.9461. \quad \text{(using three nonzero terms)}$$

Note: We are using $\displaystyle\lim_{x\to 0^+} \frac{\sin x}{x} = 1.$

56. $\displaystyle\int_0^{1/2} \frac{\arctan x}{x}\,dx = \int_0^{1/2} \left(1 - \frac{x^2}{3} + \frac{x^4}{5} - \frac{x^6}{7} + \cdots\right) dx = \left[x - \frac{x^3}{3^2} + \frac{x^5}{5^2} - \frac{x^7}{7^2} + \cdots\right]_0^{1/2}$

Since $1/(9^2 2^9) < 0.0001$, we have

$$\int_0^{1/2} \frac{\arctan x}{x}\,dx \approx \left(\frac{1}{2} - \frac{1}{3^2 2^3} + \frac{1}{5^2 2^5} - \frac{1}{7^2 2^7} + \frac{1}{9^2 2^9}\right) \approx 0.4872.$$

Note: We are using $\displaystyle\lim_{x\to 0^+} \frac{\arctan x}{x} = 1.$

57. $\displaystyle\int_{0.1}^{0.3} \sqrt{1+x^3}\,dx = \int_{0.1}^{0.3} \left(1 + \frac{x^3}{2} - \frac{x^6}{8} + \frac{x^9}{16} - \frac{5x^{12}}{128} + \cdots\right) dx = \left[x + \frac{x^4}{8} - \frac{x^7}{56} + \frac{x^{10}}{160} - \frac{5x^{13}}{1664} + \cdots\right]_{0.1}^{0.3}$

Since $\frac{1}{56}(0.3^7 - 0.1^7) < 0.0001$, we need two terms.

$$\int_{0.1}^{0.3} \sqrt{1+x^3}\,dx = \left[(0.3 - 0.1) + \frac{1}{8}(0.3^4 - 0.1^4)\right] \approx 0.201.$$

58. $\displaystyle\int_0^{1/4} x\ln(x+1)\,dx = \int_0^{1/4} \left(x^2 - \frac{x^3}{2} + \frac{x^4}{3} - \frac{x^5}{4} + \cdots\right) dx = \left[\frac{x^3}{3} - \frac{x^4}{4 \cdot 2} + \frac{x^5}{5 \cdot 3} - \frac{x^6}{6 \cdot 4} + \cdots\right]_0^{1/4}$

Since $\displaystyle\frac{(1/4)^5}{15} < 0.0001$, $\displaystyle\int_0^{1/4} x\ln(x+1)\,dx \approx \frac{(1/4)^3}{3} - \frac{(1/4)^4}{8} \approx 0.00472.$

59. $\displaystyle\int_0^{\pi/2} \sqrt{x}\cos x\,dx = \int_0^{\pi/2} \left[\sum_{n=0}^{\infty} \frac{(-1)^n x^{(4n+1)/2}}{(2n)!}\right] dx = \left[\sum_{n=0}^{\infty} \frac{(-1)^n x^{(4n+3)/2}}{\left(\frac{4n+3}{2}\right)(2n)!}\right]_0^{\pi/2} = \left[\sum_{n=0}^{\infty} \frac{(-1)^n 2x^{(4n+3)/2}}{(4n+3)(2n)!}\right]_0^{\pi/2}$

Since $2(\pi/2)^{23/2}/(23 \cdot 10!) < 0.0001$, we need five terms.

$$\int_0^1 \sqrt{x}\cos x\,dx = 2\left[\frac{(\pi/2)^{3/2}}{3} - \frac{(\pi/2)^{7/2}}{14} + \frac{(\pi/2)^{11/2}}{264} - \frac{(\pi/2)^{15/2}}{10{,}800} + \frac{(\pi/2)^{19/2}}{766{,}080}\right] \approx 0.7040.$$

60. $\displaystyle\int_{0.5}^1 \cos\sqrt{x}\,dx = \int_{0.5}^1 \left(1 - \frac{x}{2!} + \frac{x^2}{4!} - \frac{x^3}{6!} + \frac{x^4}{8!} - \cdots\right) dx = \left[x - \frac{x^2}{2(2!)} + \frac{x^3}{3(4!)} - \frac{x^4}{4(6!)} + \frac{x^5}{5(8!)} - \cdots\right]_{0.5}^1$

Since $\frac{1}{201{,}600}(1 - 0.5^5) < 0.0001$, we have

$$\int_{0.5}^1 \cos\sqrt{x}\,dx \approx \left[(1 - 0.5) - \frac{1}{4}(1 - 0.5^2) + \frac{1}{72}(1 - 0.5^3) - \frac{1}{2880}(1 - 0.5^4) + \frac{1}{201{,}600}(1 - 0.5)^5\right] \approx 0.3243.$$

61. From Exercise 21, we have

$$\frac{1}{\sqrt{2\pi}}\int_0^1 e^{-x^2/2}\,dx = \frac{1}{\sqrt{2\pi}}\int_0^1 \sum_{n=0}^{\infty} \frac{(-1)^n x^{2n}}{2^n n!}\,dx = \frac{1}{\sqrt{2\pi}}\left[\sum_{n=0}^{\infty} \frac{(-1)^n x^{2n+1}}{2^n n!(2n+1)}\right]_0^1 = \frac{1}{\sqrt{2\pi}}\sum_{n=0}^{\infty} \frac{(-1)^n}{2^n n!(2n+1)}$$

$$\approx \frac{1}{\sqrt{2\pi}}\left[1 - \frac{1}{2 \cdot 1 \cdot 3} + \frac{1}{2^2 \cdot 2! \cdot 5} - \frac{1}{2^3 \cdot 3! \cdot 7}\right] \approx 0.3414.$$

62. From Exercise 21, we have

$$\frac{1}{\sqrt{2\pi}}\int_1^2 e^{-x^2/2}dx = \frac{1}{\sqrt{2\pi}}\int_1^2 \sum_{n=0}^{\infty}\frac{(-1)^n x^{2n}}{2^n n!}\,dx = \frac{1}{\sqrt{2\pi}}\left[\sum_{n=0}^{\infty}\frac{(-1)^n x^{2n+1}}{2^n n!(2n+1)}\right]_1^2$$

$$= \frac{1}{\sqrt{2\pi}}\sum_{n=0}^{\infty}\frac{(-1)^n(2^{n+1}-1)}{2^n n!(2n+1)}$$

$$\approx \frac{1}{\sqrt{2\pi}}\left[1 - \frac{7}{2\cdot 1\cdot 3} + \frac{31}{2^2\cdot 2!\cdot 5} - \frac{127}{2^3\cdot 3!\cdot 7} + \frac{511}{2^4\cdot 4!\cdot 9} - \frac{2047}{2^5\cdot 5!\cdot 11}\right.$$

$$\left. + \frac{8191}{2^6\cdot 6!\cdot 13} - \frac{32{,}767}{2^7\cdot 7!\cdot 15} + \frac{131{,}071}{2^8\cdot 8!\cdot 17} - \frac{524{,}287}{2^9\cdot 9!\cdot 19}\right] \approx 0.1359.$$

63. $f(x) = x\cos 2x = \sum_{n=0}^{\infty}\frac{(-1)^n 4^n x^{2n+1}}{(2n)!}$

$P_5(x) = x - 2x^3 + \frac{2x^5}{3}$

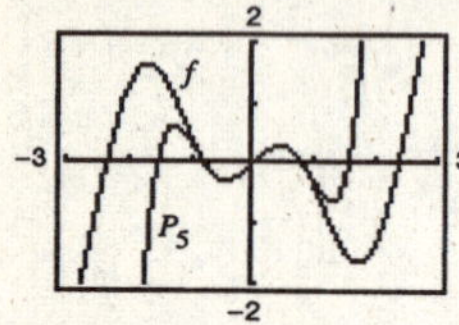

The polynomial is a reasonable approximation on the interval $\left[-\frac{3}{4}, \frac{3}{4}\right]$.

64. 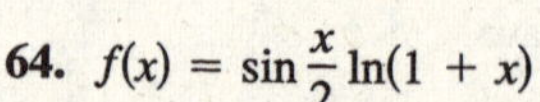 $f(x) = \sin\frac{x}{2}\ln(1+x)$

$P_5(x) = \frac{x^2}{2} - \frac{x^3}{4} + \frac{7x^4}{48} - \frac{11x^5}{96}$

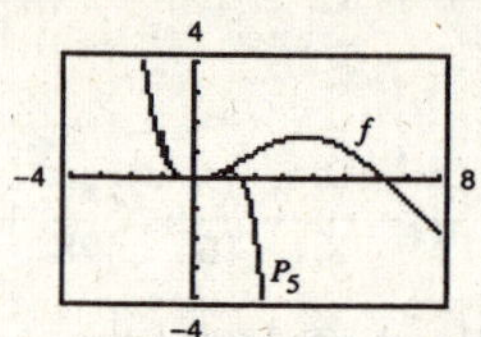

The polynomial is a reasonable approximation on the interval $(-0.60, 0.73)$.

65. $f(x) = \sqrt{x}\ln x,\ c = 1$

$P_5(x) = (x-1) - \frac{(x-1)^3}{24} + \frac{(x-1)^4}{24} - \frac{71(x-1)^5}{1920}$

The polynomial is a reasonable approximation on the interval $\left[\frac{1}{4}, 2\right]$.

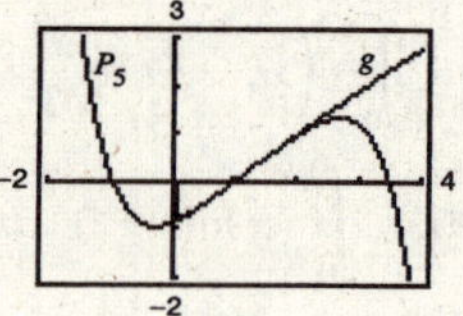

66. $f(x) = \sqrt[3]{x}\cdot\arctan x,\ c = 1$

$$P_5(x) \approx 0.7854 + 0.7618(x-1) - 0.3412\left[\frac{(x-1)^2}{2!}\right] - 0.0424\left[\frac{(x-1)^3}{3!}\right]$$

$$+ 1.3025\left[\frac{(x-1)^4}{4!}\right] - 5.5913\left[\frac{(x-1)^5}{5!}\right]$$

The polynomial is a reasonable approximation on the interval $(0.48, 1.75)$.

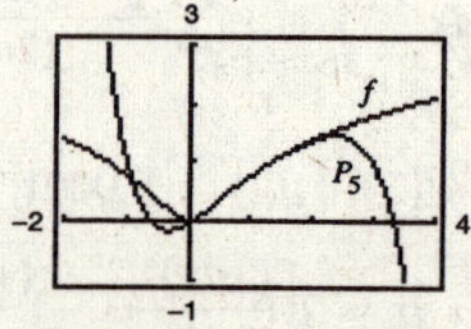

67. See Guidelines, page 680.

68. $a_{2n+1} = 0$ (odd coefficients are zero)

69. (a) Replace x with $(-x)$.

(b) Replace x with $3x$.

(c) Multiply series by x.

(d) Replace x with $2x$, then replace x with $-2x$, and add the two together.

70. The binomial series is $(1+x)^k = 1 + kx + \frac{k(k-1)}{2!}x^2 + \frac{k(k-1)(k-2)}{3!}x^3 + \cdots$. The radius of convergence is $R = 1$.

71. $y = \left(\tan\theta - \dfrac{g}{kv_0\cos\theta}\right)x - \dfrac{g}{k^2}\ln\left(1 - \dfrac{kx}{v_0\cos\theta}\right)$

$$= (\tan\theta)x - \frac{gx}{kv_0\cos\theta} - \frac{g}{k^2}\left[-\frac{kx}{v_0\cos\theta} - \frac{1}{2}\left(\frac{kx}{v_0\cos\theta}\right)^2 - \frac{1}{3}\left(\frac{kx}{v_0\cos\theta}\right)^3 - \frac{1}{4}\left(\frac{kx}{v_0\cos\theta}\right)^4 - \cdots\right]$$

$$= (\tan\theta)x - \frac{gx}{kv_0\cos\theta} + \frac{gx}{kv_0\cos\theta} + \frac{gx^2}{2v_0^2\cos^2\theta} + \frac{gkx^3}{3v_0^3\cos^3\theta} + \frac{gk^2x^4}{4v_0^4\cos^4\theta} + \cdots\Bigg]$$

$$= (\tan\theta)x + \frac{gx^2}{2v_0^2\cos^2\theta} + \frac{kgx^3}{3v_0^3\cos^3\theta} + \frac{k^2gx^4}{4v_0^4\cos^4\theta} + \cdots$$

72. $\theta = 60°,\ v_0 = 64,\ k = \dfrac{1}{16},\ g = -32$

$$y = \sqrt{3}x - \frac{32x^2}{2(64)^2(1/2)^2} - \frac{(1/16)(32)x^3}{3(64)^3(1/2)^3} - \frac{(1/16)^2(32)x^4}{4(64)^4(1/2)^4} - \cdots$$

$$= \sqrt{3}x - 32\left[\frac{2^2x^2}{2(64)^2} + \frac{2^3x^3}{3(64)^3 16} + \frac{2^4x^4}{4(64)^4(16)^2} + \cdots\right]$$

$$= \sqrt{3}x - 32\sum_{n=2}^{\infty}\frac{2^nx^n}{n(64)^n(16)^{n-2}} = \sqrt{3}x - 32\sum_{n=2}^{\infty}\frac{x^n}{n(32)^n(16)^{n-2}}$$

73. $f(x) = \begin{cases} e^{-1/x^2}, & x \neq 0 \\ 0, & x = 0 \end{cases}$

(a)

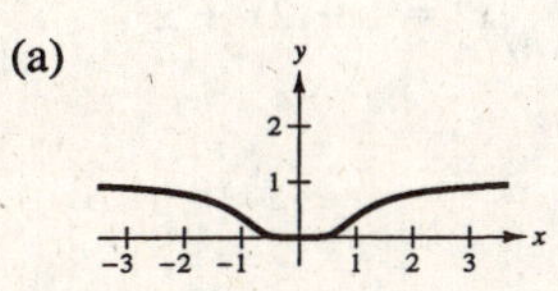

(b) $f'(0) = \lim\limits_{x\to 0}\dfrac{f(x) - f(0)}{x - 0} = \lim\limits_{x\to 0}\dfrac{e^{-1/x^2} - 0}{x}$

Let $y = \lim\limits_{x\to 0}\dfrac{e^{-1/x^2}}{x}$. Then

$$\ln y = \lim_{x\to 0}\ln\left(\frac{e^{-1/x^2}}{x}\right) = \lim_{x\to 0^+}\left[-\frac{1}{x^2} - \ln x\right] = \lim_{x\to 0^+}\left[\frac{-1 - x^2\ln x}{x^2}\right] = -\infty.$$

Thus, $y = e^{-\infty} = 0$ and we have $f'(0) = 0$.

(c) $\displaystyle\sum_{n=0}^{\infty}\frac{f^{(n)}(0)}{n!}x^n = f(0) + \frac{f'(0)x}{1!} + \frac{f''(0)x^2}{2!} + \cdots = 0 \neq f(x)$ This series converges to f at $x = 0$ only.

74. (a) $f(x) = \dfrac{\ln(x^2 + 1)}{x^2}$.

From Exercise 8, you obtain:

$$P = \frac{1}{x^2}\sum_{n=0}^{\infty}\frac{(-1)^nx^{2n+2}}{n+1} = \sum_{n=0}^{\infty}\frac{(-1)^nx^{2n}}{n+1}$$

$$P_8 = 1 - \frac{x^2}{2} + \frac{x^4}{3} - \frac{x^6}{4} + \frac{x^8}{5}.$$

(b)

1.5
0
2
-0.5

(c) $F(x) = \displaystyle\int_0^x \frac{\ln(t^2 + 1)}{t^2}\,dt$

$G(x) = \displaystyle\int_0^x P_8(t)\,dt$

x	0.25	0.50	0.75	1.00	1.50	2.00
$F(x)$	0.2475	0.4810	0.6920	0.8776	1.1798	1.4096
$G(x)$	0.2475	0.4810	0.6924	0.8865	1.6878	9.6063

(d) The curves are nearly identical for $0 < x < 1$. Hence, the integrals nearly agree on that interval.

75. By the Ratio Test: $\lim_{n\to\infty}\left|\frac{x^{n+1}}{(n+1)!}\cdot\frac{n!}{x^n}\right| = \lim_{n\to\infty}\frac{|x|}{n+1} = 0$ which shows that $\sum_{n=0}^{\infty}\frac{x^n}{n!}$ converges for all x.

76. $\ln\left(\frac{1+x}{1-x}\right) = \ln(1+x) - \ln(1-x)$

$$= \left(x - \frac{x^2}{2} + \frac{x^3}{3} - \cdots\right) - \left(-x - \frac{x^2}{2} - \frac{x^3}{3} - \cdots\right)$$

$$= 2x + 2\frac{x^3}{3} + 2\frac{x^5}{5} + \cdots) = 2x\sum_{n=0}^{\infty}\frac{x^{2n}}{2n+1},\ R = 1$$

$$\ln 3 = \ln\left(\frac{1+1/2}{1-1/2}\right) \approx 2\left(\frac{1}{2}\right)\left[1 + \frac{(1/2)^2}{3} + \frac{(1/2)^4}{5} + \frac{(1/2)^6}{7}\right] = 1 + \frac{1}{12} + \frac{1}{80} + \frac{1}{448} \approx 1.098065$$

($\ln 3 \approx 1.098612$)

77. $\binom{5}{3} = \frac{5\cdot 4\cdot 3}{3!} = \frac{60}{6} = 10$

78. $\binom{-2}{2} = \frac{(-2)(-3)}{2!} = 3$

79. $\binom{0.5}{4} = \frac{(0.5)(-0.5)(-1.5)(-2.5)}{4!}$

$$= -0.0390625 = -\frac{5}{128}$$

80. $\binom{-1/3}{5} = \frac{(-1/3)(-4/3)(-7/3)(-10/3)(-13/3)}{5!}$

$$= \frac{-91}{729} \approx -0.12483$$

81. $(1+x)^k = \sum_{n=0}^{\infty}\binom{k}{n}x^n$

Example: $(1+x)^2 = \sum_{n=0}^{\infty}\binom{2}{n}x^n = 1 + 2x + x^2$

82. Assume $e = p/q$ is rational. Let $N > q$ and form the following.

$$e - \left[1 + 1 + \frac{1}{2!} + \cdots + \frac{1}{N!}\right] = \frac{1}{(N+1)!} + \frac{1}{(N+2)!} + \cdots$$

Set $a = N!\left[e - \left(1 + 1 + \ldots + \frac{1}{N!}\right)\right]$, a positive integer. But,

$$a = N!\left[\frac{1}{(N+1)!} + \frac{1}{(N+2)!} + \cdots\right] = \frac{1}{N+1} + \frac{1}{(N+1)(N+2)} + \cdots < \frac{1}{N+1} + \frac{1}{(N+1)^2} + \cdots$$

$$= \frac{1}{N+1}\left[1 + \frac{1}{N+1} + \frac{1}{(N+1)^2} + \ldots\right] = \frac{1}{N+1}\left[\frac{1}{1 - \left(\frac{1}{N+1}\right)}\right] = \frac{1}{N},$$ a contradiction.

83. $g(x) = \frac{x}{1 - x - x^2} = a_0 + a_1x + a_2x^2 + \cdots$

$x = (1 - x - x^2)(a_0 + a_1x + a_2x^2 + \cdots)$

$x = a_0 + (a_1 - a_0)x + (a_2 - a_1 - a_0)x^2 + (a_3 - a_2 - a_1)x^3 + \cdots$

Equating coefficients,

$a_0 = 0$

$a_1 - a_0 = 1 \Rightarrow a_1 = 1$

$a_2 - a_1 - a_0 = 0 \Rightarrow a_2 = 1$

$a_3 - a_2 - a_1 = 0 \Rightarrow a_3 = 2$

$a_4 = a_3 + a_2 = 3$, etc.

In general, $a_n = a_{n-1} + a_{n-2}$. The coefficients are the Fibonacci numbers.

84. Assume the interval is $[-1, 1]$. Let $x \in [-1, 1]$,

$$f(1) = f(x) + (1 - x)f'(x) + \tfrac{1}{2}(1 - x)^2 f''(c),\ c \in (x, 1)$$

$$f(-1) = f(x) + (-1 - x)f'(x) + \tfrac{1}{2}(-1 - x)^2 f''(d),\ d \in (-1, x).$$

Hence, $f(1) - f(-1) = 2f'(x) + \tfrac{1}{2}(1 - x)^2 f''(c) - \tfrac{1}{2}(1 + x)^2 f''(d)$

$$2f'(x) = f(1) - f(-1) - \tfrac{1}{2}(1 - x)^2 f''(c) + \tfrac{1}{2}(1 + x)^2 f''(d).$$

Since $|f(x)| \le 1$ and $|f''(x)| \le 1$,

$$\begin{aligned} 2|f'(x)| &\le |f(1)| + |f(-1)| + \tfrac{1}{2}(1 - x)^2|f''(c)| + \tfrac{1}{2}(1 + x)^2|f''(d)| \\ &\le 1 + 1 + \tfrac{1}{2}(1 - x^2) + \tfrac{1}{2}(1 + x)^2 \\ &= 3 + x^2 \le 4. \end{aligned}$$

Thus, $|f'(x)| \le 2$.

Note: Let $f(x) = \tfrac{1}{2}(x + 1)^2 - 1$. Then $|f(x)| \le 1$, $|f''(x)| = 1$ and $f'(1) = 2$.

Review Exercises for Chapter 9

1. $a_n = \dfrac{1}{n!}$

2. $a_n = \dfrac{n}{n^2 + 1}$

3. $a_n = 4 + \dfrac{2}{n}$: 6, 5, 4.67, . . .

Matches (a)

4. $a_n = 4 - \dfrac{n}{2}$: 3.5, 3, . . .

Matches (c)

5. $a_n = 10(0.3)^{n-1}$: 10, 3, . . .

Matches (d)

6. $a_n = 6\left(-\dfrac{2}{3}\right)^{n-1}$: 6, −4, . . .

Matches (b)

7. $a_n = \dfrac{5n + 2}{n}$

The sequence seems to converge to 5.

$$\lim_{n\to\infty} a_n = \lim_{n\to\infty} \frac{5n + 2}{n} = \lim_{n\to\infty}\left(5 + \frac{2}{n}\right) = 5$$

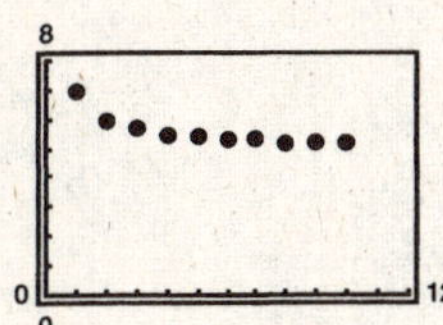

8. $a_n = \sin\dfrac{n\pi}{2}$

The sequence seems to diverge (oscillates).

$\sin\dfrac{n\pi}{2}$: 1, 0, −1, 0, 1, 0, . . .

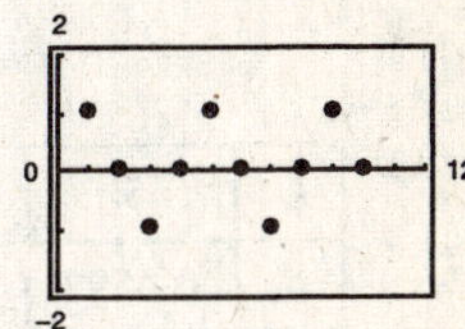

9. $\displaystyle\lim_{n\to\infty} \frac{n + 1}{n^2} = 0$

Converges

10. $\displaystyle\lim_{n\to\infty} \frac{1}{\sqrt{n}} = 0$

Converges

11. $\displaystyle\lim_{n\to\infty} \frac{n^3}{n^2 + 1} = \infty$

12. $\displaystyle\lim_{n\to\infty} \frac{n}{\ln(n)} = \lim_{n\to\infty} \frac{1}{1/n} = \infty$

Diverges

13. $\displaystyle\lim_{n\to\infty} \left(\sqrt{n + 1} - \sqrt{n}\right) = \lim_{n\to\infty} \left(\sqrt{n + 1} - \sqrt{n}\right)\frac{\sqrt{n + 1} + \sqrt{n}}{\sqrt{n + 1} + \sqrt{n}} = \lim_{n\to\infty} \frac{1}{\sqrt{n + 1} + \sqrt{n}} = 0$ Converges

14. $\displaystyle\lim_{n\to\infty} \left(1 + \frac{1}{2n}\right)^n = \lim_{k\to\infty}\left[\left(1 + \frac{1}{k}\right)^k\right]^{1/2} = e^{1/2}$

Converges; $k = 2n$

15. $\displaystyle\lim_{n\to\infty} \frac{\sin\sqrt{n}}{\sqrt{n}} = 0$

Converges

16. Let $y = (b^n + c^n)^{1/n}$

$$\ln y = \frac{\ln(b^n + c^n)}{n}$$

$$\lim_{n\to\infty} \ln y = \lim_{n\to\infty} \frac{1}{b^n + c^n}(b^n \ln b + c^n \ln c).$$

Assume $b \geq c$ and note that the terms

$$\frac{b^n \ln b + c^n \ln c}{b^n + c^n} = \frac{b^n \ln b}{b^n + c^n} + \frac{c^n \ln c}{b^n + c^n}$$

converge as $n \to \infty$. Hence a_n converges.

17. $A_n = 5000\left(1 + \frac{0.05}{4}\right)^n = 5000(1.0125)^n$

$n = 1, 2, 3$

(a) $A_1 = 5062.50$ $\quad A_5 \approx 5320.41$

$A_2 \approx 5125.78$ $\quad A_6 \approx 5386.92$

$A_3 \approx 5189.85$ $\quad A_7 \approx 5454.25$

$A_4 \approx 5254.73$ $\quad A_8 \approx 5522.43$

(b) $A_{40} \approx 8218.10$

18. (a) $V_n = 120{,}000(0.70)^n$, $n = 1, 2, 3, 4, 5$

(b) $V_5 = 120{,}000(0.70)^5 = \$20{,}168.40$

19. (a)

k	5	10	15	20	25
S_k	13.2	113.3	873.8	6448.5	50,500.3

The series diverges (geometric $r = \frac{3}{2} > 1$).

(b)

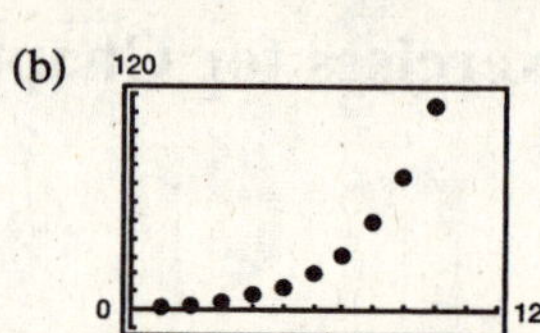

20. (a)

k	5	10	15	20	25
S_k	0.3917	0.3228	0.3627	0.3344	0.3564

The series converges by the Alternating Series Test.

(b)

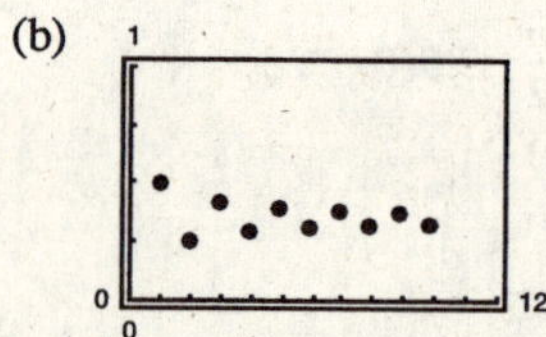

21. (a)

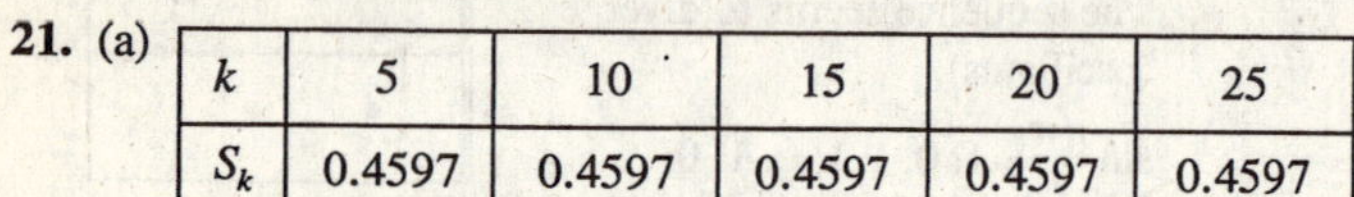

k	5	10	15	20	25
S_k	0.4597	0.4597	0.4597	0.4597	0.4597

The series converges by the Alternating Series Test.

(b)

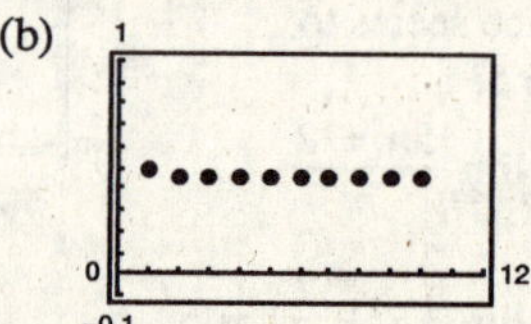

22. (a)

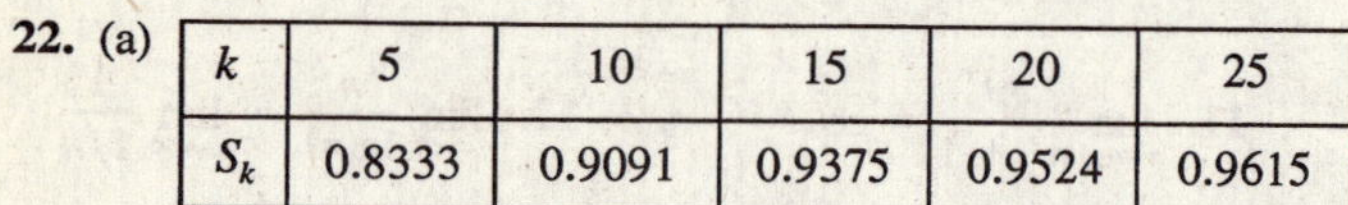

k	5	10	15	20	25
S_k	0.8333	0.9091	0.9375	0.9524	0.9615

The series converges, by the Limit Comparison Test with $\sum \frac{1}{n^2}$.

(b)

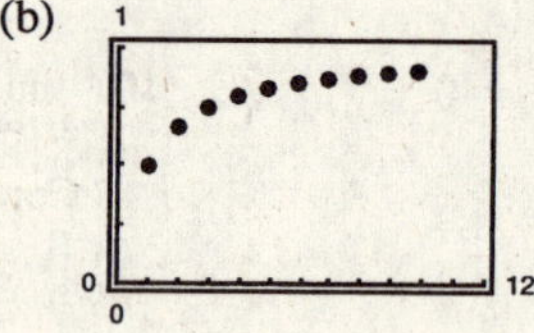

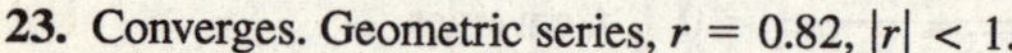

23. Converges. Geometric series, $r = 0.82$, $|r| < 1$.

24. Diverges. Geometric series, $r = 1.82 > 1$.

25. Diverges. nth-Term Test. $\lim_{n\to\infty} a_n \neq 0$.

26. Diverges. nth-Term Test, $\lim_{n\to\infty} a_n = \frac{2}{3}$.

27. $\sum_{n=0}^{\infty} \left(\frac{2}{3}\right)^n$ Geometric series with $a = 1$ and $r = \frac{2}{3}$.

$$S = \frac{a}{1-r} = \frac{1}{1-(2/3)} = \frac{1}{1/3} = 3$$

28. $\sum_{n=0}^{\infty} \frac{2^{n+2}}{3^n} = 4\sum_{n=0}^{\infty}\left(\frac{2}{3}\right)^n = 4(3) = 12$

See Exercise 27.

29. $\sum_{n=0}^{\infty}\left(\frac{1}{2^n}-\frac{1}{3^n}\right)=\sum_{n=0}^{\infty}\left(\frac{1}{2}\right)^n-\sum_{n=0}^{\infty}\left(\frac{1}{3}\right)^n=\frac{1}{1-(1/2)}-\frac{1}{1-(1/3)}=2-\frac{3}{2}=\frac{1}{2}$

30. $\sum_{n=0}^{\infty}\left[\left(\frac{2}{3}\right)^n-\frac{1}{(n+1)(n+2)}\right]=\sum_{n=0}^{\infty}\left(\frac{2}{3}\right)^n-\sum_{n=0}^{\infty}\left(\frac{1}{n+1}-\frac{1}{n+2}\right)$

$=\frac{1}{1-(2/3)}-\left[\left(1-\frac{1}{2}\right)+\left(\frac{1}{2}-\frac{1}{3}\right)+\left(\frac{1}{3}-\frac{1}{4}\right)+\cdots\right]=3-1=2$

31. (a) $0.\overline{09}=0.09+0.0009+0.000009+\cdots=0.09(1+0.01+0.0001+\cdots)=\sum_{n=0}^{\infty}(0.09)(0.01)^n$

(b) $0.\overline{09}=\frac{0.09}{1-0.01}=\frac{1}{11}$

32. (a) $0.\overline{923076}=0.923076[1+0.000001+(0.000001)^2+\cdots]=\sum_{n=0}^{\infty}(0.923076)(0.000001)^n$

(b) $0.\overline{923076}=\frac{0.923076}{1-0.000001}=\frac{923{,}076}{999{,}999}=\frac{12(76{,}923)}{13(76{,}923)}=\frac{12}{13}$

33. $D_1=8$

$D_2=0.7(8)+0.7(8)=16(0.7)$

$\vdots$

$D=8+16(0.7)+16(0.7)^2+\cdots+16(0.7)^n+\cdots$

$=-8+\sum_{n=0}^{\infty}16(0.7)^n=-8+\frac{16}{1-0.7}=45\frac{1}{3}$ meters

34. $S=\sum_{n=0}^{39}32{,}000(1.055)^n=\frac{32{,}000(1-1.055^{40})}{1-1.055}$

$\approx \$4{,}371{,}379.65$

35. See Exercise 110 in Section 9.2.

$A=\frac{P(e^{rt}-1)}{e^{r/12}-1}$

$=\frac{200(e^{(0.06)(2)}-1)}{e^{0.06/12}-1}$

$\approx \$5087.14$

36. See Exercise 110 in Section 9.2.

$A=P\left(\frac{12}{r}\right)\left[\left(1+\frac{r}{12}\right)^{12t}-1\right]$

$=100\left(\frac{12}{0.035}\right)\left[\left(1+\frac{0.035}{12}\right)^{120}-1\right]$

$\approx \$14{,}343.25$

37. $\int_1^{\infty}x^{-4}\ln(x)\,dx=\lim_{b\to\infty}\left[-\frac{\ln x}{3x^3}-\frac{1}{9x^3}\right]_1^b=0+\frac{1}{9}=\frac{1}{9}$

By the Integral Test, the series converges.

38. $\sum_{n=1}^{\infty}\frac{1}{\sqrt[4]{n^3}}=\sum_{n=1}^{\infty}\frac{1}{n^{3/4}}$

Divergent p-series, $p=\frac{3}{4}<1$

39. $\sum_{n=1}^{\infty}\left(\frac{1}{n^2}-\frac{1}{n}\right)=\sum_{n=1}^{\infty}\frac{1}{n^2}-\sum_{n=1}^{\infty}\frac{1}{n}$

Since the second series is a divergent p-series while the first series is a convergent p-series, the difference diverges.

40. $\sum_{n=1}^{\infty}\left(\frac{1}{n^2}-\frac{1}{2^n}\right)=\sum_{n=1}^{\infty}\frac{1}{n^2}-\sum_{n=1}^{\infty}\frac{1}{2^n}$

The first series is a convergent p-series and the second series is a convergent geometric series. Therefore, their difference converges.

41. $\sum_{n=1}^{\infty}\frac{1}{\sqrt{n^3+2n}}$

$\lim_{n\to\infty}\frac{1/\sqrt{n^3+2n}}{1/(n^{3/2})}=\lim_{n\to\infty}\frac{n^{3/2}}{\sqrt{n^3+2n}}=1$

By a limit comparison test with the convergent p-series $\sum_{n=1}^{\infty}\frac{1}{n^{3/2}}$, the series converges.

42. $\sum_{n=1}^{\infty}\frac{n+1}{n(n+2)}$

$\lim_{n\to\infty}\frac{(n+1)/n(n+2)}{1/n}=\lim_{n\to\infty}\frac{n+1}{n+2}=1$

By a limit comparison test with $\sum_{n=1}^{\infty}\frac{1}{n}$, the series diverges.

43. $\displaystyle\sum_{n=1}^{\infty} \frac{1 \cdot 3 \cdot 5 \cdots (2n-1)}{2 \cdot 4 \cdot 6 \cdots (2n)}$

$$a_n = \frac{1 \cdot 3 \cdot 5 \cdots (2n-1)}{2 \cdot 4 \cdot 6 \cdots (2n)}$$

$$= \left(\frac{3}{2} \cdot \frac{5}{4} \cdots \frac{2n-1}{2n-2}\right)\frac{1}{2n} > \frac{1}{2n}$$

Since $\displaystyle\sum_{n=1}^{\infty} \frac{1}{2n} = \frac{1}{2}\sum_{n=1}^{\infty} \frac{1}{n}$ diverges (harmonic series), so does the original series.

44. Since $\displaystyle\sum_{n=1}^{\infty} \frac{1}{3^n}$ converges, $\displaystyle\sum_{n=1}^{\infty} \frac{1}{3^n - 5}$ converges by the Limit Comparison Test.

45. Converges by the Alternating Series Test. (Conditional convergence)

46. $\displaystyle\sum_{n=1}^{\infty} \frac{(-1)^n \sqrt{n}}{n+1}$

$$a_{n+1} = \frac{\sqrt{n+1}}{n+2} \le \frac{\sqrt{n}}{n+1} = a_n$$

$$\lim_{n\to\infty} \frac{\sqrt{n}}{n+1} = 0$$

By the Alternating Series Test, the series converges.

47. Diverges by the nth-Term Test.

48. Converges by the Alternating Series Test.

$$a_{n+1} = \frac{3\ln(n+1)}{n+1} < \frac{3\ln n}{n} = a_n, \quad \lim_{n\to\infty} \frac{3\ln n}{n} = 0$$

49. $\displaystyle\sum_{n=1}^{\infty} \frac{n}{e^{n^2}}$

$$\lim_{n\to\infty} \left|\frac{a_{n+1}}{a_n}\right| = \lim_{n\to\infty} \left|\frac{n+1}{e^{(n+1)^2}} \cdot \frac{e^{n^2}}{n}\right|$$

$$= \lim_{n\to\infty} \left|\frac{e^{n^2}(n+1)}{e^{n^2+2n+1}n}\right|$$

$$= \lim_{n\to\infty} \left(\frac{1}{e^{2n+1}}\right)\left(\frac{n+1}{n}\right)$$

$$= (0)(1) = 0 < 1$$

By the Ratio Test, the series converges.

50. $\displaystyle\sum_{n=1}^{\infty} \frac{n!}{e^n}$

$$\lim_{n\to\infty} \left|\frac{a_{n+1}}{a_n}\right| = \lim_{n\to\infty} \left|\frac{(n+1)!}{e^{n+1}} \cdot \frac{e^n}{n!}\right|$$

$$= \lim_{n\to\infty} \frac{n+1}{e} = \infty$$

By the Ratio Test, the series diverges.

51. $\displaystyle\sum_{n=1}^{\infty} \frac{2^n}{n^3}$

$$\lim_{n\to\infty} \left|\frac{a_{n+1}}{a_n}\right| = \lim_{n\to\infty} \left|\frac{2^{n+1}}{(n+1)^3} \cdot \frac{n^3}{2^n}\right| = \lim_{n\to\infty} \frac{2n^3}{(n+1)^3} = 2$$

Therefore, by the Ratio Test, the series diverges.

52. $\displaystyle\sum_{n=1}^{\infty} \frac{1 \cdot 3 \cdot 5 \cdots (2n-1)}{2 \cdot 5 \cdot 8 \cdots (3n-1)}$

$$\lim_{n\to\infty} \left|\frac{a_{n+1}}{a_n}\right| = \lim_{n\to\infty} \left|\frac{1 \cdot 3 \cdots (2n-1)(2n+1)}{2 \cdot 5 \cdots (3n-1)(3n+2)} \cdot \frac{2 \cdot 5 \cdots (3n-1)}{1 \cdot 3 \cdots (2n-1)}\right| = \lim_{n\to\infty} \frac{2n+1}{3n+2} = \frac{2}{3}$$

By the Ratio Test, the series converges.

53. (a) Ratio Test: $\lim_{n\to\infty}\left|\frac{a_{n+1}}{a_n}\right| = \lim_{n\to\infty}\frac{(n+1)(3/5)^{n+1}}{n(3/5)^n} = \lim_{n\to\infty}\left(\frac{n+1}{n}\right)\left(\frac{3}{5}\right) = \frac{3}{5} < 1$, Converges

(b)

x	5	10	15	20	25
S_n	2.8752	3.6366	3.7377	3.7488	3.7499

(c)

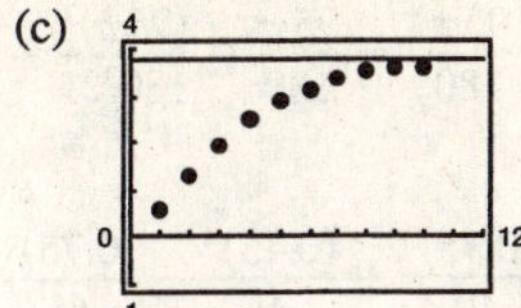

(d) The sum is approximately 3.75.

54. (a) The series converges by the Alternating Series Test.

(b)

x	5	10	15	20	25
S_n	0.0871	0.0669	0.0734	0.0702	0.0721

(c)

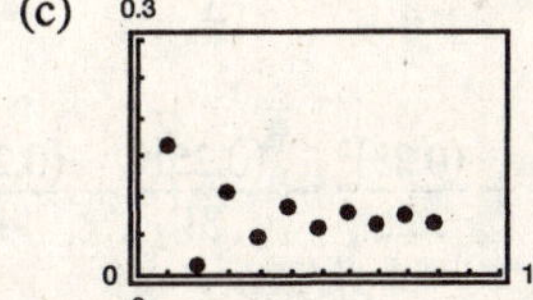

(d) The sum is approximately 0.0714.

55. (a) $\int_N^\infty \frac{1}{x^2}\,dx = \left[-\frac{1}{x}\right]_N^\infty = \frac{1}{N}$

N	5	10	20	30	40
$\sum_{n=1}^{N}\frac{1}{n^2}$	1.4636	1.5498	1.5962	1.6122	1.6202
$\int_N^\infty \frac{1}{x^2}\,dx$	0.2000	0.1000	0.0500	0.0333	0.0250

(b) $\int_N^\infty \frac{1}{x^5}\,dx = \left[-\frac{1}{4x^4}\right]_N^\infty = \frac{1}{4N^4}$

N	5	10	20	30	40
$\sum_{n=1}^{N}\frac{1}{n^5}$	1.0367	1.0369	1.0369	1.0369	1.0369
$\int_N^\infty \frac{1}{x^5}\,dx$	0.0004	0.0000	0.0000	0.0000	0.0000

The series in part (b) converges more rapidly. The integral values represent the remainders of the partial sums.

56. No. Let $a_n = \frac{3937.5}{n^2}$, then $a_{75} = 0.7$. The series $\sum_{n=1}^{\infty}\frac{3937.5}{n^2}$ is a convergent p-series.

57.

$f(x) = e^{-x/2}$ $\qquad f(0) = 1$

$f'(x) = -\frac{1}{2}e^{-x/2}$ $\qquad f'(0) = -\frac{1}{2}$

$f''(x) = \frac{1}{4}e^{-x/2}$ $\qquad f''(0) = \frac{1}{4}$

$f'''(x) = -\frac{1}{8}e^{-x/2}$ $\qquad f'''(0) = -\frac{1}{8}$

$$P_3(x) = f(0) + f'(0)x + f''(0)\frac{x^2}{2!} + f'''(0)\frac{x^3}{3!}$$
$$= 1 - \frac{1}{2}x + \frac{1}{4}\frac{x^2}{2!} - \frac{1}{8}\frac{x^3}{3!}$$
$$= 1 - \frac{1}{2}x + \frac{1}{8}x^2 - \frac{1}{48}x^3$$

58.

$f(x) = \tan x$ $\qquad f\left(-\frac{\pi}{4}\right) = -1$

$f'(x) = \sec^2 x$ $\qquad f'\left(-\frac{\pi}{4}\right) = 2$

$f''(x) = 2\sec^2 x\tan x$ $\qquad f''\left(-\frac{\pi}{4}\right) = -4$

$f'''(x) = 4\sec^2 x\tan^2 x + 2\sec^4 x$ $\qquad f'''\left(-\frac{\pi}{4}\right) = 16$

$$P_3(x) = -1 + 2\left(x + \frac{\pi}{4}\right) - 2\left(x + \frac{\pi}{4}\right)^2 + \frac{8}{3}\left(x + \frac{\pi}{4}\right)^3$$

59. Since $\dfrac{(95\pi)^9}{180^9 \cdot 9!} < 0.001$, use four terms.

$$\sin 95^\circ = \sin\left(\frac{95\pi}{180}\right) \approx \frac{95\pi}{180} - \frac{(95\pi)^3}{180^3 3!} + \frac{(95\pi)^5}{180^5 5!} - \frac{(95\pi)^7}{180^7 7!} \approx 0.99594$$

60. $\cos(0.75) \approx 1 - \dfrac{(0.75)^2}{2!} + \dfrac{(0.75)^4}{4!} - \dfrac{(0.75)^6}{6!} \approx 0.7317$

61. $\ln(1.75) \approx (0.75) - \dfrac{(0.75)^2}{2} + \dfrac{(0.75)^3}{3} - \dfrac{(0.75)^4}{4} + \dfrac{(0.75)^5}{5} - \dfrac{(0.75)^6}{6} + \cdots - \dfrac{(0.75)^{14}}{14} \approx 0.559062$

62. $e^{-0.25} \approx 1 - 0.25 + \dfrac{(0.25)^2}{2!} - \dfrac{(0.25)^3}{3!} + \dfrac{(0.25)^4}{4!} \approx 0.779$

63. $f(x) = \cos x,\ c = 0$

$$R_n(x) = \frac{f^{(n+1)}(z)}{(n+1)!}x^{n+1}$$

$$|f^{(n+1)}(z)| \le 1 \Rightarrow R_n(x) \le \frac{x^{n+1}}{(n+1)!}$$

(a) $R_n(x) \le \dfrac{(0.5)^{n+1}}{(n+1)!} < 0.001$

This inequality is true for $n = 4$.

(b) $R_n(x) \le \dfrac{(1)^{n+1}}{(n+1)!} < 0.001$

This inequality is true for $n = 6$.

(c) $R_n(x) \le \dfrac{(0.5)^{n+1}}{(n+1)!} < 0.0001$

This inequality is true for $n = 5$.

(d) $R_n(x) \le \dfrac{2^{n+1}}{(n+1)!} < 0.0001$

This inequality is true for $n = 10$.

64. $f(x) = \cos x$

$$P_4(x) = 1 - \frac{x^2}{2!} + \frac{x^4}{4!}$$

$$P_6(x) = 1 - \frac{x^2}{2!} + \frac{x^4}{4!} - \frac{x^6}{6!}$$

$$P_{10}(x) = 1 - \frac{x^2}{2!} + \frac{x^4}{4!} - \frac{x^6}{6!} + \frac{x^8}{8!} - \frac{x^{10}}{10!}$$

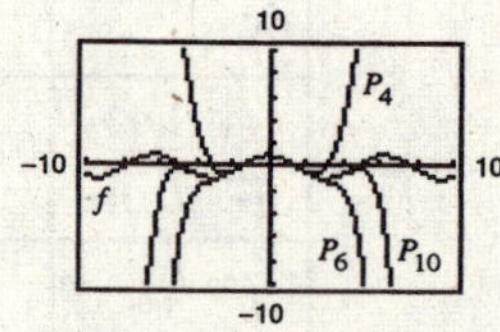

65. $\displaystyle\sum_{n=0}^{\infty}\left(\frac{x}{10}\right)^n$

Geometric series which converges only if $|x/10| < 1$ or $-10 < x < 10$.

66. $\displaystyle\sum_{n=0}^{\infty}(2x)^n$

Geometric series which converges only if $|2x| < 1$ or $-\frac{1}{2} < x < \frac{1}{2}$.

67. $\displaystyle\sum_{n=0}^{\infty}\frac{(-1)^n(x-2)^n}{(n+1)^2}$

$$\lim_{n\to\infty}\left|\frac{u_{n+1}}{u_n}\right| = \lim_{n\to\infty}\left|\frac{(-1)^{n+1}(x-2)^{n+1}}{(n+2)^2} \cdot \frac{(n+1)^2}{(-1)^n(x-2)^n}\right|$$
$$= |x-2|$$

$R = 1$

Center: 2

Since the series converges when $x = 1$ and when $x = 3$, the interval of convergence is $1 \le x \le 3$.

68. $\displaystyle\sum_{n=1}^{\infty}\frac{3^n(x-2)^n}{n}$

$$\lim_{n\to\infty}\left|\frac{u_{n+1}}{u_n}\right| = \lim_{n\to\infty}\left|\frac{3^{n+1}(x-2)^{n+1}}{n+1} \cdot \frac{n}{3^n(x-2)^n}\right|$$
$$= 3|x-2|$$

$R = \dfrac{1}{3}$

Center: 2

Since the series converges at $\frac{5}{3}$ and diverges at $\frac{7}{3}$, the interval of convergence is $\frac{5}{3} \le x < \frac{7}{3}$.

69. $\sum_{n=0}^{\infty} n!(x-2)^n$

$$\lim_{n\to\infty}\left|\frac{u_{n+1}}{u_n}\right| = \lim_{n\to\infty}\left|\frac{(n+1)!(x-2)^{n+1}}{n!(x-2)^n}\right| = \infty$$

which implies that the series converges only at the center $x = 2$.

70. $\sum_{n=0}^{\infty} \frac{(x-2)^n}{2^n} = \sum_{n=0}^{\infty}\left(\frac{x-2}{2}\right)^n$

Geometric series which converges only if

$$\left|\frac{x-2}{2}\right| < 1 \quad \text{or} \quad 0 < x < 4.$$

71.

$$y = \sum_{n=0}^{\infty}(-1)^n\frac{x^{2n}}{4^n(n!)^2}$$

$$y' = \sum_{n=1}^{\infty}\frac{(-1)^n(2n)x^{2n-1}}{4^n(n!)^2} = \sum_{n=0}^{\infty}\frac{(-1)^{n+1}(2n+2)x^{2n+1}}{4^{n+1}[(n+1)!]^2}$$

$$y'' = \sum_{n=0}^{\infty}\frac{(-1)^{n+1}(2n+2)(2n+1)x^{2n}}{4^{n+1}[(n+1)!]^2}$$

$$x^2y'' + xy' + x^2y = \sum_{n=0}^{\infty}\frac{(-1)^{n+1}(2n+2)(2n+1)x^{2n+2}}{4^{n+1}[(n+1)!]^2} + \sum_{n=0}^{\infty}\frac{(-1)^{n+1}(2n+2)x^{2n+2}}{4^{n+1}[(n+1)!]^2} + \sum_{n=0}^{\infty}(-1)^n\frac{x^{2n+2}}{4^n(n!)^2}$$

$$= \sum_{n=0}^{\infty}\left[(-1)^{n+1}\frac{(2n+2)(2n+1)}{4^{n+1}[(n+1)!]^2} + \frac{(-1)^{n+1}(2n+2)}{4^{n+1}[(n+1)!]^2} + \frac{(-1)^n}{4^n(n!)^2}\right]x^{2n+2}$$

$$= \sum_{n=0}^{\infty}\left[\frac{(-1)^{n+1}(2n+2)(2n+1+1)}{4^{n+1}[(n+1)!]^2} + (-1)^n\frac{1}{4^n(n!)^2}\right]x^{2n+2}$$

$$= \sum_{n=0}^{\infty}\left[\frac{(-1)^{n+1}4(n+1)^2}{4^{n+1}[(n+1)!]^2} + (-1)^n\frac{1}{4^n(n!)^2}\right]x^{2n+2}$$

$$= \sum_{n=0}^{\infty}\left[\frac{(-1)^{n+1}1}{4^n(n!)^2} + (-1)^n\frac{1}{4^n(n!)^2}\right]x^{2n+2} = 0$$

72.

$$y = \sum_{n=0}^{\infty}\frac{(-3)^nx^{2n}}{2^nn!}$$

$$y' = \sum_{n=1}^{\infty}\frac{(-3)^n(2n)x^{2n-1}}{2^nn!} = \sum_{n=0}^{\infty}\frac{(-3)^{n+1}(2n+2)x^{2n+1}}{2^{n+1}(n+1)!}$$

$$y'' = \sum_{n=0}^{\infty}\frac{(-3)^{n+1}(2n+2)(2n+1)x^{2n}}{2^{n+1}(n+1)!}$$

$$y'' + 3xy' + 3y = \sum_{n=0}^{\infty}\frac{(-3)^{n+1}(2n+2)(2n+1)x^{2n}}{2^{n+1}(n+1)!} + \sum_{n=0}^{\infty}\frac{(-1)^{n+1}3^{n+2}(2n+2)x^{2n+2}}{2^{n+1}(n+1)!} + \sum_{n=0}^{\infty}\frac{(-1)^n3^{n+1}x^{2n}}{2^nn!}$$

$$= \sum_{n=0}^{\infty}\frac{(-1)^{n+1}3^{n+1}(2n+2)x^{2n}}{2^nn!} + \sum_{n=0}^{\infty}\frac{(-1)^{n+1}3^{n+2}x^{2n+2}}{2^nn!} + \sum_{n=0}^{\infty}\frac{(-1)^n3^{n+1}x^{2n}}{2^nn!}$$

$$= \sum_{n=0}^{\infty}\frac{(-1)^n3^{n+1}x^{2n}}{2^nn!}[-(2n+1)+1] + \sum_{n=0}^{\infty}\frac{(-1)^{n+1}3^{n+2}x^{2n+2}}{2^nn!}$$

$$= \sum_{n=0}^{\infty}\frac{(-1)^n3^{n+1}x^{2n}}{2^nn!}(-2n) + \sum_{n=0}^{\infty}\frac{(-1)^{n+1}3^{n+2}x^{2n+2}}{2^nn!}$$

$$= \sum_{n=1}^{\infty}\frac{(-1)^{n+1}3^{n+1}x^{2n}}{2^nn!}(2n) + \sum_{n=1}^{\infty}\frac{(-1)^n3^{n+1}x^{2n}}{2^{n-1}(n-1)!}\cdot\frac{2n}{2n}$$

$$= \sum_{n=1}^{\infty}\frac{(-1)^n3^{n+1}x^{2n}}{2^nn!}[-2n+2n] = 0$$

73. $\frac{2}{3-x} = \frac{2/3}{1-(x/3)} = \frac{a}{1-r}$

$$\sum_{n=0}^{\infty}\frac{2}{3}\left(\frac{x}{3}\right)^n = \sum_{n=0}^{\infty}\frac{2x^n}{3^{n+1}}$$

74. $\frac{3}{2+x} = \frac{3/2}{1+(x/2)} = \frac{3/2}{1-(-x/2)} = \frac{a}{1-r}$

$$\sum_{n=0}^{\infty}\frac{3}{2}\left(-\frac{x}{2}\right)^n = \sum_{n=0}^{\infty}\frac{(-1)^n3x^n}{2^{n+1}}$$

75. $g(x) = \dfrac{2}{3-x}$. Power series $\displaystyle\sum_{n=0}^{\infty} \frac{2}{3}\left(\frac{x}{3}\right)^n$

Derivative: $\displaystyle\sum_{n=1}^{\infty} \frac{2}{3} n\left(\frac{x}{3}\right)^{n-1}\left(\frac{1}{3}\right) = \sum_{n=1}^{\infty} \frac{2}{9} n\left(\frac{x}{3}\right)^{n-1}$

$$= \sum_{n=0}^{\infty} \frac{2}{9}(n+1)\left(\frac{x}{3}\right)^n$$

76. Integral: $\displaystyle\sum_{n=0}^{\infty} \frac{(-1)^n 3x^{n+1}}{(n+1)2^{n+1}}$

77. $1 + \frac{2}{3}x + \frac{4}{9}x^2 + \frac{8}{27}x^3 + \cdots = \displaystyle\sum_{n=0}^{\infty}\left(\frac{2x}{3}\right)^n = \frac{1}{1-(2x/3)} = \frac{3}{3-2x}, \quad -\frac{3}{2} < x < \frac{3}{2}$

78. $8 - 2(x-3) + \frac{1}{2}(x-3)^2 - \frac{1}{8}(x-3)^3 + \cdots = \displaystyle\sum_{n=0}^{\infty} 8\left[\frac{-(x-3)}{4}\right]^n = \frac{8}{1-[-(x-3)/4]}$

$$= \frac{32}{4+(x-3)} = \frac{32}{1+x}, \quad -1 < x < 7$$

79. $f(x) = \sin x$

$f'(x) = \cos x$

$f''(x) = -\sin x$

$f'''(x) = -\cos x, \cdots$

$\sin(x) = \displaystyle\sum_{n=0}^{\infty} \frac{f^{(n)}(x)[x-(3\pi/4)]^n}{n!}$

$= \dfrac{\sqrt{2}}{2} - \dfrac{\sqrt{2}}{2}\left(x - \dfrac{3\pi}{4}\right) - \dfrac{\sqrt{2}}{2\cdot 2!}\left(x - \dfrac{3\pi}{4}\right)^2 + \cdots = \dfrac{\sqrt{2}}{2}\displaystyle\sum_{n=0}^{\infty} \frac{(-1)^{n(n+1)/2}[x-(3\pi/4)]^n}{n!}$

80. $f(x) = \cos x$

$f'(x) = -\sin x$

$f''(x) = -\cos x$

$f'''(x) = \sin x$

$\cos x = \displaystyle\sum_{n=0}^{\infty} \frac{f^{(n)}(-\pi/4)[x+(\pi/4)]^n}{n!} = \frac{\sqrt{2}}{2} + \frac{\sqrt{2}}{2}\left(x+\frac{\pi}{4}\right) - \frac{\sqrt{2}}{2\cdot 2!}\left(x+\frac{\pi}{4}\right)^2 - \frac{\sqrt{2}}{2\cdot 3!}\left(x+\frac{\pi}{4}\right)^3 + \frac{\sqrt{2}}{2\cdot 4!}\left(x+\frac{\pi}{4}\right)^4 + \cdots$

$= \dfrac{\sqrt{2}}{2}\left[1 + \left(x+\dfrac{\pi}{4}\right) + \displaystyle\sum_{n=1}^{\infty} \frac{(-1)^{[n(n+1)]/2}[x+(\pi/4)]^{n+1}}{(n+1)!}\right]$

81. $3^x = (e^{\ln(3)})^x = e^{x\ln(3)}$ and since $e^x = \displaystyle\sum_{n=0}^{\infty}\frac{x^n}{n!}$, we have $3^x = \displaystyle\sum_{n=0}^{\infty}\frac{(x\ln 3)^n}{n!} = 1 + x\ln 3 + \frac{x^2[\ln 3]^2}{2!} + \frac{x^3[\ln 3]^3}{3!} + \frac{x^4[\ln 3]^4}{4!} + \cdots.$

82. $f(x) = \csc(x)$

$f'(x) = -\csc(x)\cot(x)$

$f''(x) = \csc^3(x) + \csc(x)\cot^2(x)$

$f'''(x) = -5\csc^3(x)\cot(x) - \csc(x)\cot^3(x)$

$f^{(4)}(x) = 5\csc^5(x) + 15\csc^3(x)\cot^2(x) + \csc(x)\cot^4(x)$

$\csc(x) = \displaystyle\sum_{n=0}^{\infty}\frac{f^{(n)}(\pi/2)[x-(\pi/2)]^n}{n!} = 1 + \frac{1}{2!}\left(x-\frac{\pi}{2}\right)^2 + \frac{5}{4!}\left(x-\frac{\pi}{2}\right)^4 + \cdots$

83. $f(x) = \frac{1}{x}$

$f'(x) = -\frac{1}{x^2}$

$f''(x) = \frac{2}{x^3}$

$f'''(x) = -\frac{6}{x^4}, \ldots$

$$\frac{1}{x} = \sum_{n=0}^{\infty} \frac{f^{(n)}(-1)(x+1)^n}{n!} = \sum_{n=0}^{\infty} \frac{-n!(x+1)^n}{n!} = -\sum_{n=0}^{\infty} (x+1)^n, \; -2 < x < 0$$

84. $f(x) = x^{1/2}$

$f'(x) = \frac{1}{2}x^{-1/2}$

$f''(x) = -\left(\frac{1}{2}\right)\left(\frac{1}{2}\right)x^{-3/2}$

$f'''(x) = \left(\frac{1}{2}\right)\left(\frac{1}{2}\right)\left(\frac{3}{2}\right)x^{-5/2}$

$f^{(4)}(x) = -\left(\frac{1}{2}\right)\left(\frac{1}{2}\right)\left(\frac{3}{2}\right)\left(\frac{5}{2}\right)x^{-7/2}, \ldots$

$$\sqrt{x} = \sum_{n=0}^{\infty} \frac{f^{(n)}(4)(x-4)^n}{n!} = 2 + \frac{(x-4)}{2^2} - \frac{(x-4)^2}{2^5 2!} + \frac{1 \cdot 3(x-4)^3}{2^8 3!} - \frac{1 \cdot 3 \cdot 5(x-4)^4}{2^{11} 4!} + \cdots$$

$$= 2 + \frac{(x-4)}{2^2} + \sum_{n=2}^{\infty} \frac{(-1)^{n+1} 1 \cdot 3 \cdot 5 \cdots (2n-3)(x-4)^n}{2^{3n-1} n!}$$

85. $(1+x)^k = 1 + kx + \frac{k(k-1)x^2}{2!} + \frac{k(k-1)(k-2)x^3}{3!} + \cdots$

$$(1+x)^{1/5} = 1 + \frac{x}{5} + \frac{(1/5)(-4/5)x^2}{2!} + \frac{1/5(-4/5)(-9/5)x^3}{3!} + \cdots$$

$$= 1 + \frac{1}{5}x - \frac{1 \cdot 4x^2}{5^2 2!} + \frac{1 \cdot 4 \cdot 9x^3}{5^3 3!} - \cdots$$

$$= 1 + \frac{x}{5} + \sum_{n=2}^{\infty} \frac{(-1)^{n+1} 4 \cdot 9 \cdot 14 \cdots (5n-6)x^n}{5^n n!}$$

$$= 1 + \frac{x}{5} - \frac{2}{25}x^2 + \frac{6}{125}x^3 - \cdots$$

86. $h(x) = (1+x)^{-3}$

$h'(x) = -3(1+x)^{-4}$

$h''(x) = 12(1+x)^{-5}$

$h'''(x) = -60(1+x)^{-6}$

$h^{(4)}(x) = 360(1+x)^{-7}$

$h^{(5)}(x) = -2520(1+x)^{-8}$

$$\frac{1}{(1+x)^3} = 1 - 3x + \frac{12x^2}{2!} - \frac{60x^3}{3!} + \frac{360x^4}{4!} - \frac{2520x^5}{5!} + \cdots = \sum_{n=0}^{\infty} \frac{(-1)^n(n+2)!x^n}{2n!} = \sum_{n=0}^{\infty} \frac{(-1)^n(n+2)(n+1)x^n}{2}$$

87. $\ln x = \sum_{n=1}^{\infty} (-1)^{n+1} \frac{(x-1)^n}{n}, \quad 0 < x \le 2$

$$\ln\left(\frac{5}{4}\right) = \sum_{n=1}^{\infty} (-1)^{n+1}\left(\frac{(5/4)-1}{n}\right)^n$$

$$= \sum_{n=1}^{\infty} (-1)^{n+1} \frac{1}{4^n n} \approx 0.2231$$

88. $\ln x = \sum_{n=1}^{\infty} (-1)^{n+1} \frac{(x-1)^n}{n}, \quad 0 < x \le 2$

$$\ln\left(\frac{6}{5}\right) = \sum_{n=1}^{\infty} (-1)^{n+1}\left(\frac{(6/5)-1}{n}\right)^n$$

$$= \sum_{n=1}^{\infty} (-1)^{n+1} \frac{1}{5^n n} \approx 0.1823$$

89. $e^x = \sum_{n=0}^{\infty} \frac{x^n}{n!}, \quad -\infty < x < \infty$

$$e^{1/2} = \sum_{n=0}^{\infty} \frac{(1/2)^n}{n!} = \sum_{n=0}^{\infty} \frac{1}{2^n n!} \approx 1.6487$$

90. $e^x = \sum_{n=0}^{\infty} \frac{x^n}{n!}, \quad -\infty < x < \infty$

$$e^{2/3} = \sum_{n=0}^{\infty} \frac{(2/3)^n}{n!} = \sum_{n=0}^{\infty} \frac{2^n}{3^n n!} \approx 1.9477$$

91. $\cos x = \sum_{n=0}^{\infty} (-1)^n \frac{x^{2n}}{(2n)!}, \quad -\infty < x < \infty$

$$\cos\left(\frac{2}{3}\right) = \sum_{n=0}^{\infty} (-1)^n \frac{2^{2n}}{3^{2n}(2n)!} \approx 0.7859$$

92. $\sin x = \sum_{n=0}^{\infty} (-1)^n \frac{x^{2n+1}}{(2n+1)!}, \quad -\infty < x < \infty$

$$\sin\left(\frac{1}{3}\right) = \sum_{n=0}^{\infty} (-1)^n \frac{1}{3^{2n+1}(2n+1)!} \approx 0.3272$$

93. The series for Exercise 41 converges very slowly because the terms approach 0 at a slow rate.

94. $\frac{1}{\sqrt{1+x^3}} = (1+x^3)^{-1/2}, k = -\frac{1}{2}$

$$= 1 - \frac{1}{2}(x^3) + \frac{(-1/2)(-3/2)}{2!}x^6 + \frac{(-1/2)(-3/2)(-5/2)}{3!}x^9 + \cdots$$

$$= 1 - \frac{x^3}{2} + \sum_{n=2}^{\infty} \frac{(-1)^n 1 \cdot 3 \cdots (2n-1)}{2^n n!} x^{3n}$$

95. (a) $f(x) = e^{2x}$ $\quad f(0) = 1$

$f'(x) = 2e^{2x}$ $\quad f'(0) = 2$

$f''(x) = 4e^{2x}$ $\quad f''(0) = 4$

$f'''(x) = 8e^{2x}$ $\quad f'''(0) = 8$

$$P(x) = 1 + 2x + \frac{4x^2}{2!} + \frac{8x^3}{3!} = 1 + 2x + 2x^2 + \frac{4}{3}x^3$$

(b) $e^x = \sum_{n=0}^{\infty} \frac{x^n}{n!}, e^{2x} = \sum_{n=0}^{\infty} \frac{(2x)^n}{n!}$

$$P(x) = 1 + 2x + 2x^2 + \frac{4}{3}x^3$$

(c) $e^x \cdot e^x = \left(1 + x + \frac{x^2}{2!} + \cdots\right)\left(1 + x + \frac{x^2}{2!} + \cdots\right)$

$$P = 1 + 2x + 2x^2 + \frac{4}{3}x^3$$

96. (a) $f(x) = \sin 2x$ $\quad f(0) = 0$

$f'(x) = 2\cos 2x$ $\quad f'(0) = 2$

$f''(x) = -4\sin 2x$ $\quad f''(0) = 0$

$f'''(x) = -8\cos 2x$ $\quad f'''(0) = -8$

$f^{(4)}(x) = 16\sin 2x$ $\quad f^{(4)}(0) = 0$

$f^{(5)}(x) = 32\cos 2x$ $\quad f^{(5)}(0) = 32$

$f^{(6)}(x) = -64\sin 2x$ $\quad f^{(6)}(0) = 0$

$f^{(7)}(x) = -128\cos 2x$ $\quad f^{(7)}(0) = -128$

$$\sin 2x = 0 + 2x + \frac{0x^2}{2!} - \frac{8x^3}{3!} + \frac{0x^4}{4!} + \frac{32x^5}{5!} + \frac{0x^6}{6!} - \frac{128x^7}{7!} + \cdots = 2x - \frac{4}{3}x^3 + \frac{4}{15}x^5 - \frac{8}{315}x^7 + \cdots$$

—CONTINUED—

96. —CONTINUED—

(b) $\sin x = \sum_{n=0}^{\infty} \frac{(-1)^n x^{2n+1}}{(2n+1)!}$

$$\sin 2x = \sum_{n=0}^{\infty} \frac{(-1)^n (2x)^{2n+1}}{(2n+1)!} = 2x - \frac{(2x)^3}{3!} + \frac{(2x)^5}{5!} - \frac{(2x)^7}{7!} + \cdots$$

$$= 2x - \frac{8x^3}{6} + \frac{32x^5}{120} - \frac{128x^7}{5040} + \cdots = 2x - \frac{4}{3}x^3 + \frac{4}{15}x^5 - \frac{8}{315}x^7 + \cdots$$

(c) $\sin 2x = 2 \sin x \cos x$

$$= 2\left(x - \frac{x^3}{6} + \frac{x^5}{120} - \frac{x^7}{5040} + \cdots\right)\left(1 - \frac{x^2}{2} + \frac{x^4}{24} - \frac{x^6}{720} + \cdots\right)$$

$$= 2\left[x + \left(-\frac{x^3}{2} - \frac{x^3}{6}\right) + \left(\frac{x^5}{24} + \frac{x^5}{12} + \frac{x^5}{120}\right) + \left(-\frac{x^7}{720} - \frac{x^7}{144} - \frac{x^7}{240} - \frac{x^7}{5040}\right) + \cdots\right]$$

$$= 2\left[x - \frac{2x^3}{3} + \frac{2x^5}{15} - \frac{4x^7}{315} + \cdots\right] = 2x - \frac{4}{3}x^3 + \frac{4}{15}x^5 - \frac{8}{315}x^7 + \cdots$$

97. $\sin t = \sum_{n=0}^{\infty} \frac{(-1)^n t^{2n+1}}{(2n+1)!}$

$$\frac{\sin t}{t} = \sum_{n=0}^{\infty} \frac{(-1)^n t^{2n}}{(2n+1)!}$$

$$\int_0^x \frac{\sin t}{t}\,dt = \left[\sum_{n=0}^{\infty} \frac{(-1)^n t^{2n+1}}{(2n+1)(2n+1)!}\right]_0^x = \sum_{n=0}^{\infty} \frac{(-1)^n x^{2n+1}}{(2n+1)(2n+1)!}$$

98. $\cos t = \sum_{n=0}^{\infty} \frac{(-1)^n t^{2n}}{(2n)!}$

$$\cos \frac{\sqrt{t}}{2} = \sum_{n=0}^{\infty} \frac{(-1)^n t^n}{2^{2n}(2n)!}$$

$$\int_0^x \cos \frac{\sqrt{t}}{2}\,dt = \left[\sum_{n=0}^{\infty} \frac{(-1)^n t^{n+1}}{2^{2n}(2n)!(n+1)}\right]_0^x = \sum_{n=0}^{\infty} \frac{(-1)^n x^{n+1}}{2^{2n}(2n)!(n+1)}$$

99. $\frac{1}{1+t} = \sum_{n=0}^{\infty} (-1)^n t^n$

$$\ln(1+t) = \int \frac{1}{1+t}\,dt = \sum_{n=0}^{\infty} \frac{(-1)^n t^{n+1}}{n+1}$$

$$\frac{\ln(t+1)}{t} = \sum_{n=0}^{\infty} \frac{(-1)^n t^n}{n+1}$$

$$\int_0^x \frac{\ln(t+1)}{t}\,dt = \left[\sum_{n=0}^{\infty} \frac{(-1)^n t^{n+1}}{(n+1)^2}\right]_0^x = \sum_{n=0}^{\infty} \frac{(-1)^n x^{n+1}}{(n+1)^2}$$

100. $e^t = \sum_{n=0}^{\infty} \frac{t^n}{n!}$

$$e^t - 1 = \sum_{n=1}^{\infty} \frac{t^n}{n!}$$

$$\frac{e^t - 1}{t} = \sum_{n=1}^{\infty} \frac{t^{n-1}}{n!}$$

$$\int_0^x \frac{e^t - 1}{t}\,dt = \left[\sum_{n=1}^{\infty} \frac{t^n}{n \cdot n!}\right]_0^x = \sum_{n=1}^{\infty} \frac{x^n}{n \cdot n!}$$

101. $\arctan x = x - \frac{x^3}{3} + \frac{x^5}{5} - \frac{x^7}{7} + \frac{x^9}{9} - \cdots$

$$\frac{\arctan x}{\sqrt{x}} = \sqrt{x} - \frac{x^{5/2}}{3} + \frac{x^{9/2}}{5} - \frac{x^{13/2}}{7} + \frac{x^{17/2}}{9} - \cdots$$

$$\lim_{x \to 0^+} \frac{\arctan x}{\sqrt{x}} = 0$$

By L'Hôpital's Rule, $\lim_{x \to 0^+} \frac{\arctan x}{\sqrt{x}} = \lim_{x \to 0^+} \frac{\left(\frac{1}{1+x^2}\right)}{\left(\frac{1}{2\sqrt{x}}\right)} = \lim_{x \to 0^+} \frac{2\sqrt{x}}{1+x^2} = 0.$

102. $\arcsin x = x + \dfrac{x^3}{2 \cdot 3} + \dfrac{1 \cdot 3x^5}{2 \cdot 4 \cdot 5} + \dfrac{1 \cdot 3 \cdot 5x^7}{2 \cdot 4 \cdot 6 \cdot 7} + \cdots$

$$\frac{\arcsin x}{x} = 1 + \frac{x^2}{2 \cdot 3} + \frac{1 \cdot 3x^4}{2 \cdot 4 \cdot 5} + \frac{1 \cdot 3 \cdot 5x^6}{2 \cdot 4 \cdot 6 \cdot 7} + \cdots$$

$$\lim_{x \to 0} \frac{\arcsin x}{x} = 1$$

By L'Hôpital's Rule, $\displaystyle\lim_{x \to 0} \frac{\arcsin x}{x} = \lim_{x \to 0} \frac{\left(\dfrac{1}{\sqrt{1 - x^2}}\right)}{1} = 1.$

Problem Solving for Chapter 9

1. (a) $1\left(\dfrac{1}{3}\right) + 2\left(\dfrac{1}{9}\right) + 4\left(\dfrac{1}{27}\right) + \cdots = \displaystyle\sum_{n=0}^{\infty} \frac{1}{3}\left(\frac{2}{3}\right)^n$

$$= \frac{1/3}{1 - (2/3)} = 1$$

(b) $0, \dfrac{1}{3}, \dfrac{2}{3}, 1$, etc.

(c) $\displaystyle\lim_{n \to \infty} C_n = 1 - \sum_{n=0}^{\infty} \frac{1}{3}\left(\frac{2}{3}\right)^n = 1 - 1 = 0$

2. Let $S = \displaystyle\sum_{n=1}^{\infty} \frac{1}{(2n - 1)^2} = \frac{1}{1^2} + \frac{1}{3^2} + \frac{1}{5^2} + \cdots.$

Then $\dfrac{\pi^2}{6} = \dfrac{1}{1^2} + \dfrac{1}{2^2} + \dfrac{1}{3^2} + \dfrac{1}{4^2} + \cdots$

$$= S + \frac{1}{2^2} + \frac{1}{4^2} + \cdots$$

$$= S + \frac{1}{2^2}\left[1 + \frac{1}{2^2} + \frac{1}{3^2} + \cdots\right] = S + \frac{1}{2^2}\left(\frac{\pi^2}{6}\right).$$

Thus, $S = \dfrac{\pi^2}{6} - \dfrac{1}{4}\dfrac{\pi^2}{6} = \dfrac{\pi^2}{6}\left(\dfrac{3}{4}\right) = \dfrac{\pi^2}{8}.$

3. If there are n rows, then $a_n = \dfrac{n(n + 1)}{2}.$

For one circle,

$$a_1 = 1 \text{ and } r_1 = \frac{1}{3}\left(\frac{\sqrt{3}}{2}\right) = \frac{\sqrt{3}}{6} = \frac{1}{2\sqrt{3}}.$$

For three circles,

$$a_2 = 3 \text{ and } 1 = 2\sqrt{3}r_2 + 2r_2$$

$$r_2 = \frac{1}{2 + 2\sqrt{3}}.$$

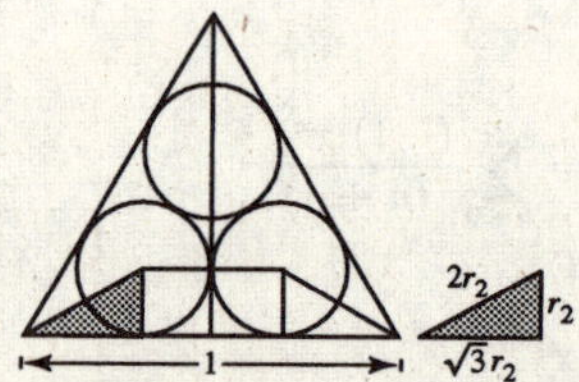

For six circles,

$$a_3 = 6 \text{ and } 1 = 2\sqrt{3}r_3 + 4r_3$$

$$r_3 = \frac{1}{2\sqrt{3} + 4}.$$

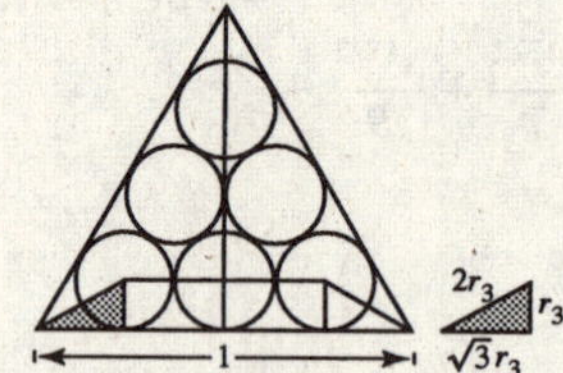

Continuing this pattern, $r_n = \dfrac{1}{2\sqrt{3} + 2(n - 1)}.$

$$\text{Total Area} = (\pi r_n^2)a_n = \pi\left(\frac{1}{2\sqrt{3} + 2(n - 1)}\right)^2 \frac{n(n + 1)}{2}$$

$$A_n = \frac{\pi}{2} \frac{n(n + 1)}{\left[2\sqrt{3} + 2(n - 1)\right]^2}$$

$$\lim_{n \to \infty} A_n = \frac{\pi}{2} \cdot \frac{1}{4} = \frac{\pi}{8}$$

4. (a) Position the three blocks as indicated in the figure. The bottom block extends 1/6 over the edge of the table, the middle block extends 1/4 over the edge of the bottom block, and the top block extends 1/2 over the edge of the middle block.

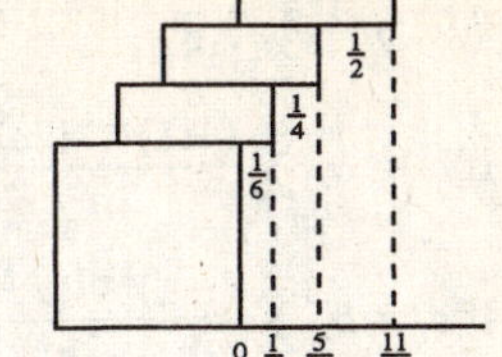

The centers of gravity are located at

bottom block: $\frac{1}{6} - \frac{1}{2} = -\frac{1}{3}$

middle block: $\frac{1}{6} + \frac{1}{4} - \frac{1}{2} = -\frac{1}{12}$

top block: $\frac{1}{6} + \frac{1}{4} + \frac{1}{2} - \frac{1}{2} = \frac{5}{12}$.

The center of gravity of the top 2 blocks is

$$\left(-\frac{1}{12} + \frac{5}{12}\right)/2 = \frac{1}{6},$$

which lies over the bottom block. The center of gravity of the 3 blocks is

$$\left(-\frac{1}{3} - \frac{1}{12} + \frac{5}{12}\right)/3 = 0$$

which lies over the table. Hence, the far edge of the top block lies

$$\frac{1}{6} + \frac{1}{4} + \frac{1}{2} = \frac{11}{12}$$

beyond the edge of the table.

(b) Yes. If there are n blocks, then the edge of the top block lies $\sum_{i=1}^{n} \frac{1}{2i}$ from the edge of the table. Using 4 blocks,

$$\sum_{i=1}^{4} \frac{1}{2i} = \frac{1}{2} + \frac{1}{4} + \frac{1}{6} + \frac{1}{8} = \frac{25}{24}$$

which shows that the top block extends beyond the table.

(c) The blocks can extend any distance beyond the table because the series diverges:

$$\sum_{i=1}^{\infty} \frac{1}{2i} = \frac{1}{2}\sum_{i=1}^{\infty} \frac{1}{i} = \infty.$$

5. (a) $\sum a_n x^n = 1 + 2x + 3x^2 + x^3 + 2x^4 + 3x^5 + \cdots$

$$= (1 + x^3 + x^6 + \cdots) + 2(x + x^4 + x^7 + \cdots) + 3(x^2 + x^5 + x^8 + \cdots)$$

$$= (1 + x^3 + x^6 + \cdots)[1 + 2x + 3x^2]$$

$$= (1 + 2x + 3x^2)\frac{1}{1 - x^3}$$

$R = 1$ because each series in the second line has $R = 1$.

(b) $\sum a_n x^n = (a_0 + a_1 x + \cdots + a_{p-1}x^{p-1}) + (a_0 x^p + a_1 x^{p+1} + \cdots) + \cdots$

$$= a_0(1 + x^p + \cdots) + a_1 x(1 + x^p + \cdots) + \cdots + a_{p-1}x^{p-1}(1 + x^p + \cdots)$$

$$= (a_0 + a_1 x + \cdots + a_{p-1}x^{p-1})(1 + x^p + \cdots)$$

$$= (a_0 + a_1 x + \cdots + a_{p-1}x^{p-1})\frac{1}{1 - x^p}$$

$R = 1$

(Assume all $a_n > 0$.)

6. $a - \frac{b}{2} + \frac{a}{3} - \frac{b}{4} + \cdots = \sum_{n=1}^{\infty} \frac{(-1)^{n+1}(a+b) + (a-b)}{2n}$

If $a = b$, $\sum_{n=1}^{\infty} \frac{(-1)^{n+1}(2a)}{2n} = a\sum_{n=1}^{\infty} \frac{(-1)^{n+1}}{n}$ converges conditionally.

If $a \neq b$, $\sum_{n=1}^{\infty} \frac{(-1)^{n+1}(a+b)}{2n} + \sum_{n=1}^{\infty} \frac{a-b}{2n}$ diverges.

No values of a and b give absolute convergence. $a = b$ implies conditional convergence.

7. (a)

$$e^x = 1 + x + \frac{x^2}{2!} + \cdots = \sum_{n=0}^{\infty} \frac{x^n}{n!}$$

$$xe^x = \sum_{n=0}^{\infty} \frac{x^{n+1}}{n!}$$

$$\int xe^x\,dx = xe^x - e^x + C = \sum_{n=0}^{\infty} \frac{x^{n+2}}{(n+2)n!}$$

Letting $x = 0$, you have $C = 1$. Letting $x = 1$,

$$e - e + 1 = \sum_{n=0}^{\infty} \frac{1}{(n+2)n!} = \frac{1}{2} + \sum_{n=1}^{\infty} \frac{1}{(n+2)n!}.$$

Thus, $\sum_{n=1}^{\infty} \frac{1}{(n+2)n!} = \frac{1}{2}$.

(b) Differentiating,

$$xe^x + e^x = \sum_{n=0}^{\infty} \frac{(n+1)x^n}{n!}.$$

Letting $x = 1$,

$$2e = \sum_{n=0}^{\infty} \frac{n+1}{n!} \approx 5.4366.$$

8.

$$e^x = 1 + x + \frac{x^2}{2!} + \cdots$$

$$e^{x^2} = 1 + x^2 + \frac{x^4}{2!} + \cdots + \frac{x^{12}}{6!} + \cdots$$

$$\frac{f^{(12)}(0)}{12!} = \frac{1}{6!} \Rightarrow f^{(12)}(0) = \frac{12!}{6!} = 665{,}280$$

9. Let $a_1 = \int_0^{\pi} \frac{\sin x}{x}\,dx$, $a_2 = -\int_{\pi}^{2\pi} \frac{\sin x}{x}\,dx$, $a_3 = \int_{2\pi}^{3\pi} \frac{\sin x}{x}\,dx$, etc.

Then,

$$\int_0^{\infty} \frac{\sin x}{x}\,dx = a_1 - a_2 + a_3 - a_4 + \cdots.$$

Since $\lim_{n\to\infty} a_n = 0$ and $a_{n+1} < a_n$, this series converges.

10. (a) If $p = 1$, $\int_2^{\infty} \frac{1}{x \ln x}\,dx = \ln \ln x\Big]_2^{\infty}$ diverges.

If $p > 1$, $\int_2^{\infty} \frac{1}{x(\ln x)^p}\,dx = \lim_{b\to\infty}\left[\frac{(\ln b)^{1-p}}{1-p} - \frac{(\ln 2)^{1-p}}{1-p}\right]$ converges.

If $p < 1$, diverges.

(b) $\sum_{n=4}^{\infty} \frac{1}{n \ln(n^2)} = \frac{1}{2}\sum_{n=4}^{\infty} \frac{1}{n \ln n}$ diverges by part (a).

11. (a) $a_1 = 3.0$

$a_2 \approx 1.73205$

$a_3 \approx 2.17533$

$a_4 \approx 2.27493$

$a_5 \approx 2.29672$

$a_6 \approx 2.30146$

$\lim_{n\to\infty} a_n = \dfrac{1+\sqrt{13}}{2}$ [See part (b) for proof.]

(b) Use mathematical induction to show the sequence is increasing. Clearly, $a_2 = \sqrt{a + a_1} = \sqrt{a\sqrt{a}} > \sqrt{a} = a_1$.

Now assume $a_n > a_{n-1}$. Then

$$a_n + a > a_{n-1} + a$$
$$\sqrt{a_n + a} > \sqrt{a_{n-1} + a}$$
$$a_{n+1} > a_n.$$

Use mathematical induction to show that the sequence is bounded above by a. Clearly, $a_1 = \sqrt{a} < a$.

Now assume $a_n < a$. Then $a > a_n$ and $a - 1 > 1$ implies

$$a(a-1) > a_n(1)$$
$$a^2 - a > a_n$$
$$a^2 > a_n + a$$
$$a > \sqrt{a_n + a} = a_{n+1}.$$

Hence, the sequence converges to some number L. To find L, assume $a_{n+1} \approx a_n \approx L$:

$$L = \sqrt{a + L} \Rightarrow L^2 = a + L \Rightarrow L^2 - L - a = 0$$
$$L = \frac{1 \pm \sqrt{1+4a}}{2}.$$

Hence, $L = \dfrac{1+\sqrt{1+4a}}{2}$.

12. Let $b_n = a_n r^n$.

$(b_n)^{1/n} = (a_n r^n)^{1/n} = a_n^{1/n} \cdot r \to Lr$ as $n \to \infty$.

$Lr < \dfrac{1}{r}r = 1.$

By the Root Test, $\sum b_n$ converges $\Rightarrow \sum a_n r^n$ converges.

13. (a) $\displaystyle\sum_{n=1}^{\infty} \frac{1}{2^{n+(-1)^n}} = \frac{1}{2^{1-1}} + \frac{1}{2^{2+1}} + \frac{1}{2^{3-1}} + \frac{1}{2^{4+1}} + \frac{1}{2^{5-1}} + \cdots$

$$S_1 = \frac{1}{2^0} = 1$$
$$S_1 = 1 + \frac{1}{8} = \frac{9}{8}$$
$$S_3 = \frac{9}{8} + \frac{1}{4} = \frac{11}{8}$$
$$S_4 = \frac{11}{8} + \frac{1}{32} = \frac{45}{32}$$
$$S_5 = \frac{45}{32} + \frac{1}{16} = \frac{47}{32}$$

—CONTINUED—

13. —CONTINUED—

(b) $\dfrac{a_{n+1}}{a_n} = \dfrac{2^{n+(-1)^n}}{2^{(n+1)+(-1)^{n+1}}} = \dfrac{2^{(-1)^n}}{2^{1+(-1)^{n+1}}}$

This sequence is $\frac{1}{8}, 2, \frac{1}{8}, 2, \ldots$ which diverges.

(c) $\sqrt[n]{\dfrac{1}{2^{n+(-1)^n}}} = \left(\dfrac{1}{2^n \cdot 2^{(-1)^n}}\right)^{1/n}$

$= \dfrac{1}{2 \cdot \sqrt[n]{2^{(-1)^n}}} \to \dfrac{1}{2} < 1$ converges because $\{2^{(-1)^n}\} = \frac{1}{2}, 2, \frac{1}{2}, 2, \ldots$ and $\sqrt[n]{1/2} \to 1$ and $\sqrt[n]{2} \to 1$.

14. (a) $\dfrac{1}{0.99} = \dfrac{1}{1-0.01} = \displaystyle\sum_{n=0}^{\infty} (0.01)^n$

$= 1 + 0.01 + (0.01)^2 + \cdots$

$= 1.010101\ldots$

(b) $\dfrac{1}{0.98} = \dfrac{1}{1-0.02} = \displaystyle\sum_{n=0}^{\infty} (0.02)^n$

$= 1 + 0.02 + (0.02)^2 + \cdots$

$= 1 + 0.02 + 0.0004 + \cdots$

$= 1.0204081632\ldots$

15. $S_6 = 130 + 70 + 40 = 240$

$S_7 = 240 + 130 + 70 = 440$

$S_8 = 440 + 240 + 130 = 810$

$S_9 = 810 + 440 + 240 = 1490$

$S_{10} = 1490 + 810 + 440 = 2740$

16. (a) Height $= 2\left[1 + \dfrac{1}{\sqrt{2}} + \dfrac{1}{\sqrt{3}} + \cdots\right]$

$= 2\displaystyle\sum_{n=1}^{\infty} \frac{1}{n^{1/2}} = \infty \quad \left(p\text{-series}, p = \frac{1}{2} < 1\right)$

(b) $S = 4\pi\left[1 + \dfrac{1}{2} + \dfrac{1}{3} + \cdots\right] = 4\pi\displaystyle\sum_{n=1}^{\infty}\frac{1}{n} = \infty$

(c) $W = \dfrac{4}{3}\pi\left[1 + \dfrac{1}{2^{3/2}} + \dfrac{1}{3^{3/2}} + \cdots\right]$

$= \dfrac{4}{3}\pi\displaystyle\sum_{n=1}^{\infty}\frac{1}{n^{3/2}}$ converges.

17. (a) $\displaystyle\sum_{n=1}^{\infty}\frac{1}{2n} = \frac{1}{2}\sum_{n=1}^{\infty}\frac{1}{n}$ diverges (harmonic)

(b) Let $f(x) = \sin x$. By the Mean Value Theorem,

$|f(x) - f(y)| = f'(c)|x - y| = \cos(c)|x - y| \le |x - y|,$

where c is between x and y. Thus,

$0 \le \left|\sin\left(\dfrac{1}{2n}\right) - \sin\left(\dfrac{1}{2n+1}\right)\right|$

$\le \left|\dfrac{1}{2n} - \dfrac{1}{2n+1}\right|$

$= \dfrac{1}{2n(2n+1)}$

Because $\displaystyle\sum_{n=1}^{\infty}\frac{1}{2n(2n+1)}$ converges, the Comparison Theorem

tells us that $\displaystyle\sum_{n=1}^{\infty}\left[\sin\left(\frac{1}{2n}\right) - \sin\left(\frac{1}{2n+1}\right)\right]$ converges.

CHAPTER 10
Conics, Parametric Equations, and Polar Coordinates

CHAPTER 10
Conics, Parametric Equations, and Polar Coordinates

Section 10.1 Conics and Calculus

1. $y^2 = 4x$

Vertex: $(0, 0)$

$p = 1 > 0$

Opens to the right
Matches graph (h).

2. $x^2 = 8y$

Vertex: $(0, 0)$

$p = 2 > 0$

Opens upward
Matches graph (a).

3. $(x + 3)^2 = -2(y - 2)$

Vertex: $(-3, 2)$

$p = -\frac{1}{2} < 0$

Opens downward
Matches graph (e).

4. $\dfrac{(x-2)^2}{16} + \dfrac{(y+1)^2}{4} = 1$

Center: $(2, -1)$
Ellipse
Matches (b)

5. $\dfrac{x^2}{9} + \dfrac{y^2}{4} = 1$

Center: $(0, 0)$
Ellipse
Matches (f)

6. $\dfrac{x^2}{9} + \dfrac{y^2}{9} = 1$

Circle radius 3.
Matches (g)

7. $\dfrac{y^2}{16} - \dfrac{x^2}{1} = 1$

Hyperbola
Center: $(0, 0)$
Vertical transverse axis
Matches (c)

8. $\dfrac{(x-2)^2}{9} - \dfrac{y^2}{4} = 1$

Hyperbola
Center: $(-2, 0)$
Horizontal transverse axis
Matches (d)

9. $y^2 = -6x = 4\left(-\frac{3}{2}\right)x$

Vertex: $(0, 0)$
Focus: $\left(-\frac{3}{2}, 0\right)$
Directrix: $x = \frac{3}{2}$

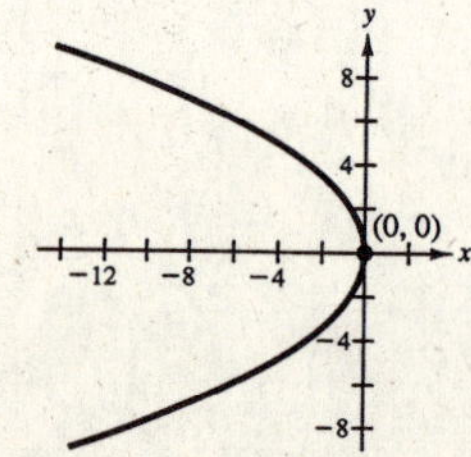

10. $x^2 + 8y = 0$

$x^2 = 4(-2)y$

Vertex: $(0, 0)$
Focus: $(0, -2)$
Directrix: $y = 2$

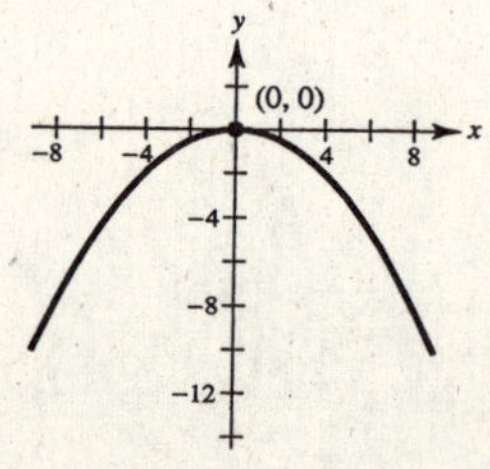

11. $(x + 3) + (y - 2)^2 = 0$

$(y - 2)^2 = 4\left(-\frac{1}{4}\right)(x + 3)$

Vertex: $(-3, 2)$
Focus: $(-3.25, 2)$
Directrix: $x = -2.75$

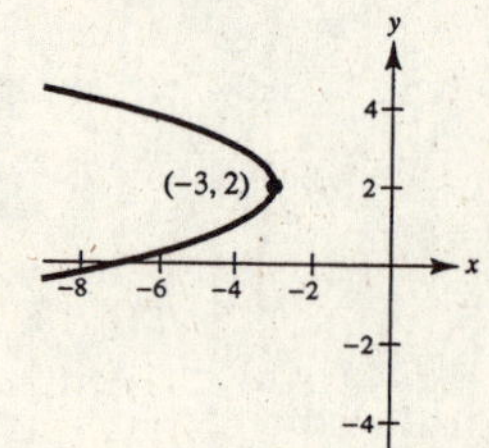

12. $(x - 1)^2 + 8(y + 2) = 0$

$(x - 1)^2 = 4(-2)(y + 2)$

Vertex: $(1, -2)$
Focus: $(1, -4)$
Directrix: $y = 0$

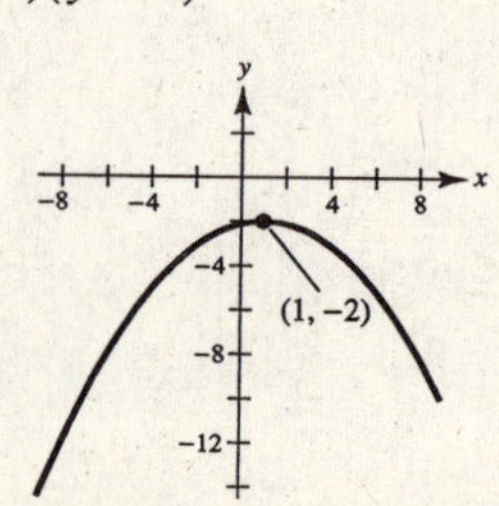

13. $y^2 - 4y - 4x = 0$

$$y^2 - 4y + 4 = 4x + 4$$

$$(y - 2)^2 = 4(1)(x + 1)$$

Vertex: $(-1, 2)$

Focus: $(0, 2)$

Directrix: $x = -2$

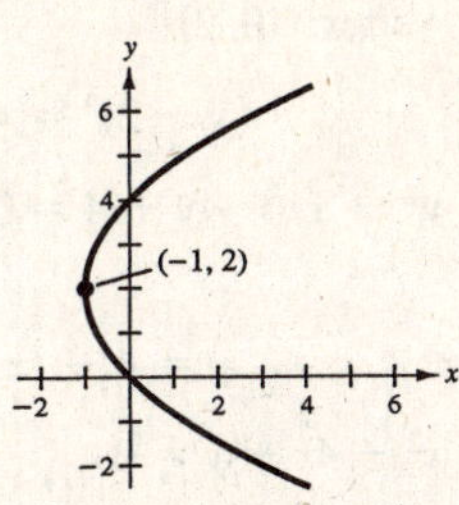

14. $y^2 + 6y + 8x + 25 = 0$

$$y^2 + 6y + 9 = -8x - 25 + 9$$

$$(y + 3)^2 = 4(-2)(x + 2)$$

Vertex: $(-2, -3)$

Focus: $(-4, -3)$

Directrix: $x = 0$

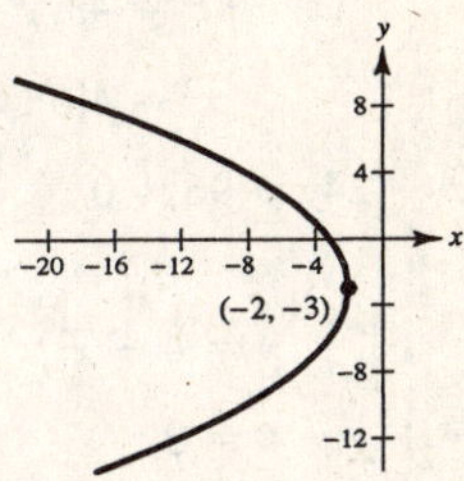

15. $x^2 + 4x + 4y - 4 = 0$

$$x^2 + 4x + 4 = -4y + 4 + 4$$

$$(x + 2)^2 = 4(-1)(y - 2)$$

Vertex: $(-2, 2)$

Focus: $(-2, 1)$

Directrix: $y = 3$

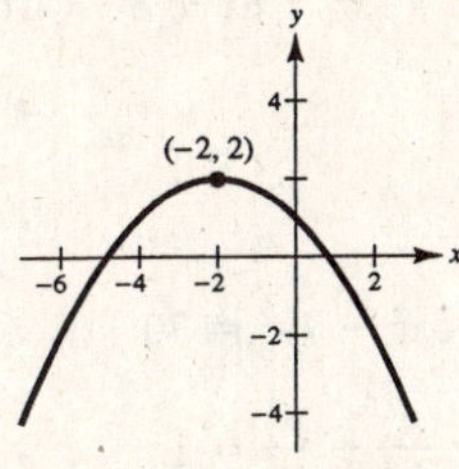

16. $y^2 + 4y + 8x - 12 = 0$

$$y^2 + 4y + 4 = -8x + 12 + 4$$

$$(y + 2)^2 = 4(-2)(x - 2)$$

Vertex: $(2, -2)$

Focus: $(0, -2)$

Directrix: $x = 4$

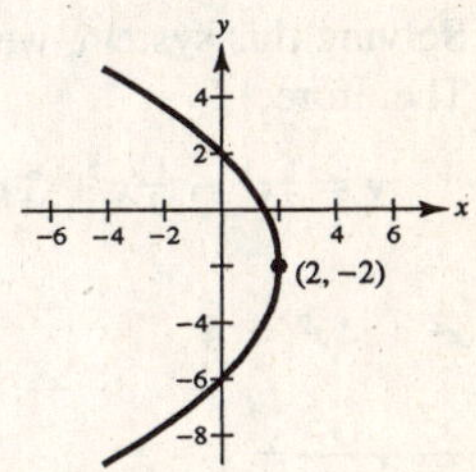

17. $y^2 + x + y = 0$

$$y^2 + y + \frac{1}{4} = -x + \frac{1}{4}$$

$$\left(y + \frac{1}{2}\right)^2 = 4\left(-\frac{1}{4}\right)\left(x - \frac{1}{4}\right)$$

Vertex: $\left(\frac{1}{4}, -\frac{1}{2}\right)$

Focus: $\left(0, -\frac{1}{2}\right)$

Directrix: $x = \frac{1}{2}$

$y_1 = -\frac{1}{2} \pm \sqrt{\frac{1}{4} - x}$

$y_2 = -\frac{1}{2} - \sqrt{\frac{1}{4} - x}$

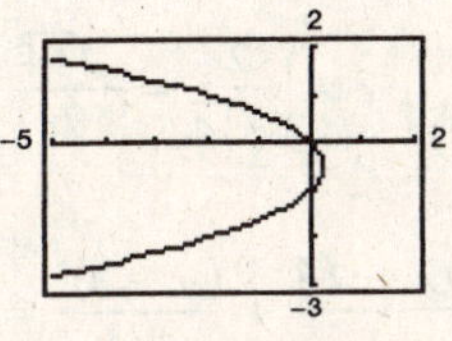

18. $y = -\frac{1}{6}(x^2 - 8x + 6) = -\frac{1}{6}(x^2 - 8x + 16 - 10)$

$$-6y = (x - 4)^2 - 10$$

$$-6y + 10 = (x - 4)^2$$

$$(x - 4)^2 = -6\left(y - \frac{5}{3}\right)$$

$$(x - 4)^2 = 4\left(-\frac{3}{2}\right)\left(y - \frac{5}{3}\right)$$

Vertex: $\left(4, \frac{5}{3}\right)$

Focus: $\left(4, \frac{1}{6}\right)$

Directrix: $y = \frac{19}{6}$

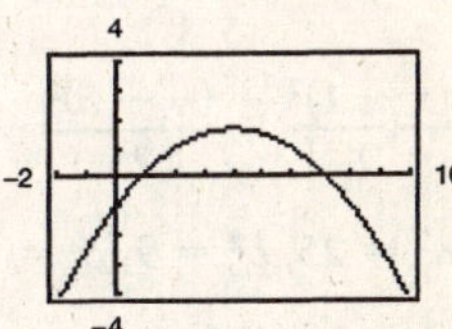

19. $y^2 - 4x - 4 = 0$

$$y^2 = 4x + 4$$

$$= 4(1)(x + 1)$$

Vertex: $(-1, 0)$

Focus: $(0, 0)$

Directrix: $x = -2$

$y_1 = 2\sqrt{x + 1}$

$y_2 = -2\sqrt{x + 1}$

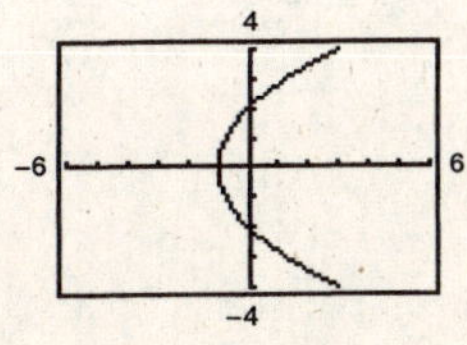

20. $x^2 - 2x + 8y + 9 = 0$

$$x^2 - 2x + 1 = -8y - 9 + 1$$

$$(x - 1)^2 = 4(-2)(y + 1)$$

Vertex: $(1, -1)$

Focus: $(1, -3)$

Directrix: $y = 1$

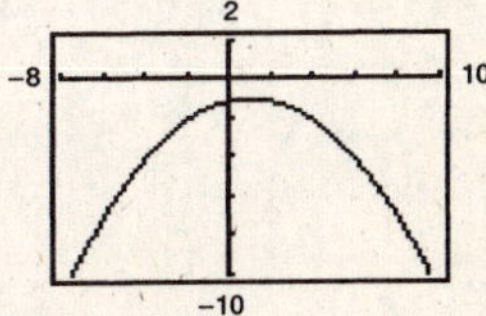

21. $(y-2)^2 = 4(-2)(x-3)$

$y^2 - 4y + 8x - 20 = 0$

22. $(x+1)^2 = 4(-2)(y-2)$

$x^2 + 2x + 8y - 15 = 0$

23. $(x-h)^2 = 4p(y-k)$

$x^2 = 4(6)(y-4)$

$x^2 - 24y + 96 = 0$

24. Vertex: $(0, 2)$

$(y-2)^2 = 4(2)(x-0)$

$y^2 - 8x - 4y + 4 = 0$

25. $y = 4 - x^2$

$x^2 + y - 4 = 0$

26. $y = 4 - (x-2)^2 = 4x - x^2$

$x^2 - 4x + y = 0$

27. Since the axis of the parabola is vertical, the form of the equation is $y = ax^2 + bx + c$. Now, substituting the values of the given coordinates into this equation, we obtain

$$3 = c,\ 4 = 9a + 3b + c,\ 11 = 16a + 4b + c.$$

Solving this system, we have $a = \frac{5}{3}$, $b = -\frac{14}{3}$, $c = 3$. Therefore,

$$y = \tfrac{5}{3}x^2 - \tfrac{14}{3}x + 3 \text{ or } 5x^2 - 14x - 3y + 9 = 0.$$

28. From Example 2: $4p = 8$ or $p = 2$

Vertex: $(4, 0)$

$(x-4)^2 = 8(y-0)$

$x^2 - 8x - 8y + 16 = 0$

29. $x^2 + 4y^2 = 4$

$$\frac{x^2}{4} + \frac{y^2}{1} = 1$$

$a^2 = 4,\ b^2 = 1,\ c^2 = 3$

Center: $(0, 0)$

Foci: $\left(\pm\sqrt{3}, 0\right)$

Vertices: $(\pm 2, 0)$

$$e = \frac{\sqrt{3}}{2}$$

30. $5x^2 + 7y^2 = 70$

$$\frac{x^2}{14} + \frac{y^2}{10} = 1$$

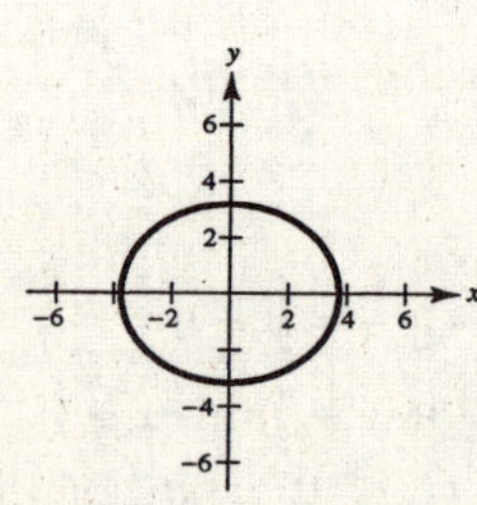

$a^2 = 14,\ b^2 = 10,\ c^2 = 4$

Center: $(0, 0)$

Foci: $(\pm 2, 0)$

Vertices: $\left(\pm\sqrt{14}, 0\right)$

$$e = \frac{2}{\sqrt{14}} = \frac{\sqrt{14}}{7}$$

31. $$\frac{(x-1)^2}{9} + \frac{(y-5)^2}{25} = 1$$

$a^2 = 25,\ b^2 = 9,\ c^2 = 16$

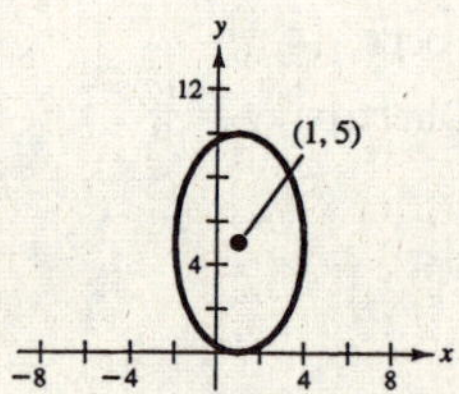

Center: $(1, 5)$

Foci: $(1, 9), (1, 1)$

Vertices: $(1, 10), (1, 0)$

$$e = \frac{4}{5}$$

32. $$\frac{(x+2)^2}{1} + \frac{(y+4)^2}{1/4} = 1$$

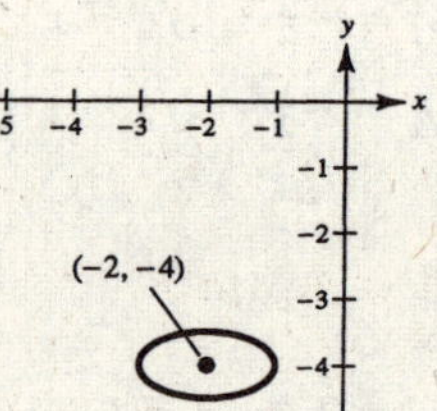

$a^2 = 1,\ b^2 = \dfrac{1}{4},\ c^2 = \dfrac{3}{4}$

Center: $(-2, -4)$

Foci: $\left(-2 \pm \dfrac{\sqrt{3}}{2}, -4\right)$

Vertices: $(-1, -4), (-3, -4)$

$$e = \frac{\sqrt{3}}{2}$$

33. $9x^2 + 4y^2 + 36x - 24y + 36 = 0$

$$9(x^2 + 4x + 4) + 4(y^2 - 6y + 9) = -36 + 36 + 36$$
$$= 36$$
$$\frac{(x+2)^2}{4} + \frac{(y-3)^2}{9} = 1$$

$a^2 = 9,\ b^2 = 4,\ c^2 = 5$

Center: $(-2, 3)$

Foci: $\left(-2, 3 \pm \sqrt{5}\right)$

Vertices: $(-2, 6), (-2, 0)$

$$e = \frac{\sqrt{5}}{3}$$

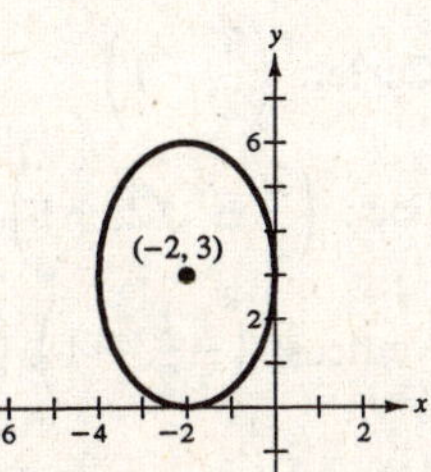

34. $16x^2 + 25y^2 - 64x + 150y + 279 = 0$

$$16(x^2 - 4x + 4) + 25(y^2 + 6y + 0) = -279 + 64 + 225$$
$$= 10$$
$$\frac{(x-2)^2}{(5/8)} + \frac{(y+3)^2}{(2/5)} = 1$$

$a^2, \frac{5}{8}, b^2 = \frac{2}{5}, c^2 = a^2 - b^2 = \frac{9}{40}$

Center: $(2, -3)$

Foci: $\left(2 \pm \frac{3\sqrt{10}}{20}, -3\right)$

Vertices: $\left(2 \pm \frac{\sqrt{10}}{4}, -3\right)$

$$e = \frac{c}{a} = \frac{3}{5}$$

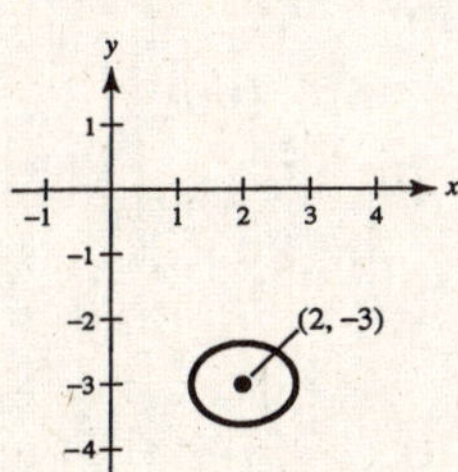

35. $12x^2 + 20y^2 - 12x + 40y - 37 = 0$

$$12\left(x^2 - x + \frac{1}{4}\right) + 20(y^2 + 2y + 1) = 37 + 3 + 20$$
$$= 60$$
$$\frac{[x - (1/2)]^2}{5} + \frac{(y+1)^2}{3} = 1$$

$a^2 = 5,\ b^2 = 3,\ c^2 = 2$

Center: $\left(\frac{1}{2}, -1\right)$

Foci: $\left(\frac{1}{2} \pm \sqrt{2}, -1\right)$

Vertices: $\left(\frac{1}{2} \pm \sqrt{5}, -1\right)$

Solve for y:

$$20(y^2 + 2y + 1) = -12x^2 + 12x + 37 + 20$$
$$(y+1)^2 = \frac{57 + 12x - 12x^2}{20}$$
$$y = -1 \pm \sqrt{\frac{57 + 12x - 12x^2}{20}}$$

(Graph each of these separately.)

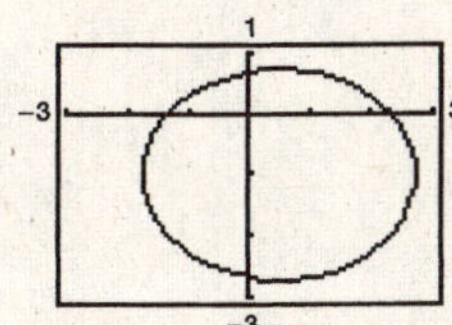

36. $36x^2 + 9y^2 + 48x - 36y + 43 = 0$

$$36\left(x^2 + \frac{4}{3}x + \frac{4}{9}\right) + 9(y^2 - 4y + 4) = -43 + 16 + 36$$

$$= 9$$

$$\frac{[x + (2/3)]^2}{1/4} + \frac{(y - 2)^2}{1} = 1$$

$a^2 = 1,\ b^2 = \frac{1}{4},\ c^2 = \frac{3}{4}$

Center: $\left(-\frac{2}{3}, 2\right)$

Foci: $\left(-\frac{2}{3}, 2 \pm \frac{\sqrt{3}}{2}\right)$

Vertices: $\left(-\frac{2}{3}, 3\right), \left(-\frac{2}{3}, 1\right)$

Solve for y:

$$9(y^2 - 4y + 4) = -36x^2 - 48x - 43 + 36$$

$$(y - 2)^2 = \frac{-(36x^2 + 48x + 7)}{9}$$

$$y = 2 \pm \frac{1}{3}\sqrt{-(36x^2 + 48x + 7)}$$

(Graph each of these separately.)

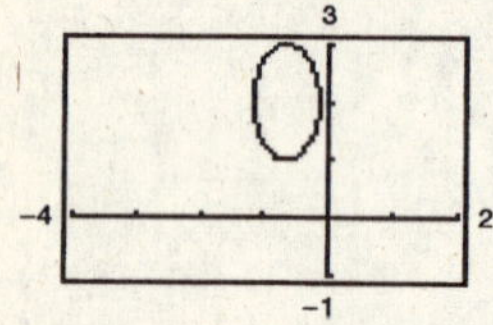

37. $x^2 + 2y^2 - 3x + 4y + 0.25 = 0$

$$\left(x^2 - 3x + \frac{9}{4}\right) + 2(y^2 + 2y + 1) = -\frac{1}{4} + \frac{9}{4} + 2 = 4$$

$$\frac{[x - (3/2)]^2}{4} + \frac{(y + 1)^2}{2} = 1$$

$a^2 = 4,\ b^2 = 2,\ c^2 = 2$

Center: $\left(\frac{3}{2}, -1\right)$

Foci: $\left(\frac{3}{2} \pm \sqrt{2}, -1\right)$

Vertices: $\left(-\frac{1}{2}, -1\right), \left(\frac{7}{2}, -1\right)$

Solve for y: $2(y^2 + 2y + 1) = -x^2 + 3x - \frac{1}{4} + 2$

$$(y + 1)^2 = \frac{1}{2}\left(\frac{7}{4} + 3x - x^2\right)$$

$$y = -1 \pm \sqrt{\frac{7 + 12x - 4x^2}{8}}$$

(Graph each of these separately.)

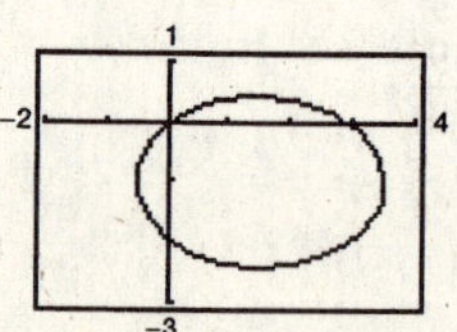

38. $2x^2 + y^2 + 4.8x - 6.4y + 3.12 = 0$

$$50x^2 + 25y^2 + 120x - 160y + 78 = 0$$

$$50\left(x^2 + \frac{12}{5}x + \frac{36}{25}\right) + 25\left(y^2 - \frac{32}{5}y + \frac{256}{25}\right) = -78 + 72 + 256 = 250$$

$$\frac{[x + (6/5)]^2}{5} + \frac{[y - (16/5)]^2}{10} = 1$$

$a^2 = 10,\ b^2 = 5,\ c^2 = 5$

Center: $\left(-\frac{6}{5}, \frac{16}{5}\right)$

Foci: $\left(-\frac{6}{5}, \frac{16}{5} \pm \sqrt{5}\right)$

Vertices: $\left(-\frac{6}{5}, \frac{16}{5} \pm \sqrt{10}\right)$

Solve for y: $(y^2 - 6.4y + 10.24) = -2x^2 - 4.8x - 3.12 + 10.24$

$$(y - 3.2)^2 = 7.12 - 4x - 2x^2$$

$$y = 3.2 \pm \sqrt{7.12 - 4x - 2x^2}$$

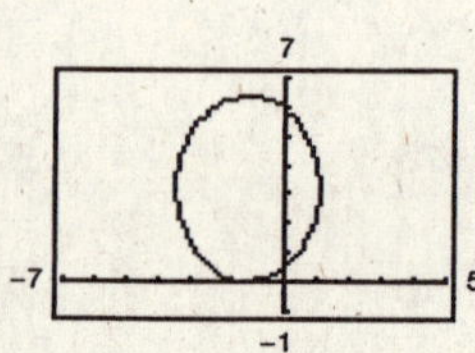

(Graph each of these separately.)

39. Center: $(0, 0)$

Focus: $(2, 0)$

Vertex: $(3, 0)$

Horizontal major axis

$a = 3, c = 2 \Rightarrow b = \sqrt{5}$

$$\frac{x^2}{9} + \frac{y^2}{5} = 1$$

40. Vertices: $(0, 2), (4, 2)$

Eccentricity: $\frac{1}{2}$

Horizontal major axis

Center: $(2, 2)$

$a = 2, c = 1 \Rightarrow b = \sqrt{3}$

$$\frac{(x-2)^2}{4} + \frac{(y-2)^2}{3} = 1$$

41. Vertices: $(3, 1), (3, 9)$

Minor axis length: 6

Vertical major axis

Center: $(3, 5)$

$a = 4, b = 3$

$$\frac{(x-3)^2}{9} + \frac{(y-5)^2}{16} = 1$$

42 Foci: $(0, \pm 5)$

Major axis length: 14

Vertical major axis

Center: $(0, 0)$

$c = 5, a = 7 \Rightarrow b = \sqrt{24}$

$$\frac{x^2}{24} + \frac{y^2}{49} = 1$$

43. Center: $(0, 0)$

Horizontal major axis

Points on ellipse: $(3, 1), (4, 0)$

Since the major axis is horizontal,

$$\left(\frac{x^2}{a^2}\right) + \left(\frac{y^2}{b^2}\right) = 1.$$

Substituting the values of the coordinates of the given points into this equation, we have

$$\left(\frac{9}{a^2}\right) + \left(\frac{1}{b^2}\right) = 1, \text{ and } \frac{16}{a^2} = 1.$$

The solution to this system is $a^2 = 16, b^2 = 16/7$.

Therefore,

$$\frac{x^2}{16} + \frac{y^2}{16/7} = 1, \frac{x^2}{16} + \frac{7y^2}{16} = 1.$$

44. Center: $(1, 2)$

Vertical major axis

Points on ellipse: $(1, 6), (3, 2)$

From the sketch, we can see that $h = 1, k = 2, a = 4, b = 2$

$$\frac{(x-1)^2}{4} + \frac{(y-2)^2}{16} = 1.$$

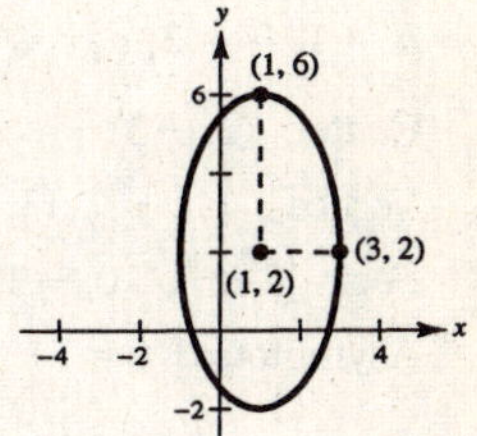

45. $\frac{y^2}{1} - \frac{x^2}{4} = 1$

$a = 1, b = 2, c = \sqrt{5}$

Center: $(0, 0)$

Vertices: $(0, \pm 1)$

Foci: $(0, \pm\sqrt{5})$

Asymptotes: $y = \pm\frac{1}{2}x$

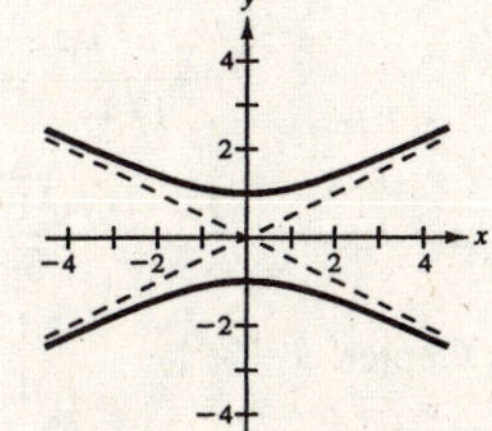

46. $\frac{x^2}{25} - \frac{y^2}{9} = 1$

$a = 5, b = 3, c = \sqrt{a^2 + b^2} = \sqrt{34}$

Center: $(0, 0)$

Vertices: $(\pm 5, 0)$

Foci: $(\pm\sqrt{34}, 0)$

Asymptotes: $y = \pm\frac{3}{5}x$

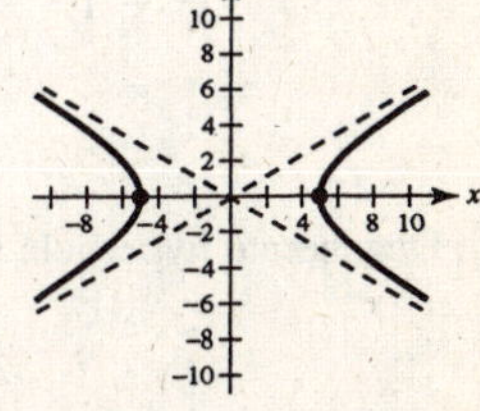

47. $\dfrac{(x-1)^2}{4} - \dfrac{(y+2)^2}{1} = 1$

$a = 2, b = 1, c = \sqrt{5}$

Center: $(1, -2)$

Vertices: $(-1, -2), (3, -2)$

Foci: $\left(1 \pm \sqrt{5}, -2\right)$

Asymptotes: $y = -2 \pm \dfrac{1}{2}(x - 1)$

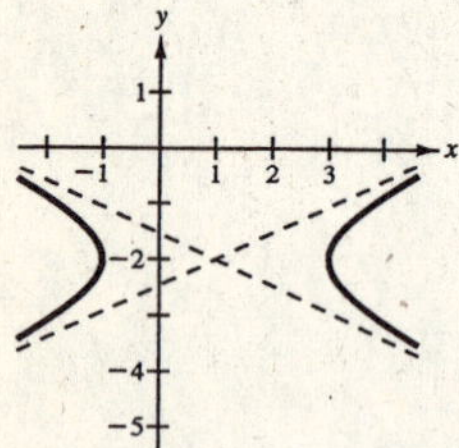

48. $\dfrac{(y+1)^2}{12^2} - \dfrac{(x-4)^2}{5^2} = 1$

$a = 12, b = 5, c = \sqrt{a^2 + b^2} = 13$

Center: $(4, -1)$

Vertices: $(4, 11), (4, -13)$

Foci: $(4, -14), (4, 12)$

Asymptotes: $y = -1 \pm \dfrac{12}{5}(x - 4)$

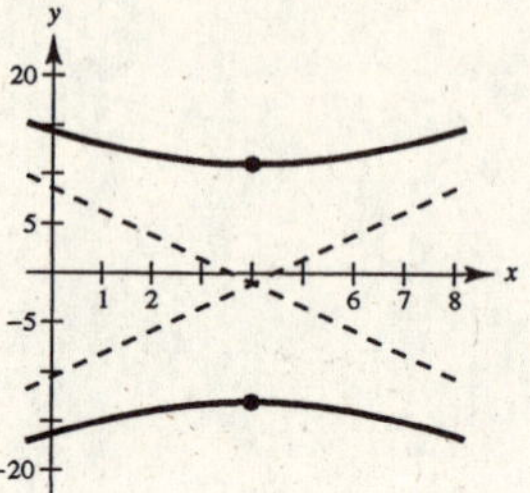

49. $9x^2 - y^2 - 36x - 6y + 18 = 0$

$9(x^2 - 4x + 4) - (y^2 + 6y + 9) = -18 + 36 - 9$

$\dfrac{(x-2)^2}{1} - \dfrac{(y+3)^2}{9} = 1$

$a = 1, b = 3, c = \sqrt{10}$

Center: $(2, -3)$

Vertices: $(1, -3), (3, -3)$

Foci: $\left(2 \pm \sqrt{10}, -3\right)$

Asymptotes: $y = -3 \pm 3(x - 2)$

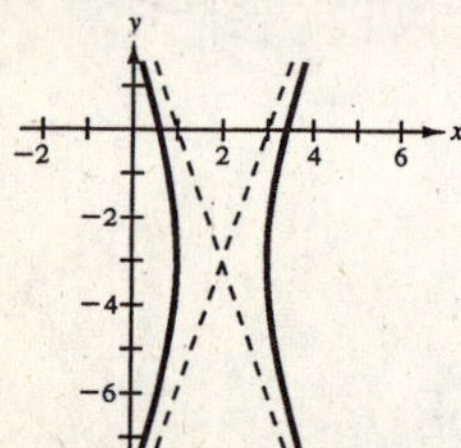

50. $y^2 - 9x^2 + 36x - 72 = 0$

$y^2 - 9(x^2 - 4x + 4) = 72 - 36 = 36$

$\dfrac{y^2}{36} - \dfrac{(x-2)^2}{4} = 1$

$a = 6, b = 2, c = \sqrt{a^2 + b^2} = 2\sqrt{10}$

Center: $(2, 0)$

Vertices: $(2, 6), (2, -6)$

Foci: $\left(2, 2\sqrt{10}\right), \left(2, -2\sqrt{10}\right)$

Asymptotes: $y = \pm 3(x - 2)$

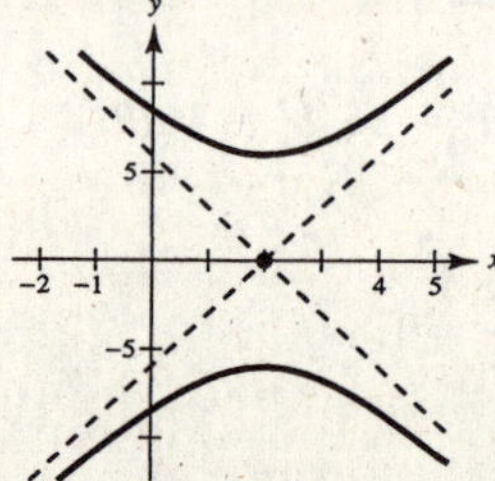

51. $x^2 - 9y^2 + 2x - 54y - 80 = 0$

$(x^2 + 2x + 1) - 9(y^2 + 6y + 9) = 80 + 1 - 81 = 0$

$(x + 1)^2 - 9(y + 3)^2 = 0$

$y + 3 = \pm\dfrac{1}{3}(x + 1)$

Degenerate hyperbola is two lines intersecting at $(-1, -3)$.

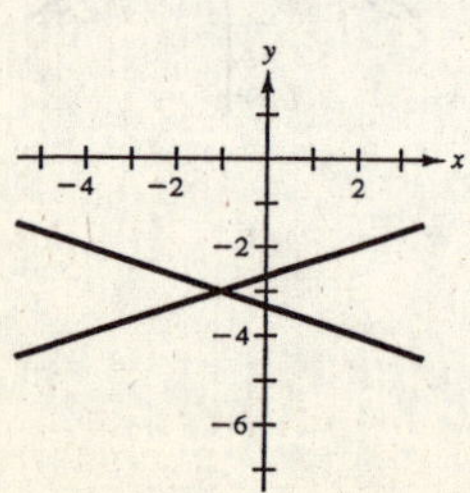

52. $9(x^2 + 6x + 9) - 4(y^2 - 2y + 1) = -78 + 81 - 4 = -1$

$9(x + 3)^2 - 4(y - 1)^2 = -1$

$\dfrac{(y-1)^2}{1/4} - \dfrac{(x+3)^2}{1/9} = 1$

$a = \dfrac{1}{2}, b = \dfrac{1}{3}, c = \dfrac{\sqrt{13}}{6}$

Center: $(-3, 1)$

Vertices: $\left(-3, \dfrac{1}{2}\right), \left(-3, \dfrac{3}{2}\right)$

Foci: $\left(-3, 1 \pm \dfrac{1}{6}\sqrt{13}\right)$

Asymptotes: $y = 1 \pm \dfrac{3}{2}(x + 3)$

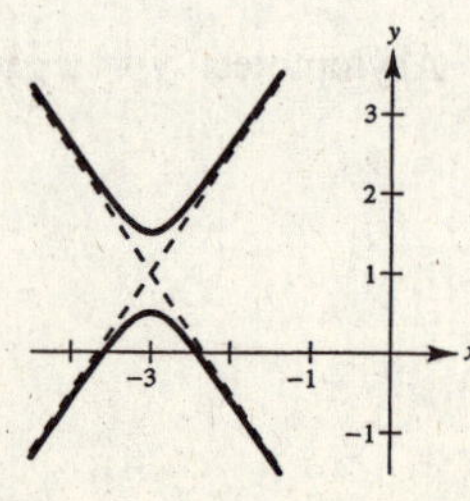

53. $9y^2 - x^2 + 2x + 54y + 62 = 0$

$$9(y^2 + 6y + 9) - (x^2 - 2x + 1) = -62 - 1 + 81 = 18$$

$$\frac{(y+3)^2}{2} - \frac{(x-1)^2}{18} = 1$$

$a = \sqrt{2}, b = 3\sqrt{2}, c = 2\sqrt{5}$

Center: $(1, -3)$

Vertices: $\left(1, -3 \pm \sqrt{2}\right)$

Foci: $\left(1, -3 \pm 2\sqrt{5}\right)$

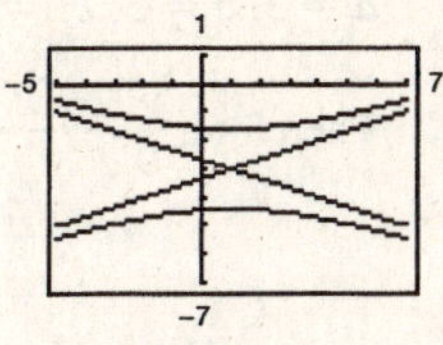

Solve for y:

$$9(y^2 + 6y + 9) = x^2 - 2x - 62 + 81$$

$$(y+3)^2 = \frac{x^2 - 2x + 19}{9}$$

$$y = -3 \pm \frac{1}{3}\sqrt{x^2 - 2x + 19}$$

(Graph each curve separately.)

54. $9x^2 - y^2 + 54x + 10y + 55 = 0$

$$9(x^2 + 6x + 9) - (y^2 - 10y + 25) = -55 + 81 - 25 = 1$$

$$\frac{(x+3)^2}{1/9} - \frac{(y-5)^2}{1} = 1$$

$a = \frac{1}{3}, b = 1, c = \frac{\sqrt{10}}{3}$

Center: $(-3, 5)$

Vertices: $\left(-3 \pm \frac{1}{3}, 5\right)$

Foci: $\left(-3 \pm \frac{\sqrt{10}}{3}, 5\right)$

Solve for y:

$$y^2 - 10y + 25 = 9x^2 + 54x + 55 + 25$$

$$(y-5)^2 = 9x^2 + 54x + 80$$

$$y = 5 \pm \sqrt{9x^2 + 54x + 80}$$

(Graph each curve separately.)

55. $3x^2 - 2y^2 - 6x - 12y - 27 = 0$

$$3(x^2 - 2x + 1) - 2(y^2 + 6y + 9) = 27 + 3 - 18 = 12$$

$$\frac{(x-1)^2}{4} - \frac{(y+3)^2}{6} = 1$$

$a = 2, b = \sqrt{6}, c = \sqrt{10}$

Center: $(1, -3)$

Vertices: $(-1, -3), (3, -3)$

Foci: $\left(1 \pm \sqrt{10}, -3\right)$

Solve for y:

$$2(y^2 + 6y + 9) = 3x^2 - 6x - 27 + 18$$

$$(y+3)^2 = \frac{3x^2 - 6x - 9}{2}$$

$$y = -3 \pm \sqrt{\frac{3(x^2 - 2x - 3)}{2}}$$

(Graph each curve separately.)

56. $3y^2 - x^2 + 6x - 12y = 0$

$$3(y^2 - 4y + 4) - (x^2 - 6x + 9) = 0 + 12 - 9 = 3$$

$$\frac{(y-2)^2}{1} - \frac{(x-3)^2}{3} = 1$$

$a = 1, b = \sqrt{3}, c = 2$

Center: $(3, 2)$

Vertices: $(3, 1), (3, 3)$

Foci: $(3, 0), (3, 4)$

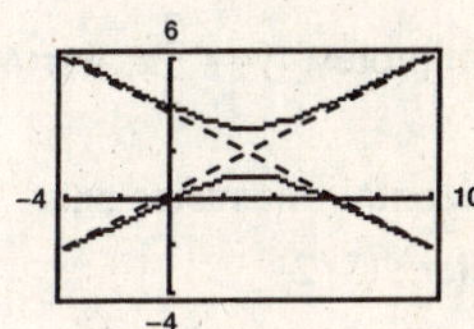

Solve for y:

$$3(y^2 - 4y + 4) = x^2 - 6x + 12$$

$$(y-2)^2 = \frac{x^2 - 6x + 12}{3}$$

$$y = 2 \pm \sqrt{\frac{x^2 - 6x + 12}{3}}$$

(Graph each curve separately.)

57. Vertices: $(\pm 1, 0)$

Asymptotes: $y = \pm 3x$

Horizontal transverse axis

Center: $(0, 0)$

$a = 1, \pm\frac{b}{a} = \pm\frac{b}{1} = \pm 3 \Rightarrow b = 3$

Therefore, $\frac{x^2}{1} - \frac{y^2}{9} = 1$.

58. Vertices: $(0, \pm 3)$

Asymptotes: $y = \pm 3x$

Vertical transverse axis

$a = 3$

Slopes of asymptotes: $\pm\frac{a}{b} = \pm 3$

Thus, $b = 1$. Therefore,

$$\frac{y^2}{9} - \frac{x^2}{1} = 1.$$

59. Vertices: $(2, \pm 3)$

Point on graph: $(0, 5)$

Vertical transverse axis

Center: $(2, 0)$

$a = 3$

Therefore, the equation is of the form

$$\frac{y^2}{9} - \frac{(x-2)^2}{b^2} = 1.$$

Substituting the coordinates of the point $(0, 5)$, we have

$$\frac{25}{9} - \frac{4}{b^2} = 1 \quad \text{or} \quad b^2 = \frac{9}{4}.$$

Therefore, the equation is $\dfrac{y^2}{9} - \dfrac{(x-2)^2}{9/4} = 1.$

60. Vertices: $(2, \pm 3)$

Foci: $(2, \pm 5)$

Vertical transverse axis

Center: $(2, 0)$

$a = 3, c = 5, b^2 = c^2 - a^2 = 16$

Therefore, $\dfrac{y^2}{9} - \dfrac{(x-2)^2}{16} = 1.$

61. Center: $(0, 0)$

Vertex: $(0, 2)$

Focus: $(0, 4)$

Vertical transverse axis

$a = 2, c = 4, b^2 = c^2 - a^2 = 12$

Therefore, $\dfrac{y^2}{4} - \dfrac{x^2}{12} = 1.$

62. Center: $(0, 0)$

Vertex: $(3, 0)$

Focus: $(5, 0)$

Horizontal transverse axis

$a = 3, c = 5, b^2 = c^2 - a^2 = 16$

Therefore, $\dfrac{x^2}{9} - \dfrac{y^2}{16} = 1.$

63. Vertices: $(0, 2), (6, 2)$

Asymptotes: $y = \frac{2}{3}x, y = 4 - \frac{2}{3}x$

Horizontal transverse axis

Center: $(3, 2)$

$a = 3$

Slopes of asymptotes: $\pm\frac{b}{a} = \pm\frac{2}{3}$

Thus, $b = 2$. Therefore,

$$\frac{(x-3)^2}{9} - \frac{(y-2)^2}{4} = 1.$$

64. Focus: $(10, 0)$

Asymptotes: $y = \pm\frac{3}{4}x$

Horizontal transverse axis

Center: $(0, 0)$ since asymptotes intersect at the origin.

$c = 10$

Slopes of asymptotes: $\pm\frac{b}{a} = \pm\frac{3}{4}$ and $b = \frac{3}{4}a$

$c^2 = a^2 + b^2 = 100$

Solving these equations, we have $a^2 = 64$ and $b^2 = 36$. Therefore, the equation is

$$\frac{x^2}{64} - \frac{y^2}{36} = 1.$$

65. (a) $\dfrac{x^2}{9} - y^2 = 1, \dfrac{2x}{9} - 2yy' = 0, \dfrac{x}{9y} = y'$

At $x = 6$: $y = \pm\sqrt{3}, y' = \dfrac{\pm 6}{9\sqrt{3}} = \dfrac{\pm 2\sqrt{3}}{9}$

At $(6, \sqrt{3})$: $y - \sqrt{3} = \dfrac{2\sqrt{3}}{9}(x - 6)$

or $2x - 3\sqrt{3}y - 3 = 0$

At $(6, -\sqrt{3})$: $y + \sqrt{3} = \dfrac{-2\sqrt{3}}{9}(x - 6)$

or $2x + 3\sqrt{3}y - 3 = 0$

(b) From part (a) we know that the slopes of the normal lines must be $\mp 9/(2\sqrt{3})$.

At $(6, \sqrt{3})$: $y - \sqrt{3} = -\dfrac{9}{2\sqrt{3}}(x - 6)$

or $9x + 2\sqrt{3}y - 60 = 0$

At $(6, -\sqrt{3})$: $y + \sqrt{3} = \dfrac{9}{2\sqrt{3}}(x - 6)$

or $9x - 2\sqrt{3}y - 60 = 0$

66. (a) $\dfrac{y^2}{4} - \dfrac{x^2}{2} = 1,\ y^2 - 2x^2 = 4,\ 2yy' - 4x = 0,$

$$y' = \frac{4x}{2y} = \frac{2x}{y}$$

At $x = 4$: $y = \pm 6,\ y' = \dfrac{\pm 2(4)}{6} = \pm\dfrac{4}{3}$

At $(4, 6)$: $y - 6 = -\dfrac{4}{3}(x - 4)$ or $4x + 3y - 34 = 0$

At $(4, -6)$: $y + 6 = -\dfrac{4}{3}(x - 4)$ or $4x + 3y + 2 = 0$

(b) From part (a) we know that the slopes of the normal lines must be $\mp 3/4$.

At $(4, 6)$: $y - 6 = -\dfrac{3}{4}(x - 4)$ or $3x + 4y - 36 = 0$

At $(4, -6)$: $y + 6 = \dfrac{3}{4}(x - 4)$ or $3x - 4y - 36 = 0$

67.
$$x^2 + 4y^2 - 6x + 16y + 21 = 0$$
$$(x^2 - 6x + 9) + 4(y^2 + 4y + 4) = -21 + 9 + 16$$
$$(x - 3)^2 + 4(y + 2)^2 = 4$$

Ellipse

68.
$$4x^2 - y^2 - 4x - 3 = 0$$
$$4\left(x^2 - x + \tfrac{1}{4}\right) - y^2 = 3 + 1$$
$$4\left(x - \tfrac{1}{2}\right)^2 - y^2 = 4$$

Hyperbola

69.
$$y^2 - 4y - 4x = 0$$
$$y^2 - 4y + 4 = 4x + 4$$
$$(y - 2)^2 = 4(x + 1)$$

Parabola

70.
$$25x^2 - 10x - 200y - 119 = 0$$
$$25\left(x^2 - \tfrac{2}{5}x + \tfrac{1}{25}\right) = 200y + 119 + 1$$
$$25\left(x - \tfrac{1}{5}\right)^2 = 200(y + 1)$$

Parabola

71.
$$4x^2 + 4y^2 - 16y + 15 = 0$$
$$4x^2 + 4(y^2 - 4y + 4) = -15 + 16$$
$$4x^2 + 4(y - 2)^2 = 1$$

Circle (Ellipse)

72.
$$y^2 - 4y = x + 5$$
$$y^2 - 4y + 4 = x + 5 + 4$$
$$(y - 2)^2 = x + 9$$

Parabola

73.
$$9x^2 + 9y^2 - 36x + 6y + 34 = 0$$
$$9(x^2 - 4x + 4) + 9\left(y^2 + \tfrac{2}{3}y + \tfrac{1}{9}\right) = -34 + 36 + 1$$
$$9(x - 2)^2 + 9\left(y + \tfrac{1}{3}\right)^2 = 3$$

Circle (Ellipse)

74.
$$2x(x - y) = y(3 - y - 2x)$$
$$2x^2 - 2xy = 3y - y^2 - 2xy$$
$$2x^2 + y^2 - 3y = 0$$
$$2x^2 + \left(y - \frac{3}{2}\right)^2 = \frac{9}{4}$$

Ellipse

75.
$$3(x - 1)^2 = 6 + 2(y + 1)^2$$
$$3(x - 1)^2 - 2(y + 1)^2 = 6$$
$$\frac{(x - 1)^2}{2} - \frac{(y + 1)^2}{3} = 1$$

Hyperbola

76.
$$9(x + 3)^2 = 36 - 4(y - 2)^2$$
$$9(x + 3)^2 + 4(y - 2)^2 = 36$$
$$\frac{(x + 3)^2}{4} + \frac{(y - 2)^2}{9} = 1$$

Ellipse

77. (a) A parabola is the set of all points (x, y) that are equidistant from a fixed line (directrix) and a fixed point (focus) not on the line.

(b) $(x - h)^2 = 4p(y - k)$ or $(y - k)^2 = 4p(x - h)$

(c) See Theorem 10.2.

78. (a) An ellipse is the set of all points (x, y), the sum of whose distance from two distinct fixed points (foci) is constant.

(b) $\dfrac{(x-h)^2}{a^2} + \dfrac{(y-k)^2}{b^2} = 1$ or $\dfrac{(x-h)^2}{b^2} + \dfrac{(y-k)^2}{a^2} = 1$

79. (a) A hyperbola is the set of all points (x, y) for which the absolute value of the difference between the distances from two distance fixed points (foci) is constant.

(b) $\dfrac{(x-h)^2}{a^2} - \dfrac{(y-k)^2}{b^2} = 1$ or $\dfrac{(y-k)^2}{a^2} - \dfrac{(x-h)^2}{b^2} = 1$

(c) $y = k \pm \dfrac{b}{a}(x-h)$ or $y = k \pm \dfrac{a}{b}(x-h)$

80. $e = \dfrac{c}{a}$, $c = \sqrt{a^2 - b^2}$, $0 < e < 1$

For $e \approx 0$, the ellipse is nearly circular.

For $e \approx 1$, the ellipse is elongated.

81. Assume that the vertex is at the origin.

$$x^2 = 4py$$
$$(3)^2 = 4p(1)$$
$$\frac{9}{4} = p$$

The pipe is located $\frac{9}{4}$ meters from the vertex.

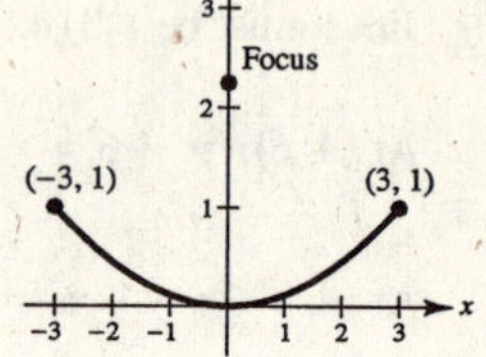

82. Assume that the vertex is at the origin.

(a) $x^2 = 4py$

$$8^2 = 4p\left(\frac{3}{100}\right)$$
$$\frac{1600}{3} = p$$
$$x^2 = 4\left(\frac{1600}{3}\right)y = \frac{6400}{3}y$$

(b) The deflection is 1 cm when

$$y = \frac{2}{100} \implies x = \pm\sqrt{\frac{128}{3}} \approx \pm 6.53 \text{ meters.}$$

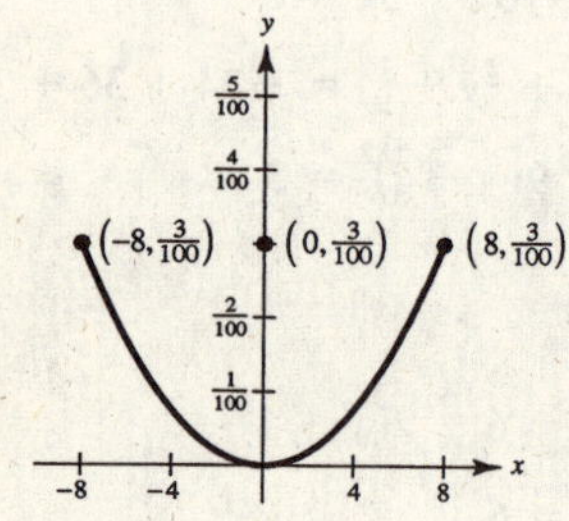

83. $y = ax^2$

$y' = 2ax$

The equation of the tangent line is

$$y - ax_0^2 = 2ax_0(x - x_0)$$
or $y = 2ax_0x - ax_0^2$.

Let $y = 0$. Then:

$$-ax_0^2 = 2ax_0x - 2ax_0^2$$
$$ax_0^2 = 2ax_0x$$
$$x = \frac{x_0}{2}$$

Therefore, $\left(\dfrac{x_0}{2}, 0\right)$ is the x-intercept.

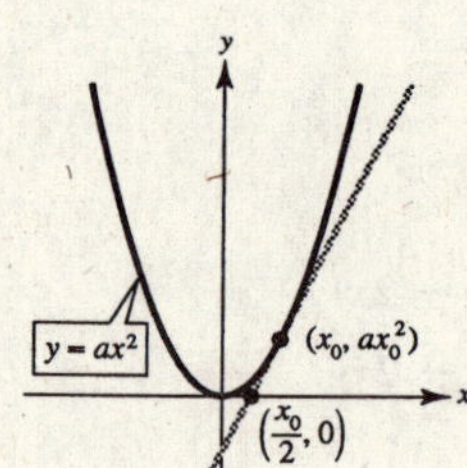

84. (a) Without loss of generality, place the coordinate system so that the equation of the parabola is $x^2 = 4py$ and, hence,

$$y' = \left(\frac{1}{2p}\right)x.$$

Therefore, for distinct tangent lines, the slopes are unequal and the lines intersect.

(b) $x^2 - 4x - 4y = 0$

$$2x - 4 - 4\frac{dy}{dx} = 0$$

$$\frac{dy}{dx} = \frac{1}{2}x - 1$$

At $(0, 0)$, the slope is -1: $y = -x$. At $(6, 3)$, the slope is 2: $y = 2x - 9$. Solving for x,

$$-x = 2x - 9$$

$$-3x = -9$$

$$x = 3$$

$$y = -3.$$

Point of intersection: $(3, -3)$

85. (a) Consider the parabola $x^2 = 4py$. Let m_0 be the slope of the one tangent line at (x_1, y_1) and therefore, $-1/m_0$ is the slope of the second at (x_2, y_2). Differentiating, $2x = 4py'$ or $y' = \dfrac{x}{2p}$, and we have:

$$m_0 = \frac{1}{2p}x_1 \text{ or } x_1 = 2pm_0$$

$$\frac{-1}{m_0} = \frac{1}{2p}x_2 \text{ or } x_2 = \frac{-2p}{m_0}.$$

Substituting these values of x into the equation $x^2 = 4py$, we have the coordinates of the points of tangency $(2pm_0, pm_0^2)$ and $(-2p/m_0, p/m_0^2)$ and the equations of the tangent lines are

$$(y - pm_0^2) = m_0(x - 2pm_0) \quad \text{and}$$

$$\left(y - \frac{p}{m_0^2}\right) = \frac{-1}{m_0}\left(x + \frac{2p}{m_0}\right).$$

The point of intersection of these lines is

$\left(\dfrac{p(m_0^2 - 1)}{m_0}, -p\right)$ and is on the directrix, $y = -p$.

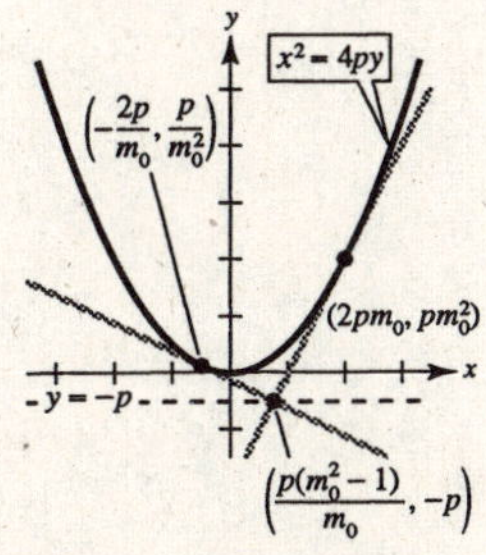

(b) $x^2 - 4x - 4y + 8 = 0$

$$(x - 2)^2 = 4(y - 1)$$

Vertex: $(2, 1)$

$$2x - 4 - 4\frac{dy}{dx} = 0$$

$$\frac{dy}{dx} = \frac{1}{2}x - 1$$

At $(-2, 5)$, $\dfrac{dy}{dx} = -2$. At $\left(3, \dfrac{5}{4}\right)$, $\dfrac{dy}{dx} = \dfrac{1}{2}$.

Tangent line at $(-2, 5)$:

$$y - 5 = -2(x + 2) \Rightarrow 2x + y - 1 = 0.$$

Tangent line at $\left(3, \dfrac{5}{4}\right)$:

$$y - \frac{5}{4} = \frac{1}{2}(x - 3) \Rightarrow 2x - 4y - 1 = 0.$$

Since $m_1m_2 = (-2)\left(\dfrac{1}{2}\right) = -1$, the lines are perpendicular.

Point of intersection: $-2x + 1 = \dfrac{1}{2}x - \dfrac{1}{4}$

$$-\frac{5}{2}x = -\frac{5}{4}$$

$$x = \frac{1}{2}$$

$$y = 0$$

Directrix: $y = 0$ and the point of intersection $\left(\dfrac{1}{2}, 0\right)$ lies on this line.

86. The focus of $x^2 = 8y = 4(2)y$ is $(0, 2)$. The distance from a point on the parabola, $(x, x^2/8)$, and the focus, $(0, 2)$, is

$$d = \sqrt{(x - 0)^2 + \left(\frac{x^2}{8} - 2\right)^2}.$$

Since d is minimized when d^2 is minimized, it is sufficient to minimize the function

$$f(x) = x^2 + \left(\frac{x^2}{8} - 2\right)^2.$$

$$f'(x) = 2x + 2\left(\frac{x^2}{8} - 2\right)\left(\frac{x}{4}\right) = \frac{x^3}{16} + x.$$

$f'(x) = 0$ implies that $\dfrac{x^3}{16} + x = x\left(\dfrac{x^2}{16} + 1\right) = 0 \Rightarrow x = 0.$

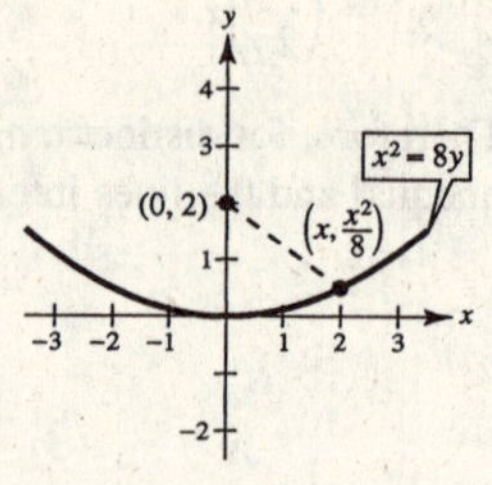

This is a minimum by the First Derivative Test. Hence, the closest point to the focus is the vertex, $(0, 0)$.

87. $y = x - x^2$

$$\frac{dy}{dx} = 1 - 2x$$

At the point of tangency (x_1, y_1) on the mountain, $m = 1 - 2x_1$. Also, $m = \dfrac{y_1 - 1}{x_1 + 1}$.

$$\frac{y_1 - 1}{x_1 + 1} = 1 - 2x_1$$

$$(x_1 - x_1^2) - 1 = (1 - 2x_1)(x_1 + 1)$$

$$-x_1^2 + x_1 - 1 = -2x_1^2 - x_1 + 1$$

$$x_1^2 + 2x_1 - 2 = 0$$

$$x_1 = \frac{-2 \pm \sqrt{2^2 - 4(1)(-2)}}{2(1)} = \frac{-2 \pm 2\sqrt{3}}{2} = -1 \pm \sqrt{3}$$

Choosing the positive value for x_1, we have $x_1 = -1 + \sqrt{3}$.

$$m = 1 - 2\left(-1 + \sqrt{3}\right) = 3 - 2\sqrt{3}$$

$$m = \frac{0 - 1}{x_0 + 1} = -\frac{1}{x_0 + 1}$$

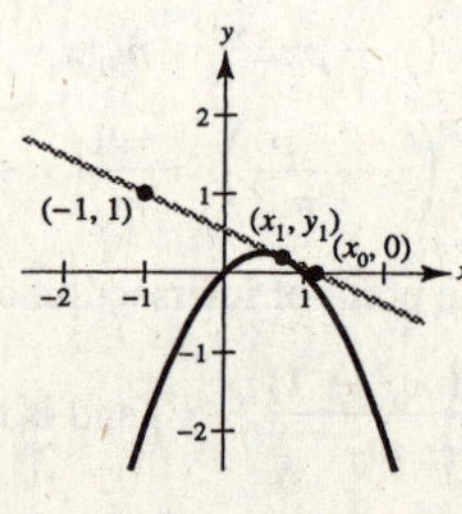

Thus, $-\dfrac{1}{x_0 + 1} = 3 - 2\sqrt{3}$

$$\frac{-1}{3 - 2\sqrt{3}} = x_0 + 1$$

$$\frac{3 + 2\sqrt{3}}{3} - 1 = x_0$$

$$\frac{2\sqrt{3}}{3} = x_0.$$

The closest the receiver can be to the hill is $\left(2\sqrt{3}/3\right) - 1 \approx 0.155$.

88. (a) $A = 0.75t^2 - 9.2t + 301$

(b)

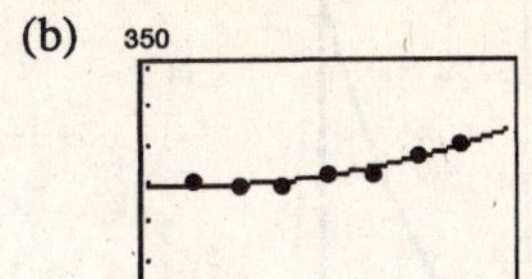

(c) $\dfrac{dA}{dt} = 1.5t - 9.2, \quad 6 \le t \le 12$

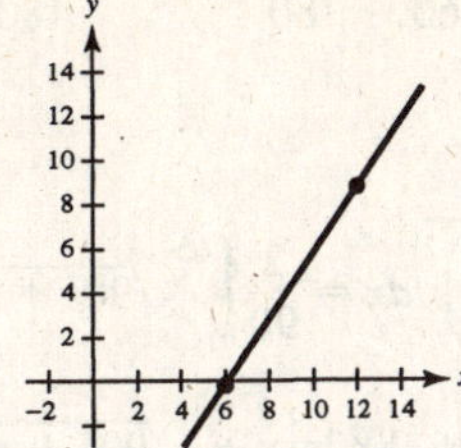

Women are spending more time watching TV each year.

89. Parabola

Vertex: $(0, 4)$

$$x^2 = 4p(y - 4)$$

$$4^2 = 4p(0 - 4)$$

$$p = -1$$

$$x^2 = -4(y - 4)$$

$$y = 4 - \frac{x^2}{4}$$

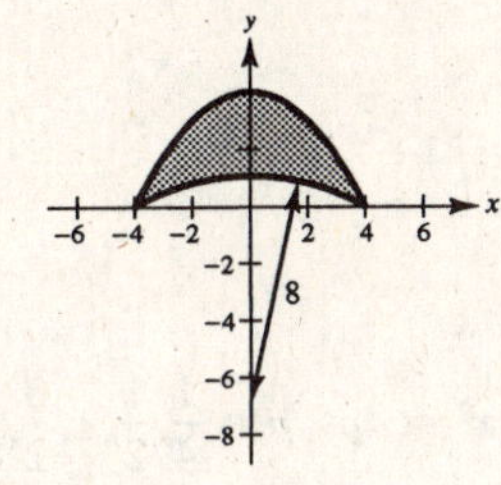

Circle

Center: $(0, k)$

Radius: 8

$$x^2 + (y - k)^2 = 64$$

$$4^2 + (0 - k)^2 = 64$$

$$k^2 = 48$$

$$k = -4\sqrt{3} \quad \text{(Center is on the negative } y\text{-axis.)}$$

$$x^2 + \left(y + 4\sqrt{3}\right)^2 = 64$$

$$y = -4\sqrt{3} \pm \sqrt{64 - x^2}$$

Since the y-value is positive when $x = 0$, we have $y = -4\sqrt{3} + \sqrt{64 - x^2}$.

$$A = 2\int_0^4 \left[\left(4 - \frac{x^2}{4}\right) - \left(-4\sqrt{3} + \sqrt{64 - x^2}\right)\right] dx$$

$$= 2\left[4x - \frac{x^3}{12} + 4\sqrt{3}x - \frac{1}{2}\left(x\sqrt{64 - x^2} + 64 \arcsin\frac{x}{8}\right)\right]_0^4$$

$$= 2\left[16 - \frac{64}{12} + 16\sqrt{3} - 2\sqrt{48} - 32 \arcsin\frac{1}{2}\right]$$

$$= \frac{16\left(4 + 3\sqrt{3} - 2\pi\right)}{3} \approx 15.536 \text{ square feet}$$

90.

$$x = \frac{1}{4}y^2$$

$$x' = \frac{1}{2}y$$

$$1 + (x')^2 = 1 + \frac{y^2}{4}$$

$$s = \int_0^4 \sqrt{1 + \left(\frac{y^2}{4}\right)}\, dy = \frac{1}{2}\int_0^4 \sqrt{4 + y^2}\, dy$$

$$= \frac{1}{4}\left[y\sqrt{4 + y^2} + 4 \ln\left|y + \sqrt{4 + y^2}\right|\right]_0^4$$

$$= \frac{1}{4}\left[4\sqrt{20} + 4 \ln\left|4 + \sqrt{20}\right| - 4 \ln 2\right]$$

$$= 2\sqrt{5} + \ln\left(2 + \sqrt{5}\right) \approx 5.916$$

91. (a) Assume that $y = ax^2$.

$$20 = a(60)^2 \Rightarrow a = \frac{2}{360} = \frac{1}{180} \Rightarrow y = \frac{1}{180}x^2$$

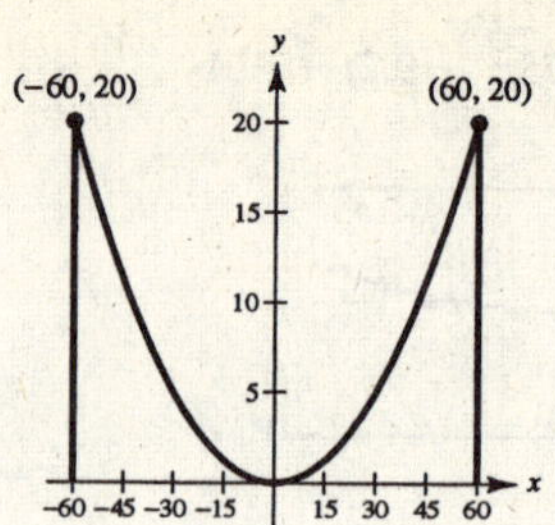

(b) $f(x) = \frac{1}{180}x^2, f'(x) = \frac{1}{90}x$

$$S = 2\int_0^{60}\sqrt{1 + \left(\frac{1}{90}x\right)^2}\,dx = \frac{2}{90}\int_0^{60}\sqrt{90^2 + x^2}\,dx$$

$$= \frac{2}{90}\,\frac{1}{2}\left[x\sqrt{90^2 + x^2} + 90^2\ln\left|x + \sqrt{90^2 + x^2}\right|\right]_0^{60} \quad \text{(Formula 26)}$$

$$= \frac{1}{90}\left[60\sqrt{11{,}700} + 90^2\ln\left(60 + \sqrt{11{,}700}\right) - 90^2\ln 90\right]$$

$$= \frac{1}{90}\left[1800\sqrt{13} + 90^2\ln\left(60 + 30\sqrt{13}\right) - 90^2\ln 90\right]$$

$$= 20\sqrt{13} + 90\ln\left(\frac{60 + 30\sqrt{13}}{90}\right)$$

$$= 10\left[2\sqrt{13} + 9\ln\left(\frac{2 + \sqrt{13}}{3}\right)\right] \approx 128.4 \text{ m}$$

92. $x^2 = 20y$

$$y = \frac{x^2}{20}$$

$$y' = \frac{x}{10}$$

$$S = 2\pi\int_0^r x\sqrt{1 + \left(\frac{x}{10}\right)^2}\,dx = 2\pi\int_0^r \frac{x\sqrt{100 + x^2}}{10}\,dx$$

$$= \left[\frac{\pi}{10}\cdot\frac{2}{3}(100 + x^2)^{3/2}\right]_0^r = \frac{\pi}{15}\left[(100 + r^2)^{3/2} - 1000\right]$$

93. $x^2 = 4py, p = \frac{1}{4}, \frac{1}{2}, 1, \frac{3}{2}, 2$

As p increases, the graph becomes wider.

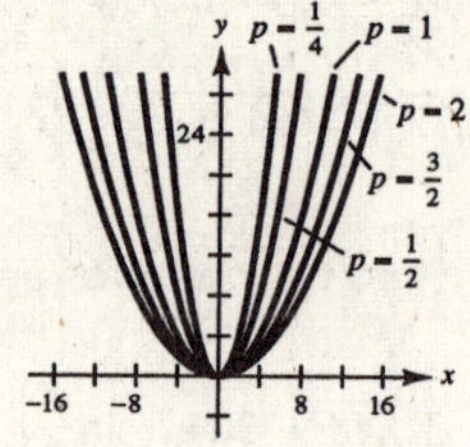

94. $A = 2\int_0^h \sqrt{4py}\,dy$

$$= 4\sqrt{p}\int_0^h y^{1/2}\,dy$$

$$= \left[4\sqrt{p}\left(\frac{2}{3}\right)y^{3/2}\right]_0^h$$

$$= \frac{8}{3}\sqrt{p}h^{3/2}$$

95. (a) At the vertices we notice that the string is horizontal and has a length of $2a$.

(b) The thumbtacks are located at the foci and the length of string is the constant sum of the distances from the foci.

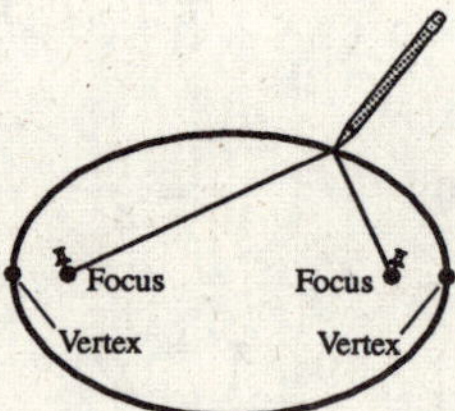

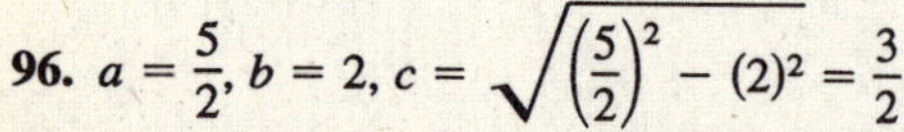

96. $a = \frac{5}{2}, b = 2, c = \sqrt{\left(\frac{5}{2}\right)^2 - (2)^2} = \frac{3}{2}$

The tacks should be placed 1.5 feet from the center. The string should be $2a = 5$ feet long.

97.

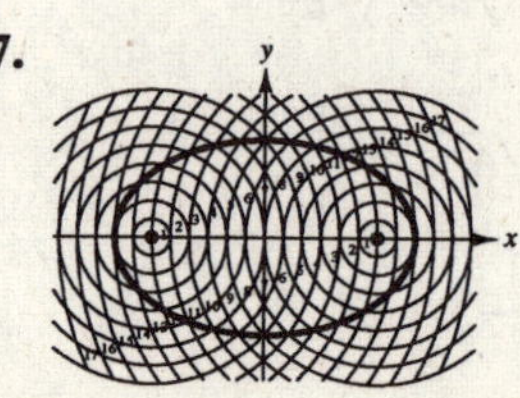

98. $e = \dfrac{c}{a}$

$0.0167 = \dfrac{c}{149{,}598{,}000}$

$c \approx 2{,}498{,}286.6$

Least distance: $a - c = 147{,}099{,}713.4$ km

Greatest distance: $a + c = 152{,}096{,}286.6$ km

99. $e = \dfrac{c}{a}$

$A + P = 2a$

$a = \dfrac{A + P}{2}$

$c = a - P = \dfrac{A + P}{2} - P = \dfrac{A - P}{2}$

$e = \dfrac{c}{a} = \dfrac{(A - P)/2}{(A + P)/2} = \dfrac{A - P}{A + P}$

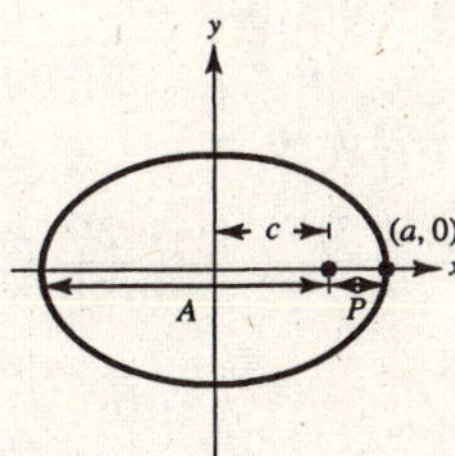

100. $e = \dfrac{A - P}{A + P}$

$= \dfrac{(123{,}000 + 4000) - (119 + 4000)}{(123{,}000 + 4000) + (119 + 4000)}$

$= \dfrac{122{,}881}{131{,}119} \approx 0.9372$

101. $e = \dfrac{A - P}{A + P} = \dfrac{35.29 - 0.59}{35.29 + 0.59} = 0.9671$

102. $\dfrac{x^2}{a^2} + \dfrac{y^2}{b^2} = 1$

$\dfrac{x^2}{a^2} + \dfrac{y^2}{a^2(b^2/a^2)} = 1$

$\dfrac{x^2}{a^2} + \dfrac{y^2}{a^2(a^2 - c^2)/a^2} = 1$

$\dfrac{x^2}{a^2} + \dfrac{y^2}{a^2(1 - e^2)} = 1$

As $e \rightarrow 0$, $1 - e^2 \rightarrow 1$ and we have

$\dfrac{x^2}{a^2} + \dfrac{y^2}{a^2} = 1$ or the circle $x^2 + y^2 = a^2$.

103. $\dfrac{x^2}{10^2} + \dfrac{y^2}{5^2} = 1$

$\dfrac{2x}{10^2} + \dfrac{2yy'}{5^2} = 0$

$y' = \dfrac{-5^2x}{10^2y} = \dfrac{-x}{4y}$

At $(-8, 3)$: $y' = \dfrac{8}{12} = \dfrac{2}{3}$

The equation of the tangent line is $y - 3 = \frac{2}{3}(x + 8)$. It will cross the y-axis when $x = 0$ and $y = \frac{2}{3}(8) + 3 = \frac{25}{3}$.

104. $\dfrac{x^2}{(4.5)^2} + \dfrac{y^2}{(2.5)^2} = 1$

$x^2 = (4.5)^2\left[1 - \dfrac{y^2}{(2.5)^2}\right]$

$x = \pm\dfrac{9}{5}\sqrt{(2.5)^2 - y^2}$

$V = $ (Area of bottom)(Length) + (Area of top)(Length)

$V = \left[\dfrac{\pi(4.5)(2.5)}{2}\right](16) + 16\displaystyle\int_0^{0.5} 2\frac{9}{5}\sqrt{(2.5)^2 - y^2}\,dy$ (Recall: Area of ellipse is πab.)

$= 90\pi + \dfrac{144}{5} \cdot \left[y\sqrt{(2.5)^2 - y^2} + (2.5)^2 \arcsin \dfrac{y}{2.5}\right]_0^{0.5} = 90\pi + \dfrac{144}{5}\left[0.5\sqrt{6} + (2.5)^2 \arcsin \dfrac{1}{5}\right] \approx 354.3 \text{ ft}^3$

105. $16x^2 + 9y^2 + 96x + 36y + 36 = 0$

$$32x + 18yy' + 96 + 36y' = 0$$

$$y'(18y + 36) = -(32x + 96)$$

$$y' = \frac{-(32x + 96)}{18y + 36}$$

$y' = 0$ when $x = -3$. y' is undefined when $y = -2$.

At $x = -3$, $y = 2$ or -6.

Endpoints of major axis: $(-3, 2), (-3, -6)$

At $y = -2$, $x = 0$ or -6.

Endpoints of minor axis: $(0, -2), (-6, -2)$

Note: Equation of ellipse is $\dfrac{(x + 3)^2}{9} + \dfrac{(y + 2)^2}{16} = 1$

106. $9x^2 + 4y^2 + 36x - 24y + 36 = 0$

$$18x + 8yy' + 36 - 24y' = 0$$

$$(8y - 24)y' = -(18x + 36)$$

$$y' = \frac{-(18x + 36)}{8y - 24}$$

$y' = 0$ when $x = -2$. y' undefined when $y = 3$.

At $x = -2$, $y = 0$ or 6.

Endpoints of major axis: $(-2, 0), (-2, 6)$

At $y = 3$, $x = 0$ or -4.

Endpoints of minor axis: $(0, 3), (-4, 3)$

Note: Equation of ellipse is $\dfrac{(x + 2)^2}{4} + \dfrac{(y - 3)^2}{9} = 1$

107. (a) $A = 4\displaystyle\int_0^2 \frac{1}{2}\sqrt{4 - x^2}\,dx = \left[x\sqrt{4 - x^2} + 4\arcsin\left(\frac{x}{2}\right)\right]_0^2 = 2\pi$ [or, $A = \pi ab = \pi(2)(1) = 2\pi$]

(b) **Disk:** $V = 2\pi\displaystyle\int_0^2 \frac{1}{4}(4 - x^2)\,dx = \frac{1}{2}\pi\left[4x - \frac{1}{3}x^3\right]_0^2 = \frac{8\pi}{3}$

$$y = \frac{1}{2}\sqrt{4 - x^2}$$

$$y' = \frac{-x}{2\sqrt{4 - x^2}}$$

$$\sqrt{1 + (y')^2} = \sqrt{1 + \frac{x^2}{16 - 4x^2}} = \sqrt{\frac{16 - 3x^2}{4y}}$$

$$S = 2(2\pi)\int_0^2 y\left(\frac{\sqrt{16 - 3x^2}}{4y}\right)dx = \pi\int_0^2 \sqrt{16 - 3x^2}\,dx$$

$$= \frac{\pi}{2\sqrt{3}}\left[\sqrt{3}x\sqrt{16 - 3x^2} + 16\arcsin\left(\frac{\sqrt{3}x}{4}\right)\right]_0^2$$

$$= \frac{2\pi}{9}(9 + 4\sqrt{3}\pi) \approx 21.48$$

(c) **Shell:** $V = 2\pi\displaystyle\int_0^2 x\sqrt{4 - x^2}\,dx = -\pi\int_0^2 -2x(4 - x^2)^{1/2}\,dx = -\frac{2\pi}{3}\left[(4 - x^2)^{3/2}\right]_0^2 = \frac{16\pi}{3}$

$$x = 2\sqrt{1 - y^2}$$

$$x' = \frac{-2y}{\sqrt{1 - y^2}}$$

$$\sqrt{1 + (x')^2} = \sqrt{1 + \frac{4y^2}{1 - y^2}} = \frac{\sqrt{1 + 3y^2}}{\sqrt{1 - y^2}}$$

$$S = 2(2\pi)\int_0^1 2\sqrt{1 - y^2}\,\frac{\sqrt{1 + 3y^2}}{\sqrt{1 - y^2}}\,dy = 8\pi\int_0^1 \sqrt{1 + 3y^2}\,dy$$

$$= \frac{8\pi}{2\sqrt{3}}\left[\sqrt{3}y\sqrt{1 + 3y^2} + \ln\left|\sqrt{3}y + \sqrt{1 + 3y^2}\right|\right]_0^1$$

$$= \frac{4\pi}{3}\left|6 + \sqrt{3}\ln(2 + \sqrt{3})\right| \approx 34.69$$

108. (a) $A = 4\int_0^4 \frac{3}{4}\sqrt{16 - x^2}\,dx = \frac{3}{2}\left[x\sqrt{16 - x^2} + 16 \arcsin \frac{x}{4}\right]_0^4 = 12\pi$

(b) **Disk:** $V = 2\pi \int_0^4 \frac{9}{16}(16 - x^2)\,dx = \frac{9\pi}{8}\left[\left(16x - \frac{1}{3}x^3\right)\right]_0^4 = 48\pi$

$$y = \frac{3}{4}\sqrt{16 - x^2}$$

$$y' = \frac{-3x}{4\sqrt{16 - x^2}}$$

$$\sqrt{1 + (y')^2} = \sqrt{1 + \frac{9x^2}{16(16 - x^2)}}$$

$$S = 2(2\pi)\int_0^4 \frac{3}{4}\sqrt{16 - x^2}\sqrt{\frac{16(16 - x^2) + 9x^2}{16(16 - x^2)}}\,dx = 4\pi \int_0^4 \frac{3}{4}\sqrt{16 - x^2}\,\frac{\sqrt{256 - 7x^2}}{4\sqrt{16 - x^2}}\,dx = \frac{3\pi}{4}\int_0^4 \sqrt{256 - 7x^2}\,dx$$

$$= \frac{3\pi}{8\sqrt{7}}\left[\sqrt{7}x\sqrt{256 - 7x^2} + 256 \arcsin \frac{\sqrt{7}x}{16}\right]_0^4 = \frac{3\pi}{8\sqrt{7}}\left(48\sqrt{7} + 256 \arcsin \frac{\sqrt{7}}{4}\right) \approx 138.93$$

(c) **Shell:** $V = 4\pi \int_0^4 x\left[\frac{3}{4}\sqrt{16 - x^2}\right]dx = 3\pi\left[\left(-\frac{1}{2}\right)\left(\frac{2}{3}\right)(16 - x^2)^{3/2}\right]_0^4 = 64\pi$

$$x = \frac{4}{3}\sqrt{9 - y^2}$$

$$x' = \frac{-4y}{3\sqrt{9 - y^2}}$$

$$\sqrt{1 + (x')^2} = \sqrt{1 + \frac{16y^2}{9(9 - y^2)}}$$

$$S = 2(2\pi)\int_0^3 \frac{4}{3}\sqrt{9 - y^2}\sqrt{\frac{9(9 - y^2) + 16y^2}{9(9 - y^2)}}\,dy$$

$$= 4\pi \int_0^3 \frac{4}{9}\sqrt{81 + 7y^2}\,dy$$

$$= \frac{16}{9}\left(\frac{\pi}{2\sqrt{7}}\right)\left[\sqrt{7}y\sqrt{81 + 7y^2} + 81 \ln\left|\sqrt{7}y + \sqrt{81 + 7y^2}\right|\right]_0^3$$

$$= \frac{8\pi}{9\sqrt{7}} 3\sqrt{7}(12) + 81 \ln\left(3\sqrt{7} + 12\right) - 81 \ln 9 \approx 168.53$$

109. From Example 5,

$$C = 4a\int_0^{\pi/2} \sqrt{1 - e^2 \sin^2 \theta}\,d\theta$$

For $\frac{x^2}{25} + \frac{y^2}{49} = 1$, we have

$a = 7, b = 5, c = \sqrt{49 - 25} = 2\sqrt{6}, e = \frac{c}{a} = \frac{2\sqrt{6}}{7}$.

$$C = 4(7)\int_0^{\pi/2} \sqrt{1 - \frac{24}{49}\sin^2 \theta}\,d\theta$$

$$\approx 28(1.3558) \approx 37.9614$$

110. (a) $\dfrac{x^2}{a^2} + \dfrac{y^2}{b^2} = 1$

$$\frac{2x}{a^2} + \frac{2yy'}{b^2} = 0$$

$$y' = -\frac{xb^2}{ya^2}$$

At P, $y' = -\dfrac{b^2}{a^2} \cdot \dfrac{x_0}{y_0} = m.$

(b) Slope of line through $(-c, 0)$ and (x_0, y_0): $m_1 = \dfrac{y_0}{x_0 + c}$

Slope of line through $(c, 0)$ and (x_0, y_0): $m_2 = \dfrac{y_0}{x_0 - c}$

(c) $$\tan\alpha = \frac{m_2 - m}{1 + m_2 m} = \frac{\dfrac{y_0}{x_0 - c} - \left(-\dfrac{b^2x_0}{a^2y_0}\right)}{1 + \left(\dfrac{y_0}{x_0 - c}\right)\left(-\dfrac{b^2x_0}{a^2y_0}\right)} = \frac{a^2y_0^2 + b^2x_0(x_0 - c)}{a^2y_0(x_0 - c) - b^2x_0y_0}$$

$$= \frac{a^2y_0^2 + b^2x_0^2 - b^2x_0c}{x_0y_0(a^2 - b^2) - a^2y_0c} = \frac{a^2b^2 - b^2x_0c}{x_0y_0c^2 - a^2y_0c} = \frac{b^2(a^2 - x_0c)}{y_0c(x_0c - a^2)} = -\frac{b^2}{y_0c}$$

$$\alpha = \arctan\left(-\frac{b^2}{y_0c}\right) = -\arctan\left(\frac{b^2}{y_0c}\right)$$

$$\tan\beta = \frac{m_1 - m}{1 + m_1 m} = \frac{\dfrac{y_0}{x_0 + c} - \left(-\dfrac{b^2x_0}{a^2y_0}\right)}{1 + \left(\dfrac{y_0}{x_0 + c}\right)\left(-\dfrac{b^2x_0}{a^2y_0}\right)} = \frac{a^2y_0^2 + b^2x_0(x_0 + c)}{a^2y_0(x_0 + c) - b^2x_0y_0}$$

$$= \frac{a^2y_0^2 + b^2x_0^2 + b^2x_0c}{a^2x_0y_0 + a^2cy_0 - b^2x_0y_0} = \frac{a^2b^2 + b^2x_0c}{x_0y_0(a^2 - b^2) + a^2cy_0} = \frac{b^2(a^2 + x_0c)}{y_0c(x_0c + a^2)} = \frac{b^2}{y_0c}$$

$$\beta = \arctan\left(\frac{b^2}{y_0c}\right)$$

Since $|\alpha| = |\beta|$, the tangent line to an ellipse at a point P makes equal angles with the lines through P and the foci.

111. Area circle $= \pi r^2 = 100\pi$

Area ellipse $= \pi ab = \pi a(10)$

$2(100\pi) = 10\pi a \Rightarrow a = 20$

Hence, the length of the major axis is $2a = 40$.

112. (a) $e = \dfrac{c}{a} = \dfrac{\sqrt{a^2 + b^2}}{a} \Rightarrow (ea)^2 - a^2 = b^2$. Hence,

$$\frac{(x - h)^2}{a^2} + \frac{(y - k)^2}{b^2} = 1$$

$$\frac{(x - h)^2}{a^2} + \frac{(y - k)^2}{a^2(1 - e^2)} = 1.$$

(b) $\dfrac{(x - 2)^2}{4} + \dfrac{(y - 3)^2}{4(1 - e^2)} = 1$

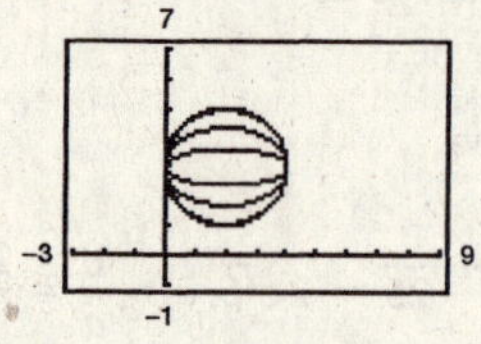

(c) As e approaches 0, the ellipse approaches a circle.

113. The transverse axis is horizontal since $(2, 2)$ and $(10, 2)$ are the foci (see definition of hyperbola).

Center: $(6, 2)$

$c = 4$, $2a = 6$, $b^2 = c^2 - a^2 = 7$

Therefore, the equation is

$$\frac{(x - 6)^2}{9} - \frac{(y - 2)^2}{7} = 1.$$

114. The transverse axis is vertical since $(-3, 0)$ and $(-3, 3)$ are the foci.

Center: $\left(-3, \dfrac{3}{2}\right)$

$c = \dfrac{3}{2}, \; 2a = 2, \; b^2 = c^2 - a^2 = \dfrac{5}{4}$

Therefore, the equation is

$$\frac{[y - (3/2)]^2}{1} - \frac{(x + 3)^2}{5/4} = 1.$$

115. $2a = 10 \Rightarrow a = 5$

$c = 6 \Rightarrow b = \sqrt{11}$

116. Center: $(0, 0)$

Horizontal transverse axis

Foci: $(\pm c, 0)$

Vertices: $(\pm a, 0)$

The difference of the distances from any point on the hyperbola is constant. At a vertex, this constant difference is

$$(a + c) - (c - a) = 2a.$$

Now, for any point (x, y) on the hyperbola, the difference of the distances between (x, y) and the two foci must also be $2a$.

$$\sqrt{(x - c)^2 + (y - 0)^2} - \sqrt{(x + c)^2 + (y - 0)^2} = 2a$$

$$\sqrt{(x - c)^2 + y^2} = 2a + \sqrt{(x + c)^2 + y^2}$$

$$(x - c)^2 + y^2 = 4a^2 + 4a\sqrt{(x + c)^2 + y^2} + (x + c)^2 + y^2$$

$$-4xc - 4a^2 = 4a\sqrt{(x + c)^2 + y^2}$$

$$-(xc + a^2) = a\sqrt{(x + c)^2 + y^2}$$

$$x^2c^2 + 2a^2cx + a^4 = a^2[x^2 + 2cx + c^2 + y^2]$$

$$x^2(c^2 - a^2) - a^2y^2 = a^2(c^2 - a^2)$$

$$\frac{x^2}{a^2} - \frac{y^2}{c^2 - a^2} = 1$$

Since $a^2 + b^2 = c^2$, we have $(x^2/a^2) - (y^2/b^2) = 1$.

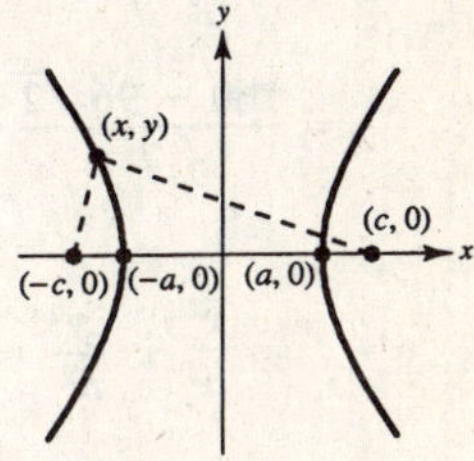

117. Time for sound of bullet hitting target to reach (x, y): $\dfrac{2c}{v_m} + \dfrac{\sqrt{(x - c)^2 + y^2}}{v_s}$

Time for sound of rifle to reach (x, y): $\dfrac{\sqrt{(x + c)^2 + y^2}}{v_s}$

Since the times are the same, we have: $\dfrac{2c}{v_m} + \dfrac{\sqrt{(x - c)^2 + y^2}}{v_s} = \dfrac{\sqrt{(x + c)^2 + y^2}}{v_s}$

$$\frac{4c^2}{v_m^2} + \frac{4c}{v_m v_s}\sqrt{(x - c)^2 + y^2} + \frac{(x - c)^2 + y^2}{v_s^2} = \frac{(x + c)^2 + y^2}{v_s^2}$$

$$\sqrt{(x - c)^2 + y^2} = \frac{v_m^2 x - v_s^2 c}{v_s v_m}$$

$$\left(1 - \frac{v_m^2}{v_s^2}\right)x^2 + y^2 = \left(\frac{v_s^2}{v_m^2} - 1\right)c^2$$

$$\frac{x^2}{c^2 v_s^2/v_m^2} - \frac{y^2}{c^2(v_m^2 - v_s^2)/v_m^2} = 1$$

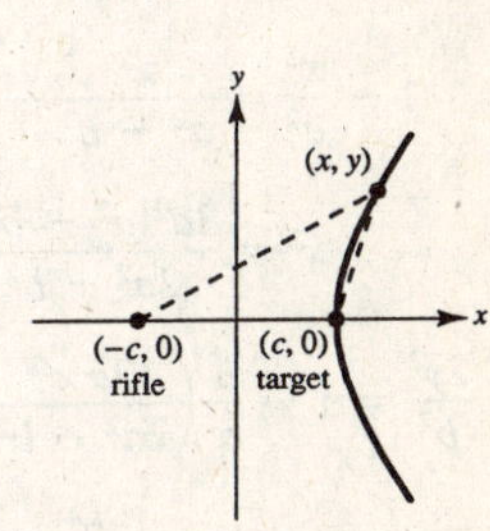

118. $c = 150$, $2a = 0.001(186{,}000)$, $a = 93$,

$b = \sqrt{150^2 - 93^2} = \sqrt{13{,}851}$

$$\frac{x^2}{93^2} - \frac{y^2}{13{,}851} = 1$$

When $y = 75$, we have

$$x^2 = 93^2\left(1 + \frac{75^2}{13{,}851}\right)$$

$$x \approx 110.3 \text{ miles.}$$

119. The point (x, y) lies on the line between $(0, 10)$ and $(10, 0)$. Thus, $y = 10 - x$. The point also lies on the hyperbola $(x^2/36) - (y^2/64) = 1$. Using substitution, we have:

$$\frac{x^2}{36} - \frac{(10 - x)^2}{64} = 1$$

$$16x^2 - 9(10 - x)^2 = 576$$

$$7x^2 + 180x - 1476 = 0$$

$$x = \frac{-180 \pm \sqrt{180^2 - 4(7)(-1476)}}{2(7)}$$

$$= \frac{-180 \pm 192\sqrt{2}}{14} = \frac{-90 \pm 96\sqrt{2}}{7}$$

Choosing the positive value for x we have:

$$x = \frac{-90 + 96\sqrt{2}}{7} \approx 6.538 \text{ and}$$

$$y = \frac{160 - 96\sqrt{2}}{7} \approx 3.462$$

120. $$\frac{x^2}{a^2} - \frac{y^2}{b^2} = 1$$

$$\frac{2x}{a^2} - \frac{2yy'}{b^2} = 0 \text{ or } y' = \frac{b^2x}{a^2y}$$

$$y - y_0 = \frac{b^2x_0}{a^2y_0}(x - x_0)$$

$$a^2y_0y - a^2y_0^2 = b^2x_0x - b^2x_0^2$$

$$b^2x_0^2 - a^2y_0^2 = b^2x_0x - a^2y_0y$$

$$a^2b^2 = b^2x_0x - a^2y_0y$$

$$\frac{x_0x}{a^2} - \frac{y_0y}{b^2} = 1$$

121. $$\frac{x^2}{a^2} + \frac{2y^2}{b^2} = 1 \Rightarrow \frac{2y^2}{b^2} = 1 - \frac{x^2}{a^2}. \text{ Let } c^2 = a^2 - b^2.$$

$$\frac{x^2}{a^2 - b^2} - \frac{2y^2}{b^2} = 1 \Rightarrow \frac{2y^2}{b^2} = \frac{x^2}{a^2 - b^2} - 1$$

$$1 - \frac{x^2}{a^2} = \frac{x^2}{a^2 - b^2} - 1 \Rightarrow 2 = x^2\left(\frac{1}{a^2} + \frac{1}{a^2 - b^2}\right)$$

$$x^2 = \frac{2a^2(a^2 - b^2)}{2a^2 - b^2} \Rightarrow x = \pm\frac{\sqrt{2}a\sqrt{a^2 - b^2}}{\sqrt{2a^2 - b^2}} = \pm\frac{\sqrt{2}ac}{\sqrt{2a^2 - b^2}}$$

$$\frac{2y^2}{b^2} = 1 - \frac{1}{a^2}\left(\frac{2a^2c^2}{2a^2 - b^2}\right) \Rightarrow \frac{2y^2}{b^2} = \frac{b^2}{2a^2 - b^2}$$

$$y^2 = \frac{b^4}{2(2a^2 - b^2)} \Rightarrow y = \pm\frac{b^2}{\sqrt{2}\sqrt{2a^2 - b^2}}$$

There are four points of intersection: $\left(\frac{\sqrt{2}ac}{\sqrt{2a^2 - b^2}}, \pm\frac{b^2}{\sqrt{2}\sqrt{2a^2 - b^2}}\right), \left(-\frac{\sqrt{2}ac}{\sqrt{2a^2 - b^2}}, \pm\frac{b^2}{\sqrt{2}\sqrt{2a^2 - b^2}}\right)$

$$\frac{x^2}{a^2} + \frac{2y^2}{b^2} = 1 \Rightarrow \frac{2x}{a^2} + \frac{4yy'}{b^2} = 0 \Rightarrow y'_e = -\frac{b^2x}{2a^2y}$$

$$\frac{x^2}{a^2 - b^2} - \frac{2y^2}{b^2} = 1 \Rightarrow \frac{2x}{c^2} - \frac{4yy'}{b^2} = 0 \Rightarrow y'_h = \frac{b^2x}{2c^2y}$$

—CONTINUED—

121. —CONTINUED—

At $\left(\dfrac{\sqrt{2}ac}{\sqrt{2a^2 - b^2}}, \dfrac{b^2}{\sqrt{2}\sqrt{2a^2 - b^2}}\right)$, the slopes of the tangent lines are:

$$y'_e = \frac{-b^2\left(\dfrac{\sqrt{2}ac}{\sqrt{2a^2 - b^2}}\right)}{2a^2\left(\dfrac{b^2}{\sqrt{2}\sqrt{2a^2 - b^2}}\right)} = -\frac{c}{a} \quad \text{and} \quad y'_h = \frac{b^2\left(\dfrac{\sqrt{2}ac}{\sqrt{2a^2 - b^2}}\right)}{2c^2\left(\dfrac{b^2}{\sqrt{2}\sqrt{2a^2 - b^2}}\right)} = \frac{a}{c}.$$

Since the slopes are negative reciprocals, the tangent lines are perpendicular. Similarly, the curves are perpendicular at the other three points of intersection.

122. $$Ax^2 + Cy^2 + Dx + Ey + F = 0 \quad \text{(Assume } A \neq 0 \text{ and } C \neq 0\text{; see (b) below)}$$

$$A\left(x^2 + \frac{D}{A}x\right) + C\left(y^2 + \frac{E}{C}y\right) = -F$$

$$A\left(x^2 + \frac{D}{A}x + \frac{D^2}{4A^2}\right) + C\left(y^2 + \frac{E}{C}y + \frac{E^2}{4C^2}\right) = -F + \frac{D^2}{4A} + \frac{E^2}{4C} = R$$

$$\frac{\left[x + \left(\dfrac{D}{2A}\right)\right]^2}{C} + \frac{\left[y + \left(\dfrac{E}{2C}\right)\right]^2}{A} = \frac{R}{AC}$$

(a) If $A = C$, we have

$$\left(x + \frac{D}{2A}\right)^2 + \left(y + \frac{E}{2C}\right)^2 = \frac{R}{A}$$

which is the standard equation of a circle.

(c) If $AC < 0$, we have

$$\frac{\left[x + \left(\dfrac{D}{2A}\right)\right]^2}{\left|\dfrac{R}{A}\right|} + \frac{\left[y + \left(\dfrac{E}{2C}\right)\right]^2}{\left|\dfrac{R}{C}\right|} = 1$$

which is the equation of an ellipse.

(b) If $C = 0$, we have

$$A\left(x + \frac{D}{2A}\right)^2 = -F - Ey + \frac{D^2}{4A}.$$

If $A = 0$, we have

$$C\left(y + \frac{E}{2C}\right)^2 = -F - Dx + \frac{E^2}{4C}.$$

These are the equations of parabolas.

(d) If $AC < 0$, we have

$$\frac{\left[x + \left(\dfrac{D}{2A}\right)\right]^2}{\left|\dfrac{R}{A}\right|} - \frac{\left[y + \left(\dfrac{E}{2C}\right)\right]^2}{\left|\dfrac{R}{C}\right|} = \pm 1$$

which is the equation of a hyperbola.

123. False. See the definition of a parabola.

124. True

125. True

126. False. $y^2 - x^2 + 2x + 2y = 0$ yields two intersecting lines:

$y + 1 = \pm(x - 1)$

127. True

128. True

129. Let $\dfrac{x^2}{a^2} + \dfrac{y^2}{b^2} = 1$ be the equation of the ellipse with $a > b > 0$. Let $(\pm c, 0)$ be the foci, $c^2 = a^2 - b^2$. Let (u, v) be a point on the tangent line at $P(x, y)$, as indicated in the figure.

$$x^2b^2 + y^2a^2 = a^2b^2$$

$$2xb^2 + 2yy'a^2 = 0$$

$$y' = -\frac{b^2x}{a^2y} \quad \text{Slope at } P(x, y)$$

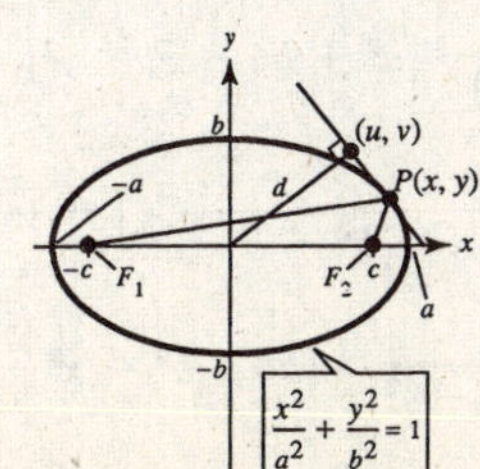

—CONTINUED—

129. —CONTINUED—

Now, $\dfrac{y - v}{x - u} = -\dfrac{b^2x}{a^2y}$

$$y^2a^2 - a^2vy = -b^2x^2 + b^2xu$$

$$y^2a^2 + x^2b^2 = a^2vy + b^2ux$$

$$a^2b^2 = a^2vy + b^2ux$$

Since there is a right angle at (u, v),

$$\frac{v}{u} = \frac{a^2y}{b^2x}$$

$$vb^2x = a^2uy.$$

We have two equations:

$$a^2vy + b^2ux = a^2b^2$$

$$a^2uy - b^2vx = 0.$$

Multiplying the first by v and the second by u, and adding,

$$a^2v^2y + a^2u^2y = a^2b^2v$$

$$y[u^2 + v^2] = b^2v$$

$$yd^2 = b^2v$$

$$v = \frac{yd^2}{b^2}.$$

Similarly, $u = \dfrac{xd^2}{a^2}$.

From the figure, $u = d\cos\theta$ and $v = d\sin\theta$. Thus, $\cos\theta = \dfrac{xd}{a^2}$ and $\sin\theta = \dfrac{yd}{b^2}$.

$$\cos^2\theta + \sin^2\theta = \frac{x^2d^2}{a^4} + \frac{y^2d^2}{b^4} = 1$$

$$x^2b^4d^2 + y^2a^4d^2 = a^4b^4$$

$$d^2 = \frac{a^4b^4}{x^2b^4 + y^2a^4}$$

Let $r_1 = PF_1$ and $r_2 = PF_2$, $r_1 + r_2 = 2a$.

$$r_1r_2 = \frac{1}{2}[(r_1 + r_2)^2 - r_1^2 - r_2^2]$$

$$= \frac{1}{2}[4a^2 - (x + c)^2 - y^2 - (x - c)^2 - y^2]$$

$$= 2a^2 - x^2 - y^2 - c^2$$

$$= a^2 + b^2 - x^2 - y^2$$

Finally, $d^2r_1r_2 = \dfrac{a^4b^4}{x^2b^4 + y^2a^4} \cdot [a^2 + b^2 - x^2 - y^2]$

$$= \frac{a^4b^4}{b^2(b^2x^2) + a^2(a^2y^2)} \cdot [a^2 + b^2 - x^2 - y^2]$$

$$= \frac{a^4b^4}{b^2(a^2b^2 - a^2y^2) + a^2(a^2b^2 - b^2x^2)} \cdot [a^2 + b^2 - x^2 - y^2]$$

$$= \frac{a^4b^4}{a^2b^2[a^2 + b^2 - x^2 - y^2]} \cdot [a^2 + b^2 - x^2 - y^2]$$

$= a^2b^2$, a constant!

130. Consider circle $x^2 + y^2 = 2$ and hyperbola $y = \dfrac{9}{x}$.

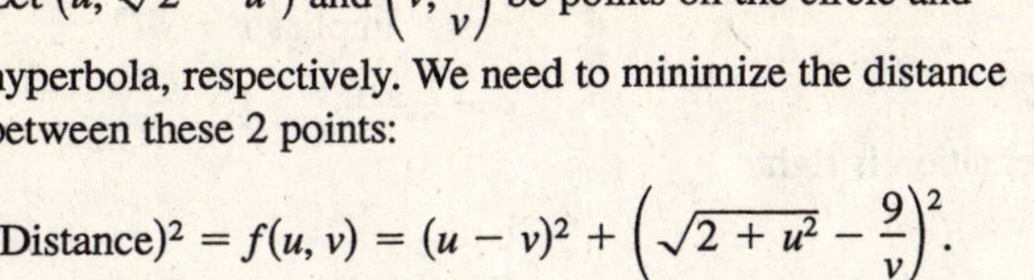

Let $\left(u, \sqrt{2 - u^2}\right)$ and $\left(v, \dfrac{9}{v}\right)$ be points on the circle and hyperbola, respectively. We need to minimize the distance between these 2 points:

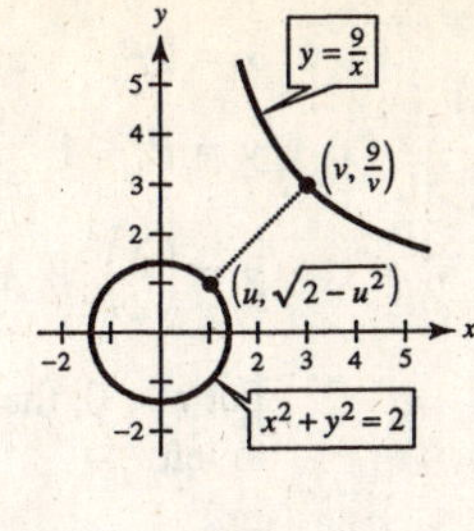

$$(\text{Distance})^2 = f(u, v) = (u - v)^2 + \left(\sqrt{2 + u^2} - \frac{9}{v}\right)^2.$$

The tangent lines at $(1, 1)$ and $(3, 3)$ are both perpendicular to $y = x$, and hence parallel.

The minimum value is $(3 - 1)^2 + (3 - 1)^2 = 8$.

Section 10.2 Plane Curves and Parametric Equations

1. $x = \sqrt{t}$, $y = 1 - t$

(a)

t	0	1	2	3	4
x	0	1	$\sqrt{2}$	$\sqrt{3}$	2
y	1	0	-1	-2	-3

(b)

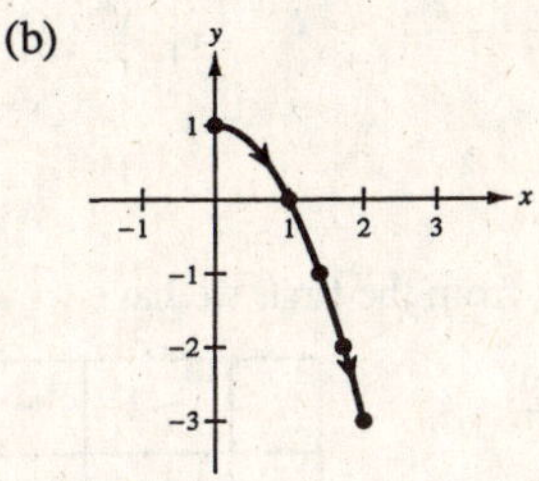

(c)

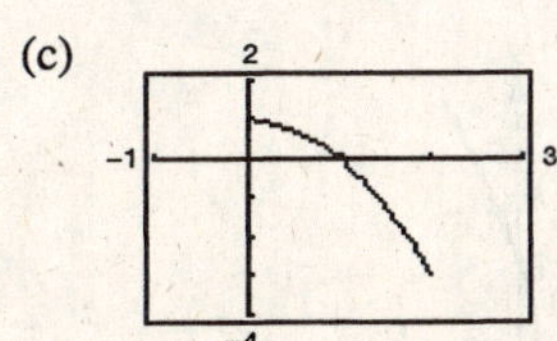

(d) $x^2 = t$

$y = 1 - x^2, x \geq 0$

2. $x = 4\cos^2\theta$ $\quad y = 2\sin\theta$

$0 \leq x \leq 4$ $\quad -2 \leq y \leq 2$

(a)

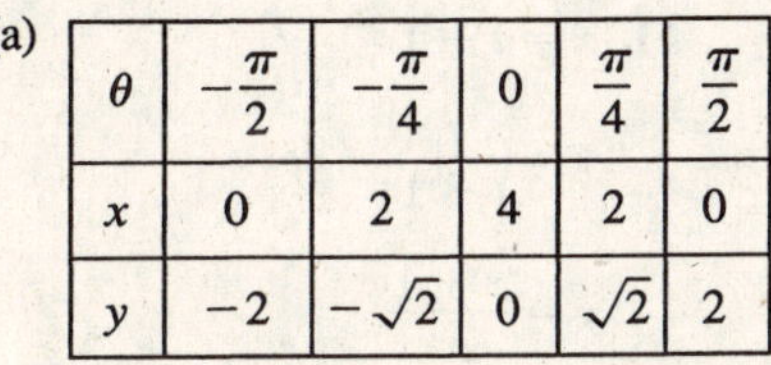

θ	$-\frac{\pi}{2}$	$-\frac{\pi}{4}$	0	$\frac{\pi}{4}$	$\frac{\pi}{2}$
x	0	2	4	2	0
y	-2	$-\sqrt{2}$	0	$\sqrt{2}$	2

(b)

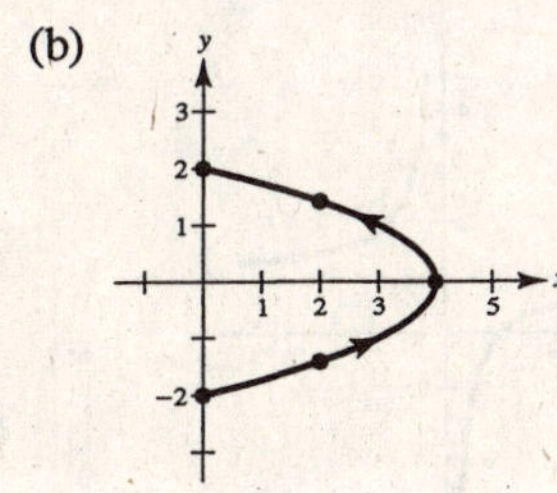

(c)

(d) $\dfrac{x}{4} = \cos^2\theta$

$\dfrac{y^2}{4} = \sin^2\theta$

$\dfrac{x}{4} + \dfrac{y^2}{4} = 1$

$x = 4 - y^2, -2 \leq y \leq 2$

(e) The graph would be oriented in the opposite direction.

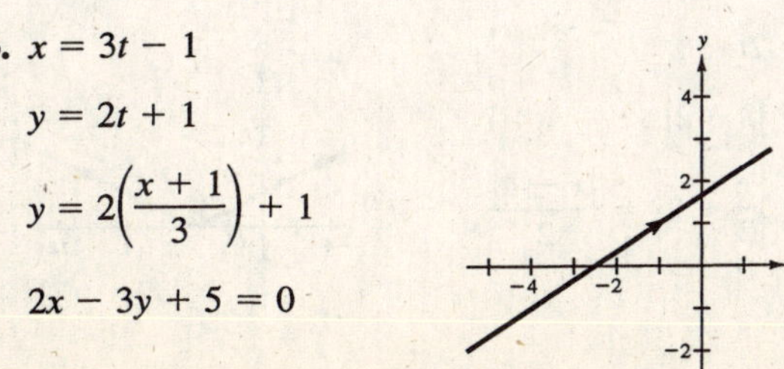

3. $x = 3t - 1$

$y = 2t + 1$

$y = 2\left(\dfrac{x + 1}{3}\right) + 1$

$2x - 3y + 5 = 0$

4. $x = 3 - 2t$

$y = 2 + 3t$

$y = 2 + 3\left(\dfrac{3 - x}{2}\right)$

$2y + 3x - 13 = 0$

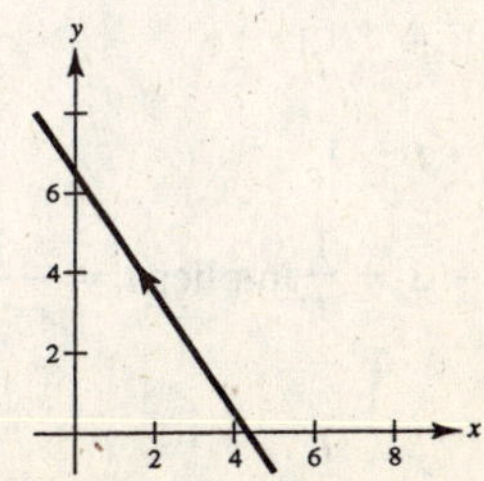

5. $x = t + 1$

$y = t^2$

$y = (x - 1)^2$

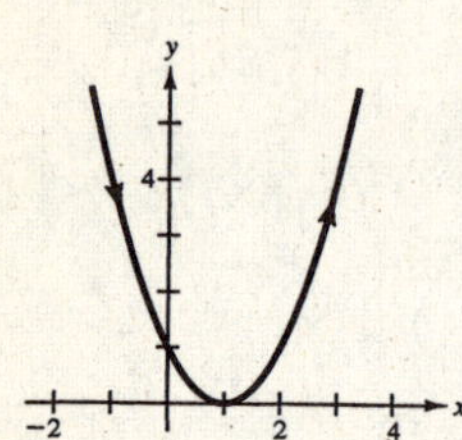

6. $x = 2t^2$

$y = t^4 + 1$

$y = \left(\frac{x}{2}\right)^2 + 1 = \frac{x^2}{4} + 1, x \geq 0$

For $t < 0$, the orientation is right to left.

For $t > 0$, the orientation is left to right.

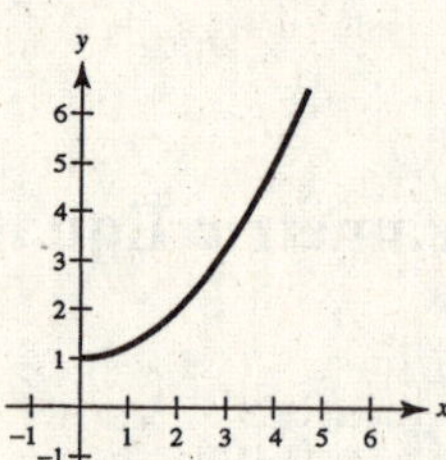

7. $x = t^3$

$y = \frac{1}{2}t^2$

$x = t^3$ implies $t = x^{1/3}$

$y = \frac{1}{2}x^{2/3}$

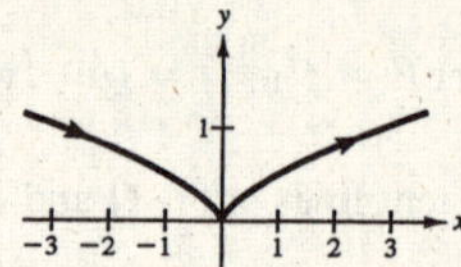

8. $x = t^2 + t,\ y = t^2 - t$

Subtracting the second equation from the first, we have

$x - y = 2t$ or $t = \frac{x - y}{2}$.

$y = \frac{(x - y)^2}{4} - \frac{x - y}{2}$

t	-2	-1	0	1	2
x	2	0	0	2	6
y	6	2	0	0	2

Since the discriminant is

$B^2 - 4AC = (-2)^2 - 4(1)(1) = 0,$

the graph is a rotated parabola.

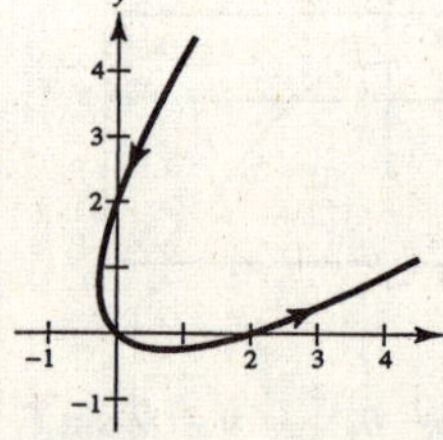

9. $x = \sqrt{t}, t \geq 0$

$y = t - 2$

$y = x^2 - 2, x \geq 0$

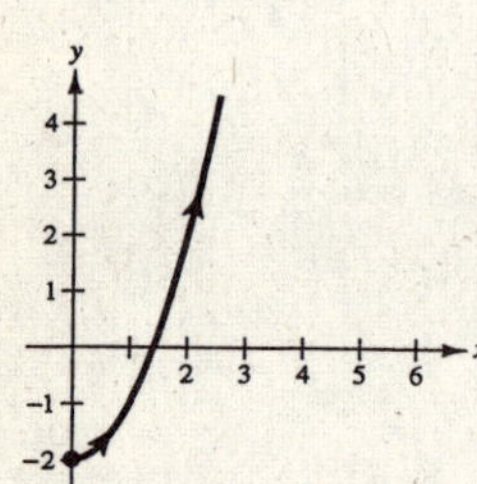

10. $x = \sqrt[4]{t}, t \geq 0$

$y = 3 - t$

$y = 3 - x^4, x \geq 0$

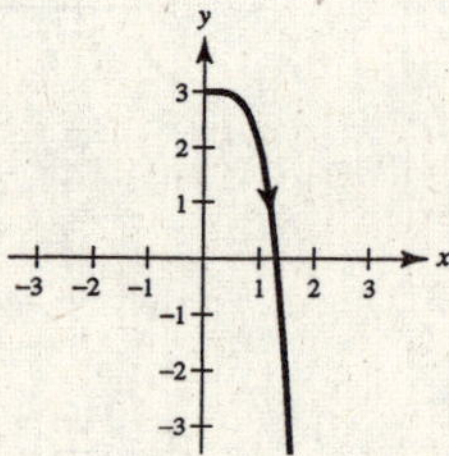

11. $x = t - 1$

$y = \frac{t}{t - 1}$

$y = \frac{x + 1}{x}$

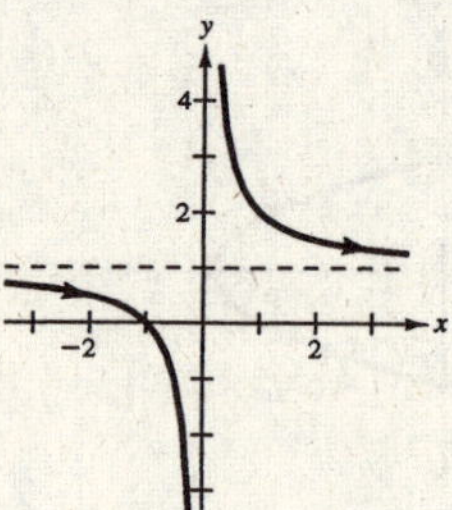

12. $x = 1 + \frac{1}{t}$

$y = t - 1$

$x = 1 + \frac{1}{t}$ implies $t = \frac{1}{x - 1}$

$y = \frac{1}{x - 1} - 1$

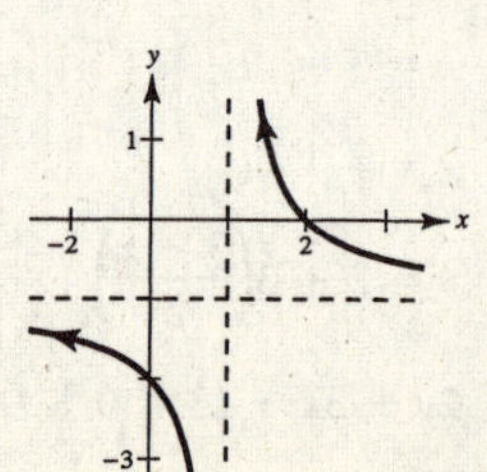

13. $x = 2t$

$y = |t - 2|$

$y = \left|\frac{x}{2} - 2\right| = \frac{|x - 4|}{2}$

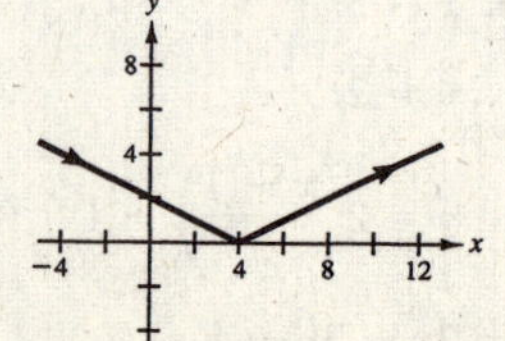

14. 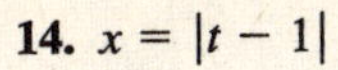 $x = |t - 1|$

$y = t + 2$

$x = |(y - 2) - 1| = |y - 3|$

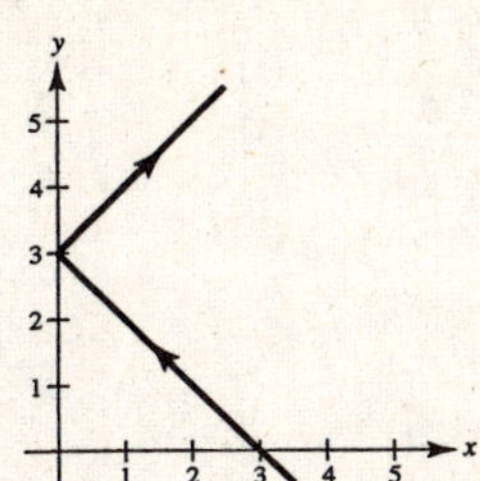

15. $x = e^t, x > 0$

$y = e^{3t} + 1$

$y = x^3 + 1, x > 0$

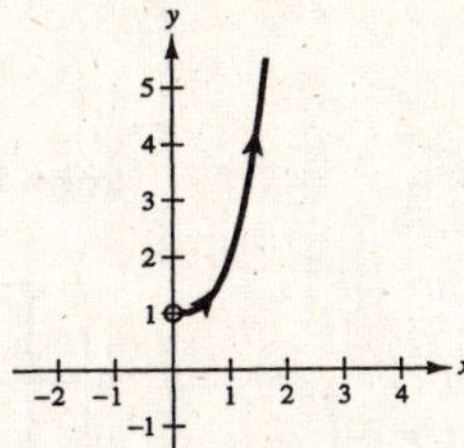

16. $x = e^{-t}, x > 0$

$y = e^{2t} - 1$

$y = x^{-2} - 1 = \dfrac{1}{x^2} - 1, x > 0$

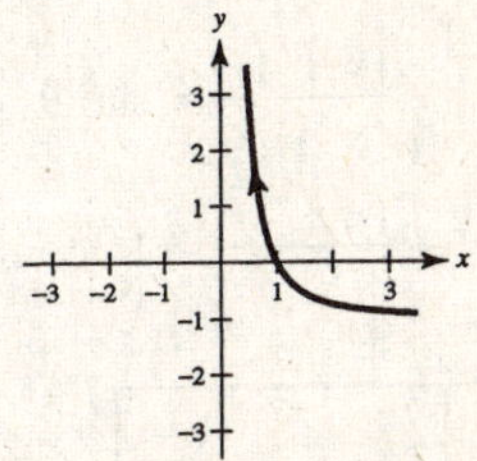

17. $x = \sec \theta$

$y = \cos \theta$

$0 \le \theta < \dfrac{\pi}{2}, \dfrac{\pi}{2} < \theta \le \pi$

$xy = 1$

$y = \dfrac{1}{x}$

$|x| \ge 1,\ |y| \le 1$

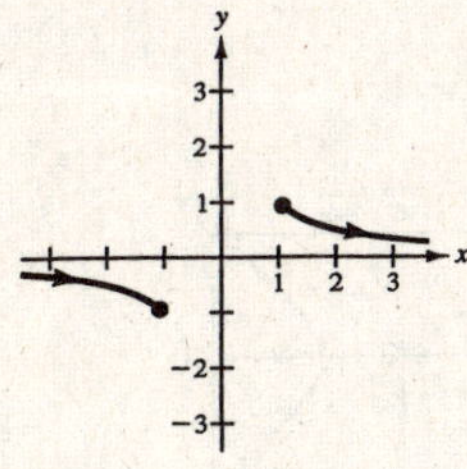

18. $x = \tan^2 \theta$

$y = \sec^2 \theta$

$\sec^2 \theta = \tan^2 \theta + 1$

$y = x + 1$

$x \ge 0$

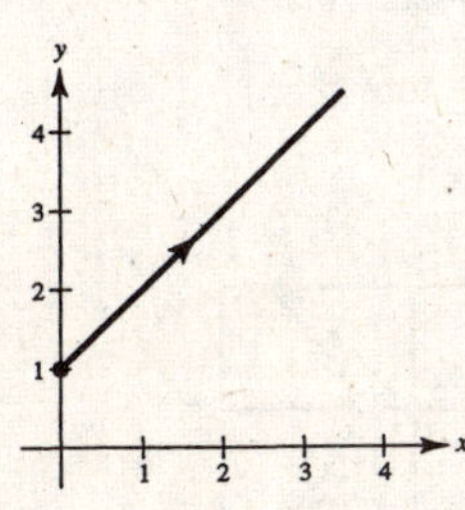

19. $x = 3 \cos \theta,\ y = 3 \sin \theta$

Squaring both equations and adding, we have

$x^2 + y^2 = 9.$

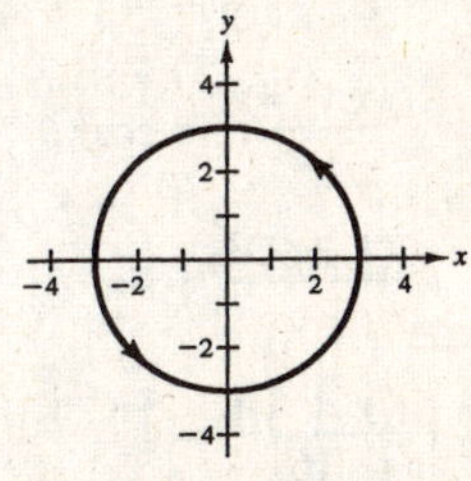

20. $x = 2 \cos \theta$

$y = 6 \sin \theta$

$\left(\dfrac{x}{2}\right)^2 + \left(\dfrac{y}{6}\right)^2 = \cos^2 + \sin^2 \theta = 1$

$\dfrac{x^2}{4} + \dfrac{y^2}{36} = 1$ ellipse

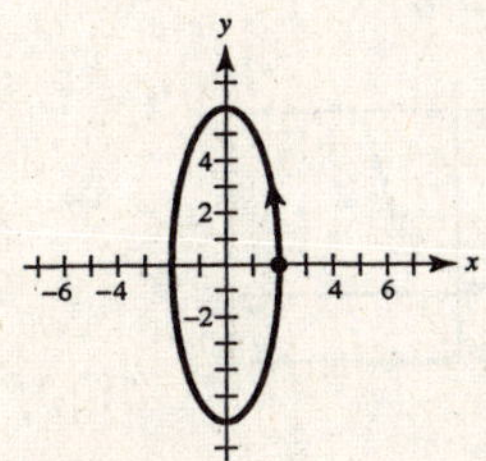

21. $x = 4 \sin 2\theta$

$y = 2 \cos 2\theta$

$\dfrac{x^2}{16} = \sin^2 2\theta$

$\dfrac{y^2}{4} = \cos^2 2\theta$

$\dfrac{x^2}{16} + \dfrac{y^2}{4} = 1$

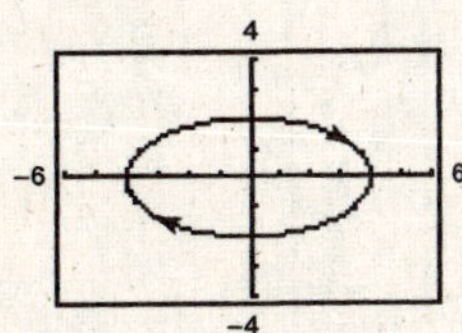

22. $x = \cos \theta$

$y = 2 \sin 2\theta$

$y = 4 \sin \theta \cos \theta$

$1 - x^2 = \sin^2 \theta$

$y = \pm 4x\sqrt{1 - x^2}$

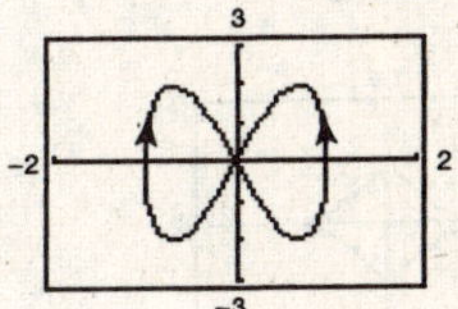

23.

$$x = 4 + 2\cos\theta$$
$$y = -1 + \sin\theta$$
$$\frac{(x-4)^2}{4} = \cos^2\theta$$
$$\frac{(y+1)^2}{1} = \sin^2\theta$$
$$\frac{(x-4)^2}{4} + \frac{(y+1)^2}{1} = 1$$

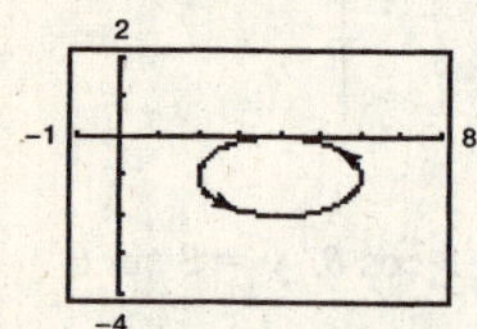

24.

$$x = 4 + 2\cos\theta$$
$$y = -1 + 2\sin\theta$$
$$(x-4)^2 = 4\cos^2\theta$$
$$(y+1)^2 = 4\sin^2\theta$$
$$(x-4)^2 + (y+1)^2 = 4$$

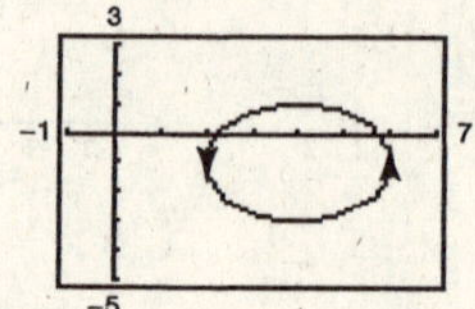

25.

$$x = 4 + 2\cos\theta$$
$$y = -1 + 4\sin\theta$$
$$\frac{(x-4)^2}{4} = \cos^2\theta$$
$$\frac{(y+1)^2}{16} = \sin^2\theta$$
$$\frac{(x-4)^2}{4} + \frac{(y+1)^2}{16} = 1$$

26.

$$x = \sec\theta$$
$$y = \tan\theta$$
$$x^2 = \sec^2\theta$$
$$y^2 = \tan^2\theta$$
$$x^2 - y^2 = 1$$

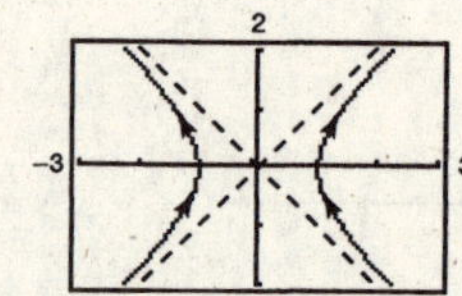

27.

$$x = 4\sec\theta$$
$$y = 3\tan\theta$$
$$\frac{x^2}{16} = \sec^2\theta$$
$$\frac{y^2}{9} = \tan^2\theta$$
$$\frac{x^2}{16} - \frac{y^2}{9} = 1$$

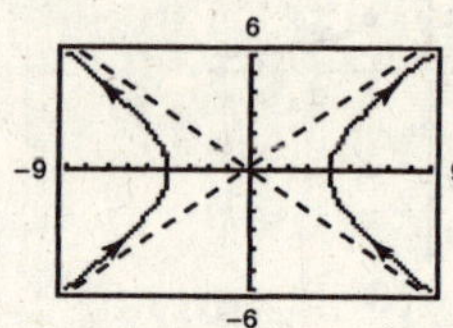

28.

$$x = \cos^3\theta$$
$$y = \sin^3\theta$$
$$x^{2/3} = \cos^2\theta$$
$$y^{2/3} = \sin^2\theta$$
$$x^{2/3} + y^{2/3} = 1$$

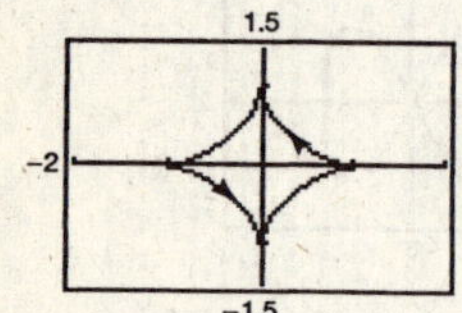

29. $x = t^3$

$$y = 3\ln t$$
$$y = 3\ln\sqrt[3]{x} = \ln x$$

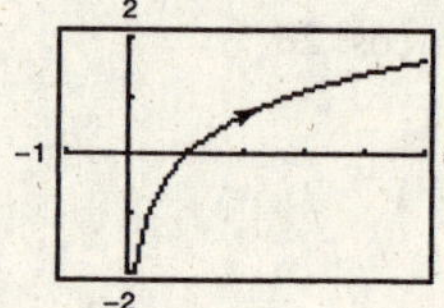

30. $x = \ln 2t$

$$y = t^2$$
$$t = \frac{e^x}{2}$$
$$y = \frac{e^{2x}}{r} = \frac{1}{4}e^{2x}$$

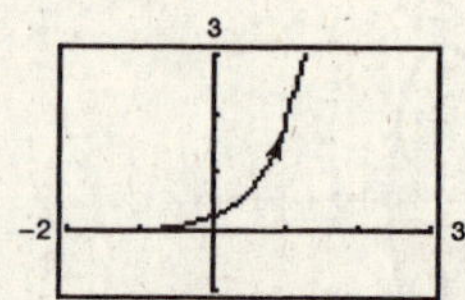

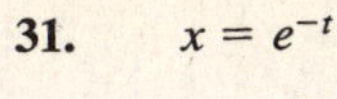

31. $x = e^{-t}$

$y = e^{3t}$

$e^t = \frac{1}{x}$

$e^t = \sqrt[3]{y}$

$\sqrt[3]{y} = \frac{1}{x}$

$y = \frac{1}{x^3}$

$x > 0$

$y > 0$

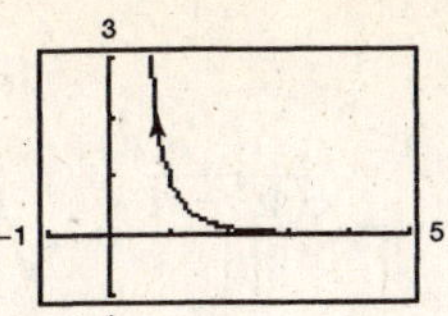

32. $x = e^{2t}$

$y = e^t$

$y^2 = x$

$y > 0$

$y = \sqrt{x}, x > 0$

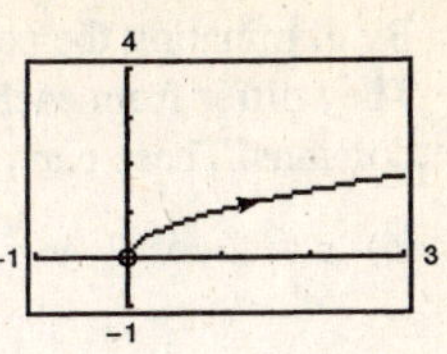

33. By eliminating the parameters in (a) – (d), we get $y = 2x + 1$. They differ from each other in orientation and in restricted domains. These curves are all smooth except for (b).

(a) $x = t,\ y = 2t + 1$

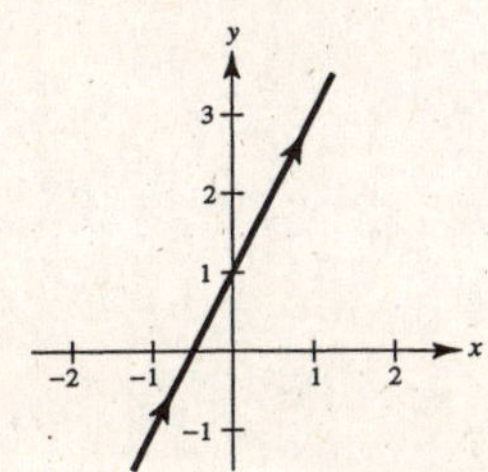

(b) $x = \cos\theta \qquad y = 2\cos\theta + 1$

$-1 \le x \le 1 \qquad -1 \le y \le 3$

$\frac{dx}{d\theta} = \frac{dy}{d\theta} = 0$ when $\theta = 0, \pm\pi, \pm 2\pi, \ldots$

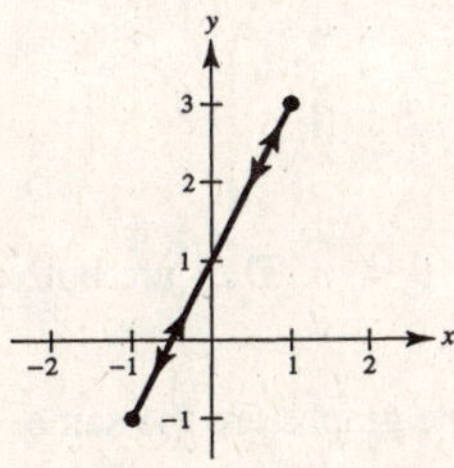

(c) $x = e^{-t} \qquad y = 2e^{-t} + 1$

$x > 0 \qquad y > 1$

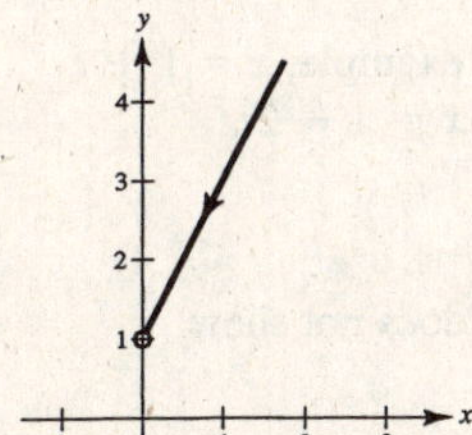

(d) $x = e^t \qquad y = 2e^t + 1$

$x > 0 \qquad y > 1$

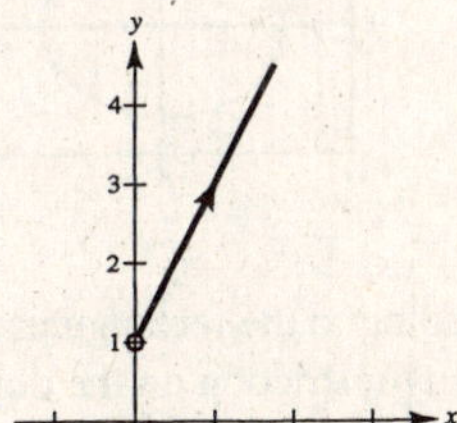

34. By eliminating the parameters in (a) – (d), we get $x^2 + y^2 = 4$. They differ from each other in orientation and in restricted domains. These curves are all smooth.

(a) $x = 2\cos\theta,\ y = 2\sin\theta$

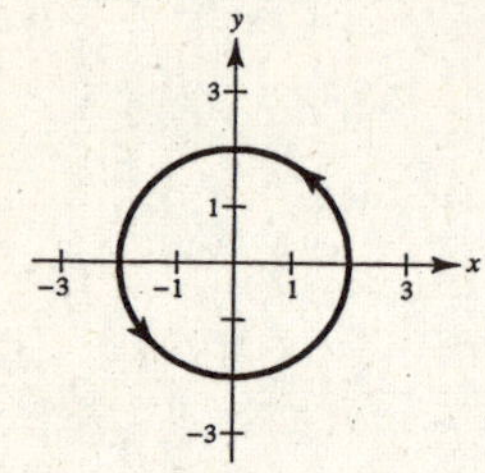

(b) $x = \dfrac{\sqrt{4t^2 - 1}}{|t|} = \sqrt{4 - \dfrac{1}{t^2}}$ $\quad y = \dfrac{1}{t}$

$x \geq 0,\ x \neq 2$ $\quad y \neq 0$

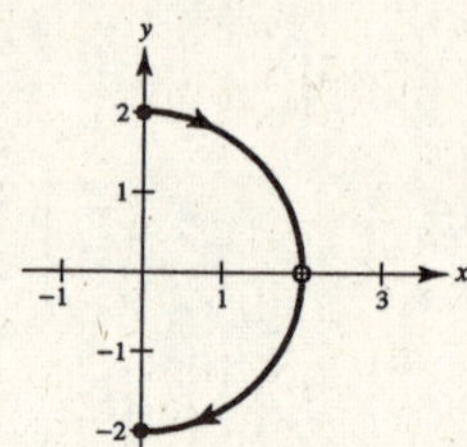

(c) $x = \sqrt{t}$ $\quad y = \sqrt{4 - t}$

$x \geq 0$ $\quad y \geq 0$

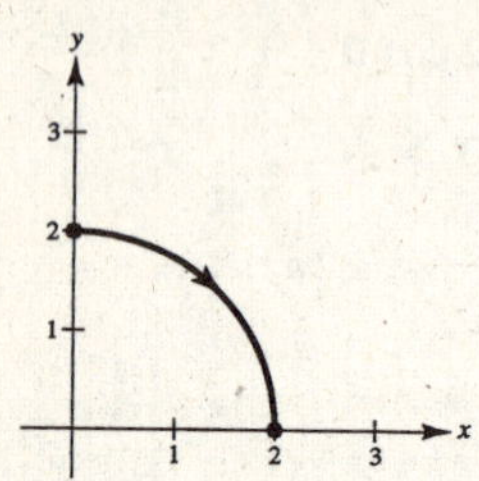

(d) $x = -\sqrt{4 - e^{2t}}$ $\quad y = e^t$

$-2 < x \leq 0$ $\quad y > 0$

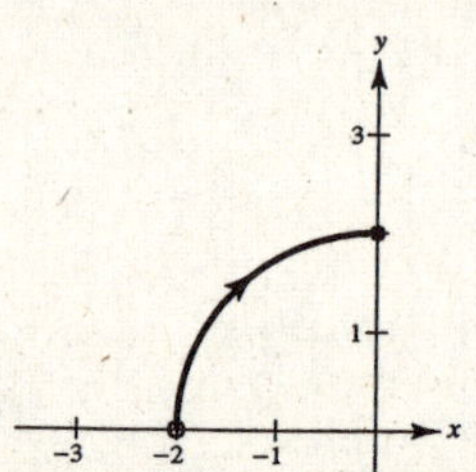

35. The curves are identical on $0 < \theta < \pi$. They are both smooth. Represent $y = 2(1 - x^2)$

36. The orientations are reversed. The graphs are the same. They are both smooth.

37. (a)

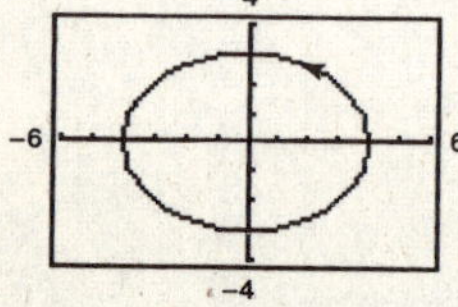

(b) The orientation of the second curve is reversed.

(c) The orientation will be reversed.

(d) Many answers possible. For example, $x = 1 + t$, $y = 1 + 2t$, and $x = 1 - t$, $x = 1 - 2t$.

38. The set of points (x, y) corresponding to the rectangular equation of a set of parametric equations does not show the orientation of the curve nor any restriction on the domain of the original parametric equations.

39.

$$x = x_1 + t(x_2 - x_1)$$

$$y = y_1 + t(y_2 - y_1)$$

$$\frac{x - x_1}{x_2 - x_1} = t$$

$$y = y_1 + \left(\frac{x - x_1}{x_2 - x_1}\right)(y_2 - y_1)$$

$$y - y_1 = \frac{y_2 - y_1}{x_2 - x_1}(x - x_1)$$

$$y - y_1 = m(x - x_1)$$

40.

$$x = h + r\cos\theta$$

$$y = k + r\sin\theta$$

$$\cos\theta = \frac{x - h}{r}$$

$$\sin\theta = \frac{y - k}{r}$$

$$\cos^2\theta + \sin^2\theta = \frac{(x - h)^2}{r^2} + \frac{(y - k)^2}{r^2} = 1$$

$$(x - h)^2 + (y - k)^2 = r^2$$

41.

$$x = h + a\cos\theta$$
$$y = k + b\sin\theta$$
$$\frac{x-h}{a} = \cos\theta$$
$$\frac{y-k}{b} = \sin\theta$$
$$\frac{(x-h)^2}{a^2} + \frac{(y-k)^2}{b^2} = 1$$

42.

$$x = h + a\sec\theta$$
$$y = k + b\tan\theta$$
$$\frac{x-h}{a} = \sec\theta$$
$$\frac{y-k}{b} = \tan\theta$$
$$\frac{(x-h)^2}{a^2} - \frac{(y-k)^2}{b^2} = 1$$

43. From Exercise 39 we have

$x = 5t$

$y = -2t.$

Solution not unique

44. From Exercise 39 we have

$x = 1 + 4t$

$y = 4 - 6t.$

Solution not unique

45. From Exercise 40 we have

$x = 2 + 4\cos\theta$

$y = 1 + 4\sin\theta.$

Solution not unique

46. From Exercise 40 we have

$x = -3 + 3\cos\theta$

$y = 1 + 3\sin\theta.$

Solution not unique

47. From Exercise 41 we have

$a = 5, c = 4 \Longrightarrow b = 3$

$x = 5\cos\theta$

$y = 3\sin\theta.$

Center: $(0, 0)$
Solution not unique

48. From Exercise 41 we have

$a = 5, c = 3 \Longrightarrow b = 4$

$x = 4 + 5\cos$

$y = 2 + 4\sin\theta.$

Center: $(4, 2)$
Solution not unique

49. From Exercise 42 we have

$a = 4, c = 5 \Longrightarrow b = 3$

$x = 4\sec\theta$

$y = 3\tan\theta.$

Center: $(0, 0)$
Solution not unique

50. From Exercise 42 we have

$a = 1, c = 2 \Longrightarrow b = \sqrt{3}$

$x = \sqrt{3}\tan\theta$

$y = \sec\theta.$

Center: $(0, 0)$
Solution not unique
The transverse axis is vertical, therefore, x and y are interchanged.

51. $y = 3x - 2$

Example

$x = t, \quad y = 3t - 2$

$x = t - 3, \quad y = 3t - 11$

52. $y = \dfrac{2}{x-1}$

Example

$x = t, y = \dfrac{2}{t-1}$

$x = -t, y = \dfrac{2}{-t-1}$

53. $y = x^3$

Example

$x = t, \quad y = t^3$

$x = \sqrt[3]{t}, \quad y = t$

$x = \tan t, \quad y = \tan^3 t$

54. $y = x^2$

Example

$x = t, \quad y = t^2$

$x = t^3, \quad y = t^6$

55. $x = 2(\theta - \sin\theta)$

$y = 2(1 - \cos\theta)$

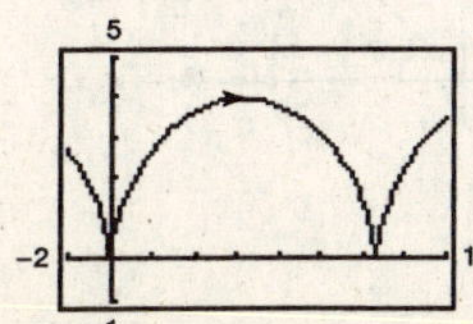

Not smooth at $\theta = 2n\pi$

56. $x = \theta + \sin\theta$

$y = 1 - \cos\theta$

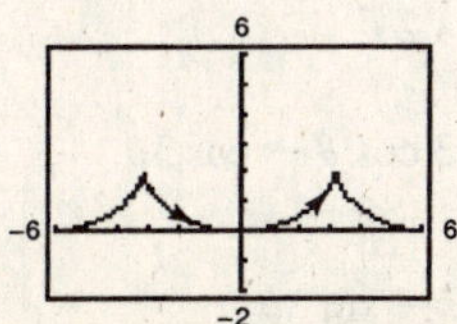

Not smooth at $x = (2n - 1)\pi$

57. $x = \theta - \frac{3}{2}\sin\theta$

$y = 1 - \frac{3}{2}\cos\theta$

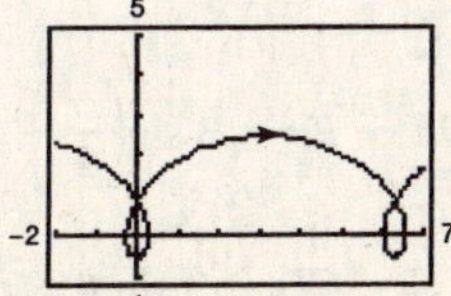

58. $x = 2\theta - 4\sin\theta$

$y = 2 - 4\cos\theta$

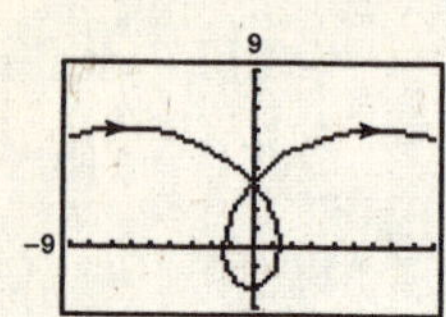

59. $x = 3\cos^3\theta$

$y = 3\sin^3\theta$

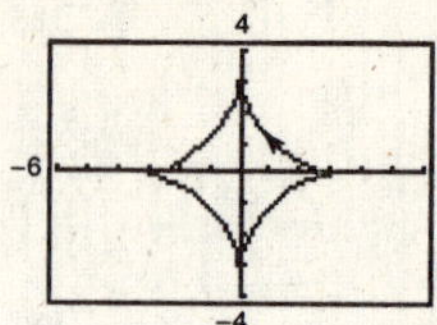

Not smooth at $(x, y) = (\pm 3, 0)$ and $(0, \pm 3)$, or $\theta = \frac{1}{2}n\pi$.

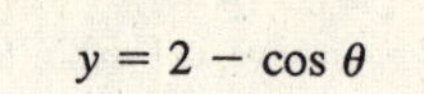

60. $x = 2\theta - \sin\theta$

$y = 2 - \cos\theta$

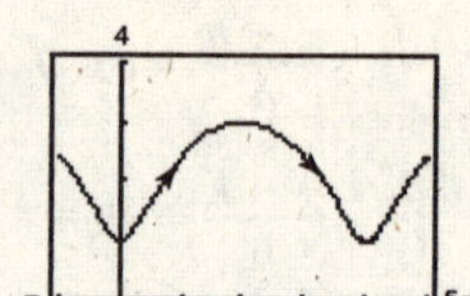

Smooth everywhere

61. $x = 2\cot\theta$

$y = 2\sin^2\theta$

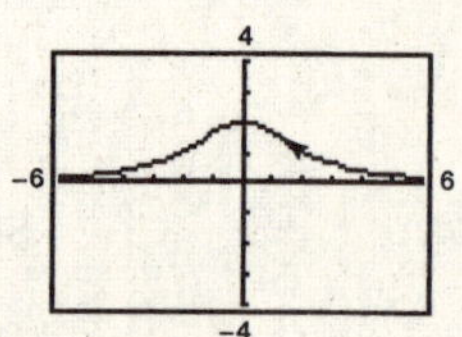

Smooth everywhere

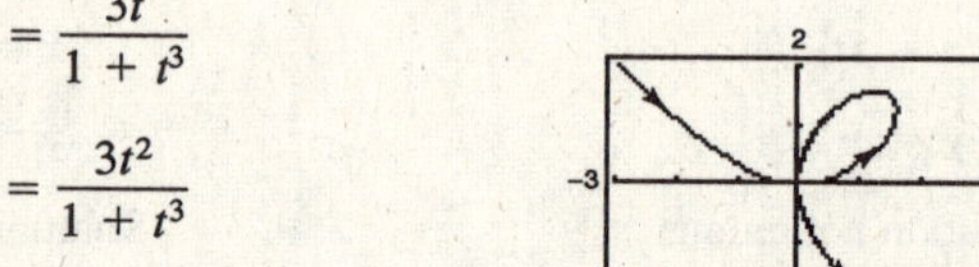

62. $x = \dfrac{3t}{1 + t^3}$

$y = \dfrac{3t^2}{1 + t^3}$

Smooth everywhere

63. See definition on page 709.

64. Each point (x, y) in the plane is determined by the plane curve $x = f(t)$, $y = g(t)$. For each t, plot (x, y). As t increases, the curve is traced out in a specific direction called the orientation of the curve.

65. A plane curve C, represented by $x = f(t)$, $y = g(t)$, is smooth if f' and g' are continuous and not simultaneously 0. See page 714.

66. (a) Matches (iv) because $(0, 2)$ is on the graph.

(b) Matches (v) because $(1, 0)$ is on the graph.

(c) Matches (ii) because $-1 \le x \le 0$ and $1 \le y \le 3$.

(d) Matches (iii) because $(4, 0)$ is on the graph.

(e) Matches (vi) because undefined at $\theta = 0$.

(f) Matches (i) because $x = (y - 2)^2 - 1$ for all y.

67. When the circle has rolled θ radians, we know that the center is at $(a\theta, a)$.

$$\sin\theta = \sin(180° - \theta) = \frac{|AC|}{b} = \frac{|BD|}{b} \quad \text{or} \quad |BD| = b\sin\theta$$

$$\cos\theta = -\cos(180° - \theta) = \frac{|AP|}{-b} \quad \text{or} \quad |AP| = -b\cos\theta$$

Therefore, $x = a\theta - b\sin\theta$ and $y = a - b\cos\theta$.

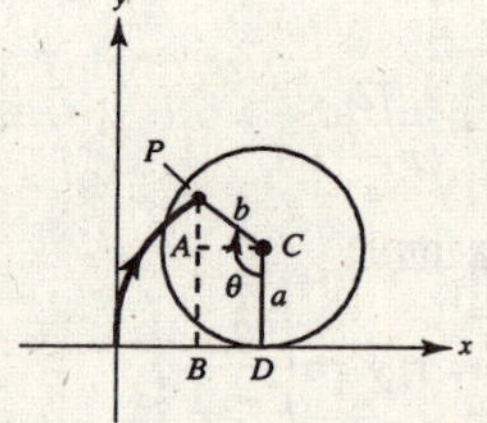

68. Let the circle of radius 1 be centered at C. A is the point of tangency on the line OC. $OA = 2$, $AC = 1$, $OC = 3$. $P = (x, y)$ is the point on the curve being traced out as the angle θ changes $\widehat{AB} = \widehat{AP}$, $\widehat{AB} = 2\theta$ and $\widehat{AP} = \alpha \Rightarrow \alpha = 2\theta$. Form the right triangle $\triangle CDP$. The angle $OCE = (\pi/2) - \theta$ and

$$\angle DCP = \alpha - \left(\frac{\pi}{2} - \theta\right) = \alpha + \theta - \left(\frac{\pi}{2}\right) = 3\theta - \left(\frac{\pi}{2}\right).$$

$$x = OE + Ex = 3\sin\left(\frac{\pi}{2} - \theta\right) + \sin\left(3\theta - \frac{\pi}{2}\right) = 3\cos\theta - \cos 3\theta$$

$$y = EC - CD = 3\sin\theta - \cos\left(3\theta - \frac{\pi}{2}\right) = 3\sin\theta - \sin 3\theta$$

Hence, $x = 3\cos\theta - \cos 3\theta$, $y = 3\sin\theta - \sin 3\theta$.

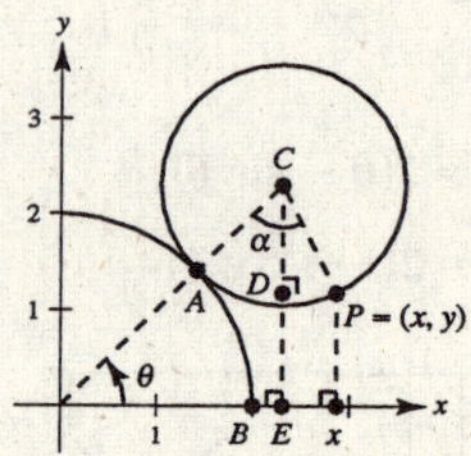

69. False

$x = t^2 \Rightarrow x \geq 0$

$y = t^2 \Rightarrow y \geq 0$

The graph of the parametric equations is only a portion of the line $y = x$.

70. False. Let $x = t^2$ and $y = t$. Then $x = y^2$ and y is not a function of x.

71. (a) $100 \text{ mi/hr} = \dfrac{(100)(5280)}{3600} = \dfrac{440}{3} \text{ ft/sec}$

$$x = (v_0 \cos\theta)t = \left(\frac{440}{3}\cos\theta\right)t$$

$$y = h + (v_0 \sin\theta)t - 16t^2$$

$$= 3 + \left(\frac{440}{3}\sin\theta\right)t - 16t^2$$

(b)

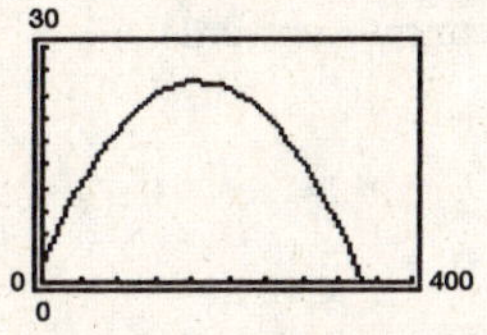

It is not a home run—when $x = 400$, $y < 10$.

(c)

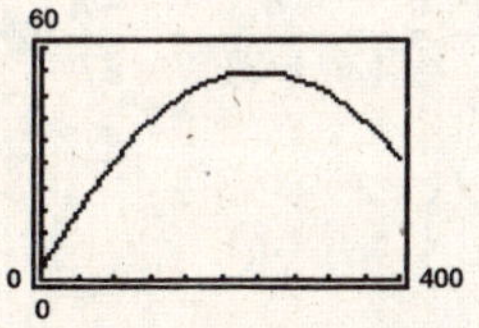

Yes, it's a home run when $x = 400$, $y > 10$.

(d) We need to find the angle θ (and time t) such that

$$x = \left(\frac{440}{3}\cos\theta\right)t = 400$$

$$y = 3 + \left(\frac{440}{3}\sin\theta\right)t - 16t^2 = 10.$$

From the first equation $t = 1200/440 \cos\theta$. Substituting into the second equation,

$$10 = 3 + \left(\frac{440}{3}\sin\theta\right)\left(\frac{1200}{440\cos\theta}\right) - 16\left(\frac{1200}{440\cos\theta}\right)^2$$

$$7 = 400\tan\theta - 16\left(\frac{120}{44}\right)^2 \sec^2\theta$$

$$= 400\tan\theta - 16\left(\frac{120}{44}\right)^2(\tan^2\theta + 1).$$

We now solve the quadratic for $\tan\theta$:

$$16\left(\frac{120}{44}\right)^2 \tan^2\theta - 400\tan\theta + 7 + 16\left(\frac{120}{44}\right)^2 = 0$$

$\tan\theta \approx 0.35185 \Rightarrow \theta \approx 19.4°$

72. (a) $x = (v_0 \cos\theta)t$

$$y = h + (v_0 \sin\theta)t - 16t^2$$

$$t = \frac{x}{v_0\cos\theta} \Rightarrow y = h + (v_0\sin\theta)\frac{x}{v_0\cos\theta} - 16\left(\frac{x}{v_0\cos\theta}\right)^2$$

$$y = h + (\tan\theta)x - \frac{16\sec^2\theta}{v_0^2}x^2$$

(b) $y = 5 + x - 0.005x^2 = h + (\tan\theta)x - \dfrac{16\sec^2\theta}{v_0^2}x^2$

$$h = 5,\ \tan\theta = 1 \Rightarrow \theta = \frac{\pi}{4}, \text{ and}$$

$$0.005 = \frac{16\sec^2(\pi/4)}{v_0^2} = \frac{16}{v_0^2}(2)$$

$$v_0^2 = \frac{32}{0.005} = 6400 \Rightarrow v_0 = 80.$$

Hence, $x = (80\cos(45°))t$

$$y = 5 + (80\sin(45°))t - 16t^2.$$

(c)

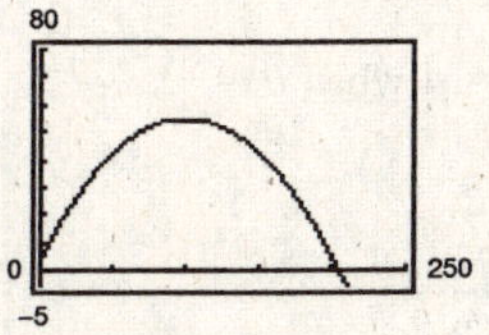

(d) Maximum height: $y = 55$ (at $x = 100$)

Range: 204.88

Section 10.3 Parametric Equations and Calculus

1. $\frac{dy}{dx} = \frac{dy/dt}{dx/dt} = \frac{-4}{2t} = \frac{-2}{t}$

2. $\frac{dy}{dx} = \frac{dy/dt}{dx/dt} = \frac{-1}{(1/3)t^{-2/3}} = -3t^{2/3}$

3. $\frac{dy}{dx} = \frac{dy/d\theta}{dx/d\theta} = \frac{-2\cos\theta\sin\theta}{2\sin\theta\cos\theta} = -1$

$\left[\text{Note: } x + y = 1 \Rightarrow y = 1 - x \text{ and } \frac{dy}{d\theta} = -1\right]$

4. $\frac{dy}{dx} = \frac{dy/d\theta}{dx/d\theta} = \frac{(-1/2)e^{-\theta/2}}{2e^{\theta}} = -\frac{1}{4}e^{-3\theta/2} = \frac{-1}{4e^{3\theta/2}}$

5. $x = 2t,\ y = 3t - 1$

$\frac{dy}{dx} = \frac{dy/dt}{dx/dt} = \frac{3}{2}$

$\frac{d^2y}{dx^2} = 0$ Line

6. $x = \sqrt{t},\ y = 3t - 1$

$\frac{dy}{dx} = \frac{3}{1/(2\sqrt{t})} = 6\sqrt{t} = 6$ when $t = 1$.

$\frac{d^2y}{dx^2} = \frac{3/\sqrt{t}}{1/(2\sqrt{t})} = 6$ concave upwards

7. $x = t + 1,\ y = t^2 + 3t$

$\frac{dy}{dx} = \frac{2t + 3}{1} = 1$ when $t = -1$.

$\frac{d^2y}{dx^2} = 2$ concave upwards

8. $x = t^2 + 3t + 2, y = 2t$

$\frac{dy}{dx} = \frac{2}{2t + 3} = \frac{2}{3}$ when $t = 0$.

$\frac{d^2y}{dx^2} = \frac{-2(2)/(2t + 3)^2}{2t + 3} = \frac{-4}{(2t + 3)^3} = \frac{-4}{27}$ when $t = 0$.

concave downward

9. $x = 2\cos\theta,\ y = 2\sin\theta$

$\frac{dy}{dx} = \frac{2\cos\theta}{-2\sin\theta} = -\cot\theta = -1$ when $\theta = \frac{\pi}{4}$.

$\frac{d^2y}{dx^2} = \frac{\csc^2\theta}{-2\sin\theta} = \frac{-\csc^3\theta}{2} = -\sqrt{2}$ when $\theta = \frac{\pi}{4}$.

concave downward

10. $x = \cos\theta,\ y = 3\sin\theta$

$\frac{dy}{dx} = \frac{3\cos\theta}{-\sin\theta} = -3\cot\theta \cdot \frac{dy}{dx}$ is undefined when $\theta = 0$.

$\frac{d^2y}{dx^2} = \frac{3\csc^2\theta}{-\sin\theta} = \frac{-3}{\sin^3\theta} \cdot \frac{d^2y}{dx^2}$ is undefined when $\theta = 0$.

11. $x = 2 + \sec\theta,\ y = 1 + 2\tan\theta$

$\frac{dy}{dx} = \frac{2\sec^2\theta}{\sec\theta\tan\theta}$

$= \frac{2\sec\theta}{\tan\theta} = 2\csc\theta = 4$ when $\theta = \frac{\pi}{6}$.

$\frac{d^2y}{dx^2} = \frac{d\left[\frac{dy}{dx}\right]}{\frac{dx}{d\theta}} = \frac{-2\csc\theta\cot\theta}{\sec\theta\tan\theta}$

$= -2\cot^3\theta = -6\sqrt{3}$ when $\theta = \frac{\pi}{6}$.

concave downward

12. $x = \sqrt{t},\ y = \sqrt{t - 1}$

$\frac{dy}{dx} = \frac{1/(2\sqrt{t - 1})}{1/(2\sqrt{t})}$

$= \frac{\sqrt{t}}{\sqrt{t - 1}} = \sqrt{2}$ when $t = 2$.

$\frac{d^2y}{dx^2} = \frac{\left[\sqrt{t - 1}/(2\sqrt{t}) - \sqrt{t}(1/2\sqrt{t - 1})\right]/(t - 1)}{1/(2\sqrt{t})}$

$= \frac{-1}{(t - 1)^{3/2}} = -1$ when $t = 2$.

concave downward

13. $x = \cos^3 \theta,\ y = \sin^3 \theta$

$$\frac{dy}{dx} = \frac{3 \sin^2 \theta \cos \theta}{-3 \cos^2 \theta \sin \theta}$$

$$= -\tan \theta = -1 \text{ when } \theta = \frac{\pi}{4}.$$

$$\frac{d^2y}{dx^2} = \frac{-\sec^2 \theta}{-3 \cos^2 \theta \sin \theta} = \frac{1}{3 \cos^4 \theta \sin \theta}$$

$$= \frac{\sec^4 \theta \csc \theta}{3} = \frac{4\sqrt{2}}{3} \text{ when } \theta = \frac{\pi}{4}.$$

concave upward

14. $x = \theta - \sin \theta,\ y = 1 - \cos \theta$

$$\frac{dy}{dx} = \frac{\sin \theta}{1 - \cos \theta} = 0 \text{ when } \theta = \pi.$$

$$\frac{d^2y}{dx^2} = \frac{\dfrac{[(1 - \cos \theta) \cos \theta - \sin^2 \theta]}{(1 - \cos \theta)^2}}{(1 - \cos \theta)}$$

$$= \frac{-1}{(1 - \cos \theta)^2} = -\frac{1}{4} \text{ when } \theta = \pi.$$

concave downward

15. $x = 2 \cot \theta, y = 2 \sin^2 \theta$

$$\frac{dy}{dx} = \frac{4 \sin \theta \cos \theta}{-2 \csc^2 \theta} = -2 \sin^3 \theta \cos \theta$$

At $\left(-\frac{2}{\sqrt{3}}, \frac{3}{2}\right)$, $\theta = \frac{2\pi}{3}$, and $\frac{dy}{dx} = \frac{3\sqrt{3}}{8}$.

Tangent line: $y - \frac{3}{2} = \frac{3\sqrt{3}}{8}\left(x + \frac{2}{\sqrt{3}}\right)$

$$3\sqrt{3}x - 8y + 18 = 0$$

At $(0, 2)$, $\theta = \frac{\pi}{2}$, and $\frac{dy}{dx} = 0$.

Tangent line: $y - 2 = 0$

At $\left(2\sqrt{3}, \frac{1}{2}\right)$, $\theta = \frac{\pi}{6}$, and $\frac{dy}{dx} = -\frac{\sqrt{3}}{8}$.

Tangent line: $y - \frac{1}{2} = -\frac{\sqrt{3}}{8}(x - 2\sqrt{3})$

$$\sqrt{3}x + 8y - 10 = 0$$

16. $x = 2 - 3 \cos \theta, y = 3 + 2 \sin \theta$

$$\frac{dy}{dx} = \frac{2 \cos \theta}{3 \sin \theta} = \frac{2}{3} \cot \theta$$

At $(-1, 3)$, $\theta = 0$, and $\frac{dy}{dx}$ is undefined.

Tangent line: $x = -1$

At $(2, 5)$, $\theta = \frac{\pi}{2}$, and $\frac{dy}{dx} = 0$.

Tangent line: $y = 5$

At $\left(\frac{4 + 3\sqrt{3}}{2}, 2\right)$, $\theta = \frac{7\pi}{6}$, and $\frac{dy}{dx} = \frac{2\sqrt{3}}{3}$.

Tangent line:

$$y - 2 = \frac{2\sqrt{3}}{3}\left(x - \frac{4 + 3\sqrt{3}}{2}\right)$$

$$2\sqrt{3}x - 3y - 4\sqrt{3} - 3 = 0$$

17. $x = 2t,\ y = t^2 - 1,\ t = 2$

(a)

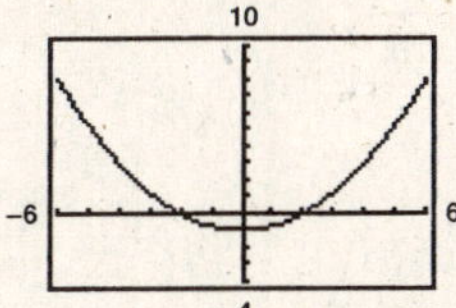

(b) At $t = 2$, $(x, y) = (4, 3)$, and

$$\frac{dx}{dt} = 2,\ \frac{dy}{dt} = 4,\ \frac{dy}{dx} = 2.$$

(c) $\frac{dy}{dx} = 2$. At $(4, 3)$, $y - 3 = 2(x - 4)$

$$y = 2x - 5.$$

(d)

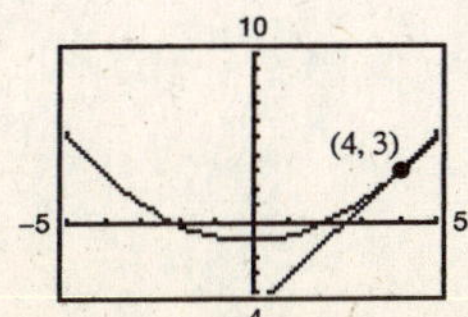

18. $x = t - 1,\ y = \frac{1}{t} + 1,\ t = 1$

(a)

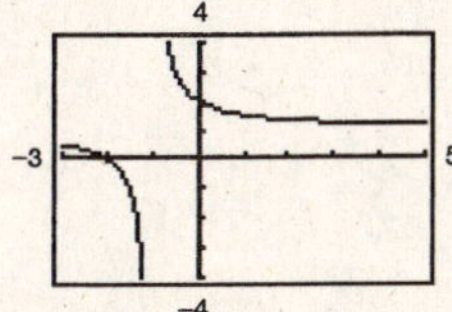

(b) At $t = 1$, $(x, y) = (0, 2)$, and

$$\frac{dx}{dt} = 1,\ \frac{dy}{dt} = -1,\ \frac{dy}{dx} = -1.$$

(c) $\frac{dy}{dx} = -1$. At $(0, 2)$, $y - 2 = -1(x - 0)$

$$y = -x + 2.$$

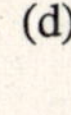

(d)

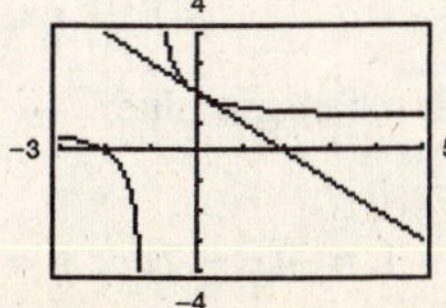

19. $x = t^2 - t + 2,\ y = t^3 - 3t,\ t = -1$

(a)

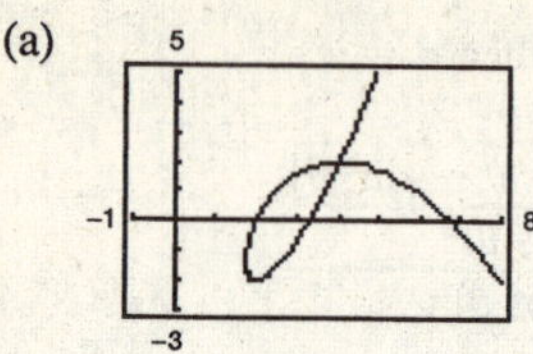

(b) At $t = -1$, $(x, y) = (4, 2)$, and

$$\frac{dx}{dt} = -3,\ \frac{dy}{dt} = 0,\ \frac{dy}{dx} = 0.$$

(c) $\frac{dy}{dx} = 0$. At $(4, 2)$, $y - 2 = 0(x - 4)$

$$y = 2$$

(d)

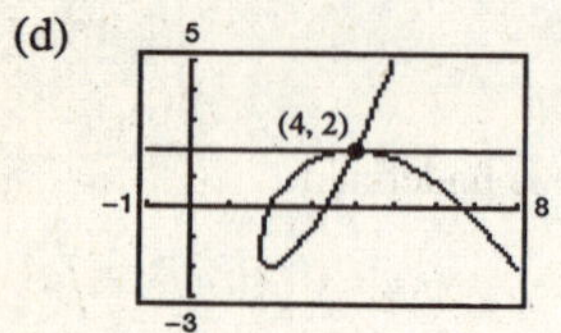

20. $x = 4\cos\theta,\ y = 3\sin\theta,\ \theta = \frac{3\pi}{4}$

(a)

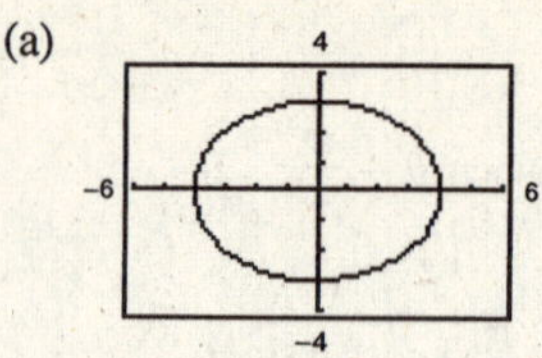

(b) At $\theta = \frac{3\pi}{4}$, $(x, y) = \left(\frac{-4}{\sqrt{2}}, \frac{3}{\sqrt{2}}\right)$, and

$$\frac{dx}{dt} = -2\sqrt{2},\ \frac{dy}{dt} = -\frac{3\sqrt{2}}{2},\ \frac{dy}{dx} = \frac{3}{4}.$$

(c) $\frac{dy}{dx} = \frac{3}{4}$. At $\left(\frac{-4}{\sqrt{2}}, \frac{3}{\sqrt{2}}\right)$, $y - \frac{3}{\sqrt{2}} = \frac{3}{4}\left(x + \frac{4}{\sqrt{2}}\right)$

$$y = \frac{3}{4}x + 3\sqrt{2}.$$

(d)

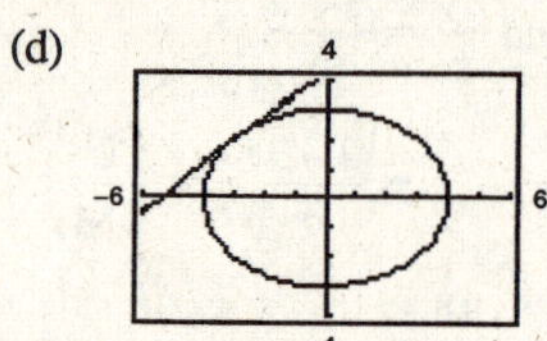

21. $x = 2\sin 2t, y = 3\sin t$ crosses itself at the origin, $(x, y) = (0, 0)$.

At this point, $t = 0$ or $t = \pi$.

$$\frac{dy}{dx} = \frac{3\cos t}{4\cos 2t}$$

At $t = 0$: $\frac{dy}{dx} = \frac{3}{4}$ and $y = \frac{3}{4}x$. Tangent Line

At $t = \pi$, $\frac{dy}{dx} = -\frac{3}{4}$ and $y = \frac{-3}{4}x$. Tangent Line

22. $x = 2 - \pi\cos t,\ y = 2t - \pi\sin t$ crosses itself at a point on the x-axis: $(2, 0)$. The corresponding t-values are $t = \pm\pi/2$.

$$\frac{dy}{dt} = 2 - \pi\cos t,\ \frac{dx}{dt} = \pi\sin t,\ \frac{dy}{dx} = \frac{2 - \pi\cos t}{\pi\sin t}$$

At $t = \frac{\pi}{2}$: $\frac{dy}{dx} = \frac{2}{\pi}$.

Tangent line: $y - 0 = \frac{2}{\pi}(x - 2)$

$$y = \frac{2}{\pi}x - \frac{4}{\pi}$$

At $t = -\frac{\pi}{2}$: $\frac{dy}{dx} = -\frac{2}{\pi}$.

Tangent line: $y - 0 = -\frac{2}{\pi}(x - 2)$

$$y = -\frac{2}{\pi}x + \frac{4}{\pi}$$

23. $x = t^2 - t, y = t^3 - 3t - 1$ crosses itself at the point $(x, y) = (2, 1)$.

At this point, $t = -1$ or $t = 2$.

$$\frac{dy}{dx} = \frac{3t^2 - 3}{2t - 1}$$

At $t = -1$, $\frac{dy}{dx} = 0$ and $y = 1$. Tangent Line

At $t = 2$, $\frac{dy}{dt} = \frac{9}{3} = 3$ and $y - 1 = 3(x - 2)$ or $y = 3x - 5$. Tangent Line

24. $x = t^3 - 6t$, $y = t^2$ crosses itself at $(0, 6)$. The corresponding t-values are $t = \pm\sqrt{6}$.

$$\frac{dy}{dx} = \frac{2t}{3t^2 - 6}$$

At $t = \sqrt{6}$, $\dfrac{dy}{dx} = \dfrac{2\sqrt{6}}{12} = \dfrac{\sqrt{6}}{6}$.

Tangent line: $y - 6 = \dfrac{\sqrt{6}}{6}(x - 0)$

$$y = \frac{\sqrt{6}}{6}x + 6$$

At $t = -\sqrt{6}$, $\dfrac{dy}{dx} = -\dfrac{2\sqrt{6}}{12} = -\dfrac{\sqrt{6}}{6}$.

Tangent line: $y = -\dfrac{\sqrt{6}}{6}x + 6$

25. $x = \cos\theta + \theta\sin\theta$, $y = \sin\theta - \theta\cos\theta$

Horizontal tangents: $\dfrac{dy}{d\theta} = \theta\sin\theta = 0$ when $\theta = \pm\pi, \pm 2\pi, \pm 3\pi, \ldots$

Points: $(-1, [2n - 1]\pi)$, $(1, 2n\pi)$ where n is an integer. Points shown: $(1, 0)$, $(-1, \pi)$, $(1, -2\pi)$

Vertical tangents: $\dfrac{dx}{d\theta} = \theta\cos\theta = 0$ when $\theta = \pm\dfrac{\pi}{2}, \pm\dfrac{3\pi}{2}, \pm\dfrac{5\pi}{2}, \ldots$

Note: $\theta = 0$ corresponds to the cusp at $(x, y) = (1, 0)$.

$$\frac{dy}{dx} = \frac{\theta\sin\theta}{\theta\cos\theta} = \tan\theta = 0 \text{ at } \theta = 0$$

Points: $\left(\dfrac{(-1)^{n+1}(2n - 1)\pi}{2}, (-1)^{n+1}\right)$ Points shown: $\left(\dfrac{\pi}{2}, 1\right)$, $\left(-\dfrac{3\pi}{2}, -1\right)$, $\left(\dfrac{5\pi}{2}, 1\right)$

26. $x = 2\theta$, $y = 2(1 - \cos\theta)$

Horizontal tangents: $\dfrac{dy}{d\theta} = 2\sin\theta = 0$ when $\theta = 0, \pm\pi, \pm 2\pi, \ldots$

Points: $(4n\pi, 0)$, $(2[2n - 1]\pi, 4)$ where n is an integer

Points shown: $(0, 0)$, $(2\pi, 4)$, $(4\pi, 0)$

Vertical tangents: $\dfrac{dx}{d\theta} = 2 \neq 0$; none

27. $x = 1 - t$, $y = t^2$

Horizontal tangents: $\dfrac{dy}{dt} = 2t = 0$ when $t = 0$

Point: $(1, 0)$

Vertical tangents: $\dfrac{dx}{dt} = -1 \neq 0$; none

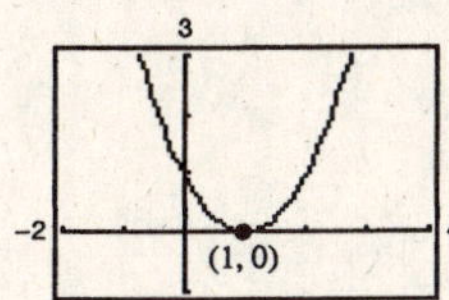

28. $x = t + 1$, $y = t^2 + 3t$

Horizontal tangents: $\dfrac{dy}{dt} = 2t + 3 = 0$ when $t = -\dfrac{3}{2}$

Point: $\left(-\dfrac{1}{2}, -\dfrac{9}{4}\right)$

Vertical tangents: $\dfrac{dx}{dt} = 1 \neq 0$; none

29. $x = 1 - t,\ y = t^3 - 3t$

Horizontal tangents: $\frac{dy}{dt} = 3t^2 - 3 = 0$ when $t = \pm 1$.

Points: $(0, -2), (2, 2)$

Vertical tangents: $\frac{dx}{dt} = -1 \neq 0$; none

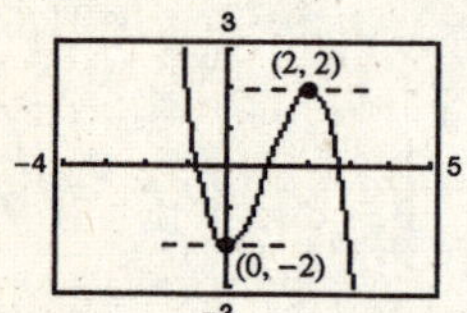

30. $x = t^2 - t + 2,\ y = t^3 - 3t$

Horizontal tangents: $\frac{dy}{dt} = 3t^2 - 3 = 0$ when $t = \pm 1$.

Points: $(2, -2), (4, 2)$

Vertical tangents: $\frac{dx}{dt} = 2t - 1 = 0$ when $t = \frac{1}{2}$.

Point: $\left(\frac{7}{4}, -\frac{11}{8}\right)$

31. $x = 3\cos\theta,\ y = 3\sin\theta$

Horizontal tangents: $\frac{dy}{d\theta} = 3\cos\theta = 0$ when $\theta = \frac{\pi}{2}, \frac{3\pi}{2}$.

Points: $(0, 3), (0, -3)$

Vertical tangents: $\frac{dx}{d\theta} = -3\sin\theta = 0$ when $\theta = 0, \pi$.

Points: $(3, 0), (-3, 0)$

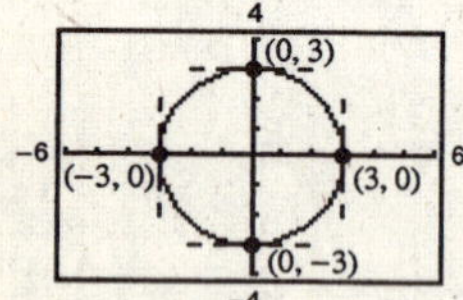

32. $x = \cos\theta,\ y = 2\sin 2\theta$

Horizontal tangents: $\frac{dy}{d\theta} = 4\cos 2\theta = 0$ when

$\theta = \frac{\pi}{4}, \frac{3\pi}{4}, \frac{5\pi}{4}, \frac{7\pi}{4}$.

Points: $\left(\frac{\sqrt{2}}{2}, 2\right), \left(-\frac{\sqrt{2}}{2}, -2\right), \left(-\frac{\sqrt{2}}{2}, 2\right), \left(\frac{\sqrt{2}}{2}, -2\right)$

Vertical tangents: $\frac{dx}{d\theta} = -\sin\theta = 0$ when $\theta = 0, \pi$.

Points: $(1, 0), (-1, 0)$

33. $x = 4 + 2\cos\theta,\ y = -1 + \sin\theta$

Horizontal tangents: $\frac{dy}{d\theta} = \cos\theta = 0$ when $\theta = \frac{\pi}{2}, \frac{3\pi}{2}$.

Points: $(4, 0), (4, -2)$

Vertical tangents: $\frac{dx}{d\theta} = -2\sin\theta = 0$ when $\theta = 0, \pi$.

Points: $(6, -1), (2, -1)$

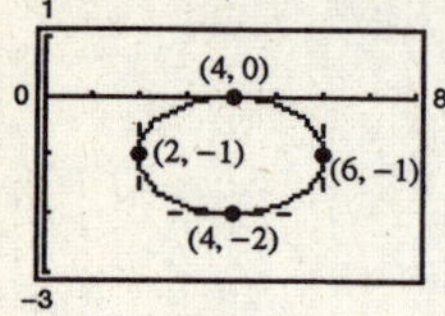

34. $x = 4\cos^2\theta,\ y = 2\sin\theta$

Horizontal tangents: $\frac{dy}{d\theta} = 2\cos\theta = 0$ when $\theta = \frac{\pi}{2}, \frac{3\pi}{2}$.

Since $dx/d\theta = 0$ at $\pi/2$ and $3\pi/2$, exclude them.

Vertical tangents: $\frac{dx}{d\theta} = -8\cos\theta\sin\theta = 0$ when

$\theta = 0, \pi$.

Point: $(4, 0)$

35. $x = \sec\theta,\ y = \tan\theta$

Horizontal tangents: $\frac{dy}{d\theta} = \sec^2\theta \neq 0$; none

Vertical tangents: $\frac{dx}{d\theta} = \sec\theta\tan\theta = 0$ when $x = 0, \pi$.

Points: $(1, 0), (-1, 0)$

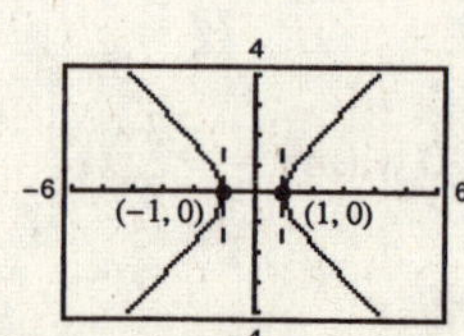

36. $x = \cos^2 \theta,\ y = \cos \theta$

Horizontal tangents: $\dfrac{dy}{d\theta} = -\sin \theta = 0$ when $x = 0, \pi$.

Since $dx/d\theta = 0$ at these values, exclude them.

Vertical tangents: $\dfrac{dx}{d\theta} = -2 \cos \theta \sin \theta = 0$ when

$\theta = \dfrac{\pi}{2}, \dfrac{3\pi}{2}$.

(Exclude $0, \pi$.)

Point: $(0, 0)$

37. $x = t^2, y = t^3 - t$

$$\frac{dy}{dx} = \frac{3t^2 - 1}{2t}$$

$$\frac{d^2y}{dx^2} = \left[\frac{\dfrac{2t(6t) - (3t^2 - 1)2}{4t^2}}{2t}\right]$$

$$= \frac{6t^2 + 2}{8t^3} = \frac{1 + 3t^2}{4t^3}$$

Concave upward for $t > 0$

Concave downward for $t < 0$

38. $x = 2 + t^2, y = t^2 + t^3$

$$\frac{dy}{dx} = \frac{2t + 3t^2}{2t} = 1 + \frac{3}{2}t$$

$$\frac{d^2y}{dx^2} = \frac{3/2}{2t} = \frac{3}{4t}$$

Concave upward for $t > 0$

Concave downward for $t < 0$

39. $x = 2t + \ln t, y = 2t - \ln t, t > 0$

$$\frac{dy}{dx} = \frac{2 - (1/t)}{2 + (1/t)} = \frac{2t - 1}{2t + 1}$$

$$\frac{d^2y}{dx^2} = \left[\frac{(2t + 1)2 - (2t - 1)2}{(2t + 1)^2}\right] \bigg/ \left(2 + \frac{1}{t}\right)$$

$$= \frac{4}{(2t + 1)^2} \cdot \frac{t}{2t + 1}$$

$$= \frac{4t}{(2t + 1)^3}$$

Because $t > 0$, $\dfrac{d^2y}{dx^2} > 0$

Concave upward for $t > 0$

40. $x = t^2, y = \ln t, t > 0$

$$\frac{dy}{dx} = \frac{1/t}{2t} = \frac{1}{2t^2}$$

$$\frac{d^2y}{dx^2} = -\frac{1/t^3}{2t} = -\frac{1}{2t^4}$$

Because $t > 0$, $\dfrac{d^2y}{dx^2} < 0$

Concave downward for $t > 0$

41. $x = \sin t, y = \cos t, 0 < t < \pi$

$$\frac{dy}{dx} = -\frac{\sin t}{\cos t} = -\tan t$$

$$\frac{d^2y}{dx^2} = -\frac{\sec^2 t}{\cos t} = -\frac{1}{\cos^3 t}$$

Concave upward on $\pi/2 < t < \pi$

Concave downward on $0 < t < \pi/2$

42. $x = 2 \cos t, y = \sin t, 0 < t < 2\pi$

$$\frac{dy}{dx} = \frac{-\cos t}{2 \sin t} = -\frac{1}{2} \cot t$$

$$\frac{d^2y}{dx^2} = \frac{(1/2) \csc^2 t}{-2 \sin t} = \frac{-1}{4 \sin^3 t}$$

Concave upward on $\pi < t < 2\pi$

Concave downward on $0 < t < \pi$

43. $x = 2t - t^2, y = 2t^{3/2}, 1 \le t \le 2$

$$\frac{dx}{dt} = 2 - 2t, \frac{dy}{dt} = 3t^{1/2}$$

$$s = \int_a^b \sqrt{\left(\frac{dx}{dt}\right)^2 + \left(\frac{dy}{dt}\right)^2}\, dt$$

$$= \int_1^2 \sqrt{(2 - 2t)^2 + (3t^{1/2})^2}\, dt$$

$$= \int_1^2 \sqrt{4 - 8t + 4t^2 + 9t}\, dt$$

$$= \int_1^2 \sqrt{4t^2 + t + 4}\, dt$$

44. $x = \ln t,\ y = t + 1,\ 1 \le t \le 6$

$$\frac{dx}{dt} = \frac{1}{t},\ \frac{dy}{dt} = 1$$

$$s = \int_a^b \sqrt{\left(\frac{dx}{dt}\right)^2 + \left(\frac{dy}{dt}\right)^2}\,dt$$

$$= \int_1^6 \sqrt{\frac{1}{t^2} + 1}\,dt$$

45. $x = e^t + 2,\ y = 2t + 1,\ -2 \le t \le 2$

$$\frac{dx}{dt} = e^t,\ \frac{dy}{dt} = 2$$

$$s = \int_a^b \sqrt{\left(\frac{dx}{dt}\right)^2 + \left(\frac{dy}{dt}\right)^2}\,dt$$

$$= \int_{-2}^2 \sqrt{e^{2t} + 4}\,dt$$

46. $x = t + \sin t,\ y = t - \cos t,\ 0 \le t \le \pi$

$$\frac{dx}{dt} = 1 + \cos t,\ \frac{dy}{dt} = 1 + \sin t$$

$$s = \int_a^b \sqrt{\left(\frac{dx}{dt}\right)^2 + \left(\frac{dy}{dt}\right)^2}\,dt$$

$$= \int_0^\pi \sqrt{(1 + \cos t)^2 + (1 + \sin t)^2}\,dt$$

$$= \int_0^\pi \sqrt{3 + 2\cos t + 2\sin t}\,dt$$

47. $x = t^2,\ y = 2t,\ 0 \le t \le 2$

$$\frac{dx}{dt} = 2t,\ \frac{dy}{dt} = 2,\ \left(\frac{dx}{dt}\right)^2 + \left(\frac{dy}{dt}\right)^2 = 4t^2 + 4 = 4(t^2 + 1)$$

$$s = 2\int_0^2 \sqrt{t^2 + 1}\,dt$$

$$= \left[t\sqrt{t^2 + 1} + \ln\left|t + \sqrt{t^2 + 1}\right|\right]_0^2$$

$$= 2\sqrt{5} + \ln\left(2 + \sqrt{5}\right) \approx 5.916$$

48. $x = t^2 + 1,\ y = 4t^3 + 3,\ -1 \le t \le 0$

$$\frac{dx}{dt} = 2t,\ \frac{dy}{dt} = 12t^2,\ \left(\frac{dx}{dt}\right)^2 + \left(\frac{dy}{dt}\right)^2 = 4t^2 + 144t^4$$

$$s = \int_{-1}^0 \sqrt{4t^2 + 144t^4}\,dt = \int_{-1}^0 -2t\sqrt{1 + 36t^2}\,dt$$

$$= \left[\frac{-(1 + 36t^2)^{3/2}}{54}\right]_{-1}^0 = \frac{-1}{54}(1 - 37^{3/2}) \approx 4.149$$

49. $x = e^{-t}\cos t,\ y = e^{-t}\sin t,\ 0 \le t \le \frac{\pi}{2}$

$$\frac{dx}{dt} = -e^{-t}(\sin t + \cos t),\ \frac{dy}{dt} = e^{-t}(\cos t - \sin t)$$

$$s = \int_0^{\pi/2} \sqrt{\left(\frac{dx}{dt}\right)^2 + \left(\frac{dy}{dt}\right)^2}\,dt$$

$$= \int_0^{\pi/2} \sqrt{2e^{-2t}}\,dt = -\sqrt{2}\int_0^{\pi/2} e^{-t}(-1)\,dt$$

$$= \left[-\sqrt{2}e^{-t}\right]_0^{\pi/2} = \sqrt{2}(1 - e^{-\pi/2}) \approx 1.12$$

50. $x = \arcsin t,\ y = \ln\sqrt{1 - t^2},\ 0 \le t \le \frac{1}{2}$

$$\frac{dx}{dt} = \frac{1}{\sqrt{1 - t^2}},\ \frac{dy}{dt} = \frac{1}{2}\left(\frac{-2t}{1 - t^2}\right) = \frac{t}{1 - t^2}$$

$$s = \int_0^{1/2} \sqrt{\left(\frac{dx}{dt}\right)^2 + \left(\frac{dy}{dt}\right)^2}\,dt$$

$$= \int_0^{1/2} \sqrt{\frac{1}{(1 - t^2)^2}}\,dt = \int_0^{1/2} \frac{1}{1 - t^2}\,dt$$

$$= \left[-\frac{1}{2}\ln\left|\frac{t - 1}{t + 1}\right|\right]_0^{1/2}$$

$$= -\frac{1}{2}\ln\left(\frac{1}{3}\right) = \frac{1}{2}\ln(3) \approx 0.549$$

51. $x = \sqrt{t},\ y = 3t - 1,\ \frac{dx}{dt} = \frac{1}{2\sqrt{t}},\ \frac{dy}{dt} = 3$

$$s = \int_0^1 \sqrt{\frac{1}{4t} + 9}\,dt = \frac{1}{2}\int_0^1 \frac{\sqrt{1 + 36t}}{\sqrt{t}}\,dt$$

$$= \frac{1}{6}\int_0^6 \sqrt{1 + u^2}\,du$$

$$= \frac{1}{12}\left[\ln\left(\sqrt{1 + u^2} + u\right) + u\sqrt{1 + u^2}\right]_0^6$$

$$= \frac{1}{12}\left[\ln\left(\sqrt{37} + 6\right) + 6\sqrt{37}\right] \approx 3.249$$

$$u = 6\sqrt{t},\ du = \frac{3}{\sqrt{t}}\,dt$$

52. $x = t,\ y = \dfrac{t^5}{10} + \dfrac{1}{6t^3},\ \dfrac{dx}{dt} = 1,\ \dfrac{dy}{dt} = \dfrac{t^4}{2} - \dfrac{1}{2t^4}$

$$S = \int_1^2 \sqrt{1 + \left(\frac{t^4}{2} - \frac{1}{2t^4}\right)^2}\,dt =$$

$$= \int_1^2 \sqrt{\left(\frac{t^4}{2} + \frac{1}{2t^4}\right)^2}\,dt$$

$$= \int_1^2 \left(\frac{t^4}{2} + \frac{1}{2t^4}\right) dt$$

$$= \left[\frac{t^5}{10} - \frac{1}{6t^3}\right]_1^2 = \frac{779}{240}$$

53. $x = a\cos^3\theta,\ y = a\sin^3\theta,\ \dfrac{dx}{d\theta} = -3a\cos^2\theta\sin\theta,$

$$\frac{dy}{d\theta} = 3a\sin^2\theta\cos\theta$$

$$s = 4\int_0^{\pi/2} \sqrt{9a^2\cos^4\theta\sin^2\theta + 9a^2\sin^4\theta\cos^2\theta}\,d\theta$$

$$= 12a\int_0^{\pi/2} \sin\theta\cos\theta\sqrt{\cos^2\theta + \sin^2\theta}\,d\theta$$

$$= 6a\int_0^{\pi/2} \sin 2\theta\,d\theta = \Big[-3a\cos 2\theta\Big]_0^{\pi/2} = 6a$$

54. $x = a\cos\theta,\ y = a\sin\theta,\ \dfrac{dx}{d\theta} = -a\sin\theta,\ \dfrac{dy}{d\theta} = a\cos\theta$

$$S = 4\int_0^{\pi/2} \sqrt{a^2\sin^2\theta + a^2\cos^2\theta}\,d\theta$$

$$= 4a\int_0^{\pi/2} d\theta = \Big[4a\theta\Big]_0^{\pi/2} = 2\pi a$$

55. $x = a(\theta - \sin\theta),\ y = a(1 - \cos\theta),$

$$\frac{dx}{d\theta} = a(1 - \cos\theta),\ \frac{dy}{d\theta} = a\sin\theta$$

$$s = 2\int_0^{\pi} \sqrt{a^2(1 - \cos\theta)^2 + a^2\sin^2\theta}\,d\theta$$

$$= 2\sqrt{2}a\int_0^{\pi} \sqrt{1 - \cos\theta}\,d\theta$$

$$= 2\sqrt{2}a\int_0^{\pi} \frac{\sin\theta}{\sqrt{1 + \cos\theta}}\,d\theta$$

$$= \Big[-4\sqrt{2}a\sqrt{1 + \cos\theta}\Big]_0^{\pi} = 8a$$

56. $x = \cos\theta + \theta\sin\theta,\ y = \sin\theta - \theta\cos\theta,\ \dfrac{dx}{d\theta} = \theta\cos\theta$

$$\frac{dy}{d\theta} = \theta\sin\theta$$

$$S = \int_0^{2\pi} \sqrt{\theta^2\cos^2\theta + \theta^2\sin^2\theta}\,d\theta$$

$$= \int_0^{2\pi} \theta\,d\theta = \left[\frac{\theta^2}{2}\right]_0^{2\pi} = 2\pi^2$$

57. $x = (90\cos 30°)t,\ y = (90\sin 30°)t - 16t^2$

(a)

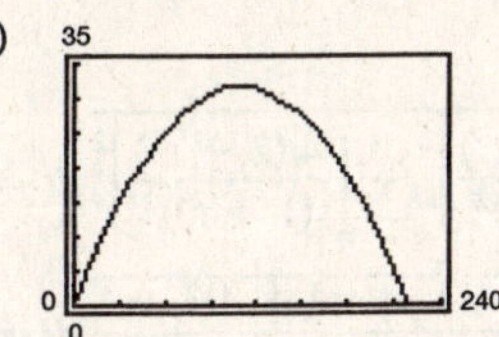

(b) Range: 219.2 ft, $\left(t = \dfrac{45}{16}\right)$

(c) $\dfrac{dx}{dt} = 90\cos 30°,\ \dfrac{dy}{dt} = 90\sin 30° - 32t$

$y = 0$ for $t = \dfrac{45}{16}$.

$$s = \int_0^{45/16} \sqrt{(90\cos 30°)^2 + (90\sin 30° - 32t)^2}\,dt$$

$$\approx 230.8 \text{ ft}$$

58. $y = 0 \Rightarrow (90 \sin \theta)t = 16t^2 \Rightarrow t = 0, \dfrac{90}{16} \sin \theta$

$$x = (90 \cos \theta)t = (90 \cos \theta)\frac{90}{16} \sin \theta$$

$$= \frac{90^2}{16} \sin \theta \cos \theta = \frac{90^2}{32} \sin 2\theta$$

$$x'(\theta) = \frac{90^2}{32} 2 \cos 2\theta = 0 \Rightarrow \theta = \frac{\pi}{4}$$

By the First Derivative Test, $\theta = \dfrac{\pi}{4}(45°)$ maximizes the range ($x = 253.125$ feet).

To maximize the arc length, we have $\dfrac{dx}{dt} = 90 \cos \theta$, $\dfrac{dy}{dt} = 90 \sin \theta - 32t$.

$$s = \int_0^{(90/16)\sin \theta} \sqrt{(90 \cos \theta)^2 + (90 \sin \theta - 32t)^2}\, dt$$

$$= \frac{2025}{8} \sin \theta + \frac{2025}{16} \cos^2 \theta \ln\left[\frac{1 + \sin \theta}{1 - \sin \theta}\right]$$

Using a graphing utility, we see that s is a maximum of approximately 303.67 feet at $\theta \approx 0.9855(56.5°)$.

59. $x = \dfrac{4t}{1 + t^3}$, $y = \dfrac{4t^2}{1 + t^3}$

(a) $x^3 + y^3 = 4xy$

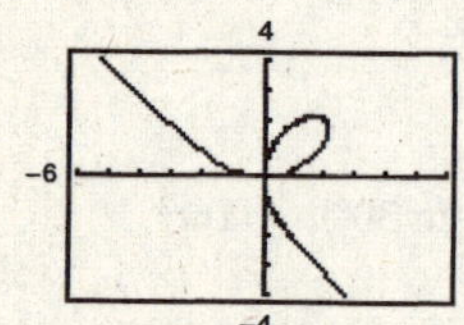

(b) $$\frac{dy}{dt} = \frac{(1 + t^3)(8t) - 4t^2(3t^2)}{(1 + t^3)^2}$$

$$= \frac{4t(2 - t^3)}{(1 + t^3)^2} = 0 \text{ when } t = 0 \text{ or } t = \sqrt[3]{2}.$$

Points: $(0, 0)$, $\left(\dfrac{4\sqrt[3]{2}}{3}, \dfrac{4\sqrt[3]{4}}{3}\right) \approx (1.6799, 2.1165)$

(c) $$s = 2\int_0^1 \sqrt{\left[\frac{4(1 - 2t^3)}{(1 + t^3)^2}\right]^2 + \left[\frac{4t(2 - t^3)}{(1 + t^3)^2}\right]^2}\, dt = 2\int_0^1 \sqrt{\frac{16}{(1 + t^3)^4}[t^8 + 4t^6 - 4t^5 - 4t^3 + 4t^2 + 1]}\, dt$$

$$= 8\int_0^1 \frac{\sqrt{t^8 + 4t^6 - 4t^5 - 4t^3 + 4t^2 + 1}}{(1 + t^3)^2}\, dt \approx 6.557$$

60. $x = 4 \cot \theta = \dfrac{4}{\tan \theta}$, $y = 4 \sin^2 \theta$, $-\dfrac{\pi}{2} \le \theta \le \dfrac{\pi}{2}$

(a)

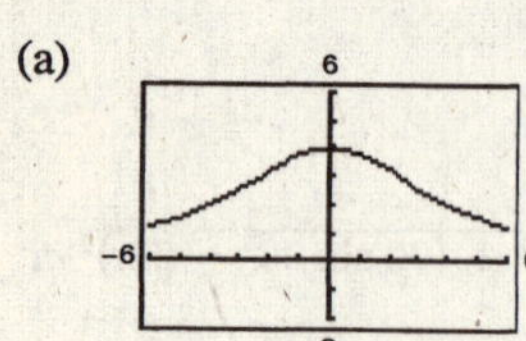

(b) $$\frac{dy}{d\theta} = 8 \sin \theta \cdot \cos \theta$$

$$\frac{dx}{d\theta} = -4 \csc^2 \theta$$

$$\frac{dy}{d\theta} = 0 \text{ for } \theta = 0, \pm\frac{\pi}{2}$$

Horizontal tangent at $(x, y) = (0, 4)\left(\theta = \pm\dfrac{\pi}{2}\right)$

(Function is not defined at $\theta = 0$)

(c) Arc length over $\dfrac{\pi}{4} \le t \le \dfrac{\pi}{2}$: 4.5183

61. (a) $x = t - \sin t$ $\quad$ $x = 2t - \sin(2t)$

$y = 1 - \cos t$ $\quad$ $y = 1 - \cos(2t)$

$0 \le t \le 2\pi$ $\quad$ $0 \le t \le \pi$

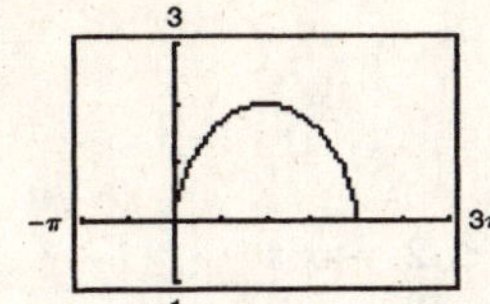

(b) The average speed of the particle on the second path is twice the average speed of a particle on the first path.

(c) $x = \frac{1}{2}t - \sin\left(\frac{1}{2}t\right)$

$y = 1 - \cos\left(\frac{1}{2}t\right)$

The time required for the particle to traverse the same path is $t = 4\pi$.

62. (a) First particle: $x = 3\cos t$, $y = 4\sin t, 0 \le t \le 2\pi$

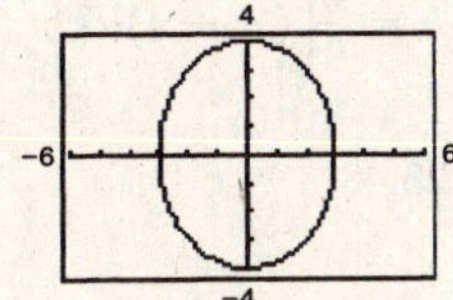

Second particle: $x = 4\sin t$, $y = 3\cos t$, $0 \le t \le 2\pi$

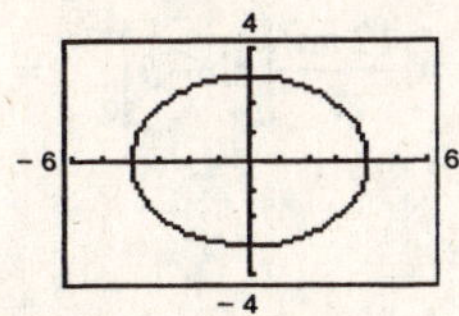

(b) There are 4 points of intersection.

(c) Suppose at time t that

$3\cos t = 4\sin t$ and $4\sin t = 3\cos t$

$\tan t = \frac{3}{4}$ and $\tan t = \frac{3}{4}$

Yes, the particles are at the same place at the same time for $\tan t = \frac{3}{4}$. $t \approx 0.6435, 3.7851$. The intersection points are $(2.4, 2.4)$ and $(-2.4, -2.4)$

(d) The curves intersect twice, but not at the same time.

63. $x = 4t, \dfrac{dx}{dt} = 4$

$y = t + 1, \dfrac{dy}{dt} = 1$

$$S = 2\pi\int_0^2 (t+1)\sqrt{4^2 + 1^2}\,dt$$

$$= 2\pi\int_0^2 \sqrt{17}(t+1)\,dt = 8\pi\sqrt{17}$$

$$\approx 103.6249$$

64. $x = \dfrac{1}{4}t^2, \dfrac{dx}{dt} = \dfrac{1}{2}t$

$y = t + 2, \dfrac{dy}{dt} = 1$

$$S = \pi\int_0^4 (t+2)\sqrt{\frac{t^2}{4} + 1}\,dt$$

$$= \pi\int_0^4 (t+2)\sqrt{t^2 + 4}\,dt$$

$$\approx 159.6264$$

65. $x = \cos^2\theta, \dfrac{dx}{d\theta} = -2\cos\theta\sin\theta$

$y = \cos\theta, \dfrac{dy}{d\theta} = -\sin\theta$

$$S = 2\pi\int_0^{\pi/2} \cos\theta\sqrt{4\cos^2\theta\sin^2\theta + \sin^2\theta}\,d\theta$$

$$\approx 5.3304$$

66. $x = \theta + \sin\theta, \dfrac{dx}{d\theta} = 1 + \cos\theta$

$y = \theta + \cos\theta, \dfrac{dy}{d\theta} = 1 - \sin\theta$

$$S = 2\pi\int_0^{\pi/2} (\theta + \cos\theta)\sqrt{(1+\cos\theta)^2 + (1-\sin\theta)^2}\,d\theta$$

$$= 2\pi\int_0^{\pi/2} (\theta + \cos\theta)\sqrt{3 + 2\cos\theta - 2\sin\theta}\,d\theta$$

$$\approx 23.2433$$

67. $x = t,\ y = 2t,\ \dfrac{dx}{dt} = 1,\ \dfrac{dy}{dt} = 2$

(a) $S = 2\pi\displaystyle\int_0^4 2t\sqrt{1+4}\,dt = 4\sqrt{5}\pi\int_0^4 t\,dt$

$= \left[2\sqrt{5}\pi t^2\right]_0^4 = 32\pi\sqrt{5}$

(b) $S = 2\pi\displaystyle\int_0^4 t\sqrt{1+4}\,dt = 2\sqrt{5}\pi\int_0^4 t\,dt$

$= \left[\sqrt{5}\pi t^2\right]_0^4 = 16\pi\sqrt{5}$

68. $x = t,\ y = 4 - 2t,\ \dfrac{dx}{dt} = 1,\ \dfrac{dy}{dt} = -2$

(a) $S = 2\pi \displaystyle\int_0^2 (4 - 2t)\sqrt{1 + 4}\, dt$

$= \left[2\sqrt{5}\pi(4t - t^2)\right]_0^2 = 8\pi\sqrt{5}$

(b) $S = 2\pi \displaystyle\int_0^2 t\sqrt{1 + 4}\, dt = \left[\sqrt{5}\pi t^2\right]_0^2 = 4\pi\sqrt{5}$

69. $x = 4\cos\theta,\ y = 4\sin\theta,\ \dfrac{dx}{d\theta} = -4\sin\theta,\ \dfrac{dy}{d\theta} = 4\cos\theta$

$$S = 2\pi \int_0^{\pi/2} 4\cos\theta\sqrt{(-4\sin\theta)^2 + (4\cos\theta)^2}\, d\theta$$

$$= 32\pi \int_0^{\pi/2} \cos\theta\, d\theta = \left[32\pi\sin\theta\right]_0^{\pi/2} = 32\pi$$

70. $x = \dfrac{1}{3}t^3,\ y = t + 1,\ 1 \le t \le 2$, y-axis

$\dfrac{dx}{dt} = t^2,\ \dfrac{dy}{dt} = 1$

$$S = 2\pi \int_1^2 \frac{1}{3}t^3\sqrt{t^4 + 1}\, dt = \frac{\pi}{9}\Big[(x^4 + 1)^{3/2}\Big]_1^2$$

$$= \frac{\pi}{9}(17^{3/2} - 2^{3/2}) \approx 23.48$$

71. $x = a\cos^3\theta,\ y = a\sin^3\theta,\ \dfrac{dx}{d\theta} = -3a\cos^2\theta\sin\theta,\ \dfrac{dy}{d\theta} = 3a\sin^2\theta\cos\theta$

$$S = 4\pi\int_0^{\pi/2} a\sin^3\theta\sqrt{9a^2\cos^4\theta\sin^2\theta + 9a^2\sin^4\theta\cos^2\theta}\, d\theta = 12a^2\pi\int_0^{\pi/2}\sin^4\theta\cos\theta\, d\theta = \frac{12\pi a^2}{5}\Big[\sin^5\theta\Big]_0^{\pi/2} = \frac{12}{5}\pi a^2$$

72. $x = a\cos\theta,\ y = b\sin\theta,\ \dfrac{dx}{d\theta} = -a\sin\theta,\ \dfrac{dy}{d\theta} = b\cos\theta$

(a) $S = 4\pi\displaystyle\int_0^{\pi/2} b\sin\theta\sqrt{a^2\sin^2\theta + b^2\cos^2\theta}\, d\theta$

$$= 4\pi\int_0^{\pi/2} ab\sin\theta\sqrt{1 - \left(\frac{a^2 - b^2}{a^2}\right)\cos^2\theta}\, d\theta = \frac{-4ab\pi}{e}\int_0^{\pi/2}(-e\sin\theta)\sqrt{1 - e^2\cos^2\theta}\, d\theta$$

$$= \frac{-2ab\pi}{e}\Big[e\cos\theta\sqrt{1 - e^2\cos^2\theta} + \arcsin(e\cos\theta)\Big]_0^{\pi/2} = \frac{2ab\pi}{e}\Big[e\sqrt{1 - e^2} + \arcsin(e)\Big]$$

$$= 2\pi b^2 + \left(\frac{2\pi a^2 b}{\sqrt{a^2 - b^2}}\right)\arcsin\left(\frac{\sqrt{a^2 - b^2}}{a}\right) = 2\pi b^2 + 2\pi\left(\frac{ab}{e}\right)\arcsin(e)$$

$$\left(e = \frac{\sqrt{a^2 - b^2}}{a} = \frac{c}{a}\text{: eccentricity}\right)$$

(b) $S = 4\pi\displaystyle\int_0^{\pi/2} a\cos\theta\sqrt{a^2\sin^2\theta + b^2\cos^2\theta}\, d\theta$

$$= 4\pi\int_0^{\pi/2} a\cos\theta\sqrt{b^2 + c^2\sin^2\theta}\, d\theta = \frac{4a\pi}{c}\int_0^{\pi/2} c\cos\theta\sqrt{b^2 + c^2\sin^2\theta}\, d\theta$$

$$= \frac{2a\pi}{c}\Big[c\sin\theta\sqrt{b^2 + c^2\sin^2\theta} + b^2\ln\left|c\sin\theta + \sqrt{b^2 + c^2\sin^2\theta}\right|\Big]_0^{\pi/2}$$

$$= \frac{2a\pi}{c}\Big[c\sqrt{b^2 + c^2} + b^2\ln\left|c + \sqrt{b^2 + c^2}\right| - b^2\ln b\Big]$$

$$= 2\pi a^2 + \frac{2\pi ab^2}{\sqrt{a^2 - b^2}}\ln\left|\frac{a + \sqrt{a^2 - b^2}}{b}\right| = 2\pi a^2 + \left(\frac{\pi b^2}{e}\right)\ln\left|\frac{1 + e}{1 - e}\right|$$

73. $\dfrac{dy}{dx} = \dfrac{dy/dt}{dx/dt}$

See Theorem 10.7.

74. (a) 0

(b) 4

75. One possible answer is the graph given by

$$x = t,\ y = -t.$$

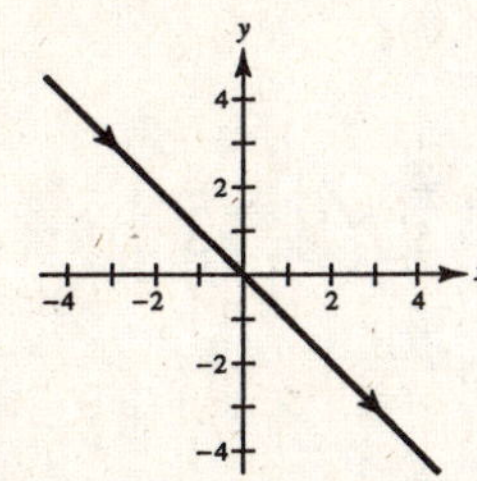

76. One possible answer is the graph given by

$$x = -t,\ y = -t.$$

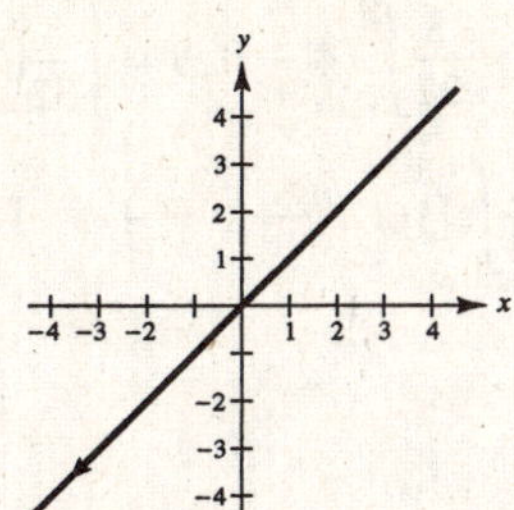

77. $s = \displaystyle\int_a^b \sqrt{\left(\frac{dx}{dt}\right)^2 + \left(\frac{dy}{dt}\right)^2}\,dt$

See Theorem 10.8.

78. (a) $S = 2\pi \displaystyle\int_a^b g(t)\sqrt{\left(\frac{dx}{dt}\right)^2 + \left(\frac{dy}{dt}\right)^2}\,dt$

(b) $S = 2\pi \displaystyle\int_a^b f(t)\sqrt{\left(\frac{dx}{dt}\right)^2 + \left(\frac{dy}{dt}\right)^2}\,dt$

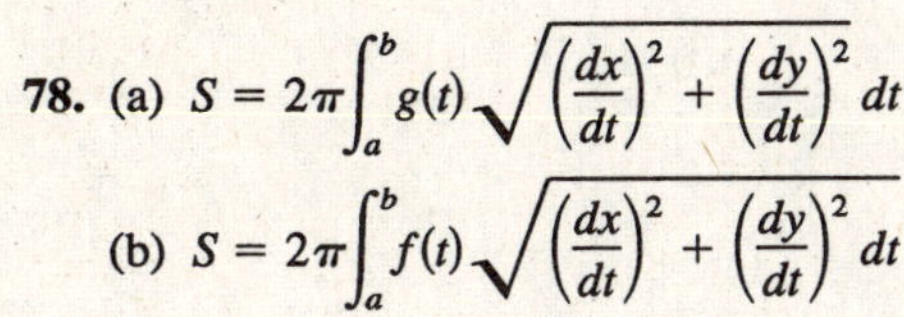

79. Let y be a continuous function of x on $a \le x \le b$. Suppose that $x = f(t)$, $y = g(t)$, and $f(t_1) = a$, $f(t_2) = b$. Then using integration by substitution, $dx = f'(t)\,dt$ and

$$\int_a^b y\,dx = \int_{t_1}^{t_2} g(t)f'(t)\,dt.$$

80. $x = r\cos\phi,\ y = r\sin\phi$

$$S = 2\pi\int_0^\theta r\sin\phi\sqrt{r^2\sin^2\phi + r^2\cos^2\phi}\,d\phi$$

$$= 2\pi r^2\int_0^\theta \sin\phi\,d\phi$$

$$= \Big[-2\pi r^2\cos\phi\Big]_0^\theta$$

$$= 2\pi r^2(1 - \cos\theta)$$

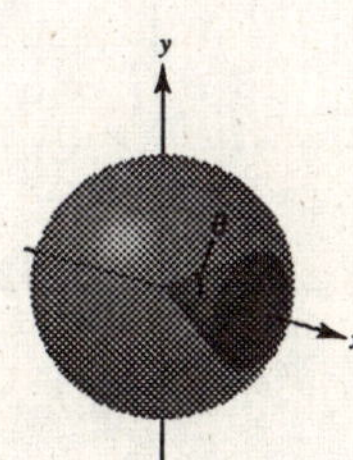

81. $x = 2\sin^2\theta$

$y = 2\sin^2\theta\tan\theta$

$\dfrac{dx}{d\theta} = 4\sin\theta\cos\theta$

$$A = \int_0^{\pi/2} 2\sin^2\theta\tan\theta(4\sin\theta\cos\theta)\,d\theta = 8\int_0^{\pi/2}\sin^4\theta\,d\theta$$

$$= 8\left[\frac{-\sin^3\theta\cos\theta}{4} - \frac{3}{8}\sin\theta\cos\theta + \frac{3}{8}\theta\right]_0^{\pi/2} = \frac{3\pi}{2}$$

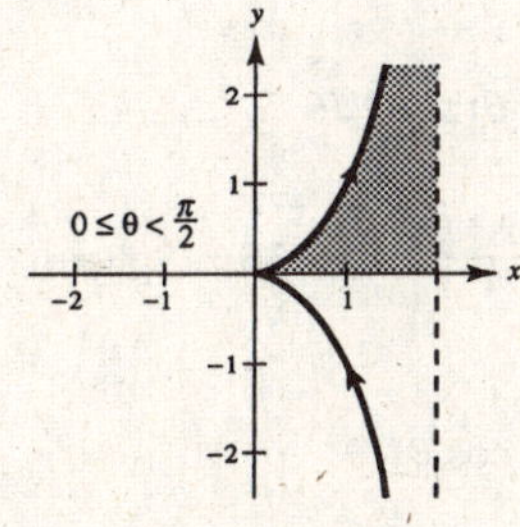

82. $x = 2\cot\theta,\ y = 2\sin^2\theta,\ \dfrac{dx}{d\theta} = -2\csc^2\theta$

$$A = 2\int_{\pi/2}^0 (2\sin^2\theta)(-2\csc^2\theta)\,d\theta = -8\int_{\pi/2}^0 d\theta = \Big[-8\theta\Big]_{\pi/2}^0 = 4\pi$$

83. πab is area of ellipse (d).

84. $\frac{3}{8}\pi a^2$ is area of asteroid (b).

85. $6\pi a^2$ is area of cardioid (f).

86. $2\pi a^2$ is area of deltoid (c).

87. $\frac{8}{3}ab$ is area of hourglass (a).

88. $2\pi ab$ is area of teardrop (e).

89. $x = \sqrt{t},\ y = 4 - t,\ 0 < t < 4$

$$A = \int_0^2 y\,dx = \int_0^4 (4-t)\frac{1}{2\sqrt{t}}\,dt = \frac{1}{2}\int_0^4 (4t^{-1/2} - t^{1/2})\,dt = \left[\frac{1}{2}\left(8\sqrt{t} - \frac{2}{3}t\sqrt{t}\right)\right]_0^4 = \frac{16}{3}$$

$$\bar{x} = \frac{1}{A}\int_0^2 yx\,dx = \frac{3}{16}\int_0^4 (4-t)\sqrt{t}\left(\frac{1}{2\sqrt{t}}\right)dt = \frac{3}{32}\int_0^4 (4-t)\,dt = \left[\frac{3}{32}\left(4t - \frac{t^2}{2}\right)\right]_0^4 = \frac{3}{4}$$

$$\bar{y} = \frac{1}{A}\int_0^2 \frac{y^2}{2}\,dx = \frac{3}{32}\int_0^4 (4-t)^2\frac{1}{2\sqrt{t}}\,dt = \frac{3}{64}\int_0^4 [16t^{-1/2} - 8t^{1/2} + t^{3/2}]\,dt = \frac{3}{64}\left[32\sqrt{t} - \frac{16}{3}t\sqrt{t} + \frac{2}{5}t^2\sqrt{t}\right]_0^4 = \frac{8}{5}$$

$$(\bar{x}, \bar{y}) = \left(\frac{3}{4}, \frac{8}{5}\right)$$

90. $x = \sqrt{4 - t},\ y = \sqrt{t},\ \dfrac{dx}{dt} = -\dfrac{1}{2\sqrt{4-t}},\quad 0 \le t \le 4$

$$A = \int_4^0 \sqrt{t}\left(-\frac{1}{2\sqrt{4-t}}\right)dt = \int_0^2 \sqrt{4-u^2}\,du = \frac{1}{2}\left[u\sqrt{4-u^2} + 4\arcsin\frac{u}{2}\right]_0^2 = \pi$$

Let $u = \sqrt{4-t}$, then $du = -1/(2\sqrt{4-t})\,dt$ and $\sqrt{t} = \sqrt{4-u^2}$.

$$\bar{x} = \frac{1}{\pi}\int_4^0 \sqrt{4-t}\sqrt{t}\left(-\frac{1}{2\sqrt{4-t}}\right)dt = -\frac{1}{2\pi}\int_4^0 \sqrt{t}\,dt = \left[-\frac{1}{2\pi}\frac{2}{3}t^{3/2}\right]_4^0 = \frac{8}{3\pi}$$

$$\bar{y} = \frac{1}{2\pi}\int_4^0 (\sqrt{t})^2\left(-\frac{1}{2\sqrt{4-t}}\right)dt = -\frac{1}{4\pi}\int_4^0 \frac{t}{\sqrt{4-t}}\,dt = -\frac{1}{4\pi}\left[\frac{-2(8+t)}{3}\sqrt{4-t}\right]_4^0 = \frac{8}{3\pi}$$

$$(\bar{x}, \bar{y}) = \left(\frac{8}{3\pi}, \frac{8}{3\pi}\right)$$

91. $x = 3\cos\theta,\ y = 3\sin\theta,\ \dfrac{dx}{d\theta} = -3\sin\theta$

$$V = 2\pi\int_{\pi/2}^0 (3\sin\theta)^2(-3\sin\theta)\,d\theta$$

$$= -54\pi\int_{\pi/2}^0 \sin^3\theta\,d\theta$$

$$= -54\pi\int_{\pi/2}^0 (1 - \cos^2\theta)\sin\theta\,d\theta$$

$$= -54\pi\left[-\cos\theta + \frac{\cos^3\theta}{3}\right]_{\pi/2}^0 = 36\pi \text{ (Sphere)}$$

92. $x = \cos\theta,\ y = 3\sin\theta,\ \dfrac{dx}{d\theta} = -\sin\theta$

$$V = 2\pi\int_{\pi/2}^0 (3\sin\theta)^2(-\sin\theta)\,d\theta$$

$$= -18\pi\int_{\pi/2}^0 \sin^3\theta\,d\theta$$

$$= -18\pi\left[-\cos\theta + \frac{\cos^3\theta}{3}\right]_{\pi/2}^0 = 12\pi$$

93. $x = a(\theta - \sin\theta),\ y = a(1 - \cos\theta)$

(a) $\dfrac{dy}{d\theta} = a\sin\theta,\ \dfrac{dx}{d\theta} = a(1 - \cos\theta)$

$$\frac{dy}{dx} = \frac{a\sin\theta}{a(1-\cos\theta)} = \frac{\sin\theta}{1-\cos\theta}$$

$$\frac{d^2y}{dx^2} = \left[\frac{(1-\cos\theta)\cos\theta - \sin\theta(\sin\theta)}{(1-\cos\theta)^2}\right]\Big/[a(1-\cos\theta)]$$

$$= \frac{\cos\theta - 1}{a(1-\cos\theta)^3} = \frac{-1}{a(\cos\theta - 1)^2}$$

(b) At $\theta = \dfrac{\pi}{6},\ x = a\left(\dfrac{\pi}{6} - \dfrac{1}{2}\right),\ y = a\left(1 - \dfrac{\sqrt{3}}{2}\right)\dfrac{dy}{dx} = \dfrac{1/2}{1 - \sqrt{3}/2} = 2 + \sqrt{3}.$

Tangent line: $y - a\left(1 - \dfrac{\sqrt{3}}{2}\right) = (2 + \sqrt{3})\left(x - a\left(\dfrac{\pi}{6} - \dfrac{1}{2}\right)\right)$

—CONTINUED—

93. —CONTINUED—

(c) $\dfrac{dy}{dx} = \dfrac{\sin\theta}{1 - \cos\theta} = 0 \Rightarrow \sin\theta = 0, 1 - \cos\theta \neq 0$

Points of horizontal tangency: $(x, y) = (a(2n + 1)\pi, 2a)$

(d) Concave downward on all open θ-intervals:
$\ldots, (-2\pi, 0), (0, 2\pi), (2\pi, 4\pi), \ldots$

(e) $s = \displaystyle\int_0^{2\pi} \sqrt{a^2\sin^2\theta + a^2(1 - \cos\theta)^2}\, d\theta$

$= a\displaystyle\int_0^{2\pi} \sqrt{2 - 2\cos\theta}\, d\theta$

$= a\displaystyle\int_0^{2\pi} \sqrt{4\sin^2\theta/2}\, d\theta$

$= 2a\displaystyle\int_0^{2\pi} \sin\frac{\theta}{2}\, d\theta$

$= -4a\cos\left(\dfrac{\theta}{2}\right)\Big]_0^{2\pi} = 8a$

94. $x = t^2\sqrt{3}, y = 3t - \frac{1}{3}t^3$

(a)

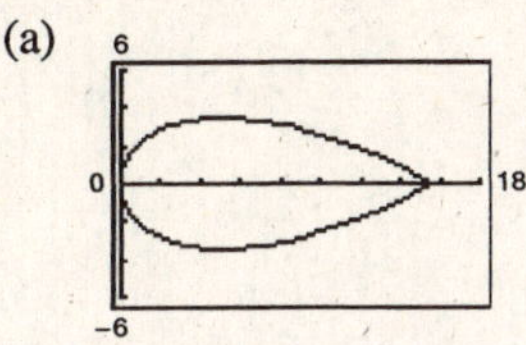

(b) $\dfrac{dx}{dt} = 2\sqrt{3}t, \dfrac{dy}{dt} = 3 - t^2, \dfrac{dy}{dx} = \dfrac{3 - t^2}{2\sqrt{3}t}$

$\dfrac{d^2y}{dx^2} = \left[\dfrac{2\sqrt{3}(t)(-2t) - (3 - t^2)2\sqrt{3}}{12t^2}\right] \Big/ \left[2\sqrt{3}t\right]$

$= \dfrac{-2\sqrt{3}t^2 - 6\sqrt{3}}{(12t^2)(2\sqrt{3}t)} = -\dfrac{t^2 + 3}{12t^3}$

(c) $(x, y) = \left(\sqrt{3}, \frac{8}{3}\right)$ at $t = 1$. $\dfrac{dy}{dx} = \dfrac{2}{2\sqrt{3}} = \dfrac{\sqrt{3}}{3}$

$y - \dfrac{8}{3} = \dfrac{\sqrt{3}}{3}(x - \sqrt{3})$

$y = \dfrac{\sqrt{3}}{3}x + \dfrac{5}{3}$

(d) $s = \displaystyle\int_{-3}^{3} \sqrt{12t^2 + (3 - t)^2}\, dt$

$= \displaystyle\int_{-3}^{3} \sqrt{t^4 - 6t^2 + 9 + 12t^2}\, dt$

$= \displaystyle\int_{-3}^{3} \sqrt{(t^2 + 3)^2}\, dt$

$= \displaystyle\int_{-3}^{3} (t^2 + 3)\, dt = 36$

(e) $S = 2\pi\displaystyle\int_0^3 \left(3t - \tfrac{1}{3}t^3\right)(t^2 + 3)\, dt = 81\pi$

95. $x = t + u = r\cos\theta + r\theta\sin\theta$

$= r(\cos\theta + \theta\sin\theta)$

$y = v - w = r\sin\theta - r\theta\cos\theta$

$= r(\sin\theta - \theta\cos\theta)$

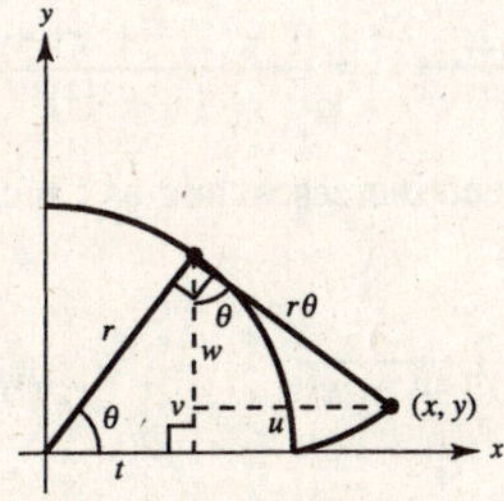

96. Let's focus on the region above the x-axis. From Exercise 95, the equation of the involute from $(1, 0)$ to $(-1, \pi)$ is

$$x = \cos\theta + \theta\sin\theta$$

$$y = \sin\theta - \theta\cos\theta$$

$$0 \le \theta \le \pi.$$

At $(-1, \pi)$, the string is fully extended and has length π. So, the area of region A is $\frac{1}{4}\pi(\pi^2) = \frac{1}{4}\pi^3$.

We now need to find the area of region B.

$$\frac{dx}{d\theta} = -\sin\theta + \sin\theta + \theta\cos\theta = \theta\cos\theta = 0 \Rightarrow \theta = \frac{\pi}{2}. \ (\theta = 0 \text{ is cusp.})$$

Hence, the far right point on the involute is $(\pi/2, 1)$.

The area of the region B + C + D is given by

$$\int_{\theta=\pi}^{\theta=\pi/2} y\,dx - \int_{\theta=0}^{\theta=\pi/2} y\,dx = \int_{\theta=\pi}^{\theta=0} y\,dx$$

where $y = \sin\theta - \theta\cos\theta$ and $dx = \theta\cos\theta\,d\theta$.

Thus, we can calculate

$$\int_{\pi}^{0} [\sin\theta - \theta\cos\theta]\theta\cos\theta\,d\theta = \frac{\pi}{6}(\pi^2 + 3).$$

Since the area of C + D is $\pi/2$, we have

$$\text{Total area covered} = 2\left[\frac{1}{4}\pi^3 + \frac{\pi}{6}(\pi^2 + 3) - \frac{\pi}{2}\right] = \frac{5}{6}\pi^3.$$

97. (a)

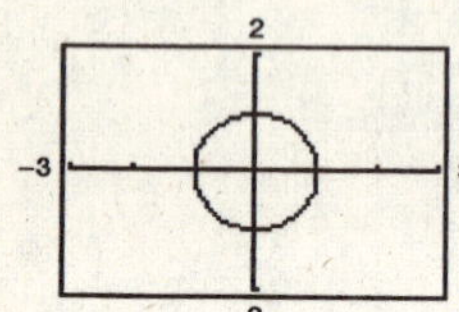

(b) $x = \dfrac{1 - t^2}{1 + t^2}, \ y = \dfrac{2t}{1 + t^2}, \ -20 \le t \le 20$

The graph (for $-\infty < t < \infty$) is the circle $x^2 + y^2 = 1$, except the point $(-1, 0)$.

Verify: $x^2 + y^2 = \left(\dfrac{1 - t^2}{1 + t^2}\right)^2 + \left(\dfrac{2t}{1 + t^2}\right)^2 = \dfrac{1 - 2t^2 + t^4 + 4t^2}{(1 + t^2)^2} = \dfrac{(1 + t^2)^2}{(1 + t^2)^2} = 1$

(c) As t increases from -20 to 0, the speed increases, and as t increases from 0 to 20, the speed decreases.

98. (a) $y = -12\ln\left(\dfrac{12 - \sqrt{144 - x^2}}{x}\right) - \sqrt{144 - x^2}$

$0 < x \le 12$

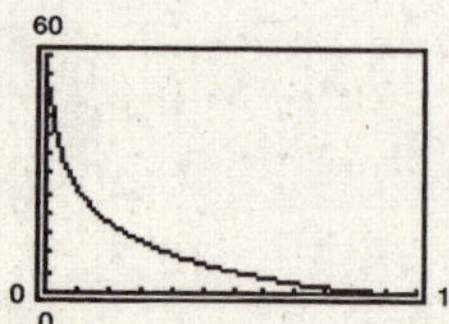

(b) $x = 12\,\text{sech}\,\dfrac{t}{12}, \ y = t - 12\tanh\dfrac{t}{12}, \ 0 \le t$

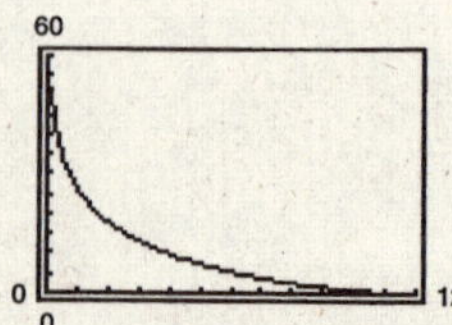

Same as the graph in (a), but has the advantage of showing the position of the object and any given time t.

—CONTINUED—

98. —CONTINUED—

(c) $\dfrac{dy}{dx} = \dfrac{1 - \operatorname{sech}^2(t/12)}{-\operatorname{sech}(t/12)\tanh(t/12)} = -\sinh\dfrac{t}{12}$

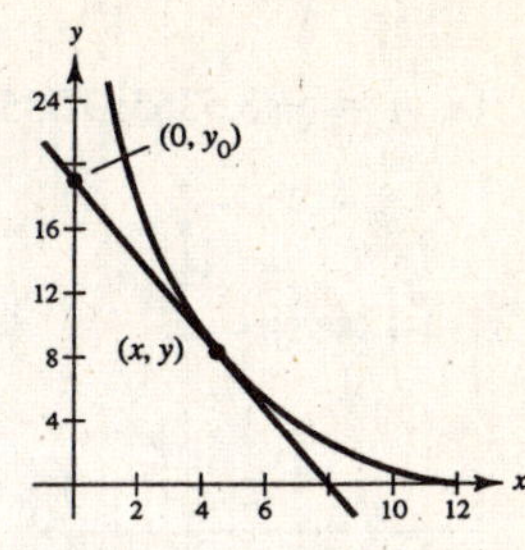

Tangent line: $y - \left(t_0 - 12\tanh\dfrac{t_0}{12}\right) = -\sinh\dfrac{t_0}{12}\left(x - 12\operatorname{sech}\dfrac{t_0}{12}\right)$

$$y = t_0 - \left(\sinh\frac{t_0}{12}\right)x$$

y-intercept: $(0, t_0)$

Distance between $(0, t_0)$ and (x, y): $d = \sqrt{\left(12\operatorname{sech}\dfrac{t_0}{12}\right)^2 + \left(-12\tanh\dfrac{t_0}{12}\right)^2} = 12$

$$d = 12 \text{ for any } t \geq 0.$$

99. False. $\dfrac{d^2y}{dx^2} = \dfrac{\dfrac{d}{dt}\left[\dfrac{g'(t)}{f'(t)}\right]}{f'(t)} = \dfrac{f'(t)g''(t) - g'(t)f''(t)}{[f'(t)]^3}$

100. False. Both dx/dt and dy/dt are zero when $t = 0$. By eliminating the parameter, we have $y = x^{2/3}$ which does not have a horizontal tangent at the origin.

Section 10.4 Polar Coordinates and Polar Graphs

1. $\left(4, \dfrac{\pi}{2}\right)$

$x = 4\cos\left(\dfrac{\pi}{2}\right) = 0$

$y = 4\sin\left(\dfrac{\pi}{2}\right) = 4$

$(x, y) = (0, 4)$

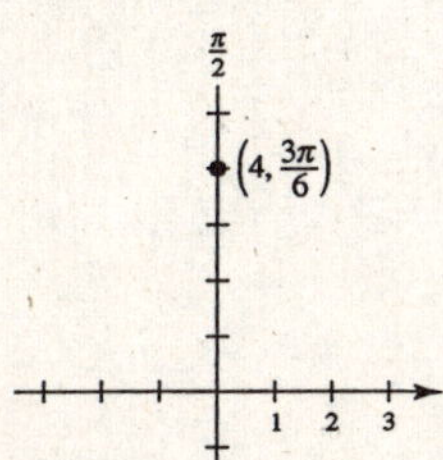

2. $\left(-2, \dfrac{7\pi}{4}\right)$

$x = -2\cos\left(\dfrac{7\pi}{4}\right) = -\sqrt{2}$

$y = -2\sin\left(\dfrac{7\pi}{4}\right) = \sqrt{2}$

$(x, y) = (-\sqrt{2}, \sqrt{2})$

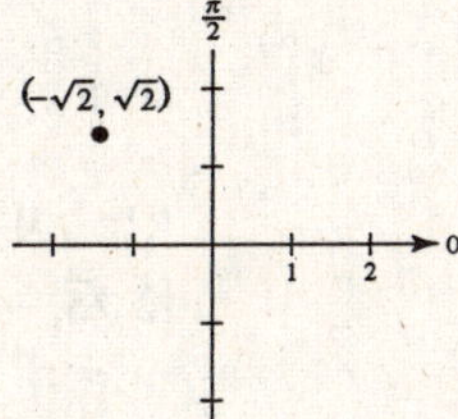

3. $\left(-4, -\dfrac{\pi}{3}\right)$

$x = -4\cos\left(-\dfrac{\pi}{3}\right) = -2$

$y = -4\sin\left(-\dfrac{\pi}{3}\right) = 2\sqrt{3}$

$(x, y) = (-2, 2\sqrt{3})$

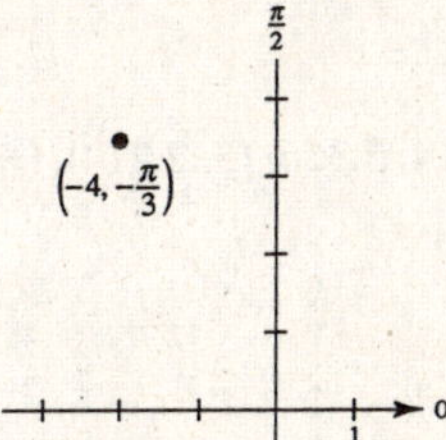

4. $\left(0, -\dfrac{7\pi}{6}\right)$

$x = 0\cos\left(-\dfrac{7\pi}{6}\right) = 0$

$y = 0\sin\left(-\dfrac{7\pi}{6}\right) = 0$

$(x, y) = (0, 0)$

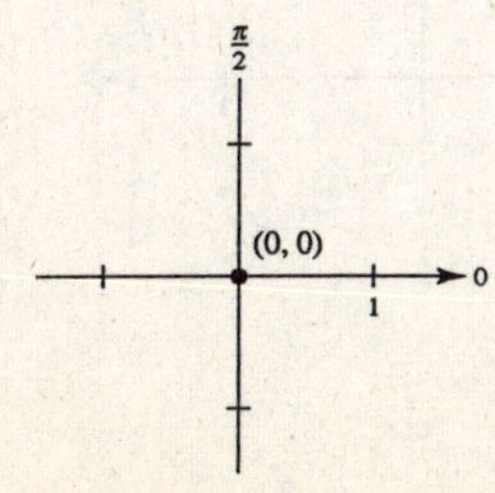

5. $(\sqrt{2}, 2.36)$

$x = \sqrt{2}\cos(2.36) \approx -1.004$

$y = \sqrt{2}\sin(2.36) \approx 0.996$

$(x, y) = (-1.004, 0.996)$

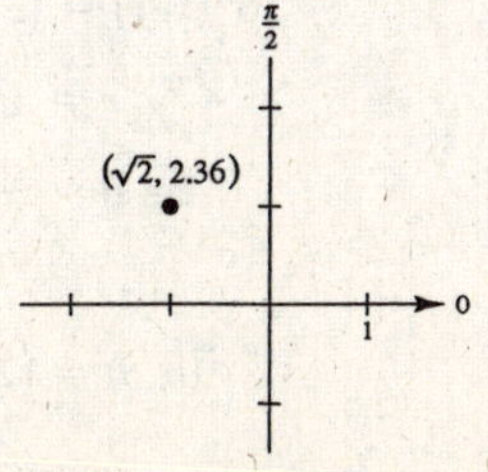

6. $(-3, -1.57)$

$x = -3\cos(-1.57) \approx -0.0024$

$y = -3\sin(-1.57) \approx 3$

$(x, y) = (-0.0024, 3)$

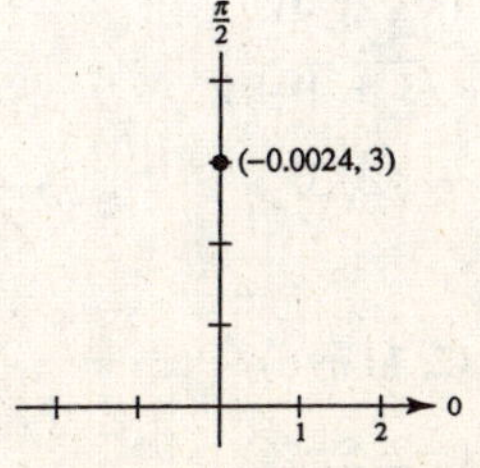

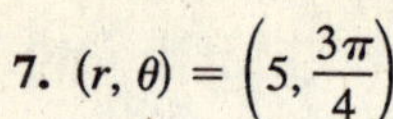

7. $(r, \theta) = \left(5, \dfrac{3\pi}{4}\right)$

$(x, y) = (-3.5355, 3.5355)$

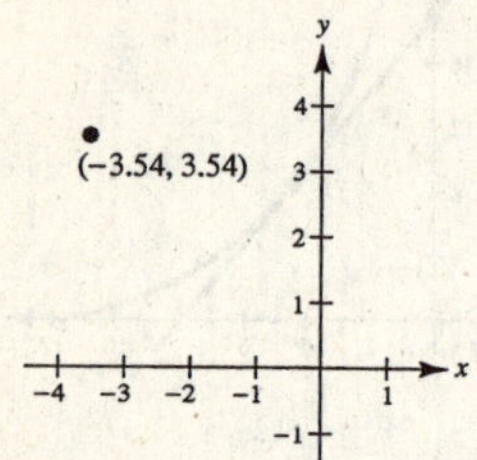

8. $(r, \theta) = \left(-2, \dfrac{11\pi}{6}\right)$

$(x, y) = (-1.7321, 1)$

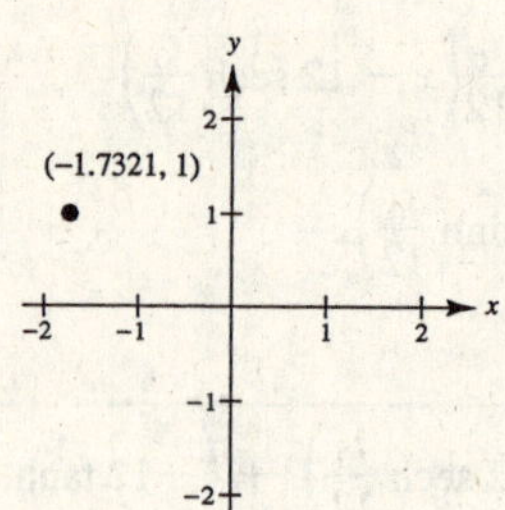

9. $(r, \theta) = (-3.5, 2.5)$

$(x, y) = (2.804, -2.095)$

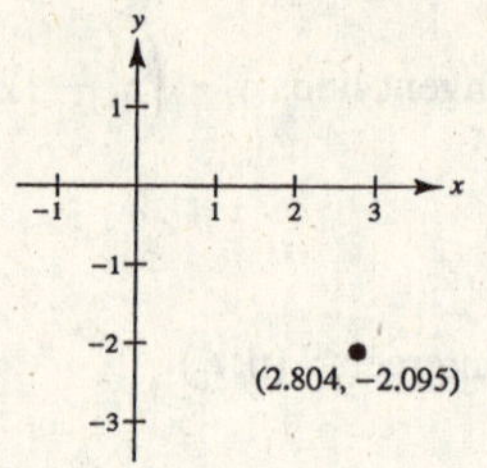

10. $(r, \theta) = (8.25, 1.3)$

$(x, y) = (2.2069, 7.9494)$

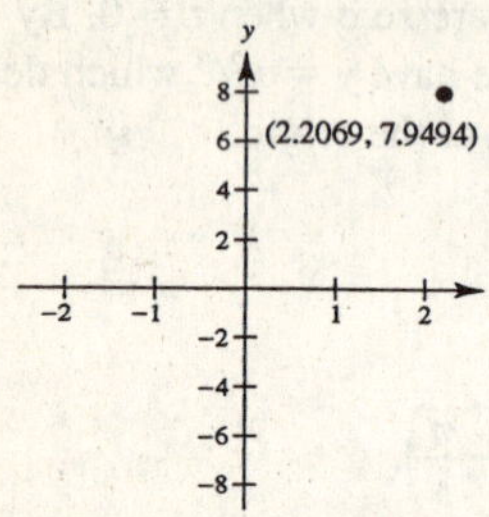

11. $(x, y) = (1, 1)$

$r = \pm\sqrt{2}$

$\tan\theta = 1$

$\theta = \dfrac{\pi}{4}, \dfrac{5\pi}{4}, \left(\sqrt{2}, \dfrac{\pi}{4}\right), \left(-\sqrt{2}, \dfrac{5\pi}{4}\right)$

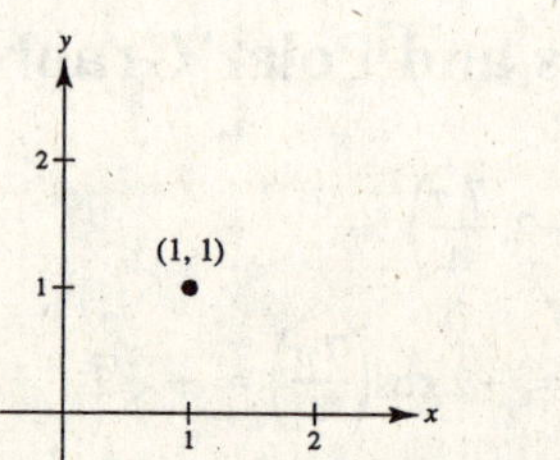

12. $(x, y) = (0, -5)$

$r = \pm 5$

$\tan\theta$ undefined

$\theta = \dfrac{\pi}{2}, \dfrac{3\pi}{2}, \left(5, \dfrac{3\pi}{2}\right), \left(-5, \dfrac{\pi}{2}\right)$

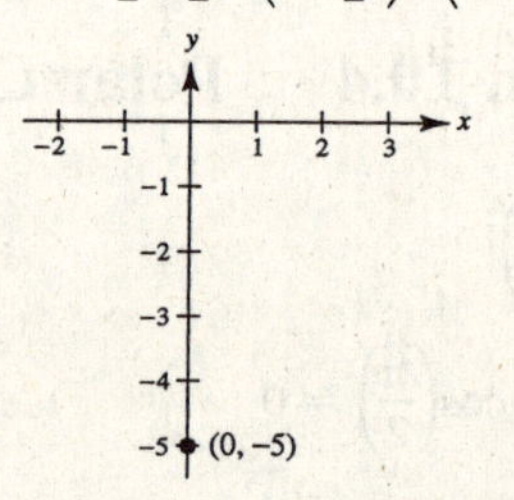

13. $(x, y) = (-3, 4)$

$r = \pm\sqrt{9 + 16} = \pm 5$

$\tan\theta = -\frac{4}{3}$

$\theta \approx 2.214, 5.356, (5, 2.214), (-5, 5.356)$

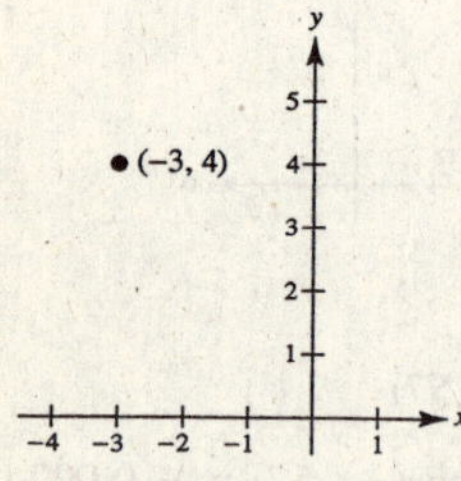

14. $(x, y) = (4, -2)$

$r = \pm\sqrt{16 + 4} = \pm 2\sqrt{5}$

$\tan\theta = -\dfrac{2}{4} = -\dfrac{1}{2}$

$\theta \approx -0.464$

$\left(2\sqrt{5}, -0.464\right), \left(-2\sqrt{5}, 2.678\right)$

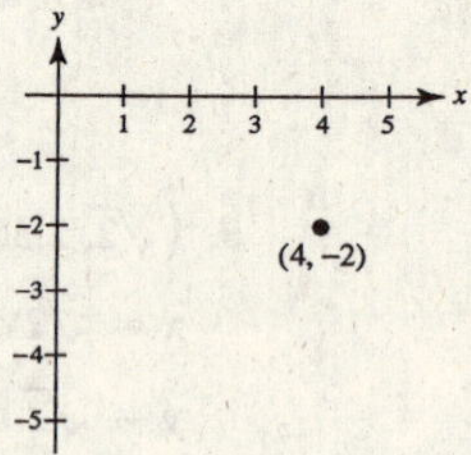

15. $(x, y) = \left(\sqrt{3}, -1\right)$

$r = \sqrt{3 + 1} = 2$

$\tan\theta = -\sqrt{3}/3$

$(r, \theta) = (2, 11\pi/6)$

$= (-2, 5\pi/6)$

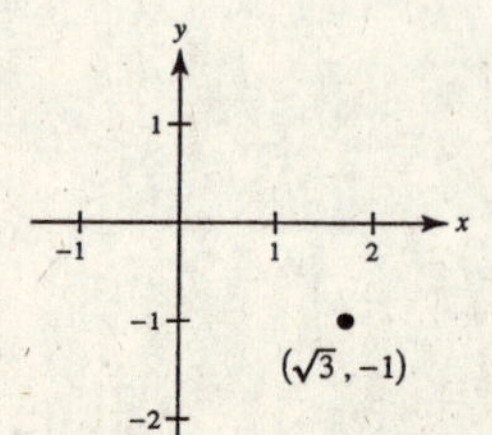

16 $(x, y) = \left(3, \sqrt{3}\right)$

$r = \sqrt{9 + 3} = 2\sqrt{3}$

$\tan\theta = -\sqrt{3}/3$

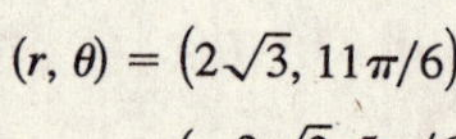

$(r, \theta) = \left(2\sqrt{3}, 11\pi/6\right)$

$= \left(-2\sqrt{3}, 5\pi/6\right)$

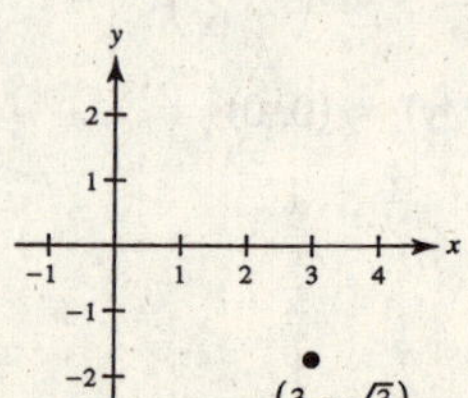

17. $(x, y) = (3, -2)$

$(r, \theta) = (3.606, -0.588)$

18. $(x, y) = (3\sqrt{2}, 3\sqrt{2})$

$(r, \theta) = (6, 0.785)$

19. $(x, y) = \left(\frac{5}{2}, \frac{4}{3}\right)$

$(r, \theta) = (2.833, 0.490)$

20. $(x, y) = (0, -5)$

$(r, \theta) = (5, -1.571)$

21. (a) $(x, y) = (4, 3.5)$

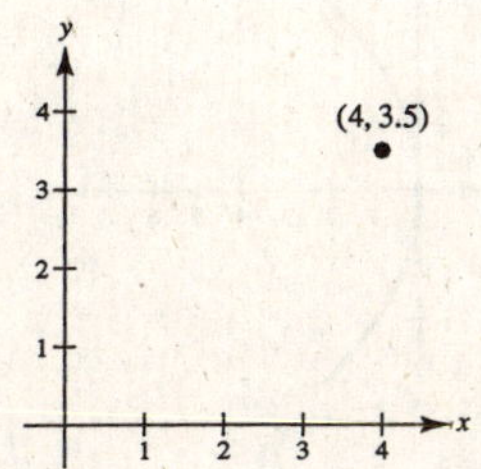

(b) $(r, \theta) = (4, 3.5)$

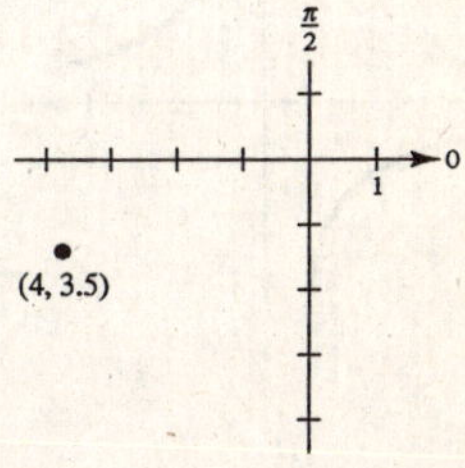

22. (a) Moving horizontally, the x-coordinate changes. Moving vertically, the y-coordinate changes.

(b) Both r and θ values change.

(c) In polar mode, horizontal (or vertical) changes result in changes in both r and θ.

23. $r = 2 \sin \theta$ circle

Matches (c)

24. $r = 4 \cos 2\theta$

Rose curve

Matches (b)

25. $r = 3(1 + \cos \theta)$

Cardioid

Matches (a)

26. $r = 2 \sec \theta$

Line

Matches (d)

27. $x^2 + y^2 = a^2$

$r = a$

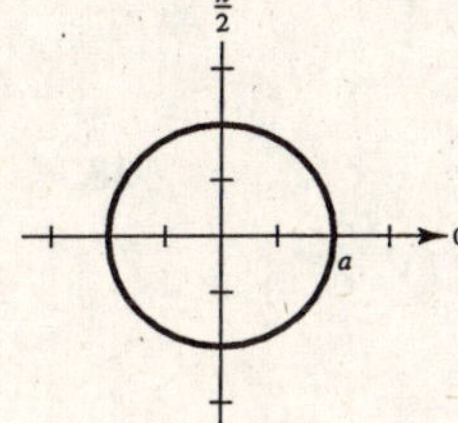

28. $x^2 + y^2 - 2ax = 0$

$r^2 - 2ar \cos \theta = 0$

$r(r - 2a \cos \theta) = 0$

$r = 2a \cos \theta$

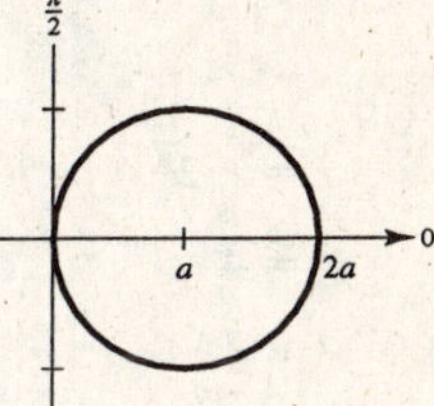

29. $y = 4$

$r \sin \theta = 4$

$r = 4 \csc \theta$

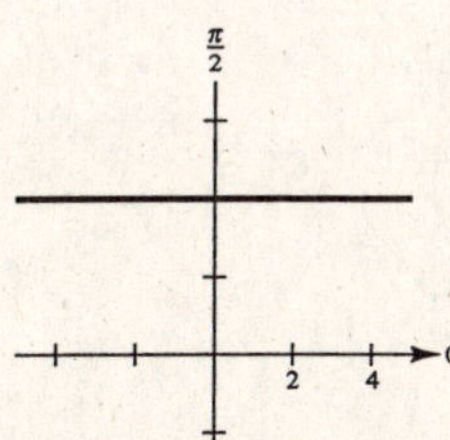

30. $x = 10$

$r \cos \theta = 10$

$r = 10 \sec \theta$

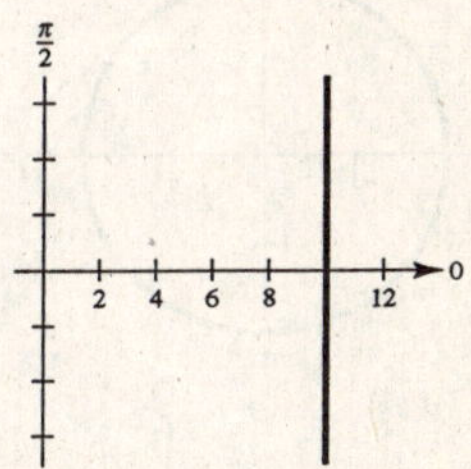

31. $3x - y + 2 = 0$

$3r \cos \theta - r \sin \theta + 2 = 0$

$r(3 \cos \theta - \sin \theta) = -2$

$$r = \frac{-2}{3 \cos \theta - \sin \theta}$$

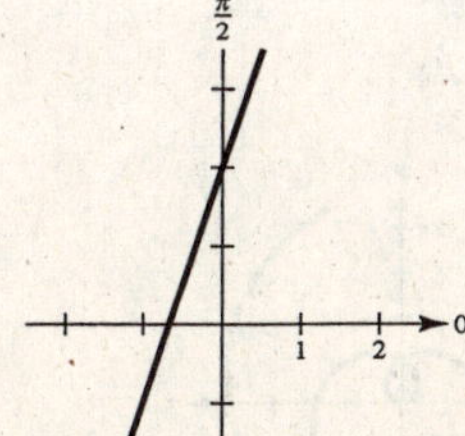

32.
$$xy = 4$$
$$(r\cos\theta)(r\sin\theta) = 4$$
$$r^2 = 4\sec\theta\csc\theta$$
$$= 8\csc 2\theta$$

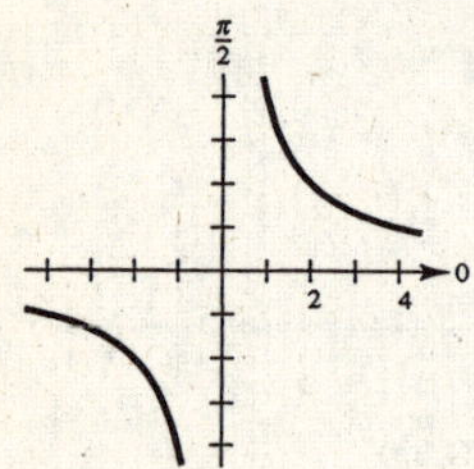

33.

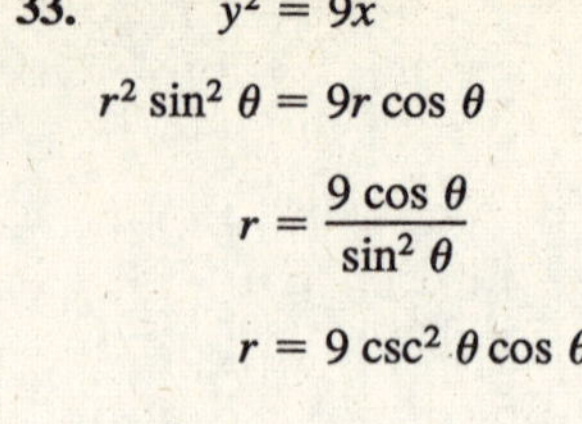

$$y^2 = 9x$$
$$r^2\sin^2\theta = 9r\cos\theta$$
$$r = \frac{9\cos\theta}{\sin^2\theta}$$
$$r = 9\csc^2\theta\cos\theta$$

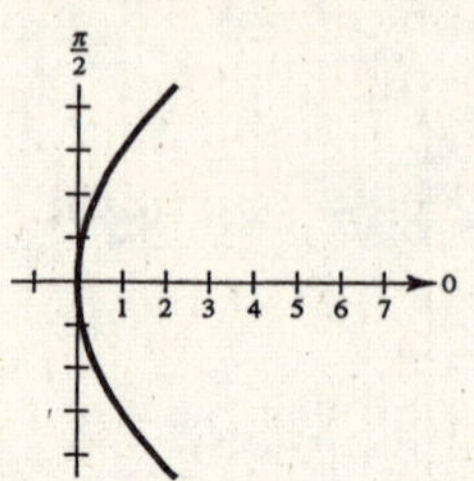

34.
$$(x^2 + y^2)^2 - 9(x^2 - y^2) = 0$$
$$(r^2)^2 - 9(r^2\cos^2\theta - r^2\sin^2\theta) = 0$$
$$r^2[r^2 - 9(\cos 2\theta)] = 0$$
$$r^2 = 9\cos 2\theta$$

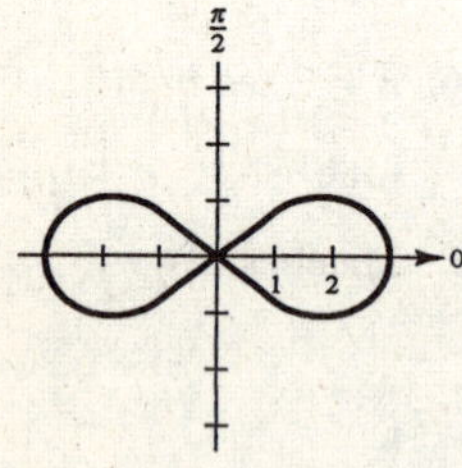

35.
$$r = 3$$
$$r^2 = 9$$
$$x^2 + y^2 = 9$$

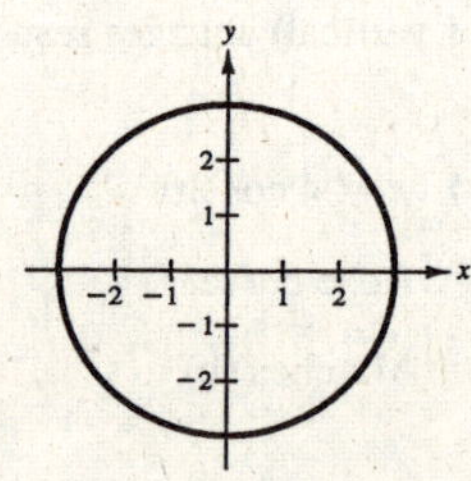

36.
$$r = -2$$
$$r^2 = 4$$
$$x^2 + y^2 = 4$$

37.
$$r = \sin\theta$$
$$r^2 = r\sin\theta$$
$$x^2 + y^2 = y$$
$$x^2 + \left(y - \frac{1}{2}\right)^2 = \frac{1}{4}$$
$$x^2 + y^2 - y = 0$$

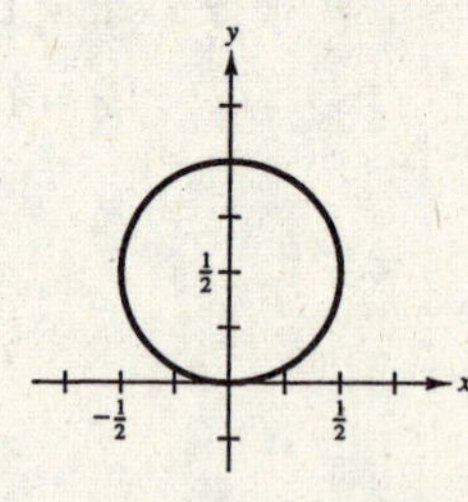

38.

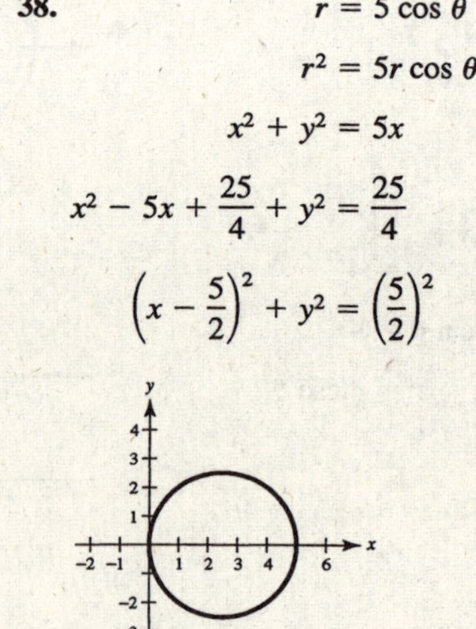

$$r = 5\cos\theta$$
$$r^2 = 5r\cos\theta$$
$$x^2 + y^2 = 5x$$
$$x^2 - 5x + \frac{25}{4} + y^2 = \frac{25}{4}$$
$$\left(x - \frac{5}{2}\right)^2 + y^2 = \left(\frac{5}{2}\right)^2$$

39.
$$r = \theta$$
$$\tan r = \tan\theta$$
$$\tan\sqrt{x^2 + y^2} = \frac{y}{x}$$
$$\sqrt{x^2 + y^2} = \arctan\frac{y}{x}$$

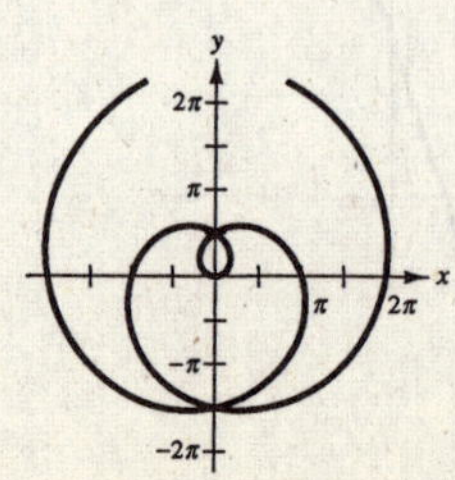

40. $\theta = \dfrac{5\pi}{6}$

$\tan\theta = \tan\dfrac{5\pi}{6}$

$\dfrac{y}{x} = -\dfrac{\sqrt{3}}{3}$

$y = -\dfrac{\sqrt{3}}{3}x$

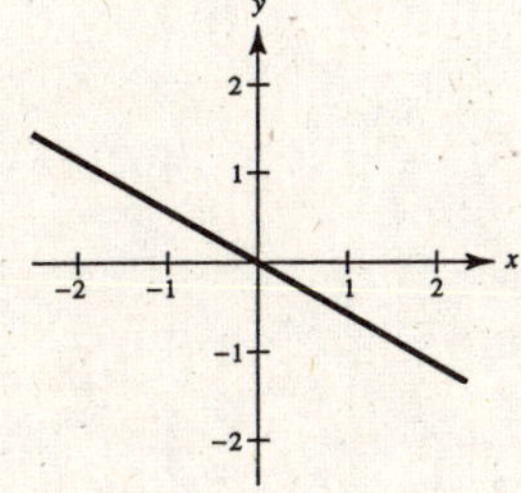

41. $r = 3\sec\theta$

$r\cos\theta = 3$

$x = 3$

$x - 3 = 0$

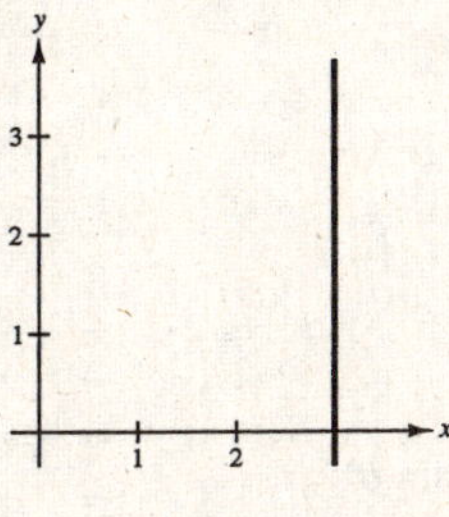

42. $r = 2\csc\theta$

$r\sin\theta = 2$

$y = 2$

$y - 2 = 0$

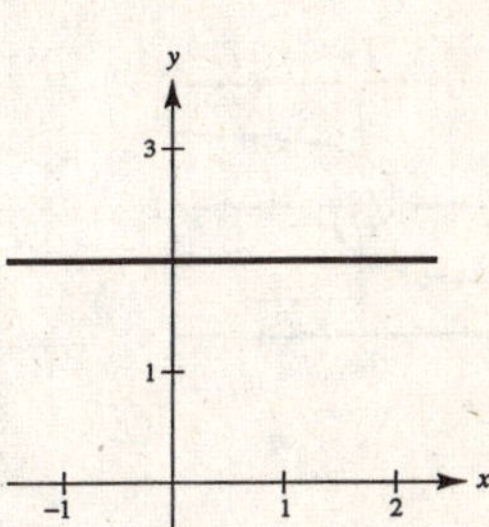

43. $r = 3 - 4\cos\theta$

$0 \le \theta < 2\pi$

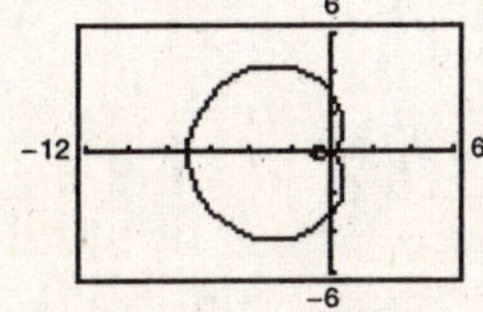

44. $r = 5(1 - 2\sin\theta)$

$0 \le \theta < 2\pi$

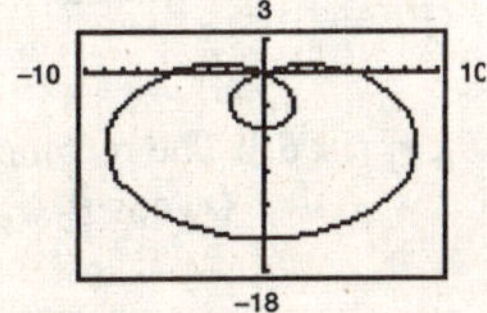

45. $r = 2 + \sin\theta$

$0 \le \theta < 2\pi$

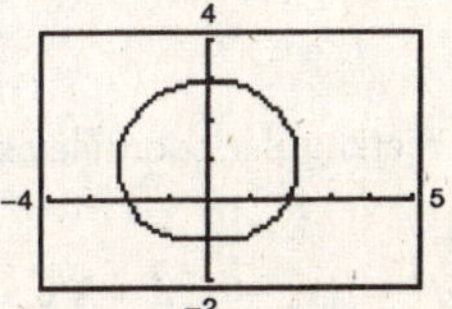

46. $r = 4 + 3\cos\theta$

$0 \le \theta < 2\pi$

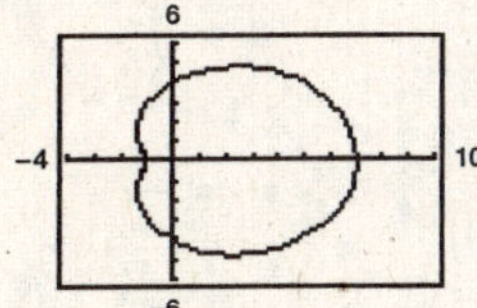

47. $r = \dfrac{2}{1 + \cos\theta}$

Traced out once on $-\pi < \theta < \pi$

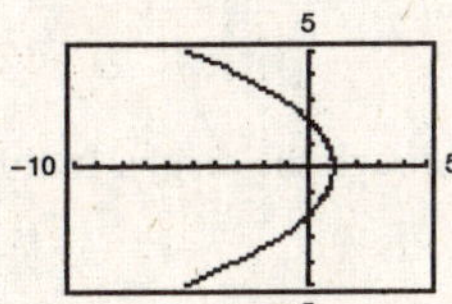

48. $r = \dfrac{2}{4 - 3\sin\theta}$

Traced out once on $0 \le \theta \le 2\pi$

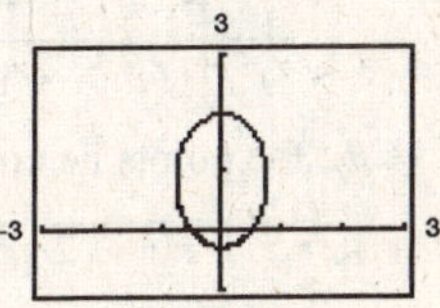

49. $r = 2\cos\left(\dfrac{3\theta}{2}\right)$

$0 \le \theta < 4\pi$

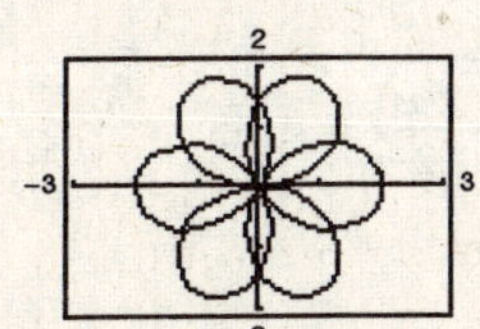

50. $r = 3\sin\left(\dfrac{5\theta}{2}\right)$

$0 \le \theta < 4\pi$

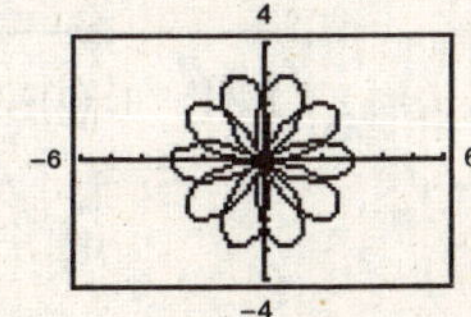

51. $r^2 = 4\sin 2\theta$

$r_1 = 2\sqrt{\sin 2\theta}$

$r_2 = -2\sqrt{\sin 2\theta}$

$0 \le \theta < \dfrac{\pi}{2}$

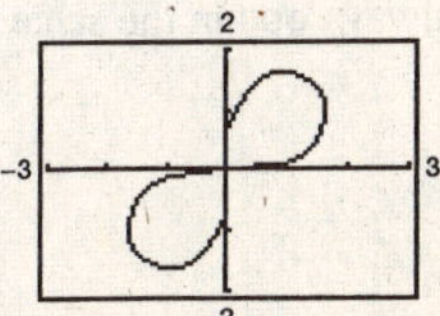

52. $r^2 = \dfrac{1}{\theta}$.

Graph as $r_1 = \dfrac{1}{\sqrt{\theta}}$, $r_2 = -\dfrac{1}{\sqrt{\theta}}$.

It is traced out once on $[0, \infty)$.

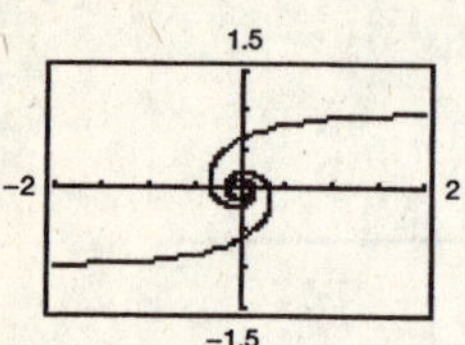

53.

$$\begin{aligned}
r &= 2(h \cos \theta + k \sin \theta)\\
r^2 &= 2r(h \cos \theta + k \sin \theta)\\
r^2 &= 2[h(r \cos \theta) + k(r \sin \theta)]\\
x^2 + y^2 &= 2(hx + ky)\\
x^2 + y^2 - 2hx - 2ky &= 0\\
(x^2 - 2hx + h^2) + (y^2 - 2ky + k^2) &= 0 + h^2 + k^2\\
(x - h)^2 + (y - k)^2 &= h^2 + k^2
\end{aligned}$$

Radius: $\sqrt{h^2 + k^2}$

Center: (h, k)

54. (a) The rectangular coordinates of (r_1, θ_1) are $(r_1 \cos \theta_1, r_1 \sin \theta_1)$. The rectangular coordinates of (r_2, θ_2) are $(r_2 \cos \theta_2, r_2 \sin \theta_2)$.

$$\begin{aligned}
d^2 &= (x_2 - x_1)^2 + (y_2 - y_1)^2\\
&= (r_2 \cos \theta_2 - r_1 \cos \theta_1)^2 + (r_2 \sin \theta_2 - r_1 \sin \theta_1)^2\\
&= r_2^2 \cos^2 \theta_2 - 2r_1r_2 \cos \theta_1 \cos \theta_2 + r_1^2 \cos^2 \theta_1 + r_2^2 \sin^2 \theta_2{}^2 - 2r_1r_2 \sin \theta_1 \sin \theta_2 + r_1^2 \sin^2 \theta_1\\
&= r_2^2 (\cos^2 \theta_2 + \sin^2 \theta_2) + r_1^2 (\cos^2 \theta_1 + \sin^2 \theta_1) - 2\,r_1r_2(\cos \theta_1 \cos \theta_2 + \sin \theta_1 \sin \theta_2)\\
&= r_1^2 + r_2^2 - 2r_1r_2 \cos(\theta_1 - \theta_2)\\
d &= \sqrt{r_1^2 + r_2^2 - 2r_1r_2 \cos(\theta_1 - \theta_2)}
\end{aligned}$$

(b) If $\theta_1 = \theta_2$, the points lie on the same line passing through the origin. In this case,

$$\begin{aligned}
d &= \sqrt{r_1^2 + r_2^2 - 2r_1r_2 \cos(0)}\\
&= \sqrt{(r_1 - r_2)^2} = |r_1 - r_2|.
\end{aligned}$$

(c) If $\theta_1 - \theta_2 = 90°$, then $\cos(\theta_1 - \theta_2) = 0$ and $d = \sqrt{r_1^2 + r_2^2}$, the Pythagorean Theorem!

(d) Many answers are possible. For example, consider the two points $(r_1, \theta_1) = (1, 0)$ and $(r_2, \theta_2) = (2, \pi/2)$.

$$d = \sqrt{1 + 2^2 - 2(1)(2) \cos\left(0 - \frac{\pi}{2}\right)} = \sqrt{5}$$

Using $(r_1, \theta_1) = (-1, \pi)$ and $(r_2, \theta_2) = [2, (5\pi/2)]$, $d = \sqrt{(-1)^2 + (2)^2 - 2(-1)(2) \cos\left(\pi - \frac{5\pi}{2}\right)} = \sqrt{5}$.

You always obtain the same distance.

55. $\left(4, \frac{2\pi}{3}\right), \left(2, \frac{\pi}{6}\right)$

$$d = \sqrt{4^2 + 2^2 - 2(4)(2)\cos\left(\frac{2\pi}{3} - \frac{\pi}{6}\right)}$$

$$= \sqrt{20 - 16\cos\frac{\pi}{2}} = 2\sqrt{5} \approx 4.5$$

56. $\left(10, \frac{7\pi}{6}\right), (3, \pi)$

$$d = \sqrt{10^2 + 3^2 - 2(10)(3)\cos\left(\frac{7\pi}{6} - \pi\right)}$$

$$= \sqrt{109 - 60\cos\frac{\pi}{6}} = \sqrt{109 - 30\sqrt{3}} \approx 7.6$$

57. $(2, 0.5), (7, 1.2)$

$$d = \sqrt{2^2 + 7^2 - 2(2)(7)\cos(0.5 - 1.2)}$$

$$= \sqrt{53 - 28\cos(-0.7)} \approx 5.6$$

58. $(4, 2.5), (12, 1)$

$$d = \sqrt{4^2 + 12^2 - 2(4)(12)\cos(2.5 - 1)}$$

$$= \sqrt{160 - 96\cos 1.5} \approx 12.3$$

59. $r = 2 + 3\sin\theta$

$$\frac{dy}{dx} = \frac{3\cos\theta\sin\theta + \cos\theta(2 + 3\sin\theta)}{3\cos\theta\cos\theta - \sin\theta(2 + 3\sin\theta)}$$

$$= \frac{2\cos\theta(3\sin\theta + 1)}{3\cos 2\theta - 2\sin\theta} = \frac{2\cos\theta(3\sin\theta + 1)}{6\cos^2\theta - 2\sin\theta - 3}$$

At $\left(5, \frac{\pi}{2}\right)$, $\frac{dy}{dx} = 0$.

At $(2, \pi)$, $\frac{dy}{dx} = -\frac{2}{3}$.

At $\left(-1, \frac{3\pi}{2}\right)$, $\frac{dy}{dx} = 0$.

60. $r = 2(1 - \sin\theta)$

$$\frac{dy}{dx} = \frac{-2\cos\theta\sin\theta + 2\cos\theta(1 - \sin\theta)}{-2\cos\theta\cos\theta - 2\sin\theta(1 - \sin\theta)}$$

At $(2, 0)$, $\frac{dy}{dx} = -1$.

At $\left(3, \frac{7\pi}{6}\right)$, $\frac{dy}{dx}$ is undefined.

At $\left(4, \frac{3\pi}{2}\right)$, $\frac{dy}{dx} = 0$.

61. (a), (b) $r = 3(1 - \cos\theta)$

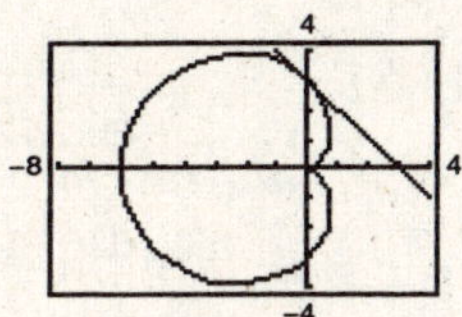

$(r, \theta) = \left(3, \frac{\pi}{2}\right) \Rightarrow (x, y) = (0, 3)$

Tangent line: $y - 3 = -1(x - 0)$

$$y = -x + 3$$

(c) At $\theta = \frac{\pi}{2}$, $\frac{dy}{dx} = -1.0$.

62. (a), (b) $r = 3 - 2\cos\theta$

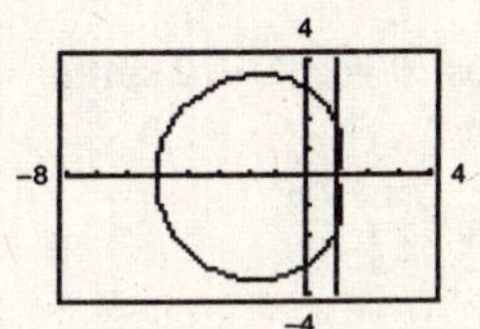

$(r, \theta) = (1, 0) \Rightarrow (x, y) = (1, 0)$

Tangent line: $x = 1$

(c) At $\theta = 0$, $\frac{dy}{dx}$ does not exist (vertical tangent).

63. (a), (b) $r = 3\sin\theta$

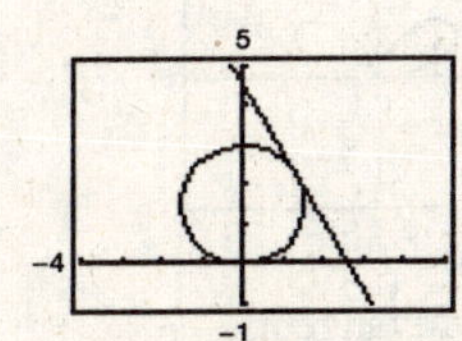

$(r, \theta) = \left(\frac{3\sqrt{3}}{2}, \frac{\pi}{3}\right) \Rightarrow (x, y) = \left(\frac{3\sqrt{3}}{4}, \frac{9}{4}\right)$

Tangent line: $y - \frac{9}{4} = -\sqrt{3}\left(x - \frac{3\sqrt{3}}{4}\right)$

$$y = -\sqrt{3}x + \frac{9}{2}$$

(c) At $\theta = \frac{\pi}{3}$, $\frac{dy}{dx} = -\sqrt{3} \approx -1.732$.

64. (a), (b) $r = 4$

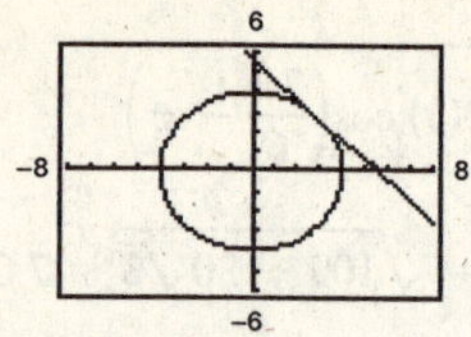

$(r, \theta) = \left(4, \frac{\pi}{4}\right) \Rightarrow (x, y) = \left(2\sqrt{2}, 2\sqrt{2}\right)$

Tangent line: $y - 2\sqrt{2} = -1\left(x - 2\sqrt{2}\right)$

$y = -x + 4\sqrt{2}$

(c) At $\theta = \frac{\pi}{4}, \frac{dy}{dx} = -1.$

65. $r = 1 - \sin\theta$

$\frac{dy}{d\theta} = (1 - \sin\theta)\cos\theta - \cos\theta\sin\theta$

$= \cos\theta(1 - 2\sin\theta) = 0$

$\cos\theta = 0 \text{ or } \sin\theta = \frac{1}{2} \Rightarrow \theta = \frac{\pi}{2}, \frac{3\pi}{2}, \frac{\pi}{6}, \frac{5\pi}{6}$

Horizontal tangents: $\left(2, \frac{3\pi}{2}\right), \left(\frac{1}{2}, \frac{\pi}{6}\right), \left(\frac{1}{2}, \frac{5\pi}{6}\right)$

$\frac{dx}{d\theta} = (-1 + \sin\theta)\sin\theta - \cos\theta\cos\theta$

$= -\sin\theta + \sin^2\theta + \sin^2\theta - 1$

$= 2\sin^2\theta - \sin\theta - 1$

$= (2\sin\theta + 1)(\sin\theta - 1) = 0$

$\sin\theta = 1 \text{ or } \sin\theta = -\frac{1}{2} \Rightarrow \theta = \frac{\pi}{2}, \frac{7\pi}{6}, \frac{11\pi}{6}$

Vertical tangents: $\left(\frac{3}{2}, \frac{7\pi}{6}\right), \left(\frac{3}{2}, \frac{11\pi}{6}\right)$

66. $r = a\sin\theta$

$\frac{dy}{d\theta} = a\sin\theta\cos\theta + a\cos\theta\sin\theta$

$= 2a\sin\theta\cos\theta = 0$

$\theta = 0, \frac{\pi}{2}, \pi, \frac{3\pi}{2}$

$\frac{dx}{d\theta} = -a\sin^2\theta + a\cos^2\theta = a(1 - 2\sin^2\theta) = 0$

$\sin\theta = \pm\frac{1}{\sqrt{2}}, \quad \theta = \frac{\pi}{4}, \frac{3\pi}{4}, \frac{5\pi}{4}, \frac{7\pi}{4}$

Horizontal: $(0, 0), \left(a, \frac{\pi}{2}\right)$

Vertical: $\left(\frac{a\sqrt{2}}{2}, \frac{\pi}{4}\right), \left(\frac{a\sqrt{2}}{2}, \frac{3\pi}{4}\right)$

67. $r = 2\csc\theta + 3$

$\frac{dy}{d\theta} = (2\csc\theta + 3)\cos\theta + (-2\csc\theta\cot\theta)\sin\theta$

$= 3\cos\theta = 0$

$\theta = \frac{\pi}{2}, \frac{3\pi}{2}$

Horizontal: $\left(5, \frac{\pi}{2}\right), \left(1, \frac{3\pi}{2}\right)$

68. $r = a\sin\theta\cos^2\theta$

$\frac{dy}{d\theta} = a\sin\theta\cos^3\theta + [-2a\sin^2\theta\cos\theta + a\cos^3\theta]\sin\theta$

$= 2a[\sin\theta\cos^3\theta - \sin^3\theta\cos\theta]$

$= 2a\sin\theta\cos\theta(\cos^2\theta - \sin^2\theta) = 0$

$\theta = 0, \ \tan^2\theta = 1, \ \theta = \frac{\pi}{4}, \frac{3\pi}{4}$

Horizontal: $\left(\frac{\sqrt{2}a}{4}, \frac{\pi}{4}\right), \left(\frac{\sqrt{2}a}{4}, \frac{3\pi}{4}\right), (0, 0)$

69. $r = 4\sin\theta\cos^2\theta$

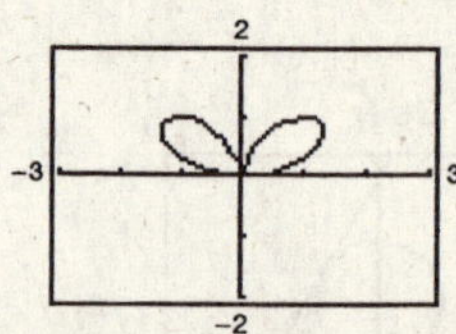

Horizontal tangents:

$(r, \theta) = (0, 0), (1.4142, 0.7854), (1.4142, 2.3562)$

70. $r = 3\cos 2\theta \sec\theta$

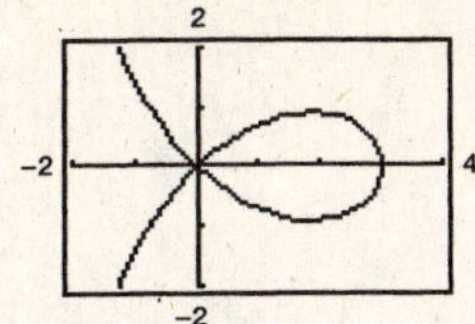

Horizontal tangents: $(r, \theta) = (2.061, \pm 0.452)$

71. $r = 2\csc\theta + 5$

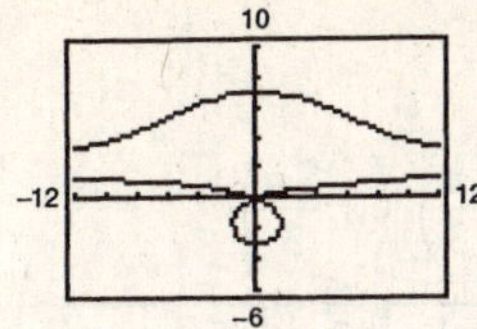

Horizontal tangents: $(r, \theta) = \left(7, \dfrac{\pi}{2}\right), \left(3, \dfrac{3\pi}{2}\right)$

72. $r = 2\cos(3\theta - 2)$

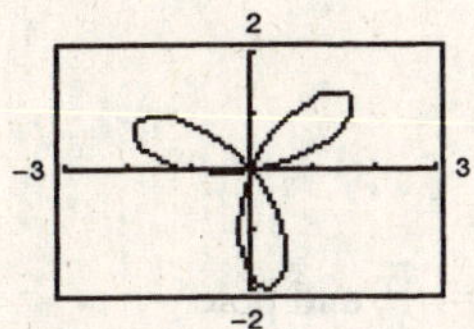

Horizontal tangents:

$(r, \theta) = (1.894, 0.776), (1.755, 2.594), (1.998, -1.442), (-0.423, 0.072)$

73.

$$r = 3\sin\theta$$
$$r^2 = 3r\sin\theta$$
$$x^2 + y^2 = 3y$$
$$x^2 + \left(y - \frac{3}{2}\right)^2 = \frac{9}{4}$$

Circle $r = \dfrac{3}{2}$

Center: $\left(0, \dfrac{3}{2}\right)$

Tangent at the pole: $\theta = 0$

74.

$$r = 3\cos\theta$$
$$r^2 = 3r\cos\theta$$
$$x^2 + y^2 = 3x$$
$$\left(x - \frac{3}{2}\right)^2 + y^2 = \frac{9}{4}$$

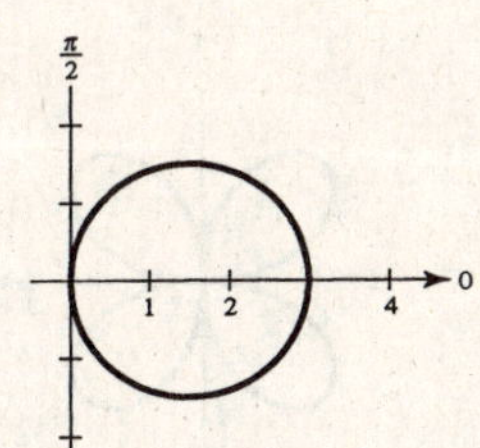

Circle: $r = \dfrac{3}{2}$

Center: $\left(\dfrac{3}{2}, 0\right)$

Tangent at pole: $\theta = \dfrac{\pi}{2}$

75. $r = 2(1 - \sin\theta)$

Cardioid

Symmetric to y-axis, $\theta = \dfrac{\pi}{2}$

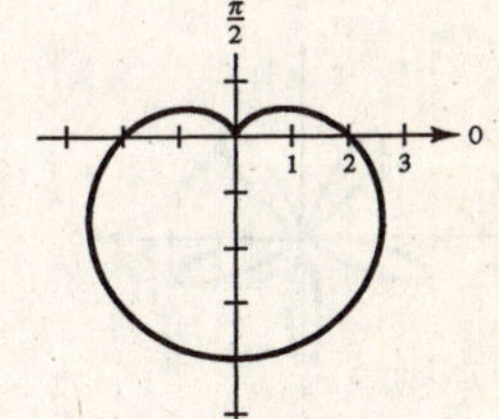

76. $r = 3(1 - \cos\theta)$

Cardioid

Symmetric to polar axis since r is a function of $\cos\theta$.

θ	0	$\frac{\pi}{3}$	$\frac{\pi}{2}$	$\frac{2\pi}{3}$	π
r	0	$\frac{3}{2}$	3	$\frac{9}{2}$	6

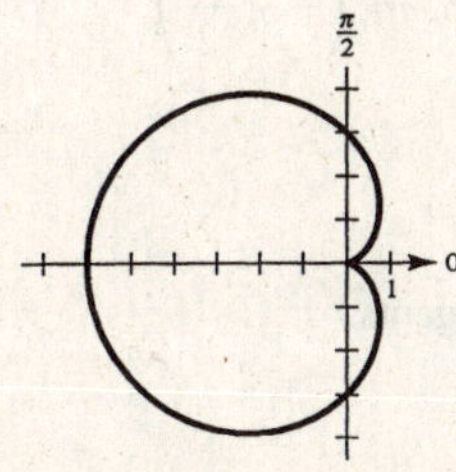

77. $r = 2\cos(3\theta)$

Rose curve with three petals

Symmetric to the polar axis

Relative extrema: $(2, 0), \left(-2, \frac{\pi}{3}\right), \left(2, \frac{2\pi}{3}\right)$

θ	0	$\frac{\pi}{6}$	$\frac{\pi}{4}$	$\frac{\pi}{3}$	$\frac{\pi}{2}$	$\frac{2\pi}{3}$	$\frac{5\pi}{6}$	π
r	2	0	$-\sqrt{2}$	-2	0	2	0	-2

Tangents at the pole: $\theta = \frac{\pi}{6}, \frac{\pi}{2}, \frac{5\pi}{6}$

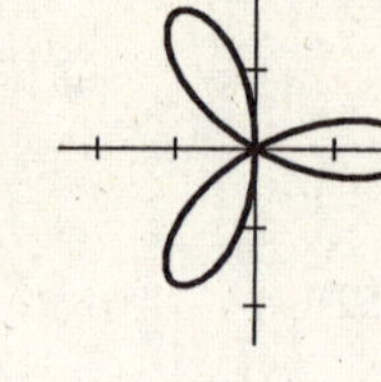
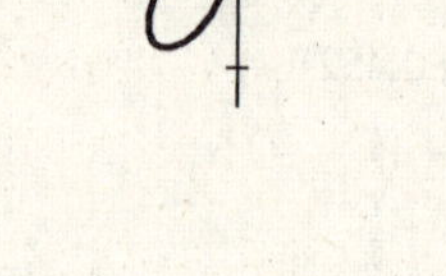

78. $r = -\sin(5\theta)$

Rose curve with five petals

Symmetric to $\theta = \frac{\pi}{2}$

Relative extrema occur when

$$\frac{dr}{d\theta} = -5\cos(5\theta) = 0 \text{ at } \theta = \frac{\pi}{10}, \frac{3\pi}{10}, \frac{5\pi}{10}, \frac{7\pi}{10}, \frac{9\pi}{10}.$$

Tangents at the pole: $\theta = 0, \frac{\pi}{5}, \frac{2\pi}{5}, \frac{3\pi}{5}, \frac{4\pi}{5}$

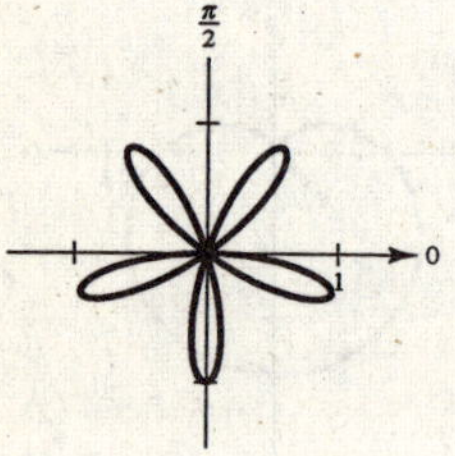

79. $r = 3\sin 2\theta$

Rose curve with four petals

Symmetric to the polar axis, $\theta = \frac{\pi}{2}$, and pole

Relative extrema: $\left(\pm 3, \frac{\pi}{4}\right), \left(\pm 3, \frac{5\pi}{4}\right)$

Tangents at the pole: $\theta = 0, \frac{\pi}{2}$

($\theta = \pi$, $3\pi/2$ give the same tangents.)

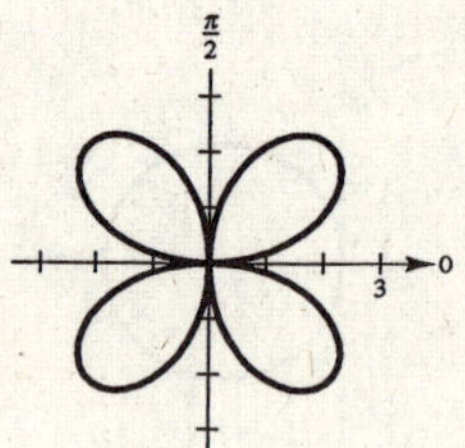

80. $r = 3\cos 2\theta$

Rose curve with four petals

Symmetric to the polar axis, $\theta = \frac{\pi}{2}$, and pole

Relative extrema: $(3, 0), \left(-3, \frac{\pi}{2}\right), (3, \pi), \left(-3, \frac{3\pi}{2}\right)$

Tangents at the pole: $\theta = \frac{\pi}{4}, \frac{3\pi}{4}$

$\theta = \frac{5\pi}{4}$ and $\frac{7\pi}{4}$ given the same tangents.

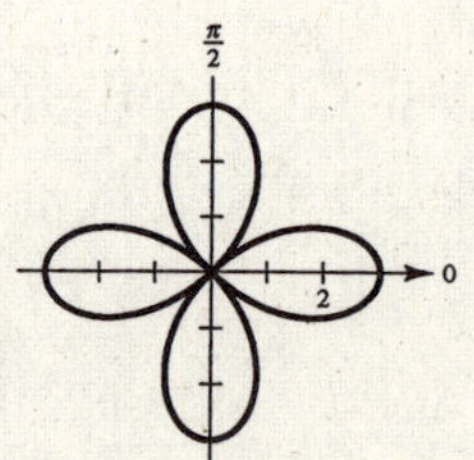

81. $r = 5$

Circle radius: 5

$x^2 + y^2 = 25$

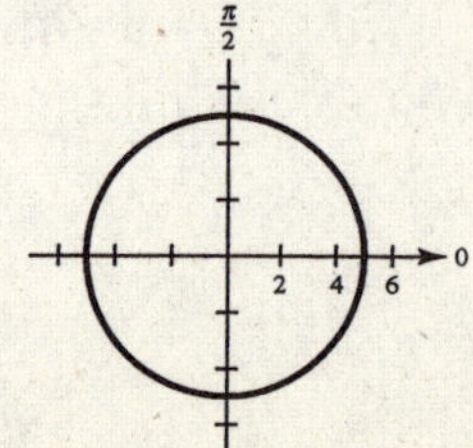

82. $r = 2$

Circle radius: 2

$x^2 + y^2 = 4$

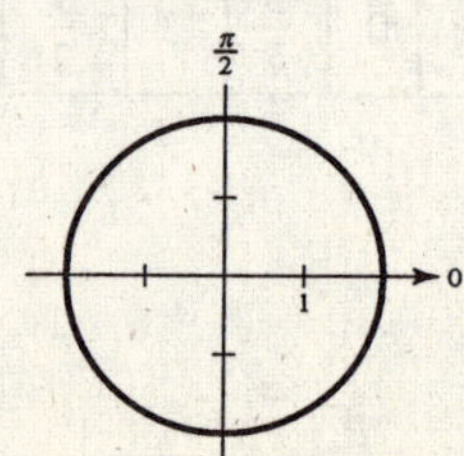

83. $r = 4(1 + \cos \theta)$

Cardioid

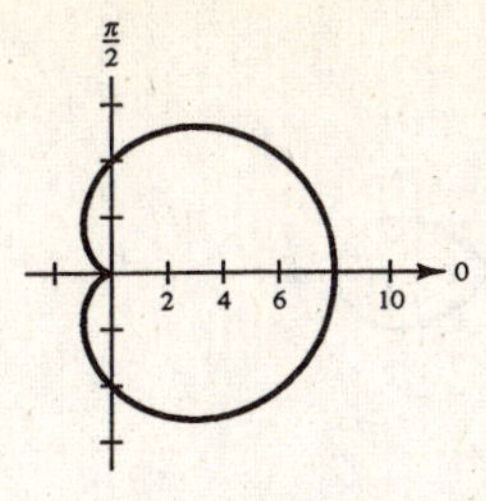

84. $r = 1 + \sin \theta$

Cardioid

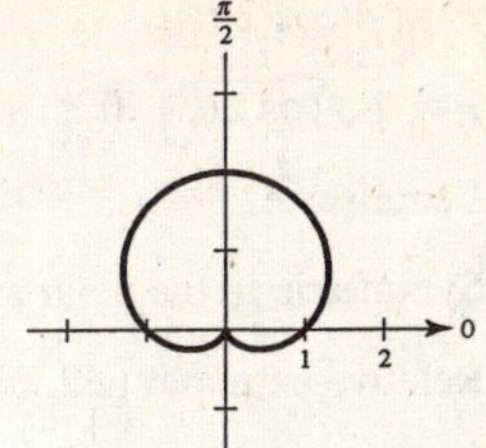

85. $r = 3 - 2 \cos \theta$

Limaçon

Symmetric to polar axis

θ	0	$\frac{\pi}{3}$	$\frac{\pi}{2}$	$\frac{2\pi}{3}$	π
r	1	2	3	4	5

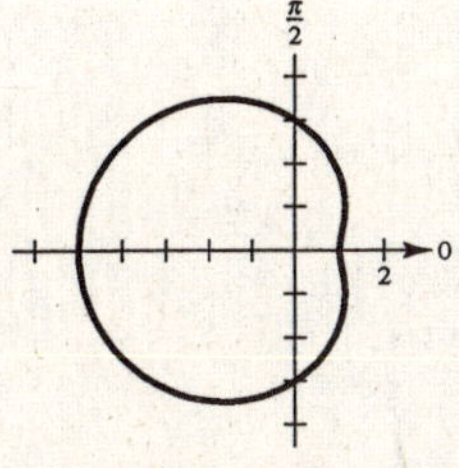

86. $r = 5 - 4 \sin \theta$

Limaçon

Symmetric to $\theta = \frac{\pi}{2}$

θ	$-\frac{\pi}{2}$	$-\frac{\pi}{6}$	0	$\frac{\pi}{6}$	$\frac{\pi}{2}$
r	9	7	5	3	1

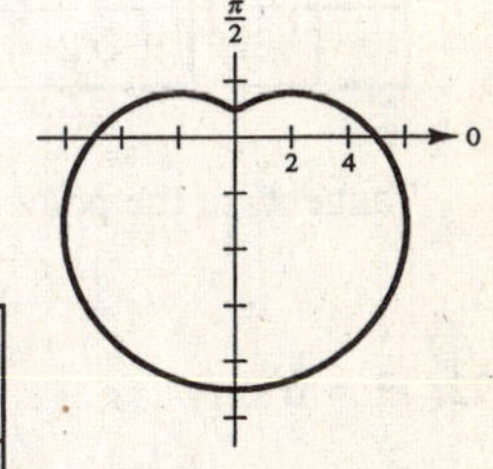

87. $r = 3 \csc \theta$

$r \sin \theta = 3$

$y = 3$

Horizontal line

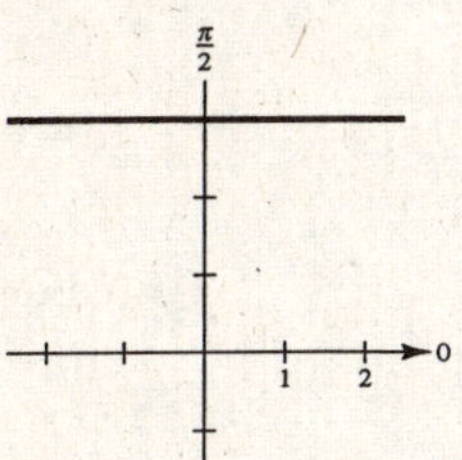

88. $$r = \frac{6}{2 \sin \theta - 3 \cos \theta}$$

$2r \sin \theta - 3r \cos \theta = 6$

$2y - 3x = 6$

Line

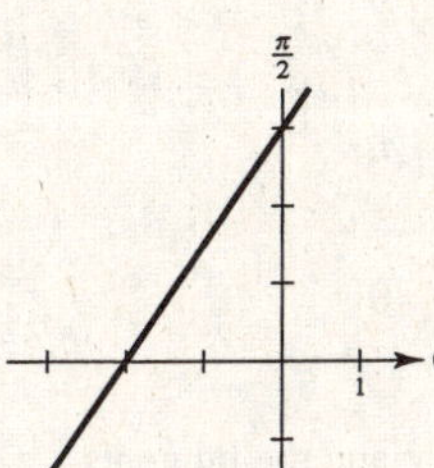

89. $r = 2\theta$

Spiral of Archimedes

Symmetric to $\theta = \frac{\pi}{2}$

θ	0	$\frac{\pi}{4}$	$\frac{\pi}{2}$	$\frac{3\pi}{4}$	π	$\frac{5\pi}{4}$	$\frac{3\pi}{2}$
r	0	$\frac{\pi}{2}$	π	$\frac{3\pi}{2}$	2π	$\frac{5\pi}{2}$	3π

Tangent at the pole: $\theta = 0$

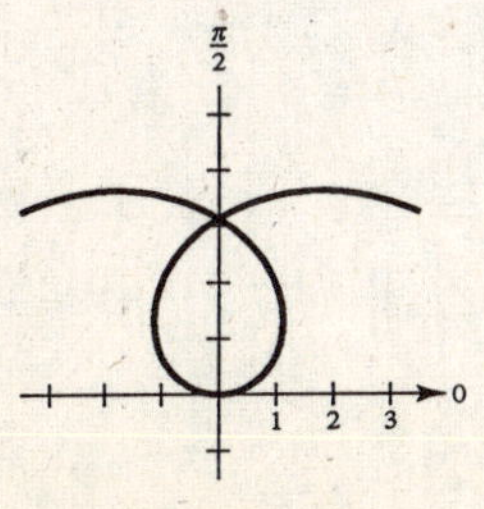

90. $r = \frac{1}{\theta}$

Hyperbolic spiral

θ	$\frac{\pi}{4}$	$\frac{\pi}{2}$	$\frac{3\pi}{4}$	π	$\frac{5\pi}{4}$	$\frac{3\pi}{2}$
r	$\frac{4}{\pi}$	$\frac{2}{\pi}$	$\frac{4}{3\pi}$	$\frac{1}{\pi}$	$\frac{4}{5\pi}$	$\frac{2}{3\pi}$

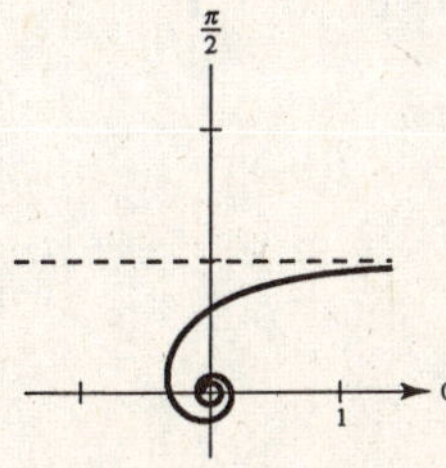

91. $r^2 = 4\cos(2\theta)$

$r = 2\sqrt{\cos 2\theta}, \quad 0 \le \theta \le 2\pi$

Lemniscate

Symmetric to the polar axis, $\theta = \dfrac{\pi}{2}$, and pole

Relative extrema: $(\pm 2, 0)$

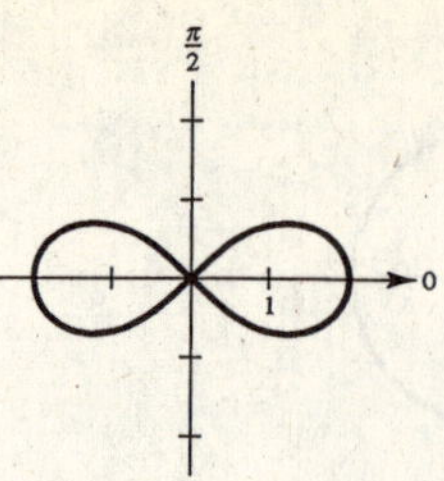

θ	0	$\dfrac{\pi}{6}$	$\dfrac{\pi}{4}$
r	± 2	$\pm\sqrt{2}$	0

Tangents at the pole: $\theta = \dfrac{\pi}{4}, \dfrac{3\pi}{4}$

92. $r^2 = 4\sin\theta$

Lemniscate

Symmetric to the polar axis, $\theta = \dfrac{\pi}{2}$, and pole

Relative extrema: $\left(\pm 2, \dfrac{\pi}{2}\right)$

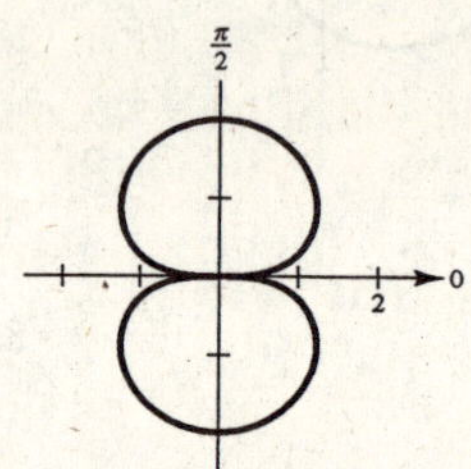
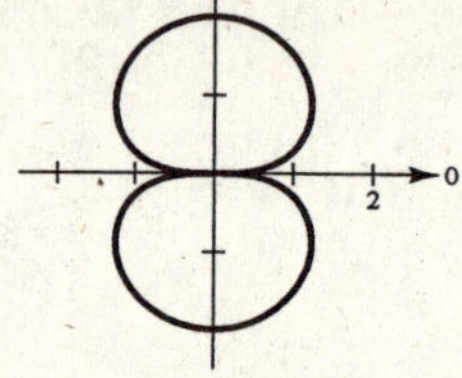

θ	0	$\dfrac{\pi}{6}$	$\dfrac{\pi}{2}$	$\dfrac{5\pi}{6}$	π
r	0	$\pm\sqrt{2}$	± 2	$\pm\sqrt{2}$	0

Tangent at the pole: $\theta = 0$

93. Since

$$r = 2 - \sec\theta = 2 - \frac{1}{\cos\theta},$$

the graph has polar axis symmetry and the tangents at the pole are

$$\theta = \frac{\pi}{3}, \frac{-\pi}{3}.$$

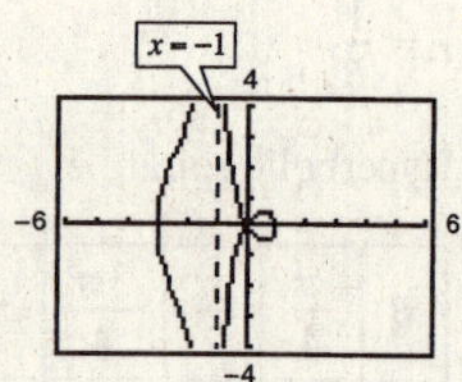

Furthermore,

$$r \Rightarrow -\infty \text{ as } \theta \Rightarrow \frac{\pi}{2^-}$$

$$r \Rightarrow \infty \text{ as } \theta \Rightarrow -\frac{\pi}{2^+}.$$

Also, $r = 2 - \dfrac{1}{\cos\theta} = 2 - \dfrac{r}{r\cos\theta} = 2 - \dfrac{r}{x}$

$$rx = 2x - r$$

$$r = \frac{2x}{1 + x}.$$

Thus, $r \Rightarrow \pm\infty$ as $x \Rightarrow -1$.

94. Since

$$r = 2 + \csc\theta = 2 + \frac{1}{\sin\theta},$$

the graphs has symmetry with respect to $\theta = \pi/2$. Furthermore,

$$r \Rightarrow \infty \text{ as } \theta \Rightarrow 0^+$$

$$r \Rightarrow \infty \text{ as } \theta \Rightarrow \pi^-.$$

Also, $r = 2 + \dfrac{1}{\sin\theta} = 2 + \dfrac{r}{\sin\theta} = 2 + \dfrac{r}{y}$

$$ry = 2y + r$$

$$r = \frac{2y}{y - 1}.$$

Thus, $r \Rightarrow \pm\infty$ as $y \Rightarrow 1$.

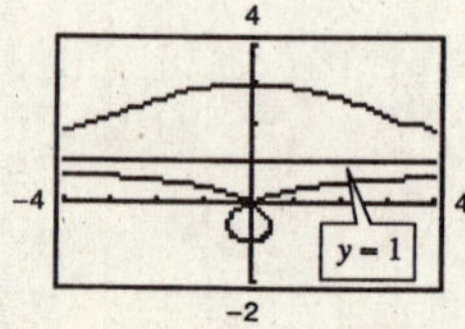

95. $r = \dfrac{2}{\theta}$

Hyperbolic spiral

$r \Rightarrow \infty$ as $\theta \Rightarrow 0$

$$r = \frac{2}{\theta} \Rightarrow \theta = \frac{2}{r} = \frac{2\sin\theta}{r\sin\theta} = \frac{2\sin\theta}{y}$$

$$y = \frac{2\sin\theta}{\theta}$$

$$\lim_{\theta\to 0}\frac{2\sin\theta}{\theta} = \lim_{\theta\to 0}\frac{2\cos\theta}{1} = 2$$

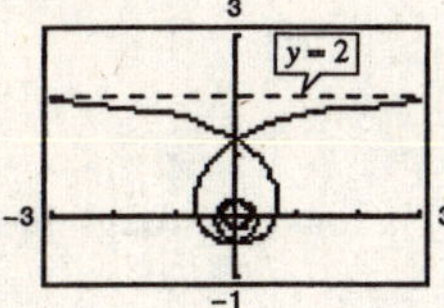

96. $r = 2\cos 2\theta \sec\theta$

Strophoid

$r \Rightarrow -\infty$ as $\theta \Rightarrow \dfrac{\pi^-}{2}$

$r \Rightarrow \infty$ as $\theta \Rightarrow \dfrac{-\pi^+}{2}$

$$r = 2\cos 2\theta \sec\theta = 2(2\cos^2\theta - 1)\sec\theta$$

$$r\cos\theta = 4\cos^2\theta - 2$$

$$x = 4\cos^2\theta - 2$$

$$\lim_{\theta\to\pm\pi/2}(4\cos^2\theta - 2) = -2$$

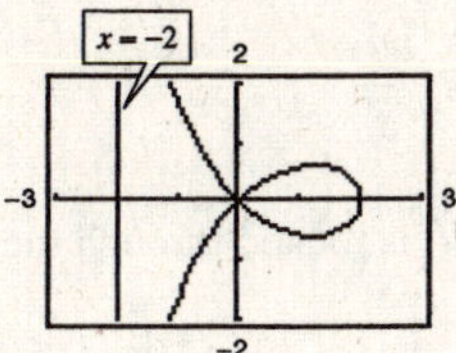

97. The rectangular coordinate system consists of all points of the form (x, y) where x is the directed distance from the y-axis to the point, and y is the directed distance from the x-axis to the point. Every point has a unique representation.

The polar coordinate system uses (r, θ) to designate the location of a point.

r is the directed distance to the origin and θ is the angle the point makes with the positive x-axis, measured clockwise.

Points do not have a unique polar representation.

98. $x = r\cos\theta,\ y = r\sin\theta$

$x^2 + y^2 = r^2,\ \tan\theta = \dfrac{y}{x}$

99. $r = a$, circle

$\theta = b$, line

100. Slope of tangent line to graph of $r = f(\theta)$ at (r, θ) is

$$\frac{dy}{dx} = \frac{f(\theta)\cos\theta + f'(\theta)\sin\theta}{-f(\theta)\sin\theta + f'(\theta)\cos\theta}.$$

If $f(\alpha) = 0$ and $f'(\alpha) \neq 0$, then $\theta = \alpha$ is tangent at the pole.

101. $r = 4\sin\theta$

(a) $0 \le \theta \le \dfrac{\pi}{2}$

(b) $\dfrac{\pi}{2} \le \theta \le \pi$

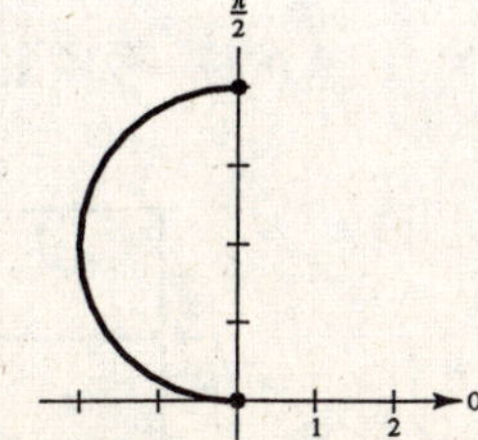

(c) $-\dfrac{\pi}{2} \le \theta \le \dfrac{\pi}{2}$

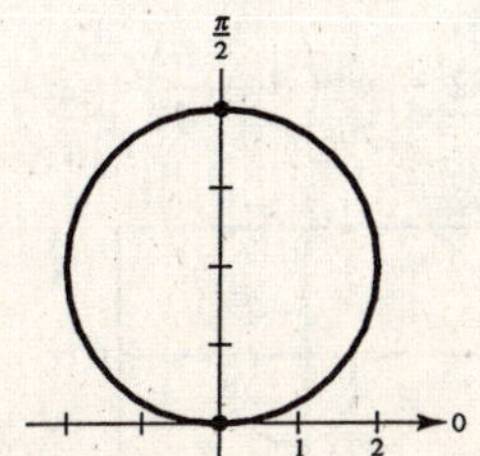

102. $r = 6[1 + \cos(\theta - \phi)]$

(a) $\phi = 0,\ r = 6[1 + \cos\theta]$

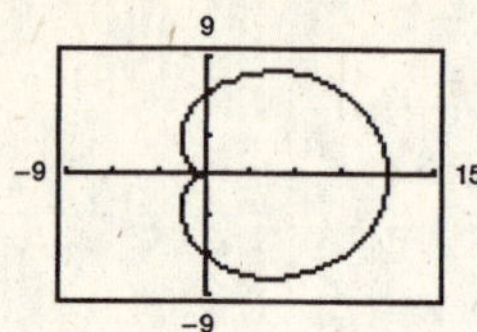

(b) $\phi = \dfrac{\pi}{4},\ r = 6\left[1 + \cos\left(\theta - \dfrac{\pi}{4}\right)\right]$

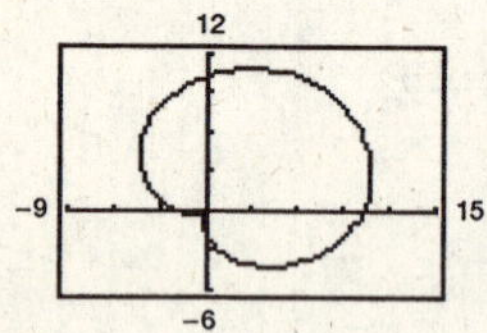

The graph of $r = 6[1 + \cos\theta]$ is rotated through the angle $\pi/4$.

(c) $\theta = \dfrac{\pi}{2}$

$$r = 6\left[1 + \cos\left(\theta - \frac{\pi}{2}\right)\right]$$

$$= 6\left[1 + \cos\theta\cos\frac{\pi}{2} + \sin\theta\sin\frac{\pi}{2}\right]$$

$$= 6[1 + \sin\theta]$$

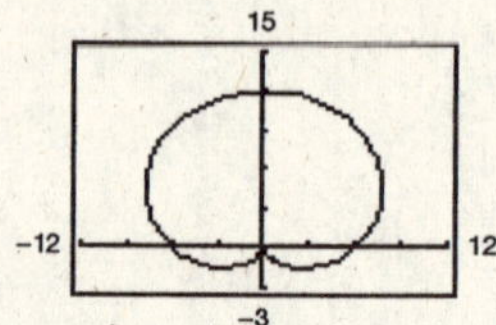

The graph of $r = 6[1 + \cos\theta]$ is rotated through the angle $\pi/2$.

103. Let the curve $r = f(\theta)$ be rotated by ϕ to form the curve $r = g(\theta)$. If (r_1, θ_1) is a point on $r = f(\theta)$, then $(r_1, \theta_1 + \phi)$ is on $r = g(\theta)$. That is,

$$g(\theta_1 + \phi) = r_1 = f(\theta_1).$$

Letting $\theta = \theta_1 + \phi$, or $\theta_1 = \theta - \phi$, we see that

$$g(\theta) = g(\theta_1 + \phi) = f(\theta_1) = f(\theta - \phi).$$

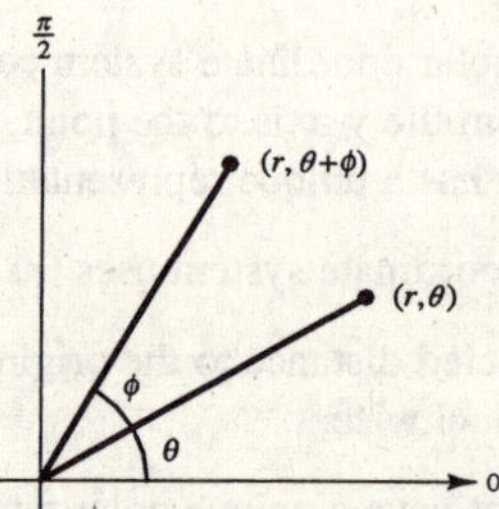

104. (a) $\sin\left(\theta - \dfrac{\pi}{2}\right) = \sin\theta\cos\left(\dfrac{\pi}{2}\right) - \cos\theta\sin\left(\dfrac{\pi}{2}\right)$

$$= -\cos\theta$$

$$r = f\left[\sin\left(\theta - \frac{\pi}{2}\right)\right]$$

$$= f(-\cos\theta)$$

(b) $\sin(\theta - \pi) = \sin\theta\cos\pi - \cos\theta\sin\pi$

$$= -\sin\theta$$

$$r = f[\sin(\theta - \pi)]$$

$$= f(-\sin\theta)$$

(c) $\sin\left(\theta - \dfrac{3\pi}{2}\right) = \sin\theta\cos\left(\dfrac{3\pi}{2}\right) - \cos\theta\sin\left(\dfrac{3\pi}{2}\right)$

$$= \cos\theta$$

$$r = f\left[\sin\left(\theta - \frac{3\pi}{2}\right)\right] = f(\cos\theta)$$

105. $r = 2 - \sin\theta$

(a) $r = 2 - \sin\left(\theta - \dfrac{\pi}{4}\right) = 2 - \dfrac{\sqrt{2}}{2}(\sin\theta - \cos\theta)$

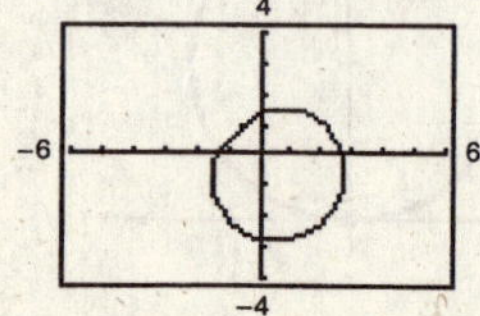

(b) $r = 2 - \sin\left(\theta - \dfrac{\pi}{2}\right) = 2 - (-\cos\theta) = 2 + \cos\theta$

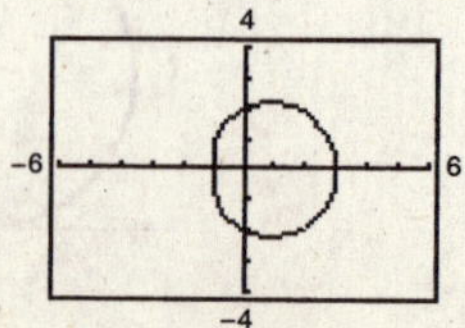

—CONTINUED—

105. —CONTINUED—

(c) $r = 2 - \sin(\theta - \pi) = 2 - (-\sin\theta) = 2 + \sin\theta$

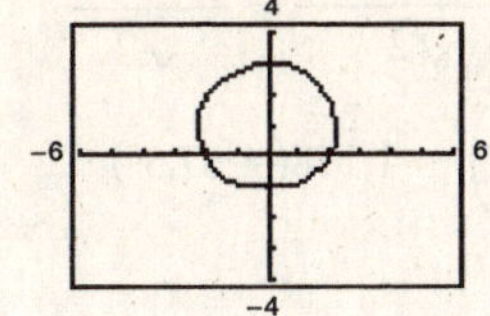

(d) $r = 2 - \sin\left(\theta - \dfrac{3\pi}{2}\right) = 2 - \cos\theta$

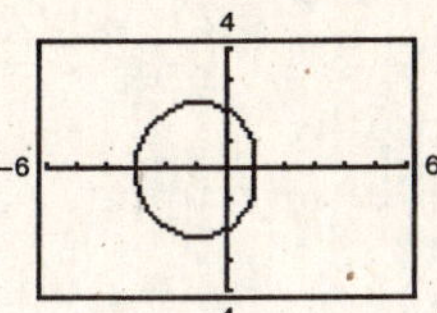

106. $r = 2\sin 2\theta = 4\sin\theta\cos\theta$

(a) $r = 4\sin\left(\theta - \dfrac{\pi}{6}\right)\cos\left(\theta - \dfrac{\pi}{6}\right)$

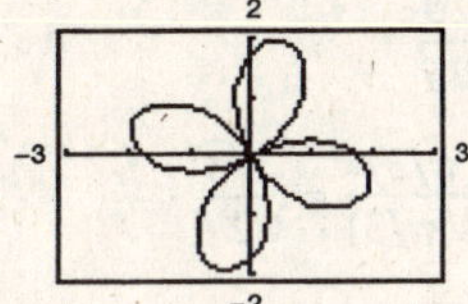

(b) $r = 4\sin\left(\theta - \dfrac{\pi}{2}\right)\cos\left(\theta - \dfrac{\pi}{2}\right) = -4\sin\theta\cos\theta$

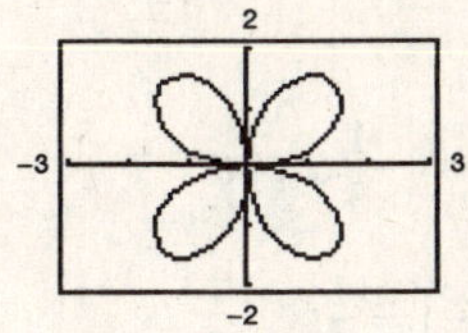

(c) $r = 4\sin\left(\theta - \dfrac{2\pi}{3}\right)\cos\left(\theta - \dfrac{2\pi}{3}\right)$

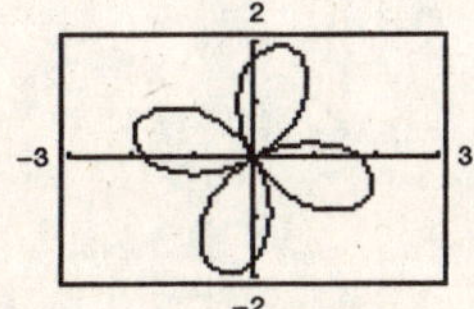

(d) $r = 4\sin(\theta - \pi)\cos(\theta - \pi) = 4\sin\theta\cos\theta$

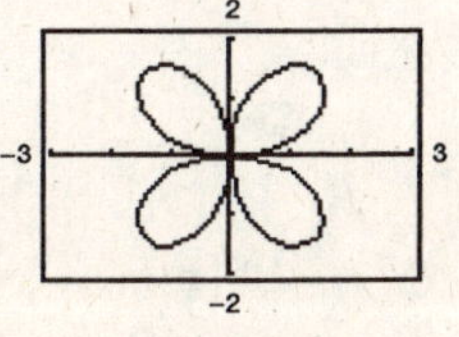

107. (a) $r = 1 - \sin\theta$

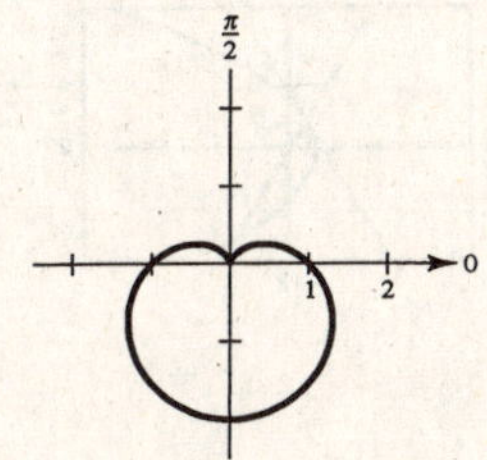

(b) $r = 1 - \sin\left(\theta - \dfrac{\pi}{4}\right)$

Rotate the graph of

$r = 1 - \sin\theta$

through the angle $\pi/4$.

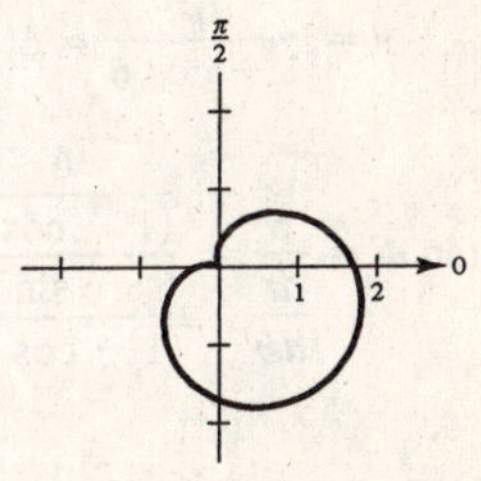

108. By Theorem 9.11, the slope of the tangent line through A and P is

$$\frac{f\cos\theta + f'\sin\theta}{-f\sin\theta + f'\cos\theta}.$$

This is equal to

$$\tan(\theta + \psi) = \frac{\tan\theta + \tan\psi}{1 - \tan\theta\tan\psi} = \frac{\sin\theta + \cos\theta\tan\psi}{\cos\theta - \sin\theta\tan\psi}.$$

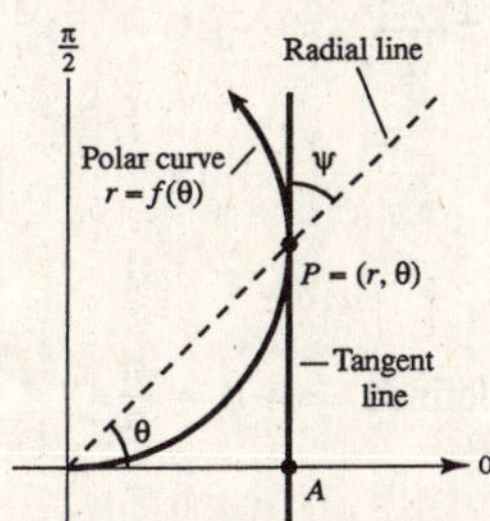

Equating the expressions and cross-multiplying, you obtain

$$(f\cos\theta + f'\sin\theta)(\cos\theta - \sin\theta\tan\psi) = (\sin\theta + \cos\theta\tan\psi)(-f\sin\theta + f'\cos\theta)$$

$$f\cos^2\theta - f\cos\theta\sin\theta\tan\psi + f'\sin\theta\cos\theta - f'\sin^2\theta\tan\psi = -f\sin^2\theta - f\sin\theta\cos\theta\tan\psi + f'\sin\theta\cos\theta + f'\cos^2\theta\tan\psi$$

$$f(\cos^2\theta + \sin^2\theta) = f'\tan\psi(\cos^2\theta + \sin^2\theta)$$

$$\tan\psi = \frac{f}{f'} = \frac{r}{dr/d\theta}.$$

109. $\tan\psi = \dfrac{r}{dr/d\theta} = \dfrac{2(1-\cos\theta)}{2\sin\theta}$

At $\theta = \pi$, $\tan\psi$ is undefined $\Rightarrow \psi = \dfrac{\pi}{2}$.

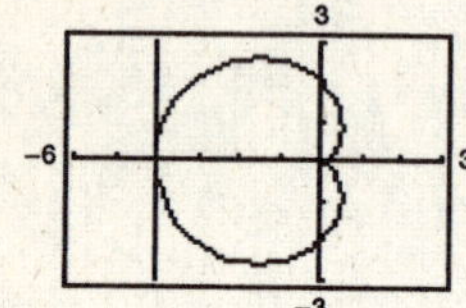

110. $\tan\psi = \dfrac{r}{dr/d\theta} = \dfrac{3(1-\cos\theta)}{3\sin\theta}$

At $\theta = \dfrac{3\pi}{4}$, $\tan\psi = \dfrac{1+(\sqrt{2}/2)}{\sqrt{2}} = \dfrac{2+\sqrt{2}}{\sqrt{2}}$.

$\psi = \arctan\left(\dfrac{2+\sqrt{2}}{\sqrt{2}}\right) \approx 1.178 (\approx 67.5^\circ)$

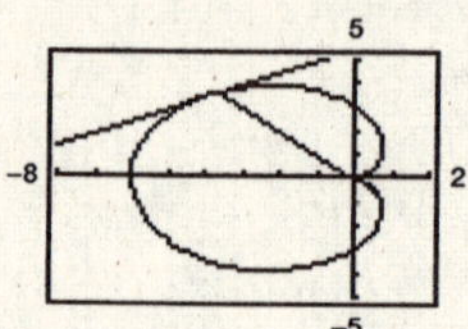

111. $r = 2\cos 3\theta$

$\tan\psi = \dfrac{r}{dr/d\theta} = \dfrac{2\cos 3\theta}{-6\sin 3\theta} = -\dfrac{1}{3}\cot 3\theta$

At $\theta = \dfrac{\pi}{4}$, $\tan\psi = -\dfrac{1}{3}\cot\left(\dfrac{3\pi}{4}\right) = \dfrac{1}{3}$.

$\psi = \arctan\left(\dfrac{1}{3}\right) \approx 18.4^\circ$

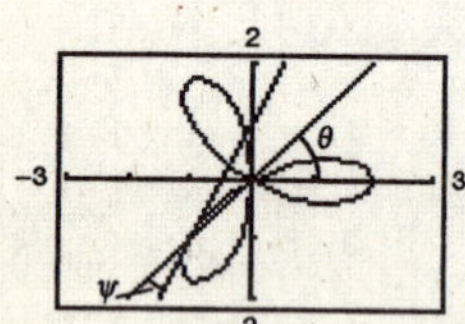

112. $\tan\psi = \dfrac{r}{dr/d\theta} = \dfrac{4\sin 2\theta}{8\cos 2\theta}$

At $\theta = \dfrac{\pi}{6}$, $\tan\psi = \dfrac{\sin(\pi/3)}{2\cos(\pi/3)} = \dfrac{\sqrt{3}}{2}$.

$\psi = \arctan\left(\dfrac{\sqrt{3}}{2}\right) \approx 0.7137 (\approx 40.89^\circ)$

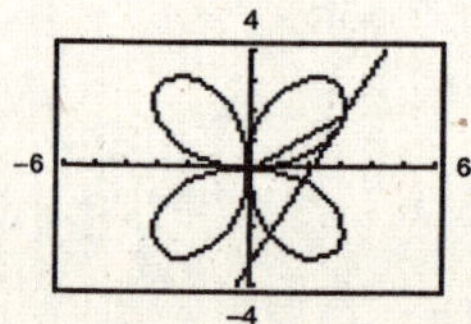

113. $r = \dfrac{6}{1-\cos\theta} = 6(1-\cos\theta)^{-1} \Rightarrow \dfrac{dr}{d\theta} = \dfrac{6\sin\theta}{(1-\cos\theta)^2}$

$\tan\psi = \dfrac{r}{\frac{dr}{d\theta}} = \dfrac{\frac{6}{(1-\cos\theta)}}{\frac{-6\sin\theta}{(1-\cos\theta)^2}} = \dfrac{1-\cos\theta}{-\sin\theta}$

At $\theta = \dfrac{2\pi}{3}$, $\tan\psi = \dfrac{1-\left(-\frac{1}{2}\right)}{-\frac{\sqrt{3}}{2}} = -\sqrt{3}$.

$\psi = \dfrac{\pi}{3}, (60^\circ)$

114. $\tan\psi = \dfrac{r}{dr/d\theta} = \dfrac{5}{0}$ undefined $\Rightarrow \psi = \dfrac{\pi}{2}$

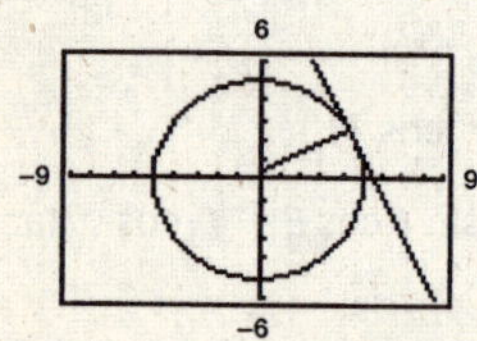

115. True

116. True

117. True

118. True

Section 10.5 Area and Arc Length in Polar Coordinates

1. $A = \frac{1}{2}\int_{\alpha}^{\beta} [f(\theta)]^2\, d\theta$

$= \frac{1}{2}\int_{\pi/2}^{\pi} (2\sin\theta)^2\, d\theta$

$= 2\int_{\pi/2}^{\pi} \sin^2\theta\, d\theta$

2. $A = \frac{1}{2}\int_{\alpha}^{\beta} [f(\theta)]^2\, d\theta$

$= \frac{1}{2}\int_{3\pi/4}^{5\pi/4} (\cos 2\theta)^2\, d\theta$

3. $A = \frac{1}{2}\int_{\alpha}^{\beta} [f(\theta)]^2\, d\theta$

$= \frac{1}{2}\int_{\pi/2}^{3\pi/2} (1 - \sin\theta)^2\, d\theta$

4. $A = \frac{1}{2}\int_{\alpha}^{\beta} [f(\theta)]^2\, d\theta$

$= \frac{1}{2}\int_{0}^{\pi} (1 - \cos 2\theta)^2\, d\theta$

5. (a) $r = 8\sin\theta$

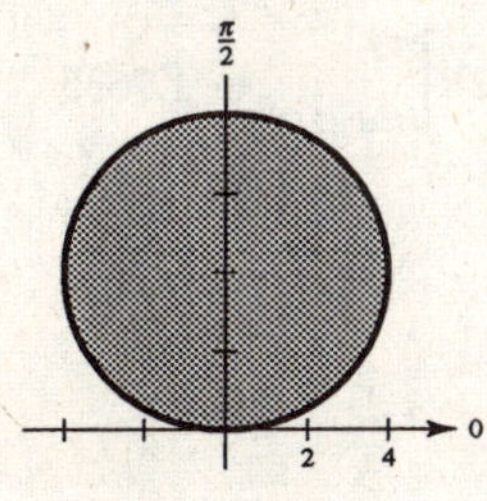

$A = \pi(4)^2 = 16\pi$

(b) $A = 2\left(\frac{1}{2}\right)\int_{0}^{\pi/2} [8\sin\theta]^2\, d\theta$

$= 64\int_{0}^{\pi/2} \sin^2\theta\, d\theta$

$= 32\int_{0}^{\pi/2} (1 - \cos 2\theta)\, d\theta$

$= 32\left[\theta - \frac{\sin 2\theta}{2}\right]_{0}^{\pi/2} = 16\pi$

6. (a) $r = 3\cos\theta$

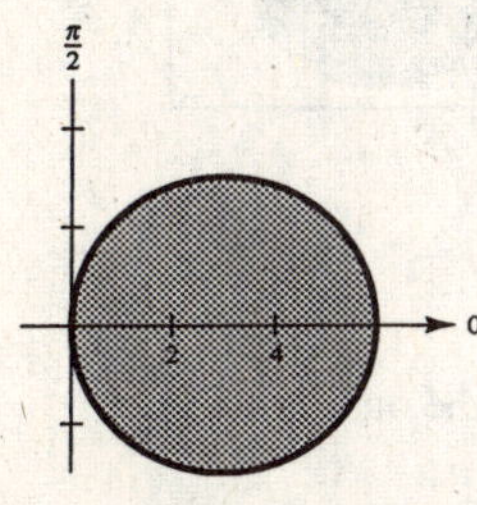

$A = \pi\left(\frac{3}{2}\right)^2 = \frac{9\pi}{4}$

(b) $A = 2\left(\frac{1}{2}\right)\int_{0}^{\pi/2} [3\cos\theta]^2\, d\theta$

$= 9\int_{0}^{\pi/2} \cos^2\theta\, d\theta$

$= \frac{9}{2}\int_{0}^{\pi/2} (1 + \cos 2\theta)\, d\theta$

$= \frac{9}{2}\left[\theta + \frac{\sin 2\theta}{2}\right]_{0}^{\pi/2} = \frac{9\pi}{4}$

7. $A = 2\left[\frac{1}{2}\int_{0}^{\pi/6} (2\cos 3\theta)^2\, d\theta\right] = 2\left[\theta + \frac{1}{6}\sin 6\theta\right]_{0}^{\pi/6} = \frac{\pi}{3}$

8. $A = 2\left[\frac{1}{2}\int_{0}^{\pi/4} (6\sin 2\theta)^2\, d\theta\right] = 36\int_{0}^{\pi/4} \sin^2 2\theta\, d\theta$

$= 36\int_{0}^{\pi/4} \frac{1 - \cos 4\theta}{2}\, d\theta$

$= 18\left[\theta - \frac{\sin 4\theta}{4}\right]_{0}^{\pi/4}$

$= 18\left[\frac{\pi}{4}\right] = \frac{9\pi}{2}$

9. $A = 2\left[\frac{1}{2}\int_0^{\pi/4} (\cos 2\theta)^2\, d\theta\right]$

$= \frac{1}{2}\left[\theta + \frac{1}{4}\sin 4\theta\right]_0^{\pi/4} = \frac{\pi}{8}$

10. $A = 2\left[\frac{1}{2}\int_0^{\pi/10} (\cos 5\theta)^2\, d\theta\right]$

$= \frac{1}{2}\left[\theta + \frac{1}{10}\sin(10\theta)\right]_0^{\pi/10} = \frac{\pi}{20}$

11. $A = 2\left[\frac{1}{2}\int_{-\pi/2}^{\pi/2} (1 - \sin\theta)^2\, d\theta\right]$

$= \left[\frac{3}{2}\theta + 2\cos\theta - \frac{1}{4}\sin 2\theta\right]_{-\pi/2}^{\pi/2} = \frac{3\pi}{2}$

12. $A = 2\left[\frac{1}{2}\int_0^{\pi/2} (1 - \sin\theta)^2\, d\theta\right]$

$= \left[\frac{3}{2}\theta + 2\cos\theta - \frac{1}{4}\sin 2\theta\right]_0^{\pi/2} = \frac{3\pi - 8}{4}$

13. $A = 2\frac{1}{2}\left[\int_{2\pi/3}^{\pi} (1 + 2\cos\theta)^2\, d\theta\right]$

$= \left[3\theta + 4\sin\theta + \sin 2\theta\right]_{2\pi/3}^{\pi} = \frac{2\pi - 3\sqrt{3}}{2}$

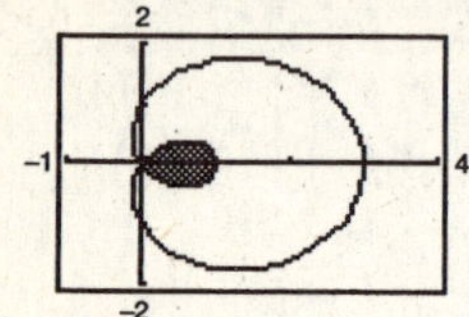

14. $A = 2\left[\frac{1}{2}\int_{\arcsin(2/3)}^{\pi/2} (4 - 6\sin\theta)^2\, d\theta\right]$

$= \int_{\arcsin(2/3)}^{\pi/2} \left[16 - 48\sin\theta + 36\sin^2\theta\right] d\theta$

$= \int_{\arcsin(2/3)}^{\pi/2} \left[16 - 48\sin\theta + 36\left(\frac{1 - \cos 2\theta}{2}\right)\right] d\theta$

$= \left[34\theta + 48\cos\theta - 9\sin 2\theta\right]_{\arcsin(2/3)}^{\pi/2} \approx 1.7635$

15. The area inside the outer loop is

$2\left[\frac{1}{2}\int_0^{2\pi/3} (1 + 2\cos\theta)^2\, d\theta\right] = \left[3\theta + 4\sin\theta + \sin 2\theta\right]_0^{2\pi/3} = \frac{4\pi + 3\sqrt{3}}{2}.$

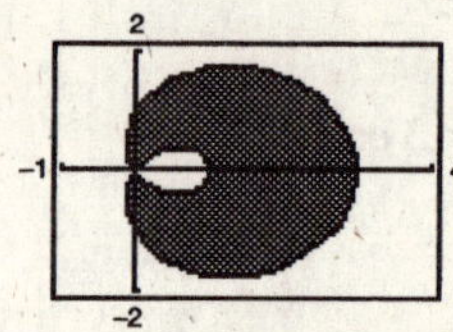

From the result of Exercise 13, the area between the loops is

$A = \left(\frac{4\pi + 3\sqrt{3}}{2}\right) - \left(\frac{2\pi - 3\sqrt{3}}{2}\right) = \pi + 3\sqrt{3}.$

16. Four times the area in Exercise 15, $A = 4\left(\pi + 3\sqrt{3}\right)$. More specifically, we see that the area inside the outer loop is

$2\left[\frac{1}{2}\int_{-\pi/6}^{\pi/2} (2(1 + 2\sin\theta))^2\, d\theta\right] = \int_{-\pi/6}^{\pi/2} (4 + 16\sin\theta + 16\sin^2\theta)\, d\theta = 8\pi + 6\sqrt{3}.$

The area inside the inner loop is

$2\frac{1}{2}\left[\int_{7\pi/6}^{3\pi/2} (2(1 + 2\sin\theta))^2\, d\theta\right] = 4\pi - 6\sqrt{3}.$

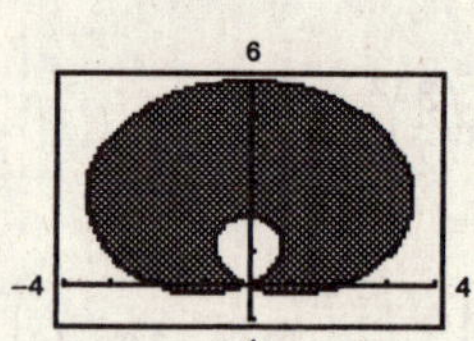

Thus, the area between the loops is $\left(8\pi + 6\sqrt{3}\right) - \left(4\pi - 6\sqrt{3}\right) = 4\pi + 12\sqrt{3}.$

17. $r = 1 + \cos\theta$

$r = 1 - \cos\theta$

Solving simultaneously,

$$1 + \cos\theta = 1 - \cos\theta$$
$$2\cos\theta = 0$$
$$\theta = \frac{\pi}{2}, \frac{3\pi}{2}.$$

Replacing r by $-r$ and θ by $\theta + \pi$ in the first equation and solving, $-1 + \cos\theta = 1 - \cos\theta$, $\cos\theta = 1$, $\theta = 0$. Both curves pass through the pole, $(0, \pi)$, and $(0, 0)$, respectively.

Points of intersection: $\left(1, \frac{\pi}{2}\right), \left(1, \frac{3\pi}{2}\right), (0, 0)$

18. $r = 3(1 + \sin\theta)$

$r = 3(1 - \sin\theta)$

Solving simultaneously,

$$3(1 + \sin\theta) = 3(1 - \sin\theta)$$
$$2\sin\theta = 0$$
$$\theta = 0, \pi.$$

Replacing r by $-r$ and θ by $\theta + \pi$ in the first equation and solving, $-3(1 - \sin\theta) = 3(1 - \sin\theta)$, $\sin\theta = 1$, $\theta = \pi/2$. Both curves pass through the pole, $(0, 3\pi/2)$, and $(0, \pi/2)$, respectively.

Points of intersection: $(3, 0), (3, \pi), (0, 0)$

19. $r = 1 + \cos\theta$

$r = 1 - \sin\theta$

Solving simultaneously,

$$1 + \cos\theta = 1 - \sin\theta$$
$$\cos\theta = -\sin\theta$$
$$\tan\theta = -1$$
$$\theta = \frac{3\pi}{4}, \frac{7\pi}{4}.$$

Replacing r by $-r$ and θ by $\theta + \pi$ in the first equation and solving, $-1 + \cos\theta = 1 - \sin\theta$, $\sin\theta + \cos\theta = 2$, which has no solution. Both curves pass through the pole, $(0, \pi)$, and $(0, \pi/2)$, respectively.

Points of intersection: $\left(\frac{2-\sqrt{2}}{2}, \frac{3\pi}{4}\right), \left(\frac{2+\sqrt{2}}{2}, \frac{7\pi}{4}\right), (0, 0)$

20. $r = 2 - 3\cos\theta$

$r = \cos\theta$

Solving simultaneously,

$$2 - 3\cos\theta = \cos\theta$$
$$\cos\theta = \frac{1}{2}$$
$$\theta = \frac{\pi}{3}, \frac{5\pi}{3}.$$

Both curves pass through the pole, $(0, \arccos 2/3)$, and $(0, \pi/2)$, respectively.

Points of intersection: $\left(\frac{1}{2}, \frac{\pi}{3}\right), \left(\frac{1}{2}, \frac{5\pi}{3}\right), (0, 0)$

21. $r = 4 - 5\sin\theta$

$r = 3\sin\theta$

Solving simultaneously,

$$4 - 5\sin\theta = 3\sin\theta$$
$$\sin\theta = \frac{1}{2}$$
$$\theta = \frac{\pi}{6}, \frac{5\pi}{6}.$$

Both curves pass through the pole, $(0, \arcsin 4/5)$, and $(0, 0)$, respectively.

Points of intersection: $\left(\frac{3}{2}, \frac{\pi}{6}\right), \left(\frac{3}{2}, \frac{5\pi}{6}\right), (0, 0)$

22. $r = 1 + \cos\theta$

$r = 3\cos\theta$

Solving simultaneously,

$$1 + \cos\theta = 3\cos\theta$$
$$\cos\theta = \frac{1}{2}$$
$$\theta = \frac{\pi}{3}, \frac{5\pi}{3}.$$

Both curves pass through the pole, $(0, \pi)$, and $(0, \pi/2)$, respectively.

Points of intersection: $\left(\frac{3}{2}, \frac{\pi}{3}\right), \left(\frac{3}{2}, \frac{5\pi}{3}\right), (0, 0)$

23. $r = \dfrac{\theta}{2}$

$r = 2$

Solving simultaneously, we have

$\theta/2 = 2, \theta = 4.$

Points of intersection:

$(2, 4), (-2, -4)$

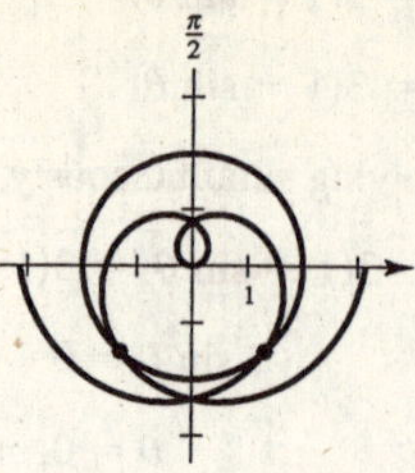

24. $\theta = \dfrac{\pi}{4}$

$r = 2$

Line of slope 1 passing through the pole and a circle of radius 2 centered at the pole.

Points of intersection:

$\left(2, \dfrac{\pi}{4}\right), \left(-2, \dfrac{\pi}{4}\right)$

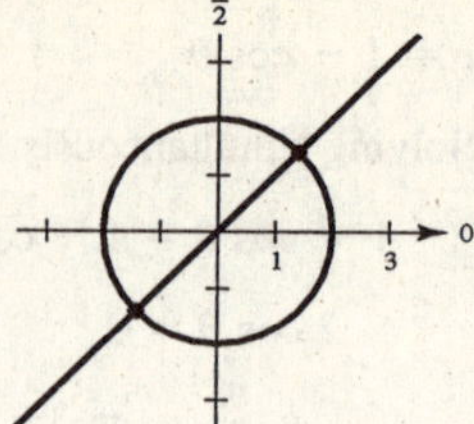

25. $r = 4 \sin 2\theta$

$r = 2$

$r = 4 \sin 2\theta$ is the equation of a rose curve with four petals and is symmetric to the polar axis, $\theta = \pi/2$, and the pole. Also, $r = 2$ is the equation of a circle of radius 2 centered at the pole. Solving simultaneously,

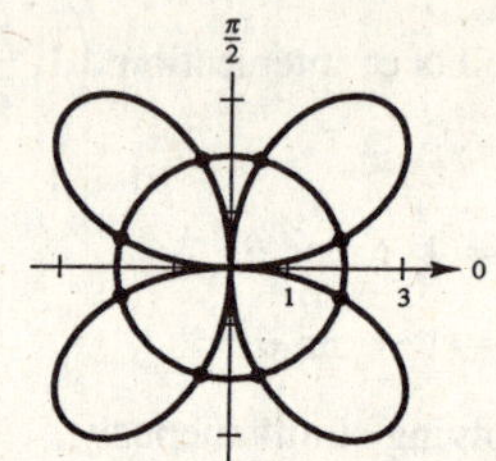

$$4 \sin 2\theta = 2$$

$$2\theta = \frac{\pi}{6}, \frac{5\pi}{6}$$

$$\theta = \frac{\pi}{12}, \frac{5\pi}{12}.$$

Therefore, the points of intersection for one petal are $(2, \pi/12)$ and $(2, 5\pi/12)$. By symmetry, the other points of intersection are $(2, 7\pi/12)$, $(2, 11\pi/12)$, $(2, 13\pi/12)$, $(2, 17\pi/12)$, $(2, 19\pi/12)$, and $(2, 23\pi/12)$.

26. $r = 3 + \sin \theta$

$r = 2 \csc \theta$

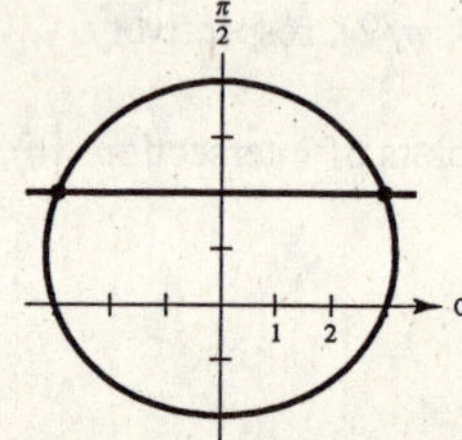

The graph of $r = 3 + \sin \theta$ is a limaçon symmetric to $\theta = \pi/2$, and the graph of $r = 2 \csc \theta$ is the horizontal line $y = 2$. Therefore, there are two points of intersection. Solving simultaneously,

$$3 + \sin \theta = 2 \csc \theta$$

$$\sin^2 \theta + 3 \sin \theta - 2 = 0$$

$$\sin \theta = \frac{-3 \pm \sqrt{17}}{2}$$

$$\theta = \arcsin\left(\frac{\sqrt{17} - 3}{2}\right) \approx 0.596.$$

Points of intersection:

$$\left(\frac{\sqrt{17} + 3}{2}, \arcsin\left(\frac{\sqrt{17} - 3}{2}\right)\right),$$

$$\left(\frac{\sqrt{17} + 3}{2}, \pi - \arcsin\left(\frac{\sqrt{17} - 3}{2}\right)\right),$$

$(3.56, 0.596), (3.56, 2.545)$

27. $r = 2 + 3\cos\theta$

$r = \dfrac{\sec\theta}{2}$

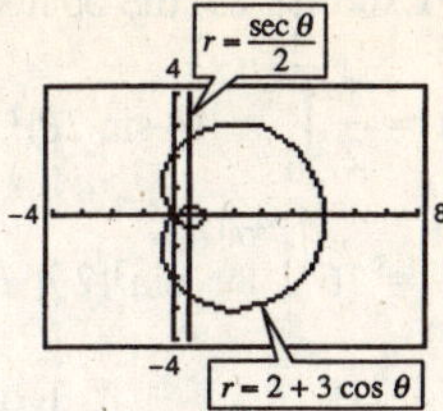

The graph of $r = 2 + 3\cos\theta$ is a limaçon with an inner loop $(b > a)$ and is symmetric to the polar axis. The graph of $r = (\sec\theta)/2$ is the vertical line $x = 1/2$. Therefore, there are four points of intersection. Solving simultaneously,

$$2 + 3\cos\theta = \frac{\sec\theta}{2}$$

$$6\cos^2\theta + 4\cos\theta - 1 = 0$$

$$\cos\theta = \frac{-2 \pm \sqrt{10}}{6}$$

$$\theta = \arccos\left(\frac{-2 + \sqrt{10}}{6}\right) \approx 1.376$$

$$\theta = \arccos\left(\frac{-2 - \sqrt{10}}{6}\right) \approx 2.6068.$$

Points of intersection: $(-0.581, \pm 2.607), (2.581, \pm 1.376)$

28. $r = 3(1 - \cos\theta)$

$r = \dfrac{6}{1 - \cos\theta}$

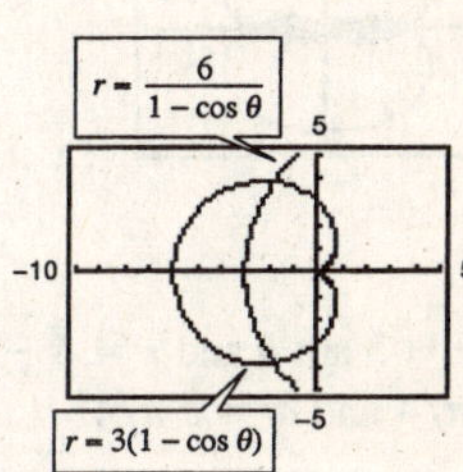

The graph of $r = 3(1 - \cos\theta)$ is a cardioid with polar axis symmetry. The graph of

$r = 6/(1 - \cos\theta)$

is a parabola with focus at the pole, vertex$(3, \pi)$, and polar axis symmetry. Therefore, there are two points of intersection. Solving simultaneously,

$$3(1 - \cos\theta) = \frac{6}{1 - \cos\theta}$$

$$(1 - \cos\theta)^2 = 2$$

$$\cos\theta = 1 \pm \sqrt{2}$$

$$\theta = \arccos\left(1 - \sqrt{2}\right).$$

Points of intersection: $\left(3\sqrt{2}, \arccos\left(1 - \sqrt{2}\right)\right) \approx (4.243, 1.998), \left(3\sqrt{2}, 2\pi - \arccos\left(1 - \sqrt{2}\right)\right) \approx (4.243, 4.285)$

29. $r = \cos\theta$

$r = 2 - 3\sin\theta$

Points of intersection:

$(0, 0), (0.935, 0.363), (0.535, -1.006)$

The graphs reach the pole at different times (θ values).

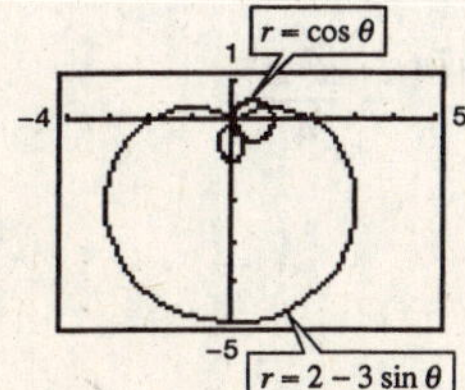

30. $r = 4\sin\theta$

$r = 2(1 + \sin\theta)$

Points of intersection: $(0, 0), \left(4, \dfrac{\pi}{2}\right)$

The graphs reach the pole at different times (θ values).

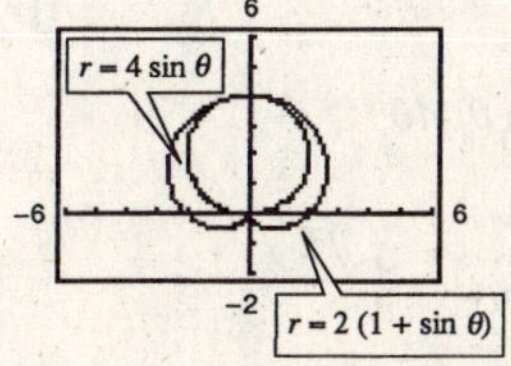

31. From Exercise 25, the points of intersection for one petal are $(2, \pi/12)$ and $(2, 5\pi/12)$. The area within one petal is

$$A = \frac{1}{2}\int_0^{\pi/12} (4\sin 2\theta)^2\,d\theta + \frac{1}{2}\int_{\pi/12}^{5\pi/12} (2)^2\,d\theta + \frac{1}{2}\int_{5\pi/12}^{\pi/2} (4\sin 2\theta)^2\,d\theta$$

$$= 16\int_0^{\pi/12} \sin^2(2\theta)\,d\theta + 2\int_{\pi/12}^{5\pi/12} d\theta \text{ (by symmetry of the petal)}$$

$$= 8\left[\theta - \frac{1}{4}\sin 4\theta\right]_0^{\pi/12} + \left[2\theta\right]_{\pi/12}^{5\pi/12} = \frac{4\pi}{3} - \sqrt{3}.$$

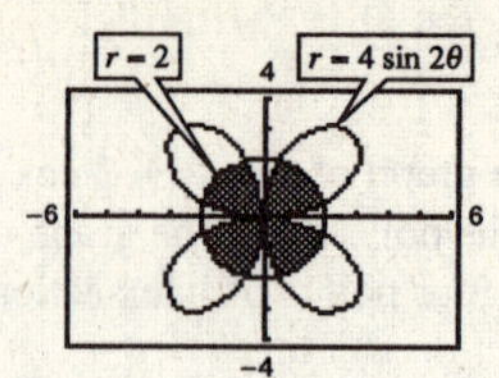

$$\text{Total area} = 4\left(\frac{4\pi}{3} - \sqrt{3}\right) = \frac{16\pi}{3} - 4\sqrt{3} = \frac{4}{3}\left(4\pi - 3\sqrt{3}\right)$$

32. $$A = 4\left[\frac{1}{2}\int_0^{\pi/2} 9(1 - \sin\theta)^2\,d\theta\right]$$

$$= 18\int_0^{\pi/2} (1 - \sin\theta)^2\,d\theta = \frac{9}{2}(3\pi - 8)$$

(from Exercise 14)

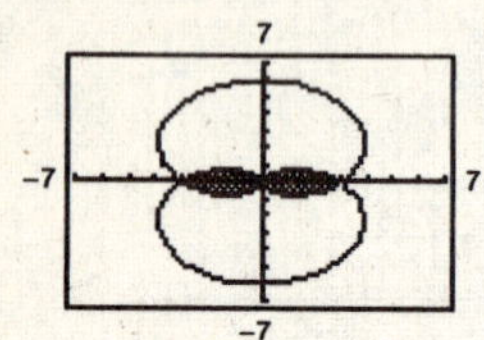

33. $$A = 4\left[\frac{1}{2}\int_0^{\pi/2} (3 - 2\sin\theta)^2\,d\theta\right]$$

$$= 2\left[11\theta + 12\cos\theta - \sin(2\theta)\right]_0^{\pi/2} = 11\pi - 24$$

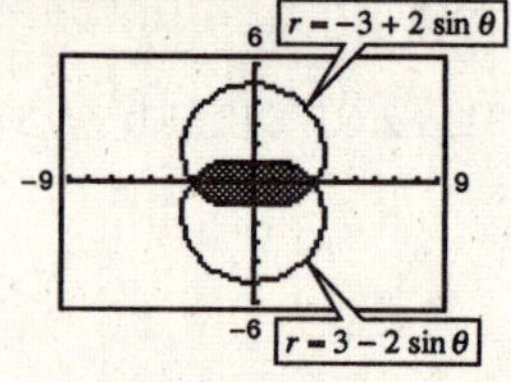

34. $r = 5 - 3\sin\theta$ and $r = 5 - 3\cos\theta$ intersect at $\theta = \pi/4$ and $\pi = 5\pi/4$.

$$A = 2\left[\frac{1}{2}\int_{\pi/4}^{5\pi/4} (5 - 3\sin\theta)^2\,d\theta\right]$$

$$= \left[\frac{59}{2}\theta + 30\cos\theta - \frac{9}{4}\sin 2\theta\right]_{\pi/4}^{5\pi/4}$$

$$= \left(\frac{59}{2}\left(\frac{5\pi}{4}\right) - 30\frac{\sqrt{2}}{2} - \frac{9}{4}\right) - \left(\frac{59}{2}\left(\frac{\pi}{4}\right) + 30\frac{\sqrt{2}}{2} - \frac{9}{4}\right)$$

$$= \frac{59\pi}{2} - 30\sqrt{2} \approx 50.251$$

35. $$A = 2\left[\frac{1}{2}\int_0^{\pi/6} (4\sin\theta)^2\,d\theta + \frac{1}{2}\int_{\pi/6}^{\pi/2} (2)^2\,d\theta\right]$$

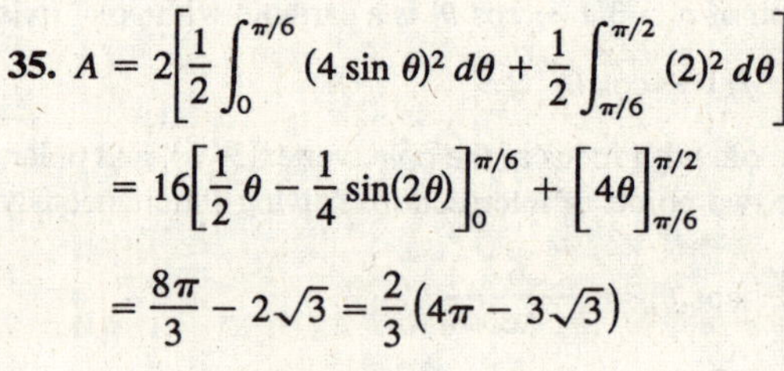

$$= 16\left[\frac{1}{2}\theta - \frac{1}{4}\sin(2\theta)\right]_0^{\pi/6} + \left[4\theta\right]_{\pi/6}^{\pi/2}$$

$$= \frac{8\pi}{3} - 2\sqrt{3} = \frac{2}{3}\left(4\pi - 3\sqrt{3}\right)$$

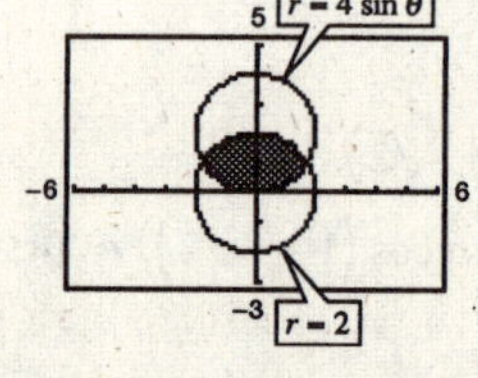

36. $$A = 2\left[\frac{1}{2}\int_{\pi/6}^{\pi/2} (3\sin\theta)^2\,d\theta - \frac{1}{2}\int_{\pi/6}^{\pi/2} (2 - \sin\theta)^2\,d\theta\right]$$

$$= \int_{\pi/6}^{\pi/2} (-4\cos 2\theta + 4\sin\theta)\,d\theta$$

$$= \left[-2\sin(2\theta) - 4\cos\theta\right]_{\pi/6}^{\pi/2} = 3\sqrt{3}$$

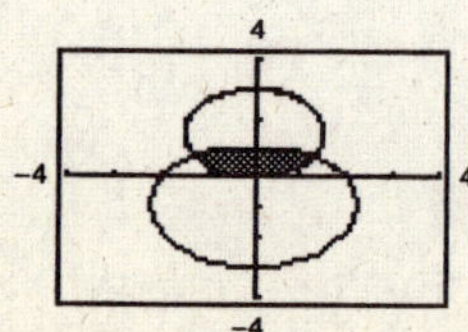

37. $$A = 2\left[\frac{1}{2}\int_0^{\pi} [a(1 + \cos\theta)]^2\,d\theta\right] - \frac{a^2\pi}{4}$$

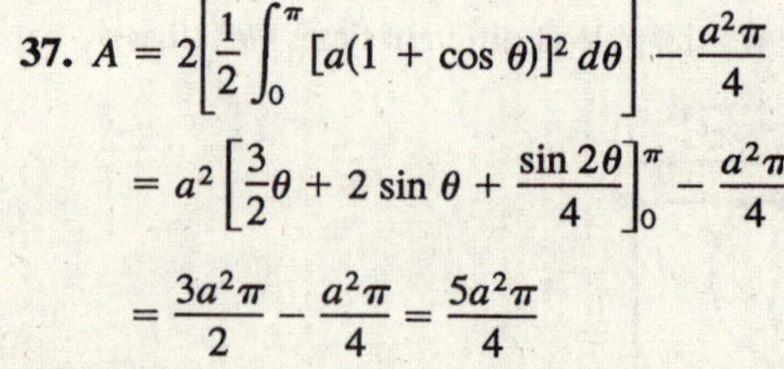

$$= a^2\left[\frac{3}{2}\theta + 2\sin\theta + \frac{\sin 2\theta}{4}\right]_0^{\pi} - \frac{a^2\pi}{4}$$

$$= \frac{3a^2\pi}{2} - \frac{a^2\pi}{4} = \frac{5a^2\pi}{4}$$

38. Area = Area of $r = 2a\cos\theta$ − Area of sector − twice area between $r = 2a\cos\theta$ and the lines

$$\theta = \frac{\pi}{3}, \theta = \frac{\pi}{2}.$$

$$A = \pi a^2 - \left(\frac{\pi}{3}\right)a^2 - 2\left[\frac{1}{2}\int_{\pi/3}^{\pi/2}(2a\cos\theta)^2\,d\theta\right]$$

$$= \frac{2\pi a^2}{3} - 2a^2\int_{\pi/3}^{\pi/2}(1+\cos 2\theta)\,d\theta$$

$$= \frac{2\pi a^2}{3} - 2a^2\left[\theta + \frac{\sin 2\theta}{2}\right]_{\pi/3}^{\pi/2}$$

$$= \frac{2\pi a^2}{3} - 2a^2\left[\frac{\pi}{2} - \frac{\pi}{3} - \frac{\sqrt{3}}{4}\right] = \frac{2\pi a^2 + 3\sqrt{3}a^2}{6}$$

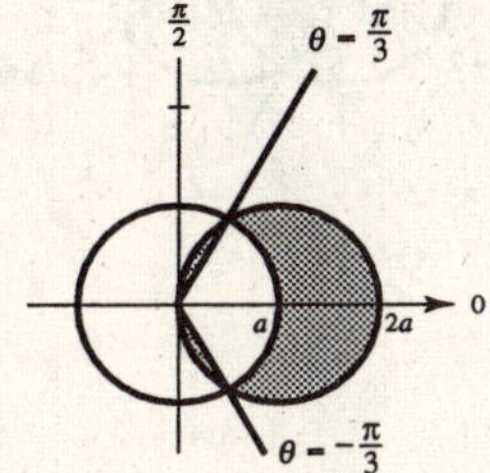

39. $$A = \frac{\pi a^2}{8} + \frac{1}{2}\int_{\pi/2}^{\pi}[a(1+\cos\theta)]^2\,d\theta$$

$$= \frac{\pi a^2}{8} + \frac{a^2}{2}\int_{\pi/2}^{\pi}\left(\frac{3}{2} + 2\cos\theta + \frac{\cos 2\theta}{2}\right)d\theta$$

$$= \frac{\pi a^2}{8} + \frac{a^2}{2}\left[\frac{3}{2}\theta + 2\sin\theta + \frac{\sin 2\theta}{4}\right]_{\pi/2}^{\pi}$$

$$= \frac{\pi a^2}{8} + \frac{a^2}{2}\left[\frac{3\pi}{2} - \frac{3\pi}{4} - 2\right] = \frac{a^2}{2}[\pi - 2]$$

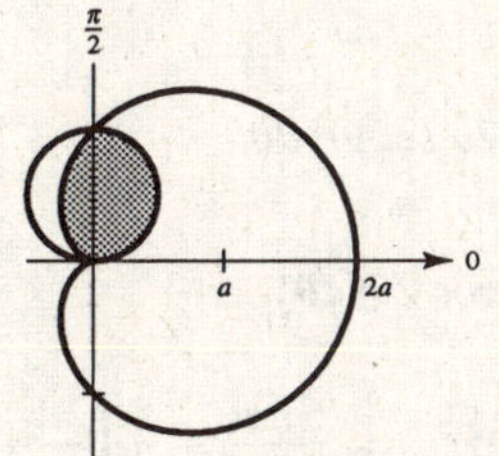

40. $r = a\cos\theta, r = a\sin\theta$

$\tan\theta = 1, \theta = \pi/4$

$$A = 2\left[\frac{1}{2}\int_0^{\pi/4}(a\sin\theta)^2\,d\theta\right]$$

$$= a^2\int_0^{\pi/4}\frac{1-\cos 2\theta}{2}\,d\theta$$

$$= \frac{1}{2}a^2\left[\theta - \frac{\sin 2\theta}{2}\right]_0^{\pi/4}$$

$$= \frac{1}{2}a^2\left[\frac{\pi}{4} - \frac{1}{2}\right]$$

$$= \frac{1}{8}a^2\pi - \frac{1}{4}a^2$$

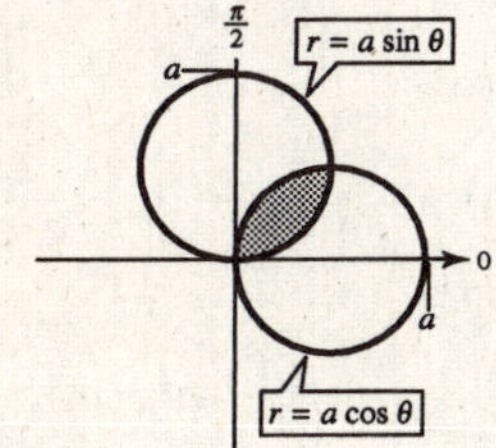

41. (a) $r = a\cos^2\theta$

$r^3 = ar^2\cos^2\theta$

$(x^2 + y^2)^{3/2} = ax^2$

(b)

a = 4
a = 6
4
−6
6
−4

(c) $$A = 4\left(\frac{1}{2}\right)\int_0^{\pi/2}[(6\cos^2\theta)^2 - (4\cos^2\theta)^2]\,d\theta$$

$$= 40\int_0^{\pi/2}\cos^4\theta\,d\theta$$

$$= 10\int_0^{\pi/2}(1+\cos 2\theta)^2\,d\theta$$

$$= 10\int_0^{\pi/2}\left(1 + 2\cos 2\theta + \frac{1-\cos 4\theta}{2}\right)d\theta$$

$$= 10\left[\frac{3}{2}\theta + \sin 2\theta + \frac{1}{8}\sin 4\theta\right]_0^{\pi/2} = \frac{15\pi}{2}$$

42. By symmetry, $A_1 = A_2$ and $A_3 = A_4$.

$$A_1 = A_2 = \frac{1}{2}\int_{-\pi/3}^{\pi/6} [(2a\cos\theta)^2 - (a)^2]\,d\theta + \frac{1}{2}\int_{\pi/6}^{\pi/4} [(2a\cos\theta)^2 - (2a\sin\theta)^2]\,d\theta$$

$$= \frac{a^2}{2}\int_{-\pi/3}^{\pi/6} (4\cos^2\theta - 1)\,d\theta + 2a^2\int_{\pi/6}^{\pi/4} \cos 2\theta\,d\theta$$

$$= \frac{a^2}{2}\Big[\theta + \sin 2\theta\Big]_{-\pi/3}^{\pi/6} + a^2\Big[\sin 2\theta\Big]_{\pi/6}^{\pi/4} = \frac{a^2}{2}\left(\frac{\pi}{2} + \sqrt{3}\right) + a^2\left(1 - \frac{\sqrt{3}}{2}\right) = a^2\left(\frac{\pi}{4} + 1\right)$$

$$A_3 = A_4 = \frac{1}{2}\left(\frac{\pi}{2}\right)a^2 = \frac{\pi a^2}{4}$$

$$A_5 = \frac{1}{2}\left(\frac{5\pi}{6}\right)a^2 - 2\left(\frac{1}{2}\right)\int_{5\pi/6}^{\pi} (2a\sin\theta)^2\,d\theta$$

$$= \frac{5\pi a^2}{12} - 2a^2\int_{5\pi/6}^{\pi} (1 - \cos 2\theta)\,d\theta$$

$$= \frac{5\pi a^2}{12} - a^2\Big[2\theta - \sin 2\theta\Big]_{5\pi/6}^{\pi} = \frac{5\pi a^2}{12} - a^2\left(\frac{\pi}{3} - \frac{\sqrt{3}}{2}\right) = a^2\left(\frac{\pi}{12} + \frac{\sqrt{3}}{2}\right)$$

$$A_6 = 2\left(\frac{1}{2}\right)\int_0^{\pi/6} (2a\sin\theta)^2\,d\theta + 2\left(\frac{1}{2}\right)\int_{\pi/6}^{\pi/4} a^2\,d\theta$$

$$= 2a^2\int_0^{\pi/6} (1 - \cos 2\theta)\,d\theta + \Big[a^2\theta\Big]_{\pi/6}^{\pi/4}$$

$$= a^2\Big[2\theta - \sin 2\theta\Big]_0^{\pi/6} + \frac{\pi a^2}{12} = a^2\left(\frac{\pi}{3} - \frac{\sqrt{3}}{2}\right) + \frac{\pi a^2}{12} = a^2\left(\frac{5\pi}{12} - \frac{\sqrt{3}}{2}\right)$$

$$A_7 = 2\left(\frac{1}{2}\right)\int_{\pi/6}^{\pi/4} [(2a\sin\theta)^2 - (a)^2]\,d\theta$$

$$= a^2\int_{\pi/6}^{\pi/4} (4\sin^2\theta - 1)\,d\theta = a^2\Big[\theta - \sin 2\theta\Big]_{\pi/6}^{\pi/4} = a^2\left(\frac{\pi}{12} - 1 + \frac{\sqrt{3}}{2}\right)$$

[**Note:** $A_1 + A_6 + A_7 + A_4 = \pi a^2$ = area of circle of radius a]

43. $r = a\cos(n\theta)$

For $n = 1$:

$r = a\cos\theta$

$$A = \pi\left(\frac{a}{2}\right)^2 = \frac{\pi a^2}{4}$$

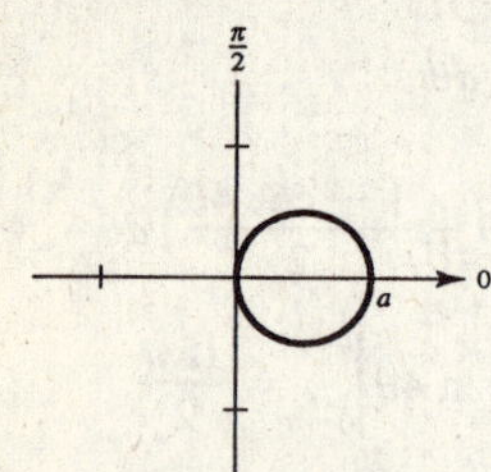

For $n = 2$:

$r = a\cos 2\theta$

$$A = 8\left(\frac{1}{2}\right)\int_0^{\pi/4} (a\cos 2\theta)^2\,d\theta = \frac{\pi a^2}{2}$$

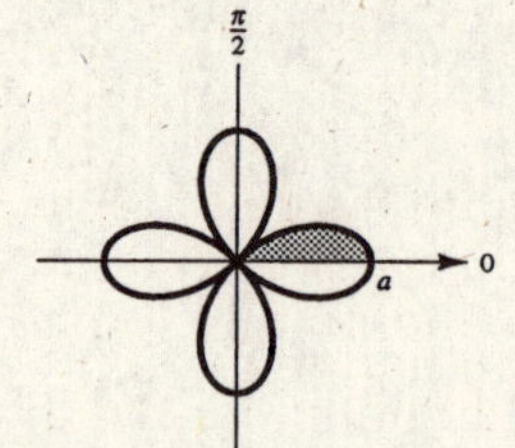

—CONTINUED—

43. —CONTINUED—

For $n = 3$:

$r = a \cos 3\theta$

$$A = 6\left(\frac{1}{2}\right)\int_0^{\pi/6} (a\cos 3\theta)^2\, d\theta = \frac{\pi a^2}{4}$$

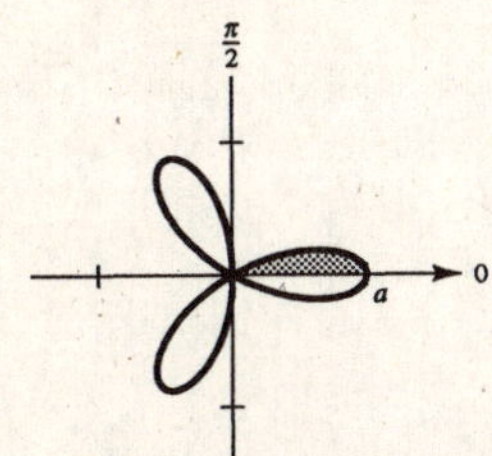

For $n = 4$:

$r = a \cos 4\theta$

$$A = 16\left(\frac{1}{2}\right)\int_0^{\pi/8} (a\cos 4\theta)^2\, d\theta = \frac{\pi a^2}{2}$$

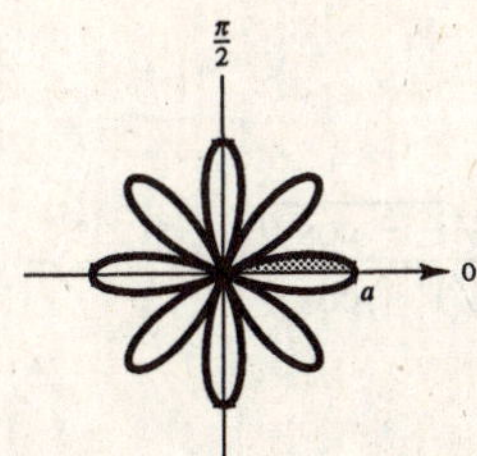

In general, the area of the region enclosed by $r = a\cos(n\theta)$ for $n = 1, 2, 3, \ldots$ is $(\pi a^2)/4$ if n is odd and is $(\pi a^2)/2$ if n is even.

44. $r = \sec\theta - 2\cos\theta,\ -\dfrac{\pi}{2} < \theta < \dfrac{\pi}{2}$

$r\cos\theta = 1 - 2\cos^2\theta$

$$x = 1 - 2\frac{r^2\cos^2\theta}{r^2} = 1 - 2\left(\frac{x^2}{x^2+y^2}\right)$$

$(x^2 + y^2)x = x^2 + y^2 - 2x^2$

$y^2(x - 1) = -x^2 - x^3$

$$y^2 = \frac{x^2(1+x)}{1-x}$$

$$A = 2\left(\frac{1}{2}\right)\int_0^{\pi/4} (\sec\theta - 2\cos\theta)^2\, d\theta$$

$$= \int_0^{\pi/4} (\sec^2\theta - 4 + 4\cos^2\theta)\, d\theta = \int_0^{\pi/4} (\sec^2\theta - 4 + 2(1 + \cos 2\theta))\, d\theta = \Big[\tan\theta - 2\theta + \sin 2\theta\Big]_0^{\pi/4} = 2 - \frac{\pi}{2}$$

45. $r = a$

$r' = 0$

$$s = \int_0^{2\pi} \sqrt{a^2 + 0^2}\, d\theta = \Big[a\theta\Big]_0^{2\pi} = 2\pi a$$

(circumference of circle of radius a)

46. $r = 2a\cos\theta$

$r' = -2a\sin\theta$

$$s = \int_{-\pi/2}^{\pi/2} \sqrt{(2a\cos\theta)^2 + (-2a\sin\theta)^2}\, d\theta$$

$$= \int_{-\pi/2}^{\pi/2} 2a\, d\theta = \Big[2a\theta\Big]_{-\pi/2}^{\pi/2} = 2\pi a$$

47. $r = 1 + \sin\theta$

$r' = \cos\theta$

$$s = 2\int_{\pi/2}^{3\pi/2} \sqrt{(1+\sin\theta)^2 + (\cos\theta)^2}\, d\theta$$

$$= 2\sqrt{2}\int_{\pi/2}^{3\pi/2} \sqrt{1+\sin\theta}\, d\theta$$

$$= 2\sqrt{2}\int_{\pi/2}^{3\pi/2} \frac{-\cos\theta}{\sqrt{1-\sin\theta}}\, d\theta$$

$$= \Big[4\sqrt{2}\sqrt{1-\sin\theta}\Big]_{\pi/2}^{3\pi/2}$$

$$= 4\sqrt{2}(\sqrt{2} - 0) = 8$$

48. $r = 8(1 + \cos\theta), 0 \le \theta \le 2\pi$

$r' = -8\sin\theta$

$$s = 2\int_0^{\pi} \sqrt{[8(1+\cos\theta)]^2 + (-8\sin\theta]^2}\,d\theta$$
$$= 16\int_0^{\pi} \sqrt{1 + 2\cos\theta + \cos^2\theta + \sin^2\theta}\,d\theta$$
$$= 16\sqrt{2}\int_0^{\pi} \sqrt{1+\cos\theta}\,d\theta$$
$$= 16\sqrt{2}\int_0^{\pi} \sqrt{1+\cos\theta}\cdot\left(\frac{\sqrt{1-\cos\theta}}{\sqrt{1-\cos\theta}}\right)d\theta$$
$$= 16\sqrt{2}\int_0^{\pi} \frac{\sin\theta}{\sqrt{1-\cos\theta}}\,d\theta$$
$$= \Big[32\sqrt{2}\sqrt{1-\cos\theta}\Big]_0^{\pi}$$
$$= 64$$

49. $r = 2\theta, 0 \le \theta \le \dfrac{\pi}{2}$

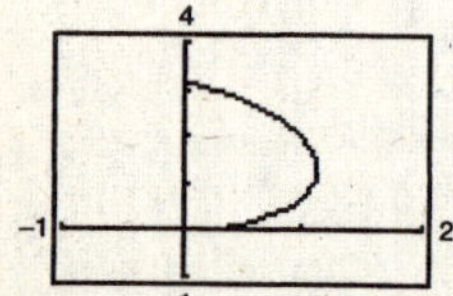

Length $\approx$ 4.16

50. $r = \sec\theta, 0 \le \theta \le \dfrac{\pi}{3}$

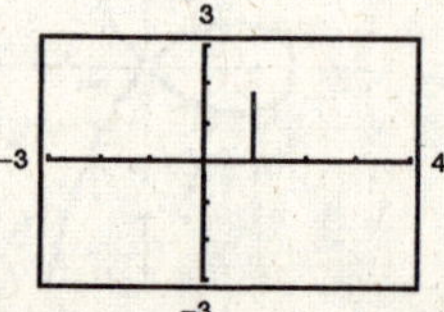

Length $\approx$ 1.73 (exact $\sqrt{3}$)

51. $r = \dfrac{1}{\theta}, \pi \le \theta \le 2\pi$

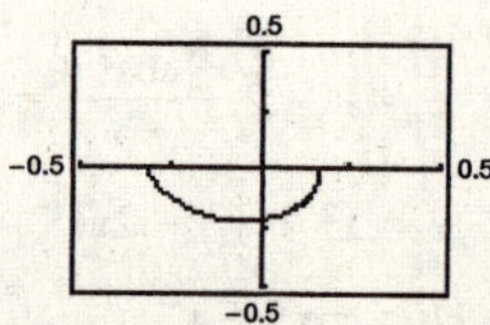

Length $\approx$ 0.71

52. $r = e^{\theta}, 0 \le \theta \le \pi$

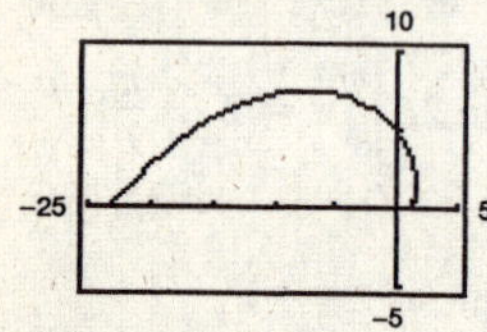

Length $\approx$ 31.31

53. $r = \sin(3\cos\theta), 0 \le \theta \le \pi$

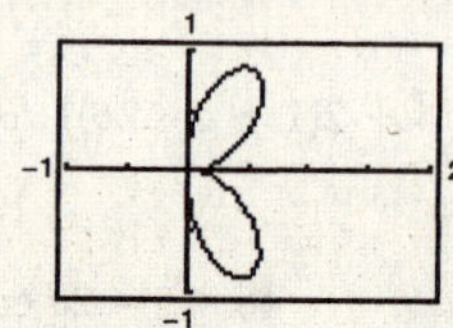

Length $\approx$ 4.39

54. $r = 2\sin(2\cos\theta),\ 0 \le \theta \le \pi$

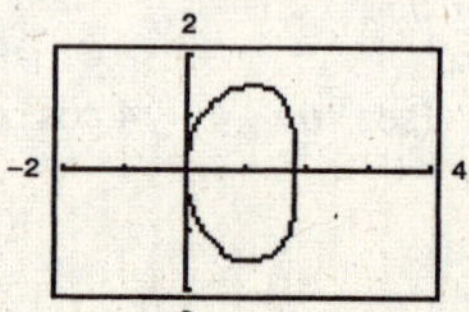

Length $\approx$ 7.78

55. $r = 6\cos\theta$

$r' = -6\sin\theta$

$$S = 2\pi\int_0^{\pi/2} 6\cos\theta\sin\theta\sqrt{36\cos^2\theta + 36\sin^2\theta}\,d\theta$$
$$= 72\pi\int_0^{\pi/2} \sin\theta\cos\theta\,d\theta$$
$$= \Big[36\pi\sin^2\theta\Big]_0^{\pi/2}$$
$$= 36\pi$$

56. $r = a\cos\theta$

$r' = -a\sin\theta$

$$S = 2\pi\int_0^{\pi/2} a\cos\theta(\cos\theta)\sqrt{a^2\cos\theta + a^2\sin^2\theta}\,d\theta$$
$$= 2\pi a^2\int_0^{\pi/2} \cos^2\theta\,d\theta = \pi a^2\int_0^{\pi/2}(1+\cos 2\theta)\,d\theta$$
$$= \left[\pi a^2\left(\theta + \frac{\sin 2\theta}{2}\right)\right]_0^{\pi/2} = \frac{\pi^2 a^2}{2}$$

57. $r = e^{a\theta}$

$r' = ae^{a\theta}$

$$S = 2\pi \int_0^{\pi/2} e^{a\theta} \cos\theta \sqrt{(e^{a\theta})^2 + (ae^{a\theta})^2}\, d\theta$$

$$= 2\pi\sqrt{1 + a^2} \int_0^{\pi/2} e^{2a\theta} \cos\theta\, d\theta$$

$$= 2\pi\sqrt{1 + a^2}\left[\frac{e^{2a\theta}}{4a^2 + 1}(2a\cos\theta + \sin\theta)\right]_0^{\pi/2}$$

$$= \frac{2\pi\sqrt{1 + a^2}}{4a^2 + 1}(e^{\pi a} - 2a)$$

58. $r = a(1 + \cos\theta)$

$r' = -a\sin\theta$

$$S = 2\pi \int_0^{\pi} a(1 + \cos\theta)\sin\theta\sqrt{a^2(1 + \cos\theta)^2 + a^2\sin^2\theta}\, d\theta = 2\pi a^2 \int_0^{\pi} \sin\theta(1 + \cos\theta)\sqrt{2 + 2\cos\theta}\, d\theta$$

$$= -2\sqrt{2}\pi a^2 \int_0^{\pi} (1 + \cos\theta)^{3/2}(-\sin\theta)\, d\theta = -\frac{4\sqrt{2}\pi a^2}{5}\left[(1 + \cos\theta)^{5/2}\right]_0^{\pi} = \frac{32\pi a^2}{5}$$

59. $r = 4\cos 2\theta$

$r' = -8\sin 2\theta$

$$S = 2\pi \int_0^{\pi/4} 4\cos 2\theta \sin\theta\sqrt{16\cos^2 2\theta + 64\sin^2 2\theta}\, d\theta$$

$$= 32\pi \int_0^{\pi/4} \cos 2\theta \sin\theta\sqrt{\cos^2 2\theta + 4\sin^2 2\theta}\, d\theta \approx 21.87$$

60. $r = \theta$

$r' = 1$

$$S = 2\pi \int_0^{\pi} \theta\sin\theta\sqrt{\theta^2 + 1}\, d\theta \approx 42.32$$

61. $$\text{Area} = \frac{1}{2}\int_\alpha^\beta [f(\theta)]^2\, d\theta = \frac{1}{2}\int_\alpha^\beta r^2\, d\theta$$

$$\text{Arc length} = \int_\alpha^\beta \sqrt{f(\theta)^2 + f'(\theta)^2}\, d\theta = \int_\alpha^\beta \sqrt{r^2 + \left(\frac{dr}{d\theta}\right)^2}\, d\theta$$

62. The curves might intersect for different values of θ:

See page 741.

63. (a) is correct: $s \approx 33.124$.

64. (a) $S = 2\pi \int_\alpha^\beta f(\theta)\sin\theta\sqrt{f(\theta)^2 + f'(\theta)^2}\, d\theta$

(b) $S = 2\pi \int_\alpha^\beta f(\theta)\cos\theta\sqrt{f(\theta)^2 + f'(\theta)^2}\, d\theta$

65. Revolve $r = 2$ about the line $r = 5\sec\theta$.

$f(\theta) = 2, f'(\theta) = 0$

$$S = 2\pi \int_0^{2\pi} (5 - 2\cos\theta)\sqrt{2^2 + 0^2}\, d\theta$$

$$= 4\pi \int_0^{2\pi} (5 - 2\cos\theta)\, d\theta$$

$$= 4\pi\left[5\theta - 2\sin\theta\right]_0^{2\pi}$$

$$= 40\pi^2$$

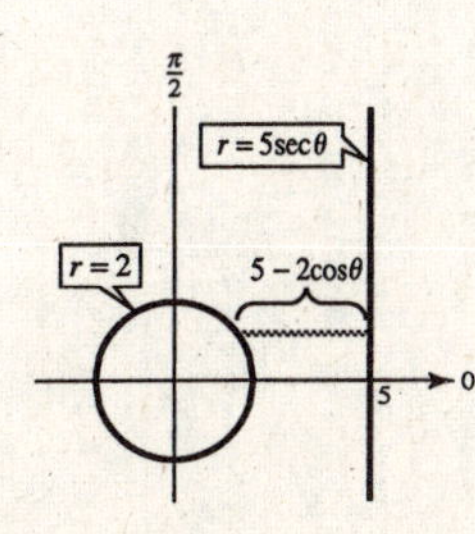

66. Revolve $r = a$ about the line $r = b \sec \theta$ where $b > a > 0$.

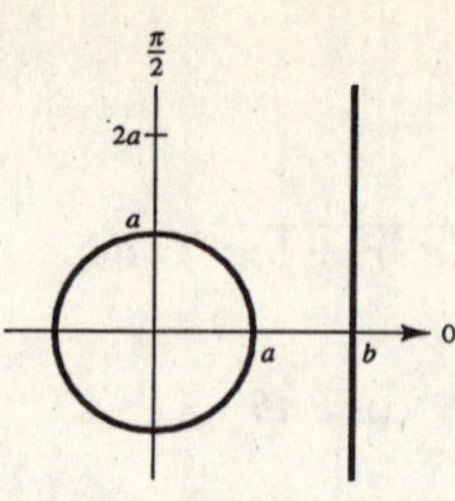

$f(\theta) = a$

$f'(\theta) = 0$

$$S = 2\pi \int_0^{2\pi} [b - a \cos \theta]\sqrt{a^2 + 0^2}\, d\theta$$

$$= 2\pi a\Big[b\theta - a \sin \theta\Big]_0^{2\pi}$$

$$= 2\pi a(2\pi b) = 4\pi^2 ab$$

67. $r = 8 \cos \theta, 0 \le \theta \le \pi$

(a) $A = \frac{1}{2}\int_0^{\pi} r^2\, d\theta = \frac{1}{2}\int_0^{\pi} 64 \cos^2 \theta\, d\theta = 32 \int_0^{\pi} \frac{1 + \cos 2\theta}{2}\, d\theta = 16\left[\theta + \frac{\sin 2\theta}{2}\right]_0^{\pi} = 16\,\pi$

(Area circle $= \pi r^2 = \pi 4^2 = 16\pi$)

(b)

θ	0.2	0.4	0.6	0.8	1.0	1.2	1.4
A	6.32	12.14	17.06	20.80	23.27	24.60	25.08

(c), (d) For $\frac{1}{4}$ of area ($4\pi \approx 12.57$): 0.42

For $\frac{1}{2}$ of area ($8\pi \approx 25.13$): 1.57 ($\pi/2$)

For $\frac{3}{4}$ of area ($12\pi \approx 37.70$): 2.73

(e) No, it does not depend on the radius.

68. $r = 3 \sin \theta, 0 \le \theta \le \pi$

(a) $A = \frac{1}{2}\int_0^{\pi} r^2\, d\theta = \frac{9}{2}\int_0^{\pi} \sin^2 \theta\, d\theta = \frac{9}{4}\int_0^{\pi} (1 - \cos 2\theta)\, d\theta = \frac{9}{4}\left[\theta - \frac{1}{2}\sin 2\theta\right]_0^{\pi} = \frac{9}{4}\pi$

$\left[\textbf{Note:}\ \text{radius of circle is } \frac{3}{2} \Rightarrow A = \pi\left(\frac{3}{2}\right)^2\right]$

(b)

θ	0.2	0.4	0.6	0.8	1.0	1.2	1.4
A	0.0119	0.0930	0.3015	0.6755	1.2270	1.9401	2.7731

(c), (d) For $\frac{1}{8}$ of area $\left(\frac{1}{8}\frac{9}{4}\pi \approx 0.8836\right)$: $\theta \approx 0.88$

For $\frac{1}{4}$ of area $\left(\frac{1}{4}\frac{9}{4}\pi \approx 1.7671\right)$: $\theta \approx 1.15$

For $\frac{1}{2}$ of area $\left(\frac{1}{2}\frac{9}{4}\pi \approx 3.5343\right)$: $\theta = \pi/2 \approx 1.57$

69.

$$r = a \sin \theta + b \cos \theta$$

$$r^2 = ar \sin \theta + br \cos \theta$$

$$x^2 + y^2 = ay + bx$$

$x^2 + y^2 - bx - ay = 0$ represents a circle.

70. $r = \sin\theta + \cos\theta$, Circle

$$A = \frac{1}{2}\int_{\alpha}^{\beta} r^2\,d\theta$$
$$= \frac{1}{2}\int_0^{\pi} (\sin\theta + \cos\theta)^2\,d\theta$$
$$= \frac{1}{2}\int_0^{\pi} (1 + 2\sin\theta\cos\theta)\,d\theta$$
$$= \frac{1}{2}\Big[\theta + \sin^2\theta\Big]_0^{\pi} = \frac{\pi}{2}$$

Converting to rectangular form:

$$r^2 = r\sin\theta + r\cos\theta$$
$$x^2 + y^2 = y + x$$
$$\left(x^2 - x + \frac{1}{4}\right) + \left(y^2 - y + \frac{1}{4}\right) = \frac{1}{2}$$
$$\left(x - \frac{1}{2}\right)^2 + \left(y - \frac{1}{2}\right)^2 = \frac{1}{2}$$

Circle of radius $\frac{1}{\sqrt{2}}$ and center $\left(\frac{1}{2}, \frac{1}{2}\right)$

$$\text{Area} = \pi\left(\frac{1}{\sqrt{2}}\right)^2 = \frac{\pi}{2}$$

71. (a) $r = \theta,\ \theta \ge 0$

As a increases, the spiral opens more rapidly. If $\theta < 0$, the spiral is reflected about the y-axis.

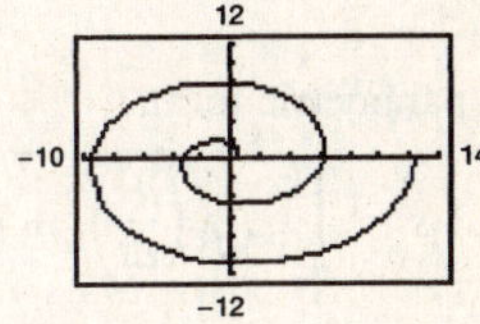

(b) $r = a\theta,\ \theta \ge 0$, crosses the polar axis for $\theta = n\pi$, n and integer. To see this

$$r = a\theta \implies r\sin\theta = y = a\theta\sin\theta = 0$$

for $\theta = n\pi$. The points are $(r, \theta) = (an\pi, n\pi)$, $n = 1, 2, 3, \ldots$.

(c) $f(\theta) = \theta, f'(\theta) = 1$

$$s = \int_0^{2\pi} \sqrt{\theta^2 + 1}\,d\theta = \frac{1}{2}\Big[\ln\big(\sqrt{x^2+1} + x\big) + x\sqrt{x^2+1}\Big]_0^{2\pi}$$
$$= \frac{1}{2}\ln\big(\sqrt{4\pi^2+1} + 2\pi\big) + \pi\sqrt{4\pi^2+1} \approx 21.2563$$

(d) $A = \frac{1}{2}\int_{\alpha}^{\beta} r^2\,dr$

$$= \frac{1}{2}\int_0^{2\pi} \theta^2\,d\theta$$
$$= \frac{\theta^3}{6}\Big]_0^{2\pi} = \frac{4}{3}\pi^3$$

72. $r = e^{\theta/6}$

$$A = \frac{1}{2}\int_0^{2\pi} (e^{\theta/6})^2\,d\theta - \frac{1}{2}\int_{-2\pi}^{0} (e^{\theta/6})^2\,d\theta$$
$$= \frac{1}{2}\int_0^{2\pi} e^{\theta/3}\,d\theta - \frac{1}{2}\int_{-2\pi}^{0} e^{\theta/3}\,d\theta$$
$$= \left[\frac{3}{2}e^{\theta/3}\right]_0^{2\pi} - \left[\frac{3}{2}e^{\theta/3}\right]_{-2\pi}^{0}$$
$$= \frac{3}{2}e^{2\pi/3} - \frac{3}{2} - \frac{3}{2} + \frac{3}{2}e^{-2\pi/3} = \frac{3}{2}\left[e^{2\pi/3} + e^{-2\pi/3} - 2\right]$$
$$\approx 9.3655$$

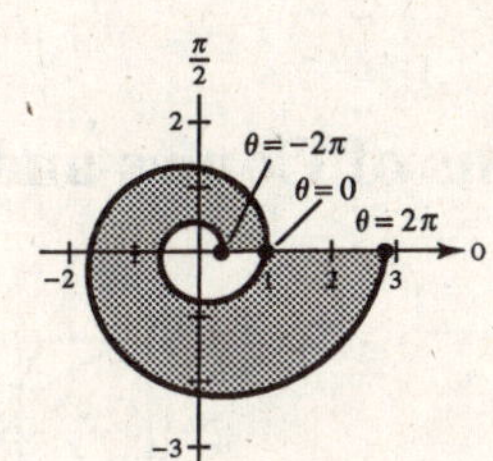

73. The smaller circle has equation $r = a\cos\theta$.

The area of the shaded lune is:

$$A = 2\left(\frac{1}{2}\right)\int_0^{\pi/4} \left[(a\cos\theta)^2 - 1\right]d\theta$$
$$= \int_0^{\pi/4} \left[\frac{a^2}{2}(1 + \cos 2\theta) - 1\right]d\theta$$
$$= \left[\frac{a^2}{2}\left(\theta + \frac{\sin 2\theta}{2}\right) - \theta\right]_0^{\pi/4}$$
$$= \frac{a^2}{2}\left(\frac{\pi}{4} + \frac{1}{2}\right) - \frac{\pi}{4}$$

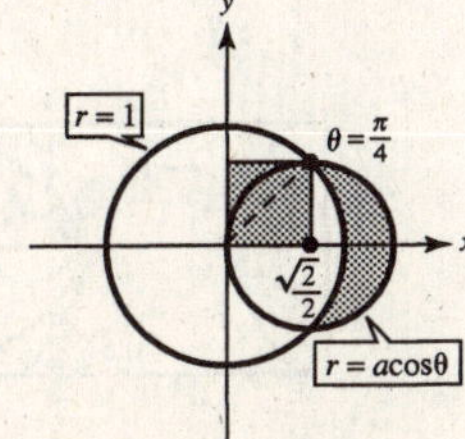

This equals the area of the square, $\left(\frac{\sqrt{2}}{2}\right)^2 = \frac{1}{2}$.

$$\frac{a^2}{2}\left(\frac{\pi}{4} + \frac{1}{2}\right) - \frac{\pi}{4} = \frac{1}{2}$$
$$\pi a^2 + 2a^2 - 2\pi - 4 = 0$$
$$a^2 = \frac{4 + 2\pi}{2 + \pi} = 2$$
$$a = \sqrt{2}$$

Smaller circle: $r = \sqrt{2}\cos\theta$

74. $x = \dfrac{3t}{1+t^3}, y = \dfrac{3t^2}{1+t^3}$

(a) $x^3 + y^3 = \dfrac{27(t^3+t^6)}{(1+t^3)^3} = \dfrac{27t^3}{(1+t^3)^2}$

$3xy = \dfrac{27t^3}{(1+t^3)^2}$

Hence, $x^3 + y^3 = 3xy$.

$(r\cos\theta)^3 + (r\sin\theta)^3 = 3(r\cos\theta)(r\sin\theta)$

$r = \dfrac{3\cos\theta\sin\theta}{\cos^3\theta + \sin^3\theta}$

(b)

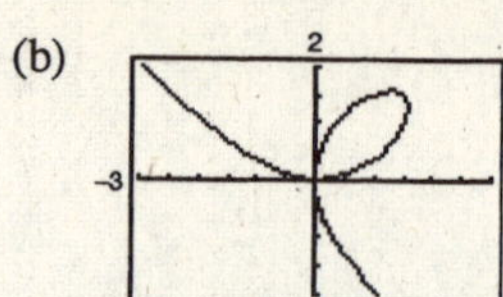

(c) $A = \dfrac{1}{2}\displaystyle\int_0^{\pi/2} r^2\, d\theta = \dfrac{3}{2}$

75. False. $f(\theta) = 1$ and $g(\theta) = -1$ have the same graphs.

76. False. $f(\theta) = 0$ and $g(\theta) = \sin 2\theta$ have only one point of intersection.

77. In parametric form,

$$s = \int_a^b \sqrt{\left(\frac{dx}{dt}\right)^2 + \left(\frac{dy}{dt}\right)^2}\, dt.$$

Using θ instead of t, we have $x = r\cos\theta = f(\theta)\cos\theta$ and $y = r\sin\theta = f(\theta)\sin\theta$. Thus,

$$\frac{dx}{d\theta} = f'(\theta)\cos\theta - f(\theta)\sin\theta \text{ and } \frac{dy}{d\theta} = f'(\theta)\sin\theta + f(\theta)\cos\theta.$$

It follows that

$$\left(\frac{dx}{d\theta}\right)^2 + \left(\frac{dy}{d\theta}\right)^2 = [f(\theta)]^2 + [f'(\theta)]^2.$$

Therefore, $s = \displaystyle\int_\alpha^\beta \sqrt{[f(\theta)]^2 + [f'(\theta)]^2}\, d\theta.$

Section 10.6 Polar Equations of Conics and Kepler's Laws

1. $r = \dfrac{2e}{1 + e\cos\theta}$

(a) $e = 1, r = \dfrac{2}{1+\cos\theta}$, parabola

(b) $e = 0.5, r = \dfrac{1}{1+0.5\cos\theta} = \dfrac{2}{2+\cos\theta}$, ellipse

(c) $e = 1.5, r = \dfrac{3}{1+1.5\cos\theta} = \dfrac{6}{2+3\cos\theta}$, hyperbola

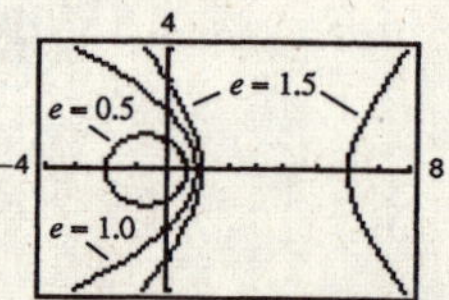

2. $r = \dfrac{2e}{1 - e\cos\theta}$

(a) $e = 1, r = \dfrac{2}{1-\cos\theta}$, parabola

(b) $e = 0.5, r = \dfrac{1}{1-0.5\cos\theta} = \dfrac{2}{2-\cos\theta}$, ellipse

(c) $e = 1.5, r = \dfrac{3}{1-1.5\cos\theta} = \dfrac{6}{2-3\cos\theta}$, hyperbola

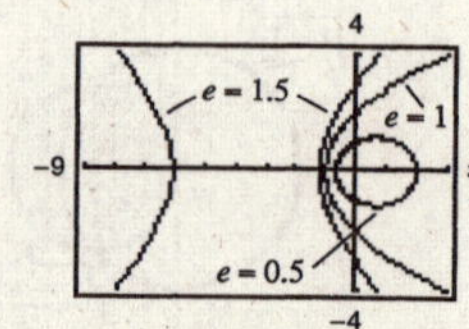

3. $r = \dfrac{2e}{1 - e \sin \theta}$

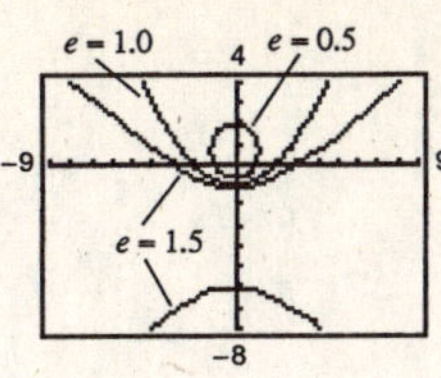

(a) $e = 1, r = \dfrac{2}{1 - \sin \theta}$, parabola

(b) $e = 0.5, r = \dfrac{1}{1 - 0.5 \sin \theta} = \dfrac{2}{2 - \sin \theta}$, ellipse

(c) $e = 1.5, r = \dfrac{3}{1 - 1.5 \sin \theta} = \dfrac{6}{2 - 3 \sin \theta}$, hyperbola

4. $r = \dfrac{2e}{1 + e \sin \theta}$

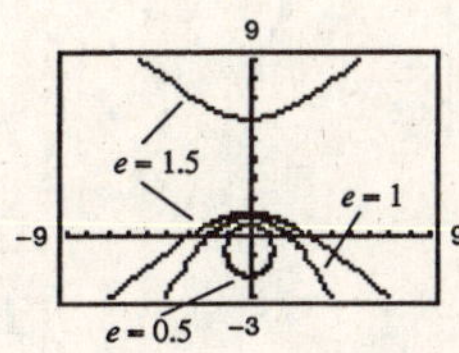

(a) $e = 1, r = \dfrac{2}{1 + \sin \theta}$, parabola

(b) $e = 0.5, r = \dfrac{1}{1 + 0.5 \sin \theta} = \dfrac{2}{2 + \sin \theta}$, ellipse

(c) $e = 1.5, r = \dfrac{3}{1 + 1.5 \sin \theta} = \dfrac{6}{2 + 3 \sin \theta}$, hyperbola

5. $r = \dfrac{4}{1 + e \sin \theta}$

(a)

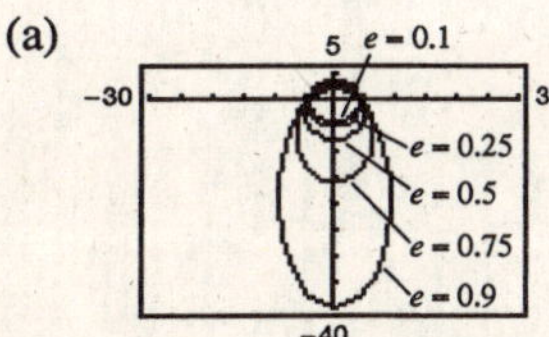

The conic is an ellipse. As $e \to 1^-$, the ellipse becomes more elliptical, and as $e \to 0^+$, it becomes more circular.

(b)

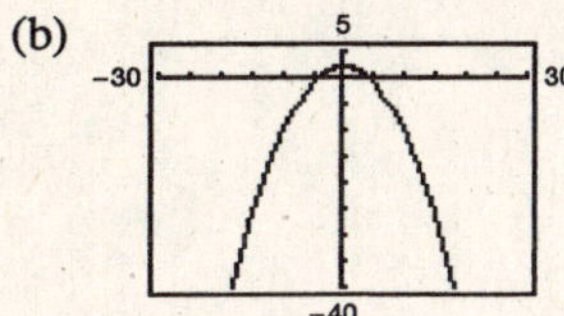

The conic is a parabola.

(c)

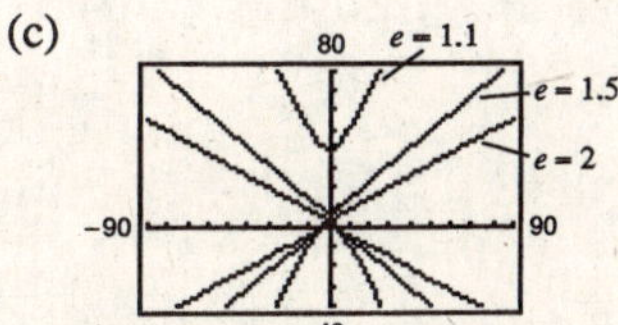

The conic is a hyperbola. As $e \to 1^+$, the hyperbolas opens more slowly, and as $e \to \infty$, they open more rapidly.

6. $r = \dfrac{4}{1 - 0.4 \cos \theta}$

(a) Because $e = 0.4 < 1$, the conic is an ellipse with vertical directrix to the left of the pole.

(b) $r = \dfrac{4}{1 + 0.4 \cos \theta}$

The ellipse is shifted to the left. The vertical directrix is to the right of the pole

$r = \dfrac{4}{1 - 0.4 \sin \theta}$.

The ellipse has a horizontal directrix below the pole.

(c)

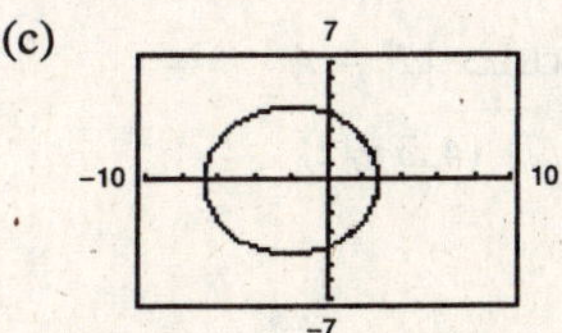

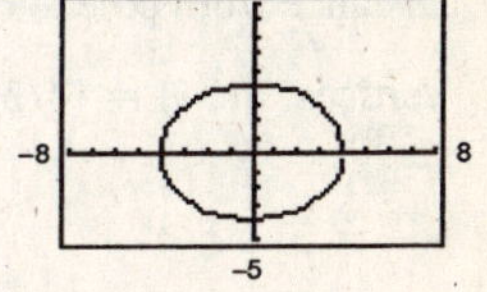

7. Parabola; Matches (c)

8. Ellipse; Matches (f)

9. Hyperbola; Matches (a)

10. Parabola; Matches (e)

11. Ellipse; Matches (b)

12. Hyperbola; Matches (d)

13. $r = \dfrac{-1}{1 - \sin\theta}$

Parabola because $e = 1$; $d = -1$

Distance from pole to directrix: $|d| = 1$

Directrix: $y = 1$

Vertex: $(r, \theta) = \left(-\dfrac{1}{2}, \dfrac{3\pi}{2}\right)$

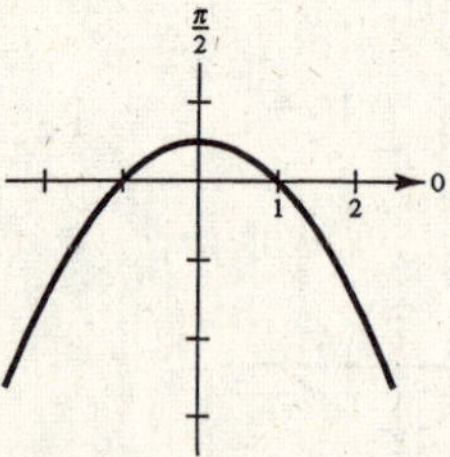

14. $r = \dfrac{6}{1 + \cos\theta}$

Parabola because $e = 1$; $d = 6$

Directrix: $x = 6$

Distance from pole to directrix: $|d| = 6$

Vertex: $(r, \theta) = (3, 0)$

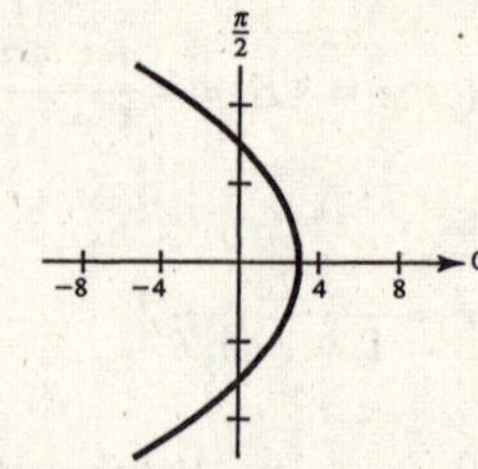

15. $r = \dfrac{6}{2 + \cos\theta} = \dfrac{3}{1 + (1/2)\cos\theta}$

Ellipse because $e = 1/2$; $d = 6$

Directrix: $x = 6$

Distance from pole to directrix: $|d| = 6$

Vertices: $(r, \theta) = (2, 0), (6, \pi)$

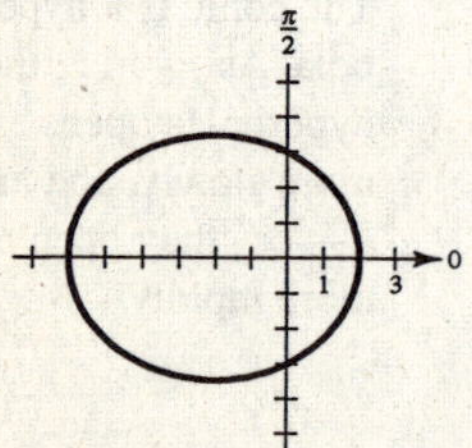

16. $r = \dfrac{5}{5 + 3\sin\theta} = \dfrac{1}{1 + (3/5)\sin\theta}$

Ellipse because $e = 3/5 < 1$; $d = 5/3$

Directrix: $y = 5/3$

Distance from pole to directrix: $|d| = 5/3$

Vertices: $(r, \theta) = (5/8, \pi/2), (5/2, 3\pi/2)$

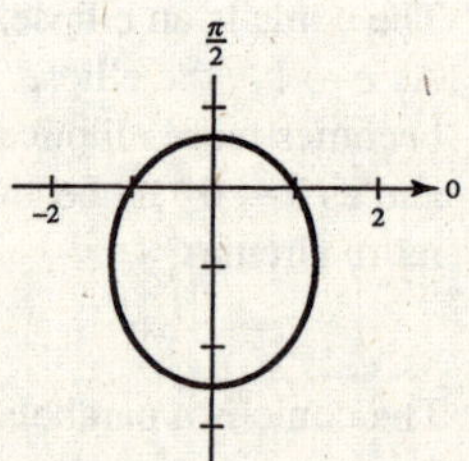

17. $r(2 + \sin\theta) = 4$

$r = \dfrac{4}{2 + \sin\theta} = \dfrac{2}{1 + (1/2)\sin\theta}$

Ellipse because $e = 1/2$; $d = 4$

Directrix: $y = 4$

Distance from pole to directrix: $|d| = 4$

Vertices: $(r, \theta) = (4/3, \pi/2), (4, 3\pi/2)$

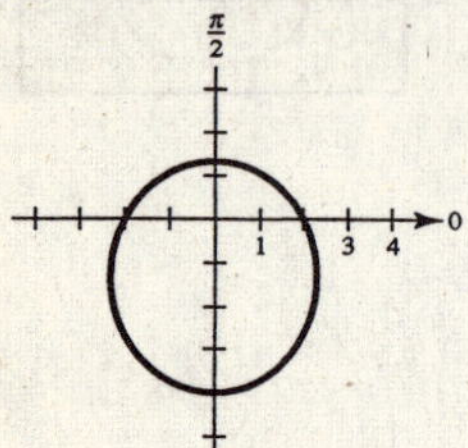

18. $r(3 - 2\cos\theta) = 6$

$r = \dfrac{6}{3 - 2\cos\theta} = \dfrac{2}{1 - (2/3)\cos\theta}$

Ellipse because $e = 2/3 < 1$; $d = 3$

Directrix: $x = -3$

Distance from pole to directrix: $|d| = 3$

Vertices: $(r, \theta) = (6, 0), (6/5, \pi)$

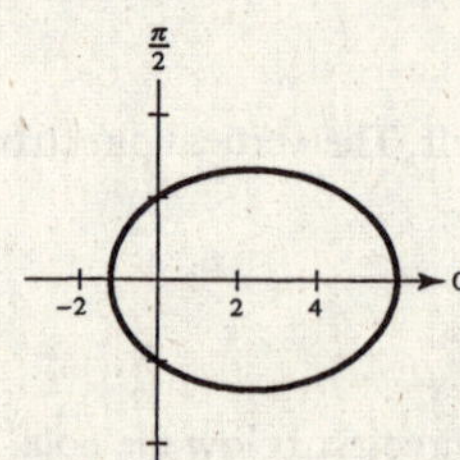

19. $r = \dfrac{5}{-1 + 2\cos\theta} = \dfrac{-5}{1 - 2\cos\theta}$

Hyperbola because $e = 2 > 1$; $d = -5/2$

Directrix: $x = 5/2$

Distance from pole to directrix: $|d| = 5/2$

Vertices: $(r, \theta) = (5, 0), (-5/3, \pi)$

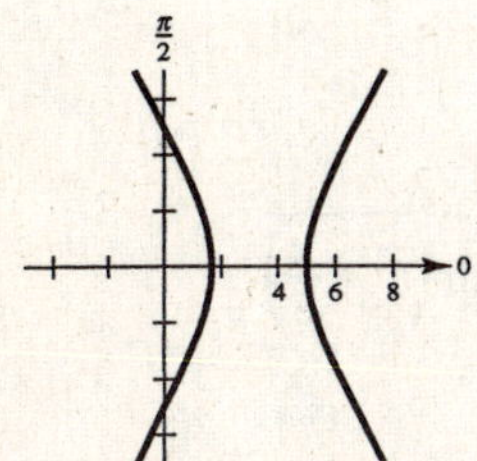

20. $r = \dfrac{-6}{3 + 7\sin\theta} = \dfrac{-2}{1 + (7/3)\sin\theta}$

Hyperbola because $e = 7/3 > 1$; $d = -6/7$

Directrix: $y = -6/7$

Distance from pole to directrix: $|d| = 6/7$

Vertices: $(r, \theta) = (-3/5, \pi/2), (3/2, 3\pi/2)$

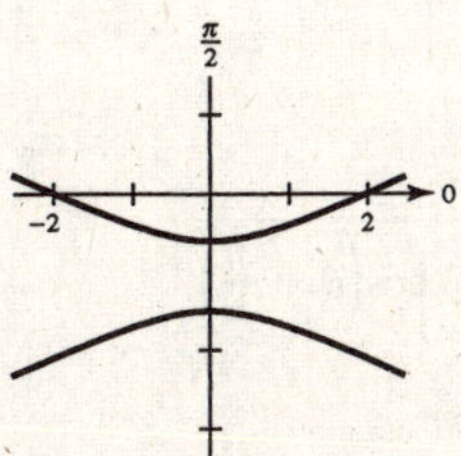

21. $r = \dfrac{3}{2 + 6\sin\theta} = \dfrac{3/2}{1 + 3\sin\theta}$

Hyperbola because $e = 3 > 0$; $d = 1/2$

Directrix: $y = 1/2$

Distance from pole to directrix: $|d| = 1/2$

Vertices: $(r, \theta) = (3/8, \pi/2), (-3/4, 3\pi/2)$

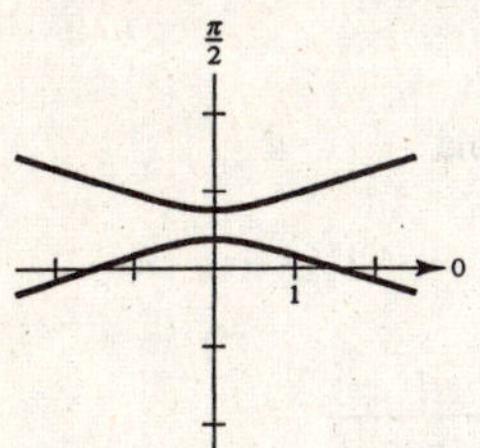

22. $r = \dfrac{4}{1 + 2\cos\theta}$

Hyperbola because $e = 2 > 1$; $d = 2$

Directrix: $x = 2$

Distance from pole to directrix: $|d| = 2$

Vertices: $(r, \theta) = (4/3, 0), (-4, \pi)$

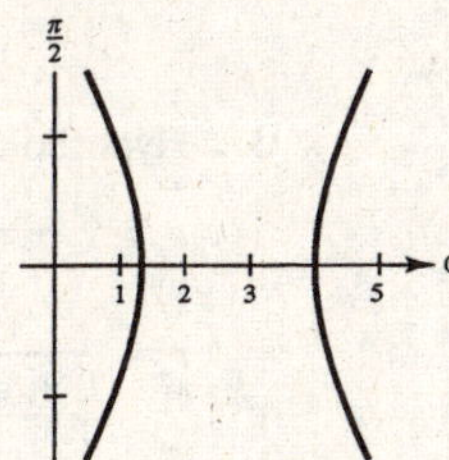

23. $r = 3/(-4 + 2\sin\theta)$

Ellipse

24. $r = -3/(2 + 4\sin\theta)$

Hyperbola

25. $r = -1/(1 - \cos\theta)$

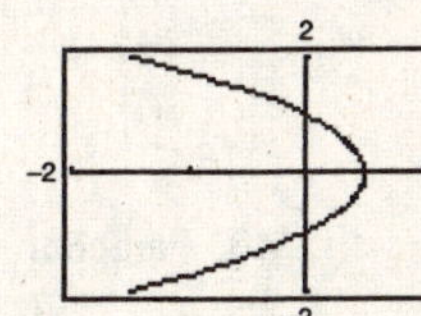

Parabola

26. $r = 2/(2 + 3\sin\theta)$

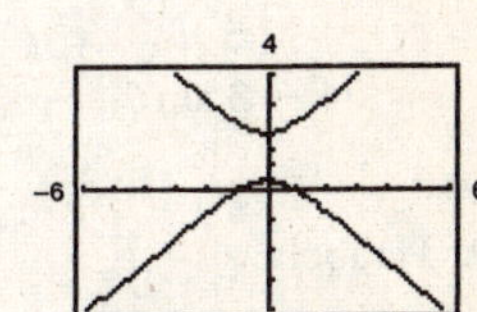

Hyperbola

27. $r = \dfrac{-1}{1 - \sin\left(\theta - \dfrac{\pi}{4}\right)}$

Rotate the graph of

$r = \dfrac{-1}{1 - \sin\theta}$

counterclockwise through the angle $\dfrac{\pi}{4}$.

28. $r = \dfrac{6}{1 + \cos\left(\theta - \dfrac{\pi}{3}\right)}$

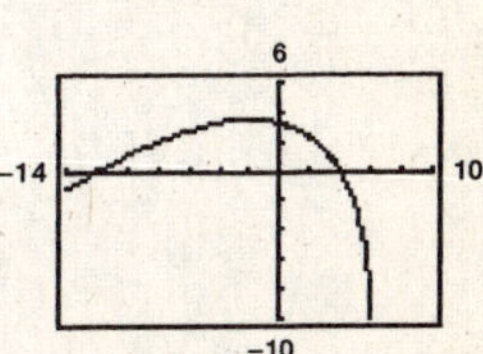

Rotate the graph of

$r = \dfrac{6}{1 + \cos\theta}$

counterclockwise through the angle $\dfrac{\pi}{3}$.

29. $r = \dfrac{6}{2 + \cos\left(\theta + \dfrac{\pi}{6}\right)}$

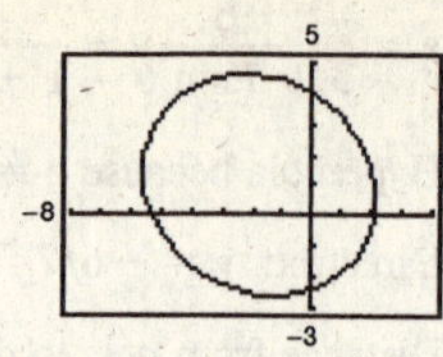

Rotate the graph of

$r = \dfrac{6}{2 + \cos\theta}$

clockwise through the angle $\dfrac{\pi}{6}$.

30. $r = \dfrac{-6}{3 + 7\sin(\theta + (2\pi/3))}$

Rotate graph of

$r = \dfrac{-6}{3 + 7\sin\theta}$

clockwise through angle of $2\pi/3$.

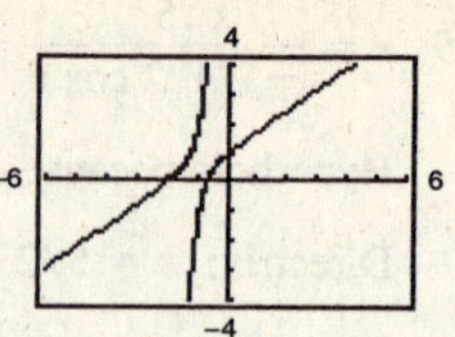

31. Change θ to $\theta + \dfrac{\pi}{4}$: $r = \dfrac{5}{5 + 3\cos\left(\theta + \dfrac{\pi}{4}\right)}$

32. Change θ to $\theta - \dfrac{\pi}{6}$: $r = \dfrac{2}{1 + \sin\left(\theta - \dfrac{\pi}{6}\right)}$

33. Parabola

$e = 1, x = -1, d = 1$

$r = \dfrac{ed}{1 - e\cos\theta} = \dfrac{1}{1 - \cos\theta}$

34. Parabola

$e = 1, y = 1, d = 1$

$r = \dfrac{ed}{1 + e\sin\theta} = \dfrac{1}{1 + \sin\theta}$

35. Ellipse

$e = \dfrac{1}{2}, y = 1, d = 1$

$r = \dfrac{ed}{1 + e\sin\theta}$

$= \dfrac{1/2}{1 + (1/2)\sin\theta}$

$= \dfrac{1}{2 + \sin\theta}$

36. Ellipse

$e = \dfrac{3}{4}, y = -2, d = 2$

$r = \dfrac{ed}{1 - e\sin\theta}$

$= \dfrac{2(3/4)}{1 - (3/4)\sin\theta}$

$= \dfrac{6}{4 - 3\sin\theta}$

37. Hyperbola

$e = 2, x = 1, d = 1$

$r = \dfrac{ed}{1 + e\cos\theta} = \dfrac{2}{1 + 2\cos\theta}$

38. Hyperbola

$e = \dfrac{3}{2}, x = -1, d = 1$

$r = \dfrac{ed}{1 - e\cos\theta}$

$= \dfrac{3/2}{1 - (3/2)\cos\theta}$

$= \dfrac{3}{2 - 3\cos\theta}$

39. Parabola

Vertex: $\left(1, -\dfrac{\pi}{2}\right)$

$e = 1, d = 2, r = \dfrac{2}{1 - \sin\theta}$

40. Parabola

Vertex: $(5, \pi)$

$e = 1, d = 10$

$r = \dfrac{ed}{1 - e\cos\theta} = \dfrac{10}{1 - \cos\theta}$

41. Ellipse

Vertices: $(2, 0), (8, \pi)$

$e = \dfrac{3}{5}, d = \dfrac{16}{3}$

$r = \dfrac{ed}{1 + e\cos\theta}$

$= \dfrac{16/5}{1 + (3/5)\cos\theta}$

$= \dfrac{16}{5 + 3\cos\theta}$

42. Ellipse

Vertices: $\left(2, \frac{\pi}{2}\right), \left(4, \frac{3\pi}{2}\right)$

$e = \frac{1}{3}, d = 8$

$$r = \frac{ed}{1 + e\sin\theta} = \frac{8/3}{1 + (1/3)\sin\theta} = \frac{8}{3 + \sin\theta}$$

43. Hyperbola

Vertices: $\left(1, \frac{3\pi}{2}\right), \left(9, \frac{3\pi}{2}\right)$

$e = \frac{5}{4}, d = \frac{9}{5}$

$$r = \frac{ed}{1 - e\sin\theta} = \frac{9/4}{1 - (5/4)\sin\theta} = \frac{9}{4 - 5\sin\theta}$$

44. Hyperbola

Vertices: $(2, 0), (10, 0)$

$e = \frac{3}{2}, d = \frac{10}{3}$

$$r = \frac{ed}{1 + e\cos\theta} = \frac{5}{1 + (3/2)\cos\theta} = \frac{10}{2 + 3\cos\theta}$$

45. Ellipse if $0 < e < 1$, parabola if $e = 1$, hyperbola if $e > 1$.

46. $r = \dfrac{4}{1 + \sin\theta}$ is a parabola with horizontal directrix above the pole.

(a) Parabola with vertical directrix to left of pole.

(b) Parabola with horizontal directrix below pole.

(c) Parabola with vertical directrix to right of pole.

(d) Parabola (b) rotated counterclockwise $\pi/4$.

47. (a) Hyperbola ($e = 2 > 1$)

(b) Ellipse $\left(e = \frac{1}{10} < 1\right)$

(c) Parabola ($e = 1$)

(d) Rotated hyperbola ($e = 3$)

48. If the foci are fixed and $e \to 0$, then $d \to \infty$. To see this, compare the ellipses

$$r = \frac{1/2}{1 + (1/2)\cos\theta},\ e = 1/2,\ d = 1$$

$$r = \frac{5/16}{1 + (1/4)\cos\theta},\ e = 1/4,\ d = 5/4.$$

49.

$$\frac{x^2}{a^2} + \frac{y^2}{b^2} = 1$$

$$x^2b^2 + y^2a^2 = a^2b^2$$

$$b^2r^2\cos^2\theta + a^2r^2\sin^2\theta = a^2b^2$$

$$r^2[b^2\cos^2\theta + a^2(1 - \cos^2\theta)] = a^2b^2$$

$$r^2[a^2 + \cos^2\theta(b^2 - a^2)] = a^2b^2$$

$$r^2 = \frac{a^2b^2}{a^2 + (b^2 - a^2)\cos^2\theta} = \frac{a^2b^2}{a^2 - c^2\cos^2\theta}$$

$$= \frac{b^2}{1 - (c/a)^2\cos^2\theta} = \frac{b^2}{1 - e^2\cos^2\theta}$$

50.

$$\frac{x^2}{a^2} - \frac{y^2}{b^2} = 1$$

$$x^2b^2 - y^2a^2 = a^2b^2$$

$$b^2r^2\cos^2\theta - a^2r^2\sin^2\theta = a^2b^2$$

$$r^2[b^2\cos^2\theta - a^2(1 - \cos^2\theta)] = a^2b^2$$

$$r^2[-a^2 + \cos^2\theta(a^2 + b^2)] = a^2b^2$$

$$r^2 = \frac{a^2b^2}{-a^2 + c^2\cos^2\theta} = \frac{b^2}{-1 + (c^2/a^2)\cos^2\theta}$$

$$= \frac{-b^2}{1 - e^2\cos^2\theta}$$

51. $a = 5, c = 4, e = \frac{4}{5}, b = 3$

$$r^2 = \frac{9}{1 - (16/25)\cos^2\theta}$$

52. $a = 4, c = 5, b = 3, e = \frac{5}{4}$

$$r^2 = \frac{-9}{1 - (25/16)\cos^2\theta}$$

53. $a = 3, b = 4, c = 5, e = \frac{5}{3}$

$$r^2 = \frac{-16}{1 - (25/9)\cos^2\theta}$$

54. $a = 2, b = 1, c = \sqrt{3}, e = \frac{\sqrt{3}}{2}$

$$r^2 = \frac{1}{1 - (3/4)\cos^2\theta}$$

55. $A = 2\left[\frac{1}{2}\int_0^{\pi}\left(\frac{3}{2 - \cos\theta}\right)^2 d\theta\right]$

$$= 9\int_0^{\pi}\frac{1}{(2 - \cos\theta)^2}\,d\theta \approx 10.88$$

56. $A = 2\left[\frac{1}{2}\int_{-\pi/2}^{\pi/2}\left(\frac{2}{3 - 2\sin\theta}\right)^2 d\theta\right] = 4\int_{-\pi/2}^{\pi/2}\frac{1}{(3 - 2\sin\theta)^2}\,d\theta \approx 3.37$

57. Vertices: $(123{,}000 + 4000, 0) = (127{,}000, 0)$

$(119 + 4000, \pi) = (4119, \pi)$

$$a = \frac{127{,}000 + 4119}{2} = 65{,}559.5$$

$$c = 65{,}559.5 - 4119 = 61{,}440.5$$

$$e = \frac{c}{a} = \frac{122{,}881}{131{,}119} \approx 0.93717$$

$$r = \frac{ed}{1 - e\cos\theta}$$

$$\theta = 0\colon\ r = \frac{ed}{1 - e},\ \theta = \pi\colon\ r = \frac{ed}{1 + e}$$

$$2a = 2(65{,}559.5) = \frac{ed}{1 - e} + \frac{ed}{1 + e}$$

$$131{,}119 = d\left(\frac{e}{1 - e} + \frac{e}{1 + e}\right) = d\left(\frac{2e}{1 - e^2}\right)$$

$$d = \frac{131{,}119(1 - e^2)}{2e} \approx 8514.1397$$

$$r = \frac{7979.21}{1 - 0.93717\cos\theta} = \frac{1{,}046{,}226{,}000}{131{,}119 - 122{,}881\cos\theta}$$

When $\theta = 60° = \frac{\pi}{3}$, $r \approx 15{,}015$.

Distance between earth and the satellite is $r - 4000 \approx 11{,}015$ miles.

58. (a) $r = \dfrac{ed}{1 - e\cos\theta}$

When $\theta = 0$, $r = c + a = ea + a = a(1 + e)$.

Therefore,

$$a(1 + e) = \frac{ed}{1 - e}$$

$$a(1 + e)(1 - e) = ed$$

$$a(1 - e^2) = ed.$$

Thus, $r = \dfrac{(1 - e^2)a}{1 - e\cos\theta}$.

(b) The perihelion distance is $a - c = a - ea = a(1 - e)$.

When $\theta = \pi$, $r = \dfrac{(1 - e^2)a}{1 + e} = a(1 - e)$.

The aphelion distance is $a + c = a + ea = a(1 + e)$.

When $\theta = 0$, $r = \dfrac{(1 - e^2)a}{1 - e} = a(1 + e)$.

59. $a = 1.496 \times 10^8$, $e = 0.0167$

$$r = \frac{(1 - e^2)a}{1 - e\cos\theta} = \frac{149{,}558{,}278.1}{1 - 0.0167\cos\theta}$$

Perihelion distance: $a(1 - e) \approx 147{,}101{,}680$ km

Aphelion distance: $a(1 + e) \approx 152{,}098{,}320$ km

60. $a = 1.427 \times 10^9$, $e = 0.0542$

$$r = \frac{(1 - e^2)a}{1 - e\cos\theta} = \frac{1{,}422{,}807{,}988}{1 - 0.0542\cos\theta}$$

Perihelion distance: $a(1 - e) \approx 1{,}349{,}656{,}600$ km

Aphelion distance: $a(1 + e) \approx 1{,}504{,}343{,}400$ km

61. $a = 5.906 \times 10^9$, $e = 0.2488$

$$r = \frac{(1 - e^2)a}{1 - e\cos\theta} = \frac{5{,}540{,}410{,}095}{1 - 0.2488\cos\theta}$$

Perihelion distance: $a(1 - e) \approx 4{,}436{,}587{,}200$ km

Aphelion distance: $a(1 + e) \approx 7{,}375{,}412{,}800$ km

62. $a = 5.791 \times 10^7$, $e = 0.2056$

$$r = \frac{(1 - e^2)a}{1 - e\cos\theta} \approx \frac{55{,}462{,}065.54}{1 - 0.2056\cos\theta}$$

Perihelion distance $\approx a(1 - e) \approx 46{,}003{,}704$ km

Aphelion distance $\approx a(1 + e) \approx 69{,}816{,}296$ km

63. $r = \dfrac{5.540 \times 10^9}{1 - 0.2488 \cos \theta}$

(a) $A = \dfrac{1}{2}\displaystyle\int_0^{\pi/9} r^2\, d\theta \approx 9.368 \times 10^{18}$ km^2

$$248\left[\frac{\frac{1}{2}\int_0^{\pi/9} r^2\, d\theta}{\frac{1}{2}\int_0^{2\pi} r^2\, d\theta}\right] \approx 21.89 \text{ yrs}$$

(b) $\dfrac{1}{2}\displaystyle\int_\pi^\alpha r^2\, d\theta = 9.368 \times 10^8$

By trial and error, $\alpha = \pi + 0.9018$

$0.9018 > \pi/9 \approx 0.3491$ because the rays in part (a) are longer than those in part (b).

(c) For part (a)

$$s = \int_0^{\pi/9} \sqrt{r^2 + (dr/d\theta)^2}\, d\theta \approx 2.563 \times 10^9 \text{ km}$$

$$\text{Average per year} = \frac{2.563 \times 10^9}{21.89} \approx 1.17 \times 10^8 \text{ km/yr}$$

For part (b)

$$s = \int_\pi^{\pi + 0.9018} \sqrt{r^2 + (dr/d\theta)^2}\, d\theta \approx 4.133 \times 10^9$$

$$\text{Average per year} = \frac{4.133 \times 10^9}{21.89} \approx 1.89 \times 10^8 \text{ km/yr}$$

64. $a = \dfrac{1}{2}(250) = 125$ au, $e = 0.995$

(a) $e = \dfrac{c}{a} \Rightarrow c = 124.375$

$b^2 = a^2 - c^2 \Rightarrow b \approx 12.4844$

length of minor axis: $2b \approx 24.97$ au

(b) $r = \dfrac{(1 - e^2)a}{1 - e \cos \theta} = \dfrac{1.246875}{1 - 0.995 \cos \theta}$

(c) Perihelion distance: $a(1 - e) = 0.625$ au

Aphelion distance: $a(1 + e) = 249.375$ au

65. $r_1 = a + c,\ r_0 = a - c,\ r_1 - r_0 = 2c,\ r_1 + r_0 = 2a$

$$e = \frac{c}{a} = \frac{r_1 - r_0}{r_1 + r_0}$$

$$\frac{1 + e}{1 - e} = \frac{1 + \frac{c}{a}}{1 - \frac{c}{a}} = \frac{a + c}{a - c} = \frac{r_1}{r_0}$$

66. For a hyperbola,

$r_0 = c - a$ and $r_1 = c + a$.

Thus $r_1 + r_0 = 2c$ and $r_1 - r_0 = 2a$.

$$e = \frac{c}{a} = \frac{r_1 + r_0}{r_1 - r_0}$$

$$\frac{e + 1}{e - 1} = \frac{c/a + 1}{c/a - 1} = \frac{c + a}{c - a} = \frac{r_1}{r_0}$$

67. $r_1 = \dfrac{ed}{1 + \sin \theta}$ and $r_2 = \dfrac{ed}{1 - \sin \theta}$

Points of intersection: $(ed, 0)$, (ed, π)

$$r_1\text{: } \frac{dy}{dx} = \frac{\left(\frac{ed}{1 + \sin \theta}\right)(\cos \theta) + \left(\frac{-ed \cos \theta}{(1 + \sin \theta)^2}\right)(\sin \theta)}{\left(\frac{-ed}{1 + \sin \theta}\right)(\sin \theta) + \left(\frac{-ed \cos \theta}{(1 + \sin \theta)^2}\right)(\cos \theta)}$$

At $(ed, 0)$, $\dfrac{dy}{dx} = -1$. At (ed, π), $\dfrac{dy}{dx} = 1$.

$$r_2\text{: } \frac{dy}{dx} = \frac{\left(\frac{ed}{1 - \sin \theta}\right)(\cos \theta) + \left(\frac{ed \cos \theta}{(1 - \sin \theta)^2}\right)(\sin \theta)}{\left(\frac{-ed}{1 - \sin \theta}\right)(\sin \theta) + \left(\frac{ed \cos \theta}{(1 - \sin \theta)^2}\right)(\cos \theta)}$$

At $(ed, 0)$, $\dfrac{dy}{dx} = 1$. At (ed, π), $\dfrac{dy}{dx} = -1$.

Therefore, at $(ed, 0)$ we have $m_1 m_2 = (-1)(1) = -1$, and at (ed, π) we have $m_1 m_2 = 1(-1) = -1$. The curves intersect at right angles.

68. $r_1 = \dfrac{c}{1 + \cos\theta}, r_2 = \dfrac{d}{1 - \cos\theta}$ (Parabolas)

To find the intersection points:

$$\frac{c}{1+\cos\theta} = \frac{d}{1-\cos\theta}$$

$$c - c \cdot \cos\theta = d + d \cdot \cos\theta$$

$$\cos\theta = \frac{c-d}{c+d}$$

$$r_1 = \frac{c}{1 + \left(\dfrac{c-d}{c+d}\right)} = \frac{c(c+d)}{2c} = \frac{c+d}{2} = r_2$$

$$\frac{dr_1}{d\theta} = \frac{c \cdot \sin\theta}{(1+\cos\theta)^2}, \frac{dr_2}{d\theta} = \frac{-d \cdot \sin\theta}{(1-\cos\theta)^2}$$

For the first parabola,

$$\frac{dy}{dx} = \frac{r_1\cos\theta + r'_1\sin\theta}{-r_1\sin\theta + r'_1\cos\theta} = \frac{c \cdot \cos\theta(1+\cos\theta) + c \cdot \sin^2\theta}{-c \cdot \sin\theta(1+\cos\theta) + c \cdot \sin\theta\cos\theta} = \frac{1+\cos\theta}{-\sin\theta}.$$

Similarly for the second parabola,

$$\frac{dy}{dx} = \frac{\sin\theta}{1+\cos\theta}.$$

Since the product of the slopes is -1, they intersect at right angles.

Review Exercises for Chapter 10

1. $4x^2 + y^2 = 4$

Ellipse

Vertex: $(1, 0)$.

Matches (e)

2. $4x^2 - y^2 = 4$

Hyperbola

Vertex: $(1, 0)$

Matches (c)

3. $y^2 = -4x$

Parabola opening to left.

Matches (b)

4. $y^2 - 4x^2 = 4$

Hyperbola

Vertex: $(0, 2)$

Matches (d)

5. $x^2 + 4y^2 = 4$

Ellipse

Vertex: $(0, 1)$

Matches (a)

6. $x^2 = 4y$

Parabola opening upward.

Matches (f)

7. $16x^2 + 16y^2 - 16x + 24y - 3 = 0$

$$\left(x^2 - x + \frac{1}{4}\right) + \left(y^2 + \frac{3}{2}y + \frac{9}{16}\right) = \frac{3}{16} + \frac{1}{4} + \frac{9}{16}$$

$$\left(x - \frac{1}{2}\right)^2 + \left(y + \frac{3}{4}\right)^2 = 1$$

Circle

Center: $\left(\frac{1}{2}, \frac{3}{4}\right)$

Radius: 1

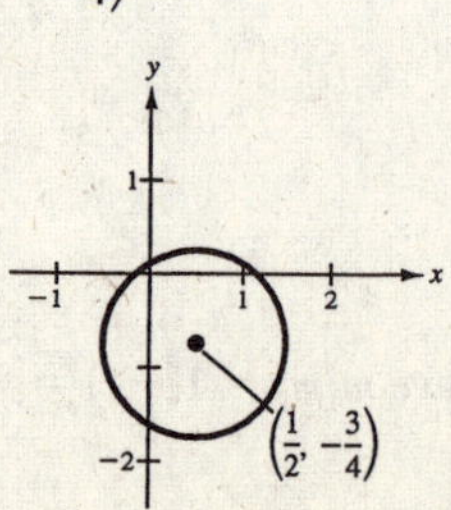

8. $y^2 - 12y - 8x + 20 = 0$

$$y^2 - 12y + 36 = 8x - 20 + 36$$

$$(y-6)^2 = 4(2)(x+2)$$

Parabola

Vertex: $(-2, 6)$

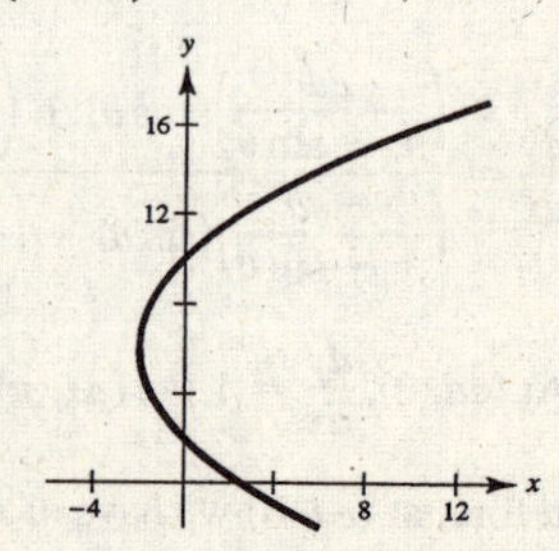

9. $3x^2 - 2y^2 + 24x + 12y + 24 = 0$

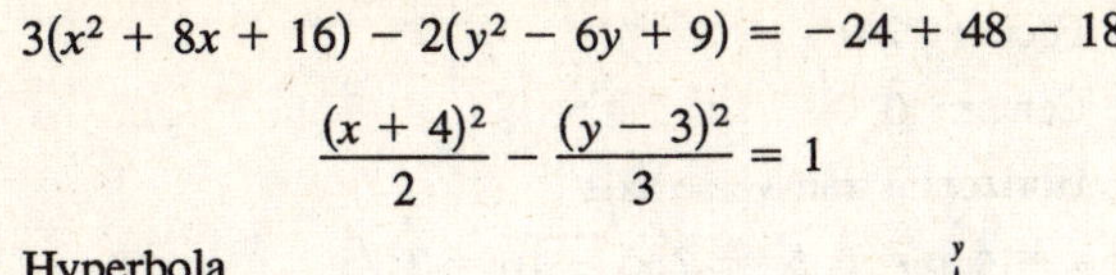

$3(x^2 + 8x + 16) - 2(y^2 - 6y + 9) = -24 + 48 - 18$

$$\frac{(x + 4)^2}{2} - \frac{(y - 3)^2}{3} = 1$$

Hyperbola

Center: $(-4, 3)$

Vertices: $(-4 \pm \sqrt{2}, 3)$

Asymptotes:

$y = 3 \pm \sqrt{\frac{3}{2}}(x + 4)$

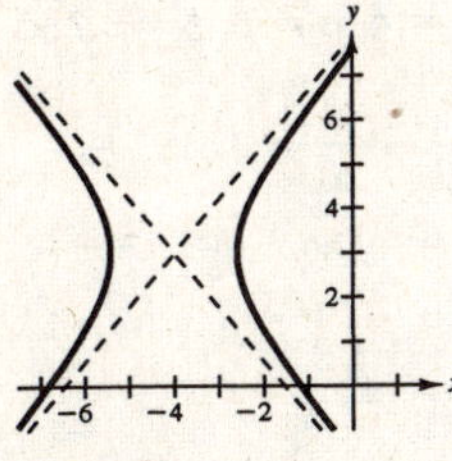

10. $4x^2 + y^2 - 16x + 15 = 0$

$4(x^2 - 4x + 4) + y^2 = -15 + 16$

$$\frac{(x - 2)^2}{1/4} + \frac{y^2}{1} = 1$$

Ellipse

Center: $(2, 0)$

Vertices: $(2, \pm 1)$

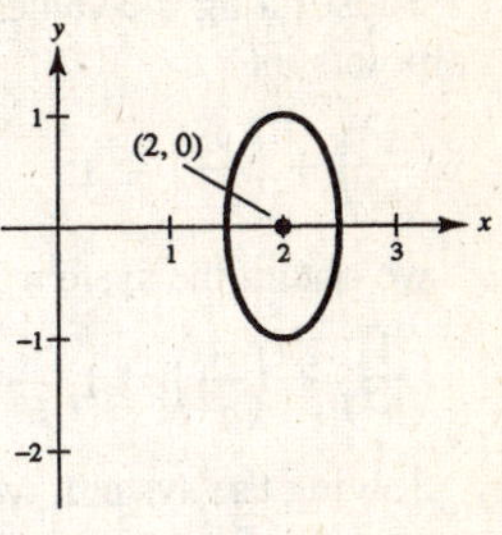

11. $3x^2 + 2y^2 - 12x + 12y + 29 = 0$

$3(x^2 - 4x + 4) + 2(y^2 + 6y + 9) = -29 + 12 + 18$

$$\frac{(x - 2)^2}{1/3} + \frac{(y + 3)^2}{1/2} = 1$$

Ellipse

Center: $(2, -3)$

Vertices: $\left(2, -3 \pm \frac{\sqrt{2}}{2}\right)$

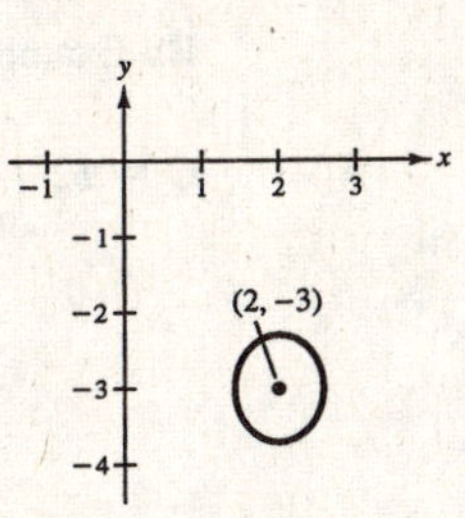

12. $4x^2 - 4y^2 - 4x + 8y - 11 = 0$

$4\left(x^2 - x + \frac{1}{4}\right) - 4(y^2 - 2y + 1) = 11 + 1 - 4$

$$\frac{[x - (1/2)]^2}{2} - \frac{(y - 1)^2}{2} = 1$$

Hyperbola

Center: $\left(\frac{1}{2}, 1\right)$

Vertices: $\left(\frac{1}{2} \pm \sqrt{2}, 1\right)$

Asymptotes: $y = 1 \pm \left(x - \frac{1}{2}\right)$

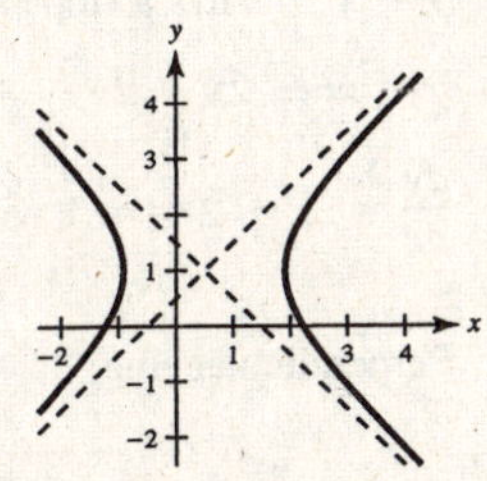

13. Vertex: $(0, 2)$

Directrix: $x = -3$

Parabola opens to the right.

$p = 3$

$(y - 2)^2 = 4(3)(x - 0)$

$y^2 - 4y - 12x + 4 = 0$

14. Vertex: $(4, 2)$

Focus: $(4, 0)$

Parabola opens downward.

$p = -2$

$(x - 4)^2 = 4(-2)(y - 2)$

$x^2 - 8x + 8y = 0$

15. Vertices: $(-3, 0), (7, 0)$

Foci: $(0, 0), (4, 0)$

Horizontal major axis

Center: $(2, 0)$

$a = 5, c = 2, b = \sqrt{21}$

$$\frac{(x - 2)^2}{25} + \frac{y^2}{21} = 1$$

16. Center: $(0, 0)$

Solution points: $(1, 2), (2, 0)$

Substituting the values of the coordinates of the given points into

$$\left(\frac{x^2}{b^2}\right) + \left(\frac{y^2}{a^2}\right) = 1,$$

we obtain the system

$$\left(\frac{1}{b^2}\right) + \left(\frac{4}{a^2}\right) = 1, \frac{4}{b^2} = 1.$$

Solving the system, we have

$$a^2 = \frac{16}{3} \text{ and } b^2 = 4, \left(\frac{x^2}{4}\right) + \left(\frac{3y^2}{16}\right) = 1.$$

17. Vertices: $(\pm 4, 0)$

Foci: $(\pm 6, 0)$

Center: $(0, 0)$

Horizontal transverse axis

$a = 4, c = 6, b = \sqrt{36 - 16} = 2\sqrt{5}$

$$\frac{x^2}{16} - \frac{y^2}{20} = 1$$

18. Foci: $(0, \pm 8)$

Asymptotes: $y = \pm 4x$

Center: $(0, 0)$

Vertical transverse axis

$c = 8$

$y = \frac{a}{b}x = 4x$ asymptote $\rightarrow a = 4b$

$b^2 = c^2 - a^2 = 64 - (4b)^2 \Rightarrow 17b^2 = 64$

$\Rightarrow b^2 = \frac{64}{17} \Rightarrow a^2 = \frac{1024}{17}$

$$\frac{y^2}{1024/17} - \frac{x^2}{64/17} = 1$$

19. $\frac{x^2}{9} + \frac{y^2}{4} = 1, a = 3, b = 2, c = \sqrt{5}, e = \frac{\sqrt{5}}{3}$

By Example 5 of Section 10.1,

$$C = 12\int_0^{\pi/2} \sqrt{1 - \left(\frac{5}{9}\right)\sin^2\theta}\, d\theta \approx 15.87.$$

20. $\frac{x^2}{4} + \frac{y^2}{25} = 1, a = 5, b = 2, c = \sqrt{21}, e = \frac{\sqrt{21}}{5}$

By Example 5 of Section 10.1,

$$C = 20\int_0^{\pi/2} \sqrt{1 - \frac{21}{25}\sin^2\theta}\, d\theta \approx 23.01.$$

21. $y = x - 2$ has a slope of 1. The perpendicular slope is -1.

$y = x^2 - 2x + 2$

$\frac{dy}{dx} = 2x - 2 = -1$ when $x = \frac{1}{2}$ and $y = \frac{5}{4}$.

Perpendicular line: $y - \frac{5}{4} = -1\left(x - \frac{1}{2}\right)$

$4x + 4y - 7 = 0$

22. $2x + y = 5$ has slope -2. The perpendicular slope is $\frac{1}{2}$.

$y = -3x^2 + x - 6$ Parabola

$y' = -6x + 1 = \frac{1}{2}$

$6x = \frac{1}{2}$

$x = \frac{1}{12}, \quad y = -\frac{95}{16}$

Perpendicular line: $y + \frac{95}{16} = \frac{1}{2}\left(x - \frac{1}{12}\right)$

$y = \frac{1}{2}x - \frac{287}{48}$

23. $y = \frac{1}{200}x^2$

(a) $x^2 = 200y$

$x^2 = 4(50)y$

Focus: $(0, 50)$

(b) $y = \frac{1}{200}x^2$

$y' = \frac{1}{100}x$

$\sqrt{1 + (y')^2} = \sqrt{1 + \frac{x^2}{10{,}000}}$

$$S = 2\pi\int_0^{100} x\sqrt{1 + \frac{x^2}{10{,}000}}\, dx \approx 38{,}294.49$$

24. (a) $V = (\pi ab)(\text{Length}) = 12\pi(16) = 192\pi \text{ ft}^3$

(b) $F = 2(62.4)\int_{-3}^{3} (3 - y)\frac{4}{3}\sqrt{9 - y^2}\, dy = \frac{8}{3}(62.4)\left[3\int_{-3}^{3} \sqrt{9 - y^2}\, dy - \int_{-3}^{3} y\sqrt{9 - y^2}\, dy\right]$

$= \frac{8}{3}(62.4)\left[\frac{3}{2}\left(y\sqrt{9 - y^2} + 9 \arcsin\frac{y}{3}\right) + \frac{1}{3}(9 - y^2)^{3/2}\right]_{-3}^{3}$

$= \frac{8}{3}(62.4)\left[\frac{3}{2}\left(\frac{9\pi}{2}\right) - \frac{3}{2}\left(-\frac{9\pi}{2}\right)\right] = \frac{8}{3}(62.4)\left(\frac{27\pi}{2}\right) \approx 7057.274$

(c) You want $\frac{3}{4}$ of the total area of 12π covered. Find h so that

$$2\int_0^h \frac{4}{3}\sqrt{9 - y^2}\, dy = 3\pi$$

$$\int_0^h \sqrt{9 - y^2}\, dy = \frac{9\pi}{8}$$

$$\frac{1}{2}\left[y\sqrt{9 - y^2} + 9 \arcsin\left(\frac{y}{3}\right)\right]_0^h = \frac{9\pi}{8}$$

$$h\sqrt{9 - h^2} + 9 \arcsin\left(\frac{h}{3}\right) = \frac{9\pi}{4}.$$

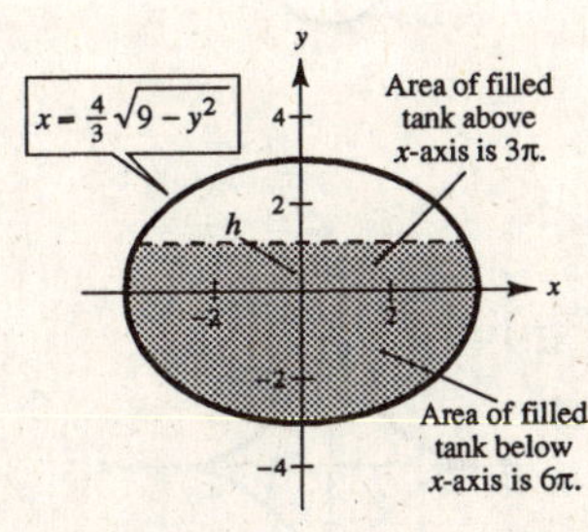

By Newton's Method, $h \approx 1.212$. Therefore, the total height of the water is $1.212 + 3 = 4.212$ ft.

(d) Area of ends $= 2(12\pi) = 24\pi$

Area of sides $=$ (Perimeter)(Length)

$= 16\int_0^{\pi/2}\left(\sqrt{1 - \left(\frac{7}{16}\right)\sin^2\theta}\right) d\theta(16)$ [from Example 5 of Section 9.1]

≈ 353.656

Total area $= 24\pi + 353.656 \approx 429.054$

25. $x = 1 + 4t, y = 2 - 3t$

$t = \frac{x - 1}{4} \Rightarrow y = 2 - 3\left(\frac{x - 1}{4}\right)$

$y = -\frac{3}{4}x + \frac{11}{4}$

$4y + 3x - 11 = 0$

Line

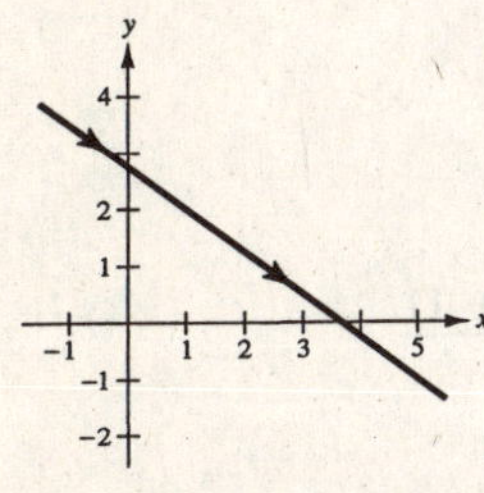

26. $x = t + 4, y = t^2$

$t = x - 4 \Rightarrow y = (x - 4)^2$

Parabola

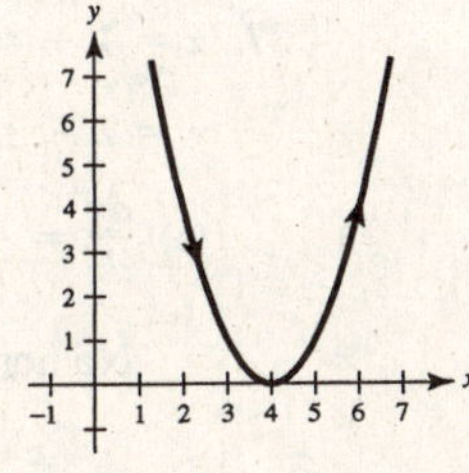

27. $x = 6\cos\theta, y = 6\sin\theta$

$\left(\frac{x}{6}\right)^2 + \left(\frac{y}{6}\right)^2 = 1$

$x^2 + y^2 = 36$

Circle

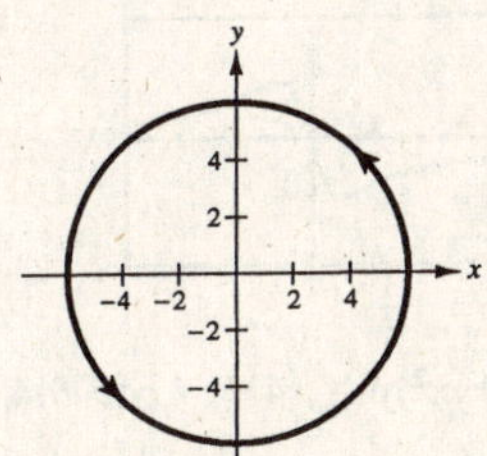

28. $x = 3 + 3\cos\theta, y = 2 + 5\sin\theta$

$$\left(\frac{x-3}{3}\right)^2 + \left(\frac{y-2}{5}\right)^2 = 1$$

$$\frac{(x-3)^2}{9} + \frac{(y-2)^2}{25} = 1$$

Ellipse

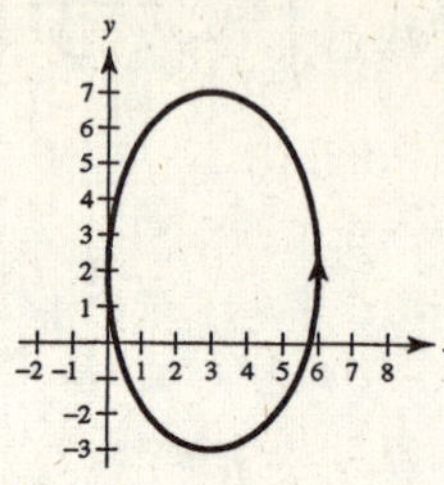

29. $x = 2 + \sec\theta, y = 3 + \tan\theta$

$(x-2)^2 = \sec^2\theta = 1 + \tan^2\theta = 1 + (y-3)^2$

$(x-2)^2 - (y-3)^2 = 1$

Hyperbola

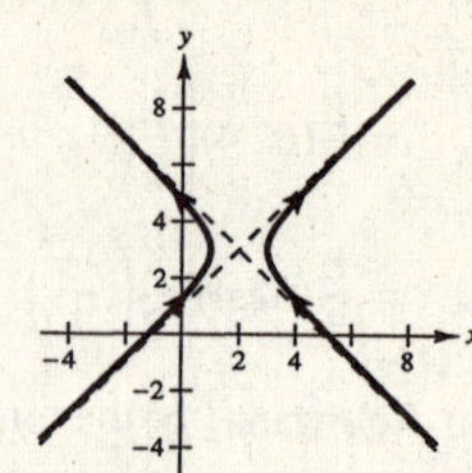

30. $x = 5\sin^3\theta, y = 5\cos^3\theta$

$$\left(\frac{x}{5}\right)^{2/3} + \left(\frac{y}{5}\right)^{2/3} = 1$$

$$x^{2/3} + y^{2/3} = 5^{2/3}$$

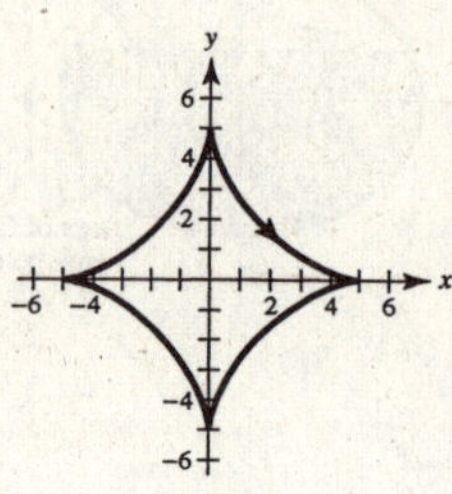

31. $x = 3 + (3 - (-2))t = 3 + 5t$

$y = 2 + (2 - 6)t = 2 - 4t$

(other answers possible)

32. $(x-h)^2 + (y-k)^2 = r^2$

$(x-5)^2 + (y-3)^2 = 2^2 = 4$

$x = 5 + 2\cos t, \quad y = 3 + 2\sin t$

33. $\dfrac{(x+3)^2}{16} + \dfrac{(y-4)^2}{9} = 1$

Let $\dfrac{(x+3)^2}{16} = \cos^2\theta$ and $\dfrac{(y-4)^2}{9} = \sin^2\theta$.

Then $x = -3 + 4\cos\theta$ and $y = 4 + 3\sin\theta$.

34. $a = 4,\ c = 5,\ b^2 = c^2 - a^2 = 9,\ \dfrac{y^2}{16} - \dfrac{x^2}{9} = 1$

Let $\dfrac{y^2}{16} = \sec^2\theta$ and $\dfrac{x^2}{9} = \tan^2\theta$.

Then $x = 3\tan\theta$ and $y = 4\sec\theta$.

35. $x = \cos 3\theta + 5\cos\theta$

$y = \sin 3\theta + 5\sin\theta$

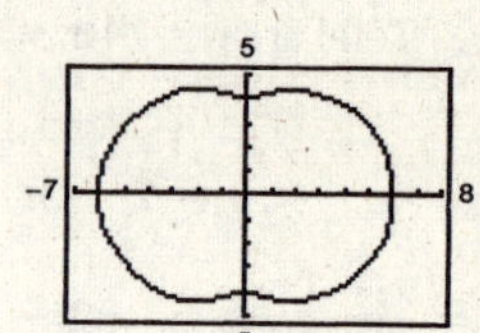

36. (a) $x = 2\cot\theta, y = 4\sin\theta\cos\theta, 0 < \theta < \pi$

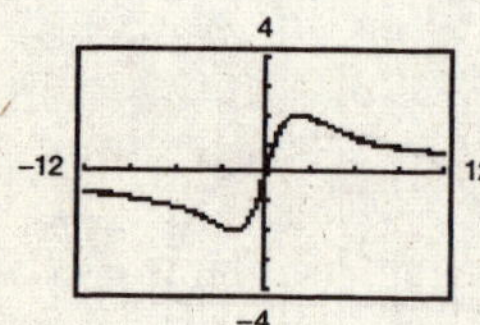

(b) $(4 + x^2)y = (4 + 4\cot^2\theta)4\sin\theta\cos\theta$

$= 16\csc^2\theta \cdot \sin\theta \cdot \cos\theta$

$= 16\dfrac{\cos\theta}{\sin\theta}$

$= 8(2\cot\theta)$

$= 8x$

37. $x = 1 + 4t$

$y = 2 - 3t$

(a) $\dfrac{dy}{dx} = -\dfrac{3}{4}$

No horizontal tangents

(b) $t = \dfrac{x-1}{4}$

$y = 2 - \dfrac{3}{4}(x-1) = \dfrac{-3x+11}{4}$

(c)

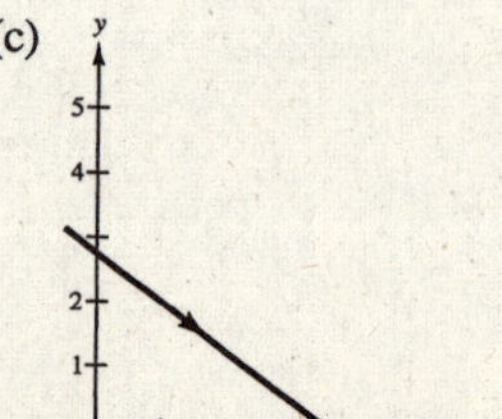

38. $x = t + 4$

$y = t^2$

(a) $\dfrac{dy}{dx} = \dfrac{2t}{1} = 2t = 0$ when $t = 0$.

Point of horizontal tangency: $(4, 0)$

(b) $t = x - 4$

$y = (x - 4)^2$

(c)

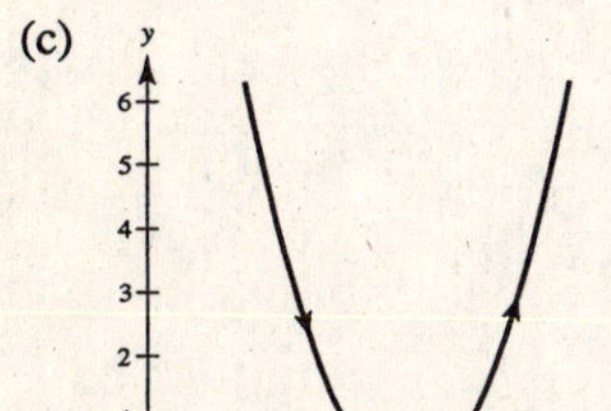

39. $x = \dfrac{1}{t}$

$y = 2t + 3$

(a) $\dfrac{dy}{dx} = \dfrac{2}{-1/t^2} = -2t^2$

No horizontal tangents, $(t \neq 0)$

(b) $t = \dfrac{1}{x}$

$y = \dfrac{2}{x} + 3$

(c)

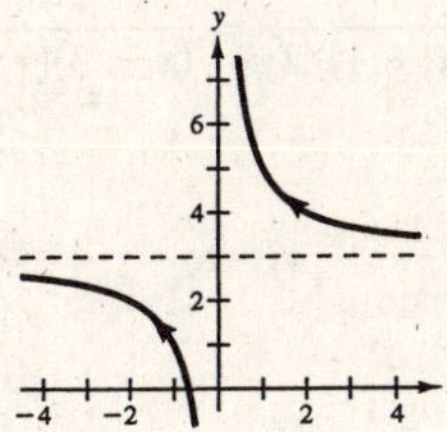

40. $x = \dfrac{1}{t}$

$y = t^2$

(a) $\dfrac{dy}{dx} = \dfrac{2t}{-1/t^2} = -2t^3$

No horizontal tangents, $(t \neq 0)$

(b) $t = \dfrac{1}{x}$

$y = \dfrac{1}{x^2}$

(c)

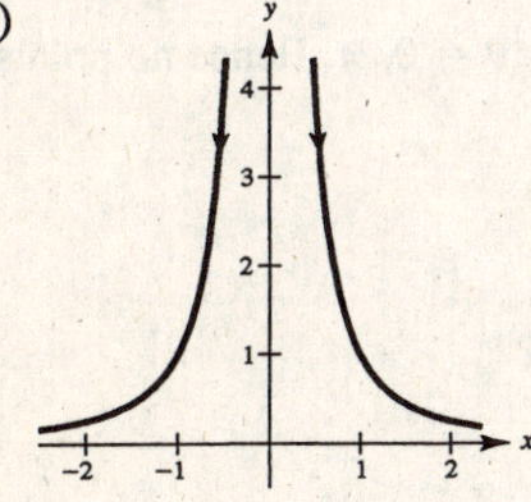

41. $x = \dfrac{1}{2t + 1}$

$y = \dfrac{1}{t^2 - 2t}$

(a) $\dfrac{dy}{dx} = \dfrac{\dfrac{-(2t - 2)}{(t^2 - 2t)^2}}{\dfrac{-2}{(2t + 1)^2}} = \dfrac{(t - 1)(2t + 1)^2}{t^2(t - 2)^2} = 0$ when $t = 1$.

Point of horizontal tangency: $\left(\dfrac{1}{3}, -1\right)$

(b) $2t + 1 = \dfrac{1}{x} \Rightarrow t = \dfrac{1}{2}\left(\dfrac{1}{x} - 1\right)$

$$y = \frac{1}{\dfrac{1}{2}\left(\dfrac{1 - x}{x}\right)\left[\dfrac{1}{2}\left(\dfrac{1 - x}{x}\right) - 2\right]}$$

$$= \frac{4x^2}{(1 - x)^2 - 4x(1 - x)} = \frac{4x^2}{(5x - 1)(x - 1)}, \quad (x \neq 0)$$

(c)

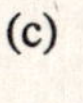

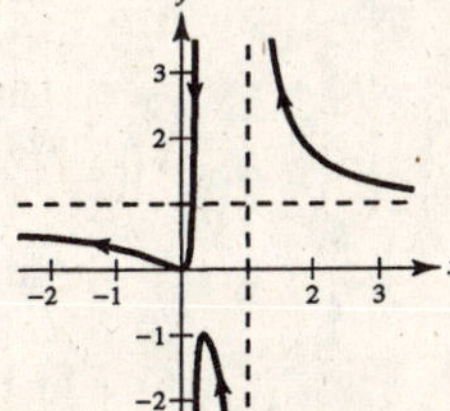

42. $x = 2t - 1$

$y = \dfrac{1}{t^2 - 2t}$

(a) $\dfrac{dy}{dx} = \dfrac{-(t^2 - 2t)^{-2}(2t - 2)}{2}$

$= \dfrac{1 - t}{t^2(t - 2)^2} = 0$ when $t = 1$.

Point of horizontal tangency: $(1, -1)$

(b) $t = \dfrac{x + 1}{2}$

$y = \dfrac{1}{[(x + 1)/2]^2 - 2[(x + 1)/2]} = \dfrac{4}{(x - 3)(x + 1)}$

(c)

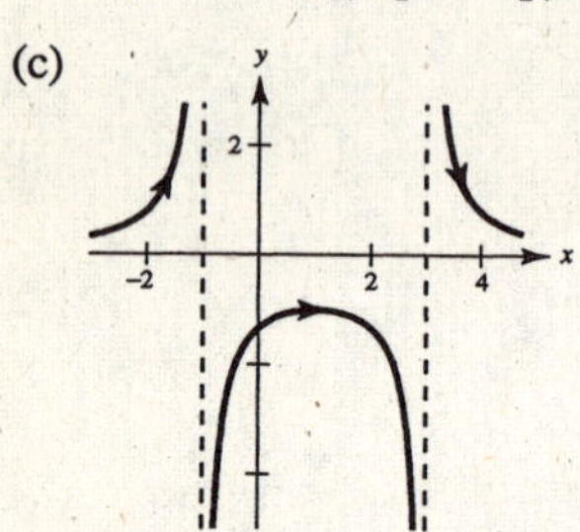

43. $x = 3 + 2\cos\theta$

$y = 2 + 5\sin\theta$

(a) $\dfrac{dy}{dx} = \dfrac{5\cos\theta}{-2\sin\theta} = -2.5\cot\theta = 0$ when $\theta = \dfrac{\pi}{2}, \dfrac{3\pi}{2}$.

Points of horizontal tangency: $(3, 7), (3, -3)$

(b) $\dfrac{(x - 3)^2}{4} + \dfrac{(y - 2)^2}{25} = 1$

(c)

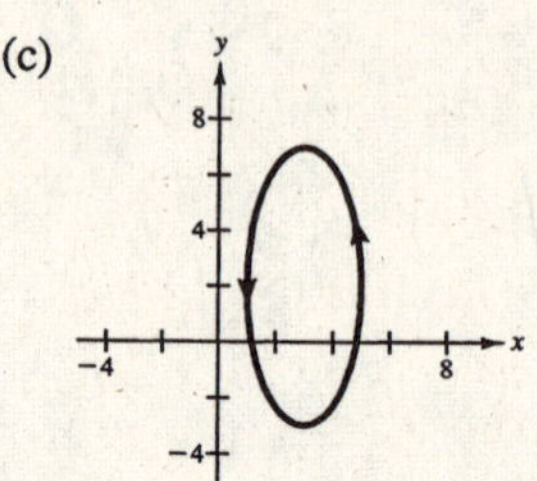

44. $x = 6\cos\theta$

$y = 6\sin\theta$

(a) $\dfrac{dy}{dx} = \dfrac{6\cos\theta}{-6\sin\theta} = -\cot\theta = 0$ when $\theta = \dfrac{\pi}{2}, \dfrac{3\pi}{2}$.

Points of horizontal tangency: $(0, 6), (0, -6)$

(b) $\left(\dfrac{x}{6}\right)^2 + \left(\dfrac{y}{6}\right)^2 = 1$

(c)

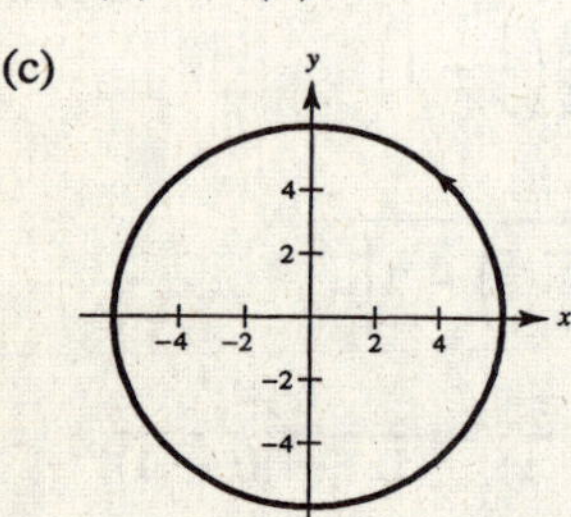

45. $x = \cos^3\theta$

$y = 4\sin^3\theta$

(a) $\dfrac{dy}{dx} = \dfrac{12\sin^2\theta\cos\theta}{3\cos^2\theta(-\sin\theta)}$

$= \dfrac{-4\sin\theta}{\cos\theta} = -4\tan\theta = 0$

when $\theta = 0, \pi$.

But, $\dfrac{dy}{dt} = \dfrac{dx}{dt} = 0$ at $\theta = 0, \pi$. Hence no points of horizontal tangency.

(b) $x^{2/3} + \left(\dfrac{y}{4}\right)^{2/3} = 1$

(c)

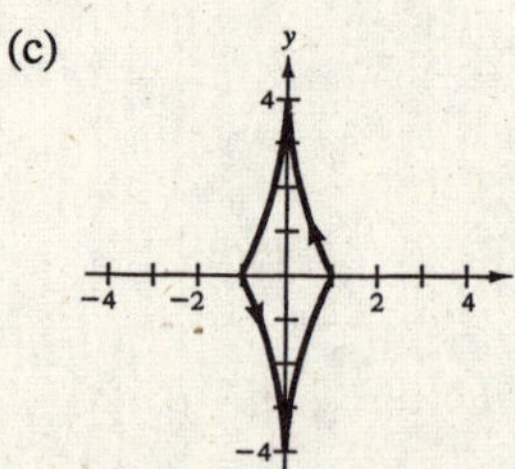

46. $x = e^t$

$y = e^{-t}$

(a) $\dfrac{dy}{dx} = \dfrac{-e^{-t}}{e^t} = -\dfrac{1}{e^{2t}} = -\dfrac{1}{x^2}$

No horizontal tangents

(b) $t = \ln x$

$y = e^{-\ln x} = e^{\ln(1/x)} = \dfrac{1}{x}, x > 0$

(c)

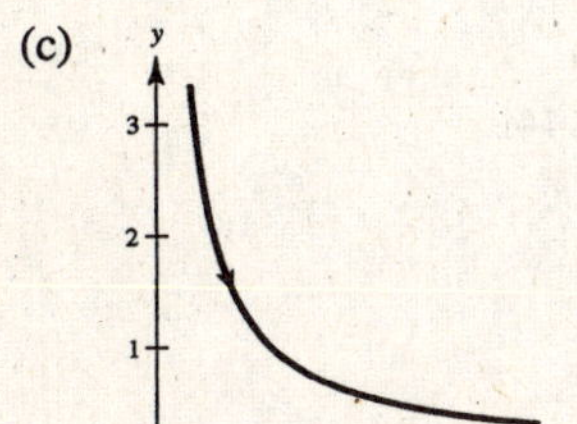

47. $x = 4 - t, \quad y = t^2$

$\dfrac{dx}{dt} = -1, \quad \dfrac{dy}{dt} = 2t$

$\dfrac{dy}{dt} = 0$ for $t = 0$.

Horizontal tangent at $t = 0$: $(x, y) = (4, 0)$

No vertical tangents

48. $x = t + 2, y = t^3 - 2t$

$\dfrac{dx}{dt} = 1, \dfrac{dy}{dt} = 3t^2 - 2$

$\dfrac{dy}{dt} = 0$ for $t = \pm\sqrt{\dfrac{2}{3}} = \dfrac{\pm\sqrt{6}}{3}$

Horizontal tangents: $t = \dfrac{\sqrt{2}}{3}$: $(x, y) = \left(\dfrac{\sqrt{6}}{3} + 2, \dfrac{2\sqrt{6}}{9} - \dfrac{2}{3}\sqrt{6}\right)$

$\approx (2.8165, -1.0887)$

$t = -\dfrac{\sqrt{6}}{3}$: $(x, y) = \left(-\dfrac{\sqrt{6}}{3} + 2, \dfrac{2}{3}\sqrt{6} - \dfrac{2\sqrt{6}}{9}\right)$

$\approx (1.1835, 1.0887)$

No vertical tangents

49. $x = 2 + 2\sin\theta, y = 1 + \cos\theta$

$\dfrac{dx}{d\theta} = 2\cos\theta, \dfrac{dy}{d\theta} = -\sin\theta$

$\dfrac{dy}{d\theta} = 0$ for $\theta = 0, \pi, 2\pi, \ldots$

Horizontal tangents: $(x, y) = (2, 2), (2, 0)$

$\dfrac{dx}{d\theta} = 0$ for $\theta = \dfrac{\pi}{2}, \dfrac{3\pi}{2}, \ldots$

Vertical tangents: $(x, y) = (4, 1), (0, 1)$

50. $x = 2 - 2\cos\theta, y = 2\sin 2\theta$

$\dfrac{dx}{d\theta} = 2\sin\theta, \dfrac{dy}{d\theta} = 4\cos 2\theta$

$\dfrac{dy}{d\theta} = 0$ for $\theta = \dfrac{\pi}{4}, \dfrac{3\pi}{4}, \dfrac{5\pi}{4}, \dfrac{7\pi}{4}, \ldots$

Horizontal tangents: $(x, y) = \left(2 \pm \sqrt{2}, 2\right),$

$\left(2 \pm \sqrt{2}, -2\right)$

$\dfrac{dx}{d\theta} = 0$ for $\theta = 0, \pi, 2\pi, \ldots$

Vertical tangents: $(x, y) = (0, 0), (4, 0)$

51. $x = \cot\theta$

$y = \sin 2\theta = 2\sin\theta\cos\theta$

(a), (c)

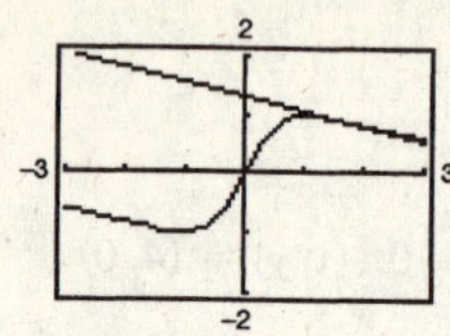

(b) At $\theta = \frac{\pi}{6}, \frac{dx}{d\theta} = -4, \frac{dy}{d\theta} = 1$, and $\frac{dy}{dx} = -\frac{1}{4}$.

52. $x = 2\theta - \sin\theta$

$y = 2 - \cos\theta$

(a), (c)

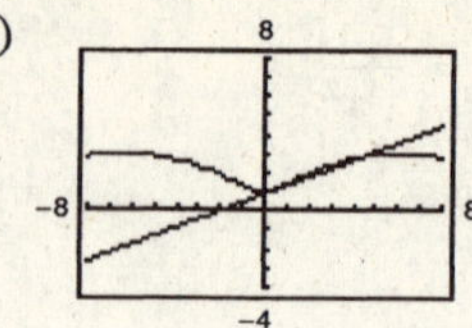

(b) At $\theta = \frac{\pi}{6}, \frac{dx}{d\theta} \approx 1.134, \left(2 - \frac{\sqrt{3}}{2}\right)$,

$\frac{dy}{dt} = 0.5$, and $\frac{dy}{dx} \approx 0.441$.

53. $x = r(\cos\theta + \theta\sin\theta)$

$y = r(\sin\theta - \theta\cos\theta)$

$\frac{dx}{d\theta} = r\theta\cos\theta$

$\frac{dy}{d\theta} = r\theta\sin\theta$

$$s = r\int_0^{\pi} \sqrt{\theta^2\cos^2\theta + \theta^2\sin^2\theta}\, d\theta$$

$$= r\int_0^{\pi} \theta\, d\theta = \frac{r}{2}\Big[\theta^2\Big]_0^{\pi} = \frac{1}{2}\pi^2 r$$

54. $x = 6\cos\theta$

$y = 6\sin\theta$

$\frac{dx}{d\theta} = -6\sin\theta$

$\frac{dy}{d\theta} = 6\cos\theta$

$$s = \int_0^{\pi} \sqrt{36\sin^2\theta + 36\cos^2\theta}\, d\theta = \Big[6\theta\Big]_0^{\pi} = 6\pi$$

(one-half circumference of circle)

55. $x = t, y = 3t, 0 \le t \le 2$

$$\frac{dx}{dt} = 1, \frac{dy}{dt} = -3, \sqrt{\left(\frac{dx}{dt}\right)^2 + \left(\frac{dy}{dt}\right)^2} = \sqrt{1+9} = \sqrt{10}$$

(a) $S = 2\pi\int_0^2 3t\sqrt{10}\, dt = 6\sqrt{10}\,\pi\left[\frac{t^2}{2}\right]_0^2 = 12\sqrt{10}\,\pi$

(b) $S = 2\pi\int_0^2 \sqrt{10}\, dt = 2\pi\sqrt{10}\Big[t\Big]_0^2 = 4\pi\sqrt{10}$

56. $x = 2\cos\theta, y = 2\sin\theta, 0 \le \theta \le \frac{\pi}{2}$

$$\frac{dx}{dt} = -2\sin\theta, \frac{dy}{dt} = 2\cos\theta, \sqrt{\left(\frac{dx}{d\theta}\right)^2 + \left(\frac{dy}{d\theta}\right)^2} = 2$$

(a) $S = 2\pi\int_0^{\pi/2} 2\sin\theta(2)\, d\theta = 8\pi\Big[-\cos\theta\Big]_0^{\pi/2} = 8\pi$

(b) $S = 2\pi\int_0^{\pi/2} 2\cos\theta(2)\, d\theta = 8\pi\Big[\sin\theta\Big]_0^{\pi/2} = 8\pi$

[**Note:** The surface is a hemisphere: $\frac{1}{2}(4\pi(2^2)) = 8\pi$]

57. $x = 3\sin\theta, y = 2\cos\theta$

$$A = \int_a^b y\, dx = \int_{-\pi/2}^{\pi/2} 2\cos\theta(3\cos\theta)\, d\theta$$

$$= 6\int_{-\pi/2}^{\pi/2} \frac{1+\cos 2\theta}{2}\, d\theta$$

$$= 3\left[\theta + \frac{\sin 2\theta}{2}\right]_{-\pi/2}^{\pi/2}$$

$$= 3\left[\frac{\pi}{2} + \frac{\pi}{2}\right] = 3\pi$$

58. $A = \int_a^b y\, dx = \int_{\pi}^{0} \sin\theta(-2\sin\theta)\, d\theta$

$$= -\int_{\pi}^{0} \frac{1-\cos 2\theta}{2}\, d\theta$$

$$= -\left[\theta - \frac{\sin 2\theta}{2}\right]_{\pi}^{0} = \pi$$

59. $(r, \theta) = \left(3, \dfrac{\pi}{2}\right)$

$(x, y) = \left(3\cos\dfrac{\pi}{2}, 3\sin\dfrac{\pi}{2}\right)$

$= (0, 3)$

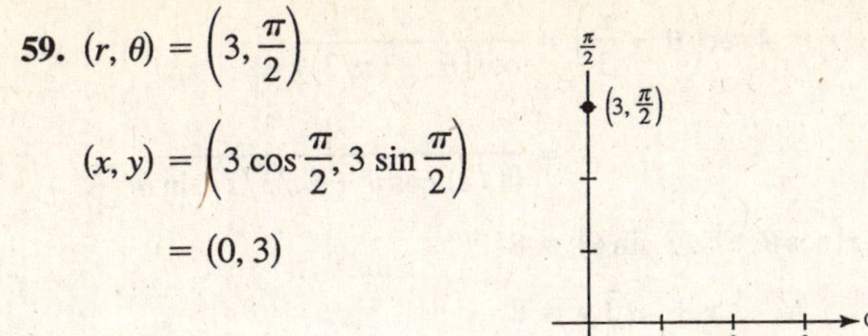

60. $(r, \theta) = \left(-4, \dfrac{11\pi}{6}\right)$

$(x, y) = \left(-4\cos\dfrac{11\pi}{6}, -4\sin\dfrac{11\pi}{6}\right)$

$= \left(-2\sqrt{3}, 2\right)$

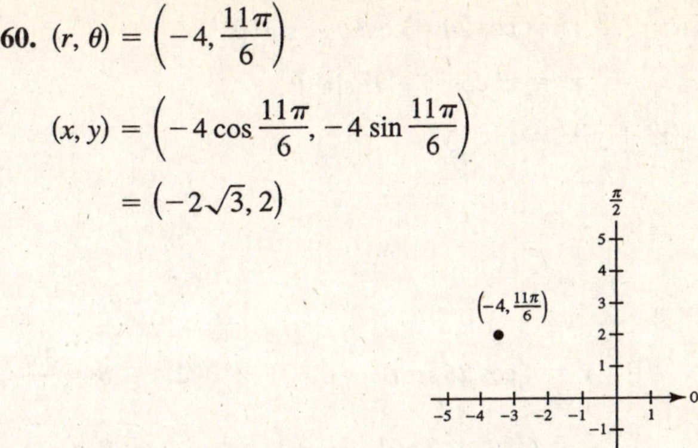

61. $(r, \theta) = \left(\sqrt{3}, 1.56\right)$

$(x, y) = \left(\sqrt{3}\cos(1.56), \sqrt{3}\sin(1.56)\right)$

$\approx (0.0187, 1.7319)$

62. $(r, \theta) = (-2, -2.45)$

$(x, y) = (-2\cos(-2.45), -\sin(-2.45))$

$\approx (1.5405, 1.2755)$

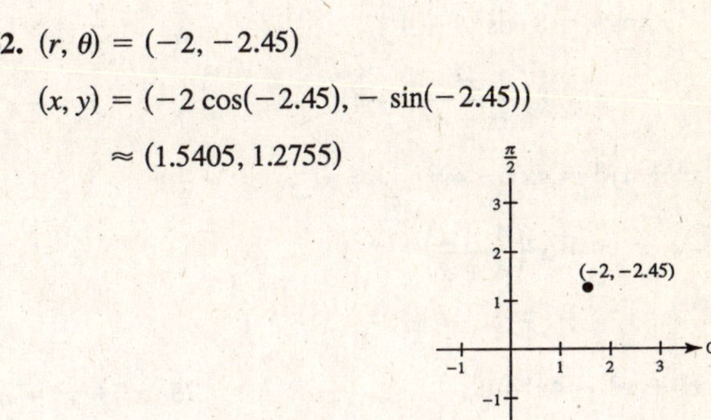

63. $(x, y) = (4, -4)$

$r = \sqrt{4^2 + (-4)^2} = 4\sqrt{2}$

$\theta = \dfrac{7\pi}{4}$

$(r, \theta) = \left(4\sqrt{2}, \dfrac{7\pi}{4}\right), \left(-4\sqrt{2}, \dfrac{3\pi}{4}\right)$

64. $(x, y) = (-1, 3)$

$r = \sqrt{(-1)^2 + 3^2} = \sqrt{10}$

$\theta = \arctan(-3) \approx 1.89(108.43°)$

$(r, \theta) = \left(\sqrt{10}, 1.89\right), \left(-\sqrt{10}, 5.03\right)$

65.

$$r = 3\cos\theta$$
$$r^2 = 3r\cos\theta$$
$$x^2 + y^2 = 3x$$
$$x^2 + y^2 - 3x = 0$$

66.

$$r = 10$$
$$r^2 = 100$$
$$x^2 + y^2 = 100$$

67.

$$r = -2(1 + \cos\theta)$$
$$r^2 = -2r(1 + \cos\theta)$$
$$x^2 + y^2 = -2\left(\pm\sqrt{x^2 + y^2}\right) - 2x$$
$$(x^2 + y^2 + 2x)^2 = 4(x^2 + y^2)$$

68.

$$r = \frac{1}{2 - \cos\theta}$$
$$2r - r\cos\theta = 1$$
$$2\left(\pm\sqrt{x^2 + y^2}\right) - x = 1$$
$$4(x^2 + y^2) = (x + 1)^2$$
$$3x^2 + 4y^2 - 2x - 1 = 0$$

69. $r^2 = \cos 2\theta = \cos^2\theta - \sin^2\theta$

$r^4 = r^2\cos^2\theta - r^2\sin^2\theta$

$(x^2 + y^2)^2 = x^2 - y^2$

70. $r = 4\sec\left(\theta - \dfrac{\pi}{3}\right) = \dfrac{4}{\cos[\theta - (\pi/3)]}$

$= \dfrac{4}{(1/2)\cos\theta + \left(\sqrt{3}/2\right)\sin\theta}$

$r\left(\cos\theta + \sqrt{3}\sin\theta\right) = 8$

$x + \sqrt{3}\,y = 8$

71. $r = 4\cos 2\theta\sec\theta$

$= 4(2\cos^2\theta - 1)\left(\dfrac{1}{\cos\theta}\right)$

$r\cos\theta = 8\cos^2\theta - 4$

$x = 8\left(\dfrac{x^2}{x^2 + y^2}\right) - 4$

$x^3 + xy^2 = 4x^2 - 4y^2$

$y^2 = x^2\left(\dfrac{4 - x}{4 + x}\right)$

72. $\theta = \dfrac{3\pi}{4}$

$\tan\theta = -1$

$\dfrac{y}{x} = -1$

$y = -x$

73. $(x^2 + y^2)^2 = ax^2y$

$r^4 = a(r^2\cos^2\theta)(r\sin\theta)$

$r = a\cos^2\theta\sin\theta$

74. $x^2 + y^2 - 4x = 0$

$r^2 - 4r\cos\theta = 0$

$r = 4\cos\theta$

75. $x^2 + y^2 = a^2\left(\arctan\dfrac{y}{x}\right)^2$

$r^2 = a^2\theta^2$

76. $(x^2 + y^2)\left(\arctan\dfrac{y}{x}\right)^2 = a^2$

$r^2\theta^2 = a^2$

77. $r = 4$

Circle of radius 4

Centered at the pole

Symmetric to polar axis,

$\theta = \pi/2$, and pole

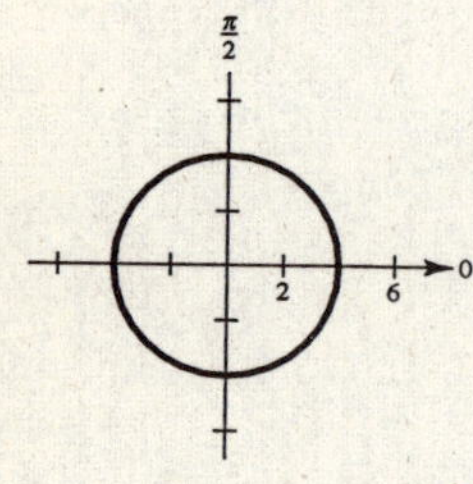

78. $\theta = \dfrac{\pi}{12}$

Line

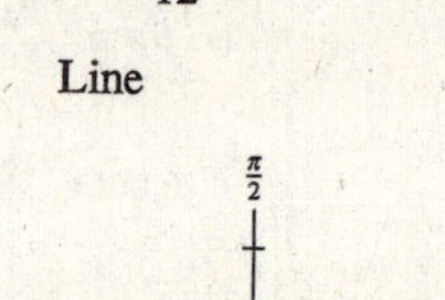

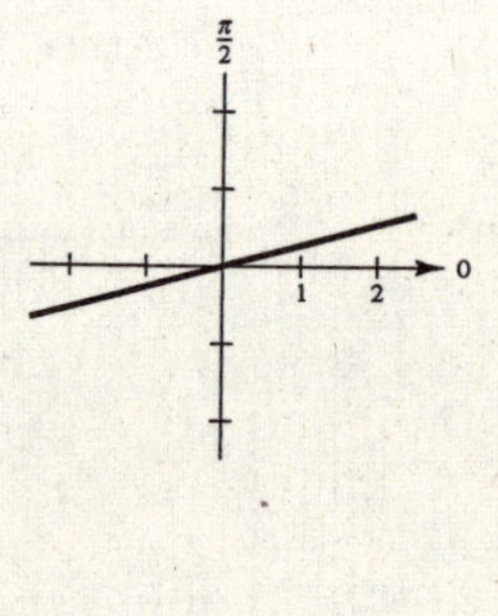

79. $r = -\sec\theta = \dfrac{-1}{\cos\theta}$

$r\cos\theta = -1, x = -1$

Vertical line

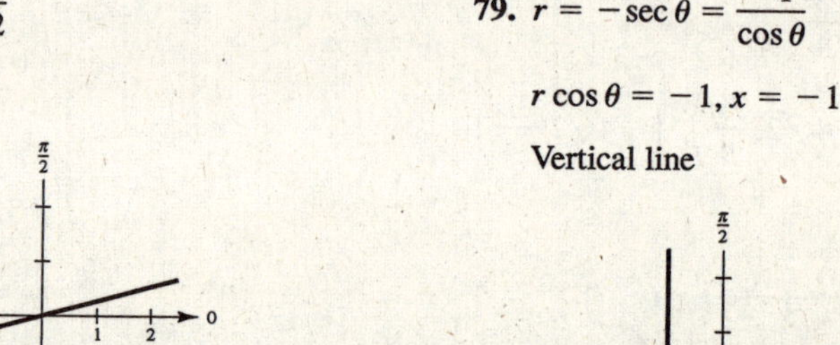

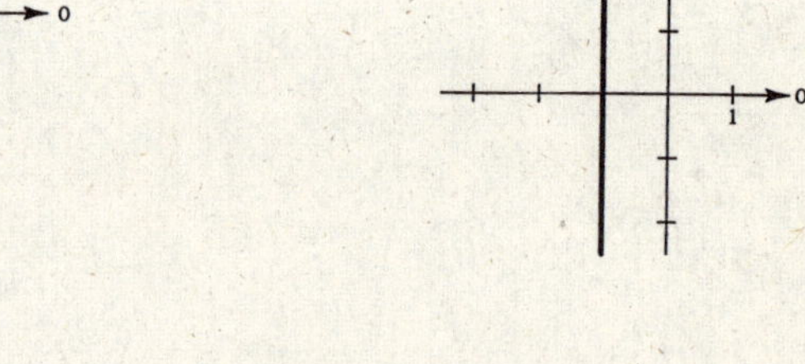

80. $r = 3\csc\theta$, $r\sin\theta = 3$, $y = 3$

Horizontal line

81. $r = -2(1 + \cos\theta)$

Cardioid

Symmetric to polar axis

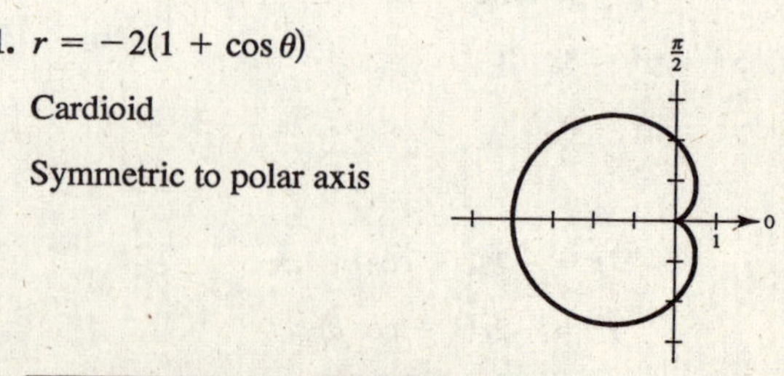

θ	0	$\frac{\pi}{3}$	$\frac{\pi}{2}$	$\frac{2\pi}{3}$	π
r	-4	-3	-2	-1	0

82. $r = 3 - 4\cos\theta$

Limaçon

Symmetric to polar axis

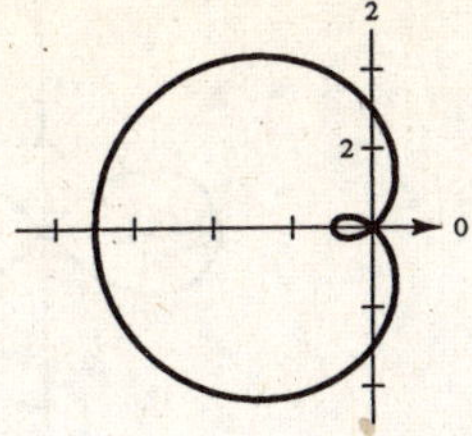

θ	0	$\frac{\pi}{3}$	$\frac{\pi}{2}$	$\frac{2\pi}{3}$	π
r	-1	1	3	5	7

83. $r = 4 - 3\cos\theta$

Limaçon

Symmetric to polar axis

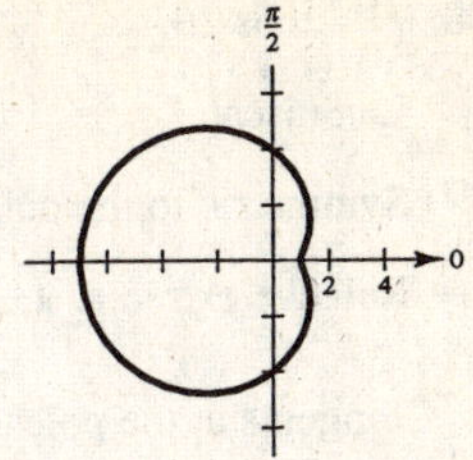

θ	0	$\frac{\pi}{3}$	$\frac{\pi}{2}$	$\frac{2\pi}{3}$	π
r	1	$\frac{5}{2}$	4	$\frac{11}{2}$	7

84. $r = 2\theta$

Spiral

Symmetric to $\theta = \pi/2$

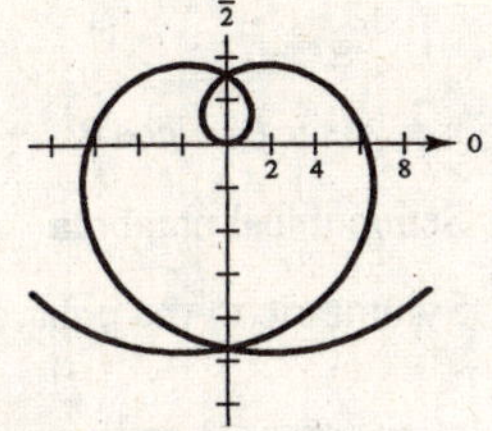

θ	0	$\frac{\pi}{4}$	$\frac{\pi}{2}$	$\frac{3\pi}{4}$	π	$\frac{5\pi}{4}$	$\frac{3\pi}{2}$
r	0	$\frac{\pi}{2}$	π	$\frac{3\pi}{2}$	2π	$\frac{5\pi}{2}$	3π

85. $r = -3\cos 2\theta$

Rose curve with four petals

Symmetric to polar axis, $\theta = \frac{\pi}{2}$, and pole

Relative extrema: $(-3, 0), \left(3, \frac{\pi}{2}\right), (-3, \pi), \left(3, \frac{3\pi}{2}\right)$

Tangents at the pole: $\theta = \frac{\pi}{4}, \frac{3\pi}{4}$

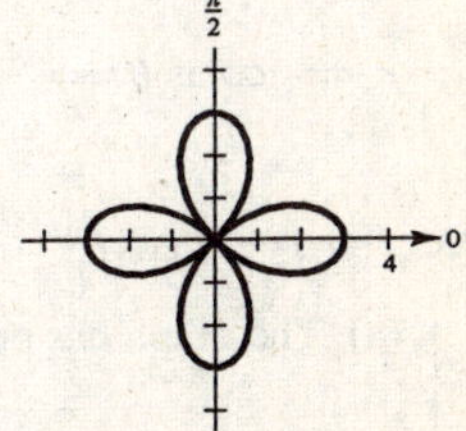

86. $r = \cos 5\theta$

Rose curve with five petals

Symmetric to polar axis

Relative extrema: $(1, 0), \left(-1, \frac{\pi}{5}\right), \left(1, \frac{2\pi}{5}\right), \left(-1, \frac{3\pi}{5}\right), \left(1, \frac{4\pi}{5}\right)$

Tangents at the pole: $\theta = \frac{\pi}{10}, \frac{3\pi}{10}, \frac{\pi}{2}, \frac{7\pi}{10}, \frac{9\pi}{10}$

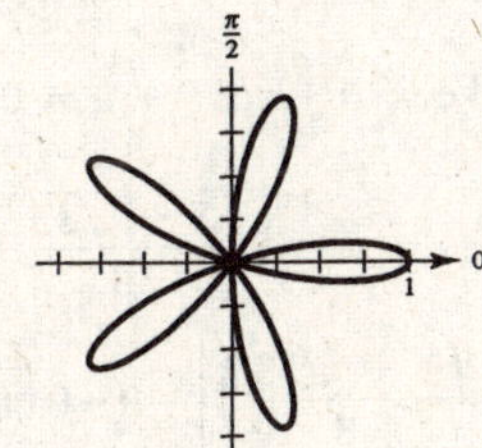

87. $r^2 = 4\sin^2 2\theta$

$r = \pm 2\sin(2\theta)$

Rose curve with four petals

Symmetric to the polar axis, $\theta = \frac{\pi}{2}$, and pole

Relative extrema: $\left(\pm 2, \frac{\pi}{4}\right), \left(\pm 2, \frac{3\pi}{4}\right)$

Tangents at the pole: $\theta = 0, \frac{\pi}{2}$

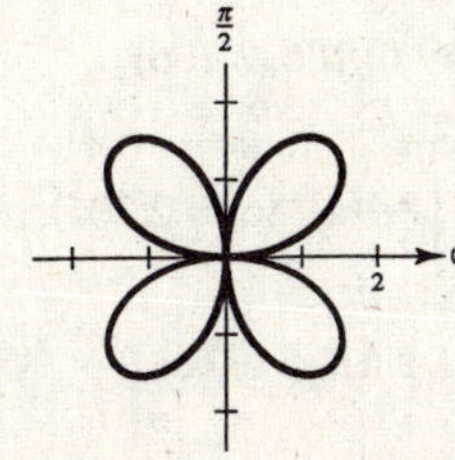

88. $r^2 = \cos 2\theta$

Lemniscate

Symmetric to the polar axis

Relative extrema: $(\pm 1, 0)$

Tangents at the pole: $\theta = \frac{\pi}{4}, \frac{3\pi}{4}$

θ	0	$\frac{\pi}{6}$	$\frac{\pi}{4}$
r	± 1	$\pm\frac{\sqrt{2}}{2}$	0

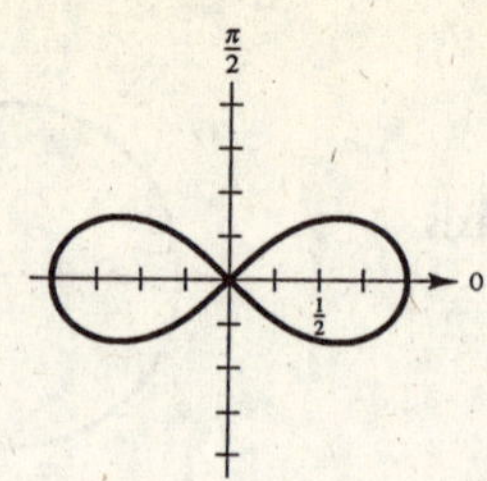

89. $r = \dfrac{3}{\cos \theta - (\pi/4)}$

Graph of $r = 3 \sec \theta$ rotated through an angle of $\pi/4$

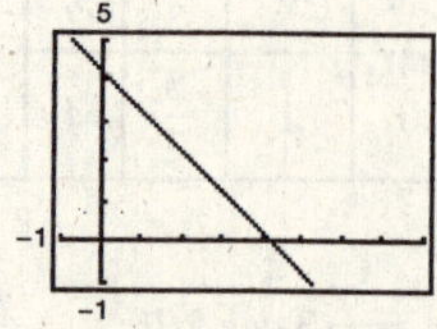

90. $r = 2 \sin \theta \cos^2 \theta$

Bifolium

Symmetric to $\theta = \pi/2$

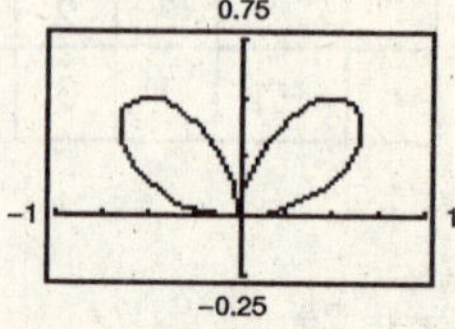

91. $r = 4 \cos 2\theta \sec \theta$

Strophoid

Symmetric to the polar axis

$r \Rightarrow \infty$ as $\theta \Rightarrow \dfrac{\pi^-}{2}$

$r \Rightarrow \infty$ as $\theta \Rightarrow \dfrac{-\pi^+}{2}$

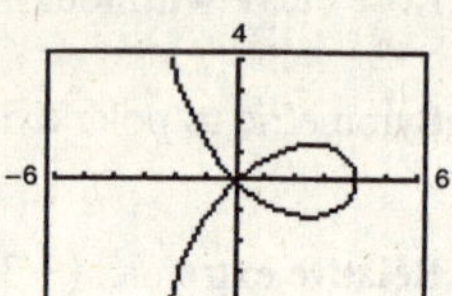

92. $r = 4(\sec \theta - \cos \theta)$

Semicubical parabola

Symmetric to the polar axis

$r \Rightarrow \infty$ as $\theta \Rightarrow \dfrac{\pi^-}{2}$

$r \Rightarrow \infty$ as $\theta \Rightarrow \dfrac{-\pi^+}{2}$

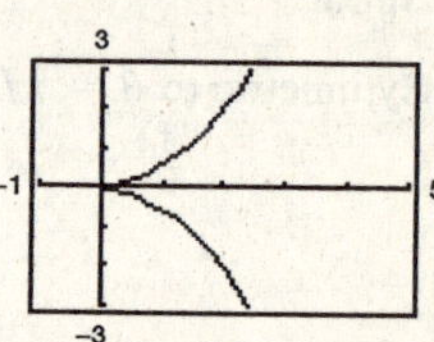

93. $r = 1 - 2 \cos \theta$

(a) The graph has polar symmetry and the tangents at the pole are $\theta = \frac{\pi}{3}, -\frac{\pi}{3}$.

(b) $\dfrac{dy}{dx} = \dfrac{2 \sin^2 \theta + (1 - 2 \cos \theta) \cos \theta}{2 \sin \theta \cos \theta - (1 - 2 \cos \theta) \sin \theta}$

Horizontal tangents: $-4 \cos^2 \theta + \cos \theta + 2 = 0, \cos \theta = \dfrac{-1 \pm \sqrt{1 + 32}}{-8} = \dfrac{1 \pm \sqrt{33}}{8}$

When $\cos \theta = \dfrac{1 \pm \sqrt{33}}{8}, r = 1 - 2\left(\dfrac{1 + \sqrt{33}}{8}\right) = \dfrac{3 \mp \sqrt{33}}{4}$,

$$\left[\frac{3 - \sqrt{33}}{4}, \arccos\left(\frac{1 + \sqrt{33}}{8}\right)\right] \approx (-0.686, 0.568)$$

$$\left[\frac{3 - \sqrt{33}}{4}, -\arccos\left(\frac{1 + \sqrt{33}}{8}\right)\right] \approx (-0.686, -0.568)$$

$$\left[\frac{3 + \sqrt{33}}{4}, \arccos\left(\frac{1 - \sqrt{33}}{8}\right)\right] \approx (2.186, 2.206)$$

$$\left[\frac{3 + \sqrt{33}}{4}, -\arccos\left(\frac{1 - \sqrt{33}}{8}\right)\right] \approx (2.186, -2.206).$$

Vertical tangents: $\sin \theta(4 \cos \theta - 1) = 0, \sin \theta = 0, \cos \theta = \frac{1}{4}, \theta = 0, \pi, \theta = \pm\arccos\left(\frac{1}{4}\right), (-1, 0), (3, \pi)$

$$\left(\frac{1}{2}, \pm\arccos \frac{1}{4}\right) \approx (0.5, \pm 1.318)$$

(c)

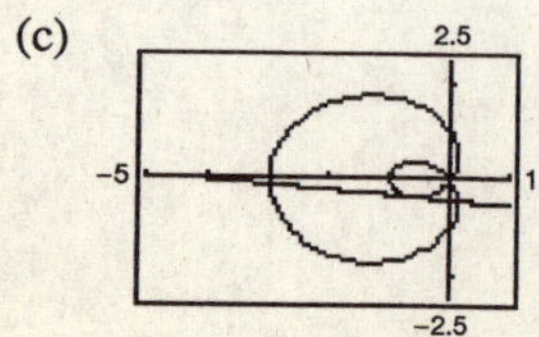

94. $r^2 = 4\sin(2\theta)$

(a) $2r\left(\dfrac{dr}{d\theta}\right) = 8\cos(2\theta)$

$$\frac{dr}{d\theta} = \frac{4\cos(2\theta)}{r}$$

Tangents at the pole: $\theta = 0, \dfrac{\pi}{2}$

(b) $$\frac{dy}{dx} = \frac{r\cos\theta + [(4\cos 2\theta\sin\theta)/r]}{-r\sin\theta + [(4\cos 2\theta\cos\theta)/r]}$$

$$= \frac{\cos(2\theta)\sin\theta + \sin(2\theta)\cos\theta}{\cos(2\theta)\cos\theta - \sin(2\theta)\sin\theta}$$

Horizontal tangents:

$\dfrac{dy}{dx} = 0$ when $\cos(2\theta)\sin\theta + \sin(2\theta)\cos\theta = 0$,

$\tan\theta = -\tan(2\theta),\ \theta = 0, \dfrac{\pi}{3}, (0, 0), \left(\pm\sqrt{2\sqrt{3}}, \dfrac{\pi}{3}\right)$

Vertical tangents when $\cos 2\theta\cos\theta - \sin 2\theta\sin\theta = 0$:

$\tan 2\theta\tan\theta = 1,\ \theta = 0, \dfrac{\pi}{6}, (0, 0), \left(\pm\sqrt{2\sqrt{3}}, \dfrac{\pi}{6}\right)$

(c)

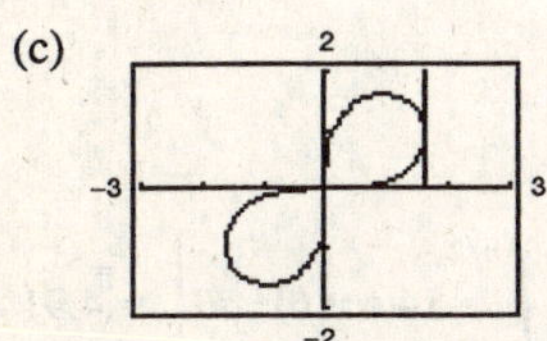

95. Circle: $r = 3\sin\theta$

$$\frac{dy}{dx} = \frac{3\cos\theta\sin\theta + 3\sin\theta\cos\theta}{3\cos\theta\cos\theta - 3\sin\theta\sin\theta} = \frac{\sin 2\theta}{\cos^2\theta - \sin^2\theta} = \tan 2\theta \text{ at } \theta = \frac{\pi}{6}, \frac{dy}{dx} = \sqrt{3}$$

Limaçon: $r = 4 - 5\sin\theta$

$$\frac{dy}{dx} = \frac{-5\cos\theta\sin\theta + (4 - 5\sin\theta)\cos\theta}{-5\cos\theta\cos\theta - (4 - 5\sin\theta)\sin\theta} \text{ at } \theta = \frac{\pi}{6}, \frac{dy}{dx} = \frac{\sqrt{3}}{9}$$

Let α be the angle between the curves:

$$\tan\alpha = \frac{\sqrt{3} - (\sqrt{3}/9)}{1 + (1/3)} = \frac{2\sqrt{3}}{3}.$$

Therefore, $\alpha = \arctan\left(\dfrac{2\sqrt{3}}{3}\right) \approx 49.1°$.

96. False. There are an infinite number of polar coordinate representations of a point. For example, the point $(x, y) = (1, 0)$ has polar representations $(r, \theta) = (1, 0), (1, 2\pi), (-1, \pi)$, etc.

97. $r = 1 + \cos\theta, r = 1 - \cos\theta$

The points $(1, \pi/2)$ and $(1, 3\pi/2)$ are the two points of intersection (other than the pole). The slope of the graph of $r = 1 + \cos\theta$ is

$$m_1 = \frac{dy}{dx} = \frac{r'\sin\theta + r\cos\theta}{r'\cos\theta - r\sin\theta} = \frac{-\sin^2\theta + \cos\theta(1 + \cos\theta)}{-\sin\theta\cos\theta - \sin\theta(1 + \cos\theta)}.$$

At $(1, \pi/2)$, $m_1 = -1/-1 = 1$ and at $(1, 3\pi/2)$, $m_1 = -1/1 = -1$. The slope of the graph of $r = 1 - \cos\theta$ is

$$m_2 = \frac{dy}{dx} = \frac{\sin^2\theta + \cos\theta(1 - \cos\theta)}{\sin\theta\cos\theta - \sin\theta(1 - \cos\theta)}.$$

At $(1, \pi/2)$, $m_2 = 1/-1 = -1$ and at $(1, 3\pi/2)$, $m_2 = 1/1 = 1$. In both cases, $m_1 = -1/m_2$ and we conclude that the graphs are orthogonal at $(1, \pi/2)$ and $(1, 3\pi/2)$.

98. $r = a\sin\theta, r = a\cos\theta$

The points of intersection are $(a/\sqrt{2}, \pi/4)$ and $(0, 0)$. For $r = a\sin\theta$,

$$m_1 = \frac{dy}{dx} = \frac{a\cos\theta\sin\theta + a\sin\theta\cos\theta}{a\cos^2\theta - a\sin^2\theta} = \frac{2\sin\theta\cos\theta}{\cos 2\theta}.$$

At $(a/\sqrt{2}, \pi/4)$, m_1 is undefined and at $(0, 0)$, $m_1 = 0$. For $r = a\cos\theta$,

$$m_2 = \frac{dy}{dx} = \frac{-a\sin^2\theta + a\cos^2\theta}{-a\sin\theta\cos\theta - a\cos\theta\sin\theta} = \frac{\cos 2\theta}{-2\sin\theta\cos\theta}.$$

At $(a/\sqrt{2}, \pi/4)$, $m_2 = 0$ and at $(0, 0)$, m_2 is undefined. Therefore, the graphs are orthogonal at $(a/\sqrt{2}, \pi/4)$ and $(0, 0)$.

99. $r = 2 + \cos\theta$

$$A = 2\left[\frac{1}{2}\int_0^{\pi} (2 + \cos\theta)^2\, d\theta\right] \approx 14.14, \left(\frac{9\pi}{2}\right)$$

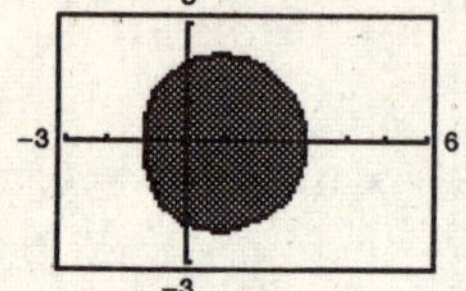

100. $r = 5(1 - \sin\theta)$

$$A = 2\left[\frac{1}{2}\int_{\pi/2}^{3\pi/2} [5(1 - \sin\theta)]^2\, d\theta\right] \approx 117.81, \left(\frac{75\pi}{2}\right)$$

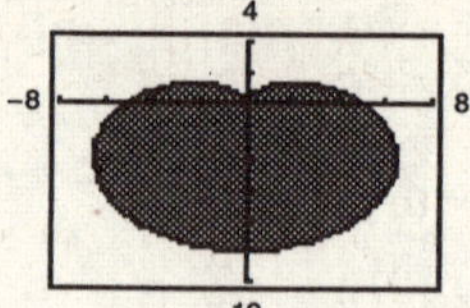

101. $r^2 = 4\sin 2\theta$

$$A = 2\left[\frac{1}{2}\int_0^{\pi/2} 4\sin 2\theta\, d\theta\right] = 4$$

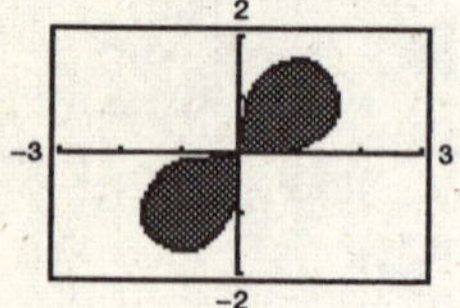

102. $r = 4\cos\theta,\ r = 2$

$$A = 2\left[\frac{1}{2}\int_0^{\pi/3} 4\, d\theta + \frac{1}{2}\int_{\pi/3}^{\pi/2} (4\cos\theta)^2\, d\theta\right] \approx 4.91$$

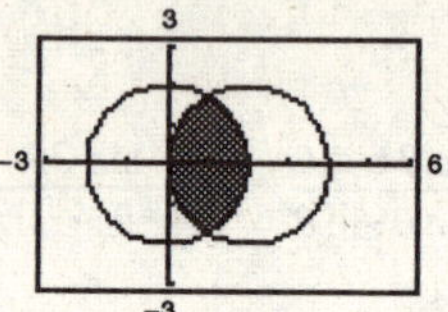

103. $r = \sin\theta\cos^2\theta$

$$A = 2\left[\frac{1}{2}\int_0^{\pi/2} (\sin\theta\cos^2\theta)^2\, d\theta\right]$$

$$\approx 0.10, \left(\frac{\pi}{32}\right)$$

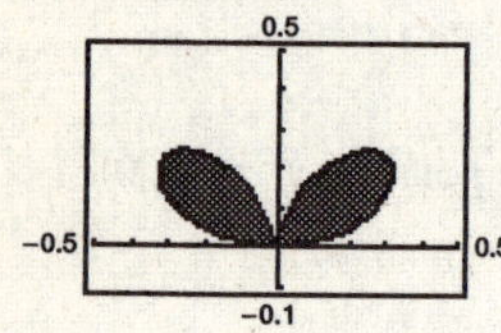

104. $r = 4\sin 3\theta$

$$A = 3\left[\frac{1}{2}\int_0^{\pi/3} (4\sin 3\theta)^2\, d\theta\right]$$

$$\approx 12.57\ (4\pi)$$

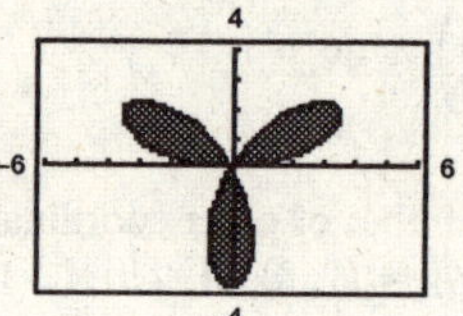

105. $r = 3,\ r^2 = 18\sin 2\theta$

$$9 = r^2 = 18\sin 2\theta$$

$$\sin 2\theta = \frac{1}{2}$$

$$\theta = \frac{\pi}{12}$$

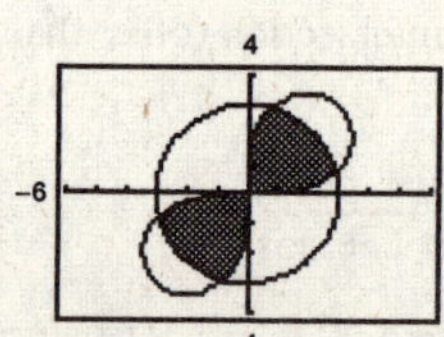

$$A = 2\left[\frac{1}{2}\int_0^{\pi/12} 18\sin 2\theta\, d\theta + \frac{1}{2}\int_{\pi/12}^{5\pi/12} 9\, d\theta + \frac{1}{2}\int_{5\pi/12}^{\pi/2} 18\sin 2\theta\, d\theta\right]$$

$$\approx 1.2058 + 9.4248 + 1.2058 \approx 11.84$$

106. $r = e^{\theta},\ 0 \le \theta \le \pi$

$$A = \frac{1}{2}\int_0^{\pi} (e^{\theta})^2\, d\theta \approx 133.62$$

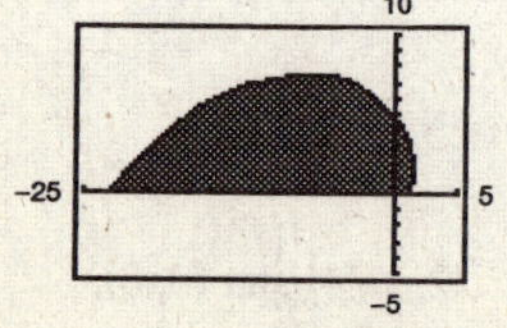

107. $r = a(1 - \cos\theta), 0 \le \theta \le \pi$

$$\frac{dr}{d\theta} = a\sin\theta$$

$$s = \int_0^{\pi} \sqrt{a^2(1-\cos\theta)^2 + a^2\sin^2\theta}\,d\theta$$

$$= \sqrt{2}a\int_0^{\pi} \sqrt{1-\cos\theta}\,d\theta$$

$$= \sqrt{2}a\int_0^{\pi} \frac{\sin\theta}{\sqrt{1+\cos\theta}}\,d\theta$$

$$= -2\sqrt{2}a\left[(1+\cos\theta)^{1/2}\right]_0^{\pi} = 4a$$

108. $r = a\cos 2\theta, -\pi/4 \le \theta \le \pi/2$

$$\frac{dr}{d\theta} = -2a\sin 2\theta$$

$$s = \int_{-\pi/2}^{\pi/2} \sqrt{a^2\cos^2 2\theta + 4a^2\sin^2 2\theta}\,d\theta$$

$$= a\int_{-\pi/2}^{\pi/2} \sqrt{1+3\sin^2 2\theta}\,d\theta$$

Using a graphing utility, $s \approx 4.8442a$.

109. $f(\theta) = 1 + 4\cos\theta$

$f'(\theta) = -4\sin\theta$

$\sqrt{f(\theta)^2 + f'(\theta)^2} = \sqrt{(1+4\cos\theta)^2 + (-4\sin\theta)^2} = \sqrt{17 + 8\cos\theta}$

$$S = 2\pi\int_0^{\pi/2} (1+4\cos\theta)\sin\theta\sqrt{17+8\cos\theta}\,d\theta$$

$$= \frac{34\pi\sqrt{17}}{5} \approx 88.08$$

110. $f(\theta) = 2\sin\theta$

$f'(\theta) = 2\cos\theta$

$\sqrt{f(\theta)^2 + f'(\theta)^2} = \sqrt{4\sin^2\theta + 4\cos^2\theta} = 2$

$$S = 2\pi\int_0^{\pi/2} 2\sin\theta\cos\theta(2)\,d\theta$$

$$= 4\pi$$

111. $r = \dfrac{2}{1-\sin\theta}, e = 1$

Parabola

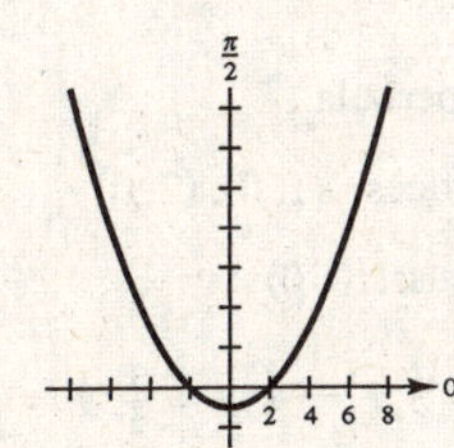

112. $r = \dfrac{2}{1+\cos\theta}, e = 1$

Parabola

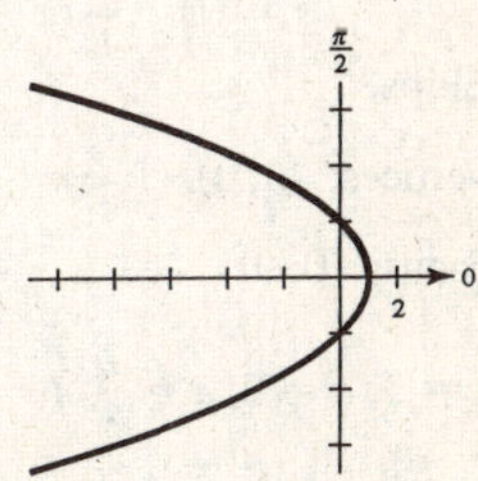

113. $r = \dfrac{6}{3+2\cos\theta} = \dfrac{2}{1+(2/3)\cos\theta}, e = \dfrac{2}{3}$

Ellipse

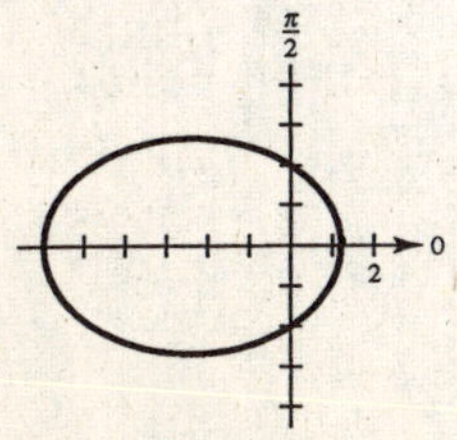

114. $r = \dfrac{4}{5-3\sin\theta} = \dfrac{4/5}{1-(3/5)\sin\theta}, e = \dfrac{3}{5}$

Ellipse

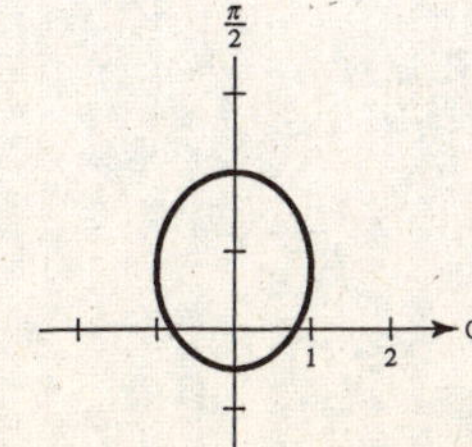

115. $r = \dfrac{4}{2 - 3\sin\theta} = \dfrac{2}{1 - (3/2)\sin\theta}, e = \dfrac{3}{2}$

Hyperbola

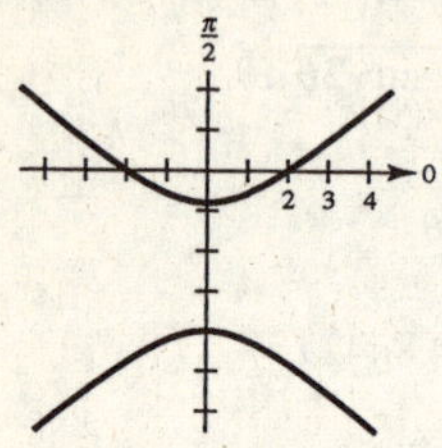

116. $r = \dfrac{8}{2 - 5\cos\theta} = \dfrac{4}{1 - (5/2)\cos\theta}, e = \dfrac{5}{2}$

Hyperbola

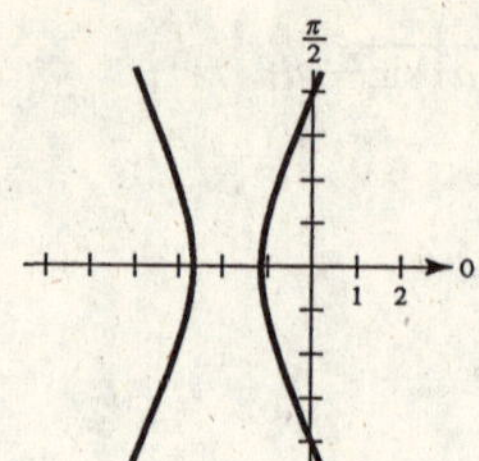

117. Circle

Center: $\left(5, \dfrac{\pi}{2}\right) = (0, 5)$ in rectangular coordinates

Solution point: $(0, 0)$

$x^2 + (y - 5)^5 = 25$

$x^2 + y^2 - 10y = 0$

$r^2 - 10r\sin\theta = 0$

$r = 10\sin\theta$

118. Line

Slope: $\sqrt{3}$

Solution point: $(0, 0)$

$y = \sqrt{3}\,x, r\sin\theta = \sqrt{3}\,r\cos\theta,$

$\tan\theta = \sqrt{3}, \theta = \dfrac{\pi}{3}$

119. Parabola

Vertex: $(2, \pi)$

Focus: $(0, 0)$

$e = 1, d = 4$

$r = \dfrac{4}{1 - \cos\theta}$

120. Parabola

Vertex: $\left(2, \dfrac{\pi}{2}\right)$

Focus: $(0, 0)$

$e = 1, d = 4$

$r = \dfrac{4}{1 + \sin\theta}$

121. Ellipse

Vertices: $(5, 0), (1, \pi)$

Focus: $(0, 0)$

$a = 3, c = 2, e = \dfrac{2}{3}, d = \dfrac{5}{2}$

$$r = \frac{\left(\frac{2}{3}\right)\left(\frac{5}{2}\right)}{1 - \left(\frac{2}{3}\right)\cos\theta} = \frac{5}{3 - 2\cos\theta}$$

122. Hyperbola

Vertices: $(1, 0), (7, 0)$

Focus: $(0, 0)$

$a = 3, c = 4, e = \dfrac{4}{3}, d = \dfrac{7}{4}$

$$r = \frac{\left(\frac{4}{3}\right)\left(\frac{7}{4}\right)}{1 + \left(\frac{4}{3}\right)\cos\theta} = \frac{7}{3 + 4\cos\theta}$$

Problem Solving for Chapter 10

1. (a)

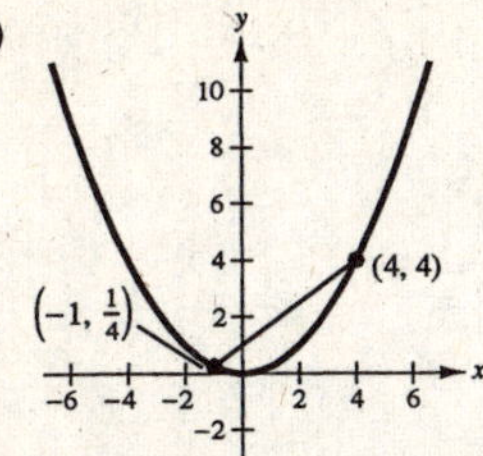

(b) $x^2 = 4y$

$2x = 4y'$

$y' = \frac{1}{2}x$

$y - 4 = 2(x - 4) \quad \Rightarrow y = 2x - 4$ Tangent line at (4, 4)

$y - \frac{1}{4} = -\frac{1}{2}(x + 1) \Rightarrow y = -\frac{1}{2}x - \frac{1}{4}$ Tangent line at $\left(-1, \frac{1}{4}\right)$

Tangent lines have slopes of 2 and $-1/2 \Rightarrow$ perpendicular.

(c) Intersection:

$$2x - 4 = -\frac{1}{2}x - \frac{1}{4}$$

$$8x - 16 = -2x - 1$$

$$10x = 15$$

$$x = \frac{3}{2} \Rightarrow \left(\frac{3}{2}, -1\right)$$

Point of intersection, $(3/2, -1)$, is on directrix $y = -1$.

2. Assume $p > 0$.

Let $y = mx + p$ be the equation of the focal chord.

First find x-coordinates of focal chord endpoints:

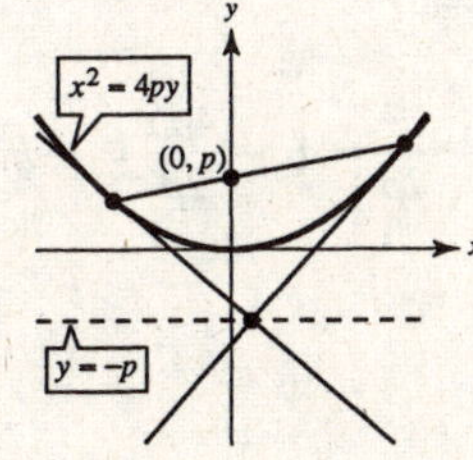

$$x^2 = 4py = 4p(mx + p)$$

$$x^2 - 4pmx - 4p^2 = 0$$

$$x = \frac{4pm \pm \sqrt{16p^2m^2 + 16p^2}}{2} = 2pm \pm 2p\sqrt{m^2 + 1}$$

$$x^2 = 4py, 2x = 4py' \Rightarrow y' = \frac{x}{2p}.$$

(a) The slopes of the tangent lines at the endpoints are perpendicular because

$$\frac{1}{2p}\left[2pm + 2p\sqrt{m^2 + 1}\right]\frac{1}{2p}\left[2pm - 2p\sqrt{m^2 + 1}\right] = \frac{1}{4p^2}[4p^2m^2 - 4p^2(m^2 + 1)] = \frac{1}{4p^2}[-4p^2] = -1$$

—CONTINUED—

2. —CONTINUED—

(b) Finally, we show that the tangent lines intersect at a point on the directrix $y = -p$.

Let $b = 2pm + 2p\sqrt{m^2 + 1}$ and $c = 2pm - 2p\sqrt{m^2 + 1}$.

$$b^2 = 8p^2m^2 + 4p^2 + 8p^2m\sqrt{m^2 + 1}$$

$$c^2 = 8p^2m^2 + 4p^2 - 8p^2m\sqrt{m^2 + 1}$$

$$\frac{b^2}{4p} = 2pm^2 + p + 2pm\sqrt{m^2 + 1}$$

$$\frac{c^2}{4p} = 2pm^2 + p - 2pm\sqrt{m^2 + 1}$$

Tangent line at $x = b$: $y - \dfrac{b^2}{4p} = \dfrac{b}{2p}(x - b) \Rightarrow y = \dfrac{bx}{2p} - \dfrac{b^2}{4p}$

Tangent line at $x = c$: $y - \dfrac{c^2}{4p} = \dfrac{c}{2p}(x - c) \Rightarrow y = \dfrac{cx}{2p} - \dfrac{c^2}{4p}$

Intersection of tangent lines: $\dfrac{bx}{2p} - \dfrac{b^2}{4p} = \dfrac{cx}{2p} - \dfrac{c^2}{4p}$

$$2bx - b^2 = 2cx - c^2$$

$$2x(b - c) = b^2 - c^2$$

$$2x\left(4p\sqrt{m^2 + 1}\right) = 16p^2m\sqrt{m^2 + 1}$$

$$x = 2pm$$

Finally, the corresponding y-value is $y - p$, which shows that the intersection point lies on the directrix.

3. Consider $x^2 = 4py$ with focus $F = (0, p)$.

Let $P(a, b)$ be point on parabola.

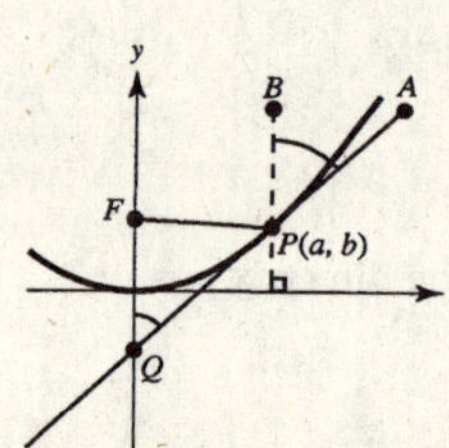

$$2x = 4py' \Rightarrow y' = \frac{x}{2p}$$

$y - b = \dfrac{a}{2p}(x - a)$ Tangent line at P

For $x = 0$, $y = b + \dfrac{a}{2p}(-a) = b - \dfrac{a^2}{2p} = b - \dfrac{4pb}{2p} = -b$.

Thus, $Q = (0, -b)$.

ΔFQP is isosceles because

$$|FQ| = p + b$$

$$|FP| = \sqrt{(a - 0)^2 + (b - p)^2} = \sqrt{a^2 + b^2 - 2bp + p^2}$$

$$= \sqrt{4pb + b^2 - 2bp + p^2}$$

$$= \sqrt{(b + p)^2}$$

$$= b + p.$$

Thus, $\measuredangle FQP = \measuredangle BPA = \measuredangle FPQ$.

4. $\dfrac{x^2}{a^2} - \dfrac{y^2}{b^2} = 1, a^2 + b^2 = c^2, MF_2 - MF_1 = 2a$

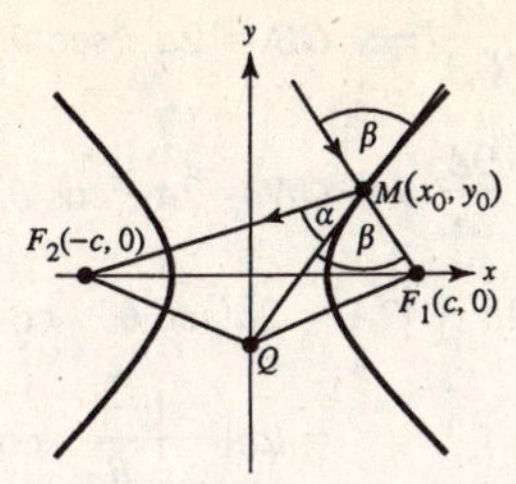

$$y' = \frac{b^2x}{a^2y}$$

Tangent line at $M(x_0, y_0)$: $\quad y - y_0 = \dfrac{b^2x_0}{a^2y_0}(x - x_0)$

$$\frac{yy_0 - y_0^2}{b^2} = \frac{x_0x - x_0^2}{a^2}$$

$$\frac{x_0x}{a^2} - \frac{y_0y}{b^2} = \frac{x_0^2}{a^2} + \frac{y_0^2}{b^2}$$

$$\frac{x_0x}{a^2} - \frac{y_0y}{b^2} = 1$$

At $x = 0, y = -\dfrac{b^2}{y_0} \Rightarrow Q = \left(0, -\dfrac{b^2}{y_0}\right)$.

$$QF_2 = QF_1 = \sqrt{c^2 + \frac{b^4}{y_0^2}} = d$$

$$MQ = \sqrt{x_0^2 + \left(y_0 + \frac{b^2}{y_0}\right)^2} = f$$

By the Law of Cosines,

$$(F_2Q)^2 = (MF_2)^2 + (MQ)^2 - 2(MF_2)(MQ)\cos\alpha$$

$$d^2 = (MF_2)^2 + f^2 - 2f(MF_2)\cos\alpha$$

$$(F_1Q)^2 = (MF_1)^2 + f^2 - 2f(MF_1)\cos\beta$$

$$d^2 = (MF_1)^2 + f^2 - 2f(MF_1)\cos\beta.$$

$$\cos\alpha = \frac{(MF_2)^2f^2 - d^2}{2f(MF_2)},\ \cos\beta = \frac{(MF_1)^2 + f^2 - d^2}{2f(MF_1)}$$

$MF_2 = MF_1 + 2a$. Let $z = MF_1$.

Slopes: MF_1: $\dfrac{y_0}{x_0 - c}$; QF_1: $\dfrac{-b^2}{y_0c}$; QF_2: $\dfrac{b^2}{y_0c}$

To show $\alpha = \beta$, consider

$$[(MF_2)^2 + f^2 - d^2][2f(MF_1)] = [(MF_1)^2 + f^2 - d^2][2f(MF_2)]$$

$$\Leftrightarrow \quad [(z + 2a)^2 + f^2 - d^2][z] = [z^2 + f^2 - d^2][z + 2a]$$

$$\Leftrightarrow \quad z^2 + 2az = f^2 - d^2$$

$$\Leftrightarrow \quad (x_0 - c)^2 + y_0^2 + 2az = \left(x_0^2 + \left(y_0 + \frac{b^2}{y_0}\right)^2\right) - \left(c^2 + \frac{b^4}{y_0^2}\right)$$

$$\Leftrightarrow \quad az - x_0c + a^2 = 0$$

$$\Leftrightarrow \quad a\sqrt{(x_0 - c)^2 + y_0^2} = x_0c - a^2$$

$$\Leftrightarrow \quad x_0^2b^2 - a^2y_0^2 = a^2b^2$$

$$\Leftrightarrow \quad \frac{x_0^2}{a^2} - \frac{y_0^2}{b^2} = 1.$$

Thus, $\alpha = \beta$ and the reflective property is verified.

5. (a) In $\triangle OCB$, $\cos\theta = \dfrac{2a}{OB} \Rightarrow OB = 2a \cdot \sec\theta$.

In $\triangle OAC$, $\cos\theta = \dfrac{OA}{2a} \Rightarrow OA = 2a \cdot \cos\theta$.

$$\begin{aligned} r = OP = AB = OB - OA &= 2a(\sec\theta - \cos\theta) \\ &= 2a\left(\frac{1}{\cos\theta} - \cos\theta\right) \\ &= 2a \cdot \frac{\sin^2\theta}{\cos\theta} \\ &= 2a \cdot \tan\theta\sin\theta \end{aligned}$$

(b) $x = r\cos\theta = (2a\tan\theta\sin\theta)\cos\theta = 2a\sin^2\theta$

$y = r\sin\theta = (2a\tan\theta\sin\theta)\sin\theta = 2a\tan\theta \cdot \sin^2\theta,\ -\dfrac{\pi}{2} < \theta < \dfrac{\pi}{2}$

Let $t = \tan\theta$, $-\infty < t < \infty$.

Then $\sin^2\theta = \dfrac{t^2}{1+t^2}$ and $x = 2a\dfrac{t^2}{1+t^2}$, $y = 2a\dfrac{t^3}{1+t^2}$.

$\sqrt{1+t^2}$ t θ 1

(c)
$$\begin{aligned} r &= 2a\tan\theta\sin\theta \\ r\cos\theta &= 2a\sin^2\theta \\ r^3\cos\theta &= 2a\,r^2\sin^2\theta \\ (x^2+y^2)x &= 2ay^2 \\ y^2 &= \frac{x^3}{(2a-x)} \end{aligned}$$

6. (a) $A = 4\displaystyle\int_0^a \frac{b}{a}\sqrt{a^2-x^2}\,dx = \frac{4b}{a}\left(\frac{1}{2}\right)\left[x\sqrt{a^2-x^2} + a^2\arcsin\left(\frac{x}{a}\right)\right]_0^a = \pi ab$

(b) **Disk:** $V = 2\pi\displaystyle\int_0^b \frac{a^2}{b^2}(b^2-y^2)\,dy = \frac{2\pi a^2}{b^2}\int_0^b (b^2-y^2)\,dy = \frac{2\pi a^2}{b^2}\left[b^2y - \frac{1}{3}y^3\right]_0^b = \frac{4}{3}\pi a^2 b$

$$\begin{aligned} S &= 4\pi\int_0^b \frac{a}{b}\sqrt{b^2-y^2}\left(\frac{\sqrt{b^4+(a^2-b^2)y^2}}{b\sqrt{b^2-y^2}}\right)dy \\ &= \frac{4\pi a}{b^2}\int_0^b \sqrt{b^4+c^2y^2}\,dy = \frac{2\pi a}{b^2c}\left[cy\sqrt{b^4+c^2y^2} + b^4\ln\left|cy+\sqrt{b^4+c^2y^2}\right|\right]_0^b \\ &= \frac{2\pi a}{b^2c}\left[b^2c\sqrt{b^2+c^2} + b^4\ln\left|cb+b\sqrt{b^2+c^2}\right| - b^4\ln(b^2)\right] \\ &= 2\pi a^2 + \frac{\pi ab^2}{c}\ln\left(\frac{c+a}{e}\right)^2 = 2\pi a^2 + \left(\frac{\pi b^2}{e}\right)\ln\left(\frac{1+e}{1-e}\right) \end{aligned}$$

(c) **Disk:** $V = 2\pi\displaystyle\int_0^a \frac{b^2}{a^2}(a^2-x^2)\,dx = \frac{2\pi b^2}{a^2}\int_0^a (a^2-x^2)\,dx = \frac{2\pi b^2}{a^2}\left[a^2x - \frac{1}{3}x^3\right]_0^a = \frac{4}{3}\pi ab^2$

$$\begin{aligned} S &= 2(2\pi)\int_0^a \frac{b}{a}\sqrt{a^2-x^2}\left(\frac{\sqrt{a^4-(a^2-b^2)x^2}}{a\sqrt{a^2-x^2}}\right)dx \\ &= \frac{4\pi b}{a^2}\int_0^a \sqrt{a^4-c^2x^2}\,dx = \frac{2\pi b}{a^2c}\left[cx\sqrt{a^4-c^2x^2} + a^4\arcsin\left(\frac{cx}{a^2}\right)\right]_0^a \\ &= \frac{a\pi b}{a^2c}\left[a^2c\sqrt{a^2-c^2} + a^4\arcsin\left(\frac{c}{a}\right)\right] = 2\pi b^2 + 2\pi\left(\frac{ab}{e}\right)\arcsin(e) \end{aligned}$$

7. (a) $y^2 = \dfrac{t^2(1-t^2)^2}{(1+t^2)^2}, x^2 = \dfrac{(1-t^2)^2}{(1+t^2)^2}.$

$$\frac{1-x}{1+x} = \frac{1-\left(\dfrac{1-t^2}{1+t^2}\right)}{1+\left(\dfrac{1-t^2}{1+t^2}\right)} = \frac{2t^2}{2} = t^2$$

Thus, $y^2 = x^2\left(\dfrac{1-x}{1+x}\right)$.

(b)
$$r^2\sin^2\theta = r^2\cos^2\theta\left(\frac{1-r\cos\theta}{1+r\cos\theta}\right)$$

$$\sin^2\theta(1+r\cos\theta) = \cos^2\theta(1-r\cos\theta)$$

$$r\cos\theta\sin^2\theta + \sin^2\theta = \cos^2\theta - r\cos^3\theta$$

$$r\cos\theta(\sin^2\theta+\cos^2\theta) = \cos^2\theta - \sin^2\theta$$

$$r\cos\theta = \cos 2\theta$$

$$r = \cos 2\theta \cdot \sec\theta$$

(c)

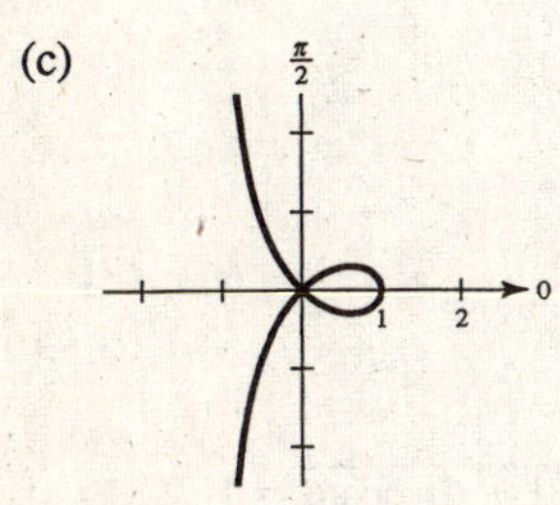

(d) $r(\theta) = 0$ for $\theta = \dfrac{\pi}{4}, \dfrac{3\pi}{4}.$

Thus, $y = x$ and $y = -x$ are tangent lines to curve at the origin.

(e) $y'(t) = \dfrac{(1+t^2)(1-3t^2)-(t-t^3)(2t)}{(1+t^2)^2} = \dfrac{1-4t^2-t^4}{(1+t^2)^2} = 0$

$$t^4 + 4t^2 - 1 = 0 \implies t^2 = -2 \pm \sqrt{5} \implies x = \frac{1-(-2\pm\sqrt{5})}{1+(-2\pm\sqrt{5})} = \frac{3\mp\sqrt{5}}{-1\pm\sqrt{5}}$$

$$= \frac{3-\sqrt{5}}{-1+\sqrt{5}} = \frac{\sqrt{5}-1}{2}$$

$$\left(\frac{\sqrt{5}-1}{2}, \pm\frac{\sqrt{5}-1}{2}\sqrt{-2+\sqrt{5}}\right)$$

8. $y = a(1-\cos\theta) \implies \cos\theta = \dfrac{a-y}{a}$

$$\theta = \arccos\left(\frac{a-y}{a}\right)$$

$$x = a(\theta - \sin\theta)$$

$$= a\left(\arccos\left(\frac{a-y}{a}\right) - \sin\left(\arccos\left(\frac{a-y}{a}\right)\right)\right)$$

$$= a\left(\arccos\left(\frac{a-y}{a}\right) - \frac{\sqrt{2ay-y^2}}{a}\right)$$

$$x = a\cdot\arccos\left(\frac{a-y}{a}\right) - \sqrt{2ay-y^2},\ 0 \le y \le 2a$$

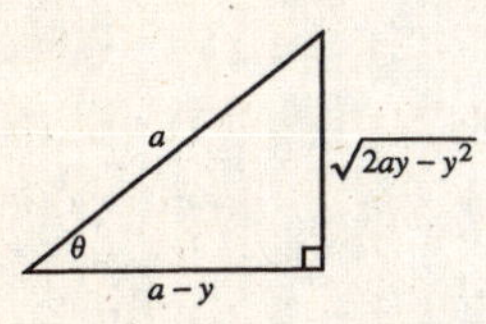

9. (a)

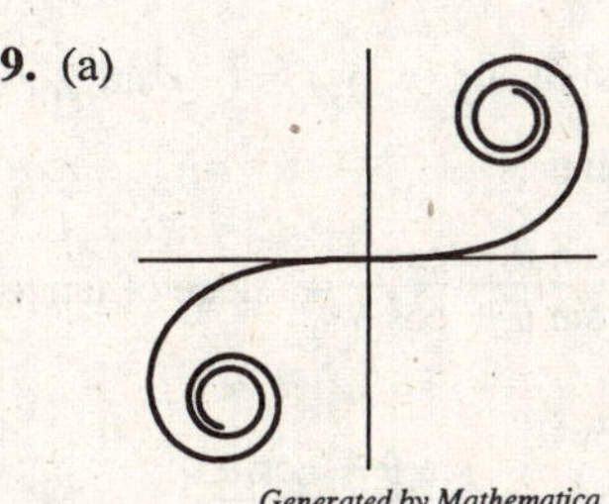

Generated by Mathematica

(b) $(-x, -y) = \left(-\displaystyle\int_0^t \cos\frac{\pi u^2}{2}\,du, -\int_0^t \sin\frac{\pi u^2}{2}\,du\right)$ is on the curve whenever (x, y) is on the curve.

(c) $x'(t) = \cos\dfrac{\pi t^2}{2}, y'(t) = \sin\dfrac{\pi t^2}{2}, x'(t)^2 + y'(t)^2 = 1$

Thus, $s = \displaystyle\int_0^a dt = a.$

On $[-\pi, \pi]$, $s = 2\pi$.

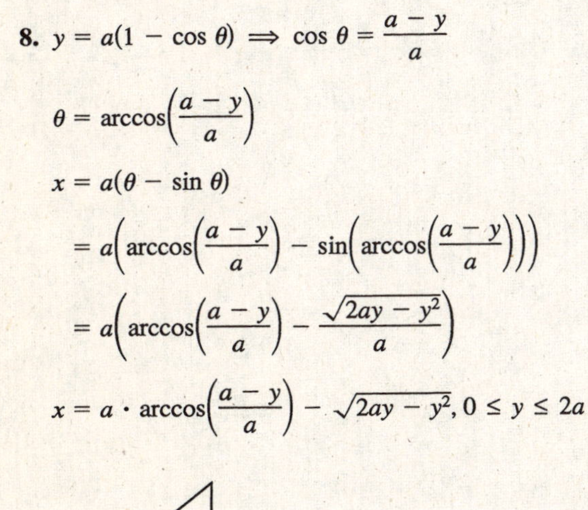

10. For $t = \frac{\pi}{2}, \frac{3}{2}, \frac{5\pi}{2}, \frac{7\pi}{2}, \ldots$

$$y = \frac{2}{\pi}, \frac{-2}{3\pi}, \frac{2}{5\pi}, \frac{-2}{7\pi}, \ldots$$

Hence, the curve has length greater that

$$S = \frac{2}{\pi} + \frac{2}{3\pi} + \frac{2}{5\pi} + \frac{2}{7\pi} + \cdots$$
$$= \frac{2}{\pi}\left(1 + \frac{1}{3} + \frac{1}{5} + \frac{1}{7} + \cdots\right)$$
$$> \frac{2}{\pi}\left(\frac{1}{2} + \frac{1}{4} + \frac{1}{6} + \frac{1}{8} + \cdots\right)$$
$$= \infty. \text{ (Harmonic series)}$$

11. $r = \dfrac{ab}{a \sin\theta + b\cos\theta}, 0 \le \theta \le \dfrac{\pi}{2}$

$$r(a\sin\theta + b\cos\theta) = ab$$
$$ay + bx = ab$$
$$\frac{y}{b} + \frac{x}{a} = 1$$

Line segment

$$\text{Area} = \frac{1}{2}ab$$

12. (a) $\text{Area} = \displaystyle\int_0^\alpha \frac{1}{2}r^2\,d\theta$

$$= \frac{1}{2}\int_0^\alpha \sec^2\theta\,d\theta$$

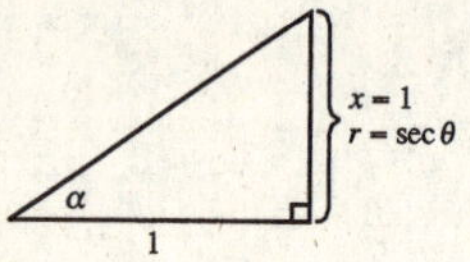

(b) $\tan\alpha = \dfrac{h}{1} \Rightarrow \text{Area} = \dfrac{1}{2}(1)\tan\alpha$

$$\Rightarrow \tan\alpha = \int_0^\alpha \sec^2\theta\,d\theta$$

(c) Differentiating, $\dfrac{d}{d\alpha}(\tan\alpha) = \sec^2\alpha.$

13. Let (r, θ) be on the graph.

$$\sqrt{r^2 + 1 + 2r\cos\theta}\sqrt{r^2 + 1 - 2r\cos\theta} = 1$$
$$(r^2 + 1)^2 - 4r^2\cos^2\theta = 1$$
$$r^4 + 2r^2 + 1 - 4r^2\cos^2\theta = 1$$
$$r^2(r^2 - 4\cos^2\theta + 2) = 0$$
$$r^2 = 4\cos^2\theta - 2$$
$$r^2 = 2(2\cos^2\theta - 1)$$
$$r^2 = 2\cos 2\theta$$

14. If a dog is located at (r, θ) in the first quadrant, then its neighbor is at $\left(r, \theta + \dfrac{\pi}{2}\right)$:

$(x_1, y_1) = (r\cos\theta, r\sin\theta)$ and $(x_2, y_2) = (-r\sin\theta, r\cos\theta)$.

The slope joining these points is

$$\frac{r\cos\theta - r\sin\theta}{-r\sin\theta - r\cos\theta} = \frac{\sin\theta - \cos\theta}{\sin\theta + \cos\theta} = \text{slope of tangent line at } (r, \theta).$$

$$\frac{dy}{dx} = \frac{\frac{dy}{dr}}{\frac{dx}{dr}} = \frac{\frac{dr}{d\theta}\sin\theta + r\cos\theta}{\frac{dr}{d\theta}\cos\theta - r\sin\theta} = \frac{\sin\theta - \cos\theta}{\sin\theta + \cos\theta}$$

$$\Rightarrow \frac{dr}{d\theta} = -r$$
$$\frac{dr}{r} = -d\theta$$
$$\ln r = -\theta + C_1$$
$$r = e^{-\theta + C_1}$$
$$r = Ce^{-\theta}$$

$$r\left(\frac{\pi}{4}\right) = \frac{d}{\sqrt{2}} \Rightarrow r = Ce^{-\pi/4} = \frac{d}{\sqrt{2}} \Rightarrow C = \frac{d}{\sqrt{2}}e^{\pi/4}$$

Finally, $r = \dfrac{d}{\sqrt{2}}e^{((\pi/4) - \theta)}, \theta \ge \dfrac{\pi}{4}.$

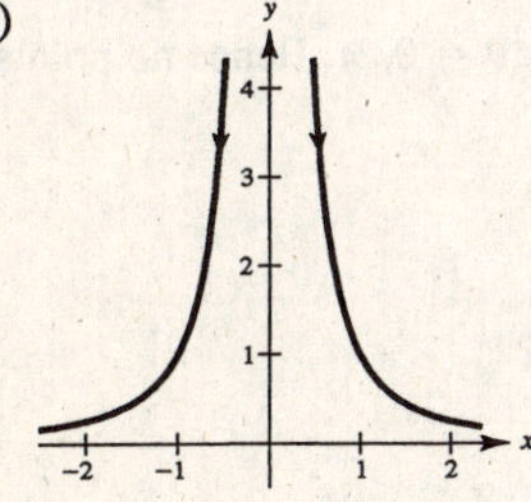

15. (a) The first plane makes an angle of 70° with the positive x-axis, and is 150 miles from P:

$$x_1 = \cos 70°(150 - 375t)$$

$$y_1 = \sin 70°(150 - 375t)$$

Similarly for the second plane,

$$x_2 = \cos 135°(190 - 450t)$$

$$= \cos 45°(-190 + 450t)$$

$$y_2 = \sin 135°(190 - 450t)$$

$$= \sin 45°(190 - 450t).$$

(b) $d = \sqrt{(x_2 - x_1)^2 + (y_2 - y_1)^2}$

$$= [[\cos 45(-190 + 450t) - \cos 70(150 - 375t)]^2 + [\sin 45(190 - 450t) - \sin 70(150 - 375t)]^2]^{1/2}$$

(c)

The minimum distance is 7.59 miles when $t = 0.4145$.

16. The curve is produced over the interval $0 \le \theta \le 10\pi$.

17.

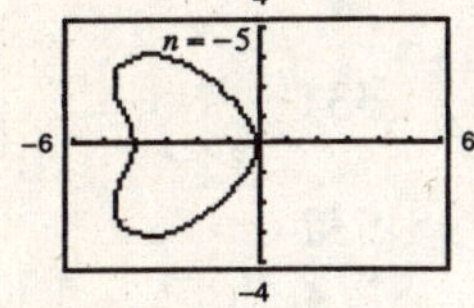

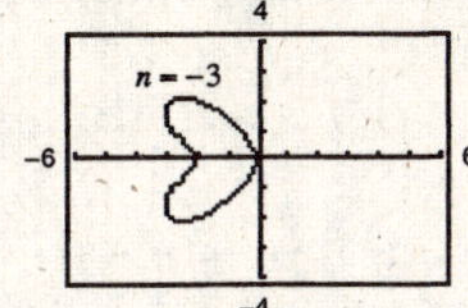

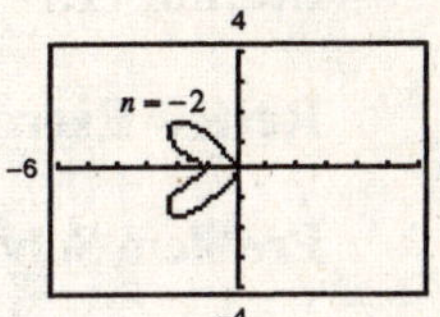

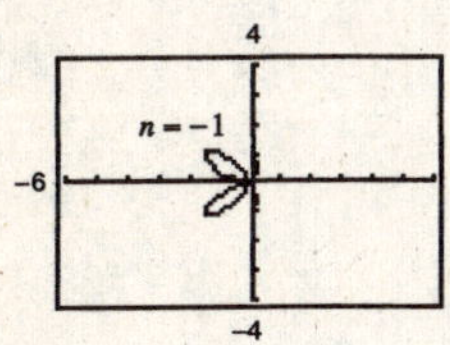

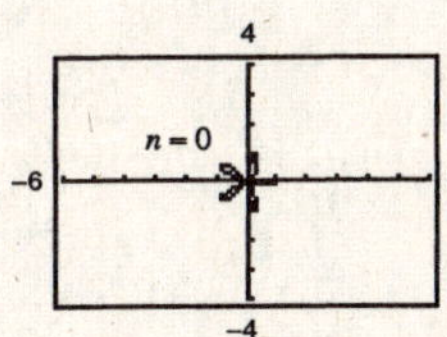

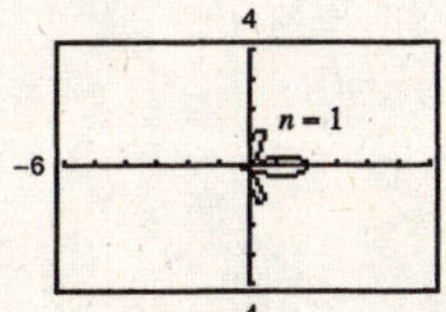

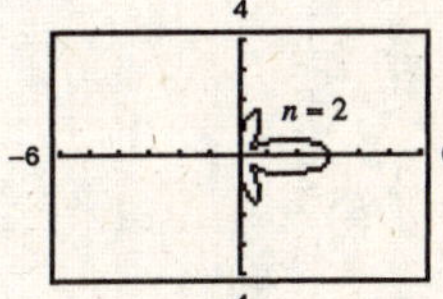

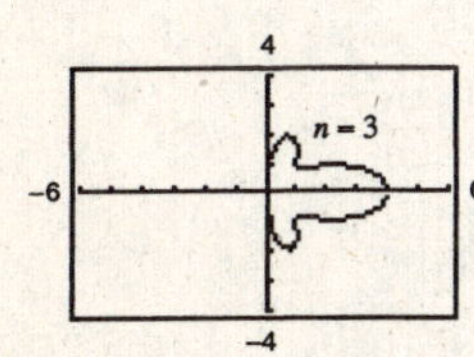

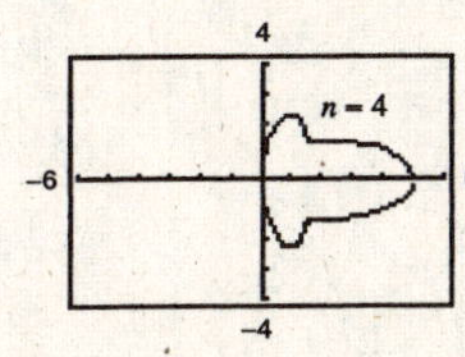

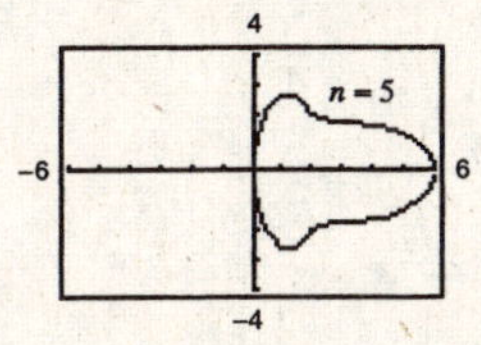

$n = 1, 2, 3, 4, 5$ produce "bells"; $n = -1, -2, -3, -4, -5$ produce "hearts".

CHAPTER 11
Vectors and the Geometry of Space

CHAPTER 11
Vectors and the Geometry of Space

Section 11.1 Vectors in the Plane

1. (a) $\mathbf{v} = \langle 5 - 1, 3 - 1 \rangle = \langle 4, 2 \rangle$

(b)

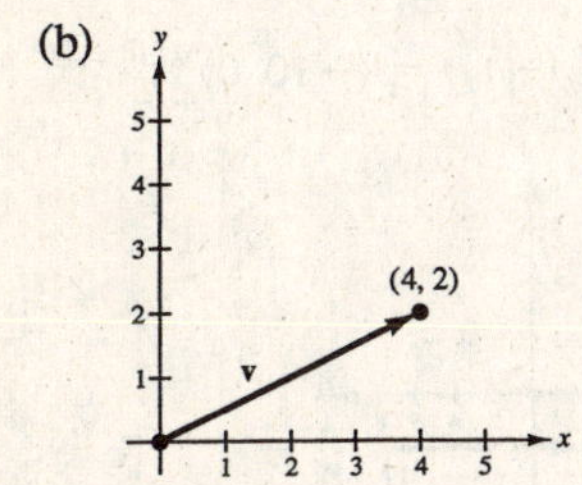

2. (a) $\mathbf{v} = \langle 3 - 3, -2 - 4 \rangle = \langle 0, -6 \rangle$

(b)

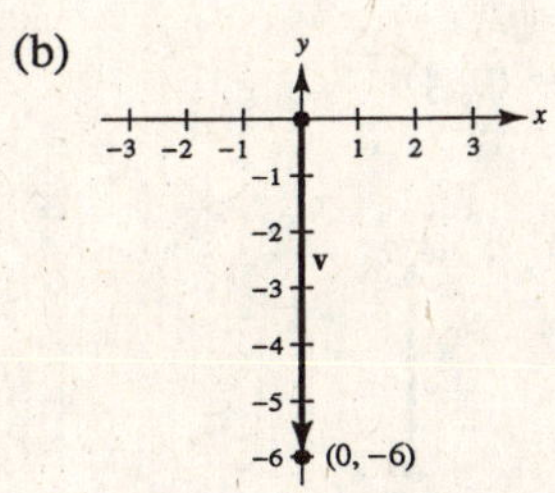

3. (a) $\mathbf{v} = \langle -4 - 3, -2 - (-2) \rangle = \langle -7, 0 \rangle$

(b)

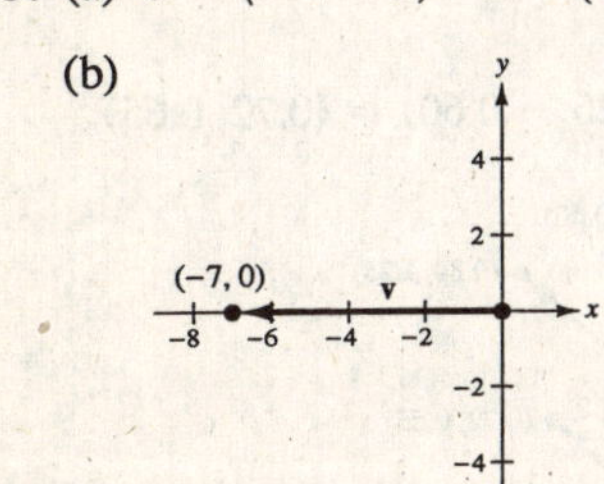

4. (a) $\mathbf{v} = \langle -1 - 2, 3 - 1 \rangle = \langle -3, 2 \rangle$

(b)

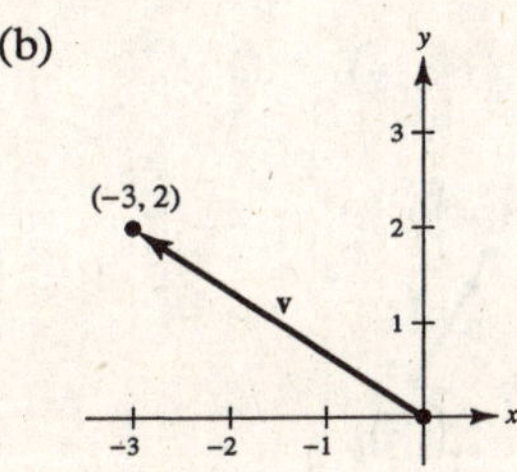

5. $\mathbf{u} = \langle 5 - 3, 6 - 2 \rangle = \langle 2, 4 \rangle$

$\mathbf{v} = \langle 1 - (-1), 8 - 4 \rangle = \langle 2, 4 \rangle$

$\mathbf{u} = \mathbf{v}$

6. $\mathbf{u} = \langle 1 - (-4), 8 - 0 \rangle = \langle 5, 8 \rangle$

$\mathbf{v} = \langle 7 - 2, 7 - (-1) \rangle = \langle 5, 8 \rangle$

$\mathbf{u} = \mathbf{v}$

7. $\mathbf{u} = \langle 6 - 0, -2 - 3 \rangle = \langle 6, -5 \rangle$

$\mathbf{v} = \langle 9 - 3, 5 - 10 \rangle = \langle 6, -5 \rangle$

$\mathbf{u} = \mathbf{v}$

8. $\mathbf{u} = \langle 11 - (-4), -4 - (-1) \rangle = \langle 15, -3 \rangle$

$\mathbf{v} = \langle 25 - 0, 10 - 13 \rangle = \langle 15, -3 \rangle$

$\mathbf{u} = \mathbf{v}$

9. (b) $\mathbf{v} = \langle 5 - 1, 5 - 2 \rangle = \langle 4, 3 \rangle$

(a) and (c).

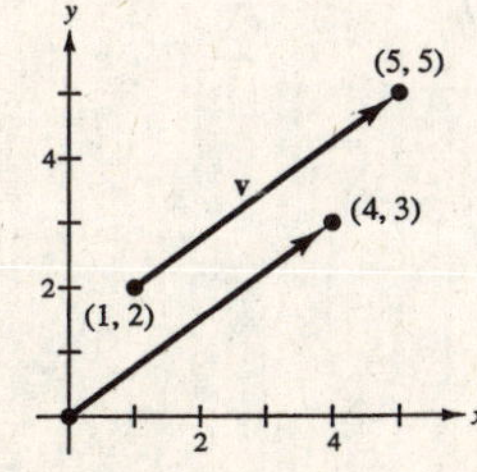

10. (b) $\mathbf{v} = \langle 3 - 2, 6 - (-6) \rangle = \langle 1, 12 \rangle$

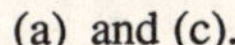
(a) and (c).

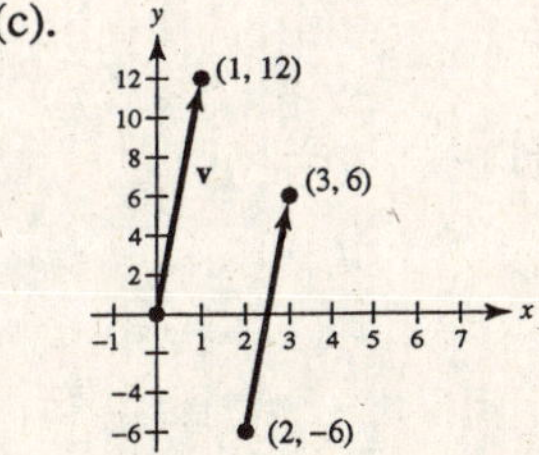

11. (b) $\mathbf{v} = \langle 6 - 10, -1 - 2 \rangle = \langle -4, -3 \rangle$

(a) and (c).

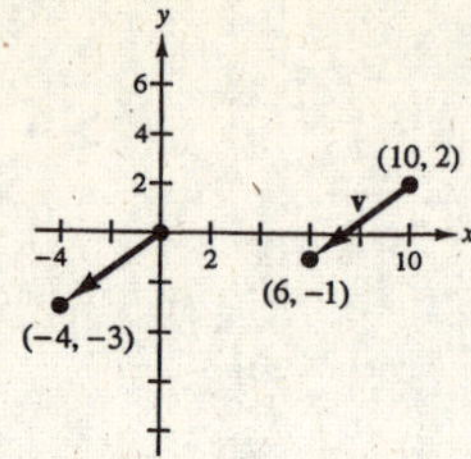

12. (b) $\mathbf{v} = \langle -5 - 0, -1 - (-4) \rangle = \langle -5, 3 \rangle$

(a) and (c).

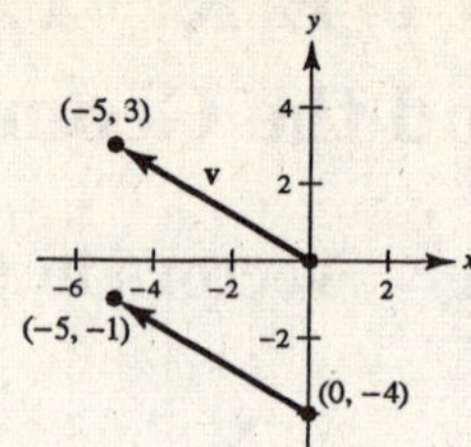

13. (b) $\mathbf{v} = \langle 6 - 6, 6 - 2 \rangle = \langle 0, 4 \rangle$

(a) and (c).

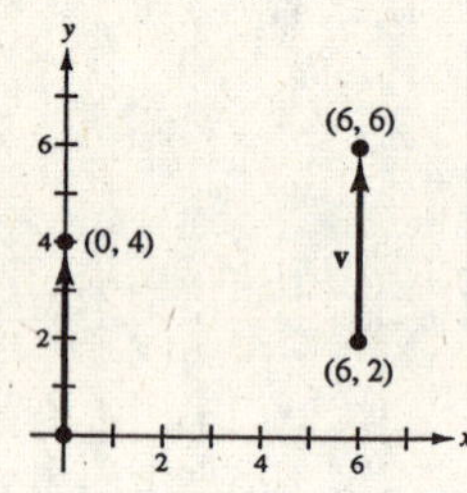

14. (b) $\mathbf{v} = \langle -3 - 7, -1 - (-1) \rangle = \langle -10, 0 \rangle$

(a) and (c).

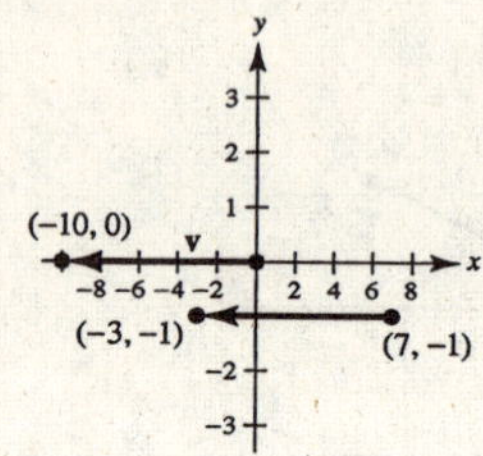

15. (b) $\mathbf{v} = \left\langle \frac{1}{2} - \frac{3}{2}, 3 - \frac{4}{3} \right\rangle = \left\langle -1, \frac{5}{3} \right\rangle$

(a) and (c).

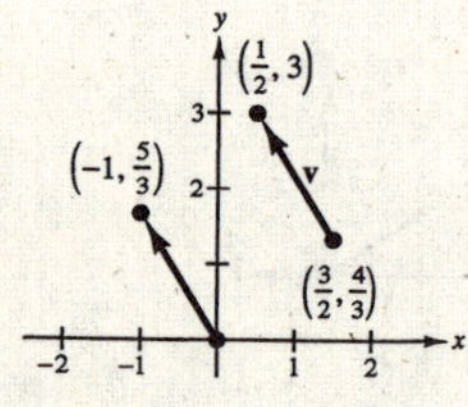

16. (b) $\mathbf{v} = \langle 0.84 - 0.12, 1.25 - 0.60 \rangle = \langle 0.72, 0.65 \rangle$

(a) and (c).

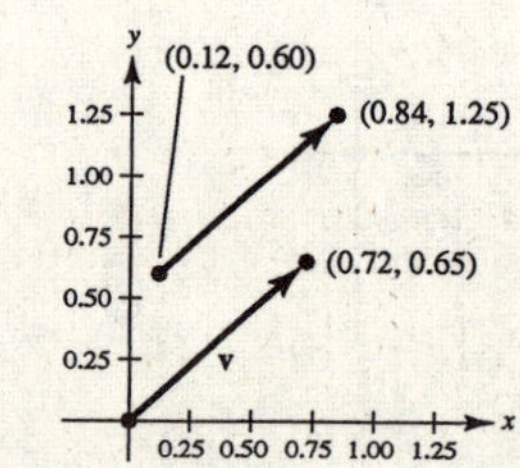

17. (a) $2\mathbf{v} = \langle 4, 6 \rangle$

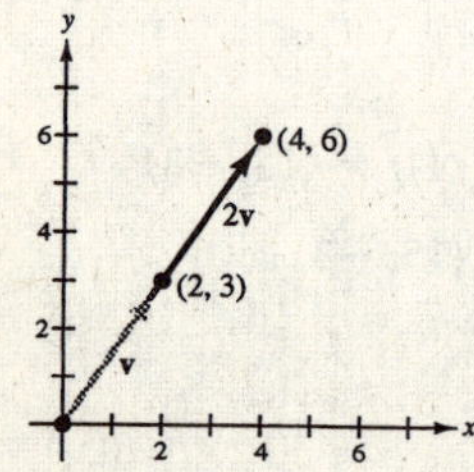

(b) $-3\mathbf{v} = \langle -6, -9 \rangle$

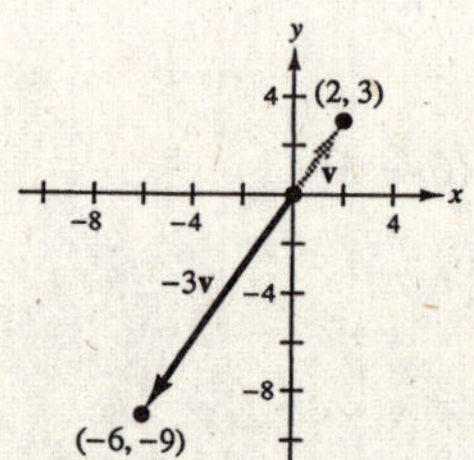

(c) $\frac{7}{2}\mathbf{v} = \left\langle 7, \frac{21}{2} \right\rangle$

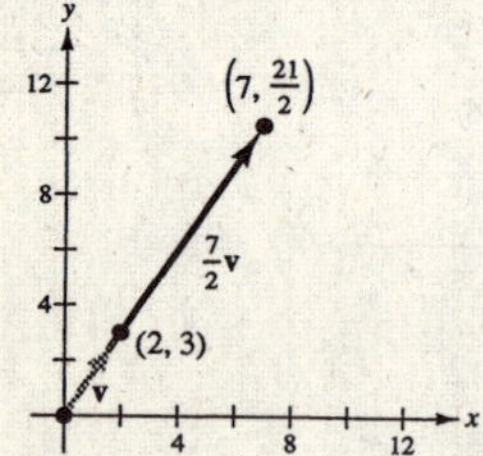

(d) $\frac{2}{3}\mathbf{v} = \left\langle \frac{4}{3}, 2 \right\rangle$

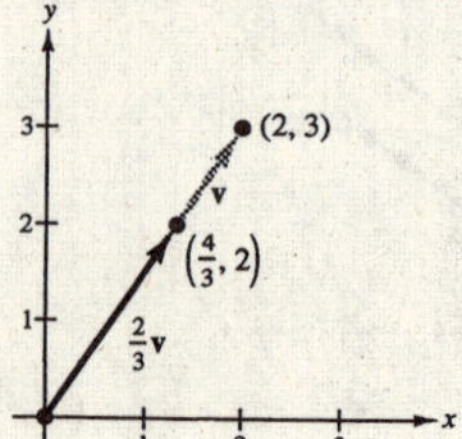

18. (a) $4\mathbf{v} = \langle -4, 20\rangle$

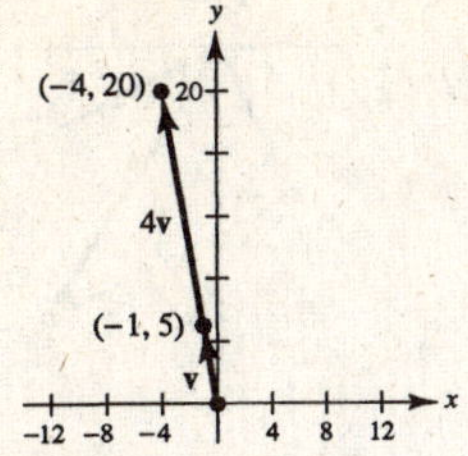

(b) $-\frac{1}{2}\mathbf{v} = \left\langle \frac{1}{2}, -\frac{5}{2}\right\rangle$

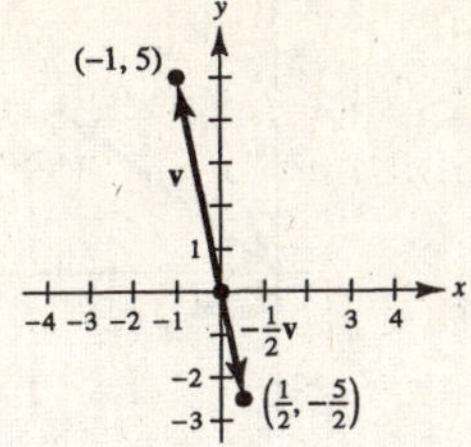

(c) $0\mathbf{v} = \langle 0, 0\rangle$

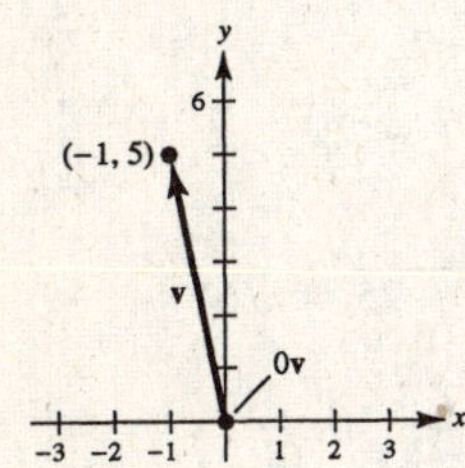

(d) $-6\mathbf{v} = \langle 6, -30\rangle$

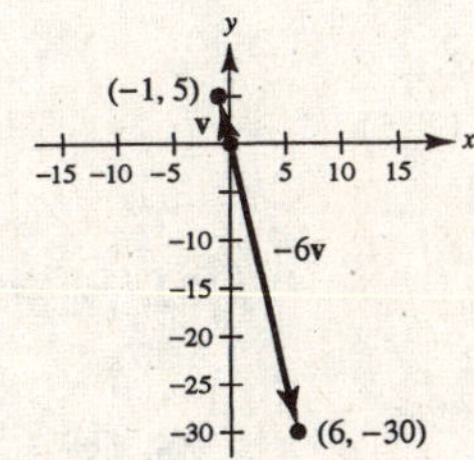

19.

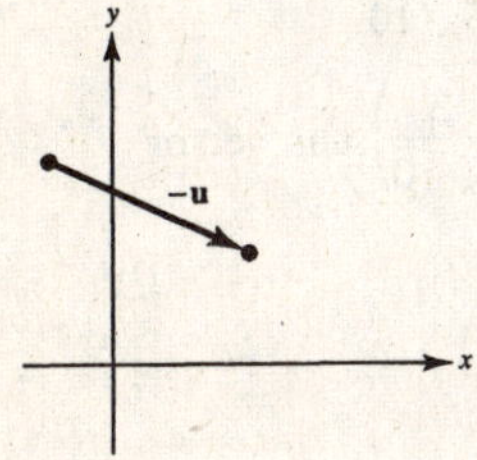

20. Twice as long as given vector **u**.

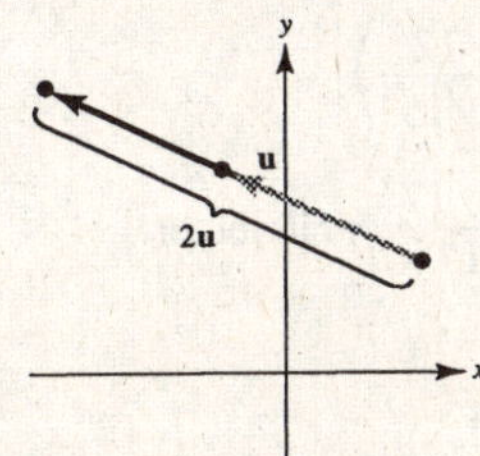

21.

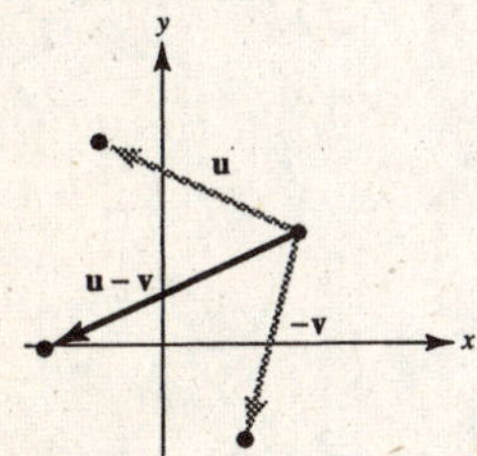

22.

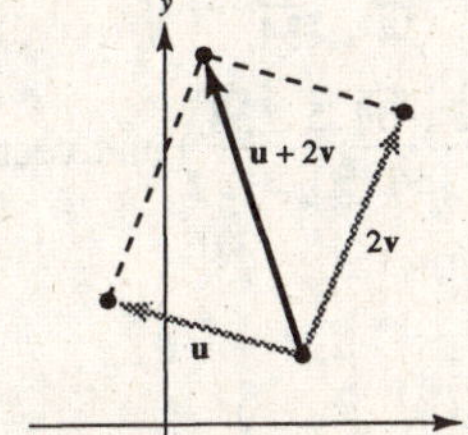

23. (a) $\frac{2}{3}\mathbf{u} = \frac{2}{3}\langle 4, 9\rangle = \left\langle \frac{8}{3}, 6\right\rangle$

(b) $\mathbf{v} - \mathbf{u} = \langle 2, -5\rangle - \langle 4, 9\rangle = \langle -2, -14\rangle$

(c) $2\mathbf{u} + 5\mathbf{v} = 2\langle 4, 9\rangle + 5\langle 2, -5\rangle = \langle 18, -7\rangle$

24. (a) $\frac{2}{3}\mathbf{u} = \frac{2}{3}\langle -3, -8\rangle = \left\langle -2, -\frac{16}{3}\right\rangle$

(b) $\mathbf{v} - \mathbf{u} = \langle 8, 25\rangle - \langle -3, -8\rangle = \langle 11, 33\rangle$

(c) $2\mathbf{u} + 5\mathbf{v} = 2\langle -3, -8\rangle + 5\langle 8, 25\rangle = \langle 34, 109\rangle$

25. $\mathbf{v} = \frac{3}{2}(2\mathbf{i} - \mathbf{j}) = 3\mathbf{i} - \frac{3}{2}\mathbf{j}$

$= \left\langle 3, -\frac{3}{2}\right\rangle$

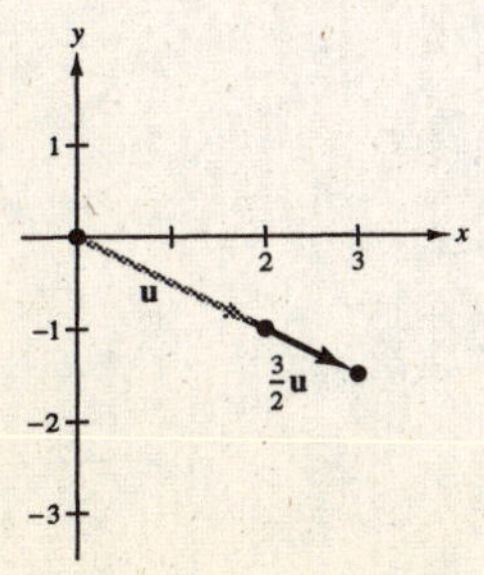

26. $\mathbf{v} = (2\mathbf{i} - \mathbf{j}) + (\mathbf{i} + 2\mathbf{j})$

$= 3\mathbf{i} + \mathbf{j} = \langle 3, 1\rangle$

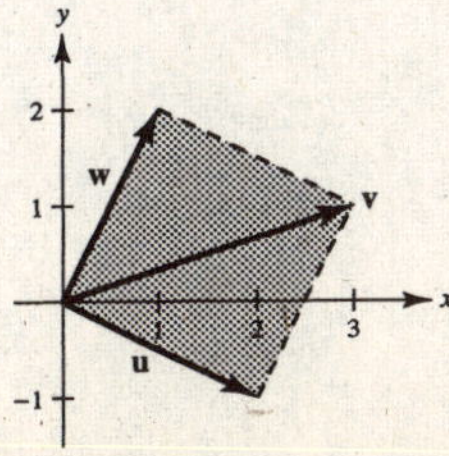

27. $\mathbf{v} = (2\mathbf{i} - \mathbf{j}) + 2(\mathbf{i} + 2\mathbf{j})$

$= 4\mathbf{i} + 3\mathbf{j} = \langle 4, 3\rangle$

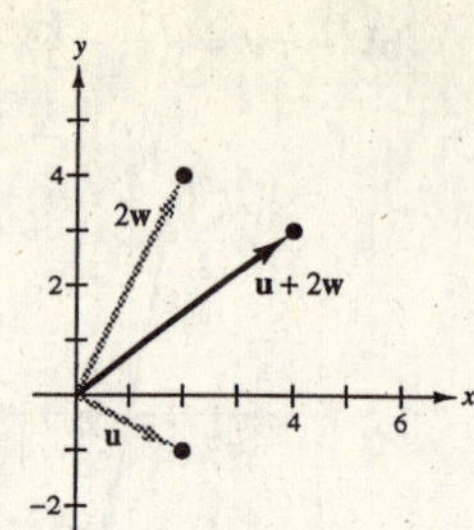

28. $\mathbf{v} = 5\mathbf{u} - 3\mathbf{w}$

$= 5\langle 2, -1\rangle - 3\langle 1, 2\rangle$

$= \langle 7, -11\rangle$

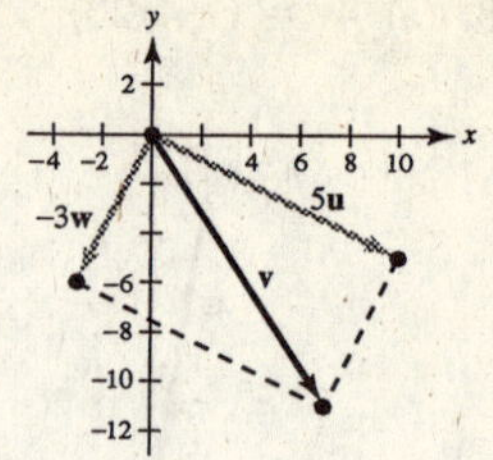

29. $u_1 - 4 = -1 \qquad u_1 = 3$

$u_2 - 2 = 3 \qquad u_2 = 5$

$Q = (3, 5)$

30. $u_1 - 3 = 4 \qquad u_1 = 7$

$u_2 - 2 = -9 \qquad u_2 = -7$

$Q = (7, -7)$

31. $\|\mathbf{v}\| = \sqrt{16 + 9} = 5$

32. $\|\mathbf{v}\| = \sqrt{144 + 25} = 13$

33. $\|\mathbf{v}\| = \sqrt{36 + 25} = \sqrt{61}$

34. $\|\mathbf{v}\| = \sqrt{100 + 9} = \sqrt{109}$

35. $\|\mathbf{v}\| = \sqrt{0 + 16} = 4$

36. $\|\mathbf{v}\| = \sqrt{1 + 1} = \sqrt{2}$

37. $\|\mathbf{u}\| = \sqrt{3^2 + 12^2} = \sqrt{153}$

$\mathbf{v} = \dfrac{\mathbf{u}}{\|\mathbf{u}\|} = \dfrac{\langle 3, 12\rangle}{\sqrt{153}} = \left\langle \dfrac{3}{\sqrt{153}}, \dfrac{12}{\sqrt{153}}\right\rangle$

$= \left\langle \dfrac{\sqrt{17}}{17}, \dfrac{4\sqrt{17}}{17}\right\rangle$ unit vector

38. $\|\mathbf{u}\| = \sqrt{5^2 + 15^2} = \sqrt{250} = 5\sqrt{10}$

$\mathbf{v} = \dfrac{\mathbf{u}}{\|\mathbf{u}\|} = \dfrac{\langle 5, 15\rangle}{5\sqrt{10}} = \left\langle \dfrac{1}{\sqrt{10}}, \dfrac{3}{\sqrt{10}}\right\rangle$ unit vector

39. $\|\mathbf{u}\| = \sqrt{\left(\dfrac{3}{2}\right)^2 + \left(\dfrac{5}{2}\right)^2} = \dfrac{\sqrt{34}}{2}$

$\mathbf{v} = \dfrac{\mathbf{u}}{\|\mathbf{u}\|} = \dfrac{\langle (3/2), (5/2)\rangle}{\sqrt{34}/2} = \left\langle \dfrac{3}{\sqrt{34}}, \dfrac{5}{\sqrt{34}}\right\rangle$

$= \left\langle \dfrac{3\sqrt{34}}{34}, \dfrac{5\sqrt{34}}{34}\right\rangle$ unit vector

40. $\|\mathbf{u}\| = \sqrt{(-6.2)^2 + (3.4)^2} = \sqrt{50} = 5\sqrt{2}$

$\mathbf{v} = \dfrac{\mathbf{u}}{\|\mathbf{u}\|} = \dfrac{\langle -6.2, 3.4\rangle}{5\sqrt{2}} = \left\langle \dfrac{-1.24}{\sqrt{2}}, \dfrac{0.68}{\sqrt{2}}\right\rangle$ unit vector

41. $\mathbf{u} = \langle 1, -1\rangle, \mathbf{v} = \langle -1, 2\rangle$

(a) $\|\mathbf{u}\| = \sqrt{1 + 1} = \sqrt{2}$

(b) $\|\mathbf{v}\| = \sqrt{1 + 4} = \sqrt{5}$

(c) $\mathbf{u} + \mathbf{v} = \langle 0, 1\rangle$

$\|\mathbf{u} + \mathbf{v}\| = \sqrt{0 + 1} = 1$

(d) $\dfrac{\mathbf{u}}{\|\mathbf{u}\|} = \dfrac{1}{\sqrt{2}}\langle 1, -1\rangle$

$\left\|\dfrac{\mathbf{u}}{\|\mathbf{u}\|}\right\| = 1$

(e) $\dfrac{\mathbf{v}}{\|\mathbf{v}\|} = \dfrac{1}{\sqrt{5}}\langle -1, 2\rangle$

$\left\|\dfrac{\mathbf{v}}{\|\mathbf{v}\|}\right\| = 1$

(f) $\dfrac{\mathbf{u} + \mathbf{v}}{\|\mathbf{u} + \mathbf{v}\|} = \langle 0, 1\rangle$

$\left\|\dfrac{\mathbf{u} + \mathbf{v}}{\|\mathbf{u} + \mathbf{v}\|}\right\| = 1$

42. $\mathbf{u} = \langle 0, 1\rangle, \mathbf{v} = \langle 3, -3\rangle$

(a) $\|\mathbf{u}\| = \sqrt{0 + 1} = 1$

(b) $\|\mathbf{v}\| = \sqrt{9 + 9} = 3\sqrt{2}$

(c) $\mathbf{u} + \mathbf{v} = \langle 3, -2\rangle$

$\|\mathbf{u} + \mathbf{v}\| = \sqrt{9 + 4} = \sqrt{13}$

(d) $\dfrac{\mathbf{u}}{\|\mathbf{u}\|} = \langle 0, 1\rangle$

$\left\|\dfrac{\mathbf{u}}{\|\mathbf{u}\|}\right\| = 1$

(e) $\dfrac{\mathbf{v}}{\|\mathbf{v}\|} = \dfrac{1}{3\sqrt{2}}\langle 3, -3\rangle$

$\left\|\dfrac{\mathbf{v}}{\|\mathbf{v}\|}\right\| = 1$

(f) $\dfrac{\mathbf{u} + \mathbf{v}}{\|\mathbf{u} + \mathbf{v}\|} = \dfrac{1}{\sqrt{13}}\langle 3, -2\rangle$

$\left\|\dfrac{\mathbf{u} + \mathbf{v}}{\|\mathbf{u} + \mathbf{v}\|}\right\| = 1$

43. $\mathbf{u} = \left\langle 1, \frac{1}{2} \right\rangle, \mathbf{v} = \langle 2, 3 \rangle$

(a) $\|\mathbf{u}\| = \sqrt{1 + \frac{1}{4}} = \frac{\sqrt{5}}{2}$

(b) $\|\mathbf{v}\| = \sqrt{4 + 9} = \sqrt{13}$

(c) $\mathbf{u} + \mathbf{v} = \left\langle 3, \frac{7}{2} \right\rangle$

$\|\mathbf{u} + \mathbf{v}\| = \sqrt{9 + \frac{49}{4}} = \frac{\sqrt{85}}{2}$

(d) $\frac{\mathbf{u}}{\|\mathbf{u}\|} = \frac{2}{\sqrt{5}}\left\langle 1, \frac{1}{2} \right\rangle$

$\left\| \frac{\mathbf{u}}{\|\mathbf{u}\|} \right\| = 1$

(e) $\frac{\mathbf{v}}{\|\mathbf{v}\|} = \frac{1}{\sqrt{13}}\langle 2, 3 \rangle$

$\left\| \frac{\mathbf{v}}{\|\mathbf{v}\|} \right\| = 1$

(f) $\frac{\mathbf{u} + \mathbf{v}}{\|\mathbf{u} + \mathbf{v}\|} = \frac{2}{\sqrt{85}}\left\langle 3, \frac{7}{2} \right\rangle$

$\left\| \frac{\mathbf{u} + \mathbf{v}}{\|\mathbf{u} + \mathbf{v}\|} \right\| = 1$

44. $\mathbf{u} = \langle 2, -4 \rangle, \mathbf{v} = \langle 5, 5 \rangle$

(a) $\|\mathbf{u}\| = \sqrt{4 + 16} = 2\sqrt{5}$

(b) $\|\mathbf{v}\| = \sqrt{25 + 25} = 5\sqrt{2}$

(c) $\mathbf{u} + \mathbf{v} = \langle 7, 1 \rangle$

$\|\mathbf{u} + \mathbf{v}\| = \sqrt{49 + 1} = 5\sqrt{2}$

(d) $\frac{\mathbf{u}}{\|\mathbf{u}\|} = \frac{1}{2\sqrt{5}}\langle 2, -4 \rangle$

$\left\| \frac{\mathbf{u}}{\|\mathbf{u}\|} \right\| = 1$

(e) $\frac{\mathbf{v}}{\|\mathbf{v}\|} = \frac{1}{5\sqrt{2}}\langle 5, 5 \rangle$

$\left\| \frac{\mathbf{v}}{\|\mathbf{v}\|} \right\| = 1$

(f) $\frac{\mathbf{u} + \mathbf{v}}{\|\mathbf{u} + \mathbf{v}\|} = \frac{1}{5\sqrt{2}}\langle 7, 1 \rangle$

$\left\| \frac{\mathbf{u} + \mathbf{v}}{\|\mathbf{u} + \mathbf{v}\|} \right\| = 1$

45. $\mathbf{u} = \langle 2, 1 \rangle$

$\|\mathbf{u}\| = \sqrt{5} \approx 2.236$

$\mathbf{v} = \langle 5, 4 \rangle$

$\|\mathbf{v}\| = \sqrt{41} \approx 6.403$

$\mathbf{u} + \mathbf{v} = \langle 7, 5 \rangle$

$\|\mathbf{u} + \mathbf{v}\| = \sqrt{74} \approx 8.602$

$\|\mathbf{u} + \mathbf{v}\| \le \|\mathbf{u}\| + \|\mathbf{v}\|$

$\sqrt{74} \le \sqrt{5} + \sqrt{41}$

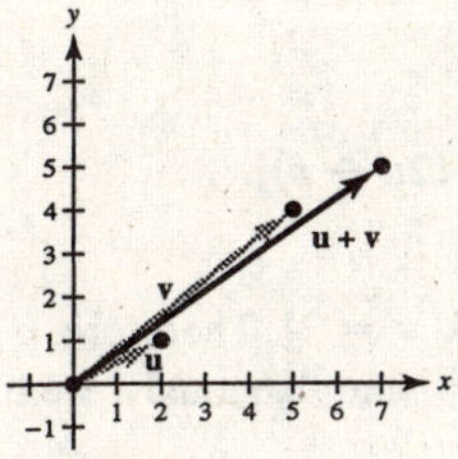

46. $\mathbf{u} = \langle -3, 2 \rangle$

$\|\mathbf{u}\| = \sqrt{13} \approx 3.606$

$\mathbf{v} = \langle 1, -2 \rangle$

$\|\mathbf{v}\| = \sqrt{5} \approx 2.236$

$\mathbf{u} + \mathbf{v} = \langle -2, 0 \rangle$

$\|\mathbf{u} + \mathbf{v}\| = 2$

$\|\mathbf{u} + \mathbf{v}\| \le \|\mathbf{u}\| + \|\mathbf{v}\|$

$2 \le \sqrt{13} + \sqrt{5}$

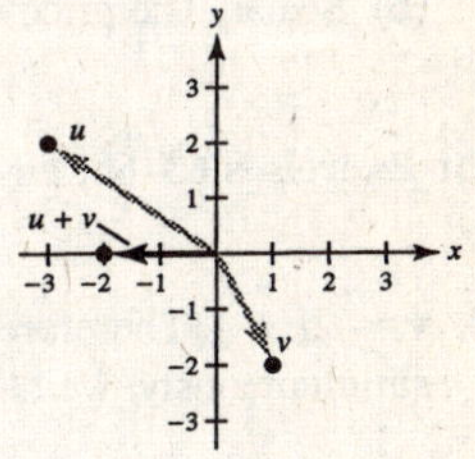

47. $\frac{\mathbf{u}}{\|\mathbf{u}\|} = \frac{1}{\sqrt{2}}\langle 1, 1 \rangle$

$4\left(\frac{\mathbf{u}}{\|\mathbf{u}\|}\right) = 2\sqrt{2}\langle 1, 1 \rangle$

$\mathbf{v} = \langle 2\sqrt{2}, 2\sqrt{2} \rangle$

48. $\frac{\mathbf{u}}{\|\mathbf{u}\|} = \frac{1}{\sqrt{2}}\langle -1, 1 \rangle$

$4\left(\frac{\mathbf{u}}{\|\mathbf{u}\|}\right) = 2\sqrt{2}\langle -1, 1 \rangle$

$\mathbf{v} = \langle -2\sqrt{2}, 2\sqrt{2} \rangle$

49. $\frac{\mathbf{u}}{\|\mathbf{u}\|} = \frac{1}{2\sqrt{3}}\langle \sqrt{3}, 3 \rangle$

$2\left(\frac{\mathbf{u}}{\|\mathbf{u}\|}\right) = \frac{1}{\sqrt{3}}\langle \sqrt{3}, 3 \rangle$

$\mathbf{v} = \langle 1, \sqrt{3} \rangle$

50. $\frac{\mathbf{u}}{\|\mathbf{u}\|} = \frac{1}{3}\langle 0, 3 \rangle$

$3\left(\frac{\mathbf{u}}{\|\mathbf{u}\|}\right) = \langle 0, 3 \rangle$

$\mathbf{v} = \langle 0, 3 \rangle$

51. $\mathbf{v} = 3[(\cos 0°)\mathbf{i} + (\sin 0°)\mathbf{j}]$

$= 3\mathbf{i} = \langle 3, 0 \rangle$

52. $\mathbf{v} = 5[(\cos 120°)\mathbf{i} + (\sin 120°)\mathbf{j}]$

$= -\frac{5}{2}\mathbf{i} + \frac{5\sqrt{3}}{2}\mathbf{j}$

53. $\mathbf{v} = 2[(\cos 150°)\mathbf{i} + (\sin 150°)\mathbf{j}]$

$= -\sqrt{3}\mathbf{i} + \mathbf{j} = \langle -\sqrt{3}, 1 \rangle$

54. $\mathbf{v} = (\cos 3.5°)\mathbf{i} + (\sin 3.5°)\mathbf{j}$

$\approx 0.9981\mathbf{i} + 0.0610\mathbf{j} = \langle 0.9981, 0.0610 \rangle$

55. $\mathbf{u} = \mathbf{i}$

$$\mathbf{v} = \frac{3\sqrt{2}}{2}\mathbf{i} + \frac{3\sqrt{2}}{2}\mathbf{j}$$

$$\mathbf{u} + \mathbf{v} = \left(\frac{2 + 3\sqrt{2}}{2}\right)\mathbf{i} + \frac{3\sqrt{2}}{2}\mathbf{j}$$

56. $\mathbf{u} = 4\mathbf{i}$

$$\mathbf{v} = \mathbf{i} + \sqrt{3}\mathbf{j}$$

$$\mathbf{u} + \mathbf{v} = 5\mathbf{i} + \sqrt{3}\mathbf{j}$$

57. $\mathbf{u} = 2(\cos 4)\mathbf{i} + 2(\sin 4)\mathbf{j}$

$$\mathbf{v} = (\cos 2)\mathbf{i} + (\sin 2)\mathbf{j}$$

$$\mathbf{u} + \mathbf{v} = (2\cos 4 + \cos 2)\mathbf{i} + (2\sin 4 + \sin 2)\mathbf{j}$$

58. $\mathbf{u} = 5[\cos(-0.5)]\mathbf{i} + 5[\sin(-0.5)]\mathbf{j}$

$$= 5[\cos(0.5)]\mathbf{i} - 5[\sin(0.5)]\mathbf{j}$$

$$\mathbf{v} = 5[\cos(0.5)]\mathbf{i} + 5[\sin(0.5)]\mathbf{j}$$

$$\mathbf{u} + \mathbf{v} = 10[\cos(0.5)]\mathbf{i}$$

59. A scalar is a real number. A vector is represented by a directed line segment. A vector has both length and direction.

60. See page 764:

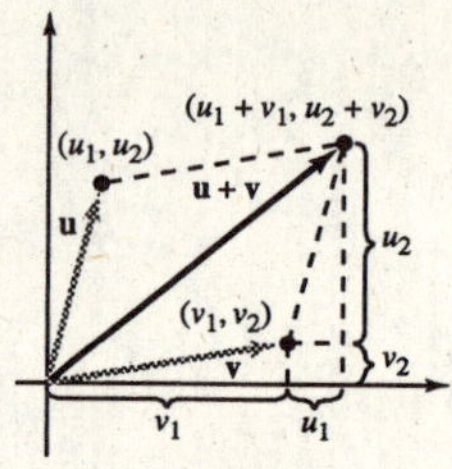

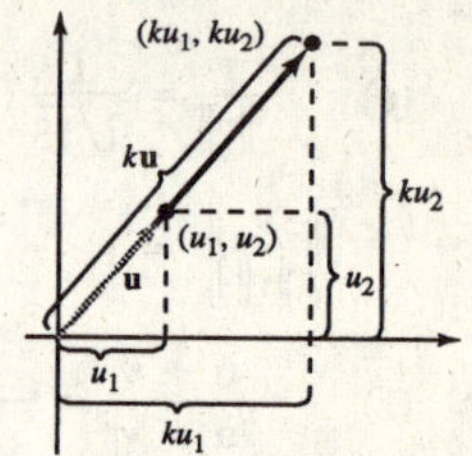

61. (a) Vector. The velocity has both magnitude and direction.

(b) Scalar. The price is a number.

62. (a) Scalar. The temperature is a number.

(b) Vector. The weight has magnitude and direction.

For Exercises 63–68, $a\mathbf{u} + b\mathbf{w} = a(\mathbf{i} + 2\mathbf{j}) + b(\mathbf{i} - \mathbf{j}) = (a + b)\mathbf{i} + (2a - b)\mathbf{j}$.

63. $\mathbf{v} = 2\mathbf{i} + \mathbf{j}$. Therefore, $a + b = 2$, $2a - b = 1$. Solving simultaneously, we have $a = 1$, $b = 1$.

64. $\mathbf{v} = 3\mathbf{j}$. Therefore, $a + b = 0$, $2a - b = 3$. Solving simultaneously, we have $a = 1$, $b = -1$.

65. $\mathbf{v} = 3\mathbf{i}$. Therefore, $a + b = 3$, $2a - b = 0$. Solving simultaneously, we have $a = 1$, $b = 2$.

66. $\mathbf{v} = 3\mathbf{i} + 3\mathbf{j}$. Therefore, $a + b = 3$, $2a - b = 3$. Solving simultaneously, we have $a = 2$, $b = 1$.

67. $\mathbf{v} = \mathbf{i} + \mathbf{j}$. Therefore, $a + b = 1$, $2a - b = 1$. Solving simultaneously, we have $a = \frac{2}{3}$, $b = \frac{1}{3}$.

68. $\mathbf{v} = -\mathbf{i} + 7\mathbf{j}$. Therefore, $a + b = -1$, $2a - b = 7$. Solving simultaneously, we have $a = 2$, $b = -3$.

69. $f(x) = x^2, f'(x) = 2x, f'(3) = 6$

(a) $m = 6$. Let $\mathbf{w} = \langle 1, 6\rangle$, $\|\mathbf{w}\| = \sqrt{37}$.

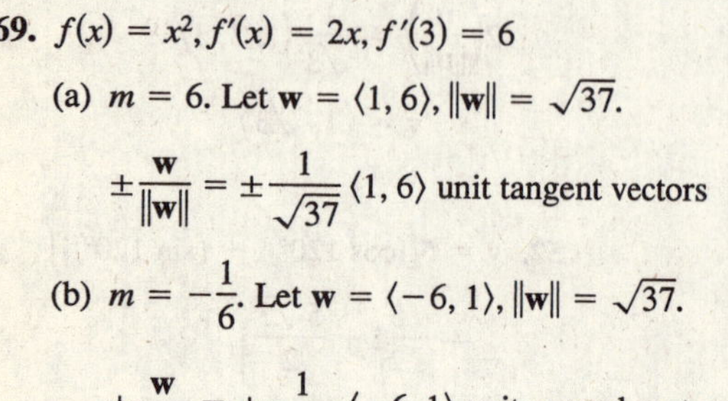

$$\pm\frac{\mathbf{w}}{\|\mathbf{w}\|} = \pm\frac{1}{\sqrt{37}}\langle 1, 6\rangle \text{ unit tangent vectors}$$

(b) $m = -\frac{1}{6}$. Let $\mathbf{w} = \langle -6, 1\rangle$, $\|\mathbf{w}\| = \sqrt{37}$.

$$\pm\frac{\mathbf{w}}{\|\mathbf{w}\|} = \pm\frac{1}{\sqrt{37}}\langle -6, 1\rangle \text{ unit normal vectors}$$

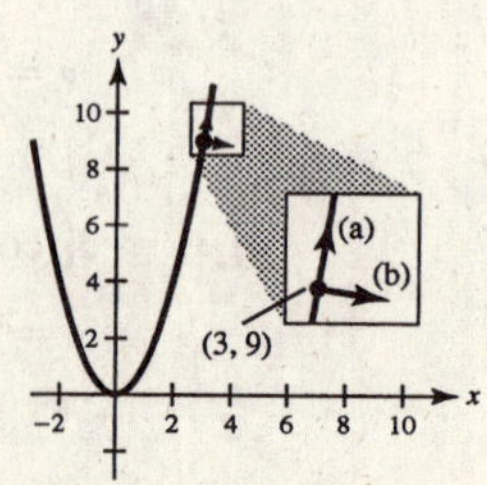

70. $f(x) = -x^2 + 5, f'(x) = -2x, f'(1) = -2$

(a) $m = -2$. Let $\mathbf{w} = \langle 1, -2\rangle, \|\mathbf{w}\| = \sqrt{5}$.

$$\pm\frac{\mathbf{w}}{\|\mathbf{w}\|} = \pm\frac{1}{\sqrt{5}}\langle 1, -2\rangle \text{ unit tangent vectors}$$

(b) $m = \frac{1}{2}$. Let $\mathbf{w} = \langle 2, 1\rangle, \|\mathbf{w}\| = \sqrt{5}$.

$$\pm\frac{\mathbf{w}}{\|\mathbf{w}\|} = \pm\frac{1}{\sqrt{5}}\langle 2, 1\rangle \text{ unit normal vectors}$$

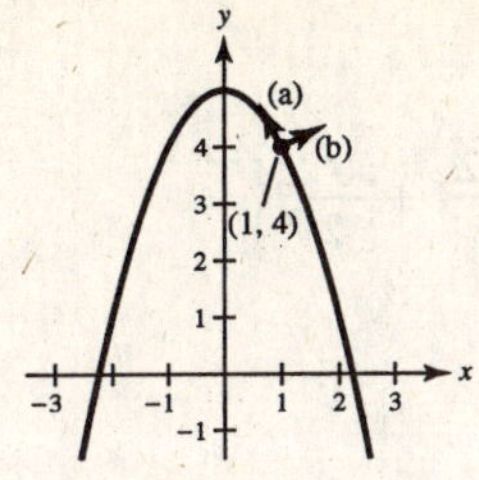

71. $f(x) = x^3, f'(x) = 3x^2 = 3$ at $x = 1$.

(a) $m = 3$. Let $\mathbf{w} = \langle 1, 3\rangle$, then

$$\frac{\mathbf{w}}{\|\mathbf{w}\|} = \pm\frac{1}{\sqrt{10}}\langle 1, 3\rangle.$$

(b) $m = -\frac{1}{3}$. Let $\mathbf{w} = \langle 3, -1\rangle$, then

$$\frac{\mathbf{w}}{\|\mathbf{w}\|} = \pm\frac{1}{\sqrt{10}}\langle 3, -1\rangle.$$

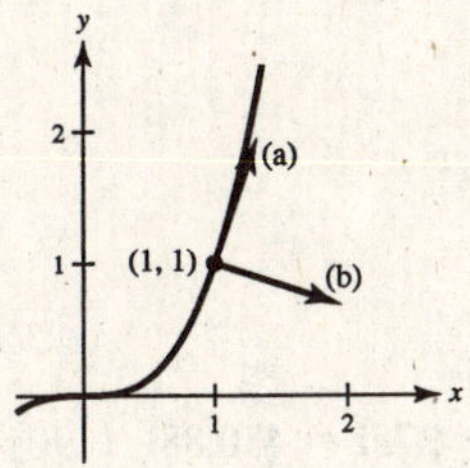

72. $f(x) = x^3, f'(x) = 3x^2 = 12$ at $x = -2$.

(a) $m = 12$. Let $\mathbf{w} = \langle 1, 12\rangle$, then

$$\frac{\mathbf{w}}{\|\mathbf{w}\|} = \pm\frac{1}{\sqrt{145}}\langle 1, 12\rangle.$$

(b) $m = -\frac{1}{12}$. Let $\mathbf{w} = \langle 12, -1\rangle$, then

$$\frac{\mathbf{w}}{\|\mathbf{w}\|} = \pm\frac{1}{\sqrt{145}}\langle 12, -1\rangle.$$

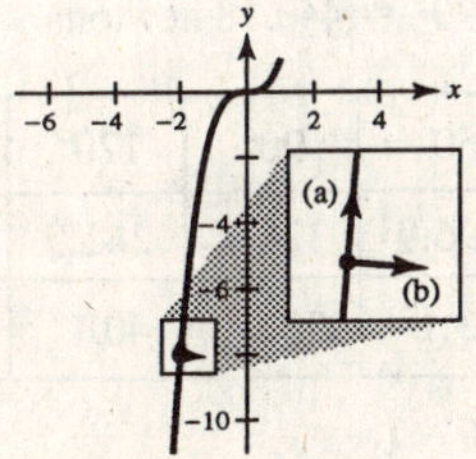

73. $f(x) = \sqrt{25 - x^2}$

$$f'(x) = \frac{-x}{\sqrt{25 - x^2}} = \frac{-3}{4} \text{ at } x = 3.$$

(a) $m = -\frac{3}{4}$. Let $\mathbf{w} = \langle -4, 3\rangle$, then

$$\frac{\mathbf{w}}{\|\mathbf{w}\|} = \pm\frac{1}{5}\langle -4, 3\rangle.$$

(b) $m = \frac{4}{3}$. Let $\mathbf{w} = \langle 3, 4\rangle$, then

$$\frac{\mathbf{w}}{\|\mathbf{w}\|} = \pm\frac{1}{5}\langle 3, 4\rangle.$$

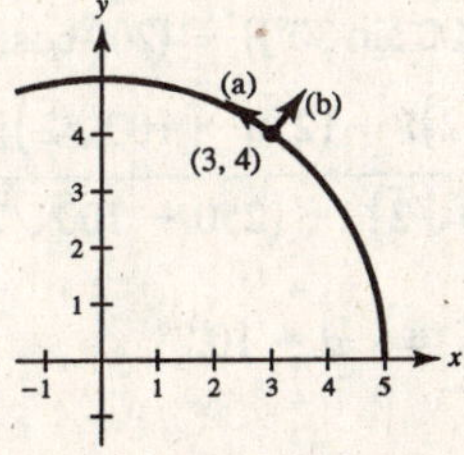

74. $f(x) = \tan x$

$$f'(x) = \sec^2 x = 2 \text{ at } x = \frac{\pi}{4}.$$

(a) $m = 2$. Let $\mathbf{w} = \langle 1, 2\rangle$, then

$$\frac{\mathbf{w}}{\|\mathbf{w}\|} = \pm\frac{1}{\sqrt{5}}\langle 1, 2\rangle.$$

(b) $m = -\frac{1}{2}$. Let $\mathbf{w} = \langle -2, 1\rangle$, then

$$\frac{\mathbf{w}}{\|\mathbf{w}\|} = \pm\frac{1}{\sqrt{5}}\langle -2, 1\rangle.$$

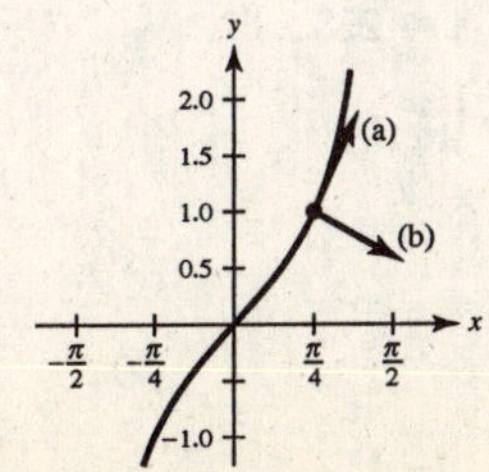

75. $\mathbf{u} = \frac{\sqrt{2}}{2}\mathbf{i} + \frac{\sqrt{2}}{2}\mathbf{j}$

$\mathbf{u} + \mathbf{v} = \sqrt{2}\mathbf{j}$

$\mathbf{v} = (\mathbf{u} + \mathbf{v}) - \mathbf{u} = -\frac{\sqrt{2}}{2}\mathbf{i} + \frac{\sqrt{2}}{2}\mathbf{j}$

76. $\mathbf{u} = 2\sqrt{3}\mathbf{i} + 2\mathbf{j}$

$\mathbf{u} + \mathbf{v} = -3\mathbf{i} + 3\sqrt{3}\mathbf{j}$

$\mathbf{v} = (\mathbf{u} + \mathbf{v}) - \mathbf{u} = \left(-3 - 2\sqrt{3}\right)\mathbf{i} + \left(3\sqrt{3} - 2\right)\mathbf{j}$

77. Programs will vary.

78. magnitude ≈ 63.5

direction $\approx -8.26°$

79. $\|\mathbf{F}_1\| = 2,\ \theta_{\mathbf{F}_1} = 33°$

$\|\mathbf{F}_2\| = 3,\ \theta_{\mathbf{F}_2} = -125°$

$\|\mathbf{F}_3\| = 2.5,\ \theta_{\mathbf{F}_3} = 110°$

$\|\mathbf{R}\| = \|\mathbf{F}_1 + \mathbf{F}_2 + \mathbf{F}_3\| \approx 1.33$

$\theta_{\mathbf{R}} = \theta_{\mathbf{F}_1+\mathbf{F}_2+\mathbf{F}_3} \approx 132.5°$

80. $\|\mathbf{F}_1\| = 2,\ \theta_{\mathbf{F}_1} = -10°$

$\|\mathbf{F}_2\| = 4,\ \theta_{\mathbf{F}_2} = 140°$

$\|\mathbf{F}_3\| = 3,\ \theta_{\mathbf{F}_3} = 200°$

$\|\mathbf{R}\| = \|\mathbf{F}_1 + \mathbf{F}_2 + \mathbf{F}_3\| \approx 4.09$

$\theta_{\mathbf{R}} = \theta_{\mathbf{F}_1+\mathbf{F}_2+\mathbf{F}_3} \approx 163.0°$

81. (a) $180(\cos 30°\mathbf{i} + \sin 30°\mathbf{j}) + 275\mathbf{i} \approx 430.88\mathbf{i} + 90\mathbf{j}$

Direction: $\alpha \approx \arctan\left(\frac{90}{430.88}\right) \approx 0.206\ (\approx 11.8°)$

Magnitude: $\sqrt{430.88^2 + 90^2} \approx 440.18$ newtons

(b) $M = \sqrt{(275 + 180\cos\theta)^2 + (180\sin\theta)^2}$

$\alpha = \arctan\left[\frac{180\sin\theta}{275 + 180\cos\theta}\right]$

(c)

θ	0°	30°	60°	90°	120°	150°	180°
M	455	440.2	396.9	328.7	241.9	149.3	95
α	0°	11.8°	23.1°	33.2°	40.1°	37.1°	0

(d)

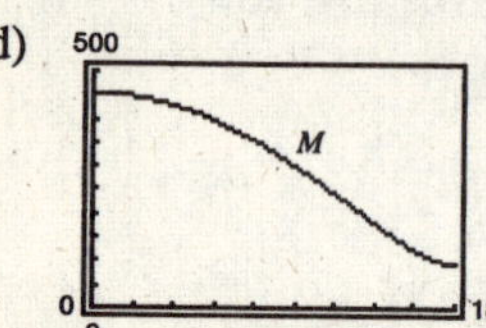

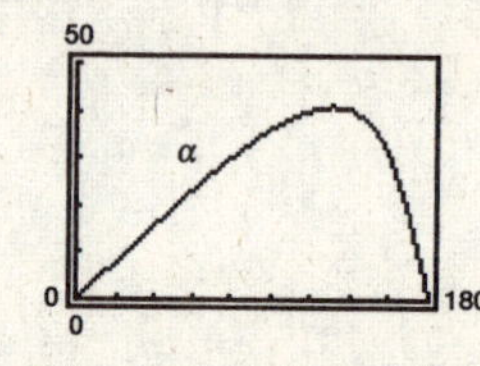

(e) M decreases because the forces change from acting in the same direction to acting in the opposite direction as θ increases from 0° to 180°.

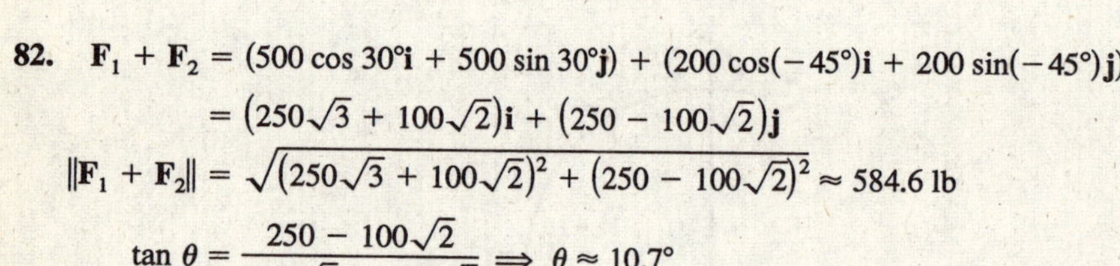

82. $\mathbf{F}_1 + \mathbf{F}_2 = (500\cos 30°\mathbf{i} + 500\sin 30°\mathbf{j}) + (200\cos(-45°)\mathbf{i} + 200\sin(-45°)\mathbf{j})$

$= \left(250\sqrt{3} + 100\sqrt{2}\right)\mathbf{i} + \left(250 - 100\sqrt{2}\right)\mathbf{j}$

$\|\mathbf{F}_1 + \mathbf{F}_2\| = \sqrt{\left(250\sqrt{3} + 100\sqrt{2}\right)^2 + \left(250 - 100\sqrt{2}\right)^2} \approx 584.6$ lb

$\tan\theta = \frac{250 - 100\sqrt{2}}{250\sqrt{3} + 100\sqrt{2}} \Rightarrow \theta \approx 10.7°$

83. $\mathbf{F}_1 + \mathbf{F}_2 + \mathbf{F}_3 = (75\cos 30°\mathbf{i} + 75\sin 30°\mathbf{j}) + (100\cos 45°\mathbf{i} + 100\sin 45°\mathbf{j}) + (125\cos 120°\mathbf{i} + 125\sin 120°\mathbf{j})$

$= \left(\frac{75}{2}\sqrt{3} + 50\sqrt{2} - \frac{125}{2}\right)\mathbf{i} + \left(\frac{75}{2} + 50\sqrt{2} + \frac{125}{2}\sqrt{3}\right)\mathbf{j}$

$\|\mathbf{R}\| = \|\mathbf{F}_1 + \mathbf{F}_2 + \mathbf{F}_3\| \approx 228.5$ lb

$\theta_{\mathbf{R}} = \theta_{\mathbf{F}_1+\mathbf{F}_2+\mathbf{F}_3} \approx 71.3°$

84. $\mathbf{F}_1 + \mathbf{F}_2 + \mathbf{F}_3 = [400(\cos(-30°)\mathbf{i} + \sin(-30°)\mathbf{j})] + [280(\cos(45°)\mathbf{i} + \sin(45°)\mathbf{j})] + [350(\cos(135°)\mathbf{i} + \sin(135°)\mathbf{j}]$

$$= \left[200\sqrt{3} + 140\sqrt{2} - 175\sqrt{2}\right]\mathbf{i} + \left[-200 + 140\sqrt{2} + 175\sqrt{2}\right]\mathbf{j}$$

$$\|\mathbf{R}\| = \sqrt{(200\sqrt{3} - 35\sqrt{2})^2 + (-200 + 315\sqrt{2})^2} \approx 385.2483 \text{ newtons}$$

$$\theta_{\mathbf{R}} = \arctan\left(\frac{-200 + 315\sqrt{2}}{200\sqrt{3} - 35\sqrt{2}}\right) \approx 0.6908 \approx 39.6°$$

85. (a) The forces act along the same direction. $\theta = 0°$.

(b) The forces cancel out each other. $\theta = 180°$.

(c) No, the magnitude of the resultant can not be greater than the sum.

86. $\mathbf{F}_1 = \langle 20, 0\rangle, \mathbf{F}_2 = 10\langle\cos\theta, \sin\theta\rangle$

(a) $\|\mathbf{F}_1 + \mathbf{F}_2\| = \|\langle 20 + 10\cos\theta, 10\sin\theta\rangle\|$

$$= \sqrt{400 + 400\cos\theta + 100\cos^2\theta + 100\sin^2\theta}$$

$$= \sqrt{500 + 400\cos\theta}$$

(b)

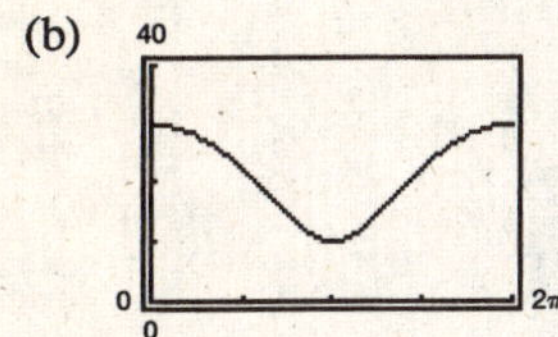

(c) The range is $10 \le \|\mathbf{F}_1 + \mathbf{F}_2\| \le 30$.

The maximum is 30, which occur at $\theta = 0$ and $\theta = 2\pi$.

The minimum is 10 at $\theta = \pi$.

(d) The minimum of the resultant is 10.

87. $(-4, -1), (6, 5), (10, 3)$

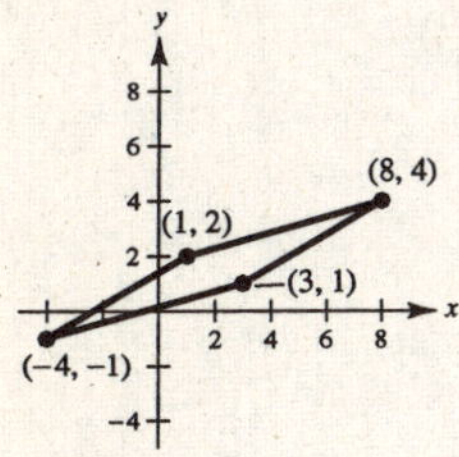

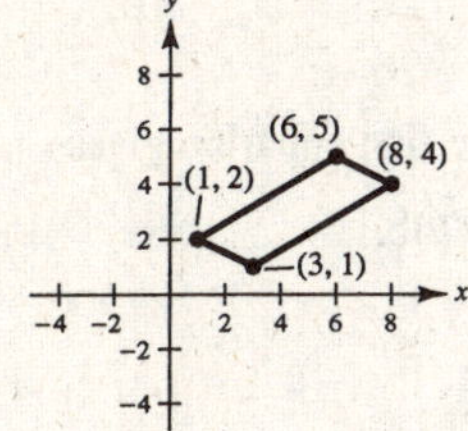

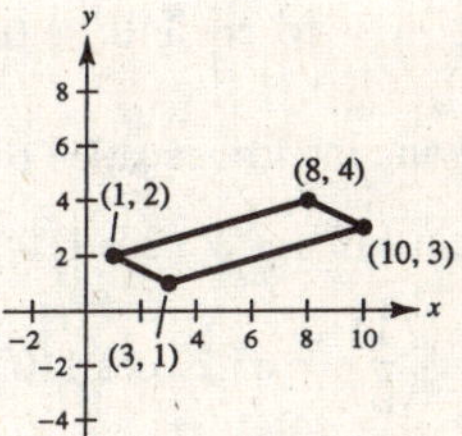

88. $\mathbf{u} = \langle 7 - 1, 5 - 2\rangle = \langle 6, 3\rangle$

$$\frac{1}{3}\mathbf{u} = \langle 2, 1\rangle$$

$$P_1 = (1, 2) + (2, 1) = (3, 3)$$

$$P_2 = (1, 2) + 2(2, 1) = (5, 4)$$

89. $\mathbf{u} = \overrightarrow{CB} = \|\mathbf{u}\|(\cos 30°\,\mathbf{i} + \sin 30°\,\mathbf{j})$

$\mathbf{v} = \overrightarrow{CA} = \|\mathbf{v}\|(\cos 130°\,\mathbf{i} + \sin 130°\,\mathbf{j})$

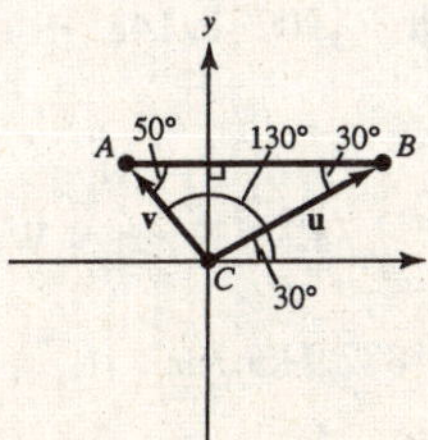

Vertical components: $\|\mathbf{u}\|\sin 30° + \|\mathbf{v}\|\sin 130° = 2000$

Horizontal components: $\|\mathbf{u}\|\cos 30° + \|\mathbf{v}\|\cos 130° = 0$

Solving this system, you obtain

$\|\mathbf{u}\| \approx 1305.5$ pounds and $\|\mathbf{v}\| \approx 1758.8$ pounds.

90. $\theta_1 = \arctan\left(\frac{24}{20}\right) \approx 0.8761$ or $50.2°$

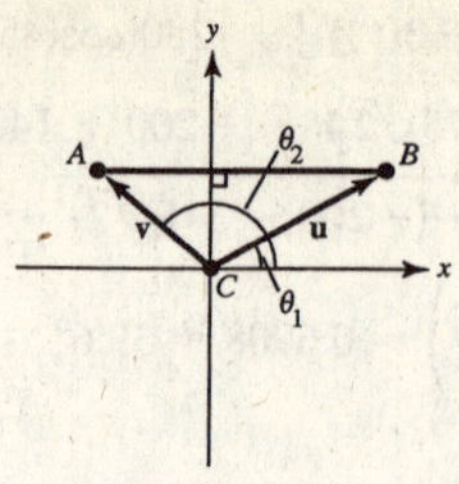

$\theta_2 = \arctan\left(\frac{24}{-10}\right) + \pi \approx 1.9656$ or $112.6°$

$\mathbf{u} = \|\mathbf{u}\|(\cos\theta_1\,\mathbf{i} + \sin\theta_1\,\mathbf{j})$

$\mathbf{v} = \|\mathbf{v}\|(\cos\theta_2\,\mathbf{i} + \sin\theta_2\,\mathbf{j})$

Vertical components: $\|\mathbf{u}\|\sin\theta_1 + \|\mathbf{v}\|\sin\theta_2 = 5000$

Horizontal components: $\|\mathbf{u}\|\cos\theta_1 + \|\mathbf{v}\|\cos\theta_2 = 0$

Solving this system, you obtain

$\|\mathbf{u}\| \approx 2169.4$ and $\|\mathbf{v}\| \approx 3611.2$.

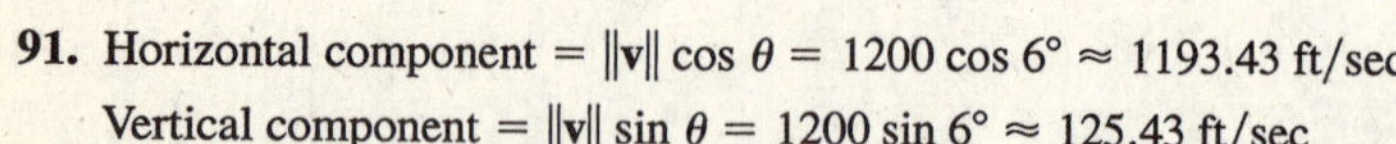

91. Horizontal component $= \|\mathbf{v}\|\cos\theta = 1200\cos 6° \approx 1193.43$ ft/sec

Vertical component $= \|\mathbf{v}\|\sin\theta = 1200\sin 6° \approx 125.43$ ft/sec

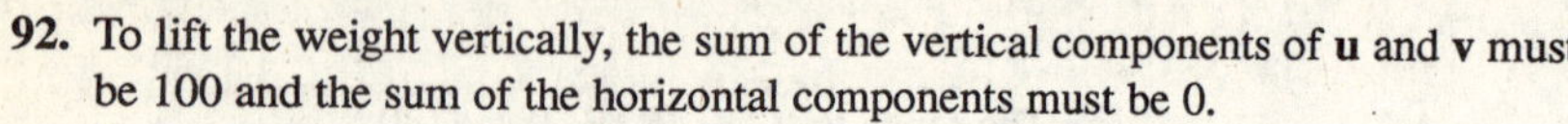

92. To lift the weight vertically, the sum of the vertical components of **u** and **v** must be 100 and the sum of the horizontal components must be 0.

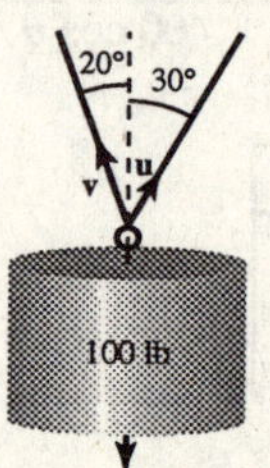

$\mathbf{u} = \|\mathbf{u}\|\,(\cos 60°\mathbf{i} + \sin 60°\mathbf{j})$

$\mathbf{v} = \|\mathbf{v}\|\,(\cos 110°\mathbf{i} + \sin 110°\mathbf{j})$

Thus, $\|\mathbf{u}\|\sin 60° + \|\mathbf{v}\|\sin 110° = 100$, or

$\|\mathbf{u}\|\left(\frac{\sqrt{3}}{2}\right) + \|\mathbf{v}\|\sin 110° = 100.$

And $\|\mathbf{u}\|\cos 60° + \|\mathbf{v}\|\cos 110° = 0$ or

$\|\mathbf{u}\|\left(\frac{1}{2}\right) + \|\mathbf{v}\|\cos 110° = 0.$

Multiplying the last equation by $(\sqrt{3})$ and adding to the first equation gives

$\|\mathbf{u}\|(\sin 110° - \sqrt{3}\cos 110°) = 100 \Longrightarrow \|\mathbf{v}\| \approx 65.27$ lb.

Then, $\|\mathbf{u}\|\left(\frac{1}{2}\right) + 65.27\cos 110° = 0$ gives

$\|\mathbf{u}\| \approx 44.65$ lb.

(a) The tension in each rope: $\|\mathbf{u}\| = 44.65$ lb, $\|\mathbf{v}\| = 65.27$ lb

(b) Vertical components: $\|\mathbf{u}\|\sin 60° \approx 38.67$ lb

$\|\mathbf{v}\|\sin 110° \approx 61.33$ lb

93. $\mathbf{u} = 900[\cos 148°\,\mathbf{i} + \sin 148°\,\mathbf{j}]$

$\mathbf{v} = 100[\cos 45°\,\mathbf{i} + \sin 45°\,\mathbf{j}]$

$\mathbf{u} + \mathbf{v} = [900\cos 148° + 100\cos 45°]\mathbf{i} + [900\sin 148° + 100\sin 45°]\mathbf{j}$

$\approx -692.53\,\mathbf{i} + 547.64\,\mathbf{j}$

$\theta \approx \arctan\left(\frac{547.64}{-692.53}\right) \approx -38.34°;\quad 38.34°$ North of West

$\|\mathbf{u} + \mathbf{v}\| \approx \sqrt{(-692.53)^2 + (547.64)^2} \approx 882.9$ km/hr

94. $\mathbf{u} = 400\mathbf{i}$(plane)

$\mathbf{v} = 50(\cos 135°\mathbf{i} + \sin 135°\mathbf{j}) = -25\sqrt{2}\mathbf{i} + 25\sqrt{2}\mathbf{j}$ (wind)

$\mathbf{u} + \mathbf{v} = \left(400 - 25\sqrt{2}\right)\mathbf{i} + 25\sqrt{2}\mathbf{j} \approx 364.64\mathbf{i} + 35.36\mathbf{j}$

$\tan\theta = \dfrac{35.36}{364.64} \Rightarrow \theta \approx 5.54°$

Direction North of East: $\approx$ N 84.46° E

Speed: $\approx$ 336.35 mph

95. True

96. True

97. True

98. False

$a = b = 0$

99. False

$\|a\mathbf{i} + b\mathbf{j}\| = \sqrt{2}|a|$

100. True

101. $\|\mathbf{u}\| = \sqrt{\cos^2\theta + \sin^2\theta} = 1,$

$\|\mathbf{v}\| = \sqrt{\sin^2\theta + \cos^2\theta} = 1$

102. Let the triangle have vertices at $(0, 0)$, $(a, 0)$, and (b, c). Let $\mathbf{u}$ be the vector joining $(0, 0)$ and (b, c), as indicated in the figure. Then $\mathbf{v}$, the vector joining the midpoints, is

$$\mathbf{v} = \left(\frac{a+b}{2} - \frac{a}{2}\right)\mathbf{i} + \frac{c}{2}\mathbf{j}$$

$$= \frac{b}{2}\mathbf{i} + \frac{c}{2}\mathbf{j} = \frac{1}{2}(b\mathbf{i} + c\mathbf{j}) = \frac{1}{2}\mathbf{u}.$$

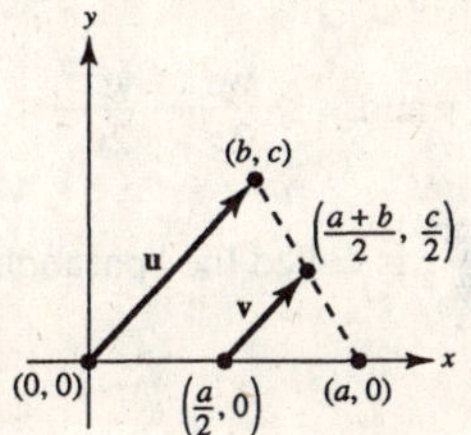

103. Let $\mathbf{u}$ and $\mathbf{v}$ be the vectors that determine the parallelogram, as indicated in the figure. The two diagonals are $\mathbf{u} + \mathbf{v}$ and $\mathbf{v} - \mathbf{u}$. Therefore, $\mathbf{r} = x(\mathbf{u} + \mathbf{v})$, $\mathbf{s} = y(\mathbf{v} - \mathbf{u})$. But,

$$\mathbf{u} = \mathbf{r} - \mathbf{s}$$

$$= x(\mathbf{u} + \mathbf{v}) - y(\mathbf{v} - \mathbf{u}) = (x + y)\mathbf{u} + (x - y)\mathbf{v}.$$

Therefore, $x + y = 1$ and $x - y = 0$. Solving we have $x = y = \frac{1}{2}$.

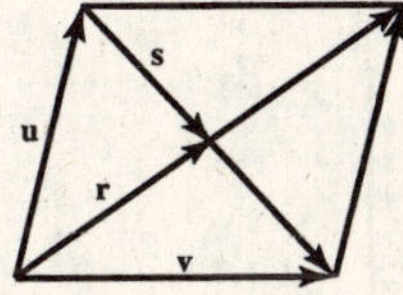

104. $\mathbf{w} = \|\mathbf{u}\|\mathbf{v} + \|\mathbf{v}\|\mathbf{u}$

$$= \|\mathbf{u}\|\left[\|\mathbf{v}\|\cos\theta_{\mathbf{v}}\mathbf{i} + \|\mathbf{v}\|\sin\theta_{\mathbf{v}}\mathbf{j}\right] + \|\mathbf{v}\|\left[\|\mathbf{u}\|\cos\theta_{\mathbf{u}}\mathbf{i} + \|\mathbf{u}\|\sin\theta_{\mathbf{u}}\mathbf{j}\right] = \|\mathbf{u}\|\,\|\mathbf{v}\|\left[(\cos\theta_{\mathbf{u}} + \cos\theta_{\mathbf{v}})\mathbf{i} + (\sin\theta_{\mathbf{u}} + \sin\theta_{\mathbf{v}})\mathbf{j}\right]$$

$$= 2\|\mathbf{u}\|\,\|\mathbf{v}\|\left[\cos\left(\frac{\theta_{\mathbf{u}} + \theta_{\mathbf{v}}}{2}\right)\cos\left(\frac{\theta_{\mathbf{u}} - \theta_{\mathbf{v}}}{2}\right)\mathbf{i} + \sin\left(\frac{\theta_{\mathbf{u}} + \theta_{\mathbf{v}}}{2}\right)\cos\left(\frac{\theta_{\mathbf{u}} - \theta_{\mathbf{v}}}{2}\right)\mathbf{j}\right]$$

$$\tan\theta_{\mathbf{w}} = \frac{\sin\left(\frac{\theta_{\mathbf{u}} + \theta_{\mathbf{v}}}{2}\right)\cos\left(\frac{\theta_{\mathbf{u}} - \theta_{\mathbf{v}}}{2}\right)}{\cos\left(\frac{\theta_{\mathbf{u}} + \theta_{\mathbf{v}}}{2}\right)\cos\left(\frac{\theta_{\mathbf{u}} - \theta_{\mathbf{v}}}{2}\right)} = \tan\left(\frac{\theta_{\mathbf{u}} + \theta_{\mathbf{v}}}{2}\right)$$

Thus, $\theta_{\mathbf{w}} = (\theta_{\mathbf{u}} + \theta_{\mathbf{v}})/2$ and $\mathbf{w}$ bisects the angle between $\mathbf{u}$ and $\mathbf{v}$.

105. The set is a circle of radius 5, centered at the origin.

$\|\mathbf{u}\| = \|\langle x, y\rangle\| = \sqrt{x^2 + y^2} = 5 \Rightarrow x^2 + y^2 = 25$

106. Let $x = v_0 t \cos\alpha$ and $y = v_0 t \sin\alpha - \frac{1}{2}gt^2$.

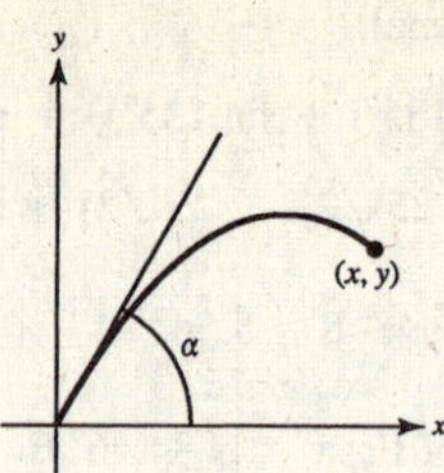

$$t = \frac{x}{v_0\cos\alpha} \Rightarrow y = v_0 \sin\alpha\left(\frac{x}{v_0\cos\alpha}\right) - \frac{1}{2}g\left(\frac{x}{v_0\cos\alpha}\right)^2$$

$$= x\tan\alpha - \frac{g}{2v_0^2}x^2\sec^2\alpha$$

$$= x\tan\alpha - \frac{gx^2}{2v_0^2}(1+\tan^2\alpha)$$

$$= \frac{v_0^2}{2g} - \frac{gx^2}{2v_0^2} - \frac{gx^2}{2v_0^2}\tan^2\alpha + x\tan\alpha - \frac{v_0^2}{2g}$$

$$= \frac{v_0^2}{2g} - \frac{gx^2}{2v_0^2} - \frac{gx^2}{2v_0^2}\left[\tan^2\alpha - 2\tan\alpha\left(\frac{v_0^2}{gx}\right) + \frac{v_0^4}{g^2x^2}\right]$$

$$= \frac{v_0^2}{2g} - \frac{gx^2}{2v_0^2} - \frac{gx^2}{2v_0^2}\left(\tan\alpha - \frac{v_0^2}{gx}\right)^2$$

If $y \le \frac{v_0^2}{2g} - \frac{gx^2}{2v_0^2}$, then α can be chosen to hit the point (x, y). To hit $(0, y)$: Let $\alpha = 90°$. Then

$$y = v_0 t - \frac{1}{2}gt^2 = \frac{v_0^2}{2g} - \frac{v_0^2}{2g}\left(\frac{g}{v_0}t - 1\right)^2, \text{ and you need } y \le \frac{v_0^2}{2g}.$$

The set H is given by $0 \le x, 0 < y$ and $y \le \frac{v_0^2}{2g} - \frac{gx^2}{2v_0^2}$

Note: The parabola $y = \frac{v_0^2}{2g} - \frac{gx^2}{2v_0^2}$ is called the "parabola of safety".

Section 11.2 Space Coordinates and Vectors in Space

1.

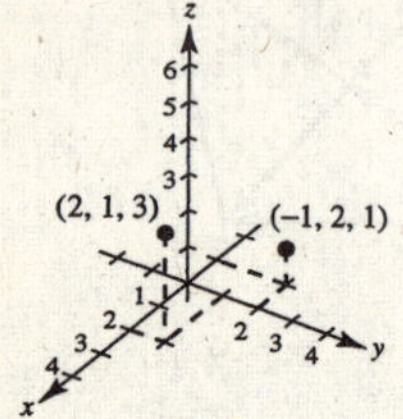

2.

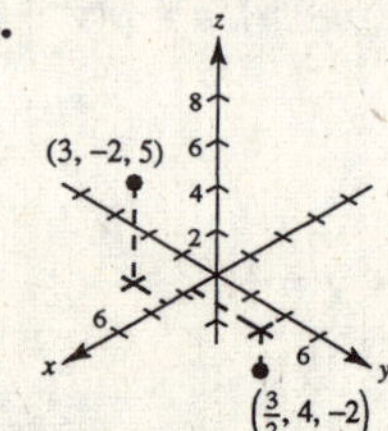

3.

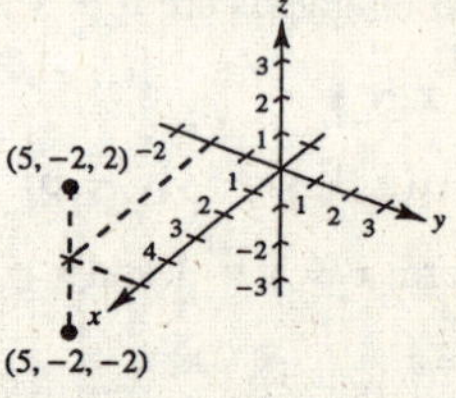

4.

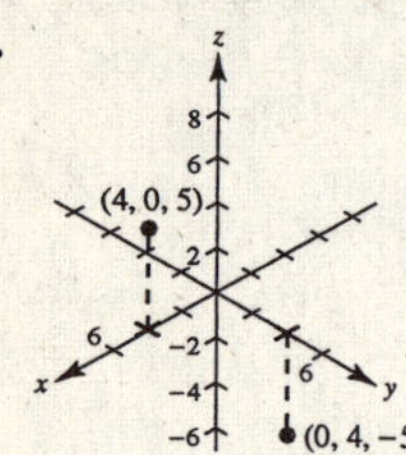

5. $A(2, 3, 4)$
$B(-1, -2, 2)$

6. $A(2, -3, -1)$
$B(-3, 1, 4)$

7. $x = -3, y = 4, z = 5$: $(-3, 4, 5)$

8. $x = 7, y = -2, z = -1$: $(7, -2, -1)$

9. $y = z = 0, x = 10$: $(10, 0, 0)$

10. $x = 0, y = 3, z = 2$: $(0, 3, 2)$

11. The z-coordinate is 0.

12. The x-coordinate is 0.

13. The point is 6 units above the xy-plane.

14. The point is 2 units in front of the xz-plane.

15. The point is on the plane parallel to the yz-plane that passes through $x = 4$.

16. The point is on the plane $z = -3$.

17. The point is to the left of the xz-plane.

18. The point is behind the yz-plane.

19. The point is on or between the planes $y = 3$ and $y = -3$.

20. The point is in front of the plane $x = 4$.

21. The point (x, y, z) is 3 units below the xy-plane, and below either quadrant I or III.

22. The point (x, y, z) is 4 units above the xy-plane, and above either quadrant II or IV.

23. The point could be above the xy-plane and thus above quadrants II or IV, or below the xy-plane, and thus below quadrants I or III.

24. The point could be above the xy-plane, and thus above quadrants I or III, or below the xy-plane, and thus below quadrants II or IV.

25. $d = \sqrt{(5-0)^2 + (2-0)^2 + (6-0)^2}$

$= \sqrt{25 + 4 + 36} = \sqrt{65}$

26. $d = \sqrt{(2-(-2))^2 + (-5-3)^2 + (-2-2)^2}$

$= \sqrt{16 + 64 + 16} = \sqrt{96} = 4\sqrt{6}$

27. $d = \sqrt{(6-1)^2 + (-2-(-2))^2 + (-2-4)^2}$

$= \sqrt{25 + 0 + 36} = \sqrt{61}$

28. $d = \sqrt{(4-2)^2 + (-5-2)^2 + (6-3)^2}$

$= \sqrt{4 + 49 + 9} = \sqrt{62}$

29. $A(0, 0, 0), B(2, 2, 1), C(2, -4, 4)$

$|AB| = \sqrt{4 + 4 + 1} = 3$

$|AC| = \sqrt{4 + 16 + 16} = 6$

$|BC| = \sqrt{0 + 36 + 9} = 3\sqrt{5}$

$|BC|^2 = |AB|^2 + |AC|^2$

Right triangle

30. $A(5, 3, 4), B(7, 1, 3), C(3, 5, 3)$

$|AB| = \sqrt{4 + 4 + 1} = 3$

$|AC| = \sqrt{4 + 4 + 1} = 3$

$|BC| = \sqrt{16 + 16 + 0} = 4\sqrt{2}$

Since $|AB| = |AC|$, the triangle is isosceles.

31. $A(1, -3, -2), B(5, -1, 2), C(-1, 1, 2)$

$|AB| = \sqrt{16 + 4 + 16} = 6$

$|AC| = \sqrt{4 + 16 + 16} = 6$

$|BC| = \sqrt{36 + 4 + 0} = 2\sqrt{10}$

Since $|AB| = |AC|$, the triangle is isosceles.

32. $A(5, 0, 0), B(0, 2, 0), C(0, 0, -3)$

$|AB| = \sqrt{25 + 4 + 0} = \sqrt{29}$

$|AC| = \sqrt{25 + 0 + 9} = \sqrt{34}$

$|BC| = \sqrt{0 + 4 + 9} = \sqrt{13}$

Neither

33. The z-coordinate is changed by 5 units:

$(0, 0, 5), (2, 2, 6), (2, -4, 9)$

34. The y-coordinate is changed by 3 units:

$(5, 6, 4), (7, 4, 3), (3, 8, 3)$

35. $\left(\frac{5 + (-2)}{2}, \frac{-9 + 3}{2}, \frac{7 + 3}{2}\right) = \left(\frac{3}{2}, -3, 5\right)$

36. $\left(\frac{4 + 8}{2}, \frac{0 + 8}{2}, \frac{-6 + 20}{2}\right) = (6, 4, 7)$

37. Center: $(0, 2, 5)$

Radius: 2

$(x-0)^2 + (y-2)^2 + (z-5)^2 = 4$

$x^2 + y^2 + z^2 - 4y - 10z + 25 = 0$

38. Center: $(4, -1, 1)$

Radius: 5

$(x-4)^2 + (y+1)^2 + (z-1)^2 = 25$

$x^2 + y^2 + z^2 - 8x + 2y - 2z - 7 = 0$

39. Center: $\dfrac{(2, 0, 0) + (0, 6, 0)}{2} = (1, 3, 0)$

Radius: $\sqrt{10}$

$$(x - 1)^2 + (y - 3)^2 + (z - 0)^2 = 10$$

$$x^2 + y^2 + z^2 - 2x - 6y = 0$$

40. Center: $(-3, 2, 4)$

$r = 3$

(tangent to yz-plane)

$$(x + 3)^2 + (y - 2)^2 + (z - 4)^2 = 9$$

41.
$$x^2 + y^2 + z^2 - 2x + 6y + 8z + 1 = 0$$
$$(x^2 - 2x + 1) + (y^2 + 6y + 9) + (z^2 + 8z + 16) = -1 + 1 + 9 + 16$$
$$(x - 1)^2 + (y + 3)^2 + (z + 4)^2 = 25$$

Center: $(1, -3, -4)$

Radius: 5

42.
$$x^2 + y^2 + z^2 + 9x - 2y + 10z + 19 = 0$$
$$\left(x^2 + 9x + \frac{81}{4}\right) + (y^2 - 2y + 1) + (z^2 + 10z + 25) = -19 + \frac{81}{4} + 1 + 25$$
$$\left(x + \frac{9}{2}\right)^2 + (y - 1)^2 + (z + 5)^2 = \frac{109}{4}$$

Center: $\left(-\dfrac{9}{2}, 1, -5\right)$

Radius: $\dfrac{\sqrt{109}}{2}$

43.
$$9x^2 + 9y^2 + 9z^2 - 6x + 18y + 1 = 0$$
$$x^2 + y^2 + z^2 - \frac{2}{3}x + 2y + \frac{1}{9} = 0$$
$$\left(x^2 - \frac{2}{3}x + \frac{1}{9}\right) + (y^2 + 2y + 1) + z^2 = -\frac{1}{9} + \frac{1}{9} + 1$$
$$\left(x - \frac{1}{3}\right)^2 + (y + 1)^2 + (z - 0)^2 = 1$$

Center: $\left(\dfrac{1}{3}, -1, 0\right)$

Radius: 1

44.
$$4x^2 + 4y^2 + 4z^2 - 4x - 32y + 8z + 33 = 0$$
$$x^2 + y^2 + z^2 - x - 8y + 2z + \frac{33}{4} = 0$$
$$\left(x^2 - x + \frac{1}{4}\right) + (y^2 - 8y + 16) + (z^2 + 2z + 1) = -\frac{33}{4} + \frac{1}{4} + 16 + 1$$
$$\left(x - \frac{1}{2}\right)^2 + (y - 4)^2 + (z + 1)^2 = 9$$

Center: $\left(\dfrac{1}{2}, 4, -1\right)$

Radius: 3

45. $x^2 + y^2 + z^2 \le 36$

Solid ball of radius 6 centered at origin.

46. $x^2 + y^2 + z^2 > 4$

Set of all points in space outside the ball of radius 2 centered at the origin.

47.
$$x^2 + y^2 + z^2 < 4x - 6y + 8z - 13$$
$$(x^2 - 4x + 4) + (y^2 + 6y + 9) + (z^2 - 8z + 16) < 4 + 9 + 16 - 13$$
$$(x - 2)^2 + (y + 3)^2 + (z - 4)^2 < 16$$

Interior of sphere of radius 4 centered at $(2, -3, 4)$.

48.
$$x^2 + y^2 + z^2 > -4x + 6y - 8z - 13$$
$$(x^2 + 4x + 4) + (y^2 - 6y + 9) + (z^2 + 8z + 16) > -13 + 4 + 9 + 16$$
$$(x + 2)^2 + (y - 3)^2 + (z + 4)^2 > 16$$

Set of all points in space outside the ball of radius 4 centered at $(-2, 3, -4)$.

49. (a) $\mathbf{v} = (2 - 4)\mathbf{i} + (4 - 2)\mathbf{j} + (3 - 1)\mathbf{k}$

$= -2\mathbf{i} + 2\mathbf{j} + 2\mathbf{k} = \langle -2, 2, 2 \rangle$

(b)

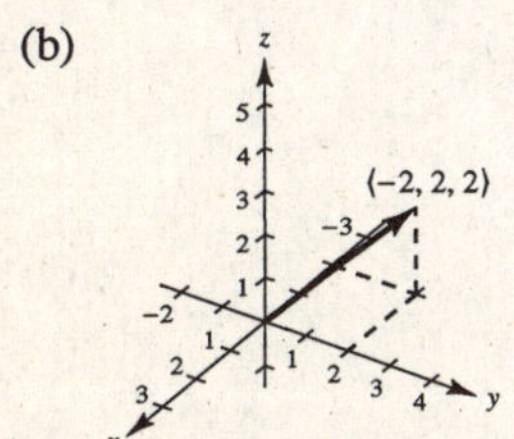

50. (a) $\mathbf{v} = (4 - 0)\mathbf{i} + (0 - 5)\mathbf{j} + (3 - 1)\mathbf{k}$

$= 4\mathbf{i} - 5\mathbf{j} + 2\mathbf{k} = \langle 4, -5, 2 \rangle$

(b)

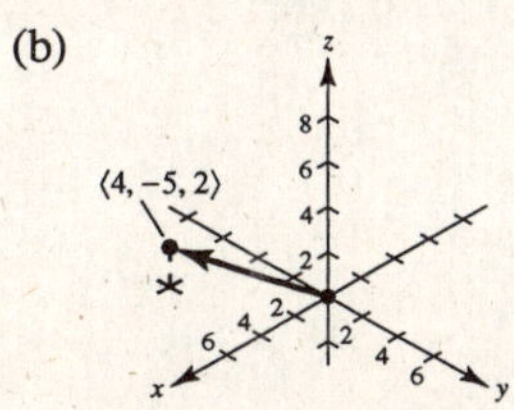

51. (a) $\mathbf{v} = (0 - 3)\mathbf{i} + (3 - 3)\mathbf{j} + (3 - 0)\mathbf{k}$

$= -3\mathbf{i} + 3\mathbf{k} = \langle -3, 0, 3 \rangle$

(b)

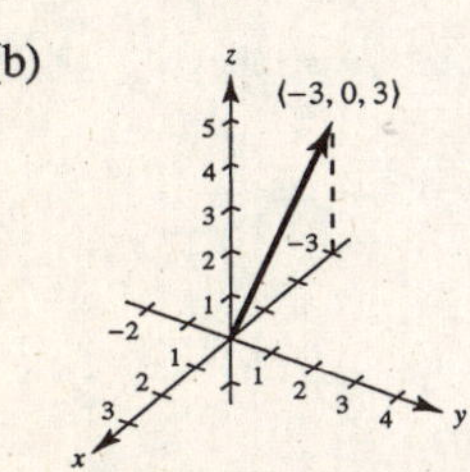

52. (a) $\mathbf{v} = (2 - 2)\mathbf{i} + (3 - 3)\mathbf{j} + (4 - 0)\mathbf{k}$

$= 4\mathbf{k} = \langle 0, 0, 4 \rangle$

(b)

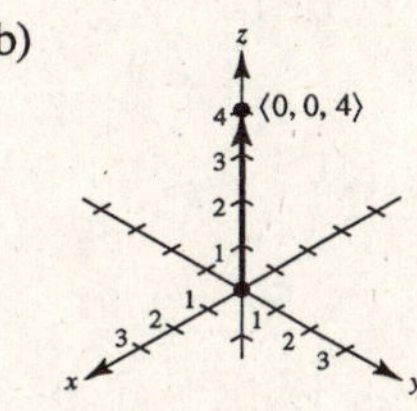

53. $\langle 4 - 3, 1 - 2, 6 - 0 \rangle = \langle 1, -1, 6 \rangle$

$\|\langle 1, -1, 6 \rangle\| = \sqrt{1 + 1 + 36} = \sqrt{38}$

Unit vector: $\dfrac{\langle 1, -1, 6 \rangle}{\sqrt{38}} = \left\langle \dfrac{1}{\sqrt{38}}, \dfrac{-1}{\sqrt{38}}, \dfrac{6}{\sqrt{38}} \right\rangle$

54. $\langle -1 - 4, 7 - (-5), -3 - 2 \rangle = \langle -5, 12, -5 \rangle$

$\|\langle -5, 12, -5 \rangle\| = \sqrt{25 + 144 + 25} = \sqrt{194}$

Unit vector: $\dfrac{\langle -5, 12, -5 \rangle}{\sqrt{194}} = \left\langle \dfrac{-5}{\sqrt{194}}, \dfrac{12}{\sqrt{194}}, \dfrac{-5}{\sqrt{194}} \right\rangle$

55. $\langle -5 - (-4), 3 - 3, 0 - 1 \rangle = \langle -1, 0, -1 \rangle$

$\|\langle -1, 0, -1 \rangle\| = \sqrt{1 + 1} = \sqrt{2}$

Unit vector: $\left\langle \dfrac{-1}{\sqrt{2}}, 0, \dfrac{-1}{\sqrt{2}} \right\rangle$

56. $\langle 2 - 1, 4 - (-2), -2 - 4 \rangle = \langle 1, 6, -6 \rangle$

$\|\langle 1, 6, -6 \rangle\| = \sqrt{1 + 36 + 36} = \sqrt{73}$

Unit vector: $\left\langle \dfrac{1}{\sqrt{73}}, \dfrac{6}{\sqrt{73}}, \dfrac{-6}{\sqrt{73}} \right\rangle$

57. (b) $\mathbf{v} = (3+1)\mathbf{i} + (3-2)\mathbf{j} + (4-3)\mathbf{k}$

$= 4\mathbf{i} + \mathbf{j} + \mathbf{k} = \langle 4, 1, 1\rangle$

(a) and (c).

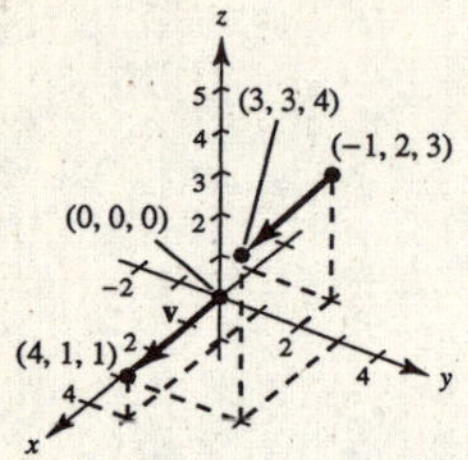

58. (b) $\mathbf{v} = (-4-2)\mathbf{i} + (3+1)\mathbf{j} + (7+2)\mathbf{k}$

$= -6\mathbf{i} + 4\mathbf{j} + 9\mathbf{k} = \langle -6, 4, 9\rangle$

(a) and (c).

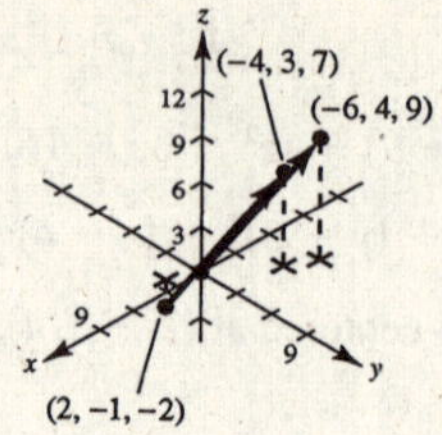

59. $(q_1, q_2, q_3) - (0, 6, 2) = (3, -5, 6)$

$Q = (3, 1, 8)$

60. $(q_1, q_2, q_3) - \left(0, 2, \frac{5}{2}\right) = \left(1, -\frac{2}{3}, \frac{1}{2}\right)$

$Q = \left(1, -\frac{4}{3}, 3\right)$

61. (a) $2\mathbf{v} = \langle 2, 4, 4\rangle$

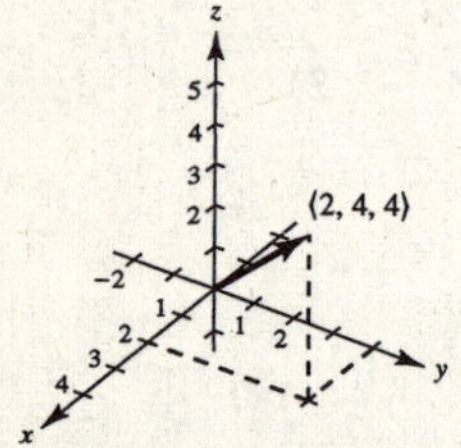

(b) $-\mathbf{v} = \langle -1, -2, -2\rangle$

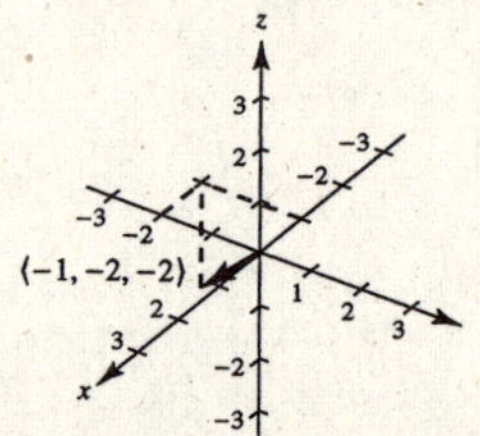

(c) $\frac{3}{2}\mathbf{v} = \left\langle \frac{3}{2}, 3, 3\right\rangle$

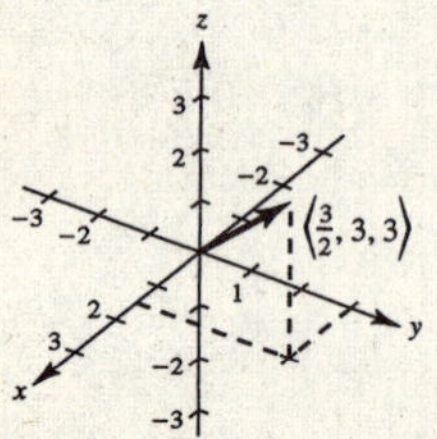

(d) $0\mathbf{v} = \langle 0, 0, 0\rangle$

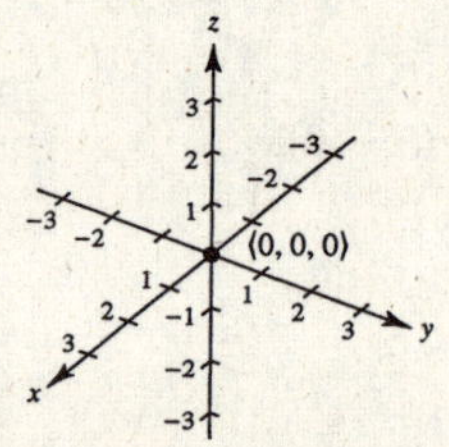

62. (a) $-\mathbf{v} = \langle -2, 2, -1\rangle$

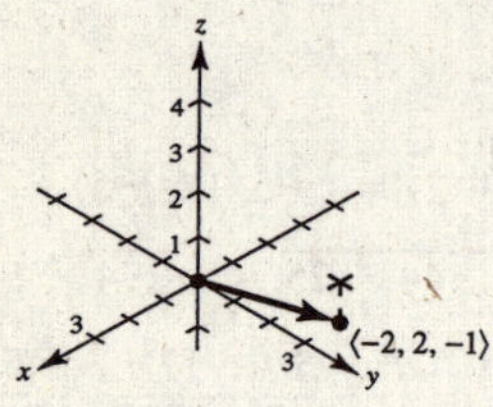

(b) $2\mathbf{v} = \langle 4, -4, 2\rangle$

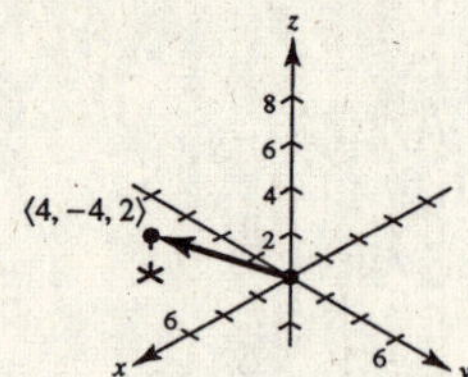

(c) $\frac{1}{2}\mathbf{v} = \left\langle 1, -1, \frac{1}{2}\right\rangle$

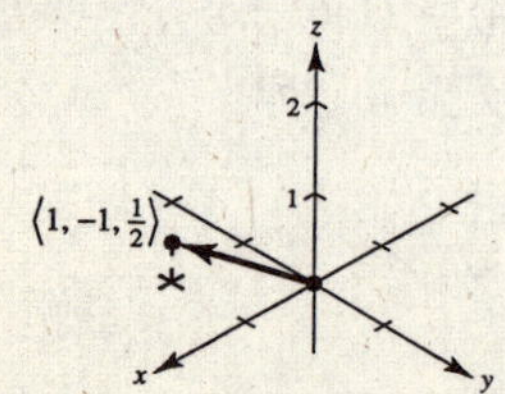

(d) $\frac{5}{2}\mathbf{v} = \left\langle 5, -5, \frac{5}{2}\right\rangle$

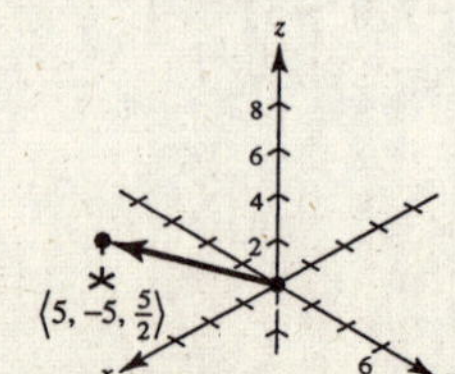

63. $\mathbf{z} = \mathbf{u} - \mathbf{v} = \langle 1, 2, 3\rangle - \langle 2, 2, -1\rangle = \langle -1, 0, 4\rangle$

64. $\mathbf{z} = \mathbf{u} - \mathbf{v} + 2\mathbf{w} = \langle 1, 2, 3\rangle - \langle 2, 2, -1\rangle + \langle 8, 0, -8\rangle = \langle 7, 0, -4\rangle$

65. $\mathbf{z} = 2\mathbf{u} + 4\mathbf{v} - \mathbf{w} = \langle 2, 4, 6\rangle + \langle 8, 8, -4\rangle - \langle 4, 0, -4\rangle = \langle 6, 12, 6\rangle$

66. $\mathbf{z} = 5\mathbf{u} - 3\mathbf{v} - \frac{1}{2}\mathbf{w} = \langle 5, 10, 15\rangle - \langle 6, 6, -3\rangle - \langle 2, 0, -2\rangle = \langle -3, 4, 20\rangle$

67. $2\mathbf{z} - 3\mathbf{u} = 2\langle z_1, z_2, z_3\rangle - 3\langle 1, 2, 3\rangle = \langle 4, 0, -4\rangle$

$2z_1 - 3 = 4 \Rightarrow z_1 = \frac{7}{2}$

$2z_2 - 6 = 0 \Rightarrow z_2 = 3$

$2z_3 - 9 = -4 \Rightarrow z_3 = \frac{5}{2}$

$\mathbf{z} = \left\langle \frac{7}{2}, 3, \frac{5}{2}\right\rangle$

68. $2\mathbf{u} + \mathbf{v} - \mathbf{w} + 3\mathbf{z} = 2\langle 1, 2, 3\rangle + \langle 2, 2, -1\rangle - \langle 4, 0, -4\rangle + 3\langle z_1, z_2, z_3\rangle = \langle 0, 0, 0\rangle$

$\langle 0, 6, 9\rangle + \langle 3z_1, 3z_2, 3z_3\rangle = \langle 0, 0, 0\rangle$

$0 + 3z_1 = 0 \Rightarrow z_1 = 0$

$6 + 3z_2 = 0 \Rightarrow z_2 = -2$

$9 + 3z_3 = 0 \Rightarrow z_3 = -3$

$\mathbf{z} = \langle 0, -2, -3\rangle$

69. (a) and (b) are parallel since $\langle -6, -4, 10\rangle = -2\langle 3, 2, -5\rangle$ and $\left\langle 2, \frac{4}{3}, -\frac{10}{3}\right\rangle = \frac{2}{3}\langle 3, 2, -5\rangle$.

70. (b) and (d) are parallel since $-\mathbf{i} + \frac{4}{3}\mathbf{j} - \frac{3}{2}\mathbf{k} = -2\left(\frac{1}{2}\mathbf{i} - \frac{2}{3}\mathbf{j} + \frac{3}{4}\mathbf{k}\right)$ and $\frac{3}{4}\mathbf{i} - \mathbf{j} + \frac{9}{8}\mathbf{k} = \frac{3}{2}\left(\frac{1}{2}\mathbf{i} - \frac{2}{3}\mathbf{j} + \frac{3}{4}\mathbf{k}\right)$.

71. $\mathbf{z} = -3\mathbf{i} + 4\mathbf{j} + 2\mathbf{k}$

(a) is parallel since $-6\mathbf{i} + 8\mathbf{j} + 4\mathbf{k} = 2\mathbf{z}$.

72. $\mathbf{z} = \langle -7, -8, 3\rangle$

(b) is parallel since $(-z)\mathbf{z} = \langle 14, 16, -6\rangle$.

73. $P(0, -2, -5), Q(3, 4, 4), R(2, 2, 1)$

$\overrightarrow{PQ} = \langle 3, 6, 9\rangle$

$\overrightarrow{PR} = \langle 2, 4, 6\rangle$

$\langle 3, 6, 9\rangle = \frac{3}{2}\langle 2, 4, 6\rangle$

Therefore, $\overrightarrow{PQ}$ and $\overrightarrow{PR}$ are parallel. The points are collinear.

74. $P(4, -2, 7), Q(-2, 0, 3), R(7, -3, 9)$

$\overrightarrow{PQ} = \langle -6, 2, -4\rangle$

$\overrightarrow{PR} = \langle 3, -1, 2\rangle$

$\langle 3, -1, 2\rangle = -\frac{1}{2}\langle -6, 2, -4\rangle$

Therefore, $\overrightarrow{PQ}$ and $\overrightarrow{PR}$ are parallel. The points are collinear.

75. $P(1, 2, 4), Q(2, 5, 0), R(0, 1, 5)$

$\overrightarrow{PQ} = \langle 1, 3, -4\rangle$

$\overrightarrow{PR} = \langle -1, -1, 1\rangle$

Since $\overrightarrow{PQ}$ and $\overrightarrow{PR}$ are not parallel, the points are not collinear.

76. $P(0, 0, 0), Q(1, 3, -2), R(2, -6, 4)$

$\overrightarrow{PQ} = \langle 1, 3, -2\rangle$

$\overrightarrow{PR} = \langle 2, -6, 4\rangle$

Since $\overrightarrow{PQ}$ and $\overrightarrow{PR}$ are not parallel, the points are not collinear.

77. $A(2, 9, 1), B(3, 11, 4), C(0, 10, 2), D(1, 12, 5)$

$\overrightarrow{AB} = \langle 1, 2, 3\rangle$

$\overrightarrow{CD} = \langle 1, 2, 3\rangle$

$\overrightarrow{AC} = \langle -2, 1, 1\rangle$

$\overrightarrow{BD} = \langle -2, 1, 1\rangle$

Since $\overrightarrow{AB} = \overrightarrow{CD}$ and $\overrightarrow{AC} = \overrightarrow{BD}$, the given points form the vertices of a parallelogram.

78. $A(1, 1, 3), B(9, -1, -2), C(11, 2, -9), D(3, 4, -4)$

$\overrightarrow{AB} = \langle 8, -2, -5\rangle$

$\overrightarrow{DC} = \langle 8, -2, -5\rangle$

$\overrightarrow{AD} = \langle 2, 3, -7\rangle$

$\overrightarrow{BC} = \langle 2, 3, -7\rangle$

Since $\overrightarrow{AB} = \overrightarrow{DC}$ and $\overrightarrow{AD} = \overrightarrow{BC}$, the given points form the vertices of a parallelogram.

79. $\|\mathbf{v}\| = 0$

80. $\|\mathbf{v}\| = \sqrt{1 + 0 + 9} = \sqrt{10}$

81. $\mathbf{v} = \langle 1, -2, -3\rangle$

$\|\mathbf{v}\| = \sqrt{1 + 4 + 9} = \sqrt{14}$

82. $\mathbf{v} = \langle -4, 3, 7\rangle$

$\|\mathbf{v}\| = \sqrt{16 + 9 + 49} = \sqrt{74}$

83. $\mathbf{v} = \langle 0, 3, -5\rangle$

$\|\mathbf{v}\| = \sqrt{0 + 9 + 25} = \sqrt{34}$

84. $\mathbf{v} = \langle 1, 3, -2\rangle$

$\|\mathbf{v}\| = \sqrt{1 + 9 + 4} = \sqrt{14}$

85. $\mathbf{u} = \langle 2, -1, 2\rangle$

$\|\mathbf{u}\| = \sqrt{4 + 1 + 4} = 3$

(a) $\dfrac{\mathbf{u}}{\|\mathbf{u}\|} = \dfrac{1}{3}\langle 2, -1, 2\rangle$

(b) $-\dfrac{\mathbf{u}}{\|\mathbf{u}\|} = -\dfrac{1}{3}\langle 2, -1, 2\rangle$

86. $\mathbf{u} = \langle 6, 0, 8\rangle$

$\|\mathbf{u}\| = \sqrt{36 + 0 + 64} = 10$

(a) $\dfrac{\mathbf{u}}{\|\mathbf{u}\|} = \dfrac{1}{10}\langle 6, 0, 8\rangle$

(b) $-\dfrac{\mathbf{u}}{\|\mathbf{u}\|} = -\dfrac{1}{10}\langle 6, 0, 8\rangle$

87. $\mathbf{u} = \langle 3, 2, -5\rangle$

$\|\mathbf{u}\| = \sqrt{9 + 4 + 25} = \sqrt{38}$

(a) $\dfrac{\mathbf{u}}{\|\mathbf{u}\|} = \dfrac{1}{\sqrt{38}}\langle 3, 2, -5\rangle$

(b) $-\dfrac{\mathbf{u}}{\|\mathbf{u}\|} = -\dfrac{1}{\sqrt{38}}\langle 3, 2, -5\rangle$

88. $\mathbf{u} = \langle 8, 0, 0\rangle$

$\|\mathbf{u}\| = 8$

(a) $\dfrac{\mathbf{u}}{\|\mathbf{u}\|} = \langle 1, 0, 0\rangle$

(b) $-\dfrac{\mathbf{u}}{\|\mathbf{u}\|} = \langle -1, 0, 0\rangle$

89. Programs will vary.

90. (a) $\mathbf{u} + \mathbf{v} = \langle 4, 7.5, -2\rangle$

(b) $\|\mathbf{u} + \mathbf{v}\| \approx 8.732$

(c) $\|\mathbf{u}\| \approx 5.099$

(d) $\|\mathbf{v}\| \approx 9.014$

91. $c\mathbf{v} = \langle 2c, 2c, -c\rangle$

$\|c\mathbf{v}\| = \sqrt{4c^2 + 4c^2 + c^2} = 5$

$9c^2 = 25$

$c = \pm\dfrac{5}{3}$

92. $c\mathbf{u} = \langle c, 2c, 3c\rangle$

$\|c\mathbf{u}\| = \sqrt{c^2 + 4c^2 + 9c^2} = 3$

$14c^2 = 9$

$c = \pm\dfrac{3\sqrt{14}}{14}$

93. $\mathbf{v} = 10\dfrac{\mathbf{u}}{\|\mathbf{u}\|} = 10\left\langle 0, \dfrac{1}{\sqrt{2}}, \dfrac{1}{\sqrt{2}}\right\rangle$

$= \left\langle 0, \dfrac{10}{\sqrt{2}}, \dfrac{10}{\sqrt{2}}\right\rangle$

94. $\mathbf{v} = 3\dfrac{\mathbf{u}}{\|\mathbf{u}\|} = 3\left\langle \dfrac{1}{\sqrt{3}}, \dfrac{1}{\sqrt{3}}, \dfrac{1}{\sqrt{3}}\right\rangle = \left\langle \dfrac{3}{\sqrt{3}}, \dfrac{3}{\sqrt{3}}, \dfrac{3}{\sqrt{3}}\right\rangle$

95. $\mathbf{v} = \dfrac{3}{2}\dfrac{\mathbf{u}}{\|\mathbf{u}\|} = \dfrac{3}{2}\left\langle \dfrac{2}{3}, \dfrac{-2}{3}, \dfrac{1}{3}\right\rangle = \left\langle 1, -1, \dfrac{1}{2}\right\rangle$

96. $\mathbf{v} = \sqrt{5}\dfrac{\mathbf{u}}{\|\mathbf{u}\|} = \sqrt{5}\left\langle \dfrac{-2}{\sqrt{14}}, \dfrac{3}{\sqrt{14}}, \dfrac{1}{\sqrt{14}}\right\rangle$

$= \left\langle \dfrac{-\sqrt{70}}{7}, \dfrac{3\sqrt{70}}{14}, \dfrac{\sqrt{70}}{14}\right\rangle$

97. $\mathbf{v} = 2[\cos(\pm 30°)\mathbf{j} + \sin(\pm 30°)\mathbf{k}]$

$= \sqrt{3}\mathbf{j} \pm \mathbf{k} = \langle 0, \sqrt{3}, \pm 1\rangle$

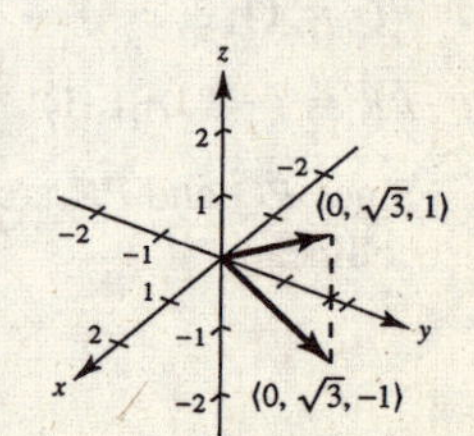

98. $\mathbf{v} = 5(\cos 45°\mathbf{i} + \sin 45°\mathbf{k}) = \dfrac{5\sqrt{2}}{2}(\mathbf{i} + \mathbf{k})$ or

$\mathbf{v} = 5(\cos 135°\mathbf{i} + \sin 135°\mathbf{k}) = \dfrac{5\sqrt{2}}{2}(-\mathbf{i} + \mathbf{k})$

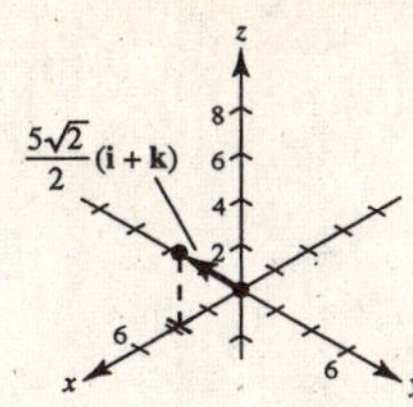

99.

$$\mathbf{v} = \langle -3, -6, 3\rangle$$
$$\tfrac{2}{3}\mathbf{v} = \langle -2, -4, 2\rangle$$
$$(4, 3, 0) + (-2, -4, 2) = (2, -1, 2)$$

100.

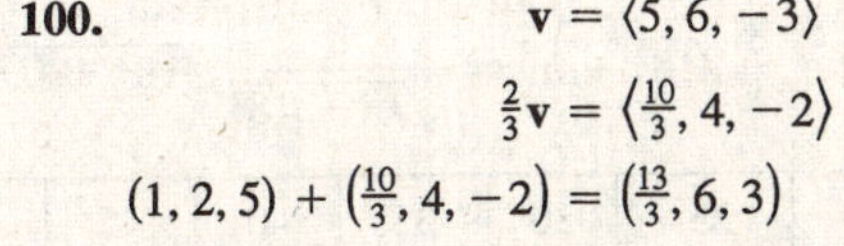

$$\mathbf{v} = \langle 5, 6, -3\rangle$$
$$\tfrac{2}{3}\mathbf{v} = \left\langle \tfrac{10}{3}, 4, -2\right\rangle$$
$$(1, 2, 5) + \left(\tfrac{10}{3}, 4, -2\right) = \left(\tfrac{13}{3}, 6, 3\right)$$

101. (a)

(b) $\mathbf{w} = a\mathbf{u} + b\mathbf{v} = a\mathbf{i} + (a + b)\mathbf{j} + b\mathbf{k} = \mathbf{0}$

$a = 0, a + b = 0, b = 0$

Thus, a and b are both zero.

(c) $a\mathbf{i} + (a + b)\mathbf{j} + b\mathbf{k} = \mathbf{i} + 2\mathbf{j} + \mathbf{k}$

$a = 1, b = 1$

$\mathbf{w} = \mathbf{u} + \mathbf{v}$

(d) $a\mathbf{i} + (a + b)\mathbf{j} + b\mathbf{k} = \mathbf{i} + 2\mathbf{j} + 3\mathbf{k}$

$a = 1, a + b = 2, b = 3$

Not possible

102. A sphere of radius 4 centered at (x_1, y_1, z_1).

$\|\mathbf{v}\| = \|\langle x - x_2, y - y_1, z - z_1\rangle\|$

$= \sqrt{(x - x_1)^2 + (y - y_1)^2 + (z - z_1)^2} = 4$

$(x - x_1)^2 + (y - y_1)^2 + (z - z_1)^2 = 16$ sphere

103. x_0 is directed distance to yz-plane.

y_0 is directed distance to xz-plane.

z_0 is directed distance to xy-plane.

104. $d = \sqrt{(x_2 - x_1)^2 + (y_2 - y_1)^2 + (z_2 - z_1)^2}$

105. $(x - x_0)^2 + (y - y_0)^2 + (z - z_0)^2 = r^2$

106. Two nonzero vectors $\mathbf{u}$ and $\mathbf{v}$ are parallel if $\mathbf{u} = c\mathbf{v}$ for some scalar c.

107.

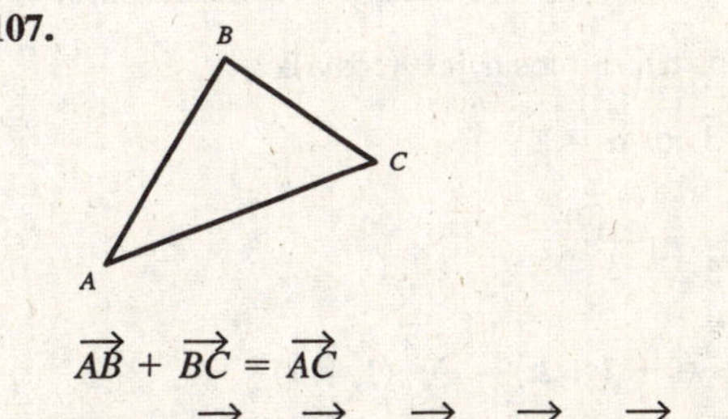

$\overrightarrow{AB} + \overrightarrow{BC} = \overrightarrow{AC}$

Hence, $\overrightarrow{AB} + \overrightarrow{BC} + \overrightarrow{CA} = \overrightarrow{AC} + \overrightarrow{CA} = \mathbf{0}$

108. $\|\mathbf{r} - \mathbf{r}_0\| = \sqrt{(x - 1)^2 + (y - 1)^2 + (z - 1)^2} = 2$

$$(x - 1)^2 + (y - 1)^2 + (z - 1)^2 = 4$$

This is a sphere of radius 2 and center $(1, 1, 1)$.

109. (a) The height of the right triangle is $h = \sqrt{L^2 - 18^2}$.
The vector $\overrightarrow{PQ}$ is given by

$$\overrightarrow{PQ} = \langle 0, -18, h \rangle.$$

The tension vector $\mathbf{T}$ in each wire is

$$\mathbf{T} = c\langle 0, -18, h \rangle \text{ where } ch = \frac{24}{3} = 8.$$

Hence, $\mathbf{T} = \dfrac{8}{h}\langle 0, -18, h \rangle$ and

$$T = \|\mathbf{T}\| = \frac{8}{h}\sqrt{18^2 + h^2} = \frac{8}{\sqrt{L^2 - 18^2}}\sqrt{18^2 + (L^2 - 18^2)} = \frac{8L}{\sqrt{L^2 - 18^2}}.$$

(b)

L	20	25	30	35	40	45	50
T	18.4	11.5	10	9.3	9.0	8.7	8.6

(c)

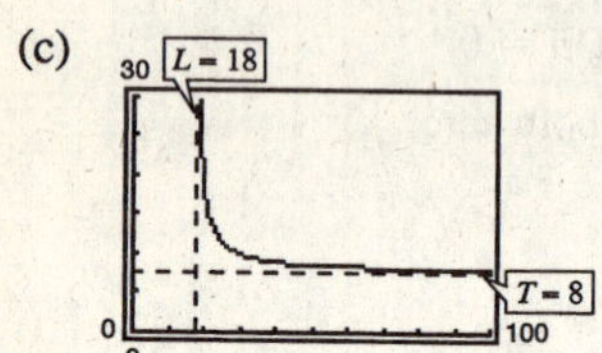

$x = 18$ is a vertical asymptote and $y = 8$ is a horizontal asymptote.

(d) $$\lim_{L \to 18^+} \frac{8L}{\sqrt{L^2 - 18^2}} = \infty$$

$$\lim_{L \to \infty} \frac{8L}{\sqrt{L^2 - 18^2}} = \lim_{L \to \infty} \frac{8}{\sqrt{1 - (18/L)^2}} = 8$$

(e) From the table, $T = 10$ implies $L = 30$ inches.

110. As in Exercise 109(c), $x = a$ will be a vertical asymptote. Hence, $\displaystyle\lim_{r_0 \to a^-} T = \infty$.

111. Let α be the angle between $\mathbf{v}$ and the coordinate axes.

$$\mathbf{v} = (\cos\alpha)\mathbf{i} + (\cos\alpha)\mathbf{j} + (\cos\alpha)\mathbf{k}$$

$$\|\mathbf{v}\| = \sqrt{3}\cos\alpha = 1$$

$$\cos\alpha = \frac{1}{\sqrt{3}} = \frac{\sqrt{3}}{3}$$

$$\mathbf{v} = \frac{\sqrt{3}}{3}(\mathbf{i} + \mathbf{j} + \mathbf{k}) = \frac{\sqrt{3}}{3}\langle 1, 1, 1 \rangle$$

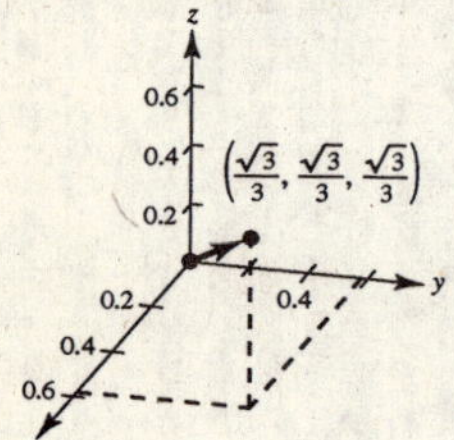

112.

$$550 = \|c(75\mathbf{i} - 50\mathbf{j} - 100\mathbf{k})\|$$

$$302{,}500 = 18{,}125c^2$$

$$c^2 = 16.689655$$

$$c \approx 4.085$$

$$\mathbf{F} \approx 4.085(75\mathbf{i} - 50\mathbf{j} - 100\mathbf{k})$$

$$\approx 306\mathbf{i} - 204\mathbf{j} - 409\mathbf{k}$$

113. $\overrightarrow{AB} = \langle 0, 70, 115 \rangle, \mathbf{F}_1 = C_1\langle 0, 70, 115 \rangle$

$\overrightarrow{AC} = \langle -60, 0, 115 \rangle, \mathbf{F}_2 = C_2\langle -60, 0, 115 \rangle$

$\overrightarrow{AD} = \langle 45, -65, 115 \rangle, \mathbf{F}_3 = C_3\langle 45, -65, 115 \rangle$

$\mathbf{F} = \mathbf{F}_1 + \mathbf{F}_2 + \mathbf{F}_3 = \langle 0, 0, 500 \rangle$

Thus:

$$\begin{aligned} -60C_2 + 45C_3 &= 0 \\ 70C_1 \qquad - 65C_3 &= 0 \\ 115(C_1 + C_2 + C_3) &= 500 \end{aligned}$$

Solving this system yields $C_1 = \dfrac{104}{69}$, $C_2 = \dfrac{28}{23}$, and $C_3 = \dfrac{112}{69}$. Thus:

$\|\mathbf{F}_1\| \approx 202.919N$

$\|\mathbf{F}_2\| \approx 157.909N$

$\|\mathbf{F}_3\| \approx 226.521N$

114. Let A lie on the y-axis and the wall on the x-axis. Then $A = (0, 10, 0)$, $B = (8, 0, 6)$, $C = (-10, 0, 6)$ and

$\overrightarrow{AB} = \langle 8, -10, 6 \rangle, \overrightarrow{AC} = \langle -10, -10, 6 \rangle.$

$\|AB\| = 10\sqrt{2}, \|AC\| = 2\sqrt{59}$

Thus, $\mathbf{F}_1 = 420\dfrac{\overrightarrow{AB}}{\|\overrightarrow{AB}\|}, \mathbf{F}_2 = 650\dfrac{\overrightarrow{AC}}{\|\overrightarrow{AC}\|}$

$$\begin{aligned} \mathbf{F} = \mathbf{F}_1 + \mathbf{F}_2 &\approx \langle 237.6, -297.0, 178.2 \rangle \\ &+ \langle -423.1, -423.1, 253.9 \rangle \\ &\approx \langle -185.5, -720.1, 432.1 \rangle \end{aligned}$$

$\|\mathbf{F}\| \approx 860.0$ lb

115. $d(AP) = 2d(BP)$

$$\sqrt{x^2 + (y+1)^2 + (z-1)^2} = 2\sqrt{(x-1)^2 + (y-2)^2 + z^2}$$

$$x^2 + y^2 + z^2 + 2y - 2z + 2 = 4(x^2 + y^2 + z^2 - 2x - 4y + 5)$$

$$0 = 3x^2 + 3y^2 + 3z^2 - 8x - 18y + 2z + 18$$

$$-6 + \frac{16}{9} + 9 + \frac{1}{9} = \left(x^2 - \frac{8}{3}x + \frac{16}{9}\right) + (y^2 - 6y + 9) + \left(z^2 + \frac{2}{3}z + \frac{1}{9}\right)$$

$$\frac{44}{9} = \left(x - \frac{4}{3}\right)^2 + (y-3)^2 + \left(z + \frac{1}{3}\right)^2$$

Sphere; center: $\left(\dfrac{4}{3}, 3, -\dfrac{1}{3}\right)$, radius: $\dfrac{2\sqrt{11}}{3}$

Section 11.3 The Dot Product of Two Vectors

1. $\mathbf{u} = \langle 3, 4 \rangle, \mathbf{v} = \langle 2, -3 \rangle$

(a) $\mathbf{u} \cdot \mathbf{v} = 3(2) + 4(-3) = -6$

(b) $\mathbf{u} \cdot \mathbf{u} = 3(3) + 4(4) = 25$

(c) $\|\mathbf{u}\|^2 = 25$

(d) $(\mathbf{u} \cdot \mathbf{v})\mathbf{v} = -6\langle 2, -3 \rangle = \langle -12, 18 \rangle$

(e) $\mathbf{u} \cdot (2\mathbf{v}) = 2(\mathbf{u} \cdot \mathbf{v}) = 2(-6) = -12$

2. $\mathbf{u} = \langle 4, 10 \rangle, \mathbf{v} = \langle -2, 3 \rangle$

(a) $\mathbf{u} \cdot \mathbf{v} = 4(-2) + 10(3) = 22$

(b) $\mathbf{u} \cdot \mathbf{u} = 4(4) + 10(10) = 116$

(c) $\|\mathbf{u}\|^2 = 116$

(d) $(\mathbf{u} \cdot \mathbf{v})\mathbf{v} = 22\langle -2, 3 \rangle = \langle -44, 66 \rangle$

(e) $\mathbf{u} \cdot (2\mathbf{v}) = 2(\mathbf{u} \cdot \mathbf{v}) = 2(22) = 44$

3. (a) $\mathbf{u} \cdot \mathbf{v} = \langle 5, -1 \rangle \cdot \langle -3, 2 \rangle = 5(-3) + (-1)(2) = -17$

(b) $\mathbf{u} \cdot \mathbf{u} = \langle 5, -1 \rangle \cdot \langle 5, -1 \rangle = 5(5) + (-1)(-1) = 26$

(c) $\|\mathbf{u}\|^2 = \mathbf{u} \cdot \mathbf{u} = 26$

(d) $(\mathbf{u} \cdot \mathbf{v})\mathbf{v} = (-17)\langle -3, 2 \rangle = \langle 51, -34 \rangle$

(e) $\mathbf{u} \cdot (2\mathbf{v}) = 2(\mathbf{u} \cdot \mathbf{v}) = 2(-17) = -34$

4. (a) $\mathbf{u} \cdot \mathbf{v} = \langle -4, 8 \rangle \cdot \langle 6, 3 \rangle = (-4)6 + 8(3) = 0$

(b) $\mathbf{u} \cdot \mathbf{u} = \langle -4, 8 \rangle \cdot \langle -4, 8 \rangle = (-4)(-4) + 8(8) = 80$

(c) $\|\mathbf{u}\|^2 = \mathbf{u} \cdot \mathbf{u} = 80$

(d) $(\mathbf{u} \cdot \mathbf{v})\mathbf{v} = 0\mathbf{v} = \mathbf{0}$

(e) $\mathbf{u} \cdot (2\mathbf{v}) = 2(\mathbf{u} \cdot \mathbf{v}) = 2(0) = 0$

5. $\mathbf{u} = \langle 2, -3, 4\rangle, \mathbf{v} = \langle 0, 6, 5\rangle$

(a) $\mathbf{u} \cdot \mathbf{v} = 2(0) + (-3)(6) + (4)(5) = 2$

(b) $\mathbf{u} \cdot \mathbf{u} = 2(2) + (-3)(-3) + 4(4) = 29$

(c) $\|\mathbf{u}\|^2 = 29$

(d) $(\mathbf{u} \cdot \mathbf{v})\mathbf{v} = 2\langle 0, 6, 5\rangle = \langle 0, 12, 10\rangle$

(e) $\mathbf{u} \cdot (2\mathbf{v}) = 2(\mathbf{u} \cdot \mathbf{v}) = 2(2) = 4$

6. $\mathbf{u} = \mathbf{i}, \mathbf{v} = \mathbf{i}$

(a) $\mathbf{u} \cdot \mathbf{v} = 1$

(b) $\mathbf{u} \cdot \mathbf{u} = 1$

(c) $\|\mathbf{u}\|^2 = 1$

(d) $(\mathbf{u} \cdot \mathbf{v})\mathbf{v} = \mathbf{i}$

(e) $\mathbf{u} \cdot (2\mathbf{v}) = 2(\mathbf{u} \cdot \mathbf{v}) = 2$

7. $\mathbf{u} = 2\mathbf{i} - \mathbf{j} + \mathbf{k}, \mathbf{v} = \mathbf{i} - \mathbf{k}$

(a) $\mathbf{u} \cdot \mathbf{v} = 2(1) + (-1)(0) + 1(-1) = 1$

(b) $\mathbf{u} \cdot \mathbf{u} = 2(2) + (-1)(-1) + (1)(1) = 6$

(c) $\|\mathbf{u}\|^2 = 6$

(d) $(\mathbf{u} \cdot \mathbf{v})\mathbf{v} = \mathbf{v} = \mathbf{i} - \mathbf{k}$

(e) $\mathbf{u} \cdot (2\mathbf{v}) = 2(\mathbf{u} \cdot \mathbf{v}) = 2$

8. $\mathbf{u} = 2\mathbf{i} + \mathbf{j} - 2\mathbf{k}, \mathbf{v} = \mathbf{i} - 3\mathbf{j} + 2\mathbf{k}$

(a) $\mathbf{u} \cdot \mathbf{v} = 2(1) + 1(-3) + (-2)(2) = -5$

(b) $\mathbf{u} \cdot \mathbf{u} = 2(2) + 1(1) + (-2)(-2) = 9$

(c) $\|\mathbf{u}\|^2 = 9$

(d) $(\mathbf{u} \cdot \mathbf{v})\mathbf{v} = -5(\mathbf{i} - 3\mathbf{j} + 2\mathbf{k}) = -5\mathbf{i} + 15\mathbf{j} - 10\mathbf{k}$

(e) $\mathbf{u} \cdot (2\mathbf{v}) = 2(\mathbf{u} \cdot \mathbf{v}) = 2(-5) = -10$

9. $\dfrac{\mathbf{u} \cdot \mathbf{v}}{\|\mathbf{u}\|\,\|\mathbf{v}\|} = \cos\theta$

$\mathbf{u} \cdot \mathbf{v} = (8)(5)\cos\dfrac{\pi}{3} = 20$

10. $\dfrac{\mathbf{u} \cdot \mathbf{v}}{\|\mathbf{u}\|\,\|\mathbf{v}\|} = \cos\theta$

$\mathbf{u} \cdot \mathbf{v} = (40)(25)\cos\dfrac{5\pi}{6} = -500\sqrt{3}$

11. $\mathbf{u} = \langle 1, 1\rangle, \mathbf{v} = \langle 2, -2\rangle$

$\cos\theta = \dfrac{\mathbf{u} \cdot \mathbf{v}}{\|\mathbf{u}\|\,\|\mathbf{v}\|} = \dfrac{0}{\sqrt{2}\sqrt{8}} = 0$

$\theta = \dfrac{\pi}{2}$

12. $\mathbf{u} = \langle 3, 1\rangle, \mathbf{v} = \langle 2, -1\rangle$

$\cos\theta = \dfrac{\mathbf{u} \cdot \mathbf{v}}{\|\mathbf{u}\|\,\|\mathbf{v}\|} = \dfrac{5}{\sqrt{10}\sqrt{5}} = \dfrac{1}{\sqrt{2}}$

$\theta = \dfrac{\pi}{4}$

13. $\mathbf{u} = 3\mathbf{i} + \mathbf{j}, \mathbf{v} = -2\mathbf{i} + 4\mathbf{j}$

$\cos\theta = \dfrac{\mathbf{u} \cdot \mathbf{v}}{\|\mathbf{u}\|\,\|\mathbf{v}\|} = \dfrac{-2}{\sqrt{10}\sqrt{20}} = \dfrac{-1}{5\sqrt{2}}$

$\theta = \arccos\left(-\dfrac{1}{5\sqrt{2}}\right) \approx 98.1°$

14. $\mathbf{u} = \cos\left(\dfrac{\pi}{6}\right)\mathbf{i} + \sin\left(\dfrac{\pi}{6}\right)\mathbf{j} = \dfrac{\sqrt{3}}{2}\mathbf{i} + \dfrac{1}{2}\mathbf{j}$

$\mathbf{v} = \cos\left(\dfrac{3\pi}{4}\right)\mathbf{i} + \sin\left(\dfrac{3\pi}{4}\right)\mathbf{j} = -\dfrac{\sqrt{2}}{2}\mathbf{i} + \dfrac{\sqrt{2}}{2}\mathbf{j}$

$$\cos\theta = \frac{\mathbf{u} \cdot \mathbf{v}}{\|\mathbf{u}\|\,\|\mathbf{v}\|}$$

$$= \frac{\sqrt{3}}{2}\left(-\frac{\sqrt{2}}{2}\right) + \frac{1}{2}\left(\frac{\sqrt{2}}{2}\right) = \frac{\sqrt{2}}{4}(1 - \sqrt{3})$$

$$\theta = \arccos\left[\frac{\sqrt{2}}{4}(1 - \sqrt{3})\right] = 105°$$

15. $\mathbf{u} = \langle 1, 1, 1\rangle, \mathbf{v} = \langle 2, 1, -1\rangle$

$\cos\theta = \dfrac{\mathbf{u} \cdot \mathbf{v}}{\|\mathbf{u}\|\,\|\mathbf{v}\|} = \dfrac{2}{\sqrt{3}\sqrt{6}} = \dfrac{\sqrt{2}}{3}$

$\theta = \arccos\dfrac{\sqrt{2}}{3} \approx 61.9°$

16. $\mathbf{u} = 3\mathbf{i} + 2\mathbf{j} + \mathbf{k}, \mathbf{v} = 2\mathbf{i} - 3\mathbf{j}$

$\cos\theta = \dfrac{\mathbf{u} \cdot \mathbf{v}}{\|\mathbf{u}\|\,\|\mathbf{v}\|} = \dfrac{3(2) + 2(-3) + 0}{\|\mathbf{u}\|\,\|\mathbf{v}\|} = 0$

$\theta = \dfrac{\pi}{2}$

17. $\mathbf{u} = 3\mathbf{i} + 4\mathbf{j}, \mathbf{v} = -2\mathbf{j} + 3\mathbf{k}$

$$\cos\theta = \frac{\mathbf{u}\cdot\mathbf{v}}{\|\mathbf{u}\|\,\|\mathbf{v}\|} = \frac{-8}{5\sqrt{13}} = \frac{-8\sqrt{13}}{65}$$

$$\theta = \arccos\left(-\frac{8\sqrt{13}}{65}\right) \approx 116.3°$$

18. $\mathbf{u} = 2\mathbf{i} - 3\mathbf{j} + \mathbf{k}, \mathbf{v} = \mathbf{i} - 2\mathbf{j} + \mathbf{k}$

$$\cos\theta = \frac{\mathbf{u}\cdot\mathbf{v}}{\|\mathbf{u}\|\,\|\mathbf{v}\|}$$

$$= \frac{9}{\sqrt{14}\sqrt{6}} = \frac{9}{2\sqrt{21}} = \frac{3\sqrt{21}}{14}$$

$$\theta = \arccos\left(\frac{3\sqrt{21}}{14}\right) \approx 10.9°$$

19. $\mathbf{u} = \langle 4, 0\rangle, \mathbf{v} = \langle 1, 1\rangle$

$\mathbf{u} \neq c\mathbf{v} \Rightarrow$ not parallel

$\mathbf{u}\cdot\mathbf{v} = 4 \neq 0 \Rightarrow$ not orthogonal

Neither

20. $\mathbf{u} = \langle 2, 18\rangle, \mathbf{v} = \left\langle \frac{3}{2}, -\frac{1}{6}\right\rangle$

$\mathbf{u} \neq c\mathbf{v} \Rightarrow$ not parallel

$\mathbf{u}\cdot\mathbf{v} = 0 \Rightarrow$ orthogonal

21. $\mathbf{u} = \langle 4, 3\rangle, \mathbf{v} = \left\langle \frac{1}{2}, -\frac{2}{3}\right\rangle$

$\mathbf{u} \neq c\mathbf{v} \Rightarrow$ not parallel

$\mathbf{u}\cdot\mathbf{v} = 0 \Rightarrow$ orthogonal

22. $\mathbf{u} = -\frac{1}{3}(\mathbf{i} - 2\mathbf{j}), \mathbf{v} = 2\mathbf{i} - 4\mathbf{j}$

$\mathbf{u} = -\frac{1}{6}\mathbf{v} \Rightarrow$ parallel

23. $\mathbf{u} = \mathbf{j} + 6\mathbf{k}, \mathbf{v} = \mathbf{i} - 2\mathbf{j} - \mathbf{k}$

$\mathbf{u} \neq c\mathbf{v} \Rightarrow$ not parallel

$\mathbf{u}\cdot\mathbf{v} = -8 \neq 0 \Rightarrow$ not orthogonal

Neither

24. $\mathbf{u} = -2\mathbf{i} + 3\mathbf{j} - \mathbf{k}, \mathbf{v} = 2\mathbf{i} + \mathbf{j} - \mathbf{k}$

$\mathbf{u} \neq c\mathbf{v} \Rightarrow$ not parallel

$\mathbf{u}\cdot\mathbf{v} = 0 \Rightarrow$ orthogonal

25. $\mathbf{u} = \langle 2, -3, 1\rangle, \mathbf{v} = \langle -1, -1, -1\rangle$

$\mathbf{u} \neq c\mathbf{v} \Rightarrow$ not parallel

$\mathbf{u}\cdot\mathbf{v} = 0 \Rightarrow$ orthogonal

26. $\mathbf{u} = \langle \cos\theta, \sin\theta, -1\rangle$,

$\mathbf{v} = \langle \sin\theta, -\cos\theta, 0\rangle$

$\mathbf{u} \neq c\mathbf{v} \Rightarrow$ not parallel

$\mathbf{u}\cdot\mathbf{v} = 0 \Rightarrow$ orthogonal

27. The vector $\langle 1, 2, 0\rangle$ joining $(1, 2, 0)$ and $(0, 0, 0)$ is perpendicular to the vector $\langle -2, 1, 0\rangle$ joining $(-2, 1, 0)$ and $(0, 0, 0)$:

$$\langle 1, 2, 0\rangle \cdot \langle -2, 1, 0\rangle = 0$$

The triangle is a right triangle.

28. Consider the vector $\langle -3, 0, 0\rangle$ joining $(0, 0, 0)$ and $(-3, 0, 0)$, and the vector $\langle 1, 2, 3\rangle$ joining $(0, 0, 0)$ and $(1, 2, 3)$:

$$\langle -3, 0, 0\rangle \cdot \langle 1, 2, 3\rangle = -3 < 0$$

The triangle has an obtuse angle.

29. The vectors forming the sides of the triangle are:

$\mathbf{u} = \langle 0 - 2, 1 + 3, 2 - 4\rangle = \langle -2, 4, -2\rangle$

$\mathbf{v} = \langle -1 - 2, 2 + 3, 0 - 4\rangle = \langle -3, 5, -4\rangle$

$\mathbf{w} = \langle -1 - 0, 2 - 1, 0 - 2\rangle = \langle -1, 1, -2\rangle$

$\mathbf{u}\cdot\mathbf{v} = 6 + 9 + 8 = 23 > 0$

$\mathbf{u}\cdot\mathbf{w} = 2 + 4 + 4 = 10 > 0$

$\mathbf{v}\cdot\mathbf{w} = 4 + 5 + 8 = 17 > 0$

The triangle has three acute angles.

30. The vectors forming the sides of the triangle are:

$\mathbf{u} = \langle -1 - 2, 5 + 7, 8 - 3\rangle = \langle -3, 12, 5\rangle$

$\mathbf{v} = \langle 4 - 2, 6 + 7, -1 - 3\rangle = \langle 2, 13, -4\rangle$

$\mathbf{w} = \langle 4 + 1, 6 - 5, -1 - 8\rangle = \langle 5, 1, -9\rangle$

$\mathbf{u}\cdot\mathbf{w} = -15 + 12 - 45 = -48 < 0$

The triangle has an obtuse angle.

31. $\mathbf{u} = \mathbf{i} + 2\mathbf{j} + 2\mathbf{k}, \|\mathbf{u}\| = 3$

$$\cos\alpha = \frac{1}{3}$$

$$\cos\beta = \frac{2}{3}$$

$$\cos\gamma = \frac{2}{3}$$

$$\cos^2\alpha + \cos^2\beta + \cos^2\gamma = \frac{1}{9} + \frac{4}{9} + \frac{4}{9} = 1$$

32. $\mathbf{u} = \langle 5, 3, -1\rangle \quad \|\mathbf{u}\| = \sqrt{35}$

$$\cos\alpha = \frac{5}{\sqrt{35}}$$

$$\cos\beta = \frac{3}{\sqrt{35}}$$

$$\cos\gamma = \frac{-1}{\sqrt{35}}$$

$$\cos^2\alpha + \cos^2\beta + \cos^2\gamma = \frac{25}{35} + \frac{9}{35} + \frac{1}{35} = 1$$

33. $\mathbf{u} = \langle 0, 6, -4\rangle, \|\mathbf{u}\| = \sqrt{52} = 2\sqrt{13}$

$$\cos\alpha = 0$$

$$\cos\beta = \frac{3}{\sqrt{13}}$$

$$\cos\gamma = -\frac{2}{\sqrt{13}}$$

$$\cos^2\alpha + \cos^2\beta + \cos^2\gamma = 0 + \frac{9}{13} + \frac{4}{13} = 1$$

34. $\mathbf{u} = \langle a, b, c\rangle, \|\mathbf{u}\| = \sqrt{a^2 + b^2 + c^2}$

$$\cos\alpha = \frac{a}{\sqrt{a^2 + b^2 + c^2}}$$

$$\cos\beta = \frac{b}{\sqrt{a^2 + b^2 + c^2}}$$

$$\cos\gamma = \frac{c}{\sqrt{a^2 + b^2 + c^2}}$$

$$\cos^2\alpha + \cos^2\beta + \cos^2\gamma = \frac{a^2}{a^2 + b^2 + c^2} + \frac{b^2}{a^2 + b^2 + c^2} + \frac{c^2}{a^2 + b^2 + c^2} = 1$$

35. $\mathbf{u} = \langle 3, 2, -2\rangle \quad \|\mathbf{u}\| = \sqrt{17}$

$$\cos\alpha = \frac{3}{\sqrt{17}} \Rightarrow \alpha \approx 0.7560 \text{ or } 43.3°$$

$$\cos\beta = \frac{2}{\sqrt{17}} \Rightarrow \beta \approx 1.0644 \text{ or } 61.0°$$

$$\cos\gamma = \frac{-2}{\sqrt{17}} \Rightarrow \gamma \approx 2.0772 \text{ or } 119.0°$$

36. $\mathbf{u} = \langle -4, 3, 5\rangle \quad \|\mathbf{u}\| = \sqrt{50} = 5\sqrt{2}$

$$\cos\alpha = \frac{-4}{5\sqrt{2}} \Rightarrow \alpha \approx 2.1721 \text{ or } 124.4°$$

$$\cos\beta = \frac{3}{5\sqrt{2}} \Rightarrow \beta \approx 1.1326 \text{ or } 64.9°$$

$$\cos\gamma = \frac{5}{5\sqrt{2}} = \frac{1}{\sqrt{2}} \Rightarrow \gamma \approx \frac{\pi}{4} \text{ or } 45°$$

37. $\mathbf{u} = \langle -1, 5, 2\rangle \quad \|\mathbf{u}\| = \sqrt{30}$

$$\cos\alpha = \frac{-1}{\sqrt{30}} \Rightarrow \alpha \approx 1.7544 \text{ or } 100.5°$$

$$\cos\beta = \frac{5}{\sqrt{30}} \Rightarrow \beta \approx 0.4205 \text{ or } 24.1°$$

$$\cos\gamma = \frac{2}{\sqrt{30}} \Rightarrow \gamma \approx 1.1970 \text{ or } 68.6°$$

38. $\mathbf{u} = \langle -2, 6, 1\rangle \quad \|\mathbf{u}\| = \sqrt{41}$

$$\cos\alpha = \frac{-2}{\sqrt{41}} \Rightarrow \alpha \approx 1.8885 \text{ or } 108.2°$$

$$\cos\beta = \frac{6}{\sqrt{41}} \Rightarrow \alpha \approx 0.3567 \text{ or } 20.4°$$

$$\cos\gamma = \frac{1}{\sqrt{41}} \Rightarrow \alpha \approx 1.4140 \text{ or } 81.0°$$

39. $\mathbf{F}_1$: $C_1 = \dfrac{50}{\|\mathbf{F}_1\|} \approx 4.3193$

$\mathbf{F}_2$: $C_2 = \dfrac{80}{\|\mathbf{F}_2\|} \approx 5.4183$

$$\mathbf{F} = \mathbf{F}_1 + \mathbf{F}_2$$

$$\approx 4.3193\langle 10, 5, 3\rangle + 5.4183\langle 12, 7, -5\rangle$$

$$= \langle 108.2126, 59.5246, -14.1336\rangle$$

$\|\mathbf{F}\| \approx 124.310$ lb

$$\cos\alpha \approx \frac{108.2126}{\|\mathbf{F}\|} \Rightarrow \alpha \approx 29.48°$$

$$\cos\beta \approx \frac{59.5246}{\|\mathbf{F}\|} \Rightarrow \beta \approx 61.39°$$

$$\cos\gamma \approx \frac{-14.1336}{\|\mathbf{F}\|} \Rightarrow \gamma \approx 96.53°$$

40. $\mathbf{F}_1$: $C_1 = \dfrac{300}{\|\mathbf{F}_1\|} \approx 13.0931$

$\mathbf{F}_2$: $C_2 = \dfrac{100}{\|\mathbf{F}_2\|} \approx 6.3246$

$$\mathbf{F} = \mathbf{F}_1 + \mathbf{F}_2$$

$$\approx 13.0931\langle -20, -10, 5\rangle + 6.3246\langle 5, 15, 0\rangle\rangle$$

$$= \langle -230.239, -36.062, 65.4655\rangle$$

$\|\mathbf{F}\| \approx 242.067$ lb

$$\cos\alpha \approx \frac{-230.239}{\|\mathbf{F}\|} \Rightarrow \alpha \approx 162.02°$$

$$\cos\beta \approx \frac{-36.062}{\|\mathbf{F}\|} \Rightarrow \beta \approx 98.57°$$

$$\cos\gamma \approx \frac{65.4655}{\|\mathbf{F}\|} \Rightarrow \gamma \approx 74.31°$$

41. $\overrightarrow{OA} = \langle 0, 10, 10 \rangle$

$$\cos \alpha = \frac{0}{\sqrt{0^2 + 10^2 + 10^2}} = 0 \Rightarrow \alpha = 90°$$

$$\cos \beta = \cos \gamma = \frac{10}{\sqrt{0^2 + 10^2 + 10^2}}$$

$$= \frac{1}{\sqrt{2}} \Rightarrow \beta = \gamma = 45°$$

42. $\mathbf{F}_1 = C_1\langle 0, 10, 10 \rangle$. $\|\mathbf{F}_1\| = 200 = C_1 10\sqrt{2} \Rightarrow C_1 = 10\sqrt{2}$

and $\mathbf{F}_1 = \langle 0, 100\sqrt{2}, 100\sqrt{2} \rangle$

$\mathbf{F}_2 = C_2\langle -4, -6, 10 \rangle$

$\mathbf{F}_3 = C_3\langle 4, -6, 10 \rangle$

$\mathbf{F} = \langle 0, 0, w \rangle$

$\mathbf{F} + \mathbf{F}_1 + \mathbf{F}_2 + \mathbf{F}_3 = \mathbf{0}$

$-4C_2 + 4C_3 = 0 \Rightarrow C_2 = C_3$

$$100\sqrt{2} - 6C_2 - 6C_3 = 0 \Rightarrow C_2 = C_3 = \frac{25\sqrt{2}}{3}N$$

$$W = 10C_2 + 10C_3 + 100\sqrt{2} = \frac{800\sqrt{2}}{3}$$

43. $\mathbf{w}_2 = \mathbf{u} - \mathbf{w}_1 = \langle 6, 7 \rangle - \langle 2, 8 \rangle = \langle 4, -1 \rangle$

44. $\mathbf{w}_2 = \mathbf{u} - \mathbf{w}_1 = \langle 9, 7 \rangle - \langle 3, 9 \rangle = \langle 6, -2 \rangle$

45. $\mathbf{w}_2 = \mathbf{u} - \mathbf{w}_1 = \langle 0, 3, 3 \rangle - \langle -2, 2, 2 \rangle = \langle 2, 1, 1 \rangle$

46. $\mathbf{w}_2 = \mathbf{u} - \mathbf{w}_1 = \langle 8, 2, 0 \rangle - \langle 6, 3, -3 \rangle = \langle 2, -1, 3 \rangle$

47. $\mathbf{u} = \langle 2, 3 \rangle, \mathbf{v} = \langle 5, 1 \rangle$

(a) $\mathbf{w}_1 = \left(\frac{\mathbf{u} \cdot \mathbf{v}}{\|\mathbf{v}\|^2}\right)\mathbf{v} = \frac{13}{26}\langle 5, 1 \rangle = \left\langle \frac{5}{2}, \frac{1}{2} \right\rangle$

(b) $\mathbf{w}_2 = \mathbf{u} - \mathbf{w}_1 = \left\langle -\frac{1}{2}, \frac{5}{2} \right\rangle$

48. $\mathbf{u} = \langle 2, -3 \rangle, \mathbf{v} = \langle 3, 2 \rangle$

(a) $\mathbf{w}_1 = \left(\frac{\mathbf{u} \cdot \mathbf{v}}{\|\mathbf{v}\|^2}\right)\mathbf{v} = 0\mathbf{v} = \langle 0, 0 \rangle$

(b) $\mathbf{w}_2 = \mathbf{u} - \mathbf{w}_1 = \langle 2, -3 \rangle$

49. $\mathbf{u} = \langle 2, 1, 2 \rangle, \mathbf{v} = \langle 0, 3, 4 \rangle$

(a) $\mathbf{w}_1 = \left(\frac{\mathbf{u} \cdot \mathbf{v}}{\|\mathbf{v}\|^2}\right)\mathbf{v}$

$= \frac{11}{25}\langle 0, 3, 4 \rangle = \left\langle 0, \frac{33}{25}, \frac{44}{25} \right\rangle$

(b) $\mathbf{w}_2 = \mathbf{u} - \mathbf{w}_1 = \left\langle 2, -\frac{8}{25}, \frac{6}{25} \right\rangle$

50. $\mathbf{u} = \langle 1, 0, 4 \rangle, \mathbf{v} = \langle 3, 0, 2 \rangle$

(a) $\mathbf{w}_1 = \left(\frac{\mathbf{u} \cdot \mathbf{v}}{\|\mathbf{v}\|^2}\right)\mathbf{v} = \frac{11}{13}\langle 3, 0, 2 \rangle = \left\langle \frac{33}{13}, 0, \frac{22}{13} \right\rangle$

(b) $\mathbf{w}_2 = \mathbf{u} - \mathbf{w}_1 = \langle 1, 0, 4 \rangle - \left\langle \frac{33}{13}, 0, \frac{22}{13} \right\rangle$

$= \left\langle -\frac{20}{13}, 0, \frac{30}{13} \right\rangle$

51. $\mathbf{u} \cdot \mathbf{v} = \langle u_1, u_2, u_3 \rangle \cdot \langle v_1, v_2, v_3 \rangle = u_1v_1 + u_2v_2 + u_3v_3$

52. The vectors **u** and **v** are orthogonal if $\mathbf{u} \cdot \mathbf{v} = 0$.

The angle θ between **u** and **v** is given by

$$\cos \theta = \frac{\mathbf{u} \cdot \mathbf{v}}{\|\mathbf{u}\| \, \|\mathbf{v}\|}.$$

53. (a) Orthogonal, $\theta = \frac{\pi}{2}$

(b) Acute, $0 < \theta < \frac{\pi}{2}$

(c) Obtuse, $\frac{\pi}{2} < \theta < \pi$

54. (a) and (b) are defined.

55. See page 784. Direction cosines of $\mathbf{v} = \langle v_1, v_2, v_3 \rangle$ are

$$\cos \alpha = \frac{v_1}{\|\mathbf{v}\|}, \cos \beta = \frac{v_2}{\|\mathbf{v}\|}, \cos \gamma = \frac{v_3}{\|\mathbf{v}\|}.$$

α, β, and γ are the direction angles. See Figure 11.26.

56. See figure 11.29, page 785.

57. (a) $\left(\dfrac{\mathbf{u}\cdot\mathbf{v}}{\|\mathbf{v}\|^2}\right)\mathbf{v} = \mathbf{u} \Rightarrow \mathbf{u} = c\mathbf{v} \Rightarrow \mathbf{u}$ and $\mathbf{v}$ are parallel.

(b) $\left(\dfrac{\mathbf{u}\cdot\mathbf{v}}{\|\mathbf{v}\|^2}\right)\mathbf{v} = \mathbf{0} \Rightarrow \mathbf{u}\cdot\mathbf{v} = 0 \Rightarrow \mathbf{u}$ and $\mathbf{v}$ are orthogonal.

58. Yes, $\left\|\dfrac{\mathbf{u}\cdot\mathbf{v}}{\|\mathbf{v}\|^2}\mathbf{v}\right\| = \left\|\dfrac{\mathbf{v}\cdot\mathbf{u}}{\|\mathbf{u}\|^2}\mathbf{u}\right\|$

$$|\mathbf{u}\cdot\mathbf{v}|\frac{\|\mathbf{v}\|}{\|\mathbf{v}\|^2} = |\mathbf{v}\cdot\mathbf{u}|\frac{\|\mathbf{u}\|}{\|\mathbf{u}\|^2}$$

$$\frac{1}{\|\mathbf{v}\|} = \frac{1}{\|\mathbf{u}\|}$$

$$\|\mathbf{u}\| = \|\mathbf{v}\|$$

59. $\mathbf{u} = \langle 3240, 1450, 2235\rangle$

$\mathbf{v} = \langle 1.35, 2.65, 1.85\rangle$

$\mathbf{u}\cdot\mathbf{v} = 3240(1.35) + 1450(2.65) + 2235(1.85)$

$= \$12{,}351.25$

This represents the total amount that the restaurant earned on its three products.

60. $\mathbf{u} = \langle 3240, 1450, 2235\rangle$

$\mathbf{v} = \langle 1.35, 2.65, 1.85\rangle$

Increase prices by 4%: $1.04\mathbf{v}$

New total amount: $1.04(\mathbf{u}\cdot\mathbf{v}) = 1.04(12{,}351.25)$

$= \$12{,}845.30$

61. Programs will vary.

62. $\|\mathbf{u}\| \approx 9.165$

$\|\mathbf{v}\| \approx 5.745$

$\theta = 90°$

63. Programs will vary.

64. $\left\langle -\dfrac{21}{26}, \dfrac{63}{26}, \dfrac{42}{13}\right\rangle$

65. Because $\mathbf{u}$ appears to be perpendicular to $\mathbf{v}$, the projection of $\mathbf{u}$ onto $\mathbf{v}$ is $\mathbf{0}$. Analytically,

$$\text{proj}_{\mathbf{v}}\mathbf{u} = \frac{\mathbf{u}\cdot\mathbf{v}}{\|\mathbf{v}\|^2}\mathbf{v} = \frac{\langle 2, -3\rangle\cdot\langle 6, 4\rangle}{\|\langle 6, 4\rangle\|^2}\langle 6, 4\rangle = 0\langle 6, 4\rangle = \mathbf{0}.$$

66. Because $\mathbf{u}$ appears to be a multiple of $\mathbf{v}$, the projection of $\mathbf{u}$ onto $\mathbf{v}$ is $\mathbf{u}$. Analytically,

$$\text{proj}_{\mathbf{v}}\mathbf{u} = \frac{\mathbf{u}\cdot\mathbf{v}}{\|\mathbf{v}\|^2}\mathbf{v} = \frac{\langle -3, -2\rangle\cdot\langle 6, 4\rangle}{\langle 6, 4\rangle\cdot\langle 6, 4\rangle}\langle 6, 4\rangle$$

$$= \frac{-26}{52}\langle 6, 4\rangle = \langle -3, -2\rangle = \mathbf{u}.$$

67. $\mathbf{u} = \dfrac{1}{2}\mathbf{i} - \dfrac{2}{3}\mathbf{j}$. Want $\mathbf{u}\cdot\mathbf{v} = 0$.

$\mathbf{v} = 8\mathbf{i} + 6\mathbf{j}$ and $-\mathbf{v} = -8\mathbf{i} - 6\mathbf{j}$ are orthogonal to $\mathbf{u}$.

68. $\mathbf{u} = -8\mathbf{i} + 3\mathbf{j}$. Want $\mathbf{u}\cdot\mathbf{v} = 0$.

$\mathbf{v} = 3\mathbf{i} + 8\mathbf{j}$ and $-\mathbf{v} = -3\mathbf{i} - 8\mathbf{j}$ are orthogonal to $\mathbf{u}$.

69. $\mathbf{u} = \langle 3, 1, -2\rangle$. Want $\mathbf{u}\cdot\mathbf{v} = 0$.

$\mathbf{v} = \langle 0, 2, 1\rangle$ and $-\mathbf{v} = \langle 0, -2, -1\rangle$ are orthogonal to $\mathbf{u}$.

70. $\mathbf{u} = \langle 0, -3, 6\rangle$. Want $\mathbf{u}\cdot\mathbf{v} = 0$.

$\mathbf{v} = \langle 0, 6, 3\rangle$ and $-\mathbf{v} = \langle 0, -6, -3\rangle$ are orthogonal to $\mathbf{u}$.

71. (a) Gravitational Force $\mathbf{F} = -48{,}000\,\mathbf{j}$

$\mathbf{v} = \cos 10°\mathbf{i} + \sin 10°\mathbf{j}$

$$\mathbf{w}_1 = \frac{\mathbf{F}\cdot\mathbf{v}}{\|\mathbf{v}\|^2}\mathbf{v} = (\mathbf{F}\cdot\mathbf{v})\mathbf{v} = (-48{,}000)(\sin 10°)\mathbf{v}$$

$$\approx -8335.1(\cos 10°\,\mathbf{i} + \sin 10°\mathbf{j})$$

$\|\mathbf{w}_1\| \approx 8335.1$ lb

(b) $\mathbf{w}_2 = \mathbf{F}\cdot\mathbf{w}_1 = -48{,}000\mathbf{j} + 8335.1(\cos 10°\mathbf{i} + \sin 10°\mathbf{j})$

$= 8208.5\mathbf{i} - 46{,}552.6\mathbf{j}$

$\|\mathbf{w}_2\| \approx 47{,}270.8$ lb

72. $\overrightarrow{OA} = \langle 10, 5, 20\rangle$, $\mathbf{v} = \langle 0, 0, 1\rangle$

$$\text{proj}_{\mathbf{v}}\overrightarrow{OA} = \frac{20}{1^2}\langle 0,0,1\rangle = \langle 0,0,20\rangle$$

$$\|\text{proj}_{\mathbf{v}}\overrightarrow{OA}\| = 20$$

73. $\mathbf{F} = 85\left(\frac{1}{2}\mathbf{i} + \frac{\sqrt{3}}{2}\mathbf{j}\right)$

$\mathbf{v} = 10\mathbf{i}$

$W = \mathbf{F}\cdot\mathbf{v} = 425$ ft · lb

74. $\mathbf{F} = 25(\cos 20°\mathbf{i} + \sin 20°\mathbf{j})$

$\mathbf{v} = 50\mathbf{i}$

$$W = \mathbf{F}\cdot\mathbf{v} = 1250\cos 20° \approx 1174.6 \text{ ft}\cdot\text{lb}$$

75. False.

Let $\mathbf{u} = \langle 2, 4\rangle$, $\mathbf{v} = \langle 1, 7\rangle$ and $\mathbf{w} = \langle 5, 5\rangle$. Then $\mathbf{u}\cdot\mathbf{v} = 2 + 28 = 30$ and $\mathbf{u}\cdot\mathbf{w} = 10 + 20 = 30$.

76. True

$$\mathbf{w}\cdot(\mathbf{u}+\mathbf{v}) = \mathbf{w}\cdot\mathbf{u} + \mathbf{w}\cdot\mathbf{v} = 0 + 0 = 0 \Rightarrow \mathbf{w}$$

and $\mathbf{u} + \mathbf{v}$ are orthogonal.

77. Let s = length of a side.

$\mathbf{v} = \langle s, s, s\rangle$

$\|\mathbf{v}\| = s\sqrt{3}$

$$\cos\alpha = \cos\beta = \cos\gamma = \frac{s}{s\sqrt{3}} = \frac{1}{\sqrt{3}}$$

$$\alpha = \beta = \gamma = \arccos\left(\frac{1}{\sqrt{3}}\right) \approx 54.7°$$

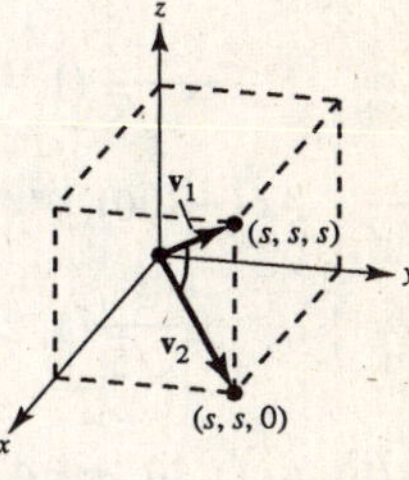

78. $\mathbf{v}_1 = \langle s, s, s\rangle$

$\|\mathbf{v}_1\| = s\sqrt{3}$

$\mathbf{v}_2 = \langle s, s, 0\rangle$

$\|\mathbf{v}_2\| = s\sqrt{2}$

$$\cos\theta = \frac{s\sqrt{2}}{s\sqrt{3}} = \frac{\sqrt{6}}{3}$$

$$\theta = \arccos\frac{\sqrt{6}}{3} \approx 35.26°$$

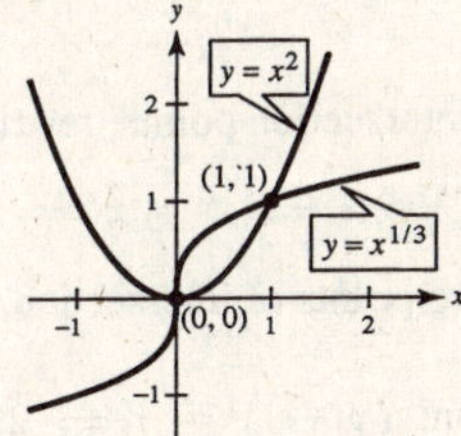

79. (a) The graphs $y_1 = x^2$ and $y_2 = x^{1/3}$ intersect at (0, 0) and (1, 1).

$y'_1 = 2x$ and $y'_2 = \dfrac{1}{3x^{2/3}}$.

At (0, 0), $\pm\langle 1, 0\rangle$ is tangent to y_1 and $\pm\langle 0, 1\rangle$ is tangent to y_2.

At (1, 1), $y'_1 = 2$ and $y'_2 = \dfrac{1}{3}$.

$\pm\dfrac{1}{\sqrt{5}}\langle 1, 2\rangle$ is tangent to y_1, $\pm\dfrac{1}{\sqrt{10}}\langle 3, 1\rangle$ is tangent to y_2.

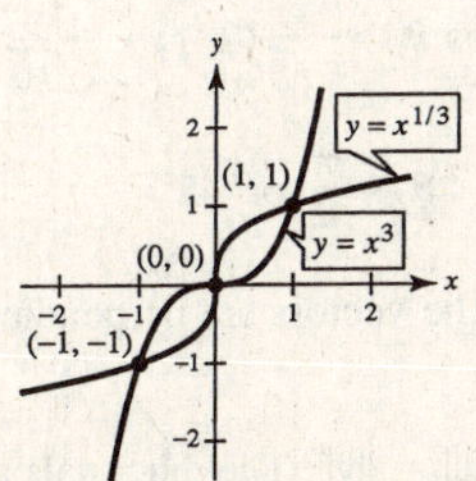

(b) At (0, 0), the vectors are perpendicular (90°).

$$\text{At } (1, 1),\ \cos\theta = \frac{\frac{1}{\sqrt{5}}\langle 1, 2\rangle\cdot\frac{1}{\sqrt{10}}\langle 3, 1\rangle}{(1)(1)} = \frac{5}{\sqrt{50}} = \frac{1}{\sqrt{2}}$$

$\theta = 45°$.

80. (a) The graphs $y_1 = x^3$ and $y_2 = x^{1/3}$ intersect at $(-1, -1)$, (0, 0) and (1, 1).

$y'_1 = 3x^2$ and $y'_2 = \dfrac{1}{3x^{2/3}}$.

At (0, 0), $\pm\langle 1, 0\rangle$ is tangent to y_1 and $\pm\langle 0, 1\rangle$ is tangent to y_2.

At (1, 1), $y'_1 = 3$ and $y'_2 = \dfrac{1}{3}$.

$\pm\dfrac{1}{\sqrt{10}}\langle 1, 3\rangle$ is tangent to y_1, $\pm\dfrac{1}{\sqrt{10}}\langle 3, 1\rangle$ is tangent to y_2.

At $(-1, -1)$, $y'_1 = 3$ and $y'_2 = \dfrac{1}{3}$.

$\pm\dfrac{1}{\sqrt{10}}\langle 1, 3\rangle$ is tangent to y_1, $\pm\dfrac{1}{\sqrt{10}}\langle 3, 1\rangle$ is tangent to y_2.

—CONTINUED—

80. —CONTINUED—

(b) At (0, 0), the vectors are perpendicular (90°).

$$\text{At } (1, 1), \cos\theta = \frac{\frac{1}{\sqrt{10}}\langle 1, 3\rangle \cdot \frac{1}{\sqrt{10}}\langle 3, 1\rangle}{(1)(1)} = \frac{6}{10} = \frac{3}{5}.$$

$\theta \approx 0.9273$ or $53.13°$

By symmetry, the angle is the same at $(-1, -1)$.

81. (a) The graphs of $y_1 = 1 - x^2$ and $y^2 = x^2 - 1$ intersect at $(1, 0)$ and $(-1, 0)$.

$y'_1 = -2x$ and $y'_2 = 2x$.

At $(1, 0)$, $y'_1 = -2$ and $y'_2 = 2$.

$\pm\frac{1}{\sqrt{5}}\langle 1, -2\rangle$ is tangent to y_1, $\pm\frac{1}{\sqrt{5}}\langle 1, 2\rangle$ is tangent to y_2.

At $(-1, 0)$, $y'_1 = 2$ and $y'_2 = -2$.

$\pm\frac{1}{\sqrt{5}}\langle 1, 2\rangle$ is tangent to y_1, $\pm\frac{1}{\sqrt{5}}\langle 1, -2\rangle$ is tangent to y_2.

(b) At $(1, 0)$, $\cos\theta = \frac{1}{\sqrt{5}}\langle 1, -2\rangle \cdot \frac{-1}{\sqrt{5}}\langle 1, -2\rangle = \frac{3}{5}$.

$\theta \approx 0.9273$ or $53.13°$

By symmetry, the angle is the same at $(-1, 0)$.

82. (a) To find the intersection points, rewrite the second equation as $y + 1 = x^3$. Substituting into the first equation

$(y + 1)^2 = x \Rightarrow x^6 = x \Rightarrow x = 0, 1.$

There are two points of intersection, $(0, -1)$ and $(1, 0)$, as indicated in the figure.

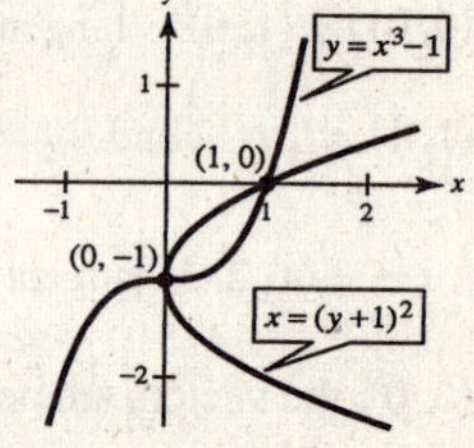

First equation: $(y + 1)^2 = x \Rightarrow 2(y + 1)y' = 1 \Rightarrow y' = \frac{1}{2(y + 1)}$

At $(1, 0)$, $y' = \frac{1}{2}$.

Second equation: $y = x^3 - 1 \Rightarrow y' = 3x^2$. At $(1, 0)$, $y' = 3$.

$\pm\frac{1}{\sqrt{5}}\langle 2, 1\rangle$ unit tangent vectors to first curve, $\pm\frac{1}{\sqrt{10}}\langle 1, 3\rangle$ unit tangent vectors to second curve

At $(0, 1)$, the unit tangent vectors to the first curve are $\pm\langle 0, 1\rangle$, and the unit tangent vectors to the second curve are $\pm\langle 1, 0\rangle$.

(b) At $(1, 0)$, $\cos\theta = \frac{1}{\sqrt{5}}\langle 2, 1\rangle \cdot \frac{1}{\sqrt{10}}\langle 1, 3\rangle = \frac{5}{\sqrt{50}} = \frac{1}{\sqrt{2}}$.

$\theta \approx \frac{\pi}{4}$ or $45°$

At $(0, -1)$ the vectors are perpendicular, $\theta = 90°$.

83. In a rhombus, $\|\mathbf{u}\| = \|\mathbf{v}\|$. The diagonals are $\mathbf{u} + \mathbf{v}$ and $\mathbf{u} - \mathbf{v}$.

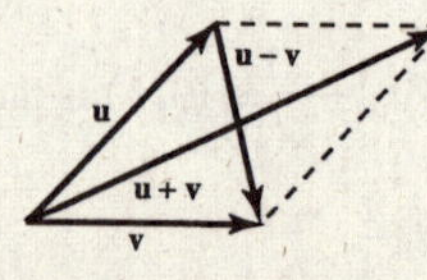

$$\begin{aligned}(\mathbf{u} + \mathbf{v}) \cdot (\mathbf{u} - \mathbf{v}) &= (\mathbf{u} + \mathbf{v}) \cdot \mathbf{u} - (\mathbf{u} + \mathbf{v}) \cdot \mathbf{v} \\ &= \mathbf{u} \cdot \mathbf{u} + \mathbf{v} \cdot \mathbf{u} - \mathbf{u} \cdot \mathbf{v} - \mathbf{v} \cdot \mathbf{v} \\ &= \|\mathbf{u}\|^2 - \|\mathbf{v}\|^2 = 0\end{aligned}$$

Therefore, the diagonals are orthogonal.

84. If **u** and **v** are the sides of the parallelogram, then the diagonals are **u** + **v** and **u** − **v**, as indicated in the figure.

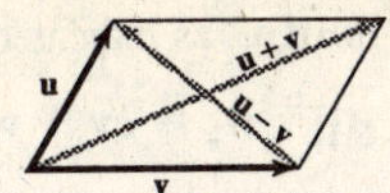

The parallelogram is a rectangle.

$\Leftrightarrow \mathbf{u} \cdot \mathbf{v} = 0$

$\Leftrightarrow 2\mathbf{u} \cdot \mathbf{v} = -2\mathbf{u} \cdot \mathbf{v}$

$\Leftrightarrow (\mathbf{u} + \mathbf{v}) \cdot (\mathbf{u} + \mathbf{v}) = (\mathbf{u} - \mathbf{v}) \cdot (\mathbf{u} - \mathbf{v})$

$\Leftrightarrow \|\mathbf{u} + \mathbf{v}\|^2 = \|\mathbf{u} - \mathbf{v}\|^2$

$\Leftrightarrow$ The diagonals are equal in length.

85. (a)

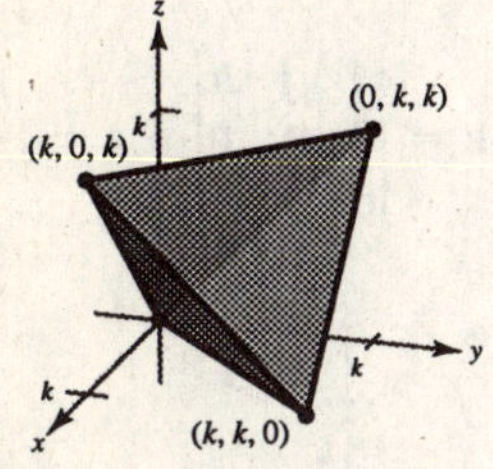

(b) Length of each edge:

$$\sqrt{k^2 + k^2 + 0^2} = k\sqrt{2}$$

(c) $\cos\theta = \dfrac{k^2}{(k\sqrt{2})(k\sqrt{2})} = \dfrac{1}{2}$

$$\theta = \arccos\left(\frac{1}{2}\right) = 60^\circ$$

(d) $\vec{r_1} = \langle k, k, 0\rangle - \left\langle \dfrac{k}{2}, \dfrac{k}{2}, \dfrac{k}{2}\right\rangle = \left\langle \dfrac{k}{2}, \dfrac{k}{2}, -\dfrac{k}{2}\right\rangle$

$$\vec{r_2} = \langle 0, 0, 0\rangle - \left\langle \frac{k}{2}, \frac{k}{2}, \frac{k}{2}\right\rangle = \left\langle -\frac{k}{2}, -\frac{k}{2}, -\frac{k}{2}\right\rangle$$

$$\cos\theta = \frac{-\dfrac{k^2}{4}}{\left(\dfrac{k}{2}\right)^2 \cdot 3} = -\frac{1}{3}$$

$$\theta = 109.5^\circ$$

86. $\mathbf{u} = \langle \cos\alpha, \sin\alpha, 0\rangle$, $\mathbf{v} = \langle \cos\beta, \sin\beta, 0\rangle$

The angle between **u** and **v** is $\alpha - \beta$. (Assuming that $\alpha > \beta$). Also,

$$\cos(\alpha - \beta) = \frac{\mathbf{u} \cdot \mathbf{v}}{\|\mathbf{u}\|\,\|\mathbf{v}\|} = \frac{\cos\alpha\cos\beta + \sin\alpha\sin\beta}{(1)(1)} = \cos\alpha\cos\beta + \sin\alpha\sin\beta.$$

87.
$$\begin{aligned}
\|\mathbf{u} - \mathbf{v}\|^2 &= (\mathbf{u} - \mathbf{v}) \cdot (\mathbf{u} - \mathbf{v})\\
&= (\mathbf{u} - \mathbf{v}) \cdot \mathbf{u} - (\mathbf{u} - \mathbf{v}) \cdot \mathbf{v}\\
&= \mathbf{u} \cdot \mathbf{u} - \mathbf{v} \cdot \mathbf{u} - \mathbf{u} \cdot \mathbf{v} + \mathbf{v} \cdot \mathbf{v}\\
&= \|\mathbf{u}\|^2 - \mathbf{u} \cdot \mathbf{v} - \mathbf{u} \cdot \mathbf{v} + \|\mathbf{v}\|^2\\
&= \|\mathbf{u}\|^2 + \|\mathbf{v}\|^2 - 2\mathbf{u} \cdot \mathbf{v}
\end{aligned}$$

88.
$$\begin{aligned}
\mathbf{u} \cdot \mathbf{v} &= \|\mathbf{u}\|\,\|\mathbf{v}\| \cos\theta\\
|\mathbf{u} \cdot \mathbf{v}| &= \big|\|\mathbf{u}\|\,\|\mathbf{v}\| \cos\theta\big|\\
&= \|\mathbf{u}\|\,\|\mathbf{v}\|\,|\cos\theta|\\
&\le \|\mathbf{u}\|\,\|\mathbf{v}\| \text{ since } |\cos\theta| \le 1.
\end{aligned}$$

89.
$$\begin{aligned}
\|\mathbf{u} + \mathbf{v}\|^2 &= (\mathbf{u} + \mathbf{v}) \cdot (\mathbf{u} + \mathbf{v})\\
&= (\mathbf{u} + \mathbf{v}) \cdot \mathbf{u} + (\mathbf{u} + \mathbf{v}) \cdot \mathbf{v}\\
&= \mathbf{u} \cdot \mathbf{u} + \mathbf{v} \cdot \mathbf{u} + \mathbf{u} \cdot \mathbf{v} + \mathbf{v} \cdot \mathbf{v}\\
&= \|\mathbf{u}\|^2 + 2\mathbf{u} \cdot \mathbf{v} + \|\mathbf{v}\|^2\\
&\le \|\mathbf{u}\|^2 + 2\|\mathbf{u}\|\,\|\mathbf{v}\| + \|\mathbf{v}\|^2\\
&\le (\|\mathbf{u}\| + \|\mathbf{v}\|)^2
\end{aligned}$$

Therefore, $\|\mathbf{u} + \mathbf{v}\| \le \|\mathbf{u}\| + \|\mathbf{v}\|$.

90. Let $\mathbf{w}_1 = \text{proj}_\mathbf{v}\mathbf{u}$, as indicated in the figure. Because $\mathbf{w}_1$ is a scalar multiple of $\mathbf{v}$, you can write

$$\mathbf{u} = \mathbf{w}_1 + \mathbf{w}_2 = c\mathbf{v} + \mathbf{w}_2.$$

Taking the dot product of both sides with $\mathbf{v}$ produces

$$\begin{aligned}\mathbf{u}\cdot\mathbf{v} &= (c\mathbf{v} + \mathbf{w}_2)\cdot\mathbf{v} = c\mathbf{v}\cdot\mathbf{v} + \mathbf{w}_2\cdot\mathbf{v}\\ &= c\|\mathbf{v}\|^2, \text{ since } \mathbf{w}_2 \text{ and } \mathbf{v} \text{ are orthogonol.}\end{aligned}$$

Thus, $\mathbf{u}\cdot\mathbf{v} = c\|\mathbf{v}\|^2 \Rightarrow c = \dfrac{\mathbf{u}\cdot\mathbf{v}}{\|\mathbf{v}\|^2}$ and $\mathbf{w}_1 = \text{proj}_\mathbf{v}\mathbf{u} = c\mathbf{v} = \dfrac{\mathbf{u}\cdot\mathbf{v}}{\|\mathbf{v}\|^2}\mathbf{v}$.

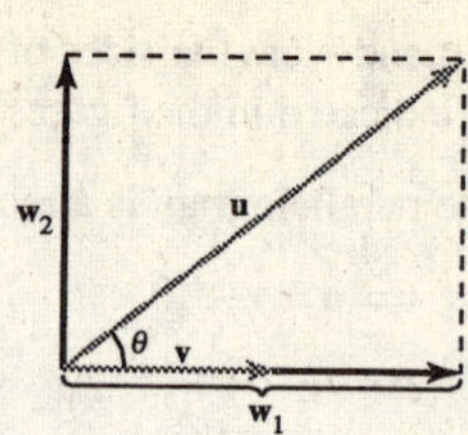

Section 11.4 The Cross Product of Two Vectors in Space

1. $\mathbf{j}\times\mathbf{i} = \begin{vmatrix}\mathbf{i} & \mathbf{j} & \mathbf{k}\\ 0 & 1 & 0\\ 1 & 0 & 0\end{vmatrix} = -\mathbf{k}$

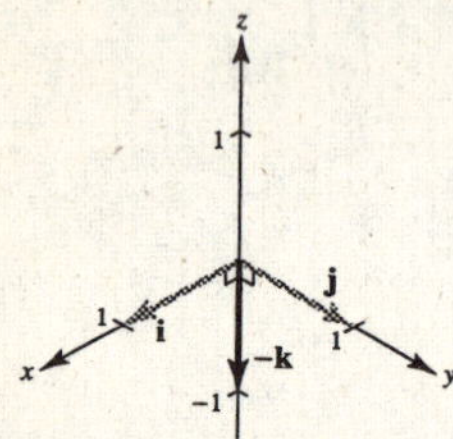

2. $\mathbf{i}\times\mathbf{j} = \begin{vmatrix}\mathbf{i} & \mathbf{j} & \mathbf{k}\\ 1 & 0 & 0\\ 0 & 1 & 0\end{vmatrix} = \mathbf{k}$

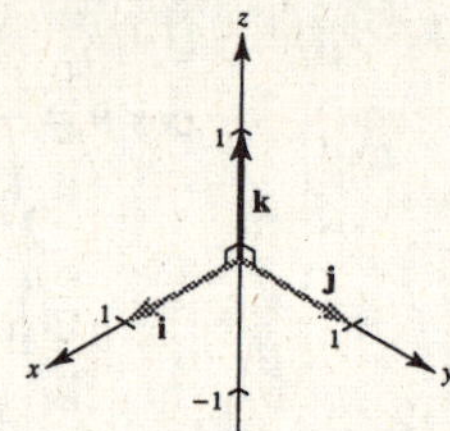

3. $\mathbf{j}\times\mathbf{k} = \begin{vmatrix}\mathbf{i} & \mathbf{j} & \mathbf{k}\\ 0 & 1 & 0\\ 0 & 0 & 1\end{vmatrix} = \mathbf{i}$

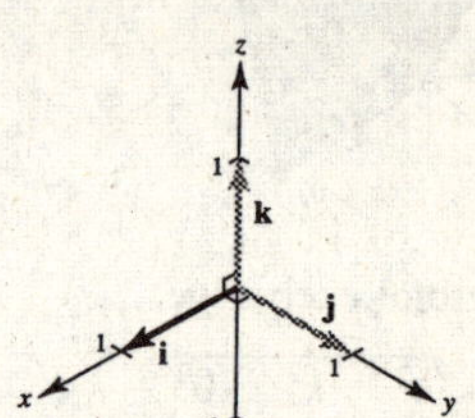

4. $\mathbf{k}\times\mathbf{j} = \begin{vmatrix}\mathbf{i} & \mathbf{j} & \mathbf{k}\\ 0 & 0 & 1\\ 0 & 1 & 0\end{vmatrix} = -\mathbf{i}$

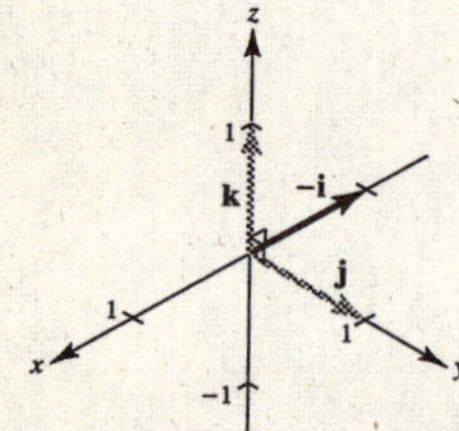

5. 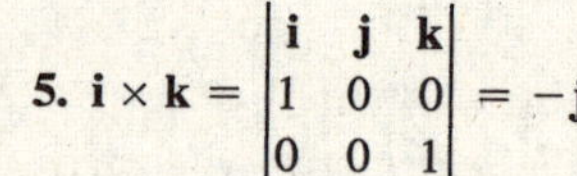
$\mathbf{i}\times\mathbf{k} = \begin{vmatrix}\mathbf{i} & \mathbf{j} & \mathbf{k}\\ 1 & 0 & 0\\ 0 & 0 & 1\end{vmatrix} = -\mathbf{j}$

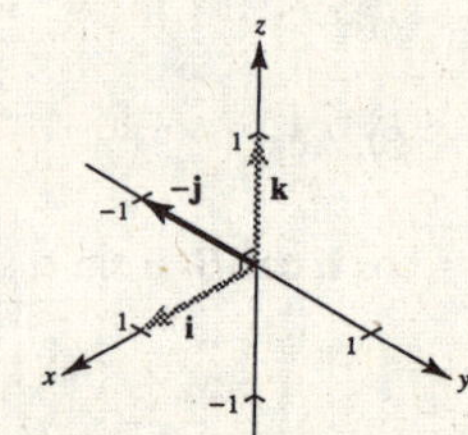

6. 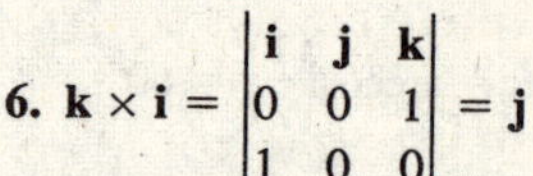
$\mathbf{k}\times\mathbf{i} = \begin{vmatrix}\mathbf{i} & \mathbf{j} & \mathbf{k}\\ 0 & 0 & 1\\ 1 & 0 & 0\end{vmatrix} = \mathbf{j}$

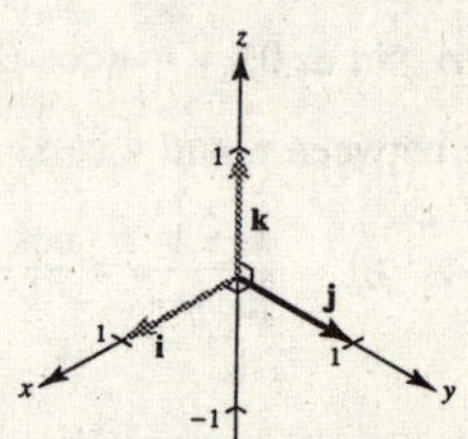

7. (a) $\mathbf{u}\times\mathbf{v} = \begin{vmatrix}\mathbf{i} & \mathbf{j} & \mathbf{k}\\ -2 & 3 & 4\\ 3 & 7 & 2\end{vmatrix} = \langle -22, 16, -23\rangle$

(b) $\mathbf{v}\times\mathbf{u} = -(\mathbf{u}\times\mathbf{v}) = \langle 22, -16, 23\rangle$

(c) $\mathbf{v}\times\mathbf{v} = \begin{vmatrix}\mathbf{i} & \mathbf{j} & \mathbf{k}\\ 3 & 7 & 2\\ 3 & 7 & 2\end{vmatrix} = \mathbf{0}$

8. (a) $\mathbf{u}\times\mathbf{v} = \begin{vmatrix}\mathbf{i} & \mathbf{j} & \mathbf{k}\\ 3 & 0 & 5\\ 2 & 3 & -2\end{vmatrix} = \langle -15, 16, 9\rangle$

(b) $\mathbf{v}\times\mathbf{u} = -(\mathbf{u}\times\mathbf{v}) = \langle 15, -16, -9\rangle$

(c) $\mathbf{v}\times\mathbf{v} = \mathbf{0}$

9. (a) $\mathbf{u}\times\mathbf{v} = \begin{vmatrix}\mathbf{i} & \mathbf{j} & \mathbf{k}\\ 7 & 3 & 2\\ 1 & -1 & 5\end{vmatrix} = \langle 17, -33, -10\rangle$

(b) $\mathbf{v}\times\mathbf{u} = -(\mathbf{u}\times\mathbf{v}) = \langle -17, 33, 10\rangle$

(c) $\mathbf{v}\times\mathbf{v} = \mathbf{0}$

10. (a) $\mathbf{u}\times\mathbf{v} = \begin{vmatrix}\mathbf{i} & \mathbf{j} & \mathbf{k}\\ 3 & -2 & -2\\ 1 & 5 & 1\end{vmatrix} = \langle 8, -5, 17\rangle$

(b) $\mathbf{v}\times\mathbf{u} = -(\mathbf{u}\times\mathbf{v}) = \langle -8, 5, -17\rangle$

(c) $\mathbf{v}\times\mathbf{v} = \mathbf{0}$

11. $\mathbf{u} = \langle 2, -3, 1\rangle, \mathbf{v} = \langle 1, -2, 1\rangle$

$$\mathbf{u}\times\mathbf{v} = \begin{vmatrix} \mathbf{i} & \mathbf{j} & \mathbf{k} \\ 2 & -3 & 1 \\ 1 & -2 & 1 \end{vmatrix} = -\mathbf{i} - \mathbf{j} - \mathbf{k} = \langle -1, -1, -1\rangle$$

$$\mathbf{u}\cdot(\mathbf{u}\times\mathbf{v}) = 2(-1) + (-3)(-1) + (1)(-1) = 0 \Rightarrow \mathbf{u}\perp\mathbf{u}\times\mathbf{v}$$

$$\mathbf{v}\cdot(\mathbf{u}\times\mathbf{v}) = 1(-1) + (-2)(-1) + (1)(-1) = 0 \Rightarrow \mathbf{v}\perp\mathbf{u}\times\mathbf{v}$$

12. $\mathbf{u} = \langle -1, 1, 2\rangle, \mathbf{v} = \langle 0, 1, 0\rangle$

$$\mathbf{u}\times\mathbf{v} = \begin{vmatrix} \mathbf{i} & \mathbf{j} & \mathbf{k} \\ -1 & 1 & 2 \\ 0 & 1 & 0 \end{vmatrix} = -2\mathbf{i} - \mathbf{k} = \langle -2, 0, -1\rangle$$

$$\mathbf{u}\cdot(\mathbf{u}\times\mathbf{v}) = (-1)(-2) + (1)(0) + (2)(-1)$$
$$= 0 \Rightarrow \mathbf{u}\perp\mathbf{u}\times\mathbf{v}$$

$$\mathbf{v}\cdot(\mathbf{u}\times\mathbf{v}) = (0)(-2) + (1)(0) + (0)(-1)$$
$$= 0 \Rightarrow \mathbf{v}\perp\mathbf{u}\times\mathbf{v}$$

13. $\mathbf{u} = \langle 12, -3, 0\rangle, \mathbf{v} = \langle -2, 5, 0\rangle$

$$\mathbf{u}\times\mathbf{v} = \begin{vmatrix} \mathbf{i} & \mathbf{j} & \mathbf{k} \\ 12 & -3 & 0 \\ -2 & 5 & 0 \end{vmatrix} = 54\mathbf{k} = \langle 0, 0, 54\rangle$$

$$\mathbf{u}\cdot(\mathbf{u}\times\mathbf{v}) = 12(0) + (-3)(0) + 0(54)$$
$$= 0 \Rightarrow \mathbf{u}\perp\mathbf{u}\times\mathbf{v}$$

$$\mathbf{v}\cdot(\mathbf{u}\times\mathbf{v}) = -2(0) + 5(0) + 0(54)$$
$$= 0 \Rightarrow \mathbf{v}\perp\mathbf{u}\times\mathbf{v}$$

14. $\mathbf{u} = \langle -10, 0, 6\rangle, \mathbf{v} = \langle 7, 0, 0\rangle$

$$\mathbf{u}\times\mathbf{v} = \begin{vmatrix} \mathbf{i} & \mathbf{j} & \mathbf{k} \\ -10 & 0 & 6 \\ 7 & 0 & 0 \end{vmatrix} = 42\mathbf{j} = \langle 0, 42, 0\rangle$$

$$\mathbf{u}\cdot(\mathbf{u}\times\mathbf{v}) = (-10)(0) + (0)(42) + 6(0)$$
$$= 0 \Rightarrow \mathbf{u}\perp\mathbf{u}\times\mathbf{v}$$

$$\mathbf{v}\cdot(\mathbf{u}\times\mathbf{v}) = 7(0) + (0)(42) + (0)(0)$$
$$= 0 \Rightarrow \mathbf{v}\perp\mathbf{u}\times\mathbf{v}$$

15. $\mathbf{u} = \mathbf{i} + \mathbf{j} + \mathbf{k}, \mathbf{v} = 2\mathbf{i} + \mathbf{j} - \mathbf{k}$

$$\mathbf{u}\times\mathbf{v} = \begin{vmatrix} \mathbf{i} & \mathbf{j} & \mathbf{k} \\ 1 & 1 & 1 \\ 2 & 1 & -1 \end{vmatrix} = -2\mathbf{i} + 3\mathbf{j} - \mathbf{k} = \langle -2, 3, -1\rangle$$

$$\mathbf{u}\cdot(\mathbf{u}\times\mathbf{v}) = 1(-2) + 1(3) + 1(-1)$$
$$= 0 \Rightarrow \mathbf{u}\perp\mathbf{u}\times\mathbf{v}$$

$$\mathbf{v}\cdot(\mathbf{u}\times\mathbf{v}) = 2(-2) + 1(3) + (-1)(-1)$$
$$= 0 \Rightarrow \mathbf{v}\perp\mathbf{u}\times\mathbf{v}$$

$$(-\mathbf{v})\times\mathbf{u} = -(\mathbf{v}\times\mathbf{u}) = \mathbf{u}\times\mathbf{v}$$

16. $$\mathbf{u}\times\mathbf{v} = \begin{vmatrix} \mathbf{i} & \mathbf{j} & \mathbf{k} \\ 1 & 6 & 0 \\ -2 & 1 & 1 \end{vmatrix} = 6\mathbf{i} - \mathbf{j} + 13\mathbf{k}$$

$$\mathbf{u}\cdot(\mathbf{u}\times\mathbf{v}) = 1(6) + 6(-1) = 0 \Rightarrow \mathbf{u}\perp(\mathbf{u}\times\mathbf{v})$$

$$\mathbf{v}\cdot(\mathbf{u}\times\mathbf{v}) = -2(6) + 1(-1) + 1(13) = 0 \Rightarrow \mathbf{v}\perp(\mathbf{u}\times\mathbf{v})$$

17.

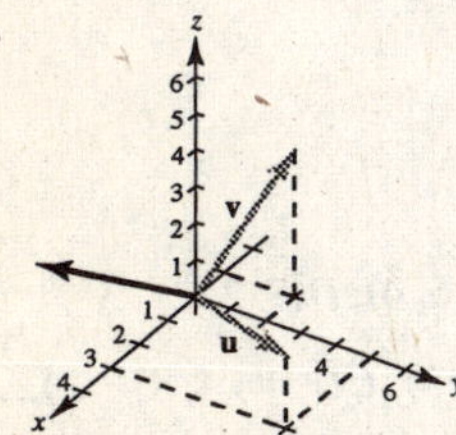

18.

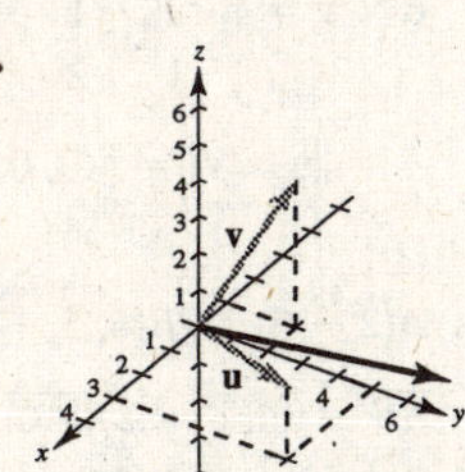

19.

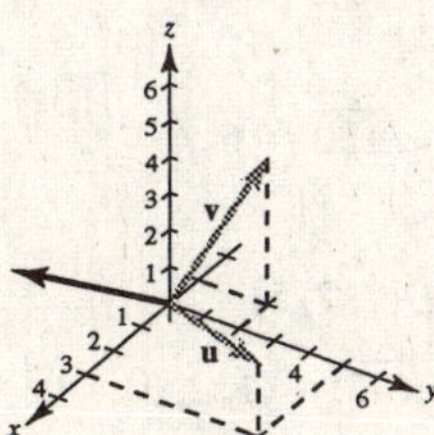

20.

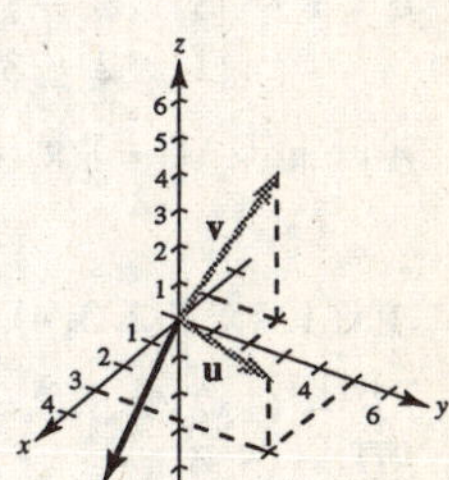

21. $\mathbf{u} = \langle 4, -3.5, 7\rangle$

$\mathbf{v} = \langle -1, 8, 4\rangle$

$$\mathbf{u}\times\mathbf{v} = \left\langle -70, -23, \frac{57}{2}\right\rangle$$

$$\frac{\mathbf{u}\times\mathbf{v}}{\|\mathbf{u}\times\mathbf{v}\|} = \left\langle \frac{-140}{\sqrt{24{,}965}}, \frac{-46}{\sqrt{24{,}965}}, \frac{57}{\sqrt{24{,}965}}\right\rangle$$

22. $\mathbf{u} = \langle -8, -6, 4\rangle$

$\mathbf{v} = \langle 10, -12, -2\rangle$

$\mathbf{u} \times \mathbf{v} = \langle 60, 24, 156\rangle$

$$\frac{\mathbf{u}\times\mathbf{v}}{\|\mathbf{u}\times\mathbf{v}\|} = \frac{1}{36\sqrt{22}}\langle 60, 24, 156\rangle = \left\langle \frac{5}{3\sqrt{22}}, \frac{2}{3\sqrt{22}}, \frac{13}{3\sqrt{22}}\right\rangle$$

23. $\mathbf{u} = -3\mathbf{i} + 2\mathbf{j} - 5\mathbf{k}$

$\mathbf{v} = \frac{1}{2}\mathbf{i} - \frac{3}{4}\mathbf{j} + \frac{1}{10}\mathbf{k}$

$\mathbf{u} \times \mathbf{v} = \left\langle -\frac{71}{20}, -\frac{11}{5}, \frac{5}{4}\right\rangle$

$$\frac{\mathbf{u}\times\mathbf{v}}{\|\mathbf{u}\times\mathbf{v}\|} = \frac{20}{\sqrt{7602}}\left\langle -\frac{71}{20}, -\frac{11}{5}, \frac{5}{4}\right\rangle = \left\langle -\frac{71}{\sqrt{7602}}, -\frac{44}{\sqrt{7602}}, \frac{25}{\sqrt{7602}}\right\rangle$$

24. $\mathbf{u} = \frac{2}{3}\mathbf{k}$

$\mathbf{v} = \frac{1}{2}\mathbf{i} + 6\mathbf{k}$

$\mathbf{u} \times \mathbf{v} = \left\langle 0, \frac{1}{3}, 0\right\rangle$

$$\frac{\mathbf{u}\times\mathbf{v}}{\|\mathbf{u}\times\mathbf{v}\|} = \langle 0, 1, 0\rangle$$

25. Programs will vary.

26. $\mathbf{u} \times \mathbf{v} = \langle -50, 40, -34\rangle$

$\|\mathbf{u}\times\mathbf{v}\| \approx 72.498$

27. $\mathbf{u} = \mathbf{j}$

$\mathbf{v} = \mathbf{j} + \mathbf{k}$

$$\mathbf{u}\times\mathbf{v} = \begin{vmatrix} \mathbf{i} & \mathbf{j} & \mathbf{k} \\ 0 & 1 & 0 \\ 0 & 1 & 1 \end{vmatrix} = \mathbf{i}$$

$A = \|\mathbf{u}\times\mathbf{v}\| = \|\mathbf{i}\| = 1$

28. $\mathbf{u} = \mathbf{i} + \mathbf{j} + \mathbf{k}$

$\mathbf{v} = \mathbf{j} + \mathbf{k}$

$$\mathbf{u}\times\mathbf{v} = \begin{vmatrix} \mathbf{i} & \mathbf{j} & \mathbf{k} \\ 1 & 1 & 1 \\ 0 & 1 & 1 \end{vmatrix} = -\mathbf{j} + \mathbf{k}$$

$A = \|\mathbf{u}\times\mathbf{v}\| = \|-\mathbf{j} + \mathbf{k}\| = \sqrt{2}$

29. $\mathbf{u} = \langle 3, 2, -1\rangle$

$\mathbf{v} = \langle 1, 2, 3\rangle$

$$\mathbf{u}\times\mathbf{v} = \begin{vmatrix} \mathbf{i} & \mathbf{j} & \mathbf{k} \\ 3 & 2 & -1 \\ 1 & 2 & 3 \end{vmatrix} = \langle 8, -10, 4\rangle$$

$A = \|\mathbf{u}\times\mathbf{v}\| = \|\langle 8, -10, 4\rangle\| = \sqrt{180} = 6\sqrt{5}$

30. $\mathbf{u} = \langle 2, -1, 0\rangle$

$\mathbf{v} = \langle -1, 2, 0\rangle$

$$\mathbf{u}\times\mathbf{v} = \begin{vmatrix} \mathbf{i} & \mathbf{j} & \mathbf{k} \\ 2 & -1 & 0 \\ -1 & 2 & 0 \end{vmatrix} = \langle 0, 0, 3\rangle$$

$A = \|\mathbf{u}\times\mathbf{v}\| = \|\langle 0, 0, 3\rangle\| = 3$

31. $A(1, 1, 1,), B(2, 3, 4), C(6, 5, 2), D(7, 7, 5)$

$\overrightarrow{AB} = \langle 1, 2, 3\rangle, \overrightarrow{AC} = \langle 5, 4, 1\rangle, \overrightarrow{CD} = \langle 1, 2, 3\rangle, \overrightarrow{BD} = \langle 5, 4, 1\rangle$

Since $\overrightarrow{AB} = \overrightarrow{CD}$ and $\overrightarrow{AC} = \overrightarrow{BD}$, the figure is a parallelogram. $\overrightarrow{AB}$ and $\overrightarrow{AC}$ are adjacent sides and

$$\overrightarrow{AB}\times\overrightarrow{AC} = \begin{vmatrix} \mathbf{i} & \mathbf{j} & \mathbf{k} \\ 1 & 2 & 3 \\ 5 & 4 & 1 \end{vmatrix} = -10\mathbf{i} + 14\mathbf{j} - 6\mathbf{k}.$$

$A = \|\overrightarrow{AB}\times\overrightarrow{AC}\| = \sqrt{332} = 2\sqrt{83}$

32. $A(2, -3, 1), B(6, 5, -1), C(3, -6, 4), D(7, 2, 2)$

$\overrightarrow{AB} = \langle 4, 8, -2\rangle, \overrightarrow{AC} = \langle 1, -3, 3\rangle, \overrightarrow{CD} = \langle 4, 8, -2\rangle, \overrightarrow{BD} = \langle 1, -3, 3\rangle$

Since $\overrightarrow{AB} = \overrightarrow{CD}$ and $\overrightarrow{AC} = \overrightarrow{BD}$, the figure is a parallelogram. $\overrightarrow{AB}$ and $\overrightarrow{AC}$ are adjacent sides and

$$\overrightarrow{AB}\times\overrightarrow{AC} = \begin{vmatrix} \mathbf{i} & \mathbf{j} & \mathbf{k} \\ 4 & 8 & -2 \\ 1 & -3 & 3 \end{vmatrix} = \langle 18, -14, -20\rangle.$$

Area $= \|\overrightarrow{AB}\times\overrightarrow{AC}\| = \sqrt{920} = 2\sqrt{230}$

33. $A(0, 0, 0), B(1, 2, 3), C(-3, 0, 0)$

$\overrightarrow{AB} = \langle 1, 2, 3\rangle, \overrightarrow{AC} = \langle -3, 0, 0\rangle$

$$\overrightarrow{AB} \times \overrightarrow{AC} = \begin{vmatrix} \mathbf{i} & \mathbf{j} & \mathbf{k} \\ 1 & 2 & 3 \\ -3 & 0 & 0 \end{vmatrix} = -9\mathbf{j} + 6\mathbf{k}$$

$$A = \frac{1}{2}\|\overrightarrow{AB} \times \overrightarrow{AC}\| = \frac{1}{2}\sqrt{117} = \frac{3}{2}\sqrt{13}$$

34. $A(2, -3, 4), B(0, 1, 2), C(-1, 2, 0)$

$\overrightarrow{AB} = \langle -2, 4, -2\rangle, \overrightarrow{AC} = \langle -3, 5, -4\rangle$

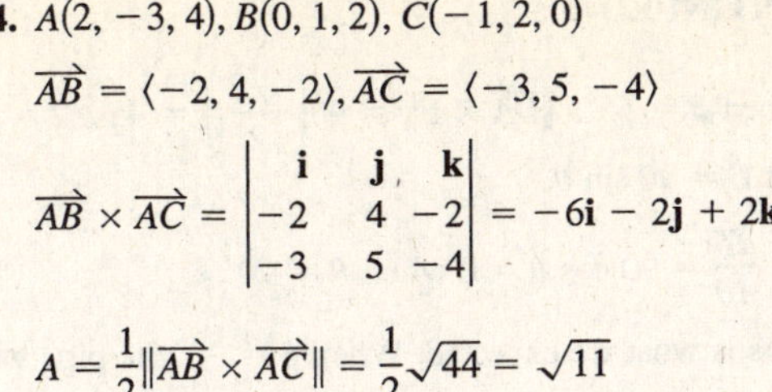

$$\overrightarrow{AB} \times \overrightarrow{AC} = \begin{vmatrix} \mathbf{i} & \mathbf{j} & \mathbf{k} \\ -2 & 4 & -2 \\ -3 & 5 & -4 \end{vmatrix} = -6\mathbf{i} - 2\mathbf{j} + 2\mathbf{k}$$

$$A = \frac{1}{2}\|\overrightarrow{AB} \times \overrightarrow{AC}\| = \frac{1}{2}\sqrt{44} = \sqrt{11}$$

35. $A(2, -7, 3), B(-1, 5, 8), C(4, 6, -1)$

$\overrightarrow{AB} = \langle -3, 12, 5\rangle, \overrightarrow{AC} = \langle 2, 13, -4\rangle$

$$\overrightarrow{AB} \times \overrightarrow{AC} = \begin{vmatrix} \mathbf{i} & \mathbf{j} & \mathbf{k} \\ -3 & 12 & 5 \\ 2 & 13 & -4 \end{vmatrix} = \langle -113, -2, -63\rangle$$

$$\text{Area} = \frac{1}{2}\|\overrightarrow{AB} \times \overrightarrow{AC}\| = \frac{1}{2}\sqrt{16{,}742}$$

36. $A(1, 2, 0), B(-2, 1, 0), C(0, 0, 0)$

$\overrightarrow{AB} = \langle -3, -1, 0\rangle, \overrightarrow{AC} = \langle -1, -2, 0\rangle$

$$\overrightarrow{AB} \times \overrightarrow{AC} = \begin{vmatrix} \mathbf{i} & \mathbf{j} & \mathbf{k} \\ -3 & -1 & 0 \\ -1 & -2 & 0 \end{vmatrix} = 5\mathbf{k}$$

$$A = \frac{1}{2}\|\overrightarrow{AB} \times \overrightarrow{AC}\| = \frac{5}{2}$$

37. $\mathbf{F} = -20\mathbf{k}$

$$\overrightarrow{PQ} = \frac{1}{2}(\cos 40^\circ \mathbf{j} + \sin 40^\circ \mathbf{k})$$

$$\overrightarrow{PQ} \times \mathbf{F} = \begin{vmatrix} \mathbf{i} & \mathbf{j} & \mathbf{k} \\ 0 & \cos 40^\circ/2 & \sin 40^\circ/2 \\ 0 & 0 & -20 \end{vmatrix} = -10\cos 40^\circ \mathbf{i}$$

$$\|\overrightarrow{PQ} \times \mathbf{F}\| = 10\cos 40^\circ \approx 7.66 \text{ ft} \cdot \text{lb}$$

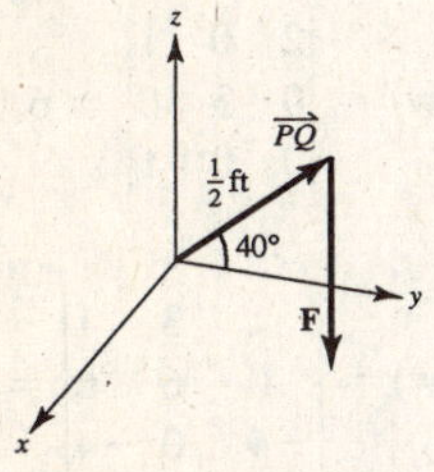

38. $\mathbf{F} = -2000(\cos 30^\circ \mathbf{j} + \sin 30^\circ \mathbf{k}) = -1000\sqrt{3}\mathbf{j} - 1000\mathbf{k}$

$\overrightarrow{PQ} = 0.16\mathbf{k}$

$$\overrightarrow{PQ} \times \mathbf{F} = \begin{vmatrix} \mathbf{i} & \mathbf{j} & \mathbf{k} \\ 0 & 0 & 0.16 \\ 0 & -1000\sqrt{3} & -1000 \end{vmatrix} = 160\sqrt{3}\mathbf{i}$$

$$\|\overrightarrow{PQ} \times \mathbf{F}\| = 160\sqrt{3} \text{ ft} \cdot \text{lb}$$

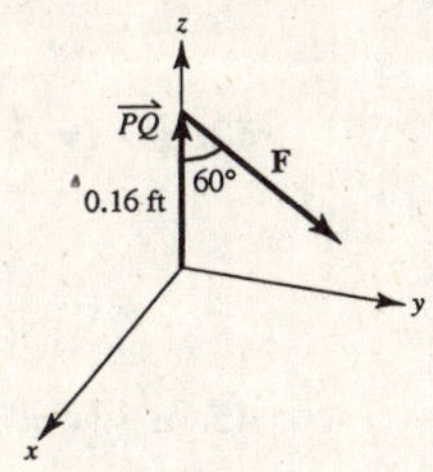

39. (a) Place the wrench in the xy-plane, as indicated in the figure.

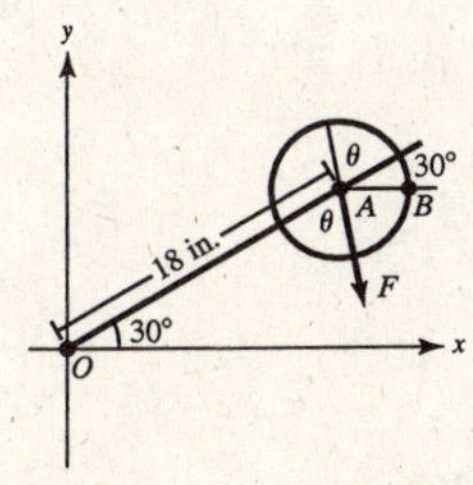

The angle from $\overrightarrow{AB}$ to $\mathbf{F}$ is $30^\circ + 180^\circ + \theta = 210^\circ + \theta$.

$\|\overrightarrow{OA}\| = 18$ inches $= 1.5$ feet

$$\overrightarrow{AB} = 1.5[\cos 30^\circ \mathbf{i} + \sin 30^\circ \mathbf{j}] = \frac{3\sqrt{3}}{4}\mathbf{i} + \frac{3}{4}\mathbf{j}$$

$$\mathbf{F} = 60[\cos(210^\circ + \theta)\mathbf{i} + \sin(210^\circ + \theta)\mathbf{j}]$$

$$\overrightarrow{OA} \times \mathbf{F} = \begin{vmatrix} \mathbf{i} & \mathbf{j} & \mathbf{k} \\ 3\sqrt{3}/4 & 3/4 & 0 \\ 60\cos(210^\circ + \theta) & 60\sin(210^\circ + \theta) & 0 \end{vmatrix}$$

$$= [45\sqrt{3}\sin(210^\circ + \theta) - 45\cos(210^\circ + \theta)]\mathbf{k}$$

$$= [45\sqrt{3}(\sin 210^\circ \cos\theta + \cos 210^\circ \sin\theta) - 45(\cos 210^\circ \cos\theta - \sin 210^\circ \sin\theta)]\mathbf{k}$$

$$= \left[45\sqrt{3}\left(-\frac{1}{2}\cos\theta - \frac{\sqrt{3}}{2}\sin\theta\right) - 45\left(-\frac{\sqrt{3}}{2}\cos\theta + \frac{1}{2}\sin\theta\right)\right]\mathbf{k}$$

$$= (-90\sin\theta)\mathbf{k}$$

Hence, $\|\overrightarrow{OA} \times \mathbf{F}\| = 90\sin\theta$.

140

y = 90 sinθ

0 180

0

—CONTINUED—

39. —CONTINUED—

(b) When $\theta = 45°$: $\|\overrightarrow{OA} \times \mathbf{F}\| = 90\left(\frac{\sqrt{2}}{2}\right) = 45\sqrt{2} \approx 63.64$.

(c) Let $T = 90 \sin \theta$.

$$\frac{dT}{d\theta} = 90 \cos \theta = 0 \text{ when } \theta = 90°.$$

This is what we expected. When $\theta = 90°$ the pipe wrench is horizontal.

40. (a) B is $-\frac{15}{12} = -\frac{5}{4}$ to the left of A, and one foot upwards:

$$\overrightarrow{AB} = \frac{-5}{4}\mathbf{j} + \mathbf{k}$$

$$\mathbf{F} = -200(\cos \theta\mathbf{j} + \sin \theta\mathbf{k})$$

(b) $$\overrightarrow{AB} \times \mathbf{F} = \begin{vmatrix} \mathbf{i} & \mathbf{j} & \mathbf{k} \\ 0 & -5/4 & 1 \\ 0 & -200\cos\theta & -200\sin\theta \end{vmatrix}$$

$$= (250 \sin \theta + 200 \cos \theta)\mathbf{i}$$

$$\|\overrightarrow{AB} \times \mathbf{F}\| = |250 \sin \theta + 200 \cos \theta|$$

$$= 25|10 \sin \theta + 8 \cos \theta|$$

(c) For $\theta = 30°$,

$$\|\overrightarrow{AB} \times \mathbf{F}\| = 25\left(10\left(\frac{1}{2}\right) + 8\left(\frac{\sqrt{3}}{2}\right)\right)$$

$$= 25(5 + 4\sqrt{3}) \approx 298.2.$$

(d) If $T = \|\overrightarrow{AB} \times \mathbf{F}\|$,

$$\frac{dT}{d\theta} = 25(10 \cos \theta - 8 \sin \theta) = 0 \Rightarrow \tan \theta = \frac{5}{4}$$

$$\Rightarrow \theta \approx 51.34°.$$

The vectors are orthogonal.

(e) The zero is $\theta \approx 141.34°$, the angle making $\overrightarrow{AB}$ parallel to $\mathbf{F}$.

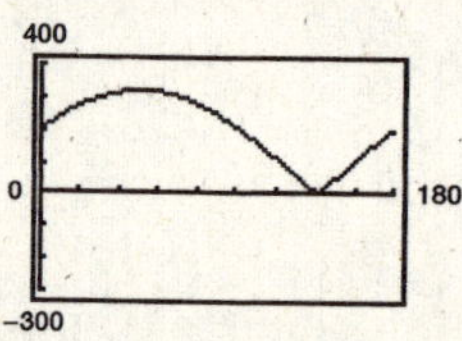

41. $\mathbf{u} \cdot (\mathbf{v} \times \mathbf{w}) = \begin{vmatrix} 1 & 0 & 0 \\ 0 & 1 & 0 \\ 0 & 0 & 1 \end{vmatrix} = 1$

42. $\mathbf{u} \cdot (\mathbf{v} \times \mathbf{w}) = \begin{vmatrix} 1 & 1 & 1 \\ 2 & 1 & 0 \\ 0 & 0 & 1 \end{vmatrix} = -1$

43. $\mathbf{u} \cdot (\mathbf{v} \times \mathbf{w}) = \begin{vmatrix} 2 & 0 & 1 \\ 0 & 3 & 0 \\ 0 & 0 & 1 \end{vmatrix} = 6$

44. $\mathbf{u} \cdot (\mathbf{v} \times \mathbf{w}) = \begin{vmatrix} 2 & 0 & 0 \\ 1 & 1 & 1 \\ 0 & 2 & 2 \end{vmatrix} = 0$

45. $\mathbf{u} \cdot (\mathbf{v} \times \mathbf{w}) = \begin{vmatrix} 1 & 1 & 0 \\ 0 & 1 & 1 \\ 1 & 0 & 1 \end{vmatrix} = 2$

$V = |\mathbf{u} \cdot (\mathbf{v} \times \mathbf{w})| = 2$

46. $\mathbf{u} \cdot (\mathbf{v} \times \mathbf{w}) = \begin{vmatrix} 1 & 3 & 1 \\ 0 & 6 & 6 \\ -4 & 0 & -4 \end{vmatrix} = -72$

$V = |\mathbf{u} \cdot (\mathbf{v} \times \mathbf{w})| = 72$

47. $\mathbf{u} = \langle 3, 0, 0 \rangle$

$\mathbf{v} = \langle 0, 5, 1 \rangle$

$\mathbf{w} = \langle 2, 0, 5 \rangle$

$\mathbf{u} \cdot (\mathbf{v} \times \mathbf{w}) = \begin{vmatrix} 3 & 0 & 0 \\ 0 & 5 & 1 \\ 2 & 0 & 5 \end{vmatrix} = 75$

$V = |\mathbf{u} \cdot (\mathbf{v} \times \mathbf{w})| = 75$

48. $\mathbf{u} = \langle 1, 1, 0 \rangle$

$\mathbf{v} = \langle 1, 0, 2 \rangle$

$\mathbf{w} = \langle 0, 1, 1 \rangle$

$\mathbf{u} \cdot (\mathbf{v} \times \mathbf{w}) = \begin{vmatrix} 1 & 1 & 0 \\ 1 & 0 & 2 \\ 0 & 1 & 1 \end{vmatrix} = -3$

$V = |\mathbf{u} \cdot (\mathbf{v} \times \mathbf{w})| = 3$

49. $\mathbf{u} \times \mathbf{v} = \langle u_1, u_2, u_3 \rangle \cdot \langle v_1, v_2, v_3 \rangle = (u_2v_3 - u_3v_2)\mathbf{i} - (u_1v_3 - u_3v_1)\mathbf{j} + (u_1v_2 - u_2v_1)\mathbf{k}$

50. See Theorem 11.8, page 792.

51. The magnitude of the cross product will increase by a factor of 4.

52. Form the vectors for two sides of the triangle, and compute their cross product:

$$\langle x_2 - x_1, y_2 - y_1, z_2 - z_1 \rangle \times \langle x_3 - x_1, y_3 - y_1, z_3 - z_1 \rangle$$

53. If the vectors are ordered pairs, then the cross product does not exist. False.

54. False, let $\mathbf{u} = \langle 1, 0, 0\rangle$, $\mathbf{v} = \langle 1, 0, 0\rangle$, $\mathbf{w} = \langle -1, 0, 0\rangle$.

Then,

$$\mathbf{u} \times \mathbf{v} = \mathbf{u} \times \mathbf{w} = \mathbf{0}, \text{ but } \mathbf{v} \neq \mathbf{w}.$$

55. True

56. $\mathbf{u} = \langle u_1, u_2, u_3\rangle$, $\mathbf{v} = \langle v_1, v_2, v_3\rangle$, $\mathbf{w} = \langle w_1, w_2, w_3\rangle$

$\mathbf{u} = u_1\mathbf{i} + u_2\mathbf{j} + u_3\mathbf{k}$

$\mathbf{v} \times \mathbf{w} = (v_2w_3 - v_3w_2)\mathbf{i} - (v_1w_3 - v_3w_1)\mathbf{j} + (v_1w_2 - v_2w_1)\mathbf{k}$

$$\mathbf{u} \cdot (\mathbf{v} + \mathbf{w}) = u_1(v_2w_3 - v_3w_2) - u_2(v_1w_3 - v_3w_1) + u_3(v_1w_2 - v_2w_1) = \begin{vmatrix} u_1 & u_2 & u_3 \\ v_1 & v_2 & v_3 \\ w_1 & w_2 & w_3 \end{vmatrix}$$

57. $\mathbf{u} = \langle u_1, u_2, u_3\rangle$, $\mathbf{v} = \langle v_1, v_2, v_3\rangle$, $\mathbf{w} = \langle w_1, w_2, w_3\rangle$

$$\mathbf{u} \times (\mathbf{v} + \mathbf{w}) = \begin{vmatrix} \mathbf{i} & \mathbf{j} & \mathbf{k} \\ u_1 & u_2 & u_3 \\ v_1 + w_1 & v_2 + w_2 & v_3 + w_3 \end{vmatrix}$$

$$= [u_2(v_3 + w_3) - u_3(v_2 + w_2)]\mathbf{i} - [u_1(v_3 + w_3) - u_3(v_1 + w_1)]\mathbf{j} + [u_1(v_2 + w_2) - u_2(v_1 + w_1)]\mathbf{k}$$

$$= (u_2v_3 - u_3v_2)\mathbf{i} - (u_1v_3 - u_3v_1)\mathbf{j} + (u_1v_2 - u_2v_1)\mathbf{k} + (u_2w_3 - u_3w_2)\mathbf{i} -$$

$$(u_1w_3 - u_3w_1)\mathbf{j} + (u_1w_2 - u_2w_1)\mathbf{k}$$

$$= (\mathbf{u} \times \mathbf{v}) + (\mathbf{u} \times \mathbf{w})$$

58. $\mathbf{u} = \langle u_1, u_2, u_3\rangle$, $\mathbf{v} = \langle v_1, v_2, v_3\rangle$, c is a scalar.

$$(c\mathbf{u}) \times \mathbf{v} = \begin{vmatrix} \mathbf{i} & \mathbf{j} & \mathbf{k} \\ cu_1 & cu_2 & cu_3 \\ v_1 & v_2 & v_3 \end{vmatrix}$$

$$= (cu_2v_3 - cu_3v_2)\mathbf{i} - (cu_1v_3 - cu_3v_1)\mathbf{j} + (cu_1v_2 - cu_2v_1)\mathbf{k}$$

$$= c[(u_2v_3 - u_3v_2)\mathbf{i} - (u_1v_3 - u_3v_1)\mathbf{j} + (u_1v_2 - u_2v_1)\mathbf{k}] = c(\mathbf{u} \times \mathbf{v})$$

59. $\mathbf{u} = \langle u_1, u_2, u_3\rangle$

$$\mathbf{u} \times \mathbf{u} = \begin{vmatrix} \mathbf{i} & \mathbf{j} & \mathbf{k} \\ u_1 & u_2 & u_3 \\ u_1 & u_2 & u_3 \end{vmatrix} = (u_2u_3 - u_3u_2)\mathbf{i} - (u_1u_3 - u_3u_1)\mathbf{j} + (u_1u_2 - u_2u_1)\mathbf{k} = \mathbf{0}$$

60. $\mathbf{u} \cdot (\mathbf{v} \times \mathbf{w}) = \begin{vmatrix} u_1 & u_2 & u_3 \\ v_1 & v_2 & v_3 \\ w_1 & w_2 & w_3 \end{vmatrix}$

$$(\mathbf{u} \times \mathbf{v}) \cdot \mathbf{w} = \mathbf{w} \cdot (\mathbf{u} \times \mathbf{v}) = \begin{vmatrix} w_1 & w_2 & w_3 \\ u_1 & u_2 & u_3 \\ v_1 & v_2 & v_3 \end{vmatrix}$$

$$= w_1(u_2v_3 - v_2u_3) - w_2(u_1v_3 - v_1u_3) + w_3(u_1v_2 - v_1u_2)$$

$$= u_1(v_2w_3 - w_2v_3) - u_2(v_1w_3 - w_1v_3) + u_3(v_1w_2 - w_1v_2)$$

$$= \mathbf{u} \cdot (\mathbf{v} \times \mathbf{w})$$

61. $\mathbf{u} \times \mathbf{v} = (u_2v_3 - u_3v_2)\mathbf{i} - (u_1v_3 - u_3v_1)\mathbf{j} + (u_1v_2 - u_2v_1)\mathbf{k}$

$(\mathbf{u} \times \mathbf{v}) \cdot \mathbf{u} = (u_2v_3 - u_3v_2)u_1 + (u_3v_1 - u_1v_3)u_2 + (u_1v_2 - u_2v_1)u_3 = 0$

$(\mathbf{u} \times \mathbf{v}) \cdot \mathbf{v} = (u_2v_3 - u_3v_2)v_1 + (u_3v_1 - u_1v_3)v_2 + (u_1v_2 - u_2v_1)v_3 = 0$

Thus, $\mathbf{u} \times \mathbf{v} \perp \mathbf{u}$ and $\mathbf{u} \times \mathbf{v} \perp \mathbf{v}$.

62. If $\mathbf{u}$ and $\mathbf{v}$ are scalar multiples of each other, $\mathbf{u} = c\mathbf{v}$ for some scalar c.

$$\mathbf{u} \times \mathbf{v} = (c\mathbf{v}) \times \mathbf{v} = c(\mathbf{v} \times \mathbf{v}) = c(\mathbf{0}) = \mathbf{0}$$

If $\mathbf{u} \times \mathbf{v} = \mathbf{0}$, then $\|\mathbf{u}\|\,\|\mathbf{v}\| \sin\theta = 0$. (Assume $\mathbf{u} \neq \mathbf{0}$, $\mathbf{v} \neq \mathbf{0}$.) Thus, $\sin\theta = 0$, $\theta = 0$, and $\mathbf{u}$ and $\mathbf{v}$ are parallel. Therefore, $\mathbf{u} = c\mathbf{v}$ for some scalar c.

63. $\|\mathbf{u} \times \mathbf{v}\| = \|\mathbf{u}\|\,\|\mathbf{v}\| \sin\theta$

If $\mathbf{u}$ and $\mathbf{v}$ are orthogonal, $\theta = \pi/2$ and $\sin\theta = 1$. Therefore, $\|\mathbf{u} \times \mathbf{v}\| = \|\mathbf{u}\|\,\|\mathbf{v}\|$.

64. $\mathbf{u} = \langle a_1, b_1, c_1\rangle$, $\mathbf{v} = \langle a_2, b_2, c_2\rangle$, $\mathbf{w} = \langle a_3, b_3, c_3\rangle$

$$\mathbf{v} \times \mathbf{w} = \begin{vmatrix} \mathbf{i} & \mathbf{j} & \mathbf{k} \\ a_2 & b_2 & c_2 \\ a_3 & b_3 & c_3 \end{vmatrix} = (b_2c_3 - b_3c_2)\mathbf{i} - (a_2c_3 - a_3c_2)\mathbf{j} + (a_2b_3 - a_3b_2)\mathbf{k}$$

$$\mathbf{u} \times (\mathbf{v} \times \mathbf{w}) = \begin{vmatrix} \mathbf{i} & \mathbf{j} & \mathbf{k} \\ a_1 & b_1 & c_1 \\ (b_2c_3 - b_3c_2) & (a_3c_2 - a_2c_3) & (a_2b_3 - a_3b_2) \end{vmatrix}$$

$$\begin{aligned}
\mathbf{u} \times (\mathbf{v} \times \mathbf{w}) &= [b_1(a_2b_3 - a_3b_2) - c_1(a_3c_2 - a_2c_3)]\mathbf{i} - [a_1(a_2b_3 - a_3b_2) - c_1(b_2c_3 - b_3c_2)]\mathbf{j} + \\
&\quad [a_1(a_3c_2 - a_2c_3) - b_1(b_2c_3 - b_3c_2)]\mathbf{k} \\
&= [a_2(a_1a_3 + b_1b_3 + c_1c_3) - a_3(a_1a_2 + b_1b_2 + c_1c_2)]\mathbf{i} + \\
&\quad [b_2(a_1a_3 + b_1b_3 + c_1c_3) - b_3(a_1a_2 + b_1b_2 + c_1c_2)]\mathbf{j} + \\
&\quad [c_2(a_1a_3 + b_1b_3 + c_1c_3) - c_3(a_1a_2 + b_1b_2 + c_1c_2)]\mathbf{k} \\
&= (a_1a_3 + b_1b_3 + c_1c_3)\langle a_2, b_2, c_2\rangle - (a_1a_2 + b_1b_2 + c_1c_2)\langle a_3, b_3, c_3\rangle \\
&= (\mathbf{u} \cdot \mathbf{w})\mathbf{v} - (\mathbf{u} \cdot \mathbf{v})\mathbf{w}
\end{aligned}$$

Section 11.5 Lines and Planes in Space

1. $x = 1 + 3t$, $y = 2 - t$, $z = 2 + 5t$

(a)

z
x
y

(b) When $t = 0$ we have $P = (1, 2, 2)$. When $t = 3$ we have $Q = (10, -1, 17)$.

$$\overrightarrow{PQ} = \langle 9, -3, 15\rangle$$

The components of the vector and the coefficients of t are proportional since the line is parallel to $\overrightarrow{PQ}$.

(c) $y = 0$ when $t = 2$. Thus, $x = 7$ and $z = 12$.
Point: $(7, 0, 12)$

$x = 0$ when $t = -\frac{1}{3}$. Point: $\left(0, \frac{7}{3}, \frac{1}{3}\right)$

$z = 0$ when $t = -\frac{2}{5}$. Point: $\left(-\frac{1}{5}, \frac{12}{5}, 0\right)$

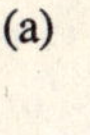

2. $x = 2 - 3t, y = 2, z = 1 - t$

(a)

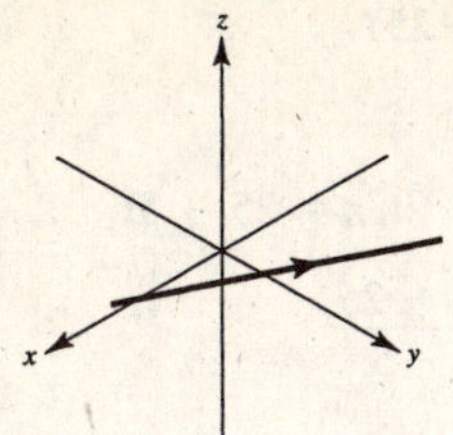

(b) When $t = 0$ we have $P = (2, 2, 1)$. When $t = 2$ we have $Q = (-4, 2, -1)$.

$$\overrightarrow{PQ} = \langle -6, 0, -2 \rangle$$

The components of the vector and the coefficients of t are proportional since the line is parallel to $\overrightarrow{PQ}$.

(c) $z = 0$ when $t = 1$. Thus, $x = -1$ and $y = 2$.

Point: $(-1, 2, 0)$

$x = 0$ when $t = \frac{2}{3}$. Point: $\left(0, 2, \frac{1}{3}\right)$

3. Point: $(0, 0, 0)$

Direction vector: $\mathbf{v} = \langle 1, 2, 3 \rangle$

Direction numbers: 1, 2, 3

(a) Parametric: $x = t, y = 2t, z = 3t$

(b) Symmetric: $x = \frac{y}{2} = \frac{z}{3}$

4. Point: $(0, 0, 0)$

Direction vector: $\mathbf{v} = \left\langle -2, \frac{5}{2}, 1 \right\rangle$

Direction numbers: $-4, 5, 2$

(a) Parametric: $x = -4t, y = 5t, z = 2t$

(b) Symmetric: $\frac{x}{-4} = \frac{y}{5} = \frac{z}{2}$

5. Point: $(-2, 0, 3)$

Direction vector: $\mathbf{v} = \langle 2, 4, -2 \rangle$

Direction numbers: $2, 4, -2$

(a) Parametric: $x = -2 + 2t, y = 4t, z = 3 - 2t$

(b) Symmetric: $\frac{x + 2}{2} = \frac{y}{4} = \frac{z - 3}{-2}$

6. Point: $(-3, 0, 2)$

Direction vector: $\mathbf{v} = \langle 0, 6, 3 \rangle$

Direction numbers: 0, 2, 1

(a) Parametric: $x = -3, y = 2t, z = 2 + t$

(b) Symmetric: $\frac{y}{2} = z - 2, x = -3$

7. Point: $(1, 0, 1)$

Direction vector: $\mathbf{v} = 3\mathbf{i} - 2\mathbf{j} + \mathbf{k}$

Direction numbers: $3, -2, 1$

(a) Parametric: $x = 1 + 3t, y = -2t, z = 1 + t$

(b) Symmetric: $\frac{x - 1}{3} = \frac{y}{-2} = \frac{z - 1}{1}$

8. Point: $(-3, 5, 4)$

Directions numbers: $3, -2, 1$

(a) Parametric: $x = -3 + 3t, y = 5 - 2t, z = 4 + t$

(b) Symmetric: $\frac{x + 3}{3} = \frac{y - 5}{-2} = z - 4$

9. Points: $(5, -3, -2), \left(\frac{-2}{3}, \frac{2}{3}, 1\right)$

Direction vector: $\mathbf{v} = \frac{17}{3}\mathbf{i} - \frac{11}{3}\mathbf{j} - 3\mathbf{k}$

Direction numbers: $17, -11, -9$

(a) Parametric: $x = 5 + 17t, y = -3 - 11t, z = -2 - 9t$

(b) Symmetric: $\frac{x - 5}{17} = \frac{y + 3}{-11} = \frac{z + 2}{-9}$

10. Points: $(2, 0, 2), (1, 4, -3)$

Direction vector: $\langle 1, -4, 5 \rangle$

Direction numbers: $1, -4, 5$

(a) Parametric: $x = 2 + t, y = -4t, z = 2 + 5t$

(b) Symmetric: $x - 2 = \frac{y}{-4} = \frac{z - 2}{5}$

11. Points: $(2, 3, 0), (10, 8, 12)$

Direction vector: $\langle 8, 5, 12 \rangle$

Direction numbers: $8, 5, 12$

(a) Parametric: $x = 2 + 8t, y = 3 + 5t, z = 12t$

(b) Symmetric: $\frac{x-2}{8} = \frac{y-3}{5} = \frac{z}{12}$

12. Points: $(0, 0, 25), (10, 10, 0)$

Direction vector: $\langle 10, 10, -25 \rangle$

Direction numbers: $2, 2, -5$

(a) Parametric: $x = 2t, y = 2t, z = 25 - 5t$

(b) Symmetric: $\frac{x}{2} = \frac{y}{2} = \frac{z-25}{-5}$

13. Point: $(2, 3, 4)$

Direction vector: $\mathbf{v} = \mathbf{k}$

Direction numbers: $0, 0, 1$

Parametric: $x = 2, y = 3, z = 4 + t$

14. Point: $(-4, 5, 2)$

Direction vector: $\mathbf{v} = \mathbf{j}$

Direction numbers: $0, 1, 0$

Parametric: $x = -4, y = 5 + t, z = 2$

15. Point: $(2, 3, 4)$

Direction vector: $\mathbf{v} = 3\mathbf{i} + 2\mathbf{j} - \mathbf{k}$

Direction numbers: $3, 2, -1$

Parametric: $x = 2 + 3t, y = 3 + 2t, z = 4 - t$

16. Point: $(-4, 5, 2)$

Direction vector: $\mathbf{v} = -\mathbf{i} + 2\mathbf{j} + \mathbf{k}$

Direction numbers: $-1, 2, 1$

Parametric: $x = -4 - t, y = 5 + 2t, z = 2 + t$

17. Point: $(5, -3, -4)$

Direction vector: $\mathbf{v} = \langle 2, -1, 3 \rangle$

Direction numbers: $2, -1, 3$

Parametric: $x = 5 + 2t, y = -3 - t, z = -4 + 3t$

18. Point: $(-1, 4, -3)$

Direction vector: $\mathbf{v} = 5\mathbf{i} - \mathbf{j}$

Direction numbers: $5, -1, 0$

Parametric: $x = -1 + 5t, y = 4 - t, z = -3$

19. Point: $(2, 1, 2)$

Direction vector: $\langle -1, 1, 1 \rangle$

Direction numbers: $-1, 1, 1$

Parametric: $x = 2 - t, y = 1 + t, z = 2 + t$

20. Point: $(-6, 0, 8)$

Direction vector: $\langle -2, 2, 0 \rangle$

Direction numbers: $-2, 2, 0$

Parametric: $x = -6 - 2t, y = 2t, z = 8$

21. Let $t = 0$: $P = (3, -1, -2)$ (other answers possible)

$\mathbf{v} = \langle -1, 2, 0 \rangle$ (any nonzero multiple of $\mathbf{v}$ is correct)

22. Let $t = 0$: $P = (0, 5, 4)$ (other answers possible)

$\mathbf{v} = \langle 4, -1, 3 \rangle$ (any nonzero multiple of $\mathbf{v}$ is correct)

23. Let each quantity equal 0: $P = (7, -6, -2)$ (other answers possible)

$\mathbf{v} = \langle 4, 2, 1 \rangle$ (any nonzero multiple of $\mathbf{v}$ is correct)

24. Let each quantity equal 0: $P = (-3, 0, 3)$ (other answers possible)

$\mathbf{v} = \langle 5, 8, 6 \rangle$ (any nonzero multiple of $\mathbf{v}$ is correct)

25. L_1: $\mathbf{v} = \langle -3, 2, 4 \rangle$ $(6, -2, 5)$ on line

L_2: $\mathbf{v} = \langle 6, -4, -8 \rangle$ $(6, -2, 5)$ on line

L_3: $\mathbf{v} = \langle -6, 4, 8 \rangle$ $(6, -2, 5)$ not on line

L_4: $\mathbf{v} = \langle 6, 4, -6 \rangle$ not parallel to L_1, L_2, nor L_3

Hence, L_1 and L_2 are identical.

$L_1 = L_2$ and L_3 are parallel.

26. L_1: $\mathbf{v} = \langle 4, -2, 3 \rangle$ $(8, -5, -9)$ on line

L_2: $\mathbf{v} = \langle 2, 1, 5 \rangle$

L_3: $\mathbf{v} = \langle -8, 4, -6 \rangle$ $(8, -5, -9)$ on line

L_4: $\mathbf{v} = \langle -2, 1, 1.5 \rangle$

L_1 and L_2 are identical.

27. At the point of intersection, the coordinates for one line equal the corresponding coordinates for the other line. Thus,

(i) $4t + 2 = 2s + 2$, (ii) $3 = 2s + 3$, and (iii) $-t + 1 = s + 1$.

From (ii), we find that $s = 0$ and consequently, from (iii), $t = 0$. Letting $s = t = 0$, we see that equation (i) is satisfied and therefore the two lines intersect. Substituting zero for s or for t, we obtain the point $(2, 3, 1)$.

$\mathbf{u} = 4\mathbf{i} - \mathbf{k}$ (First line)

$\mathbf{v} = 2\mathbf{i} + 2\mathbf{j} + \mathbf{k}$ (Second line)

$$\cos\theta = \frac{|\mathbf{u}\cdot\mathbf{v}|}{\|\mathbf{u}\|\,\|\mathbf{v}\|} = \frac{8-1}{\sqrt{17}\sqrt{9}} = \frac{7}{3\sqrt{17}} = \frac{7\sqrt{17}}{51}$$

28. By equating like variables, we have

(i) $-3t + 1 = 3s + 1$, (ii) $4t + 1 = 2s + 4$, and (iii) $2t + 4 = -s + 1$.

From (i) we have $s = -t$, and consequently from (ii), $t = \frac{1}{2}$ and from (iii), $t = -3$. The lines do not intersect.

29. Writing the equations of the lines in parametric form we have

$x = 3t$ $\quad y = 2 - t$ $\quad z = -1 + t$

$x = 1 + 4s$ $\quad y = -2 + s$ $\quad z = -3 - 3s.$

For the coordinates to be equal, $3t = 1 + 4s$ and $2 - t = -2 + s$. Solving this system yields $t = \frac{17}{7}$ and $s = \frac{11}{7}$. When using these values for s and t, the z coordinates are not equal. The lines do not intersect.

30. Writing the equations of the lines in parametric form we have

$x = 2 - 3t$ $\quad y = 2 + 6t$ $\quad z = 3 + t$

$x = 3 + 2s$ $\quad y = -5 + s$ $\quad z = -2 + 4s.$

By equating like variables, we have $2 - 3t = 3 + 2s$, $2 + 6t = -5 + s$, $3 + t = -2 + 4s$. Thus, $t = -1$, $s = 1$ and the point of intersection is $(5, -4, 2)$.

$\mathbf{u} = \langle -3, 6, 1\rangle$ (First line)

$\mathbf{v} = \langle 2, 1, 4\rangle$ (Second line)

$$\cos\theta = \frac{|\mathbf{u}\cdot\mathbf{v}|}{\|\mathbf{u}\|\,\|\mathbf{v}\|} = \frac{4}{\sqrt{46}\sqrt{21}} = \frac{4}{\sqrt{966}} = \frac{2\sqrt{966}}{483}$$

31. $x = 2t + 3$ $\quad x = -2s + 7$

$y = 5t - 2$ $\quad y = s + 8$

$z = -t + 1$ $\quad z = 2s - 1$

Point of intersection: $(7, 8, -1)$

Note: $t = 2$ and $s = 0$

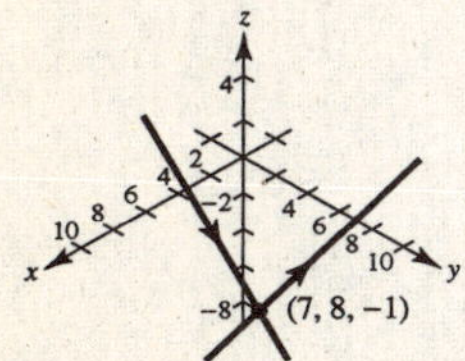

32. $x = 2t - 1$ $\quad x = -5s - 12$

$y = -4t + 10$ $\quad y = 3s + 11$

$z = t$ $\quad z = -2s - 4$

Point of intersection: $(3, 2, 2)$

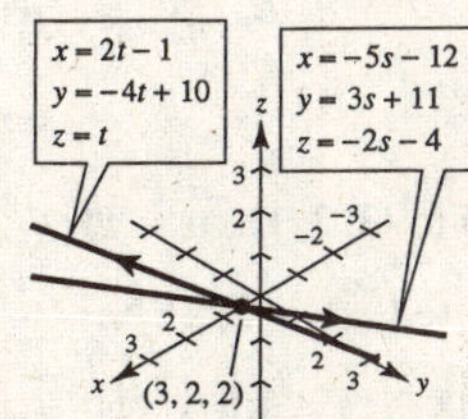

33. $4x - 3y - 6z = 6$

(a) $P = (0, 0, -1), Q = (0, -2, 0), R = (3, 4, -1)$

$\overrightarrow{PQ} = \langle 0, -2, 1\rangle, \overrightarrow{PR} = \langle 3, 4, 0\rangle$

(b) $\overrightarrow{PQ} \times \overrightarrow{PR} = \begin{vmatrix} \mathbf{i} & \mathbf{j} & \mathbf{k} \\ 0 & -2 & 1 \\ 3 & 4 & 0 \end{vmatrix} = \langle -4, 3, 6\rangle$

The components of the cross product are proportional to the coefficients of the variables in the equation. The cross product is parallel to the normal vector.

34. $2x + 3y + 4z = 4$

$P = (0, 0, 1), Q = (2, 0, 0), R = (3, 2, -2)$

(a) $\overrightarrow{PQ} = \langle 2, 0, -1\rangle, \overrightarrow{PR} = \langle 3, 2, -3\rangle$

(b) $\overrightarrow{PQ} \times \overrightarrow{PR} = \begin{vmatrix} \mathbf{i} & \mathbf{j} & \mathbf{k} \\ 2 & 0 & -1 \\ 3 & 2 & -3 \end{vmatrix} = \langle 2, 3, 4\rangle$

The components of the cross product are proportional (for this choice of P, Q, and R, they are the same) to the coefficients of the variables in the equation. The cross product is parallel to the normal vector.

35. Point: $(2, 1, 2)$

$\mathbf{n} = \mathbf{i} = \langle 1, 0, 0\rangle$

$1(x - 2) + 0(y - 1) + 0(z - 2) = 0$

$x - 2 = 0$

36. Point: $(1, 0, -3)$

$\mathbf{n} = \mathbf{k} = \langle 0, 0, 1\rangle$

$0(x - 1) + 0(y - 0) + 1[z - (-3)] = 0$

$z + 3 = 0$

37. Point: $(3, 2, 2)$

Normal vector: $\mathbf{n} = 2\mathbf{i} + 3\mathbf{j} - \mathbf{k}$

$2(x - 3) + 3(y - 2) - 1(z - 2) = 0$

$2x + 3y - z = 10$

38. Point: $(0, 0, 0)$

Normal vector: $\mathbf{n} = -3\mathbf{i} + 2\mathbf{k}$

$-3(x - 0) + 0(y - 0) + 2(z - 0) = 0$

$-3x + 2z = 0$

39. Point: $(0, 0, 6)$

Normal vector: $\mathbf{n} = -\mathbf{i} + \mathbf{j} - 2\mathbf{k}$

$-1(x - 0) + 1(y - 0) - 2(z - 6) = 0$

$-x + y - 2z + 12 = 0$

$x - y + 2z = 12$

40. Point: $(3, 2, 2)$

Normal vector: $\mathbf{v} = 4\mathbf{i} + \mathbf{j} - 3\mathbf{k}$

$4(x - 3) + (y - 2) - 3(z - 2) = 0$

$4x + y - 3z = 8$

41. Let $\mathbf{u}$ be the vector from $(0, 0, 0)$ to $(1, 2, 3)$:

$\mathbf{u} = \mathbf{i} + 2\mathbf{j} + 3\mathbf{k}$

Let $\mathbf{v}$ be the vector from $(0, 0, 0)$ to $(-2, 3, 3)$:

$\mathbf{v} = -2\mathbf{i} + 3\mathbf{j} + 3\mathbf{k}$

Normal vector: $\mathbf{u} \times \mathbf{v} = \begin{vmatrix} \mathbf{i} & \mathbf{j} & \mathbf{k} \\ 1 & 2 & 3 \\ -2 & 3 & 3 \end{vmatrix}$

$= -3\mathbf{i} + (-9)\mathbf{j} + 7\mathbf{k}$

$-3(x - 0) - 9(y - 0) + 7(z - 0) = 0$

$3x + 9y - 7z = 0$

42. Let $\mathbf{u}$ be vector from $(2, 3, -2)$ to $(3, 4, 2)$: $\langle 1, 1, 4\rangle$.

Let $\mathbf{v}$ be vector from $(2, 3, -2)$ to $(1, -1, 0)$: $\langle -1, -4, 2\rangle$.

Normal vector: $\mathbf{u} \times \mathbf{v} = \begin{vmatrix} \mathbf{i} & \mathbf{j} & \mathbf{k} \\ 1 & 1 & 4 \\ -1 & -4 & 2 \end{vmatrix} = \langle 18, -6, -3\rangle$

$= -3\langle -6, 2, 1\rangle$

$-6(x - 2) + 2(y - 3) + 1(z + 2) = 0$

$-6x + 2y + z = -8$

43. Let $\mathbf{u}$ be the vector from $(1, 2, 3)$ to $(3, 2, 1)$: $\mathbf{u} = 2\mathbf{i} - 2\mathbf{k}$

Let $\mathbf{v}$ be the vector from $(1, 2, 3)$ to $(-1, -2, 2)$: $\mathbf{v} = -2\mathbf{i} - 4\mathbf{j} - \mathbf{k}$

Normal vector: $\left(\tfrac{1}{2}\mathbf{u}\right) \times (-\mathbf{v}) = \begin{vmatrix} \mathbf{i} & \mathbf{j} & \mathbf{k} \\ 1 & 0 & -1 \\ 2 & 4 & 1 \end{vmatrix} = 4\mathbf{i} - 3\mathbf{j} + 4\mathbf{k}$

$4(x - 1) - 3(y - 2) + 4(z - 3) = 0$

$4x - 3y + 4z = 10$

44. $(1, 2, 3)$, Normal vector: $\mathbf{v} = \mathbf{i}$, $1(x - 1) = 0, x = 1$

45. $(1, 2, 3)$, Normal vector: $\mathbf{v} = \mathbf{k}$, $1(z - 3) = 0, z = 3$

46. The plane passes through the three points $(0, 0, 0)$, $(0, 1, 0)$ $(\sqrt{3}, 0, 1)$.

The vector from $(0, 0, 0)$ to $(0, 1, 0)$: $\mathbf{u} = \mathbf{j}$

The vector from $(0, 0, 0)$ to $(\sqrt{3}, 0, 1)$: $\mathbf{v} = \sqrt{3}\mathbf{i} + \mathbf{k}$

Normal vector: $\mathbf{u} \times \mathbf{v} = \begin{vmatrix} \mathbf{i} & \mathbf{j} & \mathbf{k} \\ 0 & 1 & 0 \\ \sqrt{3} & 0 & 1 \end{vmatrix} = \mathbf{i} - \sqrt{3}\mathbf{k}$

$x - \sqrt{3}z = 0$

47. The direction vectors for the lines are $\mathbf{u} = -2\mathbf{i} + \mathbf{j} + \mathbf{k}$, $\mathbf{v} = -3\mathbf{i} + 4\mathbf{j} - \mathbf{k}$.

Normal vector: $\mathbf{u} \times \mathbf{v} = \begin{vmatrix} \mathbf{i} & \mathbf{j} & \mathbf{k} \\ -2 & 1 & 1 \\ -3 & 4 & -1 \end{vmatrix} = -5(\mathbf{i} + \mathbf{j} + \mathbf{k})$

Point of intersection of the lines: $(-1, 5, 1)$

$(x + 1) + (y - 5) + (z - 1) = 0$

$x + y + z = 5$

48. The direction of the line is $\mathbf{u} = 2\mathbf{i} - \mathbf{j} + \mathbf{k}$. Choose any point on the line, [$(0, 4, 0)$, for example], and let $\mathbf{v}$ be the vector from $(0, 4, 0)$ to the given point $(2, 2, 1)$:

$\mathbf{v} = 2\mathbf{i} - 2\mathbf{j} + \mathbf{k}$

Normal vector: $\mathbf{u} \times \mathbf{v} = \begin{vmatrix} \mathbf{i} & \mathbf{j} & \mathbf{k} \\ 2 & -1 & 1 \\ 2 & -2 & 1 \end{vmatrix} = \mathbf{i} - 2\mathbf{k}$

$(x - 2) - 2(z - 1) = 0$

$x - 2z = 0$

49. Let $\mathbf{v}$ be the vector from $(-1, 1, -1)$ to $(2, 2, 1)$: $\mathbf{v} = 3\mathbf{i} + \mathbf{j} + 2\mathbf{k}$

Let $\mathbf{n}$ be a vector normal to the plane $2x - 3y + z = 3$: $\mathbf{n} = 2\mathbf{i} - 3\mathbf{j} + \mathbf{k}$

Since $\mathbf{v}$ and $\mathbf{n}$ both lie in the plane P, the normal vector to P is

$\mathbf{v} \times \mathbf{n} = \begin{vmatrix} \mathbf{i} & \mathbf{j} & \mathbf{k} \\ 3 & 1 & 2 \\ 2 & -3 & 1 \end{vmatrix} = 7\mathbf{i} + \mathbf{j} - 11\mathbf{k}$

$7(x - 2) + 1(y - 2) - 11(z - 1) = 0$

$7x + y - 11z = 5$

50. Let $\mathbf{v}$ be the vector from $(3, 2, 1)$ to $(3, 1, -5)$: $\mathbf{v} = -\mathbf{j} - 6\mathbf{k}$

Let $\mathbf{n}$ be the normal to the given plane: $\mathbf{n} = 6\mathbf{i} + 7\mathbf{j} + 2\mathbf{k}$

Since $\mathbf{v}$ and $\mathbf{n}$ both lie in the plane P, the normal vector to P is:

$\mathbf{v} \times \mathbf{n} = \begin{vmatrix} \mathbf{i} & \mathbf{j} & \mathbf{k} \\ 0 & -1 & -6 \\ 6 & 7 & 2 \end{vmatrix} = 40\mathbf{i} - 36\mathbf{j} + 6\mathbf{k}$

$= 2(20\mathbf{i} - 18\mathbf{j} + 3\mathbf{k})$

$20(x - 3) - 18(y - 2) + 3(z - 1) = 0$

$20x - 18y + 3z = 27$

51. Let $\mathbf{u} = \mathbf{i}$ and let $\mathbf{v}$ be the vector from $(1, -2, -1)$ to $(2, 5, 6)$: $\mathbf{v} = \mathbf{i} + 7\mathbf{j} + 7\mathbf{k}$

Since $\mathbf{u}$ and $\mathbf{v}$ both lie in the plane P, the normal vector to P is:

$\mathbf{u} \times \mathbf{v} = \begin{vmatrix} \mathbf{i} & \mathbf{j} & \mathbf{k} \\ 1 & 0 & 0 \\ 1 & 7 & 7 \end{vmatrix} = -7\mathbf{j} + 7\mathbf{k} = -7(\mathbf{j} - \mathbf{k})$

$[y - (-2)] - [z - (-1)] = 0$

$y - z = -1$

52. Let $\mathbf{u} = \mathbf{k}$ and let $\mathbf{v}$ be the vector from $(4, 2, 1)$ to $(-3, 5, 7)$: $\mathbf{v} = -7\mathbf{i} + 3\mathbf{j} + 6\mathbf{k}$

Since $\mathbf{u}$ and $\mathbf{v}$ both lie in the plane P, the normal vector to P is:

$\mathbf{u} \times \mathbf{v} = \begin{vmatrix} \mathbf{i} & \mathbf{j} & \mathbf{k} \\ 0 & 0 & 1 \\ -7 & 3 & 6 \end{vmatrix} = -3\mathbf{i} - 7\mathbf{j} = -(3\mathbf{i} + 7\mathbf{j})$

$3(x - 4) + 7(y - 2) = 0$

$3x + 7y = 26$

53. xy-plane: Let $z = 0$.

Then $0 = 4 - t \Rightarrow t = 4 \Rightarrow x = 1 - 2(4) = -7$ and

$y = -2 + 3(4) = 10$. Intersection: $(-7, 10, 0)$

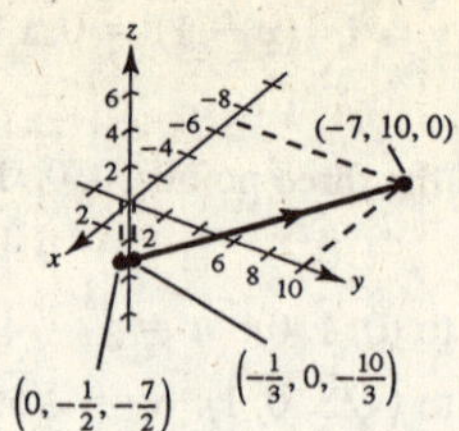

xz-plane: Let $y = 0$.

Then $0 = -2 + 3t \Rightarrow t = \frac{2}{3} \Rightarrow x = 1 - 2\left(\frac{2}{3}\right) = -\frac{1}{3}$ and

$z = -4 + \frac{2}{3} = -\frac{10}{3}$. Intersection: $\left(-\frac{1}{3}, 0, -\frac{10}{3}\right)$

yz-plane: Let $x = 0$.

Then $0 = 1 - 2t \Rightarrow t = \frac{1}{2} \Rightarrow y = -2 + 3\left(\frac{1}{2}\right) = -\frac{1}{2}$ and

$z = -4 + \frac{1}{2} = -\frac{7}{2}$. Intersection: $\left(0, -\frac{1}{2}, -\frac{7}{2}\right)$

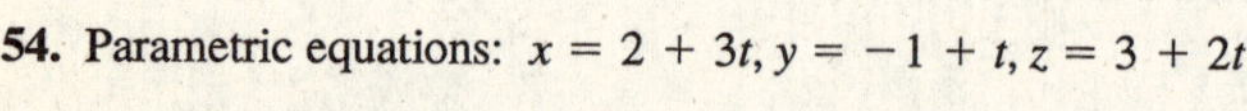

54. Parametric equations: $x = 2 + 3t, y = -1 + t, z = 3 + 2t$

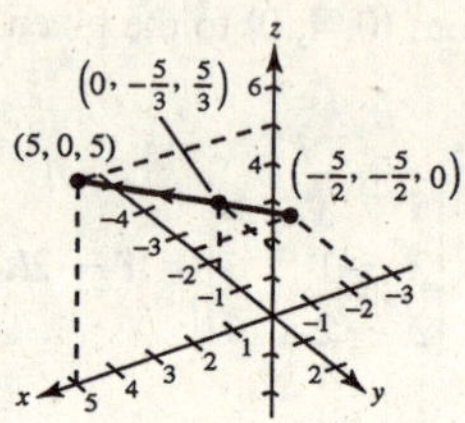

xy-plane: Let $z = 0$.

Then $3 + 2t = 0 \Rightarrow t = -\frac{3}{2} \Rightarrow x = 2 + 3\left(-\frac{3}{2}\right) = -\frac{5}{2}$ and

$y = -1 + \left(-\frac{3}{2}\right) = -\frac{5}{2}$. Intersection: $\left(-\frac{5}{2}, -\frac{5}{2}, 0\right)$

xz-plane: Let $y = 0$.

Then $t = 1 \Rightarrow x = 2 + 3(1) = 5$ and

$z = 3 + 2(1) = 5$. Intersection: $(5, 0, 5)$

yz-plane: Let $x = 0$.

Then $2 + 3t = 0 \Rightarrow t = -\frac{2}{3} \Rightarrow y = -1 - \frac{2}{3} = -\frac{5}{3}$ and

$z = 3 + 2\left(-\frac{2}{3}\right) = \frac{5}{3}$. Intersection: $\left(0, -\frac{5}{3}, \frac{5}{3}\right)$

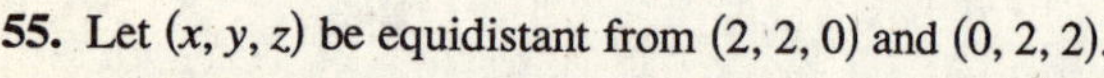

55. Let (x, y, z) be equidistant from $(2, 2, 0)$ and $(0, 2, 2)$.

$$\sqrt{(x-2)^2 + (y-2)^2 + (z-0)^2} = \sqrt{(x-0)^2 + (y-2)^2 + (z-2)^2}$$

$$x^2 - 4x + 4 + y^2 - 4y + 4 + z^2 = x^2 + y^2 - 4y + 4 + z^2 - 4z + 4$$

$$-4x + 8 = -4z + 8$$

$$x - z = 0 \quad \text{Plane}$$

56. Let (x, y, z) be equidistant from $(-3, 1, 2)$ and $(6, -2, 4)$.

$$\sqrt{(x+3)^2 + (y-1)^2 + (z-2)^2} = \sqrt{(x-6)^2 + (y+2)^2 + (z-4)^2}$$

$$x^2 + 6x + 9 + y^2 - 2y + 1 + z^2 - 4z + 4 = x^2 - 12x + 36 + y^2 + 4y + 4 + z^2 - 8z + 16$$

$$6x - 2y - 4z + 14 = -12x + 4y - 8z + 56$$

$$18x - 6y + 4z - 42 = 0$$

$$9x - 3y + 2z - 21 = 0 \quad \text{Plane}$$

57. The normal vectors to the planes are

$$\mathbf{n}_1 = \langle 5, -3, 1\rangle, \mathbf{n}_2 = \langle 1, 4, 7\rangle, \cos\theta = \frac{|\mathbf{n}_1 \cdot \mathbf{n}_2|}{\|\mathbf{n}_1\|\,\|\mathbf{n}_2\|} = 0.$$

Thus, $\theta = \pi/2$ and the planes are orthogonal.

58. The normal vectors to the planes are

$$\mathbf{n}_1 = \langle 3, 1, -4\rangle, \ \mathbf{n}_2 = \langle -9, -3, 12\rangle.$$

Since $\mathbf{n}_2 = -3\mathbf{n}_1$, the planes are parallel, but not equal.

59. The normal vectors to the planes are

$$\mathbf{n}_1 = \mathbf{i} - 3\mathbf{j} + 6\mathbf{k},\ \mathbf{n}_2 = 5\mathbf{i} + \mathbf{j} - \mathbf{k},$$

$$\cos\theta = \frac{|\mathbf{n}_1 \cdot \mathbf{n}_2|}{\|\mathbf{n}_1\|\,\|\mathbf{n}_2\|} = \frac{|5 - 3 - 6|}{\sqrt{46}\sqrt{27}} = \frac{4\sqrt{138}}{414} = \frac{2\sqrt{138}}{207}.$$

Therefore, $\theta = \arccos\left(\frac{2\sqrt{138}}{207}\right) \approx 83.5°$.

60. The normal vectors to the planes are

$$\mathbf{n}_1 = 3\mathbf{i} + 2\mathbf{j} - \mathbf{k},\ \mathbf{n}_2 = \mathbf{i} - 4\mathbf{j} + 2\mathbf{k},$$

$$\cos\theta = \frac{|\mathbf{n}_1 \cdot \mathbf{n}_2|}{\|\mathbf{n}_1\|\,\|\mathbf{n}_2\|} = \frac{|3 - 8 - 2|}{\sqrt{14}\sqrt{21}} = \frac{1}{\sqrt{6}}.$$

Therefore, $\theta = \arccos\left(\frac{1}{\sqrt{6}}\right) \approx 65.9°$.

61. The normal vectors to the planes are $\mathbf{n}_1 = \langle 1, -5, -1\rangle$ and $\mathbf{n}_2 = \langle 5, -25, -5\rangle$. Since $\mathbf{n}_2 = 5\mathbf{n}_1$, the planes are parallel, but not equal.

62. The normal vectors to the planes are

$$\mathbf{n}_1 = \langle 2, 0, -1\rangle,\ \mathbf{n}_2 = \langle 4, 1, 8\rangle,$$

$$\cos\theta = \frac{|\mathbf{n}_1 \cdot \mathbf{n}_2|}{\|\mathbf{n}_1\|\,\|\mathbf{n}_2\|} = 0.$$

Thus, $\theta = \frac{\pi}{2}$ and the planes are orthogonal.

63. $4x + 2y + 6z = 12$

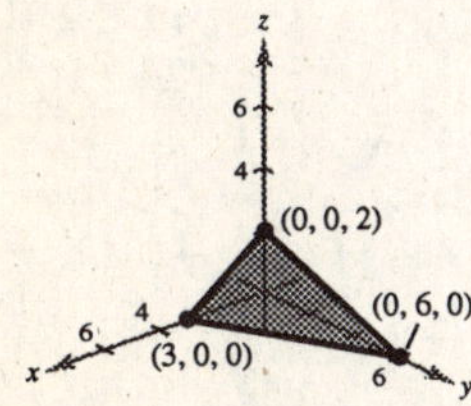

64. $3x + 6y + 2z = 6$

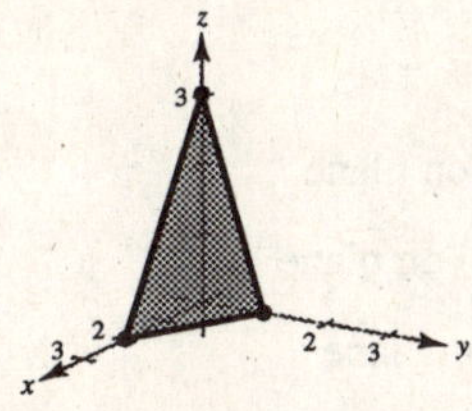

65. $2x - y + 3z = 4$

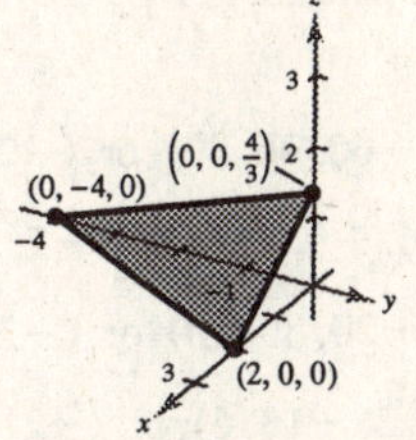

66. $2x - y + z = 4$

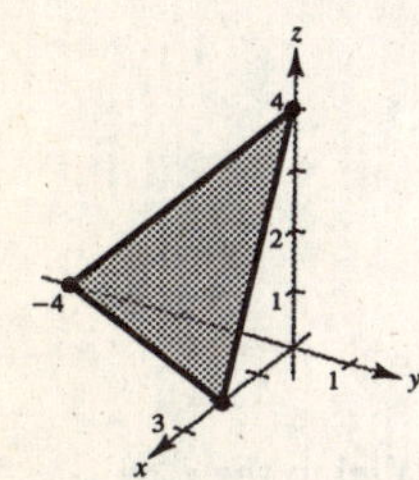

67. $y + z = 5$

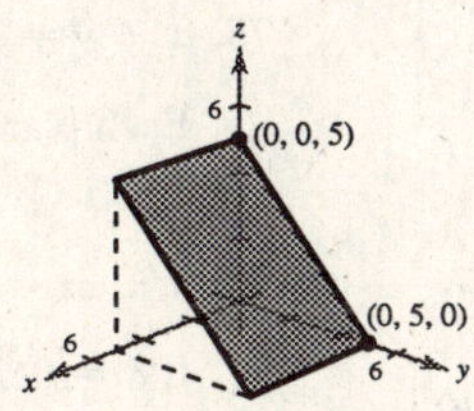

68. $x + 2y = 4$

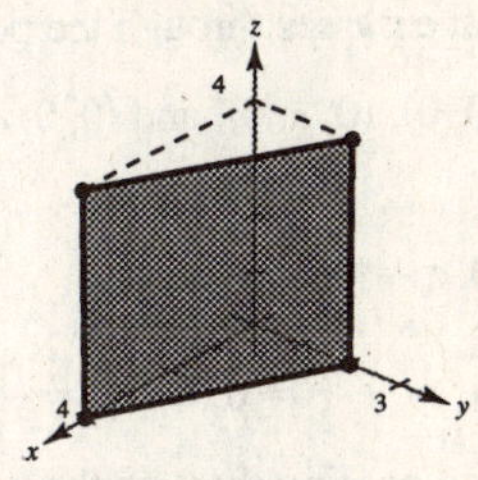

69. $x = 5$

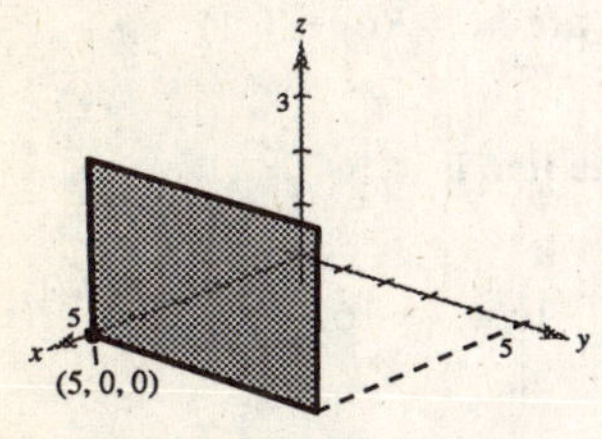

70. $z = 8$

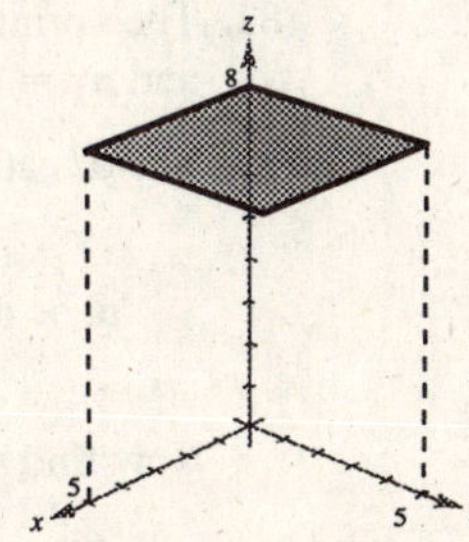

71. $2x + y - z = 6$

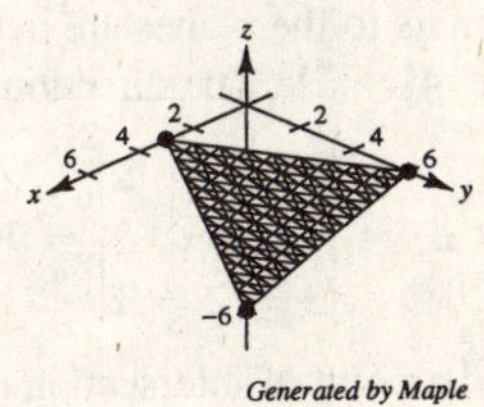

72. $x - 3z = 3$

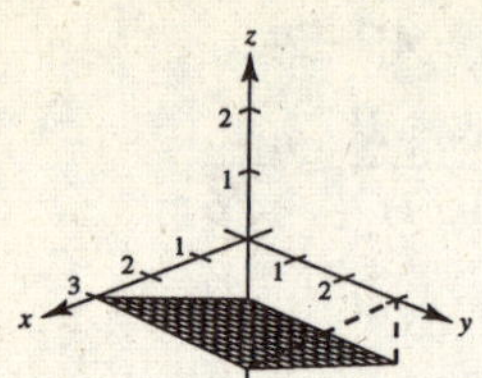

Generated by Mathematica

73. $-5x + 4y - 6z + 8 = 0$

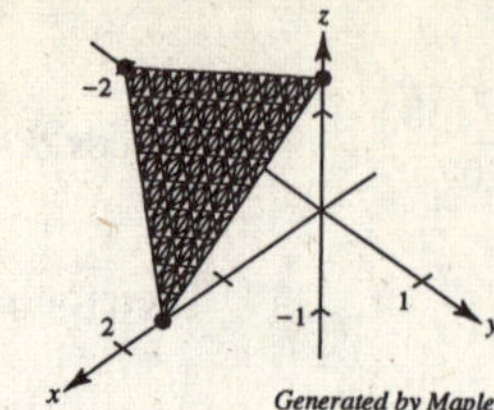

Generated by Maple

74. $2.1x - 4.7y - z + 3 = 0$

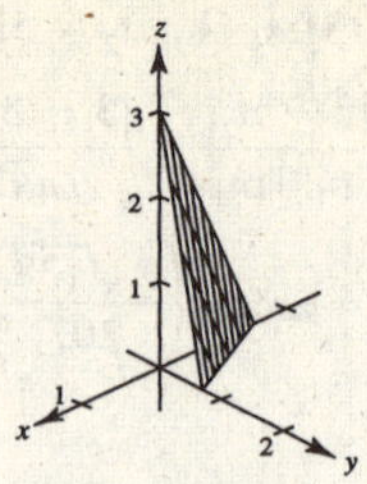

Generated by Mathematica

75. P_1: $\mathbf{n} = \langle 3, -2, 5\rangle$ $(1, -1, 1)$ on plane

P_2: $\mathbf{n} = \langle -6, 4, -10\rangle$ $(1, -1, 1)$ not on plane

P_3: $\mathbf{n} = \langle -3, 2, 5\rangle$

P_4: $\mathbf{n} = \langle 75, -50, 125\rangle$ $(1, -1, 1)$ on plane

P_1 and P_4 are identical.

$P_1 = P_4$ is parallel to P_2.

76. P_1: $\mathbf{n} = \langle -60, 90, 30\rangle$ or $\langle -2, 3, 1\rangle$ $\left(0, 0, \frac{9}{10}\right)$ on plane

P_2: $\mathbf{n} = \langle 6, -9, -3\rangle$ or $\langle -2, 3, 1\rangle$ $\left(0, 0, -\frac{2}{3}\right)$ on plane

P_3: $\mathbf{n} = \langle -20, 30, 10\rangle$ or $\langle -2, 3, 1\rangle$ $\left(0, 0, \frac{5}{6}\right)$ on plane

P_4: $\mathbf{n} = \langle 12, -18, 6\rangle$ or $\langle -2, 3, -1\rangle$

P_1, P_2, and P_3 are parallel.

77. Each plane passes through the points

$(c, 0, 0)$, $(0, c, 0)$, and $(0, 0, c)$.

78. $x + y = c$

Each plane is parallel to the z-axis.

79. If $c = 0$, $z = 0$ is xy-plane.

If $c \neq 0$, $cy + z = 0 \Rightarrow y = \dfrac{-1}{c}z$ is a plane parallel to x-axis and passing through the points $(0, 0, 0)$ and $(0, 1, -c)$.

80. $x + cz = 0$

If $c = 0$, $x = 0$ is the yz-plane.

If $c \neq 0$, $x + cz = 0$ is a plane parallel to the y-axis.

81. The normals to the planes are $\mathbf{n}_1 = 3\mathbf{i} + 2\mathbf{j} - \mathbf{k}$ and $\mathbf{n}_2 = \mathbf{i} - 4\mathbf{j} + 2\mathbf{k}$. The direction vector for the line is

$$\mathbf{n}_2 \times \mathbf{n}_1 = \begin{vmatrix} \mathbf{i} & \mathbf{j} & \mathbf{k} \\ 1 & -4 & 2 \\ 3 & 2 & -1 \end{vmatrix} = 7(\mathbf{j} + 2\mathbf{k}).$$

Now find a point of intersection of the planes.

$$\begin{aligned} 6x + 4y - 2z &= 14 \\ x - 4y + 2z &= 0 \\ 7x \quad &= 14 \\ x &= 2 \end{aligned}$$

Substituting 2 for x in the second equation, we have $-4y + 2z = -2$ or $z = 2y - 1$. Letting $y = 1$, a point of intersection is $(2, 1, 1)$.

$x = 2, y = 1 + t, z = 1 + 2t$

82. The normals to the planes are $\mathbf{n}_1 = \langle 6, -3, 1\rangle$ and $\mathbf{n}_2 = \langle -1, 1, 5\rangle$.

The direction vector for the line is

$$\mathbf{n}_1 \times \mathbf{n}_2 = \begin{vmatrix} \mathbf{i} & \mathbf{j} & \mathbf{k} \\ 6 & -3 & 1 \\ -1 & 1 & 5 \end{vmatrix} = \langle -16, -31, 3\rangle.$$

Now find a point of intersection of the planes.

$$\begin{aligned} 6x - 3y + z = 5 &\Rightarrow & 6x - 3y + z &= 5 \\ -x + y + 5z = 5 &\Rightarrow & -6x + 6y + 30z &= 30 \\ & & 3y + 31z &= 35 \end{aligned}$$

Let $y = -9, z = 2 \Rightarrow x = -4 \Rightarrow (-4, -9, 2)$.

$x = -4 - 16t, y = -9 - 31t, z = 2 + 3t$

83. Writing the equation of the line in parametric form and substituting into the equation of the plane we have:

$$x = \frac{1}{2} + t,\ y = \frac{-3}{2} - t,\ z = -1 + 2t$$

$$2\left(\frac{1}{2} + t\right) - 2\left(\frac{-3}{2} - t\right) + (-1 + 2t) = 12,\ t = \frac{3}{2}$$

Substituting $t = 3/2$ into the parametric equations for the line we have the point of intersection $(2, -3, 2)$. The line does not lie in the plane.

84. Writing the equation of the line in parametric form and substituting into the equation of the plane we have:

$$x = 1 + 4t,\ y = 2t,\ z = 3 + 6t$$

$$2(1 + 4t) + 3(2t) = -5,\ t = \frac{-1}{2}$$

Substituting $t = -\frac{1}{2}$ into the parametric equations for the line we have the point of intersection $(-1, -1, 0)$. The line does not lie in the plane.

85. Writing the equation of the line in parametric form and substituting into the equation of the plane we have:

$$x = 1 + 3t,\ y = -1 - 2t,\ z = 3 + t$$

$$2(1 + 3t) + 3(-1 - 2t) = 10,\ -1 = 10, \text{ contradiction}$$

Therefore, the line does not intersect the plane.

86. Writing the equation of the line in parametric form and substituting into the equation of the plane we have:

$$x = 4 + 2t,\ y = -1 - 3t,\ z = -2 + 5t$$

$$5(4 + 2t) + 3(-1 - 3t) = 17,\ t = 0$$

Substituting $t = 0$ into the parametric equations for the line we have the point of intersection $(4, -1, -2)$. The line does not lie in the plane.

87. Point: $Q(0, 0, 0)$

Plane: $2x + 3y + z - 12 = 0$

Normal to plane: $\mathbf{n} = \langle 2, 3, 1 \rangle$

Point in plane: $P(6, 0, 0)$

Vector $\overrightarrow{PQ} = \langle -6, 0\ 0 \rangle$

$$D = \frac{|\overrightarrow{PQ} \cdot \mathbf{n}|}{\|\mathbf{n}\|} = \frac{|-12|}{\sqrt{14}} = \frac{6\sqrt{14}}{7}$$

88. Point: $Q(0, 0, 0)$

Plane: $8x - 4y + z = 8$

Normal to plane: $\mathbf{n} = \langle 8, -4, 1 \rangle$

Point in plane: $P\langle 1, 0, 0 \rangle$

Vector: $\overrightarrow{PQ} = \langle -1, 0, 0 \rangle$

$$D = \frac{|\overrightarrow{PQ} \cdot \mathbf{n}|}{\|\mathbf{n}\|} = \frac{|-8|}{\sqrt{81}} = \frac{8}{9}$$

89. Point: $Q(2, 8, 4)$

Plane: $2x + y + z = 5$

Normal to plane: $\mathbf{n} = \langle 2, 1, 1 \rangle$

Point in plane: $P\langle 0, 0, 5 \rangle$

Vector: $\overrightarrow{PQ} = \langle 2, 8, -1 \rangle$

$$D = \frac{|\overrightarrow{PQ} \cdot \mathbf{n}|}{\|\mathbf{n}\|} = \frac{11}{\sqrt{6}} = \frac{11\sqrt{6}}{6}$$

90. Point: $Q(3, 2, 1)$

Plane: $x - y + 2z = 4$

Normal to plane: $\mathbf{n} = \langle 1, -1, 2 \rangle$

Point in plane: $P\langle 4, 0, 0 \rangle$

Vector: $\overrightarrow{PQ} = \langle -1, 2, 1 \rangle$

$$D = \frac{|\overrightarrow{PQ} \cdot \mathbf{n}|}{\|\mathbf{n}\|} = \frac{|-1|}{\sqrt{6}} = \frac{1}{\sqrt{6}} = \frac{\sqrt{6}}{6}$$

91. The normal vectors to the planes are $\mathbf{n}_1 = \langle 1, -3, 4 \rangle$ and $\mathbf{n}_2 = \langle 1, -3, 4 \rangle$. Since $\mathbf{n}_1 = \mathbf{n}_2$, the planes are parallel. Choose a point in each plane.

$P = (10, 0, 0)$ is a point in $x - 3y + 4z = 10$.

$Q = (6, 0, 0)$ is a point in $x - 3y + 4z = 6$.

$$\overrightarrow{PQ} = \langle -4, 0, 0 \rangle,\ D = \frac{|\overrightarrow{PQ} \cdot \mathbf{n}_1|}{\|\mathbf{n}_1\|} = \frac{4}{\sqrt{26}} = \frac{2\sqrt{26}}{13}$$

92. The normal vectors to the planes are $\mathbf{n}_1 = \langle 4, -4, 9 \rangle$ and $\mathbf{n}_2 = \langle 4, -4, 9 \rangle$. Since $\mathbf{n}_1 = \mathbf{n}_2$, the planes are parallel. Choose a point in each plane.

$P = (-5, 0, 3)$ is a point in $4x - 4y + 9z = 7$.

$Q = (0, 0, 2)$ is a point in $4x - 4y + 9z = 18$.

$\overrightarrow{PQ} = \langle 5, 0, -1 \rangle$

$$D = \frac{|\overrightarrow{PQ} \cdot \mathbf{n}_1|}{\|\mathbf{n}_1\|} = \frac{11}{\sqrt{113}} = \frac{11\sqrt{113}}{113}$$

93. The normal vectors to the planes are $\mathbf{n}_1 = \langle -3, 6, 7 \rangle$ and $\mathbf{n}_2 = \langle 6, -12, -14 \rangle$. Since $\mathbf{n}_2 = -2\mathbf{n}_1$, the planes are parallel. Choose a point in each plane.

$P = (0, -1, 1)$ is a point in $-3x + 6y + 7z = 1$.

$Q = \left(\frac{25}{6}, 0, 0\right)$ is a point in $6x - 12y - 14z = 25$.

$$\overrightarrow{PQ} = \left\langle \frac{25}{6}, 1, -1 \right\rangle$$

$$D = \frac{|\overrightarrow{PQ} \cdot \mathbf{n}_1|}{\|\mathbf{n}_1\|} = \frac{|-27/2|}{\sqrt{94}} = \frac{27}{2\sqrt{94}} = \frac{27\sqrt{94}}{188}$$

94. The normal vectors to the planes are $\mathbf{n}_1 = \langle 2, 0, -4\rangle$ and $\mathbf{n}_2 = \langle 2, 0, -4\rangle$. Since $\mathbf{n}_1 = \mathbf{n}_2$, the planes are parallel. Choose a point in each plane.

$P = (2, 0, 0)$ is a point in $2x - 4z = 4$.
$Q = (5, 0, 0)$ is a point in $2x - 4z = 10$.

$$\overrightarrow{PQ} = \langle 3, 0, 0\rangle, D = \frac{|\overrightarrow{PQ} \cdot \mathbf{n}_1|}{\|\mathbf{n}_1\|} = \frac{6}{\sqrt{20}} = \frac{3\sqrt{5}}{5}$$

95. $\mathbf{u} = \langle 4, 0, -1\rangle$ is the direction vector for the line. $Q(1, 5, -2)$ is the given point, and $P(-2, 3, 1)$ is on the line. Hence, $\overrightarrow{PQ} = \langle 3, 2, -3\rangle$ and

$$\overrightarrow{PQ} \times \mathbf{u} = \begin{vmatrix} \mathbf{i} & \mathbf{j} & \mathbf{k} \\ 3 & 2 & -3 \\ 4 & 0 & -1 \end{vmatrix} = \langle -2, -9, -8\rangle.$$

$$D = \frac{\|\overrightarrow{PQ} \times \mathbf{u}\|}{\|\mathbf{u}\|} = \frac{\sqrt{149}}{\sqrt{17}} = \frac{\sqrt{2533}}{17}$$

96. $\mathbf{u} = \langle 2, 1, 2\rangle$ is the direction vector for the line.

$P = (0, -3, 2)$ is a point on the line (let $t = 0$).

$\overrightarrow{PQ} = \langle 1, 1, 2\rangle$

$$\overrightarrow{PQ} \times \mathbf{u} = \begin{vmatrix} \mathbf{i} & \mathbf{j} & \mathbf{k} \\ 1 & 1 & 2 \\ 2 & 1 & 2 \end{vmatrix} = \langle 0, 2, -1\rangle$$

$$D = \frac{\|\overrightarrow{PQ} \times \mathbf{u}\|}{\|\mathbf{u}\|} = \frac{\sqrt{5}}{\sqrt{9}} = \frac{\sqrt{5}}{3}$$

97. $\mathbf{u} = \langle -1, 1, -2\rangle$ is the direction vector for the line.

$Q = (-2, 1, 3)$ is the given point, and $P = (1, 2, 0)$ is on the line (let $t = 0$ in the parametric equations for the line).

Hence, $\overrightarrow{PQ} = \langle -3, -1, 3\rangle$ and

$$\overrightarrow{PQ} \times \mathbf{u} = \begin{vmatrix} \mathbf{i} & \mathbf{j} & \mathbf{k} \\ -3 & -1 & 3 \\ -1 & 1 & -2 \end{vmatrix} = \langle -1, -9, -4\rangle.$$

$$D = \frac{\|\overrightarrow{PQ} \times \mathbf{u}\|}{\|\mathbf{u}\|} = \frac{\sqrt{1 + 81 + 16}}{\sqrt{1 + 1 + 4}} = \frac{\sqrt{98}}{6} = \frac{7}{\sqrt{3}} = \frac{7\sqrt{3}}{3}$$

98. $\mathbf{u} = \langle 0, 3, 1\rangle$ is the direction vector for the line.

$Q = (4, -1, 5)$ is the given point, and $P = (3, 1, 1)$ is on the line. Hence, $\overrightarrow{PQ} = \langle 1, -2, 4\rangle$ and

$$\overrightarrow{PQ} \times \mathbf{u} = \begin{vmatrix} \mathbf{i} & \mathbf{j} & \mathbf{k} \\ 1 & -2 & 4 \\ 0 & 3 & 1 \end{vmatrix} = \langle -14, -1, 3\rangle.$$

$$D = \frac{\|\overrightarrow{PQ} \times \mathbf{u}\|}{\|\mathbf{u}\|} = \frac{\sqrt{14^2 + 1 + 9}}{\sqrt{9 + 1}} = \sqrt{\frac{206}{10}}$$

$$= \sqrt{\frac{103}{5}} = \frac{\sqrt{515}}{5}$$

99. The direction vector for L_1 is $\mathbf{v}_1 = \langle -1, 2, 1\rangle$.

The direction vector for L_2 is $\mathbf{v}_2 = \langle 3, -6, -3\rangle$.

Since $\mathbf{v}_2 = -3\mathbf{v}_1$, the lines are parallel.

Let $Q = (2, 3, 4)$ to be a point on L_1 and $P = (0, 1, 4)$ a point on L_2. $\overrightarrow{PQ} = \langle 2, 0, 0\rangle$.
$\mathbf{u} = \mathbf{v}_2$ is the direction vector for L_2.

$$\overrightarrow{PQ} \times \mathbf{v}_2 = \begin{vmatrix} \mathbf{i} & \mathbf{j} & \mathbf{k} \\ 2 & 2 & 0 \\ 3 & -6 & -3 \end{vmatrix} = \langle -6, 6, -18\rangle$$

$$D = \frac{\|\overrightarrow{PQ} \times \mathbf{v}_2\|}{\|\mathbf{v}_2\|} = \frac{\sqrt{36 + 36 + 324}}{\sqrt{9 + 36 + 9}} = \sqrt{\frac{396}{54}}$$

$$= \sqrt{\frac{22}{3}} = \frac{\sqrt{66}}{3}$$

100. The direction vector for L_1 is $\mathbf{v}_1 = \langle 6, 9, -12\rangle$.

The direction vector for L_2 is $\mathbf{v}_2 = \langle 4, 6, -8\rangle$.

Since $\mathbf{v}_1 = \frac{3}{2}\mathbf{v}_2$, the lines are parallel.

Let $Q = (3, -2, 1)$ to be a point on L_1 and $P = (-1, 3, 0)$ a point on L_2. $\overrightarrow{PQ} = \langle 4, -5, 1\rangle$.
$\mathbf{u} = \mathbf{v}_2$ is the direction vector for L_2.

$$\overrightarrow{PQ} \times \mathbf{v}_2 = \begin{vmatrix} \mathbf{i} & \mathbf{j} & \mathbf{k} \\ 4 & -5 & 1 \\ 4 & 6 & -8 \end{vmatrix} = \langle 34, 36, 44\rangle$$

$$D = \frac{\|\overrightarrow{PQ} \times \mathbf{v}_2\|}{\|\mathbf{v}_2\|} = \frac{\sqrt{34^2 + 36^2 + 44^2}}{\sqrt{16 + 36 + 64}} = \frac{\sqrt{4388}}{\sqrt{116}}$$

$$= \sqrt{\frac{1097}{29}} = \frac{\sqrt{31813}}{29}$$

101. The parametric equations of a line L parallel to $\mathbf{v} = \langle a, b, c,\rangle$ and passing through the point $P(x_1, y_1, z_1)$ are

$$x = x_1 + at, y = y_1 + bt, z = z_1 + ct.$$

The symmetric equations are

$$\frac{x - x_1}{a} = \frac{y - y_1}{b} = \frac{z - z_1}{c}.$$

102. The equation of the plane containing $P(x_1, y_1, z_1)$ and having normal vector $\mathbf{n} = \langle a, b, c\rangle$ is

$$a(x - x_1) + b(y - y_1) + c(z - z_1) = 0.$$

You need $\mathbf{n}$ and P to find the equation.

103. Solve the two linear equations representing the planes to find two points of intersection. Then find the line determined by the two points.

104. $x = a$: plane parallel to yz-plane containing $(a, 0, 0)$

$y = b$: plane parallel to xz-plane containing $(0, b, 0)$

$z = c$: plane parallel to xy-plane containing $(0, 0, c)$

105. (a) The planes are parallel if their normal vectors are parallel:

$$\langle a_1, b_1, c_1\rangle = t\langle a_2, b_2, c_2\rangle, t \neq 0$$

(b) The planes are perpendicular if their normal vectors are perpendicular:

$$\langle a_1, b_1, c_1\rangle \cdot \langle a_2, b_2, c_2\rangle = 0$$

106. Yes. If $\mathbf{v}_1$ and $\mathbf{v}_2$ are the direction vectors for the lines L_1 and L_2, then $\mathbf{v} = \mathbf{v}_1 \times \mathbf{v}_2$ is perpendicular to both L_1 and L_2.

107. An equation for the plane is

$$\frac{x}{a} + \frac{y}{b} + \frac{z}{c} = 1.$$

For example, letting $y = z = 0$, the x-intercept is $(a, 0, 0)$.

108. (a) Sphere

$$(x - 3)^2 + (y + 2)^2 + (z - 5)^2 = 16$$
$$x^2 + y^2 + z^2 - 6x + 4y - 10z + 22 = 0$$

(b) Parallel planes

$$4x - 3y + z = 10 \pm 4\|\mathbf{n}\| = 10 \pm 4\sqrt{26}$$

109. $0.04x - 0.64y + z = 3.4 \Rightarrow z = 3.4 - 0.04x + 0.64y$

(a)

Year	1994	1995	1996	1997	1998	1999	2000
x	5.8	6.2	6.4	6.6	6.5	6.3	6.1
y	8.7	8.2	8.0	7.7	7.4	7.3	7.1
z	8.8	8.4	8.4	8.2	7.8	7.9	7.8
Model, z'	8.74	8.40	8.26	8.06	7.88	7.82	7.70

The approximations are close to the actual values.

(b) According to the model, if x and z decrease, then so will y, the consumption of reduced-fat milk.

110. On one side we have the points $(0, 0, 0)$, $(6, 0, 0)$, and $(-1, -1, 8)$.

$$\mathbf{n}_1 = \begin{vmatrix} \mathbf{i} & \mathbf{j} & \mathbf{k} \\ 6 & 0 & 0 \\ -1 & -1 & 8 \end{vmatrix} = -48\mathbf{j} - 6\mathbf{k}$$

On the adjacent side we have the points $(0, 0, 0)$, $(0, 6, 0)$, and $(-1, -1, 8)$.

$$\mathbf{n}_2 = \begin{vmatrix} \mathbf{i} & \mathbf{j} & \mathbf{k} \\ 0 & 6 & 0 \\ -1 & -1 & 8 \end{vmatrix} = 48\mathbf{i} + 6\mathbf{k}$$

$$\cos\theta = \frac{|\mathbf{n}_1 \cdot \mathbf{n}_2|}{\|\mathbf{n}_1\|\,\|\mathbf{n}_2\|} = \frac{36}{2340} = \frac{1}{65}$$

$$\theta = \arccos\frac{1}{65} \approx 89.1^\circ$$

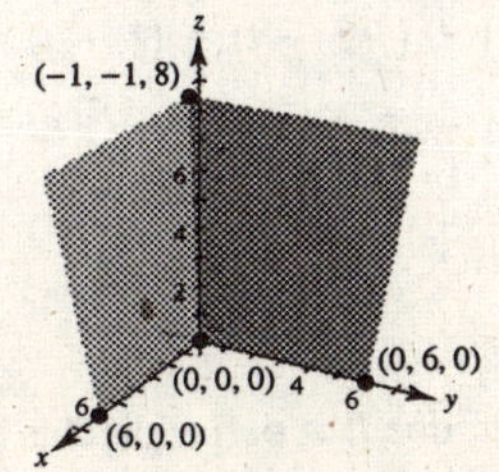

111. L_1: $x_1 = 6 + t, y_1 = 8 - t, z_1 = 3 + t$

L_2: $x_2 = 1 + t, y_2 = 2 + t, z_2 = 2t$

(a) At $t = 0$, the first insect is at $P_1 = (6, 8, 3)$ and the second insect is at $P_2 = (1, 2, 0)$.

$$\text{Distance} = \sqrt{(6-1)^2 + (8-2)^2 + (3-0)^2} = \sqrt{70} \approx 8.37 \text{ inches}$$

(b) $$\text{Distance} = \sqrt{(x_1 - x_2)^2 + (y_1 - y_2)^2 + (z_1 - z_2)^2}$$
$$= \sqrt{5^2 + (6 - 2t)^2 + (3 - t)^2}$$
$$= \sqrt{5t^2 - 30t + 70}, \quad 0 \le t \le 10$$

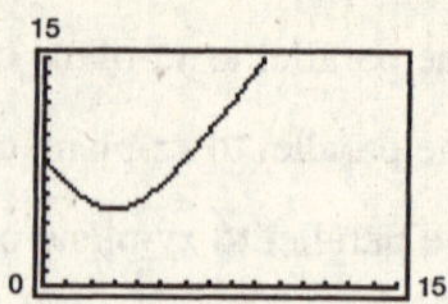

(c) The distance is never zero.

(d) Using a graphing utility, the minimum distance is 5 inches when $t = 3$ minutes.

112. First find the distance D from the point $Q = (-3, 2, 4)$ to the plane. Let $P = (4, 0, 0)$ be on the plane. $\mathbf{n} = \langle 2, 4, -3\rangle$ is the normal to the plane.

$$D = \frac{|\overrightarrow{PQ} \cdot \mathbf{n}|}{\|\mathbf{n}\|} = \frac{|\langle -7, 2, 4\rangle \cdot \langle 2, 4, -3\rangle|}{\sqrt{4 + 16 + 9}} = \frac{|-14 + 8 - 12|}{\sqrt{29}} = \frac{18}{\sqrt{29}} = \frac{18\sqrt{29}}{29}$$

The equation of the sphere with center $(-3, 2, 4)$ and radius $18\sqrt{29}/29$ is $(x + 3)^2 + (y - 2)^2 + (z - 4)^2 = \dfrac{324}{29}$.

113. The direction vector $\mathbf{v}$ of the line is the normal to the plane, $\mathbf{v} = \langle 3, -1, 4\rangle$.

The parametric equations of the line are $x = 5 + 3t, y = 4 - t, z = -3 + 4t$.

To find the point of intersection, solve for t in the following equation:

$$3(5 + 3t) - (4 - t) + 4(-3 + 4t) = 7$$
$$26t = 8$$
$$t = \frac{4}{13}$$

Point of intersection is $\left(5 + 3\left(\frac{4}{13}\right), 4 - \frac{4}{13}, -3 + 4\left(\frac{4}{13}\right)\right) = \left(\frac{77}{13}, \frac{48}{13}, -\frac{23}{13}\right)$

114. The normal to the plane, $\mathbf{n} = \langle 2, -1, -3\rangle$ is perpendicular to the direction vector $\mathbf{v} = \langle 2, 4, 0\rangle$ of the line because $\langle 2, -1, -3\rangle \cdot \langle 2, 4, 0\rangle = 0$.

Hence, the plane is parallel to the line. To find the distance between them, let $Q = (-2, -1, 4)$ be on the line and $P = (2, 0, 0)$ on the plane. $\overrightarrow{PQ} = \langle -4, -1, 4\rangle$.

$$D = \frac{|\overrightarrow{PQ} \cdot \mathbf{n}|}{\|\mathbf{n}\|} = \frac{|\langle -4, -1, 4\rangle \cdot \langle 2, -1, -3\rangle|}{\sqrt{4 + 1 + 9}} = \frac{19}{\sqrt{14}} = \frac{19\sqrt{14}}{14}$$

115. The direction vector of the line L through $(1, -3, 1)$ and $(3, -4, 2)$ is $\mathbf{v} = \langle 2, -1, 1\rangle$.

The parametric equations for L are $x = 1 + 2t, y = -3 - t, z = 1 + t$.

Substituting these equations into the equation of the plane gives

$$(1 + 2t) - (-3 - t) + (1 + t) = 2$$
$$4t = -3$$
$$t = -\frac{3}{4}.$$

Point of intersection is $\left(1 + 2\left(-\frac{3}{4}\right), -3 + \frac{3}{4}, 1 - \frac{3}{4}\right) = \left(-\frac{1}{2}, -\frac{9}{4}, \frac{1}{4}\right)$

116. The unknown line L is perpendicular to the normal vector $\mathbf{n} = \langle 1, 1, 1\rangle$ of the plane, and perpendicular to the direction vector $\mathbf{u} = \langle 1, 1, -1\rangle$. Hence, the direction vector of L is

$$\mathbf{v} = \begin{vmatrix} \mathbf{i} & \mathbf{j} & \mathbf{k} \\ 1 & 1 & 1 \\ 1 & 1 & -1 \end{vmatrix} = \langle -2, 2, 0\rangle.$$

The parametric equations for L are $x = 1 - 2t, y = 2t, z = 2$.

117. True

118. False. They may be skew lines. (See Section Project)

119. True

120. False. The lines $x = t, y = 0, z = 1$ and $x = 0, y = t, z = 1$ are both parallel to the plane $z = 0$, but the lines are not parallel.

Section 11.6 Surfaces in Space

1. Ellipsoid

Matches graph (c)

2. Hyperboloid of two sheets

Matches graph (e)

3. Hyperboloid of one sheet

Matches graph (f)

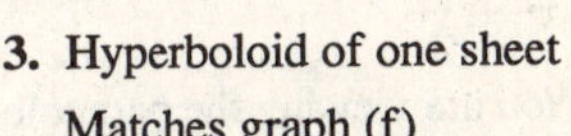

4. Elliptic cone

Matches graph (b)

5. Elliptic paraboloid

Matches graph (d)

6. Hyperbolic paraboloid

Matches graph (a)

7. $z = 3$

Plane parallel to the xy-coordinate plane

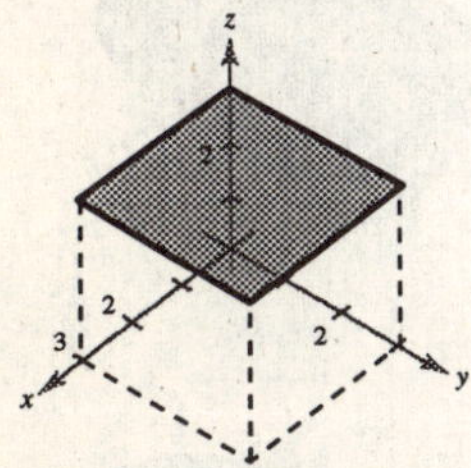

8. $x = 4$

Plane parallel to the yz-coordinate plane

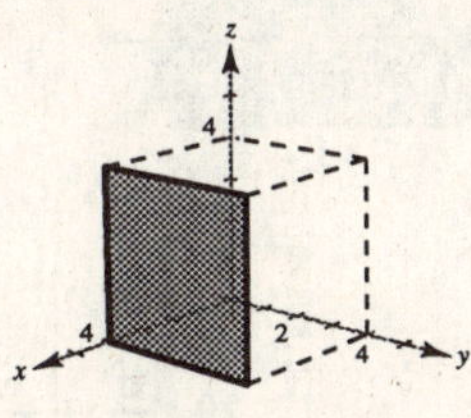

9. $y^2 + z^2 = 9$

The x-coordinate is missing so we have a cylindrical surface with rulings parallel to the x-axis. The generating curve is a circle.

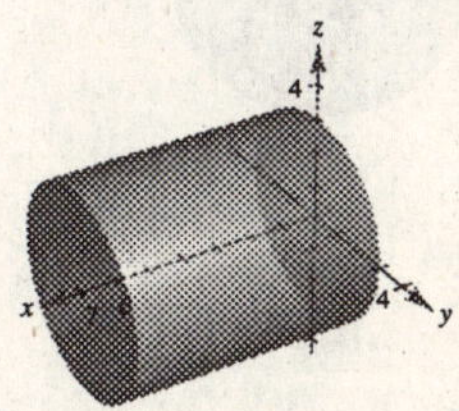

10. $x^2 + z^2 = 25$

The y-coordinate is missing so we have a cylindrical surface with rulings parallel to the y-axis. The generating curve is a circle.

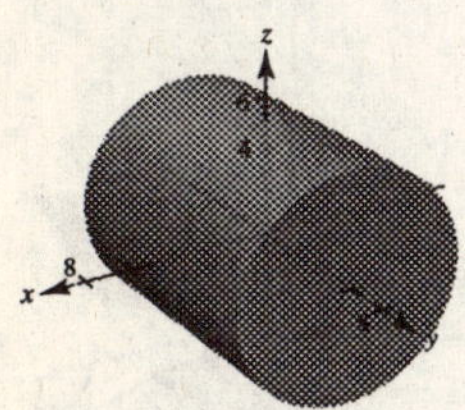

11. $y = x^2$

The z-coordinate is missing so we have a cylindrical surface with rulings parallel to the z-axis. The generating curve is a parabola.

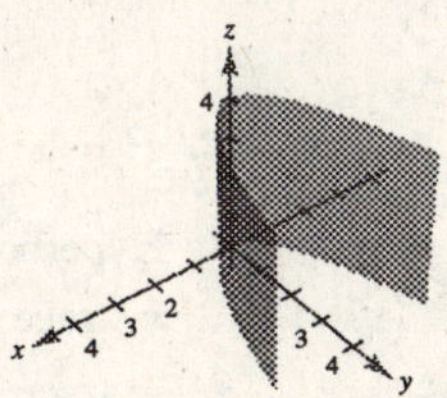

12. $z = 4 - y^2$

The x-coordinate is missing so we have a cylindrical surface with rulings parallel to the x-axis. The generating curve is a parabola.

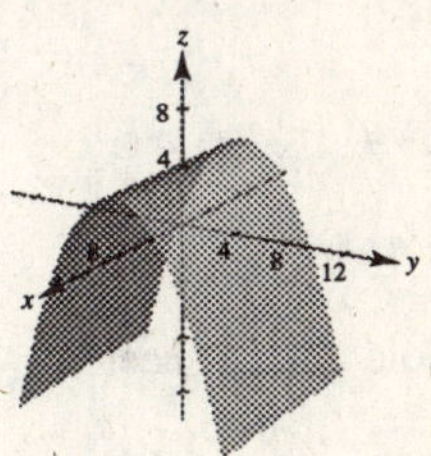

13. $4x^2 + y^2 = 4$

$$\frac{x^2}{1} + \frac{y^2}{4} = 1$$

The z-coordinate is missing so we have a cylindrical surface with rulings parallel to the z-axis. The generating curve is an ellipse.

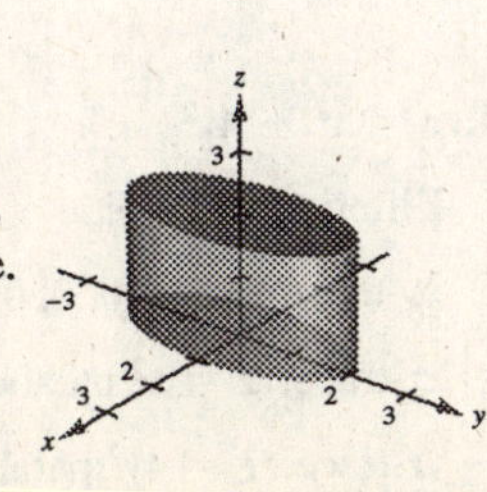

14. $y^2 - z^2 = 4$

$$\frac{y^2}{4} - \frac{z^2}{4} = 1$$

The x-coordinate is missing so we have a cylindrical surface with rulings parallel to the x-axis. The generating curve is a hyperbola.

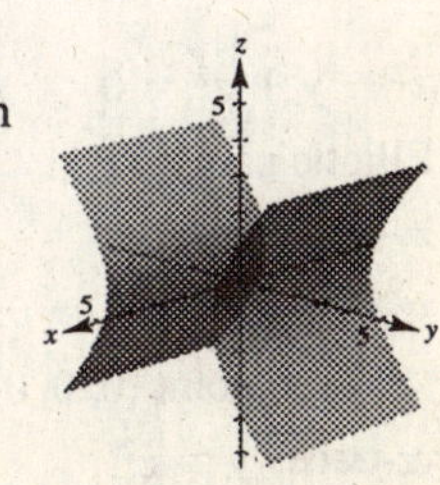

15. $z = \sin y$

The x-coordinate is missing so we have a cylindrical surface with rulings parallel to the x-axis. The generating curve is the sine curve.

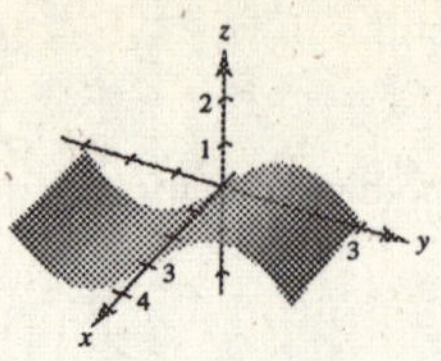

16. $z = e^y$

The x-coordinate is missing so we have a cylindrical surface with rulings parallel to the x-axis. The generating curve is the exponential curve.

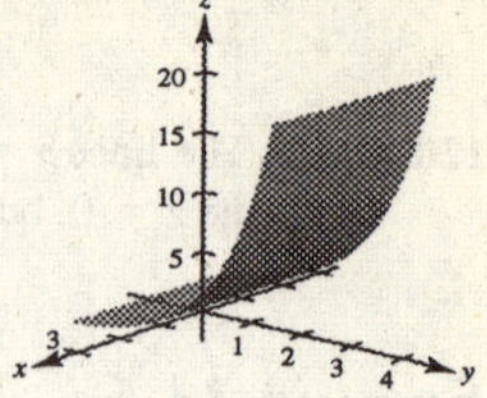

17. $z = x^2 + y^2$

(a) You are viewing the paraboloid from the x-axis: $(20, 0, 0)$

(b) You are viewing the paraboloid from above, but not on the z-axis: $(10, 10, 20)$

(c) You are viewing the paraboloid from the z-axis: $(0, 0, 20)$

(d) You are viewing the paraboloid from the y-axis: $(0, 20, 0)$

18. $y^2 + z^2 = 4$

(a) From $(10, 0, 0)$:

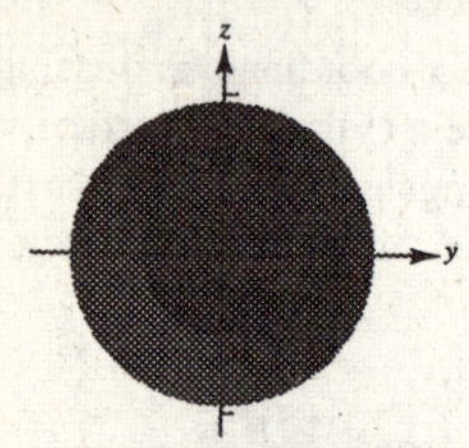

(b) From $(0, 10, 0)$:

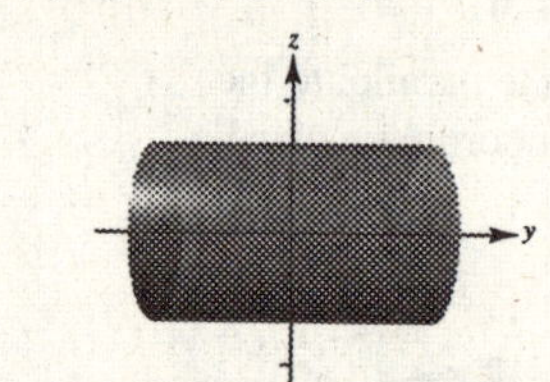

(c) From $(10, 10, 10)$:

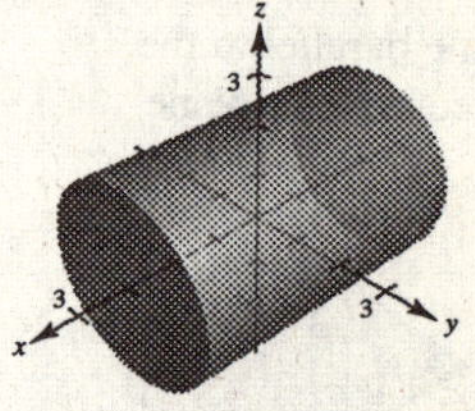

19. $\frac{x^2}{1} + \frac{y^2}{4} + \frac{z^2}{1} = 1$

Ellipsoid

xy-trace: $\frac{x^2}{1} + \frac{y^2}{4} = 1$ ellipse

xz-trace: $x^2 + z^2 = 1$ circle

yz-trace: $\frac{y^2}{4} + \frac{z^2}{1} = 1$ ellipse

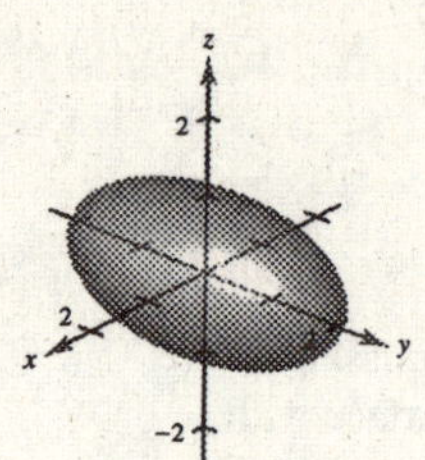

20. $\frac{x^2}{16} + \frac{y^2}{25} + \frac{z^2}{25} = 1$

Ellipsoid

xy-trace: $\frac{x^2}{16} + \frac{y^2}{25} = 1$ ellipse

xz-trace: $\frac{x^2}{16} + \frac{z^2}{25} = 1$ ellipse

yz-trace: $y^2 + z^2 = 25$ circle

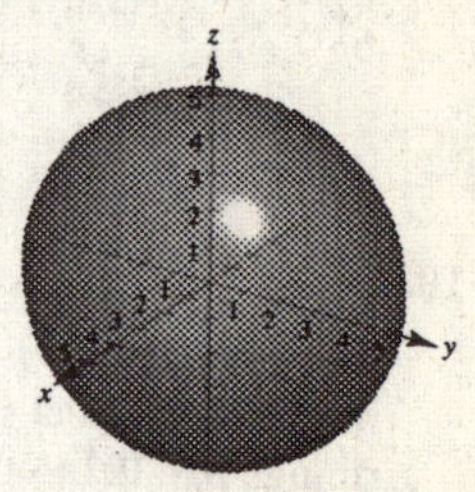

21. $16x^2 - y^2 + 16z^2 = 4$

$4x^2 - \frac{y^2}{4} + 4z^2 = 1$

Hyperboloid on one sheet

xy-trace: $4x^2 - \frac{y^2}{4} = 1$ hyperbola

xz-trace: $4(x^2 + z^2) = 1$ circle

yz-trace: $\frac{-y^2}{4} + 4z^2 = 1$ hyperbola

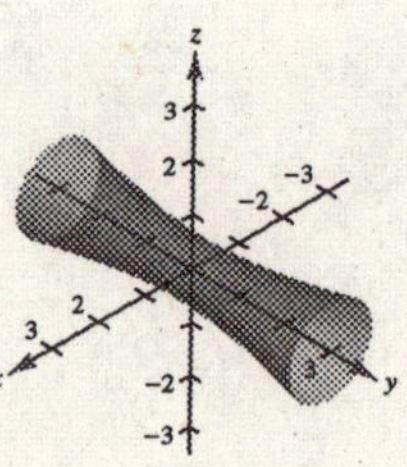

22. $z^2 - x^2 - \frac{y^2}{4} = 1$

Hyperboloid of two sheets

xy-trace: none

xz-trace: $z^2 - x^2 = 1$ hyperbola

yz-trace: $z^2 - \frac{y^2}{4} = 1$ hyperbola

$z = \pm\sqrt{10}$: $\frac{x^2}{9} + \frac{y^2}{36} = 1$ ellipse

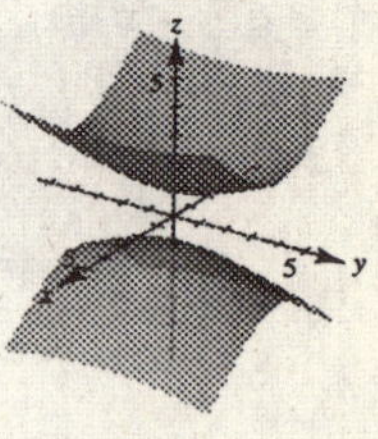

23. $x^2 - y + z^2 = 0$

Elliptic paraboloid

xy-trace: $y = x^2$

xz-trace: $x^2 + z^2 = 0$, point $(0, 0, 0)$

yz-trace: $y = z^2$

$y = 1$: $x^2 + z^2 = 1$

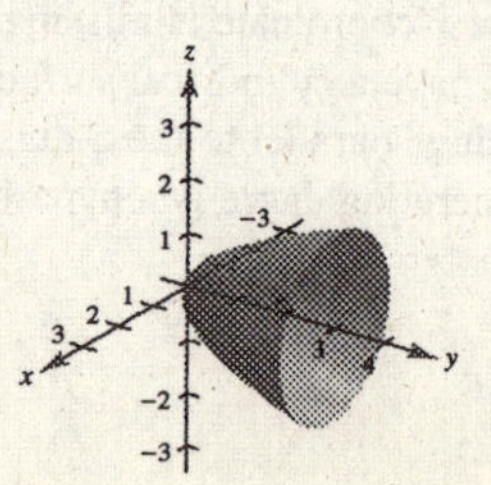

24. $z = x^2 + 4y^2$

Elliptic paraboloid

xy-trace: point $(0, 0, 0)$

xz-trace: $z = x^2$ parabola

yz-trace: $z = 4y^2$ parabola

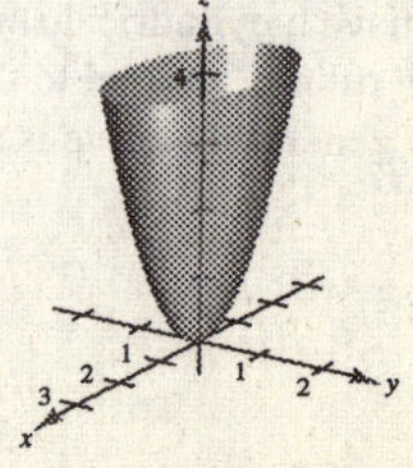

25. $x^2 - y^2 + z = 0$

Hyperbolic paraboloid

xy-trace: $y = \pm x$

xz-trace: $z = -x^2$

yz-trace: $z = y^2$

$y = \pm 1$: $z = 1 - x^2$

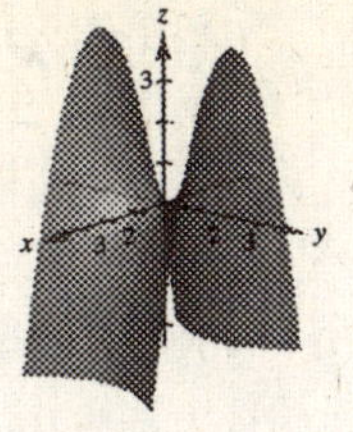
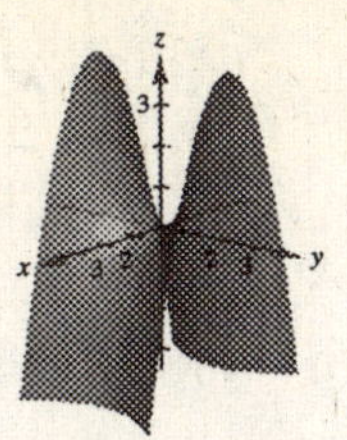

26. $3z = -y^2 + x^2$

Hyperbolic paraboloid

xy-trace: $y = \pm x$

xz-trace: $z = \frac{1}{3}x^2$

yz-trace: $z = -\frac{1}{3}y^2$

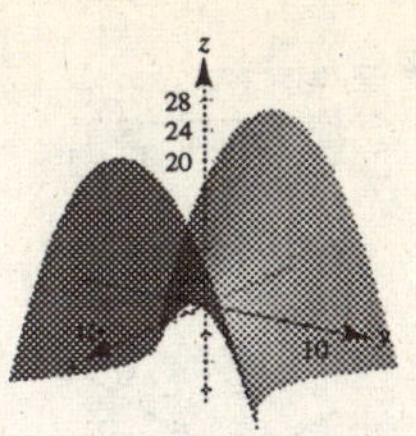

27. $z^2 = x^2 + \dfrac{y^2}{4}$

Elliptic Cone

xy-trace: point $(0, 0, 0)$

xz-trace: $z = \pm x$

yz-trace: $z = \dfrac{\pm 1}{2}y$

$z = \pm 1$: $x^2 + \dfrac{y^2}{4} = 1$

28. $x^2 = 2y^2 + 2z^2$

Elliptic Cone

xy-trace: $x = \pm\sqrt{2}y$

xz-trace: $x = \pm\sqrt{2}z$

yz-trace: point: $(0, 0, 0)$

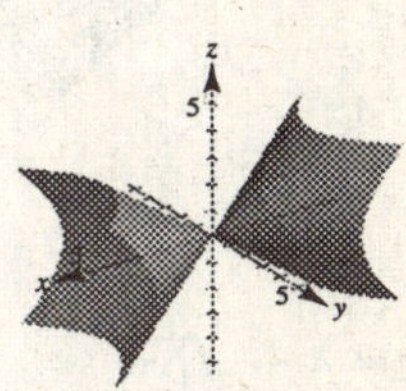

29.

$$16x^2 + 9y^2 + 16z^2 - 32x - 36y + 36 = 0$$
$$16(x^2 - 2x + 1) + 9(y^2 - 4y + 4) + 16z^2 = -36 + 16 + 36$$
$$16(x-1)^2 + 9(y-2)^2 + 16z^2 = 16$$
$$\frac{(x-1)^2}{1} + \frac{(y-2)^2}{16/9} + \frac{z^2}{1} = 1$$

Ellipsoid with center $(1, 2, 0)$.

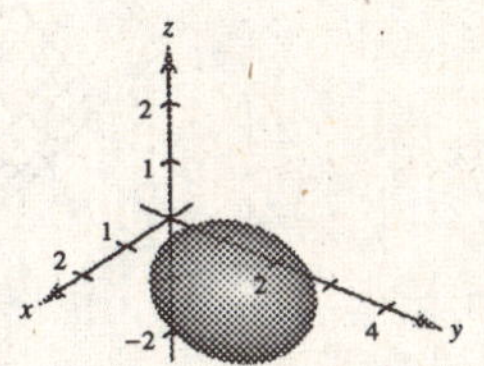

30.

$$9x^2 + y^2 - 9z^2 - 54x - 4y - 54z + 4 = 0$$
$$9(x^2 - 6x + 9) + (y^2 - 4y + 4) - 9(z^2 + 6z + 9) = 81 - 81$$
$$9(x-3)^2 + (y-2)^2 - 9(z+3)^2 = 0$$

Elliptic cone with center $(3, 2, -3)$.

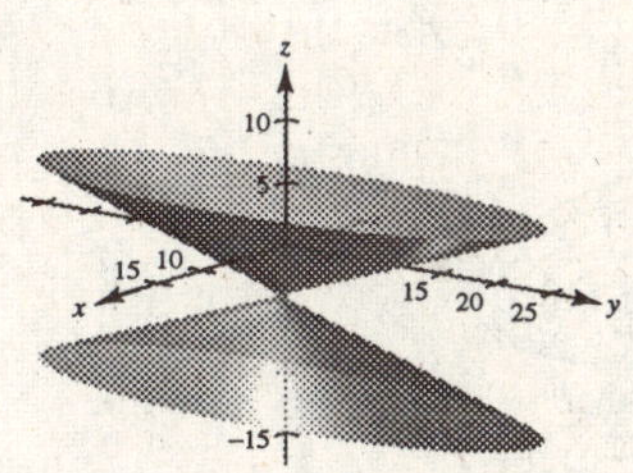

31. $z = 2 \sin x$

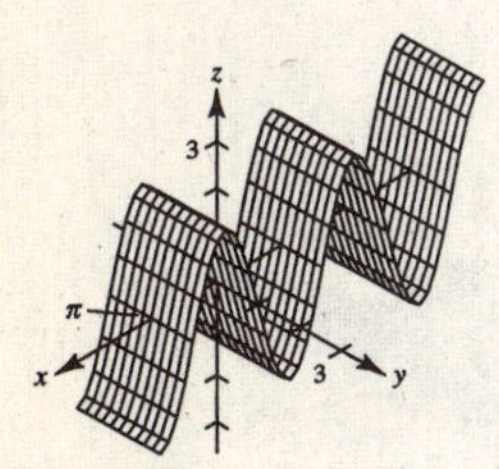

32. $z = x^2 + 0.5y^2$

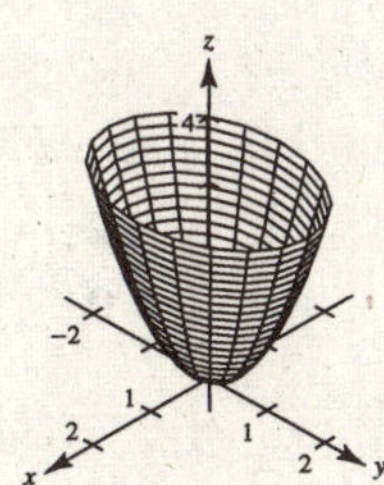

33. $z^2 = x^2 + 4y^2$

$z = \pm\sqrt{x^2 + 4y^2}$

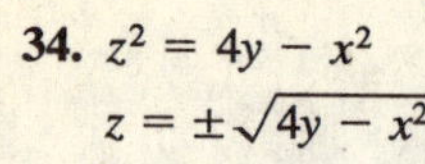

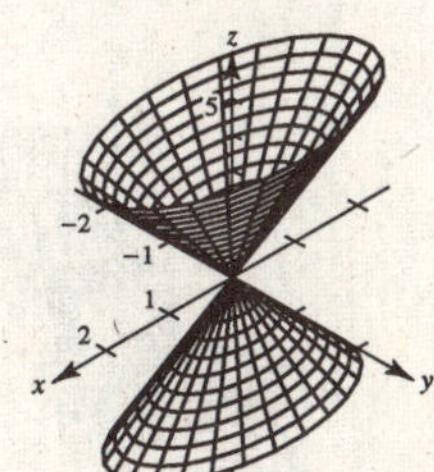

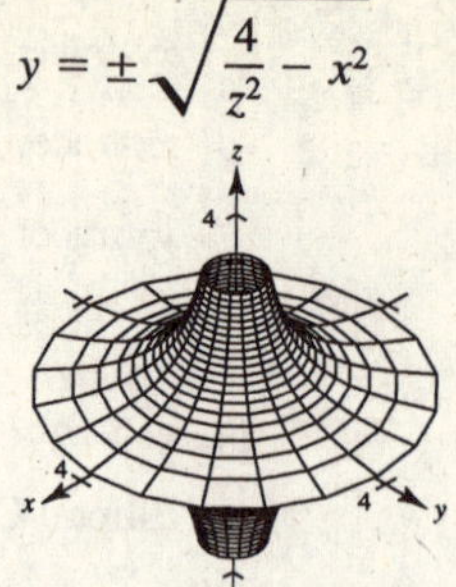

34. $z^2 = 4y - x^2$

$z = \pm\sqrt{4y - x^2}$

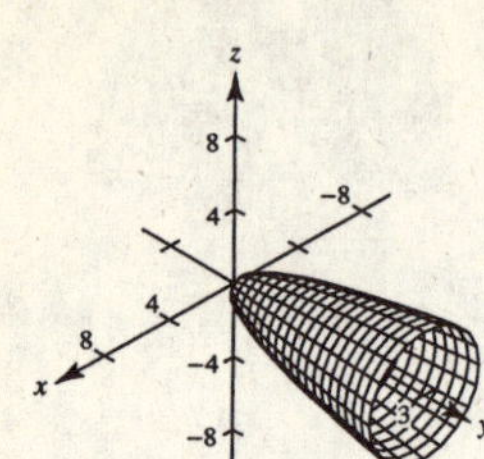

35. $x^2 + y^2 = \left(\frac{2}{z}\right)^2$

$y = \pm\sqrt{\frac{4}{z^2} - x^2}$

36. $x^2 + y^2 = e^{-z}$

$-\ln(x^2 + y^2) = z$

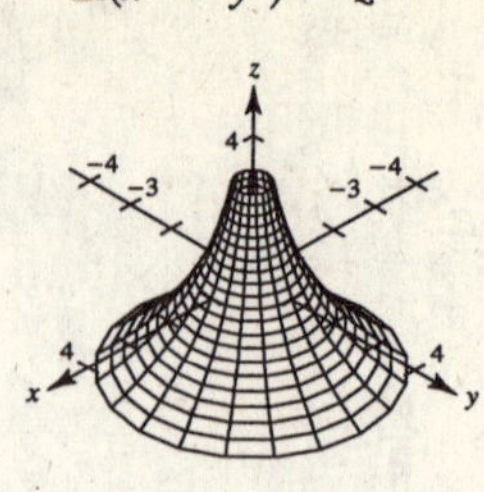

37. $z = 4 - \sqrt{|xy|}$

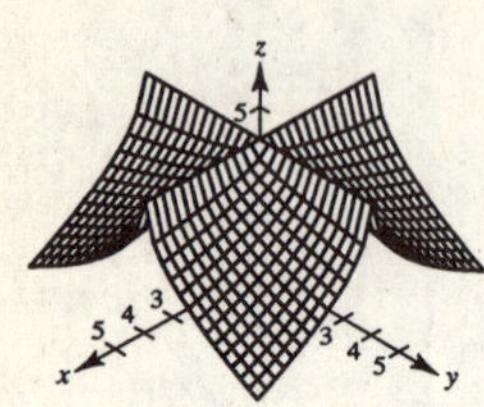

38. $z = \frac{-x}{8 + x^2 + y^2}$

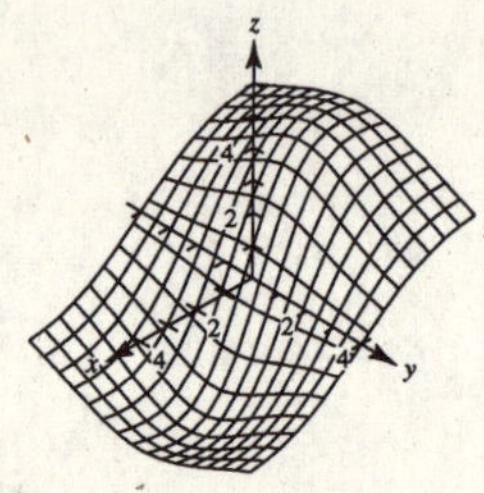

39. $4x^2 - y^2 + 4z^2 = -16$

$z = \pm\sqrt{\frac{y^2}{4} - x^2 - 4}$

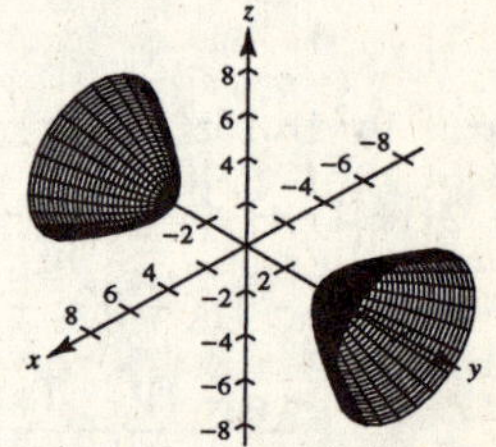

40. $9x^2 + 4y^2 - 8z^2 = 72$

$z = \pm\sqrt{\frac{9}{8}x^2 + \frac{1}{2}y^2 - 9}$

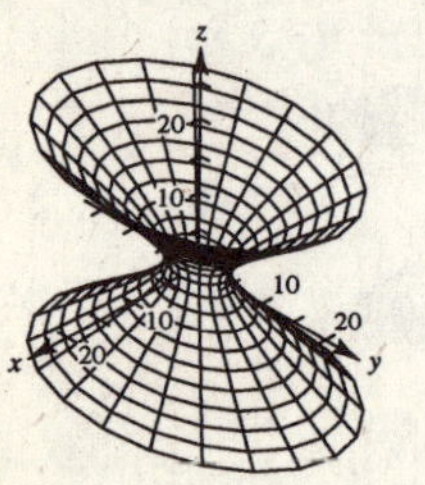

41. $z = 2\sqrt{x^2 + y^2}$

$z = 2$

$2\sqrt{x^2 + y^2} = 2$

$x^2 + y^2 = 1$

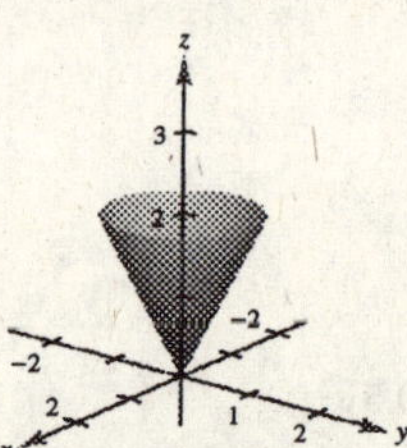

42. $z = \sqrt{4 - x^2}$

$y = \sqrt{4 - x^2}$

$x = 0, y = 0, z = 0$

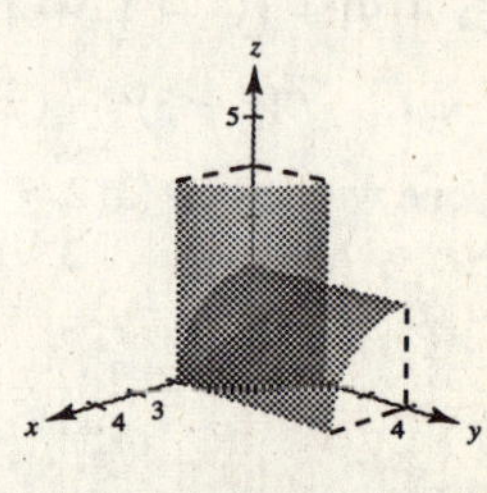

43. $x^2 + y^2 = 1$

$x + z = 2$

$z = 0$

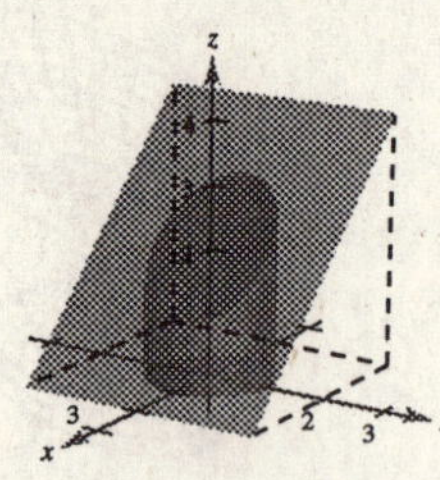

44. $z = \sqrt{4 - x^2 - y^2}$

$y = 2z$

$z = 0$

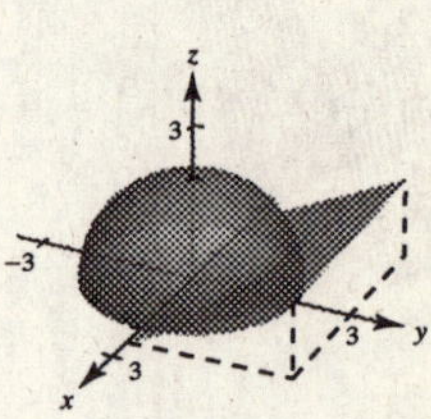

45. $x^2 + z^2 = [r(y)]^2$ and $z = r(y) = \pm 2\sqrt{y}$; therefore,

$x^2 + z^2 = 4y.$

46. $x^2 + z^2 = [r(y)]^2$ and $z = r(y) = 3y$; therefore,

$x^2 + z^2 = 9y^2.$

47. $x^2 + y^2 = [r(z)]^2$ and $y = r(z) = \frac{z}{2}$; therefore,

$x^2 + y^2 = \frac{z^2}{4}, 4x^2 + 4y^2 = z^2.$

48. $y^2 + z^2 = [r(x)]^2$ and $z = r(x) = \frac{1}{2}\sqrt{4 - x^2}$; therefore,

$y^2 + z^2 = \frac{1}{4}(4 - x^2), x^2 + 4y^2 + 4z^2 = 4.$

49. $y^2 + z^2 = [r(x)]^2$ and $y = r(x) = \dfrac{2}{x}$; therefore,

$$y^2 + z^2 = \left(\frac{2}{x}\right)^2, y^2 + z^2 = \frac{4}{x^2}.$$

50. $x^2 + y^2 = [r(z)]^2$ and $y = r(z) = e^z$; therefore,

$$x^2 + y^2 = e^{2z}.$$

51. $x^2 + y^2 - 2z = 0$

$x^2 + y^2 = \left(\sqrt{2z}\right)^2$

Equation of generating curve: $y = \sqrt{2z}$ or $x = \sqrt{2z}$

52. $x^2 + z^2 = \cos^2 y$

Equation of generating curve: $x = \cos y$ or $z = \cos y$

53. Let C be a curve in a plane and let L be a line not in a parallel plane. The set of all lines parallel to L and intersecting C is called a cylinder.

54. The trace of a surface is the intersection of the surface with a plane. You find a trace by setting one variable equal to a constant, such as $x = 0$ or $z = 2$.

55. See pages 812 and 813.

56. In the xz-plane, $z = x^2$ is a parabola.

In three-space, $z = x^2$ is a cylinder.

57. $V = 2\pi \displaystyle\int_0^4 x(4x - x^2)\,dx$

$$= 2\pi\left[\frac{4x^3}{3} - \frac{x^4}{4}\right]_0^4 = \frac{128\pi}{3}$$

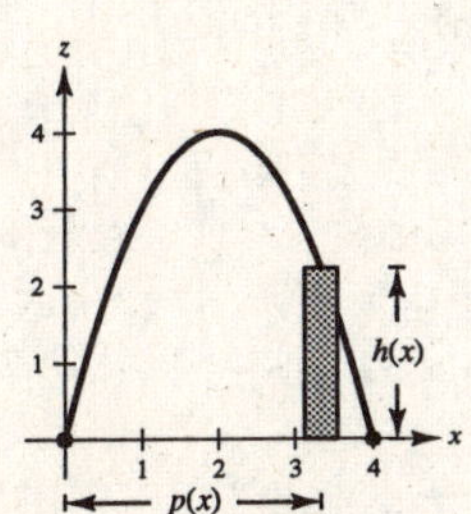

58. $V = 2\pi \displaystyle\int_0^{\pi} y \sin y\,dy$

$$= 2\pi\Big[\sin y - y\cos y\Big]_0^{\pi} = 2\pi^2$$

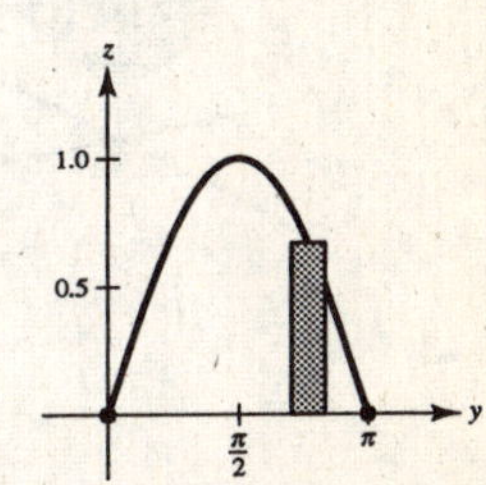

59. $z = \dfrac{x^2}{2} + \dfrac{y^2}{4}$

(a) When $z = 2$ we have $2 = \dfrac{x^2}{2} + \dfrac{y^2}{4}$, or $1 = \dfrac{x^2}{4} + \dfrac{y^2}{8}$

Major axis: $2\sqrt{8} = 4\sqrt{2}$

Minor axis: $2\sqrt{4} = 4$

$c^2 = a^2 - b^2, c^2 = 4, c = 2$

Foci: $(0, \pm 2, 2)$

(b) When $z = 8$ we have $8 = \dfrac{x^2}{2} + \dfrac{y^2}{4}$, or $1 = \dfrac{x^2}{16} + \dfrac{y^2}{32}$.

Major axis: $2\sqrt{32} = 8\sqrt{2}$

Minor axis: $2\sqrt{16} = 8$

$c^2 = 32 - 16 = 16, c = 4$

Foci: $(0, \pm 4, 8)$

60. $z = \dfrac{x^2}{2} + \dfrac{y^2}{4}$

(a) When $y = 4$ we have $z = \dfrac{x^2}{2} + 4$, $4\left(\dfrac{1}{2}\right)(z - 4) = x^2$.

Focus: $\left(0, 4, \dfrac{9}{2}\right)$

(b) When $x = 2$ we have

$z = 2 + \dfrac{y^2}{4}$, $4(z - 2) = y^2$.

Focus: $(2, 0, 3)$

61. If (x, y, z) is on the surface, then

$$(y + 2)^2 = x^2 + (y - 2)^2 + z^2$$

$$y^2 + 4y + 4 = x^2 + y^2 - 4y + 4 + z^2$$

$$x^2 + z^2 = 8y$$

Elliptic paraboloid

Traces parallel to xz-plane are circles.

62. If (x, y, z) is on the surface, then

$$z^2 = x^2 + y^2 + (z - 4)^2$$

$$z^2 = x^2 + y^2 + z^2 - 8z + 16$$

$$8z = x^2 + y^2 + 16 \Rightarrow z = \frac{x^2}{8} + \frac{y^2}{8} + 2$$

Elliptic paraboloid shifted up 2 units. Traces parallel to xy-plane are circles.

63. $$\frac{x^2}{3963^2} + \frac{y^2}{3963^2} + \frac{z^2}{3950^2} = 1$$

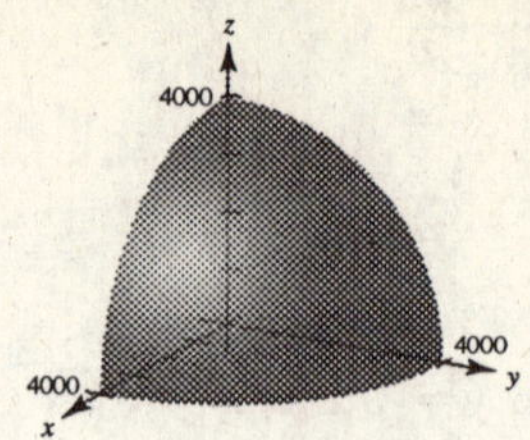

64. (a) $$x^2 + y^2 = [r(z)]^2 = \left[\sqrt{2(z-1)}\right]^2$$

$$x^2 + y^2 - 2z + 2 = 0$$

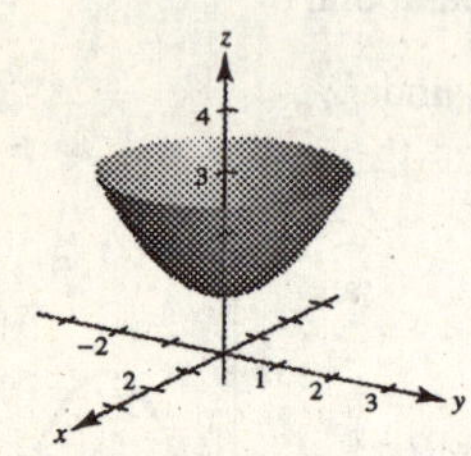

(b) $$V = 2\pi\int_0^2 x\left[3 - \left(\frac{1}{2}x^2 + 1\right)\right] dx = 2\pi\int_0^2 \left(2x - \frac{1}{2}x^3\right) dx = 2\pi\left[x^2 - \frac{x^4}{8}\right]_0^2 = 4\pi \approx 12.6 \text{ cm}^3$$

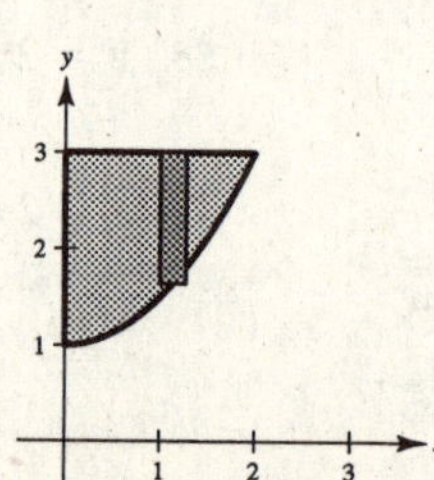

(c) $$V = 2\pi\int_{1/2}^2 x\left[3 - \left(\frac{1}{2}x^2 + 1\right)\right] dx = 2\pi\int_{1/2}^2 \left(2x - \frac{1}{2}x^3\right) dx = 2\pi\left[x^2 - \frac{x^4}{8}\right]_{1/2}^2 = 4\pi - \frac{31\pi}{64} = \frac{225\pi}{64} \approx 11.04 \text{ cm}^3$$

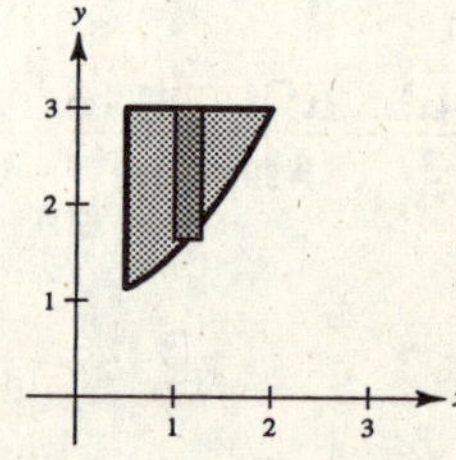

65. $z = \dfrac{y^2}{b^2} - \dfrac{x^2}{a^2}$, $z = bx + ay$

$$bx + ay = \frac{y^2}{b^2} - \frac{x^2}{a^2}$$

$$\frac{1}{a^2}\left(x^2 + a^2bx + \frac{a^4b^2}{4}\right) = \frac{1}{b^2}\left(y^2 - ab^2y + \frac{a^2b^4}{4}\right)$$

$$\frac{\left(x + \frac{a^2b}{2}\right)^2}{a^2} = \frac{\left(y - \frac{ab^2}{2}\right)^2}{b^2}$$

$$y = \pm\frac{b}{a}\left(x + \frac{a^2b}{2}\right) + \frac{ab^2}{2}$$

Letting $x = at$, you obtain the two intersecting lines $x = at$, $y = -bt$, $z = 0$ and $x = at$, $y = bt + ab^2$ $z = 2abt + a^2b^2$.

66. Equating twice the first equation with the second equation,

$$2x^2 + 6y^2 - 4z^2 + 4y - 8 = 2x^2 + 6y^2 - 4z^2 - 3x - 2$$

$$4y - 8 = -3x - 2$$

$3x + 4y = 6$, a plane

67. True. A sphere is a special case of an ellipsoid (centered at origin, for example)

$$\frac{x^2}{a^2} + \frac{y^2}{b^2} + \frac{z^2}{c^2} = 1$$

having $a = b = c$.

68. False. For example, the surface $x^2 + z^2 = e^{-2y}$ can be formed by revolving the graph of $x = e^{-y}$ about the y-axis, as the graph of $z = e^{-y}$ about the y-axis.

69. The Klein bottle *does not* have both an "inside" and an "outside." It is formed by inserting the small open end through the side of the bottle and making it contiguous with the top of the bottle.

Section 11.7 Cylindrical and Spherical Coordinates

1. $(5, 0, 2)$, cylindrical

$x = 5\cos 0 = 5$

$y = 5\sin 0 = 0$

$z = 2$

$(5, 0, 2)$, rectangular

2. $\left(4, \frac{\pi}{2}, -2\right)$, cylindrical

$x = 4\cos\frac{\pi}{2} = 0$

$y = 4\sin\frac{\pi}{2} = 4$

$z = -2$

$(0, 4, -2)$, rectangular

3. $\left(2, \frac{\pi}{3}, 2\right)$, cylindrical

$x = 2\cos\frac{\pi}{3} = 1$

$y = 2\sin\frac{\pi}{3} = \sqrt{3}$

$z = 2$

$\left(1, \sqrt{3}, 2\right)$, rectangular

4. $\left(6, -\frac{\pi}{4}, 2\right)$, cylindrical

$x = 6\cos\left(-\frac{\pi}{4}\right) = 3\sqrt{2}$

$y = 6\sin\left(-\frac{\pi}{4}\right) = -3\sqrt{2}$

$z = 2$

$\left(3\sqrt{2}, -3\sqrt{2}, 2\right)$

5. $\left(4, \frac{7\pi}{6}, 3\right)$, cylindrical

$x = 4\cos\frac{7\pi}{6} = -2\sqrt{3}$

$y = 4\sin\frac{7\pi}{6} = -2$

$z = 3$

$\left(-2\sqrt{3}, -2, 3\right)$, rectangular

6. $\left(1, \frac{3\pi}{2}, 1\right)$, cylindrical

$x = \cos\frac{3\pi}{2} = 0$

$y = \sin\frac{3\pi}{2} = -1$

$z = 1$

$(0, -1, 1)$, rectangular

7. $(0, 5, 1)$, rectangular

$r = \sqrt{(0)^2 + (5)^2} = 5$

$\theta = \arctan\frac{5}{0} = \frac{\pi}{2}$

$z = 1$

$\left(5, \frac{\pi}{2}, 1\right)$, cylindrical

8. $\left(2\sqrt{2}, -2\sqrt{2}, 4\right)$, rectangular

$r = \sqrt{\left(2\sqrt{2}\right)^2 + \left(-2\sqrt{2}\right)^2} = 4$

$\theta = \arctan(-1) = -\frac{\pi}{4}$

$z = 4$

$\left(4, -\frac{\pi}{4}, 4\right)$, cylindrical

9. $\left(1, \sqrt{3}, 4\right)$, rectangular

$r = \sqrt{1^2 + \left(\sqrt{3}\right)^2} = 2$

$\theta = \arctan\sqrt{3} = \frac{\pi}{3}$

$z = 4$

$\left(2, \frac{\pi}{3}, 4\right)$, cylindrical

10. $\left(2\sqrt{3}, -2, 6\right)$, rectangular

$r = \sqrt{12 + 4} = 4$

$\theta = \arctan\left(-\frac{1}{\sqrt{3}}\right) = \frac{5\pi}{6}$

$z = 6$

$\left(4, -\frac{\pi}{6}, 6\right)$, cylindrical

11. $(2, -2, -4)$, rectangular

$r = \sqrt{2^2 + (-2)^2} = 2\sqrt{2}$

$\theta = \arctan(-1) = -\frac{\pi}{4}$

$z = -4$

$\left(2\sqrt{2}, \frac{-\pi}{4}, -4\right)$, cylindrical

12. $(-3, 2, -1)$, rectangular

$r = \sqrt{(-3)^2 + 2^2} = \sqrt{13}$

$\theta = \arctan\left(\frac{-2}{3}\right) = -\arctan\frac{2}{3}$

$z = -1$

$\left(\sqrt{13}, -\arctan\frac{2}{3}, -1\right)$, cylindrical

13. $z = 5$ is the equation in cylindrical coordinates.

14. $x = 4,$ rectangular coordinates

$r\cos\theta = 4$

$r = 4\sec\theta,$ cylindrical coordinates

15. $x^2 + y^2 + z^2 = 10,$ rectangular equation

$r^2 + z^2 = 10,$ cylindrical equation

16. $z = x^2 + y^2 - 2,$ rectangular equation

$z = r^2 - 2,$ cylindrical equation

17. $y = x^2,$ rectangular equation

$r\sin\theta = (r\cos\theta)^2$

$\sin\theta = r\cos^2\theta$

$r = \sec\theta \cdot \tan\theta,$ cylindrical equation

18. $x^2 + y^2 = 8x,$ rectangular equation

$r^2 = 8r\cos\theta$

$r = 8\cos\theta,$ cylindrical equation

19. $y^2 = 10 - z^2$, rectangular coordinates

$(r \sin \theta)^2 = 10 - z^2$

$r^2 \sin^2 \theta + z^2 = 10$, cylindrical coordinates

20. $x^2 + y^2 + z^2 - 3z = 0$, rectangular coordinates

$r^2 + z^2 - 3z = 0$ cylindrical coordinates

21. $r = 2$

$\sqrt{x^2 + y^2} = 2$

$x^2 + y^2 = 4$

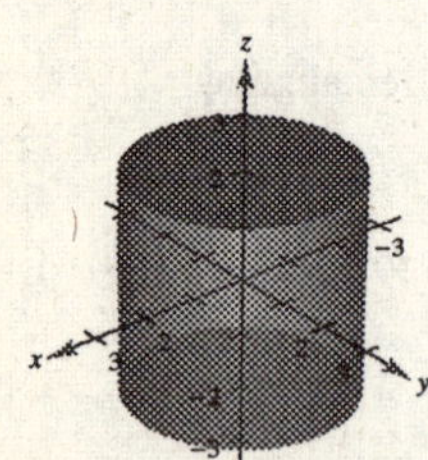

22. $z = 2$

Same

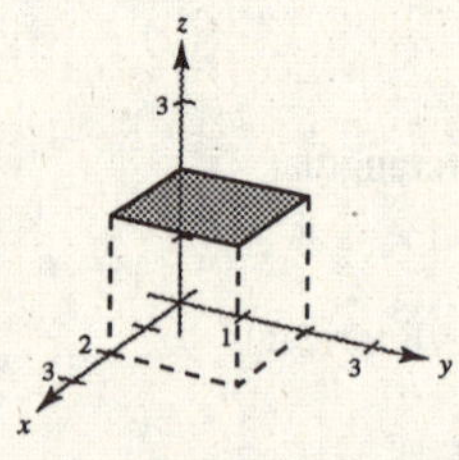

23. $\theta = \dfrac{\pi}{6}$

$\tan \dfrac{\pi}{6} = \dfrac{y}{x}$

$\dfrac{1}{\sqrt{3}} = \dfrac{y}{x}$

$x = \sqrt{3}y$

$x - \sqrt{3}y = 0$

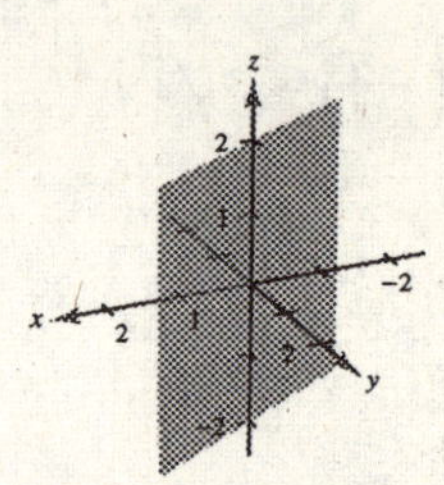

24. $r = \dfrac{z}{2}$

$\sqrt{x^2 + y^2} = \dfrac{z}{2}$

$x^2 + y^2 - \dfrac{z^2}{4} = 0$

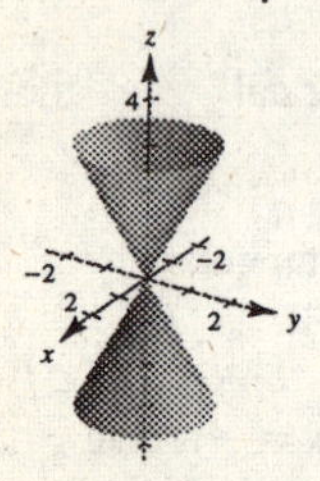

25. $r = 2 \sin \theta$

$r^2 = 2r \sin \theta$

$x^2 + y^2 = 2y$

$x^2 + y^2 - 2y = 0$

$x^2 + (y - 1)^2 = 1$

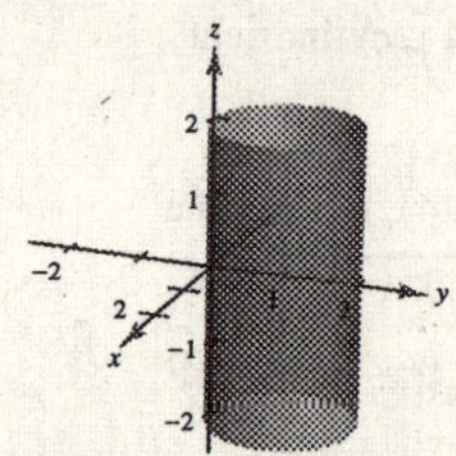

26. $r = 2 \cos \theta$

$r^2 = 2r \cos \theta$

$x^2 + y^2 = 2x$

$x^2 + y^2 - 2x = 0$

$(x - 1)^2 + y^2 = 1$

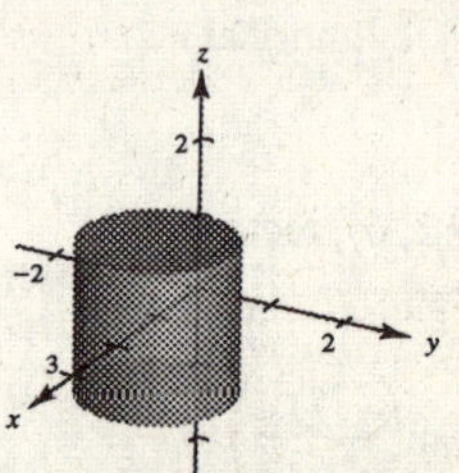

27. $r^2 + z^2 = 4$

$x^2 + y^2 + z^2 = 4$

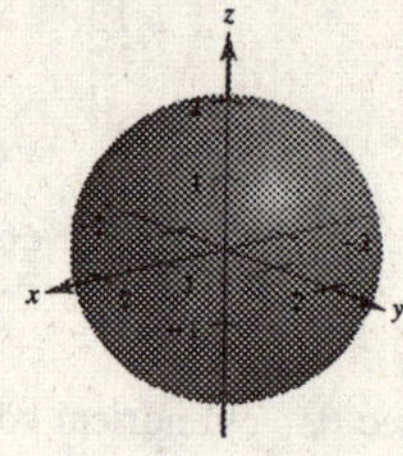

28. $z = r^2 \cos^2 \theta$

$z = x^2$

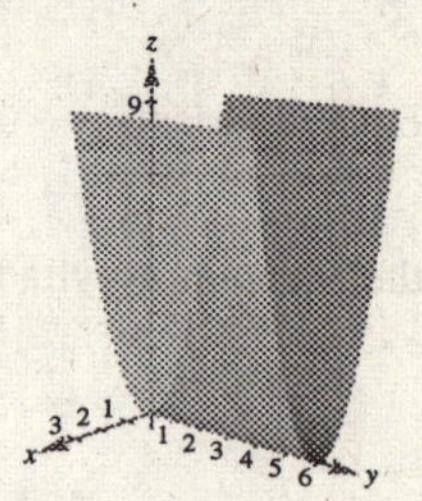

29. $(4, 0, 0)$, rectangular

$\rho = \sqrt{4^2 + 0^2 + 0^2} = 4$

$\theta = \arctan 0 = 0$

$\phi = \arccos 0 = \dfrac{\pi}{2}$

$\left(4, 0, \dfrac{\pi}{2}\right)$, spherical

30. $(1, 1, 1)$, rectangular

$\rho = \sqrt{1^2 + 1^2 + 1^2} = \sqrt{3}$

$\theta = \arctan 1 = \dfrac{\pi}{4}$

$\phi = \arccos \dfrac{1}{\sqrt{3}}$

$\left(\sqrt{3}, \dfrac{\pi}{4}, \arccos \dfrac{1}{\sqrt{3}}\right)$, spherical

31. $\left(-2, 2\sqrt{3}, 4\right)$, rectangular

$\rho = \sqrt{(-2)^2 + \left(2\sqrt{3}\right)^2 + 4^2} = 4\sqrt{2}$

$\theta = \arctan\left(-\sqrt{3}\right) = \dfrac{2\pi}{3}$

$\phi = \arccos \dfrac{1}{\sqrt{2}} = \dfrac{\pi}{4}$

$\left(4\sqrt{2}, \dfrac{2\pi}{3}, \dfrac{\pi}{4}\right)$, spherical

32. $\left(2, 2, 4\sqrt{2}\right)$, rectangular

$\rho = \sqrt{2^2 + 2^2 + \left(4\sqrt{2}\right)^2} = 2\sqrt{10}$

$\theta = \arctan 1 = \dfrac{\pi}{4}$

$\phi = \arccos \dfrac{2}{\sqrt{5}}$

$\left(2\sqrt{10}, \dfrac{\pi}{4}, \arccos \dfrac{2}{\sqrt{5}}\right)$, spherical

33. $\left(\sqrt{3}, 1, 2\sqrt{3}\right)$, rectangular

$\rho = \sqrt{3 + 1 + 12} = 4$

$\theta = \arctan \dfrac{1}{\sqrt{3}} = \dfrac{\pi}{6}$

$\phi = \arccos \dfrac{\sqrt{3}}{2} = \dfrac{\pi}{6}$

$\left(4, \dfrac{\pi}{6}, \dfrac{\pi}{6}\right)$, spherical

34. $(-4, 0, 0)$, rectangular

$\rho = \sqrt{(-4)^2 + 0^2 + 0^2} = 4$

$\theta = \pi$

$\phi = \arccos 0 = \dfrac{\pi}{2}$

$\left(4, \pi, \dfrac{\pi}{2}\right)$, spherical

35. $\left(4, \dfrac{\pi}{6}, \dfrac{\pi}{4}\right)$, spherical

$x = 4 \sin \dfrac{\pi}{4} \cos \dfrac{\pi}{6} = \sqrt{6}$

$y = 4 \sin \dfrac{\pi}{4} \sin \dfrac{\pi}{6} = \sqrt{2}$

$z = 4 \cos \dfrac{\pi}{4} = 2\sqrt{2}$

$\left(\sqrt{6}, \sqrt{2}, 2\sqrt{2}\right)$, rectangular

36. $\left(12, \dfrac{3\pi}{4}, \dfrac{\pi}{9}\right)$, spherical

$x = 12 \sin \dfrac{\pi}{9} \cos \dfrac{3\pi}{4} \approx -2.902$

$y = 12 \sin \dfrac{\pi}{9} \sin \dfrac{3\pi}{4} \approx 2.902$

$z = 12 \cos \dfrac{\pi}{9} \approx 11.276$

$(-2.902, 2.902, 11.276)$, rectangular

37. $\left(12, \dfrac{-\pi}{4}, 0\right)$, spherical

$x = 12 \sin 0 \cos\left(\dfrac{-\pi}{4}\right) = 0$

$y = 12 \sin 0 \sin\left(\dfrac{-\pi}{4}\right) = 0$

$z = 12 \cos 0 = 12$

$(0, 0, 12)$, rectangular

38. $\left(9, \dfrac{\pi}{4}, \pi\right)$, spherical

$x = 9 \sin \pi \cos \dfrac{\pi}{4} = 0$

$y = 9 \sin \pi \sin \dfrac{\pi}{4} = 0$

$z = 9 \cos \pi = -9$

$(0, 0, -9)$, rectangular

39. $\left(5, \dfrac{\pi}{4}, \dfrac{3\pi}{4}\right)$, spherical

$x = 5 \sin \dfrac{3\pi}{4} \cos \dfrac{\pi}{4} = \dfrac{5}{2}$

$y = 5 \sin \dfrac{3\pi}{4} \sin \dfrac{\pi}{4} = \dfrac{5}{2}$

$z = 5 \cos \dfrac{3\pi}{4} = -\dfrac{5\sqrt{2}}{2}$

$\left(\dfrac{5}{2}, \dfrac{5}{2}, -\dfrac{5\sqrt{2}}{2}\right)$, rectangular

40. $\left(6, \pi, \dfrac{\pi}{2}\right)$, spherical

$x = 6 \sin \dfrac{\pi}{2} \cos \pi = -6$

$y = 6 \sin \dfrac{\pi}{2} \sin \pi = 0$

$z = 6 \cos \dfrac{\pi}{2} = 0$

$(-6, 0, 0)$, rectangular

41. $y = 3$, rectangular equation

$\rho \sin \phi \sin \theta = 3$

$\rho = 3 \csc \phi \csc \theta$, spherical equation

42. $z = 2$, rectangular equation

$\rho \cos \phi = 2$

$\rho = 2 \sec \phi$, pherical equation

43. $x^2 + y^2 + z^2 = 36$, rectangular equation

$$\rho^2 = 36$$

$$\rho = 6, \quad \text{spherical equation}$$

44. $x^2 + y^2 - 3z^2 = 0$, rectangular equation

$$x^2 + y^2 + z^2 = 4z^2$$

$$\rho^2 = 4\rho^2 \cos^2 \phi$$

$$1 = 4 \cos^2 \phi$$

$$\cos \phi = \frac{1}{2}$$

$$\phi = \frac{\pi}{3}, \quad \text{(cone) spherical equation}$$

45. $x^2 + y^2 = 9$, rectangular equation

$$\rho^2 \sin^2 \phi \cos^2 \theta + \rho^2 \sin^2 \phi \sin^2 \theta = 9$$

$$\rho^2 \sin^2 \phi = 9$$

$$\rho \sin \phi = 3$$

$$\rho = 3 \csc \phi, \quad \text{spherical equation}$$

46. $x = 10$, rectangular equation

$$\rho \sin \phi \cos \theta = 10$$

$$\rho = 10 \csc \phi \sec \theta, \quad \text{spherical equation}$$

47. $x^2 + y^2 = 2z^2$ rectangular coordinates

$$\rho^2 \sin^2 \phi \cos^2 \theta + \rho^2 \sin^2 \phi \sin^2 \theta = 2\rho^2 \cos^2 \phi$$

$$\rho^2 \sin^2 \phi[\cos^2 \theta + \sin^2 \theta] = 2\rho^2 \cos^2 \phi$$

$$\rho^2 \sin^2 \phi = 2\rho^2 \cos^2 \phi$$

$$\frac{\sin^2 \phi}{\cos^2 \phi} = 2$$

$$\tan^2 \phi = 2$$

$$\tan \phi = \pm\sqrt{2} \quad \text{spherical equation}$$

48. $x^2 + y^2 + z^2 - 9z = 0$ rectangular equation

$$\rho^2 - 9\rho \cos \phi = 0$$

$$\rho = 9 \cos \phi \quad \text{spherical equation}$$

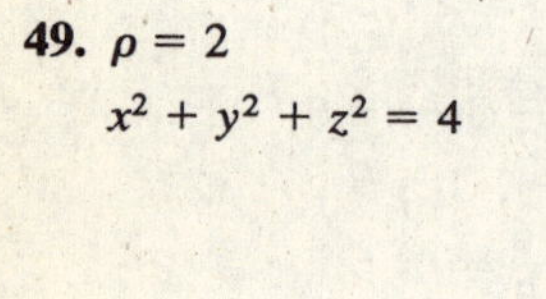

49. $\rho = 2$

$$x^2 + y^2 + z^2 = 4$$

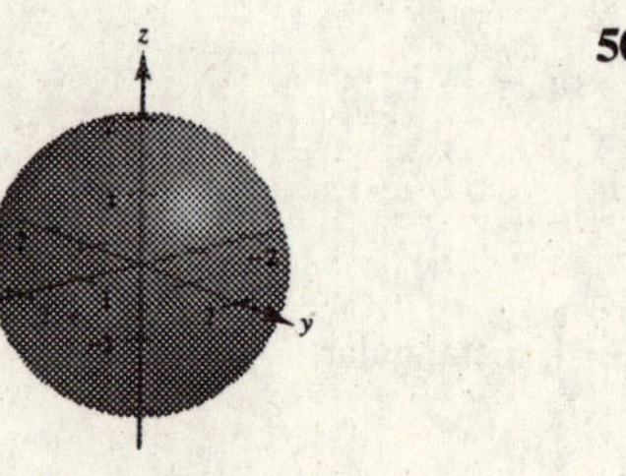

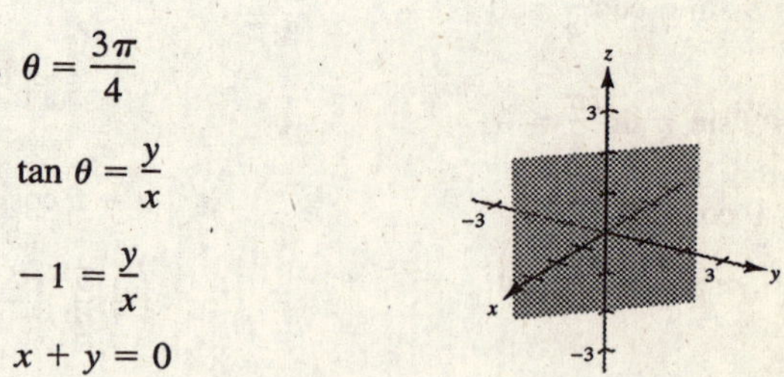

50. $\theta = \dfrac{3\pi}{4}$

$$\tan \theta = \frac{y}{x}$$

$$-1 = \frac{y}{x}$$

$$x + y = 0$$

51. $\phi = \dfrac{\pi}{6}$

$$\cos \phi = \frac{z}{\sqrt{x^2 + y^2 + z^2}}$$

$$\frac{\sqrt{3}}{2} = \frac{z}{\sqrt{x^2 + y^2 + z^2}}$$

$$\frac{3}{4} = \frac{z^2}{x^2 + y^2 + z^2}$$

$$3x^2 + 3y^2 - z^2 = 0, z \geq 0$$

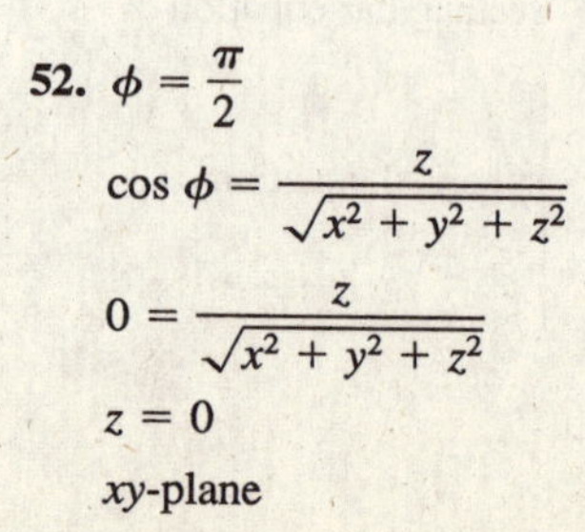

52. $\phi = \dfrac{\pi}{2}$

$$\cos \phi = \frac{z}{\sqrt{x^2 + y^2 + z^2}}$$

$$0 = \frac{z}{\sqrt{x^2 + y^2 + z^2}}$$

$$z = 0$$

xy-plane

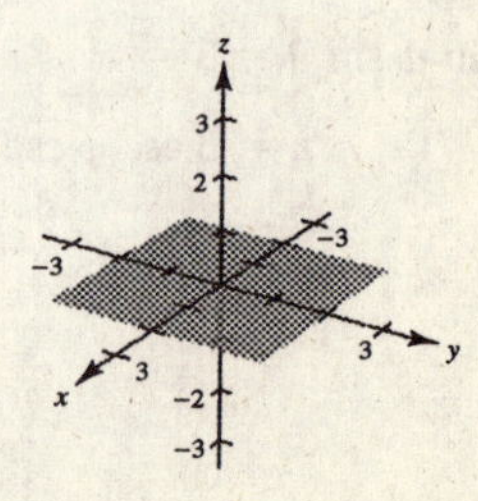

53. $\rho = 4\cos\phi$

$$\sqrt{x^2 + y^2 + z^2} = \frac{4z}{\sqrt{x^2 + y^2 + z^2}}$$

$$x^2 + y^2 + z^2 - 4z = 0$$

$$x^2 + y^2 + (z - 2)^2 = 4, z \ge 0$$

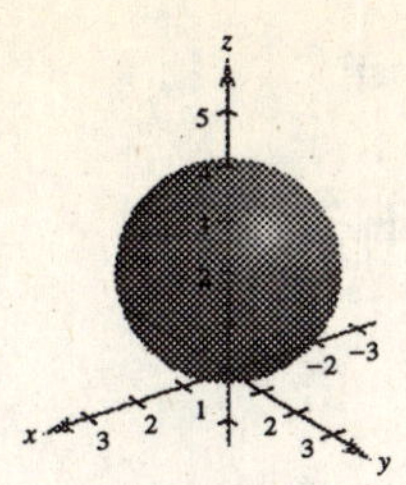

54. $\rho = 2\sec\phi$

$$\rho\cos\phi = 2$$

$$z = 2$$

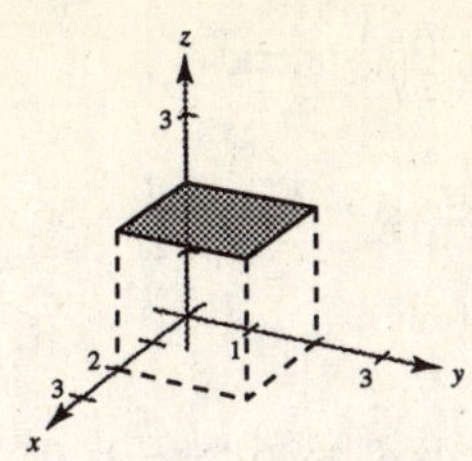

55. $\rho = \csc\phi$

$$\rho\sin\phi = 1$$

$$\sqrt{x^2 + y^2} = 1$$

$$x^2 + y^2 = 1$$

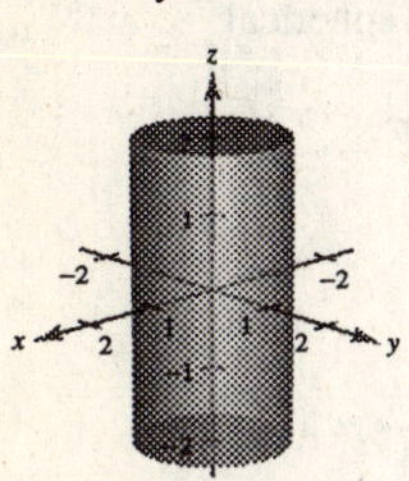

56. $\rho = 4\csc\phi\sec\phi$

$$= \frac{4}{\sin\phi\cos\theta}$$

$$\rho\sin\phi\cos\theta = 4$$

$$x = 4$$

57. $\left(4, \frac{\pi}{4}, 0\right)$, cylindrical

$$\rho = \sqrt{4^2 + 0^2} = 4$$

$$\theta = \frac{\pi}{4}$$

$$\phi = \arccos 0 = \frac{\pi}{2}$$

$\left(4, \frac{\pi}{4}, \frac{\pi}{2}\right)$, spherical

58. $\left(3, -\frac{\pi}{4}, 0\right)$, cylindrical

$$\rho = \sqrt{3^2 + 0^2} = 3$$

$$\theta = -\frac{\pi}{4}$$

$$\phi = \arccos\left(\frac{0}{9}\right) = \frac{\pi}{2}$$

$\left(3, -\frac{\pi}{4}, \frac{\pi}{2}\right)$, spherical

59. $\left(4, \frac{\pi}{2}, 4\right)$, cylindrical

$$\rho = \sqrt{4^2 + 4^2} = 4\sqrt{2}$$

$$\theta = \frac{\pi}{2}$$

$$\phi = \arccos\left(\frac{4}{4\sqrt{2}}\right) = \frac{\pi}{4}$$

$\left(4\sqrt{2}, \frac{\pi}{2}, \frac{\pi}{4}\right)$, spherical

60. $\left(2, \frac{2\pi}{3}, -2\right)$, cylindrical

$$\rho = \sqrt{2^2 + (-2)^2} = 2\sqrt{2}$$

$$\theta = \frac{2\pi}{3}$$

$$\phi = \arccos\left(\frac{-1}{\sqrt{2}}\right) = \frac{3\pi}{4}$$

$\left(2\sqrt{2}, \frac{2\pi}{3}, \frac{3\pi}{4}\right)$, spherical

61. $\left(4, \frac{-\pi}{6}, 6\right)$, cylindrical

$$\rho = \sqrt{4^2 + 6^2} = 2\sqrt{13}$$

$$\theta = \frac{-\pi}{6}$$

$$\phi = \arccos\frac{3}{\sqrt{13}}$$

$\left(2\sqrt{13}, \frac{-\pi}{6}, \arccos\frac{3}{\sqrt{13}}\right)$, spherical

62. $\left(-4, \frac{\pi}{3}, 4\right)$, cylindrical

$$\rho = \sqrt{(-4)^2 + 4^2} = 4\sqrt{2}$$

$$\theta = \frac{\pi}{3}$$

$$\phi = \arccos\frac{1}{\sqrt{2}} = \frac{\pi}{4}$$

$\left(4\sqrt{2}, \frac{\pi}{3}, \frac{\pi}{4}\right)$, spherical

63. $(12, \pi, 5)$, cylindrical

$$\rho = \sqrt{12^2 + 5^2} = 13$$

$$\theta = \pi$$

$$\phi = \arccos\frac{5}{13}$$

$\left(13, \pi, \arccos\frac{5}{13}\right)$, spherical

64. $\left(4, \frac{\pi}{2}, 3\right)$, cylindrical

$$\rho = \sqrt{4^2 + 3^2} = 5$$

$$\theta = \frac{\pi}{2}$$

$$\phi = \arccos\frac{3}{5}$$

$\left(5, \frac{\pi}{2}, \arccos\frac{3}{5}\right)$, spherical

65. $\left(10, \frac{\pi}{6}, \frac{\pi}{2}\right)$, spherical

$$r = 10\sin\frac{\pi}{2} = 10$$

$$\theta = \frac{\pi}{6}$$

$$z = 10\cos\frac{\pi}{2} = 0$$

$\left(10, \frac{\pi}{6}, 0\right)$, cylindrical

66. $\left(4, \frac{\pi}{18}, \frac{\pi}{2}\right)$, spherical

$$r = 4\sin\frac{\pi}{2} = 4$$

$$\theta = \frac{\pi}{18}$$

$$z = 4\cos\frac{\pi}{2} = 0$$

$\left(4, \frac{\pi}{18}, 0\right)$, cylindrical

67. $\left(36, \pi, \frac{\pi}{2}\right)$, spherical

$r = \rho \sin \phi = 36 \sin \frac{\pi}{2} = 36$

$\theta = \pi$

$z = \rho \cos \phi = 36 \cos \frac{\pi}{2} = 0$

$(36, \pi, 0)$, cylindrical

68. $\left(18, \frac{\pi}{3}, \frac{\pi}{3}\right)$, spherical

$r = \rho \sin \phi = 18 \sin \frac{\pi}{3} = 9$

$\theta = \frac{\pi}{3}$

$z = \rho \cos \phi = 18 \cos \frac{\pi}{3} = 9\sqrt{3}$

$\left(9, \frac{\pi}{3}, 9\sqrt{3}\right)$, cylindrical

69. $\left(6, -\frac{\pi}{6}, \frac{\pi}{3}\right)$, spherical

$r = 6 \sin \frac{\pi}{3} = 3\sqrt{3}$

$\theta = -\frac{\pi}{6}$

$z = 6 \cos \frac{\pi}{3} = 3$

$\left(3\sqrt{3}, -\frac{\pi}{6}, 3\right)$, cylindrical

70. $\left(5, -\frac{5\pi}{6}, \pi\right)$, spherical

$r = 5 \sin \pi = 0$

$\theta = -\frac{5\pi}{6}$

$z = 5 \cos \pi = -5$

$\left(0, -\frac{5\pi}{6}, -5\right)$, cylindrical

71. $\left(8, \frac{7\pi}{6}, \frac{\pi}{6}\right)$, spherical

$r = 8 \sin \frac{\pi}{6} = 4$

$\theta = \frac{7\pi}{6}$

$z = 8 \cos \frac{\pi}{6} = \frac{8\sqrt{3}}{2}$

$\left(4, \frac{7\pi}{6}, 4\sqrt{3}\right)$, cylindrical

72. $\left(7, \frac{\pi}{4}, \frac{3\pi}{4}\right)$, spherical

$r = 7 \sin \frac{3\pi}{4} = \frac{7\sqrt{2}}{2}$

$\theta = \frac{\pi}{4}$

$z = 7 \cos \frac{3\pi}{4} = -\frac{7\sqrt{2}}{2}$

$\left(\frac{7\sqrt{2}}{2}, \frac{\pi}{4}, -\frac{7\sqrt{2}}{2}\right)$, cylindrical

	Rectangular	Cylindrical	Spherical
73.	(4, 6, 3)	(7.211, 0.983, 3)	(7.810, 0.983, 1.177)
74.	(6, −2, −3)	(6.325, −0.322, −3)	(7.000, −0.322, 2.014)
75.	(4.698, 1.710, 8)	$\left(5, \frac{\pi}{9}, 8\right)$	(9.434, 0.349, 0.559)
76.	(7.317, −6.816, 6)	(10, −0.75, 6)	(11.662, −0.750, 1.030)
77.	(−7.071, 12.247, 14.142)	(14.142, 2.094, 14.142)	$\left(20, \frac{2\pi}{3}, \frac{\pi}{4}\right)$
78.	(6.115, 1.561, 4.052)	(6.311, 0.25, 4.052)	(7.5, 0.25, 1)
79.	(3, −2, 2)	(3.606, −0.588, 2)	(4.123, −0.588, 1.064)
80.	$(3\sqrt{2}, 3\sqrt{2}, -3)$	(6, 0.785, −3)	(6.708, 0.785, 2.034)
81.	$\left(\frac{5}{2}, \frac{4}{3}, \frac{-3}{2}\right)$	(2.833, 0.490, −1.5)	(3.206, 0.490, 2.058)
82.	(0, −5, 4)	(5, −1.571, 4)	(6.403, −1.571, 0.896)
83.	(−3.536, 3.536, −5)	$\left(5, \frac{3\pi}{4}, -5\right)$	(7.071, 2.356, 2.356)
84.	(−1.732, 1, 3)	$\left(-2, \frac{11\pi}{6}, 3\right)$	(3.606, 2.618, 0.588)
	[**Note:** Use the cylindrical coordinate $\left(2, \frac{5\pi}{6}, 3\right)$]		
85.	(2.804, −2.095, 6)	(−3.5, 2.5, 6)	(6.946, 5.642, 0.528)
	[**Note:** Use the cylindrical coordinates (3.5, 5.642, 6)]		
86.	(2.207, 7.949, −4)	(8.25, 1.3, −4)	(9.169, 1.3, 2.022)

87. $r = 5$

Cylinder

Matches graph (d)

88. $\theta = \dfrac{\pi}{4}$

Plane

Matches graph (e)

89. $\rho = 5$

Sphere

Matches graph (c)

90. $\phi = \dfrac{\pi}{4}$

Cone

Matches graph (a)

91. $r^2 = z, x^2 + y^2 = z$

Paraboloid

Matches graph (f)

92. $\rho = 4 \sec \phi, z = \rho \cos \phi = 4$

Plane

Matches graph (b)

93. Rectangular to cylindrical: $r^2 = x^2 + y^2$

$\tan \theta = \dfrac{y}{x}$

$z = z$

Cylindrical to rectangular: $x = r \cos \theta$

$y = r \sin \theta$

$z = z$

94. $r = a$ Cylinder with z-axis symmetry

$\theta = b$ Plane perpendicular to xy-plane

$z = c$ Plane parallel to xy-plane

95. Rectangular to spherical: $\rho^2 = x^2 + y^2 + z^2$

$\tan \theta = \dfrac{y}{x}$

$\phi = \arccos\left(\dfrac{z}{\sqrt{x^2 + y^2 + z^2}}\right)$

Spherical to rectangular: $x = \rho \sin \phi \cos \theta$

$y = \rho \sin \phi \sin \theta$

$z = \rho \cos \phi$

96. $\rho = a$ Sphere

$\theta = b$ Vertical half-plane

$\phi = c$ Half-cone

97. $x^2 + y^2 + z^2 = 16$

(a) $r^2 + z^2 = 16$

(b) $\rho^2 = 16, \rho = 4$

98. $4(x^2 + y^2) = z^2$

(a) $4r^2 = z^2, 2r = z$

(b) $4(\rho^2 \sin^2 \phi \cos^2 \theta + \rho^2 \sin^2 \phi \sin^2 \theta) = \rho^2 \cos^2 \phi,$

$4 \sin^2 \phi = \cos^2 \phi, \tan^2 \phi = \dfrac{1}{4},$

$\tan \phi = \dfrac{1}{2}, \phi = \arctan \dfrac{1}{2}$

99. $x^2 + y^2 + z^2 - 2z = 0$

(a) $r^2 + z^2 - 2z = 0, r^2 + (z - 1)^2 = 1$

(b) $\rho^2 - 2\rho \cos \phi = 0, \rho(\rho - 2 \cos \phi) = 0,$

$\rho = 2 \cos \phi$

100. $x^2 + y^2 = z$

(a) $r^2 = z$

(b) $\rho^2 \sin^2 \phi = \rho \cos \phi, \rho \sin^2 \phi = \cos \phi,$

$\rho = \dfrac{\cos \phi}{\sin^2 \phi}, \rho = \csc \phi \cot \phi$

101. $x^2 + y^2 = 4y$

(a) $r^2 = 4r \sin \theta, \ r = 4 \sin \theta$

(b) $\rho^2 \sin^2 \phi = 4\rho \sin \phi \sin \theta,$

$\rho \sin \phi(\rho \sin \phi - 4 \sin \theta) = 0,$

$\rho = \dfrac{4 \sin \theta}{\sin \phi}, \ \rho = 4 \sin \theta \csc \phi$

102. $x^2 + y^2 = 16$

(a) $r^2 = 16, r = 4$

(b) $\rho^2 \sin^2 \phi = 16, \rho^2 \sin^2 \phi - 16 = 0,$

$(\rho \sin \phi - 4)(\rho \sin \phi + 4) = 0, \rho = 4 \csc \phi$

103. $x^2 - y^2 = 9$

(a) $r^2 \cos^2 \theta - r^2 \sin^2 \theta = 9,$

$$r^2 = \frac{9}{\cos^2 \theta - \sin^2 \theta}$$

(b) $\rho^2 \sin^2 \phi \cos^2 \theta - \rho^2 \sin^2 \phi \sin^2 \theta = 9,$

$$\rho^2 \sin^2 \phi = \frac{9}{\cos^2 \theta - \sin^2 \theta},$$

$$\rho^2 = \frac{9 \csc^2 \phi}{\cos^2 \theta - \sin^2 \theta}$$

104. $y = 4$

(a) $r \sin \theta = 4, r = 4 \csc \theta$

(b) $\rho \sin \phi \sin \theta = 4, \rho = 4 \csc \phi \csc \theta$

105. $0 \le \theta \le \frac{\pi}{2}$

$0 \le r \le 2$

$0 \le z \le 4$

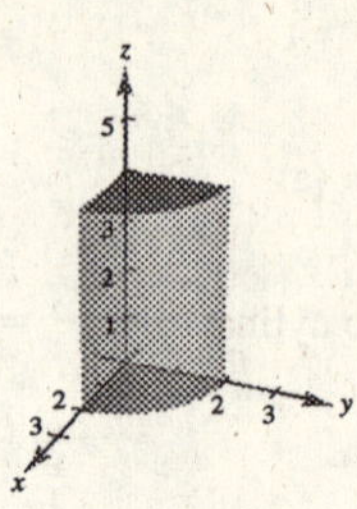

106. $-\frac{\pi}{2} \le \theta \le \frac{\pi}{2}$

$0 \le r \le 3$

$0 \le z \le r \cos \theta$

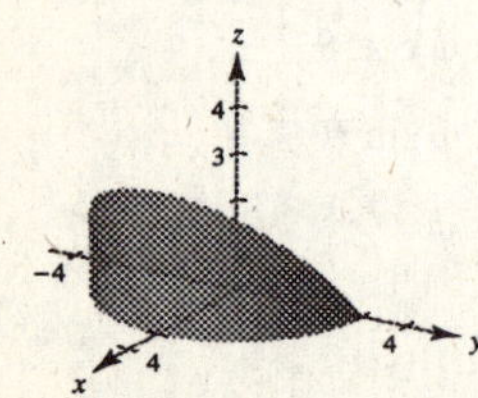

107. $0 \le \theta \le 2\pi$

$0 \le r \le a$

$r \le z \le a$

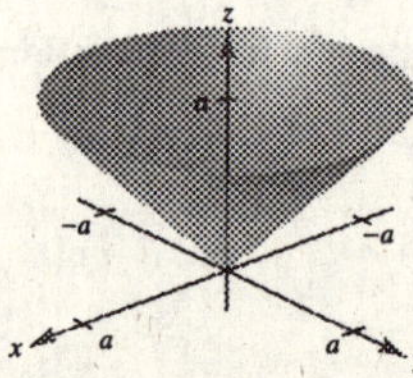

108. $0 \le \theta \le 2\pi$

$2 \le r \le 4$

$z^2 \le -r^2 + 6r - 8$

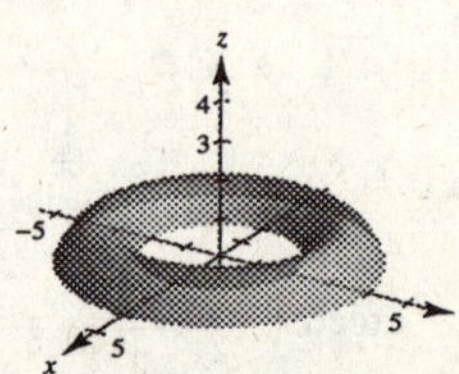

109. $0 \le \theta \le 2\pi$

$0 \le \phi \le \frac{\pi}{6}$

$0 \le \rho \le a \sec \phi$

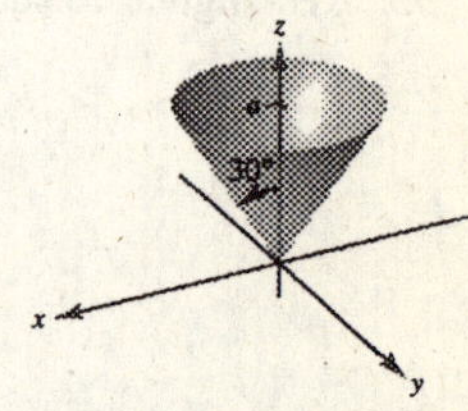

110. $0 \le \theta \le 2\pi$

$\frac{\pi}{4} \le \phi \le \frac{\pi}{2}$

$0 \le \rho \le 1$

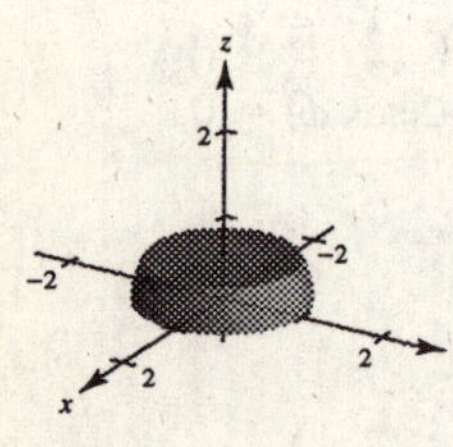

111. $0 \le \theta \le \frac{\pi}{2}$

$0 \le \phi \le \frac{\pi}{2}$

$0 \le \rho \le 2$

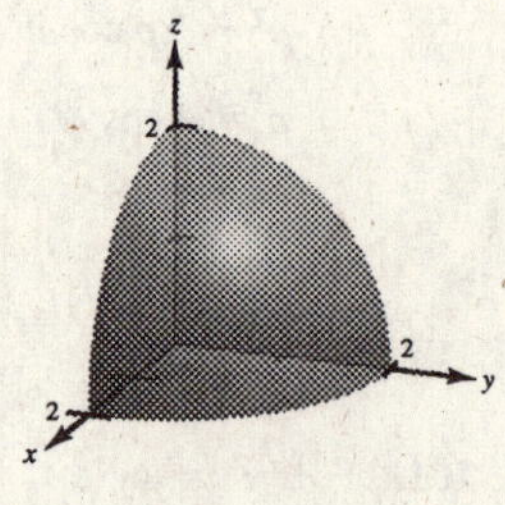

112. $0 \le \theta \le \pi$

$0 \le \phi \le \frac{\pi}{2}$

$1 \le \rho \le 3$

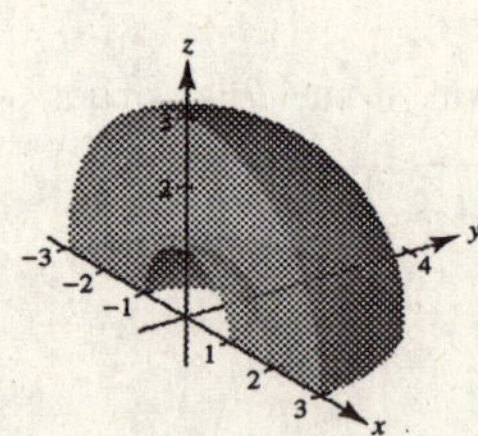

113. Rectangular

$0 \le x \le 10$

$0 \le y \le 10$

$0 \le z \le 10$

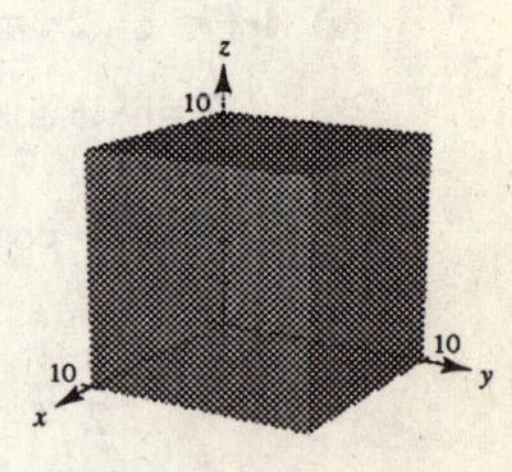

114. Cylindrical:

$0.75 \le r \le 1.25$

$0 \le z \le 8$

115. Spherical

$4 \le \rho \le 6$

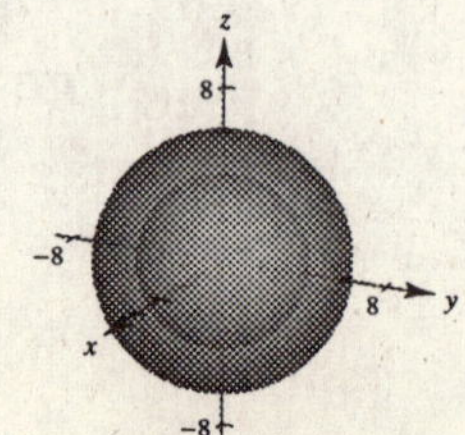

116. Cylindrical

$\frac{1}{2} \le r \le 3$

$0 \le \theta \le 2\pi$

$-\sqrt{9 - r^2} \le z \le \sqrt{9 - r^2}$

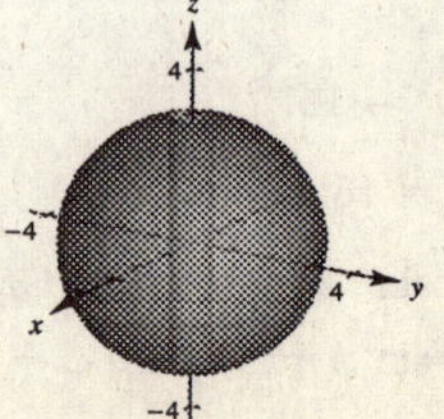

117. Cylindrical coordinates:

$r^2 + z^2 \le 9,$

$r \le 3\cos\theta, 0 \le \theta \le \pi$

118. Spherical coordinates:

$\rho \ge 2$

$\rho \le 3$

$0 \le \phi \le \dfrac{\pi}{4}$

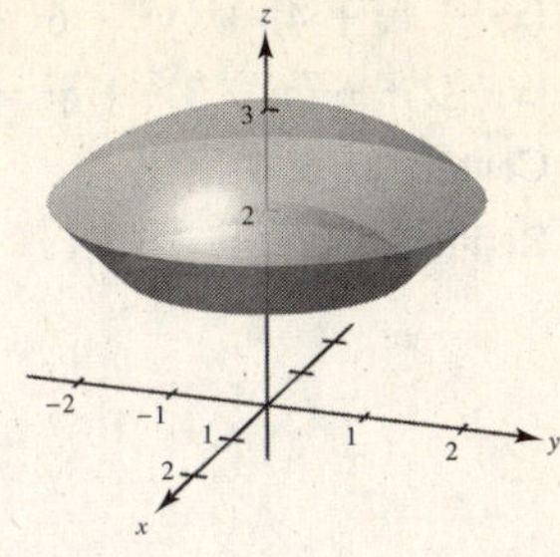

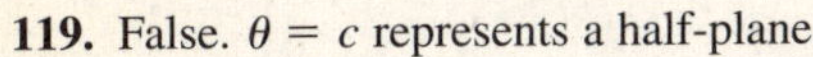

119. False. $\theta = c$ represents a half-plane.

120. True. They both represent spheres of radius 2 centered at the origin.

121. False. $(r, \theta, z) = (0, 0, 1)$ and $(r, \theta, z) = (0, \pi, 1)$ represent the same point $(x, y, z) = (0, 0, 1)$.

122. True (except for the orgin).

123. $z = \sin\theta, r = 1$

$z = \sin\theta = \dfrac{y}{r} = \dfrac{y}{1} = y$

The curve of intersection is the ellipse formed by the intersection of the plane $z = y$ and the cylinder $r = 1$.

124. $\rho = 2\sec\phi \Rightarrow \rho\cos\phi = 2 \Rightarrow z = 2$ plane

$\rho = 4$ sphere

The intersection of the plane and the sphere is a circle.

Review Exercises for Chapter 11

1. $P = (1, 2),\ Q = (4, 1),\ R = (5, 4)$

(a) $\mathbf{u} = \overrightarrow{PQ} = \langle 3, -1\rangle = 3\mathbf{i} - \mathbf{j},$

$\mathbf{v} = \overrightarrow{PR} = \langle 4, 2\rangle = 4\mathbf{i} + 2\mathbf{j}$

(b) $\|\mathbf{v}\| = \sqrt{4^2 + 2^2} = 2\sqrt{5}$

(c) $2\mathbf{u} + \mathbf{v} = \langle 6, -2\rangle + \langle 4, 2\rangle = \langle 10, 0\rangle = 10\mathbf{i}$

2. $P = (-2, -1),\ Q = (5, -1)\ R = (2, 4)$

(a) $\mathbf{u} = \overrightarrow{PQ} = \langle 7, 0\rangle = 7\mathbf{i},\ \mathbf{v} = \overrightarrow{PR} = \langle 4, 5\rangle = 4\mathbf{i} + 5\mathbf{j}$

(b) $\|\mathbf{v}\| = \sqrt{4^2 + 5^2} = \sqrt{41}$

(c) $2\mathbf{u} + \mathbf{v} = 14\mathbf{i} + (4\mathbf{i} + 5\mathbf{j}) = 18\mathbf{i} + 5\mathbf{j}$

3. $\mathbf{v} = \|\mathbf{v}\|\cos\theta\,\mathbf{i} + \|\mathbf{v}\|\sin\theta\,\mathbf{j} = 8\cos 120°\,\mathbf{i} + 8\sin 120°\,\mathbf{j}$

$= -4\mathbf{i} + 4\sqrt{3}\mathbf{j}$

4. $\mathbf{v} = \|\mathbf{v}\|\cos\theta\,\mathbf{i} + \|\mathbf{v}\|\sin\theta\,\mathbf{j} = \dfrac{1}{2}\cos 225°\,\mathbf{i} + \dfrac{1}{2}\sin 225°\,\mathbf{j}$

$= -\dfrac{\sqrt{2}}{4}\mathbf{i} - \dfrac{\sqrt{2}}{4}\mathbf{j}$

5. $z = 0,\ y = 4,\ x = -5$: $(-5, 4, 0)$

6. $x = z = 0,\ y = -7$: $(0, -7, 0)$

7. Looking down from the positive x-axis towards the yz-plane, the point is either in the first quadrant $(y > 0, z > 0)$ or in the third quadrant $(y < 0, z < 0)$. The x-coordinate can be any number.

8. Looking towards the xy-plane from the positive z-axis. The point is either in the second quadrant $(x < 0, y > 0)$ or in the fourth quadrant $(x > 0, y < 0)$. The z-coordinate can be any number.

9. $(x - 3)^2 + (y + 2)^2 + (z - 6)^2 = \left(\dfrac{15}{2}\right)^2$

10. Center: $\left(\dfrac{0 + 4}{2}, \dfrac{0 + 6}{2}, \dfrac{4 + 0}{2}\right) = (2, 3, 2)$

Radius: $\sqrt{(2 - 0)^2 + (3 - 0)^2 + (2 - 4)^2} = \sqrt{4 + 9 + 4} = \sqrt{17}$

$(x - 2)^2 + (y - 3)^2 + (z - 2)^2 = 17$

11. $(x^2 - 4x + 4) + (y^2 - 6y + 9) + z^2 = -4 + 4 + 9$

$(x - 2)^2 + (y - 3)^2 + z^2 = 9$

Center: $(2, 3, 0)$

Radius: 3

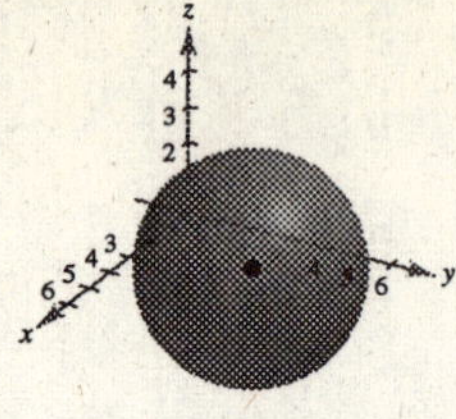

12. $(x^2 - 10x + 25) + (y^2 + 6y + 9) + (z^2 - 4z + 4) = -34 + 25 + 9 + 4$

$(x - 5)^2 + (y + 3)^2 + (z - 2)^2 = 4$

Center: $(5, -3, 2)$

Radius: 2

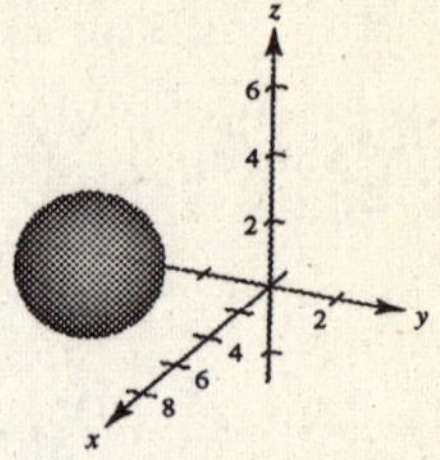

13. $\mathbf{v} = \langle 4 - 2, 4 + 1, -7 - 3\rangle = \langle 2, 5, -10\rangle$

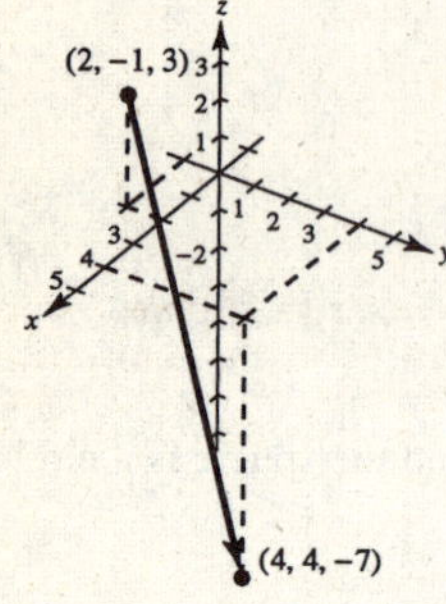

14. $\mathbf{v} = \langle 3 - 6, -3 - 2, 8 - 0\rangle = \langle -3, -5, 8\rangle$

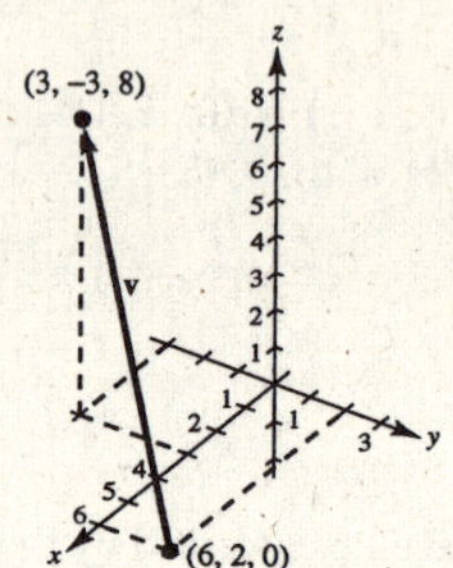

15. $\mathbf{v} = \langle -1 - 3, 6 - 4, 9 + 1\rangle = \langle -4, 2, 10\rangle$

$\mathbf{w} = \langle 5 - 3, 3 - 4, -6 + 1\rangle = \langle 2, -1, -5\rangle$

Since $-2\mathbf{w} = \mathbf{v}$, the points lie in a straight line.

16. $\mathbf{v} = \langle 8 - 5, -5 + 4, 5 - 7\rangle = \langle 3, -1, -2\rangle$

$\mathbf{w} = \langle 11 - 5, 6 + 4, 3 - 7\rangle = \langle 6, 10, -4\rangle$

Since $\mathbf{v}$ and $\mathbf{w}$ are not parallel, the points do not lie in a straight line.

17. Unit vector: $\dfrac{\mathbf{u}}{\|\mathbf{u}\|} = \dfrac{\langle 2, 3, 5\rangle}{\sqrt{38}} = \left\langle \dfrac{2}{\sqrt{38}}, \dfrac{3}{\sqrt{38}}, \dfrac{5}{\sqrt{38}}\right\rangle$

18. $8\dfrac{\langle 6, -3, 2\rangle}{\sqrt{49}} = \dfrac{8}{7}\langle 6, -3, 2\rangle = \left\langle \dfrac{48}{7}, -\dfrac{24}{7}, \dfrac{16}{7}\right\rangle$

19. $P = (5, 0, 0),\ Q = (4, 4, 0),\ R = (2, 0, 6)$

(a) $\mathbf{u} = \overrightarrow{PQ} = \langle -1, 4, 0\rangle = -\mathbf{i} + 4\mathbf{j}$,

$\mathbf{v} = \overrightarrow{PR} = \langle -3, 0, 6\rangle = -3\mathbf{i} + 6\mathbf{k}$

(b) $\mathbf{u} \cdot \mathbf{v} = (-1)(-3) + 4(0) + 0(6) = 3$

(c) $\mathbf{v} \cdot \mathbf{v} = 9 + 36 = 45$

20. $P = (2, -1, 3),\ Q = (0, 5, 1),\ R = (5, 5, 0)$

(a) $\mathbf{u} = \overrightarrow{PQ} = \langle -2, 6, -2\rangle = -2\mathbf{i} + 6\mathbf{j} - 2\mathbf{k}$,

$\mathbf{v} = \overrightarrow{PR} = \langle 3, 6, -3\rangle = 3\mathbf{i} + 6\mathbf{j} - 3\mathbf{k}$

(b) $\mathbf{u} \cdot \mathbf{v} = (-2)(3) + (6)(6) + (-2)(-3) = 36$

(c) $\mathbf{v} \cdot \mathbf{v} = 9 + 36 + 9 = 54$

21. $\mathbf{u} = \langle 7, -2, 3\rangle,\ \mathbf{v} = \langle -1, 4, 5\rangle$

Since $\mathbf{u} \cdot \mathbf{v} = 0$, the vectors are orthogonal.

22. $\mathbf{u} = \langle -4, 3, -6\rangle,\ \mathbf{v} = \langle 16, -12, 24\rangle$

Since $\mathbf{v} = -4\mathbf{u}$, the vectors are parallel.

23. $\mathbf{u} = 5\left(\cos\dfrac{3\pi}{4}\mathbf{i} + \sin\dfrac{3\pi}{4}\mathbf{j}\right) = \dfrac{5\sqrt{2}}{2}[-\mathbf{i} + \mathbf{j}]$

$\mathbf{v} = 2\left(\cos\dfrac{2\pi}{3}\mathbf{i} + \sin\dfrac{2\pi}{3}\mathbf{j}\right) = -\mathbf{i} + \sqrt{3}\mathbf{j}$

$\mathbf{u} \cdot \mathbf{v} = \dfrac{5\sqrt{2}}{2}(1 + \sqrt{3})$

$\|\mathbf{u}\| = 5 \qquad \|\mathbf{v}\| = 2$

$\cos\theta = \dfrac{|\mathbf{u} \cdot \mathbf{v}|}{\|\mathbf{u}\|\,\|\mathbf{v}\|} = \dfrac{(5\sqrt{2}/2)(1 + \sqrt{3})}{5(2)} = \dfrac{\sqrt{2} + \sqrt{6}}{4}$

$\theta = \arccos\dfrac{\sqrt{2} + \sqrt{6}}{4} = 15° \left[\text{or, } \dfrac{3\pi}{4} - \dfrac{2\pi}{3} = \dfrac{\pi}{12} \text{ or } 15°\right]$

24. $\mathbf{u} = \langle 4, -1, 5\rangle,\ \mathbf{v} = \langle 3, 2, -2\rangle$

$\mathbf{u} \cdot \mathbf{v} = 0 \Longrightarrow$ is orthogonal to $\mathbf{v}$.

$\theta = \dfrac{\pi}{2}$

25. $\mathbf{u} = \langle 10, -5, 15\rangle,\ \mathbf{v} = \langle -2, 1, -3\rangle$

$\mathbf{u} = -5\mathbf{v} \Rightarrow$ **u** is parallel to **v** and in the opposite direction.

$\theta = \pi$

26. $\mathbf{u} = \langle 1, 0, -3\rangle$

$\mathbf{v} = \langle 2, -2, 1\rangle$

$\mathbf{u}\cdot\mathbf{v} = -1$

$\|\mathbf{u}\| = \sqrt{10}$

$\|\mathbf{v}\| = 3$

$$\cos\theta = \frac{|\mathbf{u}\cdot\mathbf{v}|}{\|\mathbf{u}\|\,\|\mathbf{v}\|} = \frac{1}{3\sqrt{10}}$$

$\theta \approx 83.9°$

27. There are many correct answers.
For example: $\mathbf{v} = \pm\langle 6, -5, 0\rangle$.

28. $$W = \mathbf{F}\cdot\overrightarrow{PQ} = \|\mathbf{F}\|\,\|\overrightarrow{PQ}\|\cos\theta = (75)(8)\cos 30° = 300\sqrt{3}\text{ ft}\cdot\text{lb}$$

In Exercises 29–38, $\mathbf{u} = \langle 3, -2, 1\rangle$, $\mathbf{v} = \langle 2, -4, -3\rangle$, $\mathbf{w} = \langle -1, 2, 2\rangle$.

29. $$\mathbf{u}\cdot\mathbf{u} = 3(3) + (-2)(-2) + (1)(1) = 14 = \left(\sqrt{14}\right)^2 = \|\mathbf{u}\|^2$$

30. $$\cos\theta = \frac{|\mathbf{u}\cdot\mathbf{v}|}{\|\mathbf{u}\|\,\|\mathbf{v}\|} = \frac{11}{\sqrt{14}\sqrt{29}}$$

$$\theta = \arccos\left(\frac{11}{\sqrt{14}\sqrt{29}}\right) \approx 56.9°$$

31. $$\text{proj}_{\mathbf{u}}\mathbf{w} = \left(\frac{\mathbf{u}\cdot\mathbf{w}}{\|\mathbf{u}\|^2}\right)\mathbf{u} = -\frac{5}{14}\langle 3, -2, 1\rangle = \left\langle -\frac{15}{14}, \frac{10}{14}, -\frac{5}{14}\right\rangle = \left\langle -\frac{15}{14}, \frac{5}{7}, -\frac{5}{14}\right\rangle$$

32. Work $= \mathbf{u}\cdot\mathbf{w} = -3 - 4 + 2 = -5$

33. $$\mathbf{n} = \mathbf{v}\times\mathbf{w} = \begin{vmatrix} \mathbf{i} & \mathbf{j} & \mathbf{k} \\ 2 & -4 & -3 \\ -1 & 2 & 2 \end{vmatrix} = -2\mathbf{i} - \mathbf{j}$$

$\|\mathbf{n}\| = \sqrt{5}$

$$\frac{\mathbf{n}}{\|\mathbf{n}\|} = \frac{1}{\sqrt{5}}(-2\mathbf{i} - \mathbf{j}),\ \text{unit vector or } \frac{1}{\sqrt{5}}(2\mathbf{i} + \mathbf{j})$$

34. $$\mathbf{u}\times\mathbf{v} = \begin{vmatrix} \mathbf{i} & \mathbf{j} & \mathbf{k} \\ 3 & -2 & 1 \\ 2 & -4 & -3 \end{vmatrix} = 10\mathbf{i} + 11\mathbf{j} - 8\mathbf{k}$$

$$\mathbf{v}\times\mathbf{u} = \begin{vmatrix} \mathbf{i} & \mathbf{j} & \mathbf{k} \\ 2 & -4 & -3 \\ 3 & -2 & 1 \end{vmatrix} = -10\mathbf{i} - 11\mathbf{j} + 8\mathbf{k}$$

Thus, $\mathbf{u}\times\mathbf{v} = -(\mathbf{v}\times\mathbf{u})$.

35. $$V = |\mathbf{u}\cdot(\mathbf{v}\times\mathbf{w})| = |\langle 3, -2, 1\rangle\cdot\langle -2, -1, 0\rangle| = |-4| = 4$$

36. $$\mathbf{u}\times(\mathbf{v} + \mathbf{w}) = \langle 3, -2, 1\rangle\times\langle 1, -2, -1\rangle = \begin{vmatrix} \mathbf{i} & \mathbf{j} & \mathbf{k} \\ 3 & -2 & 1 \\ 1 & -2 & -1 \end{vmatrix} = 4\mathbf{i} + 4\mathbf{j} - 4\mathbf{k}$$

$$\mathbf{u}\times\mathbf{v} = \begin{vmatrix} \mathbf{i} & \mathbf{j} & \mathbf{k} \\ 3 & -2 & 1 \\ 2 & -4 & -3 \end{vmatrix} = 10\mathbf{i} + 11\mathbf{j} - 8\mathbf{k}$$

$$\mathbf{u}\times\mathbf{w} = \begin{vmatrix} \mathbf{i} & \mathbf{j} & \mathbf{k} \\ 3 & -2 & 1 \\ -1 & 2 & 2 \end{vmatrix} = -6\mathbf{i} - 7\mathbf{j} + 4\mathbf{k}$$

$$(\mathbf{u}\times\mathbf{v}) + (\mathbf{u}\times\mathbf{w}) = 4\mathbf{i} + 4\mathbf{j} - 4\mathbf{k} = \mathbf{u}\times(\mathbf{v} + \mathbf{w})$$

37. Area parallelogram $= \|\mathbf{u} \times \mathbf{v}\| = \|\langle 10, 11, -8\rangle\| = \sqrt{10^2 + 11^2 + (-8)^2}$ (See Exercises 34, 36)

$= \sqrt{285}$

38. Area triangle $= \frac{1}{2}\|\mathbf{v} \times \mathbf{w}\| = \frac{1}{2}\sqrt{(-2)^2 + (-1)^2} = \frac{\sqrt{5}}{2}$ (See Exercise 33)

39. $\mathbf{F} = c(\cos 20°\mathbf{j} + \sin 20°\mathbf{k})$

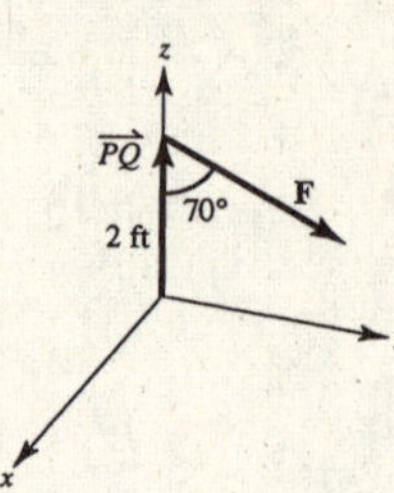

$\overrightarrow{PQ} = 2\mathbf{k}$

$$\overrightarrow{PQ} \times \mathbf{F} = \begin{vmatrix} \mathbf{i} & \mathbf{j} & \mathbf{k} \\ 0 & 0 & 2 \\ 0 & c\cos 20° & c\sin 20° \end{vmatrix} = -2c\cos 20°\mathbf{i}$$

$200 = \|\overrightarrow{PQ} \times \mathbf{F}\| = 2c\cos 20°$

$c = \dfrac{100}{\cos 20°}$

$\mathbf{F} = \dfrac{100}{\cos 20°}(\cos 20°\mathbf{j} + \sin 20°\mathbf{k}) = 100(\mathbf{j} + \tan 20°\mathbf{k})$

$\|\mathbf{F}\| = 100\sqrt{1 + \tan^2 20°} = 100 \sec 20° \approx 106.4$ lb

40. $$V = |\mathbf{u} \cdot (\mathbf{v} \times \mathbf{w})| = \begin{vmatrix} 2 & 1 & 0 \\ 0 & 2 & 1 \\ 0 & -1 & 2 \end{vmatrix} = 2(5) = 10$$

41. $\mathbf{v} = \langle 9 - 3, 11 - 0, 6 - 2\rangle = \langle 6, 11, 4\rangle$

(a) Parametric equations: $x = 3 + 6t, y = 11t, z = 2 + 4t$

(b) Symmetric equations: $\dfrac{x - 3}{6} = \dfrac{y}{11} = \dfrac{z - 2}{4}$

42. $\mathbf{v} = \langle 8 + 1, 10 - 4, 5 - 3\rangle = \langle 9, 6, 2\rangle$

(a) Parametric equations: $x = -1 + 9t, y = 4 + 6t, z = 3 + 2t$

(b) Symmetric equations: $\dfrac{x + 1}{9} = \dfrac{y - 4}{6} = \dfrac{z - 3}{2}$

43. $\mathbf{v} = \mathbf{j}$

(a) $x = 1,\ y = 2 + t,\ z = 3$

(b) None

44. Direction numbers: 1, 1, 1

(a) $x = 1 + t,\ y = 2 + t,\ z = 3 + t$

(b) $x - 1 = y - 2 = z - 3$

45. $3x - 3y - 7z = -4,\ x - y + 2z = 3$

Solving simultaneously, we have $z = 1$. Substituting $z = 1$ into the second equation we have $y = x - 1$. Substituting for x in this equation we obtain two points on the line of intersection, $(0, -1, 1)$, $(1, 0, 1)$. The direction vector of the line of intersection is $\mathbf{v} = \mathbf{i} + \mathbf{j}$.

(a) $x = t,\ y = -1 + t,\ z = 1$

(b) $x = y + 1,\ z = 1$

46. $$\mathbf{u} \times \mathbf{v} = \begin{vmatrix} \mathbf{i} & \mathbf{j} & \mathbf{k} \\ 2 & -5 & 1 \\ -3 & 1 & 4 \end{vmatrix} = -21\mathbf{i} - 11\mathbf{j} - 13\mathbf{k}$$

Direction numbers: 21, 11, 13

(a) $x = 21t,\ y = 1 + 11t,\ z = 4 + 13t$

(b) $\dfrac{x}{21} = \dfrac{y - 1}{11} = \dfrac{z - 4}{13}$

47. $P = (-3, -4, 2)$, $Q = (-3, 4, 1)$, $R = (1, 1, -2)$

$\overrightarrow{PQ} = \langle 0, 8, -1 \rangle$, $\overrightarrow{PR} = \langle 4, 5, -4 \rangle$

$$\mathbf{n} = \overrightarrow{PQ} \times \overrightarrow{PR} = \begin{vmatrix} \mathbf{i} & \mathbf{j} & \mathbf{k} \\ 0 & 8 & -1 \\ 4 & 5 & -4 \end{vmatrix} = -27\mathbf{i} - 4\mathbf{j} - 32\mathbf{k}$$

$$-27(x + 3) - 4(y + 4) - 32(z - 2) = 0$$

$$27x + 4y + 32z = -33$$

48. $\mathbf{n} = 3\mathbf{i} - \mathbf{j} + \mathbf{k}$

$$3(x + 2) - 1(y - 3) + 1(z - 1) = 0$$

$$3x - y + z + 8 = 0$$

49. The two lines are parallel as they have the same direction numbers, $-2, 1, 1$. Therefore, a vector parallel to the plane is $\mathbf{v} = -2\mathbf{i} + \mathbf{j} + \mathbf{k}$. A point on the first line is $(1, 0, -1)$ and a point on the second line is $(-1, 1, 2)$. The vector $\mathbf{u} = 2\mathbf{i} - \mathbf{j} - 3\mathbf{k}$ connecting these two points is also parallel to the plane. Therefore, a normal to the plane is

$$\mathbf{v} \times \mathbf{u} = \begin{vmatrix} \mathbf{i} & \mathbf{j} & \mathbf{k} \\ -2 & 1 & 1 \\ 2 & -1 & -3 \end{vmatrix}$$

$$= -2\mathbf{i} - 4\mathbf{j} = -2(\mathbf{i} + 2\mathbf{j}).$$

Equation of the plane: $(x - 1) + 2y = 0$

$$x + 2y = 1$$

50. Let $\mathbf{v} = \langle 5 - 2, 1 + 2, 3 - 1 \rangle = \langle 3, 3, 2 \rangle$ be the direction vector for the line through the two points. Let $\mathbf{n} = \langle 2, 1, -1 \rangle$ be the normal vector to the plane. Then

$$\mathbf{v} \times \mathbf{n} = \begin{vmatrix} \mathbf{i} & \mathbf{j} & \mathbf{k} \\ 3 & 3 & 2 \\ 2 & 1 & -1 \end{vmatrix} = \langle -5, 7, -3 \rangle$$

is the normal to the unknown plane.

$$-5(x - 5) + 7(y - 1) - 3(z - 3) = 0$$

$$-5x + 7y - 3z + 27 = 0$$

51. $Q(1, 0, 2)$ point

$2x - 3y + 6z = 6$

A point P on the plane is $(3, 0, 0)$.

$\overrightarrow{PQ} = \langle -2, 0, 2 \rangle$

$\mathbf{n} = \langle 2, -3, 6 \rangle$ normal to plane

$$D = \frac{|\overrightarrow{PQ} \cdot \mathbf{n}|}{\|\mathbf{n}\|} = \frac{8}{7}$$

52. $Q(3, -2, 4)$ point

$P(5, 0, 0)$ point on plane

$\mathbf{n} = \langle 2, -5, 1 \rangle$ normal to plane

$\overrightarrow{PQ} = \langle -2, -2, 4 \rangle$

$$D = \frac{|\overrightarrow{PQ} \cdot \mathbf{n}|}{\|\mathbf{n}\|} = \frac{10}{\sqrt{30}} = \frac{\sqrt{30}}{3}$$

53. The normal vectors to the planes are the same,

$\mathbf{n} = \langle 5, -3, 1 \rangle$.

Choose a point in the first plane, $P = (0, 0, 2)$. Choose a point in the second plane, $Q = (0, 0, -3)$.

$\overrightarrow{PQ} = \langle 0, 0, -5 \rangle$

$$D = \frac{|\overrightarrow{PQ} \cdot \mathbf{n}|}{\|\mathbf{n}\|} = \frac{|-5|}{\sqrt{35}} = \frac{5}{\sqrt{35}} = \frac{\sqrt{35}}{7}$$

54. $Q(-5, 1, 3)$ point

$\mathbf{u} = \langle 1, -2, -1 \rangle$ direction vector

$P = (1, 3, 5)$ point on line

$\overrightarrow{PQ} = \langle -6, -2, -2 \rangle$

$$\overrightarrow{PQ} \times \mathbf{u} = \begin{vmatrix} \mathbf{i} & \mathbf{j} & \mathbf{k} \\ -6 & -2 & -2 \\ 1 & -2 & -1 \end{vmatrix} = \langle -2, -8, 14 \rangle$$

$$D = \frac{\|\overrightarrow{PQ} \times \mathbf{u}\|}{\|\mathbf{u}\|} = \frac{\sqrt{264}}{\sqrt{6}} = 2\sqrt{11}$$

55. $x + 2y + 3z = 6$

Plane

Intercepts: $(6, 0, 0)$, $(0, 3, 0)$, $(0, 0, 2)$

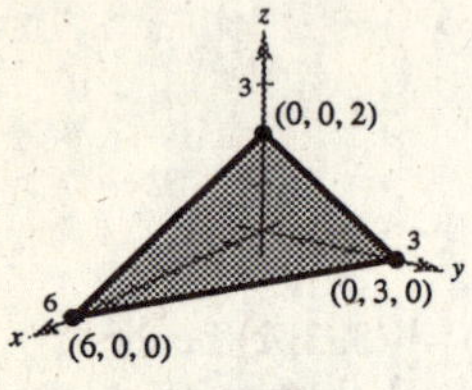

56. $y = z^2$

Since the x-coordinate is missing, we have a cylindrical surface with rulings parallel to the x-axis. The generating curve is a parabola in the yz-coordinate plane.

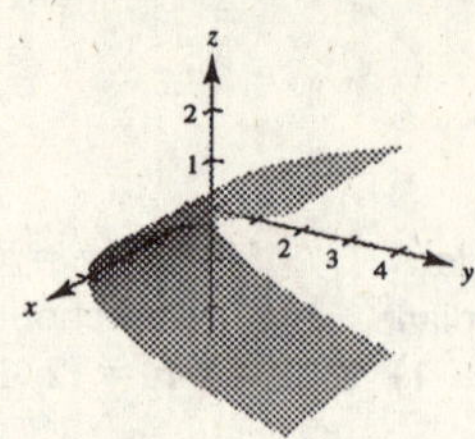

57. $y = \frac{1}{2}z$

Plane with rulings parallel to the x-axis

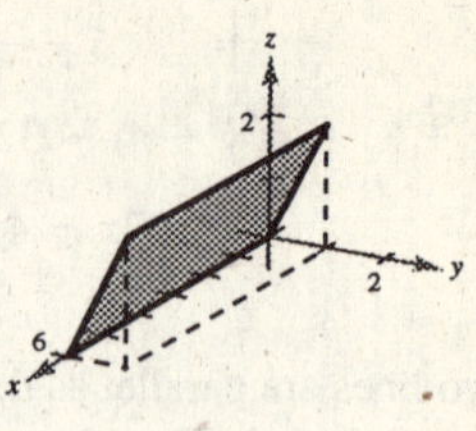

58. $y = \cos z$

Since the x-coordinate is missing, we have a cylindrical surface with rulings parallel to the x-axis. The generating curve is $y = \cos z$.

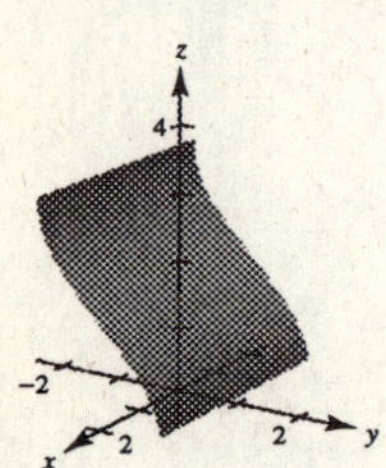

59. $\frac{x^2}{16} + \frac{y^2}{9} + z^2 = 1$

Ellipsoid

xy-trace: $\frac{x^2}{16} + \frac{y^2}{9} = 1$

xz-trace: $\frac{x^2}{16} + z^2 = 1$

yz-trace: $\frac{y^2}{9} + z^2 = 1$

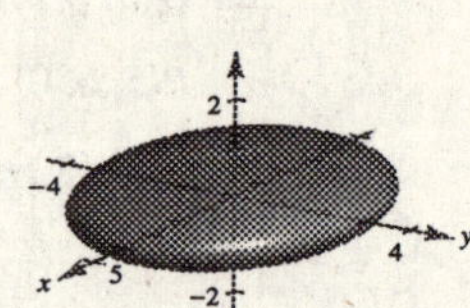

60. $16x^2 + 16y^2 - 9z^2 = 0$

Cone

xy-trace: point $(0,0, 0)$

xz-trace: $z = \pm\frac{4x}{3}$

yz-trace: $z = \pm\frac{4y}{3}$

$z = 4$, $x^2 + y^2 = 9$

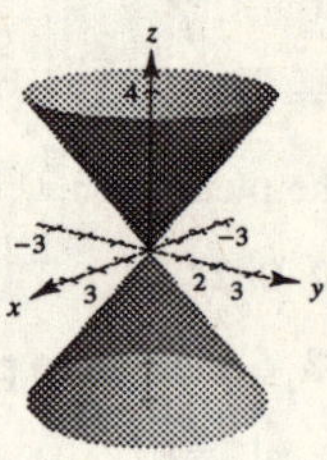

61. $\frac{x^2}{16} - \frac{y^2}{9} + z^2 = -1$

$\frac{y^2}{9} - \frac{x^2}{16} - z^2 = 1$

Hyperboloid of two sheets

xy-trace: $\frac{y^2}{9} - \frac{x^2}{16} = 1$

xz-trace: None

yz-trace: $\frac{y^2}{9} - z^2 = 1$

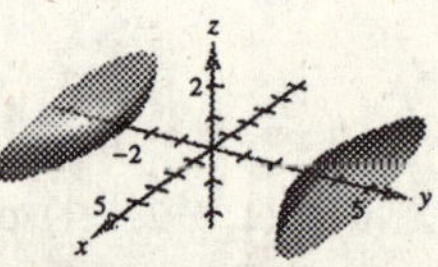

62. $\frac{x^2}{25} + \frac{y^2}{4} - \frac{z^2}{100} = 1$

Hyperboloid of one sheet

xy-trace: $\frac{x^2}{25} + \frac{y^2}{4} = 1$

xz-trace: $\frac{x^2}{25} - \frac{z^2}{100} = 1$

yz-trace: $\frac{y^2}{4} - \frac{z^2}{100} = 1$

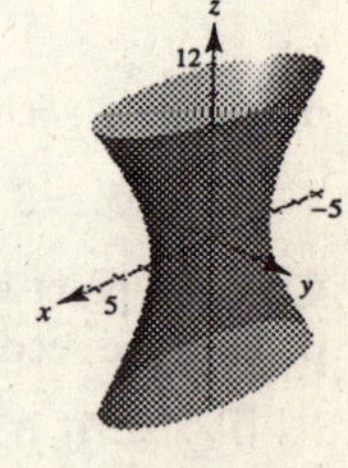

63. $x^2 + z^2 = 4$. Cylinder of radius 2 about y-axis

64. $y^2 + z^2 = 16$. Cylinder of radius 4 about x-axis

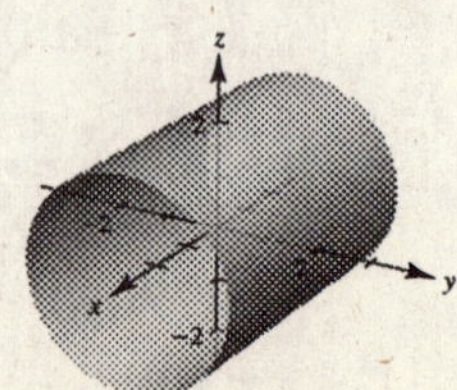

65. Let $y = r(x) = 2\sqrt{x}$ and revolve the curve about the x-axis.

66. $z^2 = 2y$ revolved about y-axis

$z = \pm\sqrt{2y}$

$x^2 + z^2 = [r(y)]^2 = 2y$

$x^2 + z^2 = 2y$

67. $(-2\sqrt{2}, 2\sqrt{2}, 2)$, rectangular

(a) $r = \sqrt{(-2\sqrt{2})^2 + (2\sqrt{2})^2} = 4,\ \theta = \arctan(-1) = \frac{3\pi}{4},\ z = 2,\ \left(4, \frac{3\pi}{4}, 2\right)$, cylindrical

(b) $\rho = \sqrt{(-2\sqrt{2})^2 + (2\sqrt{2})^2 + (2)^2} = 2\sqrt{5},\ \theta = \frac{3\pi}{4},\ \phi = \arccos\frac{2}{2\sqrt{5}} = \arccos\frac{1}{\sqrt{5}},\ \left(2\sqrt{5}, \frac{3\pi}{4}, \arccos\frac{\sqrt{5}}{5}\right)$, spherical

68. $\left(\frac{\sqrt{3}}{4}, \frac{3}{4}, \frac{3\sqrt{3}}{2}\right)$, rectangular

(a) $r = \sqrt{\left(\frac{\sqrt{3}}{4}\right)^2 + \left(\frac{3}{4}\right)^2} = \frac{\sqrt{3}}{2},\ \theta = \arctan\sqrt{3} = \frac{\pi}{3},\ z = \frac{3\sqrt{3}}{2},\ \left(\frac{\sqrt{3}}{2}, \frac{\pi}{2}, \frac{3\sqrt{3}}{2}\right)$, cylindrical

(b) $\rho = \sqrt{\left(\frac{\sqrt{3}}{4}\right)^2 + \left(\frac{3}{4}\right)^2 + \left(\frac{3\sqrt{3}}{2}\right)^2} = \frac{\sqrt{30}}{2},\ \theta = \frac{\pi}{3},\ \phi = \arccos\frac{3}{\sqrt{10}},\ \left(\frac{\sqrt{30}}{2}, \frac{\pi}{3}, \arccos\frac{3}{\sqrt{10}}\right)$, spherical

69. $\left(100, -\frac{\pi}{6}, 50\right)$, cylindrical

$\rho = \sqrt{100^2 + 50^2} = 50\sqrt{5}$

$\theta = -\frac{\pi}{6}$

$\phi = \arccos\left(\frac{50}{50\sqrt{5}}\right) = \arccos\left(\frac{1}{\sqrt{5}}\right) \approx 63.4°$ or 1.107

$\left(50\sqrt{5}, -\frac{\pi}{6}, 63.4°\right)$, spherical or $\left(50\sqrt{5}, \frac{-\pi}{6}, 1.1071\right)$

70. $\left(81, -\frac{5\pi}{6}, 27\sqrt{3}\right)$, cylindrical

$\rho = \sqrt{6561 + 2187} = 54\sqrt{3}$

$\theta = -\frac{5\pi}{6}$

$\phi = \arccos\left(\frac{27\sqrt{3}}{54\sqrt{3}}\right) = \arccos\frac{1}{2} = \frac{\pi}{3}$

$\left(54\sqrt{3}, -\frac{5\pi}{6}, \frac{\pi}{3}\right)$, spherical

71. $\left(25, -\frac{\pi}{4}, \frac{3\pi}{4}\right)$, spherical

$r^2 = \left(25\sin\left(\frac{3\pi}{4}\right)\right)^2 \Rightarrow r = 25\frac{\sqrt{2}}{2}$

$\theta = -\frac{\pi}{4}$

$z = \rho\cos\phi = 25\cos\frac{3\pi}{4} = -25\frac{\sqrt{2}}{2}$

$\left(25\frac{\sqrt{2}}{2}, -\frac{\pi}{4}, -\frac{25\sqrt{2}}{2}\right)$, cylindrical

72. $\left(12, -\frac{\pi}{2}, \frac{2\pi}{3}\right)$, spherical

$r^2 = \left(12\sin\left(\frac{2\pi}{3}\right)\right)^2 \Rightarrow r = 6\sqrt{3}$

$\theta = -\frac{\pi}{2}$

$z = \rho\cos\phi = 12\cos\left(\frac{2\pi}{3}\right) = -6$

$\left(6\sqrt{3}, -\frac{\pi}{2}, -6\right)$, cylindrical

73. $x^2 - y^2 = 2z$

(a) Cylindrical: $r^2\cos^2\theta - r^2\sin^2\theta = 2z,\ r^2\cos 2\theta = 2z$

(b) Spherical: $\rho^2\sin^2\phi\cos^2\theta - \rho^2\sin^2\phi\sin^2\theta = 2\rho\cos\phi,\ \rho\sin^2\phi\cos 2\theta - 2\cos\phi = 0,\ \rho = 2\sec 2\theta\cos\phi\csc^2\phi$

74. $x^2 + y^2 + z^2 = 16$

(a) Cylindrical: $r^2 + z^2 = 16$

(b) Spherical: $\rho = 4$

75. $r = 4\sin\theta$ cylindrical coordinates

$r^2 = 4r\sin\theta$

$x^2 + y^2 = 4y$

$x^2 + y^2 - 4y + 4 = 4$

$x^2 + (y-2)^2 = 4$ rectangular coordinates

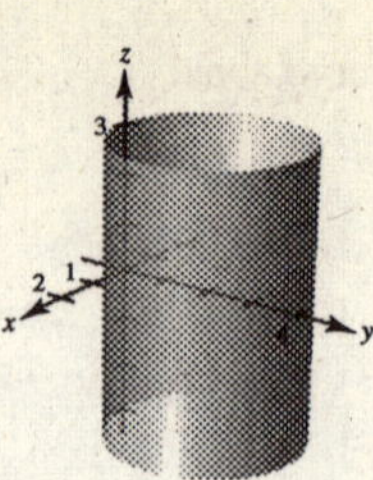

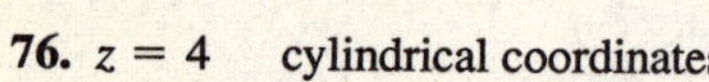

76. $z = 4$ cylindrical coordinates

$z = 4$ rectangular coordinates

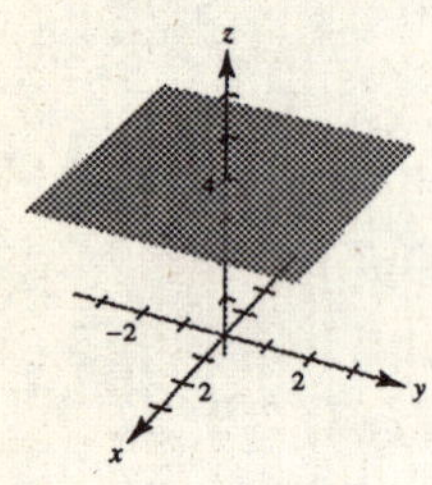

77. $\theta = \dfrac{\pi}{4}$ spherical coordinates

$\tan\theta = \tan\dfrac{\pi}{4} = 1$

$\dfrac{y}{x} = 1$

$y = x, x \geq 0$ rectangular coordinates half-plane

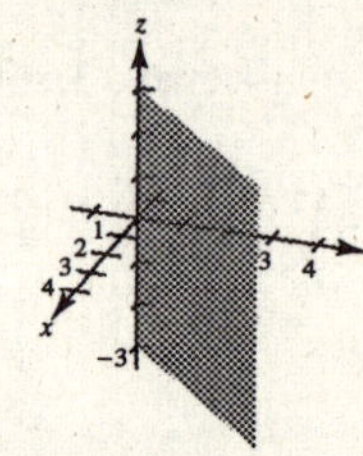

78. $\rho = 2\cos\theta$ spherical coordinates

Because $\cos\theta = \dfrac{x}{r} = \dfrac{x}{\sqrt{x^2+y^2}}$, we have

$\sqrt{x^2+y^2+z^2} = 2\dfrac{x}{\sqrt{x^2+y^2}}$ rectangular equation

Problem Solving for Chapter 11

1.

$$\mathbf{a} + \mathbf{b} + \mathbf{c} = \mathbf{0}$$
$$\mathbf{b} \times (\mathbf{a} + \mathbf{b} + \mathbf{c}) = \mathbf{0}$$
$$(\mathbf{b} \times \mathbf{a}) + (\mathbf{b} \times \mathbf{c}) = \mathbf{0}$$
$$\|\mathbf{a} \times \mathbf{b}\| = \|\mathbf{b} \times \mathbf{c}\|$$
$$\|\mathbf{b} \times \mathbf{c}\| = \|\mathbf{b}\|\,\|\mathbf{c}\| \sin A$$
$$\|\mathbf{a} \times \mathbf{b}\| = \|\mathbf{a}\|\,\|\mathbf{b}\| \sin C$$

Then,

$$\frac{\sin A}{\|\mathbf{a}\|} = \frac{\|\mathbf{b} \times \mathbf{c}\|}{\|\mathbf{a}\|\,\|\mathbf{b}\|\,\|\mathbf{c}\|} = \frac{\|\mathbf{a} \times \mathbf{b}\|}{\|\mathbf{a}\|\,\|\mathbf{b}\|\,\|\mathbf{c}\|} = \frac{\sin C}{\|\mathbf{c}\|}.$$

The other case, $\dfrac{\sin A}{\|\mathbf{a}\|} = \dfrac{\sin B}{\|\mathbf{b}\|}$ is similar.

2. $f(x) = \displaystyle\int_0^x \sqrt{t^4+1}\,dt$

(a)

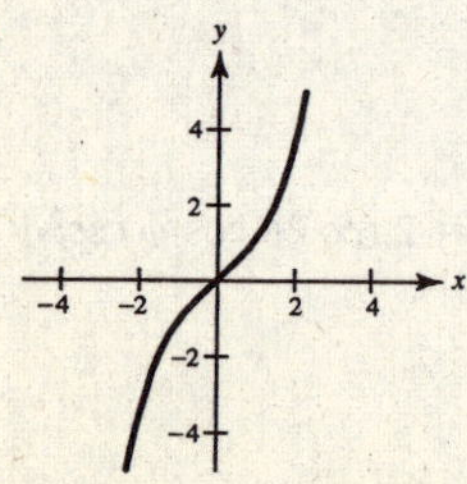

(b) $f'(x) = \sqrt{x^4+1}$

$f'(0) = 1 = \tan\theta$

$\theta = \dfrac{\pi}{4}$

$\mathbf{u} = \dfrac{1}{\sqrt{2}}(\mathbf{i} + \mathbf{j}) = \left\langle \dfrac{\sqrt{2}}{2}, \dfrac{\sqrt{2}}{2} \right\rangle$

(c) $\pm\left\langle \dfrac{\sqrt{2}}{2}, -\dfrac{\sqrt{2}}{2} \right\rangle$

(d) The line is $y = x$: $x = t, y = t$.

3. Label the figure as indicated.

From the figure, you see that

$$\overrightarrow{SP} = \frac{1}{2}\mathbf{a} - \frac{1}{2}\mathbf{b} = \overrightarrow{RQ} \quad \text{and} \quad \overrightarrow{SR} = \frac{1}{2}\mathbf{a} + \frac{1}{2}\mathbf{b} = \overrightarrow{PQ}.$$

Since $\overrightarrow{SP} = \overrightarrow{RQ}$ and $\overrightarrow{SR} = \overrightarrow{PQ}$, $PSRQ$ is a parallelogram.

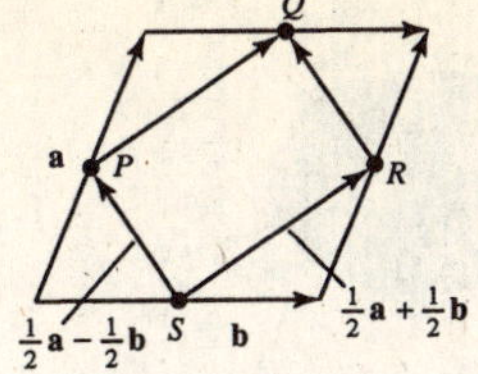

4. Label the figure as indicated.

$\overrightarrow{PR} = \mathbf{a} + \mathbf{b}$

$\overrightarrow{SQ} = \mathbf{b} - \mathbf{a}$

$(\mathbf{a} + \mathbf{b}) \cdot (\mathbf{b} - \mathbf{a}) = \|\mathbf{b}\|^2 - \|\mathbf{a}\|^2 = 0$, because

$\|\mathbf{a}\| = \|\mathbf{b}\|$ in a rhombus.

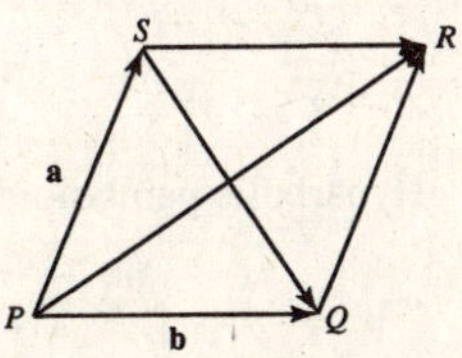

5. (a) $\mathbf{u} = \langle 0, 1, 1 \rangle$ direction vector of line determined by P_1 and P_2.

$$D = \frac{\|\overrightarrow{P_1Q} \times \mathbf{u}\|}{\|\mathbf{u}\|}$$

$$= \frac{\|\langle 2, 0, -1 \rangle \times \langle 0, 1, 1 \rangle\|}{\sqrt{2}}$$

$$= \frac{\|\langle 1, -2, 2 \rangle\|}{\sqrt{2}} = \frac{3}{\sqrt{2}} = \frac{3\sqrt{2}}{2}$$

(b) The shortest distance to the line **segment** is

$\|P_1Q\| = \|\langle 2, 0, -1 \rangle\| = \sqrt{5}$.

6. $(\mathbf{n} + \overrightarrow{PP_0}) \perp (\mathbf{n} - \overrightarrow{PP_0})$

Figure is a square.

Thus, $\|\overrightarrow{PP_0}\| = \|\mathbf{n}\|$ and the points P form a circle of radius $\|\mathbf{n}\|$ in the plane with center at P_0.

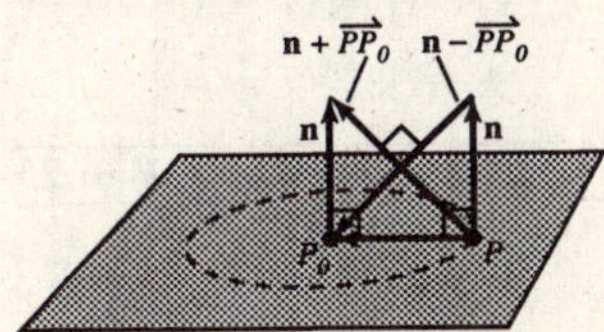

7. (a) $V = \pi \int_0^1 \left(\sqrt{z}\right)^2 dz = \left[\pi \frac{z^2}{2}\right]_0^1 = \frac{1}{2}\pi$

Note: $\frac{1}{2}$(base)(altitude) $= \frac{1}{2}\pi(1) = \frac{1}{2}\pi$

(b) $\frac{x^2}{a^2} + \frac{y^2}{b^2} = z$: (slice at $z = c$)

$$\frac{x^2}{\left(\sqrt{ca}\right)^2} + \frac{y^2}{\left(\sqrt{cb}\right)^2} = 1$$

At $z = c$, figure is ellipse of area

$\pi\left(\sqrt{ca}\right)\left(\sqrt{cb}\right) = \pi abc.$

$$V = \int_0^k \pi abc \cdot dc = \left[\frac{\pi abc^2}{2}\right]_0^k = \frac{\pi abk^2}{2}$$

(c) $V = \frac{1}{2}(\pi abk)k = \frac{1}{2}$(area of base)(height)

8. (a) $V = 2\int_0^r \pi(r^2 - x^2)\, dx = 2\pi\left[r^2x - \frac{x^3}{3}\right]_0^r = \frac{4}{3}\pi r^3$

(b) At height $z = d > 0$,

$$\frac{x^2}{a^2} + \frac{y^2}{b^2} + \frac{d^2}{c^2} = 1$$

$$\frac{x^2}{a^2} + \frac{y^2}{b^2} = 1 - \frac{d^2}{c^2} = \frac{c^2 - d^2}{c^2}$$

$$\frac{x^2}{\frac{a^2(c^2 - d^2)}{c^2}} + \frac{y^2}{\frac{b^2(c^2 - d^2)}{c^2}} = 1.$$

$$\text{Area} = \pi\sqrt{\left(\frac{a^2(c^2 - d^2)}{c^2}\right)\left(\frac{b^2(c^2 - d^2)}{c^2}\right)}$$

$$= \frac{\pi ab}{c^2}(c^2 - d^2)$$

$$V = 2\int_0^c \frac{\pi ab}{c^2}(c^2 - d^2)\, dd$$

$$= \frac{2\pi ab}{c^2}\left[c^2d - \frac{d^3}{3}\right]_0^c$$

$$= \frac{4}{3}\pi abc$$

9. (a) $\rho = 2 \sin \phi$

Torus

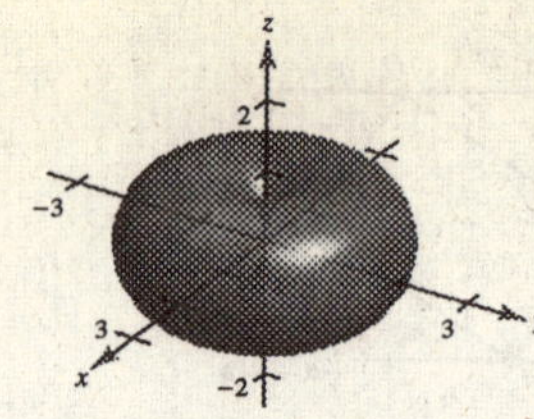

(b) $\rho = 2 \cos \phi$

Sphere

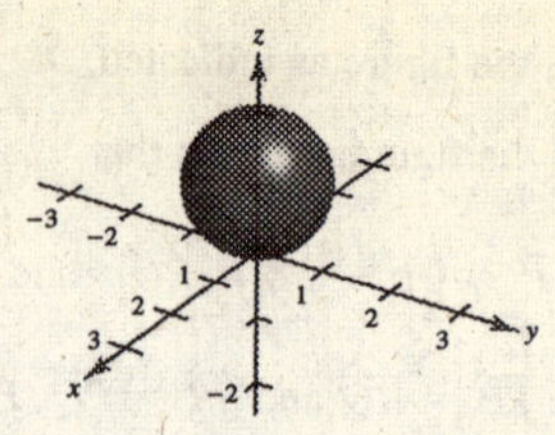

10. (a) $r = 2 \cos \theta$

Cylinder

(b) $z = r^2 \cos 2\theta$

$z^2 = x^2 - y^2$

Hyperbolic paraboloid

11. From Exercise 64, Section 11.4,

$(\mathbf{u} \times \mathbf{v}) \times (\mathbf{w} \times \mathbf{z}) = [(\mathbf{u} \times \mathbf{v}) \cdot \mathbf{z}]\mathbf{w} - [(\mathbf{u} \times \mathbf{v}) \cdot \mathbf{w}]\mathbf{z}.$

12. $x = -t + 3, y = \frac{1}{2}t + 1, z = 2t - 1; Q = (4, 3, s)$

(a) $\mathbf{u} = \langle -2, 1, 4 \rangle$ direction vector for line

$P = (3, 1, -1)$ point on line

$\overrightarrow{PQ} = \langle 1, 2, s + 1 \rangle$

$$\overrightarrow{PQ} \times \mathbf{u} = \begin{vmatrix} \mathbf{i} & \mathbf{j} & \mathbf{k} \\ 1 & 2 & s+1 \\ -2 & 1 & 4 \end{vmatrix} = (7 - s)\mathbf{i} + (-6 - 2s)\mathbf{j} + 5\mathbf{k}$$

$$D = \frac{\|\overrightarrow{PQ} \times \mathbf{u}\|}{\|\mathbf{u}\|} = \frac{\sqrt{(7 - s)^2 + (-6 - 2s)^2 + 25}}{\sqrt{21}}$$

(b)

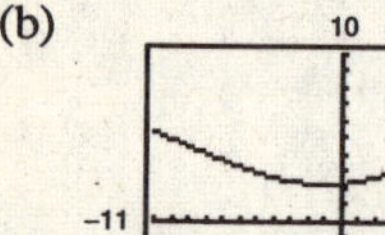

The minimum is $D \approx 2.2361$ at $s = -1$.

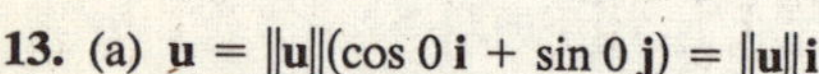

(c) Yes, there are slant asymptotes. Using $s = x$, we have

$$D(s) = \frac{1}{\sqrt{21}}\sqrt{5x^2 + 10x + 110} = \frac{\sqrt{5}}{\sqrt{21}}\sqrt{x^2 + 2x + 22}$$

$$= \frac{\sqrt{5}}{\sqrt{21}}\sqrt{(x + 1)^2 + 21} \rightarrow \pm\sqrt{\frac{5}{21}}(x + 1)$$

$y = \pm\frac{\sqrt{105}}{21}(s + 1)$ slant asymptotes.

13. (a) $\mathbf{u} = \|\mathbf{u}\|(\cos 0\,\mathbf{i} + \sin 0\,\mathbf{j}) = \|\mathbf{u}\|\mathbf{i}$

Downward force $\mathbf{w} = -\mathbf{j}$

$\mathbf{T} = \|\mathbf{T}\|(\cos(90° + \theta)\mathbf{i} + \sin(90° + \theta)\mathbf{j})$

$= \|\mathbf{T}\|(-\sin \theta\,\mathbf{i} + \cos \theta\,\mathbf{j})$

$\mathbf{0} = \mathbf{u} + \mathbf{w} + \mathbf{T} = \|\mathbf{u}\|\mathbf{i} - \mathbf{j} + \|\mathbf{T}\|(-\sin \theta\,\mathbf{i} + \cos \theta\,\mathbf{j})$

$\|\mathbf{u}\| = \sin \theta\,\|\mathbf{T}\|$

$1 = \cos \theta\,\|\mathbf{T}\|$

If $\theta = 30°$, $\|\mathbf{u}\| = (1/2)\|\mathbf{T}\|$ and $1 = (\sqrt{3}/2)\|\mathbf{T}\|$

$$\Rightarrow \|\mathbf{T}\| = \frac{2}{\sqrt{3}} \approx 1.1547 \text{ lb} \quad \text{and} \quad \|\mathbf{u}\| = \frac{1}{2}\left(\frac{2}{\sqrt{3}}\right) \approx 0.5774 \text{ lb}$$

(b) From part (a), $\|\mathbf{u}\| = \tan \theta$ and $\|\mathbf{T}\| = \sec \theta$.

Domain: $0 \leq \theta \leq 90°$

(c)

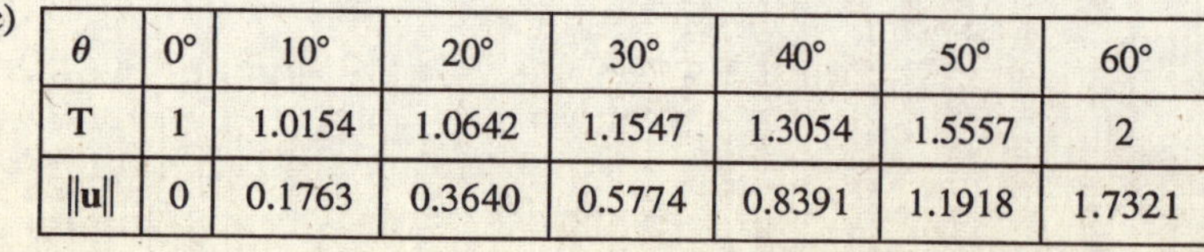

θ	0°	10°	20°	30°	40°	50°	60°
T	1	1.0154	1.0642	1.1547	1.3054	1.5557	2
$\|\mathbf{u}\|$	0	0.1763	0.3640	0.5774	0.8391	1.1918	1.7321

(d)

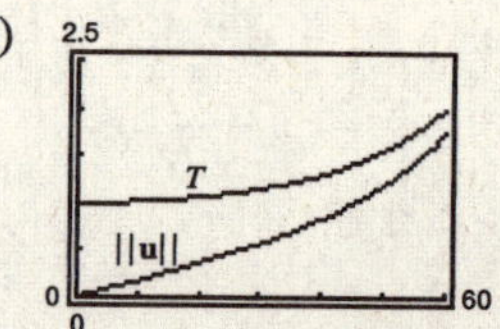

(e) Both are increasing functions.

(f) $\lim_{\theta \to \pi/2^-} T = \infty$ and $\lim_{\theta \to \pi/2^-} \|\mathbf{u}\| = \infty$.

14. (a) The tension T is the same in each tow line.

$$6000\mathbf{i} = T(\cos 20° + \cos(-20))\mathbf{i} + T(\sin 20° + \sin(-20°))\mathbf{j}$$

$$= 2T\cos 20°\mathbf{i}$$

$$\Rightarrow T = \frac{6000}{2\cos 20°} \approx 3192.5 \text{ lbs}$$

(b) As in part (a), $6000\mathbf{i} = 2T\cos\theta$

$$\Rightarrow T = \frac{3000}{\cos\theta}$$

Domain: $0 < \theta < 90°$

(c)

θ	10°	20°	30°	40°	50°	60°
T	3046.3	3192.5	3464.1	3916.2	4667.2	6000.0

(d)

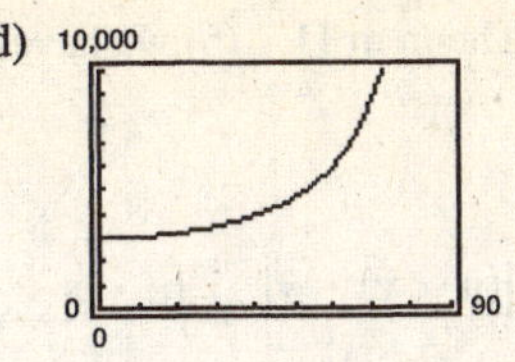

(e) As θ increases, there is less force applied in the direction of motion.

15. Let $\theta = \alpha - \beta$, the angle between **u** and **v**. Then

$$\sin(\alpha - \beta) = \frac{\|\mathbf{u}\times\mathbf{v}\|}{\|\mathbf{u}\|\,\|\mathbf{v}\|} = \frac{\|\mathbf{v}\times\mathbf{u}\|}{\|\mathbf{u}\|\,\|\mathbf{v}\|}.$$

For $\mathbf{u} = \langle\cos\alpha, \sin\alpha, 0\rangle$ and $\mathbf{v} = \langle\cos\beta, \sin\beta, 0\rangle$, $\|\mathbf{u}\| = \|\mathbf{v}\| = 1$ and

$$\mathbf{v}\times\mathbf{u} = \begin{vmatrix} \mathbf{i} & \mathbf{j} & \mathbf{k} \\ \cos\beta & \sin\beta & 0 \\ \cos\alpha & \sin\alpha & 0 \end{vmatrix} = (\sin\alpha\cos\beta - \cos\alpha\sin\beta)\mathbf{k}.$$

Thus, $\sin(\alpha - \beta) = \|\mathbf{v}\times\mathbf{u}\| = \sin\alpha\cos\beta - \cos\alpha\sin\beta$.

16. (a) Los Angeles: $(4000, -118.24°, 55.95°)$ Rio de Janeiro: $(4000, -43.23°, 112.90°)$

(b) Los Angeles: $x = 4000\sin(55.95°)\cos(-118.24°)$

$y = 4000\sin(55.95°)\sin(-118.24°)$

$z = 4000\cos(55.95°)$

$(x, y, z) \approx (-1568.2, -2919.7, 2239.7)$

Rio de Janeiro: $x = 4000\sin(112.90°)\cos(-43.23°)$

$y = 4000\sin(112.90°)\sin(-43.23°)$

$z = 4000\cos(112.90°)$

$(x, y, z) \approx (2684.7, -2523.8, -1556.5)$

(c) $$\cos\theta = \frac{\mathbf{u}\cdot\mathbf{v}}{\|\mathbf{u}\|\,\|\mathbf{v}\|} = \frac{(-1568.2)(2684.7) + (-2919.7)(-2523.8) + (2239.7)(-1556.5)}{(4000)(4000)} \approx -0.02047$$

$\theta \approx 91.17°$ or 1.59 radians

(d) $s = r\theta = 4000(1.59) \approx 6360$ miles

(e) For Boston and Honolulu:

a. Boston: $(4000, -71.06°, 47.64°)$ Honolulu: $(4000, -157.86°, 68.69°)$

b. Boston: $x = 4000\sin 47.64°\cos(-71.06°)$

$y = 4000\sin 47.64°\sin(-71.06°)$

$z = 4000\cos 47.64°$

$(959.4, -2795.7, 2695.1)$

Honolulu: $x = (4000\sin 68.69°\cos(-157.86°)$

$y = 4000\sin 68.69°\sin(-157.86°)$

$z = 4000\cos 68.69°$

$(-3451.7, -1404.4, 1453.7)$

c. $$\cos\theta = \frac{\mathbf{u}\cdot\mathbf{v}}{\|\mathbf{u}\|\,\|\mathbf{v}\|} = \frac{(959.4)(-3451.7) + (-2795.7)(-1404.4) + (2695.1)(1453.7)}{(4000)(4000)} \approx 0.28329$$

$\theta \approx 73.54°$ or 1.28 radians

d. $s = r\theta = 4000(1.28) \approx 5120$ miles

17. From Theorem 11.13 and Theorem 11.7 (6) we have

$$D = \frac{|\overrightarrow{PQ} \cdot \mathbf{n}|}{\|\mathbf{n}\|}$$

$$= \frac{|\mathbf{w} \cdot (\mathbf{u} \times \mathbf{v})|}{\|\mathbf{u} \times \mathbf{v}\|} = \frac{|(\mathbf{u} \times \mathbf{v}) \cdot \mathbf{w}|}{\|\mathbf{u} \times \mathbf{v}\|} = \frac{|\mathbf{u} \cdot (\mathbf{v} \times \mathbf{w})|}{\|\mathbf{u} \times \mathbf{v}\|}.$$

18. Assume one of a, b, c, is not zero, say a. Choose a point in the first plane such as $(-d_1/a, 0, 0)$. The distance between this point and the second plane is

$$D = \frac{|a(-d_1/a) + b(0) + c(0) + d_2|}{\sqrt{a^2 + b^2 + c^2}}$$

$$= \frac{|-d_1 + d_2|}{\sqrt{a^2 + b^2 + c^2}} = \frac{|d_1 - d_2|}{\sqrt{a^2 + b^2 + c^2}}.$$

19. $x^2 + y^2 = 1$ cylinder

$z = 2y$ plane

Introduce a coordinate system in the plane $z = 2y$.

The new u-axis is the original x-axis.

The new v-axis is the line $z = 2y, x = 0$.

Then the intersection of the cylinder and plane satisfies the equation of an ellipse:

$$x^2 + y^2 = 1$$

$$x^2 + \left(\frac{z}{2}\right)^2 = 1$$

$$x^2 + \frac{z^2}{4} = 1 \quad \text{ellipse}$$

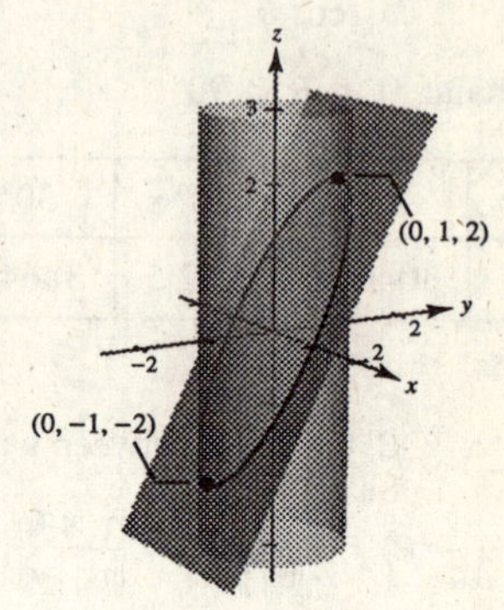

20. Essay.